.NET 开发经典名著

# ASP.NET Core 3 高级编程(第 8 版)

(上册)

[英] 亚当·弗里曼(Adam Freeman) 著
杜静芬 程凤娟 译

清华大学出版社

北 京

北京市版权局著作权合同登记号　图字：01-2021-3335
EISBN：978-1-4842-5439-4
First published in English under the title
Pro ASP.NET Core 3: Develop Cloud-Ready Web Applications Using MVC, Blazor, and Razor Pages, Eighth Edition
By Adam Freeman


**图书在版编目(CIP)数据**

ASP.NET Core 3高级编程：第8版 / (英)亚当 • 弗里曼 (Adam Freeman) 著；杜静芬，程凤娟译. —北京：清华大学出版社，2021.6

(.NET 开发经典名著)

书名原文：Pro ASP.NET Core 3: Develop Cloud-Ready Web Applications Using MVC, Blazor, and Razor Pages, Eighth Edition

ISBN 978-7-302-58271-7

Ⅰ. ①A…　Ⅱ. ①亚…　②杜…　③程…　Ⅲ. 网页制作工具—程序设计　Ⅳ. ①TP393.092.2

中国版本图书馆 CIP 数据核字(2021)第 101201 号

**责任编辑：**王　军　韩宏志
**装帧设计：**孔祥峰
**责任校对：**成凤进
**责任印制：**宋　林

**出版发行：**清华大学出版社
　　**网　址：**http://www.tup.com.cn，http://www.wqbook.com
　　**地　址：**北京清华大学学研大厦 A 座　　**邮　编：**100084
　　**社 总 机：**010-62770175　　**邮　购：**010-62786544
　　**投稿与读者服务：**010-62776969，c-service@tup.tsinghua.edu.cn
　　**质 量 反 馈：**010-62772015，zhiliang@tup.tsinghua.edu.cn
**印 装 者：**三河市国英印务有限公司
**经　销：**全国新华书店
**开　本：**170mm×240mm　　**印　张：**73　　**字　数：**2011 千字
**版　次：**2021 年 7 月第 1 版　　**印　次：**2021 年 7 月第 1 次印刷
**定　价：**268.00 元(全二册)

---

产品编号：089796-01

# 译者序

ASP.NET Core 是一款由微软创建的，用于构建 Web 应用、API、微服务的 Web 框架。它使用常见的模式，如 MVC(Model-View-Controller)、依赖注入和一个由中间件构成的请求处理管道。ASP.NET Core 是重新设计的 ASP.NET 4.x，更改了体系结构，形成了更精简的模块化框架。

ASP.NET Core 运行在微软的.NET 运行时库上，有几种语言(C#、Visual Basic 和 F#)可用来编写 ASP.NET Core 程序。C#是最常见的选择，可在 Windows、macOS 和 Linux 上构建并运行 ASP.NET Core 应用。

为什么要用 ASP.NET Core 开发应用程序？现存的 Web 框架选项已经很多了，如 Node/Express、Spring、Ruby on Rails、Django、Laravel 等，数不胜数。ASP.NET Core 具有如下优点。

- 快速：因为.NET Core 是编译运行的，执行速度远高于解释执行的语言，如 JavaScript 或 Ruby，ASP.NET Core 也已为多线程和异步任务做了专门优化。与使用 Node.js 写的代码相比，执行速度高出 5～10 倍是很正常的。
- 生态：在 NuGet(.NET 的包管理系统，如 NPM、RubyGems 或 Maven)中有成千上万的软件包。有现成的包可用来完成 JSON 反序列化、数据库连接、PDF 生成等。
- 安全：通过 ASP.NET Core，开发者可轻松配置和管理应用的安全性。ASP.NET Core 的功能包括管理身份验证、授权、数据保护、HTTPS 强制、应用机密、请求防伪保护及 CORS 管理。 通过这些安全功能，可生成安全可靠的 ASP.NET Core 应用。
- 跨平台：能在 Windows、macOS 和 Linux 上开发和运行。
- 开源：出现问题时，可阅读其源代码，来获取解决问题的方法。

本书深入浅出地介绍 ASP.NET Core 基础及实战等方面的知识，共分 4 个部分。第 I 部分介绍 ASP.NET Core。除了设置开发环境和创建第一个应用程序外，还介绍对 ASP.NET Core 开发最重要的 C#特性和如何使用 ASP.NET Core 开发工具。第 II 部分描述 ASP.NET Core 平台的主要特性，解释如何处理 HTTP 请求，如何创建和使用中间件组件，如何创建路由，如何定义和使用服务，以及如何与 Entity Framework Core 一起工作。第III部分解释如何创建不同类型的应用程序，包括 RESTful Web 服务以及使用控制器和 Razor Pages 的 HTML 应用程序。第Ⅳ部分解释如何使用 Blazor 服务器创建应用程序，如何使用实验性的 Blazor WebAssembly，以及如何使用 ASP.NET Core 验证用户身份和授予访问权限。

本书对于任何一名 C#开发者来说都是一本宝贵的指导书。书中所涉及的对于代码设计的实用建议是非常宝贵的。高效率的.NET 开发者需要对他所选择的语言有很深的理解。本书作者以令人惊叹的能力，把极其复杂的问题拆解为可消化的、易理解的一个个小问题，进行合理探讨，以一定的洞察力将知识传授给读者，并教给读者如何书写实践性强、干净简单且更容易理解的代码。无论是 C#新手还是资深开发者，都能通过阅读本书而有所收获。

本书文字简洁明快、流畅，既适合初学者及具有.NET 基础的开发者阅读，还可作为大中专院校计算机、通信、电子信息、自动化等相关专业的教材；也可供软件项目管理人员、开发团队成员学习参考。

这里要感谢清华大学出版社的编辑，他们为本书的翻译投入了巨大热情并付出了很多心血。没有他们的帮助和鼓励，本书不可能顺利付梓。

对于这本经典之作，译者本着“诚惶诚恐”的态度，在翻译过程中力求“信、达、雅”，但鉴于译者水平有限，失误在所难免，如果你有任何意见和建议，欢迎指正。

# 作者简介

Adam Freeman 是一位经验丰富的 IT 专业人士，曾在多家公司担任高级职位，后担任一家全球银行的首席技术官和首席运营官。现在退休了，他把时间花在写作和长跑上。

# 技术审校者简介

Fabio Claudio Ferracchiati 是一名使用微软技术的高级顾问和高级分析师/开发人员。他为BluArancio (www.bluarancio.com)工作。他是一名.NET 解决方案开发人员和应用程序开发人员，也是一名多产的作者和技术评论员。在过去十年里，他为意大利和国际杂志撰写文章，并与他人合著了十多本关于各种计算机主题的书籍。

# 目　录

## 第 III 部分　ASP.NET Core 应用程序

# 第 I 部分

# 介绍 ASP.NET Core

- 第 1 章　ASP.NET Core 上下文
- 第 2 章　入门
- 第 3 章　第一个 ASP.NET Core 应用程序
- 第 4 章　使用开发工具
- 第 5 章　C#的基本特点
- 第 6 章　测试 ASP.NET Core 应用程序
- 第 7 章　SportsStore：一个真正的应用程序
- 第 8 章　SportsStore：导航和购物车
- 第 9 章　SportsStore：完成购物车
- 第 10 章　SportsStore：管理
- 第 11 章　SportsStore：安全与部署

# 第 1 章

# ASP.NET Core 上下文

## 1.1 了解 ASP.NET Core

ASP.NET Core 是微软的 Web 开发平台。原来的 ASP.NET 是在 2002 年引入的，它经过了几次再创造和转世才成为 ASP.NET Core 3，这是本书的主题。

ASP.NET Core 由一个处理 HTTP 请求的平台、一系列用于创建应用程序的主要框架和提供支持特性的次要实用程序框架组成，如图 1-1 所示。

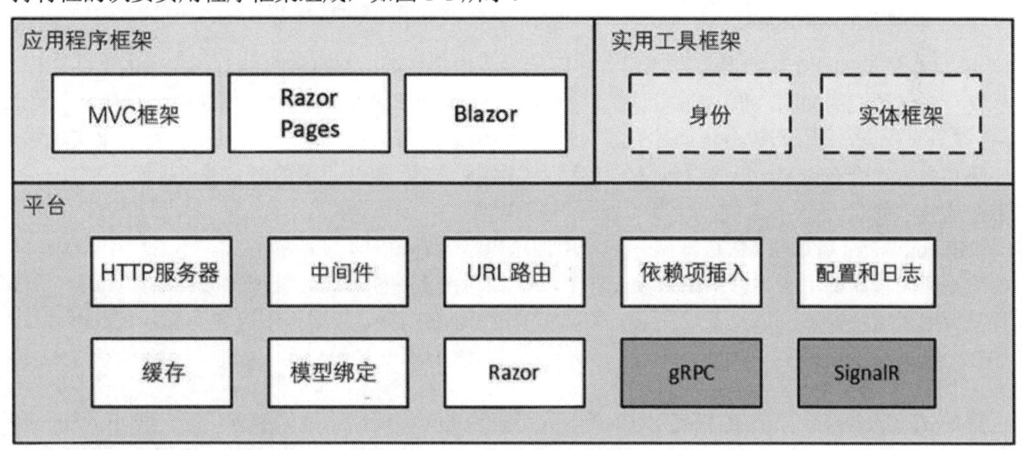


图 1-1　ASP.NET Core 的结构

### 1.1.1 理解应用程序框架

开始使用 ASP.NET Core，发现有不同的应用框架可用时，可能会感到困惑。其实，这些框架是互补的，可以解决不同的问题，或者，对于某些特性，可采用不同方式解决相同的问题。理解这些框架之间的关系意味着理解 Microsoft 所支持的不断变化的设计模式，如下面各节所述。

#### 1. 理解 MVC 框架

MVC 框架是在 ASP.NET Core 以前的时代引入的。原来的 ASP.NET 依赖于一种称为 Web Pages 的开发模型，这种模型重新创建了编写桌面应用程序的体验，但导致了无法很好扩展的、笨拙的

Web 项目。MVC 框架与 Web 页面一起引入，开发模型包含 HTTP 和 HTML 的特性，而不是试图隐藏它们。

MVC 代表 Model-View-Controller，是一种描述应用程序形状的设计模式。MVC 模式强调关注点分离，其中功能区域是独立定义的，这是对 Web 页面导致的模糊架构的有效解决方案。

早期版本的 MVC 框架建立在 ASP.NET 基础之上，最初是为 Web 页面设计的，这导致了一些笨拙的特性和解决方案。随着向.NET Core 的迁移，ASP.NET 成为 ASP.NET Core，MVC 框架是在一个开放的、可扩展的、跨平台的基础上重新构建的。

MVC 框架仍然是 ASP.NET Core 的重要组成部分。但是，随着单页应用程序(SPA)的兴起，它通常的使用方式已经发生了变化。在 SPA 中，浏览器发出一个单独的 HTTP 请求，然后接收一个 HTML 文档，该文档将提供一个富客户端，通常是用 JavaScript 客户端编写的，比如 Angular 或 React。SPA 的转变意味着 MVC 框架最初追求的清晰分离并不重要，强调遵循 MVC 模式不再是必要的，但 MVC 框架仍然是有用的(用于通过 Web 服务支持 SPA，详见第 19 章)。

#### 2. 把模式放在应有的位置上

模式只是其他人在其他项目中遇到的问题的解决方案。如果自己也面临同样的问题，了解以前是如何解决的会很有帮助。但这并不意味着必须完全遵循这个模式，或者根本不遵循，只要明白后果就行。例如，如果模式的目的是使项目可管理，而选择背离该模式，就必须接受项目可能更难管理这一事实。但是盲目地遵循模式可能比没有模式更糟糕，而且没有模式适合每个项目。

建议自由地使用模式，根据需要调整它们，忽略那些混淆模式和戒律的狂热分子。

#### 3. 理解 Razor Pages

MVC 框架的一个缺点是，在应用程序开始生成内容之前，它可能需要进行大量的准备工作。尽管存在结构上的问题，但 Web 页面的一个优点是可以在几小时内创建简单的应用程序。

Razor Pages 采用了 Web 页面的开发风格，并使用最初为 MVC 框架开发的平台特性来实现它。代码和内容混合在一起，形成自包含的页面；这重新提高了 Web 页面的开发速度，而不存在一些潜在的技术问题(尽管扩大复杂项目的规模仍然是个问题)。

Razor Pages 可以与 MVC 框架一起使用，这也是笔者倾向于使用的方式。笔者使用 MVC 框架编写应用程序的主要部分，使用 Razor Pages 编写次要功能，比如管理和报告工具。第 7～11 章就使用了这种方法，那几章开发了一个现实的 ASP.NET Core 应用程序，名为 SportsStore。

#### 4. 理解 Blazor

JavaScript 客户端框架的兴起对 C#开发人员来说是一个障碍，他们必须学习一种不同的、有些特殊的编程语言。笔者已经爱上 JavaScript，它和 C#一样流畅、富有表现力。但是，精通一种新的编程语言需要时间和投入，尤其是与 C#有根本区别的语言。

Blazor 试图通过允许将 C#用于编写客户端应用程序来弥补这一差距。Blazor 有两个版本：Blazor Server 和 Blazor WebAssembly。Blazor Server 是 ASP.NET Core 中稳定且受支持的部分。它通过使用到 ASP.NET Core 的持久 HTTP 连接工作。应用程序的 C#代码在这里执行。Blazor WebAssembly 是一个实验性版本，它更进一步，在浏览器中执行应用程序的 C#代码。如第 33 章所述，Blazor 的两种版本都不适合所有的情况，但它们都为 ASP.NET Core 开发的未来指明了方向。

### 1.1.2 理解实用程序框架

有两个框架与 ASP.NET Core 密切相关，但不直接用于生成 HTML 内容或数据。Entity Framework Core 是 Microsoft 的对象-关系映射(ORM)框架，它将存储在关系数据库中的数据表示为.NET 对象。Entity Framework Core 可以在任何.NET Core 应用程序中使用，它通常用于访问 ASP.NET Core 应用程序中的数据库。

ASP.NET Core Identity 是 Microsoft 的认证和授权框架，用于验证 ASP.NET Core 应用程序中的用户凭证，限制对应用程序特性的访问。

本书只描述了这两个框架的基本特性，重点是大多数 ASP.NET Core 应用程序所需要的功能，但它们都是复杂的框架，太大了，无法在一本关于 ASP.NET Core 的书中详细描述。

**未来版本的主题**

本书没有足够的篇幅介绍每一个 Entity Framework Core 和 ASP.NET Core Identity 特性，而只关注大多数项目都需要的方面。如果你认为某主题应该包括在本书的下一版或进行深度解析，请把建议发送到 adam@adam-freeman.com。

### 1.1.3 了解 ASP.NET Core 平台

ASP.NET Core 平台包含接收和处理 HTTP 请求以及创建响应所需的底层特性。有一个集成的 HTTP 服务器、一个处理请求的中间件组件系统和应用程序框架所依赖的核心特性，如 URL 路由和 Razor 视图引擎。

大部分开发时间都花费在应用程序框架上，但为了有效使用 ASP.NET Core，需要理解该平台提供的强大功能；没有这些功能，高层框架就无法运行。本书第 II 部分详细介绍 ASP.NET Core 平台的工作原理，解释它提供的特性是如何支撑 ASP.NET Core 开发的各个方面的。

本书没有描述两个值得注意的平台特性：SignalR 和 gRPC。SignalR 用于在应用程序之间创建低延迟通信通道，为本书第Ⅳ部分中描述的 Blazor 服务器框架提供了基础。但我们很少直接使用 SignalR，对于那些需要低延迟消息传递的项目，有更好的替代方案，如 Azure Event Grid 或 Azure Service Bus。

gRPC 是基于 HTTP 的跨平台远程过程调用(RPC)的新兴标准，最初由谷歌(gRPC 中的 g)创建，提供了效率和可伸缩性的优势。gRPC 可能是 Web 服务的未来标准，但它不能在 Web 应用程序中使用，因为它需要对它发送的 HTTP 消息进行低级控制，而这是浏览器不允许的(有一个浏览器库允许 gRPC 通过代理服务器使用，但这破坏了使用 gRPC 的好处)。直到 gRPC 可以在浏览器中使用，它才会包含在 ASP.NET Core 中，但只适用于使用它进行后端服务器之间通信的项目；对于这些项目，存在许多替代协议。当 gRPC 可以在浏览器中使用或成为主要的数据中心协议，本书的未来版本会介绍它。

## 1.2 理解本书

为从本书得到最大的收获，应该熟悉 Web 开发的基础知识，理解 HTML 和 CSS 是如何工作的，并掌握 C#的工作原理。如果没有进行过任何客户端开发(如 JavaScript)，请不要担心。本书的重点是 C#和 ASP.NET Core，随着章节的进展，读者可以获得需要知道的一切。第 5 章总结了 C#

对ASP.NET Core开发最重要的特性，如果读者从.NET Core或者.NET Framework的早期版本转到ASP.NET Core，这些特性都很有用。

### 1.2.1　需要什么软件来完成示例?

需要一个代码编辑器(Visual Studio或Visual Studio Code)、.NET核心软件开发工具包和SQL Server LocalDB。所有这些都可以从微软免费使用，第2章包含了所需软件的安装指令。

### 1.2.2　需要什么平台来完成示例?

本书是为Windows编写的，使用的是Windows 10 Pro，但是Visual Studio、Visual Studio Code和.NET Core支持的任何版本的Windows都可以工作。其他平台也支持ASP.NET Core，但本书中的示例依赖于SQL Server LocalDB特性，这是Windows特有的。如果试图使用另一个平台，可以通过adam@adam-freeman.com联系笔者，笔者会给你一些修改示例的一般指示，但有一点要注意，如果你陷入困境，笔者不能提供详细的帮助。

### 1.2.3　源代码下载

读者可访问本书的GitHub存储库(https://github.com/apress/pro-asp.net-core-3)下载源代码，也可扫描封底的二维码下载。

### 1.2.4　如果在执行这些示例时遇到问题，怎么办?

首先要做的是回到这一章的开头，重新开始。大多数问题是由于缺少一个步骤或没有完全按照清单执行而引起的。请密切关注代码清单中的重点部分，它会突出显示所需的更改。

接下来，检查该书GitHub存储库中包含的勘误表/更正列表。技术书籍是复杂的，尽管笔者和编辑尽了最大努力，错误是不可避免的。检查勘误表列表，获得已知错误列表和解决这些错误的指令。

如果仍然有问题，那么从本书的GitHub存储库(https://github.com/apress/pro-asp.net-core-3)下载正在阅读的章节的项目，并将其与项目进行比较。笔者通过遍历每一章来创建GitHub存储库的代码，因此项目中的相同文件应该具有相同的内容。

如果示例仍然不能工作，可以联系笔者(adam@adam-freeman.com)寻求帮助。请在电子邮件中说明你正在阅读的是哪本书，是哪一章/例子导致了这个问题。请记住，笔者会收到很多电子邮件，可能不会立即回复。

### 1.2.5　如果发现书中有错误，怎么办?

可以通过向adam@adam-freeman.com发电子邮件，向笔者报告错误，不过读者应先检查本书的勘误表/修正列表，本书的GitHub库可以在如下地址找到：

```
https://github.com/apress/pro-asp.net-core-3
```

你遇到的错误可能已经在该列表中列出。

### 1.2.6　本书包含的内容

本书试图涵盖大多数ASP.NET Core项目都需要的功能。它分为四个部分，每一部分都涵盖一系列相关的主题。

**第 I 部分：介绍 ASP.NET Core**

本书的这一部分——包括本章——介绍了 ASP.NET Core。除了设置开发环境和创建第一个应用程序之外，还将了解对 ASP.NET Core 开发最重要的 C#特性，和如何使用 ASP.NET Core 开发工具。但是第 I 部分的大部分内容是关于一个名为 SportsStore 的项目的开发，通过这个项目，展示了从开始到部署的实际开发过程，涉及 ASP.NET Core 的所有主要特性，并展示它们是如何结合在一起的。

**第 II 部分：ASP.NET Core 平台**

该部分的章节描述 ASP.NET Core 平台的主要特性，解释如何处理 HTTP 请求，如何创建和使用中间件组件，如何创建路由，如何定义和使用服务，以及如何与 Entity Framework Core 一起工作。解释 ASP.NET Core 的基础知识。理解它们对于有效的 ASP.NET Core 开发是至关重要的。

**第 III 部分：ASP.NET Core 应用程序**

该部分的章节解释如何创建不同类型的应用程序，包括 RESTful Web 服务以及使用控制器和 Razor Pages 的 HTML 应用程序。这些章节还描述易于生成 HTML 的特性，包括视图、视图组件和标签助手。

**第 IV 部分：高级 ASP.NET Core 功能**

本书最后一部分解释如何使用 Blazor 服务器创建应用程序，如何使用实验性的 Blazor WebAssembly，以及如何使用 ASP.NET Core 验证用户身份和授予访问权限。

### 1.2.7 本书未包含的内容

本书没有涵盖基本的 Web 开发主题，如 HTML 和 CSS，也没有详细讲述 C#(第 5 章描述了对 ASP.NET Core 开发有用的 C#特性，使用旧版本.NET 的开发人员可能不熟悉这些特性)。

虽然笔者喜欢在书中钻研细节，但不是每个 ASP.NET Core 特性在主流开发中很有用，笔者必须将本书保持适当的厚度范围内。当笔者决定去掉某个特性时，是因为认为它不重要，或者因为使用所介绍的技术可达到同样的结果。

如前所述，本书没有描述 ASP.NET Core 对 SignalR 和 gRPC 的支持，后续章节提到了没有描述的其他特性，这要么是因为它们不能广泛应用，要么是因为有更好的替代方法可用。每种情况下，笔者都会解释为什么省略了描述，并提供了关于该主题的可供参考的 Microsoft 文档。

### 1.2.8 如何联系作者?

可通过 adam@adam-freeman.com 给笔者发邮件。从第一次在书中公布电子邮件地址以来，到现在已经有好几年了。不能完全肯定这是个好主意，但很高兴笔者这么做了。笔者收到过来自世界各地的电子邮件，读者来自各个行业，大部分邮件都是积极的、礼貌的，收到邮件令人愉悦。

笔者试着迅速回复邮件，但邮件太多，有时还会积压大量邮件，尤其是当笔者埋头写书的时候。笔者总是试图帮助那些被书中的例子困住的读者，尽管读者在联系之前，应按照本章前面描述的步骤来做。

虽然笔者欢迎读者发电子邮件，但对于某些问题，笔者的回答总是否定的。笔者恐怕不会为新公司写代码，帮助完成大学作业，或参与某个开发团队的设计争议。

### 1.2.9　如果你真的喜欢本书？

请发邮件到 adam@adam-freeman.com，让笔者知道。收到快乐读者的来信总是一件让人高兴的事，笔者也很感激读者花时间发送这些邮件。写书很难，这些电子邮件是笔者前进的基本动力。

### 1.2.10　如果本书让人生气，想要抱怨该怎么办？

仍然可以通过 adam@adam-freeman.com 发邮件，笔者仍然会尽力提供帮助。请记住，只有在解释问题是什么、想让笔者怎么做的情况下，笔者才能提供帮助。有时候唯一的结果就是接受本书不是专门为每个人写的，只有放下本书，选择另一本书，事情才会结束。笔者会仔细考虑让人心烦的事情，但在写了 25 年书之后，笔者开始明白，不是每个人都喜欢读我的书。

## 1.3　小结

本章为本书其余部分做了铺垫。简要介绍了 ASP.NET Core，说明了本书的要求和内容，以及如何联系笔者。第 2 章将展示如何准备 ASP.NET Core 开发。

# 第2章 入门

欣赏软件开发框架的最好方法是直接开始使用它。本章解释如何准备 ASP.NET Core 的开发以及如何创建和运行 ASP.NET Core 应用程序。

**本书的更新**

微软对.NET Core 和 ASP.NET Core 有一个积极的开发计划，这意味着在读者阅读本书时，本书可能会有新的版本。指望读者每隔几个月就买一本新书似乎是不公平的，尤其是在大多数变化相对较小的情况下。相反，本书的 GitHub 存储库(https://github. com/apress/pro-asp.net-core-3)将免费更新，用于提供这些变化。

对笔者(和 Apress)来说，这种更新是一个正在进行的实验，而且它还在不断发展——尤其是因为不知道未来主要的 ASP.NET Core 版本将包含什么——但目标是通过补充包含的示例来延长本书的寿命。

笔者不会对更新的内容、更新的形式、将更新的内容合并到新书中的时间做任何承诺。请保持开放的思想，当新的 ASP.NET Core 版本发布时检查本书的存储库。如果有关于如何改进更新的想法，可通过 adam@adam-freeman.com 给笔者发电子邮件。

## 2.1 选择代码编辑器

微软为 ASP.NET Core 开发提供了多种工具：Visual Studio 和 Visual Studio Code。Visual Studio 是.NET 应用程序的传统开发环境，提供了用于开发各种应用程序的大量工具和特性。但它可能会消耗大量资源、速度缓慢，而且有些功能是如此有用，以至于阻碍了开发。

Visual Studio Code 是一种轻量级替代品，它没有 Visual Studio 的花哨功能，但完全能够处理 ASP.NET Core 开发。

本书中的所有示例都包括这两种编辑器的指令，Visual Studio 和 Visual Studio Code 都可以免费使用，因此可以使用任何适合自己开发风格的代码。

如果对.NET Core 开发很陌生，就从 Visual Studio 开始吧。它为创建 ASP.NET Core 开发中使用的不同类型的文件提供了更结构化的支持，有助于确保代码示例获得预期的结果。

■ **注意：**本书介绍了用于 Windows 的 ASP.NET Core 开发。可以在 Linux 和 macOS 上开发和运行 ASP.NET Core 应用程序，但大多数读者使用 Windows，所以本书选择关注它。本书中几乎所有的示例都依赖于 LocalDB，这是 SQL Server 提供的仅用于 Windows 的特性，在其他平台上是不可用的。如果想在另一个平台上阅读本书，可以使用第 1 章中的电子邮件地址联系笔者，笔者会尽力帮助你入门。

### 2.1.1　安装 Visual Studio

ASP.NET Core 3 需要 Visual Studio 2019。笔者使用的是免费的 Visual Studio 2019 社区版，可从以下地址下载它：

```
www.visualstudio.com
```

运行安装程序，显示如图 2-1 所示的提示。

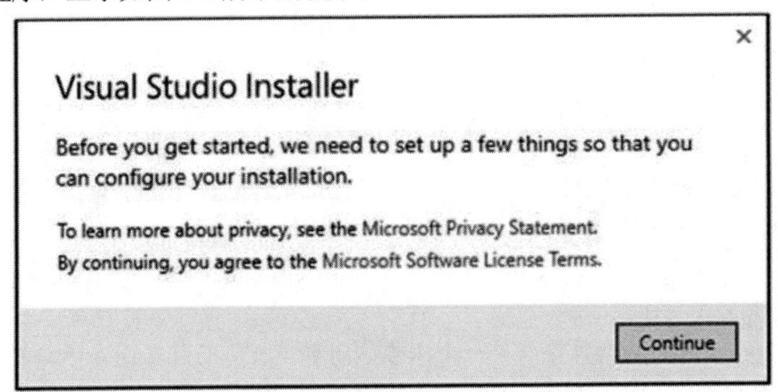


图 2-1　启动 Visual Studio 安装程序

单击 Continue 按钮，安装程序将下载安装文件，如图 2-2 所示。

Visual Studio Installer

Just a moment ... Fetching your files.

Downloading: 11.17 MB of 71.23 MB　　2.56 MB/sec

Installing

Cancel

图 2-2　下载 Visual Studio 安装程序文件

下载安装程序文件后，会显示一组按工作负载分组的安装选项。确保选中 ASP.NET and web development 工作负载，如图 2-3 所示。

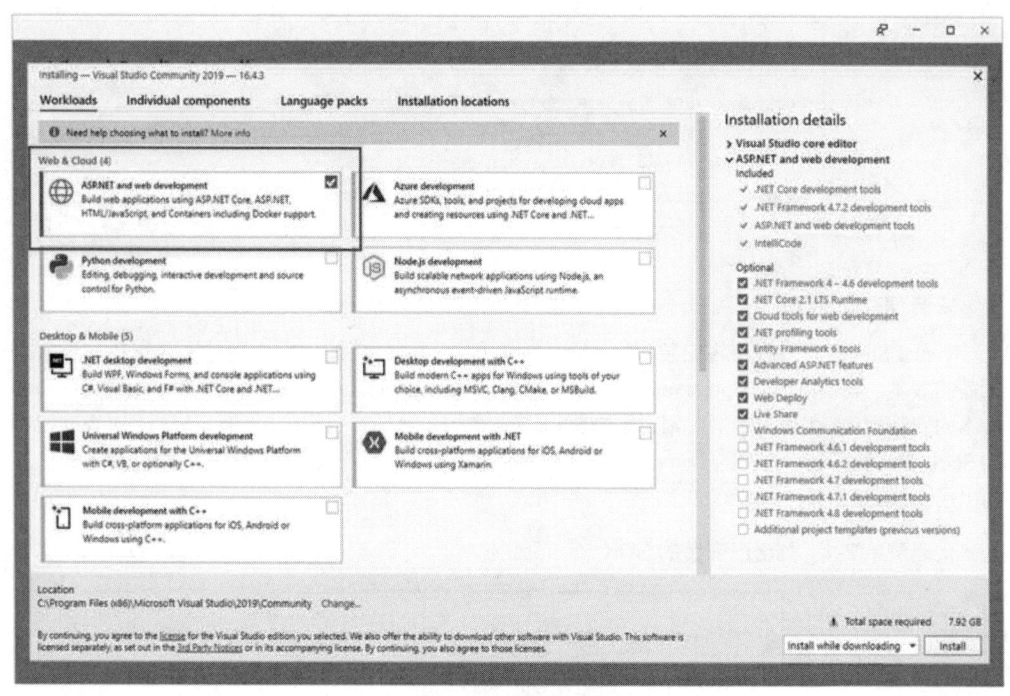

图 2-3 选择工作负载

选择窗口顶部的 Individual components 部分，并确保选中 SQL Server Express 2016 LocalDB 选项，如图 2-4 所示。后续章节将使用这个数据库组件来存储数据。

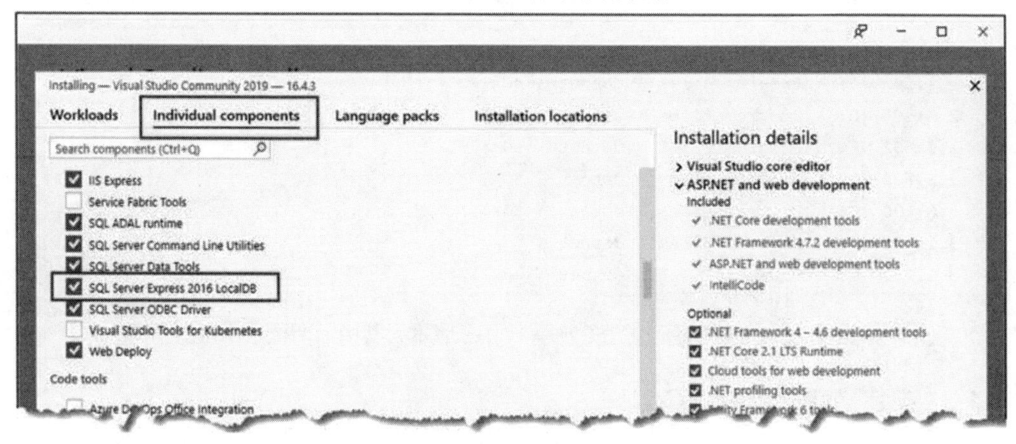

图 2-4 确保安装 LocalDB

单击 Install 按钮，就会下载和安装所选工作负载所需的文件。要完成安装，需要重新启动，如图 2-5 所示。

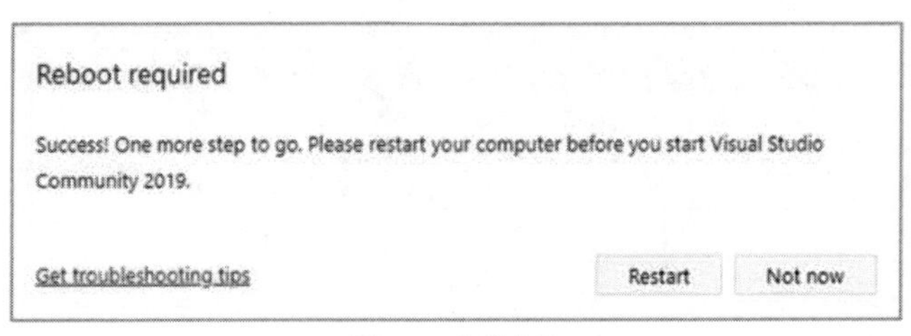


图 2-5　完成安装

#### 安装.NET Core SDK

Visual Studio 安装程序将安装.NET Core 软件开发工具包(SDK)，但它可能不会安装本书示例所需的版本。访问 https://dotnet.microsoft.com/download/dotnetcore/3.1 并下载.NET Core SDK 3.1.1 版本的安装程序，在编写本书时，该版本是长期支持版本。运行安装程序；安装完成后，从 Windows 的 Start 菜单中打开一个新的 PowerShell 命令提示符，并运行代码清单 2-1 所示的命令，该命令将列出已安装的.NET Core SDK。

**代码清单 2-1　列出已安装的 SDK**

```
dotnet --list-sdks
```

下面是在 Windows 机器上新安装的输出，该机器还没有用于.NET Core。

```
3.1.101 [C: \ Program Files \ dotnet \ sdk]
```

如果用过不同版本的.NET Core，就可能看到一个更长的列表，比如：

```
2.1.401 [C: \ Program Files \ dotnet \ sdk]
2.1.502 [C: \ Program Files \ dotnet \ sdk]
2.1.505 [C: \ Program Files \ dotnet \ sdk]
2.1.602 [C: \ Program Files \ dotnet \ sdk]
2.1.802 [C: \ Program Files \ dotnet \ sdk]
3.0.100 [C: \ Program Files \ dotnet \ sdk]
3.1.100 [C: \ Program Files \ dotnet \ sdk]
3.1.101 [C: \ Program Files \ dotnet \ sdk]
```

无论有多少项，都必须确保至少有一项是 3.1.1*xx* 版本，其中后两位数字可能不同。

### 2.1.2 安装 Visual Studio Code

如果选用 Visual Studio Code，请从 https://code.visualstudio.com 下载安装程序。不需要特定的版本，应该选择当前的稳定版本。运行安装程序，确保选中 Add to PATH 复选框，如图 2-6 所示。

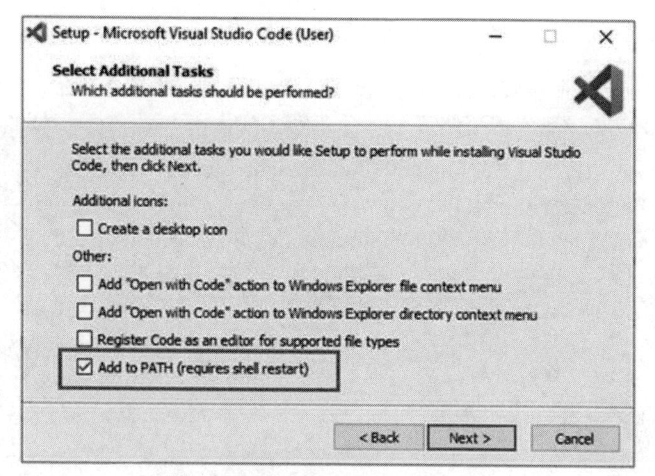


图 2-6 配置 Visual Studio Code 安装

### 1. 安装.NET Core SDK

Visual Studio 安装程序不包括.NET Core SDK，必须单独安装它。访问 https://dotnet.microsoft.com/download/dotnet-core/3.1，下载.NET Core SDK 3.1.1 版本的安装程序，该版本在撰写本书时是长期支持版本。运行安装程序；安装完成后，从 Windows“开始”菜单中打开一个新的 PowerShell 命令提示符，并运行代码清单 2-2 所示的命令，该命令将列出已安装的.NET Core SDK。

**代码清单 2-2 列出已安装的 SDK**

```
dotnet--list-sdks
```

下面是在 Windows 机器上新安装的输出，该机器还没有使用.NET Core。

```
3.1.101 [C: \ Program Files \ dotnet \ sdk]
```

如果使用过不同版本的.NET Core，可能会看到一个更长的列表，比如：

```
2.1.401 [C: \ Program Files \ dotnet \ sdk]
2.1.502 [C: \ Program Files \ dotnet \ sdk]
2.1.505 [C: \ Program Files \ dotnet \ sdk]
2.1.602 [C: \ Program Files \ dotnet \ sdk]
2.1.802 [C: \ Program Files \ dotnet \ sdk]
3.0.100 [C: \ Program Files \ dotnet \ sdk]
3.1.100 [C: \ Program Files \ dotnet \ sdk]
3.1.101 [C: \ Program Files \ dotnet \ sdk]
```

无论有多少项，都必须确保至少有一项是 3.1.1*xx* 版本，其中后两位数字可能不同。

### 2. 安装 SQL Server LocalDB

本书中的数据库示例需要LocalDB，它是SQL Server的零配置版本，可作为SQL Server Express版本的一部分安装，SQL Server Express 版本可从以下地址免费安装。

```
https://www.microsoft.com/en-in/sql-server/sql-server-downloads
```

下载并运行 Express 版本安装程序，并选择 Custom 选项，如图 2-7 所示。

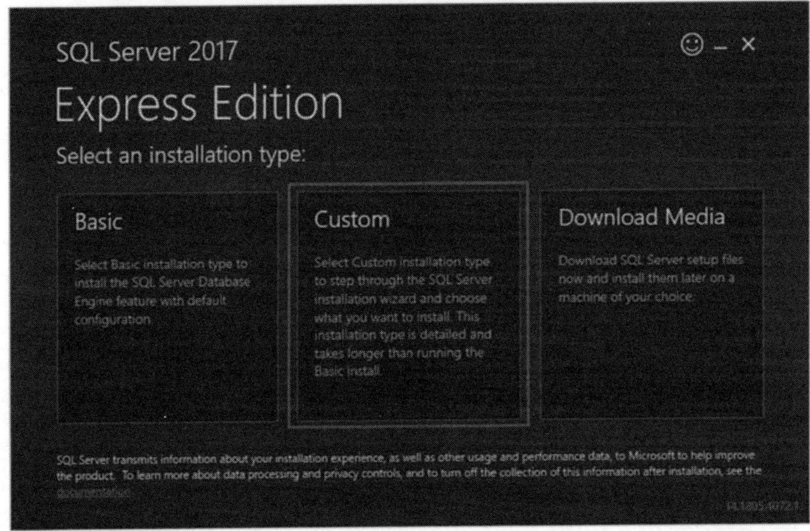


图 2-7　选择 SQL Server 的安装选项

选择自定义 Custom 后，系统会提示选择安装文件的下载位置。单击 Install 按钮，将开始下载。

当出现提示时，选择创建一个新的 SQL Server 安装选项，如图 2-8 所示。

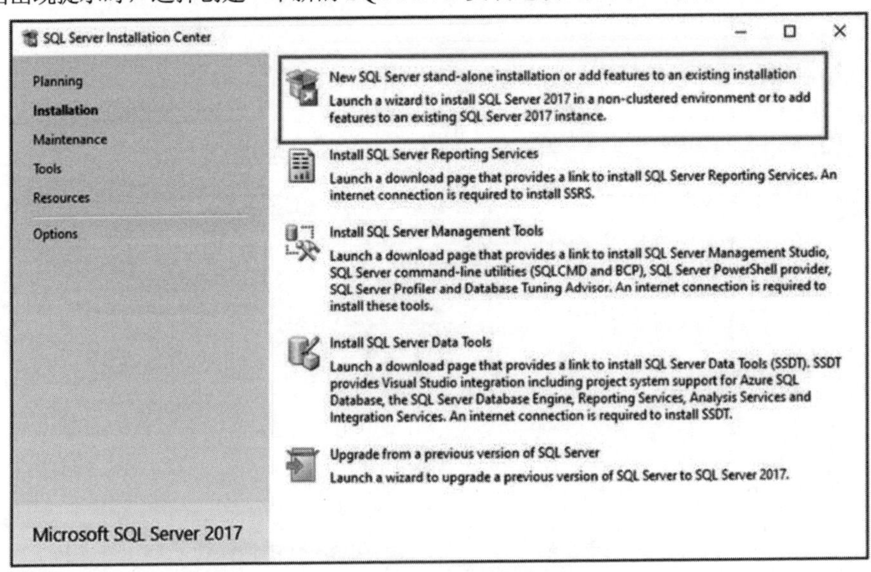


图 2-8　选择安装选项

完成安装过程，选择提供的默认选项。到达 Feature Selection 页面时，确保选中 LocalDB 选项，如图 2-9 所示(可能想取消选中 R 和 Python 选项，本书中没有使用这两个选项，下载和安装需要很长时间)。

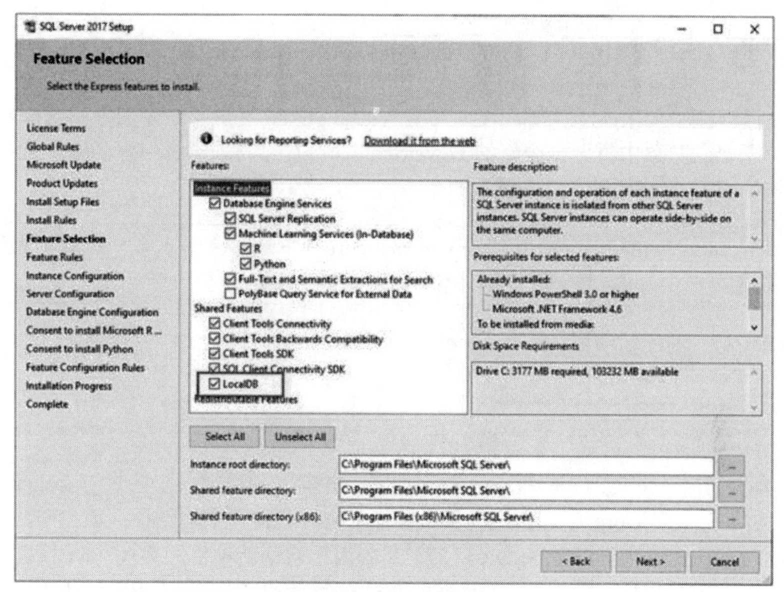


图 2-9 选中 LocalDB 特性

在 Instance Configuration 页面上，选择 Default instance 选项，如图 2-10 所示。

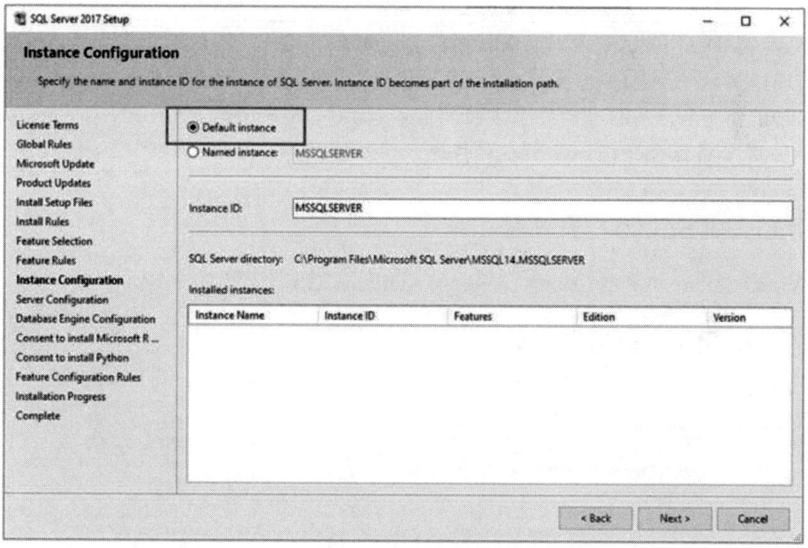

图 2-10 配置数据库

继续完成安装过程，选择默认值。安装完成后，安装 SQL Server 的最新累计更新。在撰写本书时，最新的更新可从https://support.microsoft.com/en-us/help/4527377/cumulal-update-18-forsql-server-2017 获得，尽管在阅读本章时可能已经发布了较新的更新。

**警告：**很容易跳过更新阶段，但执行这一步很重要，这样才能从本书的示例中获得预期的结果。例如，SQL Server 的基本安装有一个阻止 LocalDB 创建数据库文件的 bug，这将在第 7 章出现问题。

## 2.2　创建 ASP.NET Core 项目

创建项目最直接的方法是使用命令行，尽管 Visual Studio 提供了一个向导系统，参见第 4 章。从 Windows“开始”菜单中打开一个新的 PowerShell 命令提示符，导航到想要创建 ASP.NET Core 项目的文件夹，并运行代码清单 2-3 所示的命令。

**提示：**可以从 https://github.com/apress/pro-asp.net-core-3 下载本章和本书中所有其他章节的示例项目。如果在运行示例时遇到问题，请参阅第 1 章以获得帮助。

**代码清单 2-3　创建新项目**

```
dotnet new globaljson --sdk-version 3.1.101 --output FirstProject
dotnet new mvc --no-https --output FirstProject --framework netcoreapp3.1
```

第一个命令创建一个名为 FirstProject 的文件夹，并向其中添加一个名为 global.json 的文件，指定项目将使用的.NET Core 版本；这确保在执行示例时获得预期的结果。第二个命令创建一个新的 ASP.NET Core 项目。.NET Core SDK 包括一系列用于启动新项目的模板，mvc 模板是 ASP.NET Core 应用程序可用的选项之一。这个项目模板创建一个为 MVC 框架配置的项目，MVC 框架是 ASP.NET Core 支持的应用程序类型之一。不要被选择框架的想法吓倒，如果没有听说过 MVC，也不必担心——在本书结束时，就会了解每个框架提供的特性以及它们是如何组合在一起的。

**注意：**这是使用包含占位符内容的项目模板的少数章节之一。笔者不喜欢使用预定义的项目模板，因为它们鼓励开发人员将重要特性(如身份验证)视为黑盒。笔者写本书的目的是让读者了解和管理 ASP.NET Core 应用程序的各个方面。这就是为什么从一个空的 ASP.NET Core 应用程序开始的原因。本章非常适合介绍 mvc 模板的快速入门知识。

### 2.2.1　用 Visual Studio 打开项目

启动 Visual Studio 并单击 Open a project or solution 按钮，如图 2-11 所示。

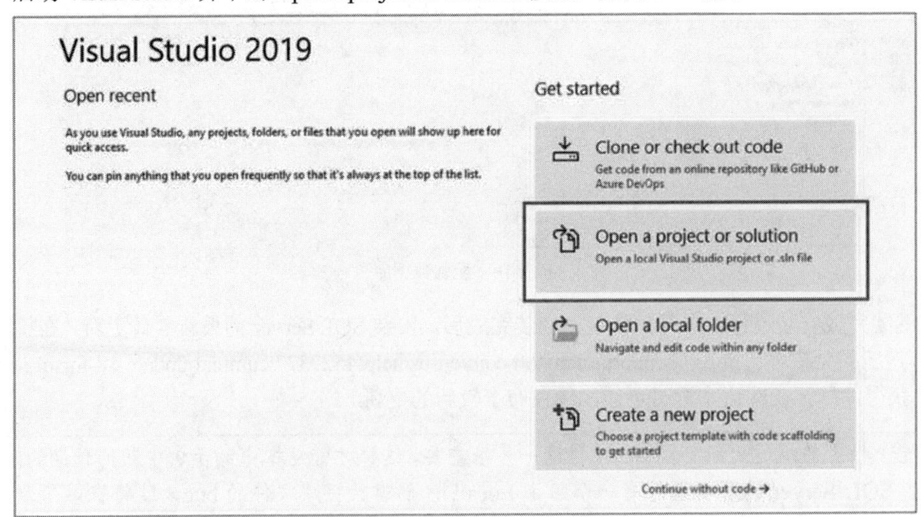


图 2-11　打开 ASP.NET Core 项目

导航到 FirstProject 文件夹，选择 FirstProject.csproj 文件，然后单击 Open 按钮。Visual Studio 将打开项目，并在 Solution Explorer 窗口中显示其内容，如图 2-12 所示。项目中的文件是由项目模板创建的。

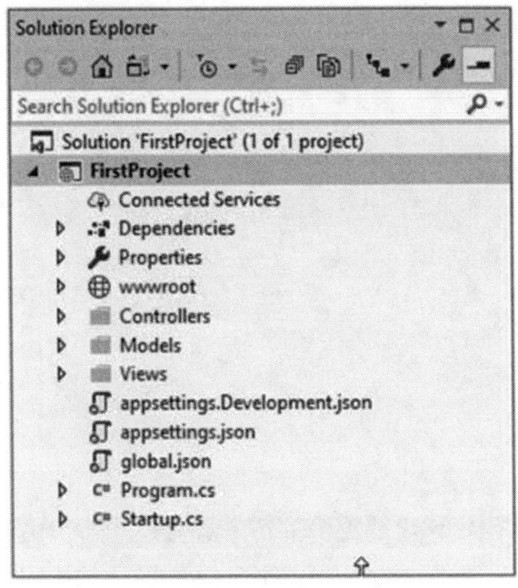


图 2-12　在 Visual Studio 中打开项目

### 选择浏览器

当项目运行时，Visual Studio 将自动打开一个浏览器窗口。要选择所使用的浏览器，单击 IIS Express 下拉菜单右边的小箭头，从 Web Browser 菜单中选择你的首选浏览器，如图 2-13 所示。本书中使用 Google Chrome。

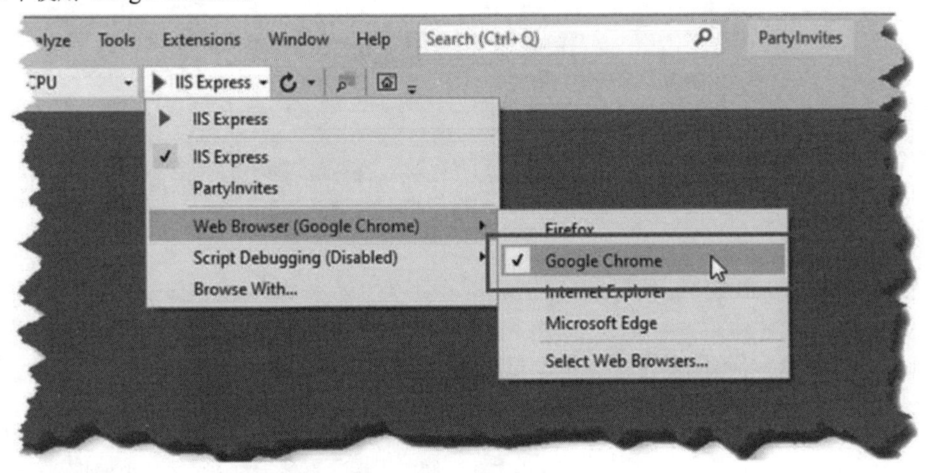


图 2-13　选择浏览器

## 2.2.2　用 Visual Studio Code 打开项目

启动 Visual Studio Code 并选择 File | Open Folder。导航到 FirstProject 文件夹并单击 Select Folder 按钮。Visual Studio Code 将打开项目，并在 Explorer 窗格中显示其内容，如图 2-14 所示(Visual

Studio Code中使用的默认暗色主题在页面上显示得不太合适，本书的截图改为浅色主题）。

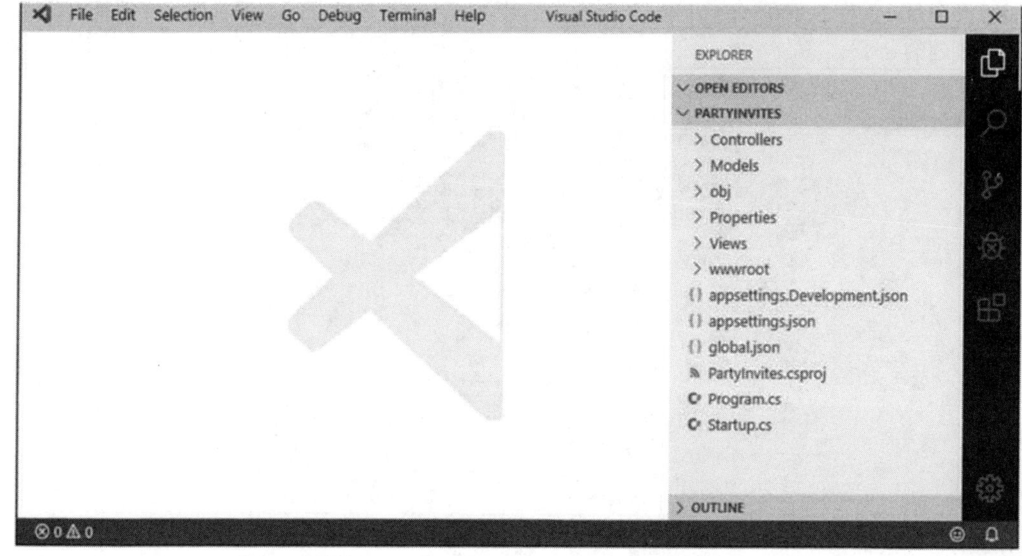


图2-14　在Visual Studio Code中打开项目

第一次在Visual Studio Code中打开.NET核心项目时，需要进行额外的配置。第一步是单击Explorer窗格中的Startup.cs文件。这将触发来自Visual Studio Code的提示来安装C#开发所需的特性，如图2-15所示。如果之前打开了一个C#项目，就将看到一个提示，提示安装所需的资源，如图2-15所示。

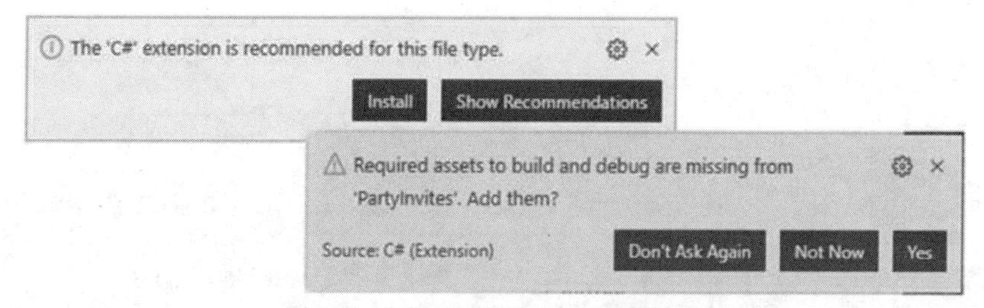


图2-15　安装Visual Studio Code C#特性

按需要单击Install或Yes按钮，Visual Studio Code将下载并安装.NET Core项目所需的特性。

## 2.3　运行ASP.NET Core应用程序

模板创建一个项目，其中包含构建和运行应用程序需要的所有内容。从Debug菜单中选择Start Without Debugging，Visual Studio将编译并启动示例应用程序，然后打开一个新的浏览器窗口，向应用程序发送HTTP请求，如图2-17所示(如果在Debug菜单中没有看到Start Without Debugging项，可在Solution Explorer窗口中单击Startup.cs文件，然后再次检查该菜单）。

如果使用的是Visual Studio Code，请在Debug菜单中选择Run Without Debugging。由于这是项目第一次启动，系统将提示选择执行环境。选择.NET Core选项，如图2-16所示。

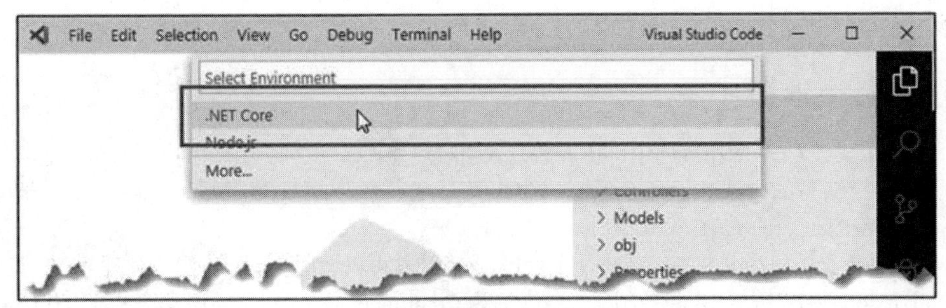


图 2-16 选择执行环境

Visual Studio Code 将创建 launch.json 文件，其中包含项目的启动设置，本书可以忽略该设置。再次从 Debug 菜单中选择 Run Without Debugging, Visual Studio Code 将编译项目，开始执行应用程序，并打开一个新的浏览器窗口，如图 2-17 所示。

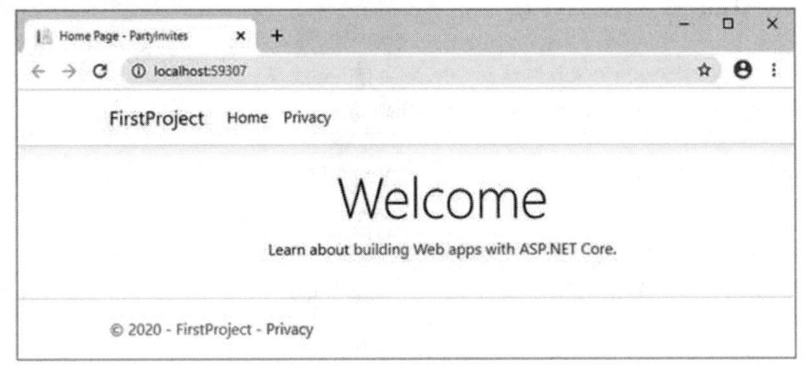


图 2-17 运行示例项目

还可以从命令行启动应用程序。从 Windows 开始菜单中打开新的 PowerShell 命令提示符，导航到 FirstProject 项目文件夹，该文件夹包含 FirstProject.csproj 文件；并运行代码清单 2-4 所示的命令。

**代码清单 2-4 启动示例应用程序**

```
dotnet run
```

一旦应用程序启动，就需要打开一个新的浏览器窗口，并请求 http://localhost:5000，这将产生如图 2-17 所示的响应。

**提示：**如果使用 Visual Studio Code 或通过命令行运行应用程序，那么使用 ASP.NET Core 在端口 5000 上监听 HTTP 请求。如果使用的是 Visual Studio，就会注意到浏览器请求一个不同的端口，该端口是在创建项目时选择的。如果查看 Windows 任务栏通知区域，会发现 IIS Express 的图标。这是完整 IIS 应用服务器的精简版，包括在 Visual Studio 中，用于在开发期间交付 ASP.NET Core 内容和服务。后续章节将展示如何更改项目配置，以使用与 Visual Studio Code 相同的 HTTP 端口。

完成后，关闭 Visual Studio 打开的浏览器窗口。如果使用的是 Visual Studio Code，则单击在代码编辑器上方弹出的窗口中显示的 stop 按钮。如果从命令行启动应用程序，则使用 Ctrl+C 组合键停止执行。

### 2.3.1 理解端点

在 ASP.NET Core 应用程序中，传入的请求由端点处理。图 2-17 中产生响应的端点是一个操作，是用 C#编写的方法。操作在控制器中定义，是一个 C#类，该类派生自内置控制器基类 Microsoft.AspNetCore.Mvc.Controller。

控制器定义的每个公共方法都是一个操作，这意味着可以调用该操作方法来处理 HTTP 请求。ASP.NET Core 项目中的约定将控制器类放在名为 Controllers 的文件夹中，该文件夹是由代码清单 2-3 中用于设置项目的模板创建的。

项目模板向 Controllers 文件夹添加一个控制器，以帮助启动开发。控制器在名为 HomeController.cs 的类文件中定义。控制器类包含一个后跟单词 Controller 的名称，意思是当看到名为 HomeController.cs 的文件时，它包含一个叫做 Home 的控制器，这是 ASP.NET Core 应用程序中使用的默认控制器。

**提示：** 如果“控制器”和“操作”这两个术语没有上述含义，不必担心。继续下面的示例，将看到浏览器发送的 HTTP 请求是如何由 C#代码处理的。

在 Solution Explorer 或 Explorer 窗格中找到 HomeController.cs 文件，单击打开它进行编辑。代码如下：

```
using System;
using System.Collections.Generic;
using System.Diagnostics;
using System.Linq;
using System.Threading.Tasks;
using Microsoft.AspNetCore.Mvc;
using Microsoft.Extensions.Logging;
using FirstProject.Models;

namespace FirstProject.Controllers {
    public class HomeController : Controller {
        private readonly ILogger<HomeController> _logger;

        public HomeController(ILogger<HomeController> logger) {
            _logger = logger;
        }

        public IActionResult Index() {
            return View();
        }

        public IActionResult Privacy() {
            return View();
        }
        [ResponseCache(Duration = 0, Location = ResponseCacheLocation.None,
```

```
            NoStore = true)]
        public IActionResult Error() {
            return View(new ErrorViewModel { RequestId = Activity.Current?.Id
              ?? HttpContext.TraceIdentifier });
        }
    }
}
```

使用代码编辑器，替换 HomeController.cs 文件的内容，使其与代码清单 2-5 匹配。这里删除了除一个方法外的所有方法，更改了结果类型及其实现，并删除了未使用名称空间的 using 语句。

**代码清单 2-5　更改 Controllers 文件夹中的 HomeController.cs 文件**

```
using Microsoft.AspNetCore.Mvc;

namespace FirstProject.Controllers {

    public class HomeController : Controller {

        public string Index() {
            return "Hello World";
        }
    }
}
```

结果是主控制器定义一个名为 Index 的操作。这些变化不会产生戏剧性效果，但可很好起到演示作用。这里改变了名为 Index 的方法，以便它返回字符串 Hello World。通过从 Debug 菜单中选择 Start Without Debugging 或 Run Without Debugging，再次运行项目。

浏览器将向 ASP.NET Core 服务器发出 HTTP 请求。代码清单 2-5 中模板创建的项目配置意味着，HTTP 请求将由主控制器定义的 Index 操作处理。换句话说，请求将由 HomeController 类定义的 Index 方法处理。Index 方法生成的字符串用于响应浏览器的 HTTP 请求，如图 2-18 所示。

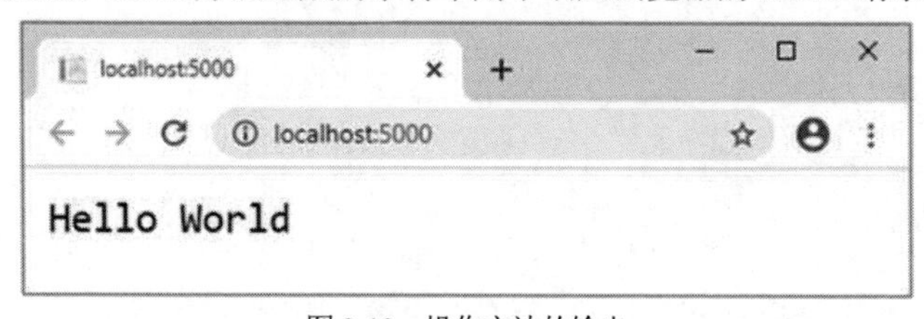


图 2-18　操作方法的输出

## 2.3.2　了解路由

ASP.NET Core 路由系统负责选择处理 HTTP 请求的端点。路由是用来决定如何处理请求的规则。当创建项目时，会创建并启动一个默认规则。可请求以下任何 URL，它们将被发送到由主控制器定义的 Index 操作：

- /
- /Home

- /Home/Index

因此，当浏览器请求 http://yoursite/或 http://yoursite/Home 时，它将从 HomeController 的 Index 方法获得输出。可以自己尝试在浏览器中更改 URL。目前，它是 http://localhost:5000/，只是如果使用 Visual Studio，端口部分可能会有所不同。如果将/Home 或/Home/Index 附加到 URL 并按 Enter 键，在应用程序中就会看到相同的 Hello World 结果。

## 2.3.3　理解 HTML 渲染

上一个示例的输出不是 HTML——只是字符串 Hello World。为了生成对浏览器请求的 HTML 响应，需要一个视图，它告诉 ASP.NET Core 如何处理由 Index 方法产生的结果，放到 HTML 响应中，发送到浏览器。

### 1. 创建和呈现视图

首先需要修改 Index 操作方法，如代码清单 2-6 所示。更改以粗体显示，这是本书遵循的惯例，以使示例更容易理解。

**代码清单 2-6　在 Controllers 文件夹中的 HomeController.cs 文件中呈现视图**

```
using Microsoft.AspNetCore.Mvc;

namespace FirstProject.Controllers {

    public class HomeController : Controller {

        public ViewResult Index() {
            return View("MyView");
        }
    }
}
```

从一个操作方法返回 ViewResult 对象时，指示 ASP.NET Core 渲染视图。通过调用 View 方法创建 ViewResult，指定要使用的视图名称，也就是 MyView。如果运行该应用程序，就可以看到 ASP.NET Core 试图找到视图，如图 2-19 中显示的错误消息所示。

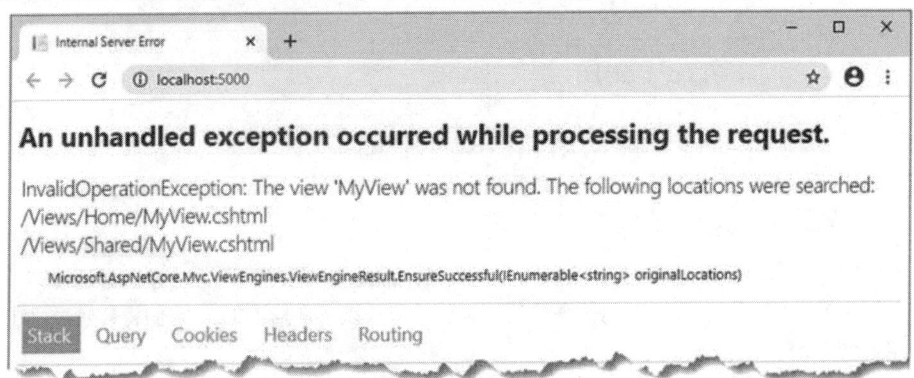


图 2-19　试图找到视图

这是一个有用的错误消息。它解释了ASP.NET Core无法找到为操作方法指定的视图，并解释了它查找的位置。视图存储在Views文件夹中，组织成子文件夹。例如，与主控制器关联的视图存储在名为Views/Home的文件夹中。不特定于单个控制器的视图存储在一个名为Views/Shared的文件夹中。用于创建项目的模板自动添加了Home和Shared文件夹，并添加了一些占位符视图来启动项目。

如果使用的是Visual Studio，右击Solution Explorer中的Views/Home文件夹，然后从弹出菜单中选择Add | New Item。Visual Studio将显示用于向项目添加项的模板列表。找到Razor View项，它可以在ASP.NET Core | Web | ASP.NET部分找到，如图2-20所示。将新文件的名称设置为MyView.cshtml，并单击Add按钮。Visual Studio会向Views/Home文件夹添加一个名为MyView.cshtml的文件，并打开它进行编辑。将文件的内容替换为代码清单2-7所示的内容。

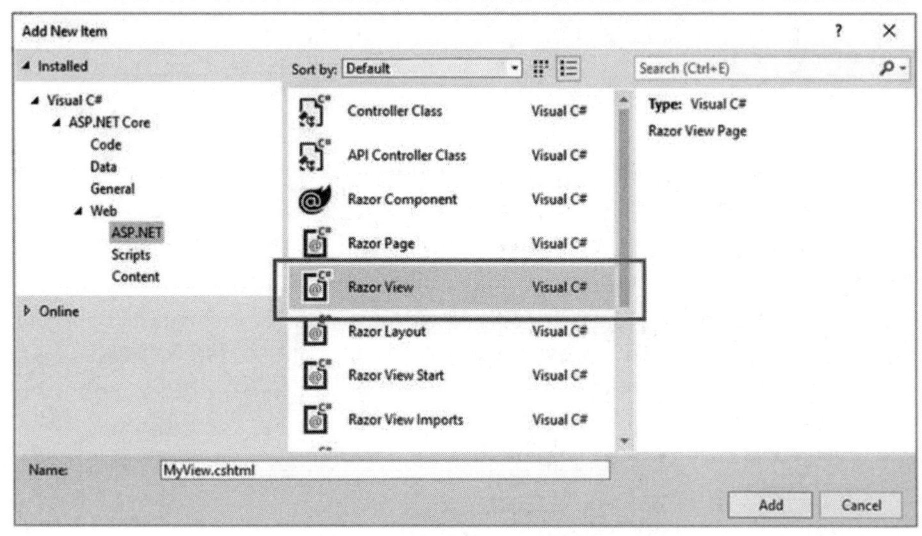


图2-20　选择Visual Studio项模板

Visual Studio Code不提供项模板。相反，在文件资源管理器窗格中右击Views/Home文件夹，并从弹出菜单中选择New File。将文件名设置为MyView.cshtml，然后按Enter键。该文件就会创建并打开，以进行编辑。添加如代码清单2-7所示的内容。

■ **提示**：很容易在错误的文件夹中创建视图文件。如果Views/Home文件夹中没有一个叫MyView.cshtml的文件，就把文件拖到正确的文件夹，或删除文件并重试。

**代码清单2-7　Views/Home文件夹中MyView.cshtml文件的内容**

```
@{
    Layout = null;
}

<!DOCTYPE html>

<html>
<head>
    <meta name="viewport" content="width=device-width" />
```

```
        <title>Index</title>
    </head>
    <body>
        <div>
            Hello World (from the view)
        </div>
    </body>
</html>
```

这个视图文件的新内容主要是 HTML，只有以下一小段除外：

```
...
@{
    Layout = null;
}
...
```

这是一个将由 Razor 解释的表达式，Razor 是处理视图内容并生成发送到浏览器的 HTML 的组件。Razor 是一个视图引擎，视图中的表达式称为 Razor 表达式。

代码清单 2-7 中的 Razor 表达式告诉 Razor：这里没有选用类似于 HTML 模板的布局，它将被发送到浏览器(详见第 22 章)。要查看创建视图的效果，请停止 ASP.NET Core(如果它正在运行)，并从 Debug 菜单中选择 Start Without Debugging (对于 Visual Studio)或 Run Without Debugging (对于 Visual Studio Code)。这会打开一个新的浏览器窗口，并生成如图 2-21 所示的响应。

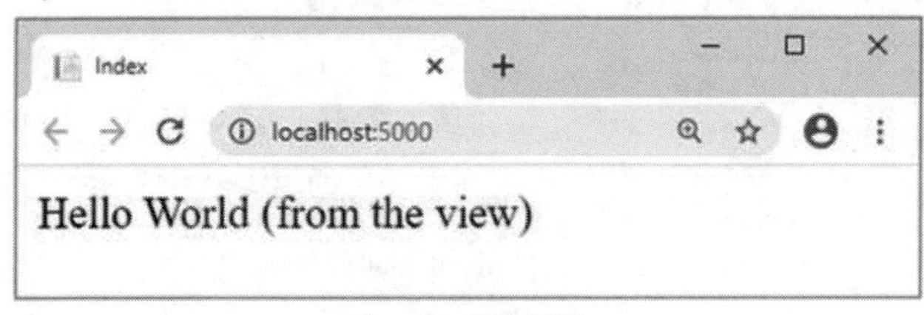


图 2-21　渲染视图

第一次编辑 Index 操作方法时，它返回一个字符串值。这意味着 ASP.NET Core 只是按原样将字符串值传递给浏览器。既然 Index 方法返回了 ViewResult，就使用 Razor 处理视图并呈现 HTML 响应。Razor 之所以能够定位视图，是因为遵循了标准的命名约定，即将视图文件放在一个名称与包含操作方法的控制器相匹配的文件夹中。在本例中，这意味着将视图文件放在 Views/Home 文件夹中，因为操作方法是由 Home 控制器定义的。

除了字符串和 ViewResult 对象外，还可以从操作方法返回其他结果。例如，如果返回一个 RedirectResult，浏览器将重定向到另一个 URL。如果返回一个 HttpUnauthorizedResult，就可以提示用户登录。这些对象统称为操作结果。操作结果系统允许在操作中封装和重用公共响应。第 19 章会介绍更多关于它们的信息，并解释它们的不同用法。

### 2. 添加动态输出

Web 应用程序的全部目的就是构造和显示动态输出。操作方法的任务是构造数据并将其传递给视图，以便使用它根据数据值创建 HTML 内容。操作方法通过向 View 方法传递参数来向视图提供数据，如代码清单 2-8 所示。提供给视图的数据称为视图模型。

**代码清单 2-8　在 Controllers 文件夹的 HomeController.cs 文件中提供一个视图模型**

```
using Microsoft.AspNetCore.Mvc;
using System;

namespace FirstProject.Controllers {

    public class HomeController : Controller {

        public ViewResult Index() {
            int hour = DateTime.Now.Hour;
            string viewModel = hour < 12 ? "Good Morning" : "Good Afternoon";
            return View("MyView", viewModel);
        }
    }
}
```

本例中的视图模型是一个字符串，它作为 View 方法的第二个参数提供给视图。代码清单 2-9 更新了视图，以便接收和使用它生成的 HTML 中的视图模型。

**代码清单 2-9　在 Views/Home 文件夹的 MyView.cshtml 文件中使用视图模型**

```
@model string
@{
    Layout = null;
}

<!DOCTYPE html>

<html>
<head>
    <meta name="viewport" content="width=device-width" />
    <title>Index</title>
</head>
<body>
    <div>
        @Model World (from the view)
    </div>
</body>
</html>
```

视图模型的类型使用@model(小写字母 m)表达式指定。视图模型值使用@Model(大写字母 M)表达式包含在 HTML 输出中。起初很难记住哪个用小写字母，哪个用大写字母，但很快就会成为第二天性)。

在呈现视图时，操作方法提供的视图模型数据将插入 HTML 响应中。选择 Start Without Debugging(使用 Visual Studio)或 Run Without Debugging(使用 Visual Studio Code)，会显示如图 2-22 所示的输出(如果在中午之后使用这个示例，可能会看到下午的问候语)。

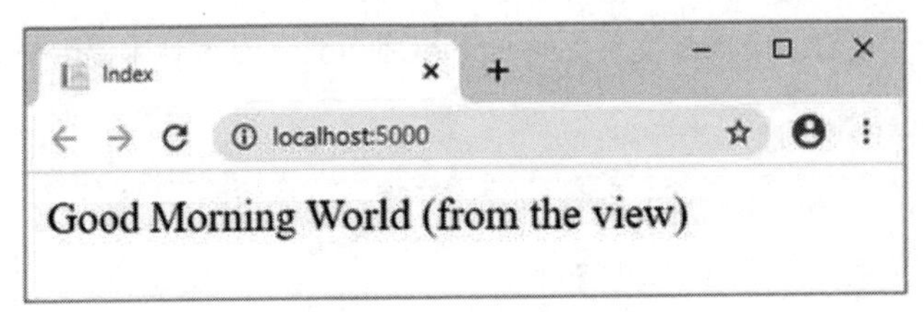


图 2-22　生成动态内容

### 2.3.4　内容综述

这是一个简单结果，但该示例揭示了创建简单 ASP.NET Core Web 应用程序并生成动态响应需要的所有构建块。ASP.NET Core 平台接收 HTTP 请求，并使用路由系统将请求 URL 匹配到一个端点。在本例中，端点是主控制器定义的 Index 操作方法。该方法被调用并生成一个 ViewResult 对象，该对象包含视图的名称和视图模型对象。Razor 视图引擎定位并处理视图，计算@Model 表达式，以将操作方法提供的数据插入响应中，响应返回给浏览器，并显示给用户。当然，还有其他许多可用的特性，但这是 ASP.NET Core 的本质，在阅读本书的其余部分时，有必要记住这个简单顺序。

## 2.4　小结

本章解释了如何通过安装 Visual Studio 或 Visual Studio Code 和.NET Core SDK，来设置 ASP.NET Core 开发。展示了如何创建简单的项目，简要说明了端点、视图和 URL 路由系统如何协同工作。第 3 章将展示如何创建简单的数据输入应用程序。

# 第3章

# 第一个ASP.NET Core应用程序

现在已经设置好了 ASP.NET Core 开发，是时候创建一个简单的应用程序了。本章将创建一个使用 ASP.NET Core 的数据输入应用程序，旨在演示 ASP.NET Core。因此，本章将加快速度，跳过一些关于幕后工作原理的解释。但别担心，后续章节将深入讨论这些主题。

## 3.1 设置场景

设想一下，一位朋友决定举办一场新年晚会，她让我创建一个 Web 应用程序，让受邀者通过电子方式回复。她提出以下四个关键特征：

- 显示聚会信息的主页
- 可用于 RSVP 的表单
- 确认 RSVP 表单，它将显示一个感谢页面
- 显示谁要来参加聚会的汇总页面

本章将创建一个 ASP.NET Core 项目，并使用它创建一个简单的应用程序，其中包含这些特征；一旦一切就绪，就应用一些样式，来改进完成的应用程序的外观。

## 3.2 创建项目

从 Windows“开始”菜单中打开 PowerShell 命令提示符，导航到一个方便的位置，运行代码清单 3-1 中的命令来创建名为 PartyInvites 的项目。

**提示：**可以从 https://github.com/apress/pro-asp.net-core-3 下载本章和本书中其他所有章节的示例项目。如果在运行示例时遇到问题，请参阅第 1 章以获得帮助。

**代码清单 3-1 创建新项目**

```
dotnet new globaljson --sdk-version 3.1.101 --output PartyInvites
dotnet new mvc --no-https --output PartyInvites --framework netcoreapp3.1
```

这些命令与第 2 章创建项目时使用的命令相同。对于 Visual Studio 用户，第 4 章将解释如何使用向导来创建项目，但是这些命令很简单，可以确保使用所需的.NET Core 版本，获得正确的

项目起点。

打开项目并编辑Controllers文件夹中的HomeController.cs文件，用代码清单3-2所示的代码替换其内容。

**代码清单3-2　替换Controllers文件夹的HomeControllers.cs文件中的内容**

```
using Microsoft.AspNetCore.Mvc;

namespace PartyInvites.Controllers {
    public class HomeController : Controller {

        public IActionResult Index() {
            return View();
        }
    }
}
```

它为新应用程序提供了一个清晰的起点，定义了一个操作方法来选择要呈现的默认视图。要向受邀者提供欢迎信息，请打开Views/Home文件夹下的Index.cshtml文件，并将其替换为代码清单3-3所示的内容。

**代码清单3-3　替换Views/Home文件夹下Index.cshtml文件的内容**

```
@{
    Layout = null;
}

<!DOCTYPE html>
<html>
<head>
    <meta name="viewport" content="width=device-width" />
    <title>Party!</title>
</head>
<body>
    <div>
        <div>
            We're going to have an exciting party.<br />
            (To do: sell it better. Add pictures or something.)
        </div>
    </div>
</body>
</html>
```

通过选择Start Without Debugging(对于Visual Studio)或Run Without Debugging (对于Visual Studio Code)来启动应用程序，就会看到聚会的详细信息(好吧，是细节的占位符，但是其想法非常明确)，如图3-1所示。

图 3-1　添加到视图 HTML

### 3.2.1 添加数据模型

数据模型是任何 ASP.NET Core 应用程序中最重要的部分。模型是定义应用程序主题(称为域)的真实对象、流程和规则的表示。模型通常称为域模型，包含 C#对象(称为域对象)，它们组成了应用程序的整体以及操作它们的方法。在大多数项目中，ASP.NET Core 应用程序的工作是为用户提供对数据模型的访问，并允许用户与之交互。

在 ASP.NET Core 应用程序的约定中，数据模型类在名为 Models 的文件夹中定义，该文件夹由代码清单 3-1 中使用的模板添加到项目中。

PartyInvites 项目不需要复杂的模型，因为它是非常简单的应用程序。实际上，只需要一个称为 GuestResponse 的域类。此对象表示来自受邀者的 RSVP。

如果使用的是 Visual Studio，则右击 Models 文件夹，然后从弹出菜单中选择 Add | Class。把类名设置为 GuestResponse .cs，然后单击 Add 按钮。如果使用的是 Visual Studio Code，则右击 Models 文件夹，选择 New File，并输入 GuestResponse.cs 作为文件名。使用新文件定义如代码清单 3-4 所示的类。

**代码清单 3-4　Models 文件夹中 GuestResponse.cs 文件的内容**

```
namespace PartyInvites.Models {

    public class GuestResponse {

        public string Name { get; set; }
        public string Email { get; set; }
        public string Phone { get; set; }
        public bool? WillAttend { get; set; }
    }
}
```

■ **提示：** WillAttend 属性是一个可空的 bool，这意味着它可以是真、假或空。本章后面的 3.2.8 节将解释这样做的基本原理。

### 3.2.2 创建第二个操作和视图

应用程序的目标之一是包含 RSVP 表单，这意味着需要定义一个操作方法来接收该表单的请求。一个控制器类可以定义多个操作方法，约定是将相关的操作分组在同一个控制器中。代码清单 3-5 向主控制器添加了一个新的操作方法。

**代码清单 3-5　在 Controllers 文件夹的 HomeController.cs 文件中添加一个操作方法**

```
using Microsoft.AspNetCore.Mvc;

namespace PartyInvites.Controllers {
    public class HomeController : Controller {

        public IActionResult Index() {
            return View();
        }

        public ViewResult RsvpForm() {
            return View();
        }
    }
}
```

这两种操作方法都调用了无参数的 View 方法，这似乎有些奇怪，但请记住，Razor 视图引擎在查找视图文件时使用了操作方法的名称。这意味着来自 Index 操作方法的结果告诉 Razor 寻找一个名为 Index.cshtml 的视图。而 RsvpForm 操作方法的结果告诉 Razor 寻找一个名为 RsvpForm.cshtml 的视图。

如果使用的是 Visual Studio，右击 Views/Home 文件夹，然后从弹出菜单中选择 Add | New Item。选择 Razor View 项，设置名称为 RsvpForm.cshtml，然后单击 Add 按钮创建文件。将内容替换为代码清单 3-6 所示的内容。

如果使用 Visual Studio Code，右击 Views/Home 文件夹，从弹出菜单中选择 New File。将文件名称设置为 RsvpForm.cshtml，并添加如代码清单 3-6 所示的内容。

**代码清单 3-6　Views/Home 文件夹中 RsvpForm.cshtml 文件的内容**

```
@{
    Layout = null;
}

<!DOCTYPE html>

<html>
<head>
    <meta name="viewport" content="width=device-width" />
    <title>RsvpForm</title>
</head>
<body>
    <div>
        This is the RsvpForm.cshtml View
    </div>
</body>
</html>
```

此内容目前只是静态HTML。要测试新的操作方法和视图，请从Debug菜单中选择Start Without Debugging或Run Without Debugging来启动应用程序。

使用打开的浏览器窗口，请求 http://localhost:5000/home/rsvpform(如果使用的是 Visual Studio，则必须将端口更改为创建项目时分配的端口)。Razor 视图引擎定位 RsvpForm.cshtml 文件，并使用它生成响应，如图 3-2 所示。

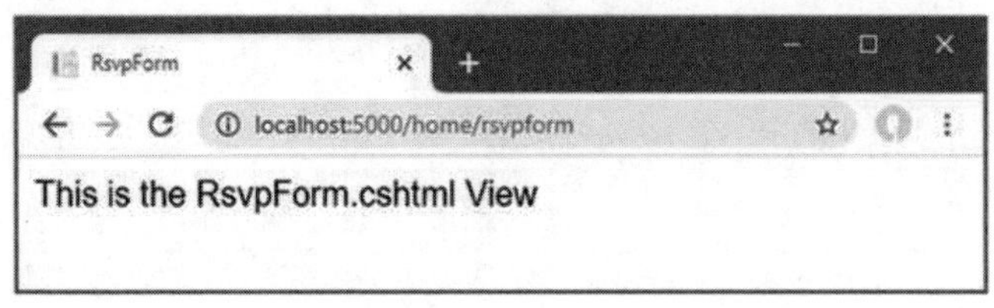


图 3-2　呈现第二个视图

## 3.2.3　连接操作方法

本例希望能够从 Index 视图创建一个链接，以便客户可以看到 RsvpForm 视图，而不必知道针对特定操作方法的 URL，如代码清单 3-7 所示。

**代码清单 3-7　在 Views/Home 文件夹的 Index.cshtml 文件中添加链接**

```
@{
    Layout = null;
}

<!DOCTYPE html>
<html>
<head>
    <meta name="viewport" content="width=device-width" />
    <title>Party!</title>
</head>
<body>
    <div>
        <div>
            We're going to have an exciting party.<br />
            (To do: sell it better. Add pictures or something.)
        </div>
        <a asp-action="RsvpForm">RSVP Now</a>
    </div>
</body>
</html>
```

添加到代码清单中的是一个具有 asp-action 属性的 a 元素。属性是标签助手属性的一个示例，是 Razor 的指令，在呈现视图时执行。asp-action 属性是向包含操作方法 URL 的 a 元素添加 href 属性的指令。第 25～27 章将解释标签助手是如何工作的，但是这个标签助手告诉 Razor 为呈现当前视图的控制器定义的操作方法插入一个 URL。可以看到助手通过运行项目创建的链接，如图 3-3 所示。

图 3-3 操作方法之间的链接

将鼠标滚动到浏览器中的 RSVP Now 链接上。链接指向以下 URL(考虑到 Visual Studio 分配给项目的不同端口号)：

```
http://localhost:5000/Home/RsvpForm
```

这里有一个重要原则，即应该使用 ASP.NET Core 提供的特性生成 URL，而不是把它们硬编码到视图中。当标签助手为 a 元素创建 href 属性时，它检查应用程序的配置，以确定 URL 应该是什么。这允许更改应用程序的配置，以支持不同的 URL 格式，而不必更新任何视图。

### 3.2.4 构建表单

前面创建了视图，并可以从 Index 视图中访问它，下面构建 RsvpForm.cshtml 文件的内容，将其转换为编辑 GuestResponse 对象的 HTML 表单，如代码清单 3-8 所示。

**代码清单 3-8 在 Views/Home 文件夹的 RsvpForm.cshtml 文件中创建表单视图**

```
@model PartyInvites.Models.GuestResponse
@{
    Layout = null;
}

<!DOCTYPE html>

<html>
<head>
    <meta name="viewport" content="width=device-width" />
    <title>RsvpForm</title>
</head>
<body>
    <form asp-action="RsvpForm" method="post">
        <div>
            <label asp-for="Name">Your name:</label>
            <input asp-for="Name" />
        </div>
        <div>
            <label asp-for="Email">Your email:</label>
            <input asp-for="Email" />
        </div>
        <div>
            <label asp-for="Phone">Your phone:</label>
```

```
            <input asp-for="Phone" />
        </div>
        <div>
            <label>Will you attend?</label>
            <select asp-for="WillAttend">
                <option value="">Choose an option</option>
                <option value="true">Yes, I'll be there</option>
                <option value="false">No, I can't come</option>
            </select>
        </div>
        <button type="submit">Submit RSVP</button>
    </form>
</body>
</html>
```

@model表达式指定视图期望接收到一个GuestResponse对象作为它的视图模型。前面为GuestResponse模型类的每个属性定义了一个标签和输入元素(或者，对于WillAttend属性，是一个选择元素)。每个元素都使用asp-for属性与模型属性相关联，这是另一个标签助手属性。标签助手属性配置元素，以将它们绑定到视图模型对象。下面是生成的一个HTML例子：

```
<p>
    <label for="Name">Your name:</label>
    <input type="text" id="Name" name="Name" value="">
</p>
```

label 元素上的 asp-for 属性设置 for 属性的值。输入元素上的 asp-for 属性设置 id 和 name 元素。这看起来可能不是特别有用，但在定义应用程序功能时，将元素与模型属性关联提供了额外的优势。

更直接的用途是应用于表单元素的 asp-action 属性，它使用应用程序的 URL 路由配置来设置 action 属性为一个 URL，这个 URL 指向一个特定的 action 方法，如下所示：

```
<form method="post" action="/Home/RsvpForm">
```

与应用于 a 元素的助手属性一样，这种方法的好处是可以更改应用程序使用的 URL 系统，由标签助手生成的内容将自动反映更改。

运行应用程序并单击 RSVP Now 链接，就可以看到这个表单，如图 3-4 所示。

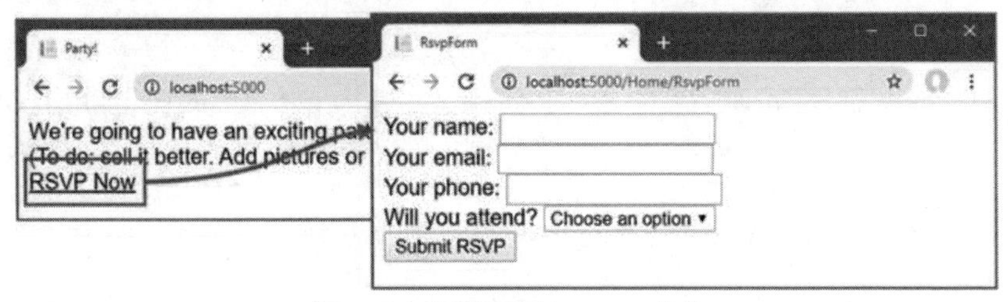


图 3-4　向应用程序添加 HTML 表单

## 3.2.5　接收表单数据

现在还没有告诉 ASP.NET Core，当表单发布到服务器时要做什么。目前，单击 Submit RSVP 按钮只会清除在表单中输入的任何值。这是因为表单返回到主控制器中的 RsvpForm 操作方法，后者只是再次呈现视图。为了接收和处理提交的表单数据，需要使用控制器的一个重要特性。添加第二个 RsvpForm 操作方法，来创建以下内容。

- 响应 HTTP GET 请求的方法：GET 请求通常是浏览器在每次单击链接时发出的。这个版本的操作将负责在有人第一次访问/Home/RsvpForm 时显示初始空白表单。
- 响应 HTTP POST 请求的方法：默认情况下，使用 Html.BeginForm()呈现的表单由浏览器作为 POST 请求提交。这个版本的操作将负责接收提交的数据，并决定如何处理它。

用不同的 C#方法处理 GET 和 POST 请求有助于保持控制器代码的整洁，因为这两个方法有不同的职责。这两个操作方法都是由同一个 URL 调用的。ASP.NET Core 根据处理的是 GET 还是 POST 请求，确保调用了适当的方法。代码清单 3-9 显示了对 HomeController 类的更改。

**代码清单 3-9　在 Controllers 文件夹的 HomeController.cs 文件中添加一个方法**

```
using Microsoft.AspNetCore.Mvc;
using PartyInvites.Models;

namespace PartyInvites.Controllers {
    public class HomeController : Controller {

        public IActionResult Index() {
            return View();
        }

        [HttpGet]
        public ViewResult RsvpForm() {
            return View();
        }
        [HttpPost]
        public ViewResult RsvpForm(GuestResponse guestResponse) {
            // TODO: store response from guest
            return View();
        }
    }
}
```

前面将 HttpGet 属性添加到现有的 RsvpForm 操作方法中，该方法声明它只能用于 GET 请求。然后添加了一个重载的 RsvpForm 方法，它接收一个 GuestResponse 对象。将 HttpPost 属性应用到这个方法，它声明新方法将处理 POST 请求。下面将解释这些添加到代码清单中的内容是如何工作的。还导入了 PartyInvites.Models 名称空间——这样就可以引用 GuestResponse 模型类型而不需要限定类名。

### 1. 理解模型绑定

RsvpForm 操作方法的第一个重载会呈现与之前相同的视图——RsvpForm.cshtml 文件，生成如图 3-4 所示的表单。由于参数的关系，第二个重载更有趣，但是考虑到调用操作方法来响应 HTTP POST 请求，并且 GuestResponse 类型是一个 C#类，这两者是如何连接的?

答案是模型绑定，一个有用的 ASP.NET Core 特性，它解析传入数据，使用 HTTP 请求中的键/值对来填充域模型类型的属性。

模型绑定是一个功能强大且可定制的特性，它消除了直接处理 HTTP 请求的繁杂工作，并允许使用 C#对象，而不是处理浏览器发送的单个数据值。GuestResponse 对象作为参数传递给操作方法，自动填充表单字段中的数据。第 28 章将深入讨论模型绑定的细节。

为了演示模型绑定如何工作，需要做一些准备工作。应用程序的目标之一是显示一个汇总页面，其中包含谁参加了聚会的详细信息，这意味着需要跟踪收到的响应。为此需要创建一个内存中的对象集合。这在真实的应用程序中没有用，因为当应用程序停止或重新启动时，响应数据将丢失，但这种方法将允许把重点放在 ASP.NET Core 上，并创建一个可以轻松重置到初始状态的应用程序。后续章节将演示持久数据存储。

在 Models 文件夹中添加一个名为 Repository.cs 的类文件，并使用它定义如代码清单 3-10 所示的类。

**代码清单 3-10　Models 文件夹中 Repository.cs 文件的内容**

```
using System.Collections.Generic;

namespace PartyInvites.Models {
   public static class Repository {
      private static List<GuestResponse> responses = new List<GuestResponse>();

      public static IEnumerable<GuestResponse> Responses => responses;

      public static void AddResponse(GuestResponse response) {
         responses.Add(response);
      }
   }
}
```

Repository 类及其成员是静态的，因此可很容易地在应用程序的不同位置存储和检索数据。ASP.NET Core 提供一种更复杂的方法来定义常见功能，称为依赖注入(dependency injection)，详见第 14 章，但是对于像这样的简单应用程序来说，静态类是一个很好的入门方法。

### 2. 存储响应

现在，有了存储数据的地方，就可以更新接收 HTTP POST 请求的操作方法，如代码清单 3-11 所示。

**代码清单 3-11　更新 Controllers 文件夹的 HomeController.cs 文件中的操作方法**

```
using Microsoft.AspNetCore.Mvc;
using PartyInvites.Models;
```

```
namespace PartyInvites.Controllers {
    public class HomeController : Controller {

        public IActionResult Index() {
            return View();
        }

        [HttpGet]
        public ViewResult RsvpForm() {
            return View();
        }

        [HttpPost]
        public ViewResult RsvpForm(GuestResponse guestResponse) {
            Repository.AddResponse(guestResponse);
            return View("Thanks", guestResponse);
        }
    }
}
```

在调用RsvpForm方法的POST版本之前，ASP.NET Core模型绑定特性从HTML表单中提取值，并将它们分配给GuestResponse对象的属性。调用这个方法时，结果作为参数用来处理HTTP请求；要处理请求中的表单数据，只需要处理传递给操作方法的GuestResponse对象。在本例中，将它作为参数传递给Repository.AddResponse方法，以便存储响应。

### 3.2.6 添加Thanks视图

在RsvpForm操作方法中对View方法的调用创建了一个ViewResult，它选择名为Thanks的视图，并使用由模型绑定器创建的GuestResponse对象作为视图模型。将一个名为Thanks.cshtml的Razor视图添加到Views/Home文件夹，其中的内容如代码清单3-12所示，以向用户显示响应。

**代码清单3-12 Views/Home文件夹中Thanks.cshtml文件的内容**

```
@model PartyInvites.Models.GuestResponse
@{
    Layout = null;
}

<!DOCTYPE html>

<html>
<head>
    <meta name="viewport" content="width=device-width" />
    <title>Thanks</title>
</head>
<body>
```

```
    <div>
        <h1>Thank you, @Model.Name!</h1>
        @if (Model.WillAttend == true) {
            @:It's great that you're coming. The drinks are already in the fridge!
        } else {
            @:Sorry to hear that you can't make it, but thanks for letting us know.
        }
    </div>
    Click <a asp-action="ListResponses">here</a> to see who is coming.
</body>
</html>
```

Thanks.cshtml 视图所生成的 HTML 依赖于分配给 GuestResponse 视图模型的值(由 RsvpForm 操作方法提供)。为访问域对象中的属性值，使用了@Model.<*PropertyName*>表达式。例如，为了获得 Name 属性的值，使用了@Model.Name 表达式。如果 Razor 语法没有任何意义，请不要担心——详见第 21 章。

现在，已经创建了 Thanks 视图，有了一个处理表单的基本工作示例。启动应用程序，单击 RSVP Now 链接，向表单添加一些数据，然后单击 Submit RSVP 按钮。就会显示如图 3-5 所示的响应(如果名字不是 Joe 或客人不能出席，将有所不同)。

图 3-5 Thanks 视图

## 3.2.7 显示响应

在 Thanks.cshtml 视图的最后，添加了一个 a 元素来创建链接，以显示即将参加聚会的人员列表。使用 asp-action 标签助手属性创建 URL，目标是一个名为 ListResponses 的操作方法，如下所示：

```
...
<div>Click <a asp-action="ListResponses">here</a> to see who is coming.</div>
...
```

如果将鼠标悬停在浏览器显示的链接上，它的目标是/Home/ListResponses URL。这与主控制器中的任何操作方法都不对应，如果单击该链接，会显示 404 Not Found error 响应。

要添加处理 URL 的端点，需要向主控制器添加另一个操作方法，如代码清单 3-13 所示。

#### 代码清单 3-13 在 Controllers 文件夹的 HomeController.cs 文件中添加一个操作方法

```
using Microsoft.AspNetCore.Mvc;
using PartyInvites.Models;
```

```
using System.Linq;

namespace PartyInvites.Controllers {
    public class HomeController : Controller {

        public IActionResult Index() {
            return View();
        }
        [HttpGet]
        public ViewResult RsvpForm() {
            return View();
        }

        [HttpPost]
        public ViewResult RsvpForm(GuestResponse guestResponse) {
            Repository.AddResponse(guestResponse);
            return View("Thanks", guestResponse);
        }

        public ViewResult ListResponses() {
            return View(Repository.Responses.Where(r => r.WillAttend == true));
        }
    }
}
```

新的操作方法称为 ListResponses，它使用 Repository.Responses 属性作为参数，调用 View 方法，这将导致 Razor 呈现默认视图，使用操作方法名作为视图文件的名称，并使用存储库中的数据作为视图模型。视图模型数据使用 LINQ 进行过滤，因此只将积极的响应提供给视图。

在 Views/Home 文件夹中添加一个名为 ListResponses.cshtml 的 Razor 视图，内容如代码清单 3-14 所示。

#### 代码清单 3-14　在 Views/Home 文件夹的 ListResponses.cshtml 文件中显示接收邀请的人员

```
@model IEnumerable<PartyInvites.Models.GuestResponse>
@{
    Layout = null;
}

<!DOCTYPE html>

<html>
<head>
    <meta name="viewport" content="width=device-width" />
    <title>Responses</title>
</head>
<body>
    <h2>Here is the list of people attending the party</h2>
```

```
    <table>
        <thead>
            <tr><th>Name</th><th>Email</th><th>Phone</th></tr>
        </thead>
        <tbody>
            @foreach (PartyInvites.Models.GuestResponse r in Model) {
                <tr>
                    <td>@r.Name</td>
                    <td>@r.Email</td>
                    <td>@r.Phone</td>
                </tr>
            }
        </tbody>
    </table>
</body>
</html>
```

Razor视图文件拥有.cshtml文件扩展名，因为它们混合了C#代码和HTML元素，如代码清单 3-14 所示，其中使用@foreach表达式处理操作方法通过view方法传递给视图的每个GuestResponse对象。与普通的C# foreach循环不同，Razor @foreach表达式的主体包含HTML元素，该元素添加到响应中，发送回浏览器。在这个视图中，每个GuestResponse对象生成一个tr元素，该元素包含用对象属性的值填充的td元素。

启动应用程序，提交一些表单数据，然后单击链接，查看响应列表。可以看到自应用程序启动以来输入的数据汇总，如图 3-6 所示。视图并没有以吸引人的方式显示数据，但目前已经足够了，本章后面将讨论应用程序的样式。

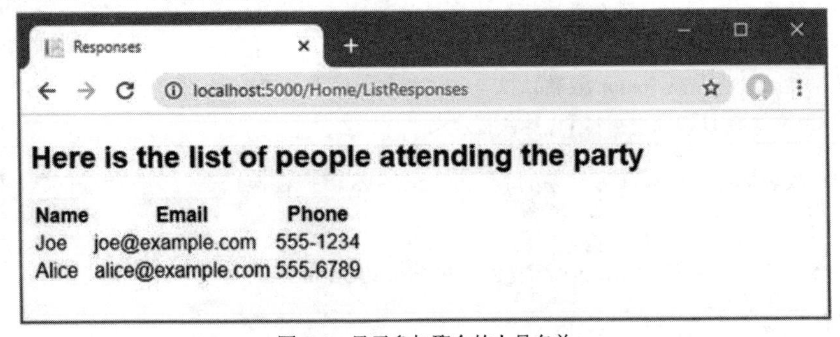


图 3-6　显示参加聚会的人员名单

## 3.2.8　添加验证

现在可以向应用程序添加数据验证。没有验证，用户可以输入无意义的数据，甚至提交空表单。在ASP.NET Core中，验证规则是通过向模型类应用属性来定义的，这意味着相同的验证规则可以应用于使用该类的任何表单。ASP.NET Core依赖于System.ComponentModel. DataAnnotations名称空间中的属性。在代码清单 3-15 中，将其应用于GuestResponse类。

**代码清单 3-15　在 Models 文件夹的 GuestResponse.cs 文件中应用验证**

```
using System.ComponentModel.DataAnnotations;

namespace PartyInvites.Models {

    public class GuestResponse {

        [Required(ErrorMessage = "Please enter your name")]
        public string Name { get; set; }

        [Required(ErrorMessage = "Please enter your email address")]
        [EmailAddress]
        public string Email { get; set; }

        [Required(ErrorMessage = "Please enter your phone number")]
        public string Phone { get; set; }

        [Required(ErrorMessage = "Please specify whether you'll attend")]
        public bool? WillAttend { get; set; }
    }
}
```

ASP.NET Core 检测属性，并在模型绑定过程中使用它们验证数据。

---

**提示：**如前所述，为 WillAttend 属性使用了一个可空的 bool。这样做是为了应用 Required 验证属性。如果使用常规 bool，那么通过模型绑定接收到的值只能是 true 或 false，且无法判断用户是否选择了一个值。可空的 bool 有三个可能的值：true、false 和 null。如果用户没有选择值，浏览器将发送空值，这将导致 Required 属性报告验证错误。这是一个很好的例子，说明 ASP.NET Core 优雅地将 C#特性与 HTML 和 HTTP 融合在一起。

---

使用接收表单数据的操作方法中的 ModelState.IsValid 属性检查是否存在验证问题，如代码清单 3-16 所示。

**代码清单 3-16　检查 Controllers 文件夹中 HomeController.cs 文件中的验证错误**

```
using Microsoft.AspNetCore.Mvc;
using PartyInvites.Models;
using System.Linq;

namespace PartyInvites.Controllers {
    public class HomeController : Controller {

        public IActionResult Index() {
            return View();
        }

        [HttpGet]
```

```
        public ViewResult RsvpForm() {
            return View();
        }

        [HttpPost]
        public ViewResult RsvpForm(GuestResponse guestResponse) {
            if (ModelState.IsValid) {
                Repository.AddResponse(guestResponse);
                return View("Thanks", guestResponse);
            } else {
                return View();
            }
        }

        public ViewResult ListResponses() {
            return View(Repository.Responses.Where(r => r.WillAttend == true));
        }
    }
}
```

控制器基类提供了一个名为 ModelState 的属性，它提供了模型绑定过程结果的详细信息。如果 ModelState.IsValid 属性返回 true，就说明模型绑定器已能满足通过 GuestResponse 类上的属性指定的验证约束。当发生这种情况时，就将呈现 Thanks 视图，就像前面所做的那样。

如果 ModelState.IsValid 属性返回 false，就说明存在验证错误。ModelState 属性返回的对象提供了我们遇到的每个问题的细节，但不需要进入这一水平的细节，因为可以依赖于一个有用的功能：要求用户通过调用无参数的 View 方法解决任何问题的自动化过程。

在呈现视图时，Razor可访问与请求关联的任何验证错误的详细信息，而标签助手可以访问这些细节，向用户显示验证错误。代码清单3-17显示了添加到RsvpForm视图的验证标签助手属性。

**代码清单 3-17　在 Views/Home 文件夹的 RsvpForm.cshtml 文件中添加验证汇总**

```
@model PartyInvites.Models.GuestResponse
@{
    Layout = null;
}

<!DOCTYPE html>

<html>
<head>
    <meta name="viewport" content="width=device-width" />
    <title>RsvpForm</title>
</head>
<body>
    <form asp-action="RsvpForm" method="post">
        <div asp-validation-summary="All"></div>
        <div>
```

```
            <label asp-for="Name">Your name:</label>
            <input asp-for="Name" />
        </div>
        <div>
            <label asp-for="Email">Your email:</label>
            <input asp-for="Email" />
        </div>
        <div>
            <label asp-for="Phone">Your phone:</label>
            <input asp-for="Phone" />
        </div>
        <div>
            <label>Will you attend?</label>
            <select asp-for="WillAttend">
                <option value="">Choose an option</option>
                <option value="true">Yes, I'll be there</option>
                <option value="false">No, I can't come</option>
            </select>
        </div>
        <button type="submit">Submit RSVP</button>
    </form>
</body>
</html>
```

将asp-validation-summary属性应用于div元素，并在呈现视图时列出验证错误。asp-validation-summary属性的值来自一个名为ValidationSummary的枚举，它指定汇总将包含哪些类型的验证错误。这里指定了All，这对于大多数应用程序来说是一个很好的起点，第29章将描述其他值，并解释它们是如何工作的。

要查看验证汇总的工作方式，请运行应用程序，填写 Name 字段并提交表单，而不输入任何其他数据。图 3-7 显示了验证错误的汇总。

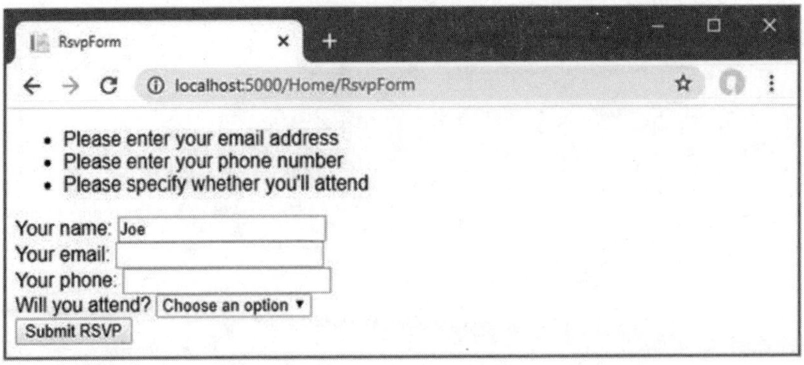

图 3-7　显示验证错误

直到满足了应用于 GuestResponse 类的所有验证约束，RsvpForm 操作方法才会呈现 Thanks 视图。注意，在 Razor 使用验证汇总呈现视图时，Name 字段中输入的数据会保留下来，并再次显示。这是模型绑定的另一个好处，它简化了对表单数据的处理。

### 高亮显示无效的字段

将模型属性与元素相关联的标签助手属性有一个方便的特性，可以与模型绑定一起使用。当模型类属性验证失败时，helper 属性将生成稍微不同的 HTML。以下是在没有验证错误时为 Phone 字段生成的输入元素：

```
<input type="text" data-val="true" data-val-required="Please enter your phone number"
id="Phone" name="Phone" value="">
```

为了比较，下面是相同的 HTML 元素，用户提交表单时没有在文本字段中输入数据(这是一个验证错误，因为 Required 属性应用到 GuestResponse 类的 Phone 属性)：

```
<input type="text" class="input-validation-error" data-val="true"
   data-val-required="Please enter your phone number" id="Phone"
   name="Phone" value="">
```

这里强调了它们的区别：asp-for标签助手属性将输入元素添加到一个名为input-validation-error的类中。可通过创建一个样式表来利用这个特性，其中包含这个类的CSS样式，以及不同HTML 助手属性使用的其他样式。

ASP.NET Core 项目中的约定是将静态内容发送给客户端，并将其放到 wwwroot 文件夹中，按照内容类型进行组织，CSS 样式表放到 wwwroot/css 文件夹中，JavaScript 文件放到 wwwroot/js 文件夹中，等等。

■ **提示**：当使用 Web 应用程序模板创建项目时，Visual Studio 会在 wwwroot/css 文件夹中创建一个 site.css 文件。可忽略这个文件，本章不使用它。

如果使用的是 Visual Studio，右击 wwwroot/css 文件夹，从弹出菜单中选择 Add | New Item。找到 Style Sheet 项模板，如图 3-8 所示；设置文件的名称为 styles.css；然后单击 Add 按钮。

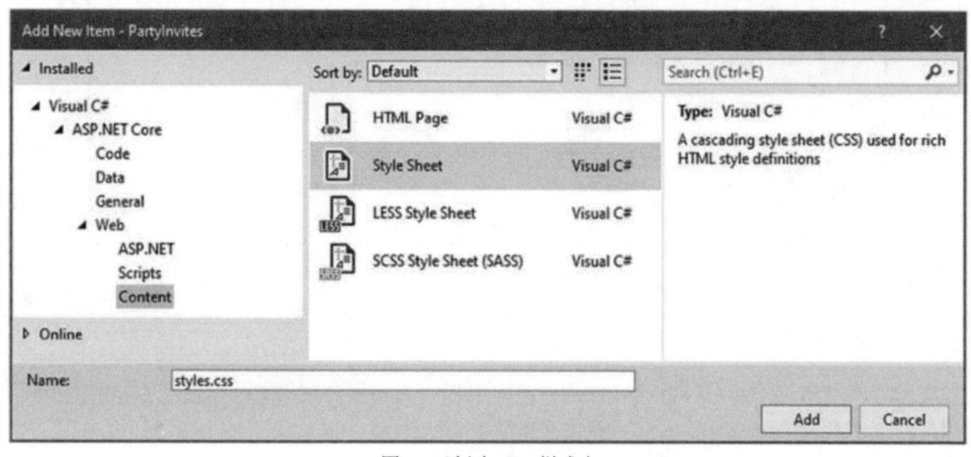


图 3-8 创建 CSS 样式表

如果使用 Visual Studio Code，右击 wwwroot/css 文件夹，从弹出菜单中选择 New File，然后使用 styles.css 作为文件名。无论使用哪种编辑器，都要使用代码清单 3-18 所示的样式替换文件内容。

**代码清单 3-18　wwwroot/css 文件夹中 styles.css 文件的内容**

```
.field-validation-error     {color: #f00;}
.field-validation-valid     { display: none;}
.input-validation-error     { border: 1px solid #f00; background-color: #fee; }
.validation-summary-errors  { font-weight: bold; color: #f00;}
.validation-summary-valid   { display: none;}
```

为了应用这个样式表，在 RsvpForm 视图的标题部分添加了一个 link 元素，如代码清单 3-19 所示。

**代码清单 3-19　在 Views/Home 文件夹的 RsvpForm.cshtml 文件中应用样式表**

```
...
<head>
    <meta name="viewport" content="width=device-width" />
    <title>RsvpForm</title>
    <link rel="stylesheet" href="/css/styles.css" />
</head>
...
```

link元素使用href属性指定样式表的位置。注意，URL中省略了wwwroot文件夹。ASP.NET的默认配置支持提供静态内容，如图像、CSS样式表和JavaScript文件，它自动将请求映射到wwwroot文件夹。应用样式表后，在提交导致验证错误的数据时，将更明显地显示验证错误，如图3-9所示。

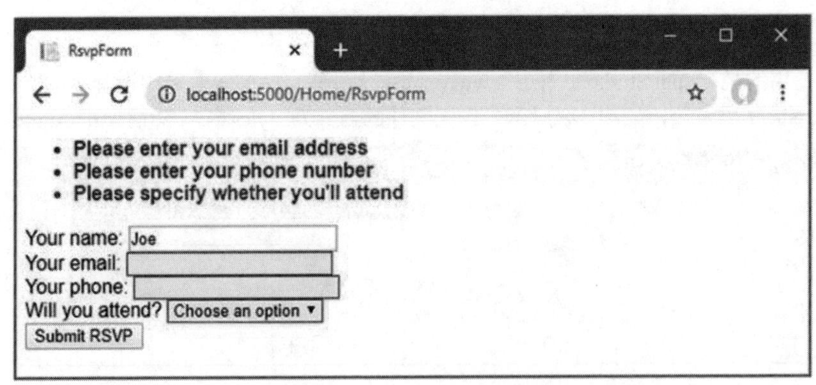


图 3-9　自动突出显示验证错误

## 3.2.9　内容的样式化

应用程序的所有功能目标都已经完成，但应用程序的整体外观很差。当使用 mvc 模板创建项目时，就像本章的示例那样，安装一些常见的客户端开发包。虽然我不喜欢使用模板项目，但喜欢微软选择的客户端库。其中一个叫做 Bootstrap，它是一个很好的 CSS 框架，最初由 Twitter 开发，现在已经成为一个主要的开源项目，成为 Web 应用程序开发的支柱。

### 1. 欢迎视图的样式化

Bootstrap的基本特性将类应用到与CSS选择器对应的元素上，这些元素定义再添加到wwwroot/lib/bootstrap文件夹的文件中。可以从http://getbootstrap.com获得Bootstrap定义的类的完整细节，还可以在代码清单3-20中看到如何将一些基本样式应用到Index.cshtml视图文件。

**代码清单 3-20　将 Bootstrap 添加到 Views/Home 文件夹的 Index.cshtml 文件**

```
@{
    Layout = null;
}

<!DOCTYPE html>

<html>
<head>
    <meta name="viewport" content="width=device-width" />
    <link rel="stylesheet" href="/lib/bootstrap/dist/css/bootstrap.css" />
    <title>Index</title>
</head>
<body>
    <div class="text-center">
        <h3> We're going to have an exciting party!</h3>
        <h4>And YOU are invited!</h4>
        <a class="btn btn-primary" asp-action="RsvpForm">RSVP Now</a>
    </div>
</body>
</html>
```

前面添加了一个链接元素，它的href属性从wwwroot/lib/bootstrap/dist/css文件夹加载bootstrap.css文件。约定是将第三方的CSS和JavaScript包安装到wwwroot/lib文件夹中，第12章将描述用于管理这些包的工具。

导入 Bootstrap 样式表后，需要对元素进行样式设置。这是一个简单示例，因此只需要使用少量 Bootstrap CSS 类：text-center、btn 和 btn-primary。

text-center 类将元素及其子元素的内容居中。btn 类将按钮、输入或元素样式化为美观的按钮，btn-primary 类指定按钮的颜色范围。运行应用程序可看到效果，如图 3-10 所示。

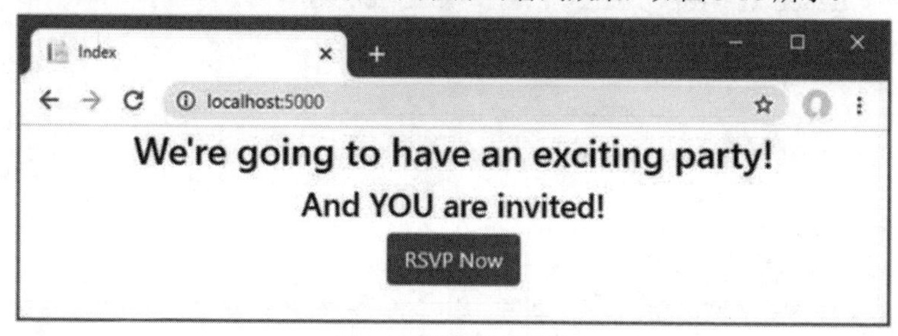


图 3-10　视图的样式化

很明显，我不是一位网页设计师。事实上，在我还是孩子的时候，就被允许不上艺术课，因为我完全没有天赋。这让我有了更多时间上数学课，但也意味着我的艺术技能并没有发展到超过 10 岁孩子的水平。对于真实的项目，我将寻求专业人员来帮助设计和指定内容的样式，但对于这个例子，我将独自完成，这意味着使用 Bootstrap 应尽可能克制，保持一致性。

### 2. 表单视图的样式化

Bootstrap 定义了可用于样式化表单的类。这里不打算详细说明，读者可在代码清单 3-21 中看到如何应用这些类。

**代码清单 3-21　向 Views/Home 文件夹中的 RsvpForm.cshtml 文件添加 Bootstrap**

```
@model PartyInvites.Models.GuestResponse
@{
    Layout = null;
}

<!DOCTYPE html>

<html>
<head>
    <meta name="viewport" content="width=device-width" />
    <title>RsvpForm</title>
    <link rel="stylesheet" href="/css/styles.css" />
    <link rel="stylesheet" href="/lib/bootstrap/dist/css/bootstrap.css" />
</head>
<body>
    <h5 class="bg-primary text-white text-center m-2 p-2">RSVP</h5>
    <form asp-action="RsvpForm" method="post" class="m-2">
        <div asp-validation-summary="All"></div>
        <div class="form-group">
            <label asp-for="Name">Your name:</label>
            <input asp-for="Name" class="form-control" />
        </div>
        <div class="form-group">
            <label asp-for="Email">Your email:</label>
            <input asp-for="Email" class="form-control" />
        </div>
        <div class="form-group">
            <label asp-for="Phone">Your phone:</label>
            <input asp-for="Phone" class="form-control" />
        </div>
        <div class="form-group">
            <label>Will you attend?</label>
            <select asp-for="WillAttend" class="form-control">
                <option value="">Choose an option</option>
                <option value="true">Yes, I'll be there</option>
                <option value="false">No, I can't come</option>
```

```
            </select>
        </div>
        <button type="submit" class="btn btn-primary">Submit RSVP</button>
    </form>
</body>
</html>
```

本例中的 Bootstrap 类创建了一个标题，只是为了给布局提供结构。为了样式化表单，使用了 form-group 类，它用于样式化包含标签和相关输入或选择元素的元素，这些元素被分配给 form-control 类。可以在图 3-11 中看到样式的效果。

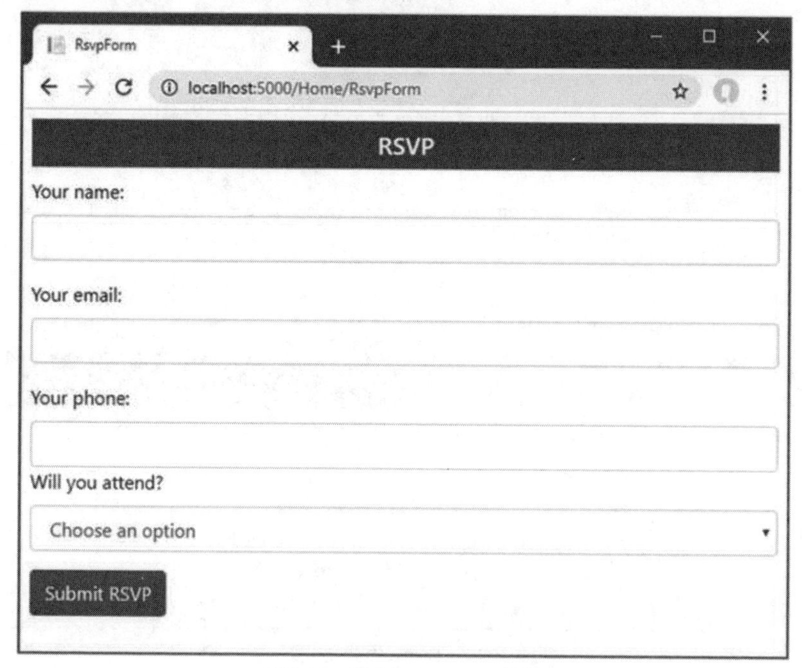


图 3-11　样式化 RsvpForm 视图

### 3. Thanks 视图的样式化

下一个要样式化的视图文件是 Thanks.cshtml。可以在代码清单 3-22 中看到如何做到这一点，其中使用的 CSS 类与用于其他视图的 CSS 类类似。为了使应用程序更易于管理，尽可能避免重复代码和标记是一个很好的原则。ASP.NET Core 提供了几个特性来帮助减少重复，详见后面的章节。这些特性包括 Razor 布局(第 23 章)、分部视图(第 22 章)和视图组件(第 24 章)。

**代码清单 3-22　对 Views/Home 文件夹中的 Thanks.cshtml 文件应用 Bootstrap**

```
@model PartyInvites.Models.GuestResponse
@{
    Layout = null;
}

<!DOCTYPE html>
```

```
<html>
<head>
    <meta name="viewport" content="width=device-width" />
    <title>Thanks</title>
    <link rel="stylesheet" href="/lib/bootstrap/dist/css/bootstrap.css" />
</head>
<body class="text-center">
    <div>
        <h1>Thank you, @Model.Name!</h1>
        @if (Model.WillAttend == true) {
            @:It's great that you're coming. The drinks are already in the fridge!
        } else {
            @:Sorry to hear that you can't make it, but thanks for letting us know.
        }
    </div>
    Click <a asp-action="ListResponses">here</a> to see who is coming.</div>
</body>
</html>
```

图 3-12 显示了样式的效果。

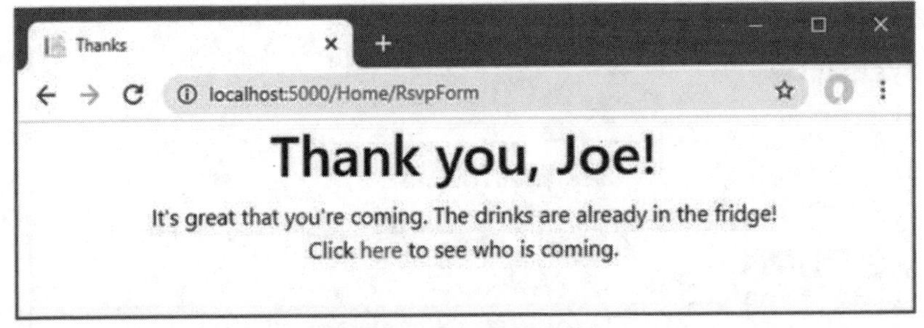


图 3-12　Thanks 视图的样式

#### 4. 列表视图的样式化

样式化的最后一个视图是 ListResponses，它列出参与者。内容的样式化遵循与其他视图相同的方法，如代码清单 3-23 所示。

**代码清单 3-23　向 Views/Home 文件夹中的 ListResponses.cshtml 文件添加 Bootstrap**

```
@model IEnumerable<PartyInvites.Models.GuestResponse>
@{
    Layout = null;
}

<!DOCTYPE html>

<html>
<head>
```

```
    <meta name="viewport" content="width=device-width" />
    <title>Responses</title>
    <link rel="stylesheet" href="/lib/bootstrap/dist/css/bootstrap.css" />
</head>
<body>
    <div class="text-center p-2">
        <h2>Here is the list of people attending the party</h2>
        <table class="table table-bordered table-striped table-sm">
            <thead>
                <tr><th>Name</th><th>Email</th><th>Phone</th></tr>
            </thead>
            <tbody>
                @foreach (PartyInvites.Models.GuestResponse r in Model) {
                    <tr>
                        <td>@r.Name</td>
                        <td>@r.Email</td>
                        <td>@r.Phone</td>
                    </tr>
                }
            </tbody>
        </table>
    </div>
</body>
</html>
```

图 3-13 显示了参与者表的呈现方式。将这些样式添加到视图中就完成了示例应用程序，该应用程序现在满足了所有开发目标，外观也得到改进。

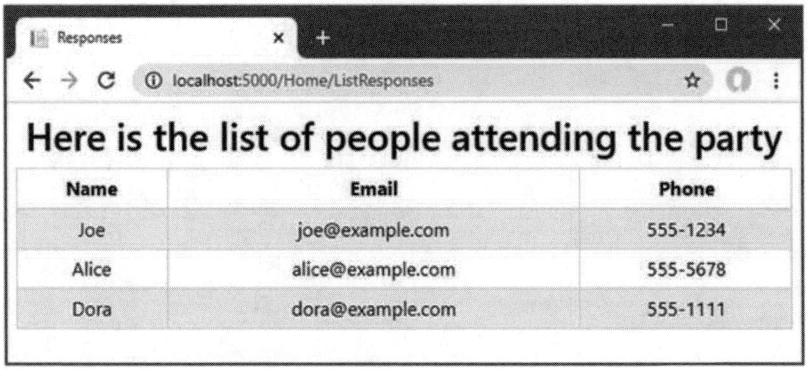

图 3-13 样式化 ListResponses 视图

## 3.3 小结

本章创建了一个新的 ASP.NET Core 项目，并用它构造一个简单的数据输入应用程序，初步论述了重要的 ASP.NET 特性，例如标签助手、模型绑定和数据验证。第 4 章将描述用于 ASP.NET Core 开发的开发工具。

# 第 4 章

# 使用开发工具

本章介绍微软为 ASP.NET Core 开发提供的工具，以及在本书中使用的工具。

与本书的早期版本不同，本书依赖.NET Core SDK 提供的命令行工具和微软发布的其他工具包。在某种程度上，这样做是为了帮助确保从示例中获得预期的结果，还因为命令行工具提供了对 ASP.NET Core 开发需要的所有特性的访问，而不管选择了哪个编辑器/IDE。

Visual Studio 以及Visual Studio Code 通过用户界面提供了对一些工具的访问，参见本章的描述，但是 Visual Studio 和 Visual Studio Code 并不支持 ASP.NET Core 开发需要的所有特性，因此有时使用命令行是不可避免的。

ASP.NET Core 已经发展了，本书逐渐过渡到只使用命令行工具，除了需要使用调试器的时候(尽管这是一种罕见的需求，如本章后面所述)。读者的首选项可能不同，特别是如果习惯完全在 IDE 中工作，那么本书的建议是尝试一下命令行工具。这些工具简洁、可预测，这对于 Visual Studio 和 Visual Studio Code 提供的所有等效功能来说是不可能的。表 4-1 总结了本章的内容。

表 4-1　本章内容摘要

| 问题 | 解决方案 | 代码清单 |
|---|---|---|
| 创建项目 | 使用 dotnet new 命令或 Visual Studio 向导 | 4-1～4-4 |
| 构建和运行项目 | 使用 dotnet build 和 dotnet run 命令或使用 Visual Studio 和 Visual Studio Code 提供的菜单 | 4-5～4-7 |
| 向项目添加包 | 使用 dotnet add package 命令或使用 Visual Studio 包管理器 | 4-8 和 4-10 |
| 安装工具命令 | 使用 dotnet tool 命令 | 4-11～4-12 |
| 管理客户端包 | 使用 libman 命令或 Visual Studio 客户端包管理器 | 4-13～4-16 |

## 4.1　创建 ASP.NET Core 项目

.NET Core SDK 包含一组用于创建、管理、构建和运行项目的命令行工具。Visual Studio 为其中一些任务提供了集成支持，但如果使用的是 Visual Studio Code，那么命令行是唯一的选择。

即使在使用 Visual Studio 时，本书也使用命令行工具，因为它们更简洁，而 Visual Studio 特性往往需要更多工作来定位需要的模板或设置。在后续章节中，将展示如何创建和使用这两套工具。无论选择哪种方法，结果都是相同的，且可以在 Visual Studio 和命令行工具之间自由切换。

■ **提示**：可以从 https://github.com/apress/pro-asp.net-core-3 下载本章和本书中其他所有章节的示例项目。如果在运行示例时遇到问题，请参阅第 1 章以获得帮助。

### 4.1.1　使用命令行创建项目

dotnet 命令提供了对.NET Core 命令行特性的访问。dotnet new 命令用于创建新项目、配置文件或解决方案文件。要查看用于创建新项的模板列表，打开 PowerShell 命令提示符，并运行代码清单 4-1 所示的命令。

**代码清单 4-1　列出.NET Core 模板**

```
dotnet new
```

每个模板都有一个简短名称，以便使用。有很多模板可用，表 4-2 描述了对创建 ASP.NET Core 项目最有用的模板。

表 4-2　有用的 ASP.NET Core 项目模板

| 名称 | 描述 |
|---|---|
| Web | 此模板创建一个项目，用 ASP.NET Core 开发所需的最少代码和内容来设置。这是本书大部分章节中使用的模板 |
| mvc | 此模板创建一个 ASP.NET Core 项目，配置为使用 MVC 框架 |
| webapp | 此模板创建一个 ASP.NET Core 项目，配置为使用 Razor Pages |
| blazorserver | 此模板创建一个 ASP.NET Core 项目，配置为使用 Razor 服务器 |
| angular | 此模板创建一个 ASP.NET Core 项目，包含使用 Angular JavaScript 框架的客户端特性 |
| react | 此模板创建一个 ASP.NET Core 项目，包含使用 React JavaScript 框架的客户端特性 |
| reactredux | 此模板创建一个 ASP.NET Core 项目，包含使用 React JavaScript 框架和流行 Redux 库的客户端特性 |

还有一些模板可创建用于配置项目的常用文件，如表 4-3 所示。

**理解项目模板的局限性**

表 4-2 中描述的项目模板旨在通过处理基本配置设置、添加占位符内容来帮助快速启动开发。这些模板可以给人一种快速进展的感觉，但是它们包含了关于应该如何配置和开发项目的假设。如果不了解这些假设的影响，就无法获得针对项目特定需求所需的结果。

这个 Web 模板创建了一个具有 ASP.NET Core 开发所需的最小配置的项目。这是本书中大多数示例使用的项目模板，这样就可以解释如何配置每个特性以及如何将这些特性一起使用。

一旦理解了 ASP.NET Core 是如何工作的，其他项目模板会很有用，因为知道了如何根据自己的需要调整它们。但在学习时，建议坚持使用 Web 模板，即使它可能需要更多努力来获得结果。

表 4-3 配置项模板

| 名称 | 描述 |
| --- | --- |
| globaljson | 此模板给项目添加了 global.json 文件，指定将使用的.NET Core 的版本 |
| sln | 这个模板创建解决方案文件，用于对多个项目进行分组，Visual Studio 经常使用这个文件。解决方案文件使用 dotnet sln add 命令填充，如下面的代码清单所示 |
| gitignore | 此模板创建.gitignore 文件，该文件从 Git 源代码控制中排除不需要的项目 |

要创建项目，打开一个新的 PowerShell 命令提示符，并运行代码清单 4-2 所示的命令。

**代码清单 4-2 创建新项目**

```
dotnet new globaljson --sdk-version 3.1.101 --output MySolution/MyProject
dotnet new web --no-https --output MySolution/MyProject --framework netcoreapp3.1
dotnet new sln -o MySolution
dotnet sln MySolution add MySolution/MyProject
```

第一个命令创建了包含 global.json 文件的 MySolution/MyProject 文件夹，它指定该项目将使用.NET Core 版本 3.1.1。顶级文件夹 MySolution 用于将多个项目分组在一起。嵌套的 MyProject 文件夹将包含单个项目。

这里使用 globaljson 模板来帮助在执行本书中的示例时获得预期的结果。微软很擅长确保与.NET Core 版本的向后兼容性，但会发生破坏性更改，最好给项目添加 global.json 文件，以便开发团队中的每个人都使用相同的版本。

第二个命令使用 Web 模板创建项目，本书中的大多数示例都使用 Web 模板。如表 4-3 所示，此模板创建一个具有 ASP.NET Core 开发所需的最小内容的项目。每个模板都有自己的一组参数，这些参数会影响所创建的项目。--no-https 参数创建不支持 HTTPS 的项目(在第 16 章解释如何使用 HTTPS)。--framework 参数选择将用于项目的.NET Core 运行时。

其他命令创建一个引用新项目的解决方案文件。解决方案文件是在同一时间打开多个相关文件的一种方便方式。MySolution.sln 文件是在 MySolution 文件夹中创建的，在 Visual Studio 中打开这个文件将加载用 Web 模板创建的项目。这不是必需的，但是当退出代码编辑器时，它会阻止 Visual Studio 提示创建文件。

### 打开项目

要打开项目，请启动 Visual Studio，选择 Open a Project or Solution，然后打开 MySolution 文件夹中的 MySolution.sln 文件。Visual Studio 将打开解决方案文件，发现对代码清单 4-2 中 final 命令添加的项目引用，并同样打开项目。

Visual Studio Code 的工作方式不同。启动 Visual Studio Code，选择 File | Open Folder，然后导航到 MySolution 文件夹。单击 Select Folder，Visual Studio Code 将打开项目。

尽管 Visual Studio Code 和 Visual Studio 处理的是同一个项目，但它们显示的内容不同。Visual Studio Code 显示了一个简单的文件列表，按字母顺序排列，如图 4-1 左侧所示。Visual Studio 隐藏一些文件，并在相关文件项中嵌入其他文件，如图 4-1 右侧所示。

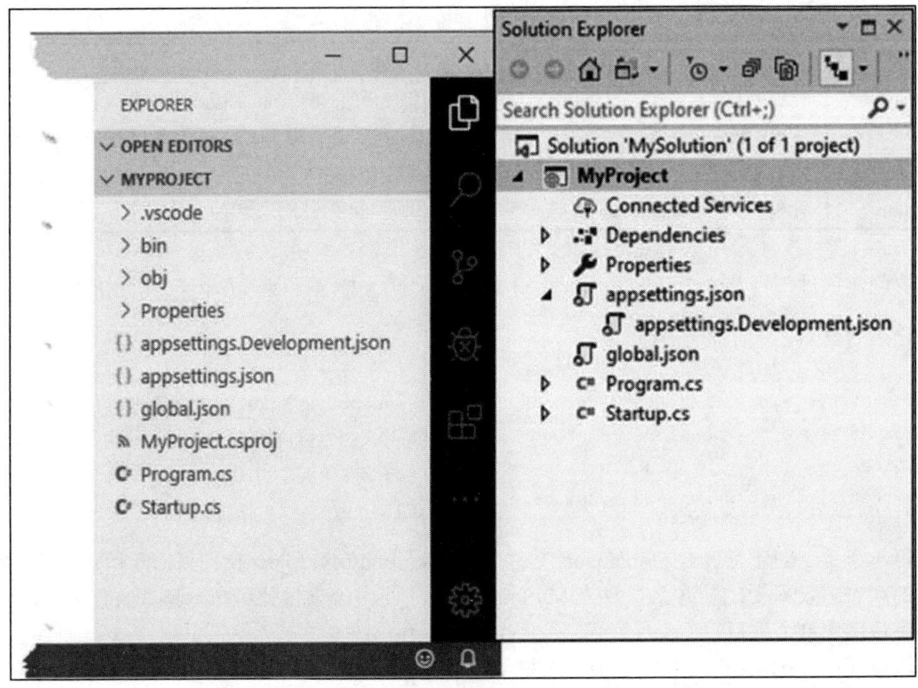


图 4-1　在 Visual Studio Code 和 Visual Studio 中打开项目

在 Visual Studio Solution Explorer 的顶部有一些按钮，它们可以禁用文件嵌套并显示项目中的隐藏项。在 Visual Studio Code 中第一次打开项目时，可能会提示添加用于构建和调试项目的资源，如图 4-2 所示。单击 Yes 按钮。

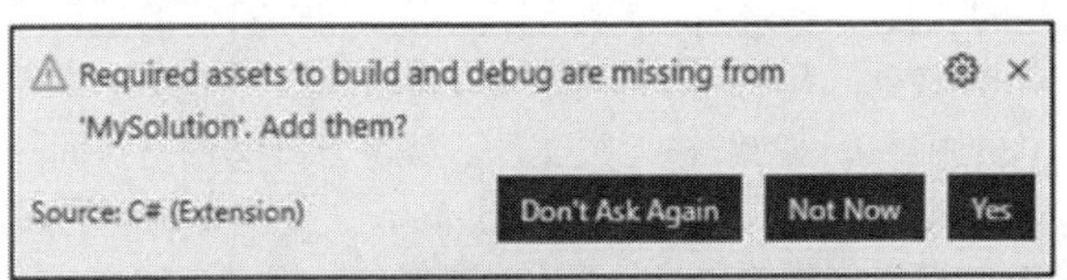


图 4-2　在 Visual Studio Code 中添加资源

## 4.1.2　使用 Visual Studio 创建项目

Visual Studio 以相同方式创建项目，使用相同的模板，但提供了一个向导。尽管本书倾向于使用命令行，但如果密切关注在此过程中选择的选项，结果是相同的。启动 Visual Studio 并单击 Create a new project，选择 ASP.NET Core Web Application 类别(如图 4-3 所示)，然后单击 Next 按钮。

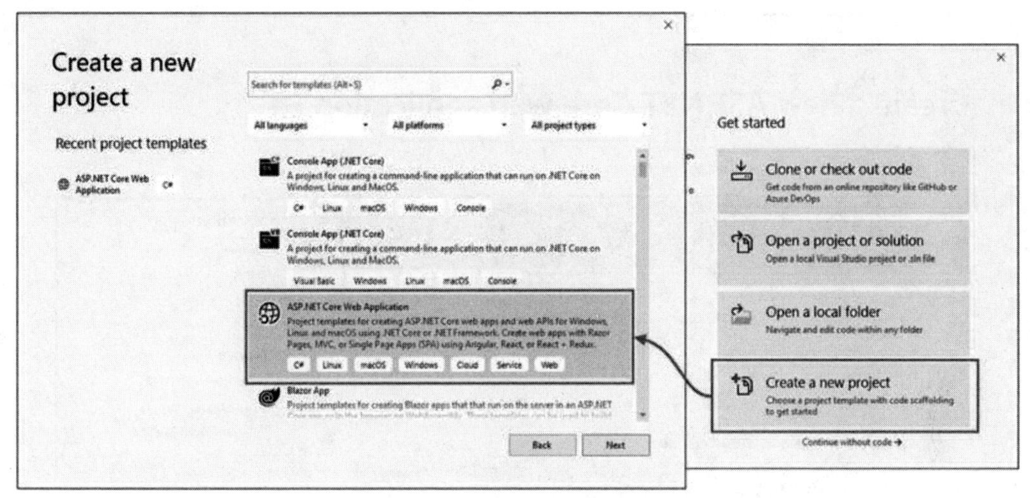


图 4-3 在 Visual Studio 中创建新项目

在 Project name 字段中输入 MyProject，在 Solution name 字段中输入 MySolution，如图 4-4 所示。使用 Location 字段选择一个在其中创建项目的文件夹，然后单击 Create 按钮。

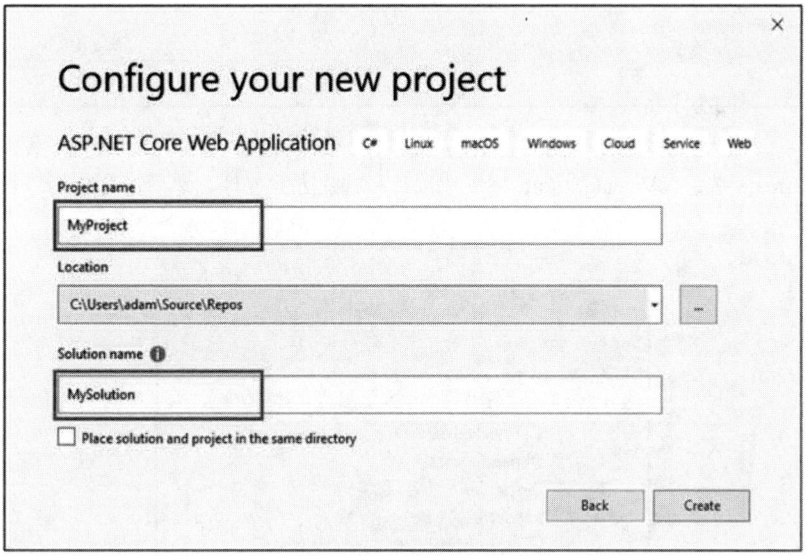

图 4-4 选择项目和解决方案名称

下一步需要密切关注，因为它很容易出错，所以本书的示例使用命令行工具。首先，使用窗口顶部的下拉框选择.NET Core 和 ASP.NET Core 3.1。接下来，从列表中选择 Empty 模板。即使使用了名称 Empty，添加到项目中的内容也与 Web 模板相对应。

取消选中 Configure for HTTPS 复选框(这相当于--no-https 命令行参数)，并确保选中 Enable Docker Support 复选框，将 Authentication 选项设置为 No Authentication，如图 4-5 所示。

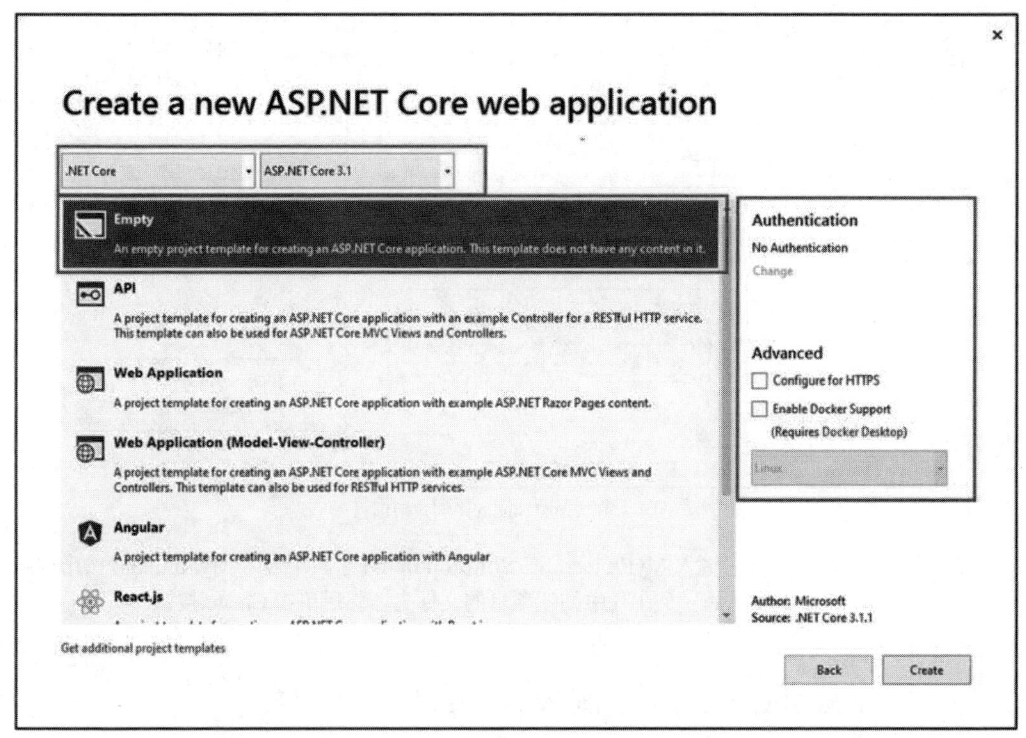


图 4-5 选择和配置项目模板

单击 Create 按钮，Visual Studio 将创建项目和解决方案文件，然后打开它们进行编辑，如图 4-6 所示。

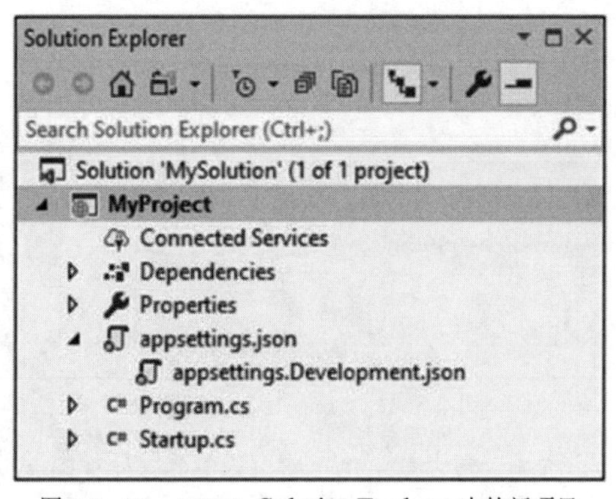


图 4-6　Visual Studio Solution Explorer 中的新项目

项目打开后，在 Solution Explorer 窗口中右击 MyProject 项，并从弹出菜单中选择 Add | New Item。从模板列表中找到 JSON 文件项，并将 Name 字段设置为 global.json。单击 Add 按钮创建文件，并将其内容替换为代码清单 4-3 所示的内容。

**代码清单 4-3　MyProject 文件夹中 global.json 文件的内容**

```
{
    "sdk": {
        "version": "3.1.101"
    }
}
```

添加 global.json 文件，确保项目使用正确版本的.NET Core SDK。

## 4.2　向项目中添加代码和内容

如果使用 Visual Studio Code，那么可以右击应该包含该文件的文件夹，并从弹出菜单中选择 New file(或者如果添加文件夹，则选择 New folder)，来向项目添加项。

---

■ **注意**：读者有责任确保文件扩展名与想添加的条目类型匹配；例如，必须添加扩展名为.html 的 html 文件。本书为添加到本书中项目的每一项提供了完整的文件名及其所在文件夹的名称，所以读者总会知道需要添加什么文件。

---

右击文件资源管理器页面中的 MyProject 项目，选择 New Folder，并将名称设置为 wwwroot，这是 ASP.NET Core 项目中存储静态内容的地方。按 Enter 键，一个名为 wwwroot 的文件夹将添加到项目中。右击新的 wwwroot 文件夹，选择 New Item，并将名称设置为 demo.html。按 Enter 键创建 HTML 文件，并添加如代码清单 4-4 所示的内容。

**代码清单 4-4　wwwroot 文件夹中 demo.html 文件的内容**

```
<!DOCTYPE html>
<html>
<head>
    <meta charset="utf-8" />
    <title></title>
</head>
<body>
    <h3>HTML File from MyProject</h3>
</body>
</html>
```

Visual Studio 提供了一种更全面的方法，这种方法只有在有选择地使用时才有用。要创建一个文件夹，在 Solution Explorer 中右击 MyProject 项目，然后从弹出菜单中选择 Add | New Folder。设置新项目的名称为 wwwroot 并按回车键；Visual Studio 将创建该文件夹。

在 Solution Explorer 中右击新的 wwwroot 条目，并从弹出菜单中选择 Add | New Item。Visual Studio 将提供大量可选择的模板，以便向项目中添加项。这些模板可以使用窗口右上角的文本字段进行搜索，或者使用窗口左侧的类别进行过滤。HTML 文件的项目模板名为 HTML Page，如图 4-7 所示。

在 Name 字段中输入 demo.html，单击 Add 按钮创建新文件，并用代码清单 4-4 所示的元素替换内容(如果省略了文件扩展名，Visual Studio 将根据所选择的项模板添加它。如果创建文件时在 Name 字段中只输入 demo，Visual Studio 将创建扩展名为.html 的文件，因为前面选择的是 HTML Page 项模板)。

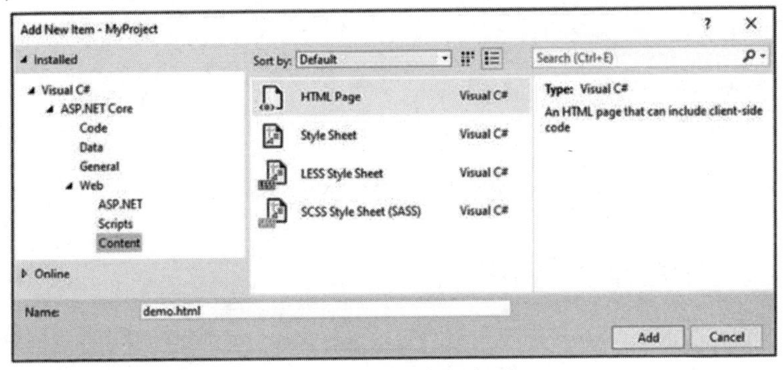


图 4-7　向示例项目添加项

**理解项的搭建**

Visual Studio 提供的项模板非常有用，特别是对于 C#类，它自动设置名称空间和类名。但是 Visual Studio 还提供 Scaffolded Item(脚手架项目)，但建议不要使用。Add | New Scaffolded Item 会引导用户选择一些项，引导完成添加更复杂项目的过程。Visual Studio 还将根据要添加项目的文件夹名称提供独立的脚手架项目。例如，如果右击一个名为 Views 的文件夹，Visual Studio 帮助将脚手架项目添加到菜单顶部，如图 4-8 所示。

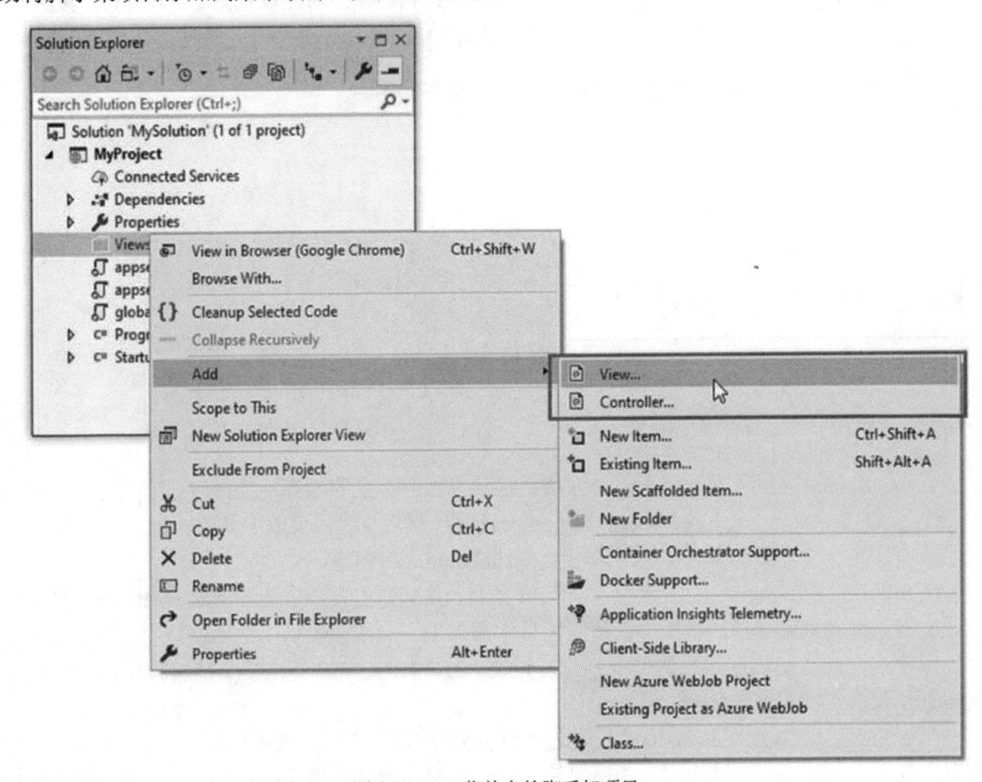


图 4-8　Add 菜单中的脚手架项目

View 和 Controller 项是搭建起来的，选择它们会呈现一些选项，决定创建的项内容。

就像项目模板一样，建议不要使用脚手架项目，至少在理解它们创建的内容之前不要使用。本书只使用 Add | New Item 菜单作为例子，并立即改变占位符的内容。

## 4.3 构建和运行项目

可以从命令行或从 Visual Studio 和 Visual Studio Code 中构建和运行项目。为此，将代码清单 4-5 所示的语句添加到 MyProject 文件夹的 Startup.cs 类文件中。

**代码清单 4-5 在 MyProject 文件夹的 Startup.cs 文件中添加语句**

```
using System;
using System.Collections.Generic;
using System.Linq;
using System.Threading.Tasks;
using Microsoft.AspNetCore.Builder;
using Microsoft.AspNetCore.Hosting;
using Microsoft.AspNetCore.Http;
using Microsoft.Extensions.DependencyInjection;
using Microsoft.Extensions.Hosting;
namespace MyProject {
   public class Startup {

      public void ConfigureServices(IServiceCollection services) {
      }

      public void Configure(IApplicationBuilder app, IWebHostEnvironment env) {

         if (env.IsDevelopment()) {
            app.UseDeveloperExceptionPage();
         }

         app.UseStaticFiles();
         app.UseRouting();

         app.UseEndpoints(endpoints => {
            endpoints.MapGet("/", async context => {
               await context.Response.WriteAsync("Hello World!");
            });
         });
      }
   }
}
```

这条语句添加了对使用 wwwroot 文件夹中的静态内容响应 HTTP 请求的支持，比如在上一节中创建的 HTML 文件(第 15 章将详细解释这个特性)。

### 4.3.1　使用命令行构建和运行项目

要构建示例项目，在 MyProject 或 MySolution 文件夹中运行代码清单 4-6 所示的命令。

**代码清单 4-6　构建项目**

```
dotnet build
```

通过在 MyProject 文件夹中运行代码清单 4-7 所示的命令，可在一个步骤中构建和运行项目。

**代码清单 4-7　构建和运行项目**

```
dotnet run
```

编译器将构建项目，然后启动集成的 ASP.NET Core HTTP 服务器监听端口 5000 上的 HTTP 请求。通过打开一个新的浏览器窗口，并请求 http://localhost:5000/demo.HTML，可以看到本章前面添加到项目中的静态 HTML 文件的内容。它将生成如图 4-9 所示的响应。

图 4-9　运行示例应用程序

### 4.3.2　使用 Visual Studio Code 构建和运行项目

如果不喜欢使用命令行，Visual Studio Code 可以构建和执行项目。选择 Terminal | Run Build Task，Visual Studio Code 将编译这个项目。

要想在一个步骤中构建和运行项目，选择 Debug Run | Without Debugging。Visual Studio Code 将编译和运行项目，并打开一个新的浏览器窗口，该窗口将向 ASP.NET Core 服务器发送一个 HTTP 请求。并生成占位符响应。请求 http://localhost:5000/demo.html，将收到如图 4-9 所示的响应。

### 4.3.3　使用 Visual Studio 构建和运行项目

Visual Studio 使用 IIS Express 作为内置 ASP.NET Core HTTP 服务器的反向代理，即在使用 dotnet run 命令时直接使用。在创建项目时，将为 IIS Express 选择一个 HTTP 端口来使用。要将 HTTP 端口更改为本书中使用的端口，请选择 Project | MyProject Properties 并选择 Debug 部分。找到 App URL 字段，将 URL 中的端口号更改为 5000，如图 4-10 所示。

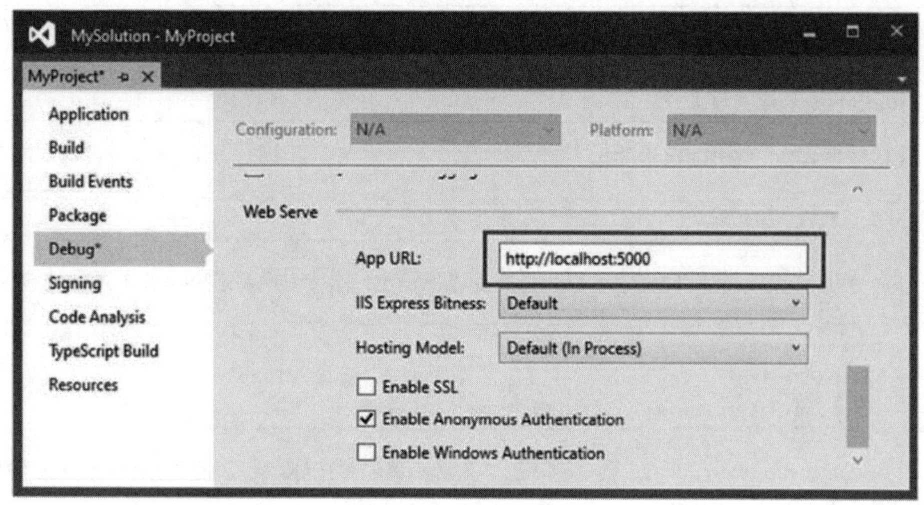


图 4-10　更改应用程序端口号

选择 File | Save All 以应用这些变化。要生成项目，请从 Build 菜单中选择 Build Solution 或 Build MyProject。要构建并运行项目，请选择 Debug | Start Without Debugging。项目编译完成后，Visual Studio 打开一个新的浏览器窗口，该窗口发送一个 HTTP 请求，IIS Express 接收该请求，并将其传递给 ASP.NET Core HTTP 服务器，生成用于创建项目的模板设置的占位符响应。请求 http://localhost:5000/demo.html，将看到如图 4-9 所示的响应。

■ **提示：** IIS Express 通常是可靠的，但如果有问题，右击 Windows 任务栏的系统托盘中的 IIS Express 图标，从弹出菜单中选择 Exit。

# 4.4　管理包

大多数项目需要除了项目模板设置的那些特性之外的其他特性，比如对访问数据库或发出 HTTP 请求的支持，这两者都没有包含在标准的 ASP.NET Core 包中(该包由用于创建示例项目的模板添加到项目中)。下一节将描述用于管理 ASP.NET Core 开发中使用的不同类型的包的工具。

## 4.4.1　管理 NuGet 包

.NET 包通过 dotnet add package 命令添加到项目中。使用 PowerShell 命令提示符在 MyProject 文件夹中运行代码清单 4-8 所示的命令，向示例项目添加一个包。

**代码清单 4-8　向示例项目添加一个包**

```
dotnet add package Microsoft.EntityFrameworkCore.SqlServer --version 3.1.1
```

这个命令将安装 Microsoft.EntityFrameworkCore.SqlServer 包的 3.1.1 版本。.NET 项目的包存储库是 nuget.org，在那里可以搜索包并看到可用的版本。例如，代码清单 4-8 中安装的包是在 https://www.nuget.org/packages/Microsoft.EntityFrameworkCore.SqlServer/3.1.1 中描述的。

通过运行代码清单 4-9 所示的命令，可看到安装在项目中的包。

■ **提示：** 项目文件(扩展名为.csproj 的文件)用于跟踪添加到项目中的包。可以在 Visual Studio Code 中打开该文件进行编辑，或者在 Visual Studio Solution Explorer 中右击项目，并从弹出菜单中选择 Edit Project File 来检查该文件。

**代码清单 4-9　列出项目中的包**

```
dotnet list package
```

该命令在 MyProject 文件夹中运行时会产生以下输出，显示在代码清单 4-8 中添加的包：

```
[netcoreapp3.1]:
Top-level Package                             Requested   Resolved
> Microsoft.EntityFrameworkCore.SqlServer     3.1.1       3.1.1
```

使用 dotnet remove package 命令删除包。要从示例项目中删除包，请在 MyProject 文件夹中运行代码清单 4-10 所示的命令。

**代码清单 4-10　从示例项目中删除一个包**

```
dotnet remove package Microsoft.EntityFrameworkCore.SqlServer
```

## 4.4.2　管理工具包

工具包安装命令可以用来从命令行在.NET Core 项目上执行操作。一个常见例子是 Entity Framework Core 工具包，它安装用于在 ASP.NET Core 项目中管理数据库的命令。工具包是使用 dotnet tool 命令管理的。要安装 Entity Framework Core 工具包，请运行代码清单 4-11 所示的命令。

**代码清单 4-11 安装工具包**

```
dotnet tool uninstall --global dotnet-ef
dotnet tool install --global dotnet-ef --version 3.1.1
```

第一个命令删除 dotnet-ef 包，它被命名为 dotnet-ef。如果软件包尚未安装，此命令将生成一个错误，但最好在安装软件包之前删除现有版本。dotnet tool uninstall 命令安装 dotnet-ef 包的 3.1.1 版本，这是本书中使用的版本。通过 dotnet 命令使用工具包安装的命令。要测试代码清单 4-11 中安装的包，在 MyProject 文件夹中运行代码清单 4-12 所示的命令。

■ **提示：** 代码清单 4-11 中的--global 参数表示包是为全局使用而安装的，而不仅仅是为特定项目。可以只将工具包安装到一个项目中，这种情况下，使用 dotnet tool run <command>访问命令。本书中使用的工具都是全局安装的。

**代码清单 4-12　运行工具包命令**

```
dotnet ef --help
```

这个工具包添加的命令是使用 dotnet ef 访问的，你将在后续章节看到依赖这些命令的例子。

### 4.4.3 管理客户端包

客户端包包含交付给客户端的内容，如图像、CSS 样式表、JavaScript 文件和静态 HTML。客户端包使用库管理器(LibMan)工具添加到 ASP.NET Core 中。要安装 LibMan 工具包，请运行代码清单 4-13 所示的命令。

**代码清单 4-13 安装 LibMan 工具包**

```
dotnet tool uninstall --global Microsoft.Web.LibraryManager.Cli
dotnet tool install --global Microsoft.Web.LibraryManager.Cli --version 2.0.96
```

这些命令将删除所有现有的 LibMan 包，并安装本书使用的版本。下一步是初始化项目，这将创建 LibMan 用于跟踪其安装的客户端包的文件。在 MyProject 文件夹中运行代码清单 4-14 所示的命令，以初始化示例项目。

**代码清单 4-14 初始化示例项目**

```
libman init -p cdnjs
```

LibMan 可从不同的存储库下载包。代码清单4-14中的-p参数指定了使用最广泛的https://cdnjs.com上的存储库。项目初始化后，就可安装客户端包了。要安装本书用于样式化HTML内容的Bootstrap CSS框架，请在MyProject文件夹中运行代码清单4-15所示的命令。

**代码清单 4-15 安装 Bootstrap CSS 框架**

```
libman install twitter-bootstrap@4.3.1 -d wwwroot/lib/twitter-bootstrap
```

该命令安装 Bootstrap 包的 4.3.1 版本，在 CDNJS 存储库中称为 twitter-bootstrap。流行的包在不同的存储库中命名的方式有些不一致，在将其添加到项目之前，应检查是否获得了期望的包。-d 参数指定包安装到的位置。ASP.NET Core 项目中的约定是将客户端包安装到 wwwroot/lib 文件夹中。

安装包之后，将代码清单 4-16 所示的类添加到 demo.html 文件的元素中。可采用这种方式应用 Bootstrap 包提供的特性。

---

■ **注意：** 本书没有详细介绍如何使用 Bootstrap CSS 框架。访问 https://getbootstrap.com 可以获取 Bootstrap 文档。

---

**代码清单 4-16 在 wwwroot 文件夹的 demo.html 文件中应用 Bootstrap 类**

```
<!DOCTYPE html>
<html>
<head>
    <meta charset="utf-8" />
    <title></title>
    <link href="/lib/twitter-bootstrap/css/bootstrap.min.css" rel="stylesheet" />
</head>
<body>
```

```
        <h3 class="bg-primary text-white text-center p-2">
            HTML File from MyProject
        </h3>
    </body>
</html>
```

启动 ASP.NET Core，请求 http://localhost:5000/demo.html，将看到如图 4-11 所示的样式化内容。

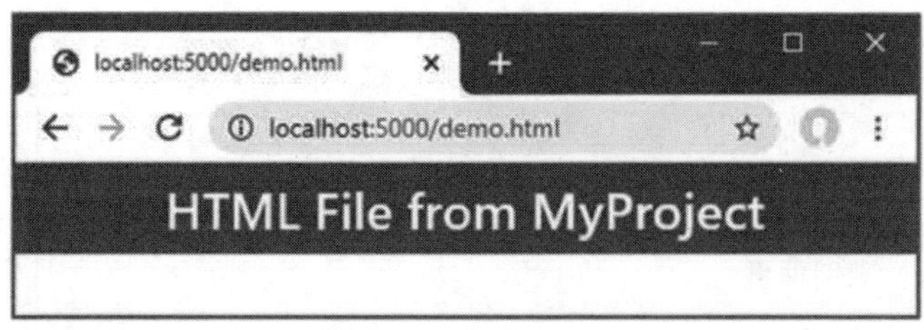


图 4-11　使用客户端包

## 4.4.4　使用 Visual Studio 管理包

Visual Studio 提供了不使用命令行管理包的工具。选择 Project | Manage NuGet Packages，Visual Studio 将打开 NuGet 包管理工具。单击 Browse 按钮，在搜索框中输入 Microsoft.EntityFrameworkCore.SqlServer，以搜索匹配的包。单击 Microsoft.EntityFrameworkCore.SqlServer 条目(它应该在列表的顶部)，就能选择版本，并安装包，如图 4-12 所示。

---

■ **警告**：Visual Studio NuGet 包管理器不能用于安装全局工具包，只能从命令行安装全局工具包。

---

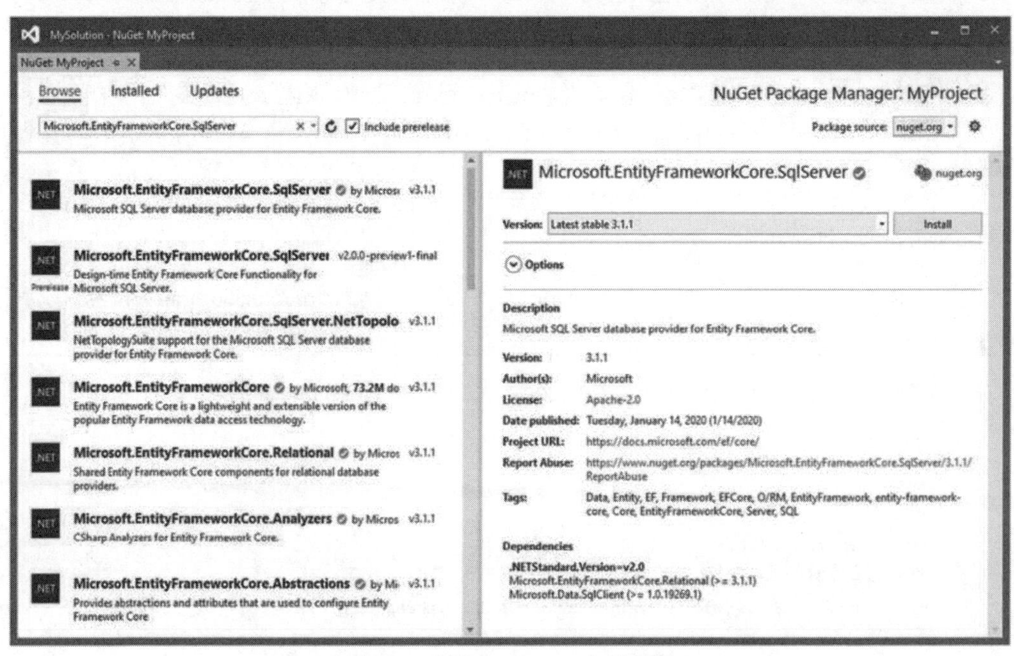


图 4-12　使用 Visual Studio 包管理器

Visual Studio NuGet 包管理器还可以用来检查已经安装在项目中的包，并检查包是否有新版本可用。

### 4.4.5 使用 Visual Studio 管理客户端包

Visual Studio 为管理客户端包提供了一个单独的工具。在 Solution Explorer 中右击 MyProject 项目，并从弹出菜单中选择 Add | Client Side Library。客户端包工具是基本的，但是它允许执行基本的搜索，选择添加到项目中的文件，并设置安装位置，如图 4-13 所示。

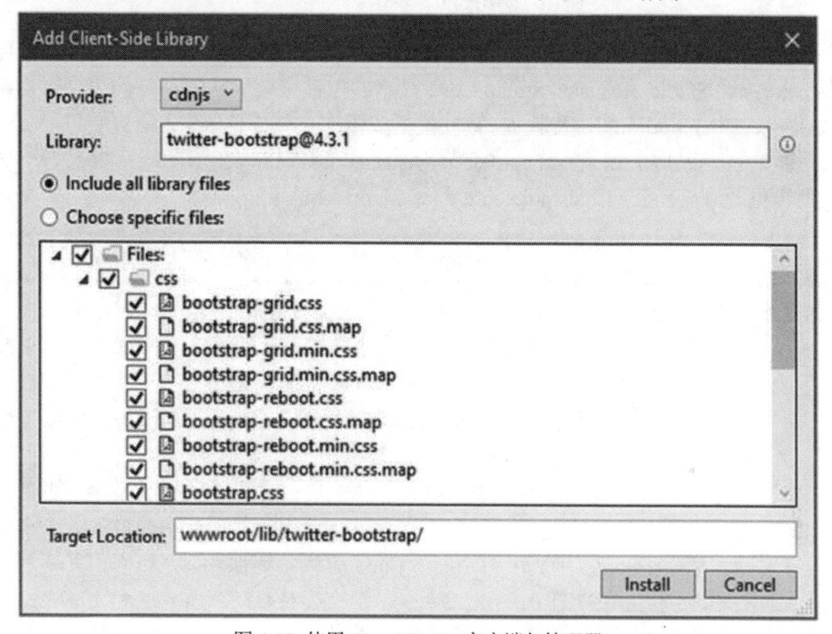


图 4-13 使用 Visual Studio 客户端包管理器

## 4.5 调试项目

Visual Studio 和 Visual Studio Code 都提供了可用于控制和检查 ASP.NET Core 应用程序执行情况的调试器。在 MyProject 文件夹中打开 Startup.cs 文件，在代码编辑器中单击下面的语句：

```
...
await context.Response.WriteAsync("Hello World!");
...
```

选择 Debug | Toggle Breakpoint，这在 Visual Studio 和 Visual Studio Code 中都有。断点在代码语句旁边显示为一个点，如图 4-14 所示，它将中断执行，并将控制权传递给用户。

通过选择 Debug | Start Debugging 来启动项目，这在 Visual Studio 和 Visual Studio Code 中都有。

如果 Visual Studio Code 提示选择环境，则选择.NET Core。然后再次选择 Start Debugging 菜单项。

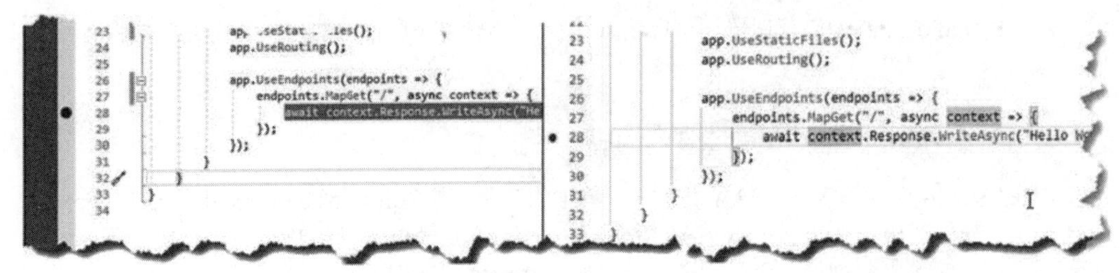

图 4-14　设置断点

应用程序将正常启动并继续运行，直到到达断点所在的语句，此时执行将停止。可以使用 Debug 菜单或 Visual Studio 和 Visual Studio Code 显示的控件来控制执行。这两个调试器都有很多特性——如果有 Visual Studio 的付费版本，特性就更多了——本书不会深入研究它们。Visual Studio 2019 调试器详见 https://docs.microsoft.com/en-us/visualstudio/debugger/?view=vs-2019，Visual Studio Code 调试器详见 https://code.visualstudio.com/docs/editor/debugging。

**如何调试代码**

调试器是功能强大的工具，但我很少使用它们。在大多数情况下，我更喜欢在代码中添加 System.Console.WriteLine 语句，以了解发生了什么，这很容易做到，因为我倾向于使用 dotnet run 命令，从命令行运行项目。这是一个基本方法，是有效的，尤其适用于代码中的大多数错误因为 if 语句中的条件无效，而没有被调用的情形。如果想详细检查一个对象，可以将其序列化为 JSON，并将结果传递给 WriteLine 方法。

对于调试器的专用用户，这可能看起来很疯狂，但是它的优点是快速、简单。当试图找出代码不能工作的原因时，常常希望快速地探索和迭代，但启动调试器所花费的时间是一个障碍。我的方法也是可靠的。Visual Studio 和 Visual Studio Code 调试器都很复杂，但它们并不总是完全可预测的，而且.NET Core 和 ASP.NET Core 变化太快，调试器特性还没有完全稳定下来。当我对某些代码的行为感到非常困惑时，希望使用最简单的诊断工具，对我来说，就是写入控制台的消息。

这里并不是建议使用这种方法，但若没有得到预期的结果，且不想与调试器斗争以找出原因时，它可以是一个很好的起点。

## 4.6　小结

本章描述了用于 ASP.NET Core 开发的工具，解释了命令行工具是处理 ASP.NET Core 项目最简洁、可靠的方法。这就是为什么在本书的示例中使用它们的原因。本章还演示了 Visual Studio 和 Visual Studio Code 提供的其他用户界面，这些界面对于某些(但不是所有)命令行工具是有用的替代。第 5 章将描述有效 ASP.NET Core 开发所需的 C#特性。

# 第5章

# C#的基本特点

本章描述 C#在 Web 应用程序开发中使用的一些特性，这些特性并没有被广泛理解，或者经常引起混淆。本书并不是一本关于 C#的书，因此我仅为每个特性提供一个简短的示例，以便读者执行本书其余部分的示例，并在自己的项目中利用这些特性。表 5-1 总结了本章的内容。

表 5-1　本章内容摘要

| 问题 | 解决方案 | 对应的代码清单 |
|---|---|---|
| 管理空值运算符 | 使用 null 条件和 null 合并运算符 | 5-7、5-10 |
| 用 getter 和 setter 创建属性 | 定义自动实现的属性 | 5-11～5-13 |
| 在字符串中混合静态和动态值 | 使用字符串插值 | 5-14 |
| 初始化和填充对象 | 使用对象和集合初始化器 | 5-15～5-18 |
| 为特定类型赋值 | 使用模式匹配 | 5-19～5-20 |
| 在不修改类的情况下扩展它的功能 | 定义扩展方法 | 5-21～5-28 |
| 简明地表达函数和方法 | 用 lambda 表达式 | 5-29～5-36 |
| 定义变量而不显式声明其类型 | 使用 var 关键字 | 5-37～5-39 |
| 修改接口而不需要更改其实现类 | 定义默认实现 | 5-40～5-44 |
| 异步执行工作 | 使用任务或 async/await 关键字 | 5-45～5-47 |
| 生成随时间变化的值序列 | 使用可枚举的异步 | 5-48～5-51 |
| 获取类或成员的名称 | 使用 nameof 表达式 | 5-52～5-53 |

## 5.1　准备工作

要创建本章的示例项目，打开一个新的 PowerShell 命令提示符，并运行代码清单 5-1 所示的命令。如果使用的是 Visual Studio，并且不喜欢使用命令行，就可以使用第 4 章描述的过程来创建项目。

■ **提示：**可以从 https://github.com/apress/pro-asp.net-core-3 下载本章和本书其他所有章节的示例项目。如果在运行示例时遇到问题，请参阅第 1 章以获得帮助。

**代码清单 5-1 创建示例项目**

```
dotnet new globaljson --sdk-version 3.1.101 --output LanguageFeatures
dotnet new web --no-https --output LanguageFeatures --framework netcoreapp3.1
dotnet new sln -o LanguageFeatures

dotnet sln LanguageFeatures add LanguageFeatures
```

### 5.1.1 打开项目

如果使用的是 Visual Studio，选择 File | Open | Project/Solution，选择 LanguageFeatures 文件夹中的 LanguageFeatures.sln 文件，然后单击 Open 按钮，以打开解决方案文件及其引用的项目。如果使用的是 Visual Studio Code，请选择 File | Open Folder，导航到 LanguageFeatures 文件夹，然后单击 Select Folder 按钮。

### 5.1.2 启用 MVC 框架

Web 项目模板创建一个包含最小 ASP.NET Core 配置的项目。这意味着第 3 章使用 mvc 模板添加的占位符内容是不可用的，需要额外的步骤才能使应用程序生成有用的输出。本节将执行设置 MVC 框架所需的更改，这是 ASP.NET Core 支持的应用程序框架之一，如第 1 章所述。首先启用 MVC 框架，对 Startup 类执行如代码清单 5-2 所示的更改。

**代码清单 5-2 在 LanguageFeatures 文件夹的 Startup.cs 文件中启用 MVC**

```
using System;
using System.Collections.Generic;
using System.Linq;
using System.Threading.Tasks;
using Microsoft.AspNetCore.Builder;
using Microsoft.AspNetCore.Hosting;
using Microsoft.AspNetCore.Http;
using Microsoft.Extensions.DependencyInjection;
using Microsoft.Extensions.Hosting;

namespace LanguageFeatures {
   public class Startup {
      public void ConfigureServices(IServiceCollection services) {
         services.AddControllersWithViews();
      }

      public void Configure(IApplicationBuilder app, IWebHostEnvironment env) {
         if (env.IsDevelopment()) {
            app.UseDeveloperExceptionPage();
         }

         app.UseRouting();
```

```
            app.UseEndpoints(endpoints => {
                //endpoints.MapGet("/", async context => {
                // await context.Response.WriteAsync("Hello World!");
                //});
                endpoints.MapDefaultControllerRoute();
            });
        }
    }
}
```

第 II 部分解释如何配置 ASP.NET Core 应用程序。但代码清单 5-2 中添加的两个语句使用默认配置提供了一个基本的 MVC 框架设置。

### 5.1.3 创建应用程序组件

现在 MVC 框架已经设置好了，可以添加将用于演示重要 C#语言特性的应用程序组件了。

#### 1. 创建数据模型

首先创建一个简单的模型类，这样就可以使用一些数据。添加一个名为 Models 的文件夹，在其中创建一个名为 Product.cs 的类文件，用于定义如代码清单 5-3 所示的类。

**代码清单 5-3 Models 文件夹中 Product.cs 文件的内容**

```
namespace LanguageFeatures.Models {
    public class Product {

        public string Name { get; set; }
        public decimal? Price { get; set; }

        public static Product[] GetProducts() {

            Product kayak = new Product {
                Name = "Kayak", Price = 275M
            };

            Product lifejacket = new Product {
                Name = "Lifejacket", Price = 48.95M
            };

            return new Product[] { kayak, lifejacket, null };
        }
    }
}
```

Product 类定义 Name 和 Price 属性，有一个名为 GetProducts 的静态方法返回一个 Product 数组。GetProducts 方法返回的数组中包含一个设置为 null 的元素，本章后面用它演示一些有用的语言特性。

#### 2. 创建控制器和视图

本章中的示例使用一个简单的控制器类来演示不同的语言特性。创建了一个 Controllers 文件夹，并添加一个名为 HomeController.cs 的类文件，其内容如代码清单 5-4 所示。

**代码清单 5-4　Controllers 文件夹中 HomeController.cs 文件的内容**

```
using Microsoft.AspNetCore.Mvc;

namespace LanguageFeatures.Controllers {
    public class HomeController : Controller {

        public ViewResult Index() {
            return View(new string[] { "C#", "Language", "Features" });
        }
    }
}
```

Index 操作方法告诉 ASP.NET Core 呈现默认视图，并为其提供一个字符串数组作为视图模型(将包含在发送给客户端的 HTML 中)。为创建视图，添加一个 Views/Home 文件夹(创建一个 Views 文件夹，然后在其中添加一个 Home 文件夹)，并添加了一个名为 Index.cshtml 的 Razor 视图，其内容如代码清单 5-5 所示。

**代码清单 5-5　Views/Home 文件夹中 Index.cshtml 文件的内容**

```
@model IEnumerable<string>
@{ Layout = null; }

<!DOCTYPE html>
<html>
<head>
    <meta name="viewport" content="width=device-width" />
    <title>Language Features</title>
</head>
<body>
    <ul>
        @foreach (string s in Model) {
            <li>@s</li>
        }
    </ul>
</body>
</html>
```

### 5.1.4　选择 HTTP 端口

如果使用的是 Visual Studio，选择 Project | LanguageFeatures 属性，然后选择 Debug 部分，并在 App URL 字段中将 HTTP 端口改为 5000，如图 5-1 所示。选择 File | Save All，以保存新端口(如果使用的是 Visual Studio Code，则不需要执行此更改)。

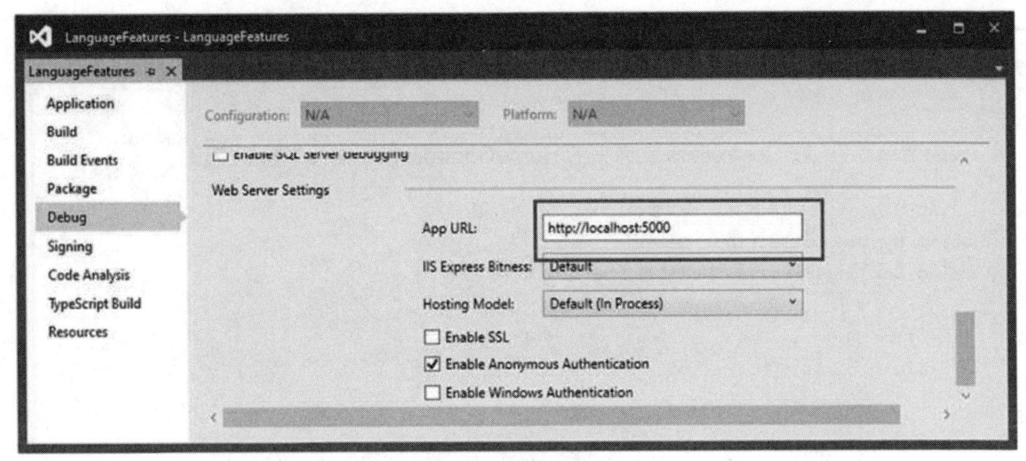


图 5-1　设置 HTTP 端口

### 5.1.5　运行示例应用程序

从 Debug 菜单中选择 Start Without Debugging (Visual Studio) 或 Run Without Debugging (Visual Studio Code)，或者在 LanguageFeatures 文件夹中运行代码清单5-6所示的命令，启动 ASP.NET Core。

**代码清单 5-6　运行示例应用程序**

```
dotnet run
```

请求 http://localhost:5000，将看到如图 5-2 所示的输出。

图 5-2　运行示例应用程序

由于本章中所有示例的输出都是文本，因此浏览器显示的消息显示如下：

```
C#
Language
Features
```

## 5.2　使用 null 条件运算符

null 条件运算符允许更优雅地检测空值。在 ASP.NET Core 开发中可能会进行很多空值检查，确定请求是否包含特定的头或值，或者模型是否包含特定的数据项。传统上，处理空值需要进行显式检查，当需要检查对象及其属性时，这种检查可能变得乏味且容易出错。null 条件运算符使这个过程更简单、更简洁，如代码清单 5-7 所示。

**代码清单 5-7 在 Controllers 文件夹的 HomeController.cs 文件中检测空值**

```
using Microsoft.AspNetCore.Mvc;
using System.Collections.Generic;
using LanguageFeatures.Models;

namespace LanguageFeatures.Controllers {
    public class HomeController : Controller {

        public ViewResult Index() {

            List<string> results = new List<string>();

            foreach (Product p in Product.GetProducts()) {
                string name = p?.Name;
                decimal? price = p?.Price;
                results.Add(string.Format("Name: {0}, Price: {1}", name, price));
            }
            return View(results);
        }
    }
}
```

Product 类定义的静态 GetProducts 方法返回一个对象数组，在 Index 操作方法中检查这些对象，以获得 Name 和 Price 值的列表。问题是数组中的对象和属性的值都可能是 null，这意味着在 foreach 循环中只引用 p.Name 或 p.Price，可能导致 NullReferenceException 异常。为避免这种情况，使用了 null 条件运算符，如下所示：

```
...
string name = p?.Name;
decimal? price = p?.Price;
...
```

null 条件运算符是一个问号(?字符)。如果 p 是 null，那么名称也设置为 null。如果 p 不为 null，那么名称就设置为 Person.Name 的值。Price 属性也这样考虑。注意，在使用 null 条件运算符时指定的变量必须能够赋予为空，这就是为什么 price 变量声明为可空的小数(decimal?)

## 5.2.1 链接 null 条件运算符

可以链接 null 条件运算符，在对象层次结构中导航，这样它就成为简化代码和允许安全导航的有效工具。在代码清单 5-8 中，向 Product 类添加了一个属性，该属性创建了一个更复杂的对象层次结构。

**代码清单 5-8　在 Models 文件夹的 Product.cs 文件中添加一个属性**

```
namespace LanguageFeatures.Models {
    public class Product {

        public string Name { get; set; }
        public decimal? Price { get; set; }
        public Product Related { get; set; }

        public static Product[] GetProducts() {

            Product kayak = new Product {
                Name = "Kayak", Price = 275M
            };
            Product lifejacket = new Product {
                Name = "Lifejacket", Price = 48.95M
            };

            kayak.Related = lifejacket;

            return new Product[] { kayak, lifejacket, null };
        }
    }
}
```

每个 Product 对象都有一个相关的属性，该属性可引用另一个 Product 对象。在 GetProducts 方法中，为表示 kayak 的 Product 对象设置了相关属性。代码清单 5-9 展示了如何链接 null 条件运算符来导航对象属性，而不会导致异常。

**代码清单 5-9　在 Controllers 文件夹的 HomeController.cs 文件中检测嵌套的空值**

```
using Microsoft.AspNetCore.Mvc;
using System.Collections.Generic;
using LanguageFeatures.Models;

namespace LanguageFeatures.Controllers {
    public class HomeController : Controller {

        public ViewResult Index() {

            List<string> results = new List<string>();

            foreach (Product p in Product.GetProducts()) {
                string name = p?.Name;
                decimal? price = p?.Price;
                string relatedName = p?.Related?.Name;
                results.Add(string.Format("Name: {0}, Price: {1}, Related: {2}",
                    name, price, relatedName));
```

```
            }
            return View(results);
        }
    }
}
```

null 条件运算符可应用到属性链的每个部分，如下所示：

```
...
string relatedName = p?.Related?.Name;
...
```

结果是，当 p 为空或当 p.Related 为空时，relatedName 变量将为空。否则，变量将被赋给 p.Related.Name 的值。重启 ASP.NET Core，并请求 http://localhost:5000，在浏览器窗口中显示以下输出：

```
Name: Kayak, Price: 275, Related: Lifejacket
Name: Lifejacket, Price: 48.95, Related:
Name: , Price: , Related:
```

### 5.2.2　结合条件运算符和合并运算符

将 null 条件运算符(一个问号)与 null 合并运算符(两个问号)结合使用，可以设置回退值，以防止在应用程序中使用 null 值，如代码清单 5-10 所示。

**代码清单 5-10　在 Controllers 文件夹的 HomeController.cs 文件中组合 Null 运算符**

```
using Microsoft.AspNetCore.Mvc;
using System.Collections.Generic;
using LanguageFeatures.Models;

namespace LanguageFeatures.Controllers {
    public class HomeController : Controller {
    public ViewResult Index() {

        List<string> results = new List<string>();

        foreach (Product p in Product.GetProducts()) {
            string name = p?.Name ?? "<No Name>";
            decimal? price = p?.Price ?? 0;
            string relatedName = p?.Related?.Name ?? "<None>";
            results.Add(string.Format("Name: {0}, Price: {1}, Related: {2}",
                name, price, relatedName));
            }

            return View(results);
        }
```

```
    }
}
```

null 条件运算符确保在浏览对象属性时不会得到 NullReferenceException 异常，null 合并运算符确保在浏览器中显示的结果中不包含空值。如果运行这个例子，浏览器窗口会显示以下结果：

```
Name: Kayak, Price: 275, Related: Lifejacket
Name: Lifejacket, Price: 48.95, Related: <None>
Name: <No Name>, Price: 0, Related: <None>
```

## 可空和不可空的引用类型

遇到意外的空值是导致错误的最常见原因之一。默认情况下，C#将 null 作为所有类型的有效值，这意味着可将 null 赋值给字符串变量，如下所示：

```
...
string product = null;
...
```

使用该变量的代码负责检查空值，当同一变量位于多个位置时，这可能特别成问题。很容易忽略其中一个检查或假设值不为 null，从而在运行时生成错误。

可空引用类型将空检查的责任转移到将值赋给变量的代码。当启用可空引用特性时，常规引用类型不能赋空值(如将空值分配给字符串)。相反，如果可能使用空值，则必须使用可空引用类型，如下所示：

```
...
string product = null; // compiler error - this is a non-nullable type
string? product = null; // no error - this is a nullable type
...
```

string?类型是可空的，而 string 不是，这意味着使用变量的代码不必担心 null 值，除非它处理的是可空的类型。要启用可空引用类型，必须在 csproj 文件中添加一个元素，如下所示：

```
...
<Project Sdk="Microsoft.NET.Sdk.Web">

  <PropertyGroup>
    <TargetFramework>netcoreapp3.1</TargetFramework>
    <Nullable>enable</Nullable>
  </PropertyGroup>
</Project>
...
```

如果使用的是 Visual Studio，则可在 Solution Explorer 中右击项目，并从弹出菜单中选择 Edit Project File，来打开项目文件。

我喜欢这个特性，但它的应用还不够广泛，还不能在本书中使用，特别是它会使一些复杂的主题更难理解。我希望一旦.NET Core 的其余部分包括这个特性，将其默认添加到 ASP.NET Core

中，也期待可以在本书的未来版本中使用可空引用类型。

## 5.3 使用自动实现的属性

C#支持自动实现的属性，上一节为 Person 类定义属性时使用了它们，如下所示：

```
namespace LanguageFeatures.Models {
    public class Product {

        public string Name { get; set; }
        public decimal? Price { get; set; }
        public Product Related { get; set; }

        public static Product[] GetProducts() {

            Product kayak = new Product {
                Name = "Kayak", Price = 275M
            };
            Product lifejacket = new Product {
                Name = "Lifejacket", Price = 48.95M
            };

            kayak.Related = lifejacket;

            return new Product[] { kayak, lifejacket, null };
        }
    }
}
```

这个特性允许在不实现 get 和 set 主体的情况下定义属性。使用自动实现的属性特性意味着，定义属性可以如下所示：

```
...
public string Name { get; set; }
...
is equivalent to the following code:
...
public string Name {
    get { return name; }
    set { name = value; }
}
...
```

这类特性被称为“语法糖”，意味着它使 C#更易于使用——这种情况下，它消除了针对每个属性重复的冗余代码——而没有实质上改变语言的行为方式。术语“糖”似乎有些贬义，但是任何使代码更容易编写和维护的增强都是有益的，在大型的复杂项目中尤其如此。

### 5.3.1 使用自动实现的属性初始化器

自 C# 3.0 以来就支持自动实现的属性。最新版本的 C#支持自动实现属性的初始化，允许不使用构造函数就可以设置初始值，如代码清单 5-11 所示。

**代码清单 5-11 在 Models 文件夹的 Product.cs 文件中使用自动实现的属性初始化器**

```
namespace LanguageFeatures.Models {
   public class Product {

      public string Name { get; set; }
      public string Category { get; set; } = "Watersports";
      public decimal? Price { get; set; }
      public Product Related { get; set; }

      public static Product[] GetProducts() {
         Product kayak = new Product {
            Name = "Kayak",
            Category = "Water Craft",
            Price = 275M
         };
         Product lifejacket = new Product {
            Name = "Lifejacket", Price = 48.95M
         };

         kayak.Related = lifejacket;

         return new Product[] { kayak, lifejacket, null };
      }
   }
}
```

将值赋给自动实现的属性并不会阻止 setter 用于以后更改属性，而只是整理简单类型的代码，这些类型最终使用包含属性赋值列表的构造函数来提供默认值。在本例中，初始化器将 Watersports 的值分配给 Category 属性。初始值可以更改，创建 kayak 对象并指定 Water Craft 值时，就会这样做。

### 5.3.2 创建自动实现的只读属性

使用初始化器，并从具有初始化器的自动实现属性中省略 set 关键字，可以创建只读属性，如代码清单 5-12 所示。

**代码清单 5-12 在 Models 文件夹的 Product.cs 文件中创建只读属性**

```
namespace LanguageFeatures.Models {
   public class Product {

      public string Name { get; set; }
```

```
        public string Category { get; set; } = "Watersports";
        public decimal? Price { get; set; }
        public Product Related { get; set; }
        public bool InStock { get; } = true;

        public static Product[] GetProducts() {

            Product kayak = new Product {
                Name = "Kayak",
                Category = "Water Craft",
                Price = 275M
            };
            Product lifejacket = new Product {
                Name = "Lifejacket", Price = 48.95M
            };

            kayak.Related = lifejacket;

            return new Product[] { kayak, lifejacket, null };
        }
    }
}
```

将 InStock 属性初始化为 true，且不能改变；但是，可以在类型的构造函数中为该值赋值，如代码清单 5-13 所示。

#### 代码清单 5-13　在 Models 文件夹的 Product.cs 文件中为只读属性分配一个值

```
namespace LanguageFeatures.Models {
    public class Product {

        public Product(bool stock = true) {
            InStock = stock;
        }

        public string Name { get; set; }
        public string Category { get; set; } = "Watersports";
        public decimal? Price { get; set; }
        public Product Related { get; set; }
        public bool InStock { get; }

        public static Product[] GetProducts() {
            Product kayak = new Product {
                Name = "Kayak",
                Category = "Water Craft",
                Price = 275M
            };
```

```
            Product lifejacket = new Product(false) {
                Name = "Lifejacket",
                Price = 48.95M
            };

            kayak.Related = lifejacket;

            return new Product[] { kayak, lifejacket, null };
        }
    }
}
```

构造函数允许将只读属性的值指定为参数，如果没有提供值，则默认为 true。构造函数一旦设置了属性值，就不能更改它。

## 5.4 使用字符串插值

string.Format 方法是传统的 C#工具，用于编写包含数据值的字符串。下面是一个来自主控制器的例子：

```
...
results.Add(string.Format("Name: {0}, Price: {1}, Related: {2}",
                    name, price, relatedName));
...
```

C#还支持一种不同的方法，称为字符串插值，这样就不必再确保字符串模板中的{0}引用与作为参数指定的变量匹配。相反，字符串插值直接使用变量名，如代码清单 5-14 所示。

**代码清单 5-14 在 Controllers 文件夹的 HomeController.cs 文件中使用字符串插值**

```
using Microsoft.AspNetCore.Mvc;
using System.Collections.Generic;
using LanguageFeatures.Models;

namespace LanguageFeatures.Controllers {
    public class HomeController : Controller {

        public ViewResult Index() {

            List<string> results = new List<string>();

            foreach (Product p in Product.GetProducts()) {
                string name = p?.Name ?? "<No Name>";
                decimal? price = p?.Price ?? 0;
                string relatedName = p?.Related?.Name ?? "<None>";
                results.Add($"Name: {name}, Price: {price}, Related: {relatedName}");
            }
```

```
            return View(results);
        }
    }
}
```

内插字符串以$字符为前缀，并包含 hole，这些 hole 是对{和}字符中包含的值的引用。在计算字符串时，使用指定的变量或常量的当前值填充 hole。

**提示：**字符串插值支持 string.Format 方法中可用的所有格式说明符。格式细节包含在 hole 中，因此$"Price: {Price: C2}"将 Price 值格式化为具有两位小数的货币值。

## 5.5　使用对象和集合初始化器

在 Product 类的静态 GetProducts 方法中创建一个对象时，使用了对象初始化器，它允许在一个步骤中创建对象，并指定它的属性值，如下所示：

```
...
Product kayak = new Product {
    Name = "Kayak",
    Category = "Water Craft",
    Price = 275M
};
...
```

这是另一个使 C#更容易使用的语法糖特性。如果没有这个特性，只能调用 Product 的构造函数，然后使用新创建的对象设置每个属性，如下所示：

```
...
Product kayak = new Product();
kayak.Name = "Kayak";
kayak.Category = "Water Craft";
kayak.Price = 275M;
...
```

一个相关的特性是集合初始化器，它允许在单个步骤中指定集合及其内容的创建。例如，在没有初始化器的情况下，创建字符串数组需要分别指定数组的大小和数组元素，如代码清单 5-15 所示。

**代码清单 5-15　在 Controllers 文件夹的 HomeController.cs 文件中初始化对象**

```
using Microsoft.AspNetCore.Mvc;
using System.Collections.Generic;
using LanguageFeatures.Models;

namespace LanguageFeatures.Controllers {
    public class HomeController : Controller {
```

```
        public ViewResult Index() {
            string[] names = new string[3];
            names[0] = "Bob";
            names[1] = "Joe";
            names[2] = "Alice";
            return View("Index", names);
        }
    }
}
```

使用集合初始化器允许将数组的内容指定为构造的一部分，这隐式地向编译器提供了数组的大小，如代码清单 5-16 所示。

**代码清单 5-16　在 Controllers 文件夹的 HomeController.cs 文件中使用集合初始化器**

```
using Microsoft.AspNetCore.Mvc;
using System.Collections.Generic;
using LanguageFeatures.Models;

namespace LanguageFeatures.Controllers {
    public class HomeController : Controller {
        public ViewResult Index() {
            return View("Index", new string[] { "Bob", "Joe", "Alice" });
        }
    }
}
```

数组元素是在{和}字符之间指定的，这样可以更简洁地定义集合，并可以在方法调用中内联地定义集合。代码清单 5-16 中的代码与代码清单 5-15 中的代码具有相同的效果。重启 ASP.NET Core，并请求 http://localhost:5000，在浏览器窗口中会显示以下输出：

```
Bob
Joe
Alice
```

### 使用索引初始化器

C#的最新版本整理了使用索引的集合(如字典)的初始化方式。代码清单 5-17 显示了重写的索引操作，以使用传统的 C#方法初始化字典，来定义集合。

**代码清单 5-17　在 Controllers 文件夹的 HomeController.cs 文件中初始化字典**

```
using Microsoft.AspNetCore.Mvc;
using System.Collections.Generic;
using LanguageFeatures.Models;

namespace LanguageFeatures.Controllers {
    public class HomeController : Controller {
```

```
        public ViewResult Index() {
            Dictionary<string, Product> products = new Dictionary<string, Product> {
                { "Kayak", new Product { Name = "Kayak", Price = 275M } },
                { "Lifejacket", new Product{ Name = "Lifejacket", Price = 48.95M } }
            };
            return View("Index", products.Keys);
        }
    }
}
```

初始化这类集合的语法过于依赖{和}字符，特别是在使用对象初始化器创建集合值时。C#的最新版本支持一种更自然的方法来初始化索引集合，这种方法与初始化集合后检索或修改值的方式一致，如代码清单 5-18 所示。

**代码清单 5-18　在 Controllers 文件夹的 HomeController.cs 文件中使用集合初始化器语法**

```
using Microsoft.AspNetCore.Mvc;
using System.Collections.Generic;
using LanguageFeatures.Models;

namespace LanguageFeatures.Controllers {
    public class HomeController : Controller {

        public ViewResult Index() {
            Dictionary<string, Product> products = new Dictionary<string, Product> {
                ["Kayak"] = new Product { Name = "Kayak", Price = 275M },
                ["Lifejacket"] = new Product { Name = "Lifejacket", Price = 48.95M }
            };

            return View("Index", products.Keys);
        }
    }
}
```

其效果是一样的——创建一个字典，其键是 Kayak 和 Lifejacket，其值是 Product 对象——但元素是使用用于其他集合操作的索引符号创建的。重启 ASP.NET Core，并请求 http://localhost:5000，在浏览器中显示以下结果：

```
Kayak
Lifejacket
```

## 5.6　模式匹配

C#最近增加的最有用的功能之一是支持模式匹配，它可以用来测试某对象是否属于特定类型或具有特定特征。这是另一种形式的语法糖，可极大地简化条件语句的复杂块。关键字 is 用于执行类型测试，如代码清单 5-19 所示。

**代码清单 5-19　在 Controllers 文件夹的 HomeController.cs 文件中执行类型测试**

```
using Microsoft.AspNetCore.Mvc;
using System.Collections.Generic;
using LanguageFeatures.Models;

namespace LanguageFeatures.Controllers {
    public class HomeController : Controller {

        public ViewResult Index() {

            object[] data = new object[] { 275M, 29.95M,
                "apple", "orange", 100, 10 };
            decimal total = 0;
            for (int i = 0; i < data.Length; i++) {
                if (data[i] is decimal d) {
                    total += d;
                }
            }

            return View("Index", new string[] { $"Total: {total:C2}" });
        }
    }
}
```

关键字 is 执行类型检查，如果值是指定的类型，将赋值给一个新变量，如下所示：

```
...
if (data[i] is decimal d) {
...
```

如果存储在 data[i]中的值是小数，则该表达式的计算结果为 true。data[i]的值被赋给变量 d，这允许它在后续语句中使用，而不需要执行任何类型转换。is 关键字只匹配指定的类型，这意味着只处理数据数组中的两个值(数组中的其他项是字符串和 int 值)。如果运行该应用程序，在浏览器窗口中将显示以下输出：

```
Total: $304.95
```

### switch 语句中的模式匹配

模式匹配也可在 switch 语句中使用，它支持 when 关键字来限制 case 语句何时匹配一个值，如代码清单 5-20 所示。

**代码清单 5-20　Controllers 文件夹的 HomeController.cs 文件中的模式匹配**

```
using Microsoft.AspNetCore.Mvc;
```

```
using System.Collections.Generic;
using LanguageFeatures.Models;

namespace LanguageFeatures.Controllers {
  public class HomeController : Controller {

    public ViewResult Index() {
      object[] data = new object[] { 275M, 29.95M,
        "apple", "orange", 100, 10 };
      decimal total = 0;
      for (int i = 0; i < data.Length; i++) {
        switch (data[i]) {
          case decimal decimalValue:
            total += decimalValue;
            break;
          case int intValue when intValue > 50:
            total += intValue;
            break;
        }
      }

      return View("Index", new string[] { $"Total: {total:C2}" });
    }
  }
}
```

要匹配任何特定类型的值，在 case 语句中使用类型和变量名，如下所示：

```
...
case decimal decimalValue:
...
```

这个 case 语句匹配任何十进制值，并将其分配给一个名为 decimalValue 的新变量。更有选择性的是 when 关键字可以包括在内，如下所示：

```
...
case int intValue when intValue > 50:
...
```

这个 case 语句匹配 int 值，并将它们分配给一个名为 intValue 的变量，但仅当值大于 50 时才这么做。重启 ASP.NET Core，并请求 http://localhost:5000，在浏览器窗口中显示以下输出：

```
Total: $404.95
```

## 5.7　使用扩展方法

扩展方法是向不能直接修改的类添加方法的一种捷径，通常是因为它们是由 Microsoft 或第三方包提供的。代码清单 5-21 显示了 ShoppingCart 类的定义，将其添加到 Models 文件夹的 ShoppingCart.cs 文件中，它表示一个 Product 对象集合。

**代码清单 5-21 Models 文件夹中 ShoppingCart.cs 文件的内容**

```
using System.Collections.Generic;

namespace LanguageFeatures.Models {

    public class ShoppingCart {
        public IEnumerable<Product> Products { get; set; }
    }
}
```

这是一个简单的类，充当 Products 对象序列的包装器(对于这个示例，只需要一个基本类)。假设需要确定 ShoppingCart 类中 Products 对象的总价值，但是不能修改这个类，因为它来自第三方，而且没有源代码。可以使用扩展方法来添加需要的功能。

在 Models 文件夹中添加一个名为 MyExtensionMethods.cs 的类文件，并用它定义如代码清单 5-22 所示的类。

**代码清单 5-22 Models 文件夹中 MyExtensionMethods.cs 文件的内容**

```
namespace LanguageFeatures.Models {

    public static class MyExtensionMethods {

        public static decimal TotalPrices(this ShoppingCart cartParam) {
            decimal total = 0;
            foreach (Product prod in cartParam.Products) {
                total += prod?.Price ?? 0;
            }
            return total;
        }
    }
}
```

扩展方法是在静态类中定义的，该静态类所在的名称空间与应用扩展方法的类所在的名称空间相同。在本例中，静态 MyExtensionMethods 类位于 LanguageFeatures.Models 名称空间，这意味着它可以包含该名称空间中类的扩展方法。

扩展方法也是静态的，代码清单 5-22 定义了一个名为 TotalPrices 的扩展方法。第一个参数前面的 this 关键字将 TotalPrices 标记为一个扩展方法。第一个参数告诉.NET，扩展方法可以应用到哪个类上——在本例中是 ShoppingCart 类。使用 cartParam 参数可以引用已应用扩展方法的 ShoppingCart 实例。此扩展方法枚举 ShoppingCart 中的 Product 对象，并返回 Product.Price 属性值的总和。代码清单 5-23 展示了如何在主控制器的操作方法中应用扩展方法。

---

**注意：** *扩展方法不能突破类为方法、字段和属性定义的访问规则。可以通过使用扩展方法来扩展类的功能，但只能使用无论如何都可以访问的类成员。*

---

**代码清单 5-23 在 Controllers 文件夹的 HomeController.cs 文件中应用扩展方法**

```
using Microsoft.AspNetCore.Mvc;
using System.Collections.Generic;
using LanguageFeatures.Models;

namespace LanguageFeatures.Controllers {

    public class HomeController : Controller {

        public ViewResult Index() {
            ShoppingCart cart
                = new ShoppingCart { Products = Product.GetProducts() };
            decimal cartTotal = cart.TotalPrices();
            return View("Index", new string[] { $"Total: {cartTotal:C2}" });
        }
    }
}
```

关键语句如下：

```
...
decimal cartTotal = cart.TotalPrices();
...
```

调用 ShoppingCart 对象上的 TotalPrices 方法，就像该方法是 ShoppingCart 类中的一部分，即使它是由一个完全不同的类定义的扩展方法。.NET 会发现扩展类是否在当前类的范围内，这意味着它们是同一个名称空间的一部分或在 using 语句指定的名称空间中。重启 ASP.NET Core，并请求 http://localhost:5000，这将在浏览器窗口中生成以下输出：

```
Total: $323.95
```

### 5.7.1 将扩展方法应用到接口

扩展方法也可应用于接口，这允许对实现该接口的所有类调用扩展方法。代码清单 5-24 显示了更新后的 ShoppingCart 类，以实现 IEnumerable<Product>接口。

**代码清单 5-24 在 Models 文件夹的 ShoppingCart.cs 文件中实现一个接口**

```
using System.Collections;
using System.Collections.Generic;

namespace LanguageFeatures.Models {

    public class ShoppingCart : IEnumerable<Product> {
        public IEnumerable<Product> Products { get; set; }

        public IEnumerator<Product> GetEnumerator() {
```

```
            return Products.GetEnumerator();
        }
        IEnumerator IEnumerable.GetEnumerator() {
            return GetEnumerator();
        }
    }
}
```

现在可以更新扩展方法，使其处理 IEnumerable<Product>，如代码清单 5-25 所示。

**代码清单 5-25 更新 Models 文件夹的 MyExtensionMethods.cs 文件中的扩展方法**

```
using System.Collections.Generic;

namespace LanguageFeatures.Models {

    public static class MyExtensionMethods {

        public static decimal TotalPrices(this IEnumerable<Product> products) {
            decimal total = 0;
            foreach (Product prod in products) {
                total += prod?.Price ?? 0;
            }
            return total;
        }
    }
}
```

第一个参数类型已更改为 IEnumerable<Product>，这意味着方法主体中的 foreach 循环直接作用于 Product 对象。使用该接口的改变意味着可以计算任何 IEnumerable<Product>枚举的 Product 对象的总价值，其中包括 ShoppingCart 的实例，也包括 Product 对象的数组，如代码清单 5-26 所示。

**代码清单 5-26 在 Controllers 文件夹的 HomeController.cs 文件中应用扩展方法**

```
using Microsoft.AspNetCore.Mvc;
using System.Collections.Generic;
using LanguageFeatures.Models;

namespace LanguageFeatures.Controllers {
    public class HomeController : Controller {

        public ViewResult Index() {

            ShoppingCart cart
                = new ShoppingCart { Products = Product.GetProducts() };

            Product[] productArray = {
```

```
                new Product {Name = "Kayak", Price = 275M},
                new Product {Name = "Lifejacket", Price = 48.95M}
            };

            decimal cartTotal = cart.TotalPrices();
            decimal arrayTotal = productArray.TotalPrices();

            return View("Index", new string[] {
                $"Cart Total: {cartTotal:C2}",
                $"Array Total: {arrayTotal:C2}" });
        }
    }
}
```

重启 ASP.NET Core，请求 http://localhost:5000，这将在浏览器中生成以下输出，说明从扩展方法得到相同的结果，不管 Product 对象是如何收集的：

```
Cart Total: $323.95
Array Total: $323.95
```

### 5.7.2 创建过滤扩展方法

关于扩展方法，要展示的最后一件事是，它们可用于过滤对象集合。对 IEnumerable<T>进行操作并返回 IEnumerable<T>的扩展方法可以使用 yield 关键字将选择条件应用于源数据中的项，从而生成一组简化的结果。代码清单 5-27 演示了这样一个方法，它已添加到 MyExtensionMethods 类中。

**代码清单 5-27　Models 文件夹的 MyExtensionMethods.cs 文件中的过滤扩展方法**

```
using System.Collections.Generic;

namespace LanguageFeatures.Models {

    public static class MyExtensionMethods {

        public static decimal TotalPrices(this IEnumerable<Product> products) {
            decimal total = 0;
            foreach (Product prod in products) {
                total += prod?.Price ?? 0;
            }
            return total;
        }

        public static IEnumerable<Product> FilterByPrice(
                this IEnumerable<Product> productEnum, decimal minimumPrice) {
            foreach (Product prod in productEnum) {
                if ((prod?.Price ?? 0) >= minimumPrice) {
```

```
                yield return prod;
            }
        }
    }
  }
}
```

这个扩展方法称为 FilterByPrice，它接收一个附加参数，允许对产品进行过滤，以便在结果中返回 Price 属性匹配或超过该参数的 Product 对象。代码清单 5-28 显示了使用的方法。

**代码清单 5-28 使用 Controllers 文件夹的 HomeController.cs 文件中的过滤扩展方法**

```
using Microsoft.AspNetCore.Mvc;
using System.Collections.Generic;
using LanguageFeatures.Models;

namespace LanguageFeatures.Controllers {
    public class HomeController : Controller {

        public ViewResult Index() {

            Product[] productArray = {
                new Product {Name = "Kayak", Price = 275M},
                new Product {Name = "Lifejacket", Price = 48.95M},
                new Product {Name = "Soccer ball", Price = 19.50M},
                new Product {Name = "Corner flag", Price = 34.95M}
            };
            decimal arrayTotal = productArray.FilterByPrice(20).TotalPrices();

            return View("Index", new string[] { $"Array Total: {arrayTotal:C2}" });
        }
    }
}
```

对 Product 对象的数组调用 FilterByPrice 方法时，TotalPrices 方法只接收那些花费超过 20 美元的对象，并用于计算总数。如果运行该应用程序，在浏览器窗口中显示以下输出：

```
Total: $358.90
```

## 5.8 使用 lambda 表达式

lambda 表达式是一个会引起很多混乱的特性，它们的简化特性尤其令人困惑。要理解正在解决的问题，请考虑上一节定义的 FilterByPrice 扩展方法。编写这个方法是为了让它能按价格过滤产品对象，这意味着必须创建第二个按名称过滤的方法，如代码清单 5-29 所示。

**代码清单 5-29 在 Models 文件夹的 MyExtensionMethods.cs 文件中添加一个过滤方法**

```
using System.Collections.Generic;
```

```
namespace LanguageFeatures.Models {

    public static class MyExtensionMethods {

        public static decimal TotalPrices(this IEnumerable<Product> products) {
            decimal total = 0;
            foreach (Product prod in products) {
                total += prod?.Price ?? 0;
            }
            return total;
        }

        public static IEnumerable<Product> FilterByPrice(
                this IEnumerable<Product> productEnum, decimal minimumPrice) {
            foreach (Product prod in productEnum) {
                if ((prod?.Price ?? 0) >= minimumPrice) {
                    yield return prod;
                }
            }
        }
        public static IEnumerable<Product> FilterByName(
                this IEnumerable<Product> productEnum, char firstLetter) {

            foreach (Product prod in productEnum) {
                if (prod?.Name?[0] == firstLetter) {
                    yield return prod;
                }
            }
        }
    }
}
```

代码清单5-30显示了在控制器中应用这两种过滤方法来创建两个不同的总数。

**代码清单5-30　使用Controllers文件夹的HomeController.cs文件中的两个过滤方法**

```
using Microsoft.AspNetCore.Mvc;
using System.Collections.Generic;
using LanguageFeatures.Models;

namespace LanguageFeatures.Controllers {
    public class HomeController : Controller {

        public ViewResult Index() {

            Product[] productArray = {
                new Product {Name = "Kayak", Price = 275M},
```

```
            new Product {Name = "Lifejacket", Price = 48.95M},
            new Product {Name = "Soccer ball", Price = 19.50M},
            new Product {Name = "Corner flag", Price = 34.95M}
        };

        decimal priceFilterTotal = productArray.FilterByPrice(20).TotalPrices();
        decimal nameFilterTotal = productArray.FilterByName('S').TotalPrices();

        return View("Index", new string[] {
            $"Price Total: {priceFilterTotal:C2}",
            $"Name Total: {nameFilterTotal:C2}" });
        }
    }
}
```

第一个过滤器选择价格在 20 美元以上的所有产品，第二个过滤器选择名称以字母 S 开头的产品。如果运行示例应用程序，在浏览器窗口中会显示如下输出：

```
Price Total: $358.90
Name Total: $19.50
```

### 5.8.1 定义函数

可以无限重复这个过程，为感兴趣的每个属性和属性组合创建过滤方法。一种更优雅的方法是将处理枚举的代码与选择标准分开。通过允许函数作为对象传递，C#使这一点变得容易。代码清单 5-31 显示了一个扩展方法，该方法过滤 Product 对象的枚举，但将结果中包含哪些对象的决定委托给一个单独的函数。

**代码清单 5-31 在 Models 文件夹的 MyExtensionMethods.cs 文件中创建通用的过滤方法**

```
using System.Collections.Generic;
using System;

namespace LanguageFeatures.Models {

    public static class MyExtensionMethods {

        public static decimal TotalPrices(this IEnumerable<Product> products) {
            decimal total = 0;
            foreach (Product prod in products) {
                total += prod?.Price ?? 0;
            }
            return total;
        }

        public static IEnumerable<Product> Filter(
                this IEnumerable<Product> productEnum,
```

```
            Func<Product, bool> selector) {
        foreach (Product prod in productEnum) {
            if (selector(prod)) {
                yield return prod;
            }
        }
    }
  }
}
```

过滤方法的第二个参数是一个接收Product对象并返回bool值的函数。过滤方法为每个Product对象调用函数，如果函数返回true，则将其包含在结果中。要使用过滤方法，可以指定一个方法或创建一个独立函数，如代码清单5-32所示。

#### 代码清单5-32　使用函数过滤Controllers文件夹的Controllers.cs文件中的对象

```
using Microsoft.AspNetCore.Mvc;
using System.Collections.Generic;
using LanguageFeatures.Models;
using System;

namespace LanguageFeatures.Controllers {
    public class HomeController : Controller {

        bool FilterByPrice(Product p) {
            return (p?.Price ?? 0) >= 20;
        }

        public ViewResult Index() {

            Product[] productArray = {
                new Product {Name = "Kayak", Price = 275M},
                new Product {Name = "Lifejacket", Price = 48.95M},
                new Product {Name = "Soccer ball", Price = 19.50M},
                new Product {Name = "Corner flag", Price = 34.95M}
            };

            Func<Product, bool> nameFilter = delegate (Product prod) {
                return prod?.Name?[0] == 'S';
            };

            decimal priceFilterTotal = productArray
                .Filter(FilterByPrice)
                .TotalPrices();
            decimal nameFilterTotal = productArray
                .Filter(nameFilter)
                .TotalPrices();
```

```
            return View("Index", new string[] {
                $"Price Total: {priceFilterTotal:C2}",
                $"Name Total: {nameFilterTotal:C2}" });
        }
    }
}
```

两种方法都不理想。定义方法(如 FilterByPrice)会使类定义混乱。创建 Func<Product, bool>对象避免了这个问题，但使用了难以理解和维护的语法。lambda 表达式通过允许以一种更优雅、更有表现力的方式定义函数来解决这个问题，如代码清单 5-33 所示。

**代码清单 5-33　在 Controllers 文件夹的 HomeController.cs 文件中使用 lambda 表达式**

```
using Microsoft.AspNetCore.Mvc;
using System.Collections.Generic;
using LanguageFeatures.Models;
using System;

namespace LanguageFeatures.Controllers {
    public class HomeController : Controller {

        public ViewResult Index() {

            Product[] productArray = {
                new Product {Name = "Kayak", Price = 275M},
                new Product {Name = "Lifejacket", Price = 48.95M},
                new Product {Name = "Soccer ball", Price = 19.50M},
                new Product {Name = "Corner flag", Price = 34.95M}
            };

            decimal priceFilterTotal = productArray
                .Filter(p => (p?.Price ?? 0) >= 20)
                .TotalPrices();
            decimal nameFilterTotal = productArray
                .Filter(p => p?.Name?[0] == 'S')
                .TotalPrices();

            return View("Index", new string[] {
                $"Price Total: {priceFilterTotal:C2}",
                $"Name Total: {nameFilterTotal:C2}" });
        }
    }
}
```

lambda 表达式以粗体显示。参数的表达没有指定类型，类型将被自动推断。=>字符读作"goes to"，并将参数链接到 lambda 表达式的结果。在本例中，一个名为 p 的产品参数将得到一个 bool 结果，如果第一个表达式中 Price 属性大于或等于 20，或者第二个表达式中 Name 属性以 S 开头，这个结果将为 true。这段代码的工作方式与单独的方法和函数委托相同，但更简洁；对于大多数人来说，更容易阅读。

**lambda 表达式的其他形式**

不需要在 lambda 表达式中表示委托的逻辑。可以很容易地调用一个方法，如下所示：

```
...
prod => EvaluateProduct(prod)
...
```

如果需要一个 lambda 表达式的委托有多个参数，就必须把参数放在括号中，如下所示：

```
...
(prod, count) => prod.Price > 20 && count > 0
...
```

最后，如果 lambda 表达式中的逻辑需要多个语句，可使用大括号({})并以一个 return 语句结束，如下所示：

```
...
(prod, count) => {
   // ...multiple code statements...
   return result;
}
...
```

不需要在代码中使用 lambda 表达式，但它们是一种表达复杂函数的简洁方式，并且具有可读性和清晰性。我非常喜欢它们，本书会介绍它们的用法。

### 5.8.2　使用 lambda 表达式方法和属性

lambda 表达式可用于实现构造函数、方法和属性。在 ASP.NET Core 开发中，方法通常包含单个语句，该语句选择要显示的数据和要呈现的视图。代码清单 5-34 重写了索引操作方法，使其遵循这个常见模式。

**代码清单 5-34　在 Controllers 文件夹的 HomeController.cs 文件中创建一个通用操作模式**

```
using Microsoft.AspNetCore.Mvc;
using System.Collections.Generic;
using LanguageFeatures.Models;
using System;
using System.Linq;

namespace LanguageFeatures.Controllers {
   public class HomeController : Controller {

      public ViewResult Index() {
         return View(Product.GetProducts().Select(p => p?.Name));
      }
```

```
    }
}
```

操作方法从静态 Product.GetProducts 方法获取 Product 对象的集合，并使用 LINQ 投影 Name 属性的值，这些值随后用作默认视图的视图模型。如果运行该应用程序，浏览器窗口就会显示以下输出：

```
Kayak
Lifejacket
```

浏览器窗口中也会有一个空列表项，因为 GetProducts 方法在其结果中包含了一个空引用，但这对本章的这一部分来说并不重要。

当构造函数或方法主体由一条语句组成时，可将其重写为 lambda 表达式，如代码清单 5-35 所示。

#### 代码清单 5-35 Controllers 文件夹中 HomeController.cs 文件的 lambda 操作方法

```
using Microsoft.AspNetCore.Mvc;
using System.Collections.Generic;
using LanguageFeatures.Models;
using System;
using System.Linq;

namespace LanguageFeatures.Controllers {
    public class HomeController : Controller {

        public ViewResult Index() =>
            View(Product.GetProducts().Select(p => p?.Name));
    }
}
```

方法的 lambda 表达式省略了 return 关键字，并使用=> (goes to)将方法签名(包括参数)与其实现关联起来。代码清单 5-35 中所示的 Index 方法与代码清单 5-34 中所示的方法工作方式相同，但表达方式更简洁。同样的基本方法也可用于定义属性。代码清单 5-36 显示在 Product 类中添加了一个使用 lambda 表达式的属性。

#### 代码清单 5-36 Models 文件夹的 Product.cs 文件中的 lambda 属性

```
namespace LanguageFeatures.Models {
    public class Product {

        public Product(bool stock = true) {
            InStock = stock;
        }

        public string Name { get; set; }
        public string Category { get; set; } = "Watersports";
        public decimal? Price { get; set; }
```

```
        public Product Related { get; set; }
        public bool InStock { get; }
        public bool NameBeginsWithS => Name?[0] == 'S';

        public static Product[] GetProducts() {

            Product kayak = new Product {
                Name = "Kayak",
                Category = "Water Craft",
                Price = 275M
            };

            Product lifejacket = new Product(false) {
                Name = "Lifejacket",
                Price = 48.95M
            };

            kayak.Related = lifejacket;

            return new Product[] { kayak, lifejacket, null };
        }
    }
}
```

## 5.9　使用类型推断和匿名类型

var 关键字允许在不显式指定变量类型的情况下定义局部变量，如代码清单 5-37 所示。这称为类型推断或隐式类型。

**代码清单 5-37　在 Controllers 文件夹的 HomeController.cs 文件中使用类型推断**

```
using Microsoft.AspNetCore.Mvc;
using System.Collections.Generic;
using LanguageFeatures.Models;
using System;
using System.Linq;

namespace LanguageFeatures.Controllers {
    public class HomeController : Controller {

        public ViewResult Index() {
            var names = new [] { "Kayak", "Lifejacket", "Soccer ball" };
            return View(names);
        }
    }
}
```

这并不是说 names 变量没有类型；相反，编译器要从代码中推断出类型。编译器检查数组声明，并确定它是一个字符串数组。运行该示例，将生成以下输出。

```
Kayak
Lifejacket
Soccer ball
```

### 使用匿名类型

通过结合对象初始化器和类型推断，可以创建简单的视图模型对象，这些对象用于在控制器和视图之间传输数据，而不必定义类或结构，如代码清单 5-38 所示。

**代码清单 5-38　在 Controllers 文件夹的 HomeController.cs 文件中创建匿名类型**

```
using Microsoft.AspNetCore.Mvc;
using System.Collections.Generic;
using LanguageFeatures.Models;
using System;
using System.Linq;

namespace LanguageFeatures.Controllers {
   public class HomeController : Controller {

      public ViewResult Index() {
         var products = new [] {
            new { Name = "Kayak", Price = 275M },
            new { Name = "Lifejacket", Price = 48.95M },
            new { Name = "Soccer ball", Price = 19.50M },
            new { Name = "Corner flag", Price = 34.95M }
         };

         return View(products.Select(p => p.Name));
      }
   }
}
```

products 数组中的每个对象都是匿名类型。这并不意味着它是动态的，因为 JavaScript 变量是动态的。这只意味着类型定义将由编译器自动创建。强类型仍然是强制的。例如，只能获取和设置在初始化器中定义的属性。重启 ASP.NET Core，并请求 http://localhost:5000，在浏览器窗口中显示以下输出：

```
Kayak
Lifejacket
Soccer ball
Corner flag
```

C#编译器根据初始化器中参数的名称和类型生成类。两个具有相同属性名称和类型的匿名类型对象将被分配给同一个自动生成的类。这意味着 products 数组中的所有对象都具有相同的类型，因为它们定义了相同的属性。

---

■ **提示：**必须使用 var 关键字来定义匿名类型对象的数组，因为直到代码编译后，才创建类型，所以不知道要使用的类型的名称。匿名类型对象数组中的元素必须定义相同的属性；否则，编译器无法计算出数组类型应该是什么。

---

为了演示这一点，更改代码清单 5-39 中的示例输出，以便它显示类型名，而不是 Name 属性的值。

**代码清单 5-39　在 Controllers 文件夹的 HomeController.cs 文件中显示类型名称**

```
using Microsoft.AspNetCore.Mvc;
using System.Collections.Generic;
using LanguageFeatures.Models;
using System;
using System.Linq;

namespace LanguageFeatures.Controllers {
    public class HomeController : Controller {

        public ViewResult Index() {
            var products = new [] {
                new { Name = "Kayak", Price = 275M },
                new { Name = "Lifejacket", Price = 48.95M },
                new { Name = "Soccer ball", Price = 19.50M },
                new { Name = "Corner flag", Price = 34.95M }
            };

            return View(products.Select(p => p.GetType().Name));
        }
    }
}
```

数组中的所有对象都被分配了相同的类型，如果运行示例，就可以看到这一点。类型名不是用户友好的，但不打算直接使用，读者看到的名称可能与下面的输出不同。

```
<>f__AnonymousType0`2
<>f__AnonymousType0`2
<>f__AnonymousType0`2
<>f__AnonymousType0`2
```

## 5.10　在接口中使用默认实现

C# 8.0 引入了新功能，允许为接口定义的属性和方法定义默认实现。这似乎是一个奇怪的特性，因为接口的目的是描述特性而没有指定实现，但是 C#的这一添加使得更新接口而不破坏现有实现成为可能。

在 Models 文件夹中添加一个名为 IProductSelect .cs 的类文件，并使用它定义接口，如代码清单 5-40 所示。

### 代码清单 5-40　Models 文件夹中 IProductSelection.cs 文件的内容

```
using System.Collections.Generic;

namespace LanguageFeatures.Models {
   public interface IProductSelection {

      IEnumerable<Product> Products { get; }
   }
}
```

更新 ShoppingCart 类以实现新接口，如代码清单 5-41 所示。

### 代码清单 5-41　在 Models 文件夹的 ShoppingCart.cs 文件中实现一个接口

```
using System.Collections;
using System.Collections.Generic;

namespace LanguageFeatures.Models {

   public class ShoppingCart : IProductSelection {
      private List<Product> products = new List<Product>();

      public ShoppingCart(params Product[] prods) {
         products.AddRange(prods);
      }

      public IEnumerable<Product> Products { get => products; }
   }
}
```

代码清单 5-42 更新了主控制器，以便它使用 ShoppingCart 类。

### 代码清单 5-42　使用 Controllers 文件夹的 HomeController.cs 文件中的接口

```
using Microsoft.AspNetCore.Mvc;
using System.Collections.Generic;
using LanguageFeatures.Models;
using System;
using System.Linq;

namespace LanguageFeatures.Controllers {
   public class HomeController : Controller {

      public ViewResult Index() {
         IProductSelection cart = new ShoppingCart(
```

```
                new Product { Name = "Kayak", Price = 275M },
                new Product { Name = "Lifejacket", Price = 48.95M },
                new Product { Name = "Soccer ball", Price = 19.50M },
                new Product { Name = "Corner flag", Price = 34.95M }
            );
            return View(cart.Products.Select(p => p.Name));
        }
    }
}
```

这是一个熟悉的使用界面，如果重新启动 ASP.NET Core 并请求 http://localhost:5000，在浏览器中会显示以下输出：

```
Kayak
Lifejacket
Soccer ball
Corner flag
```

如果想向接口添加一个新特性，就必须定位和更新实现它的所有类，这可能会很困难，特别当一个接口被其他开发团队在项目中使用时。这里可使用默认实现特性，允许向接口添加新特性，如代码清单 5-43 所示。

**代码清单 5-43　在 Models 文件夹的 IProductSelection.cs 文件中添加一个特性**

```
using System.Collections.Generic;
using System.Linq;

namespace LanguageFeatures.Models {
    public interface IProductSelection {

        IEnumerable<Product> Products { get; }

        IEnumerable<string> Names => Products.Select(p => p.Name);
    }
}
```

代码清单 5-43 定义了一个 Names 属性并提供了一个默认实现，这意味着 IProductSelection 接口可以使用 Total 属性，即使它不是由实现类定义的，如代码清单 5-44 所示。

**代码清单 5-44　使用 Controllers 文件夹的 HomeController.cs 文件中的默认实现**

```
using Microsoft.AspNetCore.Mvc;
using System.Collections.Generic;
using LanguageFeatures.Models;
using System;
using System.Linq;

namespace LanguageFeatures.Controllers {
```

```
public class HomeController : Controller {

    public ViewResult Index() {

        IProductSelection cart = new ShoppingCart(
            new Product { Name = "Kayak", Price = 275M },
            new Product { Name = "Lifejacket", Price = 48.95M },
            new Product { Name = "Soccer ball", Price = 19.50M },
            new Product { Name = "Corner flag", Price = 34.95M }
        );
        return View(cart.Names);
    }
  }
}
```

ShoppingCart 类没有修改，但是 Index 方法能够使用 Name 属性的默认实现。重启 ASP.NET Core，并请求 http://localhost:5000，在浏览器中会显示以下输出：

```
Kayak
Lifejacket
Soccer ball
Corner flag
```

## 5.11 使用异步方法

异步方法在后台执行工作，并在工作完成时发出通知，从而允许代码在执行后台工作时处理其他业务。异步方法是消除代码瓶颈的重要工具，允许应用程序利用多个处理器和处理器内核并行执行工作。

在 ASP.NET Core 中，通过允许服务器在调度和执行请求的方式上有更大的灵活性，异步方法可用来提高应用程序的整体性能。两个 C#关键字 async 和 await 用于异步执行工作。

### 5.11.1 直接处理任务

C#和.NET 对异步方法有很好的支持，但是代码往往很冗长，不习惯并行编程的开发人员经常会被不寻常的语法所困扰。要创建一个示例，向 Models 文件夹添加一个名为 MyAsyncMethods.cs 的类文件，并添加如代码清单 5-45 所示的代码。

**代码清单 5-45 Models 文件夹中 MyAsyncMethods.cs 文件的内容**

```
using System.Net.Http;
using System.Threading.Tasks;

namespace LanguageFeatures.Models {
    public class MyAsyncMethods {

        public static Task<long?> GetPageLength() {
            HttpClient client = new HttpClient();
```

```
            var httpTask = client.GetAsync("http://apress.com");
            return httpTask.ContinueWith((Task<HttpResponseMessage> antecedent) => {
               return antecedent.Result.Content.Headers.ContentLength;
            });
        }
    }
}
```

此方法使用System.Net.Http.HttpClient对象请求Apress主页的内容，返回其长度。.NET表示将作为一个任务异步完成的工作。任务对象是强类型的，基于后台工作生成的结果。因此调用HttpClient.GetAsync方法时，取回的是一个Task<HttpResponseMessage>。这说明，请求将在后台执行，请求的结果是一个HttpResponseMessage对象。

■ **提示**：使用像背景这样的词时，跳过了很多细节，只列出对ASP.NET Core世界重要的关键点。.NET对异步方法和并行编程的支持非常好，如果想创建真正的高性能应用程序，利用多核和多处理器硬件，最好更多地了解它。本书介绍不同的特性，解释ASP.NET Core是如何轻松创建异步Web应用程序的。

大多数程序员陷入困境的部分是延续，这是一种机制，通过它可以指定任务完成时希望发生什么。示例中使用了ContinueWith方法来处理从HttpClient.GetAsync方法获得的HttpResponseMessage对象，该方法使用一个lambda表达式执行，该表达式返回一个属性的值，该属性包含从Apress Web服务器获得的内容的长度。延续码如下：

```
...
return httpTask.ContinueWith((Task<HttpResponseMessage> antecedent) => {
   return antecedent.Result.Content.Headers.ContentLength;
});
...
```

注意使用了两次return关键字。这是导致混淆的部分。第一次使用return关键字指定返回一个Task<HttpResponseMessage>对象，当任务完成时，它将返回ContentLength头的长度。ContentLength头返回一个long?结果(可控的long值)，这意味着GetPageLength方法的结果是Task<long? >，如下所示：

```
...
public static Task<long?> GetPageLength() {
...
```

如果这没有任何意义，不必担心——你并不是唯一感到困惑的人。因此，微软在C#中添加了关键字来简化异步方法。

## 5.11.2 应用async和await关键字

微软在C#中引入了两个关键字，简化了使用像HttpClient.GetAsync这样的异步方法。关键字是async和await，可以在代码清单5-46中看到如何使用它们简化示例方法。

**代码清单 5-46　在 Models 文件夹的 MyAsyncMethods.cs 文件中使用 async 和 await 关键字**

```
using System.Net.Http;
using System.Threading.Tasks;

namespace LanguageFeatures.Models {

    public class MyAsyncMethods {

        public async static Task<long?> GetPageLength() {
            HttpClient client = new HttpClient();
            var httpMessage = await client.GetAsync("http://apress.com");
            return httpMessage.Content.Headers.ContentLength;
        }
    }
}
```

在调用异步方法时，使用了 await 关键字。这告诉 C#编译器，希望等待 GetAsync 方法返回的任务结果，然后继续执行相同方法的其他语句。

应用 await 关键字意味着可像对待常规方法一样对待 GetAsync 方法的结果，并将它返回的 HttpResponseMessage 对象分配给一个变量。更好的是，可以按照常规方式使用 return 关键字从另一种方法生成结果——在本例中是 ContentLength 属性的值。这是一种更自然的技术，这意味着不必担心 ContinueWith 方法和 return 关键字的多次使用。

在使用 await 关键字时，还必须向方法签名添加 async 关键字，如示例所示。方法结果类型没有改变——示例 GetPageLength 方法仍然返回一个 Task<long?>。这是因为 await 和 async 是使用一些聪明的编译技巧实现的，这意味着它们允许更自然的语法，但不会改变应用它们的方法中发生的事情。调用 GetPageLength 方法的人仍然要处理一个 Task<long? >结果，因为仍然有一个后台操作产生一个可空的长值——当然，程序员也可以选择使用 await 和 async 关键字。

此模式贯穿到控制器中，这使得编写异步操作方法更加容易，如代码清单 5-47 所示。

---

■ **注意：**也可在 lambda 表达式中使用 await 和 async 关键字，参见后面的章节。

---

**代码清单 5-47　Controllers 文件夹的 HomeController.cs 文件的一个异步操作方法**

```
using Microsoft.AspNetCore.Mvc;
using System.Collections.Generic;
using LanguageFeatures.Models;
using System;
using System.Linq;
using System.Threading.Tasks;

namespace LanguageFeatures.Controllers {
    public class HomeController : Controller {

        public async Task<ViewResult> Index() {
            long? length = await MyAsyncMethods.GetPageLength();
            return View(new string[] { $"Length: {length}" });
```

```
        }
    }
}
```

将 Index 操作方法的结果更改为 Task<ViewResult>，它声明操作方法将返回一个任务，该任务在完成时将生成 ViewResult 对象，提供应该呈现的视图的细节和它需要的数据。将 async 关键字添加到方法的定义中，这允许在调用 MyAsyncMethods.GetPathLength 方法时使用 await 关键字。.NET 方法负责处理延续，其结果是易于编写、易于阅读和易于维护的异步代码。重启 ASP.NET Core 并请求 http://localhost:5000，得到如下输出(但长度可能不同，因为 Apress 网站的内容经常变化)：

```
Length: 101868
```

### 5.11.3　使用异步枚举

异步枚举描述随时间生成的值序列。为演示该特性解决的问题，代码清单 5-48 向 MyAsyncMethods 类添加了一个方法。

**代码清单 5-48　在 Models 文件夹的 MyAsyncMethods.cs 文件中添加一个方法**

```
using System.Net.Http;
using System.Threading.Tasks;
using System.Collections.Generic;

namespace LanguageFeatures.Models {

    public class MyAsyncMethods {

        public async static Task<long?> GetPageLength() {
            HttpClient client = new HttpClient();
            var httpMessage = await client.GetAsync("http://apress.com");
            return httpMessage.Content.Headers.ContentLength;
        }

        public static async Task<IEnumerable<long?>>
                GetPageLengths(List<string> output, params string[] urls) {
            List<long?> results = new List<long?>();
            HttpClient client = new HttpClient();
            foreach (string url in urls) {
                output.Add($"Started request for {url}");
                var httpMessage = await client.GetAsync($"http://{url}");
                results.Add(httpMessage.Content.Headers.ContentLength);
                output.Add($"Completed request for {url}");
            }
            return results;
        }
    }
}
```

GetPageLengths 方法向一系列 Web 站点发出 HTTP 请求，并获取它们的长度。请求是异步执行的，但是无法在请求到达时将结果反馈给方法的调用者。相反，该方法将等待，直到所有请求完成，然后一次性返回所有结果。除了被请求的 URL 之外，该方法还接收 List<string>，在其中添加了消息，以便突出显示代码的工作方式。代码清单 5-49 更新主控制器的 Index 操作方法以使用新方法。

**代码清单 5-49　使用 Controllers 文件夹的 HomeController.cs 文件中的新方法**

```
using Microsoft.AspNetCore.Mvc;
using System.Collections.Generic;
using LanguageFeatures.Models;
using System;
using System.Linq;
using System.Threading.Tasks;

namespace LanguageFeatures.Controllers {
   public class HomeController : Controller {

      public async Task<ViewResult> Index() {
         List<string> output = new List<string>();
         foreach(long? len in await MyAsyncMethods.GetPageLengths(output,
               "apress.com", "microsoft.com", "amazon.com")) {
            output.Add($"Page length: { len}");
         }
         return View(output);
      }
   }
}
```

动作方法枚举 GetPageLengths 方法生成的序列，并将每个结果添加到 List<string>对象，生成一个有序的消息序列，显示 Index 方法中处理结果的 foreach 循环和 GetPageLengths 方法中生成它们的 foreach 循环之间的交互。重启 ASP.NET Core，请求 http://localhost:5000，会在浏览器中看到以下输出(可能需要几秒钟才能显示出来，页面长度也可能不同)：

```
Started request for apress.com
Completed request for apress.com
Started request for microsoft.com
Completed request for microsoft.com
Started request for amazon.com
Completed request for amazon.com
Page length: 101868
Page length: 159158
Page length: 91879
```

可以看到，在所有 HTTP 请求完成之前，Index 操作方法不会接收结果。这就是异步可枚举特性解决的问题，如代码清单 5-50 所示。

**代码清单 5-50　在 Models 文件夹的 MyAsyncMethods.cs 文件中使用一个异步枚举**

```
using System.Net.Http;
using System.Threading.Tasks;
using System.Collections.Generic;

namespace LanguageFeatures.Models {

    public class MyAsyncMethods {

        public async static Task<long?> GetPageLength() {
            HttpClient client = new HttpClient();
            var httpMessage = await client.GetAsync("http://apress.com");
            return httpMessage.Content.Headers.ContentLength;
        }
        public static async IAsyncEnumerable<long?>
                GetPageLengths(List<string> output, params string[] urls) {
            HttpClient client = new HttpClient();
            foreach (string url in urls) {
                output.Add($"Started request for {url}");
                var httpMessage = await client.GetAsync($"http://{url}");
                output.Add($"Completed request for {url}");
                yield return httpMessage.Content.Headers.ContentLength;
            }
        }
    }
}
```

方法的结果是 IAsyncEnumerable<long?>，表示可空的长值异步序列。这个结果类型在.NET Core 中有特殊的支持，可使用标准的 yield return 语句，没有这种支持就不可能使用该语句，因为异步方法的结果约束与 yield 关键字冲突。代码清单5-51更新控制器，以使用修改后的方法。

**代码清单 5-51　使用 Controllers 文件夹的 HomeController.cs 文件中的异步枚举**

```
using Microsoft.AspNetCore.Mvc;
using System.Collections.Generic;
using LanguageFeatures.Models;
using System;
using System.Linq;
using System.Threading.Tasks;

namespace LanguageFeatures.Controllers {
    public class HomeController : Controller {

        public async Task<ViewResult> Index() {
            List<string> output = new List<string>();
            await foreach(long? len in MyAsyncMethods.GetPageLengths(output,
                    "apress.com", "microsoft.com", "amazon.com")) {
```

```
            output.Add($"Page length: { len}");
        }
        return View(output);
    }
  }
}
```

区别在于 await 关键字在 foreach 关键字之前应用，而不是在调用 async 方法之前应用。重启 ASP.NET Core，请求 http://localhost:5000；一旦 HTTP 请求完成，响应消息的顺序就已经改变，如下所示：

```
Started request for apress.com
Completed request for apress.com
Page length: 101868
Started request for microsoft.com
Completed request for microsoft.com
Page length: 159160
Started request for amazon.com
Completed request for amazon.com
Page length: 91674
```

控制器接收序列中生成的下一个结果。如第 19 章所述，ASP.NET Core 特别支持在 Web 服务中使用 IAsyncEnumerable<T>结果，允许在生成序列中的值时序列化数据值。

## 5.12 获取名称

在 Web 应用程序开发中，有许多任务需要引用参数、变量、方法或类的名称。常见例子包括在处理用户输入时抛出异常或创建验证错误。传统方法是使用硬编码的名称字符串值，如代码清单 5-52 所示。

**代码清单 5-52 在 Controllers 文件夹的 HomeController.cs 文件中硬编码一个名称**

```
using Microsoft.AspNetCore.Mvc;
using System.Collections.Generic;
using LanguageFeatures.Models;
using System;
using System.Linq;
using System.Threading.Tasks;

namespace LanguageFeatures.Controllers {
  public class HomeController : Controller {

    public ViewResult Index() {
      var products = new[] {
        new { Name = "Kayak", Price = 275M },
        new { Name = "Lifejacket", Price = 48.95M },
        new { Name = "Soccer ball", Price = 19.50M },
```

```
                new { Name = "Corner flag", Price = 34.95M }
            };
            return View(products.Select(p => $"Name: {p.Name}, Price: {p.Price}"));
        }
    }
}
```

对 LINQ Select 方法的调用生成一系列字符串，每个字符串都包含对 Name 和 Price 属性的硬编码引用。重启 ASP.NET Core，并请求 http://localhost:5000，在浏览器窗口中显示以下输出：

```
Name: Kayak, Price: 275
Name: Lifejacket, Price: 48.95
Name: Soccer ball, Price: 19.50
Name: Corner flag, Price: 34.95
```

这种方法很容易出错，要么是因为名称输入错误，要么是因为代码被重构而字符串中的名称没有得到正确更新。C#支持 nameof 表达式，其中编译器负责生成名称字符串，如代码清单 5-53 所示。

**代码清单 5-53　在 Controllers 文件夹的 HomeController.cs 文件中使用 nameof 表达式**

```
using Microsoft.AspNetCore.Mvc;
using System.Collections.Generic;
using LanguageFeatures.Models;
using System;
using System.Linq;
using System.Threading.Tasks;

namespace LanguageFeatures.Controllers {
    public class HomeController : Controller {

        public ViewResult Index() {
            var products = new[] {
                new { Name = "Kayak", Price = 275M },
                new { Name = "Lifejacket", Price = 48.95M },
                new { Name = "Soccer ball", Price = 19.50M },
                new { Name = "Corner flag", Price = 34.95M }
            };
            return View(products.Select(p =>
                $"{nameof(p.Name)}: {p.Name}, {nameof(p.Price)}: {p.Price}"));
        }
    }
}
```

编译器处理引用，例如 p.Name，以便只包含字符串的最后一部分，生成与前面示例相同的输出。具有对 nameof 表达式的智能感知支持，因此当重构代码时，智能感知支持将提示选择引用，并正确更新表达式。由于编译器负责处理 nameof，使用无效引用会导致编译器错误，从而防止不正确或过时的引用。

## 5.13 小结

本章概述了高效的 ASP.NET Core 程序员需要知道的 C#语言的关键特性。C#是一种非常灵活的语言，通常有不同的方法来解决任何问题，但这些特性是在 Web 应用程序开发过程中最常遇到的，本书的示例中会使用它们。第 6 章将解释如何为 ASP.NET Core 建立单元测试项目。

# 第6章

# 测试 ASP.NET Core 应用程序

本章将演示如何对 ASP.NET Core 应用程序进行单元测试。单元测试是一种测试形式，在这种测试中，单个组件与应用程序的其余部分隔离，以便能够彻底验证它们的行为。ASP.NET Core 的设计使创建单元测试变得容易，并且支持广泛的单元测试框架。本章展示如何设置单元测试项目，并描述编写和运行测试的过程。表 6-1 总结了本章的内容。

**决定是否进行单元测试**

能够轻松地执行单元测试是使用 ASP.NET Core 的好处之一，但它并不适合每个人。

我喜欢单元测试，常常在项目中使用它，但并不是所有项目都使用它。我倾向于为那些很难编写、可能会在部署中产生 bug 的特性和功能编写单元测试。在这些情况下，单元测试有助于理解如何以最合理的方式实现所需功能。在开始处理实际的 bug 和缺陷之前，仅考虑需要测试的内容就有助于了解潜在问题。

也就是说，单元测试是一种工具，而不是一种信仰，只有自己知道需要多少测试。如果不认为单元测试有用，或者有更适合自己的不同方法，那么不要仅因为它流行就觉得需要进行单元测试。

如果以前没有遇到过单元测试，那么建议尝试一下，看看它是如何工作的。如果不喜欢单元测试，可以跳过这一章，转到第 7 章，从那里开始构建一个更现实的 ASP.NET Core 应用程序。

表 6-1 本章内容摘要

| 问题 | 解决方案 | 代码清单 |
|---|---|---|
| 创建单元测试项目 | 为首选测试框架使用 dotnet new 命令和项目模板 | 6-7 |
| 创建 XUnit 测试断言 | 用通过 Fact 属性修饰的方法创建一个类，并使用该类检查测试结果 | 6-9 |
| 运行单元测试 | 使用 Visual Studio 或 Visual Studio Code 测试运行器，或使用 dotnet test 命令 | 6-11 |
| 隔离组件进行测试 | 创建被测试组件需要的对象的模拟实现 | 6-12～6-19 |

## 6.1　准备工作

为了准备这一章，需要创建一个简单的 ASP.NET Core 项目。使用 Windows“开始”菜单打开一个新的 PowerShell 命令提示符，导航到一个方便的位置，并运行代码清单 6-1 所示的命令。

■ **提示：**可以从 https://github.com/apress/pro-asp.net-core-3 下载本章和本书中所有其他章节的示例项目。如果在运行示例时遇到问题，请参阅第 1 章以获得帮助。

**代码清单 6-1　创建示例项目**

```
dotnet new globaljson --sdk-version 3.1.101 --output Testing/SimpleApp
dotnet new web --no-https --output Testing/SimpleApp --framework netcoreapp3.1
dotnet new sln -o Testing

dotnet sln Testing add Testing/SimpleApp
```

这些命令使用 Web 模板创建一个名为 SimpleApp 的新项目，其中包含 ASP.NET Core 应用程序的最小配置。项目文件夹包含在解决方案文件夹 Testing 中。

### 6.1.1　打开项目

如果使用 Visual Studio，选择 File|Open|Project/Solution，然后选择 Testing.sln 文件。单击 Open 按钮，打开解决方案文件及其引用的项目。如果使用的是 Visual Studio Code，选择 File|Open Folder，导航到 Testing 文件夹，然后单击 Select Folder 按钮。

### 6.1.2　选择 HTTP 端口

如果使用的是 Visual Studio，选择 Project | SimpleApp 属性，选择 Debug 部分，并在 App URL 字段中将 HTTP 端口更改为 5000，如图 6-1 所示。选择 File|Save All，以保存新端口(如果使用的是 Visual Studio Code，则不需要进行此更改)。

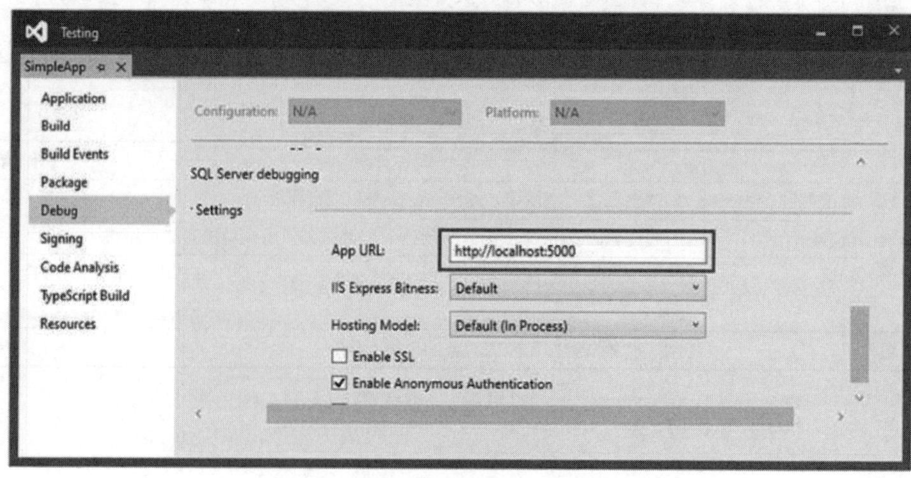


图 6-1　设置 HTTP 端口

### 6.1.3 启用 MVC 框架

如第 1 章所述，ASP.NET Core 支持不同的应用框架，但本章继续使用 MVC 框架。第 7 章开始构建 SportsStore 应用程序时会介绍其他框架，但目前，MVC 框架有助于演示如何执行与前面示例相似的单元测试。将代码清单 6-2 所示的语句添加到 SimpleApp 文件夹的 Startup.cs 文件中。

**代码清单 6-2　在 SimpleApp 文件夹的 Startup.cs 文件中启用 MVC 框架**

```
using System;
using System.Collections.Generic;
using System.Linq;
using System.Threading.Tasks;
using Microsoft.AspNetCore.Builder;
using Microsoft.AspNetCore.Hosting;
using Microsoft.AspNetCore.Http;
using Microsoft.Extensions.DependencyInjection;
using Microsoft.Extensions.Hosting;

namespace SimpleApp {
   public class Startup {

      public void ConfigureServices(IServiceCollection services) {
         services.AddControllersWithViews();
      }

      public void Configure(IApplicationBuilder app, IWebHostEnvironment env) {
         if (env.IsDevelopment()) {
            app.UseDeveloperExceptionPage();
         }

         app.UseRouting();

         app.UseEndpoints(endpoints => {
            endpoints.MapDefaultControllerRoute();
            //endpoints.MapGet("/", async context => {
            // await context.Response.WriteAsync("Hello World!");
            //});
         });
      }
   }
}
```

### 6.1.4 创建应用程序组件

现在 MVC 框架已经设置好了，可以添加用于演示重要 C#语言特性的应用程序组件了。

### 1. 创建数据模型

首先创建一个简单的模型类，这样就可以使用一些数据。添加一个名为 Models 的文件夹，并在其中创建一个名为 Product.cs 的类文件，用于定义代码清单 6-3 所示的类。

**代码清单 6-3 SimpleApp/Models 文件夹中 Product.cs 文件的内容**

```
namespace SimpleApp.Models {
    public class Product {

        public string Name { get; set; }
        public decimal? Price { get; set; }
        public static Product[] GetProducts() {

            Product kayak = new Product {
                Name = "Kayak", Price = 275M
            };

            Product lifejacket = new Product {
                Name = "Lifejacket", Price = 48.95M
            };

            return new Product[] { kayak, lifejacket };
        }
    }
}
```

Product 类定义 Name 和 Price 属性，有一个名为 GetProducts 的静态方法返回 Products 数组。

### 2. 创建控制器和视图

本章中的示例使用一个简单的控制器类来演示不同的语言特性。创建一个 Controllers 文件夹，添加一个名为 HomeController.cs 的类文件，其内容如代码清单 6-4 所示。

**代码清单 6-4 SimpleApp/Controllers 文件夹中 HomeController.cs 文件的内容**

```
using Microsoft.AspNetCore.Mvc;
using SimpleApp.Models;

namespace SimpleApp.Controllers {
    public class HomeController : Controller {

        public ViewResult Index() {
            return View(Product.GetProducts());
        }
    }
}
```

Index 操作方法告诉 ASP.NET Core 呈现默认视图，并为其提供从 Product.GetProducts 方法中获得的 Product 对象。为给操作方法创建视图，添加了一个 Views/Home 文件夹(先创建 Views

文件夹，然后在其中添加 Home 文件夹)，再添加一个名为 Index.cshtml 的 Razor 视图，内容如代码清单 6-5 所示。

**代码清单 6-5 SimpleApp/Views/Home 文件夹中 Index.cshtml 文件的内容**

```
@using SimpleApp.Models
@model IEnumerable<Product>
@{ Layout = null; }

<!DOCTYPE html>
<html>
<head>
    <meta name="viewport" content="width=device-width" />
    <title>Simple App</title>
</head>
<body>
    <ul>
        @foreach (Product p in Model) {
            <li>Name: @p.Name, Price: @p.Price</li>
        }
    </ul>
</body>
</html>
```

### 6.1.5 运行示例应用程序

启动 ASP。从 Debug 菜单中选择 Start Without Debugging (Visual Studio) 或 Run Without Debugging (Visual Studio Code)，或者在 SimpleApp 文件夹中运行代码清单 6-6 所示的命令，启动 ASP.NET Core。

**代码清单 6-6 运行示例应用程序**

```
dotnet run
```

请求 http://localhost:5000，将看到如图 6-2 所示的输出。

图 6-2 运行示例应用程序

## 6.2 创建单元测试项目

在 ASP.NET Core 应用程序中，通常创建一个单独的 Visual Studio 项目来保存单元测试，每个单元测试都定义为C#类中的一个方法。使用单独的项目意味着可在不部署测试的情况下部署应用程序。.NET Core SDK 包括使用三种流行测试工具的单元测试项目模板，如表 6-2 所示。

表 6-2　单元测试项目工具

| 名称 | 描述 |
|---|---|
| mstest | 该模板创建一个为 MS 测试框架配置的项目，该框架由 Microsoft 生成 |
| nunit | 该模板创建一个为 NUnit 框架配置的项目 |
| xunit | 该模板创建一个为 XUnit 框架配置的项目 |

这些测试框架具有基本相同的特性集，不同之处在于它们的实现方式以及如何集成到第三方测试环境中。如果没有确定的首选项，建议从 XUnit 开始，这主要是因为它是最容易使用的测试框架。

惯例是将单元测试项目命名为<ApplicationName>.Tests。在 Testing 文件夹中运行代码清单 6-7 所示的命令，创建名为 SimpleApp.Tests 的 XUnit 测试项目，将其添加到解决方案文件中，并在项目之间创建引用，以便单元测试可应用于 SimpleApp 项目中定义的类。

**代码清单 6-7　创建单元测试项目**

```
dotnet new xunit -o SimpleApp.Tests --framework netcoreapp3.1
dotnet sln add SimpleApp.Tests
dotnet add SimpleApp.Tests reference SimpleApp
```

如果使用的是 Visual Studio，系统将提示重新加载解决方案，这将导致新的单元测试项目在 Solution Explorer 中与现有项目一起显示。Visual Studio Code 不会生成新项目。如果出现这种情况，选择 Terminal | Configure Default Build Task，从列表中选择 Build；如果有提示，从环境列表中选择.NET Core。

**删除默认测试类**

项目模板向测试项目添加了一个 C#类文件，这将混淆后面示例的结果。使用 Solution Explorer 或 File Explorer 窗格删除 SimpleApp.Tests 文件夹中的 UnitTest1.cs 文件，或在 Testing 文件夹中运行代码清单 6-8 中所示的命令。

**代码清单 6-8　删除默认的测试类文件**

```
Remove-Item SimpleApp.Tests/UnitTest1.cs
```

## 6.3　编写和运行单元测试

现在已经完成了所有准备工作，可以编写一些测试了。首先，向 SimpleApp.Tests 添加一个名为 ProductTests.cs 的类文件，并使用它定义如代码清单 6-9 所示的类。这是一个简单的类，但包含了开始进行单元测试需要的所有内容。

■ **注意：** CanChangeProductPrice 方法包含一个故意的错误，本节的后面将更正这个错误。

**代码清单 6-9　SimpleApp.Tests 文件夹中 ProductTests.cs 文件的内容**

```
using SimpleApp.Models;
using Xunit;

namespace SimpleApp.Tests {

    public class ProductTests {

        [Fact]
        public void CanChangeProductName() {

            // Arrange
            var p = new Product { Name = "Test", Price = 100M };

            // Act
            p.Name = "New Name";

            //Assert
            Assert.Equal("New Name", p.Name);
        }

        [Fact]
        public void CanChangeProductPrice() {

            // Arrange
            var p = new Product { Name = "Test", Price = 100M };

            // Act
            p.Price = 200M;

            //Assert
            Assert.Equal(100M, p.Price);
        }
    }
}
```

ProductTests 类中有两个单元测试，分别用于测试 SimpleApp 项目中 Product 模型类的一个行为。一个测试项目可以包含许多类，每个类可以包含许多单元测试。

通常，测试方法的名称描述测试所做的工作，类的名称描述正在测试的内容。这很容易在项目中构造测试，理解所有测试在 Visual Studio 运行时的结果。名称 ProductTests 表示类包含针对 Product 类的测试，方法名称表示它们测试更改 Product 对象的名称和价格的能力。

将 Fact 属性应用于每个方法，以表明它是一个测试。在方法体中，单元测试遵循名为 arrange、act、assert (A / A / A)的模式。arrange 指的是为测试设置条件，act 指的是执行测试，assert 指的是验证结果是否符合预期。

这些测试的 arrange 和 act 部分是常规的 C#代码。不过，assert 部分是由 xUnit.net 处理的，它提供了一个名为 assert 的类，它的方法用于检查操作的结果是否符合预期。

■ **提示：** Fact 属性和 Asset 类是在 Xunit 名称空间中定义的，因此在每个测试类中必须有一个 using 语句。

Assert 类的方法是静态的，用于在预期结果和实际结果之间执行不同类型的比较。表 6-3 显示了常用的 Assert 方法。

表 6-3　常用的 xUnit.net Assert 方法

| 名称 | 描述 |
| --- | --- |
| Equal(expected, result) | 这个方法断言结果符合预期。此方法的重载版本用于比较不同的类型和集合。该方法还有一个版本，它接收一个对象的附加参数，该对象实现 IEqualityComparer<T>接口，用于比较对象 |
| NotEqual(expected, result) | 此方法断言结果不符合预期 |
| True(result) | 该方法断言结果为真 |
| False(result) | 该方法断言结果为假 |
| IsType(expected, result) | 这个方法断言结果是特定类型的 |
| IsNotType(expected, result) | 这个方法断言结果不是特定的类型 |
| IsNull(result) | 该方法断言结果为空 |
| IsNotNull(result) | 这个方法断言结果不是空的 |
| InRange(result, low, high) | 这个方法断言结果在 low 和 high 之间 |
| NotInRange(result, low, high) | 该方法断言结果处于 low 和 high 之外 |
| Throws(exception, expression) | 该方法断言指定的表达式抛出特定的异常类型 |

每个 Assert 方法都允许进行不同类型的比较，如果结果不同于预期，则会抛出异常。异常用于指示测试失败。在代码清单 6-9 的测试中，使用了 Equal 方法来确定属性的值是否已正确更改。

```
...
Assert.Equal("New Name", p.Name);
...
```

### 6.3.1　使用 Visual Studio Test Explorer 运行测试

Visual Studio 支持通过 Test Explorer 窗口查找和运行单元测试，这可以通过 Test | Test Explorer 菜单获得，如图 6-3 所示。

■ **提示：** 如果在 Test Explorer 窗口中看不到单元测试，请构建解决方案。编译过程会触发发现单元测试的进程。

单击 Test Explorer 窗口中的 Run All Tests 按钮来运行测试(该按钮显示两个箭头，是窗口顶部的第一个按钮)。如上所述，CanChangeProductPrice 测试包含一个导致测试失败的错误，这在图中所示的测试结果中明确指出。

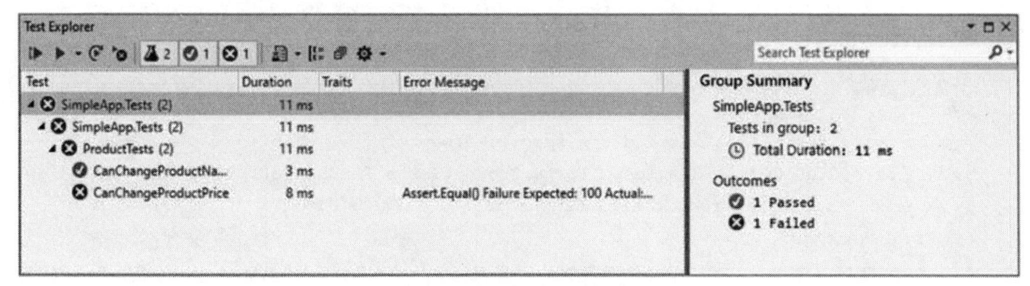

图 6-3　Visual Studio Test Explorer

## 6.3.2　使用 Visual Studio Code 运行测试

Visual Studio Code 检测测试，并允许使用 code lens 功能运行它们，该功能在编辑器中显示有关代码功能的详细信息。要运行 ProductTests 类中的所有测试，在打开单元测试类时，单击代码编辑器中的 Run All Tests，如图 6-4 所示。

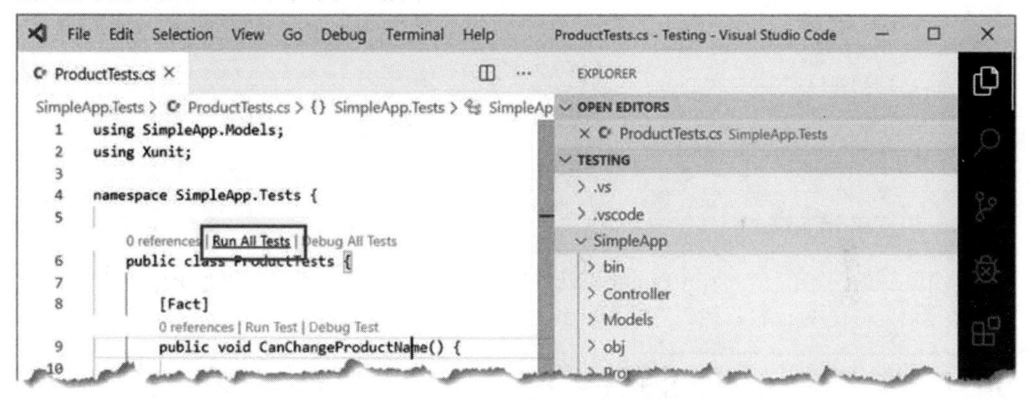

图 6-4　使用 Visual Studio Code 的 code lens 特性运行测试

■ **提示**：如果没有看到 code lens 测试特性，请关闭并在 Visual Studio Code 中重新打开 Testing 文件夹。

Visual Studio Code 使用下一节描述的命令行工具运行测试，结果将以文本形式显示在终端窗口中。

## 6.3.3　从命令行运行测试

要在项目中运行测试，请在 Testing 文件夹中运行代码清单 6-10 所示的命令。

**代码清单 6-10　运行单元测试**

```
dotnet test
```

测试被发现并执行，生成以下结果，这些结果显示了之前故意引入的错误：

```
Test run for C:\Users\adam\SimpleApp.Tests.dll(.NETCoreApp,Version=v3.1)
Microsoft (R) Test Execution Command Line Tool Version 16.3.0
```

```
Copyright (c) Microsoft Corporation. All rights reserved.
```

开始测试执行，请等待

```
A total of 1 test files matched the specified pattern.
[xUnit.net 00:00:00.83] SimpleApp.Tests.ProductTests.CanChangeProductPrice [FAIL]
   X SimpleApp.Tests.ProductTests.CanChangeProductPrice [6ms]
   Error Message:
      Assert.Equal() Failure
Expected: 100
Actual: 200
   Stack Trace:
      at SimpleApp.Tests.ProductTests.CanChangeProductPrice() in
      C:\Users\adam\Documents\Books\Pro ASP.NET Core MVC 3\Source Code\Current\Testin
g\SimpleApp.Tests\ProductTests.cs:line 31

Test Run Failed.
Total tests: 2
      Passed: 1
      Failed: 1
Total time: 1.7201 Seconds
```

### 6.3.4 纠正单元测试

单元测试的问题在于Assert.Equal方法的参数，该方法将测试结果与原始Price属性值(而不是改变后的值)进行比较。代码清单6-11纠正了这个问题。

**提示：**当测试失败时，最好先检查测试的准确性，然后查看测试的目标部件；如果测试是新的或最近修改过的，尤其应该这样做。

**代码清单6-11　纠正SimpleApp.Tests文件夹中ProductTests.cs文件的测试**

```
using SimpleApp.Models;
using Xunit;

namespace SimpleApp.Tests {

   public class ProductTests {

      [Fact]
      public void CanChangeProductName() {

         // Arrange
         var p = new Product { Name = "Test", Price = 100M };

         // Act
         p.Name = "New Name";
```

```
            //Assert
            Assert.Equal("New Name", p.Name);
        }

        [Fact]
        public void CanChangeProductPrice() {

            // Arrange
            var p = new Product { Name = "Test", Price = 100M };

            // Act
            p.Price = 200M;

            //Assert
            Assert.Equal(200M, p.Price);
        }
    }
}
```

再次运行测试，它们都通过了。如果使用 Visual Studio，可以单击 Run Failed Tests 按钮，只执行失败的测试，如图 6-5 所示。

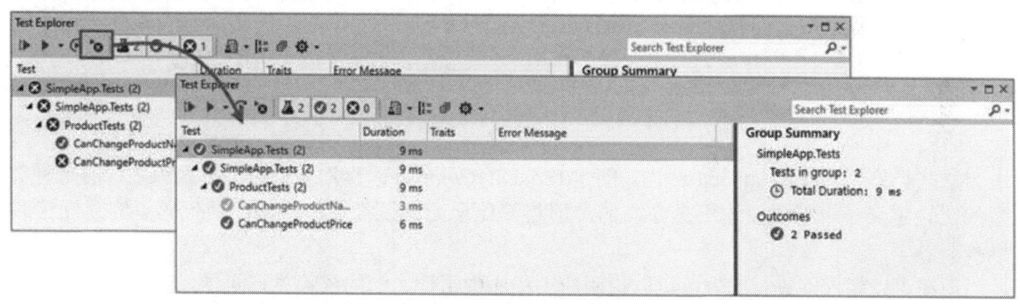


图 6-5　只运行失败的测试

## 6.3.5 为单元测试隔离组件

为像 Product 这样的模型类编写单元测试很容易。Product 类不仅简单，而且是自包含的，这意味着，对 Product 对象执行操作时，可以确信正在测试 Product 类提供的功能。

如果使用 ASP.NET Core 中的其他组件，情况就会更加复杂，因为它们之间存在依赖关系。定义的下一组测试对控制器进行操作，检查在控制器和视图之间传递的 Product 对象的序列。

在比较自定义类中实例化的对象时，需要使用 xUnit.net 的 Assert.Equal 方法。该方法接收实现 IEqualityComparer<T>接口的参数，以便比较对象。第一步是将一个名为 Comparer.cs 的类文件添加到单元测试项目中，并使用它定义如代码清单 6-12 所示的 helper 类。

**代码清单 6-12　在 SimpleApp.Tests 文件夹中比较 Comparer.cs 文件的内容**

```
using System;
using System.Collections.Generic;
```

```
namespace SimpleApp.Tests {

    public class Comparer {

        public static Comparer<U> Get<U>(Func<U, U, bool> func) {
            return new Comparer<U>(func);
        }
    }

    public class Comparer<T> : Comparer, IEqualityComparer<T> {
        private Func<T, T, bool> comparisonFunction;

        public Comparer(Func<T, T, bool> func) {
            comparisonFunction = func;
        }

        public bool Equals(T x, T y) {
            return comparisonFunction(x, y);
        }

        public int GetHashCode(T obj) {
            return obj.GetHashCode();
        }
    }
}
```

这些类将允许使用lambda表达式创建IEqualityComparer<T>对象，而不必为想进行的每种类型的比较定义一个新类。这不是必需的，但将简化单元测试类中的代码，使代码更容易阅读和维护。

现在可以轻松地进行比较，可以说明应用程序中组件之间的依赖关系问题。

在SimpleApp.Tests中添加了一个名为HomeControllerTests.cs的新类。并使用它定义单元测试，如代码清单6-13所示。

**代码清单6-13　SimpleApp.Tests文件夹中的HomeControllerTests.cs文件**

```
using Microsoft.AspNetCore.Mvc;
using System.Collections.Generic;
using SimpleApp.Controllers;
using SimpleApp.Models;
using Xunit;

namespace SimpleApp.Tests {
    public class HomeControllerTests {

        [Fact]
        public void IndexActionModelIsComplete() {
            // Arrange
```

```
            var controller = new HomeController();
            Product[] products = new Product[] {
                new Product { Name = "Kayak", Price = 275M },
                new Product { Name = "Lifejacket", Price = 48.95M}
            };

            // Act
            var model = (controller.Index() as ViewResult)?.ViewData.Model
                as IEnumerable<Product>;

            // Assert
            Assert.Equal(products, model,
                Comparer.Get<Product>((p1, p2) => p1.Name == p2.Name
                    && p1.Price == p2.Price));
        }
    }
}
```

单元测试创建一个Product对象数组，并检查它们是否与Index操作方法提供的视图模型相对应(暂时忽略测试的act部分；第21和22章将解释ViewResult类。目前，只要知道正在获取由Index操作方法返回的模型数据就足够了)。

测试通过了，但它不是一个有用的结果，因为正在测试的Product数据来自硬连接对象的Product类。例如，当有两个以上的Product对象时，或者第一个对象的Price属性具有小数部分时，就无法编写测试来确保控制器正确运行。总体效果是，测试HomeController和Product类的组合行为，并且只针对特定的硬连接对象。

当单元测试针对应用程序的小部分(如单个方法或类)时，它们是有效的。需要的能力是将主控制器与应用程序的其他部分隔离开来，这样就可以限制测试的范围，并排除存储库造成的任何影响。

### 隔离组件

隔离组件的关键是使用C#接口。为将控制器与存储库分离，在Models文件夹中添加一个名为IDataSource.cs的类，并使用它定义如代码清单6-14所示的接口。

**代码清单6-14　SimpleApp/Models文件夹中IDataSource.cs文件的内容**

```
using System.Collections.Generic;

namespace SimpleApp.Models {
    public interface IDataSource {

        IEnumerable<Product> Products { get; }
    }
}
```

在代码清单6-15中，从Product类中删除了静态方法，并创建了一个实现IDataSource接口的类。

**代码清单 6-15　在 SimpleApp/Models 文件夹的 Product.cs 文件中创建一个数据源**

```
using System.Collections.Generic;

namespace SimpleApp.Models {

    public class Product {
        public string Name { get; set; }
        public decimal? Price { get; set; }
    }

    public class ProductDataSource : IDataSource {
        public IEnumerable<Product> Products =>
            new Product[] {
                new Product { Name = "Kayak", Price = 275M },
                new Product { Name = "Lifejacket", Price = 48.95M }
            };
    }
}
```

下一步是修改控制器，使其使用 ProductDataSource 类作为其数据的源，如代码清单 6-16 所示。

---

**提示：** ASP.NET Core 支持一种更优雅的方法来解决这个问题，称为依赖注入，参见第 14 章。依赖注入通常会引起混淆，因此本章用一种更简单、更手动的方式隔离组件。

---

**代码清单 6-16　在 SimpleApp/Controllers 文件夹的 HomeController.cs 文件中添加一个属性**

```
using Microsoft.AspNetCore.Mvc;
using SimpleApp.Models;

namespace SimpleApp.Controllers {
    public class HomeController : Controller {
        public IDataSource dataSource = new ProductDataSource();

        public ViewResult Index() {
            return View(dataSource.Products);
        }
    }
}
```

这看起来可能不是一个显著变化，但它允许在测试期间更改控制器使用的数据源，这就是隔离控制器的方式。在代码清单 6-17 中，更新了控制器单元测试，以便它们使用特定版本的存储库。

**代码清单 6-17　在 SimpleApp.Tests 文件夹的 HomeControllerTests.cs 文件中隔离控制器**

```
using Microsoft.AspNetCore.Mvc;
using System.Collections.Generic;
using SimpleApp.Controllers;
```

```
using SimpleApp.Models;
using Xunit;

namespace SimpleApp.Tests {
    public class HomeControllerTests {

        class FakeDataSource : IDataSource {
            public FakeDataSource(Product[] data) => Products = data;
            public IEnumerable<Product> Products { get; set; }
        }

        [Fact]
        public void IndexActionModelIsComplete() {
            // Arrange
            Product[] testData = new Product[] {
                new Product { Name = "P1", Price = 75.10M },
                new Product { Name = "P2", Price = 120M },
                new Product { Name = "P3", Price = 110M }
            };
            IDataSource data = new FakeDataSource(testData);
            var controller = new HomeController();
            controller.dataSource = data;

            // Act
            var model = (controller.Index() as ViewResult)?.ViewData.Model
                as IEnumerable<Product>;

            // Assert
            Assert.Equal(data.Products, model,
                Comparer.Get<Product>((p1, p2) => p1.Name == p2.Name
                    && p1.Price == p2.Price));
        }
    }
}
```

我定义了IDataSource接口的一个伪实现，它允许使用控制器中的任何测试数据。

**理解测试驱动的开发**

本章遵循了最常用的单元测试风格，即编写一个应用程序特性，然后测试以确保它按要求工作。这很流行，因为大多数开发人员首先考虑应用程序代码，然后才是测试(我当然就属于这一类)。

这种方法倾向于生成单元测试，这些单元测试只关注应用程序代码中难以编写或需要进行严格调试的部分，只对特性的某些方面进行部分测试，或完全未测试。

另一种方法是测试驱动的开发(Test-Driven Development，TDD)。TDD有许多变体，但其核心思想是在实现某个特性本身之前编写该特性的测试。首先编写测试会让开发人员更仔细地考虑正在实现的规范，以及如何知道某个特性已经正确实现。TDD不需要深入实现细节，而是提前考虑成功或失败的度量。

我们编写的测试一开始都会失败，因为新特性没有实现。但是，当向应用程序添加代码时，测试将逐渐从红色转移到绿色，所有测试都将在该特性完成时通过。TDD 需要规程，但是它确实产生了一组更全面的测试，并且可以生成更健壮、可靠的代码。

### 6.3.6　使用模拟包

为 IDataSource 接口创建假实现是很容易的，但是大多数需要假实现的类更加复杂，并且不容易处理。

更合适的方法是使用模拟包，这样可以很容易地为测试创建假对象或模拟对象。有许多模仿包可用，这里使用的那个叫做 Moq(我使用了多年)。要将 Moq 添加到单元测试项目中，请在 Testing 文件夹中运行代码清单 6-18 所示的命令。

**注意：** Moq 包被添加到单元测试项目中，而非包含被测试应用程序的项目中。

**代码清单 6-18　安装模拟包**

```
dotnet add SimpleApp.Tests package Moq --version 4.13.1
```

### 6.3.7　创建模拟对象

可以使用 Moq 框架创建一个假的 IDataSource 对象，而不必定义定制的测试类，如代码清单 6-19 所示。

**代码清单 6-19　在 SimpleApp.Tests 文件夹的 HomeControllerTests.cs 文件中创建模拟对象**

```
using Microsoft.AspNetCore.Mvc;
using System.Collections.Generic;
using SimpleApp.Controllers;
using SimpleApp.Models;
using Xunit;
using Moq;

namespace SimpleApp.Tests {
   public class HomeControllerTests {

      //class FakeDataSource : IDataSource {
      //    public FakeDataSource(params Product[] data) => Products = data;
      //    public IEnumerable<Product> Products { get; set; }
      //}

      [Fact]
      public void IndexActionModelIsComplete() {

         // Arrange
         Product[] testData = new Product[] {
            new Product { Name = "P1", Price = 75.10M },
```

```
                new Product { Name = "P2", Price = 120M },
                new Product { Name = "P3", Price = 110M }
            };
            var mock = new Mock<IDataSource>();
            mock.SetupGet(m => m.Products).Returns(testData);
            var controller = new HomeController();
            controller.dataSource = mock.Object;

            // Act
            var model = (controller.Index() as ViewResult)?.ViewData.Model
                as IEnumerable<Product>;

            // Assert
            Assert.Equal(testData, model,
                Comparer.Get<Product>((p1, p2) => p1.Name == p2.Name
                    && p1.Price == p2.Price));
            mock.VerifyGet(m => m.Products, Times.Once);
        }
    }
}
```

使用 Moq 允许删除 IDataSource 接口的假实现，并用几行代码替换它。这里不打算详细介绍 Moq 支持的不同特性，但解释本例使用 Moq 的方式(可从 https://github.com/Moq/moq4 获取 Moq 的示例和文档。后续章节还会举例说明如何对不同类型的组件进行单元测试)。

第一步是创建模拟对象的新实例，指定应该实现的接口，如下所示：

```
...
var mock = new Mock<IDataSource>();
...
```

创建的模拟对象将模拟 IDataSource 接口。要创建 Product 属性的实现，使用 SetUpGet 方法，如下所示：

```
...
mock.SetupGet(m => m.Products).Returns(testData);
...
```

SetupGet 方法用于实现属性的 getter。此方法的参数是一个 lambda 表达式，它指定要实现的属性，在本例中为 Products。对 SetupGet 方法的结果调用 Returns 方法，以指定读取属性值时将返回的结果。

模拟类定义一个对象属性，该属性返回使用已定义的行为实现指定接口的对象。这里用 Object 属性来设置 HomeController 定义的 dataSource 字段，如下所示：

```
...
controller.dataSource = mock.Object;
...
```

使用的最后一个 Moq 特性是检查 Products 属性是否被调用了一次，如下所示：

```
...
mock.VerifyGet(m => m.Products, Times.Once);
...
```

VerifyGet 方法是模拟类定义的方法之一，用于在测试完成时检查模拟对象的状态。在这种情况下，VerifyGet 方法允许检查 Products 属性方法被读取的次数。Times.Once 值指定 VerifyGet 方法在属性未被准确读取一次时应抛出异常，这会导致测试失败(测试中使用的 Assert 方法通常在测试失败时抛出异常，这就是在使用模拟对象时可以使用 VerifyGet 方法替换 Assert 方法的原因)。

总体效果与伪接口实现相同，但是 mock 更灵活、更简洁，可以更深入地了解被测组件的行为。

## 6.4 小结

本章的重点是单元测试，它是提高代码质量的强大工具。单元测试并不适合每个开发人员，但值得尝试，即使只用于复杂的特性或问题诊断，它也很有用。本章描述了 xUnit.net 测试框架的使用，解释了隔离测试组件的重要性，并演示了一些简化单元测试代码的工具和技术。第 7 章开始开发一个更现实的项目，名为 SportsStore。

# 第7章

# SportsStore：一个真正的应用程序

前几章构建了快速而简单的 ASP.NET Core 应用程序。描述了 ASP.NET Core 模式、基本的 C#特性，介绍了优秀 ASP.NET Core 开发人员需要的工具。现在是时候把所有东西放在一起，构建一个简单但现实的电子商务应用程序了。

这个应用程序名为 SportsStore，它将遵循所有在线商店采用的经典方法。本章创建一个客户可以按类别和页面浏览的在线产品目录，一个用户可以添加和删除产品的购物车，以及一个客户可以输入其运输细节的结账处。还要创建一个管理区域，其中包括用于管理目录的 CRUD(创建、读取、更新和删除)工具，对它进行保护，以便只有登录的管理员才能进行更改。

本章和后续章节的目标是通过创建一个尽可能真实的示例，提供真正的 ASP.NET Core 开发过程。当然，重点是 ASP.NET Core。因此简化了与外部系统(如数据库)的集成，完全省略了其他系统，如支付处理。

构建所需的基础设施级别时，进展有点慢，但最初的投资将生成可维护、可扩展、结构良好的代码，并对单元测试提供出色的支持。

**单元测试**

本章包括了在整个开发过程中对 SportsStore 应用程序中的不同组件进行单元测试，演示了如何隔离和测试不同的 ASP.NET Core 组件。

并不是每个人都支持单元测试。如果不想进行单元测试，也可以这样做。因此，有一些纯粹关于测试的内容要提及时，就把它放在一个像这样的侧边栏中。如果对单元测试不感兴趣，可以直接跳过这些部分，这样 SportsStore 应用程序就可以正常工作了。不需要做任何类型的单元测试来获得 ASP.NET Core 的技术优势。当然，在很多项目中，对测试的支持是采用 ASP.NET Core 的一个关键原因。

在 SportsStore 应用程序中使用的大多数特性在本书后面都有独立的章节。本章没有重复这里的所有内容，只是说明示例应用程序是有意义的，并请参考另一章，以获得深入的信息。

本章将叙述构建应用程序所需的每个步骤，以便你了解 ASP.NET Core 功能如何组合在一起。创建视图时，应该特别注意。如果不紧跟示例，就会得到一些奇怪的结果。

# 7.1　创建项目

下面从一个最小的 ASP.NET Core 项目开始，添加需要的特性。从 Windows “开始”菜单中打开一个新的 PowerShell 命令提示符，并运行代码清单 7-1 所示的命令。

■ **提示：**可以从 https://github.com/apress/pro-asp.net-core-3 下载本章和本书中其他所有章节的示例项目。如果在运行示例时遇到问题，请参阅第 1 章以获得帮助。

**代码清单 7-1　创建 SportsStore 项目**

```
dotnet new globaljson --sdk-version 3.1.101 --output SportsSln/SportsStore
dotnet new web --no-https --output SportsSln/SportsStore --framework netcoreapp3.1
dotnet new sln -o SportsSln

dotnet sln SportsSln add SportsSln/SportsStore
```

这些命令创建一个 SportsSln 解决方案文件夹，其中包含一个用 Web 项目模板创建的 SportsStore 项目文件夹。SportsSln 文件夹还包含一个解决方案文件，其中添加了 SportsStore 项目。

对解决方案和项目文件夹使用不同的名称，以使示例更容易理解，但如果使用 Visual Studio 创建项目，默认情况下对两个文件夹使用相同的名称。没有“正确”的方法，可以使用适合项目的任何名称。

## 7.1.1　创建单元测试项目

要创建单元测试项目，请在用于代码清单 7-1 中命令的相同位置运行代码清单 7-2 中所示的命令。

**代码清单 7-2　创建单元测试项目**

```
dotnet new xunit -o SportsSln/SportsStore.Tests --framework netcoreapp3.1
dotnet sln SportsSln add SportsSln/SportsStore.Tests
dotnet add SportsSln/SportsStore.Tests reference SportsSln/SportsStore
```

这里使用 Moq 包来创建模拟对象。运行代码清单 7-3 所示的命令，将 Moq 包安装到单元测试项目中。在与代码清单 7-1 和 7-2 中的命令相同的位置运行此命令。

**代码清单 7-3 安装 Moq 包**

```
dotnet add SportsSln/SportsStore.Tests package Moq --version 4.13.1
```

## 7.1.2　创建应用程序项目文件夹

下一步是创建包含应用程序组件的文件夹。在 Visual Studio Solution Explorer 或 Visual Studio Code Explorer 窗格中右击 SportsStore 项目，并选择 Add | New Folder 或 New Folder 创建表 7-1 中描述的文件夹集。

表 7-1　代码清单 7-3 中创建的文件夹

| 名称 | 描述 |
| --- | --- |
| Models | 此文件夹包含数据模型和允许访问应用程序数据库中数据的类 |
| Controllers | 此文件夹将包含处理 HTTP 请求的控制器类 |
| Views | 此文件夹将包含分组到单独子文件夹中的所有 Razor 文件 |
| Views/Home | 此文件夹将包含特定于主控制器的 Razor 文件，这将在 7.1.6 节中创建 |
| Views/Shared | 此文件夹包含所有控制器共有的 Razor 文件 |

### 7.1.3　打开项目

如果使用的是 Visual Studio Code，选择 File | Open Folder，导航到 SportsSln 文件夹，然后单击 Select Folder 按钮。Visual Studio Code 将打开文件夹，并发现解决方案和项目文件。当出现提示时，如图 7-1 所示，单击 Yes 按钮，安装构建项目所需的资源。如果 Visual Studio Code 提示选择要运行的项目，请选择 SportsStore。

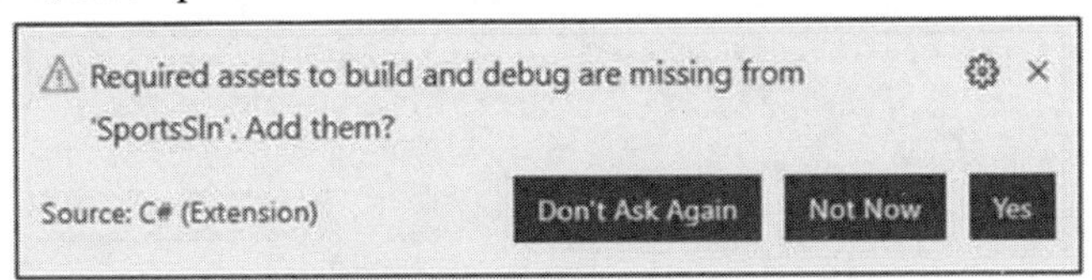


图 7-1　在 Visual Studio Code 中添加资源

如果使用的是 Visual Studio，单击屏幕上的 Open a project or solution 按钮，或者选择 File | Open | Project/Solution。在 SportsSln 文件夹中选择 SportsSln.sln 文件，然后单击 Open 按钮打开项目。项目打开后，选择 Project | SportsStore 属性，选择 Debug 部分，将 App URL 字段中的 URL 端口更改为 5000，如图 7-2 所示。选择 File | SaveAll 以保存新 URL。

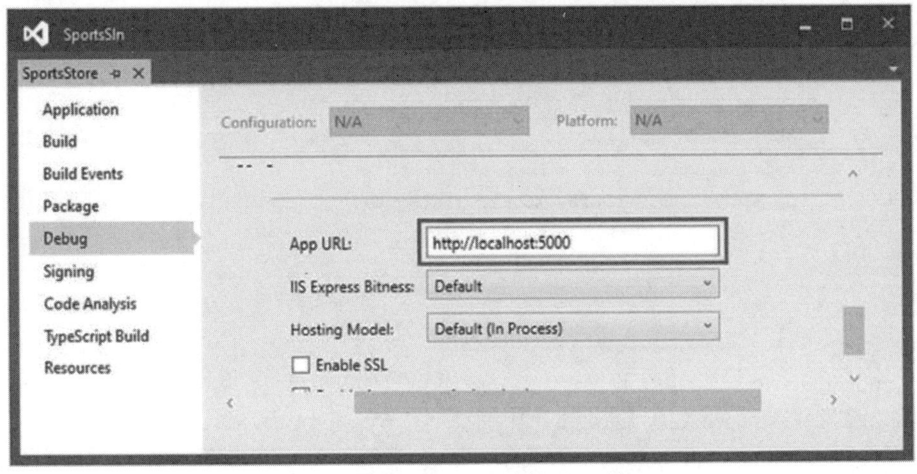


图 7-2　更改 Visual Studio 中的 HTTP 端口

### 7.1.4　准备应用程序服务和请求管道

Startup 类负责配置 ASP.NET Core 应用程序。将代码清单 7-4 中所示的更改应用于 SportsStore 项目中的 Startup 类，来配置基本应用程序特性。

**注意：** Startup 类是一个重要的 ASP.NET Core 特性。详见第 12 章。

**代码清单 7-4　在 SportsStore 文件夹的 Startup.cs 文件中配置应用程序**

```
using System;
using System.Collections.Generic;
using System.Linq;
using System.Threading.Tasks;
using Microsoft.AspNetCore.Builder;
using Microsoft.AspNetCore.Hosting;
using Microsoft.AspNetCore.Http;
using Microsoft.Extensions.DependencyInjection;
using Microsoft.Extensions.Hosting;

namespace SportsStore {
    public class Startup {

        public void ConfigureServices(IServiceCollection services) {
            services.AddControllersWithViews();
        }

        public void Configure(IApplicationBuilder app, IWebHostEnvironment env) {

            app.UseDeveloperExceptionPage();
            app.UseStatusCodePages();
            app.UseStaticFiles();

            app.UseRouting();
            app.UseEndpoints(endpoints => {
                endpoints.MapDefaultControllerRoute();
            });
        }
    }
}
```

ConfigureServices 方法用于设置对象(称为服务)，这些对象可以在整个应用程序中使用，可以通过称为依赖注入(dependency injection)的特性访问，参见第 14 章。在 ConfigureServices 方法中调用的 AddControllersWithViews 方法设置使用 MVC 框架和 Razor 视图引擎的应用程序所需的共享对象。

ASP.NET Core 接收 HTTP 请求，并通过请求管道传递它们，该管道由在 Configure 方法中注册的中间件组件填充。每个中间件组件都能够检查、修改请求、生成响应或修改其他组件生成的响应。请求管道是 ASP.NET Core 的核心，详见第 12 章，那一章还解释了如何创建自定义的中间件组件。表 7-2 描述了用于设置代码清单 7-4 中的中间件组件的方法。

表 7-2 代码清单 7-4 中使用的中间件方法

| 名称 | 描述 |
| --- | --- |
| UseDeveloperExceptionPage() | 这个扩展方法显示应用程序中发生的异常的详细信息，这在开发过程中非常有用，如第 16 章所述。在已部署的应用程序中不应该启用它，第 11 章准备部署 SportsStore 应用程序时禁用了这个特性 |
| UseStatusCodePages() | 这个扩展方法将一个简单的消息添加到 HTTP 响应中，否则该响应将没有正文，如 404-Not Found 响应。参见第 16 章 |
| UseStaticFiles() | 这个扩展方法支持从 wwwroot 文件夹提供静态内容。第 15 章将介绍对静态内容的支持 |

一个特别重要的中间件组件提供了端点路由特性，将 HTTP 请求与应用程序特性(称为端点)匹配，能为它们生成响应，这个过程详见第 13 章。端点路由特性通过 UseRouting 和 UseEndpoints 方法添加到请求管道中。要将 MVC 框架注册为端点源，代码清单 7-4 调用了 MapDefaultControllerRoute 方法。

### 7.1.5 配置 Razor 视图引擎

Razor 视图引擎负责处理扩展名为.cshtml 的视图文件来生成 HTML 响应。配置 Razor 需要进行一些初始准备，以便更容易地为应用程序创建视图。

在 Views 文件夹中添加一个名为_ViewImports.cshtml 的 Razor 视图导入文件，内容如代码清单 7-5 所示。

**警告：** 请密切注意这个文件的内容。很容易出错，导致应用程序生成不正确的 HTML 内容。

**代码清单 7-5 SportsStore/Views 文件夹中_ViewImports.cshtml 文件的内容**

```
@using SportsStore.Models
@addTagHelper *, Microsoft.AspNetCore.Mvc.TagHelpers
```

@using 语句允许在视图中使用 SportsStore.Models 名称空间中的类型，而不需要引用名称空间。@addTagHelper 语句启用了内置的标签助手，稍后将用它创建反映 SportsStore 应用程序配置的 HTML 元素，详见第 15 章。

在 SportsStore/Views 文件夹中添加一个名为_ViewStart.cshtml 的 Razor 视图启动文件，其内容如代码清单 7-6 所示(如果使用 Visual Studio 项模板创建文件，该文件已经包含此表达式)。

**代码清单 7-6 SportsStore/Views 文件夹中_ViewStart.cshtml 文件的内容**

```
@{
    Layout = "_Layout";
}
```

视图启动文件告诉 Razor，在它生成的 HTML 中使用布局文件，减少视图中的重复。要创建视图，在 Views/Shared 文件夹中添加一个名为_Layout.cshtml 的 Razor 布局，其内容如代码清单 7-7 所示。

**代码清单 7-7　SportsStore/Views/Shared 文件夹中_Layout.cshtml 文件的内容**

```
<!DOCTYPE html>
<html>
<head>
   <meta name="viewport" content="width=device-width" />
   <title>SportsStore</title>
</head>
<body>
   <div>
      @RenderBody()
   </div>
</body>
</html>
```

这个文件定义了一个简单的 HTML 文档，其他视图的内容将由@RenderBody 表达式插入其中。第 21 章将详细解释 Razor 表达式的工作原理。

### 7.1.6　创建控制器和视图

在 SportsStore/Controllers 文件夹中添加一个名为 HomeController.cs 的类文件，并使用它定义如代码清单 7-8 所示的类。这是一个最小的控制器，只包含足以生成响应的功能。

**代码清单 7-8　SportsStore/Controllers 文件夹中 HomeController.cs 文件的内容**

```
using Microsoft.AspNetCore.Mvc;

namespace SportsStore.Controllers {
   public class HomeController: Controller {

      public IActionResult Index() => View();

   }
}
```

代码清单 7-4 中使用的 MapDefaultControllerRoute 方法告诉 ASP.NET Core，如何将 URL 匹配到控制器类。该方法应用的配置声明了主控制器定义的 Index 操作方法将用于处理请求。

Index 操作方法还没有做任何有用的事情，只返回调用 View 方法的结果，该方法是从 Controller 基类继承的。这个结果告诉 ASP.NET Core，呈现与操作方法相关联的默认视图。要创建视图，在 Views/Home 文件夹中添加一个名为 Index.cshtml 的 Razor 视图文件，内容如代码清单 7-9 所示。

**代码清单 7-9　SportsStore/Views/Home 文件夹中 Index.cshtml 文件的内容**

```
<h4>Welcome to SportsStore</h4>
```

### 7.1.7 启动数据模型

几乎所有项目都有某种类型的数据模型。由于这是一个电子商务应用程序，显然，需要的模型是用于产品的。在 Models 文件夹中添加一个名为 Product.cs 的类文件，并使用它定义如代码清单 7-10 所示的类。

**代码清单 7-10 SportsStore/Models 文件夹中 Product.cs 文件的内容**

```
using System.ComponentModel.DataAnnotations.Schema;

namespace SportsStore.Models {

    public class Product {

        public long ProductID { get; set; }

        public string Name { get; set; }

        public string Description { get; set; }

        [Column(TypeName = "decimal(8, 2)")]
        public decimal Price { get; set; }

        public string Category { get; set; }
    }
}
```

Price 属性修饰为 Column 属性，以指定将用于存储此属性值的 SQL 数据类型。并不是所有 C#类型都能很好地映射到 SQL 类型，这个属性可以确保数据库为应用程序数据使用合适的类型。

### 7.1.8 检查和运行应用程序

在进一步深入之前，最好确保应用程序按照预期构建和运行。从 Debug 菜单中选择 Start Without Debugging 或 Run Without Debugging，或者在 SportsStore 文件夹中运行代码清单 7-11 所示的命令。

**代码清单 7-11 运行示例应用程序**

```
dotnet run
```

请求 http://localhost:5000，将看到如图 7-3 所示的响应。

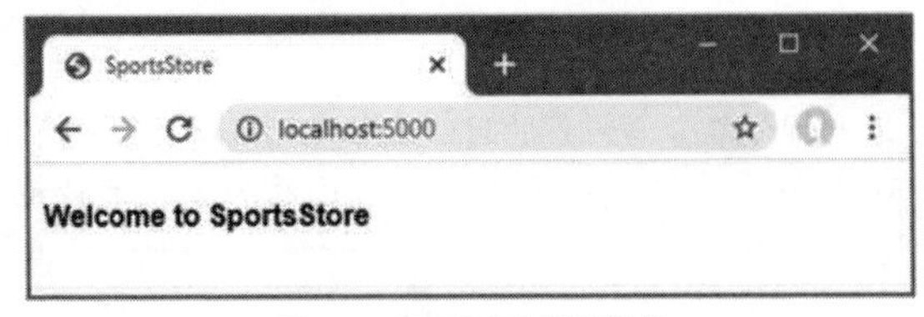


图 7-3 运行示例应用程序

## 7.2 向应用程序添加数据

既然 SportsStore 包含一些基本设置，并可以生成简单的响应，现在应该添加一些数据，以便应用程序显示更有用的内容。SportsStore 应用程序将其数据存储在一个 SQL Server LocalDB 数据库中，使用 Entity Framework Core 访问该数据库。Entity Framework Core 是 Microsoft 的 ORM(对象到关系映射)框架，是 ASP.NET Core 中使用最广泛的数据库访问方法。

---

■ **警告：** 如果在第 2 章准备开发环境时没有安装 localDB，现在就必须安装。没有数据库，SportsStore 应用程序将无法工作。

---

### 7.2.1 安装 Entity Framework Core 包

第一步是将 Entity Framework Core 添加到项目中。使用 PowerShell 命令提示符在 SportsStore 文件夹中运行代码清单 7-12 所示的命令。

**代码清单 7-12 将 Entity Framework Core 包添加到 SportsStore 项目**

```
dotnet add package Microsoft.EntityFrameworkCore.Design --version 3.1.1
dotnet add package Microsoft.EntityFrameworkCore.SqlServer --version 3.1.1
```

这些包安装 Entity Framework Core，且支持使用 SQL Server。Entity Framework Core 还需要一个工具包，其中包括为 ASP.NET Core 应用程序准备的创建数据库所需的命令行工具。运行代码清单 7-13 所示的命令，删除工具包的任何现有版本(如果有)，并安装本书中使用的版本(由于包是全局安装的，因此可在任何文件夹中运行这些命令)。

**代码清单 7-13 安装 Entity Framework Core 工具包**

```
dotnet tool uninstall --global dotnet-ef
dotnet tool install --global dotnet-ef --version 3.1.1
```

### 7.2.2 定义连接字符串

配置设置(如数据库连接字符串)存储在 JSON 配置文件中。要描述到用于 SportsStore 数据的数据库的连接，请将代码清单 7-14 中所示的条目添加到 SportsStore 文件夹的 appsettings.json 文件中。

该项目还包含 appsettings.Development.json 文件，该文件包含仅在开发中使用的配置设置。该文件由 Solution Explorer 显示为嵌套在 appsettings.json 文件中，但在 Visual Studio Code 中始终可见。appsettings.json 文件只用于开发 SportsStore 项目，第 15 章将详细解释文件之间的关系以及如何使用它们。

---

■ **提示：** 连接字符串必须表示为单个不间断的行，这在代码编辑器中没有问题，但不适合打印页面，这就是代码清单 7-14 中奇怪格式的原因。在自己的项目中定义连接字符串时，请确保 SportsStoreConnection 项的值在单行上。

---

**代码清单 7-14　在 SportsStore 文件夹的 appsettings.json 文件中添加配置设置**

```
{
  "Logging": {
    "LogLevel": {
      "Default": "Information",
      "Microsoft": "Warning",
      "Microsoft.Hosting.Lifetime": "Information"
    }
  },
  "AllowedHosts": "*",
  "ConnectionStrings": {
    "SportsStoreConnection": "Server=(localdb)\\MSSQLLocalDB;Database=
    SportsStore;MultipleActiveResultSets=true"
  }
}
```

这个配置字符串指定了一个名为 SportsStore 的 LocalDB 数据库，并启用了多活动结果集特性(MARS)，这是使用 Entity Framework Core 的 SportsStore 应用程序进行的一些数据库查询所必需的。

在添加配置设置时请密切关注。必须准确地如代码清单所示来表示 JSON 数据，这意味着必须确保正确引用属性名称和值。如果遇到困难，可以从 GitHub 存储库下载配置文件。

■ **提示：** 每个数据库服务器都需要自己的连接字符串格式。描述连接字符串的一个有用站点是 www.connectionstrings.com。

### 7.2.3　创建数据库上下文类

Entity Framework Core 通过上下文类提供对数据库的访问。在 Models 文件夹中添加一个名为 StoreDbContext.cs 的类文件，并使用它定义类，如代码清单 7-15 所示。

**代码清单 7-15　SportsStore/Models 文件夹中的 StoreDbContext.cs 文件的内容**

```
using Microsoft.EntityFrameworkCore;

namespace SportsStore.Models {
  public class StoreDbContext: DbContext {

    public StoreDbContext(DbContextOptions<StoreDbContext> options)
      : base(options) { }

    public DbSet<Product> Products { get; set; }
  }
}
```

DbContext 基类提供对 Entity Framework Core 底层功能的访问，Products 属性提供对数据库中的 Product 对象的访问。StoreDbContext 类是从 DbContext 派生出来的，并添加了用于读写应用程序数据的属性。目前只有一个属性，它提供对Product对象的访问。

### 7.2.4　配置Entity Framework Core

必须配置Entity Framework Core，以便确定连接的数据库类型、哪个连接字符串描述该连接，以及哪个上下文类在数据库中显示数据。代码清单7-16显示了Startup类所需的更改。

**代码清单7-16　在SportsStore文件夹的Startup.cs文件中配置Entity Framework Core**

```
using System;
using System.Collections.Generic;
using System.Linq;
using System.Threading.Tasks;
using Microsoft.AspNetCore.Builder;
using Microsoft.AspNetCore.Hosting;
using Microsoft.AspNetCore.Http;
using Microsoft.Extensions.DependencyInjection;
using Microsoft.Extensions.Hosting;
using Microsoft.Extensions.Configuration;
using Microsoft.EntityFrameworkCore;
using SportsStore.Models;

namespace SportsStore {
    public class Startup {

        public Startup(IConfiguration config) {
            Configuration = config;
        }

        private IConfiguration Configuration { get; set; }

        public void ConfigureServices(IServiceCollection services) {
            services.AddControllersWithViews();
                services.AddDbContext<StoreDbContext>(opts => {
                    opts.UseSqlServer(
                        Configuration["ConnectionStrings:SportsStoreConnection"]);
                });
            }

            public void Configure(IApplicationBuilder app, IWebHostEnvironment env) {
            app.UseDeveloperExceptionPage();
            app.UseStatusCodePages();
            app.UseStaticFiles();

            app.UseRouting();
            app.UseEndpoints(endpoints => {
                endpoints.MapDefaultControllerRoute();
            });
```

```
        }
    }
}
```

IConfiguration 接口提供了对 ASP.NET Core 配置系统的访问，其中包含 appsettings.json 文件的内容，详见第 15 章。构造函数接收 IConfiguration 对象，并将其分配给 Configuration 属性，该属性用于访问连接字符串。

使用 AddDbContext 方法配置 Entity Framework Core，该方法注册数据库上下文类，并配置与数据库的关系。UseSqlServer 方法声明正在使用 SQL Server，并通过 IConfiguration 对象读取连接字符串。

### 7.2.5 创建存储库

下一步是创建存储库接口和实现类。存储库模式是使用最广泛的模式之一，它提供了一致的方式来访问数据库上下文类所提供的特性。不是每个人都认为存储库有用，但经验是，它可以减少重复，并确保对数据库的操作一致地执行。在 Models 文件夹中添加一个名为 IStoreRepository.cs 的类文件，并用它定义接口，如代码清单 7-17 所示。

**代码清单 7-17 SportsStore/Models 文件夹中 IStoreRepository.cs 文件的内容**

```
using System.Linq;

namespace SportsStore.Models {
    public interface IStoreRepository {

        IQueryable<Product> Products { get; }
    }
}
```

该接口使用 IQueryable<T>来允许调用者获取 Product 对象序列。IQueryable<T>接口派生自更熟悉的 IEnumerable<T>接口，它表示可以查询的对象集合，如那些由数据库管理的对象。

依赖于 IProductRepository 接口的类可以获得 Product 对象，而不需要知道它们如何存储的细节或实现类如何交付它们的细节。

**理解 IEnumerable<T>和 IQueryable<T>接口**

IQueryable<T>接口非常有用，因为它允许高效地查询一组对象。本章的后面会添加支持功能，从数据库对象中检索一个 Product 对象子集，并使用这个 IQueryable < T >接口和标准的 LINQ 语句为需要的对象确定数据库，而不需要知道哪个数据库服务器存储了数据，或它如何处理查询。如果没有 IQueryable<T>接口，就必须从数据库中检索所有的 Product 对象，然后丢弃不想要的对象。随着应用程序使用的数据量的增加，这将成为一个昂贵的操作。正是由于这个原因，在数据库存储库接口和类中通常使用 IQueryable<T>接口，而不是 IEnumerable<T>接口。

但是，对于 IQueryable<T>接口必须小心，因为每次枚举对象集合时，该查询将再次计算，这意味着向数据库发送一个新的查询。这可能会降低使用 IQueryable<T>的效率。这种情况下，可以使用 ToList 或 ToArray 扩展方法将 IQueryable<T>接口转换为更可预测的形式。

要创建存储库接口的实现，在 Models 文件夹中添加一个名为 EFStoreRepository.cs 的类文件，并用它定义类，如代码清单 7-18 所示。

**代码清单 7-18　SportsStore/Models 文件夹中的 EFStoreRepository.cs 文件的内容**

```
using System.Linq;

namespace SportsStore.Models {
    public class EFStoreRepository : IStoreRepository {
        private StoreDbContext context;

        public EFStoreRepository(StoreDbContext ctx) {
            context = ctx;
        }

        public IQueryable<Product> Products => context.Products;
    }
}
```

在向应用程序添加特性的同时添加其他功能，但目前，存储库实现仅将 IStoreRepository 接口定义的 Products 属性映射到 StoreDbContext 类定义的 Products 属性。上下文类中的 Products 属性返回一个 DbSet<Product>对象，该对象实现 IQueryable<T>接口，这样更容易在使用 Entity Framework Core 时实现存储库接口。

本章前面解释了 ASP.NET Core 支持的允许在整个应用程序中访问对象的服务。服务的一个好处是，允许类使用接口，而不需要知道使用的是哪个实现类。详见第 14 章，但对于 SportsStore，这意味着应用程序组件可以访问实现 IStoreRepository 接口的对象，而不需要知道它使用的是 EFStoreRepository 实现类。这就更容易更改应用程序使用的实现类，而不需要更改各个组件。将代码清单 7-19 中所示的语句添加到 Startup 类中，为使用 EFStoreRepository 作为实现类的 IStoreRepository 接口创建一个服务。

---

■ **提示：** 如果现在还不明白，也不必担心。这个主题是使用 ASP.NET Core 最令人困惑的方面之一，理解它可能需要一段时间。

---

**代码清单 7-19　在 SportsStore 文件夹的 Startup.cs 文件中创建存储库服务**

```
using System;
using System.Collections.Generic;
using System.Linq;
using System.Threading.Tasks;
using Microsoft.AspNetCore.Builder;
using Microsoft.AspNetCore.Hosting;
using Microsoft.AspNetCore.Http;
using Microsoft.Extensions.DependencyInjection;
using Microsoft.Extensions.Hosting;
using Microsoft.Extensions.Configuration;
using Microsoft.EntityFrameworkCore;
```

```
using SportsStore.Models;

namespace SportsStore {
    public class Startup {
        public Startup(IConfiguration config) {
            Configuration = config;
        }

        private IConfiguration Configuration { get; set; }

        public void ConfigureServices(IServiceCollection services) {
            services.AddControllersWithViews();

            services.AddDbContext<StoreDbContext>(opts => {
                opts.UseSqlServer(
                    Configuration["ConnectionStrings:SportsStoreConnection"]);
            });
            services.AddScoped<IStoreRepository, EFStoreRepository>();
        }

        public void Configure(IApplicationBuilder app, IWebHostEnvironment env) {

            app.UseDeveloperExceptionPage();
            app.UseStatusCodePages();
            app.UseStaticFiles();

            app.UseRouting();
            app.UseEndpoints(endpoints => {
                endpoints.MapDefaultControllerRoute();
            });
        }
    }
}
```

AddScoped 方法创建一个服务，其中每个 HTTP 请求获得自己的存储库对象，这是 Entity Framework Core 通常使用的方式。

### 7.2.6 创建数据库迁移

Entity Framework Core 能通过名为 migrations 的特性，使用数据模型类生成数据库模式。当准备迁移时，Entity Framework Core 创建一个 C#类，其中包含准备数据库所需的 SQL 命令。如果需要修改模型类，那么可以创建一个包含 SQL 的新迁移命令，以反映所需的更改。这样，就不必担心手动编写和测试 SQL 命令，而只需要关注应用程序中的 C#模型类。

Entity Framework Core 命令从命令行执行。打开 PowerShell 命令提示符，并在 SportsStore 文件夹中运行代码清单 7-20 所示的命令，以创建迁移类，该迁移类为第一次使用数据库做准备。

**代码清单 7-20 创建数据库迁移**

```
dotnet ef migrations add Initial
```

当这个命令完成时，SportsStore 项目将包含 Migrations 文件夹。这是 Entity Framework Core 存储其迁移类的地方。其中一个文件名是时间戳，后面跟着_Initial.cs。这个类将用于创建数据库的初始模式。如果检查该文件的内容，可以看到如何使用 Product 模型类创建模式。

**Add-Migration 和 Update-Database 命令**

对于有经验的 EntityFramework 开发人员，可能习惯于使用 Add-Migration 命令创建数据库迁移，将 Update-Database 命令应用于数据库。

随着.NET Core 的引入，Entity Framework Core 添加了一些命令，这些命令使用 Microsoft.EntityFrameworkCore.Tools.DotNet 包添加的命令，集成到 dotnet 命令行工具中。之所以使用这些命令，是因为它们与其他.NET 命令一致，并可在任何命令提示符或 PowerShell 窗口中使用，而不像 Add-Migration 和 Update-Database 命令，它们只能在特定的 Visual Studio 窗口中工作。

### 7.2.7 创建种子数据

为了填充数据库并提供一些示例数据，在 Models 文件夹中添加了一个名为 SeedData.cs 的类文件，并定义了如代码清单 7-21 所示的类。

**代码清单 7-21 SportsStore/Models 文件夹中 SeedData.cs 文件的内容**

```
using System.Linq;
using Microsoft.AspNetCore.Builder;
using Microsoft.Extensions.DependencyInjection;
using Microsoft.EntityFrameworkCore;

namespace SportsStore.Models {

    public static class SeedData {

        public static void EnsurePopulated(IApplicationBuilder app) {
            StoreDbContext context = app.ApplicationServices
                .CreateScope().ServiceProvider.GetRequiredService<StoreDbContext>();

            if (context.Database.GetPendingMigrations().Any()) {
                context.Database.Migrate();
            }

            if (!context.Products.Any()) {
                context.Products.AddRange(
                    new Product {
                        Name = "Kayak", Description = "A boat for one person",
                        Category = "Watersports", Price = 275
                    },
```

```
                new Product {
                    Name = "Lifejacket",
                    Description = "Protective and fashionable",
                    Category = "Watersports", Price = 48.95m
                },
                new Product {
                    Name = "Soccer Ball",
                    Description = "FIFA-approved size and weight",
                    Category = "Soccer", Price = 19.50m
                },
                    new Product {
                    Name = "Corner Flags",
                    Description = "Give your playing field a professional touch",
                    Category = "Soccer", Price = 34.95m
                },
                new Product {
                    Name = "Stadium",
                    Description = "Flat-packed 35,000-seat stadium",
                    Category = "Soccer", Price = 79500
                },
                new Product {
                    Name = "Thinking Cap",
                    Description = "Improve brain efficiency by 75%",
                    Category = "Chess", Price = 16
                },
                new Product {
                    Name = "Unsteady Chair",
                    Description = "Secretly give your opponent a disadvantage",
                    Category = "Chess", Price = 29.95m
                },
                new Product {
                    Name = "Human Chess Board",
                    Description = "A fun game for the family",
                    Category = "Chess", Price = 75
                },
                new Product {
                    Name = "Bling-Bling King",
                    Description = "Gold-plated, diamond-studded King",
                    Category = "Chess", Price = 1200
                }
            );
            context.SaveChanges();
        }
    }
  }
}
```

静态 EnsurePopulated 方法接收一个 IApplicationBuilder 参数，它是 Startup 类的 Configure 方法中用于注册中间件组件以处理 HTTP 请求的接口。IApplicationBuilder 还提供对应用程序服务的访问，包括 Entity Framework Core 数据库上下文服务。

EnsurePopulated 方法通过 IApplicationBuilder 接口获得一个 StoreDbContext 对象，如果有任何挂起的迁移，则调用 Database.Migrate 方法，这意味着将创建和准备数据库，以便它能够存储 Product 对象。接下来，检查数据库中 Product 对象的数量。如果数据库中没有对象，则使用 AddRange 方法通过 Product 对象集合填充数据库，然后使用 SaveChanges 方法写入数据库。

最后的更改是在应用程序启动时为数据库添加种子，为此，添加对 Startup 类中 EnsurePopulated 方法的调用，如代码清单 7-22 所示。

**代码清单 7-22　SportsStore 文件夹的 Startup.cs 文件中的播种数据库**

```
using System;
using System.Collections.Generic;
using System.Linq;
using System.Threading.Tasks;
using Microsoft.AspNetCore.Builder;
using Microsoft.AspNetCore.Hosting;
using Microsoft.AspNetCore.Http;
using Microsoft.Extensions.DependencyInjection;
using Microsoft.Extensions.Hosting;
using Microsoft.Extensions.Configuration;
using Microsoft.EntityFrameworkCore;
using SportsStore.Models;

namespace SportsStore {
    public class Startup {

        public Startup(IConfiguration config) {
            Configuration = config;
        }

        private IConfiguration Configuration { get; set; }

        public void ConfigureServices(IServiceCollection services) {
            services.AddControllersWithViews();

            services.AddDbContext<StoreDbContext>(opts => {
                opts.UseSqlServer(
                    Configuration["ConnectionStrings:SportsStoreConnection"]);
            });
            services.AddScoped<IStoreRepository, EFStoreRepository>();
        }

        public void Configure(IApplicationBuilder app, IWebHostEnvironment env) {
```

```
            app.UseDeveloperExceptionPage();
            app.UseStatusCodePages();
            app.UseStaticFiles();

            app.UseRouting();
            app.UseEndpoints(endpoints => {
                endpoints.MapDefaultControllerRoute();
            });
            SeedData.EnsurePopulated(app);
        }
    }
}
```

**重置数据库**

如果需要重置数据库，就在 SportsStore 文件夹下运行如下命令:

```
...
dotnet ef database drop --force --context StoreDbContext
...
```

启动 ASP.NET Core，会重新创建数据库，并填充数据。

## 7.3 显示产品列表

可以看到，ASP.NET Core 项目的初始准备工作可能需要一些时间。但好消息是，一旦基础打好，速度就会提高，添加特性的速度也会更快。本节将创建一个控制器和一个操作方法，它可以显示存储库中产品的详细信息。

**使用 Visual Studio 脚手架**

如第 4 章所述，Visual Studio 支持搭建功能，可以向项目中添加项。

本书没有使用脚手架(scaffolding)。脚手架生成的代码和标记非常通用，几乎毫无用处，而且所支持的场景非常狭窄，不能解决常见的开发问题。本书的目标不仅是确保介绍如何创建 ASP.NET Core 应用程序，还解释了其幕后的工作方式(而把创建组件的责任交给脚手架功能时，就比较困难了)。

如果在 Solution Explorer 中使用 Visual Studio，从弹出菜单中选择 Add | New Item，然后从 Add New Item 窗口中选择项模板。

读者的开发风格可能与本书不同，可能更喜欢在自己的项目中使用脚手架功能，这是完全合理的，但建议花些时间理解脚手架功能的作用，这样，如果没有得到预期的结果，也知道应该去哪里查找。

### 7.3.1 准备控制器

添加代码清单 7-23 所示的语句，以准备控制器，来显示产品列表。

**代码清单 7-23　在 SportsStore/Controllers 文件夹的 HomeController.cs 文件中准备控制器**

```
using Microsoft.AspNetCore.Mvc;
using SportsStore.Models;

namespace SportsStore.Controllers {
    public class HomeController : Controller {
        private IStoreRepository repository;

        public HomeController(IStoreRepository repo) {
            repository = repo;
        }

        public IActionResult Index() => View(repository.Products);
    }
}
```

在 ASP.NET Core 需要创建 HomeController 类的一个新实例来处理 HTTP 请求时，它将检查构造函数，并查看它是否需要一个实现 IStoreRepository 接口的对象。为了确定应该使用什么实现类，ASP.NET Core 会参考 Startup 类中的配置，这说明它应该使用 EFStoreRepository，并应该为每个请求创建一个新实例。ASP.NET Core 创建一个新的 EFStoreRepository 对象，并使用它调用 HomeController 构造函数，来创建处理 HTTP 请求的控制器对象。

这称为依赖注入，它的方法允许 HomeController 对象通过 IStoreRepository 接口访问应用程序的存储库，而不需要知道配置了哪个实现类。可以把服务重新配置为使用另一个实现类(例如不使用 Entity Framework Core 的实现类)和依赖注入，则控制器将继续工作，而不必更改。

**注意：** 有些开发者不喜欢依赖注入，他们认为依赖注入会使应用程序更加复杂。这不是我的观点，但如果是依赖注入的新手，那么建议等到读完第 14 章后再做决定。

**单元测试：存储库访问**

通过创建一个模拟存储库，将其注入 HomeController 类的构造函数中，然后调用 Index 方法来获取包含产品列表的响应，可以对控制器是否正确访问存储库进行单元测试。然后将得到的 Product 对象与模拟实现中的测试数据进行比较。有关如何设置单元测试的详细信息，请参阅第 6 章。下面是为此目的在一个名为 HomeControllerTests.cs 的类文件中创建的单元测试，该文件已添加到 SportsStore.Tests 项目中。

```
using System.Collections.Generic;
using System.Linq;
using Microsoft.AspNetCore.Mvc;
using Moq;
using SportsStore.Controllers;
using SportsStore.Models;
using Xunit;

namespace SportsStore.Tests {
```

```
    public class ProductControllerTests {

        [Fact]
        public void Can_Use_Repository() {
            // Arrange
            Mock<IStoreRepository> mock = new Mock<IStoreRepository>();
            mock.Setup(m => m.Products).Returns((new Product[] {
                new Product {ProductID = 1, Name = "P1"},
                new Product {ProductID = 2, Name = "P2"}
            }).AsQueryable<Product>());

            HomeController controller = new HomeController(mock.Object);

            // Act
            IEnumerable<Product> result =
                (controller.Index() as ViewResult).ViewData.Model
                    as IEnumerable<Product>;

            // Assert
            Product[] prodArray = result.ToArray();
            Assert.True(prodArray.Length == 2);
            Assert.Equal("P1", prodArray[0].Name);
            Assert.Equal("P2", prodArray[1].Name);
        }
    }
}
```

从操作方法返回数据有点尴尬。结果是一个 ViewResult 对象，必须将其 ViewData.Model 属性值转换为期望的数据类型。第 II 部分将解释动作方法返回的不同结果类型以及如何使用它们。

## 7.3.2 更新视图

代码清单 7-23 中的 Index 操作方法将 Product 对象集合从存储库传递到 View 方法，这意味着这些对象将是 Razor 从视图生成 HTML 内容时使用的视图模型。对代码清单 7-24 所示的视图进行更改，以使用 Product 视图模型对象生成内容。

**代码清单 7-24 在 SportsStore/Views/Home 文件夹的 Index.cshtml 文件中使用产品数据**

```
@model IQueryable<Product>

@foreach (var p in Model) {
    <div>
        <h3>@p.Name</h3>
        @p.Description
        <h4>@p.Price.ToString("c")</h4>
    </div>
}
```

文件顶部的@model 表达式指定，视图期望从操作方法接收一系列 Product 对象作为其模型数据。使用@foreach 表达式遍历序列，并为接收的每个 Product 对象生成一组简单的 HTML 元素。

视图不知道 Product 对象来自何处、如何获得它们，或者它们是否表示应用程序已知的所有产品。相反，视图只处理如何使用 HTML 元素显示每个产品的详细信息。

---

■ **提示：**使用 ToString("c")方法将 Price 属性转换为字符串，该方法根据在服务器上生效的区域性设置将数值呈现为货币。例如，如果服务器设置为 en-US，那么(1002.3).ToString("c")将返回$1,002.30，但如果服务器被设置为 en-GB，那么相同的方法将返回£1,002.30。

---

### 7.3.3　运行应用程序

启动 ASP.NET Core，并请求 http://localhost:5000 查看产品列表，如图 7-4 所示。这是典型的 ASP.NET Core 开发模式。首先需要投入时间来设置一切，然后应用程序的基本特性就会快速地结合在一起。

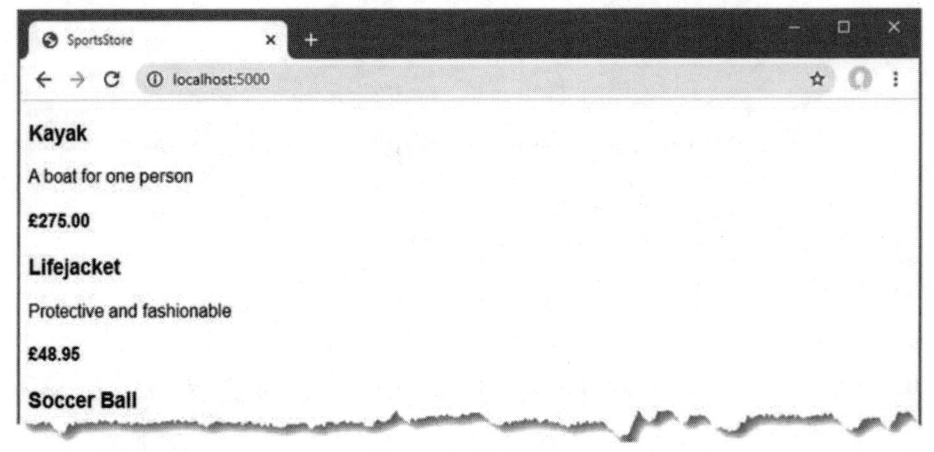


图 7-4　显示产品列表

## 7.4　添加分页

从图 7-4 可以看到，Index.cshtml 视图在单个页面上显示数据库中的产品。本节将添加对分页的支持，以便视图在一个页面上显示更少的产品，用户可以从一个页面移动到另一个页面来查看整个目录。为此，向主控制器中的 Index 方法添加一个参数，如代码清单 7-25 所示。

**代码清单 7-25　在 SportsStore/Controllers 文件夹的 HomeController.cs 文件中添加分页**

```
using Microsoft.AspNetCore.Mvc;
using SportsStore.Models;
using System.Linq;

namespace SportsStore.Controllers {
    public class HomeController : Controller {
        private IStoreRepository repository;
        public int PageSize = 4;
```

```
        public HomeController(IStoreRepository repo) {
            repository = repo;
        }

        public ViewResult Index(int productPage = 1)
            => View(repository.Products
                .OrderBy(p => p.ProductID)
                .Skip((productPage - 1) * PageSize)
                .Take(PageSize));
    }
}
```

PageSize 字段指定每个页面需要 4 个产品。向 Index 方法添加一个可选参数，这意味着，如果调用方法时没有提供参数，该调用被视为提供了参数定义中指定的值，效果是调用操作方法时若没有提供参数，则显示第一页的产品。在操作方法的主体中，获取 Product 对象，按主键对它们排序，跳过在当前页面开始之前出现的产品，并获取由 PageSize 字段指定的产品数量。

**单元测试：分页**

可以通过模拟存储库、从控制器请求特定的页面以及确保获得预期的数据子集来对分页特性进行单元测试。下面是为此目的创建的单元测试，并添加到SportsStore Tests项目的HomeControllerTests.cs文件中：

```
using System.Collections.Generic;
using System.Linq;
using Microsoft.AspNetCore.Mvc;
using Moq;
using SportsStore.Controllers;
using SportsStore.Models;
using Xunit;

namespace SportsStore.Tests {

    public class ProductControllerTests {

        [Fact]
        public void Can_Use_Repository() {
            // ...statements omitted for brevity...
        }

        [Fact]
        public void Can_Paginate() {
            // Arrange
            Mock<IStoreRepository> mock = new Mock<IStoreRepository>();
            mock.Setup(m => m.Products).Returns((new Product[] {
                new Product {ProductID = 1, Name = "P1"},
```

```
            new Product {ProductID = 2, Name = "P2"},
            new Product {ProductID = 3, Name = "P3"},
            new Product {ProductID = 4, Name = "P4"},
            new Product {ProductID = 5, Name = "P5"}
        }).AsQueryable<Product>());

        HomeController controller = new HomeController(mock.Object);
        controller.PageSize = 3;

        // Act
        IEnumerable<Product> result =
            (controller.Index(2) as ViewResult).ViewData.Model
                as IEnumerable<Product>;

        // Assert
        Product[] prodArray = result.ToArray();
        Assert.True(prodArray.Length == 2);
        Assert.Equal("P4", prodArray[0].Name);
        Assert.Equal("P5", prodArray[1].Name);
    }
  }
}
```

可以看到，新测试遵循现有测试的模式，依赖 Moq 来提供要使用的已知数据集。

## 7.4.1　显示页面的链接

重启 ASP.NET Core，并请求 http://localhost:5000，现在页面上显示了四项，如图 7-5 所示。如果想查看另一个页面，可以把查询字符串参数附加到 URL 的结尾，如下所示：

```
http://localhost:5000/?productPage=2
```

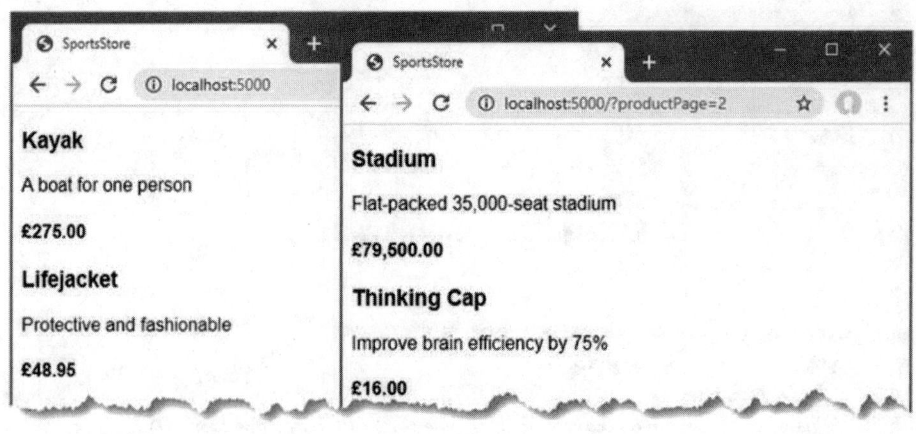


图 7-5　翻阅资料

使用这些查询字符串，可以在产品目录中导航。客户无法确定这些查询字符串参数是否存在，即使存在，客户也不希望以这种方式导航。相反，需要在每个产品列表的底部呈现一些页面链接，以便客户可以在页面之间导航。为此，创建一个标签助手，它为需要的链接生成 HTML 标记。

### 1. 添加视图模型

为了支持标签助手，向视图传递关于可用页面数量、当前页面和存储库中产品总数的信息。最简单的方法是创建一个视图模型类，它专门用于在控制器和视图之间传递数据。在 SportsStore 项目中创建 Models/ViewModels 文件夹，添加一个名为 PagingInfo.cs 的类文件。并定义如代码清单 7-26 所示的类。

**代码清单 7-26 SportsStore/Models/ViewModels 文件夹中 PagingInfo.cs 文件的内容**

```
using System;

namespace SportsStore.Models.ViewModels {

    public class PagingInfo {
        public int TotalItems { get; set; }
        public int ItemsPerPage { get; set; }
        public int CurrentPage { get; set; }

        public int TotalPages =>
            (int)Math.Ceiling((decimal)TotalItems / ItemsPerPage);
    }
}
```

### 2. 添加标签助手类

现在有了一个视图模型，该创建一个标签助手类了。在 SportsStore 项目中创建一个名为 Infrastructure 的文件夹，并向其中添加一个名为 PageLinkTagHelper.cs 的类文件，其代码如代码清单 7-27 所示。标签助手是 ASP.NET Core 开发的重要组成部分，第 25～27 章将解释它们是如何工作的，以及如何使用和创建它们。

---

■ **提示**：在 Infrastructure 文件夹中放置了一些类，它们为应用程序提供管道，但与应用程序的主要功能无关。在自己的项目中不必遵循这种约定。

---

**代码清单 7-27 SportsStore/Infrastructure 文件夹中 PageLinkTagHelper.cs 文件的内容**

```
using Microsoft.AspNetCore.Mvc;
using Microsoft.AspNetCore.Mvc.Rendering;
using Microsoft.AspNetCore.Mvc.Routing;
using Microsoft.AspNetCore.Mvc.ViewFeatures;
using Microsoft.AspNetCore.Razor.TagHelpers;
using SportsStore.Models.ViewModels;

namespace SportsStore.Infrastructure {

    [HtmlTargetElement("div", Attributes = "page-model")]
```

```
    public class PageLinkTagHelper : TagHelper {
        private IUrlHelperFactory urlHelperFactory;

        public PageLinkTagHelper(IUrlHelperFactory helperFactory) {
            urlHelperFactory = helperFactory;
        }

        [ViewContext]
        [HtmlAttributeNotBound]
        public ViewContext ViewContext { get; set; }

        public PagingInfo PageModel { get; set; }

        public string PageAction { get; set; }

        public override void Process(TagHelperContext context,
                TagHelperOutput output) {
            IUrlHelper urlHelper = urlHelperFactory.GetUrlHelper(ViewContext);
            TagBuilder result = new TagBuilder("div");
            for (int i = 1; i <= PageModel.TotalPages; i++) {
                TagBuilder tag = new TagBuilder("a");
                tag.Attributes["href"] = urlHelper.Action(PageAction,
                    new { productPage = i });
                tag.InnerHtml.Append(i.ToString());
                result.InnerHtml.AppendHtml(tag);
            }
            output.Content.AppendHtml(result.InnerHtml);
        }
    }
}
```

这个标签助手使用与产品页面对应的元素填充div元素。现在不打算详细解释标签助手；知道它们是将C#逻辑引入视图的最有用方法之一就足够了。标签助手的代码看起来十分复杂，因为C#和HTML不容易混合使用。但是使用标签助手要比在视图中包含C#代码块更好，因为标签助手可以很容易地进行单元测试。

大多数ASP.NET Core组件(如控制器和视图)是自动发现的，但标签助手必须注册。在代码清单7-28中，向_Views文件夹的_ViewImports.cshtml文件添加了一条语句，该语句告诉ASP.NET Core，在SportsStore项目中寻找标签助手类。还添加了一个@using表达式，以便在视图中引用视图模型类，而不必使用名称空间限定它们的名称。

**代码清单7-28　在SportsStore/Views文件夹的_ViewImports.cshtml文件中注册标签助手**

```
@using SportsStore.Models
@using SportsStore.Models.ViewModels
@addTagHelper *, Microsoft.AspNetCore.Mvc.TagHelpers
@addTagHelper *, SportsStore
```

**单元测试：创建页面链接**

为了测试PageLinkTagHelper标签助手类，使用测试数据调用Process方法，并提供TagHelperOutput对象，检查生成的HTML，如下所示，它是在SportsStore.Tests项目的新PageLinkTagHelperTests.cs文件中定义的：

```
using System.Collections.Generic;
using System.Threading.Tasks;
using Microsoft.AspNetCore.Mvc;
using Microsoft.AspNetCore.Mvc.Routing;
using Microsoft.AspNetCore.Razor.TagHelpers;
using Moq;
using SportsStore.Infrastructure;
using SportsStore.Models.ViewModels;
using Xunit;

namespace SportsStore.Tests {

    public class PageLinkTagHelperTests {

        [Fact]
        public void Can_Generate_Page_Links() {
            // Arrange
            var urlHelper = new Mock<IUrlHelper>();
            urlHelper.SetupSequence(x => x.Action(It.IsAny<UrlActionContext>()))
                .Returns("Test/Page1")
                .Returns("Test/Page2")
                .Returns("Test/Page3");
            var urlHelperFactory = new Mock<IUrlHelperFactory>();
            urlHelperFactory.Setup(f =>
                    f.GetUrlHelper(It.IsAny<ActionContext>()))
                        .Returns(urlHelper.Object);
            PageLinkTagHelper helper =
                    new PageLinkTagHelper(urlHelperFactory.Object) {
                PageModel = new PagingInfo {
                    CurrentPage = 2,
                    TotalItems = 28,
                ItemsPerPage = 10
                },
                PageAction = "Test"
            };
            TagHelperContext ctx = new TagHelperContext(
                new TagHelperAttributeList(),
                new Dictionary<object, object>(), "");

            var content = new Mock<TagHelperContent>();
            TagHelperOutput output = new TagHelperOutput("div",
```

```
            new TagHelperAttributeList(),
            (cache, encoder) => Task.FromResult(content.Object));
        // Act
        helper.Process(ctx, output);

        // Assert
        Assert.Equal(@"<a href=""Test/Page1"">1</a>"
            + @"<a href=""Test/Page2"">2</a>"
            + @"<a href=""Test/Page3"">3</a>",
            output.Content.GetContent());
        }
    }
}
```

这个测试的复杂性在于所创建的对象是创建和使用标签助手所需要的。标签助手使用IUrlHelperFactory对象来生成针对应用程序不同部分的url，使用Moq来创建这个接口的实现以及提供测试数据的相关IUrlHelper接口。

测试的核心部分通过使用包含双引号的字符串值来验证标签助手的输出。只要字符串以@为前缀，并且使用两组双引号("")代替一组双引号，C#就完全能够处理这些字符串。必须记住，不要将字符串断行，除非要比较的字符串被类似地断行。例如，在测试方法中使用的文字已经折开，放在几行上，因为打印页面的宽度很窄。没有添加换行符；否则，测试就会失败。

### 3. 添加视图模型数据

我们还没有完全准备好使用标签助手，因为还没有为视图提供PagingInfo视图模型类的实例。为此，向SportsStore项目的Models/ViewModels文件夹中添加了一个名为ProductsListViewModel.cs的类文件，其内容如代码清单7-29所示。

**代码清单7-29 SportsStore/Models/ViewModels文件夹中ProductsListViewModel.cs文件的内容**

```
using System.Collections.Generic;
using SportsStore.Models;

namespace SportsStore.Models.ViewModels {

    public class ProductsListViewModel {
        public IEnumerable<Product> Products { get; set; }
        public PagingInfo PagingInfo { get; set; }
    }
}
```

可更新HomeController类中的Index操作方法，以使用ProductsListViewModel类向视图提供要显示在页面上的产品的详细信息和分页的详细信息，如代码清单7-30所示。

**代码清单 7-30 更新 SportsStore/Controllers 文件夹的 HomeController.cs 文件中的操作方法**

```
using Microsoft.AspNetCore.Mvc;
using SportsStore.Models;
using System.Linq;
using SportsStore.Models.ViewModels;

namespace SportsStore.Controllers {
    public class HomeController : Controller {
        private IStoreRepository repository;
        public int PageSize = 4;

        public HomeController(IStoreRepository repo) {
            repository = repo;
        }

        public ViewResult Index(int productPage = 1)
            => View(new ProductsListViewModel {
                Products = repository.Products
                    .OrderBy(p => p.ProductID)
                    .Skip((productPage - 1) * PageSize)
                    .Take(PageSize),
                PagingInfo = new PagingInfo {
                    CurrentPage = productPage,
                    ItemsPerPage = PageSize,
                    TotalItems = repository.Products.Count()
                }
            });
    }
}
```

这些更改将一个 ProductsListViewModel 对象作为模型数据传递给视图。

**单元测试：页面模型视图数据**

需要确保控制器向视图发送正确的分页数据。下面是在测试项目的 HomeControllerTests 类中添加的单元测试，以确保：

```
...
[Fact]
public void Can_Send_Pagination_View_Model() {

    // Arrange
    Mock<IStoreRepository> mock = new Mock<IStoreRepository>();
    mock.Setup(m => m.Products).Returns((new Product[] {
        new Product {ProductID = 1, Name = "P1"},
        new Product {ProductID = 2, Name = "P2"},
        new Product {ProductID = 3, Name = "P3"},
```

```
        new Product {ProductID = 4, Name = "P4"},
        new Product {ProductID = 5, Name = "P5"}
    }).AsQueryable<Product>());

    // Arrange
    HomeController controller =
        new HomeController(mock.Object) { PageSize = 3 };

    // Act
    ProductsListViewModel result =
        controller.Index(2).ViewData.Model as ProductsListViewModel;

    // Assert
    PagingInfo pageInfo = result.PagingInfo;
    Assert.Equal(2, pageInfo.CurrentPage);
    Assert.Equal(3, pageInfo.ItemsPerPage);
    Assert.Equal(5, pageInfo.TotalItems);
    Assert.Equal(2, pageInfo.TotalPages);
}
...
```

还需要修改以前的单元测试，以反映来自 Index 操作方法的新结果。以下是修改后的测试：

```
...
[Fact]
public void Can_Use_Repository() {
    // Arrange
    Mock<IStoreRepository> mock = new Mock<IStoreRepository>();
    mock.Setup(m => m.Products).Returns((new Product[] {
        new Product {ProductID = 1, Name = "P1"},
        new Product {ProductID = 2, Name = "P2"}
    }).AsQueryable<Product>());

    HomeController controller = new HomeController(mock.Object);

    // Act
    ProductsListViewModel result =
        controller.Index().ViewData.Model as ProductsListViewModel;

    // Assert
    Product[] prodArray = result.Products.ToArray();
    Assert.True(prodArray.Length == 2);
    Assert.Equal("P1", prodArray[0].Name);
    Assert.Equal("P2", prodArray[1].Name);
}

[Fact]
public void Can_Paginate() {
```

```
    // Arrange
    Mock<IStoreRepository> mock = new Mock<IStoreRepository>();
    mock.Setup(m => m.Products).Returns((new Product[] {
        new Product {ProductID = 1, Name = "P1"},
        new Product {ProductID = 2, Name = "P2"},
        new Product {ProductID = 3, Name = "P3"},
        new Product {ProductID = 4, Name = "P4"},
        new Product {ProductID = 5, Name = "P5"}
}).AsQueryable<Product>());

HomeController controller = new HomeController(mock.Object);
controller.PageSize = 3;

// Act
ProductsListViewModel result =
    controller.Index(2).ViewData.Model as ProductsListViewModel;
    // Assert
    Product[] prodArray = result.Products.ToArray();
    Assert.True(prodArray.Length == 2);
    Assert.Equal("P4", prodArray[0].Name);
    Assert.Equal("P5", prodArray[1].Name);
}
...
```

鉴于这两种测试方法之间的重复程度，通常会创建一个公共设置方法。但由于在单独的侧边栏中交付单元测试，因此将保持所有内容独立，这样就可以单独看到每个测试。

视图当前期望得到一系列 Product 对象，因此需要更新 Index.cshtml 文件，如代码清单 7-31 所示，以处理新的视图模型类型。

**代码清单 7-31　更新 SportsStore/Views/Home 文件夹中的 Index.cshtml 文件**

```
@model ProductsListViewModel

@foreach (var p in Model.Products) {
    <div>
        <h3>@p.Name</h3>
        @p.Description
        <h4>@p.Price.ToString("c")</h4>
    </div>
}
```

前面修改了@model 指令，告诉 Razor 现在使用的是不同的数据类型。更新了 foreach 循环，以便数据源是模型数据的 Products 属性。

#### 4. 显示页面链接

目前已经准备好添加到 Index 视图的页面链接。创建了包含分页信息的视图模型，更新了控制器，以便它将此信息传递给视图，并更改了@model 指令，以匹配新的模型视图类型。剩下的就是添加一个 HTML 元素，标签助手将处理该元素来创建页面链接，如代码清单 7-32 所示。

**代码清单 7-32　在 SportsStore/Views/Home 文件夹的 Index.cshtml 文件中添加分页链接**

```
@model ProductsListViewModel

@foreach (var p in Model.Products) {
    <div>
        <h3>@p.Name</h3>
        @p.Description
        <h4>@p.Price.ToString("c")</h4>
    </div>
}

<div page-model="@Model.PagingInfo" page-action="Index"></div>
```

重启 ASP.NET Core 并请求 http://localhost:5000，将看到新的页面链接，如图 7-6 所示。样式仍然是基本的，本章的后面会修正它。目前重要的是，这些链接将用户从一个页面带到目录中的另一个页面，并允许用户对所销售的产品进行探索。当 Razor 在 div 元素上找到 page-model 属性时，它要求 PageLinkTagHelper 类转换该元素，这会生成如图所示的一组链接。

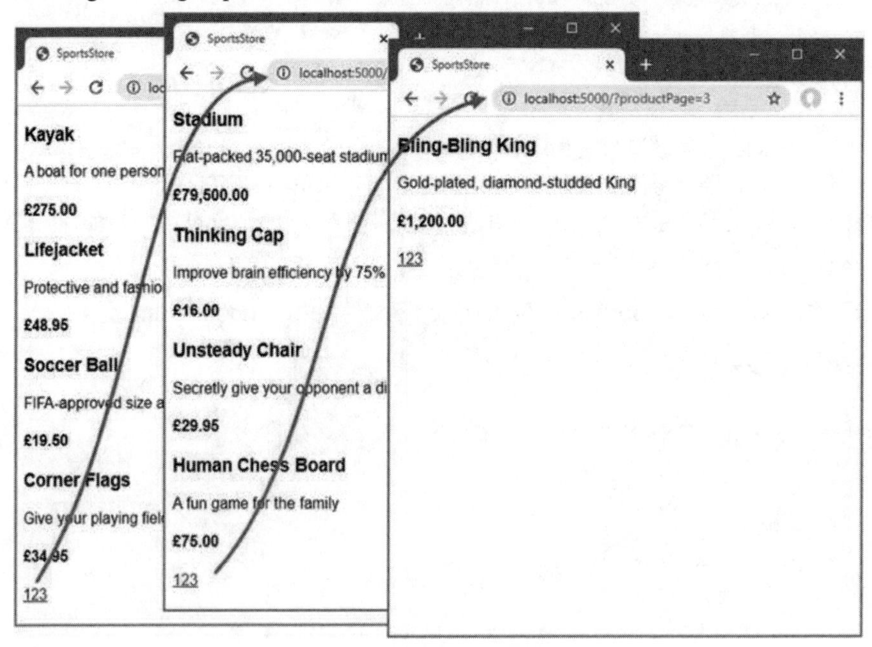


图 7-6　显示页面导航链接

## 7.4.2　改善 URL

页面链接是有效的，但它们仍然使用查询字符串将页面信息传递到服务器，如下所示：

```
http://localhost/?productPage=2
```

可以通过创建遵循可组合 URL 模式的方案来创建更吸引人的 URL。可组合的 URL 是对用户有意义的 URL，如下所示：

```
http://localhost/Page2
```

ASP.NET Core 路由特性很容易在应用程序中更改 URL 方案。只需要在 Startup 类中添加一个新路由，如代码清单 7-33 所示。

**代码清单 7-33　在 SportsStore 文件夹的 Startup.cs 文件中添加一个新路由**

```
using System;
using System.Collections.Generic;
using System.Linq;
using System.Threading.Tasks;
using Microsoft.AspNetCore.Builder;
using Microsoft.AspNetCore.Hosting;
using Microsoft.AspNetCore.Http;
using Microsoft.Extensions.DependencyInjection;
using Microsoft.Extensions.Hosting;
using Microsoft.Extensions.Configuration;
using Microsoft.EntityFrameworkCore;
using SportsStore.Models;

namespace SportsStore {
    public class Startup {

        public Startup(IConfiguration config) {
            Configuration = config;
        }

        private IConfiguration Configuration { get; set; }

        public void ConfigureServices(IServiceCollection services) {
            services.AddControllersWithViews();

            services.AddDbContext<StoreDbContext>(opts => {
                opts.UseSqlServer(
                    Configuration["ConnectionStrings:SportsStoreConnection"]);
            });
            services.AddScoped<IStoreRepository, EFStoreRepository>();
        }

        public void Configure(IApplicationBuilder app, IWebHostEnvironment env) {

            app.UseDeveloperExceptionPage();
            app.UseStatusCodePages();
            app.UseStaticFiles();
```

```
            app.UseRouting();
            app.UseEndpoints(endpoints => {
                endpoints.MapControllerRoute("pagination",
                    "Products/Page{productPage}",
                    new { Controller = "Home", action = "Index" });
                endpoints.MapDefaultControllerRoute();
            });

            SeedData.EnsurePopulated(app);
        }
    }
}
```

在调用 MapDefaultControllerRoute 方法之前添加新路由是很重要的。如第 13 章所述，路由系统按照列出的顺序处理路由，新的路由需要优先于现有的路由。

这是更改产品分页的URL方案所需的唯一更改。ASP.NET Core和路由功能是紧密集成的，因此应用程序会在应用程序使用的URL中自动反映这样的变化，包括那些由标签助手生成的URL，比如用来生成页面导航链接的那个。

重启 ASP.NET Core，请求 http://localhost:5000，并单击其中一个分页链接。浏览器将导航到一个使用新 URL 方案的 URL，如图 7-7 所示。

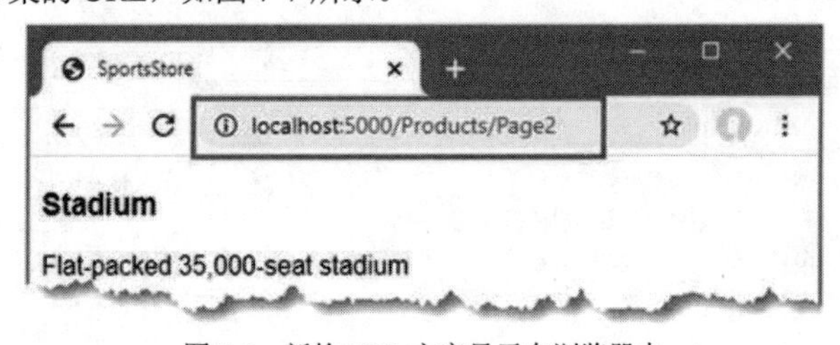


图 7-7　新的 URL 方案显示在浏览器中

## 7.5　内容的样式化

前面构建了大量的基础设施，应用程序的基本特性也开始融合在一起，但是没有考虑外观。SportsStore 应用程序设计极其平淡，破坏了自己的技术优势。虽然本书不是关于设计或 CSS 的书籍，但仍要改进设计，实现一个典型的带有标题的两列布局，如图 7-8 所示。

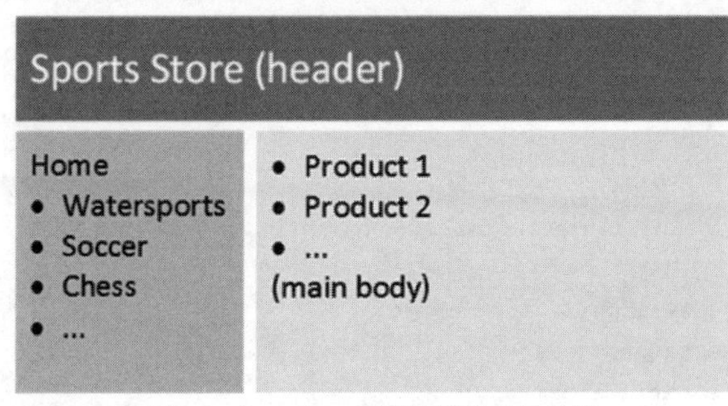


图 7-8　SportsStore 应用程序的设计目标

### 7.5.1 安装 Bootstrap 包

此处将使用 Bootstrap 包来提供将应用于应用程序的 CSS 样式。如第 4 章所述，客户端包是使用 LibMan 安装的。如果在完成第 4 章中的示例时没有安装 LibMan 包，可以使用 PowerShell 命令提示符运行代码清单 7-34 所示的命令，该命令将删除现有的 LibMan 包，并安装本书所需的版本。

**代码清单 7-34 安装 LibMan 工具包**

```
dotnet tool uninstall --global Microsoft.Web.LibraryManager.Cli
dotnet tool install --global Microsoft.Web.LibraryManager.Cli --version 2.0.96
```

安装 LibMan 后，在 SportsStore 文件夹中运行代码清单 7-35 所示的命令，以初始化示例项目，并安装 Bootstrap 包。

**代码清单 7-35 初始化示例项目**

```
libman init -p cdnjs
libman install twitter-bootstrap@4.3.1 -d wwwroot/lib/twitter-bootstrap
```

### 7.5.2 应用 Bootstrap 风格

Razor 布局提供了通用的内容，因此不必在多个视图中重复。将代码清单 7-36 所示的元素添加到 Views/Shared 文件夹的_Layout.cshtml 文件中，以便在发送到浏览器的内容中包含 Bootstrap CSS 样式表，并定义一个公共头，在整个 SportsStore 应用程序中使用。

**代码清单 7-36 在 SportsStore/Views/Shared 文件夹的_Layout .cshtml 文件中应用 Bootstrap CSS**

```
<!DOCTYPE html>
<html>
<head>
    <meta name="viewport" content="width=device-width" />
    <title>SportsStore</title>
    <link href="/lib/twitter-bootstrap/css/bootstrap.min.css" rel="stylesheet" />
</head>
<body>
    <div class="bg-dark text-white p-2">
        <span class="navbar-brand ml-2">SPORTS STORE</span>
    </div>
    <div class="row m-1 p-1">
        <div id="categories" class="col-3">
            Put something useful here later
        </div>
        <div class="col-9">
            @RenderBody()
        </div>
```

```
    </div>
</body>
</html>
```

将 Bootstrap CSS 样式表添加到布局中意味着，可以在任何依赖该布局的视图中使用它定义的样式。代码清单 7-37 显示了应用于 Index.cshtml 文件的样式。

**代码清单 7-37　在 SportsStore/Views/Home 文件夹的 Index.cshtml 文件中设置内容的样式**

```
@model ProductsListViewModel

@foreach (var p in Model.Products) {
    <div class="card card-outline-primary m-1 p-1">
        <div class="bg-faded p-1">
            <h4>
                @p.Name

                <span class="badge badge-pill badge-primary" style="float:right">
                    <small>@p.Price.ToString("c")</small>
                </span>
            </h4>
        </div>
        <div class="card-text p-1">@p.Description</div>
    </div>
}

<div page-model="@Model.PagingInfo" page-action="Index" page-classes-enabled="true"
    page-class="btn" page-class-normal="btn-outline-dark"
    page-class-selected="btn-primary" class="btn-group pull-right m-1">
</div>
```

需要设置由 PageLinkTagHelper 类生成的按钮的样式，但不想将引导类硬性链接到 C#代码中，因为这会难以在应用程序的其他地方重用标签助手或更改按钮的外观。相反，在 div 元素上定义了定制属性，用于指定需要的类，这些属性对应于添加到标签助手类的属性，这些属性随后用于对生成的 a 元素设置样式，如代码清单 7-38 所示。

**代码清单 7-38　向 SportsStore/Infrastructure 文件夹的 PageLinkTagHelper.cs 文件中的元素添加类**

```
using Microsoft.AspNetCore.Mvc;
using Microsoft.AspNetCore.Mvc.Rendering;
using Microsoft.AspNetCore.Mvc.Routing;
using Microsoft.AspNetCore.Mvc.ViewFeatures;
using Microsoft.AspNetCore.Razor.TagHelpers;
using SportsStore.Models.ViewModels;

namespace SportsStore.Infrastructure {
```

```
    [HtmlTargetElement("div", Attributes = "page-model")]
    public class PageLinkTagHelper : TagHelper {
        private IUrlHelperFactory urlHelperFactory;

        public PageLinkTagHelper(IUrlHelperFactory helperFactory) {
            urlHelperFactory = helperFactory;
        }

        [ViewContext]
        [HtmlAttributeNotBound]
        public ViewContext ViewContext { get; set; }

        public PagingInfo PageModel { get; set; }

        public string PageAction { get; set; }

        public bool PageClassesEnabled { get; set; } = false;
        public string PageClass { get; set; }
        public string PageClassNormal { get; set; }
        public string PageClassSelected { get; set; }

        public override void Process(TagHelperContext context,
                TagHelperOutput output) {
            IUrlHelper urlHelper = urlHelperFactory.GetUrlHelper(ViewContext);
            TagBuilder result = new TagBuilder("div");
            for (int i = 1; i <= PageModel.TotalPages; i++) {
                TagBuilder tag = new TagBuilder("a");
                tag.Attributes["href"] = urlHelper.Action(PageAction,
                    new { productPage = i });
                if (PageClassesEnabled) {
                    tag.AddCssClass(PageClass);
                    tag.AddCssClass(i == PageModel.CurrentPage
                        ? PageClassSelected : PageClassNormal);
                }
                tag.InnerHtml.Append(i.ToString());
                result.InnerHtml.AppendHtml(tag);
            }
            output.Content.AppendHtml(result.InnerHtml);
        }
    }
}
```

属性的值自动用于设置标签助手的属性值，并考虑 HTML 属性名格式(page-class-normal)和 C#属性名格式(PageClassNormal)之间的映射。这允许标签助手根据 HTML 元素的属性做出不同的响应，从而创建在 ASP.NETCore 应用程序中生成内容的更灵活方式。

重启 ASP.NET Core 并请求 http://localhost:5000，应用程序的外观得到了改进，如图 7-9 所示。

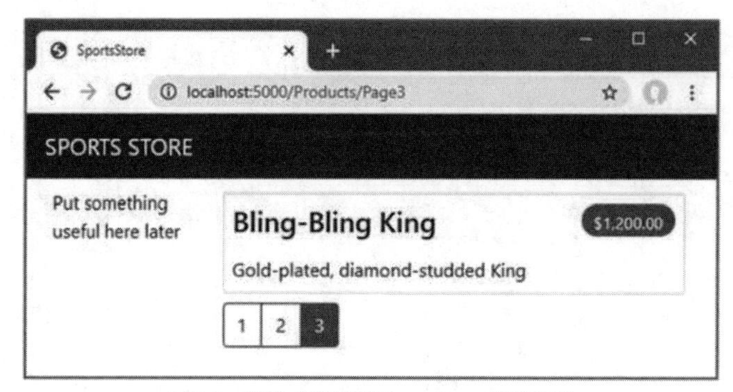


图 7-9　将样式应用到 SportsStore 应用程序

### 7.5.3　创建部分视图

作为本章的收尾工作，重构应用程序以简化 Index.cshtml 视图。创建一个部分视图；部分视图是可以嵌入另一个视图中的内容片段，类似于模板。第 22 章将详细描述部分视图；当需要相同的内容出现在应用程序的不同位置时，部分视图有助于减少重复。与其将相同的 Razor 标记复制粘贴到多个视图中，不如在局部视图中定义一次。为了创建部分视图，在 Views/Shared 文件夹中添加一个名为 ProductSummary.cshtml 的 Razor 视图，并添加如代码清单 7-39 所示的标记。

**代码清单 7-39　SportsStore/Views/Shared 文件夹中 ProductSummary.cshtml 文件的内容**

```
@model Product

<div class="card card-outline-primary m-1 p-1">
    <div class="bg-faded p-1">
        <h4>
            @Model.Name
            <span class="badge badge-pill badge-primary" style="float:right">
                <small>@Model.Price.ToString("c")</small>
            </span>
        </h4>
    </div>
    <div class="card-text p-1">@Model.Description</div>
</div>
```

现在需要更新 Views/Home 文件夹中的 Index.cshtml 文件，以便它使用部分视图，如代码清单 7-40 所示。

**代码清单 7-40　在 SportsStore/Views/Home 文件夹的 Index.cshtml 文件中使用部分视图**

```
@model ProductsListViewModel

@foreach (var p in Model.Products) {
    <partial name="ProductSummary" model="p" />
}
```

```
<div page-model="@Model.PagingInfo" page-action="Index" page-classes-enabled="true"
    page-class="btn" page-class-normal="btn-outline-dark"
    page-class-selected="btn-primary" class="btn-group pull-right m-1">
</div>
```

Index.cshtml 视图中@foreach 表达式的标记已经移动到新的部分视图。使用部分元素调用部分视图，使用名称和模型属性来指定部分视图及其视图模型的名称。使用部分视图允许将相同的标记插入需要显示产品摘要的任何视图中。

重启 ASP.NET Core，并请求 http://localhost:5000，引入部分视图不会改变应用程序的外观；只是改变 Razor 找到响应内容的位置，这些内容会发送到浏览器。

## 7.6 小结

本章为 SportsStore 应用程序构建了核心基础。此时，可以向客户端演示的特性并不多，但在幕后，已经开始使用一个域模型，该域模型有一个由 SQL Server 和 Entity Framework Core 支持的产品存储库。有一个控制器 HomeController，可以生成分页的产品列表，还建立了一个干净、友好的 URL 方案。

如果本章给人的感觉是做了很多事情，却没有什么用处，那么第 8 章将会平衡一下。既然基本结构已经就绪，就可以继续前进，添加所有面向客户的特性：按类别导航和购物车的启动。

# 第 8 章

# SportsStore：导航和购物车

本章将继续构建 SportsStore 示例应用程序。添加对导航应用程序的支持，并开始构建购物车。

■ **提示：** 可从 https://github.com/apress/pro-asp.net-core-3 下载本章和本书中其他所有章节的示例项目。如果在运行示例时遇到问题，请参阅第 1 章以获得帮助。

## 8.1 添加导航控件

如果用户可以按类别浏览产品，SportsStore 应用程序将更加有用。下面分三个阶段完成这件事。

- 增强 HomeController 类中的 Index 操作方法，以便它可以筛选存储库中的产品对象。
- 重新访问和增强 URL 方案。
- 在网站的侧边栏创建一个类别列表，突出当前类别并链接到其他类别。

### 8.1.1 筛选产品列表

下面从增强视图模型类 ProductsListViewModel 开始，它是在前一章添加到 SportsStore 项目中的。需要将当前类别传递给视图，以呈现侧边栏，这是一个很好的起点。代码清单 8-1 显示了对 Models/ViewModels 文件夹中的 ProductsListViewModel.cs 文件所做的修改。

**代码清单 8-1 修改 SportsStore/Models/ViewModels 文件夹中的 ProductsListViewModel.cs 文件**

```
using System.Collections.Generic;
using SportsStore.Models;

namespace SportsStore.Models.ViewModels {

    public class ProductsListViewModel {
        public IEnumerable<Product> Products { get; set; }
        public PagingInfo PagingInfo { get; set; }
        public string CurrentCategory { get; set; }
    }
}
```

上面添加了一个名为CurrentCategory的属性。下一步是更新主控制器，以便Index操作方法按类别筛选Product对象，并使用添加到视图模型的属性来指示选择了哪个类别，如代码清单8-2所示。

**代码清单8-2 在SportsStore/Controllers文件夹的HomeController.cs文件中添加类别支持**

```
using Microsoft.AspNetCore.Mvc;
using SportsStore.Models;
using System.Linq;
using SportsStore.Models.ViewModels;

namespace SportsStore.Controllers {
    public class HomeController : Controller {
        private IStoreRepository repository;
        public int PageSize = 4;

        public HomeController(IStoreRepository repo) {
            repository = repo;
        }

        public ViewResult Index(string category, int productPage = 1)
            => View(new ProductsListViewModel {
                Products = repository.Products
                    .Where(p => category == null || p.Category == category)
                    .OrderBy(p => p.ProductID)
                    .Skip((productPage - 1) * PageSize)
                    .Take(PageSize),
                PagingInfo = new PagingInfo {
                    CurrentPage = productPage,
                    ItemsPerPage = PageSize,
                    TotalItems = repository.Products.Count()
                },
                CurrentCategory = category
            });
    }
}
```

对操作方法做了三处更改。首先添加一个名为category的参数。这个category参数用于代码清单中的第二个更改，这是对LINQ查询的增强；如果category不是null，则只选择那些具有匹配category属性的Product对象。最后一个更改是设置添加到ProductsListViewModel类中的CurrentCategory属性的值。然而，这些变化意味着PagingInfo.TotalItems的值计算错误，因为它没有将类别筛选器考虑在内。一会儿会把它修好。

**单元测试：更新现有的单元测试**

更改Index操作方法的签名，这将阻止对一些现有的单元测试方法进行编译。为解决这个问题，需要将null作为第一个参数传递给与控制器一起工作的单元测试中的Index方法。例如，在

HomeControllerTests.cs 文件的 Can_Use_Repository 测试中，单元测试的动作部分如下：

```
...
[Fact]
public void Can_Use_Repository() {
    // Arrange
    Mock<IStoreRepository> mock = new Mock<IStoreRepository>();
    mock.Setup(m => m.Products).Returns((new Product[] {
        new Product {ProductID = 1, Name = "P1"},
        new Product {ProductID = 2, Name = "P2"}
    }).AsQueryable<Product>());

    HomeController controller = new HomeController(mock.Object);

    // Act
    ProductsListViewModel result =
        controller.Index(null).ViewData.Model as ProductsListViewModel;

    // Assert
    Product[] prodArray = result.Products.ToArray();
    Assert.True(prodArray.Length == 2);
    Assert.Equal("P1", prodArray[0].Name);
    Assert.Equal("P2", prodArray[1].Name);
}
...
```

通过对 category 参数使用 null，可接收控制器从存储库中获得的所有 Product 对象，这与添加新参数之前的情况相同。还需要对 Can_Paginate 和 Can_ Send_Pagination_View_Model 测试进行相同的更改。

```
...
[Fact]
public void Can_Paginate() {
    // Arrange
    Mock<IStoreRepository> mock = new Mock<IStoreRepository>();
    mock.Setup(m => m.Products).Returns((new Product[] {
        new Product {ProductID = 1, Name = "P1"},
        new Product {ProductID = 2, Name = "P2"},
        new Product {ProductID = 3, Name = "P3"},
        new Product {ProductID = 4, Name = "P4"},
        new Product {ProductID = 5, Name = "P5"}
    }).AsQueryable<Product>());

    HomeController controller = new HomeController(mock.Object);
    controller.PageSize = 3;

    // Act
```

```
        ProductsListViewModel result =
        controller.Index(null, 2).ViewData.Model as ProductsListViewModel;

        // Assert
        Product[] prodArray = result.Products.ToArray();

        Assert.True(prodArray.Length == 2);
        Assert.Equal("P4", prodArray[0].Name);
        Assert.Equal("P5", prodArray[1].Name);
    }

    [Fact]
    public void Can_Send_Pagination_View_Model() {

        // Arrange
        Mock<IStoreRepository> mock = new Mock<IStoreRepository>();
        mock.Setup(m => m.Products).Returns((new Product[] {
            new Product {ProductID = 1, Name = "P1"},
            new Product {ProductID = 2, Name = "P2"},
            new Product {ProductID = 3, Name = "P3"},
            new Product {ProductID = 4, Name = "P4"},
            new Product {ProductID = 5, Name = "P5"}
        }).AsQueryable<Product>());

        // Arrange
        HomeController controller =
            new HomeController(mock.Object) { PageSize = 3 };

        // Act
        ProductsListViewModel result =
            controller.Index(null, 2).ViewData.Model as ProductsListViewModel;

        // Assert
        PagingInfo pageInfo = result.PagingInfo;
        Assert.Equal(2, pageInfo.CurrentPage);
        Assert.Equal(3, pageInfo.ItemsPerPage);
        Assert.Equal(5, pageInfo.TotalItems);
        Assert.Equal(2, pageInfo.TotalPages);
    }
    ...
```

当进入测试思维模式时，保持单元测试与代码更改的同步将迅速成为第二天性。

要查看类别筛选的效果，请启动 ASP.NET Core，并使用以下网址选择一个类别，注意用大写字母 S 表示足球：

```
http://localhost:5000/?category=Soccer
```

只看到 Soccer 类别中的产品，如图 8-1 所示。

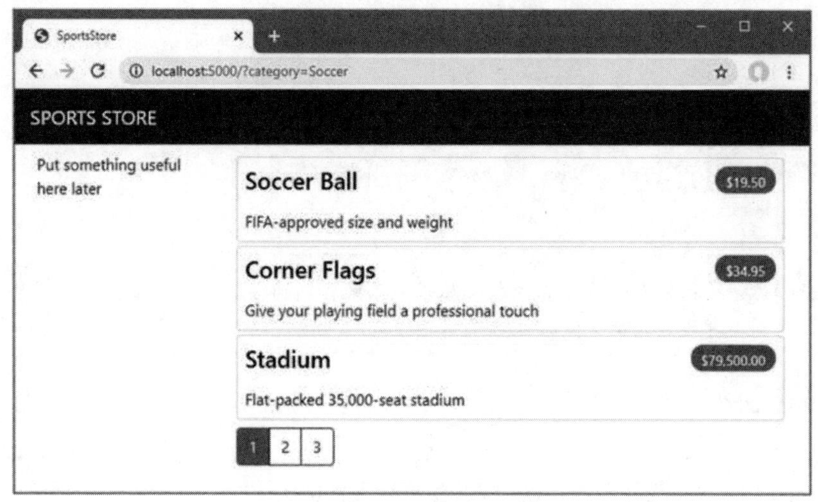


图 8-1 使用查询字符串按类别进行筛选

显然，用户不希望使用 URL 导航到类别，但是可以看到，一旦 ASP.NET Core 应用程序的基本结构已经就绪，小的变化就会产生大的影响。

**单元测试：类别筛选**

需要一个单元测试来正确地测试类别筛选函数，以确保筛选器能在指定类别中正确地生成产品。下面是添加到 ProductControllerTests 类的测试方法:

```
...
[Fact]
public void Can_Filter_Products() {

    // Arrange
    // - create the mock repository
    Mock<IStoreRepository> mock = new Mock<IStoreRepository>();
    mock.Setup(m => m.Products).Returns((new Product[] {
        new Product {ProductID = 1, Name = "P1", Category = "Cat1"},
        new Product {ProductID = 2, Name = "P2", Category = "Cat2"},
        new Product {ProductID = 3, Name = "P3", Category = "Cat1"},
        new Product {ProductID = 4, Name = "P4", Category = "Cat2"},
        new Product {ProductID = 5, Name = "P5", Category = "Cat3"}
    }).AsQueryable<Product>());

    // Arrange - create a controller and make the page size 3 items
    HomeController controller = new HomeController(mock.Object);
    controller.PageSize = 3;

    // Action
    Product[] result =
```

```
        (controller.Index("Cat2", 1).ViewData.Model as ProductsListViewModel)
            .Products.ToArray();

    // Assert
    Assert.Equal(2, result.Length);
    Assert.True(result[0].Name == "P2" && result[0].Category == "Cat2");
    Assert.True(result[1].Name == "P4" && result[1].Category == "Cat2");
}
...
```

此测试创建一个模拟存储库，其中包含属于一系列类别的 Product 对象。使用操作方法请求一个特定的类别，并检查结果，以确保结果是按正确顺序出现的正确对象。

### 8.1.2 优化 URL 方案

没有人想看到或使用像/?category=Soccer 这样难看的 URL。为解决这个问题，下面更改 Startup 类的 Configure 方法中的路由配置，以创建一组更有用的 URL，如代码清单 8-3 所示。

**警告：**在代码清单 8-3 中按显示顺序添加新路由是很重要的。路由是按照定义它们的顺序应用的；如果改变顺序，会得到一些奇怪的效果。

**代码清单 8-3 更改 SportsStore 文件夹的 Startup.cs 文件中的路由模式**

```
using System;
using System.Collections.Generic;
using System.Linq;
using System.Threading.Tasks;
using Microsoft.AspNetCore.Builder;
using Microsoft.AspNetCore.Hosting;
using Microsoft.AspNetCore.Http;
using Microsoft.Extensions.DependencyInjection;
using Microsoft.Extensions.Hosting;
using Microsoft.Extensions.Configuration;
using Microsoft.EntityFrameworkCore;
using SportsStore.Models;

namespace SportsStore {
    public class Startup {

        public Startup(IConfiguration config) {
            Configuration = config;
        }

        private IConfiguration Configuration { get; set; }

        public void ConfigureServices(IServiceCollection services) {
            services.AddControllersWithViews();
```

```
            services.AddDbContext<StoreDbContext>(opts => {
                opts.UseSqlServer(
                    Configuration["ConnectionStrings:SportsStoreConnection"]);
            });
            services.AddScoped<IStoreRepository, EFStoreRepository>();
        }

        public void Configure(IApplicationBuilder app, IWebHostEnvironment env) {
            app.UseDeveloperExceptionPage();
            app.UseStatusCodePages();
            app.UseStaticFiles();

            app.UseRouting();
            app.UseEndpoints(endpoints => {
                endpoints.MapControllerRoute("catpage",
                    "{category}/Page{productPage:int}",
                    new { Controller = "Home", action = "Index" });

                endpoints.MapControllerRoute("page", "Page{productPage:int}",
                    new { Controller = "Home", action = "Index", productPage = 1 });

                endpoints.MapControllerRoute("category", "{category}",
                    new { Controller = "Home", action = "Index", productPage = 1 });

                endpoints.MapControllerRoute("pagination",
                    "Products/Page{productPage}",
                    new { Controller = "Home", action = "Index", productPage = 1 });
                endpoints.MapDefaultControllerRoute();
            });

            SeedData.EnsurePopulated(app);
        }
    }
}
```

表 8-1 描述了这些路由表示的 URL 方案。第 13 章将详细解释路由系统。

表 8-1　路由概览

| URL | 目标 |
| --- | --- |
| / | 列出所有类别产品的第一页 |
| /Page2 | 列出指定的页面(在本例中为 Page 2)，显示来自所有类别的项目 |
| /Soccer | 显示来自指定类别的项的第一页(在本例中为 Soccer 类别) |
| /Soccer/Page2 | 显示来自指定类别(本例中为 Soccer)的项的指定页面(本例中为 Page 2) |

ASP.NET Core 路由系统处理来自客户端的传入请求，但也生成符合 URL 方案的传出 URL，可嵌入 Web 页面中。通过使用路由系统来处理传入的请求，生成传出的 URL，可以确保应用程序中的所有 URL 是一致的。

IUrlHelper 接口提供了对 URL 生成功能的访问。在前一章创建的标签助手中使用了这个接口和它定义的操作方法。现在想开始生成更复杂的 URL，需要一种方法从视图接收额外信息，而不必向标签助手类添加额外属性。幸运的是，标签助手有一个很好的特性，它允许在一个集合中接收带有公共前缀的属性，如代码清单 8-4 所示。

**代码清单 8-4　在 SportsStore/Infrastructure 文件夹的 PageLinkTagHelper.cs 文件中为值添加前缀**

```
using Microsoft.AspNetCore.Mvc;
using Microsoft.AspNetCore.Mvc.Rendering;
using Microsoft.AspNetCore.Mvc.Routing;
using Microsoft.AspNetCore.Mvc.ViewFeatures;
using Microsoft.AspNetCore.Razor.TagHelpers;
using SportsStore.Models.ViewModels;
using System.Collections.Generic;

namespace SportsStore.Infrastructure {

    [HtmlTargetElement("div", Attributes = "page-model")]
    public class PageLinkTagHelper : TagHelper {
        private IUrlHelperFactory urlHelperFactory;

        public PageLinkTagHelper(IUrlHelperFactory helperFactory) {
            urlHelperFactory = helperFactory;
        }

        [ViewContext]
        [HtmlAttributeNotBound]
        public ViewContext ViewContext { get; set; }

        public PagingInfo PageModel { get; set; }

        public string PageAction { get; set; }

        [HtmlAttributeName(DictionaryAttributePrefix = "page-url-")]
        public Dictionary<string, object> PageUrlValues { get; set; }
            = new Dictionary<string, object>();

        public bool PageClassesEnabled { get; set; } = false;
        public string PageClass { get; set; }
        public string PageClassNormal { get; set; }
        public string PageClassSelected { get; set; }

        public override void Process(TagHelperContext context,
```

```
            TagHelperOutput output) {
        IUrlHelper urlHelper = urlHelperFactory.GetUrlHelper(ViewContext);
        TagBuilder result = new TagBuilder("div");
        for (int i = 1; i <= PageModel.TotalPages; i++) {
            TagBuilder tag = new TagBuilder("a");
            PageUrlValues["productPage"] = i;
            tag.Attributes["href"] = urlHelper.Action(PageAction, PageUrlValues);
            if (PageClassesEnabled) {
                tag.AddCssClass(PageClass);
                tag.AddCssClass(i == PageModel.CurrentPage
                    ? PageClassSelected : PageClassNormal);
            }
            tag.InnerHtml.Append(i.ToString());
            result.InnerHtml.AppendHtml(tag);
        }
        output.Content.AppendHtml(result.InnerHtml);
    }
  }
}
```

使用 HtmlAttributeName 属性修饰标签助手属性，允许为元素上的属性名称指定前缀，在本例中为 page-url-。名称以这个前缀开头的任何属性的值都添加到分配给 PageUrlValues 属性的字典中，然后将该属性传递给 UrlHelper.Action 方法，生成 URL(作为标签助手生成的 a 元素的 href 属性的值)。

在代码清单 8-5 中，向标签助手处理的 div 元素添加了一个新属性，指定用于生成 URL 的类别。只向视图添加了一个新属性，但是任何具有相同前缀的属性都将添加到字典中。

#### 代码清单 8-5 在 SportsStore/Views/Home 文件夹的 Index.cshtml 文件中添加一个新属性

```
@model ProductsListViewModel

@foreach (var p in Model.Products) {
    <partial name="ProductSummary" model="p" />
}

<div page-model="@Model.PagingInfo" page-action="Index" page-classes-enabled="true"
    page-class="btn" page-class-normal="btn-outline-dark"
    page-class-selected="btn-primary" page-url-category="@Model.CurrentCategory"
    class="btn-group pull-right m-1">
</div>
```

在更改前，为分页链接生成的链接如下：

```
http://localhost:5000/Page1
```

如果用户单击这样的页面链接，则类别过滤器将丢失，应用程序将显示一个包含所有类别产品的页面。通过添加当前类别(取自视图模型)，生成如下 URL：

```
http://localhost:5000/Chess/Page1
```

当用户单击此类链接时，当前类别将传递给 Index 操作方法，过滤将被保留。要查看此更改的效果，请启动 ASP.NET Core 并请求 http://localhost:5000/Chess，这将只显示 Chess 类别的产品，如图 8-2 所示。

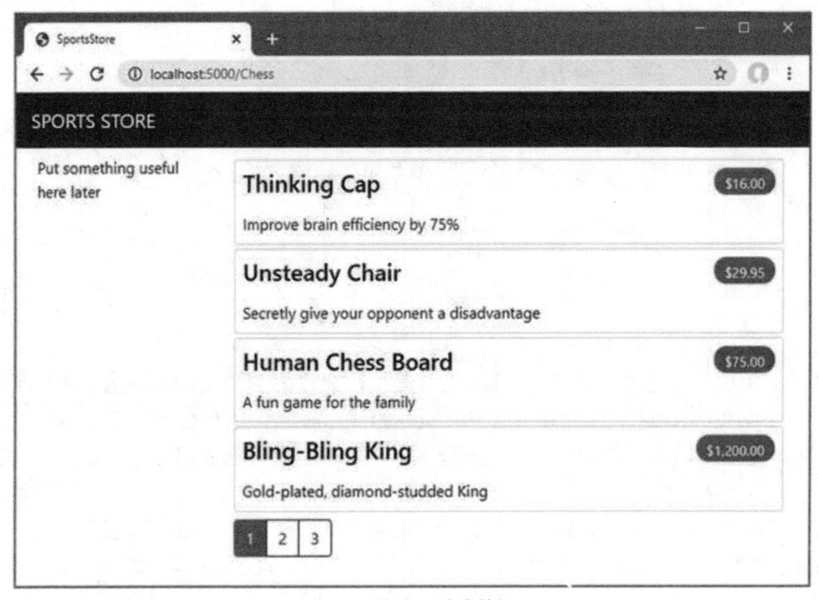


图 8-2　按类别过滤数据

## 8.1.3　构建一个类别导航菜单

需要为用户提供一种不需要输入 URL 就可以选择类别的方法。这意味着列出可用类别，并指出当前选择的类别(如果有的话)。

ASP.NET Core 有视图组件的概念，它非常适合创建可重用的导航控件等。视图组件是一个 C#类，它提供了少量可重用的应用程序逻辑，能够选择和显示 Razor 部分视图。第 24 章将详细描述视图组件。

本例将创建一个视图组件，该组件呈现导航菜单，并通过调用共享布局中的组件，将其集成到应用程序中。这种方法提供了一个常规的 C#类，可包含需要的任何应用程序逻辑，并且可以像其他类一样进行单元测试。

### 1. 创建导航视图组件

前面在 SportsStore 项目中创建了一个名为 Components 的文件夹，这是视图组件的传统主文件夹，并添加了一个名为 NavigationMenuViewComponent 的类文件。使用它定义代码清单 8-6 所示的类。

**代码清单 8-6 SportsStore/Components文件夹中NavigationMenuViewComponent.cs文件的内容**

```
using Microsoft.AspNetCore.Mvc;

namespace SportsStore.Components {

    public class NavigationMenuViewComponent : ViewComponent {

        public string Invoke() {
            return "Hello from the Nav View Component";
        }
    }
}
```

在 Razor 视图中使用组件时调用视图组件的 Invoke 方法，Invoke 方法的结果插入到发送给浏览器的 HTML 中。这里从一个返回字符串的简单视图组件开始，很快就用 HTML 替换它。

希望类别列表出现在所有页面上，因此在共享布局中使用视图组件，而不是在特定的视图中。在视图中，使用标签助手应用视图组件，如代码清单 8-7 所示。

**代码清单 8-7 在 SportsStore/Views/Shared 文件夹的_Layout.cshtml 文件中使用视图组件**

```
<!DOCTYPE html>
<html>
<head>
    <meta name="viewport" content="width=device-width" />
    <title>SportsStore</title>
    <link href="/lib/twitter-bootstrap/css/bootstrap.min.css" rel="stylesheet" />
</head>
<body>
    <div class="bg-dark text-white p-2">
        <span class="navbar-brand ml-2">SPORTS STORE</span>
    </div>
    <div class="row m-1 p-1">
        <div id="categories" class="col-3">
            <vc:navigation-menu />
        </div>
        <div class="col-9">
            @RenderBody()
        </div>
    </div>
</body>
</html>
```

删除了占位符文本，并用插入视图组件的vc:navigation-menu元素替换它。元素省略了类名的ViewComponent部分，并将其连接起来，这样vc:navigation-menu就指定了NavigationMenuViewComponent类。

重启 ASP.NET Core，并请求 http://localhost:5000，Invoke 方法的输出就包含在发送到浏览器的 HTML 中，如图 8-3 所示。

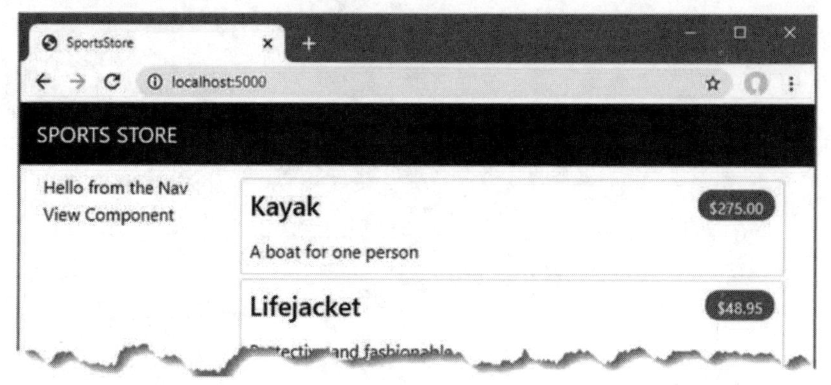


图 8-3　使用视图组件

### 2. 生成类别列表

现在可以返回到导航视图组件，并生成一组真正的类别。可通过编程方式为类别构建 HTML，就像为页面标签助手所做的那样，但是使用视图组件的好处之一是它们可以呈现 Razor 部分视图。这意味着可以使用视图组件生成组件列表，然后使用更具表达性的 Razor 语法来呈现将显示这些组件的 HTML。第一步是更新视图组件，如代码清单 8-8 所示。

**代码清单 8-8　在 SportsStore/Components 文件夹的 NavigationMenuViewComponent.cs 文件中添加类别**

```
using Microsoft.AspNetCore.Mvc;
using System.Linq;
using SportsStore.Models;

namespace SportsStore.Components {

    public class NavigationMenuViewComponent : ViewComponent {
        private IStoreRepository repository;

        public NavigationMenuViewComponent(IStoreRepository repo) {
            repository = repo;
        }

        public IViewComponentResult Invoke() {
            return View(repository.Products
                .Select(x => x.Category)
                .Distinct()
                .OrderBy(x => x));
        }
    }
}
```

代码清单 8-8 中的构造函数定义了一个 IStoreRepository 参数。在 ASP.NET Core 需要创建视图组件类的一个实例时，需要为这个参数提供一个值，并检查 Startup 类中的配置，以确定应该使用哪个实现对象。这与第 7 章的控制器中使用的依赖注入特性相同，具有相同的效果，即允许视图组件访问数据，而不知道将使用哪个存储库实现；这个特性详见第 14 章。

Invoke 方法使用 LINQ 选择存储库中的类别集，并排序，然后将它们作为参数传递给 View 方法，以显示默认的 Razor 部分视图，其细节从使用 IViewComponentResult 对象的方法返回，该过程详见第 24 章。

**单元测试：生成类别列表**

能生成类别列表的单元测试相对简单。目标是创建一个按字母顺序排序、不含重复的列表。最简单的方法是提供一些测试数据(其中有重复的类别，没有排序)，将其传递给标签助手类，声称数据已经妥善清理。下面是单元测试，是在SportsStore.Tests项目中名为NavigationMenuViewComponentTests.cs的新类文件中定义的：

```
using System.Collections.Generic;
using System.Linq;
using Microsoft.AspNetCore.Components;
using Microsoft.AspNetCore.Mvc.Rendering;
using Microsoft.AspNetCore.Mvc.ViewComponents;
using Moq;
using SportsStore.Components;
using SportsStore.Models;
using Xunit;
namespace SportsStore.Tests {

    public class NavigationMenuViewComponentTests {

        [Fact]
        public void Can_Select_Categories() {
            // Arrange
            Mock<IStoreRepository> mock = new Mock<IStoreRepository>();
            mock.Setup(m => m.Products).Returns((new Product[] {
                new Product {ProductID = 1, Name = "P1", Category = "Apples"},
                new Product {ProductID = 2, Name = "P2", Category = "Apples"},
                new Product {ProductID = 3, Name = "P3", Category = "Plums"},
                new Product {ProductID = 4, Name = "P4", Category = "Oranges"},
            }).AsQueryable<Product>());

            NavigationMenuViewComponent target =
                new NavigationMenuViewComponent(mock.Object);

            // Act = get the set of categories
            string[] results = ((IEnumerable<string>)(target.Invoke()
                as ViewViewComponentResult).ViewData.Model).ToArray();
```

```
            // Assert
            Assert.True(Enumerable.SequenceEqual(new string[] { "Apples",
                "Oranges", "Plums" }, results));
        }
    }
}
```

前面创建了一个模拟存储库实现，其中包含重复的类别和不按顺序排列的类别。断言删除重复，并强制按字母排序。

#### 3. 创建视图

Razor 使用不同的约定来定位由视图组件选择的视图。视图的默认名称和搜索视图的位置都与控制器使用的不同。为此，在 SportsStore 项目中创建了 Views/Shared/Components/NavigationMenu 文件夹，并添加了一个名为 Default.cshtml 的 Razor 视图。向其添加如代码清单 8-9 所示的内容。

**代码清单 8-9　SportsStore/Views/Shared/Components/NavigationMenu 文件夹的 Default.cshtml 文件的内容**

```
@model IEnumerable<string>

<a class="btn btn-block btn-outline-secondary"asp-action="Index"
   asp-controller="Home" asp-route-category="">
   Home
</a>

@foreach (string category in Model) {
   <a class="btn btn-block btn-outline-secondary"
      asp-action="Index" asp-controller="Home"
      asp-route-category="@category"
      asp-route-productPage="1">
      @category
   </a>
}
```

这个视图使用一个内置的标签助手(参见第 25～27 章)创建锚元素，这些锚元素的 href 属性包含一个选择不同产品类别的 URL。

重启 ASP.NET Core，并请求 http://localhost:5000，以查看类别导航按钮。如果单击一个按钮，项目列表就会更新，只显示选中类别中的项目，如图 8-4 所示。

#### 4. 突出显示当前类别

没有反馈给用户以指示选择了哪个类别。可以从列表中的项推断出其类别，也可以从一些清晰的视觉反馈中看出端倪。ASP.NET Core 组件(如控制器和视图组件)可通过请求上下文对象来接收关于当前请求的信息。大多数情况下，可依赖用于创建组件的基类来获取上下文对象，比如使用控制器基类创建控制器时。

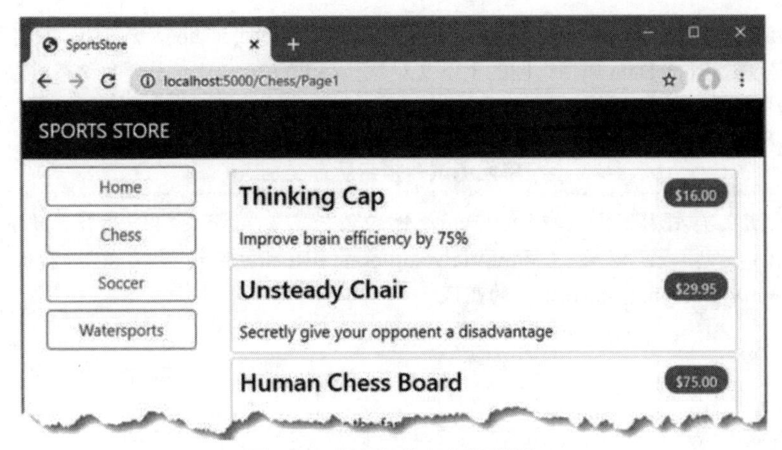


图 8-4　用视图组件生成类别链接

ViewComponent 基类也不例外，它通过一组属性提供对上下文对象的访问。其中一个属性称为 RouteData，它提供关于路由系统如何处理请求 URL 的信息。

代码清单 8-10 使用 RouteData 属性访问请求数据，以获得当前选定类别的值。可以通过创建另一个视图模型类来传递类别视图(这就是在真正的项目中执行的操作)。但为了展示多样性，此处使用视图的包功能；该功能允许将非结构化数据传递给一个视图和视图模型对象。第 22 章将详细描述这个特性的工作原理。

**代码清单 8-10　在 SportsStore/Components 文件夹的 NavigationMenuViewComponent.cs 文件中传递选定的类别**

```
using Microsoft.AspNetCore.Mvc;
using System.Linq;
using SportsStore.Models;

namespace SportsStore.Components {

    public class NavigationMenuViewComponent : ViewComponent {
        private IStoreRepository repository;

        public NavigationMenuViewComponent(IStoreRepository repo) {
            repository = repo;
        }

        public IViewComponentResult Invoke() {
            ViewBag.SelectedCategory = RouteData?.Values["category"];
            return View(repository.Products
                .Select(x => x.Category)
                .Distinct()
                .OrderBy(x => x));
        }
    }
}
```

在Invoke方法中，为ViewBag对象动态分配了一个SelectedCategory属性，并将其值设置为当前类别(是通过RouteData属性返回的上下文对象获得的)。ViewBag是一个动态对象，它允许通过给它们分配值来定义新属性。

**单元测试：报告所选类别**

可以通过在单元测试中读取ViewBag属性的值，来测试视图组件是否正确添加了所选类别的详细信息；可以通过ViewViewComponentResult类获得该值。下面是添加到NavigatioMenuViewComponentTests类的测试:

```
...
[Fact]
public void Indicates_Selected_Category() {

    // Arrange
    string categoryToSelect = "Apples";
    Mock<IStoreRepository> mock = new Mock<IStoreRepository>();
    mock.Setup(m => m.Products).Returns((new Product[] {
        new Product {ProductID = 1, Name = "P1", Category = "Apples"},
        new Product {ProductID = 4, Name = "P2", Category = "Oranges"},
    }).AsQueryable<Product>());

    NavigationMenuViewComponent target =
        new NavigationMenuViewComponent(mock.Object);
    target.ViewComponentContext = new ViewComponentContext {
        ViewContext = new ViewContext {
            RouteData = new Microsoft.AspNetCore.Routing.RouteData()
        }
    };
    target.RouteData.Values["category"] = categoryToSelect;

    // Action
    string result = (string)(target.Invoke() as
        ViewViewComponentResult).ViewData["SelectedCategory"];

    // Assert
    Assert.Equal(categoryToSelect, result);
}
...
```

这个单元测试通过ViewComponentContext属性向视图组件提供路由数据，这是视图组件接收其所有上下文数据的方式。ViewComponentContext属性通过ViewContext属性提供对特定于视图的上下文数据的访问，而ViewContext属性又通过RouteData属性提供对路由信息的访问。单元测试中的大部分代码用于创建上下文对象，这些上下文对象将提供所选类别，与应用程序运行和由ASP.NET Core MVC提供上下文数据时呈现的方式相同。

前面提供了关于所选类别的信息，可以更新视图组件所选的视图，并改变用于样式化链接的

CSS 类，以便表示当前类别的样式是不同的。代码清单 8-11 显示了对 Default.cshtml 文件所做的更改。

**代码清单 8-11　高亮显示 SportsStore/Views/Shared/Components/NavigationMenu 文件夹中的 Default.cshtml 文件**

```
@model IEnumerable<string>

<a class="btn btn-block btn-outline-secondary"asp-action="Index"
   asp-controller="Home" asp-route-category="">
   Home
</a>

@foreach (string category in Model) {
   <a class="btn btn-block
      @(category == ViewBag.SelectedCategory
         ? "btn-primary": "btn-outline-secondary")"
      asp-action="Index" asp-controller="Home"
      asp-route-category="@category"
      asp-route-productPage="1">
      @category
   </a>
}
```

在 class 属性中使用了 Razor 表达式，将 btn-primary 类应用于表示所选类别的元素，而将 btn-outline-secondary 类应用于其他元素。这些类应用不同的引导样式，并使活动按钮变得很明显，为此，可以重新启动 ASP.NET Core，请求 http://localhost:5000，单击其中一个类别按钮，如图 8-5 所示。

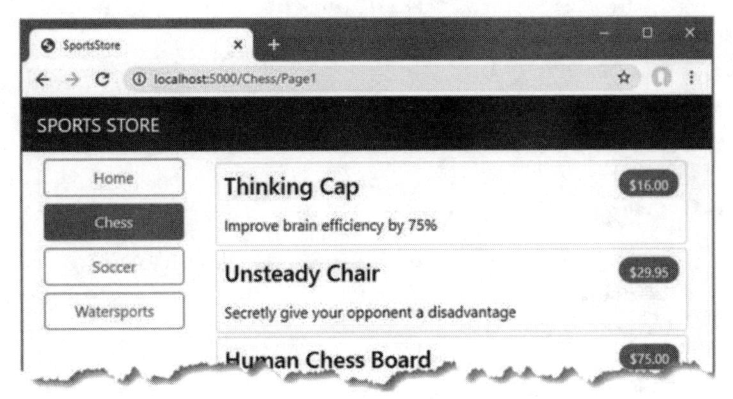


图 8-5　突出显示所选类别

## 8.1.4　更正页数

需要纠正页面链接，这样在选择类别时，链接会正确工作。目前，页面链接的数量由存储库中的产品总数决定，而不是由所选类别中的产品数量决定。这意味着客户可单击国际象棋类别第 2 页的链接，最终得到一个空页面，因为没有足够的国际象棋产品来填满两个页面。可在图 8-6

中看到这个问题。

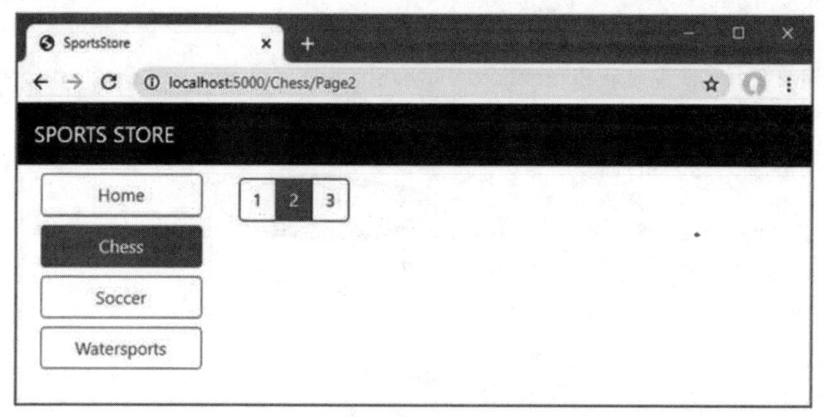


图 8-6　选择类别时显示错误的页面链接

要解决这个问题，可更新主控制器中的 Index 操作方法，以便分页信息考虑到类别，如代码清单 8-12 所示。

**代码清单 8-12　在 SportsStore/Controllers 文件夹的 HomeController.cs 文件中创建类别分页数据**

```
using Microsoft.AspNetCore.Mvc;
using SportsStore.Models;
using System.Linq;
using SportsStore.Models.ViewModels;

namespace SportsStore.Controllers {
    public class HomeController : Controller {
        private IStoreRepository repository;
        public int PageSize = 4;

        public HomeController(IStoreRepository repo) {
            repository = repo;
        }

        public ViewResult Index(string category, int productPage = 1)
            => View(new ProductsListViewModel {
                Products = repository.Products
                    .Where(p => category == null || p.Category == category)
                    .OrderBy(p => p.ProductID)
                    .Skip((productPage - 1) * PageSize)
                    .Take(PageSize),
                PagingInfo = new PagingInfo {
                    CurrentPage = productPage,
                    ItemsPerPage = PageSize,
                    TotalItems = category == null ?
                        repository.Products.Count() :
```

```
                    repository.Products.Where(e =>
                        e.Category == category).Count()
                },
                CurrentCategory = category
            });
        }
    }
```

如果选中一个类别，返回该类别的项目数量；如果没有选中，则返回产品总数。重启 ASP.NET Core 并请求 http://localhost:5000，以查看选择类别时的更改，如图 8-7 所示。

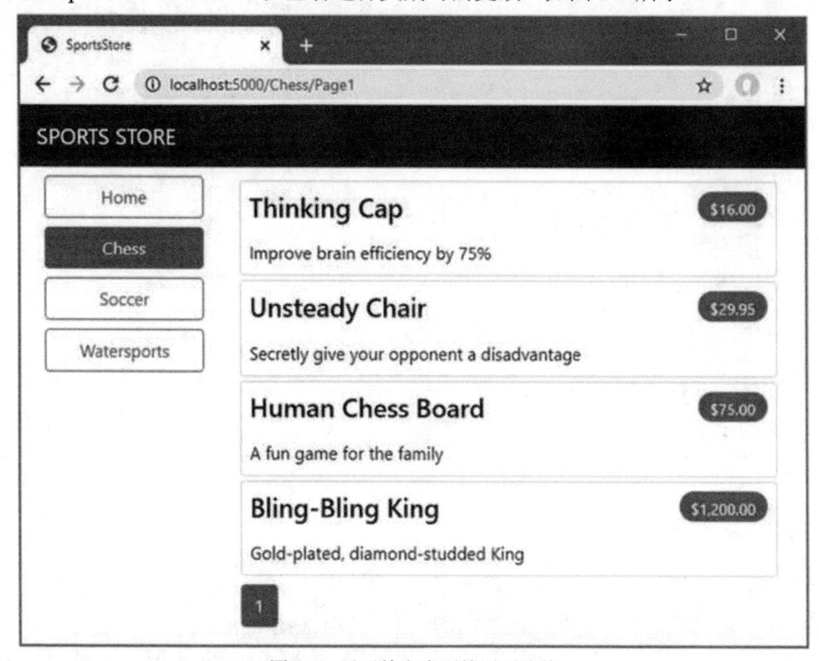


图 8-7　显示特定类别的页面计数

**单元测试：特定于类别的产品数量**

能够为不同类别生成当前产品计数的测试是很简单的。该测试创建了一个模拟存储库(其中包含一系列类别中的已知数据)，然后调用 List 操作方法依次请求每个类别。下面是添加到 HomeControllerTests 类的单元测试方法(需要为这个测试导入 System 名称空间)：

```
...
[Fact]
public void Generate_Category_Specific_Product_Count() {
    // Arrange
    Mock<IStoreRepository> mock = new Mock<IStoreRepository>();
    mock.Setup(m => m.Products).Returns((new Product[] {
        new Product {ProductID = 1, Name = "P1", Category = "Cat1"},
        new Product {ProductID = 2, Name = "P2", Category = "Cat2"},
        new Product {ProductID = 3, Name = "P3", Category = "Cat1"},
```

```
        new Product {ProductID = 4, Name = "P4", Category = "Cat2"},
        new Product {ProductID = 5, Name = "P5", Category = "Cat3"}
    }).AsQueryable<Product>());

    HomeController target = new HomeController(mock.Object);
    target.PageSize = 3;
    Func<ViewResult, ProductsListViewModel> GetModel = result =>
        result?.ViewData?.Model as ProductsListViewModel;

    // Action
    int? res1 = GetModel(target.Index("Cat1"))?.PagingInfo.TotalItems;
    int? res2 = GetModel(target.Index("Cat2"))?.PagingInfo.TotalItems;
    int? res3 = GetModel(target.Index("Cat3"))?.PagingInfo.TotalItems;
    int? resAll = GetModel(target.Index(null))?.PagingInfo.TotalItems;

    // Assert
    Assert.Equal(2, res1);
    Assert.Equal(2, res2);
    Assert.Equal(1, res3);
    Assert.Equal(5, resAll);
}
...
```

注意，还调用了 Index 方法，没有指定类别，以确保得到正确的总数。

## 8.2　构建购物车

应用程序运行得很好，但是在实现购物车之前不能销售任何产品。本节将创建如图 8-8 所示的购物车体验。这对任何在网上购买过物品的人来说都很熟悉。

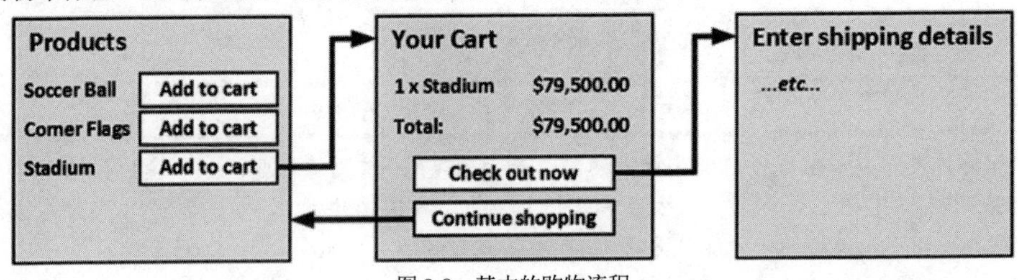


图 8-8　基本的购物流程

Add to cart 按钮显示在目录的每个产品旁边。单击此按钮将显示客户目前已选择的产品摘要，包括总价。此时，用户可以单击 Continue shopping 按钮返回产品目录，或者单击 Check out now 按钮，完成订单，并完成购物会话。

### 8.2.1　配置 Razor Pages

前面使用 MVC 框架来定义 SportsStore 项目特性。为了多样化，这里使用 Razor 页面——ASP.NET Core 支持的另一个应用程序框架——实现购物车。代码清单 8-13 配置了 Startup 类，以在 SportsStore 应用程序中启用 Razor Pages。

**代码清单 8-13　在 SportsStore 文件夹的 Startup.cs 文件中启用 Razor Pages**

```
using System;
using System.Collections.Generic;
using System.Linq;
using System.Threading.Tasks;
using Microsoft.AspNetCore.Builder;
using Microsoft.AspNetCore.Hosting;
using Microsoft.AspNetCore.Http;
using Microsoft.Extensions.DependencyInjection;
using Microsoft.Extensions.Hosting;
using Microsoft.Extensions.Configuration;
using Microsoft.EntityFrameworkCore;
using SportsStore.Models;

namespace SportsStore {
   public class Startup {

      public Startup(IConfiguration config) {
         Configuration = config;
      }

      private IConfiguration Configuration { get; set; }

      public void ConfigureServices(IServiceCollection services) {
         services.AddControllersWithViews();
         services.AddDbContext<StoreDbContext>(opts => {
            opts.UseSqlServer(
               Configuration["ConnectionStrings:SportsStoreConnection"]);
         });
         services.AddScoped<IStoreRepository, EFStoreRepository>();
         services.AddRazorPages();
      }

      public void Configure(IApplicationBuilder app, IWebHostEnvironment env) {
         app.UseDeveloperExceptionPage();
         app.UseStatusCodePages();
         app.UseStaticFiles();

         app.UseRouting();
         app.UseEndpoints(endpoints => {
            endpoints.MapControllerRoute("catpage",
               "{category}/Page{productPage:int}",
               new { Controller = "Home", action = "Index" });
```

```
                endpoints.MapControllerRoute("page", "Page{productPage:int}",
                    new { Controller = "Home", action = "Index", productPage = 1 });

                endpoints.MapControllerRoute("category", "{category}",
                    new { Controller = "Home", action = "Index", productPage = 1 });

                endpoints.MapControllerRoute("pagination",
                    "Products/Page{productPage}",
                    new { Controller = "Home", action = "Index", productPage = 1 });
                endpoints.MapDefaultControllerRoute();
                endpoints.MapRazorPages();
            });

            SeedData.EnsurePopulated(app);
        }
    }
}
```

AddRazorPages 方法设置 Razor Pages 使用的服务，MapRazorPages 方法将 Razor Pages 注册为 URL 路由系统可用来处理请求的端点。

在 SportsStore 项目中添加一个名为 Pages 的文件夹，这是 Razor Pages 的传统位置。向 Pages 文件夹添加一个名为_ViewImports.cshtml 的 Razor 视图导入文件，内容如代码清单 8-14 所示。这些表达式设置了 Razor Pages 所属的名称空间，并允许在 Razor Pages 中使用 SportsStore 类，而不必指定它们的名称空间。

#### 代码清单 8-14　SportsStore/Pages 文件夹中_ViewImports.cshtml 文件的内容

```
@namespace SportsStore.Pages
@using Microsoft.AspNetCore.Mvc.RazorPages
@using SportsStore.Models
@using SportsStore.Infrastructure
@addTagHelper *, Microsoft.AspNetCore.Mvc.TagHelpers
```

接下来，向 Pages 文件夹添加一个名为_ViewStart.cshtml 的 Razor 视图启动文件，其内容如代码清单 8-15 所示。Razor Pages 有自己的配置文件，这个文件指定了 SportsStore 项目中的 Razor Pages 默认使用一个名为_CartLayout 的布局文件。

#### 代码清单 8-15　SportsStore/Pages 文件夹中_ViewStart.cshtml 文件的内容

```
@{
    Layout = "_CartLayout";
}
```

最后，为了提供 Razor Pages 将使用的布局，向 Pages 文件夹添加一个名为_CartLayout.cshtml 的 Razor 视图，内容如代码清单 8-16 所示。

**代码清单 8-16　SportsStore/Pages 文件夹中_CartLayout.cshtml 文件的内容**

```
<!DOCTYPE html>
<html>
<head>
    <meta name="viewport" content="width=device-width" />
    <title>SportsStore</title>
    <link href="/lib/twitter-bootstrap/css/bootstrap.min.css" rel="stylesheet" />
</head>
<body>
    <div class="bg-dark text-white p-2">
        <span class="navbar-brand ml-2">SPORTS STORE</span>
    </div>
    <div class="m-1 p-1">
            @RenderBody()
    </div>
</body>
</html>
```

## 8.2.2 创建 Razor Pages

如果使用的是 Visual Studio，请使用 Razor Pages 模板项并将项名设置为 Cart.cshtml。这将创建 Cart.cshtml 文件和 Cart.cshtml.cs 类文件。将文件的内容替换为代码清单 8-17 所示的内容。如果使用的是 Visual Studio Code，只需要创建 Cart.cshtml 文件，内容如代码清单 8-17 所示。

**代码清单 8-17　SportsStore/Pages 文件夹中 Cart.cshtml 文件的内容**

```
@page

<h4>This is the Cart Page</h4>
```

重启 ASP.NET Core 并请求 http://localhost:5000/cart，以查看代码清单 8-17 中的占位符内容，如图 8-9 所示。注意，不需要注册页面，而且/cart URL 路径和 Razor Pages 之间的映射已经自动处理。

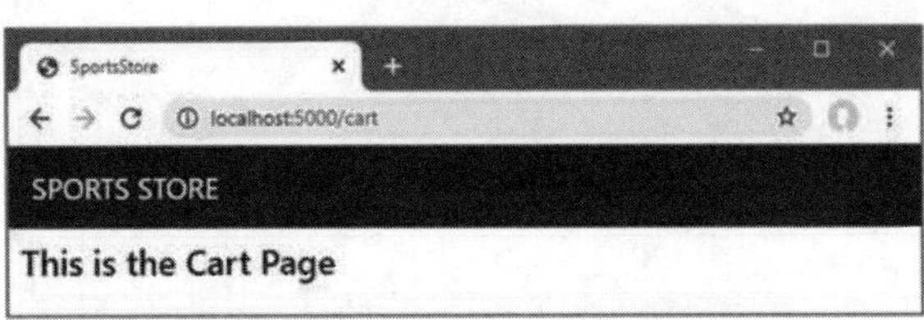


图 8-9　来自 Razor Pages 的占位符内容

## 8.2.3 创建 Add To Cart 按钮

在实现购物车功能之前，需要做一些准备工作。首先需要创建将产品添加到购物车的按钮。为此，在 Infrastructure 文件夹中添加一个名为 UrlExtensions.cs 的类文件，并定义扩展方法，如代码清单 8-18 所示。

**代码清单8-18　SportsStore/Infrastructure文件夹中UrlExtensions.cs文件的内容**

```
using Microsoft.AspNetCore.Http;

namespace SportsStore.Infrastructure {

    public static class UrlExtensions {

        public static string PathAndQuery(this HttpRequest request) =>
            request.QueryString.HasValue
                ? $"{request.Path}{request.QueryString}"
                : request.Path.ToString();
    }
}
```

PathAndQuery扩展方法操作HttpRequest类，ASP.NET Core使用它描述HTTP请求。扩展方法生成一个URL，在购物车更新后，浏览器将返回该URL，如果有查询字符串的话，则代码清单8-19将包含扩展方法的名称空间添加到视图导入文件中，以便在部分视图中使用它。

■ **注意**：这是Views文件夹中的视图导入文件，而不是添加到Pages文件夹中的视图导入文件。

**代码清单8-19　在SportsStore/Views文件夹的_ViewImports.cshtml文件中添加名称空间**

```
@using SportsStore.Models
@using SportsStore.Models.ViewModels
@using SportsStore.Infrastructure
@addTagHelper *, Microsoft.AspNetCore.Mvc.TagHelpers
@addTagHelper *, SportsStore
```

在代码清单8-20中，更新了描述每个产品的部分视图，使其包含一个Add to Cart按钮。

**代码清单8-20　在SportsStore/Views/Shared文件夹中，将按钮添加到ProductSummary.cshtml文件视图**

```
@model Product

<div class="card card-outline-primary m-1 p-1">
    <div class="bg-faded p-1">
        <h4>
            @Model.Name
            <span class="badge badge-pill badge-primary" style="float:right">
                <small>@Model.Price.ToString("c")</small>
            </span>
        </h4>
    </div>
    <form id="@Model.ProductID" asp-page="/Cart" method="post">
        <input type="hidden" asp-for="ProductID" />
        <input type="hidden" name="returnUrl"
               value="@ViewContext.HttpContext.Request.PathAndQuery()" />
```

```
        <span class="card-text p-1">
            @Model.Description
            <button type="submit"
                class="btn btn-success btn-sm pull-right" style="float:right">
                Add To Cart
            </button>
        </span>
    </form>
</div>
```

上面添加了一个表单元素，其中包含指定视图模型的 ProductID 值的隐藏输入元素，以及购物车更新后浏览器应该返回的 URL。表单元素和一个输入元素是使用内置的标签助手配置的，是生成包含模型值和目标控制器或 Razor Pages 的表单的有用方法，如第 27 章所述。另一个输入元素使用创建的扩展方法设置返回的 URL。还添加了一个按钮元素，用于向应用程序提交表单。

---

■ **注意**：将表单元素的 method 属性设置为 post，它指示浏览器使用 HTTP POST 请求提交表单数据。可以对此进行更改，以便表单使用 GET 方法，但这样做时应该慎重考虑。HTTP 规范要求 GET 请求必须是幂等的，这意味着它们不能引起更改，而将产品添加到购物车肯定是一种更改。

---

## 8.2.4 启用会话

使用会话状态存储用户购物车的详细信息，会话状态是与用户发出的一系列请求关联的数据。ASP.NET 提供了一系列不同的存储会话状态的方法，包括将其存储在内存中，这就是这里使用的方法。这样做的优点是简单，但这意味着当应用程序停止或重新启动时会话数据会丢失。启用会话需要在 Startup 类中添加服务和中间件，如代码清单 8-21 所示。

**代码清单 8-21 在 SportsStore 文件夹的 Startup.cs 文件中启用会话**

```
using System;
using System.Collections.Generic;
using System.Linq;
using System.Threading.Tasks;
using Microsoft.AspNetCore.Builder;
using Microsoft.AspNetCore.Hosting;
using Microsoft.AspNetCore.Http;
using Microsoft.Extensions.DependencyInjection;
using Microsoft.Extensions.Hosting;
using Microsoft.Extensions.Configuration;
using Microsoft.EntityFrameworkCore;
using SportsStore.Models;

namespace SportsStore {
    public class Startup {

        public Startup(IConfiguration config) {
            Configuration = config;
        }
```

```
        private IConfiguration Configuration { get; set; }

        public void ConfigureServices(IServiceCollection services) {
            services.AddControllersWithViews();
            services.AddDbContext<StoreDbContext>(opts => {
                opts.UseSqlServer(
                    Configuration["ConnectionStrings:SportsStoreConnection"]);
            });
            services.AddScoped<IStoreRepository, EFStoreRepository>();
            services.AddRazorPages();
            services.AddDistributedMemoryCache();
            services.AddSession();
        }

        public void Configure(IApplicationBuilder app, IWebHostEnvironment env) {
            app.UseDeveloperExceptionPage();
            app.UseStatusCodePages();
            app.UseStaticFiles();
            app.UseSession();
            app.UseRouting();
            app.UseEndpoints(endpoints => {
                endpoints.MapControllerRoute("catpage",
                    "{category}/Page{productPage:int}",
                    new { Controller = "Home", action = "Index" });

                endpoints.MapControllerRoute("page", "Page{productPage:int}",
                    new { Controller = "Home", action = "Index", productPage = 1 });

                endpoints.MapControllerRoute("category", "{category}",
                    new { Controller = "Home", action = "Index", productPage = 1 });

                endpoints.MapControllerRoute("pagination",
                    "Products/Page{productPage}",
                    new { Controller = "Home", action = "Index", productPage = 1 });

                endpoints.MapDefaultControllerRoute();
                endpoints.MapRazorPages();
            });

            SeedData.EnsurePopulated(app);
        }
    }
}
```

AddDistributedMemoryCache 方法调用设置内存中的数据存储。AddSession 方法注册用于访问会话数据的服务，而 UseSession 方法允许会话系统在请求从客户端到达时自动将请求与会话关联。

### 8.2.5 实现购物车功能

现在准备工作已经完成，可以实现购物车特性了。首先，将一个名为 Cart.cs 的类文件添加到 Models 文件夹中，并使用它定义如代码清单 8-22 所示的类。

**代码清单 8-22 SportsStore/Models 文件夹中 Cart.cs 文件的内容**

```
using System.Collections.Generic;
using System.Linq;

namespace SportsStore.Models {

    public class Cart {

        public List<CartLine> Lines { get; set; } = new List<CartLine>();

        public void AddItem(Product product, int quantity) {
            CartLine line = Lines
                .Where(p => p.Product.ProductID == product.ProductID)
                .FirstOrDefault();

            if (line == null) {
                Lines.Add(new CartLine {
                    Product = product,
                    Quantity = quantity
                });
            } else {
                line.Quantity += quantity;
            }
        }

        public void RemoveLine(Product product) =>
            Lines.RemoveAll(l => l.Product.ProductID == product.ProductID);

        public decimal ComputeTotalValue() =>
            Lines.Sum(e => e.Product.Price * e.Quantity);

        public void Clear() => Lines.Clear();
    }

    public class CartLine {
        public int CartLineID { get; set; }
        public Product Product { get; set; }
        public int Quantity { get; set; }
    }
}
```

Cart 类使用在同一个文件中定义的 CartLine 类来表示客户选择的产品和用户希望购买的数量。定义了向购物车中添加商品、从购物车中删除先前添加的商品、计算购物车中商品的总成本以及通过删除所有商品重置购物车的方法。

### 单元测试：测试购物车

Cart 类相对简单，但是它有一系列必须正确工作的重要行为。运行不良的购物车会破坏整个 SportsStore 应用程序。前面分解了这些特性，并分别对它们进行测试。在 SportsStore.Tests 项目中创建了一个名为 CartTests.cs 的新单元测试文件，以包含这些测试。

第一个行为与何时向购物车添加商品有关。如果这是第一次将给定的商品添加到购物车中，则需要添加一个新的 CartLine。下面是测试，包括单元测试类的定义：

```
using System.Linq;
using SportsStore.Models;
using Xunit;

namespace SportsStore.Tests {

    public class CartTests {

        [Fact]
        public void Can_Add_New_Lines() {

            // Arrange - create some test products
            Product p1 = new Product { ProductID = 1, Name = "P1" };
            Product p2 = new Product { ProductID = 2, Name = "P2" };

            // Arrange - create a new cart
            Cart target = new Cart();

            // Act
            target.AddItem(p1, 1);
            target.AddItem(p2, 1);
            CartLine[] results = target.Lines.ToArray();

            // Assert
            Assert.Equal(2, results.Length);
            Assert.Equal(p1, results[0].Product);
            Assert.Equal(p2, results[1].Product);
        }
    }
}
```

但是，如果客户已经在购物车中添加了一个产品，则需要增加对应 CartLine 的数量，而不是创建一个新的 CartLine。下面是测试：

```
...
[Fact]
public void Can_Add_Quantity_For_Existing_Lines() {
    // Arrange - create some test products
    Product p1 = new Product { ProductID = 1, Name = "P1" };
    Product p2 = new Product { ProductID = 2, Name = "P2" };

    // Arrange - create a new cart
    Cart target = new Cart();

    // Act
    target.AddItem(p1, 1);
    target.AddItem(p2, 1);
    target.AddItem(p1, 10);
    CartLine[] results = target.Lines
        .OrderBy(c => c.Product.ProductID).ToArray();

    // Assert
    Assert.Equal(2, results.Length);
    Assert.Equal(11, results[0].Quantity);
    Assert.Equal(1, results[1].Quantity);
}
...
```

还需要检查用户是否可以改变他们的想法，将产品从购物车中删除。RemoveLine 方法可用于实现该特性。下面是测试：

```
...
[Fact]
public void Can_Remove_Line() {
    // Arrange - create some test products
    Product p1 = new Product { ProductID = 1, Name = "P1" };
    Product p2 = new Product { ProductID = 2, Name = "P2" };
    Product p3 = new Product { ProductID = 3, Name = "P3" };

    // Arrange - create a new cart
    Cart target = new Cart();
    // Arrange - add some products to the cart
    target.AddItem(p1, 1);
    target.AddItem(p2, 3);
    target.AddItem(p3, 5);
    target.AddItem(p2, 1);

    // Act
    target.RemoveLine(p2);

    // Assert
```

```
        Assert.Empty(target.Lines.Where(c => c.Product == p2));
        Assert.Equal(2, target.Lines.Count());
    }
...
```

要测试的下一个行为是计算购物车中商品的总成本。下面是对这种行为的测试：

```
...
[Fact]
public void Calculate_Cart_Total() {
    // Arrange - create some test products
    Product p1 = new Product { ProductID = 1, Name = "P1", Price = 100M };
    Product p2 = new Product { ProductID = 2, Name = "P2", Price = 50M };

    // Arrange - create a new cart
    Cart target = new Cart();

    // Act
    target.AddItem(p1, 1);
    target.AddItem(p2, 1);
    target.AddItem(p1, 3);
    decimal result = target.ComputeTotalValue();

    // Assert
    Assert.Equal(450M, result);
}
...
```

最后一个测试很简单。确保购物车的内容在重置时被正确移除。下面是测试：

```
...
[Fact]
public void Can_Clear_Contents() {
    // Arrange - create some test products
    Product p1 = new Product { ProductID = 1, Name = "P1", Price = 100M };
    Product p2 = new Product { ProductID = 2, Name = "P2", Price = 50M };

    // Arrange - create a new cart
    Cart target = new Cart();

    // Arrange - add some items
    target.AddItem(p1, 1);
    target.AddItem(p2, 1);

    // Act - reset the cart
    target.Clear();

    // Assert
```

```
        Assert.Empty(target.Lines);
    }
    ...
```

有时，如本例所示，测试类功能所需的代码比类本身更长、更复杂。不要因此而放弃编写单元测试。简单类中的缺陷会产生巨大的影响，特别是那些像 Cart 一样在示例应用程序中扮演重要角色的类。

#### 1. 定义会话状态扩展方法

ASP.NET Core 中的会话状态特性只存储 int、string 和 byte[]值。因为想存储 Cart 对象，所以需要定义 ISession 接口的扩展方法，以提供对会话状态数据的访问，把 Cart 对象序列化成 JSON，然后转换回来。在 Infrastructure 文件夹中添加一个名为 SessionExtensions.cs 的类文件，并定义如代码清单 8-23 所示的扩展方法。

**代码清单 8-23　SportsStore/Infrastructure 文件夹中 SessionExtensions.cs 文件的内容**

```
using Microsoft.AspNetCore.Http;
using System.Text.Json;

namespace SportsStore.Infrastructure {

    public static class SessionExtensions {

        public static void SetJson(this ISession session, string key, object value) {
            session.SetString(key, JsonSerializer.Serialize(value));
        }

        public static T GetJson<T>(this ISession session, string key) {
            var sessionData = session.GetString(key);
            return sessionData == null
                ? default(T) : JsonSerializer.Deserialize<T>(sessionData);
        }
    }
}
```

这些方法将对象序列化为 JavaScript 对象符号格式，使得易于存储和检索 Cart 对象。

#### 2. 完成 Razor Pages

当用户单击 Add to cart 按钮时，Cart Razor Pages 将接收浏览器发送的 HTTP POST 请求。它使用请求表单数据从数据库中获取产品对象并更新用户的购物车，后者将存储为会话数据，供将来的请求使用。代码清单 8-24 实现了这些特性。

**代码清单 8-24　在 SportsStore/Pages 文件夹的 Cart.cshtml 文件中处理请求**

```
@page
@model CartModel
```

```
<h2>Your cart</h2>
<table class="table table-bordered table-striped">
    <thead>
        <tr>
            <th>Quantity</th>
            <th>Item</th>
            <th class="text-right">Price</th>
            <th class="text-right">Subtotal</th>
        </tr>
    </thead>
    <tbody>
        @foreach (var line in Model.Cart.Lines) {
            <tr>
                <td class="text-center">@line.Quantity</td>
                <td class="text-left">@line.Product.Name</td>
                <td class="text-right">@line.Product.Price.ToString("c")</td>
                <td class="text-right">
                    @((line.Quantity * line.Product.Price).ToString("c"))
                </td>
        </tr>
    }
</tbody>
<tfoot>
    <tr>
        <td colspan="3" class="text-right">Total:</td>
        <td class="text-right">
            @Model.Cart.ComputeTotalValue().ToString("c")
        </td>
        </tr>
    </tfoot>
</table>

<div class="text-center">
    <a class="btn btn-primary" href="@Model.ReturnUrl">Continue shopping</a>
</div>
```

Razor Pages允许将HTML内容、Razor表达式和代码组合在一个文件中，详见第23章，但是如果想对Razor Pages进行单元测试，则需要使用一个单独的类文件。如果使用的是Visual Studio，那么在Pages文件夹中已经有一个名为Cart.cshtml.cs的类文件(是由Razor Pages模板项创建的)。如果使用的是Visual Studio Code，就需要单独创建类文件。使用已经创建的类文件来定义如代码清单8-25所示的类。

**代码清单8-25　SportsStore/Pages文件夹中Cart.cshtml.cs文件的内容**

```
using Microsoft.AspNetCore.Mvc;
using Microsoft.AspNetCore.Mvc.RazorPages;
using SportsStore.Infrastructure;
```

```
using SportsStore.Models;
using System.Linq;

namespace SportsStore.Pages {
    public class CartModel : PageModel {
        private IStoreRepository repository;

        public CartModel(IStoreRepository repo) {
            repository = repo;
        }

        public Cart Cart { get; set; }
        public string ReturnUrl { get; set; }

        public void OnGet(string returnUrl) {
            ReturnUrl = returnUrl ?? "/";
            Cart = HttpContext.Session.GetJson<Cart>("cart") ?? new Cart();
        }

        public IActionResult OnPost(long productId, string returnUrl) {
            Product product = repository.Products
                .FirstOrDefault(p => p.ProductID == productId);
            Cart = HttpContext.Session.GetJson<Cart>("cart") ?? new Cart();
            Cart.AddItem(product, 1);
            HttpContext.Session.SetJson("cart", Cart);
            return RedirectToPage(new { returnUrl = returnUrl });
        }
    }


}
```

与 Razor Pages 关联的类称为页面模型类，它定义了为不同类型的 HTTP 请求调用的处理程序方法，这些方法在呈现视图之前更新状态。代码清单 8-25 中的页面模型类名为 CartModel，定义了一个 OnPost hander 方法，调用该方法处理 HTTP POST 请求。从数据库中检索 Product，从会话数据中检索用户的购物车，并使用该 Product 更新其内容。存储修改后的 Cart，并将浏览器重定向到相同的 Razor Pages，这将使用 GET 请求来实现(这将防止重新加载浏览器，触发重复的 POST 请求)。

GET 请求由 OnGet 处理程序方法处理，该方法设置 ReturnUrl 和 Cart 属性的值，然后呈现页面的 Razor 内容部分。HTML 内容中的表达式使用 CartModel 作为视图模型对象进行评估，这意味着可以在表达式中访问分配给 ReturnUrl 和 Cart 属性的值。Razor Pages 生成的内容详细说明了添加到用户购物车中的产品，并提供了一个按钮来导航到产品添加到购物车的位置。

处理程序方法使用与 ProductSummary.cshtml 视图生成的 HTML 表单中的输入元素匹配的参数名。这允许 ASP.NET Core 将传入的表单 POST 变量与这些参数相关联，这意味着不需要直接处理表单。这被称为模型绑定，是简化开发的强大工具，详见第 28 章。

### 理解Razor Pages

Razor Pages在第一次使用时可能令人感到有点奇怪，如果以前用过ASP.NET Core提供的MVC框架特性，这种感觉将更强烈。但是Razor Pages是对MVC框架的补充，可将它们与控制器和视图一起使用，因为它们非常适用于不需要MVC框架复杂性的自包含特性。第23章将描述Razor Pages，本书的第Ⅲ部分和第Ⅳ部分将展示它与控制器的用法。

结果是购物车的基本功能都到位了。首先，列出产品和一个按钮，将它们添加到购物车，重新启动ASP.NET Core和请求http://localhost:5000就可以看到，如图8-10所示。

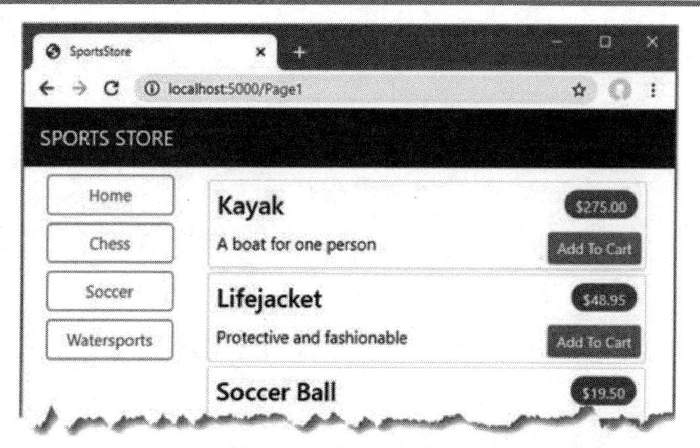


图8-10　Add To Cart按钮

其次，当用户单击Add To Cart按钮时，将适当的产品添加到他们的购物车中，并显示购物车的摘要，如图8-11所示。单击Continue shopping按钮将用户返回到他们最初的产品页面。

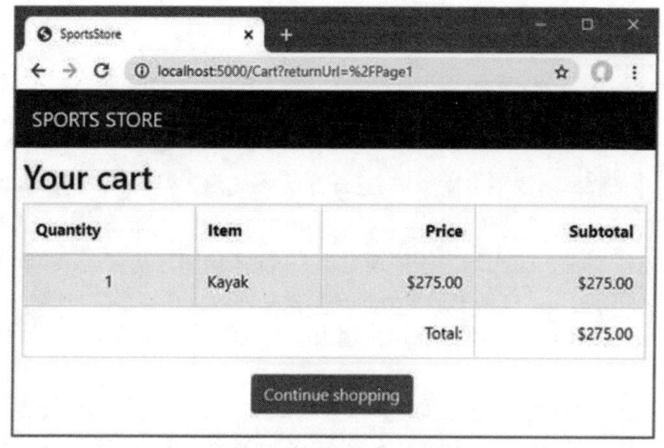


图8-11　显示购物车的内容

### 单元测试：Razor Pages

测试Razor Pages可能需要进行大量模拟，以创建页面模型类所需的上下文对象。为了测试CartModel类定义的OnGet方法的行为，向SportsStore.Tests项目添加一个名为CartPageTests.cs的类文件，并定义此测试：

```
using Microsoft.AspNetCore.Http;
using Microsoft.AspNetCore.Mvc;
using Microsoft.AspNetCore.Mvc.RazorPages;
using Microsoft.AspNetCore.Routing;
using Moq;
using SportsStore.Models;
using SportsStore.Pages;
using System.Linq;
using System.Text;
using System.Text.Json;
using Xunit;

namespace SportsStore.Tests {

    public class CartPageTests {

        [Fact]
        public void Can_Load_Cart() {

            // Arrange
            // - create a mock repository
            Product p1 = new Product { ProductID = 1, Name = "P1" };
            Product p2 = new Product { ProductID = 2, Name = "P2" };
            Mock<IStoreRepository> mockRepo = new Mock<IStoreRepository>();
            mockRepo.Setup(m => m.Products).Returns((new Product[] {
                p1, p2
            }).AsQueryable<Product>());

            // - create a cart
            Cart testCart = new Cart();
            testCart.AddItem(p1, 2);
            testCart.AddItem(p2, 1);
            // - create a mock page context and session
            Mock<ISession> mockSession = new Mock<ISession>();
            byte[] data =
                Encoding.UTF8.GetBytes(JsonSerializer.Serialize(testCart));
            mockSession.Setup(c => c.TryGetValue(It.IsAny<string>(), out data));
            Mock<HttpContext> mockContext = new Mock<HttpContext>();
            mockContext.SetupGet(c => c.Session).Returns(mockSession.Object);

            // Action
            CartModel cartModel = new CartModel(mockRepo.Object) {
                PageContext = new PageContext(new ActionContext {
                    HttpContext = mockContext.Object,
                    RouteData = new RouteData(),
                    ActionDescriptor = new PageActionDescriptor()
```

```
                })
            };
            cartModel.OnGet("myUrl");

            //Assert
            Assert.Equal(2, cartModel.Cart.Lines.Count());
            Assert.Equal("myUrl", cartModel.ReturnUrl);
        }
    }
}
```

这里不打算详述这些单元测试，因为有一种更简单的方法来执行这些测试，参见下一章。这个测试的复杂性在于模拟 ISession 接口，这样页面模型类就可以使用扩展方法检索 Cart 对象的 JSON 表示。ISession 接口只存储字节数组，获取和反序列化字符串是由扩展方法执行的。一旦定义了模拟对象，就可将它们包装在上下文对象中，并用于配置页面模型类的实例；该实例可以进行测试。

测试页面模型类的 OnPost 方法的过程意味着，捕获传递给 ISession 接口模拟的字节数组，然后对其进行反序列化，以确保其包含预期的内容。下面是添加到 CartTestsPage 类的单元测试：

```
...
[Fact]
public void Can_Update_Cart() {
    // Arrange
    // - create a mock repository
    Mock<IStoreRepository> mockRepo = new Mock<IStoreRepository>();
    mockRepo.Setup(m => m.Products).Returns((new Product[] {
        new Product { ProductID = 1, Name = "P1" }
    }).AsQueryable<Product>());

    Cart testCart = new Cart();

    Mock<ISession> mockSession = new Mock<ISession>();
    mockSession.Setup(s => s.Set(It.IsAny<string>(), It.IsAny<byte[]>()))
        .Callback<string, byte[]>((key, val) => {
            testCart =
                JsonSerializer.Deserialize<Cart>(Encoding.UTF8.GetString(val));
        });

    Mock<HttpContext> mockContext = new Mock<HttpContext>();
    mockContext.SetupGet(c => c.Session).Returns(mockSession.Object);

    // Action
    CartModel cartModel = new CartModel(mockRepo.Object) {
        PageContext = new PageContext(new ActionContext {
            HttpContext = mockContext.Object,
            RouteData = new RouteData(),
            ActionDescriptor = new PageActionDescriptor()
```

```
            })
        };
        cartModel.OnPost(1, "myUrl");

        //Assert
        Assert.Single(testCart.Lines);
        Assert.Equal("P1", testCart.Lines.First().Product.Name);
        Assert.Equal(1, testCart.Lines.First().Quantity);
    }
    ...
```

编写有效的单元测试需要耐心和少量的实验，当所测试的特性操作 ASP.NET Core 所使用的上下文对象时尤其如此。

## 8.3 小结

本章首先充实 SportsStore 应用程序面向客户的部分，提供了用户按类别导航的方法，并为添加商品到购物车提供了基本的构建块。还有更多的工作要做，第 9 章继续开发该应用程序。

# 第 9 章

# SportsStore：完成购物车

本章继续构建 SportsStore 示例应用程序。上一章添加了对购物车的基本支持，现在将改进并完成该功能。

■ **提示**：可从 https://github.com/apress/pro-asp.net-core-3 下载本章和本书中其他所有章节的示例项目。如果在运行示例时遇到问题，请参阅第 1 章以获得帮助。

## 9.1 使用服务改进 Cart 模型

上一章定义了 Cart 模型类，演示了如何使用会话特性存储它，允许用户构建一组要购买的产品。管理 Cart 类持久性的责任落到 Cart Razor Pages 上，该页面必须获取 Cart 对象，以及把它们存储为会话数据。

这种方法的问题在于，必须将获取和存储 Cart 对象的代码复制到使用它们的其他任何 Razor Pages 或控制器中。本节将使用位于 ASP.NET Core 核心的服务特性，以简化 Cart 对象的管理方式，使 Cart 控制器等单个组件不必直接处理细节。

服务通常用于向依赖它们的组件隐藏如何实现接口的细节。但是，服务还可以用来解决其他许多问题，可以用来对应用程序进行构形和重塑，甚至在使用 Cart 之类的具体类时也是如此。

### 9.1.1 创建支持存储的 Cart 类

整理使用 Cart 类的方式的第一步是创建一个子类，它知道如何使用会话状态存储自己。为此，将 virtual 关键字应用到 Cart 类，如代码清单 9-1 所示，以便可以覆盖成员。

**代码清单 9-1 在 SportsStore/Models 文件夹的 Cart.cs 文件中应用 virtual 关键字**

```
using System.Collections.Generic;
using System.Linq;

namespace SportsStore.Models {

    public class Cart {

        public List<CartLine> Lines { get; set; } = new List<CartLine>();
```

```
        public virtual void AddItem(Product product, int quantity) {
            CartLine line = Lines
                .Where(p => p.Product.ProductID == product.ProductID)
                .FirstOrDefault();
            if (line == null) {
                Lines.Add(new CartLine {
                    Product = product,
                    Quantity = quantity
                });
            } else {
                line.Quantity += quantity;
            }
        }

        public virtual void RemoveLine(Product product) =>
            Lines.RemoveAll(l => l.Product.ProductID == product.ProductID);

        public decimal ComputeTotalValue() =>
            Lines.Sum(e => e.Product.Price * e.Quantity);

        public virtual void Clear() => Lines.Clear();
    }

    public class CartLine {
        public int CartLineID { get; set; }
        public Product Product { get; set; }
        public int Quantity { get; set; }
    }
}
```

接下来，将一个名为 sessioncart.cs 的类文件添加到 Models 文件夹中，并用它定义如代码清单 9-2 所示的类。

#### 代码清单 9-2　SportsStore/Models 文件夹中 SessionCart.cs 文件的内容

```
using System;
using System.Text.Json.Serialization;
using Microsoft.AspNetCore.Http;
using Microsoft.Extensions.DependencyInjection;
using SportsStore.Infrastructure;

namespace SportsStore.Models {

    public class SessionCart : Cart {

        public static Cart GetCart(IServiceProvider services) {
            ISession session = services.GetRequiredService<IHttpContextAccessor>()?
```

```
                .HttpContext.Session;
            SessionCart cart = session?.GetJson<SessionCart>("Cart")
                ?? new SessionCart();
            cart.Session = session;
            return cart;
        }

        [JsonIgnore]
        public ISession Session { get; set; }

        public override void AddItem(Product product, int quantity) {
            base.AddItem(product, quantity);
            Session.SetJson("Cart", this);
        }

        public override void RemoveLine(Product product) {
            base.RemoveLine(product);
            Session.SetJson("Cart", this);
        }

        public override void Clear() {
            base.Clear();
            Session.Remove("Cart");
        }
    }
}
```

SessionCart 类继承 Cart 类并覆盖 AddItem、RemoveLine 和 Clear 方法，因此它们调用基本实现，然后使用 ISession 接口上的扩展方法在会话中存储更新后的状态。静态 GetCart 方法是一个工厂，用于创建 SessionCart 对象，并为它们提供 ISession 对象，以便它们能够存储自己。

获取 ISession 对象有点复杂。这里获得了 IHttpContextAccessor 服务的一个实例，它提供了对 HttpContext 对象的访问，而 HttpContext 对象又提供了 ISession。这种间接方法是必需的，因为会话不是作为常规服务提供的。

### 9.1.2 注册服务

下一步是为 Cart 类创建一个服务。目标是使用能够无缝存储自身的 SessionCart 对象来满足对 Cart 对象的请求。可在代码清单 9-3 中看到是如何创建服务的。

**代码清单 9-3 在 SportsStore 文件夹的 Startup.cs 文件中创建购物车服务**

```
using System;
using System.Collections.Generic;
using System.Linq;
using System.Threading.Tasks;
using Microsoft.AspNetCore.Builder;
using Microsoft.AspNetCore.Hosting;
```

```
using Microsoft.AspNetCore.Http;
using Microsoft.Extensions.DependencyInjection;
using Microsoft.Extensions.Hosting;
using Microsoft.Extensions.Configuration;
using Microsoft.EntityFrameworkCore;
using SportsStore.Models;

namespace SportsStore {
   public class Startup {

      public Startup(IConfiguration config) {
         Configuration = config;
      }

      private IConfiguration Configuration { get; set; }

      public void ConfigureServices(IServiceCollection services) {
         services.AddControllersWithViews();
         services.AddDbContext<StoreDbContext>(opts => {
            opts.UseSqlServer(
               Configuration["ConnectionStrings:SportsStoreConnection"]);
         });
         services.AddScoped<IStoreRepository, EFStoreRepository>();
         services.AddRazorPages();
         services.AddDistributedMemoryCache();
         services.AddSession();
         services.AddScoped<Cart>(sp => SessionCart.GetCart(sp));
      services.AddSingleton<IHttpContextAccessor, HttpContextAccessor>();
      }

      public void Configure(IApplicationBuilder app, IWebHostEnvironment env) {
         app.UseDeveloperExceptionPage();
         app.UseStatusCodePages();
         app.UseStaticFiles();
         app.UseSession();
         app.UseRouting();
         app.UseEndpoints(endpoints => {
            endpoints.MapControllerRoute("catpage",
               "{category}/Page{productPage:int}",
               new { Controller = "Home", action = "Index" });

            endpoints.MapControllerRoute("page", "Page{productPage:int}",
               new { Controller = "Home", action = "Index", productPage = 1 });

            endpoints.MapControllerRoute("category", "{category}",
               new { Controller = "Home", action = "Index", productPage = 1 });
```

```
            endpoints.MapControllerRoute("pagination",
                "Products/Page{productPage}",
                new { Controller = "Home", action = "Index", productPage = 1 });
            endpoints.MapDefaultControllerRoute();
            endpoints.MapRazorPages();
        });

        SeedData.EnsurePopulated(app);
    }
  }
}
```

AddScoped 方法指定应该使用相同的对象来满足 Cart 实例的相关请求。可以配置请求的关联方式，但默认情况下，这意味着处理相同 HTTP 请求的组件所需的任何购物车都将接收相同的对象。

与为存储库提供类型映射的 AddScoped 方法不同，这里指定了一个 lambda 表达式，该表达式将被调用以满足 Cart 请求。表达式接收已注册的服务集合，并将该集合传递给 SessionCart 类的 GetCart 方法。其结果是，通过创建 SessionCart 对象来处理对 Cart 服务的请求，这些对象在修改时将自己序列化为会话数据。

还使用 AddSingleton 方法添加了一个服务，该方法指定应该始终使用相同的对象。创建的服务告诉 ASP.NET Core，当需要实现 IHttpContextAccessor 接口时，使用 HttpContextAccessor 类。这个服务是必需的，这样就可以访问 SessionCart 类中的当前会话。

### 9.1.3 简化购物车 Razor Pages

创建这种服务的好处是，它允许简化使用 Cart 对象的代码。在代码清单 9-4 中，为 Cart Razor Pages 重新编写了页面模型类，以利用新的服务。

**代码清单 9-4 使用 SportsStore/Pages 文件夹的 Cart.cshtml.cs 文件中的 Cart 服务**

```
using Microsoft.AspNetCore.Mvc;
using Microsoft.AspNetCore.Mvc.RazorPages;
using SportsStore.Infrastructure;
using SportsStore.Models;
using System.Linq;

  public class CartModel : PageModel {
    private IStoreRepository repository;

    public CartModel(IStoreRepository repo, Cart cartService) {
        repository = repo;
        Cart = cartService;
    }

    public Cart Cart { get; set; }
    public string ReturnUrl { get; set; }
```

```
        public void OnGet(string returnUrl) {
            ReturnUrl = returnUrl ?? "/";
        }

        public IActionResult OnPost(long productId, string returnUrl) {
            Product product = repository.Products
                .FirstOrDefault(p => p.ProductID == productId);
            Cart.AddItem(product, 1);
            return RedirectToPage(new { returnUrl = returnUrl });
        }
    }
}
```

Page Model 类通过声明构造函数参数，表明它需要 Cart 对象，这允许从处理程序方法中删除加载和存储会话的语句。结果是一个更简单的页面模型类，它关注于在应用程序中的角色，而不关注如何创建或持久化 Cart 对象。而且，由于服务在整个应用程序中都可用，因此任何组件都可以使用相同的技术获得用户的购物车。

**更新单元测试**

为了简化代码清单 9-4 中的 CartModel 类，需要在单元测试项目中对 CartPageTests.cs 文件中的单元测试进行相应的更改，以便 Cart 作为构造函数参数提供，而不是通过上下文对象访问。以下是阅读购物车的测试更改:

```
...
[Fact]
public void Can_Load_Cart() {

    // Arrange
    // - create a mock repository
    Product p1 = new Product { ProductID = 1, Name = "P1" };
    Product p2 = new Product { ProductID = 2, Name = "P2" };
    Mock<IStoreRepository> mockRepo = new Mock<IStoreRepository>();
    mockRepo.Setup(m => m.Products).Returns((new Product[] {
        p1, p2
    }).AsQueryable<Product>());

    // - create a cart
    Cart testCart = new Cart();
    testCart.AddItem(p1, 2);
    testCart.AddItem(p2, 1);

    // Action
    CartModel cartModel = new CartModel(mockRepo.Object, testCart);
    cartModel.OnGet("myUrl");

    //Assert
```

```
    Assert.Equal(2, cartModel.Cart.Lines.Count());
    Assert.Equal("myUrl", cartModel.ReturnUrl);
}
...
```

对检查购物车更改的单元测试应用了相同的更改:

```
...
[Fact]
public void Can_Update_Cart() {
    // Arrange
    // - create a mock repository
    Mock<IStoreRepository> mockRepo = new Mock<IStoreRepository>();
    mockRepo.Setup(m => m.Products).Returns((new Product[] {
        new Product { ProductID = 1, Name = "P1" }
    }).AsQueryable<Product>());

    Cart testCart = new Cart();

    // Action
    CartModel cartModel = new CartModel(mockRepo.Object, testCart);
    cartModel.OnPost(1, "myUrl");

    //Assert
    Assert.Single(testCart.Lines);
    Assert.Equal("P1", testCart.Lines.First().Product.Name);
    Assert.Equal(1, testCart.Lines.First().Quantity);
}
...
```

使用服务可以简化测试过程，可以更容易地为被测试的类提供依赖项。

## 9.2 完成购物车的功能

前面介绍了 Cart 服务，现在可以通过添加两个新特性来完成 Cart 功能了。第一个将允许客户从购物车中删除一项商品。第二个特性在页面顶部显示购物车的摘要。

### 9.2.1 从购物车中删除商品

要从购物车中删除商品，需要向 Cart Razor Pages 所呈现的内容添加一个 Remove 按钮，该页面提交一个 HTTP POST 请求。更改如代码清单 9-5 所示。

**代码清单 9-5 在 SportsStore/Pages 文件夹中删除 Cart.cshtml 文件中的购物车项**

```
@page
@model CartModel

<h2>Your cart</h2>
```

```
<table class="table table-bordered table-striped">
    <thead>
        <tr>
            <th>Quantity</th>
            <th>Item</th>
            <th class="text-right">Price</th>
            <th class="text-right">Subtotal</th>
            <th></th>
        </tr>
    </thead>
    <tbody>
        @foreach (var line in Model.Cart.Lines) {
            <tr>
                <td class="text-center">@line.Quantity</td>
                <td class="text-left">@line.Product.Name</td>
                <td class="text-right">@line.Product.Price.ToString("c")</td>
                <td class="text-right">
                    @((line.Quantity * line.Product.Price).ToString("c"))
                </td>
                <td class="text-center">
                    <form asp-page-handler="Remove" method="post">
                        <input type="hidden" name="ProductID"
                                value="@line.Product.ProductID" />
                        <input type="hidden" name="returnUrl"
                                value="@Model.ReturnUrl" />
                        <button type="submit" class="btn btn-sm btn-danger">
                            Remove
                        </button>
                    </form>
                </td>
            </tr>
        }
    </tbody>
    <tfoot>
        <tr>
            <td colspan="3" class="text-right">Total:</td>
            <td class="text-right">
                @Model.Cart.ComputeTotalValue().ToString("c")
            </td>
            </tr>
        </tfoot>
    </table>

    <div class="text-center">
        <a class="btn btn-primary" href="@Model.ReturnUrl">Continue shopping</a>
</div>
```

按钮需要在页面模型类中使用一个新的处理程序方法来接收请求，并修改购物车，如代码清单 9-6 所示。

**代码清单 9-6　从 SportsStore/Pages 文件夹的 Cart.cshtml.cs 文件中删除一项**

```
using Microsoft.AspNetCore.Mvc;
using Microsoft.AspNetCore.Mvc.RazorPages;
using SportsStore.Infrastructure;
using SportsStore.Models;
using System.Linq;

namespace SportsStore.Pages {

    public class CartModel : PageModel {
        private IStoreRepository repository;

        public CartModel(IStoreRepository repo, Cart cartService) {
            repository = repo;
            Cart = cartService;
        }

        public Cart Cart { get; set; }
        public string ReturnUrl { get; set; }

        public void OnGet(string returnUrl) {
            ReturnUrl = returnUrl ?? "/";
        }

        public IActionResult OnPost(long productId, string returnUrl) {
            Product product = repository.Products
                .FirstOrDefault(p => p.ProductID == productId);
            Cart.AddItem(product, 1);
            return RedirectToPage(new { returnUrl = returnUrl });
        }

        public IActionResult OnPostRemove(long productId, string returnUrl) {
            Cart.RemoveLine(Cart.Lines.First(cl =>
                cl.Product.ProductID == productId).Product);
            return RedirectToPage(new { returnUrl = returnUrl });
        }
    }
}
```

新的 HTML 内容定义了一个 HTML 表单。接收请求的处理程序方法是用 asp-page-handler 标记 helper 属性指定的，如下所示：

```
...
<form asp-page-handler="Remove" method="post">
...
```

指定的名称以On为前缀，并提供与请求类型匹配的后缀，以便Remove的值选择OnPostRemove处理程序方法。处理程序方法使用接收到的值来定位购物车中的商品并删除它。

重启 ASP.NET Core，并请求 http://localhost:5000。单击 Add To Cart 按钮，将商品添加到购物车中，然后单击 Remove 按钮。更新购物车，以删除指定的商品，如图 9-1 所示。

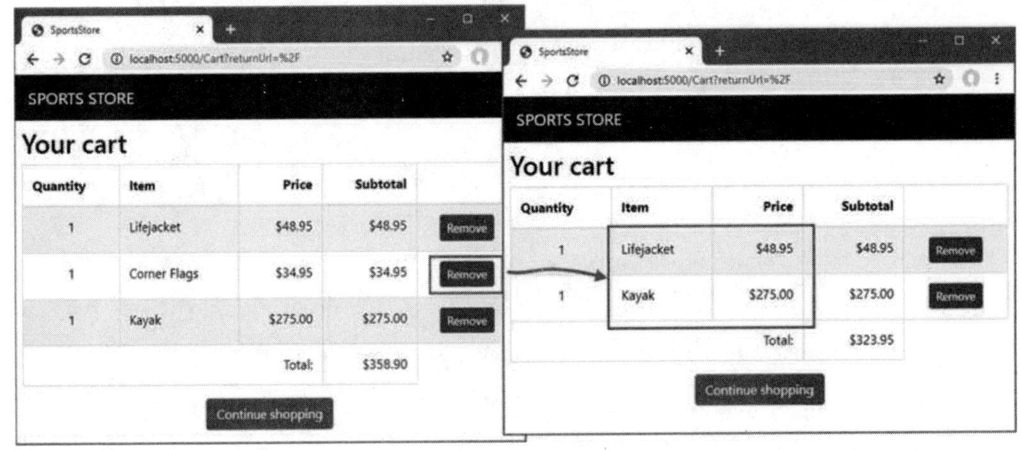


图 9-1　从购物车中删除商品

## 9.2.2　添加购物车摘要小部件

购物车的功能可能正常，但它集成到接口的方式存在一个问题。只有通过查看购物车摘要屏幕，客户才能知道购物车中有什么。而且他们只能通过向购物车中添加新项目来查看购物车摘要屏幕。

为解决这个问题，添加一个小部件，它总结购物车的内容。可以单击它，在整个应用程序中显示购物车的内容。以与添加导航小部件基本相同的方式执行此操作——作为一个视图组件，可将其输出包含在 Razor 布局中。

### 1. 添加 Font Awesome 包

作为购物车摘要的一部分，显示一个允许用户结账的按钮。这里想使用购物车符号，而不是在按钮中显示单词 checkout。由于笔者没有艺术技能，因此打算使用 Font Awesome 包，这是一组优秀的开源图标，以字体形式集成到应用程序中，字体中的每个字符都是不同的图像。可从 http://fortawesome.github.io/Font-Awesome 了解关于 Font Awesome 的更多信息，包括查看其中的图标。

要安装客户端包，使用 PowerShell 命令提示符在 SportsStore 项目中运行代码清单 9-7 所示的命令。

**代码清单 9-7　安装图标包**

```
libman install font-awesome@5.12.0 -d wwwroot/lib/font-awesome
```

### 2. 创建视图组件类和视图

在 Components 文件夹中添加一个名为 CartSummaryViewComponent.cs 的类文件，并用它定义如代码清单 9-8 所示的视图组件。

**代码清单 9-8　SportsStore/Components 文件夹中 CartSummaryViewComponent.cs 文件的内容**

```
using Microsoft.AspNetCore.Mvc;
using SportsStore.Models;

namespace SportsStore.Components {

    public class CartSummaryViewComponent : ViewComponent {
        private Cart cart;

        public CartSummaryViewComponent(Cart cartService) {
            cart = cartService;
        }

        public IViewComponentResult Invoke() {
            return View(cart);
        }
    }
}
```

这个视图组件能利用在本章前面创建的服务接收一个 Cart 对象作为构造函数参数。结果是一个简单的视图组件类，它将 Cart 传递给 View 方法，以生成将包含在布局中的 HTML 片段。为创建组件的视图，创建 Views/Shared/Components/CartSummary 文件夹，并向其中添加一个名为 Default.cshtml 的 Razor 视图，内容如代码清单 9-9 所示。

**代码清单 9-9　Views/Shared/ Components/CartSummary 文件夹中的 Default.cshtml 文件**

```
@model Cart

<div class="">
    @if (Model.Lines.Count() > 0) {
        <small class="navbar-text">
            <b>Your cart:</b>
            @Model.Lines.Sum(x => x.Quantity) item(s)
            @Model.ComputeTotalValue().ToString("c")
        </small>
    }
    <a class="btn btn-sm btn-secondary navbar-btn" asp-page="/Cart"
        asp-route-returnurl="@ViewContext.HttpContext.Request.PathAndQuery()">
        <i class="fa fa-shopping-cart"></i>
    </a>
</div>
```

该视图显示一个带有 Font Awesome 购物车图标的按钮，如果购物车中有商品，则提供一个快照，详细说明商品数量及其总价。现在有了视图组件和视图，就可以修改布局，使购物车摘要包含在主控制器生成的响应中，如代码清单 9-10 所示。

**代码清单 9-10　在 Views/Shared 文件夹的_Layout.cshtml 文件中添加购物车摘要**

```
<!DOCTYPE html>
<html>
<head>
    <meta name="viewport" content="width=device-width" />
    <title>SportsStore</title>
    <link href="/lib/twitter-bootstrap/css/bootstrap.min.css" rel="stylesheet" />
    <link href="/lib/font-awesome/css/all.min.css" rel="stylesheet" />
</head>
<body>
    <div class="bg-dark text-white p-2">
        <div class="container-fluid">
            <div class="row">
                <div class="col navbar-brand">SPORTS STORE</div>
                <div class="col-6 text-right">
                    <vc:cart-summary />
                </div>
            </div>
        </div>
    </div>
    <div class="row m-1 p-1">
        <div id="categories" class="col-3">
            <vc:navigation-menu />
        </div>
        <div class="col-9">
            @RenderBody()
        </div>
    </div>
</body>
</html>
```

启动应用程序，可查看购物车摘要。当购物车是空的，就只显示付款按钮。如果向购物车中添加商品，则会显示商品数量及其合计成本，如图 9-2 所示。有了这个附加功能，顾客就知道他们的购物车里有什么，并有一个明显的方式从商店结账。

图 9-2　显示购物车的摘要

# 9.3 提交订单

现在已经到达 SportsStore 的最终客户功能：结账和完成订单的能力。下面将扩展数据模型，以提供从用户捕获配送细节的支持，并添加应用程序支持来处理这些细节。

## 9.3.1 创建模型类

在 Models 文件夹中添加一个名为 Order.cs 的类文件，并用它定义如代码清单 9-11 所示的类。这个类用来表示客户的配送细节。

**代码清单 9-11 SportsStore/Models 文件夹中 Order.cs 文件的内容**

```
using System.Collections.Generic;
using System.ComponentModel.DataAnnotations;
using Microsoft.AspNetCore.Mvc.ModelBinding;

namespace SportsStore.Models {

    public class Order {

        [BindNever]
        public int OrderID { get; set; }
        [BindNever]
        public ICollection<CartLine> Lines { get; set; }

        [Required(ErrorMessage = "Please enter a name")]
        public string Name { get; set; }

        [Required(ErrorMessage = "Please enter the first address line")]
        public string Line1 { get; set; }
        public string Line2 { get; set; }
        public string Line3 { get; set; }

        [Required(ErrorMessage = "Please enter a city name")]
        public string City { get; set; }

        [Required(ErrorMessage = "Please enter a state name")]
        public string State { get; set; }

        public string Zip { get; set; }

        [Required(ErrorMessage = "Please enter a country name")]
        public string Country { get; set; }

        public bool GiftWrap { get; set; }
    }
}
```

前面使用了来自 System.ComponentModel.DataAnnotations 名称空间的验证属性，如第 3 章所述。第 29 章进一步描述验证。

还使用了 BindNever 属性，以阻止用户在 HTTP 请求中为这些属性提供值。这是模型绑定系统的一个特性(参见第 28 章)，可阻止 ASP.NET Core 使用来自 HTTP 请求的值填充敏感或重要的模型属性。

## 9.3.2　添加付款过程

目标是让用户能够输入配送细节并提交订单。首先向购物车视图添加一个 Checkout 按钮，如代码清单 9-12 所示。

**代码清单 9-12　在 SportsStore/Pages 文件夹的 Cart.cshtml 文件中添加按钮**

```
...
<div class="text-center">
    <a class="btn btn-primary" href="@Model.ReturnUrl">Continue shopping</a>
    <a class="btn btn-primary" asp-action="Checkout" asp-controller="Order">
        Checkout
    </a>
</div>
...
```

此更改生成一个已设计为按钮的链接，当单击该链接时，会调用 Order 控制器的 Checkout 操作方法，该操作方法将在下一节中创建。为展示 Razor Pages 和控制器如何协同工作，这里在控制器中处理订单，然后在流程结束时返回 Razor Pages。要查看 Checkout 按钮，请重新启动 ASP.NET Core，请求 http://localhost:5000，然后单击 Add To Cart 按钮之一。新按钮显示为购物车摘要的一部分，如图 9-3 所示。

图 9-3　Checkout 按钮

## 9.3.3　创建控制器和视图

现在需要定义处理订单的控制器。向 Controllers 文件夹添加一个名为 OrderController.cs 的类文件，并用它定义如代码清单 9-13 所示的类。

### 代码清单 9-13　SportsStore/Controllers 文件夹中 OrderController.cs 文件的内容

```
using Microsoft.AspNetCore.Mvc;
using SportsStore.Models;

namespace SportsStore.Controllers {

    public class OrderController : Controller {

        public ViewResult Checkout() => View(new Order());
    }
}
```

Checkout 方法返回默认视图，并传递一个新的 Order 对象作为视图模型。为创建视图，创建 Views/Order 文件夹，并向其中添加一个名为 Checkout.cshtml 的 Razor 视图，标记如代码清单 9-14 所示。

### 代码清单 9-14　SportsStore/Views/Order 文件夹中 Checkout.cshtml 文件的内容

```
@model Order

<h2>Check out now</h2>
<p>Please enter your details, and we'll ship your goods right away!</p>

<form asp-action="Checkout" method="post">
    <h3>Ship to</h3>
    <div class="form-group">
        <label>Name:</label><input asp-for="Name" class="form-control" />
    </div>
    <h3>Address</h3>
    <div class="form-group">
        <label>Line 1:</label><input asp-for="Line1" class="form-control" />
    </div>
    <div class="form-group">
        <label>Line 2:</label><input asp-for="Line2" class="form-control" />
    </div>
    <div class="form-group">
        <label>Line 3:</label><input asp-for="Line3" class="form-control" />
    </div>
    <div class="form-group">
        <label>City:</label><input asp-for="City" class="form-control" />
    </div>
    <div class="form-group">
        <label>State:</label><input asp-for="State" class="form-control" />
    </div>
    <div class="form-group">
```

```
        <label>Zip:</label><input asp-for="Zip" class="form-control" />
    </div>
    <div class="form-group">
        <label>Country:</label><input asp-for="Country" class="form-control" />
    </div>
    <h3>Options</h3>
    <div class="checkbox">
        <label>
            <input asp-for="GiftWrap" /> Gift wrap these items
        </label>
    </div>
    <div class="text-center">
        <input class="btn btn-primary" type="submit" value="Complete Order" />
    </div>
</form>
```

对于模型中的每个属性，创建了一个标签和输入元素来捕获用户输入，使用 Bootstrap 进行样式化，并用标签助手进行配置。输入元素上的 asp-for 属性由内置的标签助手处理，根据指定的模型属性生成类型、id、名称和值属性，如第 27 章所述。

为查看如图 9-4 所示的表单，重新启动 ASP.NET Core，请求 http://localhost:5000，向购物车中添加一项，并单击 Checkout 按钮。或者，更直接的做法是请求 http://localhost:5000/order/ checkout。

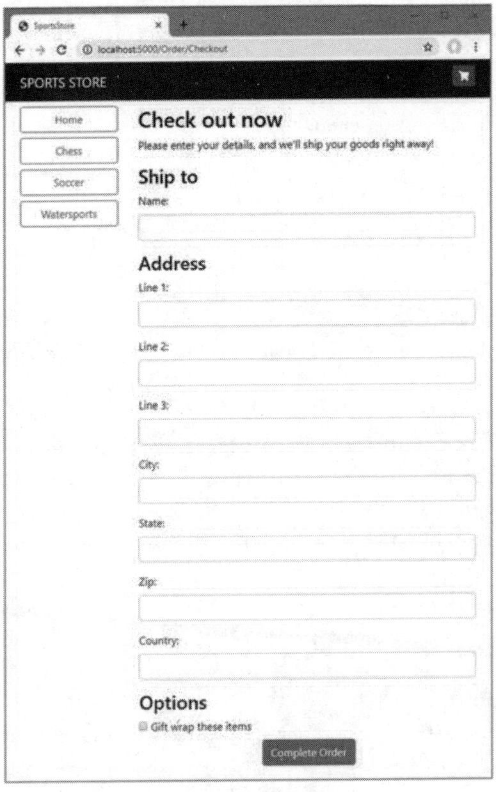

图 9-4　运输详细信息表格

### 9.3.4 实现订单处理

订单的处理是将订单写入数据库。当然，大多数电子商务网站不会就此止步，本例也没有为处理信用卡或其他支付形式提供支持。但重点是 ASP.NET Core。一个简单的数据库条目就可以了。

#### 1. 扩展数据库

向数据库中添加一种新的模型非常简单，因为第 7 章进行了初始设置。首先，向数据库上下文类添加一个新属性，如代码清单 9-15 所示。

**代码清单 9-15　在 SportsStore/Models 文件夹的 StoreDbContext.cs 文件中添加一个属性**

```
using Microsoft.EntityFrameworkCore;
namespace SportsStore.Models {
    public class StoreDbContext: DbContext {

        public StoreDbContext(DbContextOptions<StoreDbContext> options)
            : base(options) { }

        public DbSet<Product> Products { get; set; }
        public DbSet<Order> Orders { get; set; }
    }
}
```

这个更改足以让 Entity Framework Core 创建一个数据库迁移，允许 Order 对象存储在数据库中。要创建迁移，使用 PowerShell 命令提示符在 SportsStore 文件夹中运行代码清单 9-16 所示的命令。

**代码清单 9-16　创建迁移**

```
dotnet ef migrations add Orders
```

这个命令告诉 Entity Framework Core 获取应用程序数据模型的新快照，找出它与前一个数据库版本的不同之处，并生成名为 Orders 的新迁移。新的迁移将在应用程序启动时自动应用，因为 SeedData 调用 Entity Framework Core 提供的 Migrate 方法。

**重置数据库**

当频繁地对模型进行更改时，迁移和数据库模式会出现不同步的情况。最简单的方法是删除数据库并重新开始。当然，这只适用于开发期间，因为会丢失已存储的任何数据。运行此命令，会删除数据库:

```
dotnet ef database drop --force --context StoreDbContext
```

删除数据库后，在 SportsStore 文件夹中运行以下命令，重新创建数据库，并通过运行以下命令，应用前面创建的迁移:

```
dotnet ef database update --context StoreDbContext
```

如果刚启动应用程序，迁移也将由SeedData类应用。无论哪种方式，数据库都将重置，以便准确地反映数据模型，并允许返回来开发应用程序。

### 2. 创建订单存储库

遵循在产品存储库中使用的相同模式来提供对Order对象的访问。在Models文件夹中添加一个名为IOrderRepository.cs的类文件，并用它定义如代码清单9-17所示的接口。

**代码清单9-17　SportsStore/Models文件夹中IOrderRepository.cs文件的内容**

```
using System.Linq;

namespace SportsStore.Models {

    public interface IOrderRepository {

        IQueryable<Order> Orders { get; }
        void SaveOrder(Order order);
    }
}
```

为了实现订单存储库接口，在Models文件夹中添加一个名为EFOrderRepository.cs的类文件，并定义如代码清单9-18所示的类。

**代码清单9-18　SportsStore/Models文件夹中EFOrderRepository.cs文件的内容**

```
using Microsoft.EntityFrameworkCore;
using System.Linq;

namespace SportsStore.Models {

    public class EFOrderRepository : IOrderRepository {
        private StoreDbContext context;

        public EFOrderRepository(StoreDbContext ctx) {
            context = ctx;
        }

        public IQueryable<Order> Orders => context.Orders
                            .Include(o => o.Lines)
                            .ThenInclude(l => l.Product);

        public void SaveOrder(Order order) {
            context.AttachRange(order.Lines.Select(l => l.Product));
            if (order.OrderID == 0) {
                context.Orders.Add(order);
            }
            context.SaveChanges();
```

```
        }
    }
}
```

这个类使用 Entity Framework Core 实现 IOrderRepository 接口，允许检索已经存储的 Order 对象集，并允许创建或更改订单。

### 理解订单存储库

Entity Framework Core 需要指令来加载跨多个表的相关数据。在代码清单 9-18 中，使用 Include 和 ThenInclude 方法来指定，当从数据库中读取 Order 对象时，还应该加载与 Lines 属性关联的集合以及与每个集合对象关联的每个 Product 对象。

```
...
public IQueryable<Order> Orders => context.Orders
    .Include(o => o.Lines)
    .ThenInclude(l => l.Product);
...
```

这可以确保接收到需要的所有数据对象，而不必执行单独的查询，然后自己组装数据。

在数据库中存储 Order 对象时，还需要另外一个步骤。当用户的购物车数据从会话存储中反序列化时，Entity Framework Core 将创建未知的新对象，然后尝试将所有对象写入数据库。对于与 Order 关联的 Product 对象，这意味着 Entity Framework Core 试图写入已经存储的对象，这会导致错误。为避免这个问题，通知 Entity Framework Core：对象存在，且不应存储在数据库中，除非它们被修改，如下所示。

```
...
context.AttachRange(order.Lines.Select(l => l.Product));
...
```

这确保了 Entity Framework Core 不会尝试编写与 Order 对象关联的反序列化 Product 对象。

代码清单 9-19 在 Startup 类的 ConfigureServices 方法中将订单存储库注册为一个服务。

**代码清单 9-19　在 SportsStore 文件夹的 Startup.cs 文件中注册订单存储库服务**

```
...
public void ConfigureServices(IServiceCollection services) {
    services.AddControllersWithViews();
    services.AddDbContext<StoreDbContext>(opts => {
        opts.UseSqlServer(
            Configuration["ConnectionStrings:SportsStoreConnection"]);
    });
    services.AddScoped<IStoreRepository, EFStoreRepository>();
    services.AddScoped<IOrderRepository, EFOrderRepository>();
    services.AddRazorPages();
    services.AddDistributedMemoryCache();
    services.AddSession();
```

```
        services.AddScoped<Cart>(sp => SessionCart.GetCart(sp));
        services.AddSingleton<IHttpContextAccessor, HttpContextAccessor>();
    }
...
```

### 9.3.5 完成订单控制器

要完成 OrderController 类，需要修改构造函数，以便它接收处理订单所需的服务，并添加操作方法，在用户单击 Complete Order 按钮时，处理 HTTP 表单 POST 请求。代码清单 9-20 显示了这两项更改。

**代码清单 9-20　在 SportsStore/Controllers 文件夹的 OrderController.cs 文件中完成控制器**

```
using Microsoft.AspNetCore.Mvc;
using SportsStore.Models;
using System.Linq;

namespace SportsStore.Controllers {

    public class OrderController : Controller {
        private IOrderRepository repository;
        private Cart cart;

        public OrderController(IOrderRepository repoService, Cart cartService) {
            repository = repoService;
            cart = cartService;
        }

        public ViewResult Checkout() => View(new Order());

        [HttpPost]
        public IActionResult Checkout(Order order) {
            if (cart.Lines.Count() == 0) {
                ModelState.AddModelError("", "Sorry, your cart is empty!");
            }
            if (ModelState.IsValid) {
                order.Lines = cart.Lines.ToArray();
                repository.SaveOrder(order);
                cart.Clear();
                return RedirectToPage("/Completed", new { orderId = order.OrderID });
            } else {
                return View();
            }
        }
    }
}
```

Checkout 操作方法用 HttpPost 属性装饰，这意味着在本例中，当用户提交表单时，它将用于处理 POST 请求。

第 8 章使用了 ASP.NET Core 模型绑定特性从请求中接收简单的数据值。在新的操作方法中使用了相同的特性来接收已完成的 Order 对象。当处理请求时，模型绑定系统尝试查找 Order 类定义的属性值。这是在最大努力的基础上工作的，这意味着如果请求中没有相应的数据项，可能会收到缺少属性值的 Order 对象。

为了确保拥有所需的数据，对 Order 类应用了验证属性。ASP.NET Core 检查应用到 Order 类的验证约束，并通过 ModelState 属性提供结果的详细信息。可以通过检查 ModelState.IsValid 属性来查看是否有任何问题。如果购物车中没有商品，就调用 ModelState.AddModelError 方法来注册一个错误消息。后面简短地解释如何显示这些错误，第 28 和 29 章将更多地介绍模型绑定和验证。

**单元测试：订单处理**

要为 OrderController 类执行单元测试，需要测试 Checkout 方法的 POST 版本的行为。尽管该方法看起来简短，但模型绑定的使用意味着需要测试许多幕后工作。

只有购物车中有商品，客户也提供了有效的运输细节，才能处理订单。在其他所有情况下，应向客户显示错误。下面是第一个测试方法，它是在 SportsStore.Tests 项目中名为 OrderControllerTests.cs 的类文件中定义的。

```
using Microsoft.AspNetCore.Mvc;
using Moq;
using SportsStore.Controllers;
using SportsStore.Models;
using Xunit;

namespace SportsStore.Tests {

    public class OrderControllerTests {

        [Fact]
        public void Cannot_Checkout_Empty_Cart() {
            // Arrange - create a mock repository
            Mock<IOrderRepository> mock = new Mock<IOrderRepository>();
            // Arrange - create an empty cart
            Cart cart = new Cart();
            // Arrange - create the order
            Order order = new Order();
            // Arrange - create an instance of the controller
            OrderController target = new OrderController(mock.Object, cart);

            // Act
            ViewResult result = target.Checkout(order) as ViewResult;

            // Assert - check that the order hasn't been stored
            mock.Verify(m => m.SaveOrder(It.IsAny<Order>()), Times.Never);
            // Assert - check that the method is returning the default view
```

```
            Assert.True(string.IsNullOrEmpty(result.ViewName));
            // Assert - check that I am passing an invalid model to the view
            Assert.False(result.ViewData.ModelState.IsValid);
        }
    }
}
```

这个测试确保不能用空的购物车结账。为此应确保从来没有调用模拟的 IOrderRepository 实现的 SaveOrder，该方法返回的视图是默认视图(重新显示客户输入的数据，提供一个纠正的机会)，传递到视图的模型状态标记为无效。这似乎是一组非常复杂的断言，但是需要同时使用这三个断言来确保行为正确。下一个测试方法的工作方式大致相同，但是在视图模型中注入一个错误，以模拟模型绑定器报告的问题(当客户输入无效的运输数据时，会在生产中发生错误)：

```
...
[Fact]
public void Cannot_Checkout_Invalid_ShippingDetails() {
    // Arrange - create a mock order repository
    Mock<IOrderRepository> mock = new Mock<IOrderRepository>();
    // Arrange - create a cart with one item
    Cart cart = new Cart();
    cart.AddItem(new Product(), 1);
    // Arrange - create an instance of the controller
    OrderController target = new OrderController(mock.Object, cart);
    // Arrange - add an error to the model
    target.ModelState.AddModelError("error", "error");

    // Act - try to checkout
    ViewResult result = target.Checkout(new Order()) as ViewResult;

    // Assert - check that the order hasn't been passed stored
    mock.Verify(m => m.SaveOrder(It.IsAny<Order>()), Times.Never);
    // Assert - check that the method is returning the default view
    Assert.True(string.IsNullOrEmpty(result.ViewName));
    // Assert - check that I am passing an invalid model to the view
    Assert.False(result.ViewData.ModelState.IsValid);
}
...
```

确定购物车为空或详细信息无效，会阻止处理订单，需要确保在适当的时候处理订单。下面是测试：

```
...
[Fact]
public void Can_Checkout_And_Submit_Order() {
    // Arrange - create a mock order repository
    Mock<IOrderRepository> mock = new Mock<IOrderRepository>();
    // Arrange - create a cart with one item
```

```
    Cart cart = new Cart();
    cart.AddItem(new Product(), 1);
    // Arrange - create an instance of the controller
    OrderController target = new OrderController(mock.Object, cart);

    // Act - try to checkout
    RedirectToPageResult result =
          target.Checkout(new Order()) as RedirectToPageResult;

    // Assert - check that the order has been stored
    mock.Verify(m => m.SaveOrder(It.IsAny<Order>()), Times.Once);
    // Assert - check that the method is redirecting to the Completed action
    Assert.Equal("/Completed", result.PageName);
}
...
```

对于是否可以识别有效的运输细节，不必进行测试。模型绑定器使用应用到 Order 类属性的属性，自动处理这个问题。

### 9.3.6 显示验证错误

ASP.NET Core 使用应用于 Order 类的验证属性来验证用户数据，但是需要做一个简单的更改，来显示任何问题。这依赖于另一个内置的标签助手，它检查用户提供的数据的验证状态，并为发现的每个问题添加警告消息。代码清单 9-21 显示了添加的 HTML 元素，该元素由标签助手在 Checkout.cshtml 文件中进行处理。

**代码清单 9-21 在 SportsStore/Views/Order 文件夹的 Checkout.cshtml 文件中添加验证摘要**

```
@model Order

<h2>Check out now</h2>
<p>Please enter your details, and we'll ship your goods right away!</p>

<div asp-validation-summary="All" class="text-danger"></div>

<form asp-action="Checkout" method="post">
    <h3>Ship to</h3>
    <div class="form-group">
        <label>Name:</label><input asp-for="Name" class="form-control" />
    </div>
    <h3>Address</h3>
    <div class="form-group">
        <label>Line 1:</label><input asp-for="Line1" class="form-control" />
    </div>
    <div class="form-group">
        <label>Line 2:</label><input asp-for="Line2" class="form-control" />
    </div>
```

```
        <div class="form-group">
            <label>Line 3:</label><input asp-for="Line3" class="form-control" />
        </div>
        <div class="form-group">
            <label>City:</label><input asp-for="City" class="form-control" />
        </div>
        <div class="form-group">
            <label>State:</label><input asp-for="State" class="form-control" />
        </div>
        <div class="form-group">
            <label>Zip:</label><input asp-for="Zip" class="form-control" />
        </div>
        <div class="form-group">
            <label>Country:</label><input asp-for="Country" class="form-control" />
        </div>
        <h3>Options</h3>
        <div class="checkbox">
            <label>
                <input asp-for="GiftWrap" /> Gift wrap these items
            </label>
        </div>
        <div class="text-center">
            <input class="btn btn-primary" type="submit" value="Complete Order" />
        </div>
    </form>
```

通过这个简单的更改，验证错误将报告给用户。要查看效果，请重新启动 ASP.NET Core，请求 http://:/localhost:5000 /Order/Checkout，然后单击 Complete Order 按钮而不填写表格。ASP.NET Core 将处理表单数据，检测是否找到所需的值，并生成验证错误，如图 9-5 所示。

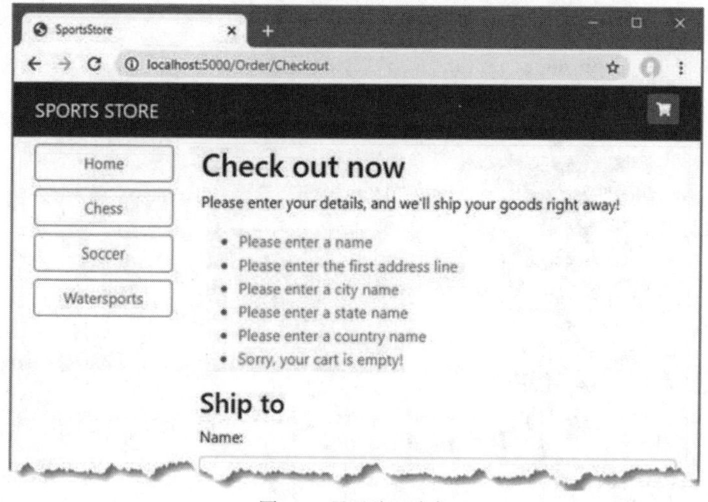


图 9-5　显示验证消息

■ **提示：**用户提交的数据在验证之前发送到服务器，这称为服务器端验证，ASP.NET Core 对这一点有很好的支持。服务器端验证的问题是，只有在数据发送到服务器、进行处理并生成结果页之后，用户才会被告知错误——在繁忙的服务器上，这可能需要几秒钟的时间。因此，服务器端验证通常用客户端验证进行补充，在客户端验证中使用 JavaScript 检查用户在将表单数据发送到服务器之前输入的值。第 29 章将描述客户端验证。

### 9.3.7 显示摘要页面

为完成结账过程，创建一个 Razor Pages，该页面显示一个感谢消息和订单摘要。向 Pages 文件夹添加一个名为 Completed.cshtml 的 Razor Pages，其内容如代码清单 9-22 所示。

**代码清单 9-22 SportsStore/Pages 文件夹中 Completed.cshtml 文件的内容**

```
@page

<div class="text-center">
    <h2>Thanks!</h2>
    <p>Thanks for placing order #@OrderId</p>
    <p>We'll ship your goods as soon as possible.</p>
    <a class="btn btn-primary" asp-controller="Home">Return to Store</a>
</div>

@functions {

    [BindProperty(SupportsGet = true)]
    public string OrderId { get; set; }
}
```

尽管 Razor Pages 通常具有页面模型类，但它们不是必需的，可以不使用它们而开发简单的特性。本例定义了一个名为 OrderId 的属性，并用 BindProperty 属性修饰它，该属性指定模型绑定系统应该从请求中获取该属性的值。

现在客户可以完成从挑选产品到结账的整个过程。如果他们提供了有效的送货细节(并且购物车中有商品)，那么单击 Complete Order 按钮时，他们将看到摘要页面，如图 9-6 所示。

图 9-6 已完成的订单摘要视图

请注意应用程序在控制器和 Razor Pages 之间移动的方式。ASP.NET Core 提供的应用程序特性是互补的，可以在项目中自由混合使用。

## 9.4 小结

本章完成了 SportsStore 面向客户的所有主要部分。这可能还不足以让亚马逊担心，但有一个可按类别和页面浏览的产品目录，一个整洁的购物车和一个简单的结账流程。

本章采用的方法意味着可以很容易地更改应用程序的任何部分，而不会在其他地方造成问题或不一致。例如，可以更改订单的存储方式，而不会对购物车、产品目录或应用程序的其他任何区域造成影响。第 10 章将添加管理 SportsStore 应用程序所需的特性。

# 第 10 章

# SportsStore：管理

本章将继续构建 SportsStore 应用程序，以便为站点管理员提供一种管理订单和产品的方法。本章使用 Blazor 创建管理特性。Blazor 是 ASP.NET Core 新添加的一个功能。它结合了客户端 JavaScript 代码和由 ASP.NET Core 执行的服务器端代码，通过持久的 HTTP 连接连接。第 32～35 章将详细描述 Blazor，但重要的是理解 Blazor 模型并不适合所有的项目(本章使用 Blazor 服务器，它是 ASP.NET Core 支持的一部分。还有，在撰写本书时，Blazor WebAssembly 是试验性的，完全在浏览器中运行。第 36 章将描述 Blazor WebAssembly)。

■ **提示：** 可以从 https://github.com/apress/pro-ASP.NET Core.net-core-3 下载本章和本书中其他所有章节的示例项目。如果在运行示例时遇到问题，请参阅第 1 章以获得帮助。

## 10.1 准备 Blazor 服务器

第一步是为 Blazor 启用服务和中间件，如代码清单 10-1 所示。

**代码清单 10-1 在 SportsStore 文件夹的 Startup.cs 文件中启用 Blazor**

```
using System;
using System.Collections.Generic;
using System.Linq;
using System.Threading.Tasks;
using Microsoft.AspNetCore.Builder;
using Microsoft.AspNetCore.Hosting;
using Microsoft.AspNetCore.Http;
using Microsoft.Extensions.DependencyInjection;
using Microsoft.Extensions.Hosting;
using Microsoft.Extensions.Configuration;
using Microsoft.EntityFrameworkCore;
using SportsStore.Models;

namespace SportsStore {
   public class Startup {
```

```
        public Startup(IConfiguration config) {
            Configuration = config;
        }

        private IConfiguration Configuration { get; set; }

        public void ConfigureServices(IServiceCollection services) {
            services.AddControllersWithViews();
            services.AddDbContext<StoreDbContext>(opts => {
            opts.UseSqlServer(
                Configuration["ConnectionStrings:SportsStoreConnection"]);
        });
        services.AddScoped<IStoreRepository, EFStoreRepository>();
        services.AddScoped<IOrderRepository, EFOrderRepository>();
        services.AddRazorPages();
        services.AddDistributedMemoryCache();
        services.AddSession();
        services.AddScoped<Cart>(sp => SessionCart.GetCart(sp));
        services.AddSingleton<IHttpContextAccessor, HttpContextAccessor>();
        services.AddServerSideBlazor();
    }

    public void Configure(IApplicationBuilder app, IWebHostEnvironment env) {
        app.UseDeveloperExceptionPage();
        app.UseStatusCodePages();
        app.UseStaticFiles();
        app.UseSession();
        app.UseRouting();

        app.UseEndpoints(endpoints => {
            endpoints.MapControllerRoute("catpage",
                "{category}/Page{productPage:int}",
                new { Controller = "Home", action = "Index" });

            endpoints.MapControllerRoute("page", "Page{productPage:int}",
                new { Controller = "Home", action = "Index", productPage = 1 });

            endpoints.MapControllerRoute("category", "{category}",
                new { Controller = "Home", action = "Index", productPage = 1 });

            endpoints.MapControllerRoute("pagination",
                "Products/Page{productPage}",
                new { Controller = "Home", action = "Index", productPage = 1 });
            endpoints.MapDefaultControllerRoute();
            endpoints.MapRazorPages();
            endpoints.MapBlazorHub();
            endpoints.MapFallbackToPage("/admin/{*catchall}", "/Admin/Index");
```

```
            });
            SeedData.EnsurePopulated(app);
        }
    }
}
```

AddServerSideBlazor 方法创建 Blazor 使用的服务，MapBlazorHub 方法注册 Blazor 中间件组件。最后要做的是调整路由系统，以确保 Blazor 与应用程序的其他部分无缝协作。

## 10.1.1 创建导入文件

Blazor 需要自己的导入文件来指定它使用的名称空间。创建 Pages/Admin 文件夹，并添加一个名为_Imports.razor 的文件，内容如代码清单 10-2 所示(如果使用的是 Visual Studio，可以使用 Razor Components 模板来创建此文件)。

---

■ **注意：** Blazor 文件通常在 Pages 文件夹中，但 Blazor 文件可以在项目的任何位置定义。例如，第Ⅳ部分使用了一个名为 Blazor 的文件夹，帮助强调 Blazor 和 Razor 页面(Razor Pages)提供了哪些特性。

---

**代码清单 10-2　SportsStore/Pages/Admin 文件夹中_Imports.razor 文件的内容**

```
@using Microsoft.AspNetCore.Components
@using Microsoft.AspNetCore.Components.Forms
@using Microsoft.AspNetCore.Components.Routing
@using Microsoft.AspNetCore.Components.Web
@using Microsoft.EntityFrameworkCore
@using SportsStore.Models
```

前四个@using 表达式用于 Blazor 所需的名称空间。最后两个表达式是为了方便在下面的示例中使用，因为它们允许使用 Entity Framework Core 和 Models 名称空间中的类。

## 10.1.2 创建 Startup Razor Pages

Blazor 依靠 Razor Pages 向浏览器提供初始内容，其中包括连接到服务器并呈现 Blazor HTML 内容的 JavaScript 代码。向 Pages/Admin 文件夹添加一个名为 Index.cshtml 的 Razor Pages，内容如代码清单 10-3 所示。

**代码清单 10-3　SportsStore/Pages/Admin 文件夹中 Index.cshtml 文件的内容**

```
@page "/admin"
@{ Layout = null; }

<!DOCTYPE html>
<html>
<head>
    <title>SportsStore Admin</title>
    <link href="/lib/twitter-bootstrap/css/bootstrap.min.css" rel="stylesheet" />
    <base href="/" />
```

```
</head>
<body>
    <component type="typeof(Routed)" render-mode="Server" />
    <script src="/_framework/blazor.server.js"></script>
</body>
</html>
```

component 元素用于在 Razor Pages 的输出中插入 Razor 组件。Razor 组件命名为 Blazor 的构造块，这令人困惑。代码清单 10-3 中应用的 component 元素命名为 Routed，将在稍后创建。Razor Pages 还包含一个 script 元素，它告诉浏览器加载 Blazor 服务器使用的 JavaScript 文件。Blazor 服务器中间件会拦截对该文件的请求，不需要显式地将 JavaScript 文件添加到项目中。

### 10.1.3 创建路由和布局组件

向 Pages/Admin 文件夹添加名为 Routed.razor 的 Razor 组件，并添加如代码清单 10-4 所示的内容。

#### 代码清单 10-4 SportsStore/Pages/Admin 文件夹中 Routed.razor 文件的内容

```
<Router AppAssembly="typeof(Startup).Assembly">
    <Found>
        <RouteView RouteData="@context" DefaultLayout="typeof(AdminLayout)" />
    </Found>
    <NotFound>
        <h4 class="bg-danger text-white text-center p-2">
            No Matching Route Found
        </h4>
    </NotFound>
</Router>
```

这个组件的内容在本书的第Ⅳ部分中有详细的描述，但是对于本章来说，知道这个组件使用浏览器的当前 URL 来定位一个可以显示给用户的 Razor 组件就足够了。如果找不到匹配的组件，则显示错误消息。

Blazor有自己的布局系统。要为管理工具创建布局，向Pages/Admin文件夹添加一个名为AdminLayout.razor的Razor组件，内容如代码清单 10-5 所示。

#### 代码清单 10-5 SportsStore/Pages/Admin 文件夹中 AdminLayout.razor 文件的内容

```
@inherits LayoutComponentBase
<div class="bg-info text-white p-2">
    <span class="navbar-brand ml-2">SPORTS STORE Administration</span>
</div>
<div class="container-fluid">
    <div class="row p-2">
        <div class="col-3">
            <NavLink class="btn btn-outline-primary btn-block"
                     href="/admin/products"
                     ActiveClass="btn-primary text-white"
```

```
                Match="NavLinkMatch.Prefix">
            Products
        </NavLink>
        <NavLink class="btn btn-outline-primary btn-block"
                href="/admin/orders"
                ActiveClass="btn-primary text-white"
                Match="NavLinkMatch.Prefix">
            Orders
        </NavLink>
    </div>
    <div class="col">
        @Body
    </div>
</div>
</div>
```

Blazor 使用 Razor 语法生成 HTML，但引入了自己的指令和特性。这种布局呈现两列显示，产品和订单导航按钮是使用 NavLink 元素创建的。这些元素应用一个内置的 Razor 组件，该组件可在不触发新的 HTTP 请求的情况下更改 URL，从而允许 Blazor 在不丢失应用程序状态的情况下响应用户交互。

### 10.1.4 创建 Razor 组件

为完成初始设置，需要添加提供管理工具的组件，尽管它们首先将包含占位符消息。向 Pages/Admin 文件夹添加名为 Products.razor 的 Razor 组件，内容如代码清单 10-6 所示。

#### 代码清单 10-6 SportsStore/Pages/Admin 文件夹中 Products.razor 文件的内容

```
@page "/admin/products"
@page "/admin"

<h4>This is the products component</h4>
```

@page 指令指定此组件显示的 URL，即/admin/products 和/admin。接下来，向 Pages/Admin 文件夹添加一个名为 Orders.razor 的 Razor 组件，内容如代码清单 10-7 所示。

#### 代码清单 10-7 SportsStore/Pages/Admin 文件夹中 Orders.razor 文件的内容

```
@page "/admin/orders"
<h4>This is the orders component</h4>
```

### 10.1.5 检查 Blazor 的设置

要确保Blazor工作正常，请启动ASP.NET Core。请求http://localhost:5000/admin。这个请求由Pages/Admin文件夹中的Index Razor Pages处理，它在发送给浏览器的内容中包含Blazor JavaScript文件。JavaScript代码将打开到ASP.NET Core服务器的持久HTTP连接，呈现初始Blazor内容，如图 10-1 所示。

■ **注意**：微软还没有发布测试 Razor 组件所需的工具，这也是本章没有单元测试示例的原因。

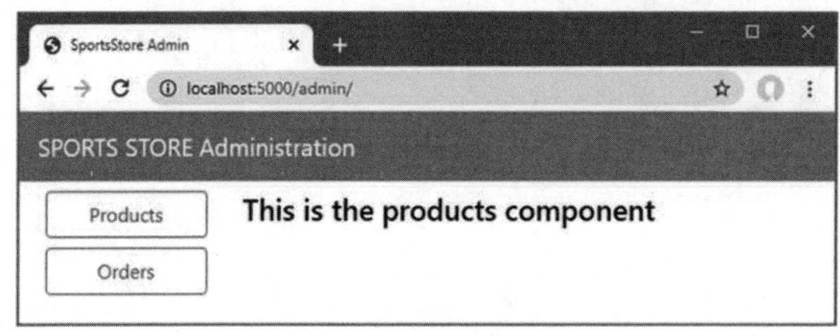


图 10-1　Blazor 应用程序

单击 Orders 按钮，显示 Orders Razor 组件生成的内容，如图 10-2 所示。不像前面章节使用的 ASP.NET Core 应用程序框架，即使浏览器显示的 URL 发生了变化，新内容的显示没有发送新的 HTTP 请求。

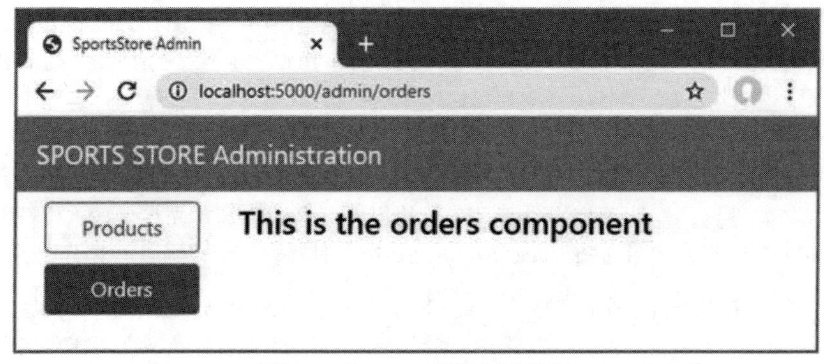


图 10-2　在 Blazor 应用程序中导航

## 10.2　管理订单

Blazor 已经设置好并进行了测试，现在开始实现管理特性。前一章增加了接收客户订单并存储在数据库中的支持功能。本节创建一个简单的管理工具，它允许查看已接收的订单并将其标记为已发货。

### 10.2.1　增强模型

首先需要更改的是增强数据模型，这样就可以记录发运了哪些订单。代码清单 10-8 显示添加到 Order 类的新属性，它定义在 Models 文件夹的 Order.cs 文件中。

**代码清单 10-8　在 SportsStore/Models 文件夹的 Order.cs 文件中添加一个属性**

```
using System.Collections.Generic;
using System.ComponentModel.DataAnnotations;
using Microsoft.AspNetCore.Mvc.ModelBinding;
```

```
namespace SportsStore.Models {

    public class Order {

        [BindNever]
        public int OrderID { get; set; }
        [BindNever]
        public ICollection<CartLine> Lines { get; set; }

        [Required(ErrorMessage = "Please enter a name")]
        public string Name { get; set; }

        [Required(ErrorMessage = "Please enter the first address line")]

        public string Line1 { get; set; }
        public string Line2 { get; set; }
        public string Line3 { get; set; }

        [Required(ErrorMessage = "Please enter a city name")]
        public string City { get; set; }

        [Required(ErrorMessage = "Please enter a state name")]
        public string State { get; set; }

        public string Zip { get; set; }

        [Required(ErrorMessage = "Please enter a country name")]
        public string Country { get; set; }

        public bool GiftWrap { get; set; }

        [BindNever]
        public bool Shipped { get; set; }
    }
}
```

这种迭代方法扩展和调整数据模型以支持不同特性，是 ASP.NET Core 开发的典型做法。在理想世界中，应能在项目的开始完全定义数据模型，在其基础上构建应用程序，但这只发生在最简单的项目中；在实践中，迭代开发是可以预料到的，因为理解什么是需要的功能是不断发展的。

Entity Framework Core 迁移使这个过程更容易，因为不必通过编写自己的 SQL 命令来手动保持数据库模式与模型类同步。要更新数据库，以反映添加到 Order 类的 Shipped 属性，请打开一个新的 PowerShell 窗口，并在 SportsStore 项目中运行代码清单 10-9 所示的命令。

**代码清单 10-9　创建一个新的迁移**

```
dotnet ef migrations add ShippedOrders
```

迁移在应用程序启动时自动应用，SeedData类调用Entity Framework Core提供的迁移方法。

## 10.2.2　向管理员显示订单

这里显示了两个表，其中一个显示等待发货的订单，另一个显示已发货的订单。每个订单都显示一个按钮，用于更改配送状态。这并不完全现实，因为订单处理通常比简单地更新数据库中的字段更复杂，但与仓库和实现系统的集成远远超出本书的范围。

为了避免重复代码和内容，本例创建一个Razor组件，它显示一个表，而不知道它处理的是哪类订单。向Pages/Admin添加一个名为OrderTable.razor的Razor组件，内容如代码清单10-10所示。

**代码清单10-10　SportsStore/Pages/Admin文件夹中OrderTable.razor文件的内容**

```
<table class="table table-sm table-striped table-bordered">
    <thead>
        <tr><th colspan="5" class="text-center">@TableTitle</th></tr>
    </thead>
    <tbody>
        @if (Orders?.Count() > 0) {
            @foreach (Order o in Orders) {
                <tr>
                    <td>@o.Name</td><td>@o.Zip</td><th>Product</th><th>Quantity</th>
                <td>
                    <button class="btn btn-sm btn-danger"
                            @onclick="@(e => OrderSelected.InvokeAsync(o.OrderID))">
                        @ButtonLabel
                    </button>
                </td>
            </tr>
                @foreach (CartLine line in o.Lines) {
                    <tr>
                        <td colspan="2"></td>
                        <td>@line.Product.Name</td><td>@line.Quantity</td>
                        <td></td>
                    </tr>
                }
            }
        } else {
            <tr><td colspan="5" class="text-center">No Orders</td></tr>
        }
    </tbody>
</table>

@code {

    [Parameter]
    public string TableTitle { get; set; } = "Orders";
```

```
    [Parameter]
    public IEnumerable<Order> Orders { get; set; }

    [Parameter]
    public string ButtonLabel { get; set; } = "Ship";

    [Parameter]
    public EventCallback<int> OrderSelected{ get; set; }
}
```

顾名思义，Razor 组件依赖 Razor 方法来注释 HTML 元素。组件的视图部分由@code 部分中的语句支持。该组件中的@code 部分定义了四个用 Parameter 属性修饰的属性，这意味着这些值将在运行时由父组件提供，稍后将创建父组件。在组件的视图部分中使用为参数提供的值来显示 Order 对象序列的详细信息。

Blazor将表达式添加到Razor语法中。这个组件的视图部分包括这个按钮元素，它有一个@onclick属性。

```
...
<button class="btn btn-sm btn-danger"
        @onclick="@(e => OrderSelected.InvokeAsync(o.OrderID))">
    @ButtonLabel
</button>
...
```

这告诉 Blazor 在用户单击按钮时如何反应。在本例中，表达式告诉 Razor 调用 OrderSelected 属性的 InvokeAsync 方法。这就是表格与 Blazor 应用程序其余部分通信的方式，在构建额外特性时这会变得更加清晰。

---

■ **提示**：本书的第Ⅳ部分深入讨论 Blazor，所以如果本章中的 Razor 组件没有直接的意义，请不要担心。SportsStore 示例的目的是展示整个开发过程，只是个别特征你暂未理解。

---

下一步是创建一个组件，该组件从数据库中获取订单数据，并使用 OrderTable 组件将其显示给用户。删除 Orders 组件中的占位符内容，并用代码清单 10-11 所示的代码和内容替换它。

#### 代码清单 10-11　SportsStore/Pages/Admin 文件夹中 Orders.razor 文件经修改的内容

```
@page "/admin/orders"
@inherits OwningComponentBase<IOrderRepository>
        <OrderTable TableTitle="Unshipped Orders"
Orders="UnshippedOrders" ButtonLabel="Ship" OrderSelected="ShipOrder" />
        <OrderTable TableTitle="Shipped Orders"
Orders="ShippedOrders" ButtonLabel="Reset" OrderSelected="ResetOrder" />
<button class="btn btn-info" @onclick="@(e => UpdateData())">Refresh Data</button>

@code {

    public IOrderRepository Repository => Service;
```

```
    public IEnumerable<Order> AllOrders { get; set; }
    public IEnumerable<Order> UnshippedOrders { get; set; }
    public IEnumerable<Order> ShippedOrders { get; set; }

    protected async override Task OnInitializedAsync() {
        await UpdateData();
  }
    public async Task UpdateData() {
        AllOrders = await Repository.Orders.ToListAsync();
        UnshippedOrders = AllOrders.Where(o => !o.Shipped);
        ShippedOrders = AllOrders.Where(o => o.Shipped);
    }

    public void ShipOrder(int id) => UpdateOrder(id, true);
    public void ResetOrder(int id) => UpdateOrder(id, false);

    private void UpdateOrder(int id, bool shipValue) {
        Order o = Repository.Orders.FirstOrDefault(o => o.OrderID == id);
        o.Shipped = shipValue;
            Repository.SaveOrder(o);
    }
  }
```

Blazor 组件与用于 SportsStore 应用程序面向用户部分的其他应用程序框架构建块不同。组件可以是长期存在的，并在较长时间内处理多个用户交互，而不是处理单个请求。这需要不同风格的开发，特别是使用 Entity Framework Core 处理数据时。@inherits 表达式确保该组件获得自己的存储库对象，从而确保其操作与显示给同一用户的其他组件执行的操作是分离的。为避免重复查询数据库(这在 Blazor 中可能是一个严重问题，参见第Ⅳ部分)，存储库仅在组件初始化、Blazor 调用 OnInitializedAsync 方法或用户单击 Refresh Data 按钮时使用。

为向用户显示它的数据，使用了 OrderTable 组件，它作为 HTML 元素应用，如下所示：

```
...
<OrderTable TableTitle="Unshipped Orders"
      Orders="UnshippedOrders" ButtonLabel="Ship" OrderSelected="ShipOrder" />
...
```

分配给 OrderTable 元素属性的值用于设置代码清单 10-10 中用 Parameter 属性装饰的属性。通过这种方式，可将单个组件配置为显示两组不同的数据，而不需要重复代码和内容。

ShipOrder 和 ResetOrder 方法用作 OrderSelected 属性的值，这意味着当用户单击 OrderTable 组件提供的一个按钮，通过存储库更新数据库中的数据时，将调用它们。

要查看新特性，请重新启动 ASP.NET Core。请求 http://localhost:5000，并创建一个订单。在数据库中至少有一个订单后，请求 http://localhost:5000/admin/orders，在 Unshipped Orders 表中会看到创建的订单的摘要。单击 Ship 按钮，订单就更新并移动到 Shipped Orders 表，如图 10-3 所示。

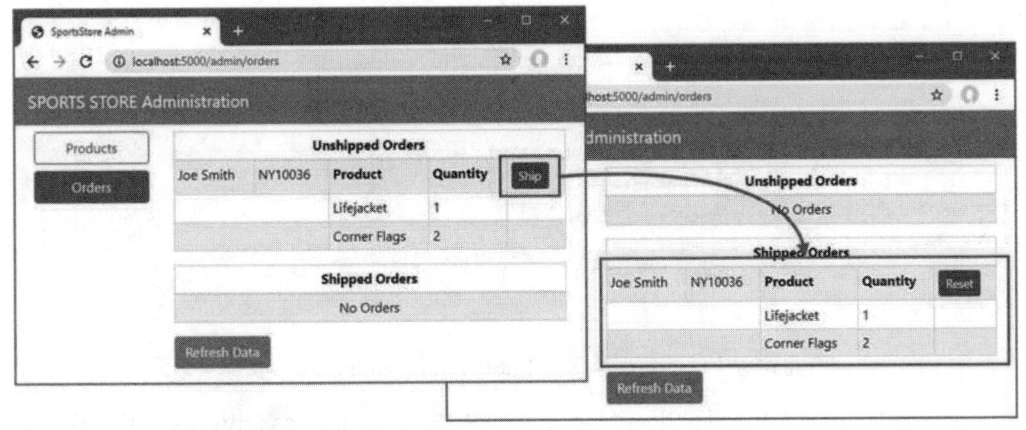


图 10-3　管理订单

# 10.3　添加目录管理

管理更复杂的项集的惯例是向用户呈现两个界面：列表界面和编辑界面，如图 10-4 所示。

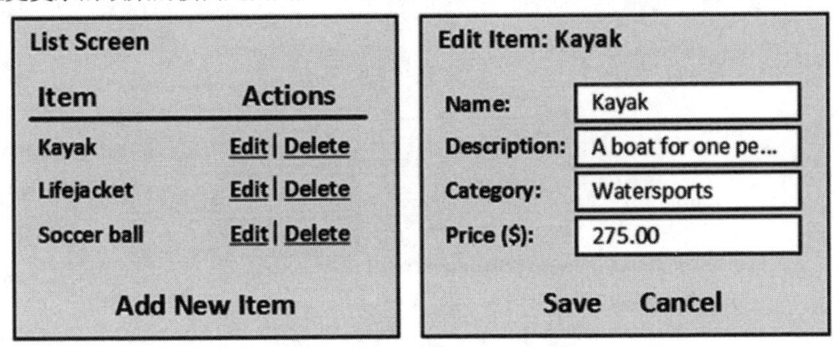


图 10-4　产品目录的 CRUD UI 草图

这些界面都允许用户创建、读取、更新和删除集合中的项。这些操作统称为 CRUD。本节将使用 Blazor 实现这些界面。

■ **提示：**开发人员经常需要实现 CRUD，因此 Visual Studio 搭建功能包括了创建 CRUD 控制器或 Razor Pages 的场景。但是，像所有的 Visual Studio 搭建功能，最好是学习如何直接创建这些特性，这就是为什么后续章节会演示所有 ASP.NET Core 应用程序框架的 CRUD 操作。

## 10.3.1　扩展存储库

第一步是向存储库添加允许创建、修改和删除 Product 对象的特性。代码清单 10-12 向 IStoreRepository 接口添加新方法。

**代码清单 10-12　在 SportsStore/Models 文件夹的 IStoreRepository.cs 文件中添加方法**

```
using System.Linq;
```

```
namespace SportsStore.Models {
    public interface IStoreRepository {

        IQueryable<Product> Products { get; }

        void SaveProduct(Product p);
        void CreateProduct(Product p);
        void DeleteProduct(Product p);
    }
}
```

代码清单 10-13 将这些方法的实现添加到 Entity Framework Core 存储库类中。

**代码清单 10-13　在 SportsStore/Models 文件夹的 EFStoreRepository.cs 文件中实现方法**

```
using System.Linq;

namespace SportsStore.Models {
    public class EFStoreRepository : IStoreRepository {
        private StoreDbContext context;

        public EFStoreRepository(StoreDbContext ctx) {
            context = ctx;
        }

        public IQueryable<Product> Products => context.Products;

        public void CreateProduct(Product p) {
            context.Add(p);
            context.SaveChanges();
        }

        public void DeleteProduct(Product p) {
            context.Remove(p);
            context.SaveChanges();
        }

        public void SaveProduct(Product p) {
            context.SaveChanges();
        }
    }
}
```

## 10.3.2　将验证属性应用到数据模型

希望验证用户在编辑或创建 Product 对象时提供的值，就像对客户付款过程所做的那样。在代码清单 10-14 中，向 Product 数据模型类添加了验证属性。

代码清单 10-14 在 SportsStore/Models 文件夹的 Product.cs 文件中添加验证属性

```
using System.ComponentModel.DataAnnotations.Schema;
using System.ComponentModel.DataAnnotations;

namespace SportsStore.Models {

    public class Product {
        public long ProductID { get; set; }

        [Required(ErrorMessage = "Please enter a product name")]
        public string Name { get; set; }

        [Required(ErrorMessage = "Please enter a description")]
        public string Description { get; set; }

        [Required]
        [Range(0.01, double.MaxValue,
            ErrorMessage = "Please enter a positive price")]
        [Column(TypeName = "decimal(8, 2)")]
        public decimal Price { get; set; }

        [Required(ErrorMessage = "Please specify a category")]
        public string Category { get; set; }
    }
}
```

Blazor 使用与其他 ASP.NET Core 相同的验证方法。但可以看到，它以不同方式应用于处理 Razor 组件的交互性更强的特性。

## 10.3.3 创建列表组件

下面首先创建一个表，它向用户显示产品表以及用于检查和编辑产品的链接。用如代码清单 10-15 所示的代码替换 Products.razor 文件的内容。

代码清单 10-15 SportsStore/Pages/Admin 文件夹中 Products.razor 文件修改后的内容

```
@page "/admin/products"
@page "/admin"
@inherits OwningComponentBase<IStoreRepository>

<table class="table table-sm table-striped table-bordered">
    <thead>
        <tr>
            <th>ID</th><th>Name</th>
            <th>Category</th><th>Price</th><td/>
        </tr>
    </thead>
```

```
    <tbody>
        @if (ProductData?.Count() > 0) {
            @foreach (Product p in ProductData) {
                <tr>
                    <td>@p.ProductID</td>
                    <td>@p.Name</td>
                    <td>@p.Category</td>
                    <td>@p.Price.ToString("c")</td>
                    <td>
                        <NavLink class="btn btn-info btn-sm"
                                href="@GetDetailsUrl(p.ProductID)">
                            Details
                        </NavLink>
                        <NavLink class="btn btn-warning btn-sm"
                                href="@GetEditUrl(p.ProductID)">
                            Edit
                        </NavLink>
                    </td>
                </tr>
            }
        } else {
            <tr>
                <td colspan="5" class="text-center">No Products</td>
            </tr>
        }
    </tbody>
</table>

<NavLink class="btn btn-primary" href="/admin/products/create">Create</NavLink>
@code {

    public IStoreRepository Repository => Service;

    public IEnumerable<Product> ProductData { get; set; }

    protected async override Task OnInitializedAsync() {
        await UpdateData();
    }

    public async Task UpdateData() {
        ProductData = await Repository.Products.ToListAsync();
    }

    public string GetDetailsUrl(long id) => $"/admin/products/details/{id}";
    public string GetEditUrl(long id) => $"/admin/products/edit/{id}";
}
```

该组件在表行中显示存储库中的每个 Product 对象，其中包含的 NavLink 组件导航到提供详细视图和编辑器的组件上。还有一个导航到组件的按钮，该组件将允许创建新 Product 对象并存储在数据库中。重启 ASP.NET Core，请求 http://localhost:5000/admin/products，将看到如图 10-5 所示的内容；但 Products 组件所显示的按钮目前都不起作用，因为还没有创建它们的目标组件。

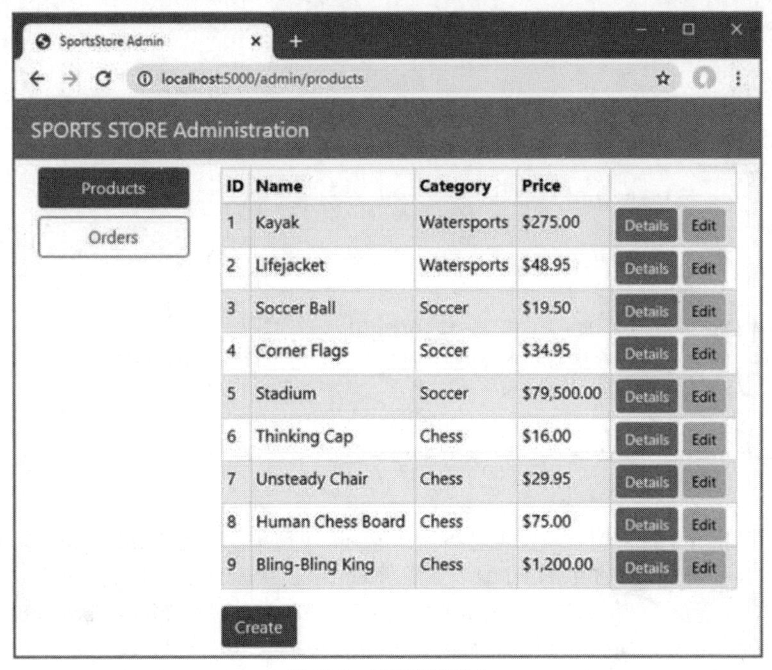


图 10-5　展示产品列表

## 10.3.4　创建细节组件

细节组件的任务是显示单个 Product 对象的所有字段。向 Pages/Admin 文件夹添加名为 Details.razor 的 Razor 组件，内容如代码清单 10-16 所示。

**代码清单 10-16　SportsStore/Pages/Admin 文件夹中 Details.razor 文件的内容**

```
@page "/admin/products/details/{id:long}"

<h3 class="bg-info text-white text-center p-1">Details</h3>

<table class="table table-sm table-bordered table-striped">
    <tbody>
        <tr><th>ID</th><td>@Product.ProductID</td></tr>
        <tr><th>Name</th><td>@Product.Name</td></tr>
        <tr><th>Description</th><td>@Product.Description</td></tr>
        <tr><th>Category</th><td>@Product.Category</td></tr>
        <tr><th>Price</th><td>@Product.Price.ToString("C")</td></tr>
    </tbody>
</table>
```

```
<NavLink class="btn btn-warning" href="@EditUrl">Edit</NavLink>
<NavLink class="btn btn-secondary" href="/admin/products">Back</NavLink>

@code {

    [Inject]
    public IStoreRepository Repository { get; set; }

    [Parameter]
    public long Id { get; set; }

    public Product Product { get; set; }

    protected override void OnParametersSet() {
        Product = Repository.Products.FirstOrDefault(p => p.ProductID == Id);
    }

    public string EditUrl => $"/admin/products/edit/{Product.ProductID}";
}
```

组件使用 Inject 属性声明它需要实现 IStoreRepository 接口，这是 Blazor 提供的访问应用程序服务的方式之一。Id 属性的值从用于导航到组件的 URL 中填充，该组件用于从数据库中检索 Product 对象。要查看详细信息视图，请重新启动 ASP.NET Core，请求 http://localhost:5000/admin/products，然后单击一个 Details 按钮，如图 10-6 所示。

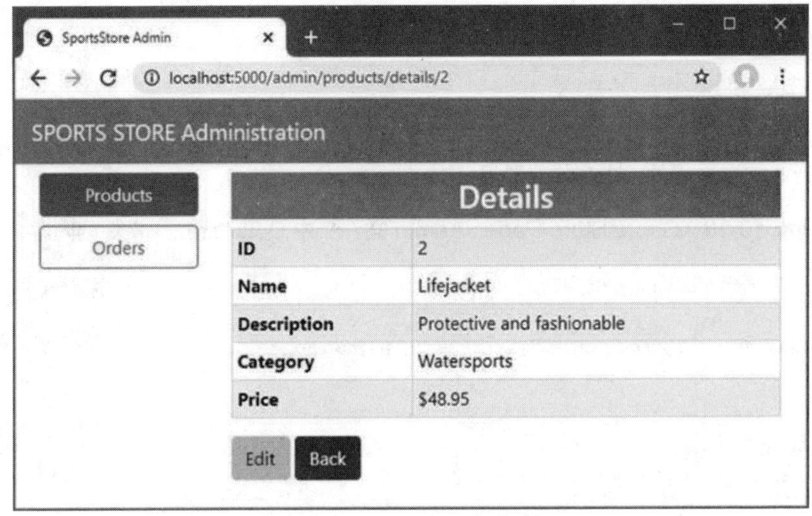


图 10-6　显示产品的详细信息

## 10.3.5　创建编辑器组件

创建和编辑数据的操作由同一个组件处理。向 Pages/Admin 文件夹添加名为 Editor 的 Razor

组件，内容如代码清单 10-17 所示。

**代码清单 10-17　SportsStore/Pages/Admin 文件夹中 Editor.razor 文件的内容**

```
@page "/admin/products/edit/{id:long}"
@page "/admin/products/create"
@inherits OwningComponentBase<IStoreRepository>

<style>
    div.validation-message { color: rgb(220, 53, 69); font-weight: 500 }
</style>

<h3 class="bg-@ThemeColor text-white text-center p-1">@TitleText a Product</h3>
<EditForm Model="Product" OnValidSubmit="SaveProduct">
    <DataAnnotationsValidator />
    @if(Product.ProductID != 0) {
        <div class="form-group">
            <label>ID</label>
            <input class="form-control" disabled value="@Product.ProductID" />
        </div>
    }
    <div class="form-group">
        <label>Name</label>
        <ValidationMessage For="@(() => Product.Name)" />
        <InputText class="form-control" @bind-Value="Product.Name" />
    </div>
    <div class="form-group">
        <label>Description</label>
        <ValidationMessage For="@(() => Product.Description)" />
        <InputText class="form-control" @bind-Value="Product.Description" />
    </div>
    <div class="form-group">
        <label>Category</label>
        <ValidationMessage For="@(() => Product.Category)" />
        <InputText class="form-control" @bind-Value="Product.Category" />
    </div>
    <div class="form-group">
        <label>Price</label>
        <ValidationMessage For="@(() => Product.Price)" />
        <InputNumber class="form-control" @bind-Value="Product.Price" />
    </div>
    <button type="submit" class="btn btn-@ThemeColor">Save</button>
    <NavLink class="btn btn-secondary" href="/admin/products">Cancel</NavLink>
</EditForm>

@code {
```

```
    public IStoreRepository Repository => Service;

    [Inject]
    public NavigationManager NavManager { get; set; }

    [Parameter]
    public long Id { get; set; } = 0;

    public Product Product { get; set; } = new Product();

    protected override void OnParametersSet() {
        if (Id != 0) {
            Product = Repository.Products.FirstOrDefault(p => p.ProductID == Id);
        }
    }

    public void SaveProduct() {
        if (Id == 0) {
            Repository.CreateProduct(Product);
        } else {
            Repository.SaveProduct(Product);
        }
        NavManager.NavigateTo("/admin/products");
    }

    public string ThemeColor => Id == 0 ? "primary" : "warning";
    public string TitleText => Id == 0 ? "Create" : "Edit";
}
```

Blazor 提供了一组内置的 Razor 组件，用于显示和验证表单，这一点很重要，因为浏览器不能在 Blazor 组件中使用 POST 请求提交数据。EditForm 组件用于呈现一个对 Blazor 友好的表单，而 InputText 和 InputNumber 组件则呈现接收字符串和数字值的输入元素，并在用户进行更改时自动更新模型属性。

数据验证集成到这些内置组件中，EditForm 组件上的 OnValidSubmit 属性用于指定一个方法，只有在输入表单中的数据符合验证属性定义的规则时，才调用该方法。

Blazor 还提供了 NavigationManager 类，用于在不触发新的 HTTP 请求的情况下，以编程方式在组件之间导航。编辑器组件使用作为服务获得的 NavigationManager，在数据库更新后返回 Products 组件。

要查看编辑器，请重新启动 ASP.NET Core，请求 http://localhost:5000/admin，然后单击 Create 按钮。单击 Save 按钮而不填写表单字段，将看到 Blazor 自动生成的验证错误，如图 10-7 所示。填写表单并再次单击 Save，将看到创建的产品显示在表中，同样如图 10-7 所示。

单击其中一个产品的 Edit 按钮，使用相同的组件编辑所选产品对象的属性。单击 Save 按钮，所做的任何更改(如果通过验证)都存储在数据库中，如图 10-8 所示。

图 10-7 使用编辑器组件

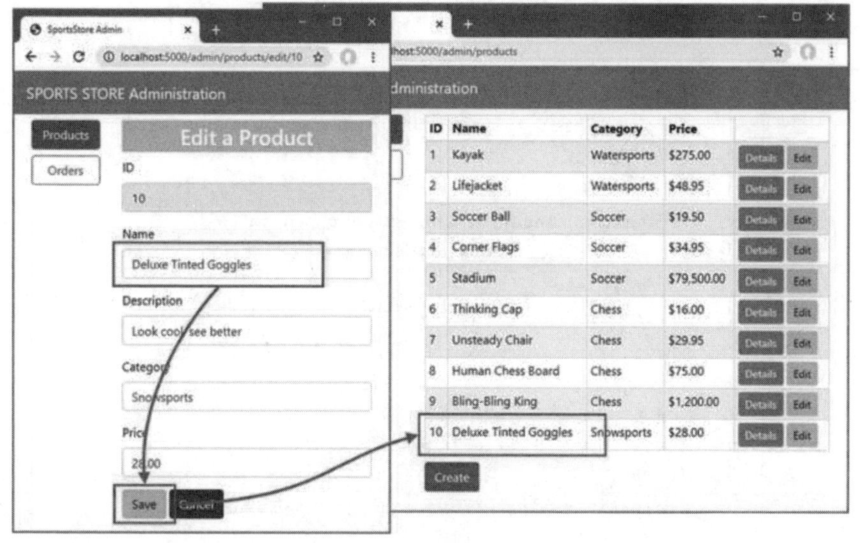


图 10-8 编辑产品

## 10.3.6 删除产品

CRUD 的最后一个特性是删除产品，这在 Products 组件中很容易实现，如代码清单 10-18 所示。

**代码清单 10-18 在 SportsStore/Pages/Admin 文件夹的 Products.razor 文件中添加删除功能**

```
@page "/admin/products"
@page "/admin"
@inherits OwningComponentBase<IStoreRepository>

<table class="table table-sm table-striped table-bordered">
    <thead>
```

```
        <tr>
            <th>ID</th><th>Name</th>
            <th>Category</th><th>Price</th><td/>
        </tr>
    </thead>
    <tbody>
        @if (ProductData?.Count() > 0) {
            @foreach (Product p in ProductData) {
                <tr>
                    <td>@p.ProductID</td>
                    <td>@p.Name</td>
                    <td>@p.Category</td>
                    <td>@p.Price.ToString("c")</td>
                    <td>
                        <NavLink class="btn btn-info btn-sm"
                                href="@GetDetailsUrl(p.ProductID)">
                            Details
                        </NavLink>
                        <NavLink class="btn btn-warning btn-sm"
                                href="@GetEditUrl(p.ProductID)">
                            Edit
                        </NavLink>
                        <button class="btn btn-danger btn-sm"
                                @onclick="@(e => DeleteProduct(p))">
                            Delete
                        </button>
                    </td>
                </tr>
            }
        } else {
            <tr>
                <td colspan="5" class="text-center">No Products</td>
            </tr>
        }
    </tbody>
</table>

<NavLink class="btn btn-primary" href="/admin/products/create">Create</NavLink>

@code {

    public IStoreRepository Repository => Service;

    public IEnumerable<Product> ProductData { get; set; }

    protected async override Task OnInitializedAsync() {
        await UpdateData();
```

```
    }

    public async Task UpdateData() {
        ProductData = await Repository.Products.ToListAsync();
    }

    public async Task DeleteProduct(Product p) {
        Repository.DeleteProduct(p);
        await UpdateData();
    }

    public string GetDetailsUrl(long id) => $"/admin/products/details/{id}";
    public string GetEditUrl(long id) => $"/admin/products/edit/{id}";
}
```

新的按钮元素使用@onclick 属性配置，该属性调用 DeleteProduct 方法。选择的产品对象从数据库中删除，更新组件显示的数据。重启 ASP.NET Core，请求 http://localhost:5000/admin/products，然后单击 Delete 按钮，从数据库中删除一个对象，如图 10-9 所示。

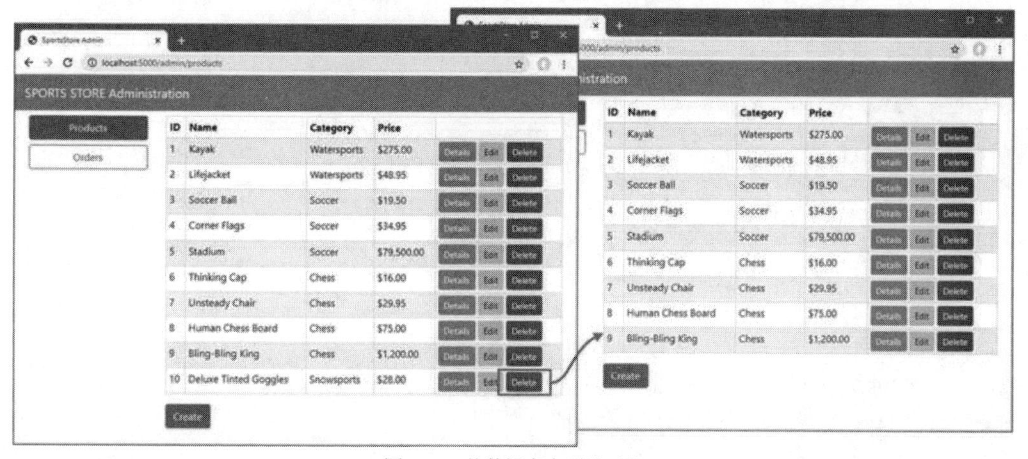


图 10-9　从数据库中删除对象

# 10.4 小结

本章介绍了管理功能，展示了如何使用 Blazor Server 实现 CRUD 操作，允许管理员创建、读取、更新和删除存储库中的产品，并将订单标记为已发货。第 11 章将展示如何保护管理功能，使它们不能对所有用户可用，并准备将 SportsStore 应用程序部署到生产环境中。

# 第 11 章

# SportsStore：安全与部署

## 11.1 确保管理功能的安全

身份验证和授权由 ASP.NET Core 身份识别系统提供，它巧妙地集成到 ASP.NET Core 平台和各个应用框架中。下面将创建一个基本的安全设置，允许名为 Admin 的用户验证和访问应用程序中的管理特性。ASP.NET Core Identity 提供了更多特性，用于验证用户的身份以及授权访问应用程序的特性和数据；参见第 37 章和第 38 章，这两章展示了如何创建和管理用户账户以及如何使用角色执行授权。但如前所述，ASP.NET Core Identity 本身就是一个大框架，本书只介绍了基本的特性。

本章的目标只是得到足够的功能，以防止客户访问 SportsStore 应用程序的敏感部分，并在这样做的同时，介绍身份验证和授权如何适合 ASP.NET Core 应用程序。

---

**提示**：可以从 https://github.com/apress/pro-asp.net-core-3 下载本章和本书中其他所有章节的示例项目。如果在运行示例时遇到问题，请参阅第 1 章以获得帮助。

---

### 11.1.1 创建身份数据库

ASP.NET Identity 系统是无限可配置和可扩展的，支持多种用户数据存储方式。

本章使用最常见的方式，就是用 Microsoft SQL Server 存储数据，用 Entity Framework Core 进行访问。

#### 1. 安装 Entity Framework Core 的 Identity 包

添加包含 ASP.NET Core Identity 的包来支持 Entity Framework Core。使用 PowerShell 命令提示符，在 SportsStore 文件夹中运行代码清单 11-1 所示的命令。

**代码清单 11-1 安装 Entity Framework Core 包**

```
dotnet add package Microsoft.AspNetCore.Identity.EntityFrameworkCore --version 3.1.0
```

### 2. 创建上下文类

需要创建一个数据库上下文文件，充当数据库和它提供访问的Identity模型对象之间的桥梁。在Models文件夹中添加一个名为AppIdentityDbContext.cs的类文件，并使用它定义如代码清单11-2所示的类。

**代码清单 11-2　SportsStore/Models 文件夹中 AppIdentityDbContext.cs 文件的内容**

```
using Microsoft.AspNetCore.Identity;
using Microsoft.AspNetCore.Identity.EntityFrameworkCore;
using Microsoft.EntityFrameworkCore;
namespace SportsStore.Models {

    public class AppIdentityDbContext : IdentityDbContext<IdentityUser> {

        public AppIdentityDbContext(DbContextOptions<AppIdentityDbContext> options)
            : base(options) { }
    }
}
```

AppIdentityDbContext 类是从 IdentityDbContext 派生的，为 Entity Framework Core 提供了特定于身份的特性。对于类型参数，使用了 IdentityUser 类，它是用于表示用户的内置类。

### 3. 定义连接字符串

下一步是定义用于数据库的连接字符串。代码清单 11-3 显示了在 SportsStore 项目的 appsettings.json 文件中定义的连接字符串，其格式与为产品数据库定义的连接字符串相同。

**代码清单 11-3　在 SportsStore 项目中 appsettings.json 文件中定义连接字符串**

```
{
    "Logging": {
        "LogLevel": {
            "Default": "Information",
            "Microsoft": "Warning",
            "Microsoft.Hosting.Lifetime": "Information"
        }
    },
    "AllowedHosts": "*",
    "ConnectionStrings": {
        "SportsStoreConnection": "Server=(localdb)\\MSSQLLocalDB;Database=
        SportsStore;MultipleActiveResultSets=true",
        "IdentityConnection": "Server=(localdb)
        \\MSSQLLocalDB;Database=Identity;MultipleActiveResultSets=true"
    }
}
```

请记住，连接字符串必须在 appsettings.json 文件中的单个不间断行中定义，然而在这个代码清单中跨多行显示，这是因为本书页面的宽度有限。代码清单中的附加部分定义了一个名为

IdentityConnection 的连接字符串，指定了一个名为 Identity 的 LocalDB 数据库。

### 4. 配置应用程序

像其他 ASP.NET Core 特性，Identity 是在 Startup 类中配置的。代码清单 11-4 显示了使用前面定义的上下文类和连接字符串，在 SportsStore 项目中设置 Identity 所添加的内容。

**代码清单 11-4 在 SportsStore 文件夹的 Startup.cs 文件中配置 Identity**

```
using System;
using System.Collections.Generic;
using System.Linq;
using System.Threading.Tasks;
using Microsoft.AspNetCore.Builder;
using Microsoft.AspNetCore.Hosting;
using Microsoft.AspNetCore.Http;
using Microsoft.Extensions.DependencyInjection;
using Microsoft.Extensions.Hosting;
using Microsoft.Extensions.Configuration;
using Microsoft.EntityFrameworkCore;
using SportsStore.Models;
using Microsoft.AspNetCore.Identity;

namespace SportsStore {
   public class Startup {

      public Startup(IConfiguration config) {
         Configuration = config;
      }

      private IConfiguration Configuration { get; set; }

      public void ConfigureServices(IServiceCollection services) {
         services.AddControllersWithViews();
         services.AddDbContext<StoreDbContext>(opts => {
            opts.UseSqlServer(
               Configuration["ConnectionStrings:SportsStoreConnection"]);
         });
         services.AddScoped<IStoreRepository, EFStoreRepository>();
         services.AddScoped<IOrderRepository, EFOrderRepository>();
         services.AddRazorPages();
         services.AddDistributedMemoryCache();
         services.AddSession();
         services.AddScoped<Cart>(sp => SessionCart.GetCart(sp));
         services.AddSingleton<IHttpContextAccessor, HttpContextAccessor>();
         services.AddServerSideBlazor();

         services.AddDbContext<AppIdentityDbContext>(options =>
```

```
            options.UseSqlServer(
                Configuration["ConnectionStrings:IdentityConnection"]));
        services.AddIdentity<IdentityUser, IdentityRole>()
            .AddEntityFrameworkStores<AppIdentityDbContext>();
    }

    public void Configure(IApplicationBuilder app, IWebHostEnvironment env) {
        app.UseDeveloperExceptionPage();
        app.UseStatusCodePages();
        app.UseStaticFiles();
        app.UseSession();
        app.UseRouting();

        app.UseAuthentication();
        app.UseAuthorization();

        app.UseEndpoints(endpoints => {
            endpoints.MapControllerRoute("catpage",
                "{category}/Page{productPage:int}",
                new { Controller = "Home", action = "Index" });

            endpoints.MapControllerRoute("page", "Page{productPage:int}",
                new { Controller = "Home", action = "Index", productPage = 1 });

            endpoints.MapControllerRoute("category", "{category}",
                new { Controller = "Home", action = "Index", productPage = 1 });
                    endpoints.MapControllerRoute("pagination",
                "Products/Page{productPage}",
                new { Controller = "Home", action = "Index", productPage = 1 });
            endpoints.MapDefaultControllerRoute();
            endpoints.MapRazorPages();
            endpoints.MapBlazorHub();
            endpoints.MapFallbackToPage("/admin/{*catchall}", "/Admin/Index");
        });

        SeedData.EnsurePopulated(app);
    }
  }
}
```

在ConfigureServices方法中，扩展了Entity Framework Core配置以注册上下文类，并使用AddIdentity方法设置Identity服务，使用内置类表示用户和角色。

在Configure方法中，调用了UseAuthentication和UseAuthorization方法来设置实现安全策略的中间件组件。这些方法必须出现在UseRouting和UseEndpoints方法之间。

#### 5. 创建和应用数据库迁移

基本配置已经就绪，现在可以使用Entity Framework Core迁移特性来定义模式，并将其应用到数据库中。打开一个新的命令提示符或 PowerShell 窗口，并在 SportsStore 文件夹中运行代码清单 11-5 所示的命令，为 Identity 数据库创建一个新的迁移。

**代码清单 11-5　创建身份迁移**

```
dotnet ef migrations add Initial --context AppIdentityDbContext
```

与前面的数据库命令的重要区别在于，使用-context 参数来指定与要处理的数据库关联的上下文类名称，即 AppIdentityDbContext。当应用程序中有多个数据库时，务必确保使用正确的上下文类。

Entity Framework Core 生成初始迁移后，在 SportsStore 文件夹中运行代码清单 11-6 所示的命令，以创建数据库并应用迁移。

**代码清单 11-6　应用身份迁移**

```
dotnet ef database update --context AppIdentityDbContext
```

结果是一个名为 Identity 的新 LocalDB 数据库，可以使用 Visual Studio SQL Server 对象资源管理器检查该数据库。

### 6. 定义种子数据

下面在应用程序启动时播种数据库，来显式地创建 Admin 用户。向 Models 文件夹添加一个类文件 IdentitySeedData.cs，并定义静态类，如代码清单 11-7 所示。

**代码清单 11-7　SportsStore/Models 文件夹中 IdentitySeedData.cs 文件的内容**

```
using Microsoft.AspNetCore.Builder;
using Microsoft.AspNetCore.Identity;
using Microsoft.Extensions.DependencyInjection;
using Microsoft.EntityFrameworkCore;
using System.Linq;

namespace SportsStore.Models {

    public static class IdentitySeedData {
        private const string adminUser = "Admin";
        private const string adminPassword = "Secret123$";

        public static async void EnsurePopulated(IApplicationBuilder app) {

        AppIdentityDbContext context = app.ApplicationServices
            .CreateScope().ServiceProvider
            .GetRequiredService<AppIdentityDbContext>();
        if (context.Database.GetPendingMigrations().Any()) {
            context.Database.Migrate();
        }
```

```
            UserManager<IdentityUser> userManager = app.ApplicationServices
                .CreateScope().ServiceProvider
                .GetRequiredService<UserManager<IdentityUser>>();

            IdentityUser user = await userManager.FindByIdAsync(adminUser);
            if (user == null) {
                user = new IdentityUser("Admin");
                user.Email = "admin@example.com";
                user.PhoneNumber = "555-1234";
                await userManager.CreateAsync(user, adminPassword);
            }
        }
    }
}
```

这段代码确保创建并更新数据库，使用UserManager<T>类；该类是由ASP.NET Core Identity为管理用户而提供的服务，如第38章所述。搜索数据库以查找Admin用户账户，如果Admin用户账户不存在，则创建该账户——密码为Secret123$。不要在本例中更改硬编码的密码，因为Identity的验证策略要求密码包含一定数量和范围的字符。有关如何更改验证设置的详细信息，请参阅第38章。

**警告：** 通常需要对管理员账户的详细信息进行硬编码，这样，一旦应用程序部署完毕，就可以登录并开始管理它。当这样做时，必须记住更改所创建账户的密码。请参阅第38章，了解如何使用身份更改密码。参见第15章，了解如何保持敏感数据(如默认密码)不受源代码控制。

为确保在应用程序启动时播种Identity数据库，将代码清单11-8所示的语句添加到Startup类的Configure方法中。

#### 代码清单11-8　在SportsStore文件夹的Startup.cs文件中播种Identity数据库

```
...
public void Configure(IApplicationBuilder app, IWebHostEnvironment env) {
    app.UseDeveloperExceptionPage();
    app.UseStatusCodePages();
    app.UseStaticFiles();
    app.UseSession();
    app.UseAuthentication();
    app.UseRouting();
    app.UseEndpoints(endpoints => {
        endpoints.MapControllerRoute("catpage",
            "{category}/Page{productPage:int}",
            new { Controller = "Home", action = "Index" });

        endpoints.MapControllerRoute("page", "Page{productPage:int}",
            new { Controller = "Home", action = "Index", productPage = 1 });

        endpoints.MapControllerRoute("category", "{category}",
            new { Controller = "Home", action = "Index", productPage = 1 });
```

```
        endpoints.MapControllerRoute("pagination",
            "Products/Page{productPage}",
            new { Controller = "Home", action = "Index", productPage = 1 });
        endpoints.MapDefaultControllerRoute();
        endpoints.MapRazorPages();
        endpoints.MapBlazorHub();
        endpoints.MapFallbackToPage("/admin/{*catchall}", "/Admin/Index");
    });

    SeedData.EnsurePopulated(app);
    IdentitySeedData.EnsurePopulated(app);
}
...
```

**删除和重新创建 ASP.NET Core Identity 数据库**

如果需要重置 Identity 数据库，则运行以下命令:

```
dotnet ef database drop --force --context AppIdentityDbContext
```

重启应用程序，重新创建数据库并填充种子数据。

### 11.1.2 添加常规的管理特性

在代码清单 11-9 中，使用 Blazor 创建管理特性，以便在 SportsStore 项目中演示广泛的 ASP.NET Core 功能。尽管 Blazor 很有用，但它并不适合所有的项目。如第Ⅳ部分所述，大多数项目都可能使用控制器或 Razor Pages 作为其管理特性。第 38 章描述了 ASP.NET Core Identity 使用所有应用程序框架的方式，但是为了平衡第 10 章创建的所有工具，下面创建一个 Razor Pages，来显示 ASP.NET Core Identity 数据库中的用户列表。第 38 章更详细地描述如何管理 Identity 数据库，这个 Razor Pages 只是为了给 SportsStore 应用程序添加一个非 Blazor 创建的敏感特性。向 SportsStore/Pages/Admin 文件夹添加一个名为 IdentityUsers.cshtml 的 Blazor 页面，内容如代码清单 11-9 所示。

**代码清单 11-9 SportsStore/Pages/Admin 文件夹中 IdentityUsers.cshtml 文件的内容**

```
@page
@model IdentityUsersModel
@using Microsoft.AspNetCore.Identity

<h3 class="bg-primary text-white text-center p-2">Admin User</h3>

<table class="table table-sm table-striped table-bordered">
    <tbody>
        <tr><th>User</th><td>@Model.AdminUser.UserName</td></tr>
        <tr><th>Email</th><td>@Model.AdminUser.Email</td></tr>
        <tr><th>Phone</th><td>@Model.AdminUser.PhoneNumber</td></tr>
    </tbody>
```

```
</table>

@functions{

    public class IdentityUsersModel: PageModel {
        private UserManager<IdentityUser> userManager;

        public IdentityUsersModel(UserManager<IdentityUser> mgr) {
            userManager = mgr;
        }

        public IdentityUser AdminUser{ get; set; }

        public async Task OnGetAsync() {
            AdminUser = await userManager.FindByNameAsync("Admin");
        }
    }
}
```

重启 ASP.NET Core，并请求 http://localhost:5000/admin/identityusers，查看 Razor Pages 生成的内容，如图 11-1 所示。

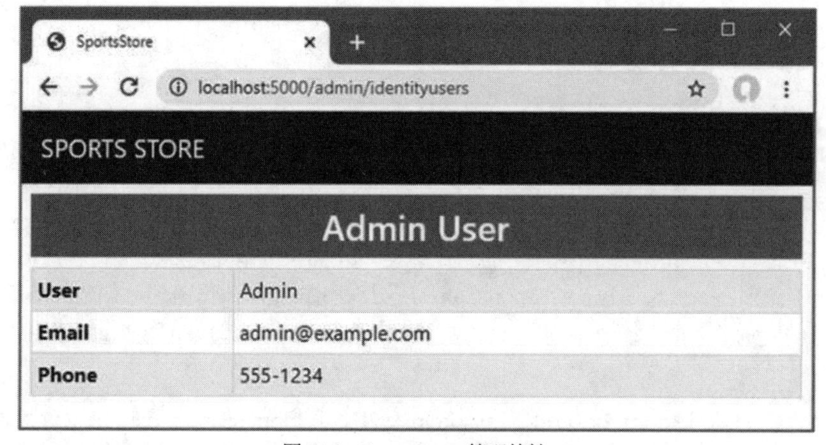


图 11-1　Razor Pages 管理特性

## 11.1.3　应用基本授权策略

现在已经配置了 ASP.NET Core Identity，可以对想要保护的应用程序部分运用授权策略。这里尽可能使用最基本的授权策略，即允许任何经过身份验证的用户访问。尽管这在实际应用程序中也是一个有用的策略，但也有一些选项可以创建粒度更细的授权控制，如第 37 章和第 38 章所述。由于 SportsStore 应用程序只有一个用户，区分匿名请求和经过身份验证的请求就足够了。

对于控制器和 Razor Pages，Authorize 属性用于限制访问，如代码清单 11-10 所示。

**代码清单 11-10 在 SportsStore/Pages/Admin 文件夹的 IdentityUsers.cshtml 文件中限制访问**

```
@page
@model IdentityUsersModel
@using Microsoft.AspNetCore.Identity
@using Microsoft.AspNetCore.Authorization

<h3 class="bg-primary text-white text-center p-2">Admin User</h3>

<table class="table table-sm table-striped table-bordered">
    <tbody>
        <tr><th>User</th><td>@Model.AdminUser.UserName</td></tr>
        <tr><th>Email</th><td>@Model.AdminUser.Email</td></tr>
        <tr><th>Phone</th><td>@Model.AdminUser.PhoneNumber</td></tr>
    </tbody>
</table>

@functions{

    [Authorize]
    public class IdentityUsersModel: PageModel {
        private UserManager<IdentityUser> userManager;

        public IdentityUsersModel(UserManager<IdentityUser> mgr) {
            userManager = mgr;
        }

        public IdentityUser AdminUser{ get; set; }

        public async Task OnGetAsync() {
            AdminUser = await userManager.FindByNameAsync("Admin");
        }
    }
}
```

当只有授权用户和未授权用户时，可将 Authorize 属性应用于充当应用程序 Blazor 部分的入口点的 Razor Pages，如代码清单 11-11 所示。

**代码清单 11-11 在 SportsStore/Pages/Admin 文件夹的 Index.cshtml 文件中应用授权**

```
@page "/admin"
@{ Layout = null; }
@using Microsoft.AspNetCore.Authorization
@attribute [Authorize]

<!DOCTYPE html>
<html>
<head>
```

```
    <title>SportsStore Admin</title>
    <link href="/lib/twitter-bootstrap/css/bootstrap.min.css" rel="stylesheet" />
    <base href="/" />
</head>
<body>
    <component type="typeof(Routed)" render-mode="Server" />
    <script src="/_framework/blazor.server.js"></script>
</body>
</html>
```

因为这个 Razor Pages 已经配置了一个页面模型类，所以可通过@attribute 表达式应用这个属性。

### 11.1.4　创建账户控制器和视图

当未经身份验证的用户发送需要授权的请求时，该用户被重定向到/Account/Login URL，应用程序可以使用该 URL 提示用户提供其凭证。代码清单 11-12 展示了如何使用 Razor Pages 处理身份验证；为了多样化，下面为 SportsStore 使用控制器和视图。在准备过程中，添加了一个视图模型来表示用户的凭证，方法是将一个名为 LoginModel.cs 的类文件添加到 Models/ViewModels 文件夹中，并使用它定义如代码清单 11-12 所示的类。

**代码清单 11-12　SportsStore/Models/ViewModels 文件夹中 LoginModel.cs 文件的内容**

```
using System.ComponentModel.DataAnnotations;

namespace SportsStore.Models.ViewModels {

    public class LoginModel {

        [Required]
        public string Name { get; set; }

        [Required]
        public string Password { get; set; }

        public string ReturnUrl { get; set; } = "/";
    }
}
```

Name 和 Password 属性已经用 Required 属性进行了修饰，该属性使用模型验证来确保提供了值。接下来，将一个名为 AccountController.cs 的类文件添加到 Controllers 文件夹，并使用它定义如代码清单 11-13 所示的控制器。这个控制器将响应对/Account/Login URL 的请求。

**代码清单 11-13　SportsStore/Controllers 文件夹中 AccountController.cs 文件的内容**

```
using System.Threading.Tasks;
using Microsoft.AspNetCore.Authorization;
using Microsoft.AspNetCore.Identity;
```

```
using Microsoft.AspNetCore.Mvc;
using SportsStore.Models.ViewModels;

namespace SportsStore.Controllers {

    public class AccountController : Controller {
        private UserManager<IdentityUser> userManager;
        private SignInManager<IdentityUser> signInManager;

        public AccountController(UserManager<IdentityUser> userMgr,
                SignInManager<IdentityUser> signInMgr) {
            userManager = userMgr;
            signInManager = signInMgr;
        }

        public ViewResult Login(string returnUrl) {
            return View(new LoginModel {
                ReturnUrl = returnUrl
            });
        }

        [HttpPost]
        [ValidateAntiForgeryToken]

        public async Task<IActionResult> Login(LoginModel loginModel) {
            if (ModelState.IsValid) {
                IdentityUser user =
                    await userManager.FindByNameAsync(loginModel.Name);
                if (user != null) {
                    await signInManager.SignOutAsync();
                    if ((await signInManager.PasswordSignInAsync(user,
                            loginModel.Password, false, false)).Succeeded) {
                        return Redirect(loginModel?.ReturnUrl ?? "/Admin");
                    }
                }
            }
            ModelState.AddModelError("", "Invalid name or password");
            return View(loginModel);
        }

        [Authorize]
        public async Task<RedirectResult> Logout(string returnUrl = "/") {
            await signInManager.SignOutAsync();
            return Redirect(returnUrl);
        }
    }
}
```

当用户重定向到/Account/Login URL 时，Login 操作方法的 GET 版本会呈现页面的默认视图，提供一个视图模型对象，其中包括如果验证请求成功，浏览器应该重定向到的 URL。

身份验证凭证提交到 Login 方法的 POST 版本中，该方法使用 UserManager<IdentityUser>和 SignInManager<IdentityUser>服务(通过控制器的构造函数接收)对用户进行身份验证并将其登录到系统中。第 37 章和第 38 章解释了这些类是如何工作的，但是现在，只要知道如果身份验证失败，就会创建一个模型验证错误，并呈现默认视图；但是，如果身份验证成功，就将用户重定向到他们想要访问的 URL，然后提示他们输入凭证。

**警告：** 一般来说，使用客户端数据验证是个好主意。它减轻了服务器上的一些工作，并为用户提供关于他们提供的数据的即时反馈。但是，不应该在客户端执行身份验证，因为这通常涉及把有效凭证发送给客户端，以检查用户输入的用户名和密码，或者至少相信客户“已经成功验证”的报告。身份验证应该始终在服务器上进行。

为给 Login 方法提供一个要呈现的视图，创建 Views/Account 文件夹，并添加了一个名为 Login.cshtml 的 Razor 视图文件，内容如代码清单 11-14 所示。

#### 代码清单 11-14　SportsStore/Views/Account 文件夹中 Login.cshtml 文件的内容

```
@model LoginModel
@{ Layout = null; }
<!DOCTYPE html>
<html>
<head>
    <meta name="viewport" content="width=device-width" />
    <title>SportsStore</title>
    <link href="/lib/twitter-bootstrap/css/bootstrap.min.css" rel="stylesheet" />
</head>
<body>
    <div class="bg-dark text-white p-2">
        <span class="navbar-brand ml-2">SPORTS STORE</span>
    </div>
    <div class="m-1 p-1">
        <div class="text-danger" asp-validation-summary="All"></div>

        <form asp-action="Login" asp-controller="Account" method="post">
            <input type="hidden" asp-for="ReturnUrl" />
            <div class="form-group">
                <label asp-for="Name"></label>
                <div asp-validation-for="Name" class="text-danger"></div>
                <input asp-for="Name" class="form-control" />
            </div>
            <div class="form-group">
                <label asp-for="Password"></label>
                <div asp-validation-for="Password" class="text-danger"></div>
                <input asp-for="Password" type="password" class="form-control" />
            </div>
            <button class="btn btn-primary" type="submit">Log In</button>
```

```
        </form>
    </div>
</body>
</html>
```

最后一步是对共享管理布局的更改，添加一个按钮，该按钮向注销操作发送请求，注销当前用户，如代码清单 11-15 所示。这是一个很有用的特性，使测试应用程序变得更容易；如果没有它，就需要清除浏览器的 cookie，以返回未经身份验证的状态。

**代码清单 11-15　在 SportsStore/Pages/Admin 文件夹的 AdminLayout.razor 文件中添加注销按钮**

```
@inherits LayoutComponentBase

<div class="bg-info text-white p-2">
    <div class="container-fluid">
        <div class="row">
            <div class="col">
                <span class="navbar-brand ml-2">SPORTS STORE Administration</span>
            </div>
            <div class="col-2 text-right">
                <a class="btn btn-sm btn-primary" href="/account/logout">Log Out</a>
            </div>
        </div>
    </div>
</div>
<div class="container-fluid">
    <div class="row p-2">
        <div class="col-3">
            <NavLink class="btn btn-outline-primary btn-block"
                     href="/admin/products"
                     ActiveClass="btn-primary text-white"
                     Match="NavLinkMatch.Prefix">
                Products
            </NavLink>
            <NavLink class="btn btn-outline-primary btn-block"
                     href="/admin/orders"
                     ActiveClass="btn-primary text-white"
                     Match="NavLinkMatch.Prefix">
                Orders
            </NavLink>
        </div>
        <div class="col">
            @Body
        </div>
    </div>
</div>
```

### 11.1.5 测试安全策略

一切就绪，重新启动 ASP.NET Core，请求 http://localhost:5000/admin 或 http://localhost:5000/admin/identityusers，可以测试安全策略。

由于目前未通过身份验证，但尝试一个需要授权的操作，因此浏览器就会重定向到/Account/Login URL。输入 Admin 和 Secret123$作为名称和密码，然后提交表单。Account 控制器会检查用户提供的凭证与添加到 Identity 数据库的种子数据，并假设输入了正确的消息，进行身份验证并重定向到请求的 URL(现在你可以访问该 URL)。图 11-2 说明了这个过程。

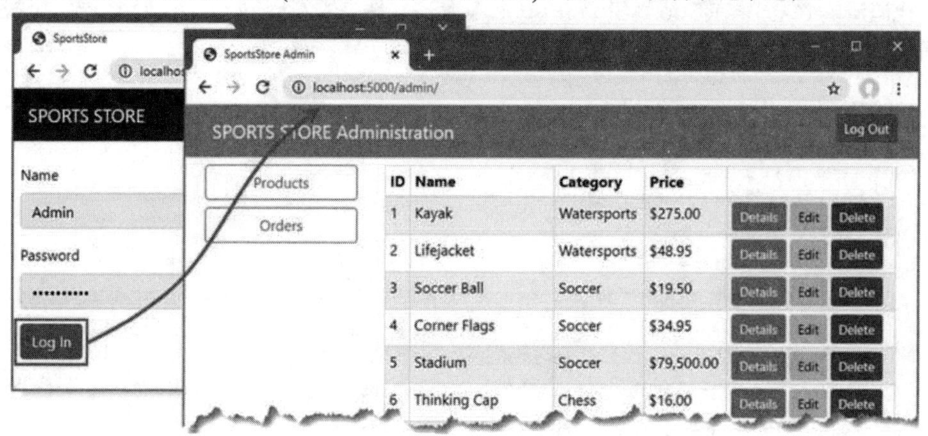


图 11-2 管理身份验证/授权过程

## 11.2 准备进行部署

本节准备 SportsStore，创建一个可以部署到生产环境中的容器。ASP.NET Core 应用程序有广泛的部署模型。但这里选择了 Docker 容器，因为它们可以在大多数托管平台上运行，或者部署到私有数据中心。这不是一个完整的部署指南，但是可以了解准备应用程序的过程。

### 11.2.1 配置错误的处理

目前，应用程序配置为使用对开发人员友好的错误页面，当出现问题时，该页面提供有用的信息。这不是终端用户应该看到的信息，因此向文件夹添加一个名为 Error.cshtml 的 Razor Pages，内容如代码清单11-16所示。

**代码清单 11-16 Pages 文件夹中 Error.cshtml 文件的内容**

```
@page "/error"
@{ Layout = null; }
<!DOCTYPE html>
<html>
<head>
    <meta name="viewport" content="width=device-width" />
    <link href="/lib/twitter-bootstrap/css/bootstrap.min.css" rel="stylesheet" />
    <title>Error</title>
```

```
</head>
<body class="text-center">
    <h2 class="text-danger">Error.</h2>
    <h3 class="text-danger">An error occurred while processing your request</h3>
</body>
</html>
```

这种类型的错误页面是最后一种手段，最好使其尽可能简单，不要依赖共享视图、视图组件或其他富特性。本例禁用了共享布局，并定义了一个简单的 HTML 文档来解释发生了错误，但没有提供任何关于发生了什么的信息。

在代码清单 11-17 中，重新配置了应用程序，以便当应用程序处于生产环境中时，错误页面用于未处理的异常。

**代码清单 11-17　在 SportsStore 文件夹的 Startup.cs 文件中配置错误处理**

```
...
public void Configure(IApplicationBuilder app, IWebHostEnvironment env) {

    if (env.IsProduction()) {
        app.UseExceptionHandler("/error");
    } else {
        app.UseDeveloperExceptionPage();
        app.UseStatusCodePages();
    }

    app.UseStaticFiles();
    app.UseSession();

    app.UseRouting();

    app.UseAuthentication();
    app.UseAuthorization();

    app.UseEndpoints(endpoints => {
        endpoints.MapControllerRoute("catpage",
            "{category}/Page{productPage:int}",
            new { Controller = "Home", action = "Index" });

        endpoints.MapControllerRoute("page", "Page{productPage:int}",
            new { Controller = "Home", action = "Index", productPage = 1 });

        endpoints.MapControllerRoute("category", "{category}",
            new { Controller = "Home", action = "Index", productPage = 1 });

        endpoints.MapControllerRoute("pagination",
            "Products/Page{productPage}",
                new { Controller = "Home", action = "Index", productPage = 1 });
```

```
            endpoints.MapDefaultControllerRoute();
            endpoints.MapRazorPages();
            endpoints.MapBlazorHub();
            endpoints.MapFallbackToPage("/admin/{*catchall}", "/Admin/Index");
        });
        SeedData.EnsurePopulated(app);
        IdentitySeedData.EnsurePopulated(app);
    }
...
```

如第 12 章所述，由 Configure 方法定义的 IWebHostEnvironment 参数描述了应用程序运行的环境。这些更改意味着在应用程序处于生产状态时调用 UseExceptionHandler 方法，但在其他情况下使用对开发人员友好的错误页面。

### 11.2.2　创建生产配置设置

可以创建用于定义设置(如连接字符串)的 JSON 配置文件，以便仅在应用程序处于特定环境(如开发、登台或生产环境)时应用它们。在第 7 章中，用来创建 SportsStore 项目的模板创建了 appsettings.json 和 appsettings.Development.json 文件，它是默认设置，被特定于开发的设置覆盖。本章将采用相反的方法，定义一个文件，其中仅包含那些特定于生产的设置。向 SportsStore 文件夹添加一个名为 appsettings.Production.json 的 JSON 文件，内容如代码清单 11-18 所示。

---

**警告：**不要在实际项目中使用这些连接字符串。必须正确地描述到生产数据库的连接，它不太可能与代码清单中的连接相同。

---

**代码清单 11-18　SportsStore 文件夹中 appsettings.Production.json 文件的内容**

```
{
    "ConnectionStrings": {
        "SportsStoreConnection": "Server=sqlserver;Database=SportsStore;
        MultipleActiveResultSets=true;User=sa;
        Password=MyDatabaseSecret123",
        "IdentityConnection": "Server=sqlserver;Database=
        Identity;MultipleActiveResultSets=true;User=sa;Password=
        MyDatabaseSecret123"
    }
}
```

这些连接字符串(每个都在单行上定义)描述到运行在 sqlserver 上的 SQL Server 的连接，sqlserver 是另一个运行 SQL Server 的 Docker 容器。

### 11.2.3　创建 Docker 映像

接下来的几节为应用程序配置和创建 Docker 映像，该应用程序可以部署到容器环境，如 Microsoft Azure 或 Amazon Web Services。请记住，容器只是一种部署样式，如果这种方法不适合，还有其他许多可用的部署样式。

■ **注意：**请记住，连接到一个运行在开发机器上的数据库，不是大多数实际应用程序的配置方式。请确保将数据库连接字符串和容器网络设置配置为与生产环境匹配。

#### 1. 安装 Docker 桌面

进入 Docker.com 下载并安装 Docker 桌面包。执行安装过程，重新启动 Windows 机器，运行代码清单 11-19 所示的命令，检查 Docker 是否已安装并位于正确的路径中(Docker 安装过程似乎经常变化，这就是为什么没有更具体地介绍这个过程)。

■ **注意：**必须在 docker.com 上创建一个账户，才能下载安装程序。

**代码清单 11-19　检查 Docker 桌面安装**

```
docker --version
```

#### 2. 创建 Docker 配置文件

Docker 是使用名为 Dockerfile 的文件配置的。该文件没有 Visual Studio 项模板，因此使用 Text File 模板给项目添加名为 Dockerfile 的文件，然后将文件重命名为 Dockerfile。如果使用的是 Visual Studio Code，可以只创建一个名为 Dockerfile 的文件，不带扩展名。使用代码清单 11-20 所示的配置设置作为新文件的内容。

**代码清单 11-20　SportsStore 文件夹中 Dockerfile 文件的内容**

```
FROM mcr.microsoft.com/dotnet/core/aspnet:3.1
FROM mcr.microsoft.com/dotnet/core/sdk:3.1

COPY /bin/Release/netcoreapp3.1/publish/ SportsStore/

ENV ASPNETCORE_ENVIRONMENT Production

EXPOSE 5000
WORKDIR /SportsStore
ENTRYPOINT ["dotnet", "SportsStore.dll", "--urls=http://0.0.0.0:5000"]
```

这些指令将 SportsStore 应用程序复制到 Docker 映像中并配置其执行。接下来，创建一个名为 docker-compose.yml 的文件，内容如代码清单 11-21 所示。Visual Studio 没有此类文件的模板，但如果选择 Text File 模板并输入完整的文件名，就会创建该文件。Visual Studio Code 用户可以简单地创建一个名为 docker-compose.yml 的文件。

**代码清单 11-21　SportsStore 文件夹中 docker-compose.yml 文件的内容**

```
version: "3"
services:
  sportsstore:
    build: .
    ports:
      - "5000:5000"
```

```
        environment:
            - ASPNETCORE_ENVIRONMENT=Production
        depends_on:
            - sqlserver
    sqlserver:
        image: "mcr.microsoft.com/mssql/server"
        environment:
            SA_PASSWORD: "MyDatabaseSecret123"
            ACCEPT_EULA: "Y"
```

YML 文件对格式和缩进特别敏感，因此按照如下所示创建该文件非常重要。如果有问题，可使用本书 GitHub 存储库中的 docker-compose.yml 文件，地址是：

```
https://github.com/apress/pro-asp.netcore3
```

### 3. 发布和映射应用程序

使用 PowerShell 提示符在 SportsStore 文件夹中运行代码清单 11-22 所示的命令，从而准备 SportsStore 应用程序。

**代码清单 11-22　准备应用程序**

```
dotnet publish -c Release
```

接下来，运行代码清单 11-23 所示的命令，为 SportsStore 应用程序创建 Docker 映像。这个命令在第一次运行时会花一些时间来完成，因为它为 ASP.NET Core 下载 Docker 图像。

**代码清单 11-23　执行 Docker Build**

```
docker-compose build
```

第一次运行该命令时，可能会提示允许 Docker 使用网络，如图 11-3 所示。

图 11-3　允许网络访问

单击 Allow access 按钮，返回 PowerShell 提示符，使用 Ctrl+C 组合键终止 Docker 容器，并再次运行代码清单 11-23 中的命令。

### 11.2.4 运行容器化应用程序

在 SportsStore 文件夹中运行代码清单 11-24 所示的命令，为 SportsStore 应用程序和 SQL Server 启动 Docker 容器。这个命令在第一次运行时需要一些时间才能完成，因为它为 SQL Server 下载 Docker 映像。

**代码清单 11-24 启动容器**

```
docker-compose up
```

两个容器启动可能需要一些时间。有很多输出，大部分来自 SQL Server，但当看到如下输出，表示应用程序已准备好：

```
...
sportsstore_1 | info: Microsoft.Hosting.Lifetime[0]
sportsstore_1 |       Now listening on: http://0.0.0.0:5000
sportsstore_1 | info: Microsoft.Hosting.Lifetime[0]
sportsstore_1 |       Application started. Press Ctrl+C to shut down.
sportsstore_1 | info: Microsoft.Hosting.Lifetime[0]
sportsstore_1 |       Hosting environment: Production
sportsstore_1 | info: Microsoft.Hosting.Lifetime[0]
sportsstore_1 |       Content root path: /SportsStore
...
```

打开一个新的浏览器窗口，请求 http://localhost:5000，将收到来自 SportsStore 的容器版本的响应，如图 11-4 所示，现在应用程序已经准备好部署了。在 PowerShell 命令提示符下使用 Ctrl+C 组合键终止 Docker 容器。

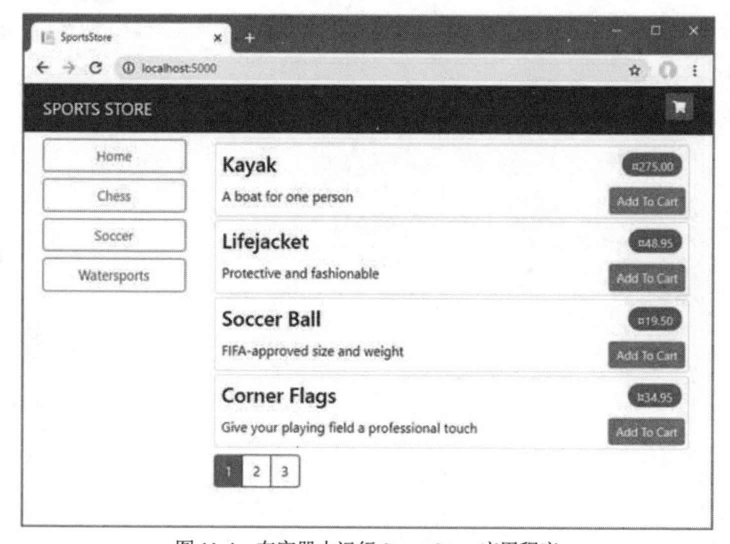


图 11-4 在容器中运行 SportsStore 应用程序

## 11.3　小结

在本章和前几章中，演示了 ASP.NET Core 如何用来创建现实的电子商务应用程序。这个扩展的示例介绍了许多关键特性：控制器、操作方法、视图、Razor Pages、Blazor、路由、验证、身份验证等。还讨论了一些关键技术是如何与 ASP.NET Core 一起使用的；其中包括 Entity Framework Core、ASP.NET Core Identity 和单元测试。这就是 SportsStore 应用程序的全部内容。本书的下一部分开始深入研究 ASP.NET Core 的细节。

# 第 II 部分

# ASP.NET Core 平台

- 第 12 章　了解 ASP.NET Core 平台
- 第 13 章　使用 URL 路由
- 第 14 章　使用依赖注入
- 第 15 章　使用平台特性(第 1 部分)
- 第 16 章　使用平台特性(第 2 部分)
- 第 17 章　处理数据

# 第 12 章

# 了解 ASP.NET Core 平台

ASP.NET Core 平台是创建 Web 应用程序的基础，提供了允许使用 MVC 和 Blazor 等框架的特性。本章解释基本 ASP. NET Core 特性的工作原理，描述文件在 ASP.NET Core 项目中的用途，解释 ASP. NET Core 请求管道如何用于处理 HTTP 请求，并演示定制它的不同方法。

本章描述的特性是 ASP. NET Core 执行所有操作的基础，了解它们是如何工作的，有助于理解日常特性，并在没有得到预期的行为时为诊断问题提供所需的知识。表 12-1 列出了 ASP.NET Core 平台的上下文。

表 12-1　ASP.NET Core 平台的上下文

| 问题 | 答案 |
| --- | --- |
| 它是什么？ | ASP.NET Core 平台是构建 Web 应用程序的基础，并提供处理 HTTP 请求的特性 |
| 它为什么有用？ | ASP.NET Core 平台处理 Web 应用程序的底层细节，以便开发人员可以专注于为最终用户提供的特性 |
| 它是如何使用的？ | 关键的构建块是服务和中间件组件，它们都可以在 Startup 类中创建 |
| 是否存在缺陷或限制？ | Startup 类的使用可能令人困惑，必须密切关注它包含的语句的顺序 |
| 还有其他选择吗？ | ASP.NET Core 平台是 ASP. NET Core 应用程序所必需的，但是可以选择不直接使用该平台，而只依赖于更高级别的 ASP.NET Core 特性，参见后续章节 |

表 12-2 总结了本章的内容。

表 12-2　本章内容摘要

| 问题 | 解决方案 | 代码清单 |
| --- | --- | --- |
| 创建中间件组件 | 调用 Use 或 UseMiddleware 方法，来向请求管道添加函数或类 | 12-4、12-6 |
| 修改响应 | 编写一个使用返回管道路径的中间件组件 | 12-7 |
| 阻止其他组件处理请求 | 使请求管道短路，或创建终端中间件 | 12-8、12-11、12-12 |
| 使用不同的中间件 | 创建管道分支 | 12-9 |
| 配置中间件组件 | 使用选项模式 | 12-13、12-16 |

## 12.1　准备工作

为了准备这一章，创建一个名为 Platform 的新项目，使用提供最小 ASP.NET Core 设置的模板。从 Windows Start 菜单中打开一个新的 PowerShell 命令提示符，并运行代码清单 12-1 所示的命令。

**提示：**可以从 https://github.com/apress/pro-asp.net-core-3 下载本章和本书中所有其他章节的示例项目。如果在运行示例时遇到问题，请参阅第 1 章以获得帮助。

**代码清单 12-1　创建项目**

```
dotnet new globaljson --sdk-version 3.1.101 --output Platform
dotnet new web --no-https --output Platform --framework netcoreapp3.1
dotnet new sln -o Platform

dotnet sln Platform add Platform
```

如果正在使用 Visual Studio，请打开 Platform 文件夹中的 Platform.sln 文件。选择 Project | Platform Properties，导航到 Debug 页面，并将 App URL 字段更改为 http://localhost:5000，如图 12-1 所示。这将更改用于接收 HTTP 请求的端口。选择 File | Save All 以保存配置变化。

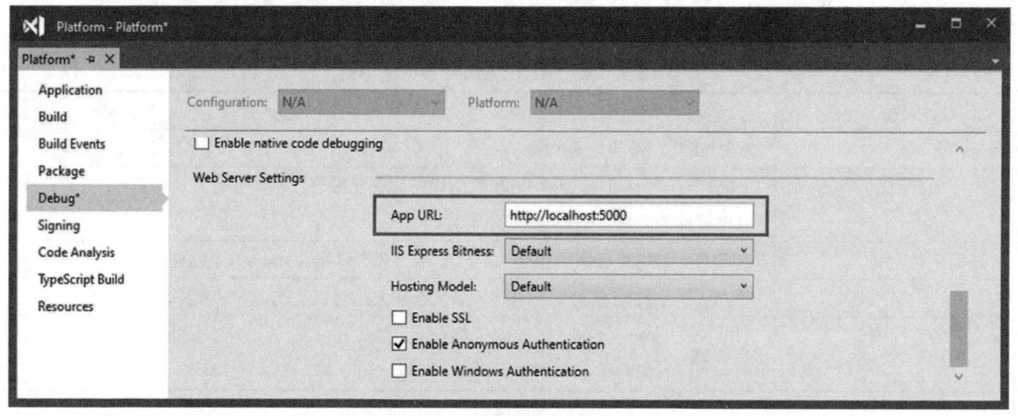


图 12-1　更改 HTTP 端口

如果使用的是 Visual Studio Code，请打开 Platform 文件夹。当提示添加构建和调试项目所需的资源时，单击 Yes 按钮，如图 12-2 所示。

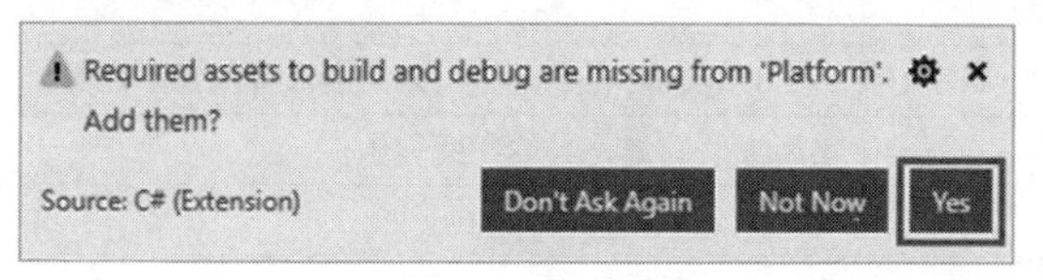


图 12-2　添加项目资源

### 运行示例应用程序

从 Debug 菜单中选择 Start Without Debugging 或Run Without Debugging，启动示例应用程序。如果正在使用 Visual Studio Code，请在提示选择环境时选择.NET Core。这是只在项目第一次启动时进行的选择。打开一个新的浏览器窗口，显示如图 12-3 所示的输出。

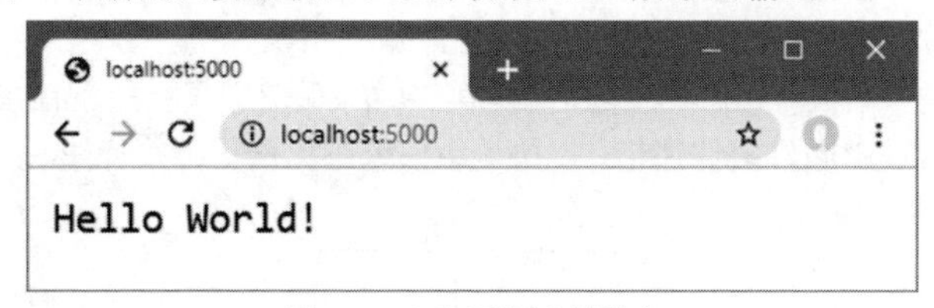


图 12-3 运行示例应用程序

还可以在 Platform 文件夹中运行代码清单 12-2 所示的命令，从命令行启动应用程序。

**代码清单 12-2 启动示例应用程序**

```
dotnet run
```

此命令不会打开新的浏览器窗口，必须手动导航到 http://localhost:5000，才能看到如图 12-3 所示的输出。

## 12.2 了解 ASP.NET Core 平台

为了解 ASP.NET Core 平台，需要关注关键特性：请求管道、中间件和服务。理解这些特性是如何组合在一起的，理解 ASP.NET Core 项目的内容和 ASP.NET Core 平台的形状的有用上下文。

### 12.2.1 理解中间件和请求管道

ASP.NET Core 的目的平台用来接收 HTTP 请求并向它们发送响应。ASP.NET Core 委托给中间件组件。中间件组件安排在一个链中，称为请求管道。

当新的 HTTP 请求到达时，ASP.NET Core 平台创建一个描述它的对象和一个对应的对象，该对象描述返回时发送的响应。这些对象传递到链中的第一个中间件组件，后者检查请求并修改响应。然后，请求传递到链中的下一个中间件组件，每个组件都检查请求并添加到响应中。一旦请求通过管道，ASP.NET Core 平台发送响应，如图 12-4 所示。

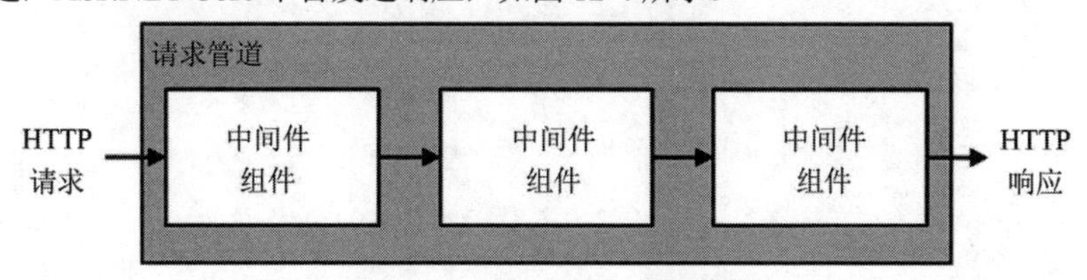


图 12-4 ASP.NETCore 请求管道

有些组件主要用于为请求生成响应，但其他组件提供支持功能，例如格式化特定数据类型或读写 cookie。ASP.NET Core 包括解决常见问题的中间件组件，如第 15 章和第 16 章所述，本章后面将展示如何创建自定义中间件组件。如果中间件组件没有生成响应，那么 ASP.NET Core 返回一个带有 HTTP 404 Not Found 状态代码的响应。

### 12.2.2 了解服务

服务是在 Web 应用程序中提供功能的对象。任何类都可以用作服务，对服务提供的特性没有任何限制。服务的特殊之处在于它们是由 ASP.NETCore 管理的，一个称为依赖注入的特性可以在应用程序的任何地方轻松访问服务，包括中间件组件。

依赖注入可能是一个很难理解的主题，详见第 14 章。现在，只要知道有些对象是由 ASP.NET Core 平台管理的，可由中间件组件共享，以协调组件或避免重复公共特性，如日志记录或加载配置数据，如图 12-5 所示。

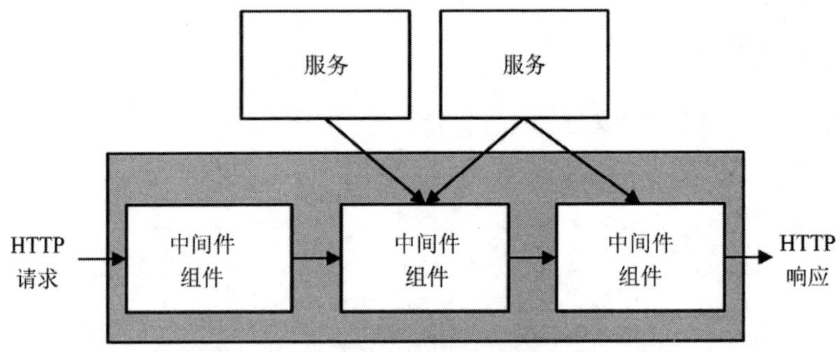


图 12-5　ASP.NET Core 平台中的服务

如图所示，中间件组件只使用它们完成工作所需的服务。如后续章节所述，ASP.NET Core 提供了一些基本服务，可以通过特定于应用程序的其他服务来补充。

## 12.3　了解 ASP.NET Core 项目

空模板生成一个项目，其中的代码和配置刚刚足够启动 ASP.NET Core 运行时，带有一些基本的服务和中间件组件。图 12-6 显示了模板添加到项目中的文件。

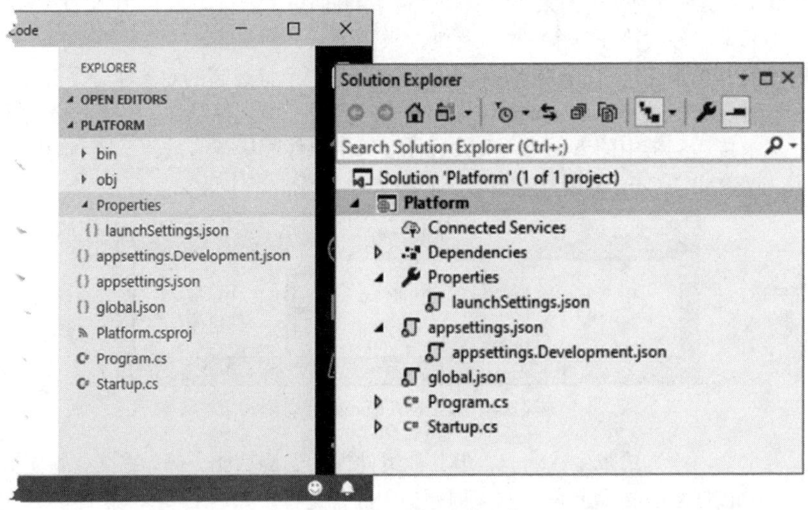


图 12-6　示例项目中的文件

Visual Studio 和 Visual Studio Code 采用不同的方法来显示文件和文件夹。Visual Studio 隐藏开发中不常用的项，并将相关项嵌套在一起，而 Visual Studio Code 则显示所有内容。

这就是图中显示的两个项目视图不同的原因：Visual Studio 隐藏了 bin 和 obj 文件夹，并在 appsettings.json 文件中嵌套了 appsettings.Development.json 文件。Solution Explorer 窗口顶部的按钮

可用于防止嵌套，显示项目中的所有文件。

尽管项目中只有很少的文件，但它们支持 ASP.NETCore 开发，参见表 12-3。

表 12-3　示例项目中的文件和文件夹

| 名称 | 描述 |
|---|---|
| appsettings.json | 这个文件用于配置应用程序，如第 15 章所述 |
| appsettings.Development.json | 这个文件用于定义特定于开发的配置设置，如第 15 章所述 |
| bin | 这个文件夹包含编译后的应用程序文件。Visual Studio 隐藏此文件夹 |
| global.json | 这个文件用于选择.NET Core SDK 的特定版本 |
| Properties/launchSettings.json | 这个文件用于在应用程序启动时配置它。Visual Studio 隐藏此文件夹和文件 |
| obj | 这个文件夹包含编译器的中间输出。Visual Studio 隐藏此文件夹 |
| Platform.csproj | 这个文件向.NETCore 工具描述了项目，包括包依赖关系和构建说明，如 12.3.3 一节所述。Visual Studio 隐藏此文件，但可在 SolutionExplorer 中右击项目项，并从弹出菜单中选择 Edit project file 来编辑该文件 |
| Program.cs | 这个文件是 ASP.NETCore 平台的入口点 |
| Startup.cs | 这个文件用于配置 ASP. NET Core 运行时及其相关框架，如 12.3.2 一节所述 |

## 12.3.1 理解入口点

.NET Core 应用程序定义了一个 main 方法，该方法在应用程序执行时调用，称为应用程序的入口点。对于 ASP.NET Core，main 方法是由 Program.cs 文件中的 Program 类定义的。下面是示例项目中 Program.cs 的内容：

```
using System;
using System.Collections.Generic;
using System.Linq;
using System.Threading.Tasks;
using Microsoft.AspNetCore.Hosting;
using Microsoft.Extensions.Configuration;
using Microsoft.Extensions.Hosting;
using Microsoft.Extensions.Logging;

namespace Platform {
    public class Program {

        public static void Main(string[] args) {
            CreateHostBuilder(args).Build().Run();
        }

        public static IHostBuilder CreateHostBuilder(string[] args) =>
            Host.CreateDefaultBuilder(args)
                .ConfigureWebHostDefaults(webBuilder => {
```

```
                webBuilder.UseStartup<Startup>();
            });
    }
}
```

.NET Core 运行时调用 Main 方法，Main 方法调用 CreateHostBuilder 方法。设置过程中的第一步是调用 Host.CreateDefaultBuilder 方法。

```
...
public static IHostBuilder CreateHostBuilder(string[] args) =>
    Host.CreateDefaultBuilder(args)
        .ConfigureWebHostDefaults(webBuilder => {
            webBuilder.UseStartup<Startup>();
    });
...
```

这个方法负责建立 ASP.NET Core 平台的基本特性，包括创建负责配置数据和日志记录的服务，这两者在第 15 章中都有描述。此方法还设置了名为 Kestrel 的 HTTP 服务器，用于接收 HTTP 请求，并添加了对使用 Internet Information Services (IIS)的支持。

来自 CreateDefaultBuilder 方法的结果传递给 ConfigureWebHostDefaults 方法，该方法选择 Startup 类作为启动过程中的下一步。

```
...
public static IHostBuilder CreateHostBuilder(string[] args) =>
    Host.CreateDefaultBuilder(args)
        .ConfigureWebHostDefaults(webBuilder => {
            webBuilder.UseStartup<Startup>();
    });
...
```

由空模板添加到 Program 类的语句适用于大多数 ASP.NET Core 项目，但是如果应用程序有特定的要求，可以更改它们。

### 12.3.2 理解 Startup 类

Startup 类使用在创建项目时添加到类中的两个方法，执行应用程序所需的大部分设置。

```
using System;
using System.Collections.Generic;
using System.Linq;
using System.Threading.Tasks;
using Microsoft.AspNetCore.Builder;
using Microsoft.AspNetCore.Hosting;
using Microsoft.AspNetCore.Http;
using Microsoft.Extensions.DependencyInjection;
using Microsoft.Extensions.Hosting;

namespace Platform {
```

```
    public class Startup {

        public void ConfigureServices(IServiceCollection services) {
        }

        public void Configure(IApplicationBuilder app, IWebHostEnvironment env) {

            if (env.IsDevelopment()) {
                app.UseDeveloperExceptionPage();
            }

            app.UseRouting();

            app.UseEndpoints(endpoints => {
                endpoints.MapGet("/", async context => {
                    await context.Response.WriteAsync("Hello World!");
                });
            });
        }
    }
}
```

ConfigureServices 方法用于定义应用程序需要的服务。默认情况下，只有由 Program 类创建的服务是可用的，但是后续章节会列举使用不同类型的服务的例子，第 14 章将详细描述使用服务的机制。

Configure 方法用于为请求管道注册中间件组件。当使用空模板创建项目时，三个中间件组件默认添加到管道中，每个中间件组件在表 12-4 中进行描述。

表 12-4 由空模板添加到管道中的中间件

| 中间件方法 | 描述 |
|---|---|
| UseDeveloperExceptionPage | 该方法添加了一个中间件组件，其中包括未处理异常的详细信息。IsDevelopment 方法用来确保这些信息不会呈现给用户，如第 16 章所述 |
| UseRouting | 此方法将端点路由中间件组件添加到管道中，该组件负责确定某些请求是如何处理的，并与 ASP.NET Core 的其他部分(如 MVC 框架)一起使用。路由详见第 13 章 |
| UseEndpoints | 此方法为 UseRouting 方法添加的端点路由中间件提供配置。配置参见第 13 章 |

微软还提供了其他中间件作为 ASP NET Core 的一部分，处理 Web 应用程序最常见的特性，参见第 15 章和第 16 章。当内置特性不符合需求时，还可以创建自己的中间件，如 12.4 节所述。

### 12.3.3 理解项目文件

Platform.csproj 文件也称为项目文件，包含.NET Core 用于构建项目和跟踪依赖关系的信息。下面是创建项目时由空模板添加到文件中的内容：

```
<Project Sdk="Microsoft.NET.Sdk.Web">
```

```
    <PropertyGroup>
        <TargetFramework>netcoreapp3.1</TargetFramework>
    </PropertyGroup>

</Project>
```

csproj 文件在使用 Visual Studio 时是隐藏的；在 Solution Explorer 中右击 Platform 项目项，并从弹出菜单中选择 Edit Project File，就可以编辑它。

项目文件包含 XML 元素，这些元素向 Microsoft 构建引擎 MSBuild 描述项目。MSBuild 可用于创建复杂的构建过程，详见 https://docs.microsoft.com/en-gb/visualstudio/msbuild/MSBuild。

在大多数项目中不需要直接编辑项目文件。对该文件最常见的更改是添加对其他.NET 包的依赖项，但这些依赖项通常是使用命令行工具或通过 Visual Studio 提供的界面添加的。

要使用命令行向项目添加包，打开一个新的 PowerShell 命令提示符，导航到 Platform 项目文件夹(包含 csproj 文件的文件夹)，并运行代码清单 12-3 所示的命令。

**代码清单 12-3　向项目添加包**

```
dotnet add package Swashbuckle.AspNetCore --version 5.0.0-rc2
```

此命令向项目添加 Swashbuckle.AspNetCore 包。第 20 章介绍如何使用这个包，但是现在，dotnet add package 命令的效果是重要的。

如果使用的是Visual Studio，为了添加包，可以在Solution Explorer中右击Platform项，并从弹出菜单中选择Manage NuGet packages。单击Browse并在搜索文本框中输入Swashbuckle.AspNetCore。从列表中选择Swashbuckle.AspNetCore包，从Version下拉列表中选择5.0.0-rc2版本，然后单击Install按钮，如图12-7所示。系统将提示接受包及其依赖项的许可。

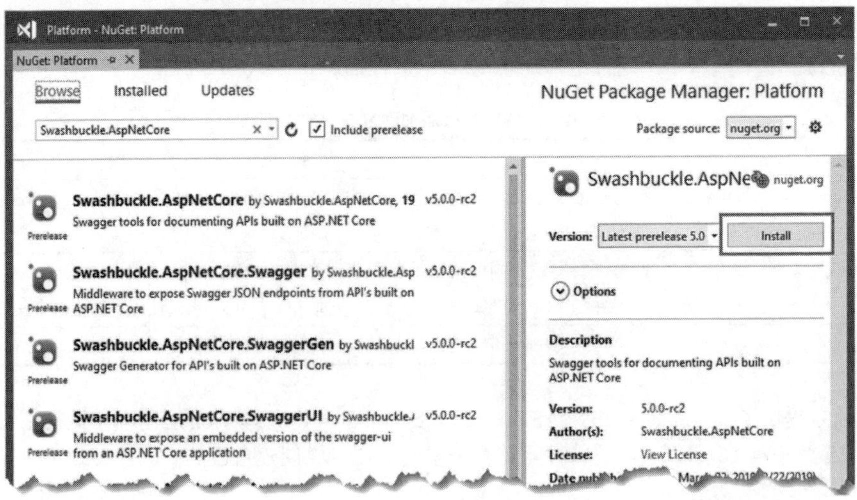

图 12-7　在 Visual Studio 中安装一个包

无论使用哪种方法来安装包，新的依赖关系显示在 Platform. csproj 文件中。

```
<Project Sdk="Microsoft.NET.Sdk.Web">
```

```
    <PropertyGroup>
        <TargetFramework>netcoreapp3.1</TargetFramework>
    </PropertyGroup>

    <ItemGroup>
        <PackageReference Include="Swashbuckle.AspNetCore" Version="5.0.0-rc2" />
    </ItemGroup>

</Project>
```

## 12.4 创建自定义中间件

如前所述，微软为 ASP.NET Core 提供了各种中间件组件。处理 Web 应用程序最常用的特性。还可以创建自己的中间件(即使在项目中只使用标准组件)，这是了解 ASP.NET Core 如何使用的一种有用方法。创建中间件的关键方法是 Use，如代码清单 12-4 所示。

**代码清单 12-4 在 Platform 文件夹的 Startup.cs 文件中创建自定义中间件**

```
using System;
using System.Collections.Generic;
using System.Linq;
using System.Threading.Tasks;
using Microsoft.AspNetCore.Builder;
using Microsoft.AspNetCore.Hosting;
using Microsoft.AspNetCore.Http;
using Microsoft.Extensions.DependencyInjection;
using Microsoft.Extensions.Hosting;

namespace Platform {
    public class Startup {

        public void ConfigureServices(IServiceCollection services) {
        }

        public void Configure(IApplicationBuilder app, IWebHostEnvironment env) {

            if (env.IsDevelopment()) {
                app.UseDeveloperExceptionPage();
            }

            app.Use(async (context, next) => {
                if (context.Request.Method == HttpMethods.Get
                        && context.Request.Query["custom"] == "true") {
                    await context.Response.WriteAsync("Custom Middleware \n");
                }
                await next();
            });
```

```
            app.UseRouting();

            app.UseEndpoints(endpoints => {
                endpoints.MapGet("/", async context => {
                    await context.Response.WriteAsync("Hello World!");
                });
            });
        }
    }
}
```

Use 方法注册一个中间件组件，该组件通常表示为一个 lambda 函数，在请求通过管道时接收请求(有另一种用于类的方法，如下一节所述)。

lambda 函数的参数是一个 HttpContext 对象和一个函数，调用它是为了告诉 ASP.NET Core，将请求传递给管道中的下一个中间件组件。

HttpContext 对象描述 HTTP 请求和 HTTP 响应，并提供附加上下文，包括与请求关联的用户的详细信息。表 12-5 描述了由 HttpContext 类提供的最有用成员，它们是在 Microsoft.AspNetCore.Http 名称空间中定义的。

表 12-5　有用的 HttpContext 成员

| 名称 | 描述 |
|---|---|
| Connection | 此属性返回一个 ConnectionInfo 对象，该对象提供有关 HTTP 请求下的网络连接的信息，包括本地和远程 IP 地址和端口的详细信息 |
| Request | 此属性返回一个 HttpRequest 对象，该对象描述正在处理的 HTTP 请求 |
| RequestServices | 此属性提供对请求可用的服务的访问，如第 14 章所述 |
| Response | 此属性返回一个 HttpResponse 对象，该对象用于创建对 HTTP 请求的响应 |
| Session | 此属性返回与请求关联的会话数据。第 16 章描述了会话数据特性 |
| User | 这个属性返回与请求关联的用户的详细信息，如第 37 和 38 章所述 |
| Features | 此属性提供对请求特性的访问，请求特性允许访问请求处理的低级方面。有关使用请求特性的示例，请参阅第 16 章 |

ASP.NET Core 平台负责处理 HTTP 请求，来创建 HttpRequest 对象，这意味着中间件和端点不必担心原始请求数据。表 12-6 描述了 HttpRequest 类中最有用的成员。

表 12-6　有用的 HttpRequest 成员

| 名称 | 描述 |
|---|---|
| Body | 此属性返回可用于读取请求主体的流 |
| ContentLength | 此属性返回 Content-Length 标题的值 |
| ContentType | 此属性返回 Content-Type 标题的值 |
| Cookies | 此属性返回请求的 cookie |
| Form | 此属性将请求主体的表示形式返回为表单 |
| Headers | 此属性返回请求标头 |

(续表)

| 名称 | 描述 |
| --- | --- |
| IsHttps | 如果使用 HTTPS 发出请求，则此属性返回 true |
| Method | 此属性返回用于请求的 HTTP 谓词 |
| Path | 此属性返回请求 URL 的路径部分 |
| Query | 此属性以键/值对的形式返回请求 URL 的查询字符串部分 |

HttpResponse 对象描述了当请求通过管道时发送回客户端的 HTTP 响应。表 12-7 描述了 HttpResponse 类中最有用的成员。ASP.NET Core 平台使响应的处理尽可能简单，自动设置报头，并便于向客户端发送内容。

表 12-7 有用的 HttpResponse 成员

| 名称 | 描述 |
| --- | --- |
| ContentLength | 这个属性设置了 Content-Length 标题的值 |
| ContentType | 此属性设置 Content-Type 标题的值 |
| Cookies | 此属性允许将 cookie 与请求关联 |
| HasStarted | 如果 ASP.NET Core 开始向客户端发送响应头，之后就不可能进行更改了，这个属性返回 true |
| Headers | 此属性允许设置响应标头 |
| StatusCode | 此属性设置响应的状态代码 |
| WriteAsync(data) | 这个异步方法向响应体写入一个数据字符串 |
| Redirect(url) | 此方法发送重定向响应 |

当创建自定义中间件时，HttpContext、HttpRequest 和 HttpResponse 对象是直接使用的，但如后面的章节所述，在使用高级 ASP.NET Core 特性，如 MVC 框架和 Razor Pages 时，这通常是不需要的。

在代码清单 12-4 中定义的中间件函数使用 HttpRequest 对象检查 HTTP 方法和查询字符串，以识别在查询字符串中有一个值为 true 的自定义参数的 GET 请求，如下所示：

```
...
if (context.Request.Method == HttpMethods.Get
    && context.Request.Query["custom"] == "true") {
...
```

HttpMethods 类为每个 HTTP 方法定义静态字符串。对于具有预期查询字符串的 GET 请求，中间件函数使用 WriteAsync 方法将字符串添加到响应体中。

```
...
await context.Response.WriteAsync("Custom Middleware \n");
...
```

▒ 注意：在本书的这一部分中，所有的例子都将简单的字符串结果发送到浏览器。第 III 部分将介绍如何创建返回 JSON 数据的 Web 服务，并介绍 ASP.NET Core 生成 HTML 结果的不同方式。

中间件的第二个参数是按惯例命名为 next 的函数，它告诉 ASP.NET Core，将请求传递给请求管道中的下一个组件。

```
...
if (context.Request.Method == HttpMethods.Get
    && context.Request.Query["custom"] == "true") {
    await context.Response.WriteAsync("Custom Middleware \n");
}
await next();
...
```

调用下一个中间件组件时不需要任何参数，因为 ASP.NET Core 负责向组件提供 HttpContext 对象和它自己的下一个函数，以便它能够处理请求。下一个函数是异步的，这就是为什么使用 await 关键字以及使用 async 关键字定义 lambda 函数的原因。

**提示：**可能会遇到调用 next. invoke()而不是 next()的中间件。它们是等价的，而 next()是由编译器提供的，以便生成简洁的代码。

重启 ASP.NET Core，导航到 http://localhost:5000/?custom=true。在将请求传递到下一个中间件组件之前，新的中间件函数将其消息附加到响应体，如图 12-8 所示。删除查询字符串，或者将 true 改为 false，中间件组件将传递请求，而不添加响应。

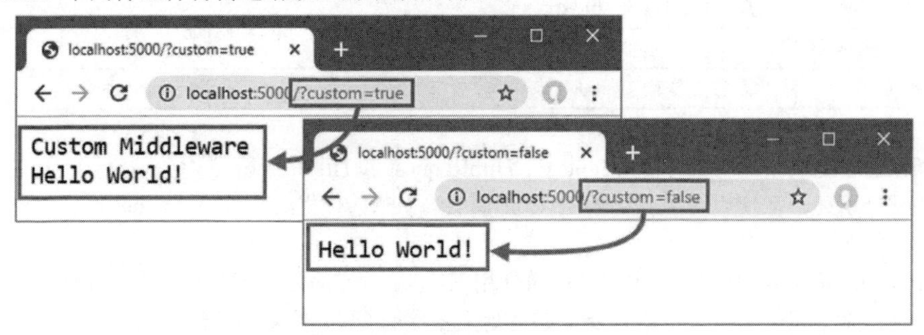


图 12-8　创建自定义中间件

## 12.4.1　使用类定义中间件

使用 lambda 函数定义中间件很方便，但是它会导致 Startup 类中的 Configure 方法很长很复杂，并且很难在不同项目中重用中间件。也可以使用类来定义中间件。将一个名为 Middleware.cs 的类文件添加到 Platform 文件夹中，并添加代码清单 12-5 所示的代码。

**代码清单 12-5　Platform 文件夹中 Middleware.cs 文件的内容**

```
using Microsoft.AspNetCore.Http;
using System.Threading.Tasks;

namespace Platform {

    public class QueryStringMiddleWare {
        private RequestDelegate next;
```

```
        public QueryStringMiddleWare(RequestDelegate nextDelegate) {
            next = nextDelegate;
        }

        public async Task Invoke(HttpContext context) {
            if (context.Request.Method == HttpMethods.Get
                    && context.Request.Query["custom"] == "true") {
                await context.Response.WriteAsync("Class-based Middleware \n");
            }
            await next(context);
        }
    }
}
```

中间件类接收 RequestDelegate 作为构造函数参数，用于将请求转发到管道中的下一个组件。接收到请求时，由 ASP.NET Core 调用 Invoke 方法。接收到的 HttpContext 对象使用 lambda 函数中间件接收到的相同类，提供对请求和响应的访问。RequestDelegate 返回一个任务，该任务允许它异步工作。

基于类的中间件的一个重要区别是，当调用 RequestDelegate 转发请求时，HttpContext 对象必须用作参数，如下所示：

```
...
await next(context);
...
```

使用 UseMiddleware 方法将基于类的中间件组件添加到管道中，该方法接收中间件作为类型参数，如代码清单 12-6 所示。

**代码清单 12-6　在 Platform 文件夹的 Startup.cs 文件中添加一个基于类的中间件组件**

```
using System;
using System.Collections.Generic;
using System.Linq;
using System.Threading.Tasks;
using Microsoft.AspNetCore.Builder;
using Microsoft.AspNetCore.Hosting;
using Microsoft.AspNetCore.Http;
using Microsoft.Extensions.DependencyInjection;
using Microsoft.Extensions.Hosting;

namespace Platform {
    public class Startup {

        public void ConfigureServices(IServiceCollection services) {
        }

        public void Configure(IApplicationBuilder app, IWebHostEnvironment env) {
```

```
            if (env.IsDevelopment()) {
                app.UseDeveloperExceptionPage();
            }

            app.Use(async (context, next) => {
                if (context.Request.Method == HttpMethods.Get
                        && context.Request.Query["custom"] == "true") {
                    await context.Response.WriteAsync("Custom Middleware \n");
                }
                await next();
            });

            app.UseMiddleware<QueryStringMiddleWare>();

            app.UseRouting();

            app.UseEndpoints(endpoints => {
                endpoints.MapGet("/", async context => {
                    await context.Response.WriteAsync("Hello World!");
                });
            });
        }
    }
}
```

启动 ASP.NET Core，实例化 QueryStringMiddleware 类，并调用其 Invoke 方法来处理收到的请求。

---

**警告：**单个中间件对象用于处理所有请求，这意味着 Invoke 方法中的代码必须是线程安全的。

---

从 Debug 菜单中选择 Start Without Debugging 或使用 dotnet run 命令重新启动 ASP.NET Core。导航到 http://localhost: 5000?custom=true，将看到来自两个中间件组件的输出，如图 12-9 所示。

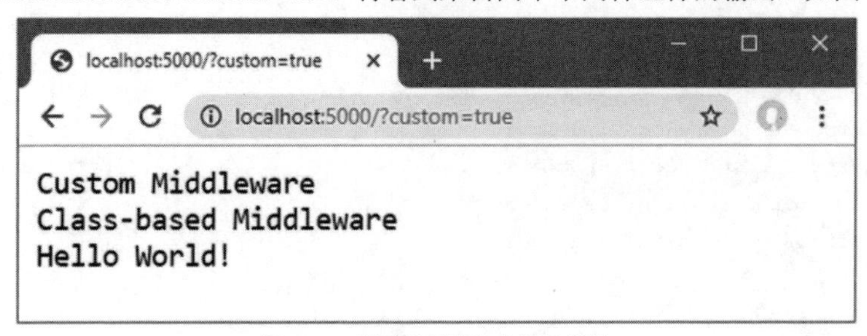


图 12-9　使用基于类的中间件组件

### 12.4.2 理解返回管道路径

中间件组件可在调用下一个函数后修改 HTTPResponse 对象，如代码清单 12-7 中的新中间件所示。

**代码清单 12-7 在 Platform 文件夹的 Startup.cs 文件中添加新的中间件**

```
using System;
using System.Collections.Generic;
using System.Linq;
using System.Threading.Tasks;
using Microsoft.AspNetCore.Builder;
using Microsoft.AspNetCore.Hosting;
using Microsoft.AspNetCore.Http;
using Microsoft.Extensions.DependencyInjection;
using Microsoft.Extensions.Hosting;

namespace Platform {
    public class Startup {

        public void ConfigureServices(IServiceCollection services) {
        }

        public void Configure(IApplicationBuilder app, IWebHostEnvironment env) {

            if (env.IsDevelopment()) {
                app.UseDeveloperExceptionPage();
            }

            app.Use(async (context, next) => {
                await next();
                await context.Response
                    .WriteAsync($"\nStatus Code: { context.Response.StatusCode}");
            });

            app.Use(async (context, next) => {
                if (context.Request.Method == HttpMethods.Get
                        && context.Request.Query["custom"] == "true") {
                    await context.Response.WriteAsync("Custom Middleware \n");
                }
                await next();
            });

            app.UseMiddleware<QueryStringMiddleWare>();

            app.UseRouting();
```

```
            app.UseEndpoints(endpoints => {
                endpoints.MapGet("/", async context => {
                    await context.Response.WriteAsync("Hello World!");
                });
            });
        }
    }
}
```

新的中间件立即调用下一个方法来沿着管道传递请求，然后使用 WriteAsync 方法向响应体添加一个字符串。这似乎是一种奇怪的方法，但是它允许中间件在响应沿着请求管道传递之前和之后对响应进行更改，方法是在调用下一个函数之前和之后定义语句，如图 12-10 所示。

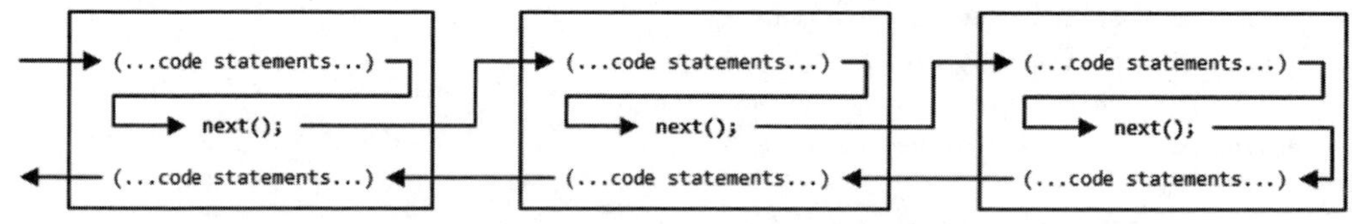


图 12-10　通过 ASP.NET Core 管道传递请求和响应

中间件可以在请求被传递之前进行操作，也可以在请求被其他组件处理之后进行操作，或者同时进行操作。结果是，多个中间件组件共同参与响应的生成，每个组件提供响应的某些方面，或者提供稍后在管道中使用的某些特性或数据。

从 Debug 菜单中选择 Start Without Debugging 或使用 dotnet 查看代码清单 12-7 中定义的中间件的效果，如图 12-11 所示。如果使用命令行，可使用 dotnet run 命令启动 ASP.NET Core，并导航到 http://localhost:5000。

图 12-11　修改返回路径中的响应

---

**注意：** *在 ASP.NET Core 开始向客户端发送响应后，中间件组件不能更改 HttpResponse 对象。可以检查表 12-7 中描述的 HasStarted 属性，以避免异常。*

---

## 12.4.3　请求管道短路

生成完整响应的组件可以选择不调用下一个函数，以便不传递请求。不传递请求的组件会导致管道短路，这就是代码清单 12-8 中显示的新中间件组件对以/short URL 为目标的请求所做的操作。

**代码清单 12-8　在 Platform 文件夹的 Startup.cs 文件中对请求管道执行短路操作**

```
using System;
using System.Collections.Generic;
using System.Linq;
using System.Threading.Tasks;
using Microsoft.AspNetCore.Builder;
using Microsoft.AspNetCore.Hosting;
using Microsoft.AspNetCore.Http;
using Microsoft.Extensions.DependencyInjection;
using Microsoft.Extensions.Hosting;

namespace Platform {
    public class Startup {

        public void ConfigureServices(IServiceCollection services) {
        }

        public void Configure(IApplicationBuilder app, IWebHostEnvironment env) {

        if (env.IsDevelopment()) {
            app.UseDeveloperExceptionPage();
        }

        app.Use(async (context, next) => {
            await next();
            await context.Response
                .WriteAsync($"\nStatus Code: { context.Response.StatusCode}");
        });

        app.Use(async (context, next) => {
            if (context.Request.Path == "/short") {
                await context.Response
                    .WriteAsync($"Request Short Circuited");
            } else {
                await next();
            }
        });

        app.Use(async (context, next) => {
            if (context.Request.Method == HttpMethods.Get
                    && context.Request.Query["custom"] == "true") {
                await context.Response.WriteAsync("Custom Middleware \n");
            }
            await next();
        });
```

```
            app.UseMiddleware<QueryStringMiddleWare>();

            app.UseRouting();

            app.UseEndpoints(endpoints => {
                endpoints.MapGet("/", async context => {
                    await context.Response.WriteAsync("Hello World!");
                });
            });
        }
    }
}
```

新的中间件检查 HttpRequest 对象的 Path 属性，以查看请求是否为/short URL；如果是，则调用 WriteAsync 方法而不调用下一个函数。要查看效果，请重新启动 ASP.NET Core，导航到 http://localhost:5000/short?，生成如图 12-12 所示的输出。

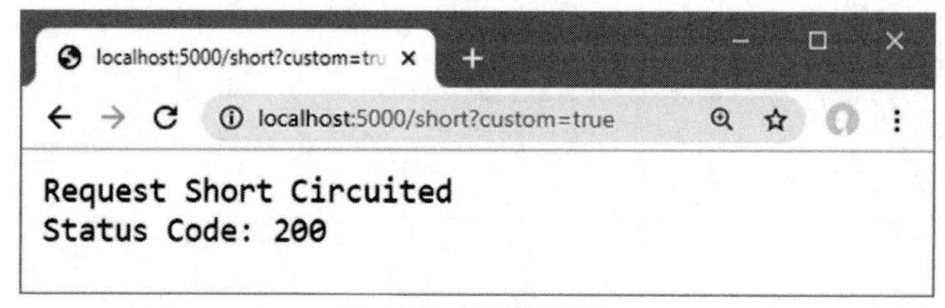


图 12-12　短路请求管道

即使 URL 具有管道中的下一个组件所期望的查询字符串参数，请求也不会转发，因此不会使用中间件。但请注意，管道中的前一个组件已经在响应中添加了消息。这是因为短路只会阻止使用管道上后期的组件，而不会影响早期的组件，如图 12-13 所示。

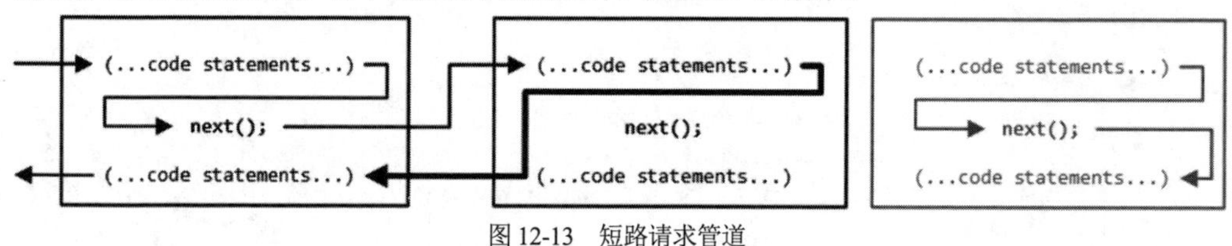


图 12-13　短路请求管道

## 12.4.4　创建管道分支

Map 方法用于创建处理特定 URL 请求的管道部分，创建一个单独的中间件组件序列，如代码清单 12-9 所示。

### 代码清单 12-9　创建管道分支

```
using System;
using System.Collections.Generic;
using System.Linq;
using System.Threading.Tasks;
using Microsoft.AspNetCore.Builder;
```

```
using Microsoft.AspNetCore.Hosting;
using Microsoft.AspNetCore.Http;
using Microsoft.Extensions.DependencyInjection;
using Microsoft.Extensions.Hosting;

namespace Platform {
   public class Startup {

      public void ConfigureServices(IServiceCollection services) {
      }

      public void Configure(IApplicationBuilder app, IWebHostEnvironment env) {

      if (env.IsDevelopment()) {
         app.UseDeveloperExceptionPage();
      }
         app.Map("/branch", branch => {

            branch.UseMiddleware<QueryStringMiddleWare>();

            branch.Use(async (context, next) => {
               await context.Response.WriteAsync($"Branch Middleware");
            });
         });

         app.UseMiddleware<QueryStringMiddleWare>();

         app.UseRouting();

         app.UseEndpoints(endpoints => {
            endpoints.MapGet("/", async context => {
               await context.Response.WriteAsync("Hello World!");
            });
         });
      }
   }
}
```

Map 方法的第一个参数指定用于匹配 URL 的字符串。第二个参数是管道的分支，中间件组件通过 Use 和 UseMiddleware 方法添加到该分支。代码清单 12-9 中的语句创建了一个分支，该分支用于以/branch 开头的 URL，并通过代码清单 12-9 中定义的 QueryStringMiddleWare 类传递请求，该语句定义了一个中间件 lambda 表达式，该表达式将消息添加到响应中。图 12-14 显示了分支对请求管道的影响。

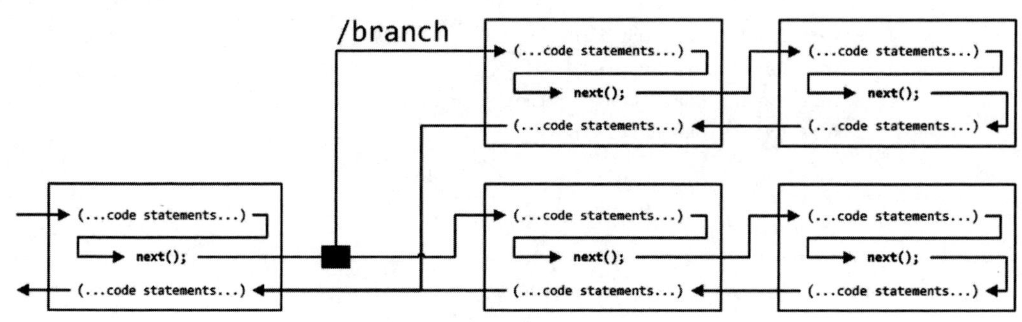


图 12-14　向请求管道添加分支

当 Map 方法匹配 URL 时，它将跟随分支，不再通过管道在主路径上的中间件组件。相同的中间件可用于管道的不同部分，如代码清单 12-9 所示，其中 QueryStringMiddleWare 类用于管道的主要部分和分支。

要查看处理请求的不同方式，请重新启动 ASP.NET Core，请求 http://localhost:5000?custom=true URL，它在管道的主要部分处理，并将生成图 12-15 左侧所示的输出。导航到 http://localhost:5000/branch?custom=true，请求转发到分支中的中间件，生成图 12-15 所示的输出。

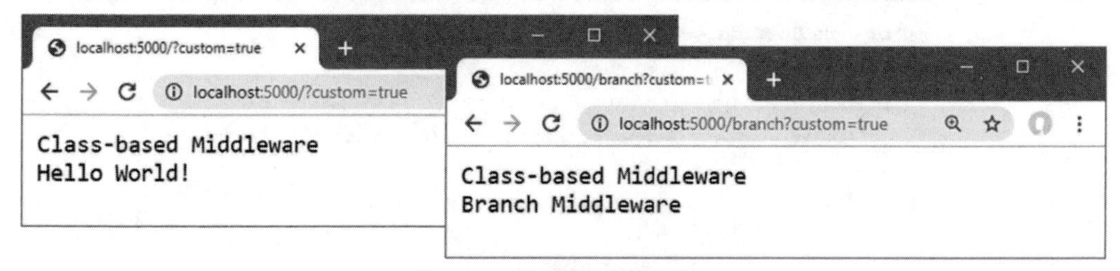


图 12-15　请求管道分支的效果

**带谓词的分支**

ASP.NET Core 还支持 MapWhen 方法，该方法可用于使用谓词匹配请求，允许根据标准为管道分支选择请求，而不仅仅是 URL。

MapWhen 方法的参数是一个谓词函数，该函数接收 HttpContext，并对跟随该分支的请求返回 true，另一个参数是一个函数，该函数接收表示管道分支的 IApplicationBuilder 对象，中间件添加到该分支。下面是一个使用 MapWhen 方法分支管道的例子:

```
...
app.MapWhen(context => context.Request.Query.Keys.Contains("branch"),
    branch => {
        // ...add middleware components here...
});
...
```

对于查询字符串中包含名为 branch 的参数的请求，谓词函数将 true 返回给分支。

## 12.4.5　创建终端中间件

终端中间件从不将请求转发给其他组件，并始终标志着请求管道的结束。Startup 类中有一个终端中间件组件，如下所示：

```
...
branch.Use(async (context, next) => {
    await context.Response.WriteAsync($"Branch Middleware");
});
...
```

ASP.NET Core 支持 Run 方法作为创建终端中间件的便利特性，该中间件组件显然不会转发请求，做出了不调用下一个函数的慎重决定。在代码清单 12-10 中，为管道分支中的终端中间件使用了 Run 方法。

**代码清单 12-10　在 Platform 文件夹的 Startup.cs 文件中使用 Run 方法**

```
...
app.Map("/branch", branch => {

    branch.UseMiddleware<QueryStringMiddleWare>();

    branch.Run(async (context) => {
        await context.Response.WriteAsync($"Branch Middleware");
    });
});
...
```

传递给 Run 方法的中间件函数只接收 HttpContext 对象，而不必定义未使用的参数。在后台，Run 方法是通过 Use 方法实现的，提供这个特性只是为了方便。

---

▒ **警告：** 中间件添加到管道后，终端组件将永远不会收到请求。如果在管道结束前添加终端组件，ASP.NET Core 不会发出警告。

---

可以编写基于类的组件，以便它们用作常规和终端中间件，如代码清单 12-11 所示。

**代码清单 12-11　在 Platform 文件夹的 Middleware.cs 文件中添加终端支持**

```
using Microsoft.AspNetCore.Http;
using System.Threading.Tasks;

namespace Platform {

    public class QueryStringMiddleWare {
        private RequestDelegate next;

        public QueryStringMiddleWare() {
            // do nothing
        }

        public QueryStringMiddleWare(RequestDelegate nextDelegate) {
            next = nextDelegate;
```

```
        }

        public async Task Invoke(HttpContext context) {
            if (context.Request.Method == HttpMethods.Get
                    && context.Request.Query["custom"] == "true") {
                await context.Response.WriteAsync("Class-based Middleware \n");
            }
            if (next != null) {
                await next(context);
            }
        }
    }
}
```

当为 nextDelegate 参数提供一个非空值时，组件才会转发请求。代码清单 12-12 以标准和终端形式显示了该组件的应用程序。

#### 代码清单 12-12　在 Platform 文件夹的 Startup.cs 文件中应用基于类的中间件

```
using System;
using System.Collections.Generic;
using System.Linq;
using System.Threading.Tasks;
using Microsoft.AspNetCore.Builder;
using Microsoft.AspNetCore.Hosting;
using Microsoft.AspNetCore.Http;
using Microsoft.Extensions.DependencyInjection;
using Microsoft.Extensions.Hosting;

namespace Platform {
    public class Startup {

        public void ConfigureServices(IServiceCollection services) {
        }

        public void Configure(IApplicationBuilder app, IWebHostEnvironment env) {

            if (env.IsDevelopment()) {
                app.UseDeveloperExceptionPage();
            }

            app.Map("/branch", branch => {
                branch.Run(new QueryStringMiddleWare().Invoke);
            });

            app.UseMiddleware<QueryStringMiddleWare>();

            app.UseRouting();
```

```
                app.UseEndpoints(endpoints => {
                endpoints.MapGet("/", async context => {
                    await context.Response.WriteAsync("Hello World!");
                });
            });
        }
    }
}
```

终端中间件没有与 UseMiddleware 等效的方法，因此必须通过创建中间件类的新实例并选择其 Invoke 方法来使用 Run 方法。使用 Run 方法不会改变中间件的输出，为了查看其效果，可以重新启动 ASP.NET Core，导航到 http://localhost:5000/branch?，生成如图 12-16 所示的内容。

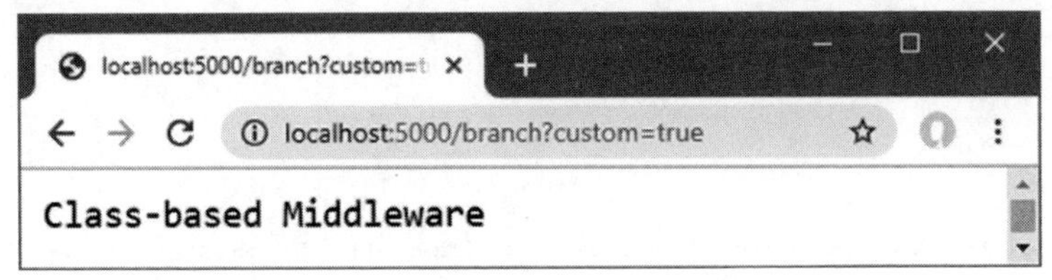


图 12-16　使用 Run 方法创建终端中间件

## 12.5　配置中间件

有一种用于配置中间件的通用模式，称为 options 模式，后续章节介绍的一些内置中间件组件使用该模式。

首先定义一个包含中间件组件配置选项的类。使用代码清单 12-13 所示的代码将名为 MessageOptions.cs 的类文件添加到 Platform 文件夹中。

#### 代码清单 12-13　Platform 文件夹中 MessageOptions.cs 文件的内容

```
namespace Platform {

    public class MessageOptions {

        public string CityName { get; set; } = "New York";
        public string CountryName{ get; set; } = "USA";
    }
}
```

MessageOptions 类定义了详细说明城市和国家/地区的属性。在代码清单 12-14 中，使用 Options 模式创建了一个自定义中间件组件，该组件的配置依赖于 MessageOptions 类。为简洁起见，还从前面的示例中删除了一些中间件。

#### 代码清单 12-14　在 Platform 文件夹的 Startup.cs 文件中使用 options 模式

```
using System;
using System.Collections.Generic;
```

```
using System.Linq;
using System.Threading.Tasks;
using Microsoft.AspNetCore.Builder;
using Microsoft.AspNetCore.Hosting;
using Microsoft.AspNetCore.Http;
using Microsoft.Extensions.DependencyInjection;
using Microsoft.Extensions.Hosting;
using Microsoft.Extensions.Options;

namespace Platform {
    public class Startup {

        public void ConfigureServices(IServiceCollection services) {
            services.Configure<MessageOptions>(options => {
                options.CityName= "Albany";
            });
        }

        public void Configure(IApplicationBuilder app, IWebHostEnvironment env,
                IOptions<MessageOptions> msgOptions) {

            if (env.IsDevelopment()) {
                app.UseDeveloperExceptionPage();
            }

            app.Use(async (context, next) => {
                if (context.Request.Path == "/location") {
                    MessageOptions opts = msgOptions.Value;
                    await context.Response
                        .WriteAsync($"{opts.CityName}, {opts.CountryName}");
                } else {
                    await next();
                }
            });

            app.UseRouting();

            app.UseEndpoints(endpoints => {
                endpoints.MapGet("/", async context => {
                    await context.Response.WriteAsync("Hello World!");
                });
            });
        }
    }
}
```

选项是在 ConfigureServices 方法中使用 IServiceCollection.Configure 设置的。泛型类型参数用

于指定 options 类，如下所示：

```
...
services.Configure<MessageOptions>(options => {
    options.CityName= "Albany";
});
...
```

该语句使用 MessageOptions 类创建选项，并更改 CityName 属性的值。当应用程序启动时，ASP.NET Core 平台将创建 options 类的一个新实例，并将其传递给作为 Configure 方法的参数提供的函数，从而允许更改默认选项值。

可以给 Startup.Configure 方法添加 IOptions<T>参数来访问这些选项。其中泛型类型参数指定 options 类，如下所示：

```
...
public void Configure(IApplicationBuilder app,IWebHostEnvironment env,
    CounterService counter, IOptions<MessageOptions> msgOptions) {
...
```

IOptions<T>接口定义了一个 Value 属性，该属性返回由 ASP.NET Core 平台创建的 options 对象，允许中间件组件和端点使用选项，如下所示：

```
...
app.Use(async (context, next) => {
    if (context.Request.Path == "/location") {
        MessageOptions opts = msgOptions.Value;
        await context.Response.WriteAsync($"{opts.CityName}, {opts.CountryName}");
    } else {
        await next();
    }
});
...
```

要查看结果，可以重新启动 ASP.NET Core，使用浏览器导航到 http://localhost:5000/location。中间件组件使用选项来生成如图 12-17 所示的输出。

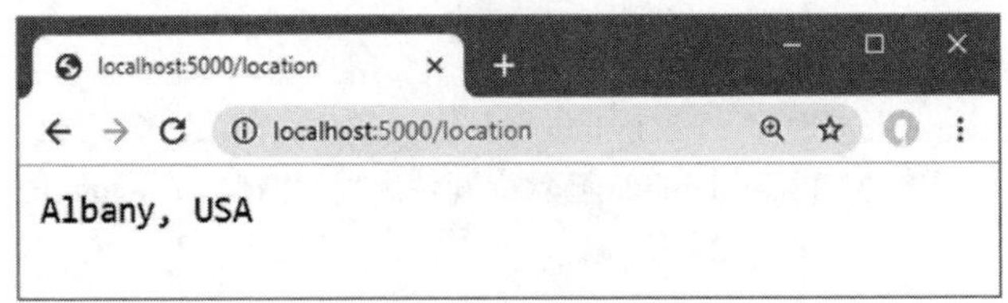


图 12-17　使用 options 模式

### 在基于类的中间件中使用 options 模式

虽然 options 模式可用于 lambda 函数中间件，但通常它会应用于基于类的中间件，如第 15 章和第 16 章中描述的内置特性。当与基于类的中间件一起使用时，不必通过 Startup.Configure 方法参数来访问配置选项，从而产生更优雅的结果。

将代码清单 12-15 中所示的语句添加到 Middleware.cs 文件中，以定义使用 MessageOptions 类进行配置的基于类的中间件组件。

**代码清单 12-15　在 Platform 文件夹的 Middleware.cs 文件中定义中间件组件**

```
using Microsoft.AspNetCore.Http;
using System.Threading.Tasks;
using Microsoft.Extensions.Options;

namespace Platform {

    public class QueryStringMiddleWare {
        private RequestDelegate? next;

        // ...statements omitted for brevity...
    }

    public class LocationMiddleware {
        private RequestDelegate next;
        private MessageOptions options;

        public LocationMiddleware(RequestDelegate nextDelegate,
                IOptions<MessageOptions> opts) {
            next = nextDelegate;
            options = opts.Value;
        }

        public async Task Invoke(HttpContext context) {
            if (context.Request.Path == "/location") {
                await context.Response
                    .WriteAsync($"{options.CityName}, {options.CountryName}");
            } else {
                await next(context);
            }
        }
    }
}
```

LocationMiddleware 类定义了一个 IOptions<MessageOptions>构造函数参数。在代码清单 12-16 中，用代码清单 12-15 中的类替换了 lambda 函数中间件组件，并删除了 IOptions <MessageOptions> 参数。

**代码清单 12-16　在 Platform 文件夹的 Startup.cs 文件中使用基于类的中间件**

```
using System;
using System.Collections.Generic;
using System.Linq;
using System.Threading.Tasks;
using Microsoft.AspNetCore.Builder;
```

```
using Microsoft.AspNetCore.Hosting;
using Microsoft.AspNetCore.Http;
using Microsoft.Extensions.DependencyInjection;
using Microsoft.Extensions.Hosting;
using Microsoft.Extensions.Options;

namespace Platform {
    public class Startup {

        public void ConfigureServices(IServiceCollection services) {
            services.Configure<MessageOptions>(options => {
                options.CityName= "Albany";
            });
        }

        public void Configure(IApplicationBuilder app, IWebHostEnvironment env) {

            if (env.IsDevelopment()) {
                app.UseDeveloperExceptionPage();
            }

            app.UseMiddleware<LocationMiddleware>();

            app.UseRouting();

            app.UseEndpoints(endpoints => {
                endpoints.MapGet("/", async context => {
                    await context.Response.WriteAsync("Hello World!");
                });
            });
        }
    }
}
```

当执行 UseMiddleware 语句时，会检查 LocationMiddleware 构造函数，它的 IOptions<MessageOptions>参数使用在ConfigureServices方法中创建的对象进行解析。这是使用第 14 章描述的依赖注入特性来完成的，但直接效果是，options模式可以用来轻松地生成配置基于类的中间件。重启ASP.NET Core，请求http://localhost:5000/location，测试新中间件，生成如图 12-17 所示的输出。

# 12.6 小结

本章重点介绍了 ASP.NET Core 平台，介绍了请求管道、中间件组件和服务，描述了在使用空模板时添加到项目中的文件，并解释了 Program 和 Startup 类的角色。第 13 章将描述 ASP.NET Core 的 URL 路由功能。

# 第 13 章

# 使用 URL 路由

URL 路由特性由一对中间件组件提供，通过将请求 URL 的处理和匹配进行合并，可更容易地生成响应。本章解释如何使用 ASP.NET Core 平台支持 URL 路由，展示它的用法，并解释为什么它比创建自己的定制中间件组件更可取。表 13-1 概述 URL 路由。

**注意：**本章主要讨论 ASP.NET Core 平台的 URL 路由，请参阅第Ⅲ部分，了解 ASP.NET Core 的高级部分如何构建在本章描述的特性之上。

表 13-1　URL 路由

| 问题 | 答案 |
|---|---|
| 它是什么？ | URL 路由合并了 URL 的处理和匹配，允许称为端点的组件生成响应 |
| 它为什么有用？ | 使用 URL 路由，每个中间件就不必分析 URL 以确定是处理请求还是沿着管道传递请求。其结果是更高效，更容易维护 |
| 它是如何使用的？ | URL 路由中间件组件添加到请求管道中，并使用一组路由进行配置。每个路由包含一个 URL 路径和一个委托，该委托在接收到具有匹配路径的请求时生成响应 |
| 是否存在缺陷或限制？ | 很难定义路由集，来匹配复杂应用程序支持的所有 URL |
| 还有其他选择吗？ | URL 路由是可选的，可以使用自定义中间件组件代替 |

表 13-2 总结了本章的内容。

表 13-2　本章内容摘要

| 问题 | 解决方案 | 代码清单 |
|---|---|---|
| 处理特定 URL 集的请求 | 定义了一个路由，该路由具有匹配所需 URL 的模式 | 13-1～13-6 |
| 从 URL 中提取值 | 使用区段变量 | 13-7～13-10、13-14 |
| 生成 URL | 使用链接生成器从路由生成 URL | 13-11～13-13、13-15 |
| 用不同数量的段匹配 URL | 在 URL 路由模式中使用可选段或综合段 | 13-16～13-18 |
| 限制匹配 | 使用 URL 路由模式中的约束 | 13-19～13-21、13-23～13-26 |
| 匹配没有以其他方式处理的请求 | 定义后备路由 | 13-22 |
| 查看哪个端点处理请求 | 使用路由上下文数据 | 13-27 |

## 13.1 准备工作

本章将继续使用第 12 章中的 Platform 项目。为了准备本章，向 Platform 文件夹添加一个名为 Population.cs 的文件，代码如代码清单 13-1 所示。

---

**提示：** 可以从 https://github.com/apress/pro-asp.net-core-3 下载本章和本书中其他所有章节的示例项目。如果在运行示例时遇到问题，请参阅第 1 章以获得帮助。

---

**代码清单 13-1 Platform 文件夹中 Population.cs 文件的内容**

```
using Microsoft.AspNetCore.Http;
using System;
using System.Threading.Tasks;

namespace Platform {
    public class Population {
        private RequestDelegate next;

        public Population() { }

        public Population(RequestDelegate nextDelegate) {
            next = nextDelegate;
        }

        public async Task Invoke(HttpContext context) {
            string[] parts = context.Request.Path.ToString()
                .Split("/", StringSplitOptions.RemoveEmptyEntries);
            if (parts.Length == 2 && parts[0] == "population") {
                string city = parts[1];
                int? pop = null;
                switch (city.ToLower()) {
                    case "london":
                        pop = 8_136_000;
                        break;
                    case "paris":
                        pop = 2_141_000;
                        break;
                    case "monaco":
                        pop = 39_000;
                        break;
                }
                if (pop.HasValue) {
                    await context.Response
                        .WriteAsync($"City: {city}, Population: {pop}");
                    return;
                }
            }
```

```
            if (next != null) {
                await next(context);
            }
        }
    }
}
```

此中间件组件响应/population/<city>的请求，其中<city>为伦敦、巴黎或摩纳哥。中间件组件拆分 URL 路径字符串，检查它是否具有预期的长度，并使用 switch 语句确定它是可以响应的 URL 请求。如果 URL 与中间件寻找的模式相匹配，则生成响应；否则，将通过管道传递请求。

用代码清单 13-2 所示的代码将一个名为 Capital.cs 的类文件添加到 Platform 文件夹中。

**代码清单13-2　Platform 文件夹中 Capital.cs 文件的内容**

```
using Microsoft.AspNetCore.Http;
using System;
using System.Threading.Tasks;

namespace Platform {
    public class Capital {
        private RequestDelegate next;

        public Capital() { }

        public Capital(RequestDelegate nextDelegate) {
            next = nextDelegate;
        }

        public async Task Invoke(HttpContext context) {
            string[] parts = context.Request.Path.ToString()
                .Split("/", StringSplitOptions.RemoveEmptyEntries);
            if (parts.Length == 2 && parts[0] == "capital") {
                string capital = null;
                string country = parts[1];
                switch (country.ToLower()) {
                    case "uk":
                        capital= "London";
                        break;
                    case "france":
                        capital = "Paris";
                        break;
                    case "monaco":
                        context.Response.Redirect($"/population/{country}");
                        return;
                }
                if (capital != null) {
                    await context.Response
```

```
                    .WriteAsync($"{capital} is the capital of {country}");
                return;
            }
        }
        if (next != null) {
            await next(context);
        }
    }
  }
}
```

此中间件组件正在寻找/capital/<country>的请求，其中<country>是英国、法国或摩纳哥。显示了英国和法国的首都，但是对摩纳哥的请求被重定向到/population/monaco，摩纳哥是一个城市，也是一个国家。

代码清单 13-3 替换了前一章中的中间件示例，并将新的中间件组件添加到请求管道中。

**代码清单 13-3　替换 Platform 文件夹中 Startup.cs 文件的内容**

```
using Microsoft.AspNetCore.Builder;
using Microsoft.AspNetCore.Hosting;
using Microsoft.AspNetCore.Http;
using Microsoft.Extensions.DependencyInjection;

namespace Platform {
    public class Startup {

        public void ConfigureServices(IServiceCollection services) {
        }

        public void Configure(IApplicationBuilder app, IWebHostEnvironment env) {
            app.UseDeveloperExceptionPage();
            app.UseMiddleware<Population>();
            app.UseMiddleware<Capital>();
            app.Use(async (context, next) => {
                await context.Response.WriteAsync("Terminal Middleware Reached");
            });
        }
    }
}
```

为了启动应用程序，从 Debug 菜单中选择 Start Without Debugging 或 Run Without Debugging，或者打开一个新的 PowerShell 命令提示符，导航到 Platform 项目文件夹(其中包含 Platform.csproj 文件)，并运行代码清单 13-4 所示的命令。

**代码清单 13-4　启动 ASP.NET Core 运行时**

```
dotnet run
```

导航到 http://localhost:5000/population/london，在图 13-1 的左侧会看到输出。导航到 http://localhost:5000/capital/France，查看来自其他中间件组件的输出，如图13-1的右侧所示。

图 13-1 运行示例应用程序

### 13.1.1 理解 URL 路由

每个中间件组件决定是否在请求通过管道传递时对其进行操作。一些组件寻找特定的标题或查询字符串值，但大多数组件——特别是终端和短路组件——则尝试匹配 URL。

当请求沿着管道运行时，每个中间件都必须重复相同的步骤集。在前面定义的中间件中可以看到这一点，其中两个组件执行相同的过程：拆分 URL、检查部件的数量、检查第一部分，等等。

这种方法效率低，难以维护，而且在更改时很容易崩溃。这是低效的，因为要重复相同的操作集来处理 URL。它很难维护，因为每个组件寻找的 URL 隐藏在其代码中。它很容易崩溃，因为更改必须在多个地方小心地完成。例如，Capital 组件将请求重定向到路径以/population 开始的 URL，该 URL 由 Population 组件处理。如果 Population 组件修改为支持/size URL，那么这个更改也必须反映在 Capital 组件中。真正的应用程序可以支持复杂的 URL 集，并且完全通过单个中间件组件进行更改可能十分困难。

URL 路由通过引入负责匹配请求 URL 的中间件，解决了这些问题，这样称为端点的组件就可以专注于响应。端点和它们需要的 URL 之间的映射用一个路由表示。路由中间件处理 URL、检查路由集，并找到处理请求的端点，这个过程称为路由。

### 13.1.2 添加路由中间件、定义端点

路由中间件使用两个单独的方法添加：UseRouting 和 UseEndpoints。UseRouting 方法将负责处理请求的中间件添加到管道中。UseEndpoints 方法用于定义将 URL 与端点匹配的路径。使用与请求 URL 路径相比较的模式匹配 URL，每个路由在一个 URL 模式和一个端点之间创建关系。代码清单 13-5 显示了路由中间件的使用，并包含一个简单的路由。

**提示：**13.3.3 节中解释了为什么有两种路由方法。

**代码清单 13-5 在 Platform 文件夹的 Startup.cs 文件中使用路由中间件**

```
using Microsoft.AspNetCore.Builder;
using Microsoft.AspNetCore.Hosting;
using Microsoft.AspNetCore.Http;
using Microsoft.Extensions.DependencyInjection;

namespace Platform {
    public class Startup {
```

```
        public void ConfigureServices(IServiceCollection services) {
        }

        public void Configure(IApplicationBuilder app, IWebHostEnvironment env) {
            app.UseDeveloperExceptionPage();
            app.UseMiddleware<Population>();
            app.UseMiddleware<Capital>();

            app.UseRouting();

            app.UseEndpoints(endpoints => {
                endpoints.MapGet("routing", async context => {
                    await context.Response.WriteAsync("Request Was Routed");
                });
            });

            app.Use(async (context, next) => {
                await context.Response.WriteAsync("Terminal Middleware Reached");
            });
        }
    }
}
```

UseRouting方法没有参数。UseEndpoints方法接收一个函数，该函数接收IEndpointRouteBuilder对象，并通过它使用表13-3中描述的扩展方法创建路由。

**提示：** 还有一些扩展方法可以为 ASP.NET Core 的其他部分(例如 MVC 框架)设置端点，如第III部分所述。

表 13-3　IEndpointRouteBuilder 扩展方法

| 名称 | 描述 |
|---|---|
| MapGet(pattern, endpoint) | 该方法将匹配 URL 模式的 HTTP GET 请求路由到该端点 |
| MapPost(pattern, endpoint) | 该方法将匹配 URL 模式的 HTTP POST 请求路由到该端点 |
| MapPut(pattern, endpoint) | 该方法将匹配 URL 模式的 HTTP PUT 请求路由到该端点 |
| MapDelete(pattern, endpoint) | 该方法将匹配 URL 模式的 HTTP DELETE 请求路由到该端点 |
| MapMethods(pattern, methods, endpoint) | 此方法将使用与 URL 模式匹配的指定 HTTP 方法之一发出的请求路由到该端点 |
| Map(pattern, endpoint) | 此方法将匹配 URL 模式的所有 HTTP 请求路由到该端点 |

端点使用 RequestDelegate 定义，它与传统中间件使用的委托相同，因此端点是接收HttpContext 对象并使用它生成响应的异步方法。这意味着前面章节中描述的所有中间件组件的特性也可以在端点中使用。

重启 ASP.NETCore，并导航到 http://localhost:5000/routing 以测试新路由。当匹配请求时，路由中间件将路由的 URL 模式应用到 URL 的 path 部分。路径与主机名通过/字符分隔，如图 13-2 所示。

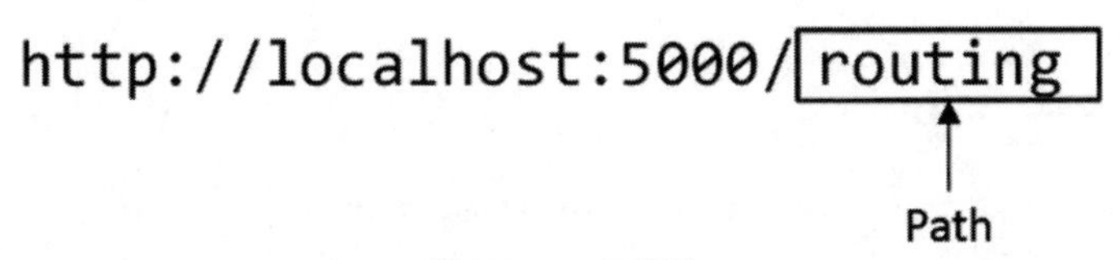


图 13-2　URL 路径

URL 中的路径与路由中指定的模式匹配。

```
...
endpoints.MapGet("routing", async context => {
...
```

URL 模式中没有前导字符/，这不是 URL 路径的一部分。当请求 URL 路径匹配 URL 模式时，请求就转发到 endpoints 函数，生成如图 13-3 所示的响应。

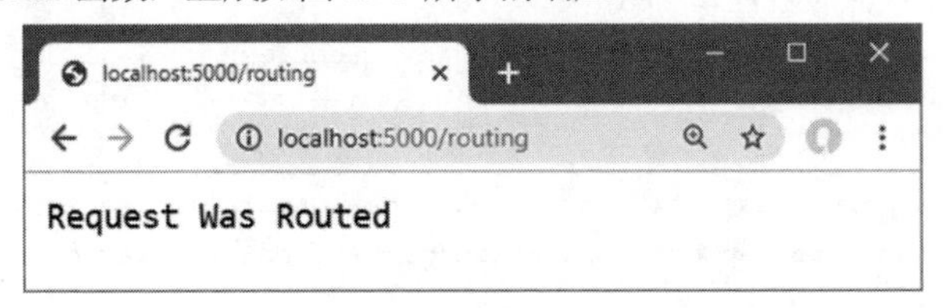


图 13-3　使用端点生成响应

当路由与 URL 匹配时，路由中间件会使管道短路，以便仅由路由的端点生成响应。请求不会转发到稍后出现在请求管道中的其他端点或中间件组件。

如果请求 URL 没有任何路由匹配，则路由中间件将请求传递到请求管道中的下一个中间件组件。要测试此行为，请求 http://localhost:5000/notrouted URL，其路径与代码清单 13-5 中定义的路由模式不匹配。

路由中间件不能将 URL 路径匹配到路由并转发请求，请求到达终端中间件，生成如图 13-4 所示的响应。

图 13-4　请求没有匹配路由的 URL

端点以与前面章节中演示的中间件组件相同的方式生成响应：它们接收 HttpContext 对象，该对象通过 HttpRequest 和 HttpResponse 对象提供对请求和响应的访问。这意味着任何中间件组件也可用作端点。代码清单 13-6 添加了一个使用 Capital 和 Population 中间件组件作为端点的路由。

**代码清单 13-6　在 Platform 文件夹的 Startup.cs 文件中使用中间件组件作为端点**

```
using Microsoft.AspNetCore.Builder;
using Microsoft.AspNetCore.Hosting;
using Microsoft.AspNetCore.Http;
using Microsoft.Extensions.DependencyInjection;
```

```
namespace Platform {
   public class Startup {

      public void ConfigureServices(IServiceCollection services) {
      }

      public void Configure(IApplicationBuilder app, IWebHostEnvironment env) {
         app.UseDeveloperExceptionPage();
         // app.UseMiddleware<Population>();
         // app.UseMiddleware<Capital>();

         app.UseRouting();

         app.UseEndpoints(endpoints => {
            endpoints.MapGet("routing", async context => {
               await context.Response.WriteAsync("Request Was Routed");
            });
            endpoints.MapGet("capital/uk", new Capital().Invoke);
            endpoints.MapGet("population/paris", new Population().Invoke);
         });
         app.Use(async (context, next) => {
            await context.Response.WriteAsync("Terminal Middleware Reached");
         });
      }
   }
}
```

使用这样的中间件组件很尴尬，因为需要创建类的新实例来选择 Invoke 方法作为端点。路由使用的 URL 模式只支持中间件组件支持的部分 URL，但是了解端点依赖于前面章节中熟悉的特性是很有用的。要测试新路由，请重新启动 ASP.NET Core，并导航到 http://localhost:5000/capital/uk 和 http://localhost:5000/population/paris，会生成如图 13-5 所示的结果。

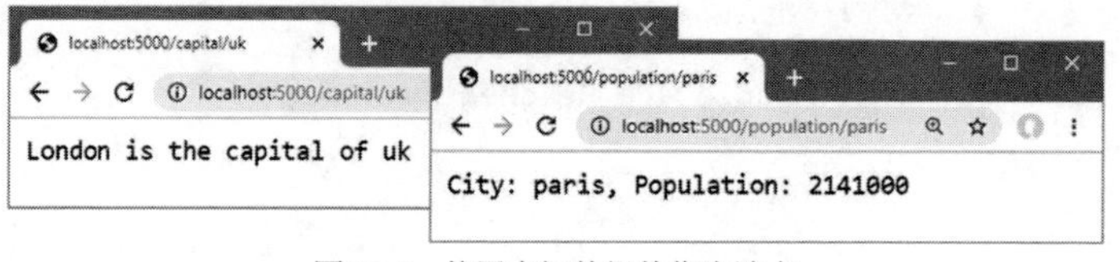


图 13-5 使用中间件组件作为端点

## 13.1.3 理解 URL 模式

使用中间件组件作为端点表明，URL 路由特性构建在标准 ASP NET Core 平台之上。虽然可以通过检查路由来查看由应用程序处理的 URL，但并不是 Capital 和 Population 类理解的所有 URL 都可以路由，这样做效率不高，因为 URL 由路由中间件处理一次，Capital 和 Population 类再次选择路由，以提取它们需要的数据值。

进行改进需要更多地了解如何使用 URL 模式。当请求到达时，路由中间件处理 URL 以从其路径中提取片段，这些片段是由/字符分隔的路径部分，如图 13-6 所示。

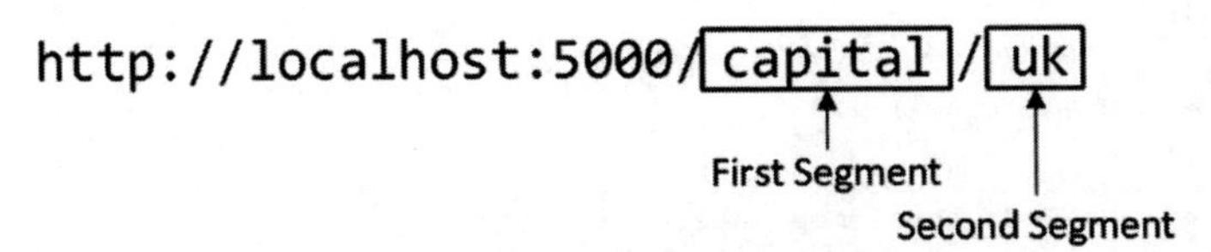


图 13-6　URL 段

路由中间件还从 URL 路由模式中提取片段，如图 13-7 所示。

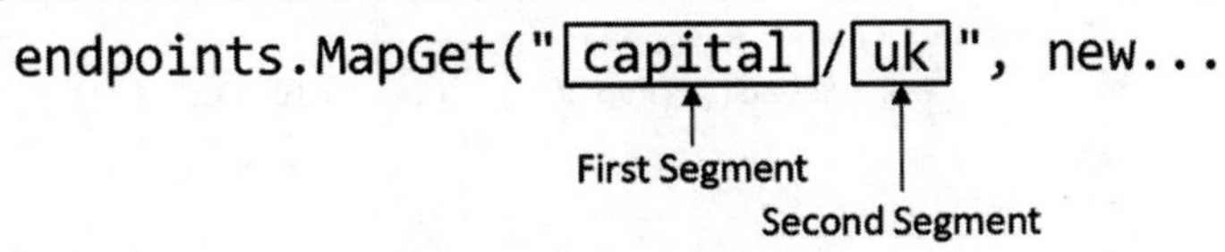


图 13-7　URL 模式段

为了路由请求，将 URL 模式中的片段与请求中的片段进行比较，看看它们是否匹配。如果请求的路径包含相同数量的段，并且每个段与 URL 模式中的内容相同，则请求被路由到端点，如表 13-4 所示。

表 13-4　匹配 URL 段

| URL 路径 | 描述 |
| --- | --- |
| /capital | 不匹配——段太少 |
| /capital/europe/uk | 不匹配——段太多 |
| /name/uk | 不匹配——第一段是 capital |
| /capital/uk | 匹配 |

### 13.1.4　在 URL 模式中使用段变量

代码清单 13-6 中使用的 URL 模式使用文字段，也称为静态段，它使用固定字符串匹配请求。例如，模式中的第一段只匹配将 capital 作为第一段的那些请求，模式中的第二段只匹配第二段为 uk 的那些请求。将它们放在一起，就可以看到为什么路由只匹配路径为/capital/uk 的请求。

段变量也称为路由参数，扩展了模式段将匹配的路径段的范围，允许更灵活的路由。段变量有一个名称，用花括号({和}字符)表示，如代码清单 13-7 所示。

**代码清单 13-7　在 Platform 文件夹的 Startup.cs 文件中使用段变量**

```
using Microsoft.AspNetCore.Builder;
using Microsoft.AspNetCore.Hosting;
using Microsoft.AspNetCore.Http;
using Microsoft.Extensions.DependencyInjection;

namespace Platform {
    public class Startup {

        public void ConfigureServices(IServiceCollection services) {
        }

        public void Configure(IApplicationBuilder app, IWebHostEnvironment env) {
```

```
            app.UseDeveloperExceptionPage();
            app.UseRouting();

            app.UseEndpoints(endpoints => {
                endpoints.MapGet("{first}/{second}/{third}", async context => {
                    await context.Response.WriteAsync("Request Was Routed\n");
                    foreach (var kvp in context.Request.RouteValues) {
                        await context.Response
                            .WriteAsync($"{kvp.Key}: {kvp.Value}\n");
                    }
                });
                endpoints.MapGet("capital/uk", new Capital().Invoke);
                endpoints.MapGet("population/paris", new Population().Invoke);
            });

            app.Use(async (context, next) => {
                await context.Response.WriteAsync("Terminal Middleware Reached");
            });
        }
    }
}
```

URL 模式{first}/{second}/{third}匹配路径包含三个段的 URL，而不管这些段包含什么。当使用段变量时，路由中间件向端点提供它们匹配的 URL 路径段的内容。该内容可以通过 HttpRequest.RouteValues 属性获得，该属性返回一个 RouteValuesDictionary 对象。表 13-5 描述了最有用的 RouteValuesDictionary 成员。

**提示：** 有些保留字不能用作段变量的名称：action、area、controller、handler 和 page。

**表 13-5　有用的 RouteValuesDictionary 成员**

| 名称 | 描述 |
| --- | --- |
| [key] | 此类定义了一个索引器，允许按键检索值 |
| Keys | 此属性返回段变量名的集合 |
| Values | 此属性返回段变量值的集合 |
| Count | 此属性返回段变量的数量 |
| ContainsKey(key) | 如果路由数据包含指定键的值，此方法返回 true |

RouteValuesDictionary 类是可枚举的，这意味着可以在 foreach 循环中使用它来生成一系列 KeyValuePair<string, object>对象，每个对象对应于从请求 URL 提取的一个段变量名称和相应的</string, object>值。代码清单 13-7 中的端点枚举 HttpRequest.RouteValues 属性，生成一个响应，该响应列出 URL 模式匹配的段变量的名称和值。

段变量的名称是 first、second 和 third，为了看到从 URL 提取的值，可以重新启动 ASP.NET Core，请求任何三段 URL，例如 http://localhost:5000/apples/oranges/cherry，这会生成如图 13-8 所示的响应。

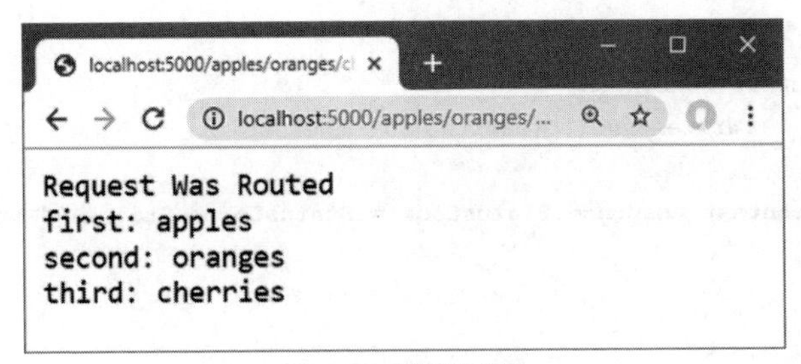


图 13-8 使用段变量

### 1. 了解路由选择

在处理请求时，中间件会找到所有能够匹配请求的路由，并给每个路由打分，然后选择得分最低的路由来处理该路由。评分过程很复杂，但效果是最具体的路由接收请求。这意味着文字段比段变量有优先权，有约束的段变量比没有约束的段变量有优先权。评分系统可能产生令人惊讶的结果，应该检查以确保应用程序支持的 URL 与期望的路由匹配。

如果两个路由具有相同的分数(这意味着它们同样适合路由请求)，就抛出一个异常，表示路由选择不明确。关于如何避免不明确路由的细节，请参阅 13.3.2 节。

### 2. 将中间件重构为端点

端点通常依赖路由中间件提供特定的段变量，而不是枚举所有的段变量。通过依赖 URL 模式提供一个特定的值，可以重构 Capital 和 Population 类以依赖路由数据，如代码清单 13-8 所示。

**代码清单 13-8 依赖 Platform 文件夹 Capital.cs 文件中的路由数据**

```
using Microsoft.AspNetCore.Http;
using System;
using System.Threading.Tasks;
using Microsoft.AspNetCore.Routing;

namespace Platform {
    public static class Capital {

        public static async Task Endpoint(HttpContext context) {
            string capital = null;
            string country = context.Request.RouteValues["country"] as string;
            switch ((country ?? "").ToLower()) {
                case "uk":
                    capital = "London";
                    break;
                case "france":
                    capital = "Paris";
                    break;
                case "monaco":
                    context.Response.Redirect($"/population/{country}");
                    return;
            }
```

```
                if (capital != null) {
                    await context.Response
                        .WriteAsync($"{capital} is the capital of {country}");
                } else {
                    context.Response.StatusCode = StatusCodes.Status404NotFound;
                }
            }
        }
    }
```

中间件组件可用作端点，但是一旦依赖于由中间件提供的数据，则相反的情况就不存在了。在代码清单 13-8 中，使用路由数据，通过由 RouteValuesDictionary 类定义的索引器获取名为 country 的段变量的值。

```
...
string country = context.Request.RouteValues["country"] as string;
...
```

索引器返回一个使用 as 关键字强制转换为字符串的对象值。代码清单删除了沿着管道传递请求的语句，路由中间件代表端点处理这些语句。

使用段变量意味着请求可能会路由到不支持的值的端点，因此添加了一条语句，该语句为端点不理解的国家返回 404 状态代码。

还删除了构造函数，并用名为 Endpoint 的静态方法替换了 Invoke 实例方法，该方法更适合在路由中使用端点的方式。代码清单 13-9 对 Population 类应用相同的更改，将其从标准中间件组件转换为依赖路由中间件处理 URL 的端点。

#### 代码清单 13-9　依赖 Platform 文件夹 Population.cs 文件中的路由数据

```
using Microsoft.AspNetCore.Http;
using System;
using System.Threading.Tasks;

namespace Platform {
    public class Population {

        public static async Task Endpoint(HttpContext context) {
            string city = context.Request.RouteValues["city"] as string;
            int? pop = null;
            switch ((city ?? "").ToLower()) {
                case "london":
                    pop = 8_136_000;
                    break;
                case "paris":
                    pop = 2_141_000;
                    break;
                case "monaco":
                    pop = 39_000;
```

```
                break;
            }
            if (pop.HasValue) {
                await context.Response
                    .WriteAsync($"City: {city}, Population: {pop}");
            } else {
                context.Response.StatusCode = StatusCodes.Status404NotFound;
            }
        }
    }
}
```

对静态方法的更改整理了定义路由时使用的端点，如代码清单 13-10 所示。

**代码清单 13-10 更新 Platform 文件夹 Startup.cs 文件中的路由**

```
...
app.UseEndpoints(endpoints => {
    endpoints.MapGet("{first}/{second}/{third}", async context => {
        await context.Response.WriteAsync("Request Was Routed\n");
        foreach (var kvp in context.Request.RouteValues) {
            await context.Response.WriteAsync($"{kvp.Key}: {kvp.Value}\n");
        }
    });
    endpoints.MapGet("capital/{country}", Capital.Endpoint);
    endpoints.MapGet("population/{city}", Population.Endpoint);
});
...
```

新的路由匹配有两段路径的 URL，第一段是 Capital 或 Population。第二段的内容分配给名为 country 和 city 的段变量，允许端点支持在本章开始时处理的全部 URL，而不需要直接处理 URL。要测试新路由，可重新启动 ASP.NET Core，请求 http://localhost:5000/capital/uk 和 http://localhost:5000/population/london，生成如图 13-9 所示的响应。

图 13-9 在端点使用段变量

这些变化解决了本章开始时描述的两个问题。由于 URL 仅由路由中间件处理一次，而不是由多个组件处理，因此效率得到提高。更容易看到每个端点支持的 URL，因为 URL 模式显示了如何匹配请求。

### 13.1.5 从路由中生成 URL

最后一个问题是很难做出改变。Capital 端点仍然对 Population 端点支持的 URL 具有硬连接依赖关系。为了打破这种依赖关系，路由系统允许通过为段变量提供数据值来生成 URL。第一步是为生成的 URL 的目标路由分配一个名称，如代码清单 13-11 所示。

**代码清单 13-11　在 Platform 文件夹的 Startup.cs 文件中命名一个路由**

```
using Microsoft.AspNetCore.Builder;
using Microsoft.AspNetCore.Hosting;
using Microsoft.AspNetCore.Http;
using Microsoft.Extensions.DependencyInjection;
using Microsoft.AspNetCore.Routing;

namespace Platform {
    public class Startup {

        public void ConfigureServices(IServiceCollection services) {
        }

        public void Configure(IApplicationBuilder app, IWebHostEnvironment env) {
            app.UseDeveloperExceptionPage();
            app.UseRouting();

            app.UseEndpoints(endpoints => {
                endpoints.MapGet("{first}/{second}/{third}", async context => {
                    await context.Response.WriteAsync("Request Was Routed\n");
                    foreach (var kvp in context.Request.RouteValues) {
                        await context.Response
                            .WriteAsync($"{kvp.Key}: {kvp.Value}\n");
                    }
                });
                endpoints.MapGet("capital/{country}", Capital.Endpoint);
                endpoints.MapGet("population/{city}", Population.Endpoint)
                    .WithMetadata(new RouteNameMetadata("population"));
            });

            app.Use(async (context, next) => {
                await context.Response.WriteAsync("Terminal Middleware Reached");
            });
        }
    }
}
```

在 MapGet 方法的结果上使用 WithMetadata 方法为路由分配元数据。生成 URL 所需的唯一元数据是名称，它是通过传递一个新的 RouteNameMetadata 对象来分配的，该对象的构造函数参数指定用于引用路由的名称。在代码清单 13-11 中，把路由命名为 population。

**提示：**命名路由有助于避免生成的链接以期望路径以外的路径为目标，但它们可以被省略；这种情况下，路由系统将尝试找到最佳匹配的路由。可以在第 17 章中看到这种方法的示例。

在代码清单 13-12 中，修改了 Capital 端点，以删除对/population URL 的直接依赖，并依赖路由特性来生成 URL。

**代码清单 13-12　在 Platform 文件夹的 Capital.cs 文件中生成 URL**

```
using Microsoft.AspNetCore.Http;
using System;
using System.Threading.Tasks;
using Microsoft.AspNetCore.Routing;
using Microsoft.Extensions.DependencyInjection;

namespace Platform {
    public static class Capital {
        public static async Task Endpoint(HttpContext context) {
            string capital = null;
            string country = context.Request.RouteValues["country"] as string;
            switch ((country ?? "").ToLower()) {
                case "uk":
                    capital= "London";
                    break;
                case "france":
                    capital = "Paris";
                    break;
                case "monaco":
                    LinkGenerator generator =
                        context.RequestServices.GetService<LinkGenerator>();
                    string url = generator.GetPathByRouteValues(context,
                        "population", new { city = country });
                    context.Response.Redirect(url);
                    return;
            }
            if (capital != null) {
                await context.Response
                    .WriteAsync($"{capital} is the capital of {country}");
            } else {
                context.Response.StatusCode = StatusCodes.Status404NotFound;
            }
        }
    }
}
```

URL 是使用 LinkGenerator 类生成的。不能仅创建一个新的 LinkGenerator 实例；必须使用第 14 章中描述的依赖注入特性来获得。对于本章的目的，知道这个语句获得端点将使用的 LinkGenerator 对象就足够了：

```
...
LinkGenerator generator = context.RequestServices.GetService<LinkGenerator>();
...
```

LinkGenerator 类提供了 GetPathByRouteValues 方法，该方法用于生成在重定向中使用的 URL。

```
...
generator.GetPathByRouteValues(context, "population", new { city = country });
...
```

GetPathByRouteValues 方法的参数是端点的 HttpContext 对象、用于生成链接的路由名称和用于为段变量提供值的对象。GetPathByRouteValues 方法返回一个 URL，该 URL 将被路由到 Population 端点。为了确认，可以重新启动 ASP.NET Core，请求 http://localhost:5000/capital/monaco URL。请求被路由到 Capital 端点，后者将生成 URL，并使用它重定向浏览器，生成如图 13-10 所示的结果。

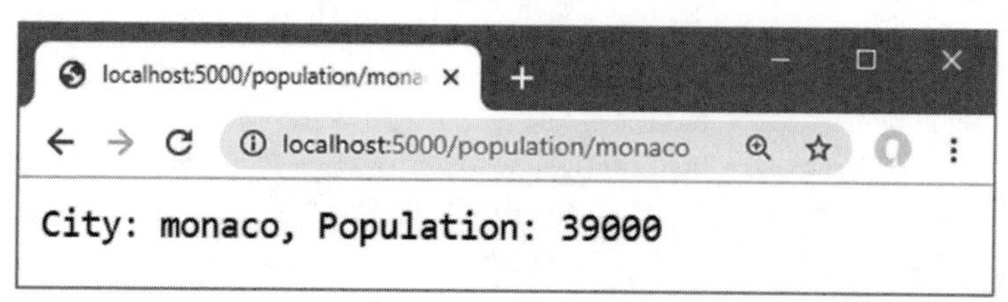


图 13-10　生成 URL

这种方法的好处是，URL 是从命名路由中的 URL 模式生成的，这意味着 URL 模式中的更改反映在生成的 URL 中，而不需要对端点进行更改。为了便于演示，代码清单 13-13 更改了 URL 模式。

**代码清单 13-13　在 Platform 文件夹的 Startup.cs 文件中更改 URL 模式**

```
...
app.UseEndpoints(endpoints => {
    endpoints.MapGet("{first}/{second}/{third}", async context => {
        await context.Response.WriteAsync("Request Was Routed\n");
        foreach (var kvp in context.Request.RouteValues) {
            await context.Response.WriteAsync($"{kvp.Key}: {kvp.Value}\n");
        }
    });
    endpoints.MapGet("capital/{country}", Capital.Endpoint);
    endpoints.MapGet("size/{city}", Population.Endpoint)
        .WithMetadata(new RouteNameMetadata("population"));
});
...
```

分配给路由的名称不变，这确保生成的 URL 针对相同的端点。要查看新模式的效果，请重新启动 ASP.NET Core，并再次请求 http://localhost:5000/capital/monaco URL。重定向到与修改后的模式匹配的 URL，如图 13-11 所示。这个特性解决了本章开始部分描述的最后一个问题，易于更改应用程序支持的 URL。

图 13-11　更改 URL 模式

**URL 路由和区域**

URL 路由系统支持一个称为“区域”的特性，它允许应用程序的不同部分拥有自己的控制器、视图和 Razor Pages。本书没有描述“区域”特性，因为它没有被广泛使用。而且当使用它时，它带来的问题往往比它解决的问题还多。如果想分解一个应用程序，那么建议创建单独的项目。

## 13.2　管理 URL 的匹配

上一节介绍了基本的 URL 路由特性，但是大多数应用程序需要更多的工作来确保正确路由 URL，以增加或限制路由匹配的 URL 的范围。接下来的各节展示调整 URL 模式以优化匹配过程的不同方法。

### 13.2.1　从一个 URL 段匹配多个值

大多数段变量直接对应于 URL 路径中的一个段，但是路由中间件能够执行更复杂的匹配，允许将单个段匹配到一个变量，同时丢弃不需要的字符。代码清单 13-14 定义了一个仅将 URL 段的一部分匹配到变量的路由。

**代码清单 13-14　在 Platform 文件夹的 Startup.cs 文件中匹配段的一部分**

```
...
app.UseEndpoints(endpoints => {
    endpoints.MapGet("files/{filename}.{ext}", async context => {
        await context.Response.WriteAsync("Request Was Routed\n");
        foreach (var kvp in context.Request.RouteValues) {
            await context.Response.WriteAsync($"{kvp.Key}: {kvp.Value}\n");
        }
    });
    endpoints.MapGet("capital/{country}", Capital.Endpoint);
    endpoints.MapGet("size/{city}", Population.Endpoint)
        .WithMetadata(new RouteNameMetadata("population"));
});
...
```

URL 模式可以包含任意多的段变量，只要它们由静态字符串分隔即可。静态分隔符的要求是让路由中间件知道一个变量的内容在哪里结束，下一个变量的内容在哪里开始。代码清单 13-14 中的模式匹配名为 filename 和 ext 的段变量，它们由句点分隔；进程文件名经常使用这种模式。要查看模式如何匹配 URL，请重新启动 ASP.NET Core，请求 http://localhost:5000 /files/myfile.txt URL，生成如图 13-12 所示的响应。

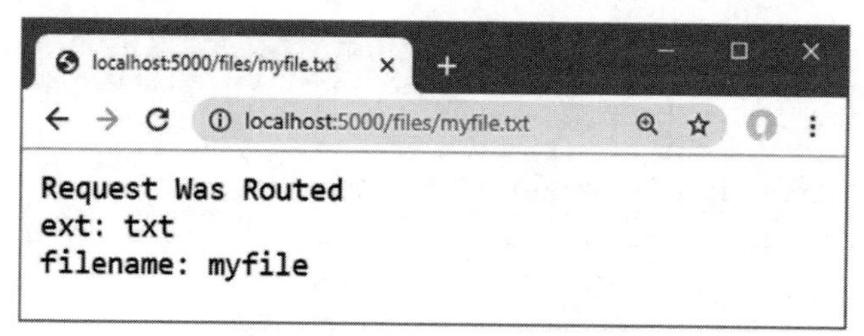


图 13-12　从单个路径段中匹配多个值

**避免复杂的模式不匹配陷阱**

如图 13-12 所示，包含多个变量的模式段是从右到左进行匹配的。这在大多数情况下并不重要，因为端点不能依赖特定的键顺序，但它确实表明复杂 URL 模式的处理方式是不同的，这反映了匹配它们的难度。

事实上，匹配过程非常困难，可能会出现意外的匹配失败。调整匹配过程，以解决问题，具体问题随着 ASP.NET Core 的每次发布而改变，但是调整往往会引入新的问题。在撰写本书时，URL 模式有一个问题，即应该由第一个变量匹配的内容也会以文字字符串的形式出现在段的开头。举个例子就更容易理解了，如下所示：

```
...
endpoints.MapGet("example/red{color}", async context => {
...
```

这个模式有一个以文字字符串 red 开头的段，后跟一个名为 color 的段变量。路由中间件根据 URL 路径示例/redgreen 正确地匹配模式，color route 变量的值是 green。但是，URL 路径示例/redredgreen 不能匹配，因为匹配过程混淆了文字内容的位置和应该分配给 color 变量内容的第一部分。阅读本书时，这个问题可能已经解决了，但是复杂模式还存在其他问题。最好让 URL 模式尽可能简单，并确保得到预期的匹配结果。

### 13.2.2　为段变量使用默认值

可以使用默认值定义模式，当 URL 不包含对应段的值时使用这些默认值，从而增加路由可以匹配的 URL 的范围。代码清单 13-15 显示了模式中默认值的使用。

**代码清单 13-15　在 Platform 文件夹的 Startup.cs 文件中使用默认值**

```
...
app.UseEndpoints(endpoints => {
    endpoints.MapGet("files/{filename}.{ext}", async context => {
        await context.Response.WriteAsync("Request Was Routed\n");
        foreach (var kvp in context.Request.RouteValues) {
            await context.Response.WriteAsync($"{kvp.Key}: {kvp.Value}\n");
        }
    });
    endpoints.MapGet("capital/{country=France}", Capital.Endpoint);
    endpoints.MapGet("size/{city}", Population.Endpoint)
        .WithMetadata(new RouteNameMetadata("population"));
```

```
});
...
```

默认值使用等号和要使用的值定义。当 URL 路径中没有第二段时，代码清单中的默认值使用值 France。结果是路由可以匹配的 URL 范围增加了，如表 13-6 所示。

表 13-6 匹配 URL

| URL 路径 | 描述 |
|---|---|
| / | 没有匹配——段太少 |
| /city | 没有匹配——第一段不是首都 |
| /capital | 匹配，国家变量为 France |
| /capital/uk | 匹配，国家变量为 uk |
| /capital/europe/italy | 没有匹配——段太多 |

要测试默认值，请重新启动 ASP.NET Core，导航到 http://localhost:5000/capital，产生如图 13-13 所示的结果。

图 13-13 为段变量使用默认值

### 13.2.3 在 URL 模式中使用可选段

默认值允许 URL 与更少的段匹配，但是对于端点来说，默认值的使用并不明显。一些端点定义自己的响应来处理省略段的 URL，这些 URL 使用了可选段。为了做好准备，代码清单 13-16 更新了 Population 端点，以便在路由数据中没有可用的城市值时使用默认值。

**代码清单 13-16 在 Platform 文件夹的 Population.cs 文件中使用默认值**

```
using Microsoft.AspNetCore.Http;
using System;
using System.Threading.Tasks;

namespace Platform {
    public class Population {

        public static async Task Endpoint(HttpContext context) {
            string city = context.Request.RouteValues["city"] as string ?? "london";
            int? pop = null;
            switch (city.ToLower()) {
                case "london":
                    pop = 8_136_000;
```

```
                    break;
                case "paris":
                    pop = 2_141_000;
                    break;
                case "monaco":
                    pop = 39_000;
                    break;
            }
            if (pop.HasValue) {
                await context.Response
                    .WriteAsync($"City: {city}, Population: {pop}");
            } else {
                context.Response.StatusCode = StatusCodes.Status404NotFound;
            }
        }
    }
}
```

改用london作为默认值，因为没有可用的city段变量。代码清单13-17更新了Population端点的路由，使第二段成为可选的。

### 代码清单13-17　在Platform文件夹的Startup.cs文件中使用可选段

```
...
app.UseEndpoints(endpoints => {
    endpoints.MapGet("files/{filename}.{ext}", async context => {
        await context.Response.WriteAsync("Request Was Routed\n");
        foreach (var kvp in context.Request.RouteValues) {
            await context.Response.WriteAsync($"{kvp.Key}: {kvp.Value}\n");
        }
    });
    endpoints.MapGet("capital/{country=France}", Capital.Endpoint);
    endpoints.MapGet("size/{city?}", Population.Endpoint)
        .WithMetadata(new RouteNameMetadata("population"));
});
...
```

可选段用变量名后面的问号(?字符)表示，并允许路由匹配没有对应路径段的URL，如表13-7所示。

表13-7　匹配的URL

| URL路径 | 描述 |
| --- | --- |
| / | 没有匹配——段太少 |
| /city | 没有匹配——第一段不是大小 |
| /size | 匹配，city变量的值不会提供给端点 |
| /size/paris | 匹配，city变量为paris |
| /size/europe/italy | 没有匹配——段太多 |

要测试可选段，请重新启动 ASP.NET Core，并导航到 http://localhost:5000/size，生成如图 13-14 所示的响应。

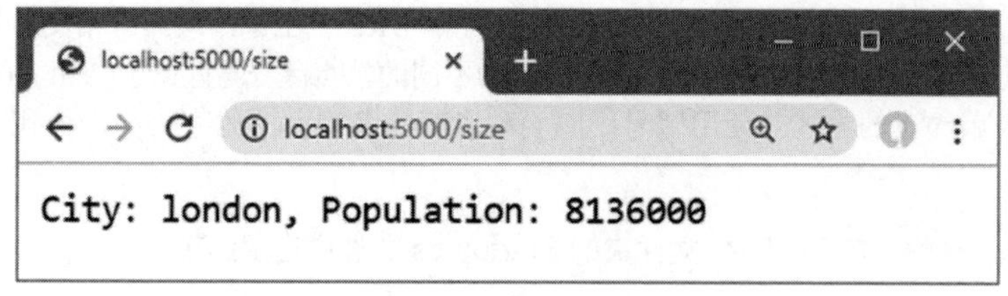


图 13-14　使用可选段

### 13.2.4　使用 catchall 段变量

可选段允许模式匹配较短的 URL 路径。catchall 段则相反，它允许路由匹配包含比模式更多段的 URL。在变量名之前用星号表示一个 catchall 段，如代码清单 13-18 所示。

**代码清单 13-18　在 Platform 文件夹的 Startup.cs 文件中使用 catchall 段**

```
...
app.UseEndpoints(endpoints => {
    endpoints.MapGet("{first}/{second}/{*catchall}", async context => {
        await context.Response.WriteAsync("Request Was Routed\n");
        foreach (var kvp in context.Request.RouteValues) {
            await context.Response.WriteAsync($"{kvp.Key}: {kvp.Value}\n");
        }
    });
    endpoints.MapGet("capital/{country=France}", Capital.Endpoint);
    endpoints.MapGet("size/{city?}", Population.Endpoint)
        .WithMetadata(new RouteNameMetadata("population"));
});
...
```

新模式包含两段变量和一个 catchall，结果是路由匹配任何路径包含两个或更多段的 URL。此路由中的 URL 模式匹配的段数量没有上限，任何其他段的内容都分配给名为 catchall 的段变量。重启 ASP.NET Core，导航到 http://localhost:5000/one/2/threes/four，生成如图 13-15 所示的响应。

> **提示：** catchall 捕捉到的段以“段/段/段”的形式呈现，端点负责处理字符串，以分离出各个段。

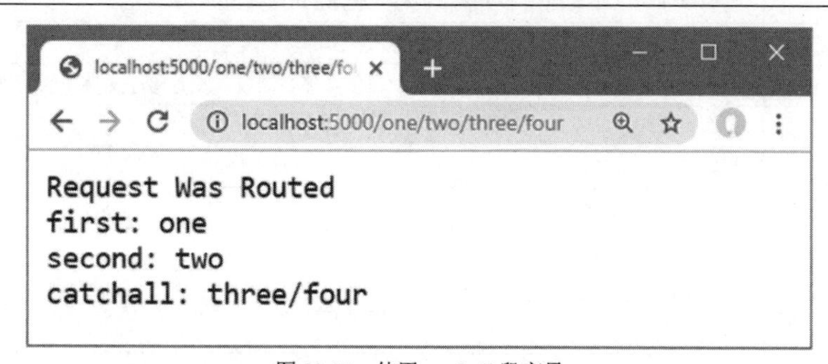


图 13-15　使用 catchall 段变量

### 13.2.5　约束段的匹配

默认值、可选段和 catchall 段都增加了路由匹配的 URL 范围。约束具有相反的效果，限制匹配。如果端点只能处理特定的段内容，或者需要为不同的端点区分与匹配密切相关的 URL，这将非常有用。应用约束的方式是使用冒号(:字符)和段变量名后面的约束类型，如代码清单 13-19 所示。

**代码清单 13-19　在 Platform 文件夹的 Startup.cs 文件中应用约束**

```
...
app.UseEndpoints(endpoints => {
    endpoints.MapGet("{first:int}/{second:bool}", async context => {
        await context.Response.WriteAsync("Request Was Routed\n");
        foreach (var kvp in context.Request.RouteValues) {
            await context.Response.WriteAsync($"{kvp.Key}: {kvp.Value}\n");
        }
    });
    endpoints.MapGet("capital/{country=France}", Capital.Endpoint);
    endpoints.MapGet("size/{city?}", Population.Endpoint)
        .WithMetadata(new RouteNameMetadata("population"));
});
...
```

这个示例限制了第一个段变量，因此它只匹配可以解析为 int 值的路径段，并且限制了第二个段，因此只匹配可以解析为 bool 的路径段。不匹配约束的值不会被路由匹配。表 13-8 描述了 URL 模式约束。

> **注意：** 有些约束匹配的类型的格式可能因地区而异。路由中间件不处理本地化格式，只匹配那些以不变区域性格式表示的值。

表 13-8　URL 模式约束

| 约束 | 描述 |
|---|---|
| alpha | 此约束匹配字母 a～z(不区分大小写) |
| bool | 此约束匹配 true 和 false(不区分大小写) |
| datetime | 此约束匹配以非本地化不变区域性格式表示的日期时间值 |
| decimal | 此约束匹配以非本地化不变区域性格式表示的十进制值 |
| double | 此约束匹配以非本地化不变区域性格式表示的双精度值 |
| file | 此约束匹配其内容在 name.ext 表单中表示文件名的段。没有验证文件是否存在 |
| float | 此约束匹配以非本地化不变区域性格式表示的浮点值 |
| guid | 此约束匹配 GUID 值 |
| int | 此约束匹配 int 值 |
| length(len) | 约束匹配具有指定字符数的路径段 |
| length(min, max) | 此约束匹配长度在指定的上下限之间的路径段 |
| long | 此约束匹配 long 值 |

(续表)

| 约束 | 描述 |
|---|---|
| max(val) | 此约束匹配可以解析为小于或等于指定值的 int 值的路径段 |
| maxlength(len) | 此约束匹配长度等于或小于指定值的路径段 |
| nonfile | 此约束匹配不表示文件名的段，例如，文件约束不匹配的值 |
| range(min, max) | 此约束匹配路径段，这些路径段可以解析为一个 int 值，该值位于指定的范围 |
| regex(expression) | 这个约束应用一个正则表达式来匹配路径段 |

要测试约束，请重新启动 ASP.NET Core，请求 http://localhost:5000/100/true，这是一个 URL，其路径段符合代码清单 13-19 中的约束，生成如图 13-16 左侧所示的结果。请求 http://localhost:5000/apples/oranges，它具有正确的段数，但包含不符合约束的值。没有一个路由可以匹配转发到终端中间件的请求，如图 13-16 所示。

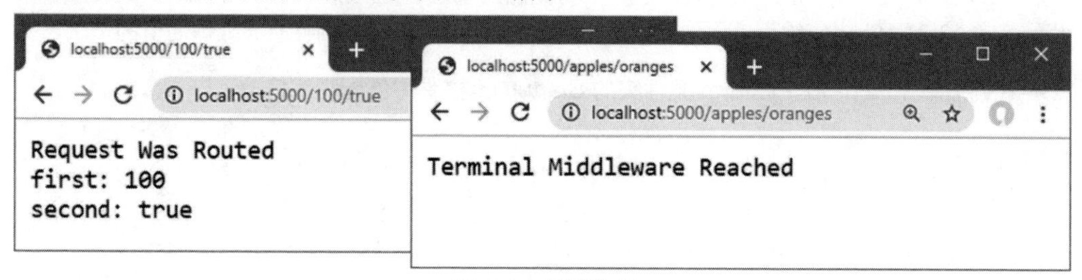


图 13-16 测试约束

可以组合约束，进一步限制匹配，如代码清单 13-20 所示。

**代码清单 13-20 在 Platform 平台文件夹的 Startup.cs 文件中合并 URL 模式约束**

```
...
app.UseEndpoints(endpoints => {
    endpoints.MapGet("{first:alpha:length(3)}/{second:bool}", async context => {
        await context.Response.WriteAsync("Request Was Routed\n");
        foreach (var kvp in context.Request.RouteValues) {
            await context.Response.WriteAsync($"{kvp.Key}: {kvp.Value}\n");
        }
    });
    endpoints.MapGet("capital/{country=France}", Capital.Endpoint);
    endpoints.MapGet("size/{city?}", Population.Endpoint)
        .WithMetadata(new RouteNameMetadata("population"));
});
...
```

合并约束，只匹配能够满足所有约束的路径段。代码清单 13-20 中的组合限制了 URL 模式，以便第一段只匹配三个字母字符。要测试该模式，请重新启动 ASP.NET Core，请求 http://localhost:5000/dog/true，生成如图 13-17 所示的输出。

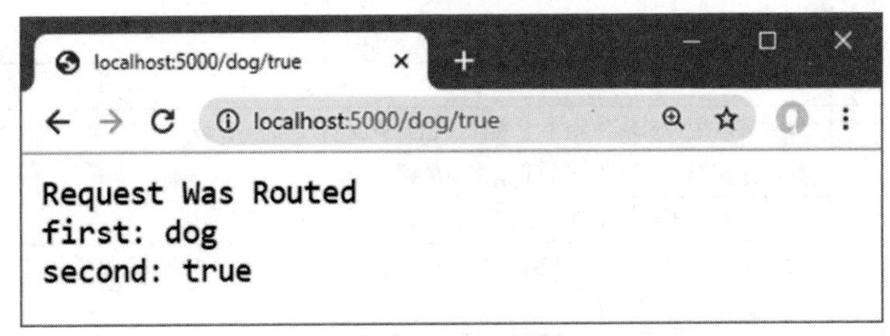


图 13-17　合并约束

请求 URL http://localhost:5000/dogs/true 不能匹配路由，因为第一段包含 4 个字符。

### 约束对特定值集的匹配

regex 约束应用一个正则表达式，它为最常见的约束之一提供了基础：只匹配一组特定的值。代码清单 13-21 将 regex 约束应用到 Capital 端点的路由，因此它只接收能够处理的值的请求。

**代码清单 13-21　在 Platform 文件夹的 Startup.cs 文件中匹配特定值**

```
...
app.UseEndpoints(endpoints => {
    endpoints.MapGet("{first:alpha:length(3)}/{second:bool}", async context => {
        await context.Response.WriteAsync("Request Was Routed\n");
        foreach (var kvp in context.Request.RouteValues) {
            await context.Response.WriteAsync($"{kvp.Key}: {kvp.Value}\n");
        }
    });

    endpoints.MapGet("capital/{country:regex(^uk|france|monaco$)}",
        Capital.Endpoint);
    endpoints.MapGet("size/{city?}", Population.Endpoint)
        .WithMetadata(new RouteNameMetadata("population"));
});
...
```

该路由仅将那些 URL 与两个段匹配。第一段必须是首都，第二段必须是英国、法国或摩纳哥。正则表达式是不区分大小写的，为了确认这一点，可以重新启动 ASP.NET Core，请求 http://localhost:5000/capital/UK，生成如图 13-18 所示的结果。

---

**提示：**浏览器可能请求/capital/uk，其中 uk 是小写。如果出现这种情况，请清除浏览器历史记录并重试。

---

图 13-18　用正则表达式匹配特定值

### 13.2.6 定义回退路由

回退路由仅在没有其他路由与请求匹配时，将请求定向到一个端点。回退路由确保路由系统始终生成响应，从而阻止请求沿着请求管道进一步传递，如代码清单 13-22 所示。

**代码清单 13-22 在 Platform 文件夹的 Startup.cs 文件中使用回退路由**

```
...
app.UseEndpoints(endpoints => {
    endpoints.MapGet("{first:alpha:length(3)}/{second:bool}", async context => {
        await context.Response.WriteAsync("Request Was Routed\n");
        foreach (var kvp in context.Request.RouteValues) {
            await context.Response.WriteAsync($"{kvp.Key}: {kvp.Value}\n");
        }
    });

    endpoints.MapGet("capital/{country:regex(^uk|france|monaco$)}",
        Capital.Endpoint);
    endpoints.MapGet("size/{city?}", Population.Endpoint)
        .WithMetadata(new RouteNameMetadata("population"));

    endpoints.MapFallback(async context => {
        await context.Response.WriteAsync("Routed to fallback endpoint");
    });
});
...
```

MapFallback 方法创建了一个路由，它用作最后的手段，匹配任何请求。表 13-9 描述了创建回退路由的方法(还有一些方法用于创建特定于 ASP.NET Core 的其他部分的回退路由。参见第III部分)。

表 13-9 用于创建回退路由的方法

| 名称 | 描述 |
|---|---|
| MapFallback(endpoint) | 这个方法创建一个回退，将请求路由到一个端点 |
| MapFallbackToFile(path) | 这个方法创建一个回退，将请求路由到一个文件 |

通过添加代码清单 13-22 中的路由，路由中间件处理所有请求，包括那些与常规路由不匹配的请求。重启 ASP.NET Core，导航到任何路由都不匹配的 URL，如 http://localhost:5000/notmatched，生成如图 13-19 所示的响应。

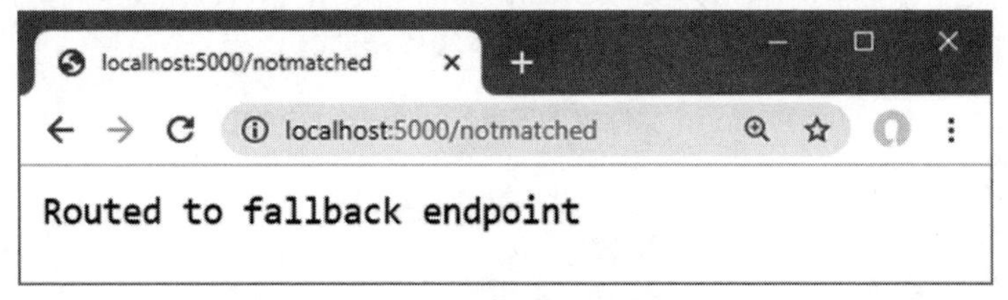


图 13-19 使用回退路由

回退路由没有魔法。回退使用的 URL 模式是{path:nofile}，它们依赖 Order 属性来确保，只有在没有其他合适路由时才使用该路由，这是 13.3.2 节中描述的特性。

# 13.3　高级路由功能

前几节中描述的路由特性满足了大多数项目的需要，特别是因为它们通常是通过第III部分中描述的 MVC 框架等高级特性访问的。对于有特殊路由需求的项目，有一些高级特性，参见下面的描述。

## 13.3.1　创建自定义约束

如果你觉得表 13-8 中描述的约束是不够的，可通过实现 IRouteConstraint 接口来定义自己的约束。要创建自定义约束，向 Platform 文件夹添加一个名为 CountryRouteConstrant.cs 的文件，并添加如代码清单 13-23 所示的代码。

**代码清单 13-23　Platform 文件夹中 CountryRouteConstraint.cs 文件的内容**

```
using System;
using Microsoft.AspNetCore.Http;
using Microsoft.AspNetCore.Routing;

namespace Platform {

    public class CountryRouteConstraint: IRouteConstraint {
        private static string[] countries = { "uk", "france", "monaco" };

        public bool Match(HttpContext httpContext, IRouter route, string routeKey,
                RouteValueDictionary values, RouteDirection routeDirection) {
            string segmentValue = values[routeKey] as string ?? "";
            return Array.IndexOf(countries, segmentValue.ToLower()) > -1;
        }
    }
}
```

IRouteConstraint 接口定义了 Match 方法，调用该方法是为了允许约束来决定路由是否匹配请求。Match 方法的参数为请求提供了 HttpContext 对象、路由、段的名称、从 URL 提取的段变量，以及请求是检查传入的 URL 还是传出的 URL。如果请求满足约束，则 Match 方法返回 true，如果不满足约束，则返回 false。代码清单 13-23 中的约束定义了一组国家/地区，这些国家/地区与应用约束的 segment 变量值进行比较。如果段匹配其中一个国家/地区，就满足约束。使用 options 模式设置定制约束，如代码清单 13-24 所示(options 模式参见第 12 章)。

**代码清单 13-24　使用 Platform 文件夹的 Startup.cs 文件中的自定义约束**

```
using Microsoft.AspNetCore.Builder;
using Microsoft.AspNetCore.Hosting;
using Microsoft.AspNetCore.Http;
using Microsoft.Extensions.DependencyInjection;
```

```
using Microsoft.AspNetCore.Routing;

namespace Platform {
    public class Startup {

        public void ConfigureServices(IServiceCollection services) {
            services.Configure<RouteOptions>(opts => {
                opts.ConstraintMap.Add("countryName",
                    typeof(CountryRouteConstraint));
            });
        }

        public void Configure(IApplicationBuilder app, IWebHostEnvironment env) {
            app.UseDeveloperExceptionPage();
            app.UseRouting();

            app.UseEndpoints(endpoints => {
                endpoints.MapGet("{first:alpha:length(3)}/{second:bool}",
                    async context => {
                        await context.Response.WriteAsync("Request Was Routed\n");
                        foreach (var kvp in context.Request.RouteValues) {
                            await context.Response
                                .WriteAsync($"{kvp.Key}: {kvp.Value}\n");
                        }
                    });

                endpoints.MapGet("capital/{country:countryName}", Capital.Endpoint);

                endpoints.MapGet("size/{city?}", Population.Endpoint)
                    .WithMetadata(new RouteNameMetadata("population"));

                endpoints.MapFallback(async context => {
                    await context.Response.WriteAsync("Routed to fallback endpoint");
                });
            });

            app.Use(async (context, next) => {
                await context.Response.WriteAsync("Terminal Middleware Reached");
            });
        }
    }
}
```

options 模式应用于 RouteOptions 类，该类定义了 ConstraintMap 属性。每个约束都使用一个键注册，该键允许在 URL 模式中应用它。在代码清单 13-24 中，CountryRouteConstraint 类的键是 countyName，它允许约束这样的路由：

```
...
endpoints.MapGet("capital/{country:countryName}", Capital.Endpoint);
...
```

只有当 URL 的第一段是 capital，而第二段是代码清单 13-23 中定义的国家/地区之一时，该路由才能匹配请求。

### 13.3.2　避免模棱两可的路由异常

当尝试路由请求时，路由中间件为每个路由分配一个分数。如本章前面所述，为更具体的路由提供优先级，而路由选择通常是一个行为可预测的直接过程，但如果没有仔细考虑和测试应用程序支持的整个 URL 范围，偶尔会有意外，

如果两条路由得分相同，路由系统无法在它们之间进行选择，并抛出一个异常，表明这两条路由是不明确的。大多数情况下，最好通过引入文字段或约束，来修改模糊路由，以增加特异性。有些情况下，这是不可能的，需要做一些额外的工作，才能让路由系统按预期工作。代码清单 13-25 用两个不明确的新路由替换了前一个示例中的路由，但仅针对某些请求。

**代码清单 13-25　在 Platform 文件夹的 Startup.cs 文件中定义不明确的路由**

```
using Microsoft.AspNetCore.Builder;
using Microsoft.AspNetCore.Hosting;
using Microsoft.AspNetCore.Http;
using Microsoft.Extensions.DependencyInjection;
using Microsoft.AspNetCore.Routing;

namespace Platform {
    public class Startup {

        public void ConfigureServices(IServiceCollection services) {
            services.Configure<RouteOptions>(opts => {
                opts.ConstraintMap.Add("countryName",
                    typeof(CountryRouteConstraint));
            });
        }

        public void Configure(IApplicationBuilder app, IWebHostEnvironment env) {
            app.UseDeveloperExceptionPage();
            app.UseRouting();

            app.UseEndpoints(endpoints => {
                endpoints.Map("{number:int}", async context => {
                    await context.Response.WriteAsync("Routed to the int endpoint");
                });
                endpoints.Map("{number:double}", async context => {
                    await context.Response
                        .WriteAsync("Routed to the double endpoint");
                });
```

```
            });

            app.Use(async (context, next) => {
                await context.Response.WriteAsync("Terminal Middleware Reached");
            });
        }
    }
}
```

这些路由仅对某些值是模糊的。如果只有一个路由匹配第一个路径段，可以解析为双精度的 URL；而如果两个路由都匹配，则可以解析为整型或双精度的 URL。要查看问题，请重新启动 ASP.NETCore，请求 http://localhost:5000/23.5。路径段 23.5 可以解析为双精度，并生成如图 13-20 左侧所示的响应。请求 http://localhost:5000/23，会看到异常显示在图 13-20 的右侧。段 23 可以被解析为 int 和 double，这意味着路由系统不能识别处理请求的单一路由。

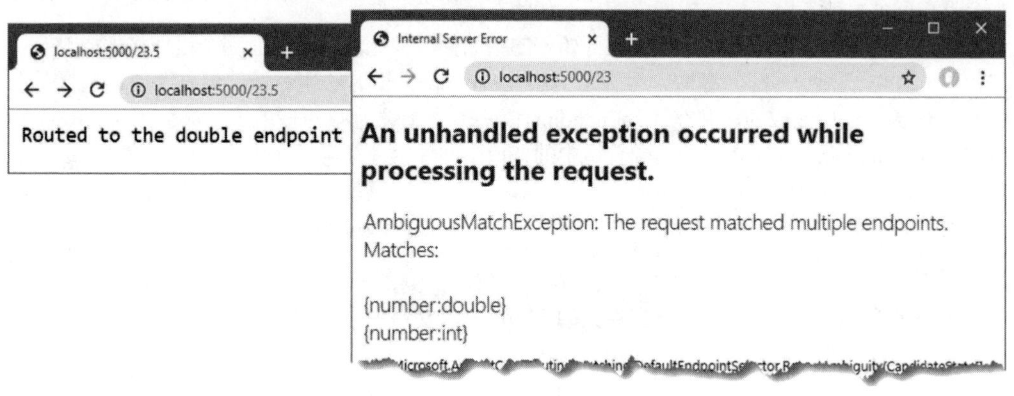


图 13-20 偶尔不明确的路由配置

对于这些情况，可以定义一个路由相对于其他匹配路由的顺序，给它指定优先级，如代码清单 13-26 所示。

### 代码清单13-26 在 Platform 文件夹的 Startup.cs 文件中消除路由歧义

```
...
app.UseEndpoints(endpoints => {
    endpoints.Map("{number:int}", async context => {
        await context.Response.WriteAsync("Routed to the int endpoint");
    }).Add(b => ((RouteEndpointBuilder)b).Order = 1);

    endpoints.Map("{number:double}", async context => {
        await context.Response.WriteAsync("Routed to the double endpoint");
    }).Add(b => ((RouteEndpointBuilder)b).Order = 2);
});
...
```

这个过程很笨拙，需要调用 Add 方法，强制转换为 RouteEndpointBuilder，并设置 Order 属性的值。具有最低 Order 值的路由具有优先级，这意味着代码清单 13-26 中的更改告诉路由系统，为两个路由都可以处理的 URL 使用第一个路由。重启 ASP.NET Core，再次请求 http://localhost:5000/23 URL，第一个路由就会处理该请求，如图 13-21 所示。

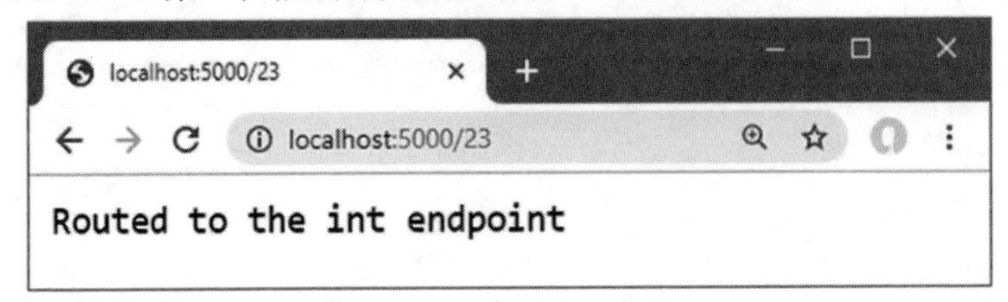


图 13-21　避免模棱两可的路由

## 13.3.3　访问中间件组件中的端点

如前面章节所述，并非所有中间件都生成响应。有些组件提供稍后在请求管道中使用的特性，如会话中间件，或者以某种方式增强响应，如状态代码中间件。

普通请求管道的一个限制是，管道开始处的中间件组件无法判断后面哪个组件将生成响应。路由中间件做了一些不同的事情。尽管路由在 UseEndpoints 方法中注册，但是路由的选择是在 UseRouting 方法中完成的，并在 UseEndpoints 方法中执行端点以生成响应。添加到 UseRouting 方法和 UseEndpoints 方法之间的请求管道中的任何中间件组件都可以在生成响应之前查看选择了哪个端点，并相应地更改其行为。

在代码清单 13-27 中，添加了一个中间件组件，它根据选择处理请求的路由，向响应添加不同的消息。

**代码清单 13-27　在 Platform 文件夹的 Startup.cs 文件中添加中间件组件**

```
using Microsoft.AspNetCore.Builder;
using Microsoft.AspNetCore.Hosting;
using Microsoft.AspNetCore.Http;
using Microsoft.Extensions.DependencyInjection;
using Microsoft.AspNetCore.Routing;

namespace Platform {
    public class Startup {

        public void ConfigureServices(IServiceCollection services) {
            services.Configure<RouteOptions>(opts => {
                opts.ConstraintMap.Add("countryName",
                    typeof(CountryRouteConstraint));
            });
        }

        public void Configure(IApplicationBuilder app, IWebHostEnvironment env) {
            app.UseDeveloperExceptionPage();
            app.UseRouting();

            app.Use(async (context, next) => {
                Endpoint end = context.GetEndpoint();
```

```
                if (end != null) {
                    await context.Response
                        .WriteAsync($"{end.DisplayName} Selected \n");
                } else {
                    await context.Response.WriteAsync("No Endpoint Selected \n");
                }
                await next();
            });

            app.UseEndpoints(endpoints => {
                endpoints.Map("{number:int}", async context => {
                    await context.Response.WriteAsync("Routed to the int endpoint");
                })
                .WithDisplayName("Int Endpoint")
                .Add(b => ((RouteEndpointBuilder)b).Order = 1);

                endpoints.Map("{number:double}", async context => {
                    await context.Response
                        .WriteAsync("Routed to the double endpoint");
                })
                .WithDisplayName("Double Endpoint")
                .Add(b => ((RouteEndpointBuilder)b).Order = 2);
            });

            app.Use(async (context, next) => {
                await context.Response.WriteAsync("Terminal Middleware Reached");
            });
        }
    }
}
```

HttpContext 类上的 GetEndpoint 扩展方法返回已选择来处理请求的端点，通过 Endpoint 对象进行描述。Endpoint 类用于定义表 13-10 中描述的属性。

■ **注意：**还有一个 SetEndpoint 方法，它允许路由中间件在生成响应之前更改所选择的端点。只是在有迫切需要干扰正常路由选择过程时，才应谨慎使用。

表 13-10　端点类定义的属性

| 名称 | 描述 |
|---|---|
| DisplayName | 此属性返回与端点关联的显示名称，可在创建路由时使用 WithDisplayName 方法进行设置 |
| Metadata | 此属性返回与端点关联的元数据集合 |
| RequestDelegate | 此属性返回用于生成响应的委托 |

为了更容易地标识路由中间件选择的端点，使用 WithDisplayName 方法为代码清单 13-27 中的路由分配名称。新的中间件组件在响应中添加消息，向已选择的端点报告。重启 ASP.NET Core，请求 http://localhost:5000/23 URL，以查看中间件的输出。该输出显示，已经在将路由中间件添加

到请求管道的两个方法之间选择端点，如图 13-22 所示。

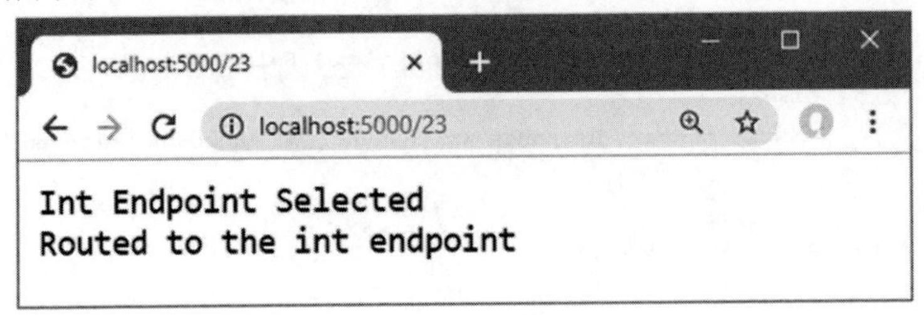


图 13-22　确定端点

## 13.4　小结

本章介绍了端点路由系统，解释了它是如何处理常规中间件中出现的一些常见问题的；展示了如何定义路由，如何匹配和生成 URL，以及如何使用约束来限制路由的使用；还介绍了路由系统的一些高级用法，包括自定义约束和避免路由歧义。第 14 章将解释 ASP.NET Core 服务的工作方式。

# 第 14 章

# 使用依赖注入

服务是中间件组件和端点之间共享的对象。服务可以提供的特性没有限制，但是它们通常用于应用程序的多个部分中需要的任务，比如日志记录或数据库访问。

ASP.NET Core 依赖注入特性来创建和使用服务。这是一个会引起混淆和难以理解的主题。本章描述依赖注入所解决的问题，并解释 ASP.NET Core 平台如何支持依赖注入。表 14-1 简要介绍依赖注入。

表 14-1　依赖注入

| 问题 | 答案 |
| --- | --- |
| 它是什么? | 依赖注入很容易创建松散耦合的组件，这通常意味着组件使用接口定义的功能，而不知道使用的是哪些实现类 |
| 它为什么有用? | 依赖注入通过更改实现接口(用于定义应用程序特性)的组件，使更改应用程序的行为变得更容易。它还会生成更容易隔离用于单元测试的组件 |
| 它是如何使用的? | Startup 类用于指定使用哪个实现类来交付应用程序使用的接口所指定的功能。可以通过 IServiceProvider 接口或声明构造函数或方法参数，显式地请求服务 |
| 是否存在缺陷或限制? | 处理中间件组件和端点的方式以及访问具有不同生命周期的服务的方式存在一些差异 |
| 还有其他选择吗? | 不必在自己的代码中使用依赖注入，但了解它是如何工作的很有帮助，因为它由 ASP.NET Core 平台为开发人员提供特性 |

表 14-2 总结了本章内容。

表 14-2　本章内容摘要

| 问题 | 解决方案 | 代码清单 |
| --- | --- | --- |
| 在 Startup 类中获取一个服务 | 将一个参数添加到 Configure 方法中 | 14-13 |
| 在中间件组件中获取服务 | 需要定义构造函数参数 | 14-14、14-33～14-35 |
| 在端点中获得服务 | 可以通过上下文对象获得 IServiceProvider 对象 | 14-15～14-18 |
| 实例化具有构造函数依赖关系的类 | 使用 ActivatorUtilities 类 | 14-19～14-21 |
| 定义为每个依赖项实例化的服务 | 定义瞬态服务 | 14-22～14-27 |
| 定义为每个请求实例化的服务 | 定义了作用域服务 | 14-28～14-32 |

(续表)

| 问题 | 解决方案 | 代码清单 |
| --- | --- | --- |
| 访问 Startup.ConfigureServices 方法中的服务 | 定义一个启动构造函数参数，并将值赋给该属性 | 14-36 |
| 管理服务实例化 | 使用服务工厂 | 14-37、14-38 |
| 定义服务的多个实现 | 定义具有相同作用域多个服务，通过 GetServices 方法使用它们 | 14-39～14-41 |
| 使用支持泛型类型参数的服务 | 使用具有未绑定类型的服务 | 14-42 |

## 14.1　为本章做准备

本章继续使用第 13 章中的 Platform 项目。需要新类为本章做准备。首先创建 Platform/Services 文件夹，向其中添加一个名为 IResponseFormatter.cs 的类文件，代码如代码清单 14-1 所示。

**提示**：可以从 https://github.com/apress/pro-asp.net-core-3 下载本章和本书中其他所有章节的示例项目。如果在运行示例时遇到问题，请参阅第 1 章以获得帮助。

**代码清单 14-1　Services 文件夹中 IResponseFormatter.cs 文件的内容**

```
using Microsoft.AspNetCore.Http;
using System.Threading.Tasks;

namespace Platform.Services {
   public interface IResponseFormatter {

      Task Format(HttpContext context, string content);
   }
}
```

IResponseFormatter 接口定义了一个接收 HttpContext 对象和字符串的方法。要创建接口的实现，将一个名为 TextResponseFormatter.cs 的类添加到 Platform/Services 文件夹中，文件的内容如代码清单 14-2 所示。

**代码清单 14-2　Services 文件夹中 TextResponseFormatter.cs 文件的内容**

```
using System.Threading.Tasks;
using Microsoft.AspNetCore.Http;

namespace Platform.Services {
   public class TextResponseFormatter : IResponseFormatter {
      private int responseCounter = 0;

      public async Task Format(HttpContext context, string content) {
         await context.Response.
            WriteAsync($"Response {++responseCounter}:\n{content}");
```

```
        }
    }
}
```

TextResponseFormatter 类实现接口，将内容作为一个简单的字符串写入响应，并带有前缀，以便在使用类时更明显。

### 14.1.1 创建中间件组件和端点

本章中的一些示例展示了在使用中间件和端点时，如何以不同的方式应用特性。将一个名为 WeatherMiddleware.cs 的文件添加到 Platform 文件夹，代码如代码清单 14-3 所示。

**代码清单 14-3 Platform 文件夹中 WeatherMiddleware.cs 文件的内容**

```
using Microsoft.AspNetCore.Http;
using System.Threading.Tasks;

namespace Platform {
    public class WeatherMiddleware {
        private RequestDelegate next;

        public WeatherMiddleware(RequestDelegate nextDelegate) {
            next = nextDelegate;
        }

        public async Task Invoke(HttpContext context) {
            if (context.Request.Path == "/middleware/class") {
                await context.Response
                    .WriteAsync("Middleware Class: It is raining in London");
            } else {
                await next(context);
            }
        }
    }
}
```

要创建与中间件组件产生类似结果的端点，请将名为 WeatherEndpoint.cs 的文件添加到 Platform 文件夹，代码如代码清单 14-4 所示。

**代码清单 14-4 Platform 文件夹中 WeatherEndpoint.cs 文件的内容**

```
using Microsoft.AspNetCore.Http;
using System.Threading.Tasks;

namespace Platform {
    public class WeatherEndpoint {

        public static async Task Endpoint(HttpContext context) {
            await context.Response
```

```
                .WriteAsync("Endpoint Class: It is cloudy in Milan");
        }
    }
}
```

### 14.1.2 配置请求管道

将 Startup.cs 文件的内容替换为代码清单 14-5 所示的内容。上一节定义的类与 lambda 函数一起应用，可以生成类似的结果。

**代码清单 14-5 替换 Platform 文件夹中 Startup.cs 文件的内容**

```
using Microsoft.AspNetCore.Builder;
using Microsoft.AspNetCore.Hosting;
using Microsoft.AspNetCore.Http;
using Microsoft.Extensions.DependencyInjection;
using Microsoft.AspNetCore.Routing;
using Platform.Services;
namespace Platform {
    public class Startup {

        public void ConfigureServices(IServiceCollection services) {
        }

        public void Configure(IApplicationBuilder app, IWebHostEnvironment env) {
            app.UseDeveloperExceptionPage();
            app.UseRouting();
            app.UseMiddleware<WeatherMiddleware>();

            IResponseFormatter formatter = new TextResponseFormatter();
            app.Use(async (context, next) => {
                if (context.Request.Path == "/middleware/function") {
                    await formatter.Format(context,
                        "Middleware Function: It is snowing in Chicago");
                } else {
                    await next();
                }
            });

            app.UseEndpoints(endpoints => {

            endpoints.MapGet("/endpoint/class", WeatherEndpoint.Endpoint);

            endpoints.MapGet("/endpoint/function", async context => {
                await context.Response
                    .WriteAsync("Endpoint Function: It is sunny in LA");
                });
```

```
            });
        }
    }
}
```

为了启动应用程序，从 Debug 菜单中选择 Start Without Debugging 或 Run Without Debugging，或者打开一个新的 PowerShell 命令提示符，导航到 Platform 项目文件夹(其中包含 Platform.csproj 文件)，并运行代码清单 14-6 所示的命令。

**代码清单 14-6 启动 ASP.NET Core 运行时**

```
dotnet run
```

使用浏览器请求 http://localhost:5000/middleware/function，显示如图 14-1 所示的响应。每次重新加载浏览器时，响应中显示的计数器都会增加。

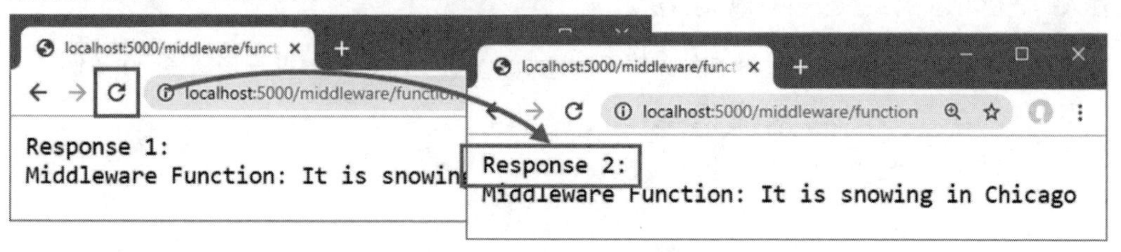


图 14-1 运行示例应用程序

## 14.2 理解服务位置和紧密耦合

要理解依赖注入，首先要了解它所解决的两个问题。接下来的几节将描述依赖项注入解决的两个问题。

**查看依赖注入**

依赖注入是读者联系我时最多谈及的话题之一。大约有一半的邮件抱怨我“强迫”他们使用依赖注入。奇怪的是，另一半人抱怨我没有足够强烈地强调依赖注入的好处，而其他读者可能没有意识到它的用处。

依赖注入可能是一个难以理解的主题，其价值也存在争议。依赖注入可能是一个有用的工具，但不是每个人都喜欢它或者需要它。

如果不做单元测试，或者在一个自包含的、稳定的小项目上工作，依赖注入的好处是有限的。理解依赖注入是如何工作的仍然很有帮助，因为依赖注入用于访问前几章描述的一些重要的 ASP.NET Core 特性。但并不总是需要在编写的自定义类中包含依赖注入。还有其他创建共享特性的方法——稍后将介绍其中的两种——如果不喜欢依赖注入，那么使用这些方法是完全可以接受的。

我在自己的应用程序中依赖于依赖注入，因为项目经常朝着意想不到的方向发展，而且能够用新的实现轻松地替换组件，可以省去许多繁杂且容易出错的更改。我宁愿在项目开始时付出一些努力，而不是在以后进行一组复杂的编辑。

但我对依赖注入并不固执己见，你也不应该固执己见。依赖注入解决了一个不是在每个项目中都会出现的问题，并且只有你才能确定项目是否需要依赖注入。

### 14.2.1　理解服务位置问题

大多数项目都有需要在应用程序的不同部分(称为服务)中使用的特性。常见的示例包括日志记录工具和配置设置，但可以扩展到任何共享特性，包括代码清单 14-2 中定义的 TextResponseFormatter 类(由一个中间件组件使用)。

每个 TextResponseFormatter 对象维护一个计数器，该计数器包含在发送到浏览器的响应中，如果想将相同的计数器合并到其他端点生成的响应中，就需要有一种方法来创建一个 TextResponseFormatter 对象，在生成响应的每个地方都可以很容易地找到并使用它。

有很多方法可以使服务可定位，但是除了本章的主题之外，主要有两种方法。第一种方法是创建一个对象，并将其用作构造函数或方法参数，传递给应用程序中需要它的部分。另一种方法是向直接访问共享实例的服务类添加静态属性，如代码清单 14-7 所示。这称为单例模式，在依赖注入广泛使用之前，它是一种常见的方法。

**代码清单 14-7　在 Services 文件夹的 TextResponseFormatter.cs 文件中创建一个单例**

```
using System.Threading.Tasks;
using Microsoft.AspNetCore.Http;

namespace Platform.Services {
   public class TextResponseFormatter : IResponseFormatter {
      private int responseCounter = 0;
      private static TextResponseFormatter shared;

      public async Task Format(HttpContext context, string content) {
         await context.Response.
            WriteAsync($"Response {++responseCounter}:\n{content}");
      }

      public static TextResponseFormatter Singleton {
         get {
            if (shared == null) {
               shared = new TextResponseFormatter();
            }
            return shared;
         }
      }
   }
}
```

这是单例模式的基本实现，还有许多变体更加关注安全并发访问等问题。对于本章来说，重要的是代码清单 14-7 中的更改依赖于 TextResponseFormatter 服务的使用者通过静态 Singleton 属性获取共享对象，如代码清单 14-8 所示。

**代码清单 14-8 在 Platform 文件夹的 Startup.cs 文件中使用服务**

```
using Microsoft.AspNetCore.Builder;
using Microsoft.AspNetCore.Hosting;
using Microsoft.AspNetCore.Http;
using Microsoft.Extensions.DependencyInjection;
using Microsoft.AspNetCore.Routing;
using Platform.Services;
namespace Platform {
    public class Startup {

        public void ConfigureServices(IServiceCollection services) {
        }

        public void Configure(IApplicationBuilder app, IWebHostEnvironment env) {
            app.UseDeveloperExceptionPage();
            app.UseRouting();
            app.UseMiddleware<WeatherMiddleware>();

            app.Use(async (context, next) => {
                if (context.Request.Path == "/middleware/function") {
                    await TextResponseFormatter.Singleton.Format(context,
                        "Middleware Function: It is snowing in Chicago");
                } else {
                    await next();
                }
            });

            app.UseEndpoints(endpoints => {

                endpoints.MapGet("/endpoint/class", WeatherEndpoint.Endpoint);

                endpoints.MapGet("/endpoint/function", async context => {
                    await TextResponseFormatter.Singleton.Format(context,
                        "Endpoint Function: It is sunny in LA");
                });
            });
        }
    }
}
```

Singleton 模式允许共享单个 TextResponseFormatter，因此由中间件组件和端点使用，效果是单个计数器通过请求两个不同的 URL 递增。要查看单例模式的效果，请重新启动 ASP.NET Core。请求 http://localhost:5000/middleware/function 和 http://localhost:5000/endpoint/function URL。为两个 URL 更新一个计数器，如图 14-2 所示。

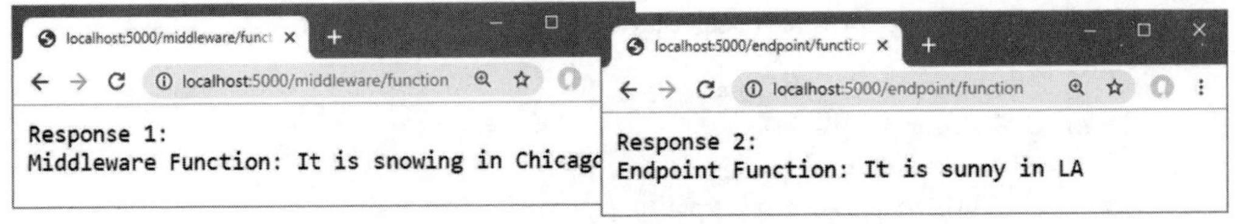


图 14-2　实现单例模式来创建共享服务

单例模式易于理解和使用，但是关于服务如何定位的知识在整个应用程序中传播，每个服务类和每个服务使用者都需要了解如何访问共享对象。当创建新服务时，这会导致单例模式发生变化，并在代码中创建许多必须在发生更改时更新的点。此模式也可以是刚性的，不允许在如何管理服务方面有任何灵活性，因为每个使用者总是共享单个服务对象。

### 14.2.2　理解紧密耦合组件的问题

尽管在代码清单 14-1 中定义了一个接口，但是使用单例模式的方式意味着消费者总是知道正在使用的实现类，因为这个类的静态属性用于获取共享对象。如果想切换到 IResponseFormatter 接口的不同实现，必须定位服务的每次使用，并用新的实现类替换现有的实现类。也有一些模式可以解决这个问题，比如 type broker 模式，其中类通过单例对象的接口提供对它们的访问。在 Platform/ Services 文件夹中添加一个名为 TypeBroker.cs 的类文件，并使用它定义如代码清单 14-9 所示的代码。

#### 代码清单 14-9　Services 文件夹中 TypeBroker.cs 文件的内容

```
namespace Platform.Services {
    public static class TypeBroker {
        private static IResponseFormatter formatter = new TextResponseFormatter();

        public static IResponseFormatter Formatter => formatter;
    }
}
```

Formatter 属性提供对实现 IResponseFormatter 接口的共享服务对象的访问。服务的使用者需要知道是由 TypeBroker 类负责选择将要使用的实现，但是这个模式意味着服务使用者可以通过接口而不是具体类工作，如代码清单 14-10 所示。

#### 代码清单 14-10　在 Platform 文件夹的 Startup.cs 文件中使用的 TypeBroker

```
using Microsoft.AspNetCore.Builder;
using Microsoft.AspNetCore.Hosting;
using Microsoft.AspNetCore.Http;
using Microsoft.Extensions.DependencyInjection;
using Microsoft.AspNetCore.Routing;
using Platform.Services;

namespace Platform {
    public class Startup {

        public void ConfigureServices(IServiceCollection services) {
```

```
        }

        public void Configure(IApplicationBuilder app, IWebHostEnvironment env) {
            app.UseDeveloperExceptionPage();
            app.UseRouting();
            app.UseMiddleware<WeatherMiddleware>();

            app.Use(async (context, next) => {
                if (context.Request.Path == "/middleware/function") {
                    await TypeBroker.Formatter.Format(context,
                "Middleware Function: It is snowing in Chicago");
                } else {
                    await next();
                }
            });

            app.UseEndpoints(endpoints => {

                endpoints.MapGet("/endpoint/class", WeatherEndpoint.Endpoint);

                endpoints.MapGet("/endpoint/function", async context => {
                    await TypeBroker.Formatter.Format(context,
                        "Endpoint Function: It is sunny in LA");
                });
            });
        }
    }
}
```

通过只更改 TypeBroker 类，这种方法可以很容易地切换到不同的实现类，并防止服务使用者创建对特定实现的依赖。还意味着服务类可以只关注它们提供的特性，不需要考虑如何定位这些特性。为了便于演示，向 Platform/Services 文件夹添加一个类文件，称为 HtmlResponseFormatter.cs，代码如代码清单 14-11 所示。

#### 代码清单 14-11　Services 文件夹中 HtmlResponseFormatter.cs 文件的内容

```
using System.Threading.Tasks;
using Microsoft.AspNetCore.Http;

namespace Platform.Services {
    public class HtmlResponseFormatter : IResponseFormatter {

        public async Task Format(HttpContext context, string content) {
            context.Response.ContentType = "text/html";
            await context.Response.WriteAsync($@"
                <!DOCTYPE html>
                <html lang=""en"">
                <head><title>Response</title></head>
```

```
            <body>
                <h2>Formatted Response</h2>
                <span>{content}</span>
            </body>
            </html>");
        }
    }
}
```

IResponseFormatter 的这个实现设置了 HttpResponse 对象的 ContentType 属性，并将内容插入 HTML 模板字符串中。要使用新的 formatter 类，只需要更改 TypeBroker，如代码清单 14-12 所示。

**代码清单 14-12　在 Platform/Services 文件夹的 TypeBroker.cs 文件中更改实现**

```
namespace Platform.Services {
    public static class TypeBroker {
        private static IResponseFormatter formatter = new HtmlResponseFormatter();

        public static IResponseFormatter Formatter => formatter;
    }
}
```

要确保新的格式化程序工作，请重新启动 ASP.NET Core，请求 http://localhost:5000/endpoint/function，生成如图 14-3 所示的结果。

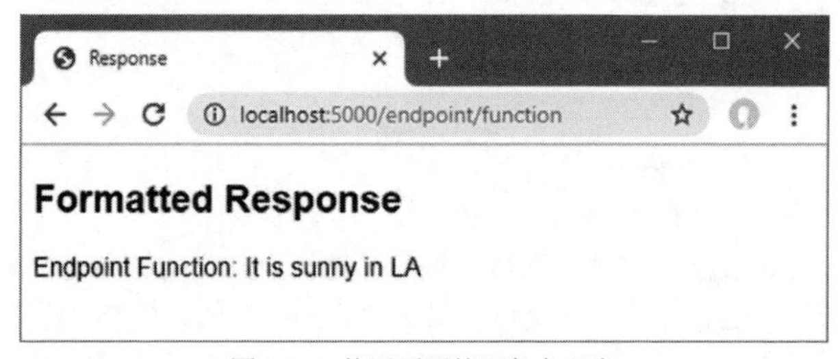


图 14-3　使用不同的服务实现类

## 14.3　使用依赖注入

依赖注入提供了另一种提供服务的方法，这些服务整理了单例和类型代理模式中出现的粗糙之处，并且与其他 ASP.NET Core 功能集成。代码清单 14-13 显示了使用 ASP.NET Core 的依赖项注入，以替换上一节中的类型代理。

**代码清单 14-13　在 Platform 文件夹的 Startup.cs 文件中使用依赖注入**

```
using Microsoft.AspNetCore.Builder;
using Microsoft.AspNetCore.Hosting;
using Microsoft.AspNetCore.Http;
using Microsoft.Extensions.DependencyInjection;
using Microsoft.AspNetCore.Routing;
using Platform.Services;
```

```
namespace Platform {
    public class Startup {

        public void ConfigureServices(IServiceCollection services) {
            services.AddSingleton<IResponseFormatter, HtmlResponseFormatter>();
        }

        public void Configure(IApplicationBuilder app, IWebHostEnvironment env,
                IResponseFormatter formatter) {
            app.UseDeveloperExceptionPage();
            app.UseRouting();
            app.UseMiddleware<WeatherMiddleware>();

            app.Use(async (context, next) => {
                if (context.Request.Path == "/middleware/function") {
                    await formatter.Format(context,
                        "Middleware Function: It is snowing in Chicago");
                } else {
                    await next();
                }
            });

            app.UseEndpoints(endpoints => {

                endpoints.MapGet("/endpoint/class", WeatherEndpoint.Endpoint);

                endpoints.MapGet("/endpoint/function", async context => {
                    await formatter.Format(context,
                        "Endpoint Function: It is sunny in LA");
                });
            });
        }
    }
}
```

使用 IServiceCollection 参数上的扩展方法，服务在 Startup 类的 ConfigureServices 方法中注册。在代码清单 14-13 中，使用一个扩展方法为 IResponseFormatter 接口创建一个服务。

```
...
public void ConfigureServices(IServiceCollection services) {
    services.AddSingleton<IResponseFormatter, HtmlResponseFormatter>();
}
...
```

AddSingleton 方法是服务可用的扩展方法之一，它告诉 ASP.NET Core.，应该使用单个对象来满足服务需求(其他扩展方法参见 14.4 节)。接口和实现类指定为泛型类型参数。为了使用该服务，向 Configure 方法添加了一个参数。

```
...
public void Configure(IApplicationBuilder app, IWebHostEnvironment env,
   IResponseFormatter formatter) {
...
```

新参数声明了对 IResponseFormatter 接口的依赖关系，该方法称为依赖于该接口。在调用 Configure 方法之前，将检查其参数，检测依赖项，并检查应用程序的服务，以确定是否可以解决依赖项。ConfigureServices 方法中的语句告诉依赖项注入系统，可以使用 HtmlResponseFormatter 对象解析 IResponseFormatter 接口上的依赖项。创建对象并用作调用方法的参数。因为解决依赖关系的对象是从使用它的类或函数外部提供的，所以称为注入，这就是该过程称为依赖关系注入的原因。

### 14.3.1　在中间件类中使用服务

在 ConfigureServices 方法中定义服务，并在 Configure 方法中使用它可能看起来不那么令人印象深刻，但是一旦定义了服务，就可以在 ASP.NET Core 应用程序中的几乎任何地方使用它。代码清单 14-14 在本章开头定义的中间件类中声明了对 IResponseFormatter 接口的依赖。

**代码清单 14-14　在 Services 文件夹的 WeatherMiddleware.cs 文件中声明一个依赖项**

```
using Microsoft.AspNetCore.Http;
using System.Threading.Tasks;
using Platform.Services;

namespace Platform {
   public class WeatherMiddleware {
      private RequestDelegate next;
      private IResponseFormatter formatter;

      public WeatherMiddleware(RequestDelegate nextDelegate,
            IResponseFormatter respFormatter) {
         next = nextDelegate;
         formatter = respFormatter;
      }

      public async Task Invoke(HttpContext context) {
         if (context.Request.Path == "/middleware/class") {
            await formatter.Format(context,
               "Middleware Class: It is raining in London");
         } else {
            await next(context);
         }
      }
   }
}
```

为了声明依赖项，添加了一个构造函数参数。要查看结果，请重新启动 ASP.NET Core 并请求 http://localhost:5000/middleware/class URL，生成如图 14-4 所示的响应。

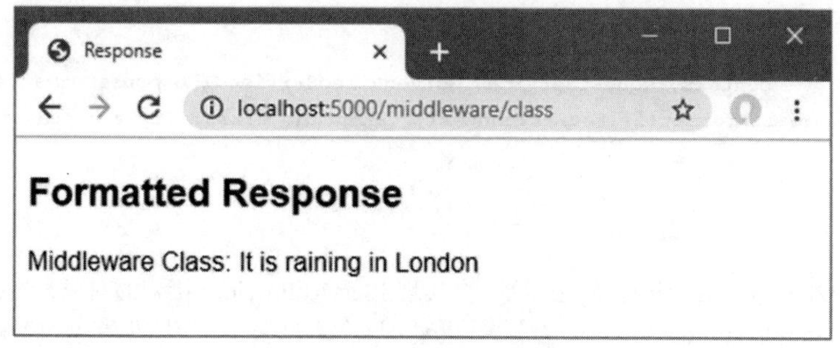


图 14-4 在中间件类中声明依赖项

设置请求管道，ASP.NET Core平台执行Configure方法中的语句，该语句将WeatherMiddleware类添加为组件。

```
...
app.UseMiddleware<WeatherMiddleware>();
...
```

平台理解它需要创建 WeatherMiddleware 类的实例，并检查构造函数。检测对 IResponseFormatter接口的依赖关系，检查服务，以查看是否可以解析依赖关系，并在调用构造函数时使用共享服务对象。

关于这个示例，有两点需要理解。首先，WeatherMiddleware 不知道使用哪个实现类来解析它对 IResponseFormatter 接口的依赖——它只知道通过构造函数参数接收到符合接口的对象。其次，WeatherMiddleware 类不知道依赖项是如何解析的——它只是声明了一个构造函数参数。这是一种比本章前面实现的单例模式和类型代理模式更优雅的方法，而且可以通过更改 Startup.ConfigureServices 方法中使用的泛型类型参数来更改用于解析服务的实现类。

## 14.3.2 在端点中使用服务

WeatherEndpoint 类中的情况更复杂，是静态的，没有可以用来声明依赖项的构造函数。有几种方法可用于解析端点类的依赖关系，参见下面的几节。

### 1. 从 HttpContext 对象中获取服务

可以通过 HttpContext 对象访问服务，该对象是在请求被路由到端点时接收到的，如代码清单 14-15 所示。

**代码清单 14-15 在 Platform 文件夹的 WeatherEndpoint.cs 文件中使用服务**

```
using Microsoft.AspNetCore.Http;
using System.Threading.Tasks;
using Platform.Services;
using Microsoft.Extensions.DependencyInjection;

namespace Platform {
```

```
public class WeatherEndpoint {

    public static async Task Endpoint(HttpContext context) {
        IResponseFormatter formatter =
            context.RequestServices.GetRequiredService<IResponseFormatter>();
        await formatter.Format(context, "Endpoint Class: It is cloudy in Milan");
    }
  }
}
```

HttpContext.RequestServices 属性返回一个实现 IServiceProvider 接口的对象，该对象可以访问在应用程序 Start.ConfigureServices 方法中启动时配置的服务。代码清单 14-15 中使用的 Microsoft.Extensions.DependencyInjection 名称空间包含 IServiceProvider 接口的扩展方法，这些方法允许获得单个服务，如表 14-3 所述。

表 14-3　获取服务的 IServiceProvider 扩展方法

| 名称 | 描述 |
| --- | --- |
| GetService<T>() | 该方法返回泛型类型参数指定的类型的服务；如果没有定义这样的服务，则返回 null |
| GetService(type) | 此方法返回指定类型的服务；如果没有定义这样的服务，则返回 null |
| GetRequiredService<T>() | 此方法返回由泛型类型参数指定的服务；如果服务不可用，则抛出异常 |
| GetRequiredService(type) | 此方法返回指定类型的服务；如果服务不可用，则抛出异常 |

当调用 Endpoint 方法时，使用 GetRequiredService<T> 方法获取用于格式化响应的 IResponseFormatter对象。要查看效果，请重新启动ASP.NET Core，使用浏览器请求http:/localhost:5000/endpoint/class，生成格式化响应，如图 14-5 所示。

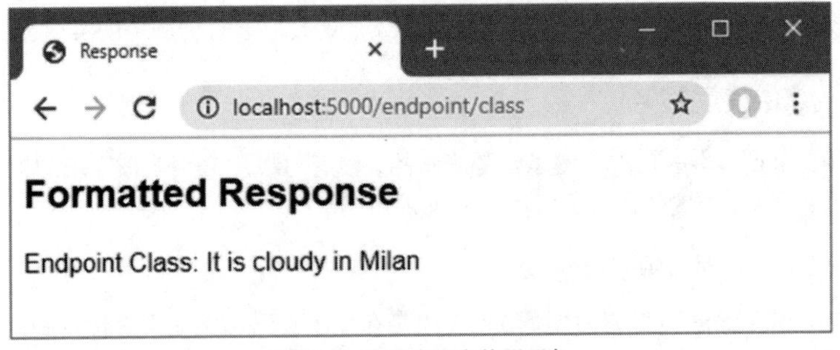


图 14-5　在端点类中使用服务

### 2. 使用适配器函数

使用 HttpContext.RequestServices 的缺点是必须为路由到端点的每个请求解析服务。如本章后面所述，有些服务需要这样做，因为它们提供特定于单个请求或响应的特性。这与 IResponseFormatter 服务不同，在该服务中，可以使用单个对象格式化多个响应。

更优雅的方法是在创建端点路由时获取服务，而不是针对每个请求获取服务。代码清单 14-16 更改静态 Endpoint 方法，使其声明对 IResponseFormatter 接口的依赖。

**代码清单 14-16 在 Platform 文件夹的 WeatherEndpoint.cs 文件中定义一个适配器函数**

```
using Microsoft.AspNetCore.Http;
using System.Threading.Tasks;
using Platform.Services;

namespace Platform {
    public class WeatherEndpoint {

        public static async Task Endpoint(HttpContext context,
                IResponseFormatter formatter) {
            await formatter.Format(context, "Endpoint Class: It is cloudy in Milan");
        }
    }
}
```

在Platform/Services文件夹中添加一个名为EndpointExtensions.cs的文件，并添加如代码清单14-17所示的代码。

**代码清单 14-17 Services 文件夹中 EndpointExtension.cs 文件的内容**

```
using Microsoft.AspNetCore.Routing;
using Microsoft.Extensions.DependencyInjection;
using Platform.Services;

namespace Microsoft.AspNetCore.Builder {

    public static class EndpointExtensions {
        public static void MapWeather(this IEndpointRouteBuilder app, string path) {
            IResponseFormatter formatter =
                app.ServiceProvider.GetService<IResponseFormatter>();
            app.MapGet(path, context => Platform.WeatherEndpoint
                .Endpoint(context, formatter));
        }
    }
}
```

新文件为IEndpointRouteBuilder接口创建一个扩展方法，该方法用于在Startup类中创建路由。该接口定义了ServiceProvider属性，该属性返回IServiceProvider对象，通过该对象可以获得服务。扩展方法获取服务，使用MapGet方法注册一个RequestDelegate，该RequestDelegate将HttpContext对象和IResponseFormatter对象传递给WeatherEndpoint.Endpoint方法。在代码清单14-18中，使用扩展方法为端点创建路由。

**代码清单 14-18 在 Platform 文件夹的 Startup.cs 文件中创建一个路由**

```
...
app.UseEndpoints(endpoints => {
```

```
        //endpoints.MapGet("/endpoint/class", WeatherEndpoint.Endpoint);
        endpoints.MapWeather("/endpoint/class");

        endpoints.MapGet("/endpoint/function", async context => {
            await formatter.Format(context,
            "Endpoint Function: It is sunny in LA");
        });
    });
    ...
```

MapWeather 扩展方法设置路由并围绕端点类创建适配器。要查看结果，请重新启动 ASP.NET Core，请求 http://localhost:5000/endpoint/class URL，生成与前面示例相同的结果，如图 14-5 所示。

### 3. 使用激活实用程序类

第 13 章对端点类使用了静态方法，因为它们在创建路由时更容易使用。但是对于需要服务的端点，使用可以实例化的类通常更容易，因为它允许使用更通用的方法来处理服务。代码清单 14-19 用构造函数修改了端点，并从 Endpoint 方法中删除了 static 关键字。

#### 代码清单 14-19　修改 Platform 文件夹的 WeatherEndpoint.cs 文件中的端点

```
using Microsoft.AspNetCore.Http;
using System.Threading.Tasks;
using Platform.Services;

namespace Platform {
    public class WeatherEndpoint {
        private IResponseFormatter formatter;

        public WeatherEndpoint(IResponseFormatter responseFormatter) {
            formatter = responseFormatter;
        }

        public async Task Endpoint(HttpContext context) {
            await formatter.Format(context, "Endpoint Class: It is cloudy in Milan");
        }
    }
}
```

在 ASP.NET Core 应用程序中，依赖注入最常用于类构造函数。通过方法进行注入(如为中间件类执行的注入)是一个需要重新创建的复杂过程，但是有一些有用的内置工具可以检查构造函数，并使用服务解析依赖关系，如代码清单 14-20 所示。

#### 代码清单 14-20　在 Services 文件夹的 EndpointExtensions.cs 文件中解析依赖关系

```
using Microsoft.AspNetCore.Http;
using Microsoft.AspNetCore.Routing;
using Microsoft.Extensions.DependencyInjection;
```

```
using Platform.Services;
using System.Reflection;
using System.Threading.Tasks;

namespace Microsoft.AspNetCore.Builder {

    public static class EndpointExtensions {

        public static void MapEndpoint<T> (this IEndpointRouteBuilder app,
                string path, string methodName = "Endpoint") {
            MethodInfo methodInfo = typeof(T).GetMethod(methodName);
            if (methodInfo == null || methodInfo.ReturnType != typeof(Task)) {
                throw new System.Exception("Method cannot be used");
            }
            T endpointInstance =
                ActivatorUtilities.CreateInstance<T>(app.ServiceProvider);
            app.MapGet(path, (RequestDelegate)methodInfo
                .CreateDelegate(typeof(RequestDelegate), endpointInstance));
        }
    }
}
```

扩展方法接收一个泛型类型参数，该参数指定要使用的端点类。其他参数是用于创建路由的路径和处理请求的端点类方法的名称。

创建端点类的新实例，并使用指定方法的委托创建路由。与任何使用.NET 反射的代码一样，代码清单 14-20 中的扩展方法可能难以阅读，但本章的关键语句是：

```
...
T endpointInstance = ActivatorUtilities.CreateInstance<T>(app.ServiceProvider);
...
```

ActivatorUtilities 类在 Microsoft.Extensions.DependencyInjection 名称空间中定义，提供方法来实例化通过构造函数声明依赖关系的类。表 14-4 显示了最有用的 ActivatorUtilities 方法。

表 14-4 ActivatorUtilities 方法

| 名称 | 描述 |
| --- | --- |
| CreateInstance<T>(services, args) | 此方法为类型参数指定的类创建一个新实例，使用服务和其他可选参数解析依赖关系 |
| CreateInstance(services, type, args) | 此方法创建参数指定的类的新实例，使用服务和附加可选参数解析依赖关系 |
| GetServiceOrCreateInstance<T>(services, args) | 如果可用，此方法返回指定类型的服务，如果没有服务，则创建新实例 |
| GetServiceOrCreateInstance(services, type, args) | 如果可用，此方法返回指定类型的服务，如果没有服务，则创建新实例 |

这两种方法都通过IServiceProvider对象和一个可选参数数组来解析使用服务的构造函数依赖关系，这些参数用于非服务的依赖关系。这些方法很容易将依赖注入应用到自定义类，而CreateInstance方法的使用会生成一个扩展方法，该方法可以创建路由，路由带有使用服务的端点类。代码清单14-21使用新的扩展方法创建一个路由。

**代码清单14-21　在Platform文件夹的Startup.cs文件中使用扩展方法创建路由**

```
...
app.UseEndpoints(endpoints => {

    endpoints.MapEndpoint<WeatherEndpoint>("/endpoint/class");

    endpoints.MapGet("/endpoint/function", async context => {
        await formatter.Format(context,
            "Endpoint Function: It is sunny in LA");
    });
});
...
```

这种类型的扩展方法更容易使用端点类，并提供了与第12章描述的UseMiddleware方法类似的体验。要确保请求被路由到端点，请重新启动ASP.NET Core，请求http://localhost:5000/endpoint/class URL，生成如图14-5所示的响应。

## 14.4　使用服务生命周期

上一节创建服务时，使用了AddSingleton扩展方法，如下所示：

```
...
public void ConfigureServices(IServiceCollection services) {
    services.AddSingleton<IResponseFormatter, HtmlResponseFormatter>();
}
...
```

AddSingleton方法生成一个服务，该服务在第一次用于解析依赖项时实例化，然后供每个后续依赖项重用。这意味着对IResponseFormatter对象的任何依赖都使用相同的HtmlResponseFormatter对象解析。

单例是开始使用服务的好方法，但是存在一些它们不适合处理的问题。所以ASP.NET Core支持范围服务，为解决依赖关系创建的对象提供了生命周期。表14-5描述了用于创建服务的一组方法。这组方法的一些版本接收类型作为常规参数，如14.5.5节所述。

表14-5中有一些具有单一类型参数的方法版本，它们允许创建一个服务来解决服务位置问题，而不解决紧耦合问题。可以在第24章中看到此类服务的示例，其中共享了一个简单的数据源(它不是通过接口访问的)。

表 14-5 用于创建服务的扩展方法

| 名称 | 描述 |
| --- | --- |
| AddSingleton<T, U>() | 这个方法创建一个 U 类型的对象来解析 T 类型的所有依赖 |
| AddTransient<T, U>() | 这个方法创建一个 U 类型的新对象来解析每个 T 类型的依赖 |
| AddScoped<T, U>() | 这个方法创建一个 U 类型的新对象，用于解析单个范围内对 T 的依赖关系，如 request |

### 14.4.1 创建临时服务

AddTransient 方法的作用与 AddSingleton 方法相反，为每个解析的依赖项创建实现类的新实例。要创建一个演示服务生命周期用法的服务，使用代码清单 14-22 所示的代码，向 Platform/Services 文件夹添加一个名为 GuidService.cs 的文件。

**代码清单 14-22 Services 文件夹中 GuidService.cs 文件的内容**

```
using System;
using System.Threading.Tasks;
using Microsoft.AspNetCore.Http;

namespace Platform.Services {

    public class GuidService : IResponseFormatter {
        private Guid guid = Guid.NewGuid();

        public async Task Format(HttpContext context, string content) {
            await context.Response.WriteAsync($"Guid: {guid}\n{content}");
        }
    }
}
```

Guid 结构生成唯一标识符，当使用不同的实例来解析对 IResponseFormatter 接口的依赖时，该标识符将非常明显。在代码清单 14-23 中，更改了创建 IResponseFormatter 服务的语句，以使用 AddTransient 方法和 GuidService 实现类。

**代码清单 14-23 在 Platform 文件夹的 Startup.cs 文件中创建临时服务**

```
...
public void ConfigureServices(IServiceCollection services) {
    services.AddTransient<IResponseFormatter, GuidService>();
}
...
```

如果重新启动 ASP.NET Core，请求 http://localhost:5000/endpoint/class 和 http://localhost:5000/middleware/class URL，会收到如图 14-6 所示的响应。每个响应将显示一个不同的 GUID 值，以确认已使用临时服务对象，来解析端点和中间件组件对 IResponseFormatter 服务的依赖关系(GUID 的性质意味着，会在响应中看到不同的值。重要的是，不会看到两种响应使用相同的值)。

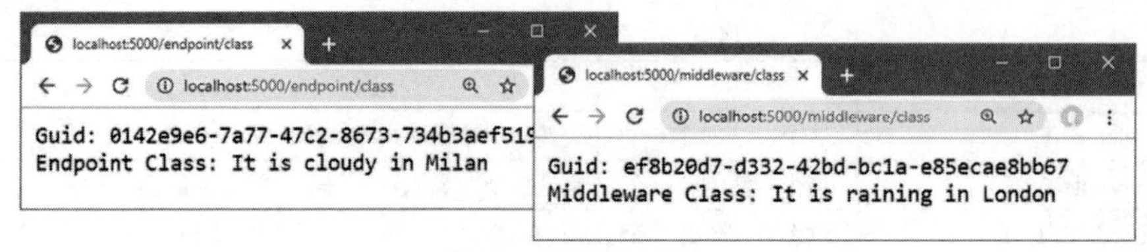


图 14-6　使用瞬态服务

## 14.4.2　避免临时服务重用陷阱

前面的示例演示了在创建不同的服务对象时，效果与预期的不太一样，可以通过单击 Reload 按钮看到这一点。响应包含相同的值，而不是新的 GUID 值，如图 14-7 所示。

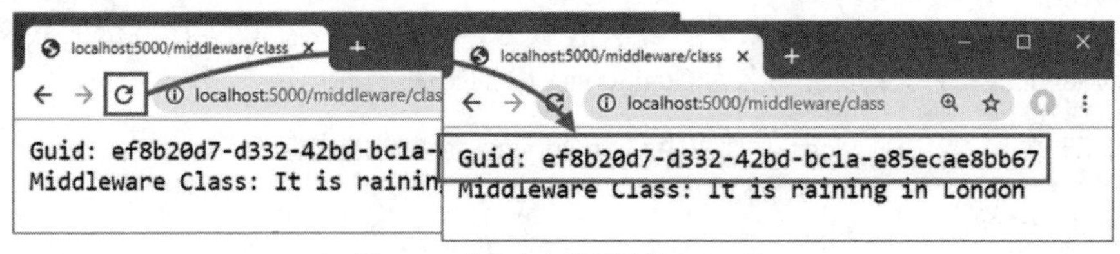


图 14-7　响应中出现相同的 GUID 值

只有在解析依赖关系时才会创建新的服务对象，而不是在使用服务时创建。只有在应用程序启动，调用 Startup.Configure 方法时，才解析示例应用程序中组件和端点的依赖关系。每个服务对象接收一个单独的服务对象，然后对处理的每个请求重用该对象。

要解决中间件组件的这个问题，可以将对服务的依赖转移到 Invoke 方法，如代码清单 14-24 所示。

### 代码清单 14-24　移动 Platform 文件夹的 WeatherMiddleware.cs 文件中的依赖项

```
using Microsoft.AspNetCore.Http;
using System.Threading.Tasks;
using Platform.Services;

namespace Platform {
    public class WeatherMiddleware {
        private RequestDelegate next;
        //private IResponseFormatter formatter;

        public WeatherMiddleware(RequestDelegate nextDelegate) {
            next = nextDelegate;
            //formatter = respFormatter;
        }

        public async Task Invoke(HttpContext context, IResponseFormatter formatter) {
            if (context.Request.Path == "/middleware/class") {
                await formatter.Format(context,
                    "Middleware Class: It is raining in London");
            } else {
```

```
                await next(context);
            }
        }
    }
}
```

ASP.NET Core 平台在每次处理请求时解析 Invoke 方法声明的依赖关系，这确保创建一个新的临时服务对象。

ActivatorUtilities 类不负责解析方法的依赖关系，而 ASP.NET Core 仅为中间件组件包含此特性。端点解决这个问题的最简单方法是在处理每个请求时显式地请求服务，这就是在前面展示如何使用服务时使用的方法。还可增强扩展方法，以代表端点请求服务，如代码清单 14-25 所示。

### 代码清单 14-25　在 Services 文件夹的 EndpointExtensions.cs 文件中请求服务

```
using Microsoft.AspNetCore.Http;
using Microsoft.AspNetCore.Routing;
using Microsoft.Extensions.DependencyInjection;
using Platform.Services;
using System.Reflection;
using System.Threading.Tasks;
using System.Linq;

namespace Microsoft.AspNetCore.Builder {

    public static class EndpointExtensions {

        public static void MapEndpoint<T>(this IEndpointRouteBuilder app,
                string path, string methodName = "Endpoint") {

            MethodInfo methodInfo = typeof(T).GetMethod(methodName);
            if (methodInfo == null || methodInfo.ReturnType != typeof(Task)) {
                throw new System.Exception("Method cannot be used");
            }

            T endpointInstance =
                ActivatorUtilities.CreateInstance<T>(app.ServiceProvider);

            ParameterInfo[] methodParams = methodInfo.GetParameters();
            app.MapGet(path, context => (Task)methodInfo.Invoke(endpointInstance,
                methodParams.Select(p => p.ParameterType == typeof(HttpContext)
                ? context
                : app.ServiceProvider.GetService(p.ParameterType)).ToArray()));
        }
    }
}
```

代码清单 14-25 中的代码不如 ASP.NET Core 平台为中间件组件采用的方法有效。处理请求的方法定义的所有参数都被视为要解析的服务，HttpContext 参数除外。路由是由一个委托创建的，该委托为每个请求解析服务，并调用处理该请求的方法。代码清单 14-26 修改了 WeatherEndpoint 类，将 IResponseFormatter 上的依赖项移动到 Endpoint 方法，以便为每个请求接收一个新的服务对象。

**代码清单 14-26　在 Platform 文件夹的 WeatherEndpoint.cs 文件中移动依赖项**

```
using Microsoft.AspNetCore.Http;
using System.Threading.Tasks;
using Platform.Services;

namespace Platform {
    public class WeatherEndpoint {
        //private IResponseFormatter formatter;

        //public WeatherEndpoint(IResponseFormatter responseFormatter) {
        // formatter = responseFormatter;
        //}

        public async Task Endpoint(HttpContext context,
                IResponseFormatter formatter) {
            await formatter.Format(context, "Endpoint Class: It is cloudy in Milan");
        }
    }
}
```

代码清单 14-24 到代码清单 14-26 中的更改确保为每个请求解析临时服务，这意味着创建一个新的 GuidService 对象，并且每个响应都包含一个唯一的 ID。

对于定义为 lambda 表达式的中间件和端点，必须在处理每个请求时获得服务，因为 Configure 方法参数声明的依赖项只在配置请求管道时解析一次。代码清单 14-27 显示了获取新服务对象所需的更改。

**代码清单 14-27　在 Platform 文件夹的 Startup.cs 文件中使用临时服务**

```
using Microsoft.AspNetCore.Builder;
using Microsoft.AspNetCore.Hosting;
using Microsoft.AspNetCore.Http;
using Microsoft.Extensions.DependencyInjection;
using Microsoft.AspNetCore.Routing;
using Platform.Services;

namespace Platform {
    public class Startup {

        public void ConfigureServices(IServiceCollection services) {
            services.AddTransient<IResponseFormatter, GuidService>();
        }
```

```
        public void Configure(IApplicationBuilder app, IWebHostEnvironment env) {
            app.UseDeveloperExceptionPage();
            app.UseRouting();
            app.UseMiddleware<WeatherMiddleware>();

            app.Use(async (context, next) => {
                if (context.Request.Path == "/middleware/function") {
                    IResponseFormatter formatter
                        = app.ApplicationServices.GetService<IResponseFormatter>();
                    await formatter.Format(context,
                        "Middleware Function: It is snowing in Chicago");
                } else {
                    await next();
                }
            });

            app.UseEndpoints(endpoints => {

                endpoints.MapEndpoint<WeatherEndpoint>("/endpoint/class");

                endpoints.MapGet("/endpoint/function", async context => {
                    IResponseFormatter formatter
                        = app.ApplicationServices.GetService<IResponseFormatter>();
                    await formatter.Format(context,
                        "Endpoint Function: It is sunny in LA");
                });
            });
        }
    }
}
```

重启 ASP.NET Core，导航到应用程序支持的四个 URL(http://localhost:5000/middleware/class、/middleware/function、/endpoint/class 和/endpoint/function)，然后单击浏览器的 Reload 按钮。每次重新加载时，都把一个新的请求发送到 ASP.NET Core。处理请求的组件或端点接收一个新的服务对象，这样在每个响应中显示不同的 GUID，如图 14-8 所示。

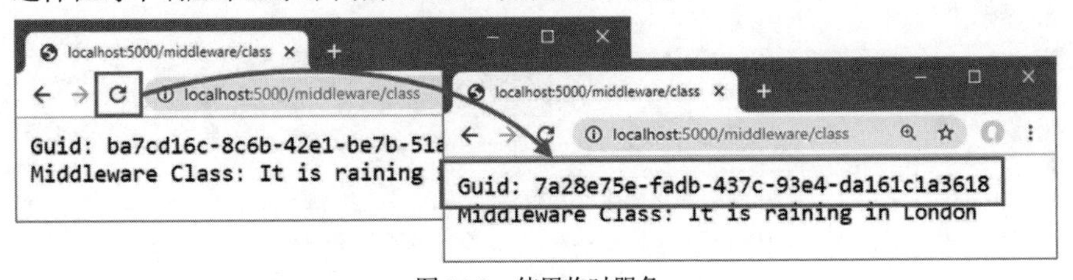


图 14-8 使用临时服务

## 14.4.3 使用有作用域的服务

有作用域的服务在单例服务和瞬态服务之间取得了平衡。在作用域内，依赖关系由同一对象解析。为每个 HTTP 请求启动一个新的作用域，这意味着处理该请求的所有组件将共享一个服务对象。为了准备有作用域的服务，代码清单 14-28 更改了 WeatherMiddleware 类，以声明对同一服务的三个依赖项。

**代码清单 14-28　在 Platform 文件夹的 WeatherMiddleware.cs 文件中添加依赖项**

```
using Microsoft.AspNetCore.Http;
using System.Threading.Tasks;
using Platform.Services;

namespace Platform {
    public class WeatherMiddleware {
        private RequestDelegate next;

        public WeatherMiddleware(RequestDelegate nextDelegate) {
            next = nextDelegate;
        }

        public async Task Invoke(HttpContext context, IResponseFormatter formatter1,
                IResponseFormatter formatter2, IResponseFormatter formatter3) {
            if (context.Request.Path == "/middleware/class") {
                await formatter1.Format(context, string.Empty);
                await formatter2.Format(context, string.Empty);
                await formatter3.Format(context, string.Empty);
            } else {
                await next(context);
            }
        }
    }
}
```

在实际项目中，并不需要对同一个服务声明几个依赖项，但是对于本例来说很有用，因为每个依赖项都是独立解析的。因为 IResponseFormatter 服务是用 AddTransient 方法创建的，所以每个依赖项都用不同的对象解析。重启 ASP.NET Core，请求 http://localhost:5000/middleware/class，对于写入响应的三条消息中的每一条，都使用了不同的 GUID，如图 14-9 所示。当重新加载浏览器时，将显示新组的三个 GUID。

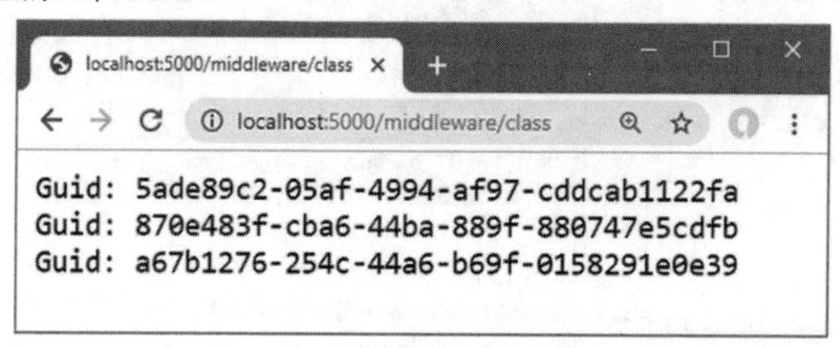


图 14-9　解析对临时服务的依赖关系

代码清单 14-29 通过 AddScoped 方法将 IResponseFormatter 服务更改为使用作用域生命周期。

---

■ **提示：**可以通过 IServiceProvider 接口的 CreateScope 扩展方法创建自己的作用域。结果是与新作用域相关联的 IServiceProvider，它让有作用域的服务拥有自己的实现对象。

---

**代码清单 14-29　在 Platform 文件夹的 Startup.cs 文件中使用有作用域的服务**

```
...
public void ConfigureServices(IServiceCollection services) {
    services.AddScoped<IResponseFormatter, GuidService>();
}
...
```

重启 ASP.NET Core，再次请求 http://localhost:5000/middleware/class，就会使用相同的 GUID 解析中间件组件声明的所有三个依赖项，如图 14-10 所示。当浏览器重新加载时，HTTP 请求发送给 ASP.NET Core，创建一个新的作用域和一个新的服务对象。

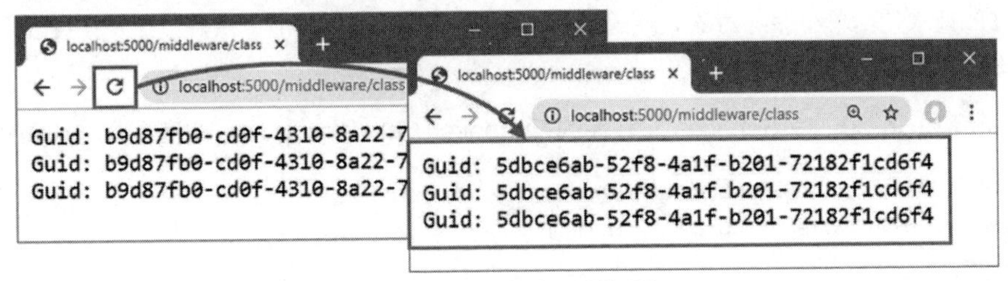


图 14-10　使用有作用域的服务

### 1. 避免有作用域的服务验证陷阱

服务使用者不知道为单例和瞬态服务选择的生命周期：会声明依赖项或请求服务并获得所需的对象。

有作用域的服务只能在一个作用域内使用。为收到的每个请求自动创建一个新的作用域。在作用域之外请求有作用域的服务会导致异常。要查看问题，请求 http://localhost:5000/endpoint/class，会生成异常响应，如图 14-11 所示。

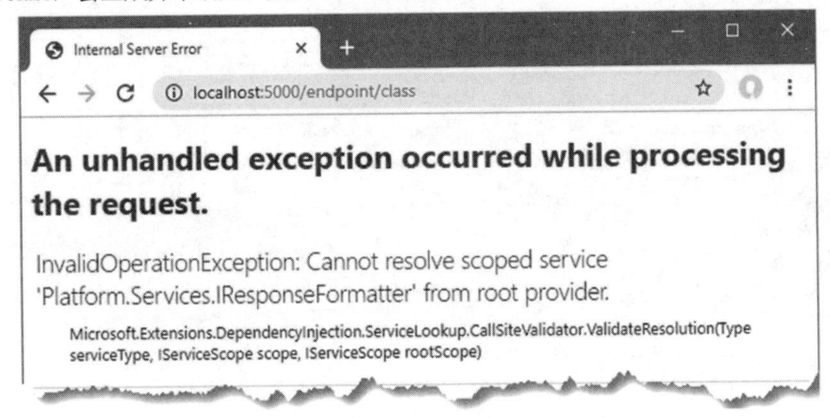


图 14-11　请求有作用域的服务

配置端点的扩展方法通过从路由中间件获得的 IServiceProvider 对象解析服务，如下所示：

```
...
app.MapGet(path, context => (Task)methodInfo.Invoke(endpointInstance,
   methodParams.Select(p => p.ParameterType == typeof(HttpContext)
   ? context : app.ServiceProvider.GetService(p.ParameterType)).ToArray()));
...
```

这称为根提供程序，它不提供对有作用域的服务的访问，以防止意外地使用不打算在作用域之外使用的服务。

### 2. 通过上下文对象访问有作用域的服务

HttpContext 类定义了 RequestServices 属性，该属性返回 IServiceProvider 对象，该对象允许访问有作用域的服务，以及单例和瞬态服务。这非常适合有作用域的服务的最常见用法，即对每个 HTTP 请求使用单个服务对象。代码清单 14-30 修改了端点扩展方法，以便使用 HttpContext 提供的服务来解析依赖关系。

**代码清单 14-30　在 Services 文件夹的 EndpointExtensions.cs 文件中使用有作用域的服务**

```
using Microsoft.AspNetCore.Http;
using Microsoft.AspNetCore.Routing;
using Microsoft.Extensions.DependencyInjection;
using Platform.Services;
using System.Reflection;
using System.Threading.Tasks;
using System.Linq;

namespace Microsoft.AspNetCore.Builder {

   public static class EndpointExtensions {

      public static void MapEndpoint<T>(this IEndpointRouteBuilder app,
            string path, string methodName = "Endpoint") {
         MethodInfo methodInfo = typeof(T).GetMethod(methodName);
         if (methodInfo == null || methodInfo.ReturnType != typeof(Task)) {
            throw new System.Exception("Method cannot be used");
         }

         T endpointInstance =
            ActivatorUtilities.CreateInstance<T>(app.ServiceProvider);

         ParameterInfo[] methodParams = methodInfo.GetParameters();
         app.MapGet(path, context => (Task)methodInfo.Invoke(endpointInstance,
            methodParams.Select(p => p.ParameterType == typeof(HttpContext)
            ? context
            : context.RequestServices.GetService(p.ParameterType)).ToArray()));
      }
```

```
    }
}
```

只有处理请求的方法所声明的依赖项才使用 HttpContext.RequestServices 属性进行解析。由端点类构造函数声明的服务仍然使用 IEndpointRouteBuilder.ServiceProvider 属性解析，该属性确保端点不会不适当地使用有作用域的服务。

### 3. 为每个请求创建新的处理程序

扩展方法的问题在于，它要求端点类知道它们所依赖的服务的生命周期。WeatherEndpoint 类依赖 IResponseFormatter 服务，必须知道依赖项只能通过 Endpoint 方法声明，而不能通过构造函数声明。

如果不愿意采用以上方式，可以创建一个端点类的新实例来处理每个请求，如代码清单 14-31 所示，它允许解析构造函数和方法依赖关系，而不需要知道哪些服务指定了作用域。

**代码清单 14-31　在 Services 文件夹的 EndpointExtensions.cs 文件中实例化端点**

```
using Microsoft.AspNetCore.Http;
using Microsoft.AspNetCore.Routing;
using Microsoft.Extensions.DependencyInjection;
using Platform.Services;
using System.Reflection;
using System.Threading.Tasks;
using System.Linq;

namespace Microsoft.AspNetCore.Builder {

    public static class EndpointExtensions {

        public static void MapEndpoint<T>(this IEndpointRouteBuilder app,
                string path, string methodName = "Endpoint") {
            MethodInfo methodInfo = typeof(T).GetMethod(methodName);
            if (methodInfo == null || methodInfo.ReturnType != typeof(Task)) {
                throw new System.Exception("Method cannot be used");
            }

            ParameterInfo[] methodParams = methodInfo.GetParameters();
            app.MapGet(path, context => {
                T endpointInstance =
                ActivatorUtilities.CreateInstance<T>(context.RequestServices);
                return (Task)methodInfo.Invoke(endpointInstance,
                    methodParams.Select(p =>
                    p.ParameterType == typeof(HttpContext) ? context :
                    context.RequestServices.GetService(p.ParameterType)).ToArray());
            });
        }
    }
}
```

这种方法需要端点类的一个新实例来处理每个请求，但是它确保不需要了解服务的生命周期。

#### 4. 在 lambda 表达式中使用有作用域的服务

HttpContext 类还可以用于定义为 lambda 表达式的中间件组件和端点，如代码清单 14-32 所示。

**代码清单 14-32　在 Platform 文件夹的 Startup.cs 文件中使用有作用域的服务**

```
using Microsoft.AspNetCore.Builder;
using Microsoft.AspNetCore.Hosting;
using Microsoft.AspNetCore.Http;
using Microsoft.Extensions.DependencyInjection;
using Microsoft.AspNetCore.Routing;
using Platform.Services;

namespace Platform {
    public class Startup {

        public void ConfigureServices(IServiceCollection services) {
            services.AddScoped<IResponseFormatter, GuidService>();
        }

        public void Configure(IApplicationBuilder app, IWebHostEnvironment env) {
            app.UseDeveloperExceptionPage();
            app.UseRouting();
            app.UseMiddleware<WeatherMiddleware>();

            app.Use(async (context, next) => {
                if (context.Request.Path == "/middleware/function") {
                    IResponseFormatter formatter
                        = context.RequestServices.GetService<IResponseFormatter>();
                    await formatter.Format(context,
                        "Middleware Function: It is snowing in Chicago");
                } else {
                    await next();
                }
            });

            app.UseEndpoints(endpoints => {

                endpoints.MapEndpoint<WeatherEndpoint>("/endpoint/class");

                endpoints.MapGet("/endpoint/function", async context => {
                    IResponseFormatter formatter
                        = context.RequestServices.GetService<IResponseFormatter>();
                    await formatter.Format(context,
                        "Endpoint Function: It is sunny in LA");
                });
```

```
            });
        }
    }
}
```

重启 ASP.NET Core，请求指向 lambda 函数的 URL:http://localhost:5000/endpoint/function 和/middleware/function。将毫无例外地获得有作用域的服务，生成如图 14-12 所示的响应。

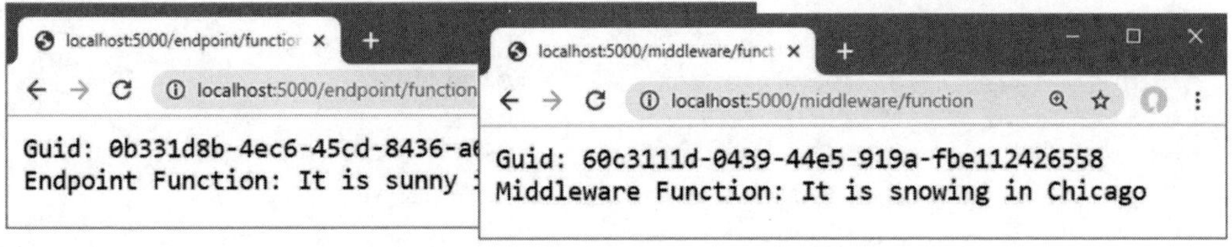


图 14-12 在 lambda 函数中使用有作用域的服务

# 14.5 其他依赖注入特性

接下来描述在使用依赖项注入时可用的一些附加特性。这些并不是所有项目都需要的，但值得理解，因为它们为依赖注入的工作方式提供了上下文。当标准特性不是项目所需要的功能时，它们会很有帮助。

## 14.5.1 创建依赖关系链

当实例化类以解析服务依赖项时，将检查其构造函数，并解析服务的任何依赖项。这允许一个服务声明对另一个服务的依赖，从而创建一个自动解析的链。为方便演示，向 Platform/Services 文件夹添加一个名为 TimeStamping.cs 的类文件，代码如代码清单 14-33 所示。

**代码清单14-33 Services 文件夹中 TimeStamping.cs 文件的内容**

```
using System;

namespace Platform.Services {

    public interface ITimeStamper {
        string TimeStamp { get; }
    }

    public class DefaultTimeStamper : ITimeStamper {

        public string TimeStamp {
            get => DateTime.Now.ToShortTimeString();
        }
    }
}
```

类文件定义了一个名为ITimeStamper的接口和一个名为DefaultTimeStamper的实现类。接下来，使用代码清单 14-34 所示的代码，将名为 TimeResponseFormatter.cs 的文件添加到 Platform/Services文件夹中。

**代码清单 14-34　Services 文件夹中 TimeResponseFormatter.cs 文件的内容**

```
using System.Threading.Tasks;
using Microsoft.AspNetCore.Http;

namespace Platform.Services {
    public class TimeResponseFormatter : IResponseFormatter {
        private ITimeStamper stamper;

        public TimeResponseFormatter(ITimeStamper timeStamper) {
            stamper = timeStamper;
        }

        public async Task Format(HttpContext context, string content) {
            await context.Response.WriteAsync($"{stamper.TimeStamp}: {content}");
        }
    }
}
```

TimeResponseFormatter 类是 IResponseFormatter 接口的实现，该接口使用构造函数参数声明对 ITimeStamper 接口的依赖。代码清单 14-35 在 Startup 类的 ConfigureServices 方法中为这两个接口定义了服务。

---

**注意：** 服务不需要与它们的依赖项有相同的生命周期，但如果把两个生命周期混在一起，可能会产生奇怪的效果。生命周期仅在解析依赖项时应用，这意味着如果一个有作用域的服务依赖于一个瞬态服务，那么瞬态对象的行为就像它分配了有作用域的生命周期一样。

---

**代码清单 14-35　在 Platform 文件夹的 Startup.cs 文件中配置服务**

```
...
public void ConfigureServices(IServiceCollection services) {
    services.AddScoped<IResponseFormatter, TimeResponseFormatter>();
    services.AddScoped<ITimeStamper, DefaultTimeStamper>();
}
...
```

当解析对 IResponseFormatter 服务的依赖时，检查 TimeResponseFormatter 构造函数，并检测其对 ITimeStamper 服务的依赖。创建一个 DefaultTimeStamper 对象并注入 TimeResponseFormatter 构造函数中，该构造函数允许解析原始依赖项。要查看依赖项链的运行情况，请重新启动 ASP.NET Core，请求 http://localhost:5000/middleware/function。可以看到，DefaultTimeStamper 类生成的时间戳被包含到 TimeResponseFormatter 类生成的响应中，如图 14-13 所示。

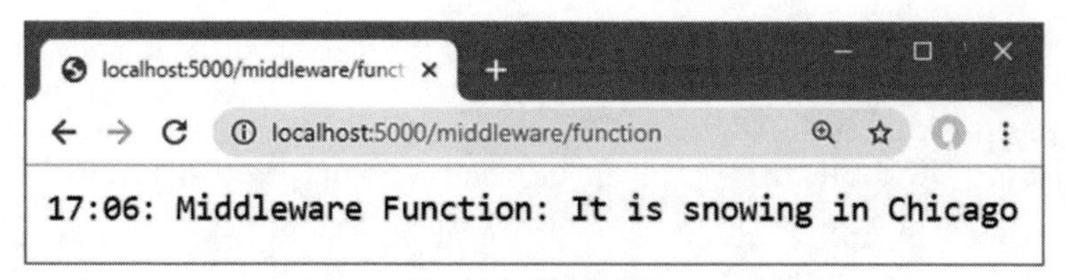


图 14-13　创建依赖链

## 14.5.2　访问 ConfigureServices 方法中的服务

依赖注入特性是由 Platform 类在实例化 Startup 类、调用 ConfigureServices 方法之前设置的。在设置过程中，对于创建 ASP.NET Core 需要的服务，以及为应用程序提供的基本服务(如配置数据、日志记录和环境设置)，都可以参阅前面的章节。

Startup 类可以通过定义构造函数来声明对这些服务的依赖关系。在应用程序启动期间实例化 Startup 类时，检查构造函数，并解析它声明的依赖项。注入构造函数中的对象可以分配属性，并在 ConfigureService 方法中访问，该方法添加特定于应用程序的服务。然后，组合的服务集对 Configure 方法、中间件组件和处理请求的端点可用。

---

■ **注意：** Startup 类定义的方法不以同样的方式对待。Configure 方法可以使用参数声明依赖关系，但如果向 ConfigureServices 方法添加参数，则会收到异常。本节中描述的技术是在 ConfigureServices 方法中声明对服务的依赖关系的唯一方法。

---

Startup 构造函数最常见的用法是声明对 IConfiguration 服务的依赖，后者提供对应用程序配置数据的访问，如代码清单 14-36 所示。配置数据详见第 15 章。

**代码清单 14-36　在 Platform 文件夹的 Startup.cs 文件中声明一个依赖项**

```
using Microsoft.AspNetCore.Builder;
using Microsoft.AspNetCore.Hosting;
using Microsoft.AspNetCore.Routing;
using Microsoft.Extensions.DependencyInjection;
using Platform.Services;
using Microsoft.Extensions.Configuration;

namespace Platform {

    public class Startup {

        public Startup(IConfiguration config) {
            Configuration = config;
        }

        private IConfiguration Configuration;

        public void ConfigureServices(IServiceCollection services) {
            services.AddScoped<IResponseFormatter, TimeResponseFormatter>();
            services.AddScoped<ITimeStamper, DefaultTimeStamper>();
```

```
        }

        public void Configure(IApplicationBuilder app, IWebHostEnvironment env) {

            // ...statements omitted for brevity...
        }
    }
}
```

IConfiguration 服务通过构造函数接收，并将其分配给名为 Configuration 的属性，然后可以由 ConfigureServices 方法使用该属性。此示例没有更改应用程序生成的响应，但下一节介绍如何使用配置数据配置服务。

### 14.5.3　使用服务工厂函数

工厂函数允许控制如何创建服务实现对象，而不是依赖ASP.NET Core来创建实例。存在AddSingleton、AddTransient和AddScoped方法的工厂版本，所有这些方法都与接收IServiceProvider对象并返回服务实现对象的函数一起使用。

工厂函数的一种用途是将服务的实现类定义为配置设置，通过 IConfguration 服务读取配置设置。这需要上一节中描述的模式，以便在 ConfigureServices 方法中访问配置数据。代码清单 14-37 为 IResponseFormatter 服务添加了一个工厂函数，该服务从配置数据中获取实现类。

**代码清单 14-37　在 Platform 文件夹的 Startup.cs 文件中使用工厂函数**

```
using Microsoft.AspNetCore.Builder;
using Microsoft.AspNetCore.Hosting;
using Microsoft.AspNetCore.Http;
using Microsoft.Extensions.DependencyInjection;
using Microsoft.AspNetCore.Routing;
using Platform.Services;
using Microsoft.Extensions.Configuration;
using System;

namespace Platform {
    public class Startup {

        public Startup(IConfiguration config) {
            Configuration = config;
        }

        private IConfiguration Configuration;

        public void ConfigureServices(IServiceCollection services) {
            services.AddScoped<IResponseFormatter>(serviceProvider => {
                string typeName = Configuration["services:IResponseFormatter"];
                return (IResponseFormatter)ActivatorUtilities
                    .CreateInstance(serviceProvider, typeName == null
```

```
                    ? typeof(GuidService) : Type.GetType(typeName, true));
            });
            services.AddScoped<ITimeStamper, DefaultTimeStamper>();
        }

        public void Configure(IApplicationBuilder app, IWebHostEnvironment env) {

            // ...statements omitted for brevity...
        }
    }
}
```

工厂函数从配置数据中读取一个值，该值转换为一种类型并传递给 ActivatorUtilities. CreateInstance 方法。代码清单 14-38 向 appsettings.Development. json 文件添加了一个配置设置。该文件选择 HtmlResponseFormatter 类作为 IResponseFormatter 服务的实现。JSON 配置文件详见第 15 章。

**代码清单 14-38　在 Platform 文件夹的 appsettings.Development.json 文件中定义设置**

```
{
  "Logging": {
    "LogLevel": {
      "Default": "Information",
      "Microsoft": "Warning",
      "Microsoft.Hosting.Lifetime": "Information"
    }
  },
  "services": {
    "IResponseFormatter": "Platform.Services.HtmlResponseFormatter"
  }
}
```

当解析对 IResponseFormatter 服务的依赖时，工厂函数创建配置文件中指定的类型的实例。重启 ASP.NET Core，请求 http://localhost:5000/middleware/function URL，生成如图 14-14 所示的响应。

图 14-14　使用服务工厂

## 14.5.4　创建具有多个实现的服务

可以用多个实现来定义服务，这允许使用者选择最适合特定问题的实现。当服务接口提供对每个实现类的洞察时，这是一个最合适的特性。要提供关于 IResponseFormatter 实现类的功能的信息，请向接口添加如代码清单 14-39 所示的默认属性。

**代码清单 14-39 在 Services 文件夹的 IResponseFormatter.cs 文件中添加一个属性**

```
using Microsoft.AspNetCore.Http;
using System.Threading.Tasks;

namespace Platform.Services {
    public interface IResponseFormatter {

        Task Format(HttpContext context, string content);

        public bool RichOutput => false;
    }
}
```

对于没有覆盖默认值的实现类，这个 RichOutput 属性将为 false。为了确保有一个返回 true 的实现，将代码清单 14-40 中所示的属性添加到 HtmlResponseFormatter 类中。

**代码清单 14-40 在 Services 文件夹的 HtmlResponseFormatter.cs 文件中重写属性**

```
using System.Threading.Tasks;
using Microsoft.AspNetCore.Http;

namespace Platform.Services {
    public class HtmlResponseFormatter : IResponseFormatter {

        public async Task Format(HttpContext context, string content) {
            context.Response.ContentType = "text/html";
            await context.Response.WriteAsync($@"
                <!DOCTYPE html>
                <html lang=""en"">
                <head><title>Response</title></head>
                <body>
                    <h2>Formatted Response</h2>
                    <span>{content}</span>
                </body>
                </html>");
        }
        public bool RichOutput => true;
    }
}
```

代码清单 14-41 为 IResponseFormatter 服务注册了多个实现，这是通过重复调用 Add<lifecycle> 方法完成的。代码清单还将现有的请求管道替换为两个路由，演示了如何使用服务。

代码清单 14-41　在 Platform 文件夹的 Startup.cs 文件中定义和使用一个服务

```
using Microsoft.AspNetCore.Builder;
using Microsoft.AspNetCore.Hosting;
using Microsoft.AspNetCore.Http;
using Microsoft.Extensions.DependencyInjection;
using Microsoft.AspNetCore.Routing;
using Platform.Services;
using Microsoft.Extensions.Configuration;
using System;
using System.Linq;

namespace Platform {
    public class Startup {

        public Startup(IConfiguration config) {
            Configuration = config;
        }

        private IConfiguration Configuration;

        public void ConfigureServices(IServiceCollection services) {
            services.AddScoped<ITimeStamper, DefaultTimeStamper>();
            services.AddScoped<IResponseFormatter, TextResponseFormatter>();
            services.AddScoped<IResponseFormatter, HtmlResponseFormatter>();
            services.AddScoped<IResponseFormatter, GuidService>();
        }
        public void Configure(IApplicationBuilder app, IWebHostEnvironment env) {
            app.UseDeveloperExceptionPage();
            app.UseRouting();
            app.UseEndpoints(endpoints => {
                endpoints.MapGet("/single", async context => {
                    IResponseFormatter formatter = context.RequestServices
                        .GetService<IResponseFormatter>();
                    await formatter.Format(context, "Single service");
                });

                endpoints.MapGet("/", async context => {
                    IResponseFormatter formatter = context.RequestServices
                        .GetServices<IResponseFormatter>().First(f => f.RichOutput);
                    await formatter.Format(context, "Multiple services");
                });
            });
        }
    }
}
```

AddScoped 语句为 IResponseFormatter 接口注册了三个服务，每个服务都有不同的实现类。单个 URL 的路由使用 IServiceProvider.GetService<T>方法请求服务，如下所示：

```
...
context.RequestServices.GetService<IResponseFormatter>();
...
```

这是一个不知道有多种可用实现的服务使用者。使用最近注册的实现(即 GuidService 类)解析服务。重启 ASP.NET Core 并请求 http://localhost:5000/single，在图 14-15 的左侧会看到输出。

图 14-15　使用多个服务实现

另一个端点是服务使用者，它知道可能有多个实现可用，并使用 IServiceProvider.GetServices<T>方法请求服务。

```
...
context.RequestServices.GetServices<IResponseFormatter>().First(f => f.RichOutput);
...
```

该方法返回一个 IEnumerable<IResponseFormatter>，它枚举可用的实现。使用 LINQ First 方法过滤它们，以选择 RichOutput 属性返回 true 的实现。如果请求 http://localhost:5000，会显示如图 14-15 右侧所示的输出，显示端点选择了最适合其需求的服务实现。

## 14.5.5　在服务中使用未绑定类型

可使用泛型类型参数定义服务，当请求服务时，泛型类型参数绑定到特定类型，如代码清单 14-42 所示。

#### 代码清单 14-42　在 Platform 文件夹的 Startup.cs 文件中使用未绑定类型

```
using Microsoft.AspNetCore.Builder;
using Microsoft.AspNetCore.Hosting;
using Microsoft.AspNetCore.Http;
using Microsoft.Extensions.DependencyInjection;
using Microsoft.AspNetCore.Routing;
using Platform.Services;
using Microsoft.Extensions.Configuration;
using System;
using System.Linq;
using System.Collections.Generic;

namespace Platform {
   public class Startup {
```

```
        public Startup(IConfiguration config) {
            Configuration = config;
        }

        private IConfiguration Configuration;

        public void ConfigureServices(IServiceCollection services) {
            services.AddSingleton(typeof(ICollection<>), typeof(List<>));
        }

        public void Configure(IApplicationBuilder app, IWebHostEnvironment env) {
            app.UseDeveloperExceptionPage();
            app.UseRouting();
            app.UseEndpoints(endpoints => {
                endpoints.MapGet("/string", async context => {
                    ICollection<string> collection
                        = context.RequestServices.GetService<ICollection<string>>();
                    collection.Add($"Request: { DateTime.Now.ToLongTimeString() }");
                    foreach (string str in collection) {
                        await context.Response.WriteAsync($"String: {str}\n");
                    }
                });

                endpoints.MapGet("/int", async context => {
                    ICollection<int> collection
                        = context.RequestServices.GetService<ICollection<int>>();
                    collection.Add(collection.Count() + 1);
                    foreach (int val in collection) {
                        await context.Response.WriteAsync($"Int: {val}\n");
                    }
                });
            });
        }
    }
}
```

此特性依赖 AddSingleton、AddScoped 和 AddTransient 方法的版本，这些方法接收类型作为常规参数，不能使用泛型类型参数执行。代码清单 14-42 中的服务是用未绑定类型创建的，如下所示：

```
...
services.AddSingleton(typeof(ICollection<>), typeof(List<>));
...
```

当解析对 ICollection<T>服务的依赖时，创建 List<T>对象，以便使用 List<string>对象解析对 ICollection<string>的依赖。未绑定服务不需要为每种类型提供单独的服务，而是允许创建所有泛型类型的映射。

代码清单14-42中的两个端点请求ICollection<string>和ICollection<int>服务，每个服务都使用不同的List<T>对象进行解析。要锁定端点，请重新启动ASP.NET Core，并请求http://localhost:5000/string和http://localhost:5000/int。该服务定义为一个单例，这意味着使用相同的List<string>和List<int>对象来解析所有针对ICollection<string>和ICollection<int>的请求。每个请求都向集合中添加一个新项，可以通过重新加载Web浏览器显示出来，如图14-16所示。

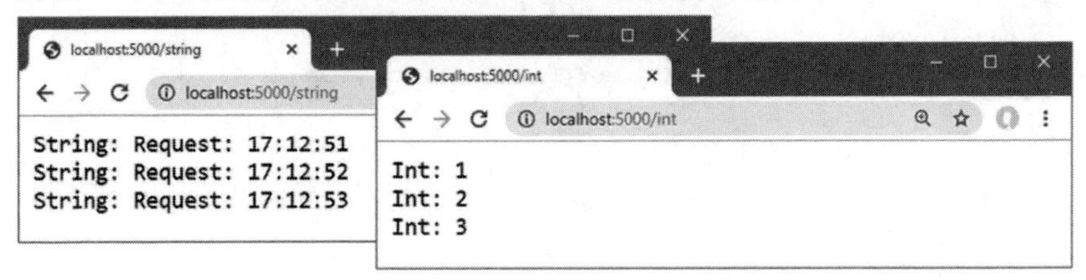


图 14-16　使用带有未绑定类型的单例服务

## 14.6　小结

本章描述了依赖注入，它用于定义易于使用、更改和消费的服务。展示了使用服务的不同方式，解释了可以提供服务的不同生命周期，探讨了一些不太常用的特性，如依赖链和未绑定的服务类型。下一章描述由 ASP.NET Core 平台提供的内置特性。

# 第 15 章

# 使用平台特性(第 1 部分)

ASP.NET Core 包括一组内置的服务和中间件组件，它们提供 Web 应用程序通常需要的特性。本章将描述三个最重要且广泛使用的特性：应用程序配置、日志记录和服务静态内容。第 16 章将继续描述平台的特性，重点关注更高级的内置服务和中间件。表 15-1 给出了本章的上下文。

**表 15-1　平台特性的相关知识**

| 问题 | 答案 |
|---|---|
| 它们是什么? | 平台特性处理常见的 Web 应用程序需求，如配置、日志记录、静态文件、会话、身份验证和数据库访问 |
| 它们为什么有用? | 使用这些特性意味着不必在自己的项目中重新创建它们的功能 |
| 它们是如何使用的? | 使用名称以 Use 开头的扩展方法，在 Startup.Configure 方法中把内置中间件组件添加到请求管道中。在 Startup.Configures 方法中，使用以 Add 开头的扩展方法设置服务 |
| 是否存在缺陷或限制? | 最常见的问题与中间件组件添加到请求管道的顺序有关。记住中间件组件形成了请求通过的链，如第 12 章所述 |
| 还有其他选择吗? | 不需要使用 ASP.NET Core 提供的任何服务或中间件组件 |

表 15-2 总结了本章的内容。

**表 15-2　本章内容摘要**

| 问题 | 解决方案 | 代码清单 |
|---|---|---|
| 访问配置数据 | 使用 IConfiguration 服务 | 15-4～15-7 |
| 设置应用程序环境 | 使用启动设置文件 | 15-11 |
| 确定应用程序环境 | 使用 IWebHostEnvironment 服务 | 15-12 |
| 在项目之外保存敏感数据 | 创建用户机密 | 15-13～15-19 |
| 日志消息 | 使用 ILogger<T>服务 | 15-20～15-22 |
| 交付静态内容 | 启用了静态内容中间件 | 15-23～15-26 |
| 交付客户端包 | 使用 LibMan 安装包并使用静态内容的中间件交付包 | 15-27～15-30 |

## 15.1　准备工作

本章将继续使用第 14 章创建的 Platform 项目。为了准备本章，更新 Startup 类以删除中间件和服务，如代码清单 15-1 所示。

**代码清单 15-1　删除 Platform 文件夹的 Startup.cs 文件中的中间件和服务**

```
using System;
using System.Collections.Generic;
using System.Linq;
using System.Threading.Tasks;
using Microsoft.AspNetCore.Builder;
using Microsoft.AspNetCore.Hosting;
using Microsoft.AspNetCore.Http;
using Microsoft.Extensions.DependencyInjection;
using Microsoft.Extensions.Hosting;
using Microsoft.Extensions.Options;

namespace Platform {
    public class Startup {

        public void ConfigureServices(IServiceCollection services) {
        }

        public void Configure(IApplicationBuilder app, IWebHostEnvironment env) {
            app.UseDeveloperExceptionPage();
            app.UseRouting();
            app.UseEndpoints(endpoints => {
                endpoints.MapGet("/", async context => {
                    await context.Response.WriteAsync("Hello World!");
                });
            });
        }
    }
}
```

本章的重要主题之一是配置数据。用代码清单 15-2 的内容替换 appsettings.Development.json 文件的内容，以删除第 14 章添加的设置。

**代码清单 15-2　替换 Platform 文件夹中 appsettings.Development.json 文件的内容**

```
{
    "Logging": {
        "LogLevel": {
            "Default": "Debug",
            "System": "Information",
            "Microsoft": "Information"
```

```
        }
    }
}
```

为启动应用程序，从 Debug 菜单中选择 Start Without Debugging 或 Run Without Debugging，或者打开一个新的 PowerShell 命令提示符，导航到 Platform 项目文件夹(其中包含 Platform.csproj 文件)，运行代码清单 15-3 所示的命令。

**提示：**可从 https://github.com/apress/pro-asp.net-core-3 下载本章和本书中其他所有章节的示例项目。如果在运行示例时遇到问题，请参阅第 1 章以获得帮助。

**代码清单 15-3 启动 ASP.NET Core 运行时**

```
dotnet run
```

如果使用 Visual Studio 或 Visual Studio Code 启动应用程序，就打开一个新的浏览器窗口并显示如图 15-1 所示的内容。如果应用程序是从命令行启动的，那么打开一个新的浏览器选项卡并导航到 http://localhost:5000；显示如图 15-1 所示的内容。

图 15-1 运行示例应用程序

## 15.2 使用配置服务

ASP.NET Core 提供的内置特性之一是访问应用程序的配置设置，然后将其作为服务呈现。

配置数据的主要来源是 appsettings.json 文件。由空模板创建的 appsettings.json 文件包含以下设置：

```
{
    "Logging": {
        "LogLevel": {
            "Default": "Information",
            "Microsoft": "Warning",
            "Microsoft.Hosting.Lifetime": "Information"
        }
    },
    "AllowedHosts": "*"
}
```

配置服务处理 JSON 配置文件，创建包含单独设置的嵌套配置节。在示例应用程序的 appsettings.json 文件中，配置服务将创建一个日志配置节，其中包含一个 LogLevel 节。LogLevel

节包含 Default、Microsoft 和 Microsoft.Hosting.Lifetime 的设置。还有一个 AllowedHosts 设置，它不是配置部分的一部分，它的值是星号(*字符)。

配置服务不理解 appsettings.json 文件中的配置部分或设置的含义，只负责处理 JSON 数据文件，将配置设置与从其他来源(如环境变量或命令行参数)获得的值合并。结果是一组层次结构的配置属性，如图 15-2 所示。

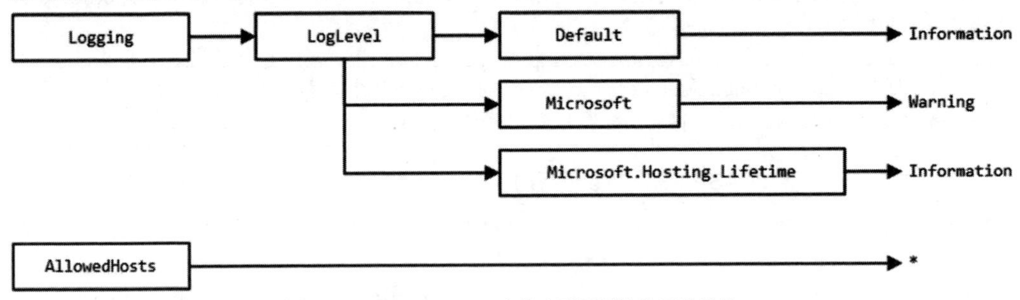


图 15-2　appsettings.json 文件中配置属性的层次结构

### 15.2.1　理解特定于环境的配置文件

大多数项目包含多个 JSON 配置文件，允许为开发周期的不同部分定义不同的设置。有三个预定义的环境，分别名为 Development、Staging 和 Production，每个环境都对应一个常用的开发阶段。在启动期间，配置服务查找名称包含当前环境的 JSON 文件。默认环境是 Development，这意味着配置服务将加载 appsets.Development.json 文件，并使用其内容补充主 appsettings.json 文件的内容。

---

**注意：** Visual Studio 的 Solution Explorer 在 appsettings.json 项目中嵌套了 appsettings.Development.json 文件。可以展开 appsettings.json 文件，来查看和编辑嵌套条目，或者单击 Solution Explorer 顶部禁用嵌套特性的按钮。

---

下面是添加到 appsettings.Ddevelopment.json 文件中的配置设置，如代码清单 15-2 所示。

```
{
  "Logging": {
    "LogLevel": {
      "Default": "Debug",
      "System": "Information",
      "Microsoft": "Information"
    }
  }
}
```

当在两个文件中定义相同的设置时，appsettings.Development.json 文件中的值替换 appsettings.json 文件中的值，这意味着这两个 JSON 文件的内容生成如图 15-3 所示的配置设置层次结构。

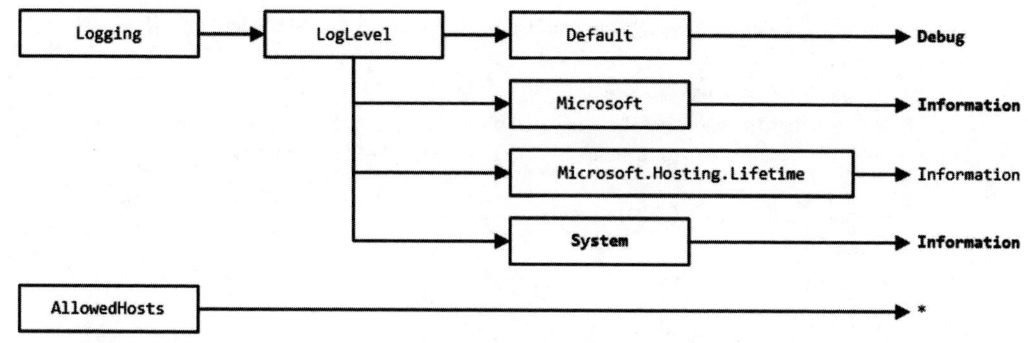


图 15-3　合并 JSON 配置设置

附加配置设置的效果是增加日志消息的详细级别，详见 15.3 一节。

### 15.2.2　访问配置设置

通过服务访问配置数据。如果只需要配置数据来配置中间件，那么可以使用 configure 方法的参数来声明对配置服务的依赖，如代码清单 15-4 所示。

#### 代码清单 15-4　在 Platform 文件夹的 Startup.cs 文件中访问配置数据

```
using System;
using System.Collections.Generic;
using System.Linq;
using System.Threading.Tasks;
using Microsoft.AspNetCore.Builder;
using Microsoft.AspNetCore.Hosting;
using Microsoft.AspNetCore.Http;
using Microsoft.Extensions.DependencyInjection;
using Microsoft.Extensions.Hosting;
using Microsoft.Extensions.Options;
using Microsoft.Extensions.Configuration;

namespace Platform {
   public class Startup {

      public void ConfigureServices(IServiceCollection services) {
      }

      public void Configure(IApplicationBuilder app, IWebHostEnvironment env,
            IConfiguration config ) {

         app.UseDeveloperExceptionPage();
         app.UseRouting();

         app.Use(async (context, next) => {
            string defaultDebug = config["Logging:LogLevel:Default"];
            await context.Response
```

```
                .WriteAsync($"The config setting is: {defaultDebug}");
            });
            app.UseEndpoints(endpoints => {
                endpoints.MapGet("/", async context => {
                    await context.Response.WriteAsync("Hello World!");
                });
            });
        }
    }
}
```

通过 IConfiguration 接口提供配置数据；这个接口在 Microsoft.Extensions.Configuration 名称空间中定义，提供用于导航配置层次结构和读取配置设置的 API。若要接收 Startup 类中的配置数据，向 Configure 方法添加一个 IConfiguration 参数。可以通过指定配置部分的路径来读取配置设置，如下所示：

```
...
string defaultDebug = config["Logging:LogLevel:Default"];
...
```

该语句读取默认设置的值，该值在配置的 Logging 下的 LogLevel 部分定义。配置部分的名称和配置设置由冒号(:字符)分隔。

代码清单 15-4 中读取的配置设置的值用于为处理/config URL 的中间件组件提供结果。为了重新启动 ASP.NET Core，需要选择 Degug | Start Without Debugging，或者在命令提示符下使用 Ctrl+C 组合键并在 Platform 文件夹中运行如代码清单 15-5 所示的命令。

**代码清单 15-5　启动 ASP.NET Core 平台**

```
dotnet run
```

运行时重启后，导航到 http://localhost:5000/config URL，在 Browser 选项卡中显示配置设置值，如图 15-4 所示。

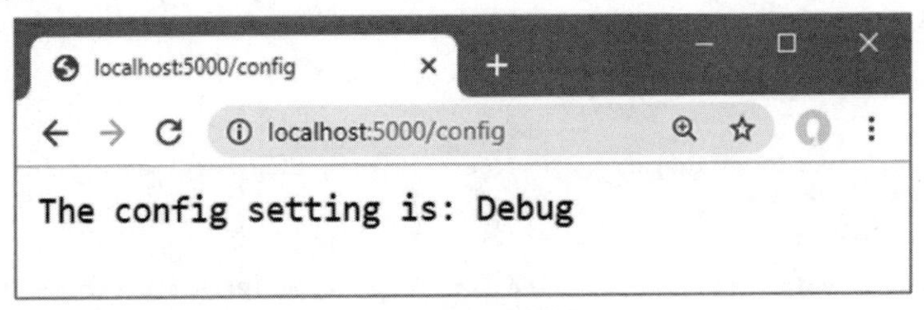


图 15-4　阅读配置数据

### 15.2.3　在服务中使用配置数据

在 ConfigureServices 方法中访问配置数据需要一种稍微不同的方法，其中不支持添加参数来接收 IConfiguration 对象。如第 14 章所述，一些服务是在实例化 Startup 类之前创建的，这允许使用构造函数参数来声明对它们的依赖。代码清单 15-6 向 Startup 类添加了一个构造函数参数，它声明对 IConfiguration 服务的依赖，并将用于解析该依赖关系的对象分配给一个可以在 ConfigureServices 和 Configure 方法中访问的属性。

**代码清单 15-6 在 Platform 文件夹的 Startup.cs 文件中使用配置数据**

```
using System;
using System.Collections.Generic;
using System.Linq;
using System.Threading.Tasks;
using Microsoft.AspNetCore.Builder;
using Microsoft.AspNetCore.Hosting;
using Microsoft.AspNetCore.Http;
using Microsoft.Extensions.DependencyInjection;
using Microsoft.Extensions.Hosting;
using Microsoft.Extensions.Options;
using Microsoft.Extensions.Configuration;

namespace Platform {
    public class Startup {

        public Startup(IConfiguration configService) {
            Configuration = configService;
        }

        private IConfiguration Configuration { get; set; }

        public void ConfigureServices(IServiceCollection services) {

            // configuration data can be accessed here
        }

        public void Configure(IApplicationBuilder app, IWebHostEnvironment env) {
            app.UseDeveloperExceptionPage();
            app.UseRouting();

            app.Use(async (context, next) => {
                string defaultDebug = Configuration["Logging:LogLevel:Default"];
                await context.Response
                    .WriteAsync($"The config setting is: {defaultDebug}");
            });

            app.UseEndpoints(endpoints => {
                endpoints.MapGet("/", async context => {
                    await context.Response.WriteAsync("Hello World!");
                });
            });
        }
```

```
    }
}
```

构造函数通过 IConfiguration 参数接收的对象分配给名为 Configuration 的属性，该属性允许访问 Configure 和 ConfigureServices 方法中的配置数据。

#### 使用 options 模式的配置数据

ConfigureServices 方法中的配置数据与第 14 章描述的 options 模式一起使用。在那一章中，展示了如何使用 lambda 函数更改默认选项值。另一种方法是使用配置设置来设置选项。

在准备时，将代码清单 15-7 所示的配置设置添加到 appsettings.json 文件。

**代码清单 15-7　在 Platform 文件夹的 appsettings.json 文件中添加配置数据**

```
{
  "Location": {
        "CityName": "Buffalo"
  },
  "Logging": {
    "LogLevel": {
      "Default": "Information",
      "Microsoft": "Warning",
      "Microsoft.Hosting.Lifetime": "Information"
    }
  },
  "AllowedHosts": "*"
}
```

在代码清单 15-8 中，使用代码清单 15-7 中定义的配置数据以及 options 模式，配置在第 14 章中创建的 LocationMiddleware 组件。

**代码清单 15-8　在 Platform 文件夹的 Startup.cs 文件中使用配置数据**

```
using System;
using System.Collections.Generic;
using System.Linq;
using System.Threading.Tasks;
using Microsoft.AspNetCore.Builder;
using Microsoft.AspNetCore.Hosting;
using Microsoft.AspNetCore.Http;
using Microsoft.Extensions.DependencyInjection;
using Microsoft.Extensions.Hosting;
using Microsoft.Extensions.Options;
using Microsoft.Extensions.Configuration;

namespace Platform {
  public class Startup {

    public Startup(IConfiguration configService) {
```

```
            Configuration = configService;
        }

        private IConfiguration Configuration { get; set; }

        public void ConfigureServices(IServiceCollection services) {
            services.Configure<MessageOptions>(Configuration.GetSection("Location"));
        }

        public void Configure(IApplicationBuilder app, IWebHostEnvironment env) {
            app.UseDeveloperExceptionPage();
            app.UseRouting();

            app.UseMiddleware<LocationMiddleware>();

            app.Use(async (context, next) => {
                string defaultDebug = Configuration["Logging:LogLevel:Default"];
                await context.Response
                    .WriteAsync($"The config setting is: {defaultDebug}");
            });
            app.UseEndpoints(endpoints => {
                endpoints.MapGet("/", async context => {
                    await context.Response.WriteAsync("Hello World!");
                });
            });
        }
    }
}
```

配置数据的部分使用 GetSection 方法获得，并在创建选项时传递给 Configure 方法。检查选中部分中的配置值，并用 options 类中的相同名称替换默认值。要查看效果，请重新启动 ASP.NET Core，并使用浏览器导航到 http://localhost:5000/location URL。会显示如图 15-5 所示的结果，其中 CityName 选项取自配置数据，CountryName 选项取自 options 类中的默认值。

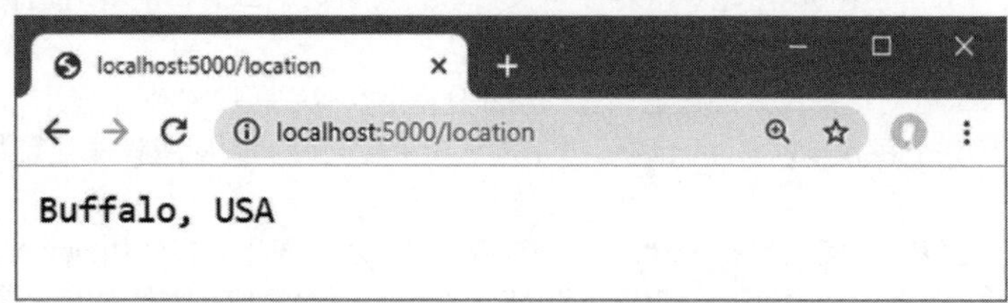


图 15-5 在 options 模式中使用配置数据

## 15.2.4 理解启动设置文件

Properties 文件夹中的 launchSettings.json 文件包含启动 ASP.NET Core 平台的配置设置，包括用于侦听 HTTP 和 HTTPS 请求的 TCP 端口，以及用于选择附加 JSON 配置文件的环境。

**提示：** Visual Studio 通常在默认情况下隐藏 Properties 文件夹。单击 Solution Explorer 顶部的 Show All Files 按钮，以显示文件夹和 launchSettings.json 文件。

下面是使用空模板创建项目时，添加到 launchSettings.json 文件的内容：

```
{
  "iisSettings": {
    "windowsAuthentication": false,
    "anonymousAuthentication": true,
    "iisExpress": {
      "applicationUrl": "http://localhost:5000",
      "sslPort": 0
    }
  },
  "profiles": {
    "IIS Express": {
      "commandName": "IISExpress",
      "launchBrowser": true,
      "environmentVariables": {
        "ASPNETCORE_ENVIRONMENT": "Development"
      }
    },
    "Platform": {
      "commandName": "Project",
      "launchBrowser": true,
      "applicationUrl": "http://localhost:5000",
      "environmentVariables": {
        "ASPNETCORE_ENVIRONMENT": "Development"
      }
    }
  }
}
```

iisSettings 部分用于在 ASP.NET Core 平台通过 IIS Express 启动时，配置 HTTP 和 HTTPS 端口。这在使用 Visual Studio 时发生。

IIS Express 部分对这些设置进行了补充，该部分指定在应用程序启动时是否应该打开一个新的浏览器窗口，其中包含 environmentVariables，它用于定义添加到应用程序配置数据中的环境变量。

Platform 部分包含在使用 Visual Studio Code 或在命令提示符下直接使用 dotnet run 命令启动应用程序时使用的配置，并且该部分指定 ASP.NET Core Kestrel HTTP 服务器的设置。

这两个部分中最重要的部分是 environmentVariables，它定义了 ASPNETCORE_ENVIRONMENT 设置。在启动期间，ASPNETCORE_ENVIRONMENT 设置的值用于选择附加的 JSON 配置文件，例如，Development 的值将导致加载 appsettings.Development.json 文件。

如果使用的是 Visual Studio Code，那么 ASPNETCORE_ENVIRONMENT 是在另一个文件中设置的。选择 Debug | Open Configurations，打开 vscode 文件夹中的 launch.json 文件，它是在使用 Visual Studio Code 编辑项目时创建的。下面是示例项目的默认配置，显示当前的

ASPNETCORE_ENVIRONMENT 值:

```
{
    "version": "0.2.0",
    "configurations": [
        {
            "name": ".NET Core Launch (web)",
            "type": "coreclr",
            "request": "launch",
            "preLaunchTask": "build",
            "program": "${workspaceFolder}/bin/Debug/netcoreapp3.0/Platform.dll",
            "args": [],
            "cwd": "${workspaceFolder}",
            "stopAtEntry": false,
            "serverReadyAction": {
                "action": "openExternally",
                "pattern": "^\\s*Now listening on:\\s+(https?://\\S+)"
            },
            "env": {
                "ASPNETCORE_ENVIRONMENT": "Development"
            },
            "sourceFileMap": {
                "/Views": "${workspaceFolder}/Views"
            }
        },
        {
            "name": ".NET Core Attach",
            "type": "coreclr",
            "request": "attach",
            "processId": "${command:pickProcess}"
        }
    ]
}
```

要显示 ASPNETCORE_ENVIRONMENT 设置的值，将语句添加到响应/config URL 的中间件组件，如代码清单 15-9 所示。

#### 代码清单 15-9 在 Platform 文件夹的 Startup.cs 文件中显示配置设置

```
using System;
using System.Collections.Generic;
using System.Linq;
using System.Threading.Tasks;
using Microsoft.AspNetCore.Builder;
using Microsoft.AspNetCore.Hosting;
using Microsoft.AspNetCore.Http;
using Microsoft.Extensions.DependencyInjection;
using Microsoft.Extensions.Hosting;
```

```
using Microsoft.Extensions.Options;
using Microsoft.Extensions.Configuration;

namespace Platform {
    public class Startup {

        public Startup(IConfiguration configService) {
            Configuration = configService;
        }

        private IConfiguration Configuration { get; set; }

        public void ConfigureServices(IServiceCollection services) {
            services.Configure<MessageOptions>(Configuration.GetSection("Location"));
        }

        public void Configure(IApplicationBuilder app, IWebHostEnvironment env) {
            app.UseDeveloperExceptionPage();
            app.UseRouting();

            app.UseMiddleware<LocationMiddleware>();

            app.Use(async (context, next) => {
                string defaultDebug = Configuration["Logging:LogLevel:Default"];
                await context.Response
                    .WriteAsync($"The config setting is: {defaultDebug}");
                string environ = Configuration["ASPNETCORE_ENVIRONMENT"];
                await context.Response
                    .WriteAsync($"\nThe env setting is: {environ}");
            });

            app.UseEndpoints(endpoints => {
                endpoints.MapGet("/", async context => {
                    await context.Response.WriteAsync("Hello World!");
                });
            });
        }
    }
}
```

重启ASP.NET Core，导航到http://localhost:5000/config，会显示ASPNETCORE_ENVIRONMENT设置的值，如图15-6所示。

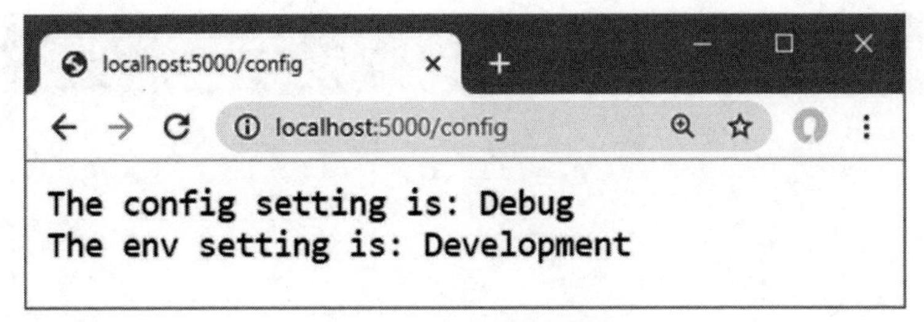


图 15-6　显示环境配置设置

要查看 ASPNETCORE_ENVIRONMENT 设置对总体配置的影响，请更改 launchSettings.json 属性中的值，如代码清单 15-10 所示。

**代码清单 15-10　在 Platform/Properties 文件夹的 launchSettings.json 文件中更改环境**

```
{
  "iisSettings": {
    "windowsAuthentication": false,
    "anonymousAuthentication": true,
    "iisExpress": {
      "applicationUrl": "http://localhost:5000",
      "sslPort": 0
    }
  },
  "profiles": {
    "IIS Express": {
      "commandName": "IISExpress",
      "launchBrowser": true,
      "environmentVariables": {
        "ASPNETCORE_ENVIRONMENT": "Production"
      }
    },
    "Platform": {
      "commandName": "Project",
      "launchBrowser": true,
      "applicationUrl": "http://localhost:5000",
      "environmentVariables": {
        "ASPNETCORE_ENVIRONMENT": "Production"
      }
    }
  }
}
```

如果使用的是 Visual Studio，为改变环境变量，可选择 Project | Platform Properties，导航到 Debug 选项卡。双击 ASPNETCORE_ENVIRONMENT 变量的值，就能将该值更改为 Production，如图 15-7 所示。

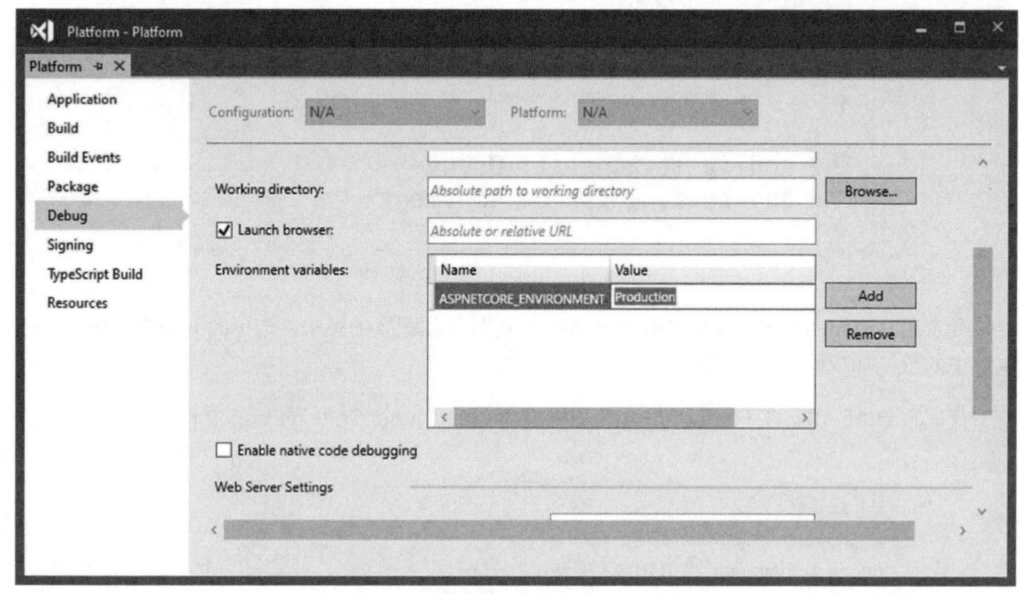


图 15-7　使用 Visual Studio 更改环境变量

如果使用的是 Visual Studio Code，请选择 Debug | Open Configurations，并更改 env 部分中的值，如代码清单 15-11 所示。

#### 代码清单 15-11　在 Platform/.vscode 文件夹的 launch.json 文件中改变环境

```
{
   "version": "0.2.0",
   "configurations": [
      {
         "name": ".NET Core Launch (web)",
         "type": "coreclr",
         "request": "launch",
         "preLaunchTask": "build",
         "program": "${workspaceFolder}/bin/Debug/netcoreapp3.0/Platform.dll",
         "args": [],
         "cwd": "${workspaceFolder}",
         "stopAtEntry": false,
         "serverReadyAction": {
            "action": "openExternally",
            "pattern": "^\\s*Now listening on:\\s+(https?://\\S+)"
         },
         "env": {
            "ASPNETCORE_ENVIRONMENT": "Production"
         },
         "sourceFileMap": {
            "/Views": "${workspaceFolder}/Views"
         }
```

```
        },
        {
            "name": ".NET Core Attach",
            "type": "coreclr",
            "request": "attach",
            "processId": "${command:pickProcess}"
        }
    ]
}
```

将更改保存到属性页或配置文件中，并重新启动 ASP.NET Core，导航到 http://localhost:5000/config，将显示环境更改的效果，如图 15-8 所示。

图 15-8 更改环境配置设置的效果

注意，浏览器中显示的两个配置值都发生了变化。appsettings.Development.json 文件不再加载，项目中没有 appsettings.Production.json 文件，所以只使用 appsettings.json 文件中的配置设置。

## 15.2.5 确定启动类中的环境

ASP.NET Core 平台提供了用于确定当前环境的 IWebHostEnvironment 服务，从而避免了手动获取配置设置的需要。IWebHostEnvironment 服务定义了表 15-3 中所示的方法。这些方法是在 Microsoft.Extensions.Hosting 名称空间中定义的扩展方法，必须使用 using 语句导入该名称空间。

表 15-3 IWebHostEnvironment 扩展方法

| 名称 | 描述 |
| --- | --- |
| IsDevelopment() | 该方法在选择 Development 环境时返回 true |
| IsStaging() | 该方法当选择 Staging 环境时返回 true |
| IsProduction() | 该方法在选择 Production 环境时返回 true |
| IsEnvironment(env) | 当参数指定的环境被选中时，此方法返回 true |

Startup 类中的 Configure 方法已经有一个 IWebHostEnvironment 参数，它是第 14 章使用空模板创建项目时添加的。这通常与表 15-3 中描述的方法一起使用，以选择添加到请求管道的中间件组件。最常见的例子是确定是否使用了 UseDeveloperExceptionPage 方法，因为 UseDeveloperExceptionPage 添加了中间件，生成只应在开发中使用的响应。为了确保中间件只在开发期间使用，代码清单 15-12 使用了 IWebHostEnvironment 服务和 IsDevelopment 方法。

**代码清单 15-12　在 Platform 文件夹的 Startup.cs 文件中选择中间件**

```
...
public void Configure(IApplicationBuilder app, IWebHostEnvironment env) {

    if (env.IsDevelopment()) {
        app.UseDeveloperExceptionPage();
    }
    app.UseRouting();

    app.UseMiddleware<LocationMiddleware>();

    app.Use(async (context, next) => {
        string defaultDebug = Configuration["Logging:LogLevel:Default"];
        await context.Response
            .WriteAsync($"The config setting is: {defaultDebug}");
        string environ = Configuration["ASPNETCORE_ENVIRONMENT"];
        await context.Response
            .WriteAsync($"\nThe env setting is: {environ}");
    });

    app.UseEndpoints(endpoints => {
        endpoints.MapGet("/", async context => {
            await context.Response.WriteAsync("Hello World!");
        });
    });
}
...
```

此中间件捕获处理请求时抛出的异常，并显示详细的堆栈跟踪，这是在部署应用程序后不应该看到的内容。第 16 章将演示错误处理中间件。

---

**提示：** IWebHostEnvironment 服务还定义了一些属性，这些属性可以用来获取应用程序文件位置的详细信息，如第 16 章所示。

---

### 15.2.6　存储用户的秘密

在开发期间，通常需要使用敏感数据来处理应用程序所依赖的服务。这些数据可以包括 API 密钥、数据库连接密码或默认的管理账户，用于访问服务和重新初始化服务，以便使用新的数据库或用户配置测试应用程序更改。

如果敏感数据包含在 C#类或 JSON 配置文件中，就检查源代码版本控制存储库，敏感数据对所有开发人员和其他可以看到代码的人而言是可见的——这可能意味着，对项目可见，是打开了不安全的仓库或存储库。

用户机密服务允许将敏感数据存储在不属于项目的文件中，且不会版本控制，从而允许每个开发人员拥有自己的敏感数据，而不会意外地通过版本控制检入暴露出来。

#### 1. 存储用户的秘密

第一步是准备用于存储敏感数据的文件。打开一个新的 PowerShell 命令提示符，并在 Platform 文件夹(包含 Platform.csproj 文件的文件夹)中运行代码清单 15-13 所示的命令。

**代码清单 15-13　安装用户机密工具包**

```
dotnet tool uninstall --global dotnet-user-secrets
dotnet tool install --global dotnet-user-secrets --version 3.0.0-preview-18579-0056
```

如果之前已经安装了包，这些命令会删除它，并安装本章要求的版本。接下来，在 Platform 文件夹中运行代码清单 15-14 所示的命令。

> ■ **注意：** 如果在运行本节中的命令时遇到问题，请在安装全局工具包后，关闭 PowerShell 提示，并打开一个新的提示。

**代码清单 15-14　初始化用户的秘密**

```
dotnet user-secrets init
```

该命令向 Platform.csproj 项目文件添加一个元素，其中包含项目的唯一 ID，该 ID 将与每个开发人员机器上的秘密相关联。接下来，在 Platform 文件夹中运行代码清单 15-15 所示的命令。

**代码清单 15-15　存储用户机密**

```
dotnet user-secrets set "WebService:Id" "MyAccount"
dotnet user-secrets set "WebService:Key" "MySecret123$"
```

每个秘密都有一个键和一个值，相关的秘密可以使用公共前缀、冒号(:字符)和秘密名称组合在一起。代码清单 15-15 中的命令创建具有 WebService 前缀的相关 Id 和密钥。

在每个命令之后，会显示一条消息，确认已将一个秘密添加到秘密存储中。要检查项目的秘密，使用命令提示符，在 Platform 文件夹中运行代码清单 15-16 所示的命令。

**代码清单 15-16　列出用户机密**

```
dotnet user-secrets list
```

该命令生成以下输出:

```
WebService:Key = MySecret123$
WebService:Id = MyAccount
```

后台在%APPDATA%\Microsoft\UserSecrets 文件夹(或用于 Linux 的~/microsoft/usersecrets 文件夹)中创建一个 JSON 文件，来存储秘密。每个项目都有自己的文件夹(其名称对应于代码清单 15-14 中 init 命令创建的唯一 ID)。

> ■ **提示：** 如果使用的是 Visual Studio，就可在 Solution Explorer 中右击项目，然后从弹出菜单中选择 Manage User Secrets，直接创建和编辑 JSON 文件。

### 2. 读取用户的机密

用户机密与正常配置设置合并，并以相同的方式访问。在代码清单 15-17 中，添加了一条语句，它显示处理/config URL 的中间件组件的秘密。

**代码清单 15-17　使用 Platform 文件夹的 Startup.cs 文件中的用户机密**

```
...
app.Use(async (context, next) => {
    string defaultDebug = Configuration["Logging:LogLevel:Default"];
    await context.Response
        .WriteAsync($"The config setting is: {defaultDebug}");
    string environ = Configuration["ASPNETCORE_ENVIRONMENT"];
    await context.Response
        .WriteAsync($"\nThe env setting is: {environ}");
    string wsID = Configuration["WebService:Id"];
    string wsKey = Configuration["WebService:Key"];
    await context.Response.WriteAsync($"\nThe secret ID is: {wsID}");
    await context.Response.WriteAsync($"\nThe secret Key is: {wsKey}");
});
...
```

只有当应用程序被设置为 Development 环境时，才会加载用户秘密。编辑 launchSettings. json 文件，将环境更改为 Development，如代码清单 15-18 所示(如果喜欢，可以选择 Project | Platform Properties，来使用 Visual Studio 界面)。

**代码清单 15-18　在 Platform/Properties 文件夹的 launchSettings.json 文件中更改环境**

```
{
    "iisSettings": {
        "windowsAuthentication": false,
        "anonymousAuthentication": true,
        "iisExpress": {
            "applicationUrl": "http://localhost:5000",
            "sslPort": 0
        }
    },
    "profiles": {
        "IIS Express": {
            "commandName": "IISExpress",
            "launchBrowser": true,
            "environmentVariables": {
                "ASPNETCORE_ENVIRONMENT": "Development"
            }
        },
        "Platform": {
            "commandName": "Project",
            "launchBrowser": true,
```

```
            "applicationUrl": "http://localhost:5000",
            "environmentVariables": {
                "ASPNETCORE_ENVIRONMENT": "Development"
            }
        }
    }
}
```

如果是从 Visual Studio Code 中启动应用程序，还必须选择 Debug | Open Configurations，更改 env 部分中的值，来编辑环境，如代码清单 15-19 所示。

**代码清单 15-19　在.vscode 文件夹的 launch.json 文件中更改环境**

```
{
    "version": "0.2.0",
    "configurations": [
        {
            "name": ".NET Core Launch (web)",
            "type": "coreclr",
            "request": "launch",
            "preLaunchTask": "build",
            "program": "${workspaceFolder}/bin/Debug/netcoreapp3.0/Platform.dll",
            "args": [],
            "cwd": "${workspaceFolder}",
            "stopAtEntry": false,
            "serverReadyAction": {
                "action": "openExternally",
                "pattern": "^\\s*Now listening on:\\s+(https?://\\S+)"
            },
            "env": {
                "ASPNETCORE_ENVIRONMENT": "Development"
            },
            "sourceFileMap": {
                "/Views": "${workspaceFolder}/Views"
            }
        },
        {
            "name": ".NET Core Attach",
            "type": "coreclr",
            "request": "attach",
            "processId": "${command:pickProcess}"
        }
    ]
}
```

保存更改，重新启动 ASP.NET Core 运行时，请求 http://localhost:5000/config URL 以查看用户机密，如图 15-9 所示。

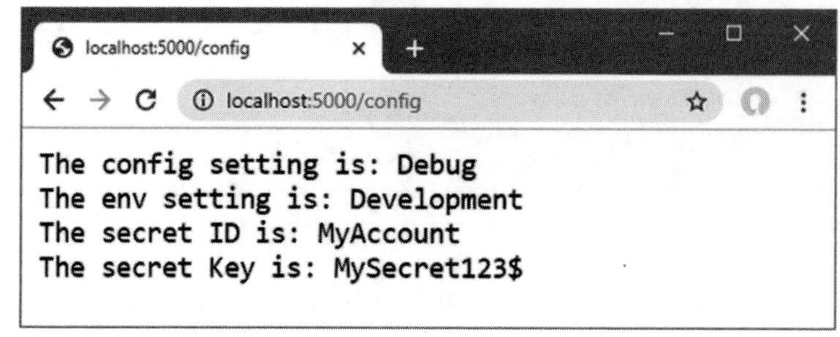


图 15-9　显示用户的机密

## 15.3　使用日志服务

ASP.NET Core 提供了一个日志服务，可用于记录描述应用程序状态的消息，以跟踪错误、监视性能并帮助诊断问题。

日志消息发送到日志提供者，后者负责将消息转发到可以看到、存储和处理它们的地方。有用于基本日志记录的内置提供程序，还有一系列第三方提供程序可用于将消息输入日志记录框架，从而允许对消息进行整理和分析。

默认情况下启用了三个内置提供程序：控制台提供程序、调试提供程序和 EventSource 提供程序。调试提供程序转发消息，以便通过 System.Diagnostics.Debug 类处理它们。Event Source 提供程序为事件跟踪工具转发消息，例如 PerfView(https://github.com/Microsoft/perfview)。

本章使用控制台提供程序，因为它很简单，不需要任何额外的配置来显示日志信息。

---

**提示：**可在 https://docs.microsoft.com/en-gb/aspnet/core/fundamentals/logging 查看可用的提供商列表和启用它们的说明。

---

### 15.3.1　生成日志消息

日志消息是使用未绑定的 ILogger<T>服务生成的。代码清单 15-20 在 Startup.Configure 方法中声明了对服务的依赖，并从前面的示例中删除中间件，以便看起来更加简洁。

**代码清单 15-20　在 Platform 文件夹的 Startup.cs 文件中生成日志消息**

```
using System;
using System.Collections.Generic;
using System.Linq;
using System.Threading.Tasks;
using Microsoft.AspNetCore.Builder;
using Microsoft.AspNetCore.Hosting;
using Microsoft.AspNetCore.Http;
using Microsoft.Extensions.DependencyInjection;
using Microsoft.Extensions.Hosting;
using Microsoft.Extensions.Options;
using Microsoft.Extensions.Configuration;
using Microsoft.Extensions.Logging;
```

```
namespace Platform {
    public class Startup {

        public Startup(IConfiguration configService) {
            Configuration = configService;
        }

        private IConfiguration Configuration { get; set; }

        public void ConfigureServices(IServiceCollection services) {
            services.Configure<MessageOptions>(Configuration.GetSection("Location"));
        }

        public void Configure(IApplicationBuilder app,
                IWebHostEnvironment env, ILogger<Startup> logger) {
            if (env.IsDevelopment()) {
                app.UseDeveloperExceptionPage();
            }

            app.UseRouting();

            app.UseMiddleware<LocationMiddleware>();

            app.UseEndpoints(endpoints => {
                endpoints.MapGet("/", async context => {
                    logger.LogDebug("Response for / started");
                    await context.Response.WriteAsync("Hello World!");
                    logger.LogDebug("Response for / completed");
                });
            });
        }
    }
}
```

日志服务根据分配给消息的类别将日志消息分组在一起。日志消息是使用 ILogger<T>接口写入的，其中通用参数 T 用于指定类别。约定是使用生成消息的类的类型作为类别类型，这就是为什么代码清单 15-20 使用 Startup 作为类型参数来声明对未绑定服务的依赖，如下所示：

```
...
public void Configure(IApplicationBuilder app,
IWebHostEnvironment env, ILogger<Startup> logger) {
...
```

这确保了在 Configure 方法中生成的日志消息在类别启动时分配。日志消息是使用表 15-4 所示的扩展方法创建的。

表 15-4　ILogger<T>扩展方法

| 名称 | 描述 |
| --- | --- |
| LogTrace | 此方法生成一个跟踪级消息，用于开发期间的低级调试 |
| LogDebug | 此方法生成一个调试级消息，用于在解决开发或生产问题期间进行低级调试 |
| LogInformation | 此方法生成一个信息级消息，用于提供有关应用程序一般状态的信息 |
| LogError | 此方法生成一个错误级消息，用于记录应用程序未处理的异常或错误 |
| LogCritical | 此方法生成一个临界级消息，用于记录严重故障 |

给日志消息分配了一个反映其重要性和细节的级别。级别从 Trace(用于详细诊断)到 Critical(用于需要立即响应的最重要信息)不等。每个方法的重载版本都允许使用字符串或异常生成日志消息。在代码清单 15-20 中，使用 LogDebug 方法在处理请求时生成日志消息。

```
...
logger.LogDebug("Response for / started");
...
```

结果是在启动和完成响应时生成调试级别的日志消息。要查看日志消息，请选择 Debug | Start Without Debugging。如果使用命令行，请使用 Ctrl+C 组合键来停止 ASP.NET Core 运行时，并在 Platform 文件夹中使用代码清单 15-21 所示的命令再次启动它。

**代码清单 15-21　启动示例应用程序**

```
dotnet run
```

应用程序启动后，使用浏览器选项卡请求 http://localhost:5000 URL。如果使用命令行，日志消息就显示在应用程序的输出中，如下所示：

```
...
info: Microsoft.AspNetCore.Hosting.Diagnostics[1]
     Request starting HTTP/1.1 GET http://localhost:5000/
info: Microsoft.AspNetCore.Routing.EndpointMiddleware[0]
     Executing endpoint '/ HTTP: GET'
dbug: Platform.Startup[0]
     Response for / started
dbug: Platform.Startup[0]
     Response for / completed
info: Microsoft.AspNetCore.Routing.EndpointMiddleware[1]
     Executed endpoint '/ HTTP: GET'
info: Microsoft.AspNetCore.Hosting.Diagnostics[2]
     Request finished in 23.022100000000002ms 200
...
```

如果使用的是 Visual Studio，则应用程序的输出显示在 Output 窗口中，可以从 View 菜单中选择该窗口。从下拉菜单中选择 ASP.NET Core Web Server，查看输出，如图 15-10 所示。

```
Output
Show output from: Platform - ASP.NET Core Web Server
info: Microsoft.AspNetCore.Hosting.Diagnostics[1]
      Request starting HTTP/1.1 GET http://localhost:5000/
info: Microsoft.AspNetCore.Routing.EndpointMiddleware[0]
      Executing endpoint '/ HTTP: GET'
dbug: Platform.Startup[0]
      Response for / started
dbug: Platform.Startup[0]
      Response for / completed
info: Microsoft.AspNetCore.Routing.EndpointMiddleware[1]
      Executed endpoint '/ HTTP: GET'
info: Microsoft.AspNetCore.Hosting.Diagnostics[2]
      Request finished in 1.2216ms 200
```

图 15-10 在 Visual Studio Output 窗口中选择日志消息

如果使用的是 Visual Studio Code，那么日志消息将示在调试控制台窗口中，如图 15-11 所示，可以从 View 菜单中选择它。

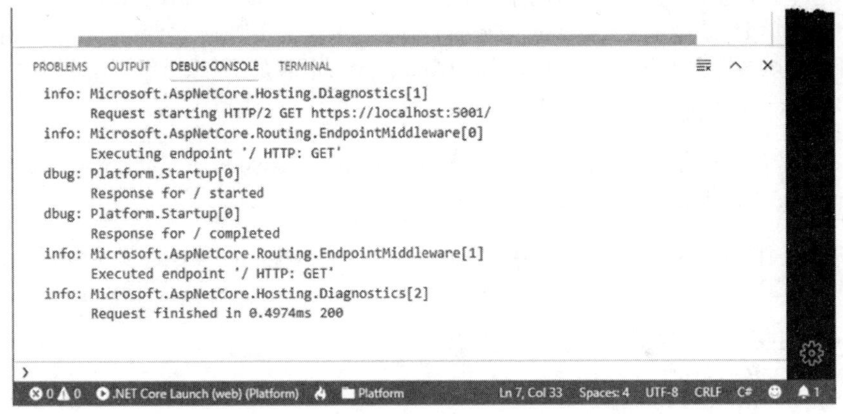


图 15-11 查看 Visual Studio Code 调试控制台窗口中的日志消息

## 15.3.2 配置最小日志级别

第 14 章展示了 appsettings.json 和 appsettings.Development.json 的默认内容，解释了如何合并它们来创建应用程序的配置设置。JSON 文件中的设置用于配置日志服务。ASP.NET Core 提供有关应用程序状态的消息。

appsettings.json 文件的 Logging:LogLevel 部分用于设置记录消息的最低级别。低于最低级别的日志消息将被丢弃。appsettings.json 文件包含以下级别:

```
...
"Default": "Information",
"Microsoft": "Warning",
"Microsoft.Hosting.Lifetime": "Information",
...
```

日志消息的类别(使用代码清单 15-20 中的泛型类型参数设置)用于选择最小过滤器级别。例如，对于 Startup 类生成的日志消息，类别是 Platform.Startup。这意味着要匹配它们，可以直接给 appsettings.json 文件添加 Platform.Startup 条目，或间接指定 Platform 名称空间。没有最小日志级别的任何类别都由 Default 条目匹配，Default 条目设置为 Information。

增加开发期间显示的日志消息的详细信息是常见的，这就是 appsettings. Development.json 文件指定更详细日志级别的原因，例如：

```
...
"Default": "Debug",
"System": "Information",
"Microsoft": "Information"
...
```

当为 Development 环境配置应用程序时，默认日志记录级别为 Debug。System 和 Microsoft 类别的级别设置为 Information，这将影响 ASP.NET Core 和微软提供的其他包和框架生成的日志消息。

通过设置级别为 Trace、Debug、Information、Error 或 Critical，可调整日志级别，使日志集中于应用程序中感兴趣的那些部分。对于使用 None 值的类别，可以禁用日志消息。

在代码清单 15-22 中，向 appsettings.Development.json 文件添加了一个新的日志记录级别，该文件禁用报告响应请求所需时间的消息。还为 Microsoft 设置增加了 Trace 级别，这增加了显示在 ASP.NET Core 平台中的类的详细信息。

---

**提示：** 如果使用的是 Visual Studio，可能需要在 Solution Explorer 中扩展 appsettings.json 项，查看 appsettings.Development.json 文件。

---

**代码清单 15-22　在 Platform 文件夹的 appsettings.Development.json 文件中选择日志级别**

```
{
  "Logging": {
    "LogLevel": {
      "Default": "Debug",
      "System": "Information",
      "Microsoft": "Trace",
      "Microsoft.AspnetCore.Hosting.Diagnostics": "None"
    }
  }
}
```

重启 ASP.NET Core，请求 http://localhost:5000 URL，为每个请求显示的详细信息量就会改变。细节的数量取决于应用程序如何运行。如果正在使用 Visual Studio Code 或使用 dotnet run 命令，可能会显示关于生成 HTTP/2 响应的其他详细信息。如果使用 Visual Studio，则不会看到这些消息，因为应用程序使用代表 ASP.NETCore 平台处理传入 HTTP 请求的 IIS Express。然而，总有一些信息会显示出来：

```
...
dbug: Microsoft.AspNetCore.HostFiltering.HostFilteringMiddleware[0]
      Wildcard detected, all requests with hosts will be allowed.
trce: Microsoft.AspNetCore.HostFiltering.HostFilteringMiddleware[2]
      All hosts are allowed.
...
```

这些消息是由 Host Filter 中间件生成的，它用于基于主机头来限制请求，如第 16 章所述。

# 15.4 使用静态内容和客户端包

大多数 Web 应用程序依赖于动态生成内容和静态内容的混合。动态内容(如购物车的内容或特定产品的详细信息)是由应用程序根据用户的身份和操作生成的，并且是为每个请求生成的。第Ⅲ部分描述使用 ASP.NET Core 创建动态内容的不同方法。

静态内容不会改变，用于提供图像、CSS 样式表、JavaScript 文件以及应用程序所依赖的其他任何内容，但不必为每个请求生成这些内容。ASP.NET Core 项目中静态内容的常规位置是 wwwroot 文件夹。

要准备在本节示例中使用的静态内容，请创建 Platform/wwwroot 文件夹并向其中添加一个名为 static.html 的文件，其内容如代码清单 15-23 所示。如果使用 Visual Studio，可以使用 HTML 页面模板创建文件。

**代码清单 15-23　wwwroot 文件夹中 static.html 文件的内容**

```
<!DOCTYPE html>
<html lang="en">
<head>
    <title>Static Content</title>
</head>
<body>
    <h3>This is static content</h3>
</body>
</html>
```

该文件包含一个基本的 HTML 文档，其中只有在浏览器中显示消息所需的基本元素。

## 15.4.1 添加静态内容中间件

ASP.NET Core 提供了一个处理静态内容请求的中间件组件，静态内容添加到请求管道中，如代码清单 15-24 所示。

**代码清单 15-24　在 Platform 文件夹的 Startup.cs 文件中添加中间件**

```
using System;
using System.Collections.Generic;
using System.Linq;
using System.Threading.Tasks;
using Microsoft.AspNetCore.Builder;
using Microsoft.AspNetCore.Hosting;
using Microsoft.AspNetCore.Http;
using Microsoft.Extensions.DependencyInjection;
using Microsoft.Extensions.Hosting;
using Microsoft.Extensions.Options;
using Microsoft.Extensions.Configuration;
using Microsoft.Extensions.Logging;
```

```
namespace Platform {
    public class Startup {

        public Startup(IConfiguration configService) {
            Configuration = configService;
        }

        private IConfiguration Configuration { get; set; }

        public void ConfigureServices(IServiceCollection services) {
            services.Configure<MessageOptions>(Configuration.GetSection("Location"));
        }

        public void Configure(IApplicationBuilder app,
                IWebHostEnvironment env, ILogger<Startup> logger) {
            if (env.IsDevelopment()) {
                app.UseDeveloperExceptionPage();
            }
            app.UseStaticFiles();
            app.UseRouting();

            app.UseMiddleware<LocationMiddleware>();
            app.UseEndpoints(endpoints => {
                endpoints.MapGet("/", async context => {
                    logger.LogDebug("Response for / started");
                    await context.Response.WriteAsync("Hello World!");
                    logger.LogDebug("Response for / completed");
                });
            });
        }
    }
}
```

UseStaticFiles 扩展方法将静态文件中间件添加到请求管道中。此中间件响应与磁盘文件名称对应的请求，并将其他所有请求传递到管道中的下一个组件。这个中间件通常添加到请求管道的开始处，以防止其他组件处理静态文件的请求。

重启 ASP.NET Core，导航到 http://localhost:5000/static.html。静态文件中间件接收请求，使用 wwwroot 文件夹中 static.html 文件的内容进行响应，如图 15-12 所示。

图 15-12 提供静态内容

中间件组件返回所请求文件的内容并设置响应头，例如Content-Type和Content-Length，用于向浏览器描述内容。

### 更改静态内容中间件的默认选项

在没有参数的情况下调用 UseStaticFiles 方法时，中间件使用 wwwroot 文件夹来定位与所请求的 URL 路径匹配的文件。

将 StaticFileOptions 对象传递给 UseStaticFiles 方法，可以调整此行为。表 15-5 描述了 StaticFileOptions 类定义的属性。

表 15-5 StaticFileOptions 类定义的属性

| 名称 | 描述 |
| --- | --- |
| ContentTypeProvider | 此属性用于获取或设置 IContentTypeProvider 对象，该对象负责生成文件的 MIME 类型。接口的默认实现使用文件扩展名来确定内容类型，并支持最常见的文件类型 |
| DefaultContentType | 如果 IContentTypeProvider 不能确定文件的类型，此属性用于设置默认的内容类型 |
| FileProvider | 此属性用于定位请求的内容，如下所述 |
| OnPrepareResponse | 此属性用于注册将在生成静态内容响应之前调用的操作 |
| RequestPath | 此属性用于指定中间件响应的 URL 路径，如下所述 |
| ServeUnknownFileTypes | 默认情况下，静态内容中间件不会提供内容类型不能由 IContentTypeProvider 确定的文件。将此属性设置为 true，可以更改此行为 |

FileProvider 和 RequestPath 属性是最常用的。FileProvider 属性用于为静态内容选择不同的位置，而 RequestPath 属性用于指定表示对静态上下文的请求的 URL 前缀。代码清单 15-25 使用这两个属性在 Startup 类中配置静态文件中间件。

■ **提示**：还有一个版本的 UseStaticFiles 方法，它接收单个字符串参数，用于设置 RequestPath 配置属性。这是一种添加支持 URL 的方便方法，而不需要创建 options 对象。

**代码清单 15-25 在 Platform 文件夹的 Startup.cs 文件中配置静态文件中间件**

```
using System;
using System.Collections.Generic;
using System.Linq;
using System.Threading.Tasks;
using Microsoft.AspNetCore.Builder;
using Microsoft.AspNetCore.Hosting;
using Microsoft.AspNetCore.Http;
using Microsoft.Extensions.DependencyInjection;
using Microsoft.Extensions.Hosting;
using Microsoft.Extensions.Options;
using Microsoft.Extensions.Configuration;
using Microsoft.Extensions.Logging;
using Microsoft.Extensions.FileProviders;

namespace Platform {
    public class Startup {
```

```
        public Startup(IConfiguration configService) {
            Configuration = configService;
        }

        private IConfiguration Configuration { get; set; }

        public void ConfigureServices(IServiceCollection services) {
            services.Configure<MessageOptions>(Configuration.GetSection("Location"));
        }

        public void Configure(IApplicationBuilder app,
                IWebHostEnvironment env, ILogger<Startup> logger) {
            if (env.IsDevelopment()) {
                app.UseDeveloperExceptionPage();
            }
            app.UseStaticFiles();

            app.UseStaticFiles(new StaticFileOptions {
                FileProvider = new
                    PhysicalFileProvider($"{env.ContentRootPath}/staticfiles"),
                RequestPath = "/files"
            });

            app.UseRouting();

            app.UseMiddleware<LocationMiddleware>();

            app.UseEndpoints(endpoints => {
                endpoints.MapGet("/", async context => {
                    logger.LogDebug("Response for / started");
                    await context.Response.WriteAsync("Hello World!");
                    logger.LogDebug("Response for / completed");
                });
            });
        }
    }
}
```

可将中间件组件的多个实例添加到管道中，每个实例处理 URL 和文件位置之间的单独映射。在代码清单中，静态文件中间件的第二个实例添加到请求管道中，以便使用名为 staticfiles 的文件夹中的文件来处理以/files 开头的 URL 请求。通过 PhysicalFileProvider 类的实例从文件夹中读取文件，该类负责读取磁盘文件。

PhysicalFileProvider 类需要使用一个绝对路径，它基于 IWebHostEnvironment 接口定义的 ContentRootPath 属性的值，该接口用于确定应用程序是在 Development 环境还是 Production 环境中运行。

要为新的中间件组件提供要使用的内容，请创建 Platform/staticfiles 文件夹，并向其中添加一个名为 hello.html 的 HTML 文件，其内容如代码清单 15-26 所示。

**代码清单 15-26　Platform/staticfiles 文件夹中 hello.html 文件的内容**

```
<!DOCTYPE html>
<html lang="en">
<head>
    <title>Static Content</title>
</head>
<body>
    <h3>This is additional static content</h3>
</body>
</html>
```

重启 ASP.NET Core，使用浏览器请求 http://localhost:5000/files/hello.html URL。对于以/files 开头并且对应于 staticfiles 文件夹中文件的 URL 的请求由新中间件处理，如图 15-13 所示。

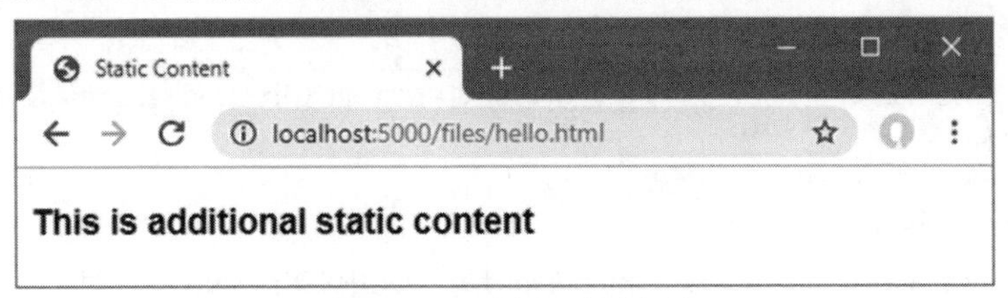


图 15-13　配置静态文件中间件

## 15.4.2　使用客户端包

大多数 Web 应用程序依赖客户端包来支持它们生成的内容，使用 CSS 框架对内容进行样式设置，或使用 JavaScript 包在浏览器中创建丰富的功能。微软提供了库管理器工具，称为 LibMan，用于下载和管理客户端包。

### 1. Bower 怎么了?

早期版本的 ASP.NET Core 依赖一个名为 Bower 的工具来管理客户端包。Bower 在 JavaScript 开发中广泛使用，直到它的创建者终止了该项目，并推荐移植到其他工具。微软引入的.NET Core 应用程序 Library Manager 应该有助于为处理客户端包提供一致的体验，并避免在 Javascript 开发的动态世界中可能出现的工具混乱状况。

### 2. 为客户端包准备项目

使用命令提示符运行代码清单 15-27 所示的命令，该命令将 LibMan 安装为一个全局的.NET Core 工具(尽管 LibMan 可能已经安装好了)。

**代码清单 15-27　安装 LibMan**

```
dotnet tool install --global Microsoft.Web.LibraryManager.Cli --version 2.0.96
```

下一步是创建 LibMan 配置文件，该文件指定用于获取客户端包的存储库和下载包的目录。打开 PowerShell 命令提示符，在 Platform 文件夹中运行代码清单 15-28 所示的命令。

代码清单 15-28　初始化 LibMan

```
libman init -p cdnjs
```

p 参数指定获取包的提供者。这里使用了 cdnjs，它选择 cdnjs.com。另一个选项是 unpkg，它选择 unpkg.com。如果没有使用包存储库的经验，那么应该从 cdnjs 选项开始。

代码清单 15-28 中的命令在 Platform 文件夹中创建一个名为 libman.json 的文件，该文件包含以下设置：

```
...
{
   "version": "1.0",
   "defaultProvider": "cdnjs",
   "libraries": []
}
...
```

如果使用的是 Visual Studio，则可以选择 Project | Manage Client-Side Libraries，来创建和编辑 libman.json 文件。

### 3. 安装客户端包

如果使用的是 Visual Studio，可以在 Solution Explorer 中右击 Platform 项目，从弹出菜单中选择 Add | Client-Side Library 来安装客户端包。Visual Studio 提供一个简单的用户界面，允许将包放在存储库中，并安装到项目中。在 Library 文本字段中输入时，存储库将查询名称匹配的包。在文本字段中输入 twitter-bootstrap，就会选中流行的 Bootstrap CSS 框架，如图 15-14 所示。

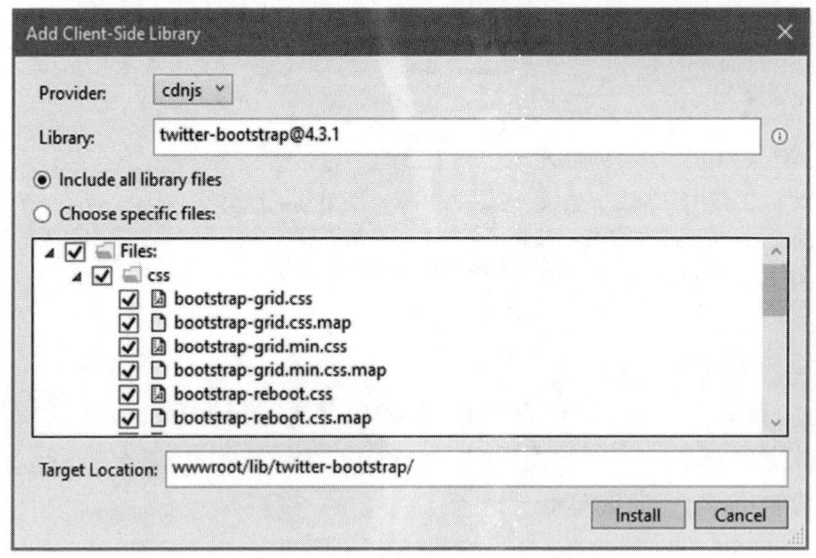


图 15-14　在 Visual Studio 中选择一个客户端包

选择包的最新版本，在撰写本书时为 4.3.1。单击 Install 按钮，下载并安装引导程序包。

**包名和存储库**

同一个包在不同的存储库中可能有不同的名称。如果习惯使用 Bower 或节点包管理器工具 NPM)来安装包，那么可能习惯于将引导 CSS 框架称为 Bootstrap。CDNJS 存储库引用与 twitter-bootstrap 相同的包，尽管包及其内容是相同的。确定需要什么包的最好方法是访问存储库网站并直接执行搜索。对于在本章中使用的 libman 配置，这意味着要访问 cdnjs.com。

也可从命令行安装包。如果正在使用 Visual Studio Code(或者喜欢使用命令行)，那么在 Platform 文件夹中运行代码清单 15-29 所示的命令，来安装 Bootstrap 包。

**代码清单 15-29　安装 Bootstrap 包**

```
libman install twitter-bootstrap@4.3.1 -d wwwroot/lib/twitter-bootstrap
```

所需的版本通过@字符与包名分离，而-d 参数用于指定安装包的位置。wwwroot/lib 文件夹是在 ASP.NETCore 项目中安装客户端包的常规位置。

不管采用哪种方法来安装客户端包，结果都是 wwwroot/lib/twitter-boostrap 文件夹包含由引导 CSS 框架提供的 CSS 样式表和 JavaScript 文件。

#### 4. 使用客户端包

一旦安装了客户端包，它的文件可以通过脚本或链接 HTML 元素或使用高级 ASP.NET Core 提供的特性来引用。参见后续章节。

为了保持简单性，代码清单 15-30 在本节前面创建的静态 HTML 文件中添加了一个 link 元素。

**代码清单 15-30　在 Platform/wwwroot 文件夹中的 static.html 文件中使用客户端包**

```
<!DOCTYPE html>
<html lang="en">
<head>
    <link rel="stylesheet" href="/lib/twitter-bootstrap/css/bootstrap.min.css" />
    <title>Static Content</title>
</head>
<body>
    <h3 class="p-2 bg-primary text-white">This is static content</h3>
</body>
</html>
```

重启 ASP.NET Core，请求 http://localhost:5000/static.html。当浏览器接收并处理 static.html 文件的内容时，它会遇到 link 元素，并向 ASP.NET Core 运行时发送一个 HTTP 请求 /lib/twitter-bootstrap/css/bootstrap.min.css URL。在代码清单 15-24 中添加的原始静态文件中间件组件接收此请求，确定它对应于 wwwroot 文件夹中的一个文件，并返回其内容，向浏览器提供启动 CSS 样式表。通过分配给 h3 元素的类应用引导样式，生成如图 15-15 所示的结果。

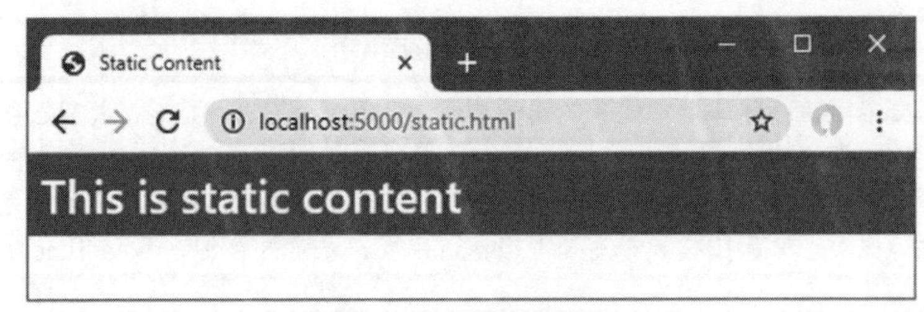


图 15-15　使用客户端包

## 15.5　小结

本章描述了 ASP.NET Core 为 Web 应用程序提供的一些最重要的有用功能，展示了如何配置 ASP.NET Core 应用程序，如何生成和管理日志消息，以及如何提供静态内容(包括客户端包的内容)。第 16 章将继续描述 ASP.NET Core 平台特性。

# 第 16 章

# 使用平台特性(第 2 部分)

本章将继续描述 ASP.NET Core 平台提供的基本特性。解释如何使用 cookie，以及如何管理 cookie consent 的跟踪。介绍会话如何提供基本 cookie 的健壮替代品，如何使用和强制 HTTPS 请求，如何处理错误，以及如何基于 Host 头过滤请求。表 16-1 总结了本章的内容。

表 16-1　本章内容摘要

| 问题 | 解决方案 | 代码清单 |
|---|---|---|
| 使用 cookie | 使用上下文对象来读写 cookie | 16-1～16-3 |
| 管理 cookie consent | 使用 consent 中间件 | 16-4～16-6 |
| 跨请求存储数据 | 使用会话 | 16-7～16-8 |
| 保护 HTTP 请求 | 使用 HTTPS 中间件 | 16-9～16-13 |
| 处理错误 | 使用错误和状态代码中间件 | 16-14～16-19 |
| 使用 Host 头限制请求 | 设置 AllowedHosts 配置设置 | 16-20 |

## 16.1　准备工作

本章将继续使用第 15 章中的 Platform 项目。为准备本章，用代码清单 16-1 的内容替换 Startup.cs 文件的内容，删除前一章中的中间件和服务。

### 代码清单 16-1　替换 Platform 文件夹中 Startup.cs 文件的内容

```
using Microsoft.AspNetCore.Builder;
using Microsoft.AspNetCore.Http;
using Microsoft.Extensions.DependencyInjection;

namespace Platform {
   public class Startup {

      public void ConfigureServices(IServiceCollection services) {
      }
```

```
        public void Configure(IApplicationBuilder app) {
            app.UseDeveloperExceptionPage();
            app.UseRouting();
            app.UseEndpoints(endpoints => {
                endpoints.MapFallback(async context =>
                    await context.Response.WriteAsync("Hello World!"));
            });
        }
    }
}
```

从 Debug 菜单中选择 Start Without Debugging 或 Run Without Debugging 启动应用程序，或者打开一个新的 PowerShell 命令提示符，导航到包含 Platform.csproj 文件的文件夹，并运行代码清单 16-2 所示的命令。

**提示：** 可从 https://github.com/apress/pro-asp.net-core-3 下载本章和本书中其他所有章节的示例项目。如果在运行示例时遇到问题，请参阅第 1 章以获得帮助。

**代码清单 16-2　启动 ASP.NET Core 运行时**

```
dotnet run
```

如果使用 Visual Studio 或 Visual Studio Code 启动应用程序，打开一个新的浏览器窗口，显示如图 16-1 所示的内容。如果应用程序是从命令行启动的，那么打开一个新的浏览器选项卡，导航到 http://localhost:5000；显示如图 16-1 所示的内容。

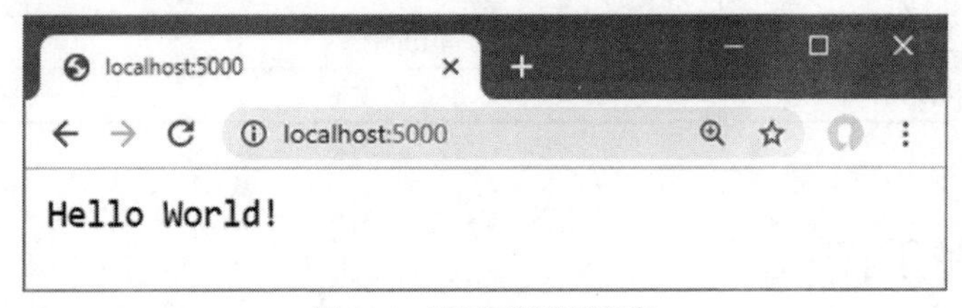


图 16-1　运行示例应用程序

# 16.2　使用 cookie

cookie 是添加到响应中的少量文本，浏览器在后续请求中会包含这些文本。cookie 对于 Web 应用程序很重要，因为它们允许开发跨越一系列 HTTP 请求的特性，每个请求都可以通过浏览器发送给服务器的 cookie 进行识别。

ASP.NET Core 通过提供给中间件组件的 HttpRequest 和 HttpResponse 对象，提供了对使用 cookie 的支持。为便于演示，代码清单 16-3 更改了示例应用程序中的路由配置，以添加实现计数器的端点。

**代码清单 16-3　在 Platform 文件夹的 Startup.cs 文件中使用 cookie**

```
using Microsoft.AspNetCore.Builder;
using Microsoft.AspNetCore.Http;
```

```
using Microsoft.Extensions.DependencyInjection;
using System;
using System.Threading.Tasks;

namespace Platform {
    public class Startup {

        public void ConfigureServices(IServiceCollection services) {
        }

        public void Configure(IApplicationBuilder app) {
            app.UseDeveloperExceptionPage();
            app.UseRouting();

            app.UseEndpoints(endpoints => {

                endpoints.MapGet("/cookie", async context => {
                    int counter1 =
                        int.Parse(context.Request.Cookies["counter1"] ?? "0") + 1;
                    context.Response.Cookies.Append("counter1", counter1.ToString(),
                        new CookieOptions {
                            MaxAge = TimeSpan.FromMinutes(30)
                        });
                    int counter2 =
                        int.Parse(context.Request.Cookies["counter2"] ?? "0") + 1;
                    context.Response.Cookies.Append("counter2", counter1.ToString(),
                        new CookieOptions {
                            MaxAge = TimeSpan.FromMinutes(30)
                        });
                    await context.Response
                        .WriteAsync($"Counter1: {counter1}, Counter2: {counter2}");
                });

                endpoints.MapGet("clear", context => {
                    context.Response.Cookies.Delete("counter1");
                    context.Response.Cookies.Delete("counter2");
                    context.Response.Redirect("/");
                    return Task.CompletedTask;
                });

                endpoints.MapFallback(async context =>
                    await context.Response.WriteAsync("Hello World!"));
            });
        }
    }
}
```

新的端点依赖名为 counter1 和 counter2 的 cookie。当请求/cookie URL 时，中间件查找 cookie 并将其解析为 int 值。如果没有 cookie，则使用回退零。

```
...
int counter1 = int.Parse(context.Request.Cookies["counter1"] ?? "0") + 1;
...
```

cookie 通过 HttpRequest.Cookies 属性访问，其中 cookie 的名称用作键。从 cookie 中获取的值是递增的，用于在响应中设置 cookie，如下所示：

```
...
context.Response.Cookies.Append("counter1", counter1.ToString(),
    new CookieOptions {
        MaxAge = TimeSpan.FromMinutes(30)
});
...
```

通过 HttpResponse.Cookies 属性设置 cookie。Append 方法在响应中创建或替换 cookie。Append 方法的参数是 cookie 的名称、值和用于配置 cookie 的 CookieOptions 对象。CookieOptions 类定义表 16-2 中描述的属性，每个属性对应一个 cookie 字段。

表 16-2　CookieOptions 属性

| 名称 | 描述 |
| --- | --- |
| Domain | 此属性指定浏览器向其发送 cookie 的主机。默认情况下，cookie 只发送给创建 cookie 的主机 |
| Expires | 该属性设置 cookie 的有效期 |
| HttpOnly | 为 true 时，该属性告诉浏览器不要在 JavaScript 代码发出的请求中包含 cookie |
| IsEssential | 此属性用于表示 cookie 是必不可少的，如 16.2.1 节所述 |
| MaxAge | 此属性指定 cookie 过期的秒数。旧的浏览器不支持有此设置的 cookie |
| Path | 此属性用于设置 URL 路径，该路径必须在浏览器发送 cookie 之前出现在请求中 |
| SameSite | 此属性用于指定是否在跨站点请求中包含 cookie。这些值是 Lax、Strict 和 None(默认值) |
| Secure | 为 true 时，该属性告诉浏览器仅使用 HTTPS 发送 cookie |

代码清单 16-3 中设置的唯一 cookie 选项是 MaxAge，它告诉浏览器 cookie 在 30 分钟后过期。代码清单 16-3 中的中间件在请求/clear URL 时删除 cookie，这是使用 HttpResponse.Cookie 完成的。删除方法，之后浏览器重定向到/ URL。

```
...
} else if (context.Request.Path == "/clear") {
    context.Response.Cookies.Delete("counter1");
    context.Response.Cookies.Delete("counter2");
    context.Response.Redirect("/");
}
...
```

重启 ASP.NETCore，导航到 http://localhost:5000/cookie。响应包含随后的请求中包含的 cookie，每次重新加载浏览器时，计数器都会增加，如图 16-2 所示。对 http://localhost:5000/clear 的请求将

删除 cookie，并重置计数器。

图 16-2 使用 cookie

## 16.2.1 启用 cookie consent 检查

欧盟一般数据保护条例(GDPR)要求在使用非必要的 cookie 前，必须征得用户同意。ASP.NET Core 支持获得同意，并防止在未获得同意的情况下向浏览器发送非必要的 cookie。options 模式用于为 cookie 创建策略，该策略由中间件组件应用，如代码清单 16-4 所示。

■ **警告**：cookie consent 只是 GDPR 的一部分。请参见 https://en.wikipedia.org/wiki/General_Data_Protection_Regulation，它包含这些规则的一个很好的概述。

#### 代码清单16-4 在 Platform 文件夹的 Startup.cs 文件中启用 cookie consent

```
using Microsoft.AspNetCore.Builder;
using Microsoft.AspNetCore.Http;
using Microsoft.Extensions.DependencyInjection;
using System;
using System.Threading.Tasks;

namespace Platform {
    public class Startup {

        public void ConfigureServices(IServiceCollection services) {
            services.Configure<CookiePolicyOptions>(opts => {
                opts.CheckConsentNeeded = context => true;
            });
        }

        public void Configure(IApplicationBuilder app) {
            app.UseDeveloperExceptionPage();
            app.UseCookiePolicy();
            app.UseRouting();

            app.UseEndpoints(endpoints => {

                endpoints.MapGet("/cookie", async context => {
                    int counter1 =
                        int.Parse(context.Request.Cookies["counter1"] ?? "0") + 1;
```

```
                context.Response.Cookies.Append("counter1", counter1.ToString(),
                    new CookieOptions {
                        MaxAge = TimeSpan.FromMinutes(30),
                        IsEssential = true
                    });
                int counter2 =
                    int.Parse(context.Request.Cookies["counter2"] ?? "0") + 1;
                context.Response.Cookies.Append("counter2", counter1.ToString(),
                    new CookieOptions {
                        MaxAge = TimeSpan.FromMinutes(30)
                    });
                await context.Response
                    .WriteAsync($"Counter1: {counter1}, Counter2: {counter2}");
            });

            endpoints.MapGet("clear", context => {
                context.Response.Cookies.Delete("counter1");
                context.Response.Cookies.Delete("counter2");
                context.Response.Redirect("/");
                return Task.CompletedTask;
            });

            endpoints.MapFallback(async context =>
                await context.Response.WriteAsync("Hello World!"));
        });
    }
  }
}
```

options 模式用于配置 CookiePolicyOptions 对象，该对象使用表 16-3 中描述的属性设置应用程序中 cookie 的总体策略。

表 16-3　CookiePolicyOptions 属性

| 名称 | 描述 |
|---|---|
| CheckConsentNeeded | 此属性分配给一个函数，该函数接收 HttpContext 对象，如果它表示需要 cookie 同意的请求，则返回 true。每个请求都会调用该函数，默认函数总是返回 false |
| ConsentCookie | 此属性返回一个对象，用于配置发送到浏览器的 cookie 以记录用户的 cookie consent |
| HttpOnly | 此属性设置 HttpOnly 属性的默认值，如表 16-2 所示 |
| MinimumSameSitePolicy | 此属性设置 SameSite 属性的最低安全级别，如表 16-2 所示 |
| Secure | 此属性设置 Secure 属性的默认值，如表 16-2 所示 |

为启用同意检查，给 CheckConsentNeeded 属性分配了一个新函数，该属性总是返回 true。为 ASP.NET Core 接收的每一个请求调用该函数，这意味着可以定义复杂的规则来选择需要获得同意的请求。对于这次应用，采取最谨慎的态度，所有应用都必须得到同意。

使用 UseCookiePolicy 方法将执行 cookie 策略的中间件添加到请求管道中。结果是只有

IsEssential 属性为 true 的 cookie 才会添加到响应中。代码清单 16-4 只设置了 cookie1 上的 IsEssential 属性，为了看到效果，可以重新启动 ASP.NET Core，请求 http://localhost:5000/cookie，并重新加载浏览器。只有计数器的 cookie 标记为彻底更新，如图 16-3 所示。

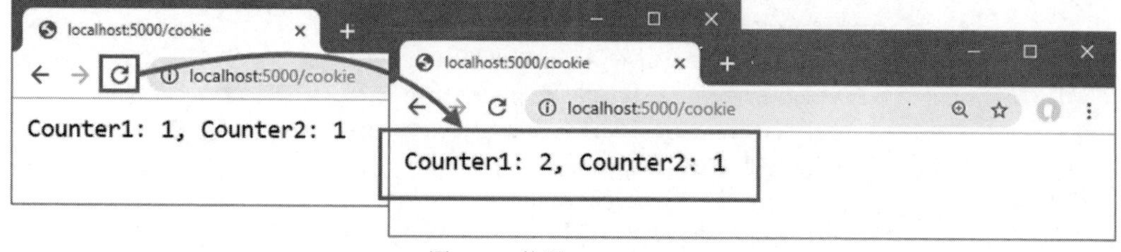


图 16-3 使用 cookie consent

## 16.2.2 管理 cookie consent

除非用户同意，否则只允许对 Web 应用程序的核心功能至关重要的 cookie。同意(consent)是通过请求特性来管理的，它为中间件组件提供了 ASP.NET Core 如何处理请求和响应的实现细节的访问。通过 HttpRequest 访问特性。每个特性都由一个接口表示，该接口的属性和方法处理低级请求处理的一个方面。

特性处理很少需要更改的请求处理方面，比如响应的结构。唯一的例外是 cookie consent 的管理，它通过 ITrackingConsentFeature 接口来处理，该接口定义了表 16-4 中描述的方法和属性。

表 16-4 ITrackingConsentFeature 成员

| 名称 | 描述 |
|---|---|
| CanTrack | 如果可将非必要的 cookie 添加到当前请求中，或者因为用户已经同意，或者因为不需要同意，则此属性返回 true |
| CreateConsentCookie() | 该方法返回一个 cookie，JavaScript 客户端可以使用该 cookie 表示同意 |
| GrantConsent() | 调用此方法，向响应添加一个 cookie，该 cookie 授予对非必需 cookie 的同意 |
| HasConsent | 如果用户同意了非必要的 cookie，这个属性返回 true |
| IsConsentNeeded | 如果当前请求需要对非必要 cookie 的同意，则此属性返回 true |
| WithdrawConsent() | 此方法删除同意 cookie |

为处理同意问题，在 Platform 文件夹中添加一个名为 ConsentMiddleware.cs 的类文件，代码如代码清单 16-5 所示。可以使用 lambda 表达式来管理 cookie consent，但是这个示例使用了一个类来保持 Configure 方法的整洁。

**代码清单 16-5 Platform 文件夹中 ConsentMiddleware.cs 文件的内容**

```
using Microsoft.AspNetCore.Http;
using Microsoft.AspNetCore.Http.Features;
using System.Threading.Tasks;

namespace Platform {
    public class ConsentMiddleware {
        private RequestDelegate next;
```

```
        public ConsentMiddleware(RequestDelegate nextDelgate) {
            next = nextDelgate;
        }

        public async Task Invoke(HttpContext context) {
            if (context.Request.Path == "/consent") {
                ITrackingConsentFeature consentFeature
                    = context.Features.Get<ITrackingConsentFeature>();
                if (!consentFeature.HasConsent) {
                    consentFeature.GrantConsent();
                } else {
                    consentFeature.WithdrawConsent();
                }
                await context.Response.WriteAsync(consentFeature.HasConsent
                    ? "Consent Granted \n" : "Consent Withdrawn\n");
            }
            await next(context);
        }
    }
}
```

使用 Get 方法获取请求特性，其中泛型类型参数指定所需的特性接口，如下所示：

```
...
ITrackingConsentFeature consentFeature
    = context.Features.Get<ITrackingConsentFeature>();
...
```

使用表 16-4 中描述的属性和方法，新的中间件组件响应/ consent URL，以确定和更改 cookie 同意。代码清单 16-6 将新的中间件添加到请求管道中。

**代码清单 16-6　在 Platform 文件夹的 Startup.cs 文件中添加中间件**

```
...
public void Configure(IApplicationBuilder app) {
    app.UseDeveloperExceptionPage();
    app.UseCookiePolicy();
    app.UseMiddleware<ConsentMiddleware>();
    app.UseRouting();

    // ...statments omitted for brevity...
}
...
```

要查看效果，请重新启动ASP.NET Core，请求http://localhost:5000/consent，然后请求http://localhost:5000/cookie。当授予同意时，允许非必要的cookie，并且示例中的两个计数器都将工作，如图 16-4 所示。重复此过程以撤销“同意”，则只有cookie标记为essential的计数器有效。

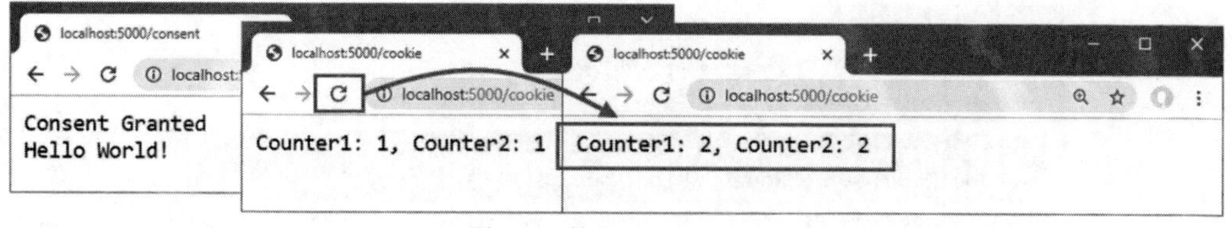


图 16-4 管理 cookie consent

# 16.3 使用会话

上一节中的示例使用 cookie 存储应用程序的状态数据，为中间件组件提供所需的数据。这种方法的问题是，cookie 的内容存储在客户端上，可以在客户端上操纵并使用它来改变应用程序的行为。

更好的方法是使用 ASP.NET Core 会话特性。会话中间件向响应添加一个 cookie，它允许识别相关的请求，还与存储在服务器上的数据相关联。

当接收到包含会话 cookie 的请求时，会话中间件组件检索与会话关联的服务器端数据，并通过 HttpContext 对象使其对其他中间件组件可用。使用会话意味着应用程序的数据保留在服务器上，只有会话的标识符被发送到浏览器。

## 16.3.1 配置会话服务和中间件

设置会话需要配置服务，并向请求管道添加中间件组件。代码清单 16-7 将语句添加到 Startup 类中，以设置示例应用程序的会话，并删除上一节中的端点。

**代码清单 16-7 在 Platform 文件夹的 Startup.cs 文件中配置会话**

```
using Microsoft.AspNetCore.Builder;
using Microsoft.AspNetCore.Http;
using Microsoft.Extensions.DependencyInjection;
using System;
using System.Threading.Tasks;

namespace Platform {
   public class Startup {

      public void ConfigureServices(IServiceCollection services) {
         services.Configure<CookiePolicyOptions>(opts => {
            opts.CheckConsentNeeded = context => true;
         });

         services.AddDistributedMemoryCache();

         services.AddSession(options => {
            options.IdleTimeout = TimeSpan.FromMinutes(30);
            options.Cookie.IsEssential = true;
         });
      }
```

```
        public void Configure(IApplicationBuilder app) {
            app.UseDeveloperExceptionPage();
            app.UseCookiePolicy();
            app.UseMiddleware<ConsentMiddleware>();
            app.UseSession();
            app.UseRouting();

            app.UseEndpoints(endpoints => {
                endpoints.MapFallback(async context =>
                    await context.Response.WriteAsync("Hello World!"));
            });
        }
    }
}
```

在使用会话时，必须决定如何存储相关数据。ASP.NET Core 为会话数据存储提供了三个选项，每个选项都有自己的方法在 Startup 类的 ConfigureServices 方法中注册服务，如表 16-5 所述。

**表 16-5　会话存储方法**

| 名称 | 描述 |
|---|---|
| AddDistributedMemoryCache | 此方法设置一个内存中的缓存。尽管有这样的名称，缓存并不是分布式的，只负责为创建它的 ASP.NET Core 运行时的实例存储数据 |
| AddDistributedSqlServerCache | 此方法设置一个缓存，用于在安装 Microsoft.Extensions.Caching.SqlServer 包时，在 SQL Server 中存储数据，这个缓存在第 17 章中使用 |
| AddStackExchangeRedisCache | 此方法设置一个 Redis 缓存，安装 Microsoft.Extensions.Caching.Redis 包时可用 |

缓存详见第 17 章，但本章使用了内存中的缓存，它在 ConfigureServices 方法中建立:

```
...
services.AddDistributedMemoryCache();
...
```

由 AddDistributedMemoryCache 方法创建的缓存服务并不是分布式的，而是为 ASP.NET Core 运行时的单个实例存储会话数据。如果通过部署多个运行时实例来扩展应用程序，那么应该使用其他缓存之一，如 SQL Server 缓存，参见第 17 章。

下一步是使用 options 模式配置会话中间件，如下所示：

```
...
services.AddSession(options => {
    options.IdleTimeout = TimeSpan.FromMinutes(30);
    options.Cookie.IsEssential = true;
});
...
```

表 16-6 显示了会话的 options 类是 SessionOptions，并描述了它定义的关键属性。

表 16-6　由 SessionOptions 类定义的属性

| 名称 | 描述 |
| --- | --- |
| Cookie | 此属性用于配置会话 cookie |
| IdleTimeout | 此属性用于配置会话过期的时间跨度 |

cookie 属性返回一个可用于配置会话 cookie 的对象。表 16-7 描述了会话数据最有用的 cookie 配置属性。

表 16-7　cookie 配置属性

| 名称 | 描述 |
| --- | --- |
| HttpOnly | 此属性指定浏览器是否会阻止 cookie 包含在 JavaScript 代码发送的 HTTP 请求中。对于使用 JavaScript 应用程序的项目，其请求应该包含在会话中，该属性应该设置为 true。默认值为 true |
| IsEssential | 此属性指定应用程序是否需要 cookie 来运行，即使用户已经指定他们不希望应用程序使用 cookie，也应该使用 cookie。默认值为 false |
| SecurityPolicy | 此属性使用来自 CookieSecurePolicy 枚举的值设置 cookie 的安全策略。值是 Always(将 cookie 限制为 HTTPS 请求)、SameAsRequest(如果最初请求是使用 HTTPS 发出的，则将 cookie 限制为 HTTPS 请求)和 None(允许 cookie 用于 HTTP 和 HTTPS 请求)。默认值是 None |

代码清单 16-7 中设置的选项允许将会话 cookie 包含在由 JavaScript 启动的请求中，并将 cookie 标记为“必需”，这样，即使用户表示不想使用 cookie，也可以使用它(参见 16.2.2 节，以获得有关基本 cookie 的更多细节)。设置了 IdleTimeout 选项，以便在 30 分钟内没有收到包含会话 cookie 的请求时，会话过期。

■ **警告：** 默认情况下会话 cookie 没有标记为“必需”，这可能会在使用 cookie consent 时造成问题。代码清单 16-7 将 IsEssential 属性设置为 true，以确保会话始终工作。如果会话不能像预期的那样工作，那么这就是可能的原因，必须将 IsEssential 设置为 true，或者调整应用程序，来处理不同意和不接受会话 cookie 的用户。

最后一步是将会话中间件组件添加到请求管道，这是使用 UseSession 方法完成的。当中间件处理包含会话 cookie 的请求时，它从缓存中检索会话数据，并通过 HttpContext 对象使其可用，然后沿着请求管道传递请求，并将其提供给其他中间件组件。当没有会话 cookie 的请求到达时，启动一个新会话，并向响应添加 cookie，以便将后续请求标识为会话的一部分。

## 16.3.2　使用会话数据

会话中间件通过 HttpContext 对象的 Session 属性，来访问与请求关联的会话的详细信息。Session 属性返回一个实现 ISession 接口的对象，该接口提供了表 16-8 中所示的用于访问会话数据的方法。

表 16-8　有用的 ISession 方法和扩展方法

| 名称 | 描述 |
| --- | --- |
| Clear() | 此方法删除会话中的所有数据 |
| CommitAsync() | 此异步方法将已更改的会话数据提交到缓存 |

(续表)

| 名称 | 描述 |
|---|---|
| GetString(key) | 此方法使用指定的键检索字符串值 |
| GetInt32(key) | 此方法使用指定的键检索一个整数值 |
| Id | 此属性返回会话的唯一标识符 |
| IsAvailable | 加载会话数据时返回 true |
| Keys | 枚举会话数据项的键 |
| Remove(key) | 此方法删除与指定键相关联的值 |
| SetString(key, val) | 此方法使用指定的键存储字符串 |
| SetInt32(key, val) | 此方法使用指定的键存储一个整数 |

会话数据存储在键/值对中，其中键是字符串，值是字符串或整数。这个简单的数据结构允许表 16-6 中列出的每个缓存轻松地存储会话数据。需要存储更复杂数据的应用程序可以使用序列化，这就是在 SportsStore 中采用的方法。代码清单 16-8 使用会话数据重新创建反例。

**代码清单 16-8　在 Platform 文件夹的 Startup.cs 文件中使用会话数据**

```
...
public void Configure(IApplicationBuilder app) {
    app.UseDeveloperExceptionPage();
    app.UseCookiePolicy();
    app.UseMiddleware<ConsentMiddleware>();
    app.UseSession();
    app.UseRouting();

    app.UseEndpoints(endpoints => {

        endpoints.MapGet("/cookie", async context => {
            int counter1 = (context.Session.GetInt32("counter1") ?? 0) + 1;
            int counter2 = (context.Session.GetInt32("counter2") ?? 0) + 1;
            context.Session.SetInt32("counter1", counter1);
            context.Session.SetInt32("counter2", counter2);
            await context.Session.CommitAsync();
            await context.Response
                .WriteAsync($"Counter1: {counter1}, Counter2: {counter2}");
        });

        endpoints.MapFallback(async context =>
            await context.Response.WriteAsync("Hello World!"));
    });
}
...
```

GetInt32 方法用于读取与键 counter1 和 counter2 关联的值。如果这是会话中的第一个请求，则没有值可用，使用 null coalescing 运算符来提供初始值。值递增后，使用 SetInt32 方法存储，用于为客户端生成一个简单结果。

使用 CommitAsync 方法是可选的，但是最好使用它，因为如果会话数据不能存储在缓存中，就抛出异常。默认情况下，如果存在缓存问题，则不会报告错误，而缓存问题可能导致不可预测和令人困惑的行为。

对会话数据的所有更改必须在响应发送到客户端之前进行，这就是为什么代码清单 16-8 在调用 Response.WriteAsync 方法之前读取、更新和存储会话数据。

注意，代码清单 16-8 中的新语句不必处理会话 cookie、检测过期会话或从缓存加载会话数据。所有这些工作都由会话中间件自动完成，它通过 HttpContext.Session 属性显示结果。这种方法的一个结果是 HttpContext.Session 属性在会话中间件处理请求之后才会填充数据，这意味着应该仅尝试在中间件或端点中访问会话数据，中间件或端点是调用 UseSession 方法后添加到请求管道中的。

重启 ASP.NET Core，导航到 http://localhost:5000/cookie URL，会看到计数器的值。重新加载浏览器，计数器值将增加，如图16-5所示。ASP.NET Core 停止时，会话和会话数据将丢失。因为本例选择了内存缓存。其他存储选项在 ASP.NET Core 运行时外部操作，一直生存到应用程序重启为止。

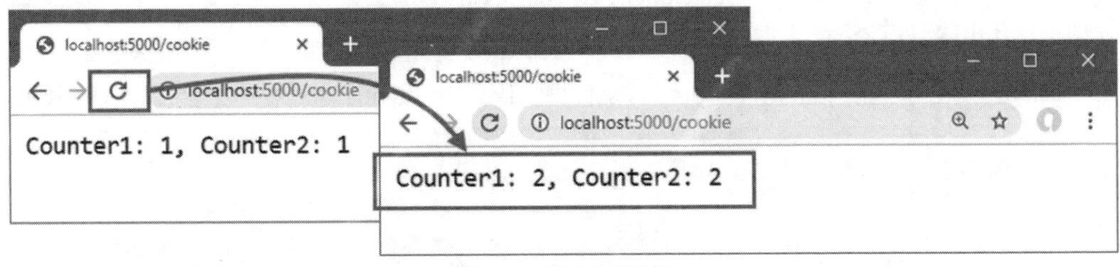


图16-5 使用会话数据

## 16.4 使用 HTTPS 连接

用户越来越希望 Web 应用程序使用 HTTPS 连接，即使是不包含或返回敏感数据的请求。ASP.NET Core 同时支持 HTTP 和 HTTPS 连接，并提供可以强制 HTTP 客户端使用 HTTPS 的中间件。

**HTTPS、SSL、TLS**

HTTPS 是 HTTP 和传输层安全(TLS)或安全套接字层(SSL)的组合。TLS 已经取代了过时的 SSL 协议，但是术语 SSL 已经成为安全网络的同义词，经常在实际使用 TLS 时使用。如果对安全性和密码学感兴趣，那么 HTTPS 的细节值得探索，https://en.wikipedia.org/wiki/HTTPS 是一个很好的起点。

### 16.4.1 启用 HTTP 连接

如果正在使用 Visual Studio，选择 Project | Platform Properties，导航到 Debug 部分，并选中 Enable SSL 选项，如图 16-6 所示。一旦选中了这个选项，选择 File | Save All，以保存配置更改。

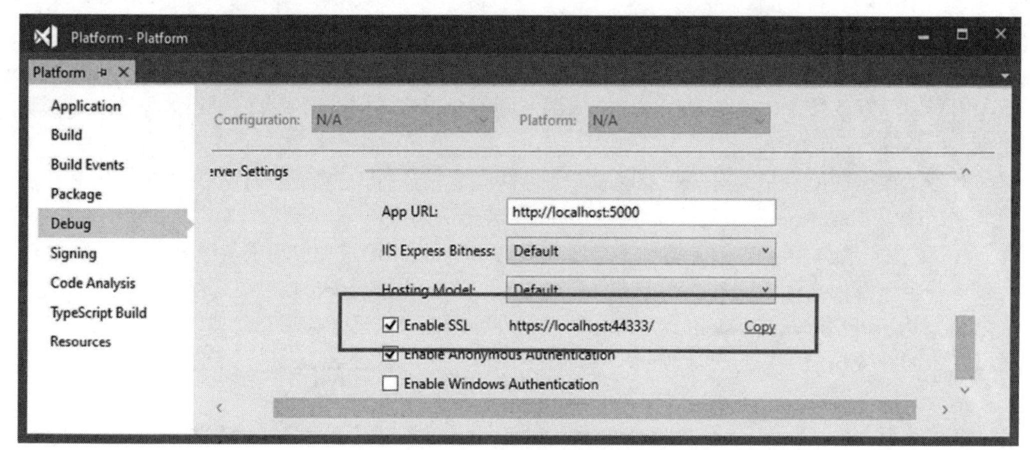


图 16-6　在 Visual Studio 中启用 HTTPS

用于接收 HTTPS 的端口是随机分配的。要更改端口，单击 Solution Explorer 顶部的 Show All 按钮，以显示 Properties 文件夹和 launchSettings.json 文件(如果它们不可见)。编辑该文件，以更改应用程序使用的 HTTPS 端口，如代码清单 16-9 所示。

**代码清单 16-9　在 Platform/Properties 文件夹的 launchSettings.json 文件中更改 HTTPS 端口**

```
{
  "iisSettings": {
    "windowsAuthentication": false,
    "anonymousAuthentication": true,
    "iisExpress": {
      "applicationUrl": "http://localhost:5000",
      "sslPort": 44350
    }
  },
  "profiles": {
    "IIS Express": {
      "commandName": "IISExpress",
      "launchBrowser": true,
      "environmentVariables": {
        "ASPNETCORE_ENVIRONMENT": "Development"
      }
    },
    "Platform": {
      "commandName": "Project",
      "launchBrowser": true,
      "environmentVariables": {
        "ASPNETCORE_ENVIRONMENT": "Development"
      },
      "applicationUrl": "https://localhost:44350;http://localhost:5000"
    }
  }
```

```
}
```

sslPort 设置的新值更改在 Visual Studio 中启动应用程序时使用的端口。新 applicationUrl 设置从命令行或使用 Visual Studio 启动应用程序时使用的端口。

---

**注意：** IIS Express 只支持端口号在 4400 到 44 399 之间的 HTTPS。

---

.NET Core 运行时包含一个用于 HTTPS 请求的测试证书。在 Platform 文件夹中运行代码清单 16-10 所示的命令，以重新生成测试证书并信任它。

**代码清单 16-10 重新生成开发证书**

```
dotnet dev-certs https --clean
dotnet dev-certs https --trust
```

在提示中选择 Yes，以删除它已经信任的现有证书，并选择 Yes 以信任新证书，如图 16-7 所示。

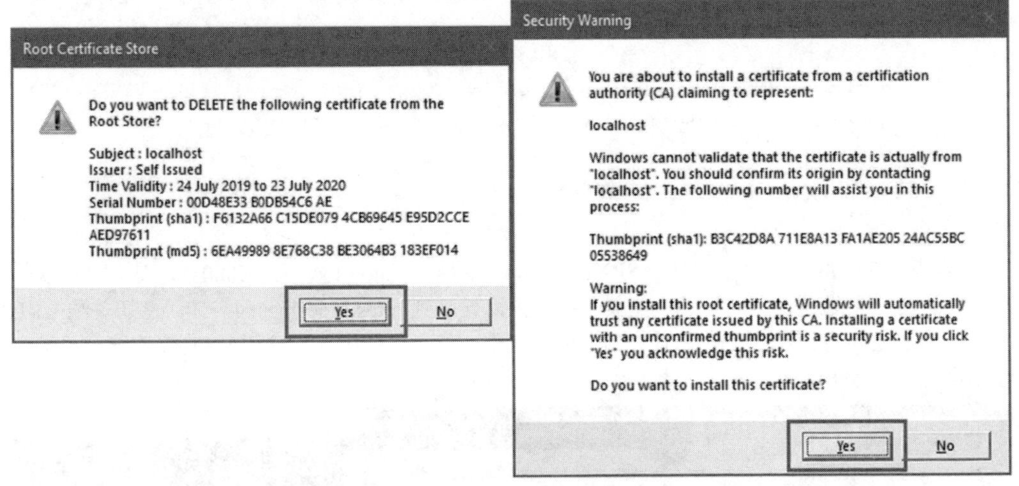


图 16-7 重新生成 HTTPS 证书

## 16.4.2 检测 HTTPS 请求

使用 HTTPS 发出的请求可通过 HttpRequest.IsHttps 属性来检测。在代码清单 16-11 中，向请求管道添加了一个新的中间件组件，该组件在使用 HTTPS 发出请求时向响应添加消息。

**代码清单 16-11 在 Platform 文件夹的 Startup.cs 文件中检测 HTTPS**

```
...
public void Configure(IApplicationBuilder app) {
    app.UseDeveloperExceptionPage();
    app.UseCookiePolicy();
    app.UseMiddleware<ConsentMiddleware>();
    app.UseSession();
    app.UseRouting();
```

```
        app.Use(async (context, next) => {
            await context.Response
                .WriteAsync($"HTTPS Request: {context.Request.IsHttps} \n");
            await next();
        });

        app.UseEndpoints(endpoints => {

            endpoints.MapGet("/cookie", async context => {
                int counter1 = (context.Session.GetInt32("counter1") ?? 0) + 1;
                int counter2 = (context.Session.GetInt32("counter2") ?? 0) + 1;
                context.Session.SetInt32("counter1", counter1);
                context.Session.SetInt32("counter2", counter2);
                await context.Session.CommitAsync();
                await context.Response
                    .WriteAsync($"Counter1: {counter1}, Counter2: {counter2}");
            });

            endpoints.MapFallback(async context =>
                await context.Response.WriteAsync("Hello World!"));
        });
    }
...
```

要测试 HTTPS，请重新启动 ASP.NET Core，导航到 http://localhost:5000。这是一个常规的 HTTP 请求，生成如图 16-8 所示的结果。接下来，导航到 https://localhost:44350，密切关注 URL 方案，它是 https 而不是 http，就像在前面的示例中一样。新的中间件检测 HTTPS 连接，并生成图 16-8 右侧的输出。

图 16-8　发出 HTTPS 请求

## 16.4.3　执行 HTTPS 请求

ASP.NET Core 提供一个中间件组件，为通过 HTTP 到达的请求发送重定向响应，来强制使用 HTTPS。代码清单 16-12 将此中间件添加到请求管道中。

**代码清单 16-12　在 Platform 文件夹的 Startup.cs 文件中强制执行 HTTPS**

```
...
public void Configure(IApplicationBuilder app) {
    app.UseDeveloperExceptionPage();
    app.UseHttpsRedirection();
    app.UseCookiePolicy();
    app.UseMiddleware<ConsentMiddleware>();
    app.UseSession();
    app.UseRouting();
    app.Use(async (context, next) => {
        await context.Response
            .WriteAsync($"HTTPS Request: {context.Request.IsHttps} \n");
        await next();
    });

    app.UseEndpoints(endpoints => {

        endpoints.MapGet("/cookie", async context => {
            int counter1 = (context.Session.GetInt32("counter1") ?? 0) + 1;
            int counter2 = (context.Session.GetInt32("counter2") ?? 0) + 1;
            context.Session.SetInt32("counter1", counter1);
            context.Session.SetInt32("counter2", counter2);
            await context.Session.CommitAsync();
            await context.Response
                .WriteAsync($"Counter1: {counter1}, Counter2: {counter2}");
        });

        endpoints.MapFallback(async context =>
            await context.Response.WriteAsync("Hello World!"));
    });
}
...
```

UseHttpsRedirection 方法添加了中间件组件，该组件出现在请求管道的开始处，以便可以在任何其他组件对管道执行短路操作，并使用常规 HTTP 生成响应之前，发生对 HTTPS 的重定向。

**配置 HTTPS 重定向**

通过调用ConfigureServices方法中的AddHttpsRedirection方法，options模式可用于配置HTTPS重定向中间件，如下所示:

```
...
services.AddHttpsRedirection(opts => {
    opts.RedirectStatusCode = StatusCodes.Status307TemporaryRedirect;
    opts.HttpsPort = 443;
});
...
```

此片段中显示了仅有的两个配置选项，它们设置重定向中使用的状态代码，响应重定向到的客户端的端口，覆盖从配置文件加载的值。在部署应用程序时，指定 HTTPS 端口可能很有用，但是在更改重定向状态代码时应该小心。

重启 ASP.NET Core，请求 http://localhost:5000，这是应用程序的 HTTP URL。HTTPS 重定向中间件将拦截请求，将浏览器重定向到 HTTPS URL，如图 16-9 所示。

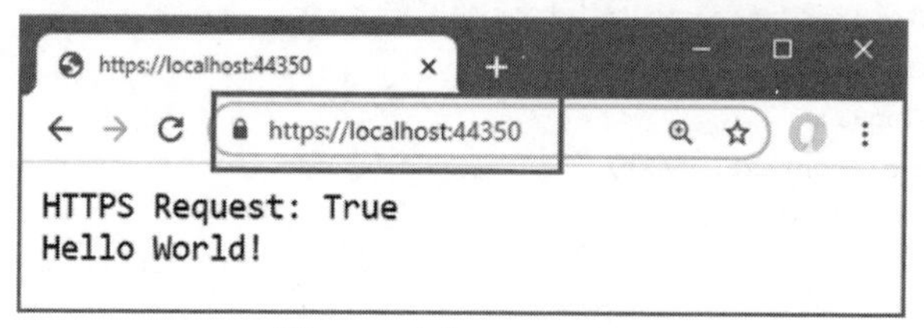


图 16-9　迫使 HTTPS 请求

**提示：**现代浏览器通常会隐藏 URL 方案，这就是为什么应该注意显示的端口号。为在图中显示完整的 URL 方案，必须单击 URL 栏。

### 16.4.4　启用 HTTP 严格传输安全性

HTTPS 重定向的一个限制是，用户在被重定向到安全连接之前，可以使用 HTTP 发出初始请求，这就存在安全风险。

HTTP 严格传输安全(HSTS)协议旨在帮助降低这种风险，它在响应中包含一个标题，告诉浏览器，只有在向 Web 应用程序的主机发送请求时才使用 HTTPS。接收到 HSTS 头之后，即使用户指定了 HTTP URL，支持 HSTS 的浏览器也会使用 HTTPS 向应用程序发送请求。代码清单 16-13 显示了向请求管道添加 HSTS 中间件。

**代码清单 16-13　在 Platform 文件夹的 Startup.cs 文件中启用 HSTS**

```
using Microsoft.AspNetCore.Builder;
using Microsoft.AspNetCore.Http;
using Microsoft.Extensions.DependencyInjection;
using System;
using System.Threading.Tasks;
using Microsoft.AspNetCore.Hosting;
using Microsoft.Extensions.Hosting;

namespace Platform {
   public class Startup {

      public void ConfigureServices(IServiceCollection services) {
         services.Configure<CookiePolicyOptions>(opts => {
            opts.CheckConsentNeeded = context => true;
         });

         services.AddDistributedMemoryCache();
```

```
        services.AddSession(options => {
            options.IdleTimeout = TimeSpan.FromMinutes(30);
            options.Cookie.IsEssential = true;
        });

        services.AddHsts(opts => {
            opts.MaxAge = TimeSpan.FromDays(1);
            opts.IncludeSubDomains = true;
        });
    }

    public void Configure(IApplicationBuilder app, IWebHostEnvironment env) {
        app.UseDeveloperExceptionPage();
        if (env.IsProduction()) {
            app.UseHsts();
        }
        app.UseHttpsRedirection();
        app.UseCookiePolicy();
        app.UseMiddleware<ConsentMiddleware>();
        app.UseSession();
        app.UseRouting();

        app.Use(async (context, next) => {
            await context.Response
                .WriteAsync($"HTTPS Request: {context.Request.IsHttps} \n");
            await next();
        });

        app.UseEndpoints(endpoints => {

            endpoints.MapGet("/cookie", async context => {
                int counter1 = (context.Session.GetInt32("counter1") ?? 0) + 1;
                int counter2 = (context.Session.GetInt32("counter2") ?? 0) + 1;
                context.Session.SetInt32("counter1", counter1);
                context.Session.SetInt32("counter2", counter2);
                await context.Session.CommitAsync();
                await context.Response
                    .WriteAsync($"Counter1: {counter1}, Counter2: {counter2}");
            });

            endpoints.MapFallback(async context =>
                await context.Response.WriteAsync("Hello World!"));
        });
    }
  }
}
```

中间件使用 UseHsts 方法添加到请求管道中。通过表 16-9 中描述的属性，可以使用 AddHsts 方法配置 HSTS 中间件。

表 16-9　HSTS 配置属性

| 名称 | 描述 |
| --- | --- |
| ExcludeHosts | 此属性返回一个 List <string>，其中包含中间件不会为其发送 HSTS 头的主机。默认设置排除了 IP 版本 4 和版本 6 的本地主机和环回地址 |
| IncludeSubDomains | 当为 true 时，浏览器将应用 HSTS 设置到子域。默认值为 false |
| MaxAge | 此属性指定浏览器仅发出 HTTPS 请求的时间段。默认值是 30 天 |
| Preload | 对于属于 HSTS 预加载方案一部分的域，此属性设置为 true。这些域硬编码到浏览器中，从而避免了初始的不安全请求，并确保只使用 HTTPS。更多细节请参见 hstspreload.org |

HSTS 在开发期间禁用，仅在生产中启用，这就是只在该环境中调用 UseHsts 方法的原因。

```
...
if (env.IsProduction()) {
    app.UseHsts();
}
...
```

必须小心使用 HSTS，因为很容易造成客户端无法访问应用程序的情况，特别是在使用 HTTP 和 HTTPS 的非标准端口时。

例如，如果示例应用程序部署到名为 myhost 的服务器上，且用户请求 http://myhost:5000，则给浏览器发送 HSTS 头并重定向到 https://myhost:5001，应用程序按预期工作。但是下次用户请求 http://myhost:5000 时，会收到一个错误，说明无法建立安全连接。

出现这个问题是因为浏览器对 HSTS 采用了一种简单的方法，假定 HTTP 请求在端口 80 上处理，HTTPS 请求在端口 443 上处理。

当用户请求 http://myhost:5000 时，浏览器检查它的 HSTS 数据，发现之前收到一个 myhost 的 HSTS 头。于是，浏览器将一个请求发送到 https://myhost:5000，而不是用户输入的 HTTP URL。ASP.NET Core 不在其用于 HTTP 的端口上处理 HTTPS，因此请求失败。浏览器不记得或不理解以前为端口 5001 接收的重定向。

如果端口 80 用于 HTTP，而端口 443 用于 HTTPS，这就不是问题了。URL http://myhost 相当于 http://myhost:80，而 https://myhost:443 相当于 https://myhost:443，这意味着更改方案的目标是正确的端口。

一旦浏览器收到一个 HSTS 标头，就在标头的 MaxAge 属性持续期间继续支持它。在首次部署应用程序时，最好将 HSTS MaxAge 属性设置为较短的持续时间，直到确信 HTTPS 基础设施能正常工作为止，这就是在代码清单 16-13 中将 MaxAge 设置为一天的原因。一旦确定客户端不需要发出 HTTP 请求，就可增加 MaxAge 属性。MaxAge 属性的常用值是一年。

**提示：** 如果正在用谷歌 Chrome 测试 HSTS，就可通过导航到 Chrome://net-internals/#hsts，检查和编辑应用 HSTS 的域列表。

## 16.5 处理异常和错误

如果在处理请求时发生异常，ASP.NET Core 使用状态码 500 返回一个响应，它告诉浏览器出现了错误。

普通的500结果在开发期间是没有用的，因为它没有揭示问题的情况。UseDeveloperExceptionPage 方法添加了一个中间件组件，该组件拦截异常，并提供更有用的响应。

为了演示异常处理的方式，代码清单 16-14 用一个有意抛出异常的新组件替换了前面示例中使用的中间件和端点。

**代码清单 16-14 在 Platform 文件夹的 Startup.cs 文件中添加中间件组件**

```
using Microsoft.AspNetCore.Builder;
using Microsoft.AspNetCore.Http;
using Microsoft.Extensions.DependencyInjection;
using System;
using System.Threading.Tasks;
using Microsoft.AspNetCore.Hosting;
using Microsoft.Extensions.Hosting;

namespace Platform {
    public class Startup {

        public void ConfigureServices(IServiceCollection services) {
            services.Configure<CookiePolicyOptions>(opts => {
                opts.CheckConsentNeeded = context => true;
            });

            services.AddDistributedMemoryCache();

            services.AddSession(options => {
                options.IdleTimeout = TimeSpan.FromMinutes(30);
                options.Cookie.IsEssential = true;
            });
            services.AddHsts(opts => {
                opts.MaxAge = TimeSpan.FromDays(1);
                opts.IncludeSubDomains = true;
            });
        }

        public void Configure(IApplicationBuilder app, IWebHostEnvironment env) {
            app.UseDeveloperExceptionPage();
            if (env.IsProduction()) {
                app.UseHsts();
            }
            app.UseHttpsRedirection();
            app.UseCookiePolicy();
```

```
            app.UseMiddleware<ConsentMiddleware>();
            app.UseSession();

            app.Run(context => {
                throw new Exception("Something has gone wrong");
            });
        }
    }
}
```

重启 ASP.NET Core，导航到 https://localhost:44350，查看中间件组件生成的响应，如图 16-10 所示。该页面显示了关于请求的堆栈跟踪和详细信息，包括请求中包含的标头和 cookie 的详细信息。

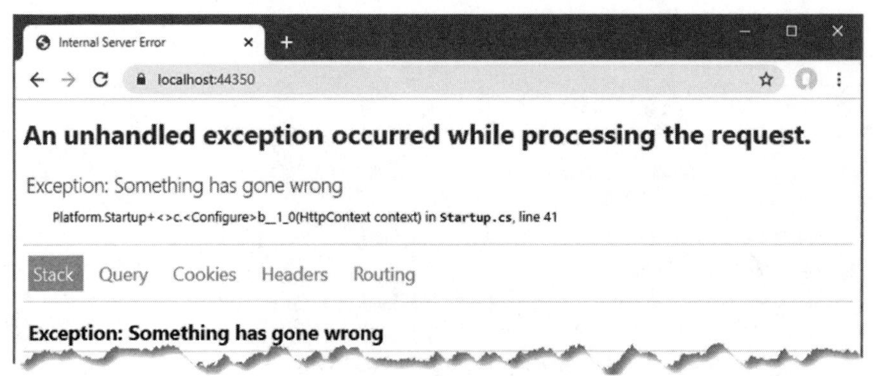


图 16-10　开发异常页面

## 管理开发和生产中间件

图 16-10 所示的详细信息应该只向开发人员显示，这就是为什么中间件只在开发期间添加到管道中。一个常见的模式是使用 IWebHostEnvironment 服务来检查环境，在开发环境中调用 UseDeveloperExceptionPage 方法，在其他环境中调用 UseHsts 方法，如:

```
...
public void Configure(IApplicationBuilder app, IWebHostEnvironment env) {
    if (env.IsDevelopment()) {
        app.UseDeveloperExceptionPage();
    } else {
        app.UseHsts();
    }
    app.UseHttpsRedirection();
    app.UseCookiePolicy();
    app.UseMiddleware<ConsentMiddleware>();
    app.UseSession();

    app.Run(context => {
        throw new Exception("Something has gone wrong");
    });
```

```
    }
...
```

这种组合确保异常处理中间件只在开发中添加，而 HSTS 中间件只在开发环境之外添加。

### 16.5.1 返回 HTML 错误响应

当禁用开发异常中间件时(应用程序在生产环境中应该如此)，ASP.NET Core 发送一个只包含错误代码的响应，来处理未处理的异常。代码清单 16-15 禁用了开发异常中间件，以便看到默认行为。

**代码清单 16-15 在 Platform 文件夹的 Startup.cs 文件中禁用中间件**

```
...
public void Configure(IApplicationBuilder app, IWebHostEnvironment env) {
    //app.UseDeveloperExceptionPage();
    if (env.IsProduction()) {
        app.UseHsts();
    }
    app.UseHttpsRedirection();
    app.UseCookiePolicy();
    app.UseMiddleware<ConsentMiddleware>();
    app.UseSession();

    app.Run(context => {
        throw new Exception("Something has gone wrong");
    });
}
...
```

重启 ASP.NET Core，导航到 https://localhost:44350。显示的响应取决于浏览器，因为 ASP.NET Core 只向它提供了一个包含状态码 500 的响应，而没有显示任何内容。图 16-11 显示了谷歌 Chrome 是如何处理这个问题的。

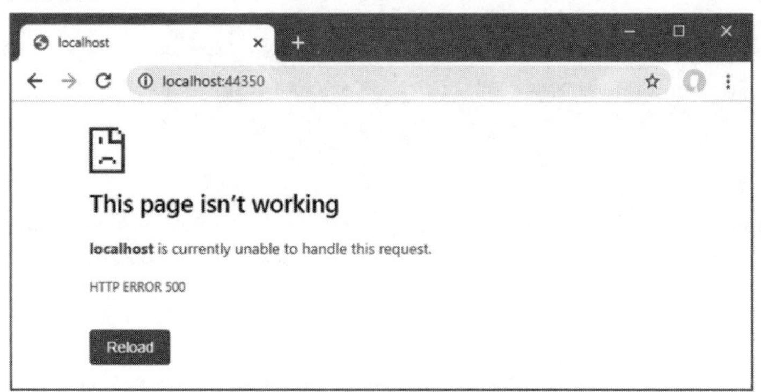


图 16-11 返回错误响应

作为只返回状态代码的替代方法，ASP.NET Core 提供了中间件，来拦截未处理的异常，向浏览器发送重定向，这可用来显示比原始状态代码更友好的响应。异常重定向中间件用 UseExceptionHandler 方法添加，如代码清单 16-16 所示。

**代码清单 16-16　在 Platform 文件夹的 Startup.cs 文件中返回一个 HTML 错误响应**

```
...
public void Configure(IApplicationBuilder app, IWebHostEnvironment env) {
    //app.UseDeveloperExceptionPage();
    app.UseExceptionHandler("/error.html");
    if (env.IsProduction()) {
        app.UseHsts();
    }
    app.UseHttpsRedirection();
    app.UseCookiePolicy();
    app.UseStaticFiles();
    app.UseMiddleware<ConsentMiddleware>();
    app.UseSession();

    app.Run(context => {
        throw new Exception("Something has gone wrong");
    });
}
...
```

当抛出异常时，异常处理程序中间件将拦截响应，将浏览器重定向到作为 UseExceptionHandler 方法的参数提供的 URL。对于本例，重定向到由静态文件处理的 URL，因此 UseStaticFiles 中间件也添加到管道中。

要添加浏览器接收的文件，在 wwwroot 文件夹中创建一个名为 error.html 的 HTML 文件，并添加如代码清单 16-17 所示的内容。

**代码清单 16-17　Platform/wwwroot 文件夹中 error.html 文件的内容**

```
<!DOCTYPE html>
<html lang="en">
<head>
    <link rel="stylesheet" href="/lib/twitter-bootstrap/css/bootstrap.min.css" />
    <title>Error</title>
</head>
<body class="text-center">
    <h3 class="p-2">Something went wrong...</h3>
    <h6>You can go back to the <a href="/">homepage</a> and try again</h6>
</body>
</html>
```

重启 ASP.NET Core，导航到 https://localhost:44350，查看新中间件的效果。浏览器发送一个重定向到/error.html URL，而不是原始状态代码，如图 16-12 所示。

图 16-12 显示 HTML 错误

有些 UseExceptionHandler 方法版本允许组成更复杂的响应，但建议保持尽可能简单的错误处理，因为不可能预测应用程序可能遇到的所有问题，当试图处理处理程序触发的异常时，可能遇到另一个异常，导致一个令人迷惑的响应或根本没有响应。

## 16.5.2 富集状态码响应

并非所有错误响应都是未捕获异常的结果。有些请求由于软件缺陷以外的原因无法处理，比如对 URL 的请求不受支持或需要身份验证。对于这类问题，将客户端重定向到另一个 URL 可能会有问题，因为有些客户端依赖错误代码来检测问题。后续章节将展示如何创建和使用 RESTful Web 应用程序时，会看到这样的示例。

ASP.NET Core 提供了中间件，可在不需要重定向的情况下向错误响应添加用户友好的内容。这样可保留错误状态代码，同时提供可读的消息，帮助用户理解问题。

最简单的方法是定义一个字符串，用作响应的主体。这比简单地指向文件要笨拙得多，但它是一种更可靠的技术，而且在处理错误时，简单可靠的技术通常更可取。要为示例项目创建字符串响应，使用代码清单 16-18 所示的代码将名为 ResponseStrings.cs 的类文件添加到 Platform 文件夹中。

**代码清单 16-18 Platform 文件夹中 ResponseStrings.cs 文件的内容**

```
namespace Platform {

public static class Responses {

    public static string DefaultResponse = @"
        <!DOCTYPE html>
            <html lang=""en"">
            <head>
                <link rel=""stylesheet""
                    href=""/lib/twitter-bootstrap/css/bootstrap.min.css"" />
                <title>Error</title>
            </head>
            <body class=""text-center"">
                <h3 class=""p-2"">Error {0}</h3>
                <h6>
                    You can go back to the <a href=""/"">homepage</a> and try again
                </h6>
```

```
        </body>
      </html>";
    }
  }
```

Response 类定义了一个 DefaultResponse 属性，为它分配了一个包含简单 HTML 文档的多行字符串。有一个占位符{0}，当响应被发送到客户端时，响应状态代码将插入其中。

代码清单 16-19 将状态代码中间件添加到请求管道，并添加一个新的中间件组件，该组件返回 404 状态代码，表示未找到所请求的 URL。

**代码清单 16-19　在 Platform 文件夹的 Startup.cs 文件中添加状态代码中间件**

```
...
public void Configure(IApplicationBuilder app, IWebHostEnvironment env) {
    //app.UseDeveloperExceptionPage();
    app.UseExceptionHandler("/error.html");
    if (env.IsProduction()) {
        app.UseHsts();
    }
    app.UseStaticFiles();
    app.UseHttpsRedirection();
    app.UseStatusCodePages("text/html", Responses.DefaultResponse);
    app.UseCookiePolicy();
    app.UseMiddleware<ConsentMiddleware>();
    app.UseSession();

    app.Use(async (context, next) => {
        if (context.Request.Path == "/error") {
            context.Response.StatusCode = StatusCodes.Status404NotFound;
            await Task.CompletedTask;
        } else {
            await next();
        }
    });

    app.Run(context => {
        throw new Exception("Something has gone wrong");
    });
}
...
```

UseStatusCodePages 方法将响应丰富的中间件添加到请求管道中。第一个参数是用于响应的 Content-Type 头的值，在本例中为 text/html。第二个参数是用作响应主体的字符串，是代码清单 16-18 中的 HTML 字符串。

定制中间件组件设置 HttpResponse.StatusCode 属性，使用 StatusCode 类定义的值指定响应的状态代码。中间件组件需要返回一个任务，所以使用了 Task.CompletedTask 属性，因为此中间件组件没有工作要做。

要查看如何处理 404 状态代码，请重新启动 ASP.NET Core，请求 https://localhost:44350/error。状态代码中间件将拦截结果，并将如图 16-13 所示的内容添加到响应中。使用状态代码插入用作 UseStatusCodePages 的第二个参数的字符串，来解析占位符。

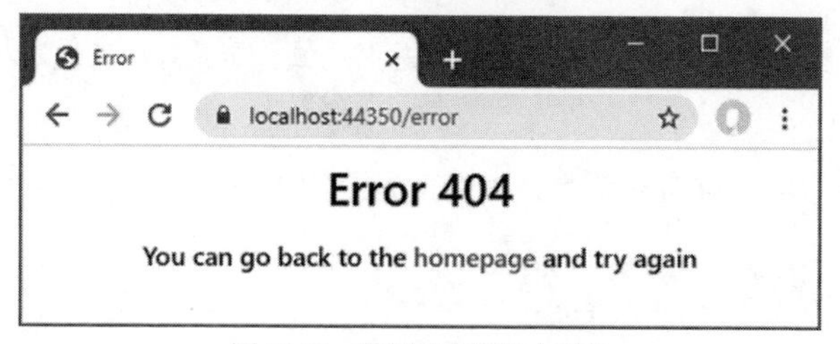


图 16-13 使用状态代码中间件

状态码中间件只响应 400 到 600 之间的状态码，不会改变已经包含内容的响应，这意味着如果在另一个中间件组件开始生成响应后出现错误，就不会看到图中的响应。状态代码中间件不会响应未处理的异常，因为异常中断了通过管道的请求流，这意味着状态代码中间件在响应发送到客户端之前没有机会检查响应。因此，UseStatusCodePages 方法通常与 UseExceptionHandler 或 UseDeveloperExceptionPage 方法一起使用。

■ **注意：** 有两个相关的方法，UseStatusCodePagesWithreRedirect 和 UseStatusCodePagesWithReExecute，它们把客户端重定向到一个不同的 URL，或者通过管道用一个不同的 URL 重新运行请求。这两种情况下，原始状态码都可能丢失。

## 16.6 使用 Host 头过滤请求

HTTP 规范要求，请求包含指定请求的主机名的 Host 头，所以可以支持虚拟服务器，其中一个 HTTP 服务器在单个端口上接收请求，并根据请求的主机名以不同方式处理它们。

由 Program 类添加到请求管道的默认中间件集包括基于 Host 头过滤请求的中间件，以便仅处理以已批准主机名列表为目标的请求，而其他所有请求都被拒绝。

Host 头中间件的默认配置包含在 appsettings.json 文件中，如下：

```
...
{
    "Location": {
        "CityName": "Buffalo"
    },
    "Logging": {
        "LogLevel": {
            "Default": "Information",
            "Microsoft": "Warning",
            "Microsoft.Hosting.Lifetime": "Information"
        }
    },
    "AllowedHosts": "*"
}
...
```

创建项目时，AllowedHosts 配置属性添加到 JSON 文件中，默认值接受请求，而不考虑 Host 头的值。可以通过编辑 JSON 文件来更改配置。还可以使用 options 模式更改配置，如代码清单 16-20 所示。

**注意：** 中间件已经通过 Program 类添加到管道中，但是如果需要显式添加中间件，可以使用 UseHostFiltering 方法。

#### 代码清单 16-20　在 Platform 文件夹的 Startup.cs 文件中配置 Host 头过滤

```
using Microsoft.AspNetCore.Builder;
using Microsoft.AspNetCore.Http;
using Microsoft.Extensions.DependencyInjection;
using System;
using System.Threading.Tasks;
using Microsoft.AspNetCore.Hosting;
using Microsoft.Extensions.Hosting;
using Microsoft.AspNetCore.HostFiltering;

namespace Platform {
    public class Startup {

        public void ConfigureServices(IServiceCollection services) {
            services.Configure<CookiePolicyOptions>(opts => {
                opts.CheckConsentNeeded = context => true;
            });

            services.AddDistributedMemoryCache();

            services.AddSession(options => {
                options.IdleTimeout = TimeSpan.FromMinutes(30);
                options.Cookie.IsEssential = true;
            });

            services.AddHsts(opts => {
                opts.MaxAge = TimeSpan.FromDays(1);
                opts.IncludeSubDomains = true;
            });

            services.Configure<HostFilteringOptions>(opts => {
                opts.AllowedHosts.Clear();
                opts.AllowedHosts.Add("*.example.com");
            });
        }

        public void Configure(IApplicationBuilder app, IWebHostEnvironment env) {

            // ...statements omitted for brevity...
```

```
            }
        }
    }
```

HostFilteringOptions 类使用表 16-10 中描述的属性，来配置主机过滤中间件。

表 16-10　HostFilteringOptions 属性

| 名称 | 描述 |
| --- | --- |
| AllowedHosts | 此属性返回一个 List<string>，其中包含允许请求的域<br>允许使用通配符，以便*.example.com 接受 example.com 域中的所有名称，*接受所有标题值 |
| AllowEmptyHosts | 为 false 时，此属性告诉中间件拒绝不包含 Host 头的请求，默认值为 true |
| IncludeFailureMessage | 为 true 时，此属性在响应中包含指示错误原因的消息，默认值为 true |

在代码清单 16-20 中，调用了 Clear 方法来删除已从 appsettings.json 文件中加载的通配符条目，然后调用 Add 方法来接受 example.com 域中的所有主机。从浏览器发送到本地主机的请求不再包含可接受的 Host 头。为了看到发生了什么，可以重新启动 ASP.NET Core，使用浏览器请求 http://localhost:5000。Host 头中间件检查请求中的 Host 头，确定请求主机名与 AllowedHosts 列表不匹配，并用 400 状态码终止请求，这表明请求失败。图 16-14 显示了错误消息。

▒ **提示：**浏览器没有重定向为使用 HTTPS，因为 Host 头中间件在 HTTPS 重定向中间件之前添加到请求管道中。

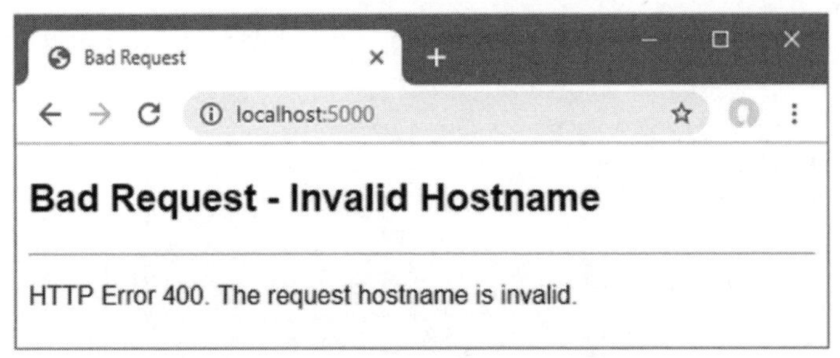


图 16-14　基于 Host 头的请求被拒绝

## 16.7　小结

本章继续描述 ASP.NET Core 平台提供的基本特性。展示了如何管理 cookie 和跟踪 cookie consent，如何使用会话，如何使用和强制 HTTPS，如何处理未处理的异常和状态代码响应，最后介绍了如何基于 Host 头过滤请求。第 17 章将解释 ASP.NET Core 如何支持使用数据。

# 第 17 章

# 处理数据

本部分前几章中的所有示例都为每个请求生成了新的响应，这在处理简单的字符串或 HTML 片段时很容易做到。大多数真实项目处理的数据生成起来很昂贵，因此需要尽可能有效地使用数据。本章描述 ASP.NET Core 提供的缓存数据和缓存整个响应的特性。介绍使用 Entity Framework Core 创建和配置用于访问数据库中的数据所需的服务。表 17-1 列出了用于处理数据的 ASP.NET Core 特性。

■ **注意：** 本章中的示例依赖第 2 章中安装的 SQL Server LocalDB 特性。如果没有安装 LocalDB 和必需的更新，就会遇到错误。

表 17-1　ASP. NET Core 数据特性

| 问题 | 答案 |
| --- | --- |
| 它们是什么? | 本章中描述的特性允许使用先前创建的数据生成响应，这些数据要么是为先前的请求创建的，要么存储在数据库中 |
| 它们为什么有用? | 大多数 Web 应用程序为每个请求重新创建数据的成本都很高。本章的特性允许以更少的资源更有效地生成响应 |
| 它们是如何使用的? | 使用服务缓存数据值。中间件组件基于 Cache-Control 头来缓存响应。通过将 LINQ 查询转换为 SQL 语句的服务来访问数据库 |
| 是否存在缺陷或限制? | 对于缓存，重要的是在部署应用程序之前测试缓存策略的效果，以确保在效率和响应能力之间找到正确的平衡。对于 Entity Framework Core，重要的是要注意发送到数据库的查询，以确保它们不会检索被应用程序处理后丢弃的大量数据 |
| 还有其他选择吗? | 本章描述的所有特性都是可选的。可选择不缓存数据或响应，也可选择使用外部缓存。可选择不使用数据库，或者使用 Entity Framework Core 以外的框架访问数据库 |

表 17-2 总结了本章的内容。

表 17-2　本章内容摘要

| 问题 | 解决方案 | 代码清单 |
| --- | --- | --- |
| 缓存数据值 | 设置一个缓存服务，并在端点和中间件组件中使用它来存储数据值 | 17-7、17-8 |
| 创建持久缓存 | 使用数据库支持的缓存 | 17-9～17-14 |

(续表)

| 问题 | 解决方案 | 代码清单 |
| --- | --- | --- |
| 缓存整个响应 | 启用缓存中间件，并在响应中设置 Cache-Control 头 | 17-15、17-16 |
| 存储应用程序数据 | 使用 Entity Framework Core | 17-17～17-23、17-26～17-28 |
| 创建数据库模式 | 创建和应用迁移 | 17-24、17-25 |
| 访问端点中的数据 | 使用数据库上下文服务 | 17-29 |
| 在日志消息中包括所有请求细节 | 启用敏感数据日志功能 | 17-30 |

## 17.1　准备工作

本章继续使用第 16 章中的 Platform 项目。为准备本章，用代码清单 17-1 所示的代码替换 Startup.cs 文件的内容。

**提示：**可从 https://github.com/apress/pro-asp.net-core-3 下载本章和本书中其他所有章节的示例项目。如果在运行示例时遇到问题，请参阅第 1 章以获得帮助。

**代码清单 17-1　替换 Platform 文件夹中 Startup.cs 文件的内容**

```
using Microsoft.AspNetCore.Builder;
using Microsoft.AspNetCore.Http;
using Microsoft.Extensions.DependencyInjection;

namespace Platform {
    public class Startup {

        public void ConfigureServices(IServiceCollection services) {
        }

        public void Configure(IApplicationBuilder app) {
            app.UseDeveloperExceptionPage();
            app.UseStaticFiles();
            app.UseRouting();
            app.UseEndpoints(endpoints => {
                endpoints.MapGet("/", async context => {
                    await context.Response.WriteAsync("Hello World!");
                });
            });
        }
    }
}
```

要减少应用程序显示的日志消息的细节级别，请更改代码清单 17-2 所示的 appsettings.Development.json 文件。

**代码清单 17-2 Platform 文件夹中 appsettings.Development.json 文件的日志细节**

```
{
  "Logging": {
    "LogLevel": {
      "Default": "Debug",
      "System": "Information",
      "Microsoft": "Information",
      "Microsoft.AspnetCore.Hosting.Diagnostics": "None"
    }
  }
}
```

如果使用 Visual Studio，选择 Project | Platform Properties，导航到 Debug 选项卡，取消选中 Enable SSL 选项(如图 17-1 所示)，然后选择 File | Save All。

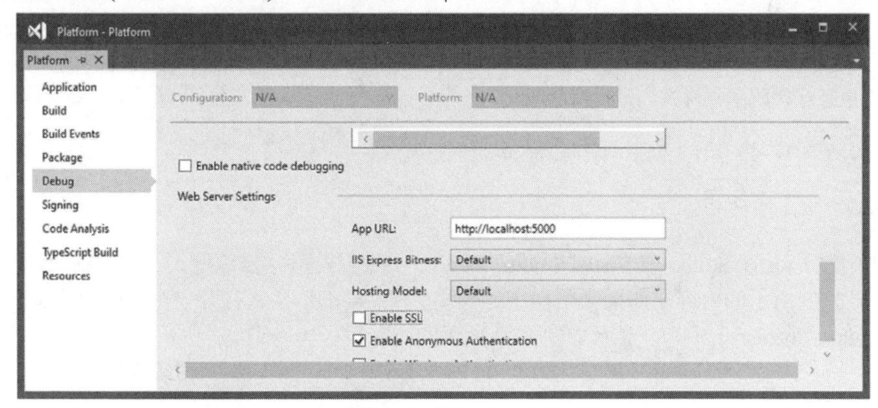


图 17-1 禁用 HTTPS

如果使用的是 Visual Studio Code，请打开 Properties/launchSettings.json 文件，删除 HTTPS URL，如代码清单 17-3 所示。

**代码清单 17-3 在 Platform/Properties 文件夹的 launchSettings.json 文件中禁用 SSL**

```
{
  "iisSettings": {
    "windowsAuthentication": false,
    "anonymousAuthentication": true,
    "iisExpress": {
      "applicationUrl": "http://localhost:5000",
      "sslPort": 0
    }
  },
  "profiles": {
    "IIS Express": {
      "commandName": "IISExpress",
      "launchBrowser": true,
```

```
            "environmentVariables": {
               "ASPNETCORE_ENVIRONMENT": "Development"
            }
         },
         "Platform": {
            "commandName": "Project",
            "launchBrowser": true,
            "environmentVariables": {
               "ASPNETCORE_ENVIRONMENT": "Development"
            },
            "applicationUrl": "http://localhost:5000"
         }
      }
   }
```

为启动应用程序，从 Debug 菜单中选择 Start Without Debugging 或 Run Without Debugging，或者打开一个新的 PowerShell 命令提示符，导航到 Platform 项目文件夹(其中包含 Platform.csproj 文件)，并运行代码清单 17-4 所示的命令。

**代码清单 17-4　启动 ASP.NET Core 运行时**

```
dotnet run
```

如果使用 Visual Studio 或 Visual Studio Code 启动应用程序，将打开一个新的浏览器窗口，并显示如图 17-2 所示的内容。如果应用程序是从命令行启动的，那么打开一个新的浏览器选项卡，导航到 https://localhost:5000；显示如图 17-2 所示的内容。

图 17-2　运行示例应用程序

## 17.2　缓存数据

在大多数 Web 应用程序中，有些数据的生成成本较高，但需要重复使用。数据的确切性质是特定于每个项目的，但是重复执行相同的计算集会增加托管应用程序所需的资源，并提高托管成本。要表示昂贵的响应，向 Platform 文件夹添加一个名为 SumEndpoint.cs 的类文件，代码如代码清单 17-5 所示。

**代码清单 17-5　Platform 文件夹中 SumEndpoint.cs 文件的内容**

```
using Microsoft.AspNetCore.Http;
using System;
using System.Threading.Tasks;
```

```
namespace Platform {

    public class SumEndpoint {

        public async Task Endpoint(HttpContext context) {
            int count = int.Parse((string)context.Request.RouteValues["count"]);
            long total = 0;
            for (int i = 1; i <= count; i++) {
                total += i;
            }
            string totalString = $"({ DateTime.Now.ToLongTimeString() }) {total}";
            await context.Response.WriteAsync(
                $"({DateTime.Now.ToLongTimeString()}) Total for {count}"
                + $" values:\n{totalString}\n");
        }
    }
}
```

代码清单 17-6 创建了一个使用端点的路由，使用第 13 章创建的 MapEndpoint 扩展方法来应用该端点。

### 代码清单 17-6　在 Platform 文件夹的 Startup.cs 文件中添加一个端点

```
using Microsoft.AspNetCore.Builder;
using Microsoft.AspNetCore.Http;
using Microsoft.Extensions.DependencyInjection;

namespace Platform {
    public class Startup {

        public void ConfigureServices(IServiceCollection services) {
        }

        public void Configure(IApplicationBuilder app) {
            app.UseDeveloperExceptionPage();
            app.UseStaticFiles();
            app.UseRouting();
            app.UseEndpoints(endpoints => {

                endpoints.MapEndpoint<SumEndpoint>("/sum/{count:int=1000000000}");

                endpoints.MapGet("/", async context => {
                    await context.Response.WriteAsync("Hello World!");
                });
            });
        }
    }
}
```

重启 ASP.NET Core，使用浏览器请求 https://localhost:5000/sum。端点将计算 1 000 000 000 个整数值的总和，生成如图 17-3 所示的结果。

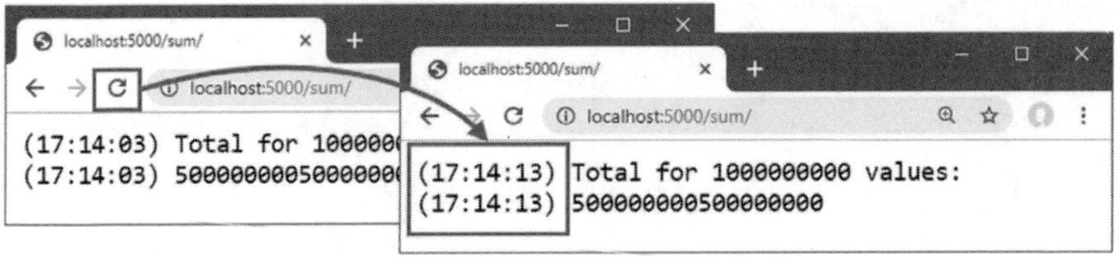


图 17-3　一个昂贵的响应

重新加载浏览器窗口，将重复计算端点。如图所示，响应中的时间戳都发生了变化，表明响应的每个部分都是为每个请求生成的。

**提示：**可能需要根据机器的性能增加或减少 route 参数的默认值。尝试找到一个需要两到三秒钟才能生成结果的值——只要时间足够长，就可以知道何时正在执行计算，但不要太长；可以在计算时出去喝杯咖啡。

### 17.2.1　缓存数据值

ASP.NET Core 提供了一个可通过 IDistributedCache 接口缓存数据值的服务。代码清单 17-7 修改了端点以声明对服务的依赖，并使用它缓存计算值。

**代码清单 17-7　在 Platform 文件夹的 SumEndpoint.cs 文件中使用缓存服务**

```
using Microsoft.AspNetCore.Http;
using System;
using System.Threading.Tasks;
using Microsoft.Extensions.Caching.Distributed;

namespace Platform {

    public class SumEndpoint {

        public async Task Endpoint(HttpContext context, IDistributedCache cache) {
            int count = int.Parse((string)context.Request.RouteValues["count"]);
            string cacheKey = $"sum_{count}";
            string totalString = await cache.GetStringAsync(cacheKey);
            if (totalString == null) {
                long total = 0;
                for (int i = 1; i <= count; i++) {
                    total += i;
                }
                totalString = $"({ DateTime.Now.ToLongTimeString() }) {total}";
                await cache.SetStringAsync(cacheKey, totalString,
                    new DistributedCacheEntryOptions {
                        AbsoluteExpirationRelativeToNow = TimeSpan.FromMinutes(2)
                    });
```

```
            }
            await context.Response.WriteAsync(
                $"({DateTime.Now.ToLongTimeString()}) Total for {count}"
                + $" values:\n{totalString}\n");
        }
    }
}
```

缓存服务只能存储字节数组，字节数组有一定的限制，但允许使用一系列 IDistributedCache 实现。有一些扩展方法允许使用字符串，这是缓存大多数数据的更简便方法。表 17-3 描述了使用缓存的最有用方法。

表 17-3 有用的 IDistributedCache 方法

| 名称 | 描述 |
|---|---|
| GetString(key) | 此方法返回与指定键相关联的缓存字符串，如果没有这样的项，则返回 null |
| GetStringAsync(key) | 此方法返回一个 Task <string>，它生成与键相关联的缓存字符串，如果没有这样的项目，则返回 null |
| SetString(key, value, options) | 此方法使用指定的键在缓存中存储字符串。缓存条目可以配置一个可选的 DistributedCacheEntryOptions 对象 |
| SetStringAsync(key, value, options) | 此方法使用指定的键异步地将字符串存储在缓存中。缓存条目可用一个可选的 DistributedCacheEntryOptions 对象配置 |
| Refresh(key) | 此方法重置与键关联的值的到期时间间隔，防止从缓存中刷新该值 |
| RefreshAsync(key) | 此方法异步重置与键关联的值的过期时间间隔，防止它从缓存中刷新 |
| Remove(key) | 此方法删除与该键关联的缓存项 |
| RemoveAsync(key) | 此方法异步删除与该键关联的缓存项 |

默认情况下，条目无限期地保留在缓存中，但是 SetString 和 SetStringAsync 方法接收一个可选的 DistributedCacheEntryOptions 参数，该参数用于设置一个过期策略，该策略指出缓存何时退出条目。表 17-4 显示了 DistributedCacheEntryOptions 类定义的属性。

表 17-4 DistributedCacheEntryOptions 属性

| 名称 | 描述 |
|---|---|
| AbsoluteExpiration | 此属性用于指定绝对过期日期 |
| AbsoluteExpirationRelativeToNow | 此属性用于指定相对过期日期 |
| SlidingExpiration | 此属性用于指定一个不活动的时间段，在此之后，如果项尚未读取，则将从缓存中弹出 |

在代码清单 17-7 中，端点使用 GetStringAsync 查看前一个请求是否有缓存的结果可用。如果没有缓存的值，端点执行计算，使用 SetStringAsync 方法缓存结果，使用 AbsoluteExpiration-RelativeToNow 属性指出缓存在两分钟后退出条目。

```
...
await cache.SetStringAsync(cacheKey, totalStr,
    new DistributedCacheEntryOptions {
```

```
        AbsoluteExpirationRelativeToNow = TimeSpan.FromMinutes(2)
    });
...
```

下一步是在 Startup 类中设置缓存服务，如代码清单 17-8 所示。

**代码清单 17-8　在 Platform 文件夹的 Startup.cs 文件中添加一个服务**

```
using Microsoft.AspNetCore.Builder;
using Microsoft.AspNetCore.Http;
using Microsoft.Extensions.DependencyInjection;

namespace Platform {
    public class Startup {
        public void ConfigureServices(IServiceCollection services) {
            services.AddDistributedMemoryCache(opts => {
                opts.SizeLimit = 200;
            });
        }

        public void Configure(IApplicationBuilder app) {
            app.UseDeveloperExceptionPage();
            app.UseStaticFiles();
            app.UseRouting();
            app.UseEndpoints(endpoints => {

                endpoints.MapEndpoint<SumEndpoint>("/sum/{count:int=1000000000}");

                endpoints.MapGet("/", async context => {
                    await context.Response.WriteAsync("Hello World!");
                });
            });
        }
    }
}
```

AddDistributedMemoryCache 与第 16 章使用的、为会话数据提供数据存储的方法相同。这是用于为 IDistributedCache 服务选择实现的三个方法之一，如表 17-5 所示。

**表 17-5　缓存服务实现方法**

| 名称 | 描述 |
| --- | --- |
| AddDistributedMemoryCache | 此方法设置一个内存中的缓存 |
| AddDistributedSqlServerCache | 此方法设置一个缓存，用于在 SQL Server 中存储数据，安装了 Microsoft.Extensions.Caching.SqlServer 包，它就是可用的。详见 17.3 节 |
| AddStackExchangeRedisCache | 此方法设置一个 Redis 缓存；安装 Microsoft.Extensions.Caching.Redis 包之后，它就是可用的 |

代码清单 17-8 使用 AddDistributedMemoryCache 方法创建一个内存中缓存，作为 IDistributedCache服务的实现。这个缓存使用MemoryCacheOptions类配置，表17-6描述了该类最有用的属性。

表 17-6 有用的 MemoryCacheOptions 属性

| 名称 | 描述 |
| --- | --- |
| ExpirationScanFrequency | 此属性用于设置 TimeSpan，确定缓存扫描过期项的频率 |
| SizeLimit | 此属性指定缓存中的最大项数。达到该大小时，缓存将弹出项 |
| CompactionPercentage | 此属性指定到达 SizeLimit 时缓存大小减少的百分比 |

代码清单 17-8 中的语句使用 SizeLimit 属性将缓存限制为 200 项。在使用内存缓存时，必须注意在分配足够的内存以使缓存有效而不耗尽服务器资源之间找到正确的平衡。

要查看缓存的效果，请重新启动 ASP.NET Core，请求 http://localhost:5000/sum URL。重新加载浏览器，只有一个时间戳会更改，如图 17-4 所示。这是因为缓存提供了计算响应，允许端点生成结果，而不必重复计算。

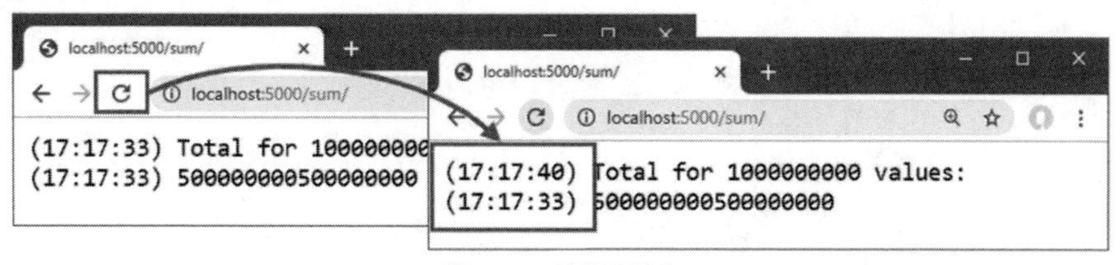


图 17-4 缓存数据值

如果等待两分钟，然后重新加载浏览器，那么两个时间戳都会发生变化，因为缓存的结果已经被弹出，端点必须执行计算来生成结果。

## 17.2.2 使用共享和持久的数据缓存

由 AddDistributedMemoryCache 方法创建的缓存并不是分布式的，尽管名称不同。这些项作为 ASP.NET Core 进程的一部分存储在内存中，这意味着运行在多个服务器或容器上的应用程序不共享缓存数据。这也意味着当 ASP. NET Core 停止时，缓存的内容会丢失。

AddDistributedSqlServerCache 方法将缓存数据存储在 SQL Server 数据库中，该数据库可以在多个 ASP.NET Core 服务器之间共享，持久化存储数据。

第一步是创建一个用于存储缓存数据的数据库。可将缓存的数据与应用程序的其他数据一起存储，但本章使用一个单独的数据库，命名为 CacheDb。可以使用 Azure Data Studio 或 SQL Server Management Studio 创建数据库，这两种工具都可从微软免费获得。还可以使用 sqlcmd 从命令行创建数据库。打开一个新的 PowerShell 命令提示符，运行代码清单 17-9 所示的命令来连接 LocalDB 服务器。

---

■ **提示：**sqlcmd 工具应该作为 Visual Studio 工作负载的一部分安装，或者作为 SQL Server Express 安装的一部分。如果还没有安装，那么可从 https://docs.microsoft.com/en-us/sql/tools/sqlcmd-utility?view=sql-server-2017 下载安装程序。

---

**代码清单 17-9　连接到数据库**

```
sqlcmd -S "(localdb)\MSSQLLocalDB"
```

密切关注指定数据库的参数。有一个反斜杠，后面跟着 MSSQLLocalDB。很难发现重复的字母：MS-SQL-LocalDB(但没有连字符)。

当连接建立后，会看到 1>提示符。输入代码清单 17-10 所示的命令，并在每个命令之后按 Enter 键。

> **注意：** 如果正在使用 Visual Studio，必须应用第 2 章描述的 SQL Server 更新。随 Visual Studio 默认安装的 SQL Server 版本无法创建 LocalDB 数据库。

**代码清单 17-10　创建数据库**

```
CREATE  DATABASE  CacheDb
GO
```

如果没有报告错误，就输入 exit 并按 Enter 键终止连接。下一步是运行代码清单 17-11 所示的命令，在新数据库中创建一个表，它使用一个全局.NET Core 工具准备数据库。

> 提示：如果需要重置缓存数据库，请使用代码清单 17-9 中的命令打开连接，并使用命令 DROP DATABASE CacheDB。然后可以使用代码清单 17-10 中的命令重新创建数据库。

**代码清单 17-11　创建缓存数据库表**

```
dotnet sql-cache create "Server=(localdb)\MSSQLLocalDB;Database=CacheDb" dbo DataCache
```

该命令的参数是连接字符串，它指定了数据库、模式和将用于存储缓存数据的表名。在一行上输入命令并按 Enter 键。该工具将花费几秒钟时间连接到数据库。如果该过程成功，会看到以下消息：

```
Table and index were created successfully.
```

### 创建持久缓存服务

既然数据库已经准备好了，就可以创建使用它来存储缓存数据的服务。要添加支持 SQL Server 缓存所需的 NuGet 包，打开一个新的 PowerShell 命令提示符，导航到 Platform 项目文件夹，并运行代码清单 17-12 所示的命令(如果使用的是 Visual Studio，可以通过选择 Project | Manage Nuget Packages 来添加这个包)。

**代码清单 17-12　向项目添加包**

```
dotnet add package Microsoft.Extensions.Caching.SqlServer --version 3.1.1
```

下一步是定义一个连接字符串，它在 JSON 配置文件中描述数据库连接，如代码清单 17-13 所示。

■ **注意：**AddDistributedSqlServerCache 方法创建的缓存是分布式的，这意味着多个应用程序可以使用同一个数据库并共享缓存数据。如果将同一个应用程序部署到多个服务器或容器，那么所有实例都将能够共享缓存的数据。如果在不同的应用程序之间共享缓存，那么应该密切注意使用的键，以确保应用程序接收到它们期望的数据类型。

**代码清单 17-13 在 Platform 文件夹的 appsettings.json 文件中定义连接字符串**

```
{
  "Location": {
    "CityName": "Buffalo"
  },
  "Logging": {
    "LogLevel": {
      "Default": "Information",
      "Microsoft": "Warning",
      "Microsoft.Hosting.Lifetime": "Information"
    }
  },
  "AllowedHosts": "*",
  "ConnectionStrings": {
    "CacheConnection": "Server=(localdb)\\MSSQLLocalDB;Database=CacheDb"
  }
}
```

注意，连接字符串使用两个反斜杠字符(\\)转义 JSON 文件中的字符。代码清单 17-14 将 Startup 类中的缓存服务实现更改为使用 SQL Server 和代码清单 17-13 中的连接字符串。

**代码清单 17-14 在 Platform 文件夹的 Startup.cs 文件中使用持久数据缓存**

```
using Microsoft.AspNetCore.Builder;
using Microsoft.AspNetCore.Http;
using Microsoft.Extensions.DependencyInjection;
using Microsoft.Extensions.Configuration;

namespace Platform {
  public class Startup {

    public Startup(IConfiguration config) {
      Configuration = config;
    }

    private IConfiguration Configuration {get; set;}

    public void ConfigureServices(IServiceCollection services) {

      services.AddDistributedSqlServerCache(opts => {
        opts.ConnectionString
          = Configuration["ConnectionStrings:CacheConnection"];
```

```
                opts.SchemaName = "dbo";
                opts.TableName = "DataCache";
            });
        }

        public void Configure(IApplicationBuilder app) {
            app.UseDeveloperExceptionPage();
            app.UseStaticFiles();
            app.UseRouting();
            app.UseEndpoints(endpoints => {

                endpoints.MapEndpoint<SumEndpoint>("/sum/{count:int=1000000000}");

                endpoints.MapGet("/", async context => {
                    await context.Response.WriteAsync("Hello World!");
                });
            });
        }
    }
}
```

IConfiguration 服务用于从应用程序的配置数据中访问连接字符串。缓存服务是使用 AddDistributedSqlServerCache 方法创建的，并使用 SqlServerCacheOptions 类的实例进行配置，表 17-7 描述了该类最有用的属性。

表 17-7 有用的 SqlServerCacheOptions 属性

| 名称 | 描述 |
| --- | --- |
| ConnectionString | 此属性指定连接字符串，它通常存储在 JSON 配置文件中，并通过 IConfguration 服务访问 |
| SchemaName | 此属性指定高速缓存表的架构名称 |
| TableName | 此属性指定高速缓存表的名称 |
| ExpiredItemsDeletionInterval | 此属性指定扫描过期项的频率。默认值是 30 分钟 |
| DefaultSlidingExpiration | 此属性指定一项在缓存中保持未读的时间有多长，直到过期。默认值是 20 分钟 |

代码清单使用 ConnectionString、SchemaName 和 TableName 属性来配置缓存中间件，以使用数据库表。重启 ASP.NET Core，使用浏览器请求 http://localhost:5000/sum URL。应用程序生成的响应没有变化，如图 17-4 所示，但是缓存的响应是持久的，甚至在重新启动 ASP.NET Core 时也会使用。

**缓存特定于会话的数据值**

使用 IDistributedCache 服务时，数据值在所有请求之间共享。如果希望为每个用户缓存不同的数据值，就可以使用第 16 章中描述的会话中间件。会话中间件依赖 IDistributedCache 服务来存储其数据，这意味着会持久存储会话数据，并在使用 AddDistributedSqlServer Cache 方法时供分布式应用程序使用。

## 17.3 缓存响应

缓存单个数据项的另一种方法是缓存整个响应，如果编写响应的开销很大，并且很可能会重复，那么这是一种有用的方法。缓存响应需要添加服务和中间件组件，如代码清单 17-15 所示。

**代码清单 17-15 在 Platform 文件夹的 Startup.cs 文件中配置响应缓存**

```
using Microsoft.AspNetCore.Builder;
using Microsoft.AspNetCore.Http;
using Microsoft.Extensions.DependencyInjection;
using Microsoft.Extensions.Configuration;
using Platform.Services;

namespace Platform {
   public class Startup {

      public Startup(IConfiguration config) {
         Configuration = config;
      }

      private IConfiguration Configuration {get; set;}
      public void ConfigureServices(IServiceCollection services) {
         services.AddDistributedSqlServerCache(opts => {
            opts.ConnectionString
               = Configuration["ConnectionStrings:CacheConnection"];
            opts.SchemaName = "dbo";
            opts.TableName = "DataCache";
         });
         services.AddResponseCaching();
         services.AddSingleton<IResponseFormatter, HtmlResponseFormatter>();
      }

      public void Configure(IApplicationBuilder app) {
         app.UseDeveloperExceptionPage();
         app.UseResponseCaching();
         app.UseStaticFiles();
         app.UseRouting();
         app.UseEndpoints(endpoints => {

            endpoints.MapEndpoint<SumEndpoint>("/sum/{count:int=1000000000}");

            endpoints.MapGet("/", async context => {
               await context.Response.WriteAsync("Hello World!");
            });
         });
      }
   }
}
```

在 ConfigureServices 方法中使用 AddResponseCaching 方法来设置缓存所使用的服务。中间件组件是通过 UseResponseCaching 方法添加的，应该在需要缓存其响应的任何端点或中间件之前调用该方法。

还定义了 IResponseFormatter 服务，第 14 章用它来解释依赖注入是如何工作的。响应缓存仅在某些情况下使用，如稍后所述，演示该特性需要 HTML 响应。

**注意：**响应缓存功能不使用 IDistributedCache 服务。响应缓存在内存中，不是分布式的。

在代码清单 17-16 中，更新了 SumEndpoint 类，以便它请求响应缓存，而不是只缓存数据值。

**代码清单 17-16　在 Platform 文件夹的 SumEndpoint.cs 文件中使用响应缓存**

```
using Microsoft.AspNetCore.Http;
using System;
using System.Threading.Tasks;
using Microsoft.Extensions.Caching.Distributed;
using Microsoft.AspNetCore.Routing;
using Platform.Services;

namespace Platform {

    public class SumEndpoint {

        public async Task Endpoint(HttpContext context, IDistributedCache cache,
                IResponseFormatter formatter, LinkGenerator generator) {
            int count = int.Parse((string)context.Request.RouteValues["count"]);
            long total = 0;
            for (int i = 1; i <= count; i++) {
                total += i;
            }
            string totalString = $"({ DateTime.Now.ToLongTimeString() }) {total}";

            context.Response.Headers["Cache-Control"] = "public, max-age=120";

            string url = generator.GetPathByRouteValues(context, null,
                new { count = count });

            await formatter.Format(context,
                $"<div>({DateTime.Now.ToLongTimeString()}) Total for {count}"
                + $" values:</div><div>{totalString}</div>"
                + $"<a href={url}>Reload</a>");
        }
    }
}
```

对端点的一些更改启用了响应缓存，但其他更改只是为了演示它是如何工作的。为了启用响应缓存，重要的语句是向响应添加头的语句，如下所示：

```
...
context.Response.Headers["Cache-Control"] = "public, max-age=120";
...
```

Cache-Control 头用于控制响应缓存。中间件只缓存具有 Cache-Control 头的响应，Cache-Control 头包含公共指令。max-age 指令用于指定可以缓存响应的周期，以秒为单位。代码清单 17-16 中使用的 Cache-Control 头支持缓存，并指定可以将响应缓存两分钟。

启用响应缓存很简单，但是检查它是否工作需要小心。重新加载浏览器窗口或在 URL 栏中按 Return 时，浏览器在请求中包含一个 Cache-Control 头，该头把 max-age 指令设置为 0，从而绕过响应缓存，并导致端点生成一个新的响应。请求 URL 而没有 Cache-Control 头的唯一可靠的方式是使用 HTML 锚元素导航，这就是为什么代码清单 17-16 中的端点使用 IResponseFormatter 服务生成一个 HTML 响应，并使用 LinkGenerator 服务创建一个 URL，该 URL 可以用于锚元素的 href 属性。

要检查响应缓存，请重新启动 ASP.NET Core，使用浏览器请求 http://localhost:5000/sum。生成响应后，单击 Reload 链接，请求相同的 URL。到响应中的两个时间戳都没有更改，这表明整个响应已被缓存，如图 17-5 所示。

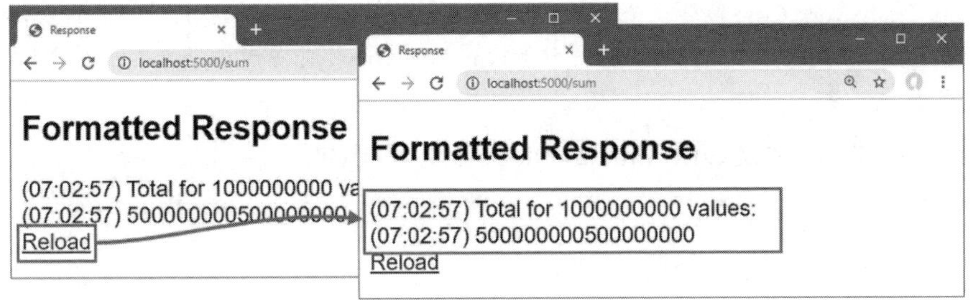


图 17-5　缓存响应

可将Cache-Control头与Vary头结合使用，以提供对缓存哪些请求的细粒度控制。参见 https://developer.mozilla.org/en-US/docs/Web/HTTP/Headers/Cache-Control和https://developer.mozilla。通过org/en-US/docs/Web/HTTP/Headers/Vary，了解两个标头提供的特性的详细信息。

**压缩响应**

ASP.NET Core 包含中间件，可为那些表示可以处理压缩数据的浏览器压缩响应。中间件通过 UseResponseCompression 方法添加到管道中。压缩是在压缩所需的服务器资源和向客户端交付内容所需的带宽之间进行权衡，在没有进行测试以确定性能影响之前，不应该启用压缩。

## 17.4 使用 Entity Framework Core

并不是所有的数据值都是由应用程序直接生成的，大多数项目都需要访问数据库中的数据。Entity Framework Core 很好地集成到 ASP.NET Core 平台中，支持用 C#类创建数据库，也支持用 C#类来表示现有数据库。在接下来的几节中，将演示创建简单数据模型、使用它创建数据库、并在端点中查询该数据库的过程。

**使用 Entity Framework Core 工作**

对 Entity Framework Core 最常见的抱怨是性能差。审查存在 Entity Framework Core 性能问题的项目时，问题几乎总是因为开发团队将 Entity Framework Core 视为一个黑箱，而没有注意发送到数据库的 SQL 查询。并不是所有的 LINQ 特性都可以转换成 SQL，最常见的问题是从数据库中检索大量数据的查询，在减少数据以生成单个值后将其丢弃。

使用 Entity Framework Core 要求你深刻理解 SQL，并确保应用程序生成的 LINQ 查询转换为高效的 SQL 查询。很少有应用程序具有 Entity Framework Core 无法满足的高性能数据需求，但这不是大多数典型 Web 应用程序的情况。

这并不是说 Entity Framework Core 是完美的。它有自己的怪癖，需要投入时间才能精通。如果不喜欢 Entity Framework Core 的工作方式，则可能更喜欢使用另一种方法，比如 Dapper (https://github.com/StackExchange/Dapper)。但是如果问题是查询执行的速度不够快，就应该花一些时间研究如何处理这些查询，使用本章其余部分描述的技术来完成。

### 17.4.1　安装 Entity Framework Core

Entity Framework Core 需要一个全局工具包，用于从命令行管理数据库，并管理提供数据访问的项目的包。要安装这个工具包，打开一个新的 PowerShell 命令提示符，运行代码清单 17-17 所示的命令。

**代码清单 17-17　安装 Entity Framework Core 全局工具包**

```
dotnet tool uninstall --global dotnet-ef
dotnet tool install --global dotnet-ef --version 3.1.1
```

第一个命令删除 dotnet-ef 包的任何现有版本，第二个命令安装本书示例所需的版本。这个包提供了在后面的示例中看到的 dotnet ef 命令。为了确保包按预期工作，运行代码清单 17-18 所示的命令。

**代码清单 17-18　测试 Entity Framework Core 全局工具**

```
dotnet ef -help
```

此命令显示全局工具的帮助消息，并生成以下输出:

```
Entity Framework Core .NET Command-line Tools 3.1.1
Usage: dotnet ef [options] [command]
Options:
    --version Show version information
    -h|--help Show help information
    -v|--verbose Show verbose output.
    --no-color Don't colorize output.
    --prefix-output Prefix output with level.
Commands:
    database Commands to manage the database.
    dbcontext Commands to manage DbContext types.
```

```
   migrations Commands to manage migrations.
Use "dotnet ef [command] --help" for more information about a command.
```

Entity Framework Core 还需要将包添加到项目中。如果正在使用 Visual Studio Code 或喜欢从命令行工作，请导航到 Platform 项目文件夹(包含 Platform.csproj 文件的文件夹)。并运行代码清单 17-19 中所示的命令。

**代码清单 17-19 向项目中添加 Entity Framework Core 包**

```
dotnet add package Microsoft.EntityFrameworkCore.Design --version 3.1.1
dotnet add package Microsoft.EntityFrameworkCore.SqlServer --version 3.1.1
```

### 17.4.2 创建数据模型

本章使用 C#类定义数据模型，使用 Entity Framework Core 创建数据库和模式。创建 Platform/Models 文件夹，并向其中添加一个名为 calculation.cs 的类文件，其内容如代码清单 17-20 所示。

**代码清单 17-20 Platform/Models 文件夹中 calculation.cs 文件的内容**

```
namespace Platform.Models {

    public class Calculation {
        public long Id { get; set; }
        public int Count { get; set; }
        public long Result { get; set; }
    }
}
```

可以在其他章节中看到更复杂的数据模型，但是对于这个示例，继续本章的主题，并对前面示例中执行的计算进行建模。Id 属性用于为存储在数据库中的每个对象创建唯一的键，Count 和 Result 属性将描述计算及其结果。

Entity Framework Core 使用上下文类提供对数据库的访问。向 Platform/Models 文件夹添加一个名为 CalculationContext.cs 的文件，其内容如代码清单 17-21 所示。

**代码清单 17-21 Platform/Models 文件夹中 CalculationContext.cs 文件的内容**

```
using Microsoft.EntityFrameworkCore;

namespace Platform.Models {

    public class CalculationContext: DbContext {

        public CalculationContext(DbContextOptions<CalculationContext> opts)
            : base(opts) {}

        public DbSet<Calculation> Calculations { get; set; }
    }
}
```

CalculationContext 类定义了一个构造函数，用于接收传递给基构造函数的 options 对象。Calculations 属性提供 Entity Framework Core 从数据库检索 Calculation 对象的访问。

### 17.4.3　配置数据库服务

通过在 Startup 类中配置的服务提供对数据库的访问，如代码清单 17-22 所示。

**代码清单 17-22　在 Platform 文件夹的 Startup.cs 文件中配置数据服务**

```
using Microsoft.AspNetCore.Builder;
using Microsoft.AspNetCore.Http;
using Microsoft.Extensions.DependencyInjection;
using Microsoft.Extensions.Configuration;
using Platform.Services;
using Microsoft.EntityFrameworkCore;
using Platform.Models;

namespace Platform {
    public class Startup {

        public Startup(IConfiguration config) {
            Configuration = config;
        }

        private IConfiguration Configuration {get; set;}

        public void ConfigureServices(IServiceCollection services) {
            services.AddDistributedSqlServerCache(opts => {
                opts.ConnectionString
                    = Configuration["ConnectionStrings:CacheConnection"];
                opts.SchemaName = "dbo";
                opts.TableName = "DataCache";
            });
            services.AddResponseCaching();
            services.AddSingleton<IResponseFormatter, HtmlResponseFormatter>();

            services.AddDbContext<CalculationContext>(opts => {
                opts.UseSqlServer(Configuration["ConnectionStrings:CalcConnection"]);
            });
        }

        public void Configure(IApplicationBuilder app) {
            app.UseDeveloperExceptionPage();
            app.UseResponseCaching();
            app.UseStaticFiles();
            app.UseRouting();
            app.UseEndpoints(endpoints => {
```

```
            endpoints.MapEndpoint<SumEndpoint>("/sum/{count:int=1000000000}");

            endpoints.MapGet("/", async context => {
                await context.Response.WriteAsync("Hello World!");
            });
        });
    }
  }
}
```

AddDbContext 方法为 Entity Framework Core 上下文类创建一个服务。该方法接收一个选项对象，该对象用于选择数据库提供程序，这是通过 UseSqlServer 方法完成的。IConfiguration 服务用于获取数据库的连接字符串，它在代码清单 17-23 中定义。

**代码清单 17-23　在 Platform 文件夹的 appsettings.json 文件中定义连接字符串**

```
{
    "Location": {
        "CityName": "Buffalo"
    },
    "Logging": {
        "LogLevel": {
            "Default": "Information",
            "Microsoft": "Warning",
            "Microsoft.Hosting.Lifetime": "Information",
            "Microsoft.EntityFrameworkCore": "Information"
        }
    },
    "AllowedHosts": "*",
    "ConnectionStrings": {
        "CacheConnection": "Server=(localdb)\\MSSQLLocalDB;Database=CacheDb",
        "CalcConnection": "Server=(localdb)\\MSSQLLocalDB;Database=CalcDb"
    }
}
```

该代码清单还设置了 Microsoft.EntityFrameworkCore 类的日志级别，它会显示 Entity Framework Core 用于查询数据库的 SQL 语句。

---

**提示：**对于对多个结果集进行查询的连接字符串，将 MultipleActiveResultSets 选项设置为 True，可以在第 7 章的 SportsStore 项目的连接字符串中看到设置此选项的示例。

---

## 17.4.4 创建和应用数据库迁移

Entity Framework Core 使用一种称为迁移的特性来管理数据模型类和数据库之间的关系。对模型类进行更改时，将创建一个新的迁移，修改数据库以匹配这些更改。为了创建初始迁移(它将创建一个新的数据库，并准备存储计算对象)，打开一个新的 PowerShell 命令提示符，导航到包含 Platform.csproj 文件的文件夹，并运行代码清单 17-24 所示的命令。

**代码清单17-24　创建一个迁移**

```
dotnet ef migrations add Initial
```

dotnet ef 命令与 Entity Framework Core 相关。代码清单 17-24 中的命令创建一个名为 Initial 的新迁移，这是项目的第一次迁移通常使用的名称。项目中已添加了 Migrations 文件夹，它包含类文件，这些类文件的语句准备数据库，以便它在数据模型中存储对象。要应用迁移，在 Platform 项目文件夹中运行代码清单 17-25 所示的命令。

**代码清单17-25　应用迁移**

```
dotnet ef database update
```

该命令执行代码清单17-24中创建的迁移中的命令，并使用它们准备数据库，可以在命令提示符写入的 SQL 语句中看到这一点。

### 17.4.5　播种数据库

大多数应用程序需要一些种子数据，特别是在开发期间。Entity Framework Core 确实提供了一个数据库种子化特性，但它在大多数项目中用途有限，因为它不允许在数据库为其存储的对象分配唯一键的地方播种数据。这是大部分数据模型中的一个重要特性，因为这意味着应用程序不必考虑分配唯一的键值。

更灵活的方法是使用常规 Entity Framework Core 特性向数据库添加种子数据。使用代码清单 17-26 所示的代码在 Platform/Models 文件夹中创建一个名为 SeedData.cs 的文件。

**代码清单17-26　Platform/Models 文件夹中 SeedData.cs 文件的内容**

```
using Microsoft.EntityFrameworkCore;
using Microsoft.Extensions.Logging;
using System.Collections.Generic;
using System.Linq;

namespace Platform.Models {
    public class SeedData {
        private CalculationContext context;
        private ILogger<SeedData> logger;

        private static Dictionary<int, long> data
            = new Dictionary<int, long>() {
                {1, 1}, {2, 3}, {3, 6}, {4, 10}, {5, 15},
                {6, 21}, {7, 28}, {8, 36}, {9, 45}, {10, 55}
            };

        public SeedData(CalculationContext dataContext, ILogger<SeedData> log) {
            context = dataContext;
            logger = log;
        }
```

```
        public void SeedDatabase() {
            context.Database.Migrate();
            if (context.Calculations.Count() == 0) {
                logger.LogInformation("Preparing to seed database");
                context.Calculations!.AddRange(data.Select(kvp => new Calculaton() {
                    Count = kvp.Key, Result = kvp.Value
                }));
                context.SaveChanges();
                logger.LogInformation("Database seeded");
            } else {
                logger.LogInformation("Database not seeded");
            }
        }
    }
}
```

SeedData 类声明了 CalculationContext 和 ILogger<T>类型的构造函数依赖关系，这些类型在 SeedDatabase 方法中用于准备数据库。上下文的 Database.Migrate 方法用于将任何挂起的迁移应用到数据库，而 calculate 属性用于通过 AddRange 方法存储新数据，该方法接收一系列 Calculation 对象。

新对象使用 SaveChanges 方法存储在数据库中。要使用 SeedData 类，对 Startup 类进行如代码清单 17-27 所示的更改。

### 代码清单 17-27 在 Platform 文件夹的 Startup.cs 文件中启用数据库播种

```
using Microsoft.AspNetCore.Builder;
using Microsoft.AspNetCore.Http;
using Microsoft.Extensions.DependencyInjection;
using Microsoft.Extensions.Configuration;
using Platform.Services;
using Microsoft.EntityFrameworkCore;
using Platform.Models;
using Microsoft.Extensions.Hosting;
using Microsoft.AspNetCore.Hosting;

namespace Platform {
    public class Startup {

        public Startup(IConfiguration config) {
            Configuration = config;
        }

        private IConfiguration Configuration {get; set;}

        public void ConfigureServices(IServiceCollection services) {
            services.AddDistributedSqlServerCache(opts => {
                opts.ConnectionString
```

```
                = Configuration["ConnectionStrings:CacheConnection"];
            opts.SchemaName = "dbo";
            opts.TableName = "DataCache";
        });
        services.AddResponseCaching();
        services.AddSingleton<IResponseFormatter, HtmlResponseFormatter>();

        services.AddDbContext<CalculationContext>(opts => {
            opts.UseSqlServer(Configuration["ConnectionStrings:CalcConnection"]);
        });
        services.AddTransient<SeedData>();
    }

    public void Configure(IApplicationBuilder app,
            IHostApplicationLifetime lifetime, IWebHostEnvironment env,
            SeedData seedData) {
        app.UseDeveloperExceptionPage();
        app.UseResponseCaching();
        app.UseStaticFiles();
        app.UseRouting();
        app.UseEndpoints(endpoints => {

            endpoints.MapEndpoint<SumEndpoint>("/sum/{count:int=1000000000}");

            endpoints.MapGet("/", async context => {
                await context.Response.WriteAsync("Hello World!");
            });
        });

        bool cmdLineInit = (Configuration["INITDB"] ?? "false") == "true";
        if (env.IsDevelopment() || cmdLineInit) {
            seedData.SeedDatabase();
            if (cmdLineInit) {
                lifetime.StopApplication();
            }
        }
    }
  }
}
```

ConfigureServices 方法中的语句创建了一个 SeedData 服务。尽管服务通常是使用接口定义的，如第 14 章所述，但也有不同版本的 AddSingleton、AddScoped 和 AddTransient 方法，它们使用单一类型参数创建服务。这便于实例化具有构造函数依赖关系、允许使用服务生命周期的类。代码清单 17-27 中的语句创建了一个临时服务，这意味着创建一个新的 SeedData 对象，来解析对该服务的每个依赖。

对 Configure 方法的添加允许在两种情况下播种数据库。如果宿主环境是 Development，数据库将在应用程序启动时自动播种。显式地为数据库播种也很有用，特别是在为登台或生产测试设置应用程序时。该语句检查名为 INITDB 的配置设置：

```
...
bool cmdLineInit = (Configuration["INITDB"] ?? "false") == "true";
...
```

可在命令行上提供此设置为数据库提供种子，之后使用 IHostApplicationLifetime 服务终止应用程序。这是较少使用的服务之一，但它提供了在应用程序启动和停止时触发的事件，并提供了终止应用程序的 StopApplication 方法。要查看数据库，打开一个新的 PowerShell 命令提示符，导航到项目文件夹，运行代码清单 17-28 所示的命令。

**代码清单 17-28 播种数据库**

```
dotnet run INITDB=true
```

应用程序启动，数据库播种到 SeedData 类定义的 10 次计算的结果，之后应用程序将终止。在播种过程中，会看到发送到数据库的 SQL 语句，这些语句检查是否有任何未决迁移，计算用于存储计算数据的表中的行数，如果表为空，则添加种子数据。

如果不喜欢使用命令行，可通过选择 Debug | Start Without Debugging 播种数据库。

只要环境是 Development，数据库将种子作为正常 ASP.NET Core 启动序列的一部分。

■ **注意：** 如果需要重置数据库，可使用 dotnet ef database drop -force 命令。然后使用 dotnet run INITDB=true 重新创建数据库并再次播种。

## 17.4.6 在端点中使用数据

端点和中间件组件通过声明对上下文类的依赖，并使用其 DbSet<t>属性执行 LINQ 查询，来访问 Entity Framework Core 数据。LINQ 查询转换为 SQL，并发送到数据库。从数据库接收到的表格数据用于创建生成响应的数据模型对象。代码清单 17-29 更新了 SumEndpoint 类，以使用 Entity Framework Core。

**代码清单 17-29 在 Platform 文件夹的 SumEndpoint.cs 文件中使用数据库**

```
using Microsoft.AspNetCore.Http;
using System;
using System.Threading.Tasks;
using Microsoft.Extensions.Caching.Distributed;
using Microsoft.AspNetCore.Routing;
using Platform.Services;
using Platform.Models;
using System.Linq;

namespace Platform {

    public class SumEndpoint {
```

```
        public async Task Endpoint(HttpContext context,
                CalculationContext dataContext) {
            int count = int.Parse((string)context.Request.RouteValues["count"]);
            long total = dataContext.Calculations
                .FirstOrDefault(c => c.Count == count)?.Result ?? 0;
            if (total == 0) {
                for (int i = 1; i <= count; i++) {
                    total += i;
                }
                dataContext.Calculations!
                    .Add(new Calculaton() { Count = count, Result = total});
                await dataContext.SaveChangesAsync();
            }
            string totalString = $"({ DateTime.Now.ToLongTimeString() }) {total}";
            await context.Response.WriteAsync(
                $"({DateTime.Now.ToLongTimeString()}) Total for {count}"
                + $" values:\n{totalString}\n");
        }
    }
}
```

端点使用 LINQ FirstOrDefault 来搜索一个存储的 Calculation 对象，计算请求如下：

```
...
dataContext.Calculations.FirstOrDefault(c => c.Count == count)?.Result ?? 0;
...
```

如果存储了对象，则使用它准备响应。如果没有，则执行计算，新的 Calculation 对象由以下语句存储：

```
...
dataContext.Calculations!.Add(new Calculaton() { Count = count, Result = total});
await dataContext.SaveChangesAsync();
...
```

Add 方法用于告诉 Entity Framework Core 应该存储对象，但是在调用 SaveChangesAsync 方法之前不会执行更新。要查看更改的效果，请重新启动 ASP.NET Core MVC(如果使用命令行，则不带 INITDB 参数)，并请求 http://localhost:5000/sum/10 URL。这是数据库已播种的计算之一，能够在应用程序生成的日志消息中看到发送到数据库的查询。

```
...
Executing DbCommand [Parameters=[@__count_0='?' (DbType = Int32)],
   CommandType='Text', CommandTimeout='30']
SELECT TOP(1) [c].[Id], [c].[Count], [c].[Result]
FROM [Calculations] AS [c]
WHERE ([c].[Count] = @__count_0) AND @__count_0 IS NOT NULL
...
```

如果请求 http://localhost:5000/sum/100，将查询数据库，但不会找到结果。端点执行计算，并将结果存储在数据库中，然后生成图 17-6 所示的结果。

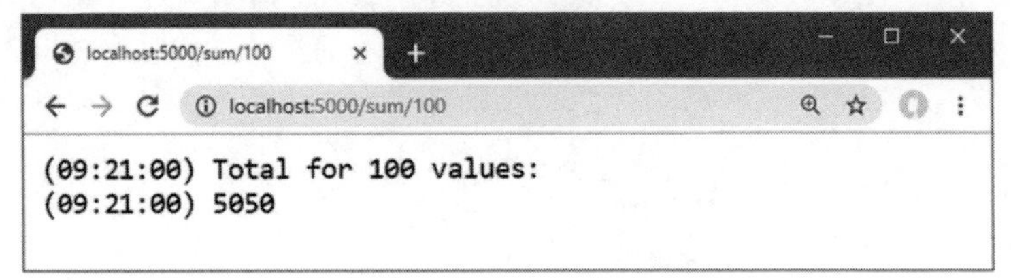


图 17-6 执行计算

一旦结果存储在数据库中，后续对相同 URL 的请求将使用存储的数据得到满足。在 Entity Framework Core 生成的日志输出中，SQL 语句用于存储数据。

```
...
Executing DbCommand [Parameters=[@p0='?' (DbType = Int32), @p1='?' (DbType = Int64)],
    CommandType='Text', CommandTimeout='30']
SET NOCOUNT ON;
INSERT INTO [Calculations] ([Count], [Result])
VALUES (@p0, @p1);
SELECT [Id]
FROM [Calculations]
WHERE @@ROWCOUNT = 1 AND [Id] = scope_identity();
...
```

**注意：** 从数据库中检索的数据不会缓存，并且每个请求都会导致一个新的 SQL 查询。根据所需查询的频率和复杂性，可能希望使用本章前面描述的技术缓存数据值或响应。

### 启用敏感数据记录

Entity Framework Core 在它生成的日志消息中不包含参数值，这就是为什么日志输出包含问号的原因，如下所示：

```
...
Executing DbCommand [Parameters=[@__count_0='?' (DbType = Int32)], CommandType='Text',
CommandTimeout='30']
...
```

为了防止敏感数据存储在日志中，数据被省略了。如果查询有问题，需要查看发送到数据库的值，那么可以在配置数据库上下文时使用 EnableSensitiveDataLogging 方法，如代码清单 17-30 所示。

**代码清单 17-30 在 Platform 文件夹的 Startup.cs 文件中启用敏感数据日志记录**

```
...
services.AddDbContext<CalculationContext>(opts => {
    opts.UseSqlServer(Configuration["ConnectionStrings:CalcConnection"]);
    opts.EnableSensitiveDataLogging(true);
});
```

```
...
```

重启 ASP.NET Core MVC，再次请求 http://localhost:5000/sum/100 URL。处理请求时，Entity Framework Core 会在它创建的日志消息中包含参数值来显示 SQL 查询，如下所示：

```
...
Executing DbCommand [Parameters=[@__count_0='100'], CommandType='Text',
   CommandTimeout='30']
SELECT TOP(1) [c].[Id], [c].[Count], [c].[Result]
FROM [Calculations] AS [c]
WHERE ([c].[Count] = @__count_0) AND @__count_0 IS NOT NULL
...
```

这是一个应该谨慎使用的特性，因为通常无法访问应用程序处理的敏感数据(如信用卡号和账户详细信息)的人可以访问日志。

## 17.5　小结

本章演示了对处理数据很有用的 ASP.NET Core 平台特性，展示了如何在本地和共享数据库中缓存单个数据值。还展示了如何缓存响应以及如何使用 Entity Framework Core 在数据库中读写数据。本书第III部分将解释如何根据 ASP.NET Core 平台创建 Web 应用程序。

# 第 III 部分

# ASP.NET Core 应用程序

- 第 18 章　创建示例项目
- 第 19 章　创建 RESTful Web 服务
- 第 20 章　高级 Web 服务特性
- 第 21 章　使用控制器和视图(第 1 部分)
- 第 22 章　使用控制器和视图(第 2 部分)
- 第 23 章　使用 Razor Pages
- 第 24 章　使用视图组件
- 第 25 章　使用标签助手
- 第 26 章　使用内置的标签助手
- 第 27 章　使用表单标签助手
- 第 28 章　使用模型绑定
- 第 29 章　使用模型验证
- 第 30 章　使用过滤器
- 第 31 章　创建表单应用程序

# 第 18 章

# 创建示例项目

本章将创建贯穿本书这一部分的示例项目。该项目包含一个简单的数据模型、一个用于格式化 HTML 内容的客户端包和一个简单的请求管道。

## 18.1 创建项目

从 Windows 的 Start 菜单中打开一个新的 PowerShell 命令提示符，并运行代码清单 18-1 所示的命令。

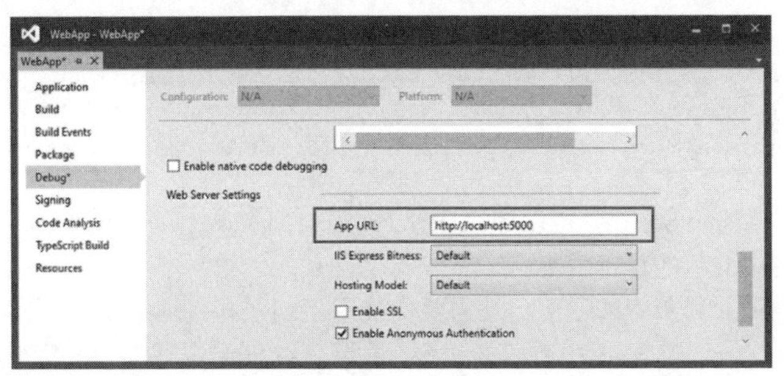


图 18-1 更改 HTTP 端口

■ **提示：** 可以从 https://github.com/apress/pro-asp.net-core-3 下载本章和本书中其他所有章节的示例项目。如果在运行示例时遇到问题，请参阅第 1 章以获得帮助。

**代码清单 18-1 创建项目**

```
dotnet new globaljson --sdk-version 3.1.101 --output WebApp
dotnet new web --no-https --output WebApp --framework netcoreapp3.1
dotnet new sln -o WebApp

dotnet sln WebApp add WebApp
```

如果使用的是 Visual Studio，打开 WebApp 文件夹中的 WebApp.sln 文件。选择 Project | Platform Properties，导航到 Debug 页面，并将 App URL 字段更改为 http://localhost:5000，如图 18-1 所示。这将更改用于接收 HTTP 请求的端口。选择 File | Save All 以保存配置变化。

如果使用的是 Visual Studio Code，请打开 WebApp 文件夹。当提示添加构建和调试项目所需的资源时，单击 Yes 按钮，如图 18-2 所示。

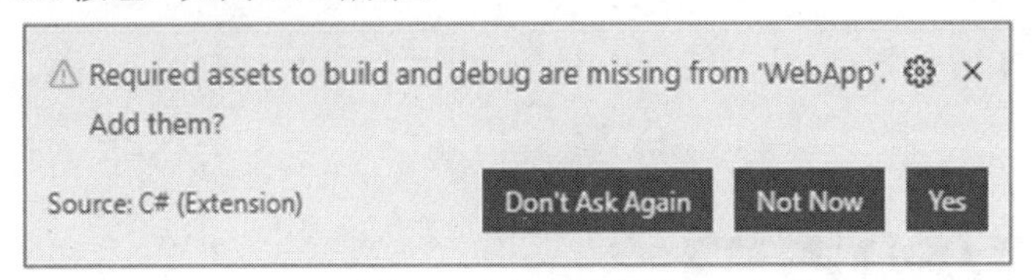


图 18-2　添加项目资源

## 18.2　添加数据模型

数据模型有助于演示使用 ASP.NET Core 构建 Web 应用程序的不同方法，展示如何组合复杂的响应，以及用户如何提交数据。在接下来的几节中，创建了一个简单的数据模型，并使用它创建用于存储应用程序数据的数据库模式。

### 18.2.1　向项目中添加 NuGet 包

数据模型使用 Entity Framework Core 在 SQL Server LocalDB 数据库中存储和查询数据。要为 Entity Framework Core 添加 NuGet 包，使用 PowerShell 命令提示符在 WebApp 项目文件夹中运行代码清单 18-2 所示的命令。

**代码清单 18-2　向项目中添加包**

```
dotnet add package Microsoft.EntityFrameworkCore.Design --version 3.1.1
dotnet add package Microsoft.EntityFrameworkCore.SqlServer --version 3.1.1
```

如果使用的是 Visual Studio，可以通过选择 Project | Manage NuGet Packages 来添加包。请注意选择要添加到项目中的包的正确版本。

如果没有遵循前面章节中的示例，就需要安装用于创建和管理 Entity Framework Core 迁移的全局工具包。运行代码清单 18-3 中所示的命令，删除包的任何现有版本，并安装本书所需的版本(如果在前面的章节中安装了这个版本的工具包，可以跳过这些命令)。

**代码清单 18-3　安装全局工具包**

```
dotnet tool uninstall --global dotnet-ef
dotnet tool install --global dotnet-ef --version 3.1.1
```

### 18.2.2　创建数据模型

本书这一部分的数据模型由三个相关的类组成：Product、Supplier 和 Category。创建一个名为 Models 的新文件夹，并向其中添加一个名为 Category.cs 的类文件，其内容如代码清单 18-4 所示。

**代码清单 18-4　Model 文件夹中 Category.cs 文件的内容**

```
using System.Collections.Generic;
```

```
namespace WebApp.Models {
    public class Category {

        public long CategoryId { get; set; }
        public string Name { get; set; }

        public IEnumerable<Product> Products { get; set; }
    }
}
```

将一个名为 Supplier.cs 的类添加到 Models 文件夹中，并使用它定义如代码清单 18-5 所示的类。

#### 代码清单 18-5　Models 文件夹中 Supplier.cs 文件的内容

```
using System.Collections.Generic;

namespace WebApp.Models {
    public class Supplier {

        public long SupplierId { get; set; }
        public string Name { get; set; }
        public string City { get; set; }

        public IEnumerable<Product> Products { get; set; }
    }
}
```

接下来，将一个名为 Product.cs 的类添加到 Models 文件夹中，并使用它定义如代码清单 18-6 所示的类。

#### 代码清单 18-6　Models 文件夹中 Product.cs 文件的内容

```
using System.ComponentModel.DataAnnotations.Schema;

namespace WebApp.Models {
    public class Product {

        public long ProductId { get; set; }

        public string Name { get; set; }
        [Column(TypeName = "decimal(8, 2)")]
        public decimal Price { get; set; }

        public long CategoryId { get; set; }
        public Category Category { get; set; }
```

```
        public long SupplierId { get; set; }
        public Supplier Supplier { get; set; }
    }
}
```

这三个数据模型类中的每个类都定义了一个键属性，数据库在存储新对象时分配该键属性的值。还有一些导航属性用于查询相关数据，这样就可以查询特定类别中的所有产品。

Price 属性用 Column 属性进行修饰，该属性指定存储在数据库中的值的精度。C#和 SQL 数值类型之间没有一对一的映射，Column 属性告诉 Entity Framework Core，数据库中应该使用哪种 SQL 类型来存储 Price 值。在本例中，decimal(8,2)类型允许有 8 位数字，包括小数点后的两位。

要创建能访问数据库的 Entity Framework Core 上下文类，请将一个名为 DataContext.cs 的文件添加到 Models 文件夹，并添加如代码清单 18-7 所示的代码。

**代码清单 18-7　Models 文件夹中 DataContext.cs 文件的内容**

```
using Microsoft.EntityFrameworkCore;

namespace WebApp.Models {
    public class DataContext: DbContext {

        public DataContext(DbContextOptions<DataContext> opts)
            : base(opts) { }

        public DbSet<Product> Products { get; set; }
        public DbSet<Category> Categories { get; set; }
        public DbSet<Supplier> Suppliers { get; set; }
    }
}
```

上下文类定义了用于查询数据库中的 Product、Category 和 Supplier 数据的属性。

### 18.2.3　准备种子数据

在 Models 文件夹中添加一个名为 SeedData.cs 的类，并添加如代码清单 18-8 所示的代码，以定义用于填充数据库的种子数据。

**代码清单 18-8　Models 文件夹中 SeedData.cs 文件的内容**

```
using Microsoft.EntityFrameworkCore;
using System.Linq;

namespace WebApp.Models {
    public static class SeedData {

        public static void SeedDatabase(DataContext context) {
            context.Database.Migrate();
            if (context.Products.Count() == 0 && context.Suppliers.Count() == 0
                    && context.Categories.Count() == 0) {
```

```
                Supplier s1 = new Supplier
                   { Name = "Splash Dudes", City = "San Jose"};
                Supplier s2 = new Supplier
                   { Name = "Soccer Town", City = "Chicago"};
                Supplier s3 = new Supplier
                   { Name = "Chess Co", City = "New York"};

                Category c1 = new Category { Name = "Watersports" };
                Category c2 = new Category { Name = "Soccer" };
                Category c3 = new Category { Name = "Chess" };

                context.Products.AddRange(
                    new Product { Name = "Kayak", Price = 275,
                       Category = c1, Supplier = s1},
                    new Product { Name = "Lifejacket", Price = 48.95m,
                       Category = c1, Supplier = s1},
                    new Product { Name = "Soccer Ball", Price = 19.50m,
                       Category = c2, Supplier = s2},
                    new Product { Name = "Corner Flags", Price = 34.95m,
                       Category = c2, Supplier = s2},
                    new Product { Name = "Stadium", Price = 79500,
                       Category = c2, Supplier = s2},
                    new Product { Name = "Thinking Cap", Price = 16,
                       Category = c3, Supplier = s3},
                    new Product { Name = "Unsteady Chair", Price = 29.95m,
                       Category = c3, Supplier = s3},
                    new Product { Name = "Human Chess Board", Price = 75,
                       Category = c3, Supplier = s3},
                    new Product { Name = "Bling-Bling King", Price = 1200,
                       Category = c3, Supplier = s3}
                );
                context.SaveChanges();
            }
        }
    }
}
```

静态 SeedDatabase 方法确保所有未决迁移都已应用到数据库。如果数据库是空的，那么用类别、供应商和产品填充它。Entity Framework Core 负责将对象映射到数据库的表中，并在存储数据时自动分配键属性。

### 18.2.4 配置 Entity Framework Core 服务和中间件

对代码清单18-9所示的Startup类进行更改，该类配置Entity Framework Core并设置DataContext服务；本书第II部分将一直使用该服务来访问数据库。

代码清单 18-9　在 WebApp 文件夹的 Startup.cs 文件中准备服务和中间件

```
using System;
using System.Collections.Generic;
using System.Linq;
using System.Threading.Tasks;
using Microsoft.AspNetCore.Builder;
using Microsoft.AspNetCore.Hosting;
using Microsoft.AspNetCore.Http;
using Microsoft.Extensions.DependencyInjection;
using Microsoft.Extensions.Hosting;
using Microsoft.Extensions.Configuration;
using Microsoft.EntityFrameworkCore;
using WebApp.Models;

namespace WebApp {
    public class Startup {

        public Startup(IConfiguration config) {
            Configuration = config;
        }

        public IConfiguration Configuration { get; set; }

        public void ConfigureServices(IServiceCollection services) {
            services.AddDbContext<DataContext>(opts => {
                opts.UseSqlServer(Configuration[
                    "ConnectionStrings:ProductConnection"]);
                opts.EnableSensitiveDataLogging(true);
            });
        }

        public void Configure(IApplicationBuilder app, DataContext context) {

            app.UseDeveloperExceptionPage();
            app.UseRouting();

            app.UseEndpoints(endpoints => {
                endpoints.MapGet("/", async context => {
                    await context.Response.WriteAsync("Hello World!");
                });
            });

            SeedData.SeedDatabase(context);
        }
    }
}
```

要定义用于应用程序数据的连接字符串，在 appsettings.json 文件中添加如代码清单 18-10 所示的配置设置。连接字符串应该在一行中输入。

**代码清单 18-10 在 WebApp 文件夹的 appsettings.json 文件中定义连接字符串**

```
{
  "Logging": {
    "LogLevel": {
      "Default": "Information",
      "Microsoft": "Warning",
      "Microsoft.Hosting.Lifetime": "Information",
      "Microsoft.EntityFrameworkCore": "Information"
    }
  },
  "AllowedHosts": "*",
  "ConnectionStrings": {
    "ProductConnection": "Server=(localdb)\\MSSQLLocalDB;Database=Products;
MultipleActiveResultSets=True"
  }
}
```

除了连接字符串，代码清单 18-10 增加了 Entity Framework Core 的日志细节，以便记录发送到数据库的 SQL 查询。

### 18.2.5 创建和应用迁移

要创建建立数据库模式的迁移，使用 PowerShell 命令提示符在 WebApp 项目文件夹中运行代码清单 18-11 所示的命令。

**代码清单 18-11 创建 Entity Framework Core 迁移**

```
dotnet ef migrations add Initial
```

创建迁移后，使用代码清单 18-12 所示的命令将其应用到数据库。

**代码清单 18-12 将迁移应用到数据库**

```
dotnet ef database update
```

应用程序显示的日志消息将显示发送到数据库的 SQL 命令。

■ **注意：**如果需要重置数据库，那么运行 dotnet ef database drop -force 命令，然后运行代码清单 18-12 中的命令。

## 18.3 添加 CSS 框架

后续章节将演示生成 HTML 响应的不同方法。运行代码清单 18-13 所示的命令，删除 LibMan 包的任何现有版本，并安装本书中使用的版本(如果在前面的章节中安装了 LibMan 的这个版本，可以跳过这些命令)。

**代码清单 18-13　安装 LibMan 工具包**

```
dotnet tool uninstall --global Microsoft.Web.LibraryManager.Cli
dotnet tool install --global Microsoft.Web.LibraryManager.Cli --version 2.0.96
```

要添加引导 CSS 框架，以便对 HTML 响应进行样式化，请在 WebApp 项目文件夹中运行代码清单 18-14 所示的命令。

**代码清单 18-14　安装引导 CSS 框架**

```
libman init -p cdnjs
libman install twitter-bootstrap@4.3.1 -d wwwroot/lib/twitter-bootstrap
```

## 18.4　配置请求管道

要定义一个简单的中间件组件，来确保示例项目已经正确设置，请将一个名为 TestMiddleware.cs 的类文件添加到 WebApp 文件夹，并添加如代码清单 18-15 所示的代码。

**代码清单 18-15　WebApp 文件夹中 TestMiddleware.cs 文件的内容**

```
using Microsoft.AspNetCore.Http;
using System.Linq;
using System.Threading.Tasks;
using WebApp.Models;

namespace WebApp {
    public class TestMiddleware {
        private RequestDelegate nextDelegate;

        public TestMiddleware(RequestDelegate next) {
            nextDelegate = next;
        }

        public async Task Invoke(HttpContext context, DataContext dataContext) {
            if (context.Request.Path == "/test") {
                await context.Response.WriteAsync(
                    $"There are {dataContext.Products.Count()} products\n");
                await context.Response.WriteAsync(
                    $"There are {dataContext.Categories.Count()} categories\n");
                await context.Response.WriteAsync(
                    $"There are {dataContext.Suppliers.Count()} suppliers\n");
            } else {
                await nextDelegate(context);
            }
        }
```

```
    }
}
```

将中间件组件添加到 Startup 类的请求管道中，如代码清单 18-16 所示。

**代码清单 18-16　在 WebApp 文件夹的 Startup.cs 文件中添加一个中间件组件**

```
using System;
using System.Collections.Generic;
using System.Linq;
using System.Threading.Tasks;
using Microsoft.AspNetCore.Builder;
using Microsoft.AspNetCore.Hosting;
using Microsoft.AspNetCore.Http;
using Microsoft.Extensions.DependencyInjection;
using Microsoft.Extensions.Hosting;
using Microsoft.Extensions.Configuration;
using Microsoft.EntityFrameworkCore;
using WebApp.Models;

namespace WebApp {
    public class Startup {

        public Startup(IConfiguration config) {
            Configuration = config;
        }

        public IConfiguration Configuration { get; set; }

        public void ConfigureServices(IServiceCollection services) {
            services.AddDbContext<DataContext>(opts => {
                opts.UseSqlServer(Configuration[
                    "ConnectionStrings:ProductConnection"]);
                opts.EnableSensitiveDataLogging(true);
            });
        }

        public void Configure(IApplicationBuilder app, DataContext context) {

            app.UseDeveloperExceptionPage();
            app.UseStaticFiles();
            app.UseRouting();
            app.UseMiddleware<TestMiddleware>();
            app.UseEndpoints(endpoints => {
                endpoints.MapGet("/", async context => {
                    await context.Response.WriteAsync("Hello World!");
                });
            });
```

```
                SeedData.SeedDatabase(context);
            }
        }
    }
```

## 18.5 运行示例应用程序

启动应用程序，从 Debug 菜单中选择 Start Without Debugging 或 Run Without Debugging，或在 WebApp 项目文件夹中运行代码清单 18-17 所示的命令。

**代码清单 18-17　运行示例应用程序**

```
dotnet run
```

使用新的浏览器选项卡，请求 http://localhost:5000/test，显示如图 18-3 所示的响应。

图 18-3　运行示例应用程序

## 18.6 小结

本章建了贯穿全书这一部分的示例应用程序。该项目使用空模板创建，包含一个依赖于 Entity Framework Core 的数据模型，使用包含简单测试中间件组件的请求管道进行配置。第 19 章将展示如何使用 ASP.NETCore 创建 Web 服务。

# 第 19 章

# 创建 RESTful Web 服务

Web 服务接收 HTTP 请求，生成包含数据的响应。本章解释 MVC 框架所提供的特性(它是 ASP.NET Core 的一个组成部分)如何用于构建第Ⅱ部分描述的功能，以创建 Web 服务。

Web 服务的本质意味着，本章中的一些示例是使用 PowerShell 提供的命令行工具进行测试的，按照所示的方式输入命令很重要。第 20 章介绍了使用 Web 服务的更复杂工具，但命令行方法更适合本书章节中的示例，即使它们在输入时可能令人感到有些别扭。表 19-1 列出 RESTful Web 服务的相关知识。

表 19-1　RESTful Web 服务的相关知识

| 问题 | 答案 |
|---|---|
| 它们是什么? | Web 服务提供对应用程序数据的访问，这些数据通常以 JSON 格式表示 |
| 它们为什么有用? | Web 服务通常用于为富客户端应用程序提供数据 |
| 它们是如何使用的? | URL 和 HTTP 方法的组合描述了由 ASP.NET Core 控制器定义的操作方法处理的操作 |
| 是否存在缺陷或限制? | 关于 Web 服务应该如何实现还没有广泛共识，必须注意只生成客户端期望的数据 |
| 还有其他选择吗? | 有许多不同的方法可以向客户端提供数据，尽管 RESTful Web 服务是最常见的方法 |

表 19-2 总结了本章的内容。

表 19-2　本章内容摘要

| 问题 | 解决方案 | 代码清单 |
|---|---|---|
| 定义 Web 服务 | 创建具有操作方法的控制器，操作方法对应于需要的操作 | 19-1～19-14 |
| 随着时间的推移生成数据序列 | 使用 IAsyncEnumerable<T>响应，这将阻止请求线程在生成结果时出现阻塞 | 19-15 |
| 防止将请求值用于敏感数据属性 | 使用绑定目标将模型绑定过程限制为仅安全属性 | 19-16～19-18 |
| 表示非数据结果 | 使用操作结果来描述 ASP.NET Core 应该发送的响应 | 19-19～19-24 |
| 验证数据 | 使用 ASP.NET Core 模型绑定和模型验证特性 | 19-25～19-27 |
| 自动验证请求 | 使用 ApiController 属性 | 19-28 |
| 从数据响应中省略空值 | 映射数据对象以过滤掉属性，或者配置 JSON 序列化器以忽略空属性 | 19-29～19-31 |

# 19.1　准备工作

本章继续使用第 18 章创建的 WebApp 项目。为了准备这一章，打开一个新的 PowerShell 命令提示符，导航到包含 WebApp.csproj 文件的文件夹，并运行代码清单 19-1 所示的命令，来删除数据库。

**提示：** 可从 https://github.com/apress/pro-asp.net-core-3 下载本章和本书中其他所有章节的示例项目。如果你在运行示例时遇到问题，请参阅第 1 章以获得帮助。

**代码清单 19-1　删除数据库**

```
dotnet ef database drop --force
```

为启动应用程序，从 Debug 菜单中选择 Start Without Debugging 或 Run Without Debugging，或者在项目文件夹中运行代码清单 19-2 所示的命令。

**代码清单 19-2　启动示例应用程序**

```
dotnet run
```

ASP.NET Core 启动后，请求 URL http://localhost:5000/test once，显示如图 19-1 所示的响应。

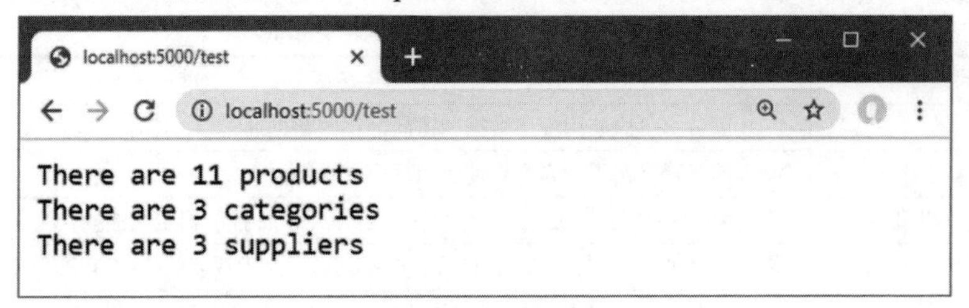


图 19-1　运行示例应用程序

# 19.2　理解 RESTful Web 服务

Web 服务通过客户端(如 JavaScript 应用程序)可以使用的数据响应 HTTP 请求。对于 Web 服务应该如何工作没有硬性规则，但最常见的方法是采用具象状态传输(Representational State Transfer, REST)模式。对于 REST 没有权威的规范，对于 REST 式 Web 服务的组成也没有共识，但是有一些普遍用于 Web 服务的主题。由于缺乏详细的规范，关于 REST 的含义和应该如何创建 RESTful Web 服务的争论会没完没了，如果为项目创建的 Web 服务有效，那么可以安全地忽略所有这些问题。

## 19.2.1　理解请求 URL 和方法

REST 的核心前提是 Web 服务通过 URL 和 HTTP 方法(如 GET 和 POST)的组合定义 API，这也是得到广泛认同的唯一方面，这些方法也称为 HTTP 动词。方法指定操作的类型，而 URL 指定应用操作的一个或多个数据对象。

作为一个例子，下面是一个 URL，它可能在示例应用程序中标识一个 Product 对象。

```
/api/products/1
```

这个 URL 可以标识 ProductId 属性值为 1 的 Product 对象。URL 标识产品，HTTP 方法指定应该对其做什么。表 19-3 列出 Web 服务中常用的 HTTP 方法以及它们通常表示的操作。

表 19-3 HTTP 方法和操作

| HTTP 方法 | 描述 |
| --- | --- |
| GET | 此方法用于检索一个或多个数据对象 |
| POST | 此方法用于创建一个新对象 |
| PUT | 此方法用于更新现有对象 |
| PATCH | 此方法用于更新现有对象的一部分 |
| DELETE | 此方法用于删除对象 |

### 19.2.2 理解 JSON

大多数基于 REST 的 Web 服务使用 JSON(JavaScript 对象符号)格式化响应数据。JSON 之所以流行，是因为它很简单，很容易被 JavaScript 客户端使用。JSON 详见 www.json.org，但是创建 Web 服务不需要了解 JSON 的每个方面，因为 ASP.NET Core 提供了创建 JSON 响应需要的所有特性。

**理解 RESTful Web 服务的替代品**

REST 不是设计 Web 服务的唯一方法，还有一些流行的替代方法。GraphQL 与 React JavaScript 框架关系最密切，但可以更广泛地使用。REST Web 服务通过 URL 和 HTTP 方法的单独组合提供特定的查询；与 REST Web 服务不同，GraphQL 提供对应用程序的所有数据的访问，并允许客户端以所需的格式查询所需的数据。GraphQL 的设置可能很复杂，可能需要更高级的客户端，但其结果是一个更灵活的 Web 服务，使客户端的开发人员能控制他们消费的数据。ASP.NET Core 并不直接支持 GraphQL，但是有一些.NET 实现可用。更多细节请参见 https://graphql.org。

一个新的替代方案是 gRPC，它是一个完整的远程过程调用框架，关注速度和效率。在撰写本书时，gRPC 还不能在 Web 浏览器(如 Angular 或 React 框架)中使用，因为浏览器不提供 gRPC 需要的细粒度访问，而这正是 gRPC 制定其 HTTP 请求所需要的。

## 19.3 使用自定义端点创建 Web 服务

当理解 ASP.NET Core 为 Web 服务提供的设施时，很容易忘记它们是基于第 II 部分中描述的特性构建的。要创建一个简单的 Web 服务，向 WebApp 文件夹添加一个名为 WebServiceEndpoint.cs 的文件，并用它定义如代码清单 19-3 所示的类。

**代码清单 19-3 WebApp 文件夹中 WebServiceEndpoint.cs 文件的内容**

```
using Microsoft.AspNetCore.Http;
using Microsoft.AspNetCore.Routing;
using Microsoft.Extensions.DependencyInjection;
```

```
using System.Collections.Generic;
using System.Text.Json;
using WebApp.Models;

namespace Microsoft.AspNetCore.Builder {

    public static class WebServiceEndpoint {
        private static string BASEURL = "api/products";

        public static void MapWebService(this IEndpointRouteBuilder app) {
            app.MapGet($"{BASEURL}/{{id}}", async context => {
                long key = long.Parse(context.Request.RouteValues["id"] as string);
                DataContext data = context.RequestServices.GetService<DataContext>();
                Product p = data.Products.Find(key);
                if (p == null) {
                    context.Response.StatusCode = StatusCodes.Status404NotFound;
                } else {
                    context.Response.ContentType = "application/json";
                    await context.Response
                        .WriteAsync(JsonSerializer.Serialize<Product>(p));
                }
            });

            app.MapGet(BASEURL, async context => {
                DataContext data = context.RequestServices.GetService<DataContext>();
                context.Response.ContentType = "application/json";
                await context.Response.WriteAsync(JsonSerializer
                    .Serialize<IEnumerable<Product>>(data.Products));
            });

            app.MapPost(BASEURL, async context => {
                DataContext data = context.RequestServices.GetService<DataContext>();
                Product p = await
                    JsonSerializer.DeserializeAsync<Product>(context.Request.Body);
                await data.AddAsync(p);
                await data.SaveChangesAsync();
                context.Response.StatusCode = StatusCodes.Status200OK;
            });
        }
    }
}
```

MapWebService 扩展方法创建了三条路由，它们只使用前面章节中描述的特性来形成一个基本的 Web 服务。这些路由匹配以/api 开头的 URL，这是 Web 服务的传统 URL 前缀。第一个路由的端点从段变量接收一个值，该值用于在数据库中定位单个 Product 对象。第二个路由的端点检索数据库中的所有 Product 对象。第三个端点处理 POST 请求并读取请求主体，以获得要添加到数据库的新对象的 JSON 表示。

创建 Web 服务有更好的 ASP.NET Core 特性，但是代码清单 19-3 中的代码显示了如何组合 HTTP 方法和 URL 来描述操作。代码清单 19-4 使用 MapWebService 扩展方法将端点添加到示例应用程序的路由配置中。

**代码清单 19-4 在 WebApp 文件夹的 Startup.cs 文件中添加路由**

```
...
public void Configure(IApplicationBuilder app, DataContext context) {
    app.UseDeveloperExceptionPage();
    app.UseRouting();
    app.UseMiddleware<TestMiddleware>();
    app.UseEndpoints(endpoints => {
        endpoints.MapGet("/", async context => {
            await context.Response.WriteAsync("Hello World!");
        });
        endpoints.MapWebService();
    });
    SeedData.SeedDatabase(context);
}
...
```

要测试 Web 服务，请重新启动 ASP.NET Core，请求 http://localhost:5000/api/products/1。该请求将与代码清单 19-4 中定义的第一个路由匹配，并生成图 19-2 左侧所示的响应。接下来，请求 http://localhost:5000/api/products，它与第二个路由匹配，并生成图 19-2 右侧所示的响应。

■ **注意：**由于 LINQ 查询不包含相关数据，图中显示的响应包含 Supplier 和 Category 属性的空值。详见第 20 章。

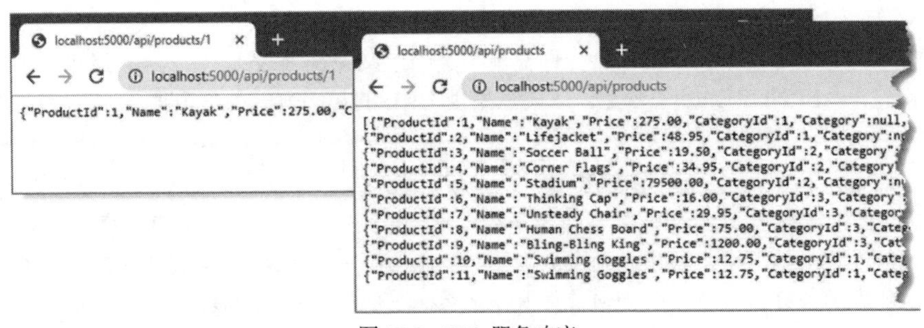


图 19-2 Web 服务响应

测试第三种路由需要一种不同的方法，因为不可能使用浏览器发送 HTTP POST 请求。打开一个新的 PowerShell 命令提示符，运行代码清单 19-5 所示的命令。按照如下所示输入命令很重要，因为 Invoke-RestMethod 命令对其参数的语法很挑剔。

■ **提示：**如果没有执行 Microsoft Edge 或 Internet Explorer 的初始安装，在使用 Invoke-RestMethod 或 Invoke-WebRequest 命令测试本章中的示例时，可能收到一个错误。可通过运行 IE 并选择所需的初始配置来解决该问题。

**代码清单 19-5　发送 POST 请求**

```
Invoke-RestMethod http://localhost:5000/api/products -Method POST -Body (@{ Name="Swimming Goggles";
Price=12.75; CategoryId=1; SupplierId=1} | ConvertTo-Json) -ContentType "application/json"
```

该命令发送一个 HTTP POST 命令，与代码清单 19-5 中定义的第三个路由匹配。请求的主体是一个 JSON 格式的对象，解析该对象以创建 Product，然后存储在数据库中。请求中包含的 JSON 对象包含 Name、Price、CategoryId 和 SupplierId 属性的值。对象的唯一键与 ProductId 属性相关联，在存储对象时由数据库分配。使用浏览器再次请求 http://localhost:5000/api/products URL，JSON 响应就包含新对象，如图 19-3 所示。

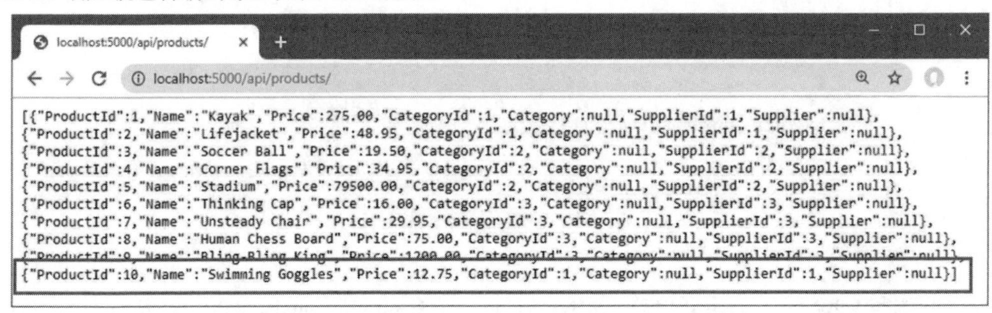


图 19-3　使用 Web 服务存储新数据

## 19.4　使用控制器创建 Web 服务

使用端点创建 Web 服务的缺点是，每个端点都要复制一套类似的步骤生成一个响应：获取 Entity Framework Core 服务，以便它可以查询数据库，为响应设置 Content-Type 头，把对象序列化到 JSON 等。因此，使用端点创建的 Web 服务很难理解，维护起来也很笨拙。

更优雅的方法是使用控制器，它允许在单个类中定义 Web 服务。控制器是 MVC 框架的一部分，它构建在 ASP.NET Core 平台上，以端点处理 URL 的相同方式处理数据。

**MVC 模式在 ASP.NET Core 中的兴衰**

MVC 框架是模型-视图-控制器模式的一种实现，它描述了一种构建应用程序的方法。本章中的例子使用了模式的三个支柱中的两个：数据模型(MVC 中的 M)和控制器(MVC 中的 C)。第 21 章提供了缺失的部分，并解释了如何使用 Razor 和视图创建 HTML 响应。

MVC 模式是 ASP.NET 发展的一个重要步骤，它允许该平台脱离之前的 Web 表单模型。Web 表单应用程序很容易启动，但很快就很难管理和隐藏 HTTP 请求和开发人员响应的细节。相比之下，遵循 MVC 模式为使用 MVC 框架编写的应用程序提供了强大的可伸缩结构，并且不会对开发人员隐藏任何东西。MVC 框架重振了 ASP.NET，并为后来的 ASP.NET Core 提供了基础，ASP.NET Core 放弃了对 Web 表单的支持，而只专注于使用 MVC 模式。

随着 ASP.NET Core 的发展，其他风格的 Web 应用程序也被包括进来，MVC 框架只是创建应用程序的方法之一。这并没有削弱 MVC 模式的效用，但是它在 ASP.NET Core 开发中不再扮演核心角色，过去 MVC 框架所独有的特性现在可以通过其他方法访问，比如 Razor 页面(Razor Pages)和 Blazor。

这种发展的结果是，理解 MVC 模式不再是有效的 ASP.NET Core 开发的先决条件。如果对理解 MVC 模式感兴趣，那么 https://en.wikipedia.org/wiki/model-view-controller 是一个很好的起点。但是对于本书来说，理解 MVC 框架提供的特性如何在 ASP.NET Core 平台上构建即可。

### 19.4.1 启用 MVC 框架

使用控制器创建 Web 服务的第一步是配置 MVC 框架，它需要服务和端点，如代码清单 19-6 所示。

**代码清单 19-6 在 WebApp 文件夹的 Startup.cs 文件中启用 MVC 框架**

```
using System;
using System.Collections.Generic;
using System.Linq;
using System.Threading.Tasks;
using Microsoft.AspNetCore.Builder;
using Microsoft.AspNetCore.Hosting;
using Microsoft.AspNetCore.Http;
using Microsoft.Extensions.DependencyInjection;
using Microsoft.Extensions.Hosting;
using Microsoft.Extensions.Configuration;
using Microsoft.EntityFrameworkCore;
using WebApp.Models;

namespace WebApp {
    public class Startup {

        public Startup(IConfiguration config) {
            Configuration = config;
        }

        public IConfiguration Configuration { get; set; }

        public void ConfigureServices(IServiceCollection services) {
            services.AddDbContext<DataContext>(opts => {
                opts.UseSqlServer(Configuration[
                    "ConnectionStrings:ProductConnection"]);
                opts.EnableSensitiveDataLogging(true);
            });

            services.AddControllers();
        }

        public void Configure(IApplicationBuilder app, DataContext context) {
            app.UseDeveloperExceptionPage();
            app.UseRouting();
            app.UseMiddleware<TestMiddleware>();
```

```
            app.UseEndpoints(endpoints => {
                endpoints.MapGet("/", async context => {
                    await context.Response.WriteAsync("Hello World!");
                });
                //endpoints.MapWebService();
                endpoints.MapControllers();
            });
            SeedData.SeedDatabase(context);
        }
    }
}
```

AddControllers 方法定义了 MVC 框架需要的服务，MapControllers 方法定义了允许控制器处理请求的路由。后续章节会介绍用于配置 MVC 框架的其他方法，这些方法提供对不同特性的访问，但是代码清单 19-6 中使用的方法是为 Web 服务配置 MVC 框架的方法。

## 19.4.2　创建控制器

控制器是类，其方法称为操作，可以处理 HTTP 请求。当应用程序启动时自动发现控制器。基本的发现过程很简单：名称以 Controller 结尾的任何公共类都是控制器，控制器定义的任何公共方法都是操作。要演示控制器有多简单，请创建 WebApp/Controllers 文件夹，并用代码清单 19-7 所示的代码向其中添加一个名为 ProductsController.cs 的文件。

**提示：** 控制器通常在 Controllers 文件夹中定义，但它们可以在项目的任何地方定义，而且仍然会被发现。

**代码清单 19-7　Controllers 文件夹中 ProductsController.cs 文件的内容**

```
using Microsoft.AspNetCore.Mvc;
using System.Collections.Generic;
using WebApp.Models;

namespace WebApp.Controllers {

    [Route("api/[controller]")]
    public class ProductsController: ControllerBase {

        [HttpGet]
        public IEnumerable<Product> GetProducts() {
            return new Product[] {
                new Product() { Name = "Product #1" },
                new Product() { Name = "Product #2" },
            };
        }

        [HttpGet("{id}")]
        public Product GetProduct() {
```

```
            return new Product() {
                ProductId = 1, Name = "Test Product"
            };
        }
    }
}
```

ProductsController 类满足 MVC 框架在控制器中寻找的标准。它定义了名为 GetProducts 和 GetProduct 的公共方法，它们被视为操作。

### 1. 理解基类

控制器是从 ControllerBase 类派生的，该类提供对 MVC 框架和底层 ASP.NET Core 平台特性的访问。表 19-4 描述了 ControllerBase 类提供的最有用属性。

> ▒ **注意：** 虽然控制器通常来源于 ControllerBase 或 Controller 类(参见第 21 章)，但 MVC 框架接受任何名称以 Controller 结尾的类，以 Controller 结尾的类的派生类，或者用 Controller 属性装饰的类。将 NonController 属性应用于满足这些条件但不应该接收 HTTP 请求的类。

表 19-4　有用的 ControllerBase 属性

| 名　称 | 描　述 |
| --- | --- |
| HttpContext | 此属性返回当前请求的 HttpContext 对象 |
| ModelState | 此属性返回数据验证过程的详细信息，详见本章后面的 19.5.4 节和第 29 章 |
| Request | 此属性返回当前请求的 HttpRequest 对象 |
| Response | 此属性返回当前响应的 HttpResponse 对象 |
| RouteData | 此属性返回路由中间件从请求 URL 中提取的数据，如第 13 章所述 |
| User | 此属性返回一个对象，描述与当前请求关联的用户，如第 38 章所述 |

每次使用控制器类的一个操作处理请求时，都会创建一个控制器类的新实例，这意味着表 19-4 中的属性只描述当前请求。

### 2. 理解控制器属性

操作方法支持的 HTTP 方法和 URL 由应用到控制器的属性组合决定。控制器的 URL 由 Route 属性指定，它应用于类，如下所示：

```
...
[Route("api/[controller]")]
public class ProductsController: ControllerBase {
...
```

属性参数的[controller]部分用于从控制器类的名称派生 URL。类名的 Controller 部分被删除，这意味着代码清单 19-7 中的属性将控制器的 URL 设置为/api/products。

每个操作都用一个属性修饰，这个属性指定了它所支持的 HTTP 方法，如下所示。

```
...
[HttpGet]
```

```
public Product[] GetProducts() {
...
```

在用于 Web 服务的控制器中，操作方法的名称并不重要。第 21 章描述了控制器的其他用途，其中名称确实很重要，但对于 Web 服务来说，重要的是 HTTP 方法属性和路由模式。

HttpGet 属性告诉 MVC 框架，GetProducts 操作方法处理 HTTP GET 请求。表 19-5 描述可应用于操作以指定 HTTP 方法的全部属性。

表 19-5　HTTP 方法属性

| 名称 | 描述 |
| --- | --- |
| HttpGet | 此属性指定该操作只能由使用 GET 谓词的 HTTP 请求调用 |
| HttpPost | 此属性指定该操作只能由使用 POST 谓词的 HTTP 请求调用 |
| HttpDelete | 此属性指定该操作只能由使用 DELETE 谓词的 HTTP 请求调用 |
| HttpPut | 此属性指定该操作只能由使用 PUT 谓词的 HTTP 请求调用 |
| HttpPatch | 此属性指定该操作只能由使用 PATCH 谓词的 HTTP 请求调用 |
| HttpHead | 此属性指定该操作只能由使用 HEAD 谓词的 HTTP 请求调用 |
| AcceptVerbs | 此属性用于指定多个 HTTP 谓词 |

应用于指定 HTTP 方法的操作的属性也可以用于构建控制器的基 URL。

```
...
[HttpGet("{id}")]
public Product GetProduct() {
...
```

这个属性告诉 MVC 框架，GetProduct 操作方法处理 URL 模式 api/ products/{id}的 GET 请求。在发现过程中，应用于控制器的属性用于构建控制器可以处理的一组 URL 模式，如表 19-6 所示。

表 19-6　URL 模式

| HTTP 方法 | URL 模式 | 操作方法名称 |
| --- | --- | --- |
| GET | api/products | GetProducts |
| GET | api/products/{id} | GetProduct |

**提示：**在编写控制器时，务必确保控制器支持的每个 HTTP 方法和 URL 模式的组合只映射到一个操作方法。当一个请求可以由多个操作处理时，由于 MVC 框架无法决定使用哪个操作，将引发一个异常。

属性的组合与本章前面使用端点创建 Web 服务时为相同 URL 模式所使用的 MapGet 方法是等效的。

**GET 和 POST：选择正确的一个**

经验法则是，GET 请求应该用于所有只读信息的检索，而 POST 请求应该用于任何更改应用程序状态的操作。按照符合标准的术语，GET 请求用于安全的交互(除了信息检索之外没有任何副作用)，而 POST 请求用于不安全的交互(做出决策或更改某些内容)。这些约定由万维网联盟

(W3C)制定，网址为 www.w3.org/protocols/rfc2616 /rfc2616-sec9.html。

GET 请求是可寻址的：所有信息都包含在 URL 中，因此可将这些地址添加为书签，并链接到它们。不要对更改状态的操作使用 GET 请求。许多 Web 开发人员在 2005 年谷歌 Web 加速器向公众发布时经历了惨痛的教训。此应用程序预取每个页面链接的所有内容，这在 HTTP 中是合法的，因为 GET 请求应该是安全的。遗憾的是，许多 Web 开发人员忽略了 HTTP 约定，在他们的应用程序中放置了“删除项目”或“添加到购物车”的简单链接。混乱接踵而至。

### 3. 理解行动方法的结果

控制器提供的主要好处之一是，MVC 框架负责设置响应头，并序列化发送到客户端的数据对象。可以在操作方法定义的结果中看到，如下所示：

```
...
[HttpGet("{id}")]
public Product GetProduct() {
...
```

使用端点时，必须直接使用 JSON 序列化器来创建一个可写入响应的字符串，并设置 Content-Type 头来告诉客户端，响应包含 JSON 数据。操作方法返回一个 Product 对象，该对象是自动处理的。

要查看操作方法的结果是如何处理的，请重新启动 ASP.NET Core，请求 http://localhost:5000/api/products，生成图 19-4 左侧所示的响应，该响应是由 GetProducts 操作方法生成的。接下来，请求 http://localhost:5000/api/products/1，它由 GetProduct 方法处理，并生成图 19-4 右侧所示的结果。

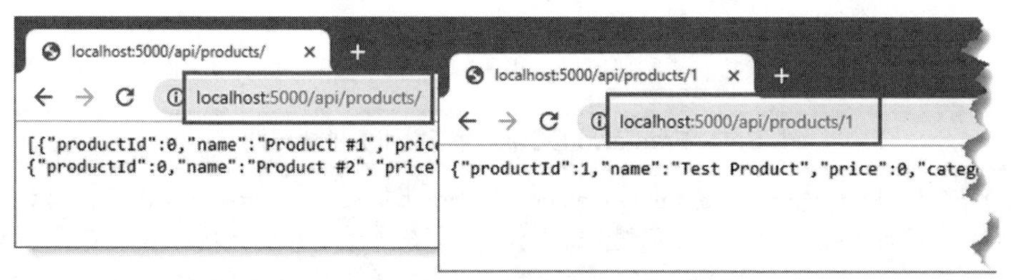


图 19-4　使用控制器

### 4. 在控制器中使用依赖注入

每次使用控制器类的一个操作处理请求时，都会创建一个控制器类的新实例。应用程序的服务用于解析控制器通过其构造函数声明的任何依赖项和操作方法定义的任何依赖项。这允许服务通过构造函数处理所有操作，同时仍然允许单个操作声明它们自己的依赖项，如代码清单 19-8 所示。

**代码清单 19-8　在 Controllers 文件夹的 ProductsController.cs 文件中使用服务**

```
using Microsoft.AspNetCore.Mvc;
using WebApp.Models;
using System.Collections.Generic;
using Microsoft.Extensions.Logging;
using System.Linq;
```

```
namespace WebApp.Controllers {

    [Route("api/[controller]")]
    public class ProductsController: ControllerBase {
        private DataContext context;

        public ProductsController(DataContext ctx) {
            context = ctx;
        }

        [HttpGet]
        public IEnumerable<Product> GetProducts() {
            return context.Products;
        }

        [HttpGet("{id}")]
        public Product GetProduct([FromServices]
                ILogger<ProductsController> logger) {
            logger.LogDebug("GetProduct Action Invoked");
            return context.Products.FirstOrDefault();
        }
    }
}
```

构造函数声明了对 DataContext 服务的依赖，该服务提供对应用程序数据的访问。使用有作用域的请求解析服务，这意味着控制器可以请求所有服务，而不需要了解它们的生命周期。

**Entity Framework Core 上下文服务生命周期**

为每个控制器创建一个新的 Entity Framework Core 上下文对象。一些开发人员会试图拒绝上下文对象来提高性能，但这会导致问题，因为来自一个查询的数据可能影响后续的查询，如第 20 章所述。在后台，Entity Framework Core 有效地管理到数据库的连接，不应该试图在为其创建的控制器之外存储或重用上下文对象。

GetProducts 操作方法使用 DataContext 请求数据库中的所有 Product 对象。GetProduct 方法也使用 DataContext 服务，但是它声明了对 ILogger<T>的依赖，ILogger<T>是第 15 章描述的日志服务。由操作方法声明的依赖项必须用 FromServices 属性修饰，如下所示：

```
...
public Product GetProduct([FromServices] ILogger<ProductsController> logger) {
...
```

默认情况下，MVC 框架尝试从请求 URL 中查找操作方法参数的值，FromServices 属性会覆盖此行为。要查看控制器中服务的使用情况，请重新启动 ASP.NET Core，请求 http://localhost:5000/api/products/1，生成如图 19-5 所示的响应。在应用程序的输出中还会显示以下日志消息：

```
...
dbug: WebApp.Controllers.ProductsController[0]
      GetProduct Action Invoked
...
```

```
{"productId":1,"name":"Kayak","price":275.00,"categoryId":1,"
category":null,"supplierId":1,"supplier":null}
```

图 19-5　在控制器中使用服务

■ **警告**：控制器生命周期的一个后果是，不能依赖以特定顺序调用方法所造成的副作用。例如，不能将代码清单 19-8 中 GetProduct 方法接收到的 ILogger<T>对象分配给可以在以后的请求中由 GetProducts 操作读取的属性。每个控制器对象用于处理一个请求，MVC 框架只会为每个对象调用一个操作方法。

### 5. 使用模型绑定访问路由数据

上一节注意到 MVC 框架使用请求 URL 来查找操作方法参数的值，这个过程称为模型绑定。模型绑定详见第 28 章，但代码清单 19-9 显示了一个简单示例。

**代码清单 19-9　在 Controllers 文件夹的 ProductsController.cs 文件中使用模型绑定**

```
using Microsoft.AspNetCore.Mvc;
using WebApp.Models;
using System.Collections.Generic;
using Microsoft.Extensions.Logging;
using System.Linq;

namespace WebApp.Controllers {

    [Route("api/[controller]")]
    public class ProductsController: ControllerBase {
        private DataContext context;

        public ProductsController(DataContext ctx) {
            context = ctx;
        }

        [HttpGet]
        public IEnumerable<Product> GetProducts() {
            return context.Products;
        }

        [HttpGet("{id}")]
```

```
        public Product GetProduct(long id,
                [FromServices] ILogger<ProductsController> logger) {
            logger.LogDebug("GetProduct Action Invoked");
            return context.Products.Find(id);
        }
    }
}
```

代码清单将一个名为 id 的 long 参数添加到 GetProduct 方法中。调用该操作方法时，MVC 框架从路由数据中注入同名值，自动将其转换为 long 值，该操作使用该 long 值来通过 LINQ Find 方法查询数据库。结果是操作方法响应 URL，要看到这个 URL，可以重新启动 ASP.NET Core，请求 http://localhost:5000/api/products/5，生成如图 19-6 所示的响应。

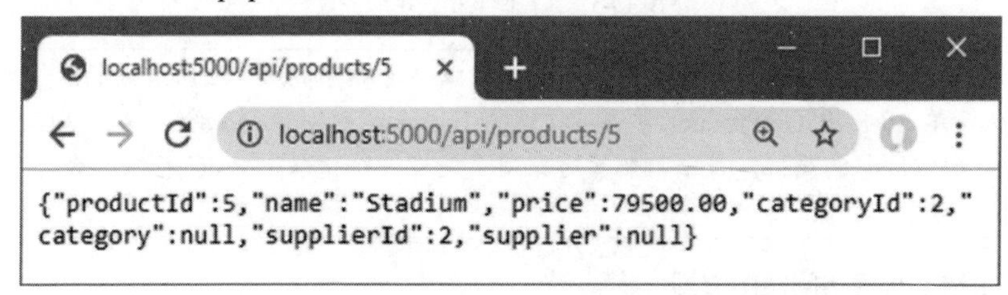


图 19-6　在操作中使用模型绑定

### 6. 在请求主体中进行模型绑定

模型绑定特性也可用于请求体中的数据，它允许客户端发送容易由操作方法接收的数据。代码清单 19-10 添加了一个新的操作方法，它响应 POST 请求，并允许客户端在请求主体中提供 Product 对象的 JSON 表示。

**代码清单 19-10　在 Controllers 文件夹的 ProductsController.cs 文件中添加一个操作**

```
using Microsoft.AspNetCore.Mvc;
using WebApp.Models;
using System.Collections.Generic;
using Microsoft.Extensions.Logging;
using System.Linq;

namespace WebApp.Controllers {

    [Route("api/[controller]")]
    public class ProductsController: ControllerBase {
        private DataContext context;

        public ProductsController(DataContext ctx) {
            context = ctx;
        }

        [HttpGet]
        public IEnumerable<Product> GetProducts() {
            return context.Products;
```

```
        }

        [HttpGet("{id}")]
        public Product GetProduct(long id,
                [FromServices] ILogger<ProductsController> logger) {
            logger.LogDebug("GetProduct Action Invoked");
            return context.Products.Find(id);
        }

        [HttpPost]
        public void SaveProduct([FromBody]Product product) {
            context.Products.Add(product);
            context.SaveChanges();
        }
    }
}
```

新操作依赖两个属性。HttpPost 属性应用于操作方法，并告诉 MVC 框架该操作可处理 POST 请求。FromBody 属性应用于操作的参数，它指定应该通过解析请求主体获得该参数的值。调用操作方法时，MVC 框架会创建一个新的 Product 对象，并用请求主体中的值填充其属性。模型绑定过程可能很复杂，通常与数据验证结合在一起，如第 29 章所述。打开一个新的 PowerShell 命令提示符，运行代码清单 19-11 所示的命令。

**代码清单 19-11　向示例应用程序发送 POST 请求**

```
Invoke-RestMethod http://localhost:5000/api/products -Method POST -Body (@{ Name="Soccer Boots"; Price=89.99;
CategoryId=2; SupplierId=2} | ConvertTo-Json) -ContentType "application/json"
```

命令执行后，使用 Web 浏览器请求 http://localhost:5000/api/products，会看到存储在数据库中的新对象，如图 19-7 所示。

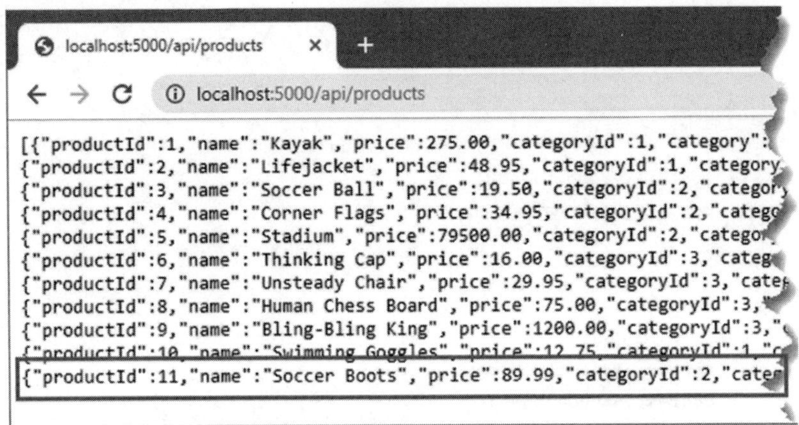


图 19-7　使用控制器存储新数据

#### 7. 添加额外的操作

现在基本特性已经就绪，可以添加允许客户端使用 HTTP PUT 和 DELETE 方法替换和删除 Product 对象的操作，如代码清单 19-12 所示。

**代码清单 19-12　在 Controllers 文件夹的 ProductsController.cs 文件中添加操作**

```
using Microsoft.AspNetCore.Mvc;
using WebApp.Models;
using System.Collections.Generic;
using Microsoft.Extensions.Logging;
using System.Linq;

namespace WebApp.Controllers {

    [Route("api/[controller]")]
    public class ProductsController: ControllerBase {
        private DataContext context;

        public ProductsController(DataContext ctx) {
            context = ctx;
        }

        [HttpGet]
        public IEnumerable<Product> GetProducts() {
            return context.Products;
        }

        [HttpGet("{id}")]
        public Product GetProduct(long id,
                [FromServices] ILogger<ProductsController> logger) {
            logger.LogDebug("GetProduct Action Invoked");
            return context.Products.Find(id);
        }

        [HttpPost]
        public void SaveProduct([FromBody]Product product) {
            context.Products.Add(product);
            context.SaveChanges();
        }

        [HttpPut]
        public void UpdateProduct([FromBody]Product product) {
            context.Products.Update(product);
            context.SaveChanges();
        }

        [HttpDelete("{id}")]
```

```
        public void DeleteProduct(long id) {
            context.Products.Remove(new Product() { ProductId = id });
            context.SaveChanges();
        }
    }
}
```

UpdateProduct 操作类似于 SaveProduct 操作，并使用模型绑定从请求主体接收 Product 对象。DeleteProduct 操作接收一个 URL 的主键值，并使用它创建一个 ProductId 属性具有特定值的 Product，这是必需的，因为 Entity Framework Core 只处理对象，但是 Web 服务客户端通常希望能够只使用一个键值删除对象。

重启 ASP.NET Core，然后使用另一个 PowerShell 命令提示符，运行代码清单 19-13 所示的命令，该命令测试 UpdateProduct 操作。

**代码清单 19-13 更新对象**

```
Invoke-RestMethod http://localhost:5000/api/products -Method PUT -Body (@{ ProductId=1;
Name="Green Kayak";
Price=275; CategoryId=1; SupplierId=1} | ConvertTo-Json) -ContentType "application/json"
```

该命令发送一个 HTTP PUT 请求，该请求的主体包含一个替换对象。操作方法通过模型绑定特性接收对象，并更新数据库。接下来，运行代码清单 19-14 所示的命令，来测试 DeleteProduct 操作。

**代码清单 19-14 删除对象**

```
Invoke-RestMethod http://localhost:5000/api/products/2 -Method DELETE
```

该命令发送一个 HTTP DELETE 请求，该请求删除 ProductId 属性为 2 的对象。要查看更改的效果，请使用浏览器请求 http://localhost:5000/api/products，发送一个 GET 请求，该请求由 GetProducts 操作处理，生成图 19-8 所示的响应。

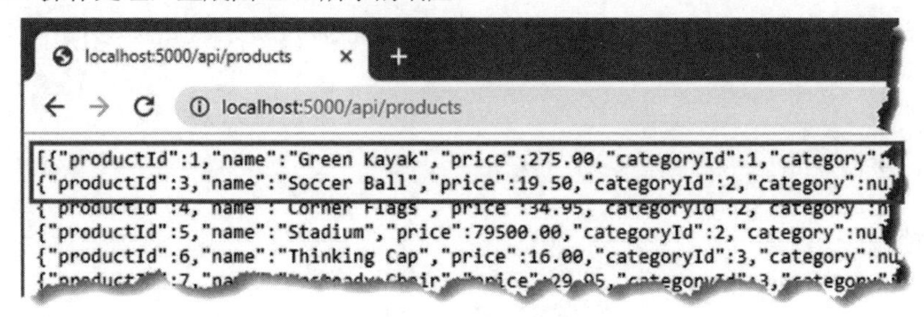


图 19-8 更新和删除对象

## 19.5 改进 Web 服务

代码清单 19-14 中的控制器重新创建了单独端点提供的所有功能，但是仍然可以进行改进，如下面几节所述。

**支持跨源请求**

如果支持第三方 JavaScript 客户端，那么可能需要启用对跨源请求(CORS)的支持。浏览器只允许 JavaScript 代码在相同的源内发出 HTTP 请求，来保护用户，这意味着 URL 具有与用于加载 JavaScript 代码的 URL 相同的方案、主机和端口。CORS 执行初始 HTTP 请求，来检查服务器是否允许来自特定 URL 的请求，帮助防止未经用户同意就使用服务的恶意代码，从而放松了这一限制。

ASP.NET Core 提供了一个处理 CORS 的内置服务，通过向 Startup 类添加 ConfigureServices 方法，以启用该服务:

```
...
Services.AddCors ();
...
```

options 模式使用在 Microsoft.AspNetCore. Cors.Infrastructure 名称空间中定义的 CorsOptions 类配置 CORS。详情请参见 https://docs.microsoft.com/en-gb/aspnet/core/security/cors?view=aspnetcore-3.1。

### 19.5.1 使用异步操作

ASP.NET Core 平台通过从池中分配一个线程来处理每个请求。池的大小决定着可以并发处理的请求数量，在等待操作产生结果时，线程不能用于处理其他任何请求。

依赖外部资源的操作可能导致请求线程等待更长时间。例如，数据库服务器可能有自己的并发性限制，并可能将查询排队等待执行。ASP.NET Core 请求线程无法处理其他任何请求，直到数据库为该操作生成结果，然后该结果生成可发送到 HTTP 客户端的响应。

这个问题可以通过定义异步操作来解决，异步操作允许 ASP.NET Core 线程处理其他可能被阻塞的请求，这增加了应用程序可以同时处理的 HTTP 请求的数量。代码清单 19-15 修改了控制器，以使用异步操作。

■ **注意：**异步操作不会产生更快的响应，其好处只是增加了可以同时处理的请求数量。

**代码清单 19-15 Controllers 文件夹的 ProductsController.cs 文件中的异步操作**

```
using Microsoft.AspNetCore.Mvc;
using WebApp.Models;
using System.Collections.Generic;
using Microsoft.Extensions.Logging;
using System.Linq;
using System.Threading.Tasks;

namespace WebApp.Controllers {

    [Route("api/[controller]")]
    public class ProductsController: ControllerBase {
        private DataContext context;

        public ProductsController(DataContext ctx) {
```

```
            context = ctx;
        }

        [HttpGet]
        public IAsyncEnumerable<Product> GetProducts() {
            return context.Products;
        }

        [HttpGet("{id}")]
        public async Task<Product> GetProduct(long id) {
            return await context.Products.FindAsync(id);
        }

        [HttpPost]
        public async Task SaveProduct([FromBody]Product product) {
            await context.Products.AddAsync(product);
            await context.SaveChangesAsync();
        }

        [HttpPut]
        public async Task UpdateProduct([FromBody]Product product) {
            context.Update(product);
            await context.SaveChangesAsync();
        }

        [HttpDelete("{id}")]
        public async Task DeleteProduct(long id) {
            context.Products.Remove(new Product() { ProductId = id });
            await context.SaveChangesAsync();
        }
    }
}
```

Entity Framework Core 提供了一些方法的异步版本，比如 FindAsync、AddAsync 和 SaveChangesAsync，它们与 await 关键字一起使用。并不是所有操作都可以异步执行，这就是 Update 和 Remove 方法不变的原因。

对于某些操作(包括对数据库的 LINQ 查询)，可以使用 IAsyncEnumerable<T>接口，该接口表示应该异步枚举的对象序列，并防止 ASP.NET Core 请求线程等待数据库生成的每个对象，如第 5 章所述。

控制器生成的响应没有变化，但是 ASP.NET Core 处理每个请求的线程并不一定会被操作方法阻止。

## 19.5.2 防止过度绑定

一些操作方法使用模型绑定特性从响应体中获取数据，以便它可以用于执行数据库操作。SaveProduct 操作有一个问题，使用 PowerShell 提示符，运行代码清单 19-16 所示的命令，就可以看到这个问题。

**代码清单 19-16　存储产品**

```
Invoke-RestMethod http://localhost:5000/api/products -Method POST -Body (@{ ProductId=100;
Name="Swim Buoy";
    Price=19.99; CategoryId=1; SupplierId=1} | ConvertTo-Json) -ContentType "application/json"
```

与代码清单 19-11 中用于测试 POST 方法的命令不同，该命令包含 ProductId 属性的值。当 Entity Framework Core 向数据库发送数据时，抛出以下异常:

```
...
Microsoft.Data.SqlClient.SqlException (0x80131904): Cannot insert explicit value for
identity column in table 'Products' when IDENTITY_INSERT is set to OFF.
...
```

默认情况下，Entity Framework Core 配置数据库，以便在存储新对象时分配主键值。这意味着应用程序不必跟踪已经分配的键值，并允许多个应用程序共享相同的数据库，而不必协调键分配。Product 数据模型类需要一个 ProductId 属性，但模型绑定过程不理解该属性的重要性，将客户端提供的任何值添加到它创建的对象中，这将导致 SaveProduct 操作方法中的异常。

这就是所谓的过度绑定，当客户端提供了开发人员没有期望的值时，就会导致严重的问题。在最好的情况下，应用程序会出现意外行为，但是这种技术用于破坏应用程序安全性，并授予用户超出其应有权限的权限。

防止过度绑定的最安全方法是创建单独的数据模型类，这些类仅用于通过模型绑定过程接收数据。将一个名为 ProductBindingTarget.cs 的类文件添加到 WebApp/Models 文件夹中，并使用它定义如代码清单 19-17 所示的类。

**代码清单 19-17　WebApp/Models 文件夹中 ProductBindingTarget.cs 文件的内容**

```
namespace WebApp.Models {
    public class ProductBindingTarget {

        public string Name { get; set; }

        public decimal Price { get; set; }

        public long CategoryId { get; set; }

        public long SupplierId { get; set; }

        public Product ToProduct() => new Product() {
            Name = this.Name, Price = this.Price,
            CategoryId = this.CategoryId, SupplierId = this.SupplierId
        };
    }
}
```

ProductBindingTarget类只定义了应用程序在存储新对象时希望从客户端接收的属性。ToProduct方法创建一个可与应用程序的其余部分一起使用的Product，确保客户端可以仅为Name、Price、CategoryId和SupplierId属性提供值。代码清单 19-18 在SaveProduct操作中使用绑定目标类防止过度绑定。

**代码清单 19-18　在 Controllers 文件夹的 ProductsController.cs 文件中使用绑定目标**

```
...
[HttpPost]
public async Task SaveProduct([FromBody]ProductBindingTarget target) {
    await context.Products.AddAsync(target.ToProduct());
    await context.SaveChangesAsync();
}
...
```

重启 ASP.NET Core，重复代码清单 19-16 中的命令，将看到如图 19-9 所示的响应。客户端包含了 ProductId 值，但是模型绑定过程会忽略它，丢弃只读属性的值(在运行此示例时，可能会看到 ProductId 属性的不同值，这取决于在运行命令之前对数据库所做的更改)。

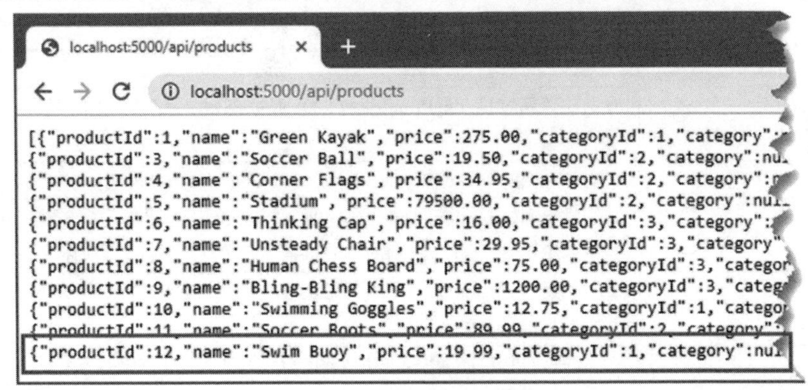


图 19-9　丢弃不想要的数据值

## 19.5.3　使用操作的结果

MVC 框架自动设置响应的状态代码，但未必能得到想要的结果，部分原因是对于 RESTful Web 服务没有严格的规则，且 Microsoft 所做的假设可能与用户的期望不一致。要查看示例，请使用 PowerShell 命令提示符，运行代码清单 19-19 所示的命令，该命令向 Web 服务发送 GET 请求。

**代码清单 19-19　发送 GET 请求**

```
Invoke-WebRequest http://localhost:5000/api/products/1000 |Select-Object StatusCode
```

Invoke-WebRequest 命令与前面示例中使用的 Invoke-RestMethod 命令类似，但是可以更容易地从响应中获取状态代码。代码清单 19-19 中请求的 URL 由 GetProduct 操作方法处理，该方法查询数据库中 ProductId 值为 1000 的对象，该命令生成以下输出：

```
StatusCode
```

```
----------
        204
```

数据库中没有匹配对象，这意味着 GetProduct 操作方法返回 null。当 MVC 框架从一个操作方法接收到 null 时，它返回 204 状态码，这表示一个没有生成任何数据的成功请求。并不是所有 Web 服务都以这种方式运行，常见的替代方法是返回 404 响应，表示未找到。

类似地，SaveProducts 操作在存储对象时返回一个 200 响应，但是由于直到存储数据时才生成主键，所以客户端不知道分配了什么键值。

**注意**：这些 Web 服务实现细节没有对错之分，应该选择最适合项目和个人偏好的方法。本节演示如何更改默认行为，而不是遵循任何特定 Web 服务样式的方向。

操作方法可以通过返回一个实现 IActionResult 接口的对象，来指示 MVC 框架，发送一个特定的响应，该对象称为操作结果。这允许操作方法指定所需的响应类型，而不必直接使用 HttpResponse 对象生成它。

ControllerBase 类提供一组用于创建操作结果对象的方法，操作结果对象可从操作方法返回。表 19-7 描述了最有用的操作结果方法。

表 19-7　有用的 ControllerBase 操作结果方法

| 名称 | 描述 |
| --- | --- |
| Ok | 此方法返回的 IActionResult 生成 200 OK 状态码，并在响应体中发送一个可选的数据对象 |
| NoContent | 此方法返回的 IActionResult 生成 204 NO CONTENT 状态码 |
| BadRequest | 此方法返回的 IActionResult 生成 400 AD REQUEST 状态码。该方法接收一个可选的模型状态对象，该对象向客户端描述问题，如 19.5.4 节所述 |
| File | 此方法返回的 IActionResult 生成 200OK 响应，为特定的类型设置 Content-Type 头，并将指定的文件发送给客户端 |
| NotFound | 此方法返回的 IActionResult 生成 404 NOT FOUND 状态码 |
| RedirectRedirectPermanent | 此方法返回的 IActionResult 将客户端重定向到指定的 URL |
| RedirectToRoute<br>RedirectToRoutePermanent | 此方法返回的 IActionResult 将客户端重定向到使用路由系统(使用约定路由)创建的指定 URL |
| LocalRedirectLocal<br>RedirectPermanent | 此方法返回的 IActionResult 将客户端重定向到应用程序本地的指定 URL |
| RedirectToActionRedirect<br>ToActionPermanent | 此方法返回的 IActionResult 将客户端重定向到一个操作方法。重定向的 URL 是使用 URL 路由系统创建的 |
| RedirectToPageRedirect<br>ToPagePermanent | 此方法返回的 IActionResult 将客户端重定向到 Razor Pages，见第 23 章 |
| StatusCode | 此方法返回的 IActionResult 会生成一个带有特定状态码的响应 |

当操作方法返回一个对象时，它等价于将对象传递给 Ok 方法并返回结果。当操作返回 null 时，它等价于从 NoContent 方法返回结果。代码清单 19-20 修改了 GetProduct 和 SaveProduct 操作的行为，以便它们使用表 19-7 中的方法来覆盖 Web 服务控制器的默认行为。

代码清单 19-20 在 Controllers 文件夹的 ProductsController.cs 文件中使用操作结果

```
using Microsoft.AspNetCore.Mvc;
using WebApp.Models;
using System.Collections.Generic;
using Microsoft.Extensions.Logging;
using System.Linq;
using System.Threading.Tasks;

namespace WebApp.Controllers {

    [Route("api/[controller]")]
    public class ProductsController : ControllerBase {
        private DataContext context;

        public ProductsController(DataContext ctx) {
            context = ctx;
        }

        [HttpGet]
        public IAsyncEnumerable<Product> GetProducts() {
            return context.Products;
        }

        [HttpGet("{id}")]
        public async Task<IActionResult> GetProduct(long id) {
            Product p = await context.Products.FindAsync(id);
            if (p == null) {
                return NotFound();

            }
            return Ok(p);
        }

        [HttpPost]
        public async Task<IActionResult>
                SaveProduct([FromBody]ProductBindingTarget target) {
            Product p = target.ToProduct();
            await context.Products.AddAsync(p);
            await context.SaveChangesAsync();
            return Ok(p);
        }

        [HttpPut]
        public async Task UpdateProduct([FromBody]Product product) {
            context.Update(product);
            await context.SaveChangesAsync();
```

```
        }

        [HttpDelete("{id}")]
        public async Task DeleteProduct(long id) {
            context.Products.Remove(new Product() { ProductId = id });
            await context.SaveChangesAsync();
        }
    }
}
```

重启ASP.NET Core，重复代码清单19-19中的命令，将看到一个异常，这就是Invoke-WebRequest命令响应错误状态代码的方式，例如GetProduct操作方法返回的404 Not Found。

要查看对 SaveProduct 操作方法的更改效果，请使用 PowerShell 命令提示符，运行代码清单 19-21 所示的命令，该命令向 Web 服务发送一个 POST 请求。

**代码清单 19-21　发送 POST 请求**

```
Invoke-RestMethod http://localhost:5000/api/products -Method POST -Body (@{Name="Boot Laces"; Price=19.99;
CategoryId=2; SupplierId=2} | ConvertTo-Json) -ContentType "application/json"
```

该命令生成以下输出，显示从 Web 服务接收的 JSON 数据中解析的值：

```
productId  : 13
name       : Boot Laces
price      : 19.99
categoryId : 2
category   :
supplierId : 2
supplier   :
```

### 1. 执行重定向

表 19-7 中的许多操作结果方法都与重定向有关，它将客户端定向到另一个 URL。执行重定向的最基本方法是调用 Redirect 方法，如代码清单 19-22 所示。

**提示：**如果控制器试图重定向任何非本地的 URL，LocalRedirect 和 LocalRedirectPermanent 方法会抛出一个异常。这在重定向到用户提供的 URL 时非常有用，这种情况下，开放重定向攻击试图将另一个用户重定向到不受信任的站点。

**代码清单 19-22　Controllers 文件夹的 ProductsController.cs 文件中的重定向**

```
using Microsoft.AspNetCore.Mvc;
using WebApp.Models;
using System.Collections.Generic;
using Microsoft.Extensions.Logging;
using System.Linq;
```

```
using System.Threading.Tasks;

namespace WebApp.Controllers {

    [Route("api/[controller]")]
    public class ProductsController : ControllerBase {
        private DataContext context;
        public ProductsController(DataContext ctx) {
            context = ctx;
        }

        // ...other action methods omitted for brevity...

        [HttpGet("redirect")]
        public IActionResult Redirect() {
            return Redirect("/api/products/1");
        }
    }
}
```

重定向 URL 表示为 Redirect 方法的字符串参数，该方法生成临时重定向。重启 ASP.NET Core，使用 PowerShell 命令提示符，运行代码清单 19-23 所示的命令，该命令发送一个 GET 请求，该请求由新的操作方法处理。

**代码清单 19-23　测试重定向**

```
Invoke-RestMethod http://localhost:5000/api/products/redirect
```

Invoke-RestMethod 命令接收来自 Web 服务的重定向响应，并向它给出的 URL 发送一个新的请求，生成以下响应：

```
productId  : 1
name       : GreenKayak
price      : 275.00
categoryId : 1
category   :
supplierId : 1
supplier   :
```

### 2. 重定向到操作方法

可以使用 RedirectToAction 方法(用于临时重定向)或 RedirectToActionPermanent 方法(用于永久重定向)重定向到另一个操作方法。代码清单 19-24 更改了 Redirect 操作方法，这样客户端将被重定向到控制器定义的另一个操作方法。

**代码清单 19-24　在 Controllers 文件夹的 ProductsController.cs 文件中重定向到操作**

```
using Microsoft.AspNetCore.Mvc;
```

```
using WebApp.Models;
using System.Collections.Generic;
using Microsoft.Extensions.Logging;
using System.Linq;
using System.Threading.Tasks;

namespace WebApp.Controllers {

    [Route("api/[controller]")]
    public class ProductsController : ControllerBase {
        private DataContext context;

        public ProductsController(DataContext ctx) {
            context = ctx;
        }

        // ...other action methods omitted for brevity...

        [HttpGet("redirect")]
        public IActionResult Redirect() {
            return RedirectToAction(nameof(GetProduct), new { Id = 1 });
        }
    }
}
```

操作方法被指定为字符串，不过 nameof 表达式可以用于选择操作方法，而不会有输入错误的风险。创建路由所需的其他任何值都使用匿名对象提供。重启 ASP.NET Core，使用 PowerShell 命令提示符，重复代码清单 19-23 中的命令。路由系统用于创建一个 URL，目标是指定的操作方法，生成以下响应：

```
productId  : 1
name       : Kayak
price      : 100.00
categoryId : 1
category   :
supplierId : 1
supplier   :
```

如果只指定一个操作方法名，则重定向以当前控制器为目标。RedirectToAction 方法的一个重载版本接收操作和控制器名称。

## 使用路由值重定向

RedirectToRoute 和 RedirectToRoutePermanent 方法将客户端重定向到一个 URL，该 URL 是通过向路由系统提供段变量的值，并允许它选择要使用的路由而创建的。这对于具有复杂路由配置的应用程序非常有用，并且应该谨慎使用，因为很容易创建到错误 URL 的重定向。下面是一个使用 RedirectToRoute 方法进行重定向的例子:

```
...
[HttpGet("redirect")]
public IActionResult Redirect() {
    return RedirectToRoute(new {
        controller = "Products", action = "GetProduct", Id = 1
    });
}
...
```

此重定向中的一组值依赖“约定路由”来选择控制器和操作方法。约定路由通常用于生成 HTML 响应的控制器，如第 21 章所述。

### 19.5.4 验证数据

接收来自客户端的数据时，必须假定许多数据是无效的，并准备过滤掉应用程序不能使用的值。为 MVC 框架控制器提供的数据验证功能详见第 29 章，但这一章只关注一个问题：确保客户端为数据库中存储数据所需的属性提供值。模型绑定的第一步是将属性应用到数据模型类的属性，如代码清单 19-25 所示。

**代码清单 19-25 在 Models 文件夹的 ProductBindingTarget.cs 文件中应用属性**

```
using System.ComponentModel.DataAnnotations;

namespace WebApp.Models {
    public class ProductBindingTarget {

        [Required]
        public string Name { get; set; }

        [Range(1, 1000)]
        public decimal Price { get; set; }

        [Range(1, long.MaxValue)]
        public long CategoryId { get; set; }

        [Range(1, long.MaxValue)]
        public long SupplierId { get; set; }

        public Product ToProduct() => new Product() {
            Name = this.Name, Price = this.Price,
            CategoryId = this.CategoryId, SupplierId = this.SupplierId
        };
    }
}
```

Required 属性表示客户端必须为其提供值的属性，可以应用于在请求中没有值时分配为 null 的属性。Range 属性需要一个介于上限和下限之间的值，用于基本类型，当请求中没有值时，该属性默认为零。

代码清单 19-26 更新了 SaveProduct 操作，以便在存储“模型绑定过程”创建的对象之前执行验证，确保只有包含所有四个属性值的对象才装饰了验证属性。

**代码清单 19-26 在 Controllers 文件夹的 ProductsController.cs 文件中应用验证**

```
...
[HttpPost]
public async Task<IActionResult> SaveProduct([FromBody]ProductBindingTarget target) {
    if (ModelState.IsValid) {
        Product p = target.ToProduct();
        await context.Products.AddAsync(p);
        await context.SaveChangesAsync();
        return Ok(p);
    }
    return BadRequest(ModelState);
}
...
```

ModelState 属性是从 ControllerBase 类继承的，如果模型绑定过程生成的数据满足验证标准，那么 IsValid 属性返回 true。如果从客户端接收的数据是有效的，则返回 Ok 方法的操作结果。如果客户端发送的数据验证检查失败，那么 IsValid 属性为 false，而使用 BadRequest 方法的操作结果。BadRequest 方法接收 ModelState 属性返回的对象，该属性用于向客户端描述验证错误(没有描述验证错误的标准方法，因此客户端可能仅依赖 400 状态码来确定存在问题)。

要测试验证，请重新启动 ASP.NET Core，使用一个新的 PowerShell 命令提示符运行代码清单 19-27 所示的命令。

**代码清单 19-27 测试验证**

```
Invoke-WebRequest http://localhost:5000/api/products -Method POST -Body (@{Name="Boot Laces"} |
ConvertTo-Json) -ContentType "application/json"
```

该命令抛出一个异常，显示 Web 服务返回一个 400 Bad Request 响应。验证错误的细节没有显示出来，因为 Invoke-WebRequest 命令和 Invoke-RestMethod 命令都不提供对错误响应体的访问。虽然看不到它，它的主体包含一个 JSON 对象，每个数据的属性已经验证失败，如下所示：

```
{
 "Price":["The field Price must be between 1 and 1000."],
 "CategoryId":["The field CategoryId must be between 1 and 9.223372036854776E+18."],
 "SupplierId":["The field SupplierId must be between 1 and 9.223372036854776E+18."]
}
```

可以在第 29 章中看到使用验证消息的示例，其中详细描述了验证特性。

### 19.5.5 应用 API 控制器属性

ApiController属性可应用于Web服务控制器类，以更改模型绑定和验证特性的行为。使用FromBody属性从请求体中选择数据并显式检查ModelState。在已使用ApiController属性修饰的控制器中不需要IsValid属性。在Web服务中，从主体获取数据和验证数据是非常普遍的要求，因此在使用属性时自动应用这些数据，从而将控制器操作中的代码焦点恢复为处理应用程序特性，如代码清单 19-28 所示。

**代码清单 19-28 在 Controllers 文件夹的 ProductsController.cs 文件中使用 ApiController**

```
using Microsoft.AspNetCore.Mvc;
using WebApp.Models;
using System.Collections.Generic;
using Microsoft.Extensions.Logging;
using System.Linq;
using System.Threading.Tasks;

namespace WebApp.Controllers {

    [ApiController]
    [Route("api/[controller]")]
    public class ProductsController : ControllerBase {
        private DataContext context;

        public ProductsController(DataContext ctx) {
            context = ctx;
        }

        [HttpGet]
        public IAsyncEnumerable<Product> GetProducts() {
            return context.Products;
        }

        [HttpGet("{id}")]
        public async Task<IActionResult> GetProduct(long id) {
            Product p = await context.Products.FindAsync(id);
            if (p == null) {
                return NotFound();
            }
            return Ok(p);
        }

        [HttpPost]
        public async Task<IActionResult> SaveProduct(ProductBindingTarget target) {
            Product p = target.ToProduct();
            await context.Products.AddAsync(p);
            await context.SaveChangesAsync();
```

```
            return Ok(p);
        }

        [HttpPut]
        public async Task UpdateProduct(Product product) {
            context.Update(product);
            await context.SaveChangesAsync();
        }

        [HttpDelete("{id}")]
        public async Task DeleteProduct(long id) {
            context.Products.Remove(new Product() { ProductId = id });
            await context.SaveChangesAsync();
        }

        [HttpGet("redirect")]
        public IActionResult Redirect() {
           return RedirectToAction(nameof(GetProduct), new { Id = 1 });
        }
    }
}
```

使用 ApiController 属性是可选的，但它有助于生成简洁的 Web 服务控制器。

### 19.5.6　忽略 Null 属性

本章要做的最后一个更改是从 Web 服务返回的数据中删除空值。数据模型类包含由 Entity Framework Core 用于在复杂查询中关联相关数据的导航属性，如第 20 章所述。对于本章执行的简单查询，这些导航属性没有赋值，这意味着客户端接收到的属性的值永远不可用。要查看这个问题，使用 PowerShell 命令提示符，运行代码清单 19-29 所示的命令。

**代码清单 19-29　发送 GET 请求**

```
Invoke-WebRequest http://localhost:5000/api/products/1 | Select-Object Content
```

该命令发送一个 GET 请求，并显示来自 Web 服务的响应体，生成以下输出:

```
Content
-------
{"productId":1,"name":"Green
Kayak","price":275.00,"categoryId":1,"category":null,"supplierId":1,"supplier":null}
```

请求是由 GetProduct 操作方法处理的，响应中的类别和供应商值始终为 null，因为操作没有要求 Entity Framework Core 填充这些属性。

#### 1. 投射选定的属性

第一种方法是只返回客户端需要的属性。这样可以完全控制每个响应，但是如果每个操作返回一组不同的值，则管理起来会很难，并使客户端开发人员感到困惑。代码清单 19-30 显示了如何投影从数据库获得的 Product 对象，以便省略导航属性。

**代码清单 19-30 在 Controllers 文件夹的 ProductsController.cs 文件中省略属性**

```
...
[HttpGet("{id}")]
public async Task<IActionResult> GetProduct(long id) {
    Product p = await context.Products.FindAsync(id);
    if (p == null) {
       return NotFound();
    }
    return Ok(new {
        ProductId = p.ProductId, Name = p.Name,
        Price = p.Price, CategoryId = p.CategoryId,
        SupplierId = p.SupplierId
    });
}
...
```

选择客户端需要的属性，并将其添加到传递给 Ok 方法的对象中。重启 ASP.NET Core 并运行代码清单 19-30 中的命令，收到一个省略导航属性及其空值的响应，如下所示：

```
Content
-------
{"productId":1,"name":"Green Kayak","price":275.00,"categoryId":1,"supplierId":1}
```

#### 2. 配置 JSON 序列化器

可将 JSON 序列化器配置为在序列化对象时省略值为 null 的属性。序列化器使用 Startup 类中的 options 模式配置，如代码清单 19-31 所示。

**代码清单 19-31 在 WebApp 文件夹的 Startup.cs 文件中配置 JSON 序列化器**

```
using System;
using System.Collections.Generic;
using System.Linq;
using System.Threading.Tasks;
using Microsoft.AspNetCore.Builder;
using Microsoft.AspNetCore.Hosting;
using Microsoft.AspNetCore.Http;
using Microsoft.Extensions.DependencyInjection;
using Microsoft.Extensions.Hosting;
using Microsoft.Extensions.Configuration;
using Microsoft.EntityFrameworkCore;
```

```
using WebApp.Models;
using Microsoft.AspNetCore.Mvc;

namespace WebApp {
    public class Startup {

        public Startup(IConfiguration config) {
            Configuration = config;
        }

        public IConfiguration Configuration { get; set; }

        public void ConfigureServices(IServiceCollection services) {
            services.AddDbContext<DataContext>(opts => {
                opts.UseSqlServer(Configuration[
                    "ConnectionStrings:ProductConnection"]);
                opts.EnableSensitiveDataLogging(true);
            });

            services.AddControllers();
            services.Configure<JsonOptions>(opts => {
                opts.JsonSerializerOptions.IgnoreNullValues = true;
            });
        }

        public void Configure(IApplicationBuilder app, DataContext context) {
            app.UseDeveloperExceptionPage();
            app.UseRouting();
            app.UseMiddleware<TestMiddleware>();
            app.UseEndpoints(endpoints => {
                endpoints.MapGet("/", async context => {
                    await context.Response.WriteAsync("Hello World!");
                });
                endpoints.MapControllers();
            });
            SeedData.SeedDatabase(context);
        }
    }
}
```

JSON 序列化器是使用 JsonOptions 类的 JsonSerializerOptions 属性配置的，当 IgnoreNullValues 属性为 true 时，null 值将被丢弃。

此配置更改会影响所有 JSON 响应，使用时应谨慎，特别是在任何数据模型类使用 null 值向客户端传递信息时。要查看更改的效果，请重新启动 ASP.NET Core，使用浏览器请求 http://localhost:5000/api/products，生成如图 19-10 所示的响应。

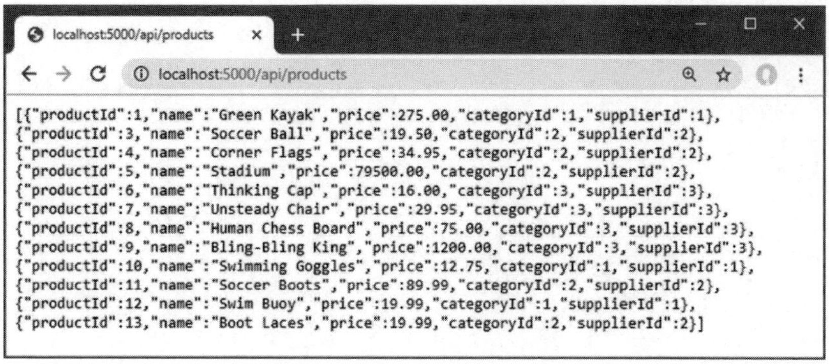
localhost:5000/api/products
localhost:5000/api/products
[{"productId":1,"name":"Green Kayak","price":275.00,"categoryId":1,"supplierId":1},
{"productId":3,"name":"Soccer Ball","price":19.50,"categoryId":2,"supplierId":2},
{"productId":4,"name":"Corner Flags","price":34.95,"categoryId":2,"supplierId":2},
{"productId":5,"name":"Stadium","price":79500.00,"categoryId":2,"supplierId":2},
{"productId":6,"name":"Thinking Cap","price":16.00,"categoryId":3,"supplierId":3},
{"productId":7,"name":"Unsteady Chair","price":29.95,"categoryId":3,"supplierId":3},
{"productId":8,"name":"Human Chess Board","price":75.00,"categoryId":3,"supplierId":3},
{"productId":9,"name":"Bling-Bling King","price":1200.00,"categoryId":3,"supplierId":3},
{"productId":10,"name":"Swimming Goggles","price":12.75,"categoryId":1,"supplierId":1},
{"productId":11,"name":"Soccer Boots","price":89.99,"categoryId":2,"supplierId":2},
{"productId":12,"name":"Swim Buoy","price":19.99,"categoryId":1,"supplierId":1},
{"productId":13,"name":"Boot Laces","price":19.99,"categoryId":2,"supplierId":2}]

图 19-10 配置 JSON 序列化器

# 19.6 小结

本章展示了如何使用 MVC 框架创建 RESTful Web 服务。解释了 MVC 框架是构建在 ASP.NET Core 平台之上的。展示了单个控制器类如何定义多个操作方法，每个操作方法都可以处理 URL 和 HTTP 方法的不同组合。论述了如何使用依赖注入来消费服务，如何使用模型绑定来访问请求数据，如何验证请求数据，以及如何控制返回给客户端的响应。第 20 章将描述为 Web 服务提供的 ASP.NET Core 高级特性。

# 第 20 章

■■■

# 高级 Web 服务特性

本章描述可用于创建 RESTful Web 服务的高级特性。解释如何在 Entity Framework Core 查询中处理相关数据，如何添加对 HTTP Patch 方法的支持，如何使用内容协商，以及如何使用 OpenAPI 来描述 Web 服务。表 20-1 给出 Web 服务特性的相关知识。

表 20-1　Web 服务特性的相关知识

| 问题 | 答案 |
| --- | --- |
| 它们是什么? | 本章描述的特性提供了对 ASP.NET Core Web 服务如何工作进行更好的控制，包括管理发送到客户端的数据和数据使用的格式 |
| 它们为什么有用? | ASP.NET Core 提供的默认行为并不能满足每个项目的需求，本章所描述的特性允许重新塑造 Web 服务，以适应特定需求 |
| 它们是如何使用的? | 本章特性的共同主题是改变操作方法产生的响应 |
| 是否存在缺陷或限制? | 很难决定如何实现 Web 服务，特别是当第三方客户端使用 Web 服务时。一旦客户端开始使用 Web 服务，Web 服务的行为就会固定下来，这意味着在使用本章描述的特性时需要仔细考虑 |
| 还有其他选择吗? | 本章描述的特性是可选的，可以依赖 ASP.NET Core Web 服务的默认行为 |

表 20-2 总结了本章的内容。

表 20-2　本章内容摘要

| 问题 | 解决方案 | 代码清单 |
| --- | --- | --- |
| 使用关系数据 | 在 LINQ 查询中使用 Include 和 ThenInclude 方法 | 20-4 |
| 打破循环引用 | 显式地将导航属性设置为 null | 20-5 |
| 允许客户端有选择地更新数据 | 支持 HTTP Patch 方法 | 20-6～20-9 |
| 支持一系列响应数据类型 | 支持内容格式化和协商 | 20-10～20-24 |
| 为 Web 服务编制文档 | 使用 OpenAPI 描述 Web 服务 | 20-25～20-29 |

## 20.1　准备工作

本章使用第 18 章创建、第 19 章修改的 WebApp 项目。为准备本章，在 WebApp/Controllers

文件夹中添加一个名为 SuppliersController.cs 的文件，其内容如代码清单 20-1 所示。

**代码清单 20-1　Controllers 文件夹中 SuppliersController.cs 文件的内容**

```
using Microsoft.AspNetCore.Mvc;
using WebApp.Models;
using System.Threading.Tasks;

namespace WebApp.Controllers {

    [ApiController]
    [Route("api/[controller]")]
    public class SuppliersController: ControllerBase {
        private DataContext context;

        public SuppliersController(DataContext ctx) {
            context = ctx;
        }

        [HttpGet("{id}")]
        public async Task<Supplier> GetSupplier(long id) {
            return await context.Suppliers.FindAsync(id);
        }
    }
}
```

控制器扩展了 ControllerBase 类，声明了对 DataContext 服务的依赖，定义了一个名为 GetSupplier 的操作来处理/api/[controller]/{id} URL 模式的 GET 请求。

### 20.1.1　删除数据库

打开一个新的 PowerShell 命令提示符，导航到包含 WebApp.csproj 文件的文件夹，并运行代码清单 20-2 所示的命令来删除数据库。

---

**提示：** 可从 https://github.com/apress/pro-asp.net-core-3 下载本章和本书中其他所有章节的示例项目。如果在运行示例时遇到问题，请参阅第 1 章以获得帮助。

---

**代码清单 20-2　删除数据库**

```
dotnet ef database drop --force
```

### 20.1.2　运行示例应用程序

删除数据库之后，从 Debug 菜单中选择 Start Without Debugging 或 Run Without Debugging，或者使用 PowerShell 命令提示符，运行代码清单 20-3 所示的命令。

**代码清单 20-3　运行示例应用程序**

```
dotnet run
```

数据库作为应用程序启动的一部分填充数据。一旦运行 ASP.NET Core，使用 Web 浏览器请求 http://localhost:5000/api/suppliers/1，生成如图 20-1 所示的响应。

图 20-1　运行示例应用程序

响应显示主键与请求 URL 的最后一部分匹配的 Supplier 对象。在第 19 章中，JSON 序列化器配置为忽略 null 值的属性，这就是为什么响应不包括由 Supplier 数据模型类定义的导航属性。

## 20.2　处理相关数据

尽管这不是一本关于 Entity Framework Core 的书，但是在数据查询方面，大多数 Web 服务都会遇到一个问题。第 18 章中定义的数据模型类包括导航属性，当使用 include 方法时，Entity Framework Core 可以通过数据库中的关系填充导航属性，如代码清单 20-4 所示。

**代码清单 20-4　在 Controllers 文件夹的 SuppliersController.cs 文件中请求相关数据**

```
using Microsoft.AspNetCore.Mvc;
using WebApp.Models;
using System.Threading.Tasks;
using Microsoft.EntityFrameworkCore;

namespace WebApp.Controllers {

    [ApiController]
    [Route("api/[controller]")]
    public class SuppliersController: ControllerBase {
        private DataContext context;

        public SuppliersController(DataContext ctx) {
            context = ctx;
        }

        [HttpGet("{id}")]
        public async Task<Supplier> GetSupplier(long id) {
            return await context.Suppliers
                .Include(s => s.Products)
                .FirstAsync(s => s.SupplierId == id);
```

```
            }
        }
    }
```

Include 方法告诉 Entity Framework Core，在数据库中遵循一个关系并加载相关数据。在本例中，Include 方法选择由 Supplier 类定义的 Products 导航属性，这将导致 Entity Framework Core 加载与所选 Supplier 关联的 Product 产品对象，并将其分配给 Products 属性。

重启 ASP.NET Core，使用浏览器请求 http://localhost:5000/api/suppliers/1，该请求针对 GetSupplier 操作方法。请求失败，显示如图 20-2 所示的异常。

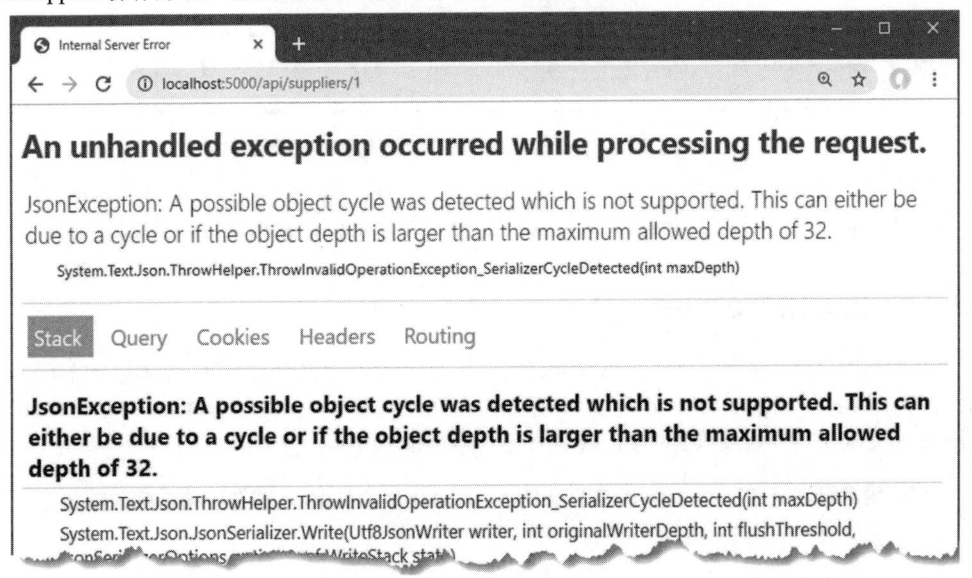


图 20-2　查询相关数据引起的异常

JSON 序列化器报告了一个“对象循环”，这意味着在为响应序列化的数据中存在一个循环引用。

查看代码清单 20-4 中的代码，可能很难理解为什么使用 Include 方法创建了一个循环引用。这个问题是由 Entity Framework Core 的一个特性引起的，该特性试图最小化从数据库读取的数据量，但这在 ASP.NETCore 应用程序中导致了问题。

当 Entity Framework Core 创建对象时，它使用已经由相同数据库上下文创建的对象填充导航属性。这在某些应用程序中可能是一个有用的特性，比如桌面应用程序，其中数据库上下文对象的生命周期很长，用于在一段时间内发出许多请求。它对 ASP.NET Core 应用程序没有用处，其中为每个 HTTP 请求创建一个新的上下文对象。

Entity Framework Core 在数据库中查询与所选 Supplier 关联的 Product 对象，并将其分配给该 Supplier. Product 导航属性。问题是 Entity Framework Core 会查看它创建的每个 Product 对象，并使用查询响应来填充 Product.Supplier 导航属性。对于 ASP.NET Core 应用程序，这个步骤没有帮助，因为它在 Supplier 和 Product 对象的导航属性之间创建了一个循环引用，如图 20-3 所示。

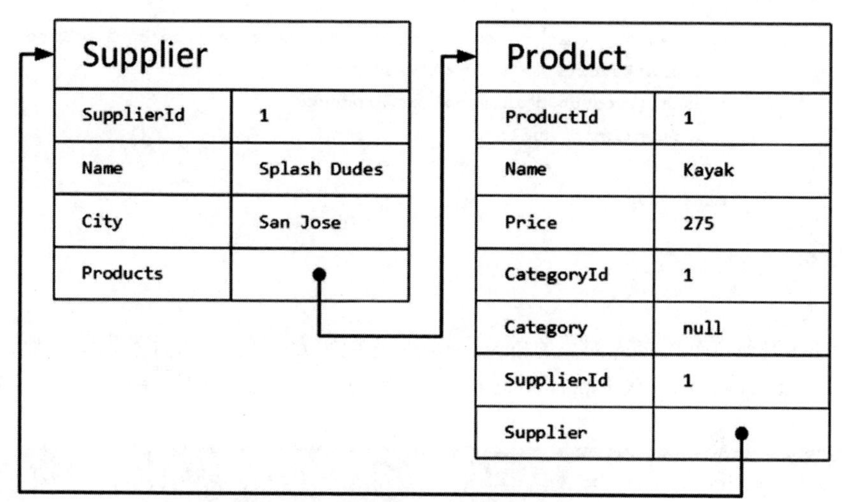


图 20-3　理解 Entity Framework Core 如何使用相关数据

当 Supplier 对象由控制器的操作方法返回时，JSON 序列化器通过属性来工作。遵循 Product 对象的引用，每个引用都有一个指向 Supplier 对象的引用，在一个循环中，直到达到最大深度、抛出异常为止，如图 20-2 所示。

## 打破相关数据中的循环引用

无法阻止 Entity Framework Core 在它加载到数据库的数据中创建循环引用。防止异常意味着为 JSON 序列化器提供不包含循环引用的数据，最简单的方法是在 Entity Framework Core 创建对象之后和序列化之前修改对象，如代码清单 20-5 所示。

### 代码清单 20-5　在 Controllers 文件夹的 SuppliersController.cs 文件中破坏引用

```
using Microsoft.AspNetCore.Mvc;
using WebApp.Models;
using System.Threading.Tasks;
using Microsoft.EntityFrameworkCore;

namespace WebApp.Controllers {

    [ApiController]
    [Route("api/[controller]")]
    public class SuppliersController: ControllerBase {
        private DataContext context;

        public SuppliersController(DataContext ctx) {
            context = ctx;
        }

        [HttpGet("{id}")]
        public async Task<Supplier> GetSupplier(long id) {
```

```
            Supplier supplier = await context.Suppliers.Include(s => s.Products)
                .FirstAsync(s => s.SupplierId == id);
            foreach (Product p in supplier.Products) {
                p.Supplier = null;
            };
            return supplier;
        }
    }
}
```

foreach 循环将每个 Product 对象的 Supplier 属性设置为 null，这会破坏循环引用。重启 ASP.NET Core，请求 http://localhost:5000/api/suppliers/1，查询供应商及其相关产品，会生成如图 20-4 所示的响应。

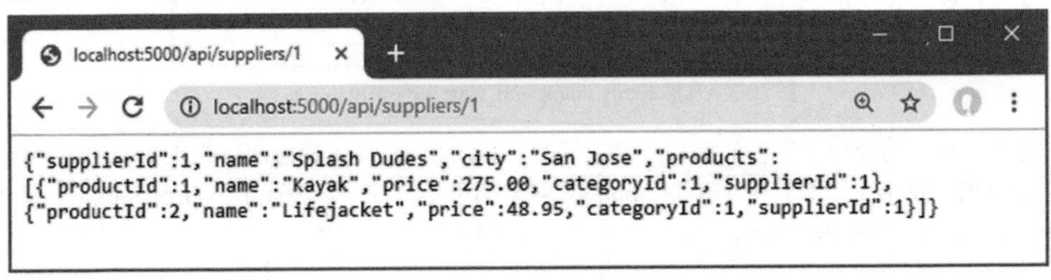


图 20-4 查询相关数据

# 20.3 支持 HTTP Patch 方法

对于简单的数据类型，可以使用 PUT 方法替换现有对象来处理编辑操作，这是第 19 章采用的方法。例如，即使只需要更改 Product 类中的单个属性值，使用 PUT 方法并包含其他所有 Product 属性的值也不会太麻烦。

并不是所有的数据类型都很容易处理，这可能是因为它们定义了太多属性，也可能是因为客户端只接收了所选属性的值。解决方案是使用 Patch 请求，只向 Web 服务发送更改，而不是完整地替换对象。

## 20.3.1 理解 JSON Patch

ASP.NET Core 支持使用 JSON Patch 标准，该标准允许以统一的方式指定更改。JSON Patch 标准允许描述一组复杂的更改，但本章只关注更改属性值的能力。

这里不打算深入讲解 JSON Patch 标准的细节，其内容参见 https://tools.ietf.org/html/ rfc6902，但是客户端会在它的 HTTP Patch 请求中发送 Web 服务 JSON 数据，如下所示：

```
[
{ "op": "replace", "path": "Name", "value": "Surf Co"},
{ "op": "replace", "path": "City", "value": "Los Angeles"},
]
```

JSON Patch 文档表示为操作数组。每个操作都有一个 op 属性(指定操作的类型)和一个 path 属性(指定操作应用于何处)。

对于这个示例应用程序以及其他大多数应用程序，只需要替换操作，该操作用于更改属性的值。这个 JSON Patch 文档为 Name 和 City 属性设置新值。不会修改 JSON Patch 文档中没有提到的 Supplier 类定义的属性。

### 20.3.2 安装和配置 JSON Patch 包

用空模板创建项目时，不会安装对 JSON Patch 的支持。要安装 JSON Patch 包，打开一个新的 PowerShell 命令提示符，导航到包含 WebApp.csproj 文件的文件夹，并运行代码清单 20-6 所示的命令。如果使用的是 Visual Studio，可以通过选择 Project | Manage NuGet Packages 来安装这个包。

**代码清单 20-6 安装 JSON Patch 包**

```
dotnet add package Microsoft.AspNetCore.Mvc.NewtonsoftJson --version 3.1.1
```

JSON Patch 的微软实现依赖第三方 Newtonsoft JSON.NET 序列化器，在 ASP.NET Core 2.x 中使用，但已被 ASP.NET Core 3.x 中定制的 JSON 序列化器所取代。将代码清单 20-7 中所示的语句添加到 Startup 类的 ConfigureServices 方法中，以启用旧的序列化器。

**代码清单 20-7 在 WebApp 文件夹的 Startup.cs 文件中支持 JSON.NET 序列化器**

```
...
public void ConfigureServices(IServiceCollection services) {
    services.AddDbContext<DataContext>(opts => {
        opts.UseSqlServer(Configuration[
            "ConnectionStrings:ProductConnection"]);
        opts.EnableSensitiveDataLogging(true);
    });

    services.AddControllers().AddNewtonsoftJson();

    services.Configure<MvcNewtonsoftJsonOptions>(opts => {
        opts.SerializerSettings.NullValueHandling
            = Newtonsoft.Json.NullValueHandling.Ignore;
    });

    //services.Configure<JsonOptions>(opts => {
    //    opts.JsonSerializerOptions.IgnoreNullValues = true;
    //});
}
...
```

AddNewtonsoftJson 方法启用 JSON.NET 序列化器，它替代标准的 ASP. NET Core 序列化器。JSON.NET 序列化器有自己的配置类 MvcNewtonsoftJsonOptions，它是通过 options 模式应用的。代码清单 20-7 设置了 NullValueHandling 值，该值告诉序列化器，丢弃具有空值的属性。

---

■ **提示：**查看 https://www.newtonsoft.com/json，了解 JSON.NET 序列化器可用的其他配置选项的详细信息。

---

### 20.3.3　定义操作方法

要添加对 Patch 方法的支持，请将代码清单 20-8 所示的操作方法添加到 SuppliersController 类中。

**代码清单 20-8　在 Controller 文件夹的 SuppliersController.cs 文件中添加一个操作**

```
using Microsoft.AspNetCore.Mvc;
using WebApp.Models;
using System.Threading.Tasks;
using Microsoft.EntityFrameworkCore;
using Microsoft.AspNetCore.JsonPatch;

namespace WebApp.Controllers {

    [ApiController]
    [Route("api/[controller]")]
    public class SuppliersController : ControllerBase {
        private DataContext context;

        public SuppliersController(DataContext ctx) {
            context = ctx;
        }

        [HttpGet("{id}")]
        public async Task<Supplier> GetSupplier(long id) {
            Supplier supplier = await context.Suppliers.Include(s => s.Products)
                .FirstAsync(s => s.SupplierId == id);
            foreach (Product p in supplier.Products) {
                p.Supplier = null;
            };
            return supplier;
        }

        [HttpPatch("{id}")]
        public async Task<Supplier> PatchSupplier(long id,
                JsonPatchDocument<Supplier> patchDoc) {
            Supplier s = await context.Suppliers.FindAsync(id);
            if (s != null) {
                patchDoc.ApplyTo(s);
                await context.SaveChangesAsync();
            }
            return s;
        }
    }
}
```

操作方法用HttpPatch属性修饰，这表示它处理HTTP请求。模型绑定特性用于通过JsonPatchDocument<T>方法参数来处理JSON Patch文档。JsonPatchDocument<T>类定义了一个ApplyTo方法，将每个操作应用于一个对象。代码清单20-8中的操作方法从数据库中检索Supplier对象，应用JSON Patch，并存储修改后的对象。

重启 ASP.NET Core，使用 PowerShell 命令提示符运行代码清单 20-9 所示的命令，该命令发送一个带有 JSON Patch 文档的 HTTP Patch 请求，将 City 属性的值更改为 Los Angeles。

**代码清单 20-9 发送 HTTP Patch 请求**

```
    Invoke-RestMethod http://localhost:5000/api/suppliers/1 -Method Patch -ContentType
"application/json"
    -Body '[{"op":"replace","path":"City","value":"Los Angeles"}]'
```

PatchSupplier 操作方法返回修改后的 Supplier 对象作为结果，该对象被序列化并在 HTTP 响应中发送到客户端。要查看更改的效果，还可使用 Web 浏览器请求 http://localhost:5000/ suppliers/1，生成如图 20-5 所示的响应。

图 20-5 使用 Patch 请求进行更新

# 20.4 理解内容的格式化

到目前为止，Web 服务示例生成 JSON 结果，但这并不是操作方法能够生成的唯一数据格式。为操作结果选择的内容格式取决于四个因素：客户端接受的格式、应用程序可以生成的格式、操作方法指定的内容策略以及操作方法返回的类型。所有事物如何相互配合可能令人生畏，但好消息是，默认策略很适合大多数应用程序，需要更改或得不到期望格式的结果时，只需要理解幕后发生了什么。

## 20.4.1 理解默认的内容策略

了解内容格式化的最好方法是了解当客户端和操作方法都没有对可使用的格式施加任何限制时会发生什么。在这种情况下，结果是简单、可预测的。

(1) 如果操作方法返回一个字符串，则把未修改的字符串发送给客户端，响应的 Content-Type 头设置为文本/纯文本。

(2) 对于所有其他数据类型，包括 int 等其他简单类型，数据格式化为 JSON，响应的 Content-Type 头设置为 application/json。

字符串得到特殊处理，因为当它们编码为 JSON 时，会导致问题。对其他简单类型进行编码时，比如 C# int 值 2，结果是一个带引号的字符串，比如" 2"。对一个字符串进行编码时，会得到两组引号，这样"Hello"就变成"Hello"。并不是所有客户端都能很好地处理这种双重编码，所以使用文本/纯文本格式并完全回避这个问题更可靠。这不是什么问题，因为很少有应用程序发送字符串值；以 JSON 格式发送对象更常见。要查看默认策略，将一个名为 ContentController.cs 的类文件添加到 WebApps/Controllers 文件夹中，如代码清单 20-10 所示。

**代码清单 20-10　Controllers 文件夹中 ContentController.cs 文件的内容**

```
using Microsoft.AspNetCore.Mvc;
using Microsoft.EntityFrameworkCore;
using System.Threading.Tasks;
using WebApp.Models;

namespace WebApp.Controllers {

    [ApiController]
    [Route("/api/[controller]")]
    public class ContentController : ControllerBase {
        private DataContext context;

        public ContentController(DataContext dataContext) {
            context = dataContext;
        }

        [HttpGet("string")]
        public string GetString() => "This is a string response";

        [HttpGet("object")]
        public async Task<Product> GetObject() {
            return await context.Products.FirstAsync();
        }
    }
}
```

控制器定义返回字符串和对象结果的操作。重启 ASP.NET Core，使用 PowerShell 提示符运行代码清单 20-11 所示的命令；这个命令发送一个调用 GetString 操作方法的请求，该方法返回一个字符串。

**代码清单 20-11　请求字符串响应**

```
    Invoke-WebRequest http://localhost:5000/api/content/string | select @{n='Content-Type';e={
$_.Headers.
    "Content-Type" }}, Content
```

该命令向//api/content/string URL 发送 GET 请求，并处理响应，以显示 Content-Type 头和响应中的内容。该命令生成以下输出，显示响应的 Content-Type 头。

```
Content-Type                Content
------------                -------
text/plain; charset=utf-8 This is a string response
```

接下来，运行代码清单 20-12 所示的命令，该命令发送一个由 GetObject 操作方法处理的请求。

**代码清单 20-12 请求对象响应**

```
Invoke-WebRequest http://localhost:5000/api/content/object | select @{n='Content-Type';e={
$_.Headers."Content-Type" }}, Content
```

这个命令生成如下输出，为清晰起见，它显示编码为 JSON 的响应：

```
Content-Type                     Content
------------                     -------
application/json; charset=utf-8 {"productId":1,"name":"Kayak",
                                  "price":275.00,"categoryId":1,"supplierId":1}
```

## 20.4.2 理解内容协商

大多数客户端在请求中包含一个 Accept 标头，指定了它们愿意在响应中接收的一组格式，表示为一组 MIME 类型。这里是谷歌 Chrome 在请求中发送的 Accept 标头：

```
Accept: text/html,application/xhtml+xml,application/xml;q=0.9,image/webp,image/apng,
        */*;q=0.8
```

这个头表示 Chrome 可以处理 HTML 和 XHTML 格式(XHTML 是一种兼容 XML 的 HTML 方言)、XML 和 WEBP 图像格式(即动态的 PNG 图像格式)。

标题中的 q 值指定相对首选项，默认值为 1.0。为 application/xml 指定一个 0.9 的 q 值告诉服务器，Chrome 接受 XML 数据，但更喜欢处理 HTML 或 XHTML。*/*项告诉服务器，Chrome 接受任何格式，但 q 值指定它是指定类型的最低首选项。把这些放在一起意味着，Chrome 发送的 Accept 标头给服务器提供以下信息：

(1) Chrome 更喜欢接收 HTML 或 XHTML 数据或 WEBP 和 APNG 图像。

(2) 如果这些格式不可用，那么接下来最受欢迎的格式是 XML。

(3) 如果没有首选格式可用，那么 Chrome 接受任何格式。

通过设置 Accept 标头，可以更改 ASP.NET Core 应用程序生成的格式。但它不是那样工作的——或者，它不是那样工作只是因为需要一些准备工作。

要查看更改 Accept 标头时会发生什么，使用 PowerShell 提示符运行代码清单 20-13 所示的命令，该命令设置 Accept 标头，告诉 ASP.NET Core，客户端只愿意接受 XML 数据。

**代码清单 20-13 请求 XML 数据**

```
Invoke-WebRequest http://localhost:5000/api/content/object -Headers @{Accept="application/xml"}
| select
@{n='Content-Type';e={ $_.Headers."Content-Type" }}, Content
```

下面是结果，显示应用程序发送了一个 application/json 响应：

```
Content-Type                     Content
------------                     -------
application/json; charset=utf-8 {"productId":1,"name":"Kayak",
                                  "price":275.00,"categoryId":1,"supplierId":1}
```

即使 ASP.NET Core 应用程序发送给客户端一个它没有指定的格式，包含 Accept 标头对格式也没有影响。问题是，默认情况下，MVC 框架配置为只使用 JSON。MVC 框架没有返回错误，而是发送 JSON 数据，希望客户端能够处理它，即使它不是请求 Accept 标头指定的格式之一。

#### 1. 启用 XML 格式

要使内容协商生效，必须对应用程序进行配置，以便可以选择使用的格式。JSON 已经成为 Web 应用程序的默认格式，但是 MVC 框架也可以支持将数据编码为 XML，如代码清单 20-14 所示。

**提示：**可通过从 Microsoft.AspNetCore.Mvc.Formatters.Outputformat 类派生来创建自己的内容格式。这种方法很少使用，因为创建自定义数据格式并不是在应用程序中公开数据的有用方法，而且已经实现了最常见的格式——JSON 和 XML。

**代码清单 20-14　在 WebApp 文件夹的 Startup.cs 文件中启用 XML 格式**

```
...
public void ConfigureServices(IServiceCollection services) {
    services.AddDbContext<DataContext>(opts => {
        opts.UseSqlServer(Configuration[
           "ConnectionStrings:ProductConnection"]);
        opts.EnableSensitiveDataLogging(true);
    });

    services.AddControllers().AddNewtonsoftJson().AddXmlSerializerFormatters();
    services.Configure<MvcNewtonsoftJsonOptions>(opts => {
        opts.SerializerSettings.NullValueHandling
           = Newtonsoft.Json.NullValueHandling.Ignore;
    });
}
...
```

XML 序列化器有一些限制，包括不能处理 Entity Framework Core 导航属性，因为它们是通过接口定义的。要创建一个可序列化的对象，代码清单 20-15 使用第 19 章定义的 ProductBindingTarget。

**代码清单 20-15　在 Controllers 文件夹的 ContentController.cs 文件中创建一个可序列化的对象**

```
using Microsoft.AspNetCore.Mvc;
using Microsoft.EntityFrameworkCore;
using System.Threading.Tasks;
using WebApp.Models;

namespace WebApp.Controllers {

    [ApiController]
    [Route("/api/[controller]")]
    public class ContentController : ControllerBase {
        private DataContext context;

        public ContentController(DataContext dataContext) {
            context = dataContext;
        }

        [HttpGet("string")]
        public string GetString() => "This is a string response";

        [HttpGet("object")]
        public async Task<ProductBindingTarget> GetObject() {
            Product p = await context.Products.FirstAsync();
            return new ProductBindingTarget() {
                Name = p.Name, Price = p.Price, CategoryId = p.CategoryId,
                SupplierId = p.SupplierId
            };
        }
    }
}
```

当 MVC 框架只有可用的 JSON 格式时，它别无选择，只能将响应编码为 JSON。现在有了选择，可以看到内容协商过程工作得更充分。重启 ASP.NET Core MVC，再次运行代码清单 20-13 中的命令，来请求 XML 数据，输出如下(为简洁起见，省略了名称空间属性)：

```
Content-Type                    Content
------------                    -------
application/xml; charset=utf-8 <ProductBindingTarget>
                                  <Name>Kayak</Name>
                                  <Price>275.00</Price>
                                  <CategoryId>1</CategoryId>
                                  <SupplierId>1</SupplierId>
                                </ProductBindingTarget>
```

### 2. 完全尊重 Accept 标头

如果 Accept 标头包含表示任何格式的*/*，MVC 框架将始终使用 JSON 格式，即使存在其他具有较高优先级的受支持格式。这是一个奇怪的特性，它旨在一致地处理来自浏览器的请求，尽管它可能造成混淆。运行代码清单 20-16 所示的命令，发送一个带有 Accept 标头的请求，该头请求 XML，但如果 XML 不可用，则接受任何其他格式。

**代码清单 20-16　请求带有回退的 XML 响应**

```
Invoke-WebRequest http://localhost:5000/api/content/object -Headers
@{Accept="application/xml,*/*;q=0.8"} |
    select @{n='Content-Type';e={ $_.Headers."Content-Type" }}, Content
```

即使 Accept 标头告诉 MVC 框架，客户端更喜欢 XML，但是出现*/*回退意味着发送了一个 JSON 响应。一个相关的问题是，当客户端请求一种 MVC 框架尚未配置生成的格式时，将发送一个 JSON 响应，可通过运行代码清单 20-17 中所示的命令看到这一点。

**代码清单 20-17　请求 PNG 响应**

```
Invoke-WebRequest http://localhost:5000/api/content/object -Headers @{Accept="img/png"}
| select @{n='Content-
    Type';e={ $_.Headers."Content-Type" }}, Content
```

代码清单 20-16 和代码清单 20-17 中的命令都会生成以下响应：

```
Content-Type                      Content
------------                      -------
application/json; charset=utf-8 {"name":"Kayak","price":275.00,
                                  "categoryId":1,"supplierId":1}
```

这两种情况下，MVC 框架都返回 JSON 数据，这可能不是客户端所期望的。两个配置设置用来告诉 MVC 框架，尊重客户端发送的 Accept 设置，默认情况下不发送 JSON 数据。要更改配置，向 Startup 类添加代码清单 20-18 所示的语句。

**代码清单 20-18　在 WebApp 文件夹的 Startup.cs 文件中配置内容协商**

```
...
public void ConfigureServices(IServiceCollection services) {
    services.AddDbContext<DataContext>(opts => {
        opts.UseSqlServer(Configuration[
            "ConnectionStrings:ProductConnection"]);
        opts.EnableSensitiveDataLogging(true);
    });

    services.AddControllers().AddNewtonsoftJson().AddXmlSerializerFormatters();

    services.Configure<MvcNewtonsoftJsonOptions>(opts => {
        opts.SerializerSettings.NullValueHandling
```

```
                = Newtonsoft.Json.NullValueHandling.Ignore;
        });

        services.Configure<MvcOptions>(opts => {
            opts.RespectBrowserAcceptHeader = true;
            opts.ReturnHttpNotAcceptable = true;
        });
    }
    ...
```

options 模式用于设置 MvcOptions 对象的属性。将 RespectBrowserAcceptHeader 设置为 true，将禁止在 Accept 标头包含*/*时回退到 JSON。将 ReturnHttpNotAcceptable 设置为 true，将禁止在客户端请求不支持的数据格式时回退到 JSON。

重启 ASP.NETCore，重复代码清单 20-16 中的命令。采用 Accept 标头指定的格式首选项，而不是 JSON 响应，并发送 XML 响应。重复代码清单 20-17 中的命令，收到一个状态码为 406 的响应。

```
...
Invoke-WebRequest : The remote server returned an error: (406) Not Acceptable.
...
```

发送 406 代码表明，客户端可处理的格式和 MVC 框架可生成的格式之间没有重叠，从而确保客户端不会接收到它不能处理的数据格式。

### 20.4.3 指定操作结果格式

对于 MVC 框架可用于“操作方法结果”的数据格式，可以使用 Produces 属性进行约束，如代码清单 20-19 所示。

**提示：** Produces 属性是一个过滤器的例子，它允许属性改变请求和响应。更多细节见第 30 章。

**代码清单 20-19 在 Controllers 文件夹的 ContentController.cs 文件中指定数据格式**

```
using Microsoft.AspNetCore.Mvc;
using Microsoft.EntityFrameworkCore;
using System.Threading.Tasks;
using WebApp.Models;

namespace WebApp.Controllers {

    [ApiController]
    [Route("/api/[controller]")]
    public class ContentController : ControllerBase {
        private DataContext context;

        public ContentController(DataContext dataContext) {
            context = dataContext;
```

```
        }

        [HttpGet("string")]
        public string GetString() => "This is a string response";

        [HttpGet("object")]
        [Produces("application/json")]
        public async Task<ProductBindingTarget> GetObject() {
            Product p = await context.Products.FirstAsync();
            return new ProductBindingTarget() {
                Name = p.Name, Price = p.Price, CategoryId = p.CategoryId,
                SupplierId = p.SupplierId
            };
        }
    }
}
```

属性的参数指定用于操作结果的格式，可以指定多种类型。Produces 属性限制了 MVC 框架在处理 Accept 标头时考虑的类型。要查看 Produces 属性的效果，使用 PowerShell 提示符运行代码清单 20-20 所示的命令。

**代码清单 20-20　请求数据**

```
Invoke-WebRequest http://localhost:5000/api/content/object -Headers
@{Accept="application/xml,application/json;q=0.8"} | select @{n='Content-Type';
e={ $_.Headers."Content-Type" }}, Content
```

Accept 标头告诉 MVC 框架，客户端更喜欢 XML 数据，但接受 JSON。Produces 属性意味着 XML 数据不能作为 GetObject 操作方法的数据格式，因此选择 JSON 序列化器，它生成以下响应：

```
Content-Type                    Content
------------                    -------
application/json; charset=utf-8 {"name":"Kayak","price":275.00,
                                 "categoryId":1,"supplierId":1}
```

## 20.4.4　在 URL 中请求格式

Accept 标头并不总是在编写客户端的程序员的控制之下。这种情况下，可考虑允许使用 URL 来请求响应的数据格式。通过使用 FormatFilter 属性修饰操作方法并确保操作方法的路由中有一个 format 段变量，来启用该特性，如代码清单 20-21 所示。

**代码清单 20-21　在 Controllers 文件夹的 ContentController.cs 文件中启用格式化**

```
using Microsoft.AspNetCore.Mvc;
using Microsoft.EntityFrameworkCore;
using System.Threading.Tasks;
using WebApp.Models;
```

```
namespace WebApp.Controllers {

    [ApiController]
    [Route("/api/[controller]")]
    public class ContentController : ControllerBase {
        private DataContext context;

        public ContentController(DataContext dataContext) {
            context = dataContext;
        }

        [HttpGet("string")]
        public string GetString() => "This is a string response";

        [HttpGet("object/{format?}")]
        [FormatFilter]
        [Produces("application/json", "application/xml")]
        public async Task<ProductBindingTarget> GetObject() {
            Product p = await context.Products.FirstAsync();
            return new ProductBindingTarget() {
                Name = p.Name, Price = p.Price, CategoryId = p.CategoryId,
                SupplierId = p.SupplierId
            };
        }
    }
}
```

FormatFilter 属性是一个过滤器的例子，是一个可以修改请求和响应的属性，如第 30 章所述。此过滤器从匹配请求的路由中获取 format 段变量的值，并使用它来覆盖客户端发送的 Accept 标头。还扩展了 Produces 属性指定的类型范围，这样操作方法就可同时返回 JSON 和 XML 响应。

应用程序支持的每种数据格式都有一个简写：XML 表示 XML 数据，JSON 表示 JSON 数据。当操作方法的目标是包含这些简写名称之一的 URL 时，将忽略 Accept 标头，并使用指定的格式。要查看效果，请重新启动 ASP.NET Core，使用浏览器请求 http://localhost:5000/api/content/object/json 和 http://localhost:5000/api/content/object/xml，生成如图 20-6 所示的响应。

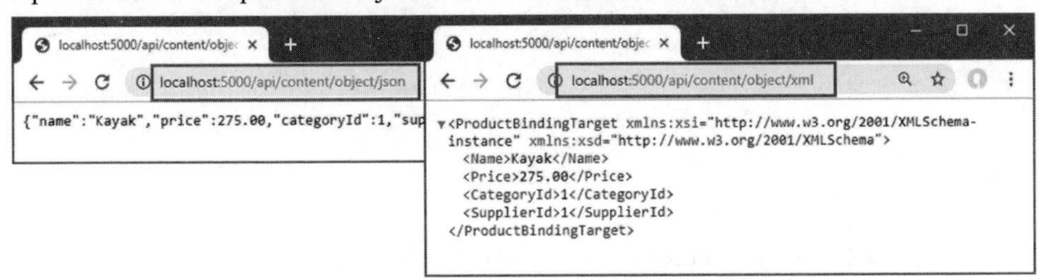


图 20-6　在 URL 中请求数据格式

### 20.4.5　限制操作方法接收的格式

大多数内容格式化决策都关注 ASP.NET Core 应用程序发送给客户端的数据格式，但使用处理结果的序列化器来反序列化客户端在请求体中发送的数据。反序列化过程是自动进行的，大多数应用程序都乐于接受它们配置的所有格式的数据。示例应用程序配置为发送 JSON 和 XML 数据，这意味着客户端可在请求中发送 JSON 和 XML 数据。

可以将 Consumes 属性应用于操作方法，以限制要处理的数据类型，如代码清单 20-22 所示。

**代码清单 20-22　在 Controllers 文件夹的 ContentController.cs 文件中添加操作方法**

```
using Microsoft.AspNetCore.Mvc;
using Microsoft.EntityFrameworkCore;
using System.Threading.Tasks;
using WebApp.Models;

namespace WebApp.Controllers {

    [ApiController]
    [Route("/api/[controller]")]
    public class ContentController : ControllerBase {
        private DataContext context;

        public ContentController(DataContext dataContext) {
            context = dataContext;
        }

        [HttpGet("string")]
        public string GetString() => "This is a string response";

        [HttpGet("object/{format?}")]
        [FormatFilter]
        [Produces("application/json", "application/xml")]
        public async Task<ProductBindingTarget> GetObject() {
            Product p = await context.Products.FirstAsync();
            return new ProductBindingTarget() {
                Name = p.Name, Price = p.Price, CategoryId = p.CategoryId,
                SupplierId = p.SupplierId
            };
        }

        [HttpPost]
        [Consumes("application/json")]
        public string SaveProductJson(ProductBindingTarget product) {
            return $"JSON: {product.Name}";
        }

        [HttpPost]
```

```
        [Consumes("application/xml")]
        public string SaveProductXml(ProductBindingTarget product) {
            return $"XML: {product.Name}";
        }
    }
}
```

新的操作方法用 Consumes 属性进行了修饰，限制了每个操作方法可以处理的数据类型。这些属性的组合意味着 Content-Type 头为 application/json 的 HTTP POST 属性由 SaveProductJson 操作方法处理。Content-Type 头为 application/xml 的 HTTP POST 请求由 SaveProductXml 操作方法处理。重启 ASP.NET Core，使用 PowerShell 命令提示符运行代码清单 20-23 所示的命令，将 JSON 数据发送给示例应用程序。

**代码清单 20-23　发送 JSON 数据**

```
Invoke-RestMethod http://localhost:5000/api/content -Method POST -Body (@{ Name="Swimming
Goggles";
Price=12.75; CategoryId=1; SupplierId=1} | ConvertTo-Json) -ContentType "application/json"
```

请求将自动路由到正确的操作方法，生成以下响应：

```
JSON: Swimming Goggles
```

运行代码清单 20-24 所示的命令，将 XML 数据发送到示例应用程序。

**代码清单 20-24　发送 XML 数据**

```
Invoke-RestMethod http://localhost:5000/api/content -Method POST -Body
"<ProductBindingTarget><Name>Kayak</Name><Price>275.00</Price><CategoryId>1</CategoryId>
<SupplierId>1</SupplierId></ProductBindingTarget>"-ContentType "application/xml"
```

请求路由到 SaveProductXml 操作方法，生成以下响应：

```
XML: Kayak
```

如果所发送请求的Content-Type头与应用程序支持的数据类型不匹配，MVC框架将发送415-Unsupported Media Type响应。

## 20.5　记录和探索 Web 服务

负责开发 Web 服务及其客户端时，每个操作的目的及结果是显而易见的，通常是同时编写的。如果你负责供第三方开发人员使用的 Web 服务，就可能需要提供描述 Web 服务如何工作的文档。OpenAPI 规范也称为 Swagger，以其他程序员可理解并通过编程方式使用的方式描述 Web 服务。本节将演示如何使用 OpenAPI 描述 Web 服务，并展示如何调整描述。

### 20.5.1　解决操作冲突

OpenAPI 发现过程要求对每个操作方法使用唯一的 HTTP 方法和 URL 模式组合。该进程不支持 Consumes 属性，因此需要对 ContentController 进行更改，以删除接收 XML 和 JSON 数据的单独操作，如代码清单 20-25 所示。

**代码清单 20-25　删除 Controllers 文件夹的 ContentController.cs 文件中的操作**

```
using Microsoft.AspNetCore.Mvc;
using Microsoft.EntityFrameworkCore;
using System.Threading.Tasks;
using WebApp.Models;

namespace WebApp.Controllers {

    [ApiController]
    [Route("/api/[controller]")]
    public class ContentController : ControllerBase {
        private DataContext context;

        public ContentController(DataContext dataContext) {
            context = dataContext;
        }

        [HttpGet("string")]
        public string GetString() => "This is a string response";

        [HttpGet("object/{format?}")]
        [FormatFilter]
        [Produces("application/json", "application/xml")]
        public async Task<ProductBindingTarget> GetObject() {
            Product p = await context.Products.FirstAsync();
            return new ProductBindingTarget() {
                Name = p.Name, Price = p.Price, CategoryId = p.CategoryId,
                SupplierId = p.SupplierId
            };
        }

        [HttpPost]
        [Consumes("application/json")]
        public string SaveProductJson(ProductBindingTarget product) {
            return $"JSON: {product.Name}";
        }

        //[HttpPost]
        //[Consumes("application/xml")]
        //public string SaveProductXml(ProductBindingTarget product) {
```

```
            // return $"XML: {product.Name}";
            //}
        }
    }
```

注释掉一个操作方法可确保每个剩余的操作都具有唯一的 HTTP 方法和 URL 组合。

### 20.5.2 安装和配置 Swashbuckle 包

Swashbuckle 包是最流行的 OpenAPI 规范的 ASP.NET Core 实现，并自动为 ASP.NET Core 应用程序中的 Web 服务生成描述。还包括使用该描述对 Web 服务进行检查和测试的工具。

打开一个新的 PowerShell 命令提示符，导航到包含 WebApp.csproj 文件的文件夹，并运行代码清单 20-26 中所示的命令来安装 NuGet 包。如果正在使用 Visual Studio，可以选择 Project | Manage Nuget Packages，通过 Visual Studio 包的用户界面安装这个包。

**代码清单 20-26 向项目添加包**

```
dotnet add package Swashbuckle.AspNetCore --version 5.0.0-rc2
```

将代码清单 20-27 中所示的语句添加到 Startup 类，以添加 Swashbuckle 包提供的服务和中间件。

**代码清单 20-27 在 WebApp 文件夹的 Startup.cs 文件中配置 Swashbuckle**

```
using System;
using System.Collections.Generic;
using System.Linq;
using System.Threading.Tasks;
using Microsoft.AspNetCore.Builder;
using Microsoft.AspNetCore.Hosting;
using Microsoft.AspNetCore.Http;
using Microsoft.Extensions.DependencyInjection;
using Microsoft.Extensions.Hosting;
using Microsoft.Extensions.Configuration;
using Microsoft.EntityFrameworkCore;
using WebApp.Models;
using Microsoft.AspNetCore.Mvc;
using Microsoft.OpenApi.Models;

namespace WebApp {
    public class Startup {

        public Startup(IConfiguration config) {
            Configuration = config;
        }

        public IConfiguration Configuration { get; set; }
        public void ConfigureServices(IServiceCollection services) {
```

```
            services.AddDbContext<DataContext>(opts => {
                opts.UseSqlServer(Configuration[
                    "ConnectionStrings:ProductConnection"]);
                opts.EnableSensitiveDataLogging(true);
            });

            services.AddControllers()
                .AddNewtonsoftJson().AddXmlSerializerFormatters();

            services.Configure<MvcNewtonsoftJsonOptions>(opts => {
                opts.SerializerSettings.NullValueHandling
                    = Newtonsoft.Json.NullValueHandling.Ignore;
            });

            services.Configure<MvcOptions>(opts => {
                opts.RespectBrowserAcceptHeader = true;
                opts.ReturnHttpNotAcceptable = true;
            });

            services.AddSwaggerGen(options => {
                options.SwaggerDoc("v1",
                    new OpenApiInfo { Title = "WebApp", Version = "v1" });
            });
        }

        public void Configure(IApplicationBuilder app, DataContext context) {
            app.UseDeveloperExceptionPage();
            app.UseRouting();
            app.UseMiddleware<TestMiddleware>();
            app.UseEndpoints(endpoints => {
                endpoints.MapGet("/", async context => {
                    await context.Response.WriteAsync("Hello World!");
                });;
                endpoints.MapControllers();
            });
            app.UseSwagger();
            app.UseSwaggerUI(options => {
               options.SwaggerEndpoint("/swagger/v1/swagger.json", "WebApp");
            });
            SeedData.SeedDatabase(context);
        }
    }
}
```

代码清单20-27中的语句设置了两个特性。该特性生成应用程序包含的Web服务的OpenAPI描述。要查看描述，可以重新启动ASP.NET Core，使用浏览器请求URL http://localhost:5000/swagger/v1/swagger，生成如图20-7所示的响应。OpenAPI格式冗长，但是可以看到Web服务控制器支持的每个 URL，以及每个控制器期望接收的数据的详细信息和它生成的响应范围。

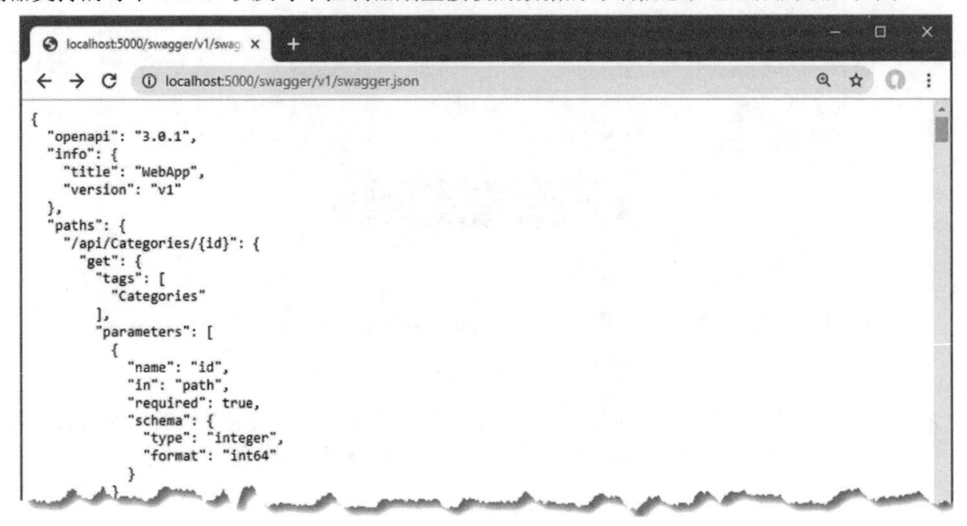


图 20-7　Web 服务的 OpenAPI 描述

第二个特性是 UI，它使用 Web 服务的 OpenAPI 描述，以更容易理解的方式显示信息显示对每个操作的测试支持。使用浏览器请求 http://localhost:5000/swagger，显示如图 20-8 所示的界面。可以展开每个操作以查看详细信息，包括请求中预期的数据和客户端预期的不同响应。

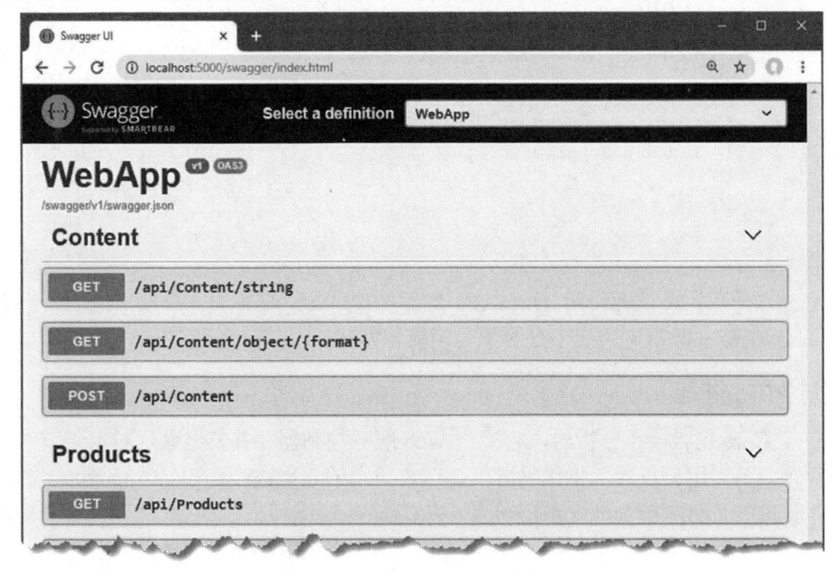


图 20-8　OpenAPI explorer 接口

## 20.5.3　微调 API 描述

依赖 API 发现过程可以生成不能真正捕获 Web 服务的结果。为此，可检查 Products 部分中的条目，该条目描述了由/api/Product/{id} URL 模式匹配的 GET 请求。展开此项并检查 Responses 部分，就会看到只返回一个状态代码响应，如图 20-9 所示。

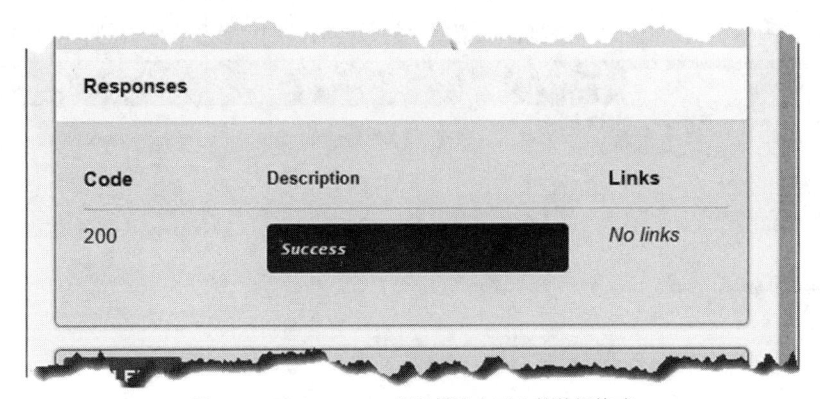


图 20-9　OpenAPI Web 服务描述中列出的数据格式

API 发现过程对操作方法产生的响应做出假设，但并不总是反映可能发生的事情。这种情况下，ProductController 类中的 GetProduct 操作方法可返回发现过程未检测到的另一个响应。

```
...
[HttpGet("{id}")]
public async Task<IActionResult> GetProduct(long id) {
    Product p = await context.Products.FindAsync(id);
    if (p == null) {
       return NotFound();
    }
    return Ok(new {
        ProductId = p.ProductId, Name = p.Name,
        Price = p.Price, CategoryId = p.CategoryId,
        SupplierId = p.SupplierId
    });
}
...
```

如果第三方开发人员试图使用 OpenAPI 数据实现 Web 服务的客户端，他们不会期望看到 404-Not Found 响应(在数据库中找不到对象时由操作发送)。

### 1. 运行 API 分析器

ASP.NET Core 包含一个分析器，它检查 Web 服务控制器，并突出显示与上一节中描述的问题类似的问题。要启用分析器，将代码清单 20-28 所示的元素添加到 WebApp.cspoj 文件中(如果使用的是 Visual Studio，在 Solution Explorer 中右击 WebApp 项目，然后从弹出菜单中选择 Edit Project File)。

**代码清单 20-28　在 WebApp 文件夹的 WebApp.csproj 文件中启用分析器**

```
<Project Sdk="Microsoft.NET.Sdk.Web">

  <PropertyGroup>
    <TargetFramework>netcoreapp3.1</TargetFramework>
  </PropertyGroup>
```

```
  <ItemGroup>
    <PackageReference Include="Microsoft.AspNetCore.Mvc.NewtonsoftJson"
        Version="3.1.1" />
    <PackageReference Include="Microsoft.EntityFrameworkCore.Design" Version="3.1.1">
      <IncludeAssets>runtime; build; native; contentfiles; analyzers;
           buildtransitive</IncludeAssets>
      <PrivateAssets>all</PrivateAssets>
    </PackageReference>
    <PackageReference Include="Microsoft.EntityFrameworkCore.SqlServer"
         Version="3.1.1" />
    <PackageReference Include="Swashbuckle.AspNetCore" Version="5.0.0-rc2" />
  </ItemGroup>

  <PropertyGroup>
    <IncludeOpenAPIAnalyzers>true</IncludeOpenAPIAnalyzers>
  </PropertyGroup>

</Project>
```

如果正在使用 Visual Studio，将在控制器类文件中看到 API 分析器检测到的任何问题，如图 20-10 所示。

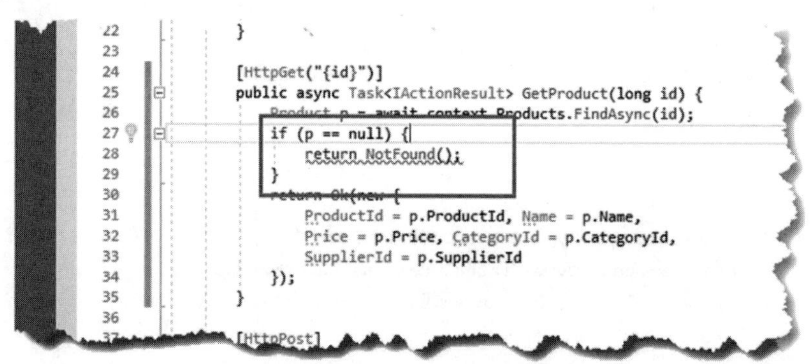

图 20-10　API 分析器检测到的问题

如果正在使用 Visual Studio Code，那么在使用 dotnet build 命令或使用 dotnet run 命令执行项目时，将看到警告消息。编译项目时，会看到这样一条消息，它描述了 ProductController 类中的问题：

```
    Controllers\ProductsController.cs(28,9): warning API1000: Action method returns
undeclared status code '404'. [C:\WebApp\WebApp.csproj]
        1 Warning(s)
        0 Error(s)
```

### 2. 声明操作方法结果类型

要修复分析器检测到的问题，可以使用 ProducesResponseType 属性声明操作方法可以生成的每个响应类型，如代码清单 20-29 所示。

**代码清单 20-29　在 Controllers 文件夹的 ProductsController.cs 文件中声明结果**

```
using Microsoft.AspNetCore.Mvc;
using WebApp.Models;
using System.Collections.Generic;
using Microsoft.Extensions.Logging;
using System.Linq;
using System.Threading.Tasks;
using Microsoft.AspNetCore.Http;

namespace WebApp.Controllers {

    [ApiController]
    [Route("api/[controller]")]
    public class ProductsController : ControllerBase {
        private DataContext context;

        public ProductsController(DataContext ctx) {
            context = ctx;
        }

        [HttpGet]
        public IAsyncEnumerable<Product> GetProducts() {
            return context.Products;
        }

        [HttpGet("{id}")]
        [ProducesResponseType(StatusCodes.Status200OK)]
        [ProducesResponseType(StatusCodes.Status404NotFound)]
        public async Task<IActionResult> GetProduct(long id) {
            Product p = await context.Products.FindAsync(id);
            if (p == null) {
                return NotFound();
            }
            return Ok(new {
                ProductId = p.ProductId, Name = p.Name,
                Price = p.Price, CategoryId = p.CategoryId,
                SupplierId = p.SupplierId
            });
        }

        // ...action methods omitted for brevity...
    }
}
```

重启 ASP.NET Core 用浏览器请求 http://localhost:5000/swagger，将看到操作方法的描述已经

更新为反映 404 响应，如图 20-11 所示。

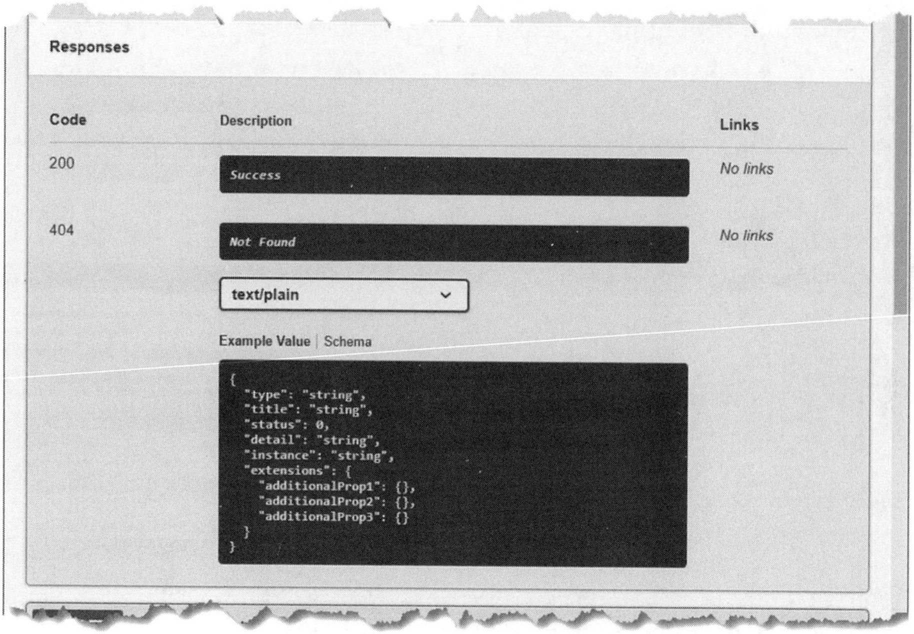

图 20-11 反映操作方法生成的所有状态代码

## 20.6 小结

本章描述了一些用于创建 Web 服务的高级特性。解释了如何在 Entity Framework Core 查询中处理相关数据，如何支持处理选择性更新的 HTTP Patch 方法，内容协商如何工作，以及如何使用 OpenAPI 描述所创建的 Web 服务。第 21 章将描述控制器如何生成 HTML 响应。

# 第 21 章

# 使用控制器和视图(第 1 部分)

本章将介绍 Razor 视图引擎，它负责生成直接显示给用户的 HTML 响应(而不是通常由其他应用程序使用的 JSON 和 XML 响应)。视图是包含 C#表达式和 HTML 片段的文件，经过视图引擎处理后，生成 HTML 响应。本章将介绍视图的工作方式，解释在操作方法中如何使用它们，并描述它们包含的不同类型的 C#表达式。第 22 章将介绍视图支持的其他功能。表 21-1 介绍了 Razor 视图的相关知识。

表 21-1　Razor 视图的相关知识

| 问题 | 答案 |
| --- | --- |
| 视图是什么? | 视图是包含静态 HTML 内容和 C#表达式的文件 |
| 视图为什么有用? | 视图用于创建 HTTP 请求的 HTML 响应。C#表达式在被计算后，其结果将与 HTML 内容合并起来，创建最终的响应 |
| 如何使用视图? | View 方法(由 Controller 类定义)创建的操作响应使用视图 |
| 使用视图时存在陷阱或者限制吗? | 可能需要一点时间才能熟悉视图文件的语法以及视图文件将代码与内容组合在一起的方式 |
| 存在替代方案吗? | 在 ASP.NET Core MVC 中可以使用一些第三方视图引擎，但对它们的使用存在一些限制 |

表 21-2 总结了本章介绍的内容。

表 21-2　本章内容摘要

| 问题 | 解决方案 | 代码清单 |
| --- | --- | --- |
| 启用视图 | 使用 AddControllersWithViews 和 MapControllerRoute 方法来设置必要的服务和端点 | 21-1～21-5 |
| 从控制器操作方法返回 HTML 响应 | 使用 View 方法创建 ViewResult | 21-6 |
| 创建动态 HTML 内容 | 创建 Razor 视图，为动态内容使用表达式 | 21-7 ～ 21-9 、21-20、21-21 |
| 按名称选择视图 | 提供视图名称作为 View 方法的实参 | 21-10、21-11 |
| 创建可被多个控制器使用的视图 | 创建一个共享视图 | 21-12～21-14 |
| 为视图指定模型类型 | 使用@model 表达式 | 21-15～21-19 |
| 有选择地生成内容 | 使用@if、@switch 或@foreach 表达式 | 21-22～21-26 |
| 在视图中包含 C#代码 | 使用代码块 | 21-27 |

## 21.1　准备工作

本章使用第 20 章创建的 WebApp 项目。为了完成准备工作，打开一个新的 PowerShell 命令提示符，在 WebApp 文件夹中运行代码清单 21-1 中的命令，以安装新包。如果使用的是 Visual Studio，则可以选择 Project | Manage NuGet Packages 来安装包。

**代码清单 21-1　在示例项目中添加包**

```
dotnet add package Microsoft.AspNetCore.Mvc.Razor.RuntimeCompilation
--version 3.1.1
```

然后，按照代码清单 21-2 所示替换 Startup 类的内容，这将移除前面章节中使用到的一些服务和中间件。

**提示：** 从 https://github.com/apress/pro-asp.net-core-3 可下载本章以及本书其他章节的示例项目。如果无法成功运行示例，请参考第 1 章的说明。

**代码清单 21-2　替换 WebApp 文件夹中 Startup.cs 文件的内容**

```
using System;
using System.Collections.Generic;
using System.Linq;
using Microsoft.AspNetCore.Builder;
using Microsoft.Extensions.DependencyInjection;
using Microsoft.Extensions.Configuration;
using Microsoft.EntityFrameworkCore;
using WebApp.Models;

namespace WebApp {
    public class Startup {

        public Startup(IConfiguration config) {
            Configuration = config;
        }

        public IConfiguration Configuration { get; set; }

        public void ConfigureServices(IServiceCollection services) {
            services.AddDbContext<DataContext>(opts => {
                opts.UseSqlServer(Configuration[
                    "ConnectionStrings:ProductConnection"]);
                opts.EnableSensitiveDataLogging(true);
            });
            services.AddControllers();
        }
```

```
        public void Configure(IApplicationBuilder app, DataContext context) {
            app.UseDeveloperExceptionPage();
            app.UseStaticFiles();
            app.UseRouting();
            app.UseEndpoints(endpoints => {
                endpoints.MapControllers();
            });
            SeedData.SeedDatabase(context);
        }
    }
}
```

### 21.1.1 删除数据库

打开一个新的 PowerShell 命令提示符，导航到 WebApp.csproj 文件所在的文件夹，然后运行代码清单 21-3 中的命令来删除数据库。

**代码清单 21-3 删除数据库**

```
dotnet ef database drop --force
```

### 21.1.2 运行示例应用程序

为运行示例应用程序，可在 Debug 菜单中选择 Start Without Debugging 或 Run Without Debugging，或者使用 PowerShell 命令提示符运行代码清单 21-4 中的命令。

**代码清单 21-4 运行示例应用程序**

```
dotnet run
```

在应用程序启动过程中，将对数据库执行 seed 操作。当 ASP.NET Core 开始运行后，使用 Web 浏览器请求 http://localhost:5000/api/products，这将得到图 21-1 中显示的响应。

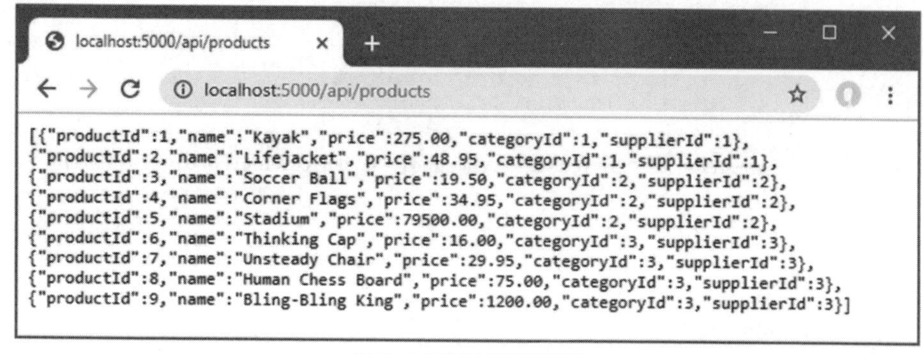


图 21-1 运行示例应用程序

# 21.2 开始使用视图

本章一开始展示了一个 Web 服务控制器，以演示其余使用视图的控制器之间的相似性。虽然很容易认为 Web 服务和视图控制器是不同的东西，但是这两种类型的响应在底层使用了相同的功能，理解这一点很重要。在接下来的小节中，将配置应用程序来支持 HTML 应用程序，并修改 Home 控制器的作用，使其生成 HTML 响应。

## 21.2.1 配置应用程序

第一步是配置 ASP.NET Core 来启用 HTML 响应，如代码清单 21-5 所示。

**代码清单 21-5　修改 WebApp 文件夹中 Startup.cs 文件的配置**

```
using System;
using System.Collections.Generic;
using System.Linq;
using Microsoft.AspNetCore.Builder;
using Microsoft.Extensions.DependencyInjection;
using Microsoft.Extensions.Configuration;
using Microsoft.EntityFrameworkCore;
using WebApp.Models;

namespace WebApp {
    public class Startup {

        public Startup(IConfiguration config) {
            Configuration = config;
        }

        public IConfiguration Configuration { get; set; }

        public void ConfigureServices(IServiceCollection services) {
            services.AddDbContext<DataContext>(opts => {
                opts.UseSqlServer(Configuration[
                    "ConnectionStrings:ProductConnection"]);
                opts.EnableSensitiveDataLogging(true);
            });
            services.AddControllersWithViews().AddRazorRuntimeCompilation();
        }

        public void Configure(IApplicationBuilder app, DataContext context) {
            app.UseDeveloperExceptionPage();
            app.UseStaticFiles();
            app.UseRouting();
            app.UseEndpoints(endpoints => {
                endpoints.MapControllers();
                endpoints.MapControllerRoute("Default",
```

```
                    "{controller=Home}/{action=Index}/{id?}");
            });
            SeedData.SeedDatabase(context);
        }
    }
}
```

HTML 响应是使用视图创建的，视图则是混合了 HTML 元素和 C#表达式的文件。第 19 章曾使用 AddControllers 方法来启用 MVC 框架，但它只支持 Web 服务控制器。为了支持视图，这里使用了 AddControllersWithViews 方法。使用 AddRazorRuntimeCompilation 方法是为了启用代码清单 21-1 安装的包所提供的功能，从而方便在开发过程中使用视图，稍后将详细介绍这方面的内容。

这里做出的第二个修改是在端点路由配置中添加了 MapControllerRoute 方法。生成 HTML 响应的控制器使用的路由特性与 Web 服务控制器不同，并且依赖约定式路由功能(下一节将进行介绍)。

### 21.2.2 创建 HTML 控制器

HTML 应用程序的控制器与 Web 服务使用的控制器类似，但存在一些重要区别。为创建一个 HTML 控制器，在 Controllers 文件夹中创建一个类文件，命名为 HomeController.cs，并添加代码 21-6 中的语句。

**代码清单 21-6 Controllers 文件夹的 HomeController.cs 文件的内容**

```
using Microsoft.AspNetCore.Mvc;
using System.Threading.Tasks;
using WebApp.Models;

namespace WebApp.Controllers {

    public class HomeController: Controller {
        private DataContext context;

        public HomeController(DataContext ctx) {
            context = ctx;
        }

        public async Task<IActionResult> Index(long id = 1) {
            return View(await context.Products.FindAsync(id));
        }
    }
}
```

HTML 控制器的基类是 Controller，它派生自 Web 服务控制器使用的 ControllerBase，并提供了一些专门用于视图的方法。

```
...
public class HomeController:Controller {
```

```
...
```

HTML 控制器中的操作方法返回的对象实现了 IActionResult 接口，第 19 章使用了相同的结果类型来返回状态码响应。Controller 基类提供了 View 方法，用于选择一个视图来创建响应。

```
...
return View(await context.Products.FindAsync(id));
...
```

**提示：** 代码清单 21-6 中的控制器没有使用特性进行装饰。ApiController 特性只用于 Web 服务控制器，不应该用于 HTML 控制器。而之所以不需要使用 Route 和 HTTP 方法特性，是因为 HTML 控制器依赖于约定式路由。代码清单 21-5 配置了约定式路由，稍后将介绍这种路由。

View 方法创建了 ViewResult 类的一个实例，该类实现了 IActionResult 接口，并告诉 MVC 框架，应该使用视图为客户端生成响应。View 方法的实参称为视图模型，它为视图提供了在生成响应时需要的数据。

目前，MVC 框架还没有视图可用，但如果重新启动 ASP.NET Core，并使用浏览器请求 http://localhost:5000，将看到一个错误消息，指出 MVC 框架响应从 Index 操作方法收到的 ViewResult，如图 21-2 所示。

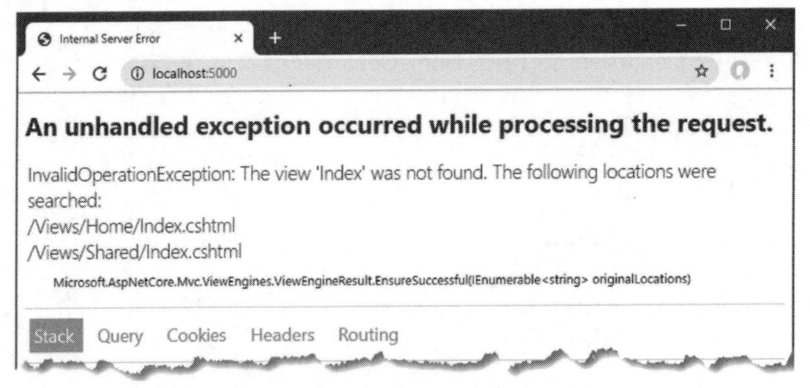


图 21-2　使用视图结果

在后台，有两个重要的约定式路由在起作用，接下来将分别进行介绍。

**注意：** 有两种功能可以增加搜索范围。第 23 章将会介绍，如果项目使用了 Razor Pages，则搜索将包括/Pages/Shared 文件夹。

### 1. 理解约定式路由

HTML 控制器依赖于约定式路由，而不是 Route 特性。“约定”指的是使用控制器类名称和操作方法名称来配置路由系统，代码清单 21-6 通过在端点路由配置中添加下面的语句来实现约定式路由：

```
...
endpoints.MapControllerRoute("Default", "{controller=Home}/{action=Index}/{id?}");
...
```

这条语句设置的路由匹配两段或三段 URL。第一个片段的值用作控制器类的名称，但不带 Controller 后缀，所以 Home 指代的是 HomeController 类。第二个片段的值是操作方法的名称，而可选的第三个片段允许操作方法收到一个名为 id 的参数。对于不包含所有片段的 URL，默认值将选择 Home 控制器的 Index 操作方法。这是一种很常用的约定，所以在设置相同的路由配置时，不需要指定 URL 模式，如代码清单 21-7 所示。

**代码清单 21-7　在 WebApp 文件夹的 Startup.cs 文件中使用默认路由约定**

```
...
public void Configure(IApplicationBuilder app, DataContext context) {
    app.UseDeveloperExceptionPage();
    app.UseStaticFiles();
    app.UseRouting();
    app.UseEndpoints(endpoints => {
        endpoints.MapControllers();
        endpoints.MapDefaultControllerRoute();
    });
    SeedData.SeedDatabase(context);
}
...
```

MapDefaultControllerRoute 方法避免了写错 URL 模式的风险，并设置了基于约定的路由。本章配置了一个路由，但应用程序可根据需要定义任意多的路由，后续章节还将扩展路由配置，使学习示例变得更加简单。

---

■ **提示：** MVC 框架认为 HTML 控制器定义的任何 public 方法是操作方法，并且操作方法支持所有 HTTP 方法。如果需要在控制器中定义一个不是操作方法的方法，则可将其声明为 private，或者如果不能将其声明为 private，则可以使用 NonAction 特性装饰该方法。通过应用特定的属性，可限制操作方法，使其只支持特定的 HTTP 方法。例如，HttpGet 特性代表操作可以处理 GET 请求，HttpPost 方法代表操作可以处理 POST 请求。

---

### 2. 理解 Razor 视图约定

调用 Home 控制器定义的 Index 操作方法时，它将使用 id 参数的值从数据库获取一个对象，然后传递给 View 方法。

```
...
public async Task<IActionResult> Index(long id = 1) {
   return View(await context.Products.FindAsync(id));
}
...
```

当操作方法调用 View 方法时，将创建一个 ViewResult，告诉 MVC 框架使用默认约定来寻找视图。Razor 视图引擎将寻找一个与操作方法同名、且带有 cshtml 文件扩展名(Razor 视图引擎使用这种文件类型)的视图。视图保存在 Views 文件夹中，按照关联的控制器进行分组。因为操作方法是由 Home 控制器定义的，所以首先搜索的位置是 Views/Home 文件夹(从控制器类名称中去掉 Controller，得到这个名称)。如果在 Views/Home 文件夹中没有找到 Index.cshtml 文件，就检查

Views/Shared 文件夹，这里包含的是控制器之间共享的视图。

虽然大部分控制器都有自己的视图，但可以把视图用作共享视图，从而避免重复创建共用的功能，如“使用共享视图”一节所述。

图 21-2 中的异常响应显示了这两种约定的结果。路由约定使用 Home 控制器定义的 Index 操作方法来处理请求，这个方法告诉 Razor 视图引擎使用视图搜索约定来寻找视图。视图引擎使用操作方法和控制器的名称来构建搜索模式，并查找 Views/Home/Index.cshtml 和 Views/Shared/Index.cshtml 文件。

## 21.2.3　创建 Razor 视图

为向 MVC 框架提供可以显示的视图，创建 Views/Home 文件夹，在其中添加一个 Index.cshtml 文件，使其包含代码清单 21-8 中的内容。如果使用的是 Visual Studio，则可以右击 Views/Home 文件夹，从弹出菜单中选择 Add | New Item，然后在 ASP.NET Core | Web 分类中选择 Razor View 来创建视图，如图 21-3 所示。

**提示：** 在弹出菜单的 Add 菜单项中，有一个用来创建视图的菜单项，但该选项依赖于 Visual Studio 的基架功能。该功能会添加模板内容，用于创建不同类型的视图。本书不依赖于基架，而将展示如何从头创建视图。

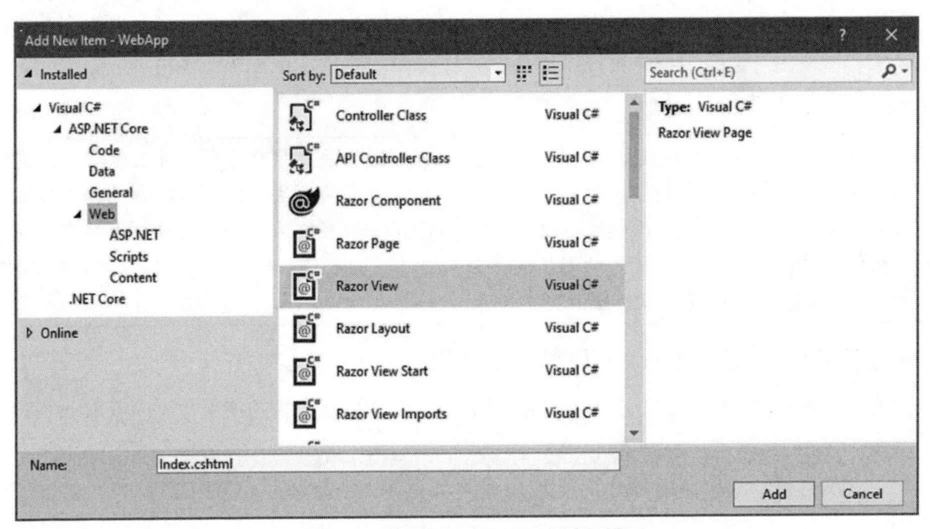


图 21-3　使用 Visual Studio 创建视图

### 代码清单 21-8　Views/Home 文件夹中 Index.cshtml 文件的内容

```
<!DOCTYPE html>
<html>
<head>
    <link href="/lib/twitter-bootstrap/css/bootstrap.min.css" rel="stylesheet" />
</head>
<body>
    <h6 class="bg-primary text-white text-center m-2 p-2">Product Table</h6>
```

```
        <div class="m-2">
            <table class="table table-sm table-striped table-bordered">
                <tbody>
                    <tr><th>Name</th><td>@Model.Name</td></tr>
                    <tr><th>Price</th><td>@Model.Price.ToString("c")</td></tr>
                </tbody>
            </table>
        </div>
    </body>
    </html>
```

视图文件包含使用 Bootstrap CSS 框架样式化的标准 HTML 元素，这些样式是通过使用 class 特性应用的。视图的关键功能是能够使用 C#表达式生成内容，如下所示：

```
...
<tr><th>Name</th><td>@Model.Name</td></tr>
<tr><th>Price</th><td>@Model.Price.ToString("c")</td></tr>
...
```

21.4 节将解释这些表达式的工作方式，现在只需要知道，这些表达式插入的是 Product 视图模型的 Name 和 Price 属性，而 Product 视图模型是由代码清单 21-6 中的操作方法传入 View 方法的。重新启动 ASP.NET Core，在浏览器中请求 http://localhost:5000，将看到如图 21-4 所示的 HTML 响应。

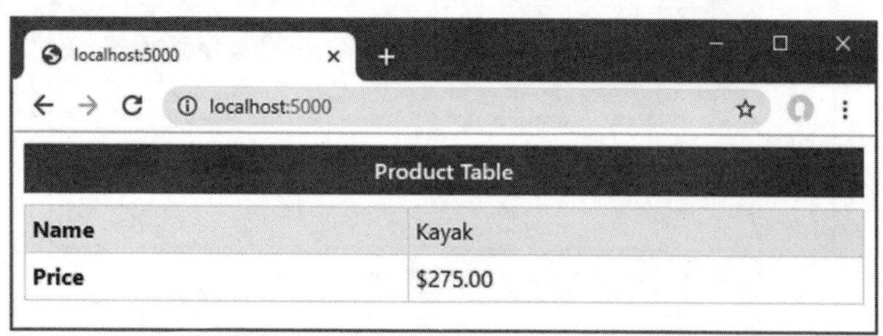


图 21-4　视图响应

## 修改 Razor 视图

在代码清单 21-1 中添加、并在代码清单 21-5 中配置的包会自动检测并重新编译 Razor 视图，这意味着并不需要重新启动 ASP.NET Core 运行时。为演示重新编译过程，代码清单 21-9 在 Index 视图中添加了新元素。

### 代码清单 21-9　在 Views/Home 文件夹的 Index.cshtml 文件中添加元素

```
<!DOCTYPE html>
<html>
<head>
    <link href="/lib/twitter-bootstrap/css/bootstrap.min.css" rel="stylesheet" />
</head>
<body>
```

```
        <h6 class="bg-primary text-white text-center m-2 p-2">Product Table</h6>
        <div class="m-2">
            <table class="table table-sm table-striped table-bordered">
                <tbody>
                    <tr><th>Name</th><td>@Model.Name</td></tr>
                    <tr><th>Price</th><td>@Model.Price.ToString("c")</td></tr>
                    <tr><th>Category ID</th><td>@Model.CategoryId</td></tr>
                </tbody>
            </table>
        </div>
    </body>
    </html>
```

在视图中保存修改，然后重新加载浏览器窗口，但不重新启动 ASP.NET Core。视图中的修改将被检测到，此时将会编译视图，所以响应会短暂停止，之后将显示图 21-5 中的响应。

**注意：**此功能只适用于视图，而不适用于项目中的 C#类。如果修改了类文件，则只有重新启动 ASP.NET Core 才能让修改生效。

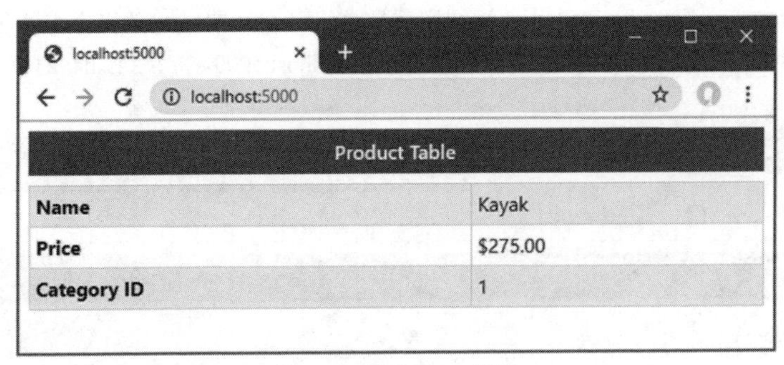


图 21-5　修改 Razor 视图

## 21.2.4　通过名称选择视图

代码清单 21-6 中的操作方法完全依赖约定，让 Razor 选择视图来生成响应。通过向 View 方法提供一个名称作为参数，操作方法可选择对应的视图，如代码清单 21-10 所示。

#### 代码清单 21-10　在 Controllers 文件夹的 HomeController.cs 文件中选择视图

```
using Microsoft.AspNetCore.Mvc;
using System.Threading.Tasks;
using WebApp.Models;

namespace WebApp.Controllers {

    public class HomeController: Controller {
        private DataContext context;
```

```
        public HomeController(DataContext ctx) {
            context = ctx;
        }

        public async Task<IActionResult> Index(long id = 1) {
            Product prod = await context.Products.FindAsync(id);
            if (prod.CategoryId == 1) {
                return View("Watersports", prod);
            } else {
                return View(prod);
            }
        }
    }
}
```

操作方法基于从数据库中检索的 Product 对象的 CategoryId 属性来选择视图。如果 CategoryId 是 1，则操作方法在调用 View 方法时将使用一个额外参数，选择名为 Watersports 的视图。

```
...
return View("Watersports", prod);
...
```

注意，操作方法并没有指定视图的文件扩展名或位置。视图引擎负责将 Watersports 翻译为一个视图文件。要创建视图，可在 Views/Home 文件夹中添加一个名为 Watersports.cshtml 的视图文件，并在其中添加代码清单 21-11 中的内容。

### 代码清单 21-11 Views/Home 文件夹的 Watersports.cshtml 文件的内容

```
<!DOCTYPE html>
<html>
<head>
    <link href="/lib/twitter-bootstrap/css/bootstrap.min.css" rel="stylesheet" />
</head>
<body>
    <h6 class="bg-secondary text-white text-center m-2 p-2">Watersports</h6>
    <div class="m-2">
        <table class="table table-sm table-striped table-bordered">
            <tbody>
                <tr><th>Name</th><td>@Model.Name</td></tr>
                <tr><th>Price</th><td>@Model.Price.ToString("c")</td></tr>
                <tr><th>Category ID</th><td>@Model.CategoryId</td></tr>
            </tbody>
        </table>
    </div>
</body>
</html>
```

新视图采用了与 Index 视图相同的模式，但表格上方有一个不同的标题。因为 HomeController

类已经被修改，所以需要重新启动 ASP.NET Core，并请求 http://localhost:5000/home/index/1 和 http://localhost:5000/home/index/4。操作方法将为第一个 URL 选择 Watersports，为第二个 URL 选择默认视图，得到的两个响应如图 21-6 所示。

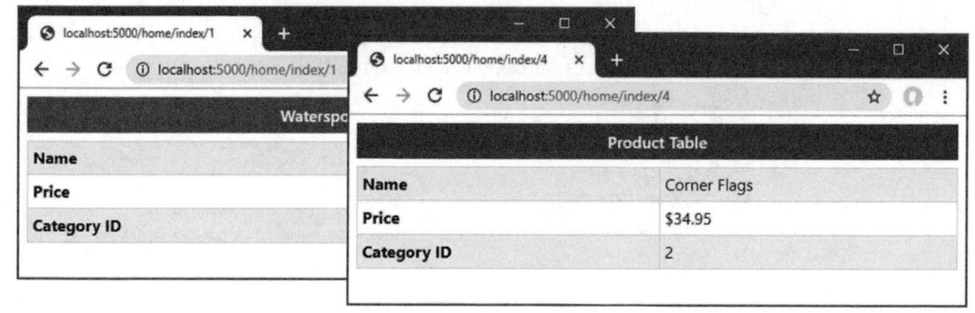


图 21-6　选择视图

### 使用共享视图

当 Razor 视图引擎寻找视图时，先在 View/[控制器]文件夹中寻找，然后在 Views/Shared 文件夹中寻找。这种搜索模式意味着可在控制器之间共享包含公共内容的视图，从而避免重复。为了解这个过程，在 Views/Shared 文件夹中添加一个名为 Common.cshtml 的 Razor 视图文件，并向其添加代码清单 21-12 中的内容。

#### 代码清单 21-12　Views/Shared 文件夹中 Common.cshtml 文件的内容

```
<!DOCTYPE html>
<html>
<head>
    <link href="/lib/twitter-bootstrap/css/bootstrap.min.css" rel="stylesheet" />
</head>
<body>
    <h6 class="bg-secondary text-white text-center m-2 p-2">Shared View</h6>
</body>
</html>
```

接下来，在 Home 控制器中添加一个操作方法来使用新视图，如代码清单 21-13 所示。

#### 代码清单 21-13　在 Controllers 文件夹的 HomeController.cs 文件中添加一个操作

```
using Microsoft.AspNetCore.Mvc;
using System.Threading.Tasks;
using WebApp.Models;

namespace WebApp.Controllers {

    public class HomeController: Controller {
        private DataContext context;

        public HomeController(DataContext ctx) {
            context = ctx;
```

```
        }

        public async Task<IActionResult> Index(long id = 1) {
            Product prod = await context.Products.FindAsync(id);
            if (prod.CategoryId == 1) {
                return View("Watersports", prod);
            } else {
                return View(prod);
            }
        }

        public IActionResult Common() {
            return View();
        }
    }
}
```

新操作依赖于“使用方法名称作为视图名称”的约定。当视图不需要向用户显示任何数据时，可以不为 View 方法指定实参。接下来，创建一个新的控制器：在 Controllers 文件夹中添加一个名为 SecondController.cs 的类文件，并为其添加代码清单 21-14 中的内容。

#### 代码清单 21-14　Controllers 文件夹中 SecondController.cs 文件的内容

```
using Microsoft.AspNetCore.Mvc;

namespace WebApp.Controllers {

    public class SecondController : Controller {

        public IActionResult Index() {
            return View("Common");
        }
    }
}
```

新控制器定义了一个名为 Index 的操作，它调用 View 方法来选择 Common 视图。重新启动 ASP.NET Core，导航到 http://localhost:5000/home/common 和 http://localhost:5000/second，它们都将呈现 Common 视图，得到如图 21-7 所示的响应。

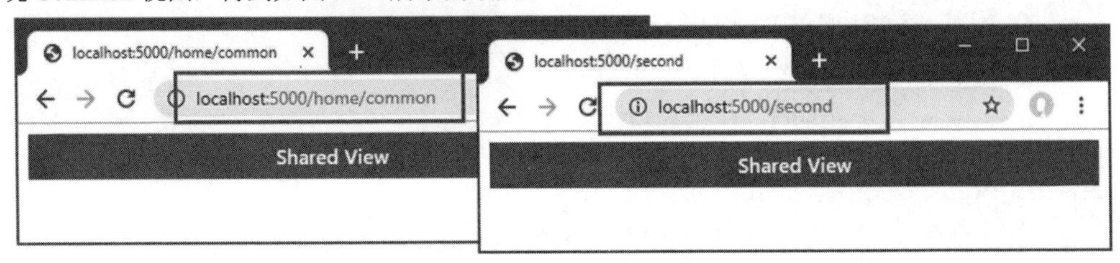


图 21-7　使用共享视图

**指定视图位置**

Razor 视图引擎首先查找特定于控制器的视图，找不到，再查找共享视图。通过指定视图文件的完整路径，可修改这种行为。如果想选择某个共享视图，但因为也存在一个特定于控制器的同名视图、导致该共享视图被忽略，就有必要指定视图文件的完整路径。

```
...
public IActionResult Index() {
    return View("/Views/Shared/Common.cshtml");
}
...
```

指定视图时，必须指定相对于项目文件夹的路径，并以/字符开头。注意，需要使用文件的完整名称，包括文件扩展名。

应该谨慎使用这种技术，因为它依赖于特定文件，而不是让视图引擎选择文件。

## 21.3　使用 Razor 视图

Razor 视图包含 HTML 表达式和 C#表达式。表达式混杂在 HTML 元素中，用@字符表示，如下所示：

```
...
<tr><th>Name</th><td>@Model.Name</td></tr>
...
```

当使用视图生成响应时，将计算表达式，并把计算结果包含到发给客户端的内容中。这里的表达式将获取操作方法提供的 Product 视图模型对象的名称，得到如下所示的输出：

```
...
<tr><th>Name</th><td>Corner Flags</td></tr>
...
```

这种转换看起来很神奇，但 Razor 比其看起来更简单。Razor 视图将被转换为 C#类，它继承自 RazorPage 类，像其他 C#类一样被编译。

**提示：** 通过在 Windows 资源管理器中查看 obj/Debug/netcoreapp3.0/Razor/Views 文件夹的内容，可看到生成的视图类。

例如，代码清单 21-11 中的视图将被转换为如下所示的类：

```
using Microsoft.AspNetCore.Mvc.Razor;
using System.Threading.Tasks;
using Microsoft.AspNetCore.Mvc.Rendering;
using Microsoft.AspNetCore.Mvc;
using Microsoft.AspNetCore.Mvc.ViewFeatures;

namespace AspNetCore {
```

```
    public class Views_Home_Watersports : RazorPage<dynamic> {

        public async override Task ExecuteAsync() {
            WriteLiteral("<!DOCTYPE html>\r\n<html>\r\n");
            WriteLiteral("<head>");
            WriteLiteral(@"<link
                href=""/lib/twitter-bootstrap/css/bootstrap.min.css""
                rel=""stylesheet"" />");
            WriteLiteral("</head>");
            WriteLiteral("<body>");
            WriteLiteral(@"<h6 class=""bg-secondary text-white text-center
                m-2 p-2"">Watersports</h6>\r\n<div class=""m-2"">\r\n<table
                class=""table table-sm table-striped table-bordered"">\r\n
                <tbody>\r\n>");
            WriteLiteral("<th>Name</th><td>");
            Write(Model.Name);
            WriteLiteral("</td></tr>");
            WriteLiteral("<tr><th>Price</th><td>");
            Write(Model.Price.ToString("c"));
            WriteLiteral("</td></tr>\r\n<tr><th>Category ID</th><td>");
            Write(Model.CategoryId);
            WriteLiteral("</td></tr>\r\n</tbody>\r\n</table>\r\n</div>");
            WriteLiteral("</body></html>");
        }

        public IUrlHelper Url { get; private set; }
        public IViewComponentHelper Component { get; private set; }
        public IJsonHelper Json { get; private set; }
        public IHtmlHelper<dynamic> Html { get; private set; }
        public IModelExpressionProvider ModelExpressionProvider { get; private set; }
    }
}
```

这里的类对生成的代码做了简化，以便能够关注对本章而言最重要的功能。首先需要注意的是，从视图生成的类继承了 RazorPage<T>类。

```
...
public class Views_Home_Watersports : RazorPage<dynamic> {
...
```

表 21-3 描述了 RazorPage<T>定义的最有用的属性和方法。

**缓存响应**

通过对操作方法应用 ResponseCache 特性，可以缓存视图的响应；还可以将该特性应用于控制器类，此时将缓存所有操作方法的响应。关于如何启用缓存的详细信息，请参阅第 17 章。

表 21-3　RazorPage<T>的成员

| 名称 | 描述 |
| --- | --- |
| Context | 此属性返回当前请求的 HttpContext 对象 |
| Layout | 此属性用于设置视图布局，如第 22 章所述 |
| Model | 此属性返回操作传递给 View 方法的视图模型 |
| RenderBody() | 此方法用来在布局中包含视图的内容，如第 22 章所述 |
| RenderSection() | 此方法用来在布局中包含视图某节的内容，如第 22 章所述 |
| TempData | 此属性用于访问临时数据功能，如第 22 章所述 |
| ViewBag | 此属性用于访问 ViewBag，如第 22 章所述 |
| ViewContext | 此属性返回一个 ViewContext 对象，用于提供上下文数据 |
| ViewData | 此属性返回视图数据，在 SportsStore 应用程序中，用来对控制器进行单元测试 |
| Write(str) | 此方法写一个字符串，该字符串将被安全编码，从而能够用在 HTML 中 |
| WriteLiteral(str) | 此方法写一个字符串，但不对其进行编码，所以在 HTML 中使用该字符串时不能保证安全 |

视图中的表达式将被转换为对 Write 方法的调用，该方法对表达式的结果进行编码，从而能够在 HTML 文档中安全地包含这个结果。WriteLiteral 方法用于视图的静态 HTML 部分，这部分不需要进一步编码。

**提示：**关于 HTML 编码的详细介绍，请参考第 22 章。

结果是 CSHTML 文件中的一个片段，如下所示：

```
...
<tr><th>Name</th><td>@Model.Name</td></tr>
...
```

这个片段在 ExecuteAsync 方法中被转换为一系列 C#语句，如下所示：

```
...
WriteLiteral("<th>Name</th><td>");
Write(Model.Name);
WriteLiteral("</td></tr>");
...
```

当调用 ExecuteAsync 方法时，将混合使用视图中包含的静态 HTML 和表达式生成响应。当执行生成类中的语句时，将把 HTML 片段和表达式的计算结果写入响应，得到如下所示的 HTML：

```
...
<th>Name</th><td>Kayak</td></tr>
...
```

除了从 RazorPage<T>类继承的属性和方法，生成的视图类还定义了表 21-4 中描述的属性，其中一些属性用于后续章节中介绍的功能。

表 21-4　更多视图类属性

| 名称 | 描述 |
| --- | --- |
| Component | 此属性返回一个使用视图组件的助手，可通过第 25 章描述的 vc 标签助手访问 |
| Html | 此属性返回 IHtmlHelper 接口的实现。此属性用于管理 HTML 编码，如第 22 章所述 |
| Json | 此属性返回 IJsonHelper 接口的实现，用于将数据编码为 JSON，如第 22 章所述 |
| ModelExpressionProvider | 此属性可用来访问从模型选择属性的表达式，通过标签助手可使用模型，如第 25～27 章所述 |
| Url | 此属性返回一个使用 URL 的助手，如第 26 章所述 |

## 设置视图模型的类型

为 Watersports.cshtml 文件生成的类派生自 RazorPage<T>，但是 Razor 不知道操作方法为视图模型使用什么类型，所以选择了 dynamic 作为泛型类型实参。这意味着@Model 表达式可用于任何属性或方法名，这个名称是在运行时生成响应的过程中确定的。为演示在表达式中使用一个不存在的成员会发生什么，将代码清单 21-15 中显示的内容添加到 Watersports.cshtml 文件中。

**代码清单 21-15　在 Views/Home 文件夹的 Watersports.cshtml 文件中添加内容**

```
<!DOCTYPE html>
<html>
<head>
   <link href="/lib/twitter-bootstrap/css/bootstrap.min.css" rel="stylesheet" />
</head>
<body>
    <h6 class="bg-secondary text-white text-center m-2 p-2">Watersports</h6>
    <div class="m-2">
        <table class="table table-sm table-striped table-bordered">
            <tbody>
                <tr><th>Name</th><td>@Model.Name</td></tr>
                <tr><th>Price</th><td>@Model.Price.ToString("c")</td></tr>
                <tr><th>Category ID</th><td>@Model.CategoryId</td></tr>
                <tr><th>Tax Rate</th><td>@Model.TaxRate</td></tr>
            </tbody>
        </table>
    </div>
</body>
</html>
```

使用浏览器请求 http://localhost:5000，将看到如图 21-8 中显示的异常。

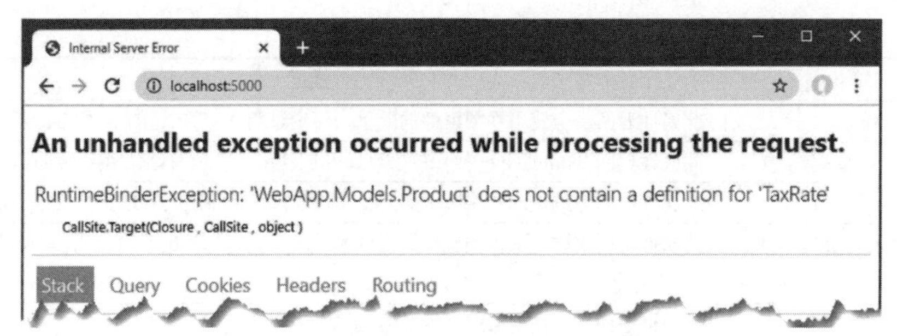


图 21-8　在视图表达式中使用不存在的属性

为在开发过程中检查表达式，可使用 model 关键字指定 Model 对象的类型，如代码清单 21-16 所示。

**提示：** 很容易混淆这两个术语。首字母大写的 Model 用于表达式中，供访问操作方法提供的视图模型对象；首字母小写的 model 则用于指定视图模型的类型。

### 代码清单 21-16　在 Views/Home 文件夹的 Watersports.cshtml 文件中声明模型类型

```
@model WebApp.Models.Product
<!DOCTYPE html>
<html>
<head>
    <link href="/lib/twitter-bootstrap/css/bootstrap.min.css" rel="stylesheet" />
</head>
<body>
    <h6 class="bg-secondary text-white text-center m-2 p-2">Watersports</h6>
    <div class="m-2">
        <table class="table table-sm table-striped table-bordered">
            <tbody>
                <tr><th>Name</th><td>@Model.Name</td></tr>
                <tr><th>Price</th><td>@Model.Price.ToString("c")</td></tr>
                <tr><th>Category ID</th><td>@Model.CategoryId</td></tr>
                <tr><th>Tax Rate</th><td>@Model.TaxRate</td></tr>
            </tbody>
        </table>
    </div>
</body>
</html>
```

几秒后，在编辑器中将显示一条警告消息，这是因为 Visual Studio 或 Visual Studio Code 会在后台检查视图，如图 21-9 所示。如果生成项目，或使用 dotnet build 或 dotnet run 命令，编译器也将报错。

```
WebApp - Watersports.cshtml
Watersports.cshtml
10  table class="table table-sm table-striped table-bordered">
11    <tbody>
12        <tr><th>Name</th><td>@Model.Name</td></tr>
13        <tr><th>Price</th><td>@Model.Price.ToString("c")</td></tr>
14        <tr><th>Category ID</th><td>@Model.CategoryId</td></tr>
15        <tr><th>Tax Rate</th><td>@Model.TaxRate</td></tr>
16    </tbody>
17  table>
18
19
20
100 %
```

图 21-9　视图文件中的错误警告

当生成视图的 C#类时，将使用视图模型的类型作为基类的泛型实参，如下所示：

```
...
public class Views_Home_Watersports : RazorPage<Product> {
...
```

指定视图模型类型使得你在编辑视图时，Visual Studio 和 Visual Studio Code 能够建议属性和方法的名称。将不存在的属性替换为代码清单 21-17 中显示的属性。

**代码清单 21-17　在 Views/Home 文件夹的 Watersports.cshtml 文件中替换属性**

```
@model WebApp.Models.Product
<!DOCTYPE html>
<html>
<head>
  <link href="/lib/twitter-bootstrap/css/bootstrap.min.css" rel="stylesheet" />
</head>
<body>
    <h6 class="bg-secondary text-white text-center m-2 p-2">Watersports</h6>
    <div class="m-2">
        <table class="table table-sm table-striped table-bordered">
            <tbody>
                <tr><th>Name</th><td>@Model.Name</td></tr>
                <tr><th>Price</th><td>@Model.Price.ToString("c")</td></tr>
                <tr><th>Category ID</th><td>@Model.CategoryId</td></tr>
                <tr><th>Supplier ID</th><td>@Model.SupplierId</td></tr>
            </tbody>
        </table>
    </div>
</body>
</html>
```

在输入过程中，编辑器将显示视图模型类定义的成员名称作为建议，如图 21-10 所示。该图显示的是 Visual Studio 的代码编译器，但 Visual Studio Code 提供了类似的功能。

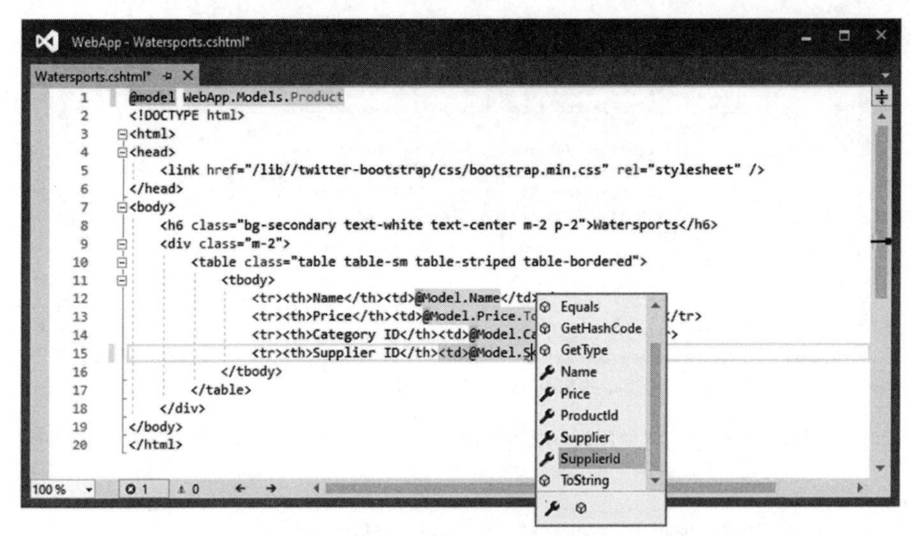


图 21-10　使用视图模型类型时，编辑器将给出建议

### 使用视图导入文件

在 Watersports.cshtml 文件开头声明视图模型对象时，必须包含该类所在的名称空间，如下所示：

```
...
@model WebApp.Models.Product
...
```

默认情况下，必须使用名称空间限定 Razor 视图中引用的所有类型。如果只引用了模型对象的类型，那么问题不大，但如果要编写更复杂的 Razor 表达式，例如本章后面将编写的那些表达式，视图就会变得难以阅读。

通过在项目中添加视图导入文件，能够指定在一组名称空间中搜索类型。视图导入文件的名称为_ViewImports.cshtml，包含在 Views 文件夹中。

---

**注意：** Views 文件夹中以下画线(_)开头的文件不会返回给用户，这就使得文件名能够区分想要展示的视图和支持这些视图的文件。视图导入文件和布局(稍后介绍)带有下画线前缀。

---

如果使用的是 Visual Studio，可在 Solution Explorer 中右击 Views 文件夹，从弹出菜单中选择 Add | New Item，然后从 ASP.NET Core 分类中选择 Razor View Imports 模板，如图 21-11 所示。

Visual Studio 将自动把文件名称设为_ViewImports.cshtml，此时单击 Add 按钮，将创建该文件，但它不包含任何内容。如果使用的是 Visual Studio Code，则只需要选择 Views 文件夹，在其中添加一个新文件，命名为_ViewImports.cshtml。

无论使用哪个编辑器，都需要添加代码清单 21-18 中显示的表达式。

#### 代码清单 21-18　Views 文件夹中_ViewImports.cshtml 文件的内容

```
@using WebApp.Models
```

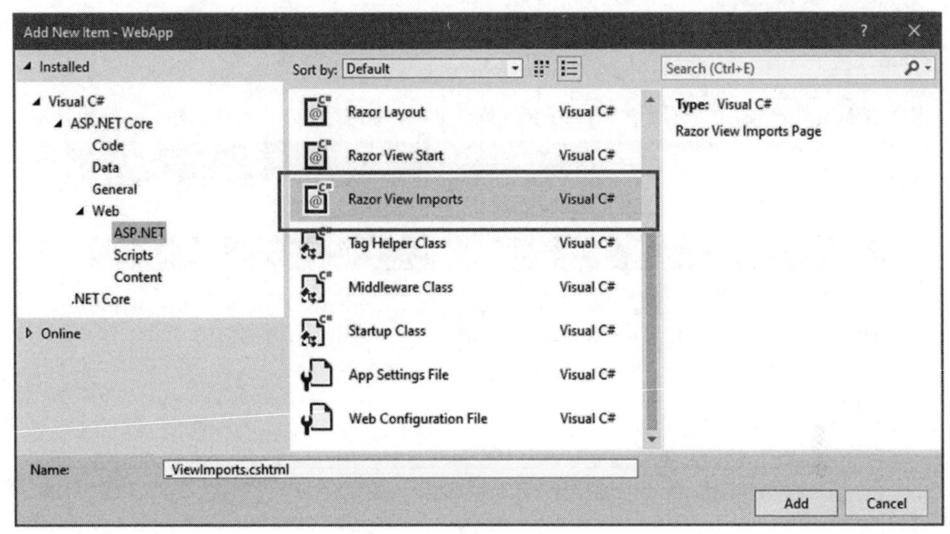


图 21-11 创建视图导入文件

通过使用@using 表达式，并在后面加上名称空间，可以指定在哪些名称空间中搜索 Razor 视图中使用的类。在代码清单 21-18 中，为 WebApp.Models 名称空间添加一个条目，该名称空间包含 Watersports.cshtml 视图中使用的视图模型类。

在视图导入文件中包含了名称空间后，就可以从视图中删除名称空间，如代码清单 21-19 所示。

---

**提示：**也可在单独的视图文件中添加@using 表达式，这样就可在视图中使用类型，而不需要添加名称空间。

---

#### 代码清单 21-19 在 Views/Home 文件夹的 Watersports.cshtml 文件中简化模型类型

```
@model Product
<!DOCTYPE html>
<html>
<head>
   <link href="/lib/twitter-bootstrap/css/bootstrap.min.css" rel="stylesheet" />
</head>
<body>
    <h6 class="bg-secondary text-white text-center m-2 p-2">Watersports</h6>
    <div class="m-2">
        <table class="table table-sm table-striped table-bordered">
            <tbody>
                <tr><th>Name</th><td>@Model.Name</td></tr>
                <tr><th>Price</th><td>@Model.Price.ToString("c")</td></tr>
                <tr><th>Category ID</th><td>@Model.CategoryId</td></tr>
                <tr><th>Supplier ID</th><td>@Model.SupplierId</td></tr>
            </tbody>
        </table>
    </div>
```

```
</body>
</html>
```

保存视图文件，使用浏览器请求 http://localhost:5000，将看到如图 21-12 所示的响应。

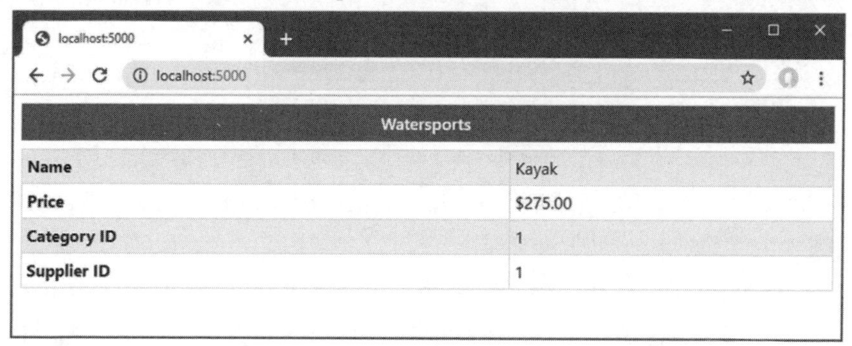


图 21-12　使用视图导入文件

## 21.4　理解 Razor 语法

Razor 编译器将 HTML 的静态部分与 C#表达式分开，然后在生成的类文件中单独处理 C#表达式。在视图中可包含几种类型的表达式，接下来将分别进行介绍。

### 21.4.1　理解指令

指令是向 Razor 视图引擎发送命令的表达式。例如，@model 表达式就是一个指令，它告诉视图引擎为视图模型使用特定类型，而@using 指令则告诉视图引擎导入一个名称空间。表 21-5 描述了最有用的 Razor 指令。

表 21-5　最有用的 Razor 指令

| 名称 | 描述 |
| --- | --- |
| @model | 此指令指定视图模型的类型 |
| @using | 此指令导入名称空间 |
| @page | 此指令指示 Razor Pages，如第 23 章所述 |
| @section | 此指令指示布局节，如第 22 章所述 |
| @addTagHelper | 此指令在视图中添加标签助手，如第 25 章所述 |
| @namespace | 此指令设置从视图生成的 C#类的名称空间 |
| @functions | 此指令向从视图生成的 C#类添加 C#属性和方法，经常用在 Razor Pages 中，如第 23 章所述 |
| @attribute | 此指令向从视图生成的 C#类添加特性。第 38 章使用此功能应用授权限制 |
| @implements | 此指令声明，从视图生成的 C#类继承某个接口，或派生自某个基类。第 33 章将演示此功能 |
| @inherits | 此指令设置了从视图生成的 C#类的基类。第 33 章将演示此功能 |
| @inject | 此指令使视图能通过依赖注入直接访问某个服务。第 23 章将演示此功能 |

## 21.4.2 理解内容表达式

Razor 内容表达式生成的内容将包含在视图生成的输出中。表 21-6 描述了最有用的内容表达式。

表 21-6 最有用的 Razor 内容表达式

| 名称 | 描述 |
|---|---|
| @<expression> | 这是基本的 Razor 表达式，其计算结果将被插入响应中 |
| @if | 此表达式用于基于表达式的结果选择内容。相关示例请参见“使用条件表达式”一节 |
| @switch | 此表达式用于基于表达式的结果选择内容。相关示例请参见“使用条件表达式”一节 |
| @foreach | 此表达式为序列中的每个元素生成相同的内容。相关示例请参见“枚举序列”一节 |
| @{ ... } | 此表达式定义了一个代码块。相关示例请参见“使用 Razor 代码块”一节 |
| @: | 此表达式指示没有包含在 HTML 元素中的一段内容。相关示例请参见“使用条件表达式”一节 |
| @try | 此表达式用于捕获异常 |
| @await | 此表达式用于执行异步操作，操作结果将被插入响应中。相关示例见第 24 章 |

## 21.4.3 设置元素内容

最简单的表达式是计算得到一个值，这个值在发送给客户端的响应中用作 HTML 元素的内容。最常见的表达式类型是插入来自视图模型对象的一个值，如 Watersports.cshtml 视图文件中的以下表达式：

```
...
<tr><th>Name</th><td>@Model.Name</td></tr>
<tr><th>Price</th><td>@Model.Price.ToString("c")</td></tr>
...
```

正如示例所示，这种类型的表达式可读取属性值或调用方法。视图可以包含更复杂的表达式，但要求把这些表达式放到括号内，以便 Razor 编译器能够区分代码和静态内容，如代码清单 21-20 所示。

**代码清单 21-20 在 Views/Home 文件夹的 Watersports.cshtml 文件中添加表达式**

```
@model Product
<!DOCTYPE html>
<html>
<head>
   <link href="/lib/twitter-bootstrap/css/bootstrap.min.css" rel="stylesheet" />
</head>
<body>
    <h6 class="bg-secondary text-white text-center m-2 p-2">Watersports</h6>
    <div class="m-2">
        <table class="table table-sm table-striped table-bordered">
            <tbody>
```

```
                <tr><th>Name</th><td>@Model.Name</td></tr>
                <tr><th>Price</th><td>@Model.Price.ToString("c")</td></tr>
                <tr><th>Tax</th><td>@Model.Price * 0.2m</td></tr>
                <tr><th>Tax</th><td>@(Model.Price * 0.2m)</td></tr>
            </tbody>
        </table>
    </div>
</body>
</html>
```

使用浏览器请求 http://localhost:5000，将得到如图 21-13 所示的响应，这个响应表明了括号的重要性。

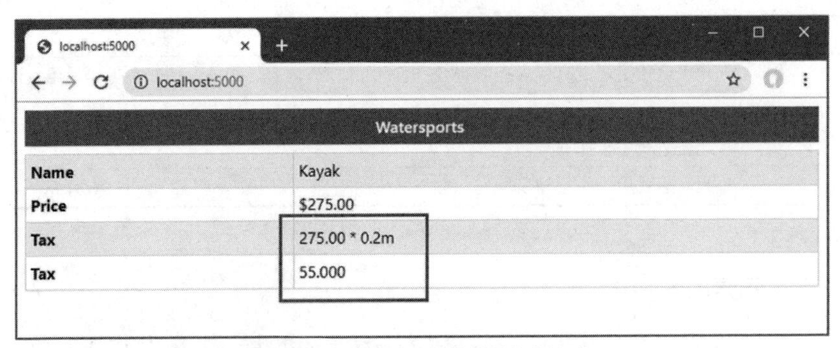


图 21-13　带括号和不带括号的表达式

Razor 视图编译器在匹配表达式时偏保守，假定第一个表达式中的星号和数值是静态内容。在第二个表达式中使用了括号来避免这个问题。

### 21.4.4　设置特性值

表达式可用于设置元素特性的值，如代码清单 21-21 所示。

**代码清单 21-21　在 Views/Home 文件夹的 Watersports.cshtml 文件中设置特性**

```
@model Product
<!DOCTYPE html>
<html>
<head>
   <link href="/lib/twitter-bootstrap/css/bootstrap.min.css" rel="stylesheet" />
</head>
<body>
    <h6 class="bg-secondary text-white text-center m-2 p-2">Watersports</h6>
    <div class="m-2">
        <table class="table table-sm table-striped table-bordered"
               data-id="@Model.ProductId">
            <tbody>
                <tr><th>Name</th><td>@Model.Name</td></tr>
                <tr><th>Price</th><td>@Model.Price.ToString("c")</td></tr>
                <tr><th>Tax</th><td>@Model.Price * 0.2m</td></tr>
```

```
                <tr><th>Tax</th><td>@(Model.Price * 0.2m)</td></tr>
            </tbody>
        </table>
    </div>
</body>
</html>
```

这里使用了 Razor 表达式来设置 table 元素的一些 data 特性的值。

> ▒ **提示：** 数据特性，即名称前带有 data-前缀的特性，在多年来都是创建自定义特性的一种非正式方法，但在 HTML5 中成为正式标准的一部分。使用数据特性，主要是为了让 JavaScript 代码能够定位特定元素，或者能在更少元素上应用 CSS 样式。

如果请求 http://localhost:5000，并查看发送给浏览器的 HTML 源，会看到 Razor 已经设置了特性的值，如下所示：

```
...
<table class="table table-sm table-striped table-bordered" data-id="1">
    <tbody>
        <tr><th>Name</th><td>Kayak</td></tr>
        <tr><th>Price</th><td>$275.00</td></tr>
        <tr><th>Tax</th><td>275.00 * 0.2m</td></tr>
        <tr><th>Tax</th><td>55.000</td></tr>
    </tbody>
</table>
...
```

### 21.4.5 使用条件表达式

Razor 支持条件表达式，这意味着可以根据视图模型调整输出。这种技术是 Razor 的核心，允许从易于阅读和维护的视图创建复杂的流式响应。在代码清单 21-22 中，向 Watersports 视图添加一个条件语句。

#### 代码清单 21-22 在 Views/Home 文件夹的 Watersports.cshtml 文件中使用 if 表达式

```
@model Product
<!DOCTYPE html>
<html>
<head>
   <link href="/lib/twitter-bootstrap/css/bootstrap.min.css" rel="stylesheet" />
</head>
<body>
    <h6 class="bg-secondary text-white text-center m-2 p-2">Watersports</h6>
    <div class="m-2">
        <table class="table table-sm table-striped table-bordered"
                data-id="@Model.ProductId">
            <tbody>
```

```
                @if (Model.Price > 200) {
                    <tr><th>Name</th><td>Luxury @Model.Name</td></tr>
                } else {
                    <tr><th>Name</th><td>Basic @Model.Name</td></tr>
                }
                <tr><th>Price</th><td>@Model.Price.ToString("c")</td></tr>
                <tr><th>Tax</th><td>@Model.Price * 0.2m</td></tr>
                <tr><th>Tax</th><td>@(Model.Price * 0.2m)</td></tr>
            </tbody>
        </table>
    </div>
</body>
</html>
```

@字符后跟 if 关键字以及在运行时计算的条件。if 表达式还支持可选的 else 和 elseif 子句，并用关闭括号(}字符)结束。如果满足条件，就将 if 子句中的内容插入响应中，否则，将 else 子句中的内容插入响应中。

注意，在条件中访问 Model 属性时，不需要使用@前缀。

```
...
@if (Model.Price > 200) {
...
```

但在 if 和 else 子句中，则需要使用@前缀，如下所示：

```
...
<tr><th>Name</th><td>Luxury @Model.Name</td></tr>
...
```

为查看条件语句的效果，使用浏览器请求 http://localhost:5000/home/index/1 和 http://localhost:5000/home/index/2。条件语句将为这两个 URL 生成不同的 HTML 元素，如图 21-14 所示。

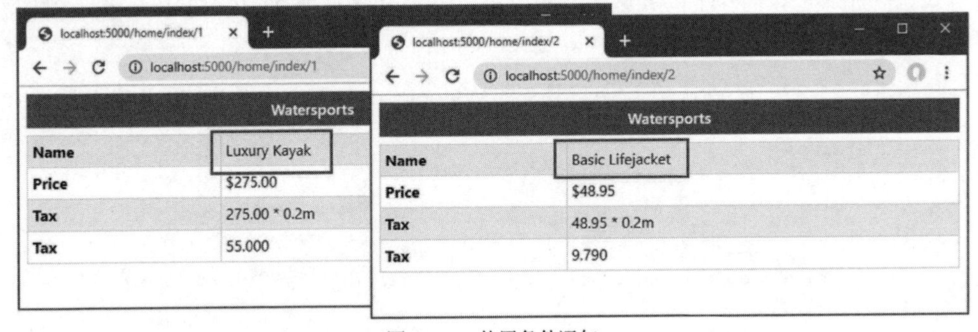


图 21-14　使用条件语句

Razor 还支持@switch 表达式，这可作为处理多个条件的一种更简洁的方式，如代码清单 21-23 所示。

**代码清单 21-23 在 Views/Home 文件夹的 Watersports.cshtml 中使用 switch 表达式**

```
@model Product
<!DOCTYPE html>
<html>
<head>
   <link href="/lib/twitter-bootstrap/css/bootstrap.min.css" rel="stylesheet" />
</head>
<body>
    <h6 class="bg-secondary text-white text-center m-2 p-2">Watersports</h6>
    <div class="m-2">
        <table class="table table-sm table-striped table-bordered"
               data-id="@Model.ProductId">
            <tbody>
                @switch (Model.Name) {
                    case "Kayak":
                        <tr><th>Name</th><td>Small Boat</td></tr>
                        break;
                    case "Lifejacket":
                        <tr><th>Name</th><td>Flotation Aid</td></tr>
                        break;
                    default:
                        <tr><th>Name</th><td>@Model.Name</td></tr>
                        break;
                }
                <tr><th>Price</th><td>@Model.Price.ToString("c")</td></tr>
                <tr><th>Tax</th><td>@Model.Price * 0.2m</td></tr>
                <tr><th>Tax</th><td>@(Model.Price * 0.2m)</td></tr>
            </tbody>
        </table>
    </div>
</body>
</html>
```

条件表达式可导致为每个结果子句重复相同的内容块。例如，在 switch 表达式中，每个 case 子句只有 td 元素的内容不同，tr 和 th 元素的内容是相同的。为删除重复，可在元素内使用条件表达式，如代码清单 21-24 所示。

**代码清单 21-24 在 Views/Home 文件夹的 Watersports.cshtml 文件中设置内容**

```
@model Product
<!DOCTYPE html>
<html>
<head>
   <link href="/lib/twitter-bootstrap/css/bootstrap.min.css" rel="stylesheet" />
</head>
<body>
    <h6 class="bg-secondary text-white text-center m-2 p-2">Watersports</h6>
```

```
    <div class="m-2">
        <table class="table table-sm table-striped table-bordered"
               data-id="@Model.ProductId">
            <tbody>
                <tr><th>Name</th><td>
                @switch (Model.Name) {
                    case "Kayak":
                        @:Small Boat
                        break;
                    case "Lifejacket":
                        @:Flotation Aid
                        break;
                    default:
                        @Model.Name
                        break;
                }
                </td></tr>
                <tr><th>Price</th><td>@Model.Price.ToString("c")</td></tr>
                <tr><th>Tax</th><td>@Model.Price * 0.2m</td></tr>
                <tr><th>Tax</th><td>@(Model.Price * 0.2m)</td></tr>
            </tbody>
        </table>
    </div>
</body>
</html>
```

Razor 编译器需要借助@:前缀来处理未包含在 HTML 元素中的字面量值，如下所示：

```
...
@:Small Boat
...
```

编译器能够检测到开始标签，所以会处理 HTML 元素，但对于文本内容，就必须获得上述额外的帮助。为了查看 switch 语句的效果，使用 Web 浏览器请求 http://localhost:5000/home/index/2，这将得到如图 21-15 所示的响应。

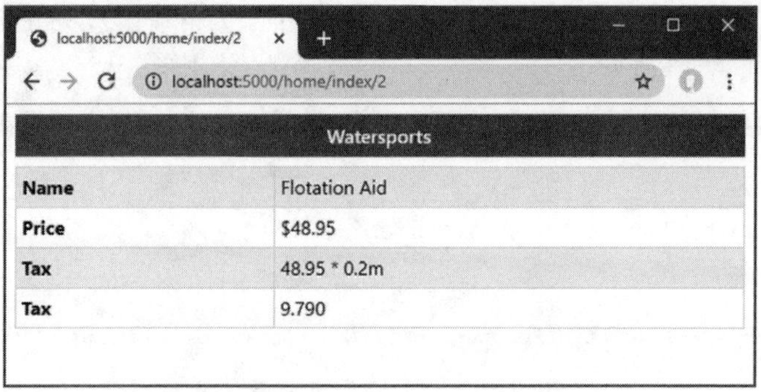

图 21-15　使用 switch 表达式和字面量内容

### 21.4.6 枚举序列

Razor 的@foreach 表达式为数组或集合中的每个对象生成内容，这在处理数据时是一种常见的需求。代码清单 21-25 在 Home 控制器中添加了一个操作方法，用于生成一系列对象。

**代码清单 21-25 在 Controllers 文件夹的 HomeController.cs 文件中添加操作**

```
using Microsoft.AspNetCore.Mvc;
using System.Threading.Tasks;
using WebApp.Models;

namespace WebApp.Controllers {

    public class HomeController : Controller {
        private DataContext context;

        public HomeController(DataContext ctx) {
            context = ctx;
        }

        public async Task<IActionResult> Index(long id = 1) {
            Product prod = await context.Products.FindAsync(id);
            if (prod.CategoryId == 1) {
                return View("Watersports", prod);
            } else {
                return View(prod);
            }
        }

        public IActionResult Common() {
            return View();
        }

        public IActionResult List() {
            return View(context.Products);
        }
    }
}
```

新操作方法被命名为 List，它为视图提供从 Entity Framework Core 数据上下文获取的一系列 Product 对象。在 Views/Home 文件夹中添加一个名为 List.cshtml 的 Razor 视图文件，并为其添加代码清单 21-26 中的内容。

**代码清单 21-26 Views/Home 文件夹中 List.cshtml 文件的内容**

```
@model IEnumerable<Product>
<!DOCTYPE html>
<html>
```

```
<head>
    <link href="/lib/twitter-bootstrap/css/bootstrap.min.css" rel="stylesheet" />
</head>
<body>
    <h6 class="bg-secondary text-white text-center m-2 p-2">Products</h6>
    <div class="m-2">
        <table class="table table-sm table-striped table-bordered">
            <thead>
                <tr><th>Name</th><th>Price</th></tr>
            </thead>
            <tbody>
                @foreach (Product p in Model) {
                    <tr><td>@p.Name</td><td>@p.Price</td></tr>
                }
            </tbody>
        </table>
    </div>
</body>
</html>
```

foreach 表达式的格式与 C#的 foreach 语句相同。在本例中，将操作方法提供的序列中的每个对象赋值给变量 p。当计算完它包含的表达式后，将为每个对象重复表达式中的内容，然后将这些内容插入响应中。在本例中，foreach 表达式中的内容将生成一个表格行，其单元格包含自己的表达式。

```
...
<tr><td>@p.Name</td><td>@p.Price</td></tr>
...
```

重新启动 ASP.NET Core，以便能够使用新的操作方法，然后使用浏览器请求 http://localhost:5000/home/list，这将得到如图 21-16 所示的结果，显示 foreach 表达式如何填充一个表格体。

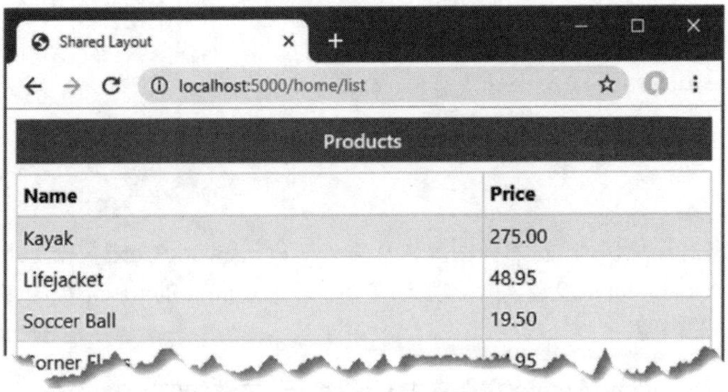

图 21-16　使用 foreach 表达式

### 21.4.7 使用 Razor 代码块

代码块是一段 C#代码，它们不生成内容，但能够为那些生成内容的表达式提供支持。代码清单 21-27 添加了一个计算平均值的代码块。

---

■ **提示：** 代码块最常见的用途是选择布局，如第 21 章所述。

---

**代码清单 21-27 在 Views/Home 文件夹的 List.cshtml 文件中使用代码块**

```
@model IEnumerable<Product>
@{
   decimal average = Model.Average(p => p.Price);
}
<!DOCTYPE html>
<html>
<head>
   <link href="/lib/twitter-bootstrap/css/bootstrap.min.css" rel="stylesheet" />
</head>
<body>
    <h6 class="bg-secondary text-white text-center m-2 p-2">Products</h6>
    <div class="m-2">
        <table class="table table-sm table-striped table-bordered">
            <thead>
                <tr><th>Name</th><th>Price</th><th></th></tr>
            </thead>
            <tbody>
                @foreach (Product p in Model) {
                    <tr>
                        <td>@p.Name</td><td>@p.Price</td>
                        <td>@((p.Price / average * 100).ToString("F1"))
                                % of average</td>
                    </tr>
                }
            </tbody>
        </table>
    </div>
</body>
</html>
```

代码块由@{和}表示，其中包含标准 C#语句。代码清单 21-27 中的代码块使用 LINQ 来计算一个值，并将其赋值给 average 变量，该变量将在表达式中用于设置表格单元格的内容，从而避免为视图模型序列中的每个对象重复计算平均值的表达式。使用浏览器请求 http://localhost:5000/home/list，将看到如图 21-17 所示的响应。

---

■ **注意：** 如果代码块中包含太多语句，会变得难以管理。对于更复杂的任务，可以考虑使用第 22 章将讨论的 ViewBag，或者在控制器中添加一个非操作方法。

---

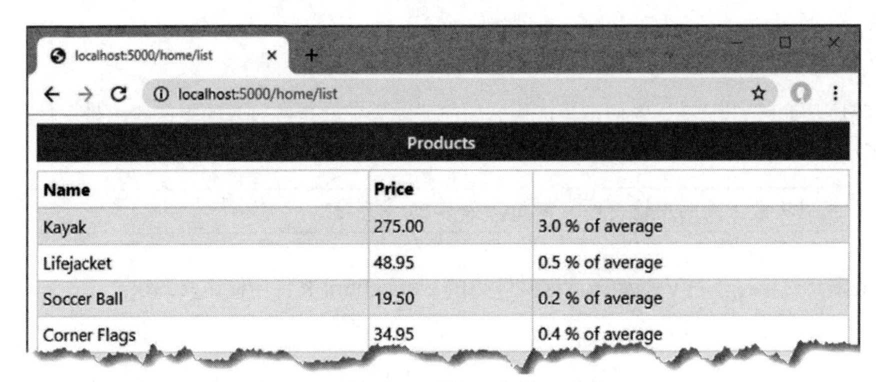


图 21-17　使用代码块

# 21.5　小结

本章介绍了 Razor 视图，它们用于从操作方法创建 HTML 响应。本章解释了如何定义视图、如何把视图转换为 C#类，以及如何使用视图包含的表达式来生成动态内容。第 22 章将继续介绍如何在视图中使用控制器。

# 第 22 章

# 使用控制器和视图(第 2 部分)

本章将介绍 Razor 视图提供的更多功能，包括使用 ViewBag 向视图传递额外的数据，以及使用布局和布局节减少重复。本章还将解释如何编码表达式的结果，以及如何禁用编码。表 22-1 总结了本章的内容。

表 22-2　本章内容摘要

| 问题 | 解决方案 | 代码清单 |
| --- | --- | --- |
| 向视图提供非结构化数据 | 使用 ViewBag | 22-5～22-6 |
| 向视图提供临时数据 | 使用临时数据 | 22-7～22-8 |
| 在多个视图中使用相同的内容 | 使用布局 | 22-9～22-12、22-15～22-18 |
| 为视图选择默认布局 | 使用 ViewStart 文件 | 22-13、22-14 |
| 穿插独特的和公共的内容 | 使用布局节 | 22-19～22-24 |
| 创建可重用的内容节 | 使用分部视图 | 22-25～22-29 |
| 使用 Razor 表达式在响应中插入 HTML | 编码 HTML | 22-30～22-32 |
| 在视图中包含 JSON | 使用 JSON 编码器 | 22-33 |

## 22.1　准备工作

本章将使用第 21 章的 WebApp 项目。为了完成准备工作，使用代码清单 22-1 中的代码替换 HomeController.cs 文件的内容。

#### 代码清单 22-1　Controller 文件夹中 HomeController.cs 文件的内容

```
using Microsoft.AspNetCore.Mvc;
using System.Threading.Tasks;
using WebApp.Models;

namespace WebApp.Controllers {

    public class HomeController: Controller {
```

```
        private DataContext context;

        public HomeController(DataContext ctx) {
            context = ctx;
        }

        public async Task<IActionResult> Index(long id = 1) {
            return View(await context.Products.FindAsync(id));
        }

        public IActionResult List() {
            return View(context.Products);
        }
    }
}
```

本章使用的功能之一需要用到第 16 章介绍的会话功能。为启用会话，需要在 Startup 类中添加代码清单 22-2 中显示的语句。

**代码清单 22-2　在 WebApp 文件夹的 Startup.cs 文件中启用会话**

```
using System;
using System.Collections.Generic;
using System.Linq;
using Microsoft.AspNetCore.Builder;
using Microsoft.Extensions.DependencyInjection;
using Microsoft.Extensions.Configuration;
using Microsoft.EntityFrameworkCore;
using WebApp.Models;

namespace WebApp {
    public class Startup {

        public Startup(IConfiguration config) {
            Configuration = config;
        }

        public IConfiguration Configuration { get; set; }

        public void ConfigureServices(IServiceCollection services) {
            services.AddDbContext<DataContext>(opts => {
                opts.UseSqlServer(Configuration[
                    "ConnectionStrings:ProductConnection"]);
                opts.EnableSensitiveDataLogging(true);
            });
            services.AddControllersWithViews().AddRazorRuntimeCompilation();

            services.AddDistributedMemoryCache();
```

```
            services.AddSession(options => {
                options.Cookie.IsEssential = true;
            });
        }

        public void Configure(IApplicationBuilder app, DataContext context) {
            app.UseDeveloperExceptionPage();
            app.UseStaticFiles();
            app.UseSession();
            app.UseRouting();
            app.UseEndpoints(endpoints => {
                endpoints.MapControllers();
                endpoints.MapDefaultControllerRoute();
            });
            SeedData.SeedDatabase(context);
        }
    }
}
```

### 22.1.1 删除数据库

打开一个新的 PowerShell 命令提示符，导航到包含 WebApp.csproj 文件的文件夹，并运行代码清单 22-3 中显示的命令来删除数据库。

> ■ **提示：** 从 https://github.com/apress/pro-asp.net-core-3 可以下载本章以及本书其他章节的示例项目。如果无法成功运行示例，请参考第 1 章的说明。

**代码清单 22-3 删除数据库**

```
dotnet ef database drop --force
```

### 22.1.2 运行示例应用程序

删除数据库后，从 Debug 菜单中选择 Start Without Debugging 或 Run Without Debugging，或者使用 PowerShell 命令提示符来运行代码清单 22-4 中显示的命令。

**代码清单 22-4 运行示例应用程序**

```
dotnet run
```

在应用程序启动过程中，将对数据库执行 seed 操作。当 ASP.NET Core 开始运行后，使用 Web 浏览器请求 http://localhost:5000，将得到如图 22-1 所示的响应。

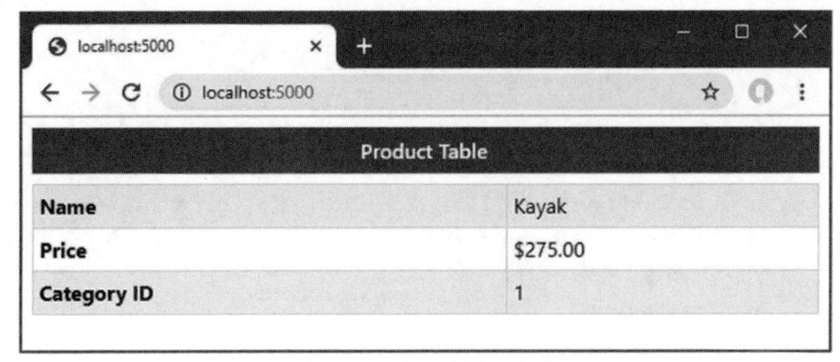


图 22-1　运行示例应用程序

## 22.2　使用 ViewBag

操作方法使用视图模型为视图提供要显示的数据，但有时候，还需要额外的信息。操作方法可以使用 ViewBag 为视图提供额外数据，如代码清单 22-5 所示。

**代码清单 22-5　在 Controllers 文件夹的 HomeControllers.cs 文件中使用 ViewBag**

```
using Microsoft.AspNetCore.Mvc;
using System.Threading.Tasks;
using WebApp.Models;
using Microsoft.EntityFrameworkCore;

namespace WebApp.Controllers {

    public class HomeController: Controller {
        private DataContext context;

        public HomeController(DataContext ctx) {
            context = ctx;
        }

        public async Task<IActionResult> Index(long id = 1) {
            ViewBag.AveragePrice = await context.Products.AverageAsync(p => p.Price);
            return View(await context.Products.FindAsync(id));
        }

        public IActionResult List() {
            return View(context.Products);
        }
    }
}
```

ViewBag 属性是从 Controller 基类继承来的，它返回一个 dynamic 对象。这就允许操作方法通过向属性赋值来创建新属性，如上面的代码清单所示。通过另外一个也叫作 ViewBag 的属性，可以在视图中访问操作方法赋给 ViewBag 属性的值，如代码清单 22-6 所示。

### 代码清单 22-6　在 Views/Home 文件夹的 Index.cshtml 文件中使用 ViewBag

```
<!DOCTYPE html>
<html>
<head>
   <link href="/lib/twitter-bootstrap/css/bootstrap.min.css" rel="stylesheet" />
</head>
<body>
    <h6 class="bg-primary text-white text-center m-2 p-2">Product Table</h6>
    <div class="m-2">
        <table class="table table-sm table-striped table-bordered">
            <tbody>
                <tr><th>Name</th><td>@Model.Name</td></tr>
                <tr>
                    <th>Price</th>
                    <td>
                        @Model.Price.ToString("c")
                        (@(((Model.Price / ViewBag.AveragePrice)
                            * 100).ToString("F2"))% of average price)
                    </td>
                </tr>
                <tr><th>Category ID</th><td>@Model.CategoryId</td></tr>
            </tbody>
        </table>
    </div>
</body>
</html>
```

ViewBag 属性将对象及视图模型对象从操作传递给视图。在代码清单中，操作方法查询数据库中的 Product.Price 属性的平均值，并将其赋值给一个名为 AveragePrice 的 ViewBag 属性，后者将用在一个表达式中。重新启动 ASP.NET Core，使用浏览器请求 http://localhost:5000，这将得到如图 22-2 所示的响应。

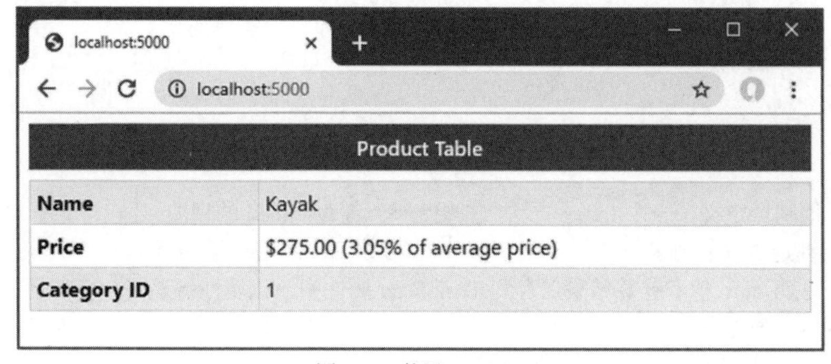


图 22-2　使用 ViewBag

**使用 ViewBag 的时机**

当需要为视图提供少量的补充数据，但不需要为每个操作方法创建新的视图模型类的时候，ViewBag 的效果最好。但 ViewBag 的问题在于，编译器不能检查动态对象上的属性，与不使用 @model 表达式的视图一样。很难判断什么时候应该使用一个新的视图模型类，我的经验是，当多个操作使用了相同的视图模型属性，或者当一个操作方法在 ViewBag 中添加了两个或三个以上的属性时，就创建一个新的视图模型类。

## 22.3　使用临时数据

临时数据功能允许控制器在请求之间保留数据，这在执行重定向时十分有用。如果启用了会话状态，将临时数据存储为会话数据；否则使用 cookie 存储临时数据。与会话数据不同，当读取完临时数据值后，将把其标记为可被删除，当处理完请求后就删除它们。

在 WebApp/Controllers 文件夹中添加一个名为 CubedController.cs 的类文件，用来定义代码清单 22-7 中显示的控制器。

**代码清单 22-7　Controllers 文件夹中 CubedController.cs 文件的内容**

```
using Microsoft.AspNetCore.Mvc;
using System;

namespace WebApp.Controllers {
    public class CubedController: Controller {

        public IActionResult Index() {
            return View("Cubed");
        }

        public IActionResult Cube(double num) {
            TempData["value"] = num.ToString();
            TempData["result"] = Math.Pow(num, 3).ToString();
            return RedirectToAction(nameof(Index));
        }
    }
}
```

Cubed 控制器定义了一个 Index 方法，它选择一个名为 Cubed 的视图。还包含一个 Cube 操作，它依赖模型绑定过程从请求中获取 num 参数的值(第 28 章将详细介绍这个过程)。Cubed 操作方法执行计算，并将 num 值和计算结果保存到 TempData 属性中，该属性将返回一个字段，用来存储键值对。因为临时数据功能建立在会话功能的基础上，所以只能存储可被序列化为字符串的值，因此在代码清单 22-7 中，将两个 double 值转换为字符串。当把值存储为临时数据后，Cube 方法将重定向到 Index 方法。为给控制器提供一个视图，在 WebApp/Views/Shared 文件夹中添加一个名为 Cubed.cshtml 的 Razor 视图文件，使其包含代码清单 22-8 中显示的内容。

### 代码清单 22-8　Views/Shared 文件夹中 Cubed.cshtml 文件的内容

```
<!DOCTYPE html>
<html>
<head>
   <link href="/lib/twitter-bootstrap/css/bootstrap.min.css" rel="stylesheet" />
</head>
<body>
    <h6 class="bg-secondary text-white text-center m-2 p-2">Cubed</h6>
    <form method="get" action="/cubed/cube" class="m-2">
        <div class="form-group">
            <label>Value</label>
            <input name="num" class="form-control" value="@(TempData["value"])" />
        </div>
        <button class="btn btn-primary" type="submit">Submit</button>
    </form>
    @if (TempData["result"] != null) {
        <div class="bg-info text-white m-2 p-2">
           The cube of @TempData["value"] is @TempData["result"]
        </div>
    }
</body>
</html>
```

为 Razor 视图使用的基类提供了 TempData 属性，用于访问临时数据，可读取表达式内的值。在本例中，使用临时数据来设置一个 input 元素的内容，并显示一个结果摘要。读取一个临时数据值并不会立即删除它，这意味着在同一个视图中可以重复读取值。只有当处理完请求后，才会删除标记的值。

为了查看效果，重新启动 ASP.NET Core，使用浏览器导航到 http://localhost:5000/cubed，在表单字段中输入一个值，然后单击 Submit 按钮。浏览器将发送一个请求来设置临时数据并触发重定向。临时数据值将保存下来，供新请求使用，结果将被显示给用户。不过，读取数据值将把它们标记为可以删除，如果重新加载浏览器，就会再显示 input 元素的内容和结果摘要，如图 22-3 所示。

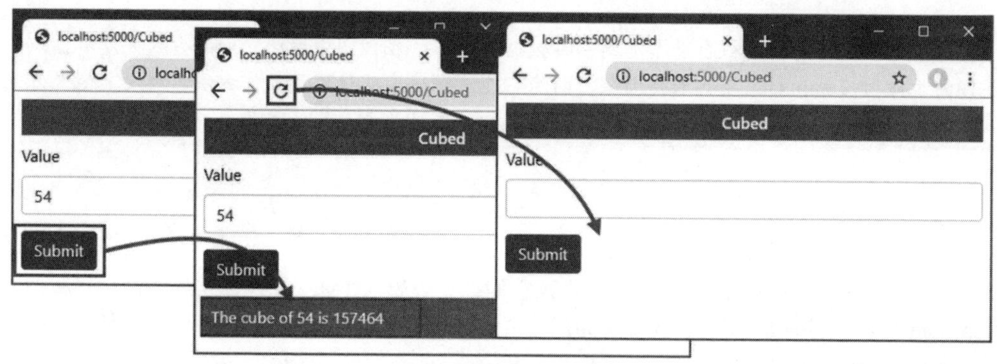


图 22-3　使用临时数据

**提示：** TempData 属性返回的对象提供了一个 Peek 方法，允许获取一个数据值而不将其标记为可以删除，还提供了一个 Keep 方法，用于防止此前读取的值被删除。Keep 方法不会永久保护值。如果再次读取该值，就会再次把它标记为可以删除。如果想存储数据项，使得在处理完请求后也不会删除它们，就需要使用会话数据。

**使用 TempData 特性**

控制器可以定义一些属性，并使用 TempData 特性来装饰它们。这种方法可以代替使用 TempData 属性，如下所示：

```
using Microsoft.AspNetCore.Mvc;
using System;

namespace WebApp.Controllers {
    public class CubedController: Controller {

        public IActionResult Index() {
            return View("Cubed");
        }

        public IActionResult Cube(double num) {
            Value = num.ToString();
            Result = Math.Pow(num, 3).ToString();
            return RedirectToAction(nameof(Index));
        }

        [TempData]
        public string Value { get; set; }

        [TempData]
        public string Result { get; set; }
    }
}
```

赋给这些属性的值将被自动添加到临时数据存储中，在视图中访问它们的方式并没有区别。我倾向于使用 TempData 字典来存储值，因为它能清晰地表明使用了操作方法，方便其他开发人员理解。但这两种方法都是有效的，选择使用哪种方法只是个人喜好问题。

## 22.4　使用布局

示例应用程序中的视图包含重复元素，它们处理设置 HTML 文档、定义 head 节和加载 Bootstrap CSS 文件等工作。Razor 支持布局，可通过将公共内容放到一个文件中，供任何视图使用，来达到避免这种重复的目的。

通常把布局存储到 Views/Shared 文件夹中，因为它们常被多个控制器的操作方法使用。如果

使用的是 Visual Studio，则右击 Views/Shared 子文件夹，从弹出菜单中选择 Add | New Item，然后选择 Razor Layout 模板，如图 22-4 所示。确保文件的名称为_Layout.cshtml，然后单击 Add 按钮来创建新文件。使用代码清单 22-9 中的元素替换 Visual Studio 在该文件中添加的内容。

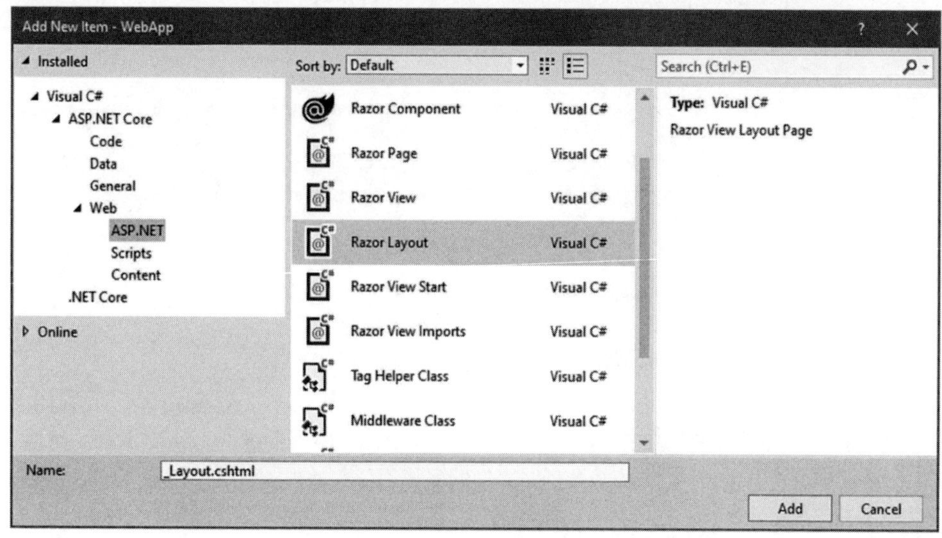


图 22-4 创建布局

如果使用的是 Visual Studio Code，则在 Views/Shared 文件夹中创建一个名为_Layout.cshtml 的文件，并添加代码清单 22-9 中的内容。

### 代码清单 22-9 Views/Shared 文件夹中_Layout.cshtml 文件的内容

```
<!DOCTYPE html>
<html>
<head>
    <link href="/lib/twitter-bootstrap/css/bootstrap.min.css" rel="stylesheet" />
</head>
<body>
    <h6 class="bg-primary text-white text-center m-2 p-2">Shared View</h6>
    @RenderBody()
</body>
</html>
```

布局包含供多个视图使用的公共内容。每个视图所特有的内容是通过调用 RenderBody 方法插入响应的，RazorPage<T>类继承了该方法，如第 21 章所述。使用布局的视图能够只关注它们特有的内容，如代码清单 22-10 所示。

### 代码清单 22-10 在 Views/Home 文件夹的 Index.cshtml 文件中使用布局

```
@model Product
@{
    Layout = "_Layout";
}
```

```
<div class="m-2">
    <table class="table table-sm table-striped table-bordered">
        <tbody>
            <tr><th>Name</th><td>@Model.Name</td></tr>
            <tr>
                <th>Price</th>
                <td>
                    @Model.Price.ToString("c")
                    (@(((Model.Price / ViewBag.AveragePrice)
                        * 100).ToString("F2"))% of average price)
                </td>
            </tr>
            <tr><th>Category ID</th><td>@Model.CategoryId</td></tr>
        </tbody>
    </table>
</div>
```

通过添加代码块(由@{和}字符表示)来设置从 RazorPage<T>类继承的 Layout 属性，可以选择布局。在本例中，将 Layout 属性设置为布局文件的名称。与普通视图一样，指定布局时不需要使用路径或文件扩展名，Razor 引擎将在/Views/[控制器]和/Views/Shared 文件夹中寻找匹配的文件。使用浏览器请求 http://localhost:5000，将看到如图 22-5 所示的响应。

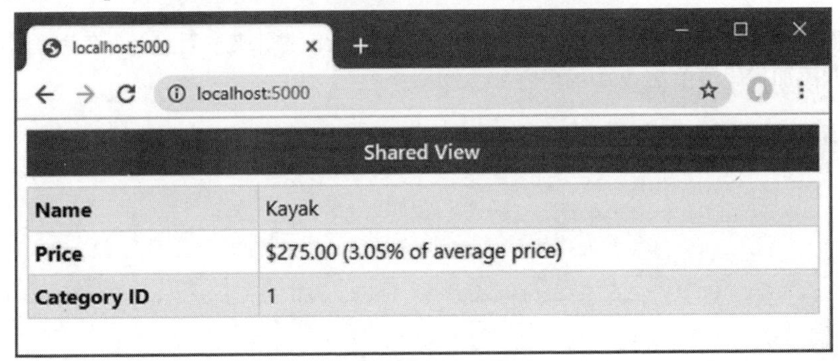


图 22-5　使用布局

### 22.4.1　使用 ViewBag 配置布局

视图可为布局提供数据值，从而允许定制视图提供的公共内容。ViewBag 属性是在选择布局的代码块中定义的，如代码清单 22-11 所示。

**代码清单 22-11　在 Views/Home 文件夹的 Index.cshtml 文件中设置 ViewBag 属性**

```
@model Product
@{
    Layout = "_Layout";
    ViewBag.Title = "Product Table";
}
```

```
<div class="m-2">
    <table class="table table-sm table-striped table-bordered">
        <tbody>
            <tr><th>Name</th><td>@Model.Name</td></tr>
            <tr>
                <th>Price</th>
                <td>
                    @Model.Price.ToString("c")
                    (@(((Model.Price / ViewBag.AveragePrice)
                        * 100).ToString("F2"))% of average price)
                </td>
            </tr>
            <tr><th>Category ID</th><td>@Model.CategoryId</td></tr>
        </tbody>
    </table>
</div>
```

视图设置了一个 Title 属性，可以把它用在布局中，如代码清单 22-12 所示。

**代码清单 22-12　在 Views/Shared 文件夹的_Layout.cshtml 文件中使用 ViewBag 属性**

```
<!DOCTYPE html>
<html>
<head>
    <title>@ViewBag.Title</title>
    <link href="/lib/twitter-bootstrap/css/bootstrap.min.css" rel="stylesheet" />
</head>
<body>
    <h6 class="bg-primary text-white text-center m-2 p-2">
       @(ViewBag.Title ?? "Layout")
    </h6>
    @RenderBody()
</body>
</html>
```

Title 属性用于设置 title 元素和 body 节的 h6 元素的内容。布局不能指望定义了 ViewBag 属性，所以 h6 元素的表达式还提供一个备用值，当视图没有定义 Title 属性时将使用这个备用值。为了查看 ViewBag 属性的效果，使用浏览器请求 http://localhost:5000，这将得到如图 22-6 所示的响应。

### 理解 ViewBag 的优先级

如果视图和操作方法定义了相同的 ViewBag 属性，则将优先采用视图定义的值。如果希望操作方法覆盖视图中定义的值，则在视图代码块中使用如下所示的语句：

```
...
@{
    Layout = "_Layout";
    ViewBag.Title = ViewBag.Title ?? "Product Table";
```

```
}
...
```

只有当操作方法没有定义 Title 属性时，这条语句才会设置 Title 属性的值。

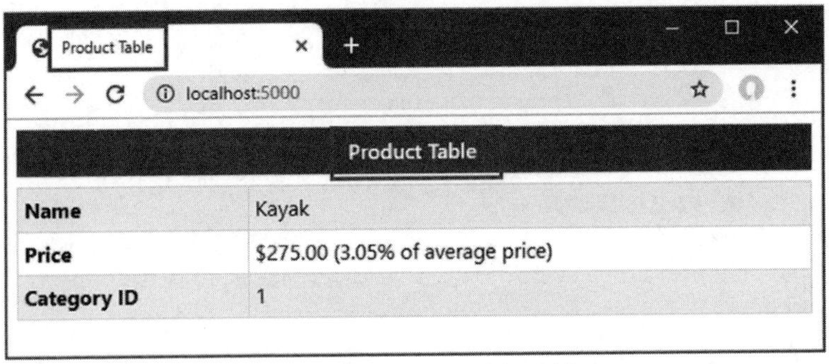

图 22-6　使用 ViewBag 属性来配置布局

## 22.4.2　使用 ViewStart 文件

不必在每个视图中设置 Layout 属性，而是可以在项目中添加一个 ViewStart 文件，用来提供默认的 Layout 值。如果使用的是 Visual Studio，则可以在 Solution Explorer 中右击 Views 文件夹，选择 Add | New Item，并定位到 Razor ViewStart 模板，如图 22-7 所示。确保将文件的名称设置为_ViewStart.cshtml，然后单击 Add 按钮来创建文件，它应该具有如代码清单 22-13 所示的内容。

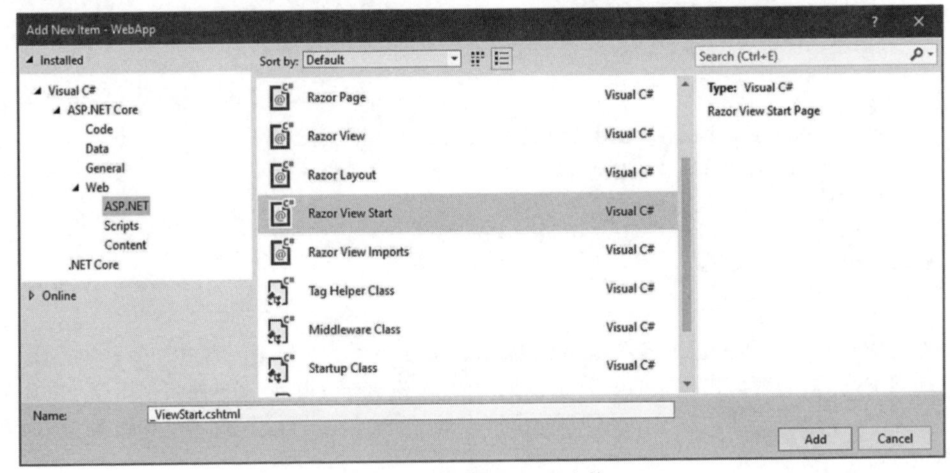


图 22-7　创建 ViewStart 文件

如果使用的是 Visual Studio Code，则在 Views 文件夹中添加一个名为_ViewStart.cshtml 的文件，并添加如代码清单 22-13 所示的内容。

**代码清单 22-13　Views 文件夹中_ViewStart.cshtml 文件的内容**

```
@{
    Layout = "_Layout";
}
```

该文件设置 Layout 属性，其值将用作 Layout 属性的默认值。代码清单 22-14 从 Common.cshtml 文件中删除了布局包含的内容。

**代码清单 22-14　从 Views/Shared 文件夹的 Common.cshtml 文件中删除内容**

```
<h6 class="bg-secondary text-white text-center m-2 p-2">Shared View</h6>
```

因为项目包含一个 ViewStart 文件，所以视图没有定义视图模型类型，也不需要设置 Layout 属性。其结果是，代码清单 22-14 中的内容将被添加到响应的 HTML 内容的 body 节中。使用浏览器导航到 http://localhost:5000/second，将看到如图 22-8 所示的响应。

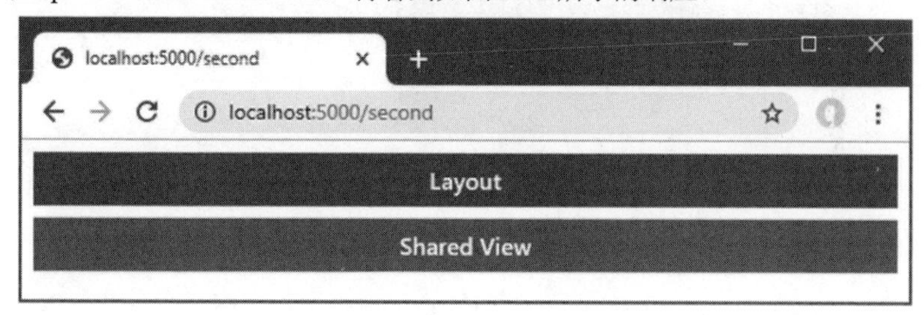


图 22-8　使用 ViewStart 文件

## 22.4.3　覆盖默认布局

在两种情况下，即使在项目中定义了 ViewStart 文件，也可能需要在视图中定义 Layout 属性。在第一种情况下，视图需要的布局与 ViewStart 文件指定的布局不同。为演示这种情况，在 Views/Shared 文件夹中添加一个名为_ImportantLayout.cshtml 的 Razor 布局文件，并为其添加代码清单 22-15 中的内容。

**代码清单 22-15　Views/Shared 文件夹中_ImportantLayout.cshtml 文件的内容**

```
<!DOCTYPE html>
<html>
<head>
    <title>@ViewBag.Title</title>
    <link href="/lib/twitter-bootstrap/css/bootstrap.min.css" rel="stylesheet" />
</head>
<body>
    <h3 class="bg-warning text-white text-center p-2 m-2">Important</h3>
    @RenderBody()
</body>
</html>
```

除了 HTML 文档结构，这个文件还包含一个标题元素，用来以大字体显示 Important 字符。通过将这个布局的名称赋给 Layout 属性，可选择这个布局，如代码清单 22-16 所示。

**提示：** 如果需要为某个控制器的全部操作使用不同的布局，则可以在 Views/[控制器]文件夹中添加一个 ViewStart 文件来选择需要的视图。Razor 引擎将使用控制器特定的 ViewStart 文件指定的布局。

**代码清单 22-16　在 Views/Home 文件夹的 Index.cshtml 文件中使用特定布局**

```
@model Product
@{
    Layout = "_ImportantLayout";
    ViewBag.Title = ViewBag.Title ?? "Product Table";
}
<div class="m-2">
    <table class="table table-sm table-striped table-bordered">
        <tbody>
            <tr><th>Name</th><td>@Model.Name</td></tr>
            <tr>
                <th>Price</th>
                <td>
                    @Model.Price.ToString("c")
                    (@(((Model.Price / ViewBag.AveragePrice)
                        * 100).ToString("F2"))% of average price)
                </td>
            </tr>
            <tr><th>Category ID</th><td>@Model.CategoryId</td></tr>
        </tbody>
    </table>
</div>
```

ViewStart 文件中的 Layout 值将被视图中的值覆盖，从而允许应用不同的布局。使用浏览器请求 http://localhost:5000，将使用新布局生成响应，如图 22-9 所示。

**使用代码选择布局**

视图赋给 Layout 属性的值可以是一个表达式的结果，从而允许视图根据表达式的结果选择不同的布局，这类似于操作方法可以选择不同的视图。下面的示例基于视图模型对象定义的一个属性来选择布局：

```
...
@model Product
@{
    Layout = Model.Price > 100 ? "_ImportantLayout" : "_Layout";
    ViewBag.Title = ViewBag.Title ?? "Product Table";
}
...
```

当视图模型对象的 Price 属性的值大于 100 时，将选择名为_ImportantLayout 的布局，否则将选择名为_Layout 的布局。

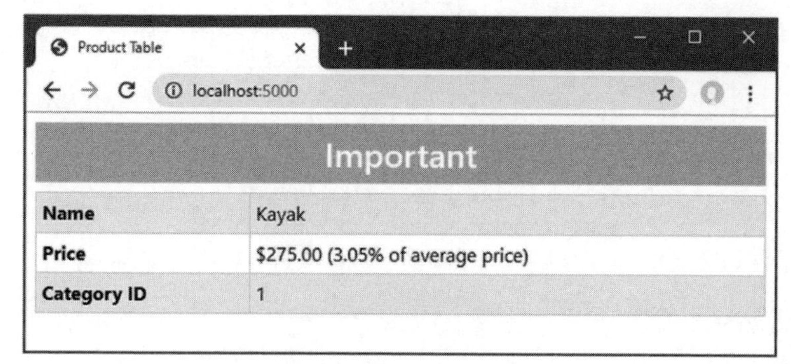


图 22-9 在视图中指定布局

需要使用 Layout 属性的第二种情况是，视图包含一个完整的 HTML 文档，并不需要一个布局。为理解这个问题，可打开一个新的 PowerShell 命令提示符，运行代码清单 22-17 中的命令。

**代码清单 22-17 发送 HTTP 请求**

```
Invoke-WebRequest http://localhost:5000/home/list | Select-Object -expand Content
```

此命令发送一个 HTTP GET 请求，其响应是使用 Views/Home 文件夹中的 List.cshtml 文件来生成的。此视图包含一个完整的 HTML 文档，它将与 ViewStart 文件指定的视图的内容合并起来，从而得到格式不正确的 HTML 文档，如下所示：

```
<!DOCTYPE html>
<html>
<head>
    <title></title>
    <link href="/lib/twitter-bootstrap/css/bootstrap.min.css" rel="stylesheet" />
</head>
<body>
    <h6 class="bg-primary text-white text-center m-2 p-2">
       Layout
    </h6>
    <!DOCTYPE html>
<html>
<head>
   <link href="/lib/twitter-bootstrap/css/bootstrap.min.css" rel="stylesheet" />
</head>
<body>
    <h6 class="bg-secondary text-white text-center m-2 p-2">Products</h6>
    <div class="m-2">
        <table class="table table-sm table-striped table-bordered">
            <thead>
               <tr><th>Name</th><th>Price</th></tr>
            </thead>
            <tbody>
                   <tr><td>Kayak</td><td>275.00</td></tr>
```

```
                        <tr><td>Lifejacket</td><td>48.95</td></tr>
                        <tr><td>Soccer Ball</td><td>19.50</td></tr>
                        <tr><td>Corner Flags</td><td>34.95</td></tr>
                        <tr><td>Stadium</td><td>79500.00</td></tr>
                        <tr><td>Thinking Cap</td><td>16.00</td></tr>
                        <tr><td>Unsteady Chair</td><td>29.95</td></tr>
                        <tr><td>Human Chess Board</td><td>75.00</td></tr>
                        <tr><td>Bling-Bling King</td><td>1200.00</td></tr>
                </tbody>
            </table>
        </div>
</body>
</html>
</body>
</html>
```

HTML 文档的结构元素重复出现，有两个 html、head、body 和 link 元素。浏览器擅长处理格式不正确的 HTML，但并非始终能够处理结构不合适的内容。当视图包含一个完整的 HTML 文档时，可将 Layout 属性设置为 null，如代码清单 22-18 所示。

#### 代码清单 22-18　在 Views/Home 文件夹的 List.cshtml 文件中禁用布局

```
@model IEnumerable<Product>
@{
    Layout = null;
}
<!DOCTYPE html>
<html>
<head>
    <link href="/lib/twitter-bootstrap/css/bootstrap.min.css" rel="stylesheet" />
</head>
<body>
    <h6 class="bg-secondary text-white text-center m-2 p-2">Products</h6>
    <div class="m-2">
        <table class="table table-sm table-striped table-bordered">
            <thead>
                <tr><th>Name</th><th>Price</th></tr>
            </thead>
            <tbody>
                @foreach (Product p in Model) {
                    <tr><td>@p.Name</td><td>@p.Price</td></tr>
                }
            </tbody>
        </table>
    </div>
</body>
</html>
```

保存视图，并再次运行代码清单 22-17 中的命令，将看到响应只包含视图中的元素，布局已被禁用。

```
<!DOCTYPE html>
<html>
<head>
   <link href="/lib/twitter-bootstrap/css/bootstrap.min.css" rel="stylesheet" />
</head>
<body>
    <h6 class="bg-secondary text-white text-center m-2 p-2">Products</h6>
    <div class="m-2">
        <table class="table table-sm table-striped table-bordered">
            <thead>
                <tr><th>Name</th><th>Price</th></tr>
            </thead>
            <tbody>
                    <tr><td>Kayak</td><td>275.00</td></tr>
                    <tr><td>Lifejacket</td><td>48.95</td></tr>
                    <tr><td>Soccer Ball</td><td>19.50</td></tr>
                    <tr><td>Corner Flags</td><td>34.95</td></tr>
                    <tr><td>Stadium</td><td>79500.00</td></tr>
                    <tr><td>Thinking Cap</td><td>16.00</td></tr>
                    <tr><td>Unsteady Chair</td><td>29.95</td></tr>
                    <tr><td>Human Chess Board</td><td>75.00</td></tr>
                    <tr><td>Bling-Bling King</td><td>1200.00</td></tr>
            </tbody>
        </table>
    </div>
</body>
</html>
```

## 22.4.4 使用布局节

Razor 视图引擎支持节的概念，从而允许在布局内提供区域内容。对于控制将视图的哪些部分插入布局中，以及把它们放到什么位置，Razor 节提供了更大程度的控制权。为了演示节，按照代码清单 22-19 所示编辑/Views/Home/Index.cshtml 文件。

**代码清单 22-19　在 Views/Home 文件夹的 Index.cshtml 文件中定义节**

```
@model Product
@{
    Layout = "_Layout";
    ViewBag.Title = ViewBag.Title ?? "Product Table";
}

@section Header {
   Product Information
```

```
}

<tr><th>Name</th><td>@Model.Name</td></tr>
<tr>
    <th>Price</th>
    <td>@Model.Price.ToString("c")</td>
</tr>
<tr><th>Category ID</th><td>@Model.CategoryId</td></tr>

@section Footer {
    @(((Model.Price / ViewBag.AveragePrice)
       * 100).ToString("F2"))% of average price
}
```

通过使用 Razor @section 表达式，并在其后添加节的名称，可以定义节。代码清单 22-19 定义了名为 Header 和 Footer 的节并且节中同样可以混合 HTML 内容和表达式，就像视图的主要部分一样。使用@RenderSection 表达式可在布局内应用节，如代码清单 22-20 所示。

#### 代码清单 22-20　在 Views/Shared 文件夹的_Layout.cshtml 文件中使用节

```
<!DOCTYPE html>
<html>
<head>
    <title>@ViewBag.Title</title>
    <link href="/lib/twitter-bootstrap/css/bootstrap.min.css" rel="stylesheet" />
</head>
<body>
    <div class="bg-info text-white m-2 p-1">
       This is part of the layout
    </div>

    <h6 class="bg-primary text-white text-center m-2 p-2">
       @RenderSection("Header")
    </h6>

    <div class="bg-info text-white m-2 p-1">
       This is part of the layout
    </div>

    <div class="m-2">
        <table class="table table-sm table-striped table-bordered">
            <tbody>
                @RenderBody()
            </tbody>
        </table>
    </div>

    <div class="bg-info text-white m-2 p-1">
```

```
        This is part of the layout
    </div>

    <h6 class="bg-primary text-white text-center m-2 p-2">
        @RenderSection("Footer")
    </h6>

    <div class="bg-info text-white m-2 p-1">
        This is part of the layout
    </div>
</body>
</html>
```

应用布局时，RenderSection 表达式将把指定节的内容插入响应中。视图中没有包含在节内的区域由 RenderBody 方法插入响应中。为了解如何应用节，使用浏览器请求 http://localhost:5000，这将得到如图 22-10 所示的响应。

■ **注意：**视图只能定义布局中引用的节。如果在视图中定义了节，但布局中没有对应的@RenderSection 表达式，视图引擎将抛出异常。

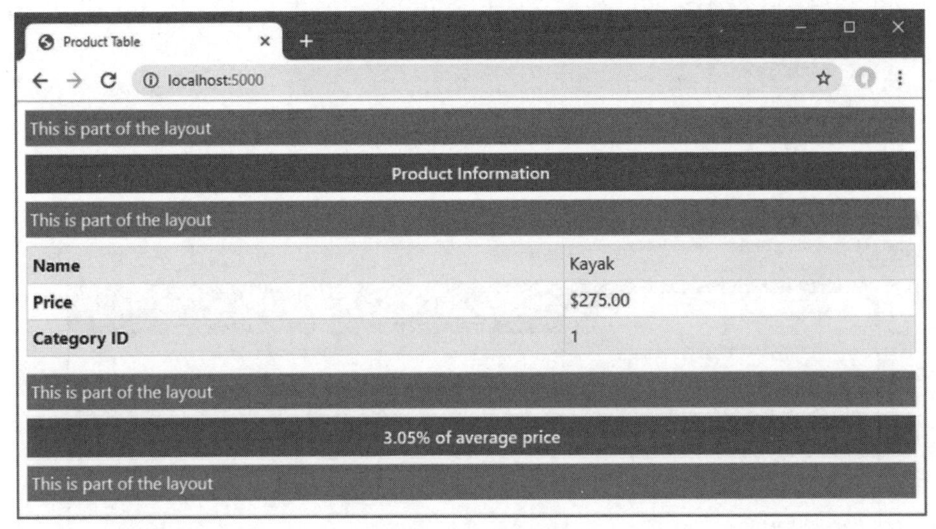


图 22-10　在布局中使用节

节允许视图向布局提供内容片段，而不指定如何使用它们。例如，代码清单 22-21 重新定义了布局，以便把 body 和节合并到一个 HTML 表格中。

#### 代码清单 22-21　在 Views/Shared 文件夹的_Layout.cshtml 文件中使用表格

```
<!DOCTYPE html>
<html>
<head>
    <title>@ViewBag.Title</title>
```

```
        <link href="/lib/twitter-bootstrap/css/bootstrap.min.css" rel="stylesheet" />
    </head>
    <body>
        <div class="m-2">
            <table class="table table-sm table-striped table-bordered">
                <thead>
                    <tr>
                        <th class="bg-primary text-white text-center" colspan="2">
                            @RenderSection("Header")
                        </th>
                    </tr>
                </thead>
                <tbody>
                    @RenderBody()
                </tbody>
                <tfoot>
                    <tr>
                        <th class="bg-primary text-white text-center" colspan="2">
                            @RenderSection("Footer")
                        </th>
                    </tr>
                </tfoot>
            </table>
        </div>
    </body>
</html>
```

为了查看对视图所做的修改的效果，使用浏览器请求 http://localhost:5000，将得到如图 22-11 所示的响应。

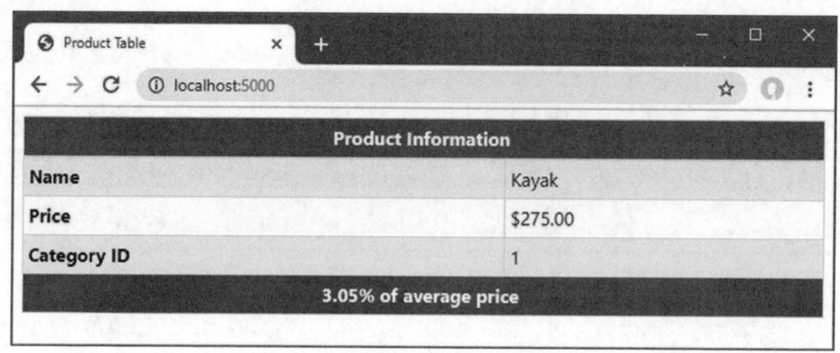


图 22-11　修改在布局中显示节的方式

### 1. 使用可选布局节

默认情况下，如果布局中存在 RenderSection 调用，就必须在视图中包含对应的节。如果布局需要一个节，但视图没有定义该节，就会抛出异常。代码清单 22-22 添加了一个对 RenderSection 方法的调用，它需要一个名为 Summary 的节。

**代码清单 22-22 在 Views/Shared 文件夹的_Layout.cshtml 文件中添加一个节**

```
<!DOCTYPE html>
<html>
<head>
    <title>@ViewBag.Title</title>
    <link href="/lib/twitter-bootstrap/css/bootstrap.min.css" rel="stylesheet" />
</head>
<body>
    <div class="m-2">
        <table class="table table-sm table-striped table-bordered">
            <thead>
                <tr>
                    <th class="bg-primary text-white text-center" colspan="2">
                        @RenderSection("Header")
                    </th>
                </tr>
            </thead>
            <tbody>
                @RenderBody()
            </tbody>
            <tfoot>
                <tr>
                    <th class="bg-primary text-white text-center" colspan="2">
                        @RenderSection("Footer")
                    </th>
                </tr>
            </tfoot>
        </table>
    </div>
    @RenderSection("Summary")
</body>
</html>
```

使用浏览器请求 http://localhost:5000，将看到如图 22-12 所示的异常。

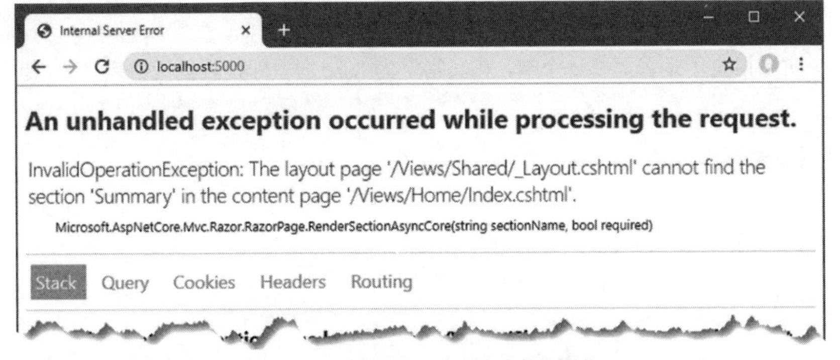


图 22-12 试图渲染一个不存在的视图节

解决这个问题有两种方法。第一种方法是创建一个可选节，只有当视图定义了该节时才渲染该节。通过向 RenderSection 传递第二个实参，可以创建可选节，如代码清单 22-23 所示。

**代码清单 22-23 在 Views/Shared 文件夹的_Layout.cshtml 文件中定义可选节**

```
<!DOCTYPE html>
<html>
<head>
    <title>@ViewBag.Title</title>
    <link href="/lib/twitter-bootstrap/css/bootstrap.min.css" rel="stylesheet" />
</head>
<body>
    <div class="m-2">
        <table class="table table-sm table-striped table-bordered">
            <thead>
                <tr>
                    <th class="bg-primary text-white text-center" colspan="2">
                        @RenderSection("Header", false)
                    </th>
                </tr>
            </thead>
            <tbody>
                @RenderBody()
            </tbody>
            <tfoot>
                <tr>
                    <th class="bg-primary text-white text-center" colspan="2">
                        @RenderSection("Footer", false)
                    </th>
                </tr>
            </tfoot>
        </table>
    </div>
    @RenderSection("Summary", false)
</body>
</html>
```

第二个实参指定了是否需要一个节，将其指定为 false 时，即使视图没有定义节，也不会抛出异常。

### 2. 测试布局节

IsSectionDefined 方法用于判断视图是否定义了指定的节，可以将其用在 if 表达式中来渲染后备内容，如代码清单 22-24 所示。

**代码清单 22-24　在 Views/Shared 文件夹的_Layout.cshtml 文件中检查节**

```
<!DOCTYPE html>
<html>
<head>
    <title>@ViewBag.Title</title>
    <link href="/lib/twitter-bootstrap/css/bootstrap.min.css" rel="stylesheet" />
</head>
<body>
    <div class="m-2">
        <table class="table table-sm table-striped table-bordered">
            <thead>
                <tr>
                    <th class="bg-primary text-white text-center" colspan="2">
                       @RenderSection("Header", false)
                    </th>
                </tr>
            </thead>
            <tbody>
               @RenderBody()
            </tbody>
            <tfoot>
                <tr>
                    <th class="bg-primary text-white text-center" colspan="2">
                       @RenderSection("Footer", false)
                    </th>
                </tr>
            </tfoot>
        </table>
    </div>
    @if (IsSectionDefined("Summary")) {
       @RenderSection("Summary", false)
    } else {
        <div class="bg-info text-center text-white m-2 p-2">
           This is the default summary
        </div>
    }
</body>
</html>
```

使用想要检查的节的名称来调用 IsSectionDefined 方法，如果视图定义了该节，这个方法将返回 true。在示例中，使用这个帮助方法检查视图是否定义了 Summary 节，当没有定义该节时，就渲染后备内容。为查看后备内容，使用浏览器请求 http://localhost:5000，将得到如图 22-13 所示的响应。

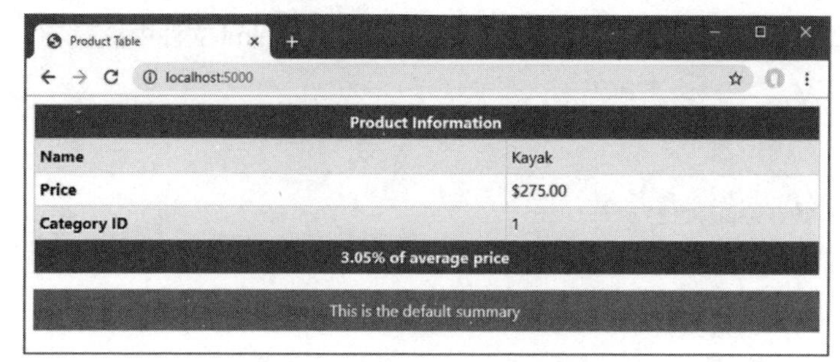


图 22-13　显示视图节的后备内容

# 22.5　使用分部视图

经常需要在几个不同位置使用相同的一组 HTML 元素和表达式。分部视图是视图，其内容是将要包含在其他视图中的内容片段，目的是生成复杂的响应，却不会造成重复。

## 22.5.1　启用分部视图

通过使用一种称为标签助手的功能(第 25 章将会详细介绍)，可以应用分部视图；在视图导入文件(在第 21 章中将其添加到项目)中配置标签助手。为了启用支持分部视图的功能，需要在_ViewImports.cshtml 文件中添加代码清单 22-25 中的语句。

**代码清单 22-25　在 Views 文件夹的_ViewImports.cshtml 文件中启用标签助手**

```
@using WebApp.Models
@addTagHelper *, Microsoft.AspNetCore.Mvc.TagHelpers
```

## 22.5.2　创建分部视图

分部视图就是普通的 CSHTML 文件，只是使用方式把它们与标准视图区分开。如果使用的是 Visual Studio，可以右击 Views/Home 文件夹，选择 Add | New Item，然后使用 Razor View 模板创建一个名为_RowPartial.cshtml 的文件。创建该文件后，将其内容替换为代码清单 22-26 中的内容。如果使用的是 Visual Studio Code，则在 Views/Home 文件夹中添加一个名为_RowPartial.cshtml 的文件，然后在其中添加如代码清单 22-26 所示的内容。

**提示：** Visual Studio 为创建预填充的分部视图提供了一些工具支持，但创建分部视图最简单的方法是使用 Razor View 模板来创建一个普通视图。

**代码清单 22-26　Views/Home 文件夹中_RowPartial.cshtml 文件的内容**

```
@model Product

<tr>
    <td>@Model.Name</td>
    <td>@Model.Price</td>
</tr>
```

model 表达式用于为分部视图定义视图模型类型；分部视图与普通视图一样，也可以混合表达式和 HTML 元素。这个分部视图的内容创建一个表格行，使用 Product 对象的 Name 和 Price 属性来填充表格单元格。

### 22.5.3 应用分部视图

通过在另一个视图或布局中添加 partial 元素，可以应用分部视图。在代码清单 22-27 中，将该元素添加到 List.cshtml 文件中，以便能够使用分部视图来生成表格中的行。

**代码清单 22-27 在 Views/Home 文件夹的 List.cshtml 文件中使用分部视图**

```
@model IEnumerable<Product>
@{
   Layout = null;
}
<!DOCTYPE html>
<html>
<head>
    <link href="/lib/twitter-bootstrap/css/bootstrap.min.css" rel="stylesheet" />
</head>
<body>
    <h6 class="bg-secondary text-white text-center m-2 p-2">Products</h6>
    <div class="m-2">
        <table class="table table-sm table-striped table-bordered">
            <thead>
                <tr><th>Name</th><th>Price</th></tr>
            </thead>
            <tbody>
                @foreach (Product p in Model) {
                    <partial name="_RowPartial" model="p" />
                }
            </tbody>
        </table>
    </div>
</body>
</html>
```

应用到 partial 元素的特性控制着分部视图的选择和配置，如表 22-2 所述。

表 22-2 partial 元素的特性

| 名称 | 描述 |
|---|---|
| name | 此特性指定了分部视图的名称，搜索该分部视图的方式与搜索普通视图相同 |
| model | 此特性指定的值用作分部视图的视图模型对象 |
| for | 此特性用于定义一个表达式，该表达式将为分部视图选择视图模型对象，接下来将进行介绍 |
| view-data | 此特性用于为分部视图提供额外数据 |

代码清单 22-27 中的 partial 元素使用 name 特性来选择_RowPartial 视图，使用 model 特性来选择 Product 对象，它将用作视图模型对象。在@foreach 表达式中应用 partial 元素，意味着将使用它来生成表格中的每一行。使用浏览器请求 http://localhost:5000/home/list，将看到如图 22-14 所示的响应，其中展示了使用 partial 元素生成的表格行。

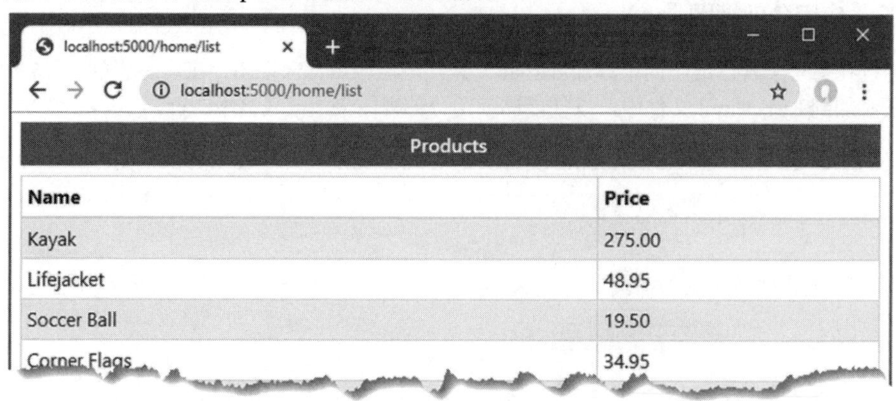


图 22-14　使用分部视图

**使用 HTML 助手来应用分部视图**

在 ASP.NET Core 的早期版本中，通过在从视图生成的 C#类(如第 21 章所述)上添加 Html 属性来应用分部视图。Html 属性返回的对象实现了 IHtmlHelper 接口，通过该接口可应用视图，如下所示:

```
...
@Html.Partial("_RowPartial")
...
```

现在仍然支持这类表达式，但 partial 元素提供了更优雅的方法，能够与视图中的其余 HTML 元素保持一致。

### 使用表达式选择分部视图模型

for 特性使用一个应用到视图模型的表达式，设置分部视图的模型。演示这种功能比描述这种功能简单得多。在 Views/Home 文件夹中添加一个名为_CellPartial.cshtml 的分部视图，并为其添加代码清单 22-28 中的内容。

**代码清单 22-28　Views/Home 文件夹中_CellPartial.cshtml 文件的内容**

```
@model string

<td class="bg-info text-white">@Model</td>
```

这个分部视图有一个字符串视图模型对象，用作表格单元格元素的内容；表格单元格元素的样式是使用 Bootstrap CSS 框架设置的。代码清单 22-29 在_RowPartial.cshtml 文件中添加了一个 partial 元素，使用_CellPartial 分部视图为 Product 对象的名称显示一个单元格。

**代码清单 22-29　在 Views/Home 文件夹的_RowPartial.cshtml 文件中使用分部视图**

```
@model Product

<tr>
    <partial name="_CellPartial" for="Name" />
    <td>@Model.Price</td>
</tr>
```

for特性选择Name属性作为_CellPartial分部视图的模型。为查看效果，使用浏览器请求http://localhost:5000/home/list，将得到如图 22-15 所示的响应。

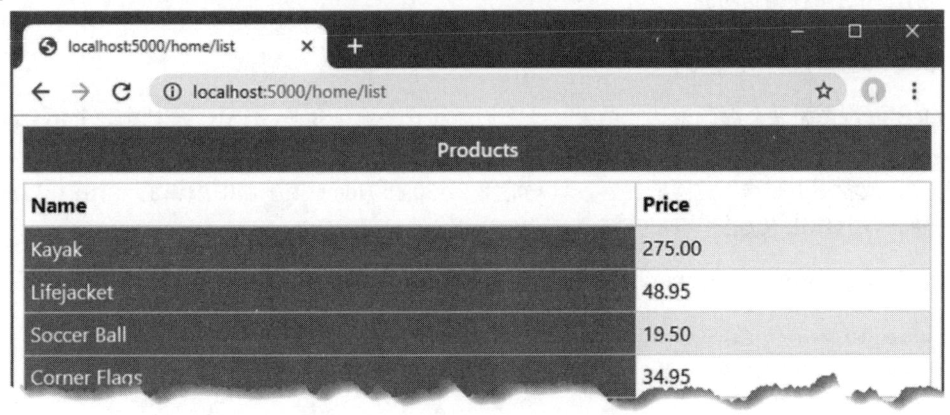


图 22-15　在分部视图中选择使用模型属性

## 使用模板化委托

模板化委托是避免在视图中发生重复的另一种方法。需要在代码块中定义模板化委托，如下所示:

```
...
@{
    Func<Product, object> row
        = @<tr><td>@item.Name</td><td>@item.Price</td></tr>;
}
...
```

模板是一个函数，它接收一个 Product 输入对象，返回一个动态结果。在模板表达式中，使用 item 来引用输入对象。模板化委托作为一个方法表达式调用，用来生成内容。

```
...
<tbody>
    @foreach (Product p in Model) {
        @row(p)
    }
</tbody>
...
```

我觉得这种功能很难使用，所以更喜欢使用分部视图，不过这是个人喜好和习惯问题，而不是说模板化委托的工作方式客观存在问题。

# 22.6 理解内容编码

Razor 视图为编码内容提供了两种有用的功能。HTML 内容编码功能确保了表达式响应不会修改发送给浏览器的响应的结构，这是一项重要的安全功能。JSON 编码功能将对象编码为 JSON，并将其插入响应中，这有助于调试，并且在把数据提供给 JavaScript 应用程序时很有用。接下来将分别介绍这两种编码功能。

## 22.6.1 理解 HTML 编码

Razor 视图引擎会编码表达式结果，使它们能够安全地包含在 HTML 文档中，并不会改变 HTML 文档的结构。在处理用户提供的内容时，这是一个重要功能，因为用户可能会试图破坏应用程序，或者不小心输入危险的内容。代码清单 22-30 在 Home 控制器中添加了一个操作方法，用来将一段 HTML 传递给 View 方法。

**代码清单 22-30 在 Controllers 文件夹的 HomeController.cs 文件中添加一个操作**

```
using Microsoft.AspNetCore.Mvc;
using System.Threading.Tasks;
using WebApp.Models;
using Microsoft.EntityFrameworkCore;

namespace WebApp.Controllers {

    public class HomeController: Controller {
        private DataContext context;

        public HomeController(DataContext ctx) {
            context = ctx;
        }

        public async Task<IActionResult> Index(long id = 1) {
            ViewBag.AveragePrice = await context.Products.AverageAsync(p => p.Price);
            return View(await context.Products.FindAsync(id));
        }

        public IActionResult List() {
            return View(context.Products);
        }

        public IActionResult Html() {
            return View((object)"This is a <h3><i>string</i></h3>");
        }
```

```
    }
}
```

新操作传递一个包含 HTML 元素的字符串。要为新操作方法创建视图，可在 Views/Home 文件夹中添加一个名为 Html.cshtml 的 Razor 视图文件，并为其添加代码清单 22-31 中的内容。

---

▒ **提示：** 这里将传递给 View 方法的字符串强制转换为一个对象，如果不这么做，将认为字符串是视图(而不是视图模型对象)的名称。

---

### 代码清单 22-31　Views/Home 文件夹中 Html.cshtml 文件的内容

```
@model string
@{
    Layout = null;
}
<!DOCTYPE html>
<html>
<head>
    <link href="/lib/twitter-bootstrap/css/bootstrap.min.css" rel="stylesheet" />
</head>
<body>
    <div class="bg-secondary text-white text-center m-2 p-2">@Model</div>
</body>
</html>
```

重新启动 ASP.NET Core，使用浏览器请求 http://localhost:5000/home/html。其响应如图 22-17 左侧所示，它显示了视图模型字符串中具有潜在风险的字符如何被转义。

要原样包含表达式的结果，而不做安全编码，可以调用Html.Raw方法。Html属性是添加到生成的视图类(如第21章所述)的属性之一，它返回的对象实现了IHtmlHelper接口，如代码清单22-32所示。

### 代码清单 22-32　在 Views/Home 文件夹的 Html.cshtml 文件中禁用编码

```
@model string
@{
    Layout = null;
}
<!DOCTYPE html>
<html>
<head>
    <link href="/lib/twitter-bootstrap/css/bootstrap.min.css" rel="stylesheet" />
</head>
<body>
    <div class="bg-secondary text-white text-center m-2 p-2">@Html.Raw(Model)</div>
</body>
</html>
```

再次请求 http://localhost:5000/home/html，将看到视图模型字符串被直接传递，并没有被编码，浏览器将其解释为 HTML 文档的一部分，如图 22-16 的右侧所示。

**警告**：除非完全确信不会向视图传递恶意内容，否则不要禁用安全编码。滥用这种功能可能为应用程序及用户带来安全风险。

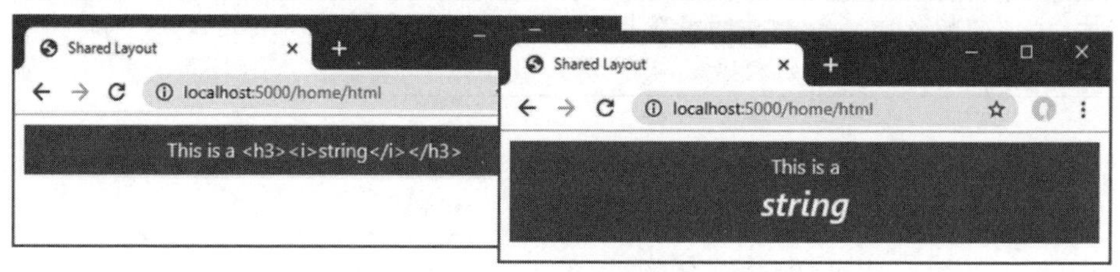


图 22-16　HTML 结果编码

## 22.6.2　理解 JSON 编码

Json 属性将被添加到从视图生成的类中，如第 21 章所述。它可以用来将对象编码为 JSON。如前面的章节所述，JSON 数据最常用于 RESTful Web 服务器中，不过，我发现在没有从视图得到期望的输出时，Razor 的 JSON 编码功能可作为一个非常有用的调试帮手。代码清单 22-33 将视图模型对象的 JSON 表示添加到 Index 视图生成的输出中。

**代码清单 22-33　在 Views/Home 文件夹的 Index.cshtml 文件中使用 JSON 编码**

```
@model Product
@{
    Layout = "_Layout";
    ViewBag.Title = ViewBag.Title ?? "Product Table";
}

@section Header {
   Product Information
}

<tr><th>Name</th><td>@Model.Name</td></tr>
<tr>
    <th>Price</th>
    <td>@Model.Price.ToString("c")</td>
</tr>
<tr><th>Category ID</th><td>@Model.CategoryId</td></tr>

@section Footer {
    @(((Model.Price / ViewBag.AveragePrice)
       * 100).ToString("F2"))% of average price
}

@section Summary {
    <div class="bg-info text-white m-2 p-2">
```

```
        @Json.Serialize(Model)
    </div>
}
```

Json 属性返回 IJsonHelper 接口的一个实现，其 Serialize 方法生成对象的 JSON 表示。使用浏览器请求 http://localhost:5000，将看到如图 22-17 所示的输出，它在视图的摘要部分包含 JSON。

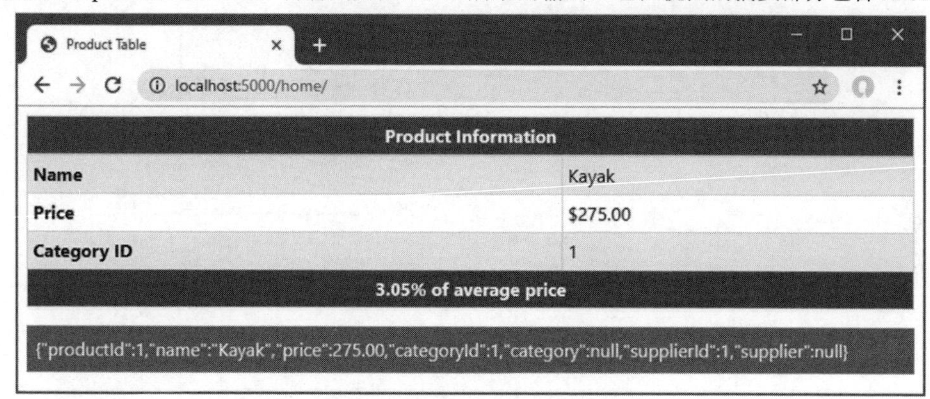


图 22-17 将表达式结果编码为 JSON

## 22.7 小结

本章继续探讨了 Razor 视图中可用的功能，展示了如何使用 ViewBag，如何使用布局和分部视图来处理公共内容，以及如何管理表达式结果的编码。第 23 章将介绍 Razor Pages，这是生成 HTML 响应的另外一种方式。

.NET 开发经典名著

# ASP.NET Core 3 高级编程(第 8 版)

## (下册)

[英] 亚当·弗里曼(Adam Freeman) 著
杜静芬 程凤娟 译

清华大学出版社
北 京

北京市版权局著作权合同登记号　图字：01-2021-3335
EISBN：978-1-4842-5439-4
First published in English under the title
Pro ASP.NET Core 3: Develop Cloud-Ready Web Applications Using MVC, Blazor, and Razor Pages, Eighth Edition
By Adam Freeman


**图书在版编目(CIP)数据**

ASP.NET Core 3高级编程：第8版 / (英)亚当•弗里曼 (Adam Freeman) 著；杜静芬，程凤娟译. —北京：清华大学出版社，2021.6
(.NET 开发经典名著)
书名原文：Pro ASP.NET Core 3: Develop Cloud-Ready Web Applications Using MVC, Blazor, and Razor Pages, Eighth Edition
ISBN 978-7-302-58271-7

Ⅰ. ①A…　Ⅱ. ①亚…　②杜…　③程…　Ⅲ. 网页制作工具—程序设计　Ⅳ. ①TP393.092.2

中国版本图书馆 CIP 数据核字(2021)第 101201 号

**责任编辑**：王　军　韩宏志
**装帧设计**：孔祥峰
**责任校对**：成凤进
**责任印制**：宋　林

**出版发行**：清华大学出版社
网　　址：http://www.tup.com.cn，http://www.wqbook.com
地　　址：北京清华大学学研大厦 A 座　　　　邮　　编：100084
社 总 机：010-62770175　　　　邮　　购：010-62786544
投稿与读者服务：010-62776969，c-service@tup.tsinghua.edu.cn
质 量 反 馈：010-62772015，zhiliang@tup.tsinghua.edu.cn
印 装 者：三河市国英印务有限公司
经　　销：全国新华书店
开　　本：170mm×240mm　　印　　张：73　　字　　数：2011 千字
版　　次：2021 年 7 月第 1 版　　印　　次：2021 年 7 月第 1 次印刷
定　　价：268.00 元(全二册)

---

产品编号：089796-01

# 第 23 章

# 使用 Razor Pages

本章将介绍 Razor Pages，这是一种更简单地生成 HTML 内容的方法，供对遗留的 ASP.NET Web Pages 框架仍然心存热情的开发人员使用。本章将解释 Razor Pages 的工作方式，它们与 MVC 框架采取的控制器和视图方法有什么区别，以及它们在更广大的 ASP.NET Core 平台中扮演什么角色。

解释 Razor Pages 工作方式的过程，能够帮助将其与前面章节介绍的控制器和视图方法的区别降到最低。你可能认为，Razor Pages 只不过是轻量级 MVC，所以对其不加重视，但这种做法并不合适。Razor Pages 之所以值得关注，原因在于开发体验，而不是它们的实现方式。

我的建议是，尝试使用 Razor Pages，如果你是经验丰富的 MVC 开发人员，就更应该给 Razor Pages 一个机会。虽然用到的技术会让你感到熟悉，但创建应用程序功能的过程并不相同，并且非常适合小型的、关注特定某个点的功能，这种功能的规模和复杂性并没有达到需要使用控制器和视图的程度。从 MVC 框架问世以来，我就开始使用它，不过必须承认，我并没有重视 Razor Pages 的早期版本。但是，现在我会在大部分项目中混合使用 Razor Pages 和 MVC 框架，就像在第 I 部分的 SportsStore 示例中所做的那样。表 23-1 介绍了 Razor Pages 的相关知识。

表 23-1　Razor Pages 的相关知识

| 问题 | 答案 |
| --- | --- |
| Razor Pages 是什么？ | Razor Pages 是生成 HTML 响应的一种简化方式 |
| Razor Pages 为什么有用？ | Razor Pages 的简单性，意味着能够比 MVC 框架更快获得结果，因为 MVC 框架需要相对复杂的准备过程。经验欠缺的 Web 开发人员也更容易理解 Razor Pages，因为在 Razor Pages 中，代码与内容的关系更明显 |
| 如何使用 Razor Pages？ | Razor Pages 将单个视图与一个类关联起来，由该类为视图提供功能，并使用基于文件的路由系统来匹配 URL |
| 存在陷阱或限制吗？ | Razor Pages 的灵活性不如 MVC 框架，所以不适合复杂的应用程序。Razor Pages 只能用于生成 HTML 响应，不能用于创建 RESTful Web 服务 |
| 存在替代方法吗？ | 可以使用 MVC 框架的控制器和视图方法替代 Razor Pages |

表 23-2 总结了本章的内容。

表 23-2　本章内容摘要

| 问题 | 解决方案 | 代码清单 |
| --- | --- | --- |
| 启用 Razor Pages | 使用 AddRazorPages 和 MapRazorPages 来设置必要的服务和中间件 | 23-3 |
| 创建自包含的端点 | 创建一个 Razor Pages | 23-4、23-26、23-27 |
| 将请求路由到 Razor Pages | 使用@page 指令使用页面的名称或指定一个路由 | 23-5～23-8 |
| 为 Razor Pages 的视图节提供逻辑支持 | 使用页面模型类 | 23-9～23-12 |
| 创建不使用 Razor Pages 渲染的结果 | 定义一个处理程序方法来返回一个操作结果 | 23-13～23-15 |
| 处理多个 HTTP 方法 | 在页面模型类中定义处理程序 | 23-16～23-18 |
| 避免内容重复 | 使用布局或分部视图 | 23-19～23-25 |

# 23.1　准备工作

本章使用第 22 章创建的 WebApp 项目。打开一个新的 PowerShell 命令提示符，导航到包含 WebApp.csproj 文件的文件夹，并运行代码清单 23-1 中显示的命令来删除数据库。

**提示：** 从 https://github.com/apress/pro-asp.net-core-3 可下载本章以及本书其他章节的示例项目。如果无法成功运行示例，请参考第 1 章的说明。

**代码清单 23-1　删除数据库**

```
dotnet ef database drop --force
```

## 运行示例应用程序

删除数据库后，从 Debug 菜单中选择 Start Without Debugging 或 Run Without Debugging，或者使用 PowerShell 命令提示符来运行代码清单 23-2 中显示的命令。

**代码清单 23-2　运行示例应用程序**

```
dotnet run
```

在应用程序启动过程中，将对数据库执行 seed 操作。当 ASP.NET Core 开始运行后，使用 Web 浏览器请求 http://localhost:5000，将得到如图 23-1 所示的响应。

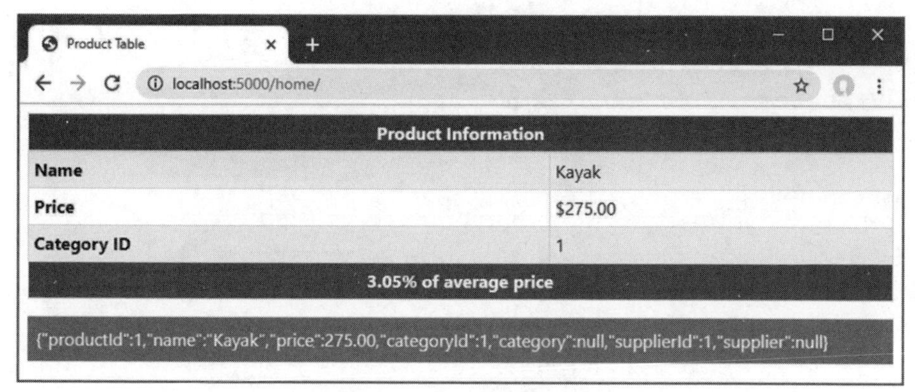


图 23-1 运行示例应用程序

## 23.2 理解 Razor Pages

在学习 Razor Pages 的工作方式的过程中，将发现它们与 MVC 框架有一些相同的功能。事实上，Razor Pages 通常被描述为 MVC 框架的简化版本，这么说并没有错，但不能帮助理解为什么 Razor Pages 很有用。

MVC 框架以相同的方式解决每个问题：控制器定义了操作方法，这些操作方法选择视图来生成响应。这种方案之所以能够执行，是因为它十分灵活：控制器可以定义多个操作方法来响应不同的请求，操作方法可以决定在处理请求时使用哪个视图，而视图则可以依赖私有或共享的分部视图来生成响应。

并不是所有 Web 应用程序功能都需要用到 MVC 框架的灵活性。许多功能使用一个操作方法来处理多个请求，并且在处理过程中使用同一个视图。Razor Pages 以牺牲灵活性来换取专注度，提供了一种更加专注的方法，将标记与 C#代码捆绑在一起。

但是，Razor Pages 也存在局限。Razor Pages 一开始通常关注一个功能，但随着不断增强，会慢慢变得难以控制。而且，与 MVC 控制器不同，不能使用 Razor Pages 来创建 Web 服务。

这两个模型并不是二选一的情况；本章将展示，MVC 框架和 Razor Pages 是可以共存的。这意味着可轻松地使用 Razor Pages 开发自包含的功能，而使用 MVC 控制器和操作来实现应用程序的更复杂的功能。

接下来将展示如何配置和使用 Razor Pages，然后将解释 Razor Pages 的工作方式，以及它们与 MVC 控制器和操作共有的基础。

### 23.2.1 配置 Razor Pages

为使应用程序能够使用 Razor Pages，必须在 Startup 类中添加语句来设置服务及配置端点路由系统，如代码清单 23-3 所示。

**代码清单 23-3 在 WebApp 文件夹的 Startup.cs 文件中配置应用程序**

```
using System;
using System.Collections.Generic;
using System.Linq;
using Microsoft.AspNetCore.Builder;
```

```
using Microsoft.Extensions.DependencyInjection;
using Microsoft.Extensions.Configuration;
using Microsoft.EntityFrameworkCore;
using WebApp.Models;

namespace WebApp {
    public class Startup {

        public Startup(IConfiguration config) {
            Configuration = config;
        }

        public IConfiguration Configuration { get; set; }

        public void ConfigureServices(IServiceCollection services) {
            services.AddDbContext<DataContext>(opts => {
                opts.UseSqlServer(Configuration[
                    "ConnectionStrings:ProductConnection"]);
                opts.EnableSensitiveDataLogging(true);
            });
            services.AddControllersWithViews().AddRazorRuntimeCompilation();
            services.AddRazorPages().AddRazorRuntimeCompilation();

            services.AddDistributedMemoryCache();
            services.AddSession(options => {
                options.Cookie.IsEssential = true;
            });
        }

        public void Configure(IApplicationBuilder app, DataContext context) {
            app.UseDeveloperExceptionPage();
            app.UseStaticFiles();
            app.UseSession();
            app.UseRouting();
            app.UseEndpoints(endpoints => {
                endpoints.MapControllers();
                endpoints.MapDefaultControllerRoute();
                endpoints.MapRazorPages();
            });
            SeedData.SeedDatabase(context);
        }
    }
}
```

AddRazorPages方法用于设置Razor Pages必须用到的服务，而可选的AddRazorRuntimeCompilation方法则启用运行时重新编译，这用到了第 21 章添加到项目的包。MapRazorPages方法创建路由配置，用于将URL匹配到页面，稍后将解释相关内容。

### 23.2.2 创建 Razor Pages

Razor Pages 是在 Pages 文件夹中定义的。如果使用的是 Visual Studio，则创建 WebApp/Pages 文件夹，然后在 Solution Explorer 中右击该文件夹，从弹出菜单中选择 Add | New Item，并选择 Razor Page 模板，如图 23-2 所示。在 Name 输入框中输入 Index.cshtml，然后单击 Add 按钮创建文件，并使用代码清单 23-4 中的内容替换该文件的内容。

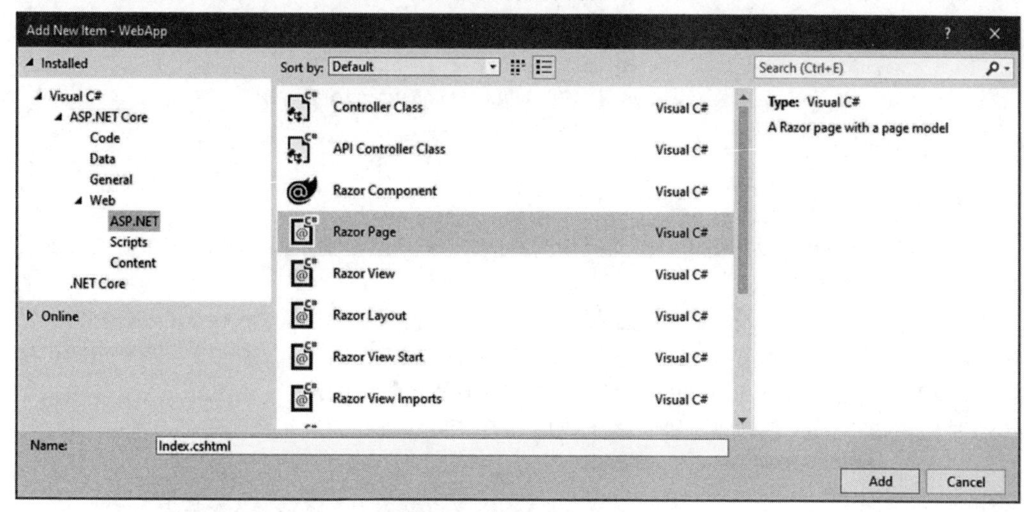


图 23-2 创建 Razor Pages

如果使用的是Visual Studio Code，则创建WebApp/Pages文件夹，在其中添加一个名为Index.cshtml的新文件，并在该文件中添加代码清单 23-4 中的内容。

#### 代码清单 23-4 Pages 文件夹的 Index.cshtml 文件的内容

```
@page
@model IndexModel
@using Microsoft.AspNetCore.Mvc.RazorPages
@using WebApp.Models;

<!DOCTYPE html>
<html>
<head>
    <link href="/lib/twitter-bootstrap/css/bootstrap.min.css" rel="stylesheet" />
</head>
<body>
    <div class="bg-primary text-white text-center m-2 p-2">@Model.Product.Name</div>
</body>
</html>

@functions {

    public class IndexModel: PageModel {
```

```
        private DataContext context;

        public Product Product { get; set; }

        public IndexModel(DataContext ctx) {
            context = ctx;
        }

        public async Task OnGetAsync(long id = 1) {
            Product = await context.Products.FindAsync(id);
        }
    }
}
```

Razor Pages 使用第 21 章和第 22 章介绍的 Razor 语法，甚至使用相同的 CSHTML 文件扩展名，但存在一些重要区别。

在 Razor Pages 中，@page 指令必须第一个出现，这确保不会将该文件误认为与控制器关联在一起的视图。不过，最重要的区别是，使用@functions 指令在同一个文件中定义支持 Razor 内容的 C#代码。稍后将介绍 Razor Pages 的工作方式，但是要查看 Razor Pages 生成的输出，可以重新启动 ASP.NET Core，使用浏览器请求 http://localhost:5000/index，这将得到如图 23-3 所示的响应。

图 23-3　使用 Razor Pages

### 1. 理解 URL 路由约定

Razor Pages 的 URL 路由基于文件名称和文件相对于 Pages 文件夹的位置。代码清单 23-4 中的 Razor Pages 包含在 Pages 文件夹中一个名为 Index.cshtml 的文件中，这意味着它将处理对/index 的请求。“理解 Razor Pages 的路由”一节将会介绍，路由约定是可被覆盖的，但默认情况下，Razor Pages 文件的位置决定了它响应的 URL。

### 2. 理解页面模型

在 Razor Pages 中，@model 指令用于选择页面模型类，而不是标识操作方法提供的对象的类型。代码清单 23-4 中的@model 指令选择了 IndexModel 类。

```
...
@model IndexModel
...
```

页面模型在@functions 指令中定义，它派生自 PageModel 类，如下所示：

```
...
@functions {
```

```
    public class IndexModel: PageModel {
...
```

当选择一个 Razor Pages 来处理 HTTP 请求时，将创建页面模型类的一个新实例，并使用依赖注入来解析使用构造函数参数声明的任何依赖，这要用到第 14 章介绍的功能。IndexModel 类声明了对第 18 章创建的 DataContext 服务的依赖，从而能够访问数据库中的数据。

```
...
public IndexModel(DataContext ctx) {
    context = ctx;
}
...
```

创建页面模型对象后，将调用一个处理程序方法。处理程序方法的名称为 On 字符后跟请求的 HTTP 方法，所以当选择 Razor Pages 来处理 HTTP GET 请求时，将调用 OnGet 方法。处理程序方法可以是异步的，此时 GET 请求将调用 OnGetAsync 方法，这是 IndexModel 类实现的一个方法。

```
...
public async Task OnGetAsync(long id = 1) {
    Product = await context.Products.FindAsync(id);
}
...
```

处理程序方法的参数值是使用模型绑定过程从 HTTP 请求获取的，第 28 章将介绍这个过程。OnGetAsync 方法从模型绑定器收到 id 参数的值，并用它来查询数据库，把结果赋值给 Product 属性。

### 3. 理解页面视图

Razor Pages 同样混合使用 HTML 片段和代码表达式来生成内容，定义展示给用户的视图。在 Razor Pages 中，通过使用@Model 表达式，可以访问页面模型的方法和属性。IndexModel 类定义的 Product 属性用于设置 HTML 元素的内容，如下所示：

```
...
<div class="bg-primary text-white text-center m-2 p-2">@Model.Product.Name</div>
...
```

@Model 表达式返回一个 IndexModel 对象，上面这个表达式读取 Product 属性返回的对象的 Name 属性。

### 4. 理解生成的 C#类

在后台，Razor Pages 被转换为 C#类，就像普通的 Razor 视图一样。下面展示了代码清单 23-4 中的 Razor Pages 生成的 C#类的简化版本：

```
using System;
using System.Collections.Generic;
using System.Linq;
```

```
using System.Threading.Tasks;
using Microsoft.AspNetCore.Mvc;
using Microsoft.AspNetCore.Mvc.Rendering;
using Microsoft.AspNetCore.Mvc.ViewFeatures;
using Microsoft.AspNetCore.Mvc.Razor.TagHelpers;
using Microsoft.AspNetCore.Mvc.ViewFeatures;
using Microsoft.AspNetCore.Mvc.Rendering;
using Microsoft.AspNetCore.Mvc.RazorPages;
using Microsoft.AspNetCore.Razor.Runtime.TagHelpers;
using Microsoft.AspNetCore.Razor.TagHelpers;
using WebApp.Models;

namespace AspNetCore {

    public class Pages_Index : Page {
        public <IndexModel> ViewData => (<IndexModel>)PageContext?.ViewData;
        public IndexModel Model => ViewData.Model;

        public async override Task ExecuteAsync() {
            WriteLiteral("\r\n<!DOCTYPE html>\r\n<html>\r\n");
            WriteLiteral("<head>");
            WriteLiteral("@<link
                href=\"/lib/twitter-bootstrap/css/bootstrap.min.css\"
                rel=\"stylesheet\" />");
            WriteLiteral("</head>");
            WriteLiteral("<body>");
            WriteLiteral("<div class=\"bg-primary text-white text-center m-2 p-2\">")
            Write(Model.Product.Name);
            WriteLiteral("</div>");
            WriteLiteral("</body></html>\r\n\r\n");
        }

        public class IndexModel: PageModel {
            private DataContext context;
            public Product Product { get; set; }

            public IndexModel(DataContext ctx) {
                context = ctx;
            }

            public async Task OnGetAsync(long id = 1) {
                Product = await context.Products.FindAsync(id);
            }
        }

        public IUrlHelper Url { get; private set; }
        public IViewComponentHelper Component { get; private set; }
```

```
        public IJsonHelper Json { get; private set; }
        public IHtmlHelper<IndexModel> Html { get; private set; }
        public IModelExpressionProvider ModelExpressionProvider { get; private set; }
    }
}
```

如果将这段代码与第 21 章的代码进行比较，可以看到 Razor Pages 同样依赖于 MVC 框架使用的功能。HTML 片段和视图表达式被转换为对 WriteLiteral 和 Write 方法的调用。

---

■ **提示：** 通过在 Windows 资源管理器中检查 obj/Debug/netcoreapp3.0/Razor/Pages 文件夹的内容，可看到生成的类。

---

## 23.3 理解 Razor Pages 的路由

Razor Pages 依赖于 CSHTML 文件的位置来进行路由，所以对 http://localhost:5000/index 的请求由 Pages/Index.cshtml 文件进行处理。要为应用程序添加更复杂的 URL 结构，可以添加文件夹，使其名称代表想要支持的 URL 片段。作为一个示例，可以创建一个 WebApp/Pages/Suppliers 文件夹，在其中添加一个名为 List.cshtml 的 Razor Pages，并在其中添加代码清单 23-5 所示的内容。

**代码清单 23-5 Pages/Suppliers 文件夹中 List.cshtml 文件的内容**

```
@page
@model ListModel
@using Microsoft.AspNetCore.Mvc.RazorPages
@using WebApp.Models;
<!DOCTYPE html>
<html>
<head>
   <link href="/lib/twitter-bootstrap/css/bootstrap.min.css" rel="stylesheet" />
</head>
<body>
    <h5 class="bg-primary text-white text-center m-2 p-2">Suppliers</h5>
    <ul class="list-group m-2">
        @foreach (string s in Model.Suppliers) {
           <li class="list-group-item">@s</li>
        }
    </ul>
</body>
</html>

@functions {

    public class ListModel : PageModel {
        private DataContext context;

        public IEnumerable<string> Suppliers { get; set; }
```

```
        public ListModel(DataContext ctx) {
            context = ctx;
        }

        public void OnGet() {
            Suppliers = context.Suppliers.Select(s => s.Name);
        }
    }
}
```

新的页面模型类定义了一个 Suppliers 属性，并使用数据库中的 Supplier 对象的 Name 值序列来设置这个属性。本例中的数据库操作是同步的，所以页面模型类定义了 OnGet 方法，而不是 OnGetAsync。使用@foreach 表达式在列表中显示供应商的名称。为使用新的 Razor Pages，使用浏览器请求 http://localhost:5000/suppliers/list，这将得到如图 23-4 所示的响应。请求 URL 中的路径片段对应于 List.cshtml Razor Pages 的文件夹和文件的名称。

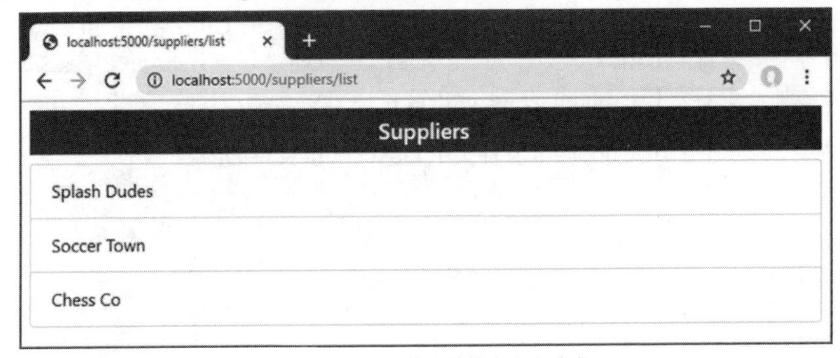


图 23-4　使用文件夹结构来路由请求

### 理解默认的 URL 处理

MapRazorPages 方法采用与 MVC 框架类似的约定，为 Index.cshtml Razor Pages 的默认 URL 设置路由。因此，添加到项目中的第一个 Razor Pages 通常被命名为 Index.cshtml。但当应用程序混合了 Razor Pages 和 MVC 框架时，先配置谁，就由谁来设置默认路由，所以在示例应用程序中，对 http://localhost:5000 的请求由 Home MVC 控制器的 Index 操作来处理。如果想让 Index.cshtml 文件来处理默认 URL，就可以修改端点路由语句的顺序，以便先设置 Razor Pages，如下所示：

```
...
app.UseEndpoints(endpoints => {
    endpoints.MapRazorPages();
    endpoints.MapControllers();
    endpoints.MapDefaultControllerRoute();
});
...
```

在我自己的项目中，会混合使用 Razor Pages 和 MVC 控制器。我倾向于依赖 MVC 框架来处理默认 URL，并避免创建 Index.cshtml Razor Pages，从而避免产生混淆。

### 23.3.1 在 Razor Pages 中指定路由模式

使用文件夹和文件结构来执行路由，意味着模型绑定过程没有可以使用的片段变量。相反，将从URL查询字符串中获取请求处理程序方法的值。为查看这一点，可以使用浏览器请求 http://localhost:5000/index?id=2，得到如图 23-5 所示的响应。

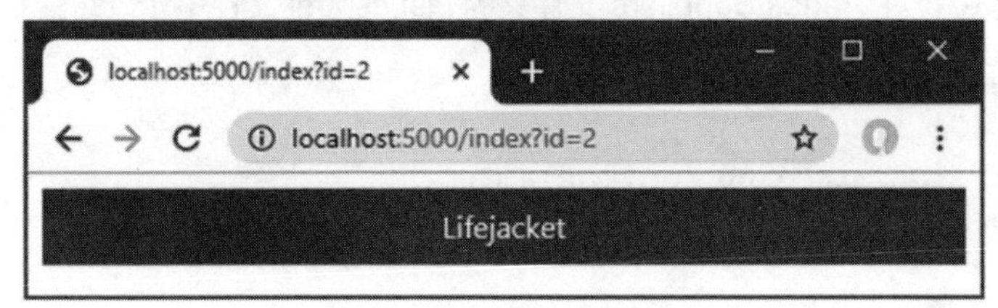


图 23-5 使用查询字符串参数

查询字符串提供了一个名为 id 的参数，模型绑定过程将使用它来作为 Index Razor Pages 的 OnGetAsync 方法中定义的 id 参数的值。

```
...
public async Task OnGetAsync(long id = 1) {
...
```

第 28 章将详细解释模型绑定的工作方式，现在只需要知道，当调用 OnGetAsync 方法来查询数据库中的产品时，请求 URL 中的查询字符串参数用于为该方法提供 id 实参。

可以在@page 指令中使用路由模式，从而能够定义片段变量，如代码清单 23-6 所示。

**代码清单 23-6 在 Pages 文件夹的 Index.cshtml 文件中定义片段变量**

```
@page "{id:long?}"
@model IndexModel
@using Microsoft.AspNetCore.Mvc.RazorPages
@using WebApp.Models;

<!DOCTYPE html>
<html>
<head>
    <link href="/lib/twitter-bootstrap/css/bootstrap.min.css" rel="stylesheet" />
</head>
<body>
    <div class="bg-primary text-white text-center m-2 p-2">@Model.Product.Name</div>
</body>
</html>

@functions {

    // ...statements omitted for brevity...
}
```

在@page 指令中，可使用第 13 章介绍的所有 URL 模式功能。代码清单 23-6 中的路由模式添加了一个可选的片段变量，名为 id，并对其施加约束，使它只匹配可被解析为一个 long 值的片段。为看到修改，需要重新启动 ASP.NET Core(自动重新编译并不会检测到路由变化)，并使用浏览器请求 http://localhost:5000/index/4，这将得到图 23-6 左侧显示的响应。

@page 指令可用于覆盖 Razor Pages 的基于文件的路由约定，如代码清单 23-7 所示。

**代码清单 23-7　在 Pages/Suppliers 文件夹的 List.cshtml 文件中修改路由**

```
@page "/lists/suppliers"
@model ListModel
@using Microsoft.AspNetCore.Mvc.RazorPages
@using WebApp.Models;

<!DOCTYPE html>
<html>
<head>
   <link href="/lib/twitter-bootstrap/css/bootstrap.min.css" rel="stylesheet" />
</head>
<body>
    <h5 class="bg-primary text-white text-center m-2 p-2">Suppliers</h5>
    <ul class="list-group m-2">
        @foreach (string s in Model.Suppliers) {
            <li class="list-group-item">@s</li>
        }
    </ul>
</body>
</html>

@functions {

   // ...statements omitted for brevity...
}
```

该指令修改了 List 页面的路由，使其匹配的 URL 的路径为/lists/suppliers。为查看修改结果，重启 ASP.NET Core，并请求 http://localhost:5000/lists/suppliers，这将得到图 23-6 右侧显示的响应。

图 23-6　使用@page 指令修改路由

## 23.3.2　为 Razor Pages 添加路由

使用@page 指令可替换 Razor Pages 默认的基于文件的路由。如果想为页面定义多个路由，则可将配置语句添加到 Startup 类中，如代码清单 23-8 所示。

**代码清单23-8 在WebApp文件夹的Startup.cs文件中添加Razor Pages路由**

```
using System;
using System.Collections.Generic;
using System.Linq;
using Microsoft.AspNetCore.Builder;
using Microsoft.Extensions.DependencyInjection;
using Microsoft.Extensions.Configuration;
using Microsoft.EntityFrameworkCore;
using WebApp.Models;
using Microsoft.AspNetCore.Mvc.RazorPages;

namespace WebApp {
    public class Startup {

        public Startup(IConfiguration config) {
            Configuration = config;
        }

        public IConfiguration Configuration { get; set; }

        public void ConfigureServices(IServiceCollection services) {
            services.AddDbContext<DataContext>(opts => {
                opts.UseSqlServer(Configuration[
                    "ConnectionStrings:ProductConnection"]);
                opts.EnableSensitiveDataLogging(true);
            });
            services.AddControllersWithViews().AddRazorRuntimeCompilation();
            services.AddRazorPages().AddRazorRuntimeCompilation();

            services.AddDistributedMemoryCache();
            services.AddSession(options => {
                options.Cookie.IsEssential = true;
            });

            services.Configure<RazorPagesOptions>(opts => {
                opts.Conventions.AddPageRoute("/Index", "/extra/page/{id:long?}");
            });
        }

        public void Configure(IApplicationBuilder app, DataContext context) {
            app.UseDeveloperExceptionPage();
            app.UseStaticFiles();
            app.UseSession();
            app.UseRouting();
            app.UseEndpoints(endpoints => {
                endpoints.MapControllers();
```

```
                endpoints.MapDefaultControllerRoute();
                endpoints.MapRazorPages();
            });
            SeedData.SeedDatabase(context);
        }
    }
}
```

这里使用 RazorPageOptions 类为 Razor Pages 添加额外的路由。在 Conventions 属性上调用 AddPageRoute 扩展方法为页面添加路由。该方法的第一个实参是相对于 Pages 文件夹的页面路径，但不包含文件扩展名。第二个实参是要添加到路由配置的 URL 模式。为了测试新路由，需要重新启动 ASP.NET Core，并使用浏览器请求 http://localhost:5000/extra/page/2，这将匹配代码清单 23-8 中添加的 URL 模式，得到图 23-7 左侧的响应。代码清单 23-8 中添加的路由补充了@page 特性定义的路由，可通过请求 http://localhost:5000/index/2 进行测试，得到的响应如图 23-7 右侧所示。

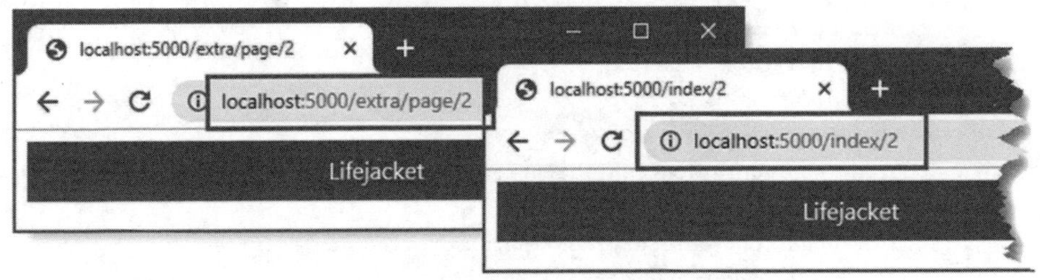


图 23-7　为 Razor Pages 添加路由

## 23.4　理解页面模型类

页面模型派生自 PageModel 类，该类将 ASP.NET Core 的其余部分与 Razor Pages 的视图部分关联起来。PageModel 类提供了一些方法来处理请求，并使用一些属性来提供上下文数据，表 23-3 列出其中最有用的一部分属性。这里为完备起见，列出这些属性，但在 Razor Pages 开发中，常常不需要使用它们，因为 Razor Pages 开发更关注于选择必要的数据来渲染页面的视图部分。

表 23-3　用于提供上下文数据的一些 PageModel 属性

| 名称 | 描述 |
| --- | --- |
| HttpContext | 此属性返回一个 HttpContext 对象，如第 12 章所述 |
| ModelState | 此属性可用于访问第 28 章和第 29 章介绍的模型绑定和验证功能 |
| PageContext | 此属性返回一个 PageContext 对象，可用于访问与 PageModel 类相同的许多属性，以及关于当前页面选择的其他信息 |
| Request | 此属性返回一个描述当前 HTTP 请求的 HttpRequest 对象，如第 12 章所述 |
| Response | 此属性返回一个提供当前响应的 HttpResponse 对象，如第 12 章所述 |
| RouteData | 此属性可用于访问路由系统匹配的数据，如第 13 章所述 |
| TempData | 此属性可用于访问临时数据功能，这种功能用于在后续请求读取数据前存储数据。详细信息请参见第 22 章 |
| User | 此属性返回的对象描述了与请求关联的用户，如第 38 章所述 |

### 23.4.1 使用代码隐藏类文件

@function 指令允许在同一个文件中定义页面背后的类和 Razor 内容，一些流行的客户端框架(如 React 或 Vue.js)就采用了这种开发方法。

在同一个文件中定义代码和标记很方便，但对于比较复杂的应用程序，可能变得难以管理。也可以把 Razor Pages 拆分为单独的视图文件和代码文件，这类似于前面章节中的 MVC 示例，并让人想起 ASP.NET Web Pages，它们在所谓的"代码隐藏文件"中定义 C#类。第一步是从 CSHTML 文件中移除页面模型类，如代码清单 23-9 所示。

**代码清单 23-9 在 Pages 文件夹的 Index.cshtml 文件中移除页面模型类**

```
@page "{id:long?}"
@model WebApp.Pages.IndexModel

<!DOCTYPE html>
<html>
<head>
    <link href="/lib/twitter-bootstrap/css/bootstrap.min.css" rel="stylesheet" />
</head>
<body>
    <div class="bg-primary text-white text-center m-2 p-2">@Model.Product.Name</div>
</body>
</html>
```

Razor Pages 代码隐藏文件的命名方式是在视图文件名称的后面加上.cs 文件扩展名。如果使用 Visual Studio，则在项目中添加 Index.cshtml 文件的时候，Razor Page 模板将创建代码隐藏文件。在 Solution Explorer 中展开 Index.cshtml 项，可以看到代码隐藏文件，如图 23-8 所示。打开文件进行编辑，将其内容替换为代码清单 23-10 中的语句。

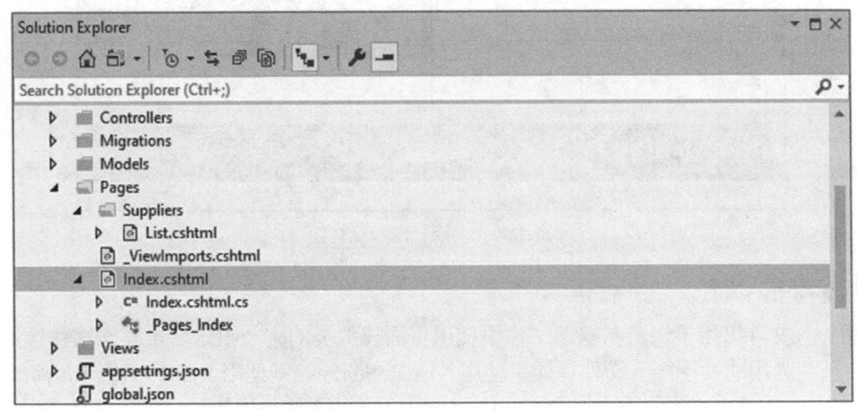


图 23-8 在 Visual Studio Solution Explorer 中显示代码隐藏文件

如果使用 Visual Studio Code，则在 WebApp/Pages 文件夹中添加一个名为 Index.cshtml.cs 的文件，并在其中添加代码清单 23-10 显示的语句。

**代码清单 23-10　Pages 文件夹中 Index.cshtml.cs 文件的内容**

```
using System.Threading.Tasks;
using Microsoft.AspNetCore.Mvc.RazorPages;
using WebApp.Models;

namespace WebApp.Pages {

    public class IndexModel: PageModel {
        private DataContext context;

        public Product Product { get; set; }

        public IndexModel(DataContext ctx) {
            context = ctx;
        }

        public async Task OnGetAsync(long id = 1) {
            Product = await context.Products.FindAsync(id);
        }
    }
}
```

在定义单独的页面模型类时，在 WebApp.Pages 名称空间中定义了该类。并非必须这么做，但这可让 C#类与应用程序的其余部分保持一致。

使用代码隐藏文件的一个缺点是，只有 CSHTML 文件会自动重新编译，这意味着在重新启动应用程序之前，不会应用对类文件做的修改。重新启动 ASP.NET Core，并请求 http://localhost:5000/index，以确保使用代码隐藏文件，得到的响应如图 23-9 所示。

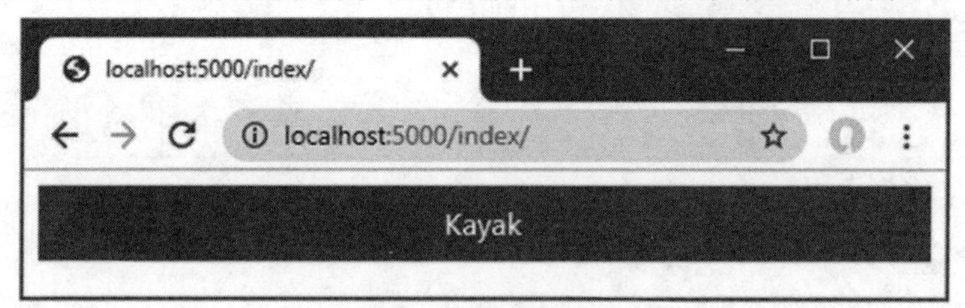


图 23-9　使用代码隐藏文件

### 添加视图导入文件

视图导入文件可用于避免在视图文件中使用页面模型类的完全限定名，其作用就像第 22 章为 MVC 框架使用的视图导入文件。如果使用 Visual Studio，则使用 Razor View Imports 模板向 WebApp/Pages 文件夹添加一个名为_ViewImports.cshtml 的文件，使其包含代码清单 23-11 中的内容。如果使用 Visual Studio Code，则直接添加该文件。

**代码清单 23-11　WebApp/Pages 文件夹中_ViewImports.cshtml 文件的内容**

```
@namespace WebApp.Pages
@using WebApp.Models
```

@namespace 指令用于为视图生成的 C#类设置名称空间，在视图导入文件中使用该指令时，将为应用程序中的所有 Razor Pages 设置默认名称空间，其效果是视图与其页面模型类位于相同的名称空间中，并且@model 指令不需要使用完全限定的类型，如代码清单 23-12 所示。

**代码清单 23-12　在 Pages 文件夹的 Index.cshtml 文件中移除页面模型名称空间**

```
@page "{id:long?}"
@model IndexModel

<!DOCTYPE html>
<html>
<head>
   <link href="/lib/twitter-bootstrap/css/bootstrap.min.css" rel="stylesheet" />
</head>
<body>
   <div class="bg-primary text-white text-center m-2 p-2">@Model.Product.Name</div>
</body>
</html>
```

使用浏览器请求 http://localhost:5000/index，这将导致重新编译视图。Razor Pages 生成的响应并没有区别，仍然如图 23-9 所示。

## 23.4.2　理解 Razor Pages 的操作结果

虽然不太明显，但 Razor Pages 处理程序方法使用相同的 IActionResult 接口来控制生成的响应。为了方便开发页面模型类，处理程序方法有一个隐含的结果，可显示页面的视图部分。代码清单 23-13 显式表达了这个结果。

**代码清单 23-13　在 Pages 文件夹的 Index.cshtml.cs 文件中使用显式结果**

```
using System.Threading.Tasks;
using Microsoft.AspNetCore.Mvc.RazorPages;
using WebApp.Models;
using Microsoft.AspNetCore.Mvc;

namespace WebApp.Pages {

    public class IndexModel : PageModel {
        private DataContext context;

        public Product Product { get; set; }

        public IndexModel(DataContext ctx) {
```

```
            context = ctx;
        }

        public async Task<IActionResult> OnGetAsync(long id = 1) {
            Product = await context.Products.FindAsync(id);
            return Page();
        }
    }
}
```

Page 方法继承自 PageModel 类，并创建一个 PageResult 对象，告诉框架渲染页面的视图部分。与 MVC 操作方法中使用的 View 方法不同，Razor Pages 的 Page 方法不接收实参，并总是渲染已被选择用来处理请求的页面的视图部分。

PageModel 类提供了其他方法，通过创建不同的操作结果来生成不同的效果，如表 23-4 所示。

表 23-4　PageModel 的操作结果方法

| 名称 | 描述 |
| --- | --- |
| Page() | 此方法返回的 IActionResult 生成 200 OK 状态码，并渲染 Razor Pages 的视图部分 |
| NotFound() | 此方法返回的 IActionResult 生成 404 NOT FOUND 状态码 |
| BadRequest(state) | 此方法返回的 IActionResult 生成 400 BAD REQUEST 状态码。此方法接收一个可选的模型状态对象，向客户端描述发生的问题，如第 19 章所述 |
| File(name, type) | 此方法返回的 IActionResult 生成 200 OK 响应，将 Content-Type 头设置为指定的类型，并将指定的文件发送到客户端 |
| Redirect(path)<br>RedirectPermanent(path) | 这些方法返回的 IActionResult 生成 302 FOUND 和 301 MOVED PERMANENTLY 响应，将客户端重定向到指定的 URL |
| RedirectToAction(name)<br>RedirectToActionPermanent(name) | 这些方法返回的 IActionResult 生成 302 FOUND 和 301 MOVED PERMANENTLY 响应，将客户端重定向到指定的操作方法。<br>用来重定向客户端的 URL 是使用第 13 章介绍的路由功能生成的 |
| RedirectToPage(name)<br>RedirectToPagePermanent(name) | 这些方法返回的 IActionResult 生成 302 FOUND 和 301 MOVED PERMANENTLY 响应，将客户端重定向到另一个 Razor Pages。如果没有提供名称，可将客户端重定向到当前页面 |
| StatusCode(code) | 此方法返回的 IActionResult 使用指定状态码生成响应 |

### 使用操作结果

除了 Page 方法外，表 23-4 中的方法与操作方法中可用的方法相同。但在使用这些方法时必须谨慎，因为在 Razor Pages 中发送状态码并没有什么用，只有当客户端期望收到视图的内容时才会使用它们。

例如，当没有找到请求的数据时，相比使用NotFound方法，更好的方法是将客户端重定向到另一个URL，为用户显示一条HTML消息。可以重定向到一个静态HTML文件、另一个Razor Pages，或者重定向到控制器定义的一个操作。在Pages文件夹中添加一个名为NotFound.cshtml的Razor Pages，并添加代码清单 23-14 中显示的内容。

### 代码清单 23-14 Pages 文件夹中 NotFound.cshtml 文件的内容

```
@page "/noid"
@model NotFoundModel
@using Microsoft.AspNetCore.Mvc.RazorPages
@using WebApp.Models;

<!DOCTYPE html>
<html>
<head>
    <link href="/lib/twitter-bootstrap/css/bootstrap.min.css" rel="stylesheet" />
    <title>Not Found</title>
</head>
<body>
    <div class="bg-primary text-white text-center m-2 p-2">No Matching ID</div>
    <ul class="list-group m-2">
        @foreach (Product p in Model.Products) {
            <li class="list-group-item">@p.Name (ID: @p.ProductId)</li>
        }
    </ul>
</body>
</html>

@functions {

    public class NotFoundModel: PageModel {
        private DataContext context;

        public IEnumerable<Product> Products { get; set; }

        public NotFoundModel(DataContext ctx) {
            context = ctx;
        }

        public void OnGetAsync(long id = 1) {
            Products = context.Products;
        }
    }
}
```

@page 指令覆盖了路由约定，所以此 Razor Pages 将处理/noid URL 路径。页面模型类使用 Entity Framework Core 上下文对象来查询数据库，并列出数据库中的产品名称和键值。

在代码清单 23-15 中，更新了 IndexModel 类的处理程序方法，当收到一个没有匹配到数据库中的 Product 对象的请求时，将用户重定向到 NotFound 页面。

### 代码清单 23-15 在 Pages 文件夹的 Index.cshtml.cs 文件中使用重定向

```
using System.Threading.Tasks;
```

```
using Microsoft.AspNetCore.Mvc.RazorPages;
using WebApp.Models;
using Microsoft.AspNetCore.Mvc;

namespace WebApp.Pages {

    public class IndexModel : PageModel {
        private DataContext context;

        public Product Product { get; set; }

        public IndexModel(DataContext ctx) {
            context = ctx;
        }

        public async Task<IActionResult> OnGetAsync(long id = 1) {
            Product = await context.Products.FindAsync(id);
            if (Product == null) {
                return RedirectToPage("NotFound");
            }
            return Page();
        }
    }
}
```

RedirectToPage 方法生成的操作结果将客户端重定向到一个不同的 Razor Pages。指定目标页面的名称时，不需要指定文件扩展名，另外，指定的任何文件夹结构都是相对于 Pages 文件夹的。为测试重定向，可重新启动 ASP.NET Core，请求 http://localhost:5000/index/500，这为 id 片段变量提供了 500 作为值，并不匹配数据库中的任何记录。浏览器将被重定向，得到如图 23-10 所示的结果。

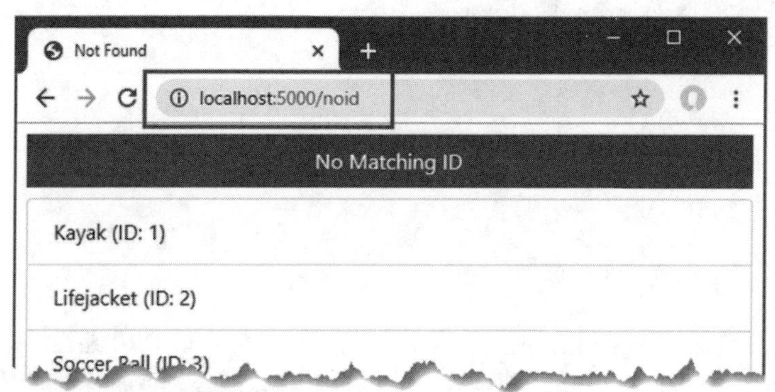


图 23-10　重定向到一个不同的 Razor Pages

注意，路由系统用于为重定向客户端生成 URL，这使用了@page 指令指定的路由模式。在本例中，使用 NotFound 作为 RedirectToPage 方法的实参，但代码清单 23-14 中的@page 指令已将其转换为到/noid 路径的重定向。

### 23.4.3 处理多个HTTP方法

Razor Pages可以定义多个处理程序方法来响应不同的HTTP方法。最常见的组合是支持GET和POST方法，以允许用户查看和编辑数据。为进行演示，可在Pages文件夹中添加一个名为Editor.cshtml的Razor Pages，并为其添加代码清单23-16中的内容。

---

■ **注意：** 我尽可能简化了这个示例，但ASP.NET Core为创建HTML表单和在提交数据后接收数据提供了出色的功能，如第31章所述。

---

**代码清单23-16 WebApps/Pages文件夹中Editor.cshtml文件的内容**

```
@page "{id:long}"
@model EditorModel

<!DOCTYPE html>
<html>
<head>
   <link href="/lib/twitter-bootstrap/css/bootstrap.min.css" rel="stylesheet" />
</head>
<body>
    <div class="bg-primary text-white text-center m-2 p-2">Editor</div>
    <div class="m-2">
        <table class="table table-sm table-striped table-bordered">
            <tbody>
                <tr><th>Name</th><td>@Model.Product.Name</td></tr>
                <tr><th>Price</th><td>@Model.Product.Price</td></tr>
            </tbody>
        </table>
        <form method="post">
            @Html.AntiForgeryToken()
            <div class="form-group">
                <label>Price</label>
                <input name="price" class="form-control"
                       value="@Model.Product.Price" />
            </div>
            <button class="btn btn-primary" type="submit">Submit</button>
        </form>
    </div>
</body>
</html>
```

此Razor Pages视图中的元素创建了一个简单的HTML表单，其包含的输入元素为用户显示了一个Product对象的Price属性的值。定义的form元素不包含action特性，这意味着当用户单击Submit按钮时，将向Razor Pages的URL发送一个POST请求。

**注意：** 代码清单23-16中的@Html.AntiForgeryToken()表达式在HTML表单中添加了一个隐藏的表单字段，ASP.NET Core使用它来防止跨站请求伪造(Cross-Site Request Forgery，CSRF)攻击。第27章将解释这种功能的工作方式，但在本章中只需要知道，不包含这个表单字段的POST请求将被拒绝。

如果使用Visual Studio，则在Solution Explorer中展开Editor.cshtml项，以显示Editor.cshtml.cs类文件，并使用代码清单23-17中的代码替换其内容。如果使用Visual Studio Code，则在WebApp/Pages文件夹中添加一个名为Editor.cshtml.cs的文件，并用它来定义代码清单23-17中显示的类。

**代码清单23-17　Pages文件夹中Editor.cshtml.cs文件的内容**

```
using System.Threading.Tasks;
using Microsoft.AspNetCore.Mvc;
using Microsoft.AspNetCore.Mvc.RazorPages;
using WebApp.Models;

namespace WebApp.Pages {
    public class EditorModel : PageModel {
        private DataContext context;

        public Product Product { get; set; }

        public EditorModel(DataContext ctx) {
            context = ctx;
        }

        public async Task OnGetAsync(long id) {
            Product = await context.Products.FindAsync(id);
        }

        public async Task<IActionResult> OnPostAsync(long id, decimal price) {
            Product p = await context.Products.FindAsync(id);
            p.Price = price;
            await context.SaveChangesAsync();
            return RedirectToPage();
        }
    }
}
```

页面模型类定义了两个处理程序方法，方法的名称告诉Razor Pages框架自己处理的HTTP方法是什么。OnGetAsync方法用于处理GET请求，当它找到Product后，将在视图中显示该Product的详细信息。

OnPostAsync方法用于处理POST请求，当用户提交HTML表单时，浏览器将发送POST请求。OnPostAsync方法的参数是从请求中获取的，id值来自URL路由，price值来自表单。第28章将介绍从表单获取数据的模型绑定功能。

**理解 POST 重定向**

注意，OnPostAsync 方法的最后一条语句调用了 RedirectToPage 方法，但没有提供实参，这将把客户端重定向到 Razor Pages 的 URL。这看起来可能有点奇怪，但它的效果是告诉浏览器，向它为 POST 请求使用的 URL 发送一个 GET 请求。这种重定向意味着如果用户重新加载了浏览器，浏览器也不会重新提交 POST 请求，从而防止了不小心多次执行相同的操作。

为了查看页面模型类如何处理不同的 HTTP 方法，可重新启动 ASP.NET Core，使用浏览器导航到 http://localhost:5000/editor/1。编辑字段，将价格设为 100，然后单击 Submit 按钮。浏览器将发送一个 POST 请求，OnPostAsync 方法将处理该请求。数据库将被更新，浏览器将被重定向，以显示更新后的数据，如图 23-11 所示。

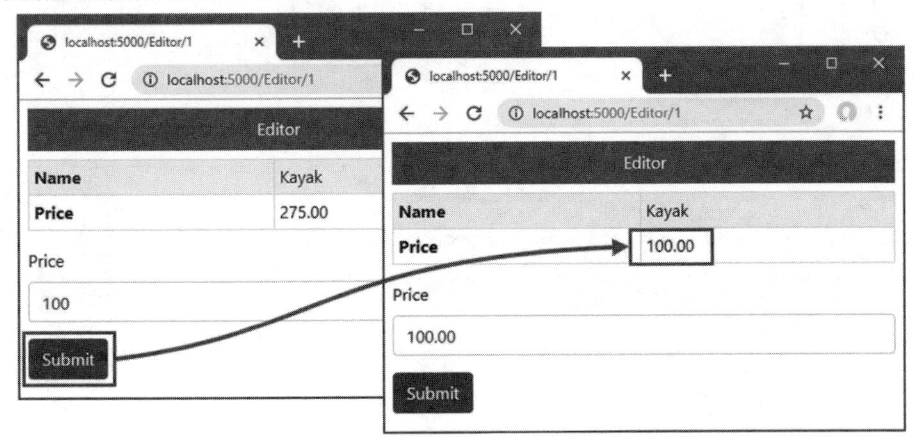


图 23-11 处理多个 HTTP 方法

### 23.4.4 选择处理程序方法

页面模型类可定义多个处理程序方法，允许请求使用 handler 查询字符串参数或者路由片段变量来选择方法。为演示这种功能，在 Pages 文件夹中添加一个名为 HandlerSelector.cshtml 的 Razor Pages 文件，并为其添加代码清单 23-18 中的内容。

**代码清单 23-18 Pages 文件夹中 HandlerSelector.cshtml 文件的内容**

```
@page
@model HandlerSelectorModel
@using Microsoft.AspNetCore.Mvc.RazorPages
@using Microsoft.EntityFrameworkCore

<!DOCTYPE html>
<html>
<head>
   <link href="/lib/twitter-bootstrap/css/bootstrap.min.css" rel="stylesheet" />
</head>
<body>
    <div class="bg-primary text-white text-center m-2 p-2">Selector</div>
```

```
    <div class="m-2">
        <table class="table table-sm table-striped table-bordered">
            <tbody>
                <tr><th>Name</th><td>@Model.Product.Name</td></tr>
                <tr><th>Price</th><td>@Model.Product.Name</td></tr>
                <tr><th>Category</th><td>@Model.Product.Category?.Name</td></tr>
                <tr><th>Supplier</th><td>@Model.Product.Supplier?.Name</td></tr>
            </tbody>
        </table>
        <a href="/handlerselector" class="btn btn-primary">Standard</a>
        <a href="/handlerselector?handler=related" class="btn btn-primary">
            Related
        </a>
    </div>
</body>
</html>

@functions{

    public class HandlerSelectorModel: PageModel {
        private DataContext context;

        public Product Product { get; set; }

        public HandlerSelectorModel(DataContext ctx) {
            context = ctx;
        }

        public async Task OnGetAsync(long id = 1) {
            Product = await context.Products.FindAsync(id);
        }

        public async Task OnGetRelatedAsync(long id = 1) {
            Product = await context.Products
                .Include(p => p.Supplier)
                .Include(p => p.Category)
                .FirstOrDefaultAsync(p => p.ProductId == id);
            Product.Supplier.Products = null;
            Product.Category.Products = null;
        }
    }
}
```

本例中的页面模型类定义了两个处理程序方法：OnGetAsync 和 OnGetRelatedAsync。默认使用 OnGetAsync 方法，通过使用浏览器请求 http://localhost:5000/handlerselector 可以看到这一点。处理程序方法查询数据库，并将结果展示给用户，如图 23-12 左侧所示。

页面渲染的锚元素之一使用一个处理程序查询字符串参数来决定目标 URL，如下所示：

```
...
<a href="/handlerselector?handler=related" class="btn btn-primary">Related</a>
...
```

在指定处理程序方法的名称时，不使用 On[method]前缀，也不使用 Async 后缀，所以使用 related 作为处理程序的值将选择 OnGetRelatedAsync 方法。这个处理程序方法在查询中包含相关数据，并把额外数据展示给用户，如图 23-12 右侧所示。

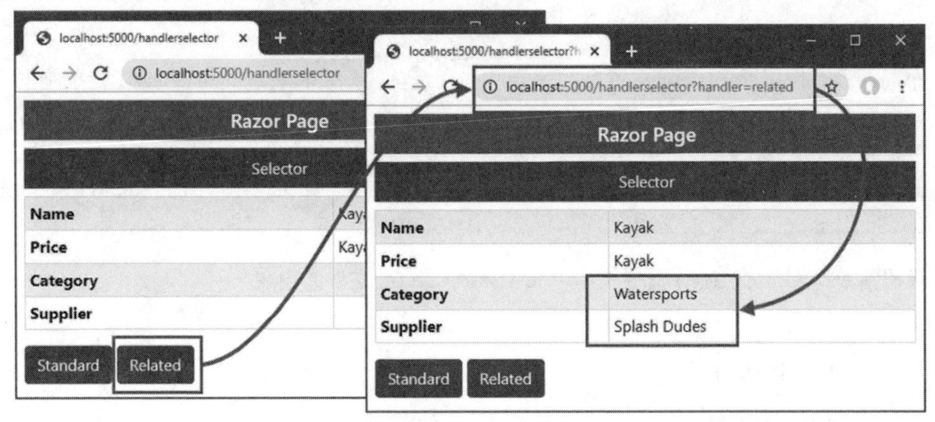


图 23-12　选择处理程序方法

# 23.5 理解 Razor Pages 视图

Razor Pages 的视图部分使用的语法和具备的功能与控制器使用的视图相同。Razor Pages 能够使用所有表达式和功能，如会话、临时数据和布局。除了使用@page 指令和页面模型类之外，仅有的区别在于需要重复一些代码来配置布局和分部视图等功能。

## 23.5.1 为 Razor Pages 创建布局

为 Razor Pages 创建布局的方式与为控制器视图创建布局相同，但需要在 Pages/Shared 文件夹中完成。如果使用 Visual Studio，则创建 Pages/Shared 文件夹，并使用 Razor Layout 模板和代码清单23-19中的内容在其中添加_Layout.cshtml 文件。如果使用 Visual Studio Code，则创建 Pages/Shared 文件夹，在该文件夹中创建_Layout.cshtml 文件，然后添加代码清单23-19中显示的内容。

■ **注意：**可在 Razor Pages 所在的文件夹中创建布局，此时将优先使用这些布局文件，而不是 Shared 文件夹中的文件。

**代码清单 23-19　Pages/Shared 文件夹中_Layout.cshtml 文件的内容**

```
<!DOCTYPE html>
<html>
<head>
    <link href="/lib/twitter-bootstrap/css/bootstrap.min.css" rel="stylesheet" />
```

```
    <title>@ViewBag.Title</title>
</head>
<body>
    <h5 class="bg-secondary text-white text-center m-2 p-2">
        Razor Page
    </h5>
    @RenderBody()
</body>
</html>
```

布局并没有使用 Razor Pages 特有的功能，并且包含的元素和表达式与第 22 章为控制器视图创建布局时使用的元素和表达式相同。

接下来使用 Razor View Start 模板，在 Pages 文件夹中添加一个名为_ViewStart.cshtml 的文件。Visual Studio 将使用代码清单 23-20 中显示的内容创建该文件。如果使用 Visual Studio Code，则创建_ViewStart.cshtml 文件，并添加代码清单 23-20 中的内容。

**代码清单 23-20 Pages 文件夹中_ViewStart.cshtml 文件的内容**

```
@{
    Layout = "_Layout";
}
```

从 Razor Pages 生成的 C#类派生自 Page 类，该类提供了一个 Layout 属性供 ViewStart 文件使用，后者的作用与控制器视图使用的 ViewStart 文件相同。代码清单 23-21 更新了 Index 页面，删除了布局将会提供的元素。

**代码清单 23-21 在 Pages 文件夹的 Index.cshtml 文件中删除元素**

```
@page "{id:long?}"
@model IndexModel

<div class="bg-primary text-white text-center m-2 p-2">@Model.Product.Name</div>
```

ViewStart 文件将布局应用到所有未覆盖 Layout 属性值的页面。代码清单 23-22 在 Editor 页面中添加了一个代码块，使其不使用布局。

**代码清单 23-22 在 Pages 文件夹的 Editor.cshtml 文件中禁用布局**

```
@page "{id:long}"
@model EditorModel
@{
    Layout = null;
}

<!DOCTYPE html>
<html>
<head>
    <link href="/lib/twitter-bootstrap/css/bootstrap.min.css" rel="stylesheet" />
</head>
```

```
<body>

   <! ...elements omitted for brevity ... />

</body>
</html>
```

使用浏览器请求 http://localhost:5000/index，将看到新布局的效果，如图 23-13 左侧所示。使用浏览器请求 http://localhost:5000/editor/1，将收到生成的内容，但没有布局，如图 23-13 右侧所示。

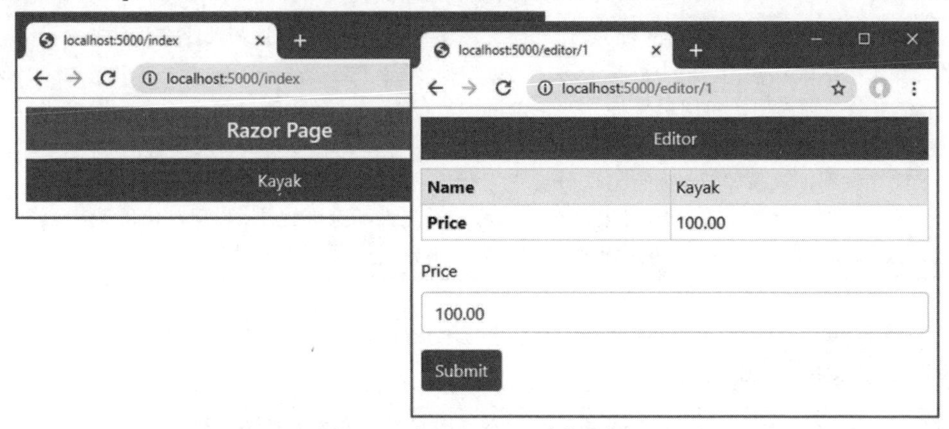


图 23-13　在 Razor Pages 中使用布局

## 23.5.2　在 Razor Pages 中使用分部视图

Razor Pages 可使用分部视图来避免重复公共内容。本节的示例依赖于第 25 章将介绍的标签助手功能。对于本章的内容，需要在 ViewImports 文件中添加代码清单 23-23 中显示的指令，以启用自定义 HTML 元素来应用分部视图。

**代码清单 23-23　在 Pages 文件夹的_ViewImports.cshtml 文件中启用标签助手**

```
@namespace WebApp.Pages
@using WebApp.Models
@addTagHelper *, Microsoft.AspNetCore.Mvc.TagHelpers
```

接下来，在 Pages/Shared 文件夹中添加一个名为_ProductPartial.cshtml 的 Razor 视图，并添加代码清单 23-24 中的内容。

**代码清单 23-24　Pages/Shared 文件夹中_ProductPartial.cshtml 文件的内容**

```
@model Product

<div class="m-2">
    <table class="table table-sm table-striped table-bordered">
        <tbody>
            <tr><th>Name</th><td>@Model.Name</td></tr>
            <tr><th>Price</th><td>@Model.Price</td></tr>
```

```
        </tbody>
    </table>
</div>
```

注意，分部视图中并没有 Razor Pages 特有的代码。分部视图使用@model 指令来接收视图模型对象，并不使用 Razor Pages 特有的@page 指令，也没有页面模型。这就使 Razor Pages 能够与 MVC 控制器共享分部视图，如侧边栏所述。

**理解分部方法的搜索路径**

Razor 视图引擎首先在 Razor Pages 所在的文件夹中寻找分部视图。如果没有找到匹配的文件，则在父目录中继续搜索，直至到达 Pages 文件夹。例如，对于 Pages/App/Data 文件夹中定义的 Razor Pages 使用的分部视图，视图引擎将首先在 Pages/App/Data 文件中进行搜索，然后是 Pages/App 文件夹，最后是 Pages 文件夹。如果没有找到文件，则将在 Pages/Shared 文件夹中继续搜索，如果仍然没有找到，就在 Views/Shared 文件夹中进行搜索。

最后这个搜索位置允许 Razor Pages 使用为控制器定义的分部视图，在同时使用 MVC 控制器和 Razor Pages 的应用程序中，这对于避免重复内容来说是一项有用的功能。

如代码清单 23-25 所示，使用 partial 元素来应用分部视图，其 name 特性指定了视图的名称，model 特性则提供了视图模型。

**警告**：分部视图通过@model 指令来接收一个视图模型，而不是页面模型。因此，model 特性的值是 Model.Product，而不只是 Model。

**代码清单 23-25　在 Pages 文件夹的 Index.cshtml 文件中使用分部视图**

```
@page "{id:long?}"
@model IndexModel

<div class="bg-primary text-white text-center m-2 p-2">@Model.Product.Name</div>
<partial name="_ProductPartial" model="Model.Product" />
```

当使用 Razor Pages 来处理响应时，分部视图的内容将包含到响应中。使用浏览器请求 http://localhost:5000/index，响应中将包含分部视图中定义的表格，如图 23-14 所示。

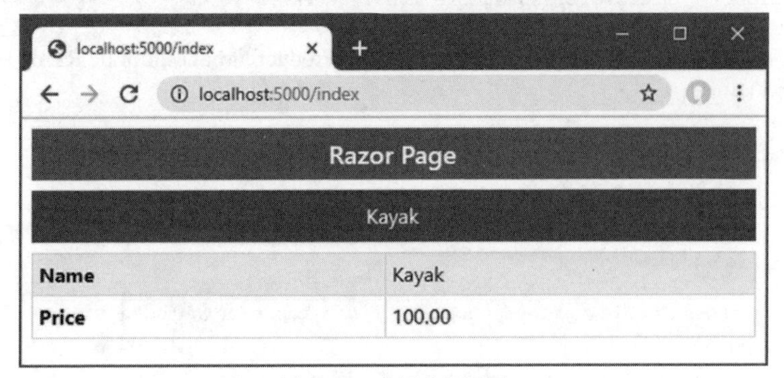


图 23-14　使用分部视图

### 23.5.3 创建没有页面模型的 Razor Pages

如果 Razor Pages 只是简单地向用户展示数据，那么其结果可以是一个页面模型类，该类简单地声明了一个构造函数依赖，用于设置视图中使用的属性。为理解这个模式，在 WebApp/Pages 文件夹中添加一个名为 Data.cshtml 的 Razor Pages，并在文件中添加代码清单 23-26 中显示的内容。

**代码清单 23-26 Pages 文件夹中 Data.cshtml 文件的内容**

```
@page
@model DataPageModel
@using Microsoft.AspNetCore.Mvc.RazorPages

<h5 class="bg-primary text-white text-center m-2 p-2">Categories</h5>
<ul class="list-group m-2">
    @foreach (Category c in Model.Categories) {
        <li class="list-group-item">@c.Name</li>
    }
</ul>

@functions {

    public class DataPageModel : PageModel {
        private DataContext context;

        public IEnumerable<Category> Categories { get; set; }

        public DataPageModel(DataContext ctx) {
            context = ctx;
        }

        public void OnGet() {
            Categories = context.Categories;
        }
    }
}
```

本例中的页面模型只是通过依赖注入让视图访问数据，并不转换数据、执行计算或做其他工作。为了避免这种只使用页面模型来访问服务的模式，可使用@inject 指令在视图中获取服务，并不需要使用页面模型，如代码清单 23-27 所示。

---

▒ **警告：** 不应该滥用@inject 指令；只有当页面模型类只用于访问服务、而不能添加其他价值的时候才使用该指令。在其他所有情况中，使用页面模型类更容易管理和维护。

---

**代码清单 23-27 在 Pages 文件夹的 Data.cshtml 文件中访问服务**

```
@page
@inject DataContext context;
```

```
<h5 class="bg-primary text-white text-center m-2 p-2">Categories</h5>
<ul class="list-group m-2">
    @foreach (Category c in context.Categories) {
        <li class="list-group-item">@c.Name</li>
    }
</ul>
```

@inject表达式指定了服务类型，以及通过什么名称来访问服务。在本例中，服务类型为DataContext，用来访问服务的名称为context。在视图内，@foreach表达式为DataContext.Categories属性返回的每个对象生成元素。本例中没有页面模型，所以移除了@page和@using指令。使用浏览器导航到http://localhost:5000/data，将看到如图23-15所示的响应。

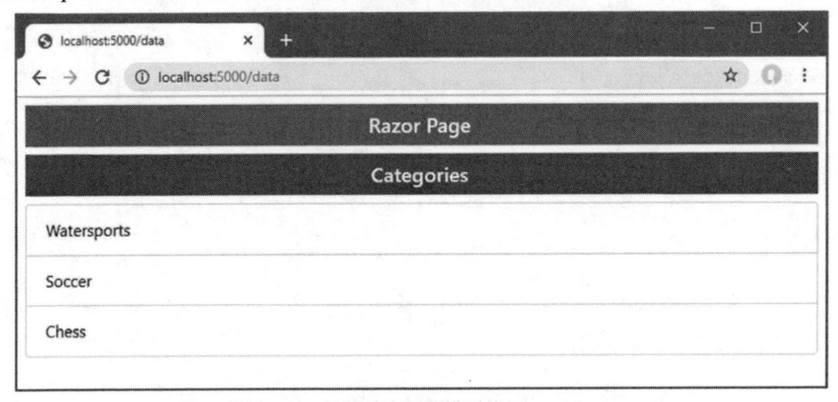


图 23-15 使用没有页面模型的 Razor Pages

## 23.6 小结

本章介绍了 Razor Pages，并解释了它们与控制器和视图的区别。本章展示了如何在同一个文件中定义内容和代码，如何使用代码隐藏文件，以及页面模型如何支持大部分重要的 Razor Pages 功能。第 24 章将介绍视图组件功能。

# 第 24 章

# 使用视图组件

本章将讨论视图组件，它们是类，提供了类似操作的逻辑来支持分部视图，这意味着视图组件提供了复杂内容，可供嵌套到视图中，同时使得支持这些内容的 C#代码变得很容易维护。表 24-1 列出视图组件的相关知识。

表 24-1　视图组件的相关知识

| 问题 | 答案 |
| --- | --- |
| 视图组件是什么？ | 视图组件是类，为支持分部视图或者在父视图中注入少量 HTML 或 JSON 数据提供了应用程序逻辑 |
| 视图组件为什么有用？ | 如果不使用视图组件，就很难创建嵌入的功能，如购物车或登录面板 |
| 如何使用视图组件？ | 视图组件一般派生自 ViewComponent 类，并使用自定义的 vc HTML 元素或者@await Component.InvokeAsync 表达式应用到父视图中 |
| 存在陷阱或局限吗？ | 视图组件是一种简单的、可预测的功能。主要陷阱是不要在使用它们后，在视图中难以测试和维护的地方包含应用程序逻辑 |
| 存在替代方案吗？ | 可将数据访问和处理逻辑直接放到分部视图中，但结果难以使用和维护 |

表 24-2 总结了本章的内容。

表 24-2　本章内容摘要

| 问题 | 解决方案 | 代码清单 |
| --- | --- | --- |
| 创建可重用的代码和内容单元 | 定义一个视图组件 | 24-7～24-13 |
| 从视图组件创建响应 | 使用一个 IViewComponentResult 实现类 | 24-14～24-18 |
| 获取上下文数据 | 使用从基类继承的属性，或者使用 Invoke 或 InvokeAsync 方法的参数 | 24-19～24-23 |
| 异步生成视图组件响应 | 重写 InvokeAsync 方法 | 24-24～24-26 |
| 将视图组件集成到另一个端点 | 创建混合的控制器或 Razor Pages | 24-27～24-34 |

## 24.1　准备工作

本章使用第 23 章创建的 WebApp 项目。为做好准备工作，在 WebApp/Models 文件夹中添加一个名为 City.cs 的类文件，在其中添加代码清单 24-1 中的内容。

---

**提示：**从 https://github.com/apress/pro-asp.net-core-3 可以下载本章以及本书其他章节的示例项目。如果无法成功运行示例，请参考第 1 章的说明。

---

**代码清单 24-1　Models 文件夹中 City.cs 文件的内容**

```
namespace WebApp.Models {

    public class City {
        public string Name { get; set; }
        public string Country { get; set; }
        public int Population { get; set; }
    }
}
```

在 WebApp/Models 文件夹中添加一个名为 CitiesData.cs 的类，并为其添加代码清单 24-2 中的内容。

**代码清单 24-2　WebApp/Models 文件夹中 CitiesData.cs 文件的内容**

```
using System.Collections.Generic;

namespace WebApp.Models {

    public class CitiesData {

        private List<City> cities = new List<City> {
            new City { Name = "London", Country = "UK", Population = 8539000},
            new City { Name = "New York", Country = "USA", Population = 8406000 },
            new City { Name = "San Jose", Country = "USA", Population = 998537 },
            new City { Name = "Paris", Country = "France", Population = 2244000 }
        };

        public IEnumerable<City> Cities => cities;

        public void AddCity(City newCity) {
            cities.Add(newCity);
        }
    }
}
```

CitiesData 类可用来访问一个 City 对象集合，还提供一个 AddCity 方法，用来将新对象添加到该集合中。在 Startup 类的 ConfigureServices 中添加代码清单 24-3 中显示的语句，为 CitiesData 类创建一个服务。

**代码清单 24-3　在 WebApp 文件夹的 Startup.cs 文件中定义一个服务**

```
using System;
```

```
using System.Collections.Generic;
using System.Linq;
using Microsoft.AspNetCore.Builder;
using Microsoft.Extensions.DependencyInjection;
using Microsoft.Extensions.Configuration;
using Microsoft.EntityFrameworkCore;
using WebApp.Models;
using Microsoft.AspNetCore.Mvc.RazorPages;

namespace WebApp {
    public class Startup {

        public Startup(IConfiguration config) {
            Configuration = config;
        }

        public IConfiguration Configuration { get; set; }

        public void ConfigureServices(IServiceCollection services) {
            services.AddDbContext<DataContext>(opts => {
                opts.UseSqlServer(Configuration[
                    "ConnectionStrings:ProductConnection"]);
                opts.EnableSensitiveDataLogging(true);
            });
            services.AddControllersWithViews().AddRazorRuntimeCompilation();
            services.AddRazorPages().AddRazorRuntimeCompilation();

            services.AddDistributedMemoryCache();
            services.AddSession(options => {
                options.Cookie.IsEssential = true;
            });

            services.Configure<RazorPagesOptions>(opts => {
                opts.Conventions.AddPageRoute("/Index", "/extra/page/{id:long?}");
            });

            services.AddSingleton<CitiesData>();
        }

        public void Configure(IApplicationBuilder app, DataContext context) {
            app.UseDeveloperExceptionPage();
            app.UseStaticFiles();
            app.UseSession();
            app.UseRouting();
            app.UseEndpoints(endpoints => {
                endpoints.MapControllers();
                endpoints.MapDefaultControllerRoute();
```

```
                endpoints.MapRazorPages();
            });
            SeedData.SeedDatabase(context);
        }
    }
}
```

新语句使用了 AddSingleton 方法来创建 CitiesData 服务。此服务没有分离接口/实现，创建它是为了便于分发共享的 CitiesData 对象。在 WebApp/Pages 文件夹中添加一个名为 Cities.cshtml 的 Razor Pages，并为其添加代码清单 24-4 中的内容。

**代码清单 24-4 Pages 文件夹中 Cities.cshtml 文件的内容**

```
@page
@inject CitiesData Data

<div class="m-2">
    <table class="table table-sm table-striped table-bordered">
        <tbody>
            @foreach (City c in Data.Cities) {
                <tr>
                    <td>@c.Name</td>
                    <td>@c.Country</td>
                    <td>@c.Population</td>
                </tr>
            }
        </tbody>
    </table>
</div>
```

### 24.1.1 删除数据库

打开一个新的 PowerShell 命令提示符，导航到包含 WebApp.csproj 文件的文件夹，并运行代码清单 24-5 中显示的命令来删除数据库。

**代码清单 24-5 删除数据库**

```
dotnet ef database drop --force
```

### 24.1.2 运行示例应用程序

从 Debug 菜单中选择 Start Without Debugging 或 Run Without Debugging，或者使用 PowerShell 命令提示符来运行代码清单 24-6 中显示的命令。

**代码清单 24-6 运行示例应用程序**

```
dotnet run
```

在应用程序启动过程中，将对数据库执行 seed 操作。当 ASP.NET Core 开始运行后，使用 Web 浏览器请求 http://localhost:5000/cities，将得到如图 24-1 所示的响应。

| London | UK | 8539000 |
| --- | --- | --- |
| New York | USA | 8406000 |
| San Jose | USA | 998537 |
| Paris | France | 2244000 |

图 24-1 运行示例应用程序

# 24.2 理解视图组件

很多时候，应用程序需要在视图中嵌入一些内容，但这些内容与应用程序的主要用途无关。常见的例子包括站点导航工具以及身份验证面板，后者使用户不必访问另一个页面就可以登录。

从操作方法或页面模型传递给视图的模型数据中不包含这类功能的数据。因此，在示例项目中，创建了两种数据源：将显示使用 City 数据生成的一些内容，在从 Entity Framework Core 存储库及其包含的 Product、Category 和 Supplier 对象获取数据的视图中，这不容易实现。

分部视图用于创建视图内必须用到的可重用标记，从而避免在应用程序的多个位置重复创建相同的内容。分部视图是有用的功能，但它们只包含 HTML 和 Razor 指令的片段，这些片段操作的数据是从父视图获取的。如果需要显示不同数据，就会出现问题。可以在分部视图中直接访问需要的数据，但这破坏了开发模型，使得到的应用程序难以理解和维护。另一方面，可以扩展应用程序使用的视图模型，使之包含需要的数据，但这意味着需要修改每个操作方法，从而难以隔离操作方法的功能，实现有效的维护和测试。

这时，视图组件就可以起到作用。视图组件是一个 C#类，可独立于操作方法或 Razor Pages，为分部视图提供需要的数据。从这方面讲，可将视图组件想象成一个专用的操作或页面，但它只用于为分部视图提供数据；它不能收到 HTTP 请求，而且提供的内容将总是被包含到父视图中。

# 24.3 创建和使用视图组件

视图组件是名称以 ViewComponent 结束、并定义了 Invoke 或 InvokeAsync 方法的类，派生自 ViewComponent 基类的类，或者使用 ViewComponent 特性装饰的类。“获取上下文数据”一节演示了特性的用法，不过本章的其他示例依赖于基类。

在项目的任何位置都可以定义视图组件，但约定做法是把它们放到一个名为 Components 的文件夹中。创建 WebApp/Components 文件夹，在其中添加一个名为 CitySummary.cs 的文件，并为该文件添加代码清单 24-7 中的内容。

**代码清单 24-7　Components 文件夹中 CitySummary.cs 文件的内容**

```
using Microsoft.AspNetCore.Mvc;
using System.Linq;
using WebApp.Models;

namespace WebApp.Components {

    public class CitySummary: ViewComponent {
        private CitiesData data;

        public CitySummary(CitiesData cdata) {
            data = cdata;
        }

        public string Invoke() {
            return $"{data.Cities.Count()} cities, "
            + $"{data.Cities.Sum(c => c.Population)} people";
        }
    }
}
```

视图组件可利用依赖注入来获取必要的服务。在本例中，视图组件声明了对 CitiesData 类的依赖，并在 Invoke 方法中使用了注入的数据来创建一个字符串，其中包含城市个数和人口总数。

#### 1. 应用视图组件

可采用两种不同方式来应用视图组件。第一种方法是使用添加到视图和 Razor Pages 生成的 C#类的 Component 属性。该属性返回的对象实现了 IViewComponentHelper 接口，后者提供了 InvokeAsync 方法。代码清单 24-8 使用了这种方法，在 Views/Home 文件夹的 Index.cshtml 文件中应用了视图组件。

**代码清单 24-8　在 Views/Index 文件夹的 Index.cshtml 文件中使用视图组件**

```
@model Product
@{
    Layout = "_Layout";
    ViewBag.Title = ViewBag.Title ?? "Product Table";
}

@section Header { Product Information }

<tr><th>Name</th><td>@Model.Name</td></tr>
<tr>
    <th>Price</th>
    <td>@Model.Price.ToString("c")</td>
</tr>
<tr><th>Category ID</th><td>@Model.CategoryId</td></tr>
```

```
@section Footer {
    @(((Model.Price / ViewBag.AveragePrice)
       * 100).ToString("F2"))% of average price
}

@section Summary {
    <div class="bg-info text-white m-2 p-2">
        @await Component.InvokeAsync("CitySummary")
    </div>
}
```

这里使用了Component.InvokeAsync方法来应用视图组件，在使用该方法时将视图组件类的名称作为实参。这种方法的语法可能让人产生混淆。取决于同步还是异步执行操作，视图组件类可能定义Invoke或InvokeAsync方法，但总会使用Component.InvokeAsync方法，即使要应用的视图组件定义了Invoke方法，且只执行同步操作，也是这种情况。

要把视图组件所在的名称空间添加到视图包含的名称空间列表中，需要在 Views 文件夹的_ViewImports.json 文件中添加代码清单 24-9 所示的语句。

**代码清单 24-9　在 Views 文件夹的_ViewImports.json 文件中添加名称空间**

```
@using WebApp.Models
@addTagHelper *, Microsoft.AspNetCore.Mvc.TagHelpers
@using WebApp.Components
```

重新启动 ASP.NET Core，使用浏览器请求 http://localhost:5000/home/index/1，这将得到如图 24-2 所示的结果。

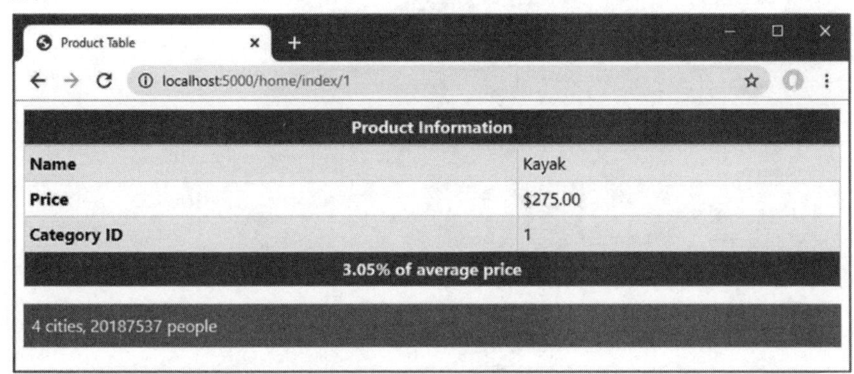


图 24-2　使用视图组件

### 2. 使用标签助手应用视图组件

Razor 视图和页面可以包含标签助手，它们是由 C#类管理的自定义 HTML 元素。第 25 章将详细介绍标签助手的工作方式，不过现在可以知道，通过使用实现为标签助手的一个 HTML 元素，可以应用视图组件。为启用这种功能，在 Views 文件夹的_ViewImports.cshtml 文件中添加代码清单 24-10 中显示的指令。

■ **注意**：只能在控制器视图或 Razor Pages 中使用视图组件，而不能使用它们直接处理请求。

**代码清单 24-10　在 Views 文件夹的_ViewImports.cshtml 文件中配置标签助手**

```
@using WebApp.Models
@addTagHelper *, Microsoft.AspNetCore.Mvc.TagHelpers
@using WebApp.Components
@addTagHelper *, WebApp
```

新指令为示例项目(通过名称指定)添加了对标签助手的支持(必须将 WebApp 改为自己的项目的名称)。在代码清单 24-11 中，使用自定义 HTML 元素来应用视图组件。

**代码清单 24-11　在 Views/Home 文件夹的 Index.cshtml 文件中应用视图组件**

```
@model Product
@{
    Layout = "_Layout";
    ViewBag.Title = ViewBag.Title ?? "Product Table";
}

@section Header { Product Information }

<tr><th>Name</th><td>@Model.Name</td></tr>
<tr>
    <th>Price</th>
    <td>@Model.Price.ToString("c")</td>
</tr>
<tr><th>Category ID</th><td>@Model.CategoryId</td></tr>

@section Footer {
    @(((Model.Price / ViewBag.AveragePrice)
       * 100).ToString("F2"))% of average price
}

@section Summary {
    <div class="bg-info text-white m-2 p-2">
       <vc:city-summary />
    </div>
}
```

自定义元素的标签是 vc，后跟一个冒号，再跟视图组件类的名称(已被转换为带短横线的形式)。类名中的每个大写字母被转换为小写字母，并用一条短横线隔开，所以 CitySummary 变成了 city-summary，将使用 vc:city-summary 元素来应用 CitySummary 视图组件。

### 3. 在 Razor Pages 中应用视图组件

Razor Pages 采用相同的方式使用视图组件，即通过 Component 属性或通过自定义 HTML 元素。因为 Razor Pages 有自己的视图导入文件，所以需要使用一个单独的@addTagHelper 指令，如代码清单 24-12 所示。

**代码清单 24-12 在 Pages 文件夹的_ViewImports.cshtml 文件中添加指令**

```
@namespace WebApp.Pages
@using WebApp.Models
@addTagHelper *, Microsoft.AspNetCore.Mvc.TagHelpers
@addTagHelper *, WebApp
```

代码清单 24-13 在 Data 页面中应用 CitySummary 视图组件。

**代码清单 24-13 在 Pages 文件夹的 Data.cshtml 文件中使用视图组件**

```
@page
@inject DataContext context;

<h5 class="bg-primary text-white text-center m-2 p-2">Categories</h5>
<ul class="list-group m-2">
    @foreach (Category c in context.Categories) {
        <li class="list-group-item">@c.Name</li>
    }
</ul>

<div class="bg-info text-white m-2 p-2">
    <vc:city-summary />
</div>
```

使用浏览器请求 http://localhost:5000/data，将看到如图 24-3 所示的响应，即数据库中的分类旁显示了城市数据。

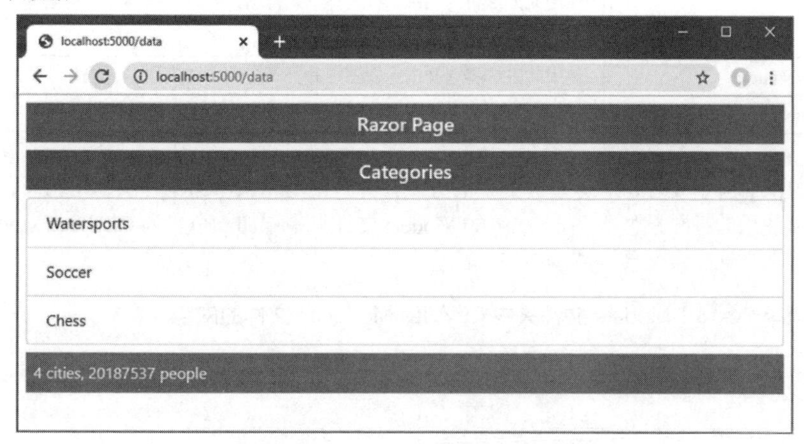


图 24-3 在 Razor Pages 中使用视图组件

## 24.4 理解视图组件的结果

将简单字符串值插入视图或页面中并不是特别有用，不过好消息是，视图组件还提供了更多功能。通过让 Invoke 或 InvokeAsync 方法返回一个实现了 IViewComponentResult 接口的对象，可以实现更复杂的效果。有 3 个内置的类实现了 IViewComponentResult 接口，表 24-3 说明了这些类，以及 ViewComponent 基类提供的、用于创建这些类的方法。接下来将介绍每种结果类型的用法。

表 24-3 实现了 IViewComponentResult 接口的内置类

| 名称 | 描述 |
| --- | --- |
| ViewViewComponentResult | 此类用于指定一个 Razor 视图，包括可选的视图模型数据。使用 View 方法可创建这个类的实例 |
| ContentViewComponentResult | 此类用于指定文本结果，该文本结果将被安全地编码，以便能够安全地包含到 HTML 文档中。使用 Content 方法可创建这个类的实例 |
| HtmlContentViewComponentResult | 此类用于指定一个 HTML 片段，该片段不会被编码，而只是直接包含到 HTML 文档中。没有 ViewComponent 方法可以创建这类结果 |

对于两种结果类型，存在特殊处理。如果视图组件返回一个字符串，则使用该字符串来创建一个 ContentViewComponentResult 对象，前面的示例依赖这种处理。如果视图组件返回一个 IHTMLContent 对象，则用它来创建一个 HtmlContentViewComponentResult 对象。

### 24.4.1 返回一个分部视图

最有用的响应是名称有些拗口的 ViewViewComponentResult，它告诉 Razor 渲染一个分部视图，并在父视图中包含渲染的结果。ViewComponent 基类提供了 View 方法，用于创建 ViewViewComponentResult 对象，这个方法有 4 个重载，如表 24-4 所述。

表 24-4 ViewComponent.View 方法

| 名称 | 说明 |
| --- | --- |
| View() | 使用此方法将为视图组件选择默认视图，并且不提供视图模型 |
| View(model) | 使用此方法将选择默认视图，并使用指定对象作为视图模型 |
| View(viewName) | 使用此方法将选择指定视图，并且不提供视图模型 |
| View(viewName, model) | 使用此方法将选择指定视图，并使用指定对象作为视图模型 |

这些方法对应于 Controller 基类提供的方法，并且使用起来与那些方法基本相同。要创建视图组件可以使用的视图模型类，在 WebApp/Models 文件夹中添加一个名为 CityViewModel.cs 的类文件，并在该文件中定义代码清单 24-14 显示的类。

**代码清单 24-14 Models 文件夹中 CityViewModel.cs 文件的内容**

```
namespace WebApp.Models {

    public class CityViewModel {
        public int Cities { get; set; }
        public int Population { get; set; }
    }
}
```

代码清单 24-15 修改了 CitySummary 视图组件的 Invoke 方法，使其使用 View 方法来选择一个分部视图，并使用 CityViewModel 对象来提供视图数据。

**代码清单 24-15　在 Components 文件夹的 CitySummary.cs 文件中选择视图**

```
using Microsoft.AspNetCore.Mvc;
using System.Linq;
using WebApp.Models;

namespace WebApp.Components {

    public class CitySummary: ViewComponent {
        private CitiesData data;

        public CitySummary(CitiesData cdata) {
           data = cdata;
        }

        public IViewComponentResult Invoke() {
            return View(new CityViewModel {
                Cities = data.Cities.Count(),
                Population = data.Cities.Sum(c => c.Population)
            });
        }
    }
}
```

目前，还没有可供视图组件使用的视图，这会得到一个错误消息。不过，该消息能够说明搜索了哪些位置。重新启动 ASP.NET Core，使用浏览器请求 http://localhost:5000/home/index/1，查看当为控制器使用该视图组件时搜索了哪些位置。请求 http://localhost:5000/data，查看当为 Razor Pages 使用该视图组件时搜索的位置。图 24-4 显示了这两个请求得到的响应。

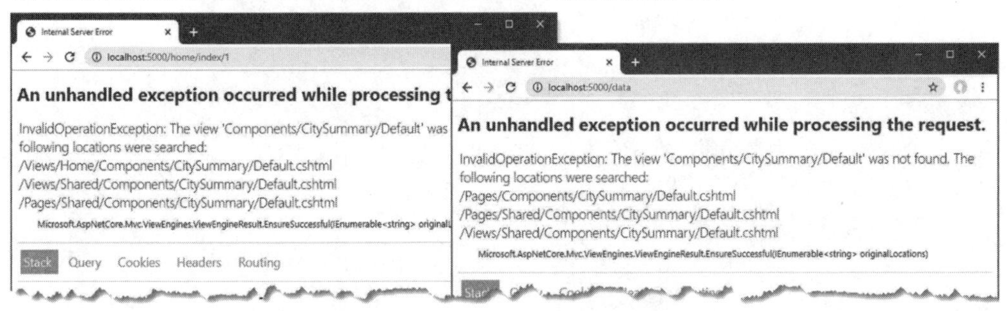


图 24-4　ViewComponent 视图的搜索位置

当视图组件调用 View 方法、但没有指定名称时，Razor 将搜索一个名为 Default.cshtml 的视图。如果为控制器使用该视图组件，则搜索位置如下所示：

- /Views/[controller]/Components/[viewcomponent]/Default.cshtml
- /Views/Shared/Components/[viewcomponent]/Default.cshtml
- /Pages/Shared/Components/[viewcomponent]/Default.cshtml

例如，当通过Home控制器选择的视图渲染CitySummary组件时，[controller]是Home，[viewcomponent]是CitySummary，这意味着第一个搜索位置是/Views/Home/Components/CitySummary/Default.cshtml。如果为Razor Pages使用视图组件，则搜索位置如下所示：

- /Pages/Components/[viewcomponent]/Default.cshtml
- /Pages/Shared/Components/[viewcomponent]/Default.cshtml
- /Views/Shared/Components/[viewcomponent]/Default.cshtml

如果 Razor Pages 的搜索路径不包含页面名称，但在子文件夹中定义了一个 Razor Pages，那么 Razor 视图引擎将在 Components/[viewcomponent]文件夹中寻找视图，相对于定义了 Razor Pages 的位置，在文件夹层次结构中向上一级级查找，直至找到视图或到达 Pages 文件夹。

---

**提示**：Razor Pages 中使用的视图组件将找到 Views/Shared/Components 文件夹中定义的视图，控制器中定义的视图组件将找到 Pages/Shared/Components 文件夹中定义的视图。这意味着当控制器和 Razor Pages 都使用一个视图组件时，不需要重复视图。

---

创建WebApp/Views/Shared/Components/CitySummary文件夹，在其中添加一个名为Default.cshtml的Razor视图，并在该视图中添加代码清单 24-16 显示的内容。

**代码清单 24-16　Views/Shared/Components/CitySummary 文件夹中的 Default.cshtml 文件**

```
@model CityViewModel

<table class="table table-sm table-bordered text-white bg-secondary">
    <thead>
        <tr><th colspan="2">Cities Summary</th></tr>
    </thead>
    <tbody>
        <tr>
            <td>Cities:</td>
            <td class="text-right">
                @Model.Cities
            </td>
        </tr>
        <tr>
            <td>Population:</td>
            <td class="text-right">
                @Model.Population.ToString("#,###")
            </td>
        </tr>
    </tbody>
</table>
```

视图组件的视图类似于分部视图，使用@model 指令来设置视图模型对象的类型。此视图将从视图组件收到一个 CityViewModel 对象，用于填充 HTML 表格中的单元格。使用浏览器请求 http://localhost:5000/home/index/1 和 http://localhost:5000/data，将看到响应中包含了视图，如图 24-5 所示。

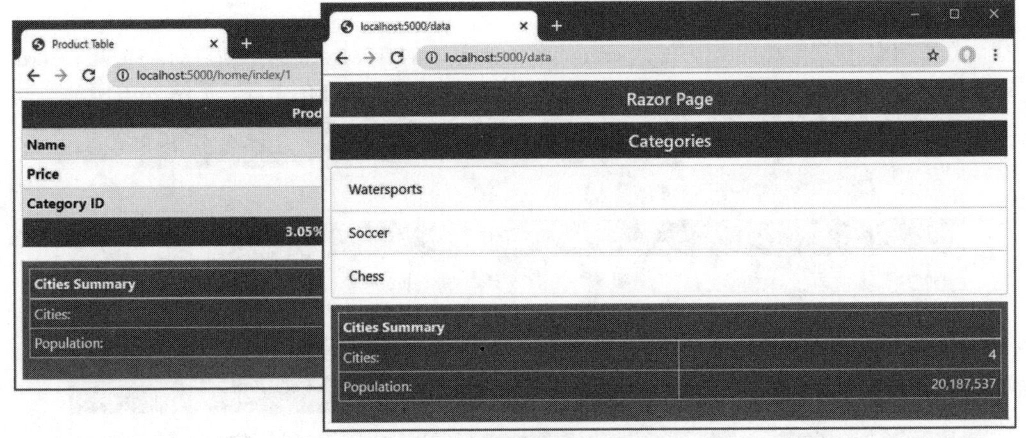


图 24-5　使用有视图组件的视图

## 24.4.2　返回 HTML 片段

ContentViewComponentResult 类用于在不使用视图的情况下，在父视图中包含 HTML 片段。使用从 ViewComponent 基类继承的 Content 方法，可以创建 ContentViewComponentResult 类的实例，该方法接收一个 string 值。代码清单 24-17 演示了 Content 方法的用法。

▇ 提示：除了 Content 方法，Invoke 方法也可返回一个 string，它将被自动转换为 ContentViewComponentResult。在前面第一次定义视图组件时，使用了这种方法。

**代码清单 24-17　在 Components 文件夹的 CitySummary.cs 文件中使用 Content 方法**

```
using Microsoft.AspNetCore.Mvc;
using System.Linq;
using WebApp.Models;

namespace WebApp.Components {

    public class CitySummary: ViewComponent {
        private CitiesData data;

        public CitySummary(CitiesData cdata) {
            data = cdata;
        }
        public IViewComponentResult Invoke() {
            return Content("This is a <h3><i>string</i></h3>");
        }
    }
}
```

Content 方法收到的字符串是编码后的字符串，以便能够安全地包含到 HTML 文档中。在处理由用户或者外部系统提供的内容时，这一点尤为重要，因为这可防止在应用程序生成的 HTML

中嵌入 JavaScript 内容。

在本例中，传递给 Content 方法的 string 包含一些基本的 HTML 标签。重新启动 ASP.NET Core，使用浏览器请求 http://localhost:5000/data。响应将包含编码后的 HTML 片段，如图 24-6 所示。

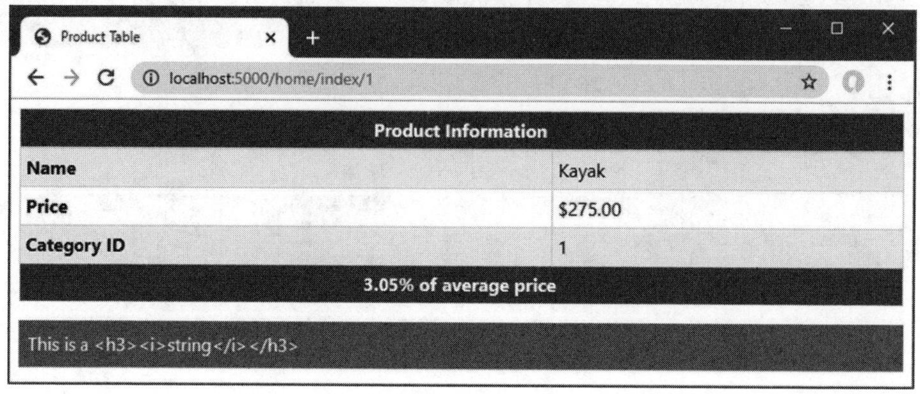


图 24-6　使用视图组件返回编码后的 HTML 片段

如果查看视图组件生成的 HTML，会发现尖括号已被替换掉，使得浏览器不会把内容解释为 HTML 元素，如下所示：

```
...
<div class="bg-info text-white m-2 p-2">
   This is a <h3><i>string</i></h3>
</div>
...
```

如果你信任内容来源，希望将内容解释为 HTML，则不需要编码内容。Content 方法总是编码其实参，所以必须直接创建 HtmlContentViewComponentResult 对象，为其构造函数提供一个 HtmlString 对象，用来代表你知道可以安全显示的一个字符串(可能是因为它来自一个可信来源，或者你确信字符串已经被编码过)，如代码清单 24-18 所示。

#### 代码清单 24-18　在 Components 文件夹的 CitySummary.cs 文件中返回一个 HTML 片段

```
using Microsoft.AspNetCore.Mvc;
using System.Linq;
using WebApp.Models;
using Microsoft.AspNetCore.Mvc.ViewComponents;
using Microsoft.AspNetCore.Html;

namespace WebApp.Components {
    public class CitySummary: ViewComponent {
        private CitiesData data;

        public CitySummary(CitiesData cdata) {
            data = cdata;
        }
```

```
        public IViewComponentResult Invoke() {
            return new HtmlContentViewComponentResult(
              new HtmlString("This is a <h3><i>string</i></h3>"));
        }
    }
}
```

在使用这种方法时应该谨慎，只将其用于不会被篡改、并且自己会进行编码的内容源。重新启动 ASP.NET Core，使用浏览器请求 http://localhost:5000/home/index/1，将看到响应未经编码，直接被解释为 HTML 元素，如图 24-7 所示。

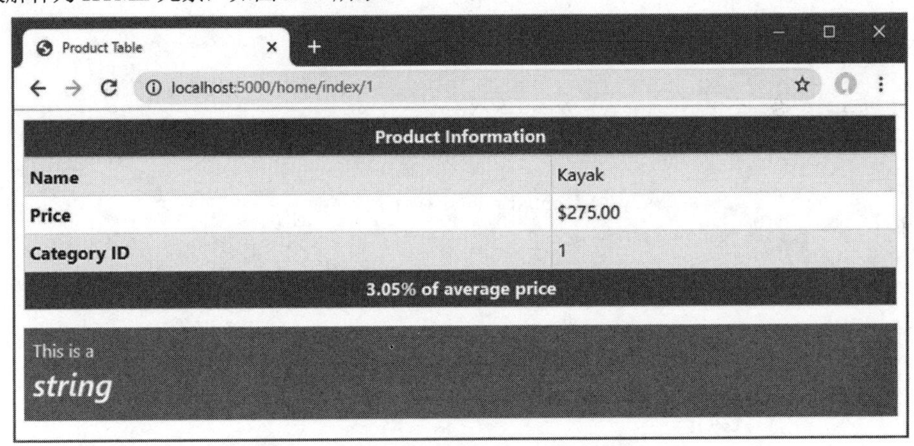


图 24-7　使用视图组件返回未编码的 HTML 片段

## 24.5　获取上下文数据

通过 ViewComponent 基类定义的属性，将关于当前请求和父视图的详细信息提供给视图组件，如表 24-5 所述。

表 24-5　ViewComponentContext 属性

| 名称 | 描述 |
|---|---|
| HttpContext | 此属性返回一个 HttpContext 对象，描述了当前请求和准备生成的响应 |
| Request | 此属性返回一个 HttpRequest 对象，描述了当前的 HTTP 请求 |
| User | 此属性返回一个 IPrincipal 对象，描述了当前用户，如第 37 章和第 38 章所述 |
| RouteData | 此属性返回一个 RouteData 对象，描述了当前请求的路由数据 |
| ViewBag | 此属性返回 dynamic ViewBag 对象，后者可用于在视图组件和视图之间传递数据，如第 22 章所述 |
| ModelState | 此属性返回一个 ModelStateDictionary，它提供了关于模型绑定过程的详细信息，如第 29 章所述 |
| ViewData | 此属性返回一个 ViewDataDictionary，用于访问为视图组件提供的视图数据 |

可以采用任何方式来使用上下文数据，帮助视图组件完成其工作，包括修改选择数据的方式，或者渲染不同的内容或视图。很难设计一个代表性的示例来演示如何在视图组件中使用上下文数

据，因为它解决的问题依赖于具体项目。在代码清单 24-19 中，检查了请求的路由数据，以判断路由模式中是否包含控制器片段变量，这可以说明请求将被控制器和视图处理。

**代码清单 24-19　在 Components 文件夹的 CitySummary.cs 文件中使用请求数据**

```
using Microsoft.AspNetCore.Mvc;
using System.Linq;
using WebApp.Models;
using Microsoft.AspNetCore.Mvc.ViewComponents;
using Microsoft.AspNetCore.Html;

namespace WebApp.Components {

    public class CitySummary: ViewComponent {
        private CitiesData data;

        public CitySummary(CitiesData cdata) {
            data = cdata;
        }

        public string Invoke() {
            if (RouteData.Values["controller"] != null) {
                return "Controller Request";
            } else {
                return "Razor Page Request";
            }
        }
    }
}
```

重新启动ASP.NET Core，使用浏览器请求http://localhost:5000/home/index/1和http://localhost:5000/data，将看到视图组件修改了其输出，如图24-8所示。

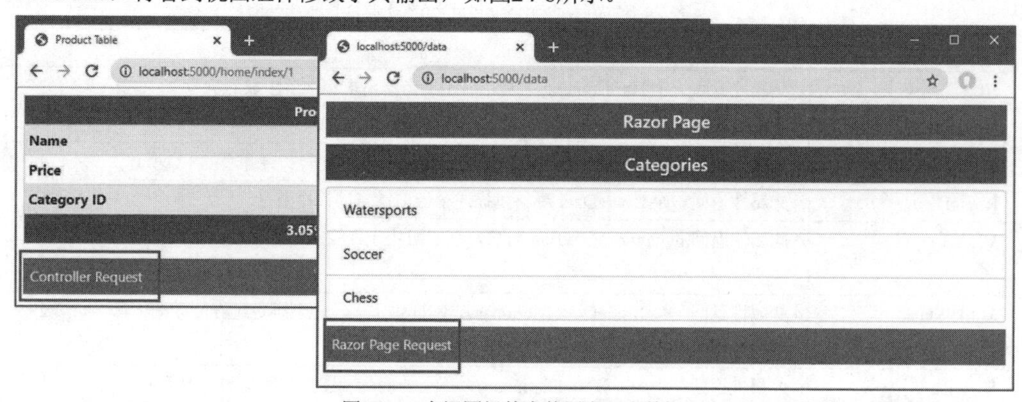


图 24-8　在视图组件中使用上下文数据

### 24.5.1 使用实参提供父视图的上下文

父视图能够为视图组件提供额外的上下文数据，包括关于应该生成的内容的数据。这些上下文数据是通过 Invoke 或 InvokeAsync 方法获取的，如代码清单 24-20 所示。

**代码清单 24-20 在 Components 文件夹的 CitySummary.cs 文件中获取值**

```
using Microsoft.AspNetCore.Mvc;
using System.Linq;
using WebApp.Models;
using Microsoft.AspNetCore.Mvc.ViewComponents;
using Microsoft.AspNetCore.Html;

namespace WebApp.Components {

    public class CitySummary: ViewComponent {
        private CitiesData data;

        public CitySummary(CitiesData cdata) {
            data = cdata;
        }

        public IViewComponentResult Invoke(string themeName) {
            ViewBag.Theme = themeName;

            return View(new CityViewModel {
                Cities = data.Cities.Count(),
                Population = data.Cities.Sum(c => c.Population)
            });
        }
    }
}
```

Invoke 方法定义了一个 themeName 参数，它将通过 ViewBag(第 22 章介绍过)传递给分部视图。代码清单 24-21 更新了 Default 视图，使其用收到的值来设置生成内容的样式。

**代码清单 24-21 在 Views/Shared/Components/CitySummary 文件夹的 Default.cshtml 文件中设置内容的样式**

```
@model CityViewModel

<table class="table table-sm table-bordered text-white bg-@ViewBag.Theme">
    <thead>
        <tr><th colspan="2">Cities Summary</th></tr>
    </thead>
    <tbody>
        <tr>
            <td>Cities:</td>
```

```
            <td class="text-right">
                @Model.Cities
            </td>
        </tr>
        <tr>
            <td>Population:</td>
            <td class="text-right">
                @Model.Population.ToString("#,###")
            </td>
        </tr>
    </tbody>
</table>
```

必须为视图组件的 Invoke 或 InvokeAsync 方法定义的所有参数提供值。代码清单 24-22 在 Home 控制器选择的视图中为 themeName 参数提供了一个值。

**提示：** 如果没有为视图组件定义的所有参数提供值，则不会使用该视图组件，也不会显示错误消息。如果没有看到视图组件生成任何内容，那么很可能是因为缺少某个参数的值。

**代码清单 24-22　在 Views/Home 文件夹的 Index.cshtml 文件中提供一个值**

```
@model Product
@{
    Layout = "_Layout";
    ViewBag.Title = ViewBag.Title ?? "Product Table";
}

@section Header { Product Information }

<tr><th>Name</th><td>@Model.Name</td></tr>
<tr>
    <th>Price</th>
    <td>@Model.Price.ToString("c")</td>
</tr>
<tr><th>Category ID</th><td>@Model.CategoryId</td></tr>

@section Footer {
    @(((Model.Price / ViewBag.AveragePrice)
        * 100).ToString("F2"))% of average price
}

@section Summary {
    <div class="bg-info text-white m-2 p-2">
        <vc:city-summary theme-name="secondary" />
    </div>
}
```

每个参数的名称被表达为用短横线形式表示的特性，所以 theme-name 特性为 themeName 参数提供了值。代码清单 24-23 在 Data.cshtml Razor Pages 中设置了一个值。

**代码清单 24-23 在 Pages 文件夹的 Data.cshtml 文件中提供值**

```
@page
@inject DataContext context;

<h5 class="bg-primary text-white text-center m-2 p-2">Categories</h5>
<ul class="list-group m-2">
    @foreach (Category c in context.Categories) {
        <li class="list-group-item">@c.Name</li>
    }
</ul>

<div class="bg-info text-white m-2 p-2">
   <vc:city-summary theme-name="danger" />
</div>
```

重新启动 ASP.NET Core，使用浏览器请求 http://localhost:5000/home/index/1 和 http://localhost:5000/home/index/1。由于为视图组件的 themeName 参数提供了不同的值，所以生成了不同的响应，如图 24-9 所示。

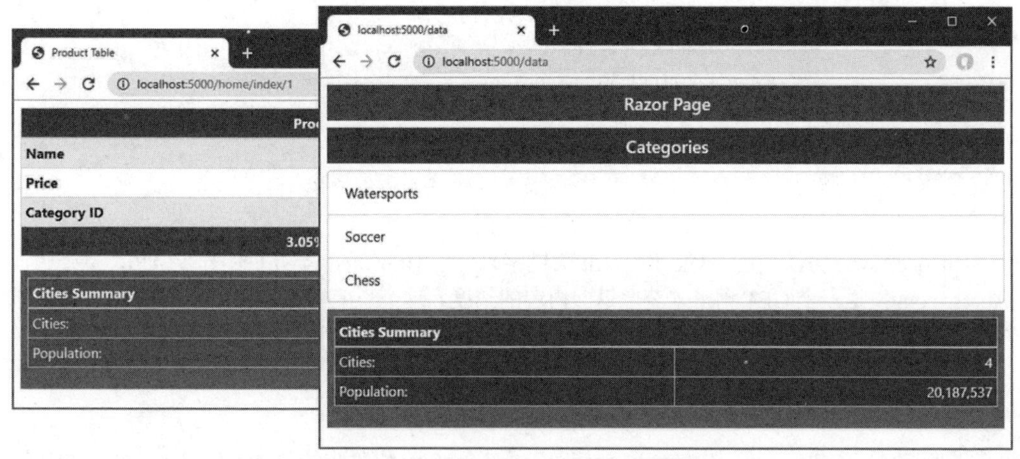


图 24-9 在视图组件中使用上下文数据

## 使用组件助手提供值

如果你喜欢使用 Component.InvokeAsync 助手来应用视图组件，则可以使用方法参数提供上下文，如下所示：

```
...
<div class="bg-info text-white m-2 p-2">
   @await Component.InvokeAsync("CitySummary", new { themeName = "danger" })
</div>
...
```

InvokeAsync 方法的第一个实参是视图组件类的名称，第二个实参是一个对象，它的名称对应于视图组件定义的形参。

## 24.5.2 创建异步视图组件

本章到目前为止的所有示例都是同步视图组件，都定义了 Invoke 方法。如果视图组件依赖于异步 API，则可通过定义一个返回 Task 的 InvokeAsync 方法，创建异步视图组件。当 Razor 从 InvokeAsync 方法收到 Task 时，会等待 Task 完成，然后把结果插入主视图。为创建一个新组件，在 Components 文件夹中添加一个名为 PageSize.cs 的类文件，用它来定义代码清单 24-24 中的类。

**代码清单 24-24　Components 文件夹中 PageSize.cs 文件的内容**

```
using Microsoft.AspNetCore.Mvc;
using System.Net.Http;
using System.Threading.Tasks;

namespace WebApp.Components {

    public class PageSize : ViewComponent {

        public async Task<IViewComponentResult> InvokeAsync() {
            HttpClient client = new HttpClient();
            HttpResponseMessage response
               = await client.GetAsync("http://apress.com");
            return View(response.Content.Headers.ContentLength);
        }
    }
}
```

InvokeAsync 方法使用 async 和 await 关键字来使用 HttpClient 类提供的异步 API，并通过向 Apress.com 发送一个 GET 请求来获取返回内容的长度。这个长度将传递给 View 方法，后者将选择与该视图组件关联的默认分部视图。

创建 Views/Shared/Components/PageSize 文件夹，在其中添加一个名为 Default.cshtml 的 Razor 视图，并在其中添加代码清单 24-25 所示的内容。

**代码清单 24-25　Views/Shared/Components/PageSize 文件夹的 Default.cshtml 文件的内容**

```
@model long
<div class="m-1 p-1 bg-light text-dark">Page size: @Model</div>
```

最后的步骤是使用组件。如代码清单 24-26 所示，在 Home 控制器的 Index 视图中使用了组件。使用异步视图组件的方式不需要修改。

**代码清单 24-26　在 Views/Home 文件夹的 Index.cshtml 文件中使用异步组件**

```
@model Product
@{
```

```
    Layout = "_Layout";
    ViewBag.Title = ViewBag.Title ?? "Product Table";
}

@section Header { Product Information }

<tr><th>Name</th><td>@Model.Name</td></tr>
<tr>
    <th>Price</th>
    <td>@Model.Price.ToString("c")</td>
</tr>
<tr><th>Category ID</th><td>@Model.CategoryId</td></tr>

@section Footer {
    @(((Model.Price / ViewBag.AveragePrice)
        * 100).ToString("F2"))% of average price
}

@section Summary {
    <div class="bg-info text-white m-2 p-2">
        <vc:city-summary theme-name="secondary" />
        <vc:page-size />
    </div>
}
```

重新启动 ASP.NET Core，使用浏览器请求 http://localhost:5000/home/index/1，得到的响应将包含 Apress.com 主页的大小，如图 24-10 所示。Apress 的网站会频繁更新，所以你看到的数字可能与图中不同。

---

■ **注意：**当需要创建几个不同的内容区域，每个区域可以独立执行时，异步视图组件很有用。直到内容都准备好后，才会将响应发送给浏览器。如果想动态更新展示给用户的内容，则可以使用 Blazor，如第 IV 部分所述。

---

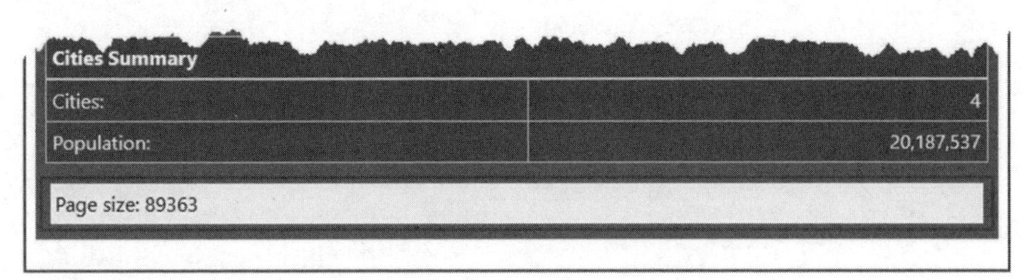


图 24-10　使用异步组件

## 24.6　创建视图组件类

视图组件常提供功能的一个汇总或快照，由控制器或 Razor Pages 来深入处理功能。例如，对于汇总购物车的一个视图组件，通常会有一个链接来访问一个控制器，该控制器提供了购物车中包含的产品的详细列表，并且可用于付款以完成购买。

这种情况下，可创建一个类，使其既作为视图组件，又作为控制器或 Razor Pages。如果使用 Visual Studio，则在 Solution Explorer 中展开 Cities.cshtml，以显示 Cities.cshtml.cs 文件，然后使用代码清单 24-27 中的代码替换其内容。如果使用的是 Visual Studio Code，则在 Pages 文件夹中添加一个名为 Cities.cshtml.cs 的文件，并在其中添加代码清单 24-27 中的内容。

**代码清单 24-27　Pages 文件夹中 Cities.cshtml.cs 文件的内容**

```
using System.Linq;
using Microsoft.AspNetCore.Mvc;
using Microsoft.AspNetCore.Mvc.RazorPages;
using Microsoft.AspNetCore.Mvc.ViewComponents;
using Microsoft.AspNetCore.Mvc.ViewFeatures;
using WebApp.Models;

namespace WebApp.Pages {

    [ViewComponent(Name = "CitiesPageHybrid")]
    public class CitiesModel : PageModel {

        public CitiesModel(CitiesData cdata) {
            Data = cdata;
        }

        public CitiesData Data { get; set; }

        [ViewComponentContext]
        public ViewComponentContext Context { get; set; }

        public IViewComponentResult Invoke() {
            return new ViewViewComponentResult() {
                ViewData = new ViewDataDictionary<CityViewModel>(
                    Context.ViewData,
                    new CityViewModel {
                        Cities = Data.Cities.Count(),
                        Population = Data.Cities.Sum(c => c.Population)
                    })
            };
        }
    }
}
```

这个页面模型类使用了 ViewComponent 进行装饰，从而允许它用作一个视图组件。Name 实参指定了使用什么名称来应用该视图组件。因为页面模型不能继承 ViewComponent 基类，所以使用 ViewComponentContext 特性来装饰一个类型为 ViewComponentContext 的属性，这表明在调用 Invoke 或 InvokeAsync 方法之前，应该把一个定义了表 24-5 中的属性的对象赋值给这个属性。View 方法不可用，所以我创建了一个 ViewViewComponentResult 对象，它依赖于通过被装饰的属性收到的上下文对象。代码清单 24-28 更新了页面的视图部分来使用新的页面模型类。

**代码清单 24-28　在 Pages 文件夹的 Cities.cshtml 文件中更新视图**

```
@page
@model WebApp.Pages.CitiesModel

<div class="m-2">
    <table class="table table-sm table-striped table-bordered">
        <tbody>
            @foreach (City c in Model.Data.Cities) {
                <tr>
                    <td>@c.Name</td>
                    <td>@c.Country</td>
                    <td>@c.Population</td>
                </tr>
            }
        </tbody>
    </table>
</div>
```

这里做的修改更新了指令，以使用页面模型类。为创建混合视图组件的视图，需要创建一个 Pages/Shared/Components/CitiesPageHybrid 文件夹，在其中添加一个名为 Default.cshtml 的 Razor 视图，并为其添加代码清单 24-29 中的内容。

**代码清单 24-29　Pages/Shared/Components/CitiesPageHybrid 文件夹中的 Default.cshtml 文件**

```
@model CityViewModel

<table class="table table-sm table-bordered text-white bg-dark">
    <thead><tr><th colspan="2">Hybrid Page Summary</th></tr></thead>
    <tbody>
        <tr>
            <td>Cities:</td>
            <td class="text-right">@Model.Cities</td>
        </tr>
        <tr>
            <td>Population:</td>
            <td class="text-right">
                @Model.Population.ToString("#,###")
            </td>
        </tr>
    </tbody>
</table>
```

代码清单 24-30 在另一个页面中应用了混合类的视图组件部分。

#### 代码清单 24-30 在 Pages 文件夹的 Data.cshtml 文件中使用视图组件

```
@page
@inject DataContext context;

<h5 class="bg-primary text-white text-center m-2 p-2">Categories</h5>
<ul class="list-group m-2">
    @foreach (Category c in context.Categories) {
        <li class="list-group-item">@c.Name</li>
    }
</ul>

<div class="bg-info text-white m-2 p-2">
    <vc:cities-page-hybrid />
</div>
```

应用混合类的方式与应用其他任何视图组件相同。重新启动 ASP.NET Core，请求 http://localhost:5000/cities 和 http://localhost:5000/data。这两个 URL 由同一个类处理。对于第一个 URL，类用作一个页面模型；对于第二个 URL，类用作一个视图组件。图 24-11 显示了这两个 URL 的输出。

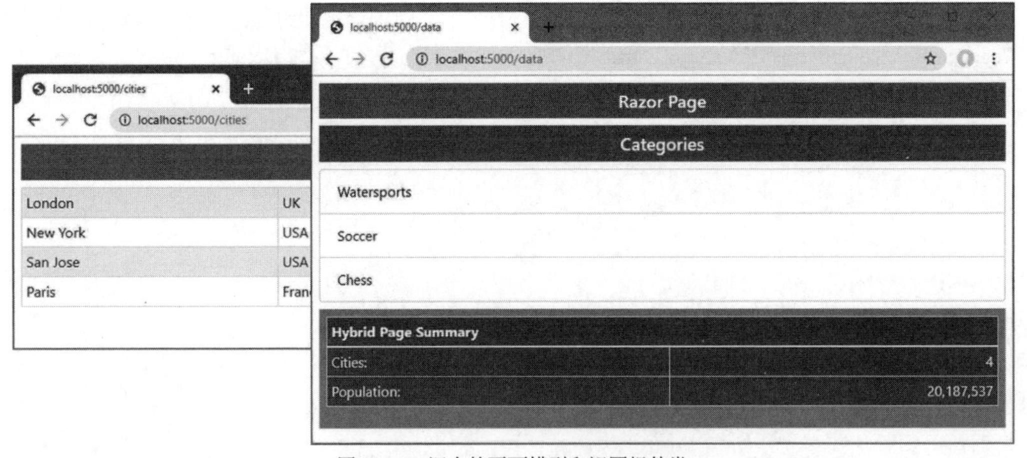


图 24-11 混合的页面模型和视图组件类

### 创建混合控制器类

对于控制器，也可以采用相同的技术。在 Controllers 文件夹中添加一个名为 CitiesController.cs 的类文件，并为其添加代码清单 24-31 的语句。

#### 代码清单 24-31 Controllers 文件夹中 CitiesController.cs 文件的内容

```
using Microsoft.AspNetCore.Mvc;
using Microsoft.AspNetCore.Mvc.ViewComponents;
using Microsoft.AspNetCore.Mvc.ViewFeatures;
using System.Linq;
```

```
using WebApp.Models;

namespace WebApp.Controllers {

    [ViewComponent(Name = "CitiesControllerHybrid")]
    public class CitiesController: Controller {
        private CitiesData data;

        public CitiesController(CitiesData cdata) {
            data = cdata;
        }

        public IActionResult Index() {
            return View(data.Cities);
        }

        public IViewComponentResult Invoke() {
            return new ViewViewComponentResult() {
                ViewData = new ViewDataDictionary<CityViewModel>(
                    ViewData,
                    new CityViewModel {
                        Cities = data.Cities.Count(),
                        Population = data.Cities.Sum(c => c.Population)
                    })
            };
        }
    }
}
```

控制器的实例化方式有一个特别之处，不需要使用 ViewComponentContext 特性装饰的属性，而可以使用从 Controller 基类继承的 ViewData 属性创建视图组件结果。

为给操作方法提供视图，创建 Views/Cities 文件夹，在其中添加一个名为 Index.cshtml 的文件，并在文件中添加代码清单 24-32 中的内容。

#### 代码清单 24-32 Views/Cities 文件夹中 Index.cshtml 文件的内容

```
@model IEnumerable<City>
@{
    Layout = "_ImportantLayout";
}

<div class="m-2">
    <table class="table table-sm table-striped table-bordered">
        <tbody>
            @foreach (City c in Model) {
                <tr>
                    <td>@c.Name</td>
```

```
                    <td>@c.Country</td>
                    <td>@c.Population</td>
                </tr>
            }
        </tbody>
    </table>
</div>
```

要为视图组件提供视图，创建 Views/Shared/Components/CitiesControllerHybrid 文件夹，在其中添加一个名为 Default.cshtml 的 Razor 视图，使其包含代码清单 24-33 中的内容。

**代码清单 24-33　Views/Shared/Components/CitiesControllerHybrid 文件夹中的 Default.cshtml 文件**

```
@model CityViewModel
<table class="table table-sm table-bordered text-white bg-dark">
    <thead><tr><th colspan="2">Hybrid Controller Summary</th></tr></thead>
    <tbody>
        <tr>
            <td>Cities:</td>
            <td class="text-right">@Model.Cities</td>
        </tr>
        <tr>
            <td>Population:</td>
            <td class="text-right">
                @Model.Population.ToString("#,###")
            </td>
        </tr>
    </tbody>
</table>
```

代码清单 24-34 在 Data.cshtml Razor Pages 中应用混合视图组件，替换了上一节创建的混合类。

**代码清单 24-34　在 Pages 文件夹的 Data.cshtml 文件中应用视图组件**

```
@page
@inject DataContext context;

<h5 class="bg-primary text-white text-center m-2 p-2">Categories</h5>
<ul class="list-group m-2">
    @foreach (Category c in context.Categories) {
        <li class="list-group-item">@c.Name</li>
    }
</ul>

<div class="bg-info text-white m-2 p-2">
    <vc:cities-controller-hybrid />
</div>
```

重新启动ASP.NET Core，使用浏览器请求http://localhost:5000/cities/index和http://localhost:5000/data。对于第一个URL，代码清单24-34中的类用作一个控制器；对于第二个URL，该类用作一个视图组件。图24-12显示了这两个URL的响应。

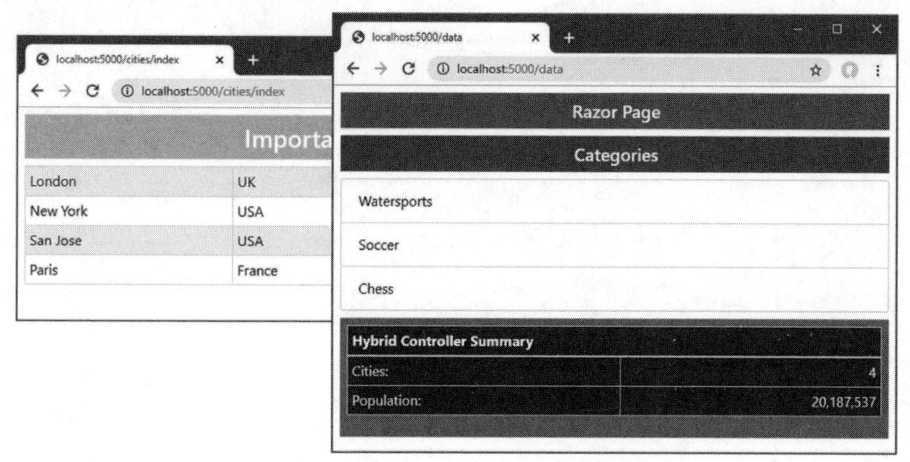


图 24-12 混合的控制器和视图组件类

## 24.7 小结

本章介绍了视图组件功能，这允许在控制器或 Razor Pages 使用的视图中包含其他功能。本章解释了视图组件的工作方式，以及如何应用它们，还演示了视图组件生成的不同类型的结果。本章最后介绍了如何创建既是视图组件又是控制器或 Razor Pages 的类。第 25 章将介绍标签助手，它们用于转换 HTML 元素。

# 第 25 章

# 使用标签助手

标签助手是转换视图或页面中的 HTML 元素的 C#类。标签助手的常见用途包括使用应用程序的路由配置为表单生成 URL，确保特定类型的元素具有一致的样式，以及用常用的内容片段替换定制的速记元素。本章将描述标签助手的工作原理，以及自定义标签助手是如何创建和应用的。第 26 章描述内置的标签助手，第 27 章使用标签助手来解释 HTML 表单是如何创建的。表 25-1 给出了标签助手的相关知识。

表 25-1　标签助手的相关知识

| 问题 | 答案 |
| --- | --- |
| 它们是什么? | 标签助手是操作 HTML 元素的类，可通过某种方式更改它们，用附加内容补充它们，或者用新内容完全替换它们 |
| 它们为什么有用? | 标签助手允许使用 C#逻辑生成或转换视图内容，确保发送到客户端的 HTML 反映了应用程序的状态 |
| 它们是如何使用的? | 应用标签助手的 HTML 元素是基于类名或使用 HTMLTargetElement 属性选择的。当呈现视图时，元素被标签助手转换，并包含在发送给客户端的 HTML 中 |
| 是否存在缺陷或限制? | 使用标签助手很容易生成复杂的 HTML 内容，而使用视图组件更容易实现这一点，详见第 24 章 |
| 还有其他选择吗? | 不必使用标签助手，但是它们很容易在 ASP.NETCore 应用程序中生成复杂的 HTML |

表 25-2 总结了本章的内容。

表 25-2　本章内容摘要

| 问题 | 解决方案 | 代码清单 |
| --- | --- | --- |
| 创建标签助手 | 定义一个从 TagHelper 类派生的类 | 25-1～25-7 |
| 控制标签助手属性的作用域 | 改变 HTMLTargetElement 指定的元素范围 | 25-8～25-11 |
| 创建被内容替换的定制 HTML 元素 | 使用速记元素 | 25-12、25-13 |
| 以编程方式创建元素 | 使用 TagBuilder 类 | 25-14 |
| 控制内容的插入位置 | 使用 prepend 和 append 特性 | 25-15～25-18 |
| 获取上下文数据 | 使用上下文对象 | 25-19、25-20 |
| 对视图模型或页面模型进行操作 | 使用模型表达式 | 25-21～25-24 |

(续表)

| 问题 | 解决方案 | 代码清单 |
| --- | --- | --- |
| 创建协调标签助手 | 使用 Items 属性 | 25-26 |
| 抑制内容 | 使用 SuppressOutput 方法 | 25-27、25-28 |
| 将标签助手定义为服务 | 创建标签助手组件 | 25-29～25-32 |

## 25.1　准备工作

本章使用了第 24 章中的 WebApp 项目。为了准备这一章，将 startup.cs 文件的内容替换为代码清单 25-1 中的内容，删除前面章节中使用的一些配置语句。

**提示：**可从 https://github.com/apress/pro-asp.net-core-3 下载本章和本书中其他所有章节的示例项目。如果在运行示例时遇到问题，请参阅第 1 章，以了解如何获得帮助。

**代码清单 25-1　WebApp 文件夹中 Startup.cs 文件的内容**

```
using Microsoft.AspNetCore.Builder;
using Microsoft.Extensions.DependencyInjection;
using Microsoft.Extensions.Configuration;
using Microsoft.EntityFrameworkCore;
using WebApp.Models;

namespace WebApp {
    public class Startup {

        public Startup(IConfiguration config) {
            Configuration = config;
        }

        public IConfiguration Configuration { get; set; }

        public void ConfigureServices(IServiceCollection services) {
            services.AddDbContext<DataContext>(opts => {
                opts.UseSqlServer(Configuration[
                    "ConnectionStrings:ProductConnection"]);
                opts.EnableSensitiveDataLogging(true);
            });
            services.AddControllersWithViews().AddRazorRuntimeCompilation();
            services.AddRazorPages().AddRazorRuntimeCompilation();
            services.AddSingleton<CitiesData>();
        }

        public void Configure(IApplicationBuilder app, DataContext context) {
            app.UseDeveloperExceptionPage();
            app.UseStaticFiles();
```

```
            app.UseRouting();
            app.UseEndpoints(endpoints => {
                endpoints.MapControllers();
                endpoints.MapDefaultControllerRoute();
                endpoints.MapRazorPages();
            });
            SeedData.SeedDatabase(context);
        }
    }
}
```

接下来，用代码清单 25-2 的内容替换 Views/Home 文件夹中 Index.cshtml 文件的内容。

**代码清单 25-2　Views/Home 文件夹中 Index.cshtml 文件的内容**

```
@model Product
@{
   Layout = "_SimpleLayout";
}

<table class="table table-striped table-bordered table-sm">
    <thead>
        <tr>
            <th colspan="2">Product Summary</th>
        </tr>
    </thead>
    <tbody>
        <tr><th>Name</th><td>@Model.Name</td></tr>
        <tr>
            <th>Price</th>
            <td>@Model.Price.ToString("c")</td>
        </tr>
        <tr><th>Category ID</th><td>@Model.CategoryId</td></tr>
    </tbody>
</table>
```

代码清单 25-2 中的视图依赖新的布局。在 Views/Shared 文件夹中添加一个名为_SimpleLayout.cshtml 的 Razor 视图文件，内容如代码清单 25-3 所示。

**代码清单 25-3　Views/Shared 文件夹中_SimpleLayout.cshtml 文件的内容**

```
<!DOCTYPE html>
<html>
<head>
    <title>@ViewBag.Title</title>
    <link href="/lib/twitter-bootstrap/css/bootstrap.min.css" rel="stylesheet" />
</head>
<body>
    <div class="m-2">
```

```
        @RenderBody()
    </div>
</body>
</html>
```

### 25.1.1　删除数据库

打开一个新的 PowerShell 命令提示符，导航到包含 WebApp.csproj 文件的文件夹，并运行代码清单 25-4 所示的命令来删除数据库。

**代码清单 25-4　删除数据库**

```
dotnet ef database drop --force
```

### 25.1.2　运行示例应用程序

从 Debug 菜单中选择 Start Without Debugging 或 Run Without Debugging，或者使用 PowerShell 命令提示符运行代码清单 25-5 所示的命令。

**代码清单 25-5　运行示例应用程序**

```
dotnet run
```

使用浏览器请求 http://localhost:5000/home，这将生成如图 25-1 所示的响应。

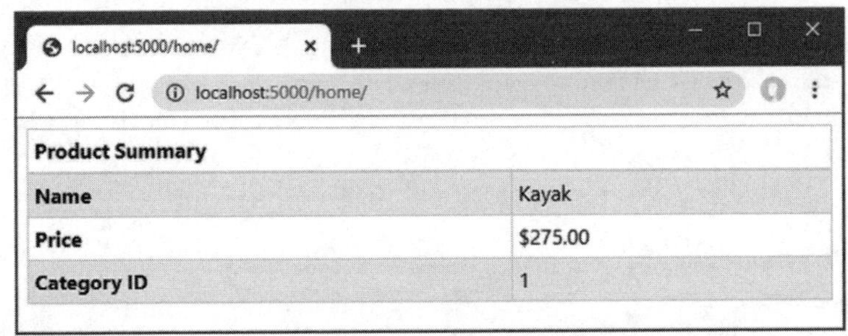


图 25-1　运行示例应用程序

## 25.2　创建标签助手

理解标签助手的最好方法是创建一个标签助手，揭示它们是如何操作的，以及它们如何用于 ASP.NET Core 应用程序。接下来将介绍创建和应用标签助手的过程，为 tr 元素设置引导 CSS 类：

```
...
<tr tr-color="primary">
    <th colspan="2">Product Summary</th>
</tr>
...
```

转化为:

```
...
<tr class="bg-primary text-white text-center">
   <th colspan="2">Product Summary</th>
</tr>
...
```

标签助手将识别 tr-color 属性，并使用其值设置发送到浏览器的元素的 class 属性。这并不是最引人注目或最有用的转变，但它为解释标签助手如何工作提供了基础。

### 25.2.1 定义标签助手类

标签助手可以在项目的任何地方定义，但是将它们放在一起是有帮助的，因为在使用之前需要先注册它们。创建 WebApp/TagHelpers 文件夹，并向其中添加一个名为 TrTagHelper.cs 的类文件，如代码清单 25-6 所示。

**代码清单 25-6 TagHelpers 文件夹中 TrTagHelper.cs 文件的内容**

```
using Microsoft.AspNetCore.Razor.TagHelpers;

namespace WebApp.TagHelpers {

    public class TrTagHelper: TagHelper {

        public string BgColor { get; set; } = "dark";
        public string TextColor { get; set; } = "white";

        public override void Process(TagHelperContext context,
                TagHelperOutput output) {
            output.Attributes.SetAttribute("class",
                $"bg-{BgColor} text-center text-{TextColor}");
        }
    }
}
```

标签助手是从 TagHelper 类派生的，该类在 Microsoft.AspNetCore.Razor.TagHelpers 名称空间中定义。TagHelper 类定义了一个 Process 方法，它被子类覆盖以实现转换元素的行为。

标签助手的名称组合了它转换的元素的名称，后跟 TagHelper。在本例中，类名 TrTagHelper 表示这是一个操作 tr 元素的标签助手。标签助手可应用到的元素范围可使用属性来扩展或缩小，如本章后面所述，但默认行为是由类名定义的。

---

▒ **提示：** 异步标签助手可通过覆盖 ProcessAsync 方法而不是 Process 方法来创建，但这对大多数助手来说不是必需的，因为它们倾向于对 HTML 元素进行小而集中的更改。可以在“高级标签助手功能”一节中看到一个异步标签助手的示例。

---

#### 1. 接收上下文数据

标签助手通过 TagHelperContext 类的实例接收关于它们正在转换的元素的信息，这个实例作为 Process 方法的参数接收，定义了表 25-3 中描述的属性。

表 25-3　TagHelperContext 属性

| 名称 | 描述 |
| --- | --- |
| AllAttributes | 此属性返回一个特性的只读字典，其中包含应用于被转换元素的特性，按名称和索引排序 |
| Items | 此属性返回一个字典，该字典用于在标签助手之间进行协调，如 25.3.6“标签助手之间的协调”所述 |
| UniqueId | 此属性返回被转换元素的唯一标识符 |

虽然可通过 AllAttributes 字典访问元素属性的详细信息，但更方便的方法是定义一个属性，其名称对应于你感兴趣的属性，如下所示：

```
...
public string BgColor { get; set; } = "dark";
public string TextColor { get; set; } = "white";
...
```

当使用标签助手时，检查它定义的属性，并将名称与应用于 HTML 元素的属性相匹配的任何属性值分配给它。作为这个过程的一部分，属性值转换为与 C#属性相匹配的类型，这样 bool 属性就可以用来接收 true 和 false 属性值，int 属性就可以用来接收属性数字值，如 1 和 2。

无对应 HTML 元素属性的属性没有设置，这意味着应该进行检查，以确保没有处理 null 或提供默认值，如代码清单 25-6 所示。

属性的名称会自动从默认的 HTML 样式 bg-color 转换为 C#样式 BgColor。除了 asp- (Microsoft 使用的)和 data-(为发送给客户端的自定义属性保留的)之外，可以使用任何属性前缀。示例标签助手使用 bg-color 和 text-color 属性配置，它们为 BgColor 和 TextColor 属性提供值，并用于在 Process 方法中配置 tr 元素，如下所示：

```
...
output.Attributes.SetAttribute("class",
   $"bg-{BgColor} text-center text-{TextColor}");
...
```

**提示：** 为标签助手属性使用 HTML 属性名并不总是导致可读或可理解的类。可以使用 HTMLAttributeName 属性中断属性名称与其所表示的属性之间的链接，该属性可用于指定所表示的 HTML 属性。

#### 2. 生成输出

Process 方法通过配置作为参数接收的 TagHelperOutput 对象来转换元素。TagHelperOuput 对象首先描述出现在视图中的 HTML 元素，然后通过表 25-4 中描述的属性和方法进行修改。

表 25-4 TagHelperOutput 属性和方法

| 名称 | 描述 |
| --- | --- |
| TagName | 此属性用于获取或设置输出元素的标记名 |
| Attributes | 此属性返回一个包含输出元素属性的字典 |
| Content | 此属性返回一个 TagHelperContent 对象，该对象用于设置元素的内容 |
| GetChildContentAsync() | 这个异步方法提供了对将要转换的元素内容的访问，如 25.3.1“创建快捷元素”一节所述 |
| PreElement | 该属性返回一个 TagHelperContext 对象，该对象用于在输出元素之前在视图中插入内容。参见 25.3.3“追加、附加内容和元素”一节 |
| PostElement | 该属性返回一个 TagHelperContext 对象，该对象用于在输出元素之后在视图中插入内容。参见 25.3.3“追加、附加内容和元素”一节 |
| PreContent | 此属性返回一个 TagHelperContext 对象，用于在输出元素的内容之前插入内容。参见 25.3.3“追加、附加内容和元素”一节 |
| PostContent | 此属性返回一个 TagHelperContext 对象，用于在输出元素的内容之后插入内容。参见 25.3.3“追加、附加内容和元素”一节 |
| TagMode | 此属性指定如何使用 TagMode 枚举的值写入输出元素。参见 25.3.1“创建快捷元素”一节 |
| SupressOuput() | 调用该方法会从视图中排除一个元素。请参阅 25.3.7“抑制输出元素”一节 |

在 TrTagHelper 类中，使用 Attributes 字典向指定引导样式的 HTML 元素添加了一个 class 属性，包括 BgColor 和 TextColor 属性的值。其效果是，可通过将 bg-color 和 text-color 属性设置为引导名称(如 primary、info 和 danger)来指定 tr 元素的背景颜色。

## 25.2.2 注册标签助手

在使用标签助手类之前，它们必须在@addTagHelper 指令中注册。标签助手可以应用到的一组视图或页面取决于使用@addTagHelper 指令的位置。

对于单个视图或页面，指令出现在 CSHTML 文件中。要使标签助手更广泛地可用，可以将其添加到视图导入文件中，该文件在控制器的 Views 文件夹和 Razor Pages 的 Pages 文件夹中定义。

希望本章创建的标签助手在应用程序的任何地方都可用，这意味着@addTagHelper 指令添加到 Views 和 Pages 文件夹的_ViewImports.cshtml 文件中。在第 24 章中用于应用视图组件的 vc 元素是一个标签助手，这就是启用标签助手所需的指令位于_ViewImports.cshtml 文件的原因。

```
@using WebApp.Models
@addTagHelper *, Microsoft.AspNetCore.Mvc.TagHelpers
@using WebApp.Components
@addTagHelper *, WebApp
```

参数的第一部分指定标签助手类的名称，并支持通配符，第二部分指定定义它们的程序集的名称。@addTagHelper 指令使用通配符选择 WebApp 程序集中的所有名称空间，其效果是在项目中任何地方定义的标签助手都可在任何控制器视图中使用。Razor Pages 文件夹的_ViewImports.cshtml 文件中有一个相同的语句。

```
@namespace WebApp.Pages
@using WebApp.Models
@addTagHelper *, Microsoft.AspNetCore.Mvc.TagHelpers
@addTagHelper *, WebApp
```

另一个@addTagHelper 指令启用了微软提供的内置标签助手，参见第 26 章。

### 25.2.3　使用标签助手

最后一步是使用标签助手转换元素。在代码清单 25-7 中，将属性添加到 tr 元素，它将应用标签助手。

**代码清单 25-7　在 Views/Home 文件夹的 Index.cshtml 文件中使用标签助手**

```
@model Product
@{
    Layout = "_SimpleLayout";
}

<table class="table table-striped table-bordered table-sm">
    <thead>
        <tr bg-color="info" text-color="white">
            <th colspan="2">Product Summary</th>
        </tr>
    </thead>
    <tbody>
        <tr><th>Name</th><td>@Model.Name</td></tr>
        <tr>
            <th>Price</th>
            <td>@Model.Price.ToString("c")</td>
        </tr>
        <tr><th>Category ID</th><td>@Model.CategoryId</td></tr>
    </tbody>
</table>
```

重启 ASP.NET Core 并使用浏览器请求 http://localhost:5000/home，这会生成如图 25-2 所示的响应。

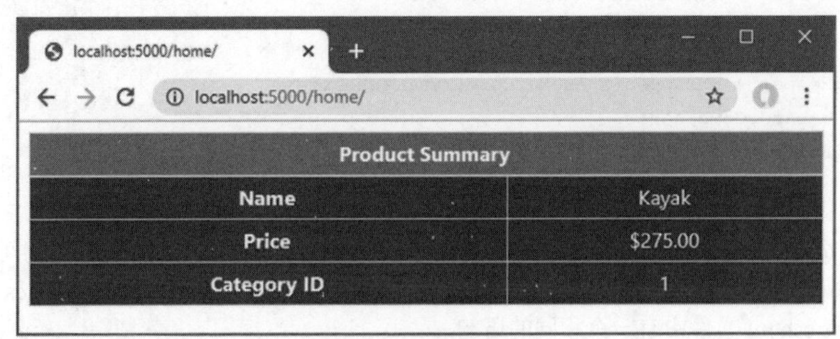


图 25-2　使用标签助手

对代码清单 25-7 中应用属性的 tr 元素进行了转换，但这不是图中显示的唯一更改。默认情况下，标签助手应用于特定类型的所有元素，这意味着视图中的所有 tr 元素都使用标签助手类中定义的默认值进行转换，因为没有定义任何属性。

事实上，这个问题更加严重，因为视图导入文件中的@addTagHelper 指令意味着示例标签助手应用于控制器和 Razor Pages 所呈现的任何视图中使用的所有 tr 元素。例如，使用浏览器请求 http://localhost:5000/cities，来自 cities Razor Pages 的响应中的 tr 元素也进行了转换，如图 25-3 所示。

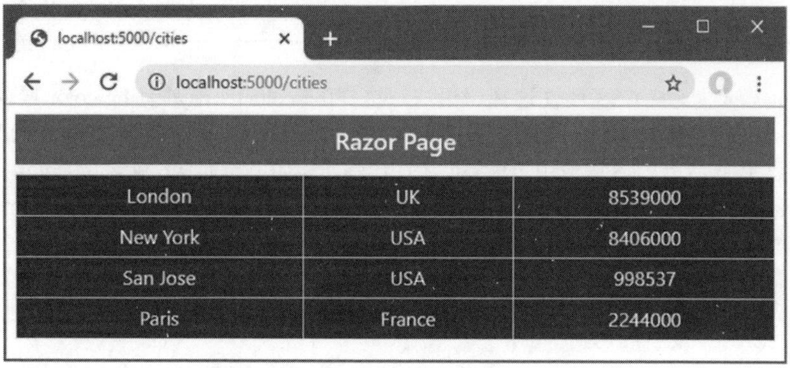

图 25-3 意外地使用标签助手修改元素

### 25.2.4 缩小标签助手的范围

可使用 HTMLTargetElement 元素控制由标签助手转换的元素范围，如代码清单 25-8 所示。

**代码清单 25-8 在 TagHelpers 文件夹的 TrTagHelper.cs 文件中缩小范围**

```
using Microsoft.AspNetCore.Razor.TagHelpers;

namespace WebApp.TagHelpers {

    [HtmlTargetElement("tr", Attributes = "bg-color,text-color", ParentTag ="thead")]
    public class TrTagHelper: TagHelper {

        public string BgColor { get; set; } = "dark";
        public string TextColor { get; set; } = "white";

        public override void Process(TagHelperContext context,
                TagHelperOutput output) {
            output.Attributes.SetAttribute("class",
                $"bg-{BgColor} text-center text-{TextColor}");
        }
    }
}
```

HTMLTargetElement 属性描述了应用标签助手的元素。第一个参数指定元素类型，并支持表 25-5 中描述的其他命名属性。

表 25-5　HTMLTargetElement 属性

| 名称 | 描述 |
| --- | --- |
| Attributes | 指定标签助手应仅应用于具有一组给定属性(以逗号分隔的列表提供)的元素。以星号结尾的属性名称将被视为前缀，这样 bg-*将匹配 bg-color、bg-size 等 |
| ParentTag | 指定标签助手应仅应用于给定类型元素中包含的元素 |
| TagStructure | 指定标签助手应仅应用于其标记结构对应于 TagStructure 枚举中给定值的元素，该枚举定义了 Unspecified、NormalOrSelfClosing 和 WithoutEndTag |

Attributes 属性支持 CSS 属性选择器语法，这样[bg-color]匹配具有 bg-color 属性的元素，[bg-color=primary]匹配 bg-color 属性值为 primary 的元素，[bg-color ^ = p]匹配 bg-color 属性值以 p 开头的元素。代码清单 25-8 中应用于标签助手的属性匹配如下 tr 元素：包含 bg-color 和 text-color 属性，父元素是 thead。重启 ASP.NET Core，并用浏览器请求 http://localhost:5000/home/index/1，标签助手的范围缩小了，如图 25-4 所示。

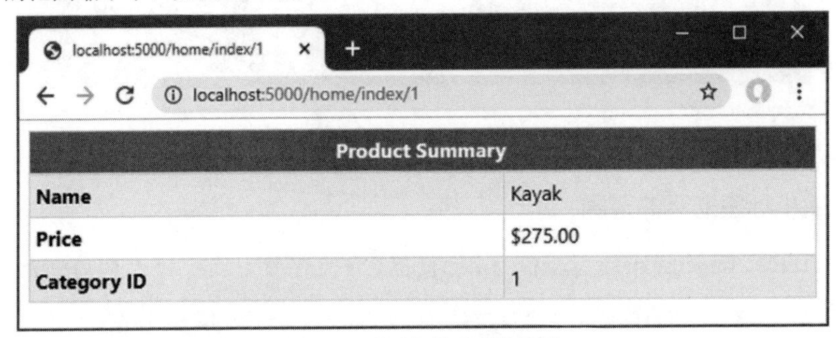


图 25-4　缩小标签助手的范围

## 25.2.5　扩展标签助手的范围

HTMLTargetElement 属性还可用于扩大标签助手的范围，以便匹配更大范围的元素。这是通过将属性的第一个参数设置为星号(*字符)来实现的，*匹配任何元素。代码清单 25-9 更改了应用于示例标签助手的属性，以便它匹配任何具有 bg-color 和 text-color 属性的元素。

**代码清单 25-9　在 TagHelpers 文件夹的 TrTagHelper.cs 文件中扩大范围**

```
using Microsoft.AspNetCore.Razor.TagHelpers;

namespace WebApp.TagHelpers {

    [HtmlTargetElement("*", Attributes = "bg-color,text-color")]
    public class TrTagHelper: TagHelper {

        public string BgColor { get; set; } = "dark";
        public string TextColor { get; set; } = "white";
        public override void Process(TagHelperContext context,
                TagHelperOutput output) {
            output.Attributes.SetAttribute("class",
```

```
                $"bg-{BgColor} text-center text-{TextColor}");
        }
    }
}
```

使用星号时必须小心，因为很容易匹配过大范围，并选择不应该转换的元素。更安全的中间方法是为每类元素应用 HTMLTargetElement 属性，如代码清单 25-10 所示。

#### 代码清单 25-10　在 TagHelpers 文件夹的 TrTagHelper.cs 文件中平衡范围

```
using Microsoft.AspNetCore.Razor.TagHelpers;

namespace WebApp.TagHelpers {

    [HtmlTargetElement("tr", Attributes = "bg-color,text-color")]
    [HtmlTargetElement("td", Attributes = "bg-color")]
    public class TrTagHelper: TagHelper {

        public string BgColor { get; set; } = "dark";
        public string TextColor { get; set; } = "white";

        public override void Process(TagHelperContext context,
                TagHelperOutput output) {
            output.Attributes.SetAttribute("class",
                $"bg-{BgColor} text-center text-{TextColor}");
        }
    }
}
```

属性的每个实例可使用不同的选择条件。标签助手使用 bg-color 和 text- color 属性匹配 tr 元素，并使用 bg-color 属性匹配 td 元素。代码清单 25-11 添加了一个要转换到 Index 视图的元素，以演示修改后的范围。

#### 代码清单 25-11　在 Views/Home 文件夹的 Index.cshtml 文件中添加属性

```
@model Product
@{
    Layout = "_SimpleLayout";
}

<table class="table table-striped table-bordered table-sm">
    <thead>
        <tr bg-color="info" text-color="white">
            <th colspan="2">Product Summary</th>
        </tr>
    </thead>
    <tbody>
        <tr><th>Name</th><td>@Model.Name</td></tr>
```

```
        <tr>
            <th>Price</th>
            <td bg-color="dark">@Model.Price.ToString("c")</td>
        </tr>
        <tr><th>Category ID</th><td>@Model.CategoryId</td></tr>
    </tbody>
</table>
```

重启 ASP.NET Core 并使用浏览器请求 http://localhost:5000/home/index/1。响应包含两个转换后的元素，如图 25-5 所示。

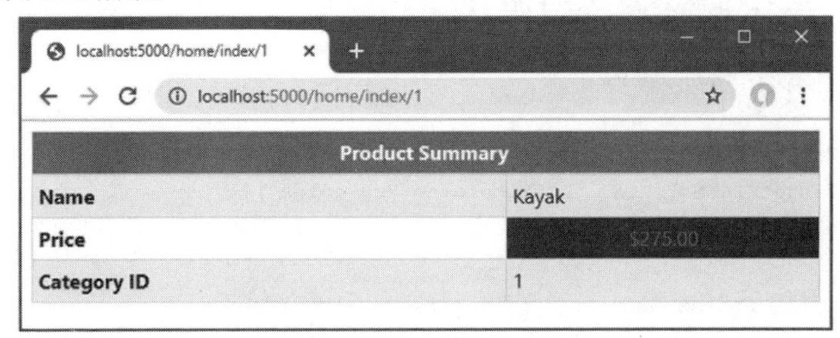


图 25-5　管理标签助手的范围

**预定标签助手的执行**

如果需要对一个元素应用多个标签助手，可通过设置 Order 属性来控制它们的执行顺序，Order 属性继承自 TagHelper 基类。管理序列可帮助最小化标签助手之间的冲突，但仍然很容易遇到问题。

# 25.3　高级标签助手功能

上一节演示了如何创建一个基本的标签助手，但这仅触及了皮毛。接下来将展示标签助手的更高级用法及其提供的特性。

## 25.3.1　创建快捷元素

标签助手并不仅限于转换标准 HTML 元素，还可用常用内容替换定制元素。这是一个有用的特性，可使视图更简洁，意图更明显。为演示这一点，代码清单 25-12 用定制的 HTML 元素替换了 Index 视图中的 thead 元素。

**代码清单 25-12　在 Views/Home 文件夹的 Index.cshtml 文件中添加定制的 HTML 元素**

```
@model Product
@{
    Layout = "_SimpleLayout";
}

<table class="table table-striped table-bordered table-sm">
    <tablehead bg-color="dark">Product Summary</tablehead>
```

```
        <tbody>
            <tr><th>Name</th><td>@Model.Name</td></tr>
            <tr>
                <th>Price</th>
                <td bg-color="dark">@Model.Price.ToString("c")</td>
            </tr>
            <tr><th>Category ID</th><td>@Model.CategoryId</td></tr>
        </tbody>
    </table>
```

tablehead 元素不是 HTML 规范的一部分，浏览器无法理解。相反，下面使用这个元素作为生成 HTML 表的 thead 元素及其内容的简写。将一个名为 TableHeadTagHelper.cs 的类添加到 TagHelpers 文件夹中，并用它定义如代码清单 25-13 所示的类。

> **提示：** 在处理不属于 HTML 规范的自定义元素时，必须应用 HTMLTargetElement 属性并指定元素名称，如代码清单 25-13 所示。根据类名对元素应用标签助手的约定只适用于标准元素名。

**代码清单 25-13　TagHelpers 文件夹中 TableHeadTagHelper.cs 文件的内容**

```
using Microsoft.AspNetCore.Razor.TagHelpers;
using System.Threading.Tasks;

namespace WebApp.TagHelpers {

    [HtmlTargetElement("tablehead")]
    public class TableHeadTagHelper: TagHelper {

        public string BgColor { get; set; } = "light";

        public override async Task ProcessAsync(TagHelperContext context,
                TagHelperOutput output) {

            output.TagName = "thead";
            output.TagMode = TagMode.StartTagAndEndTag;
            output.Attributes.SetAttribute("class",
                $"bg-{BgColor} text-white text-center");

            string content = (await output.GetChildContentAsync()).GetContent();
            output.Content
                .SetHtmlContent($"<tr><th colspan=\"2\">{content}</th></tr>");
        }
    }
}
```

这个标签助手是异步的，并且覆盖了ProcessAsync方法，这样它就可以访问它转换的元素的现有内容。ProcessAsync方法使用TagHelperOuput对象的属性来生成一个完全不同的元素：TagName属性用于指定thead元素，TagMode属性用于指定使用开始和结束标记编写元素。

Attributes.SetAttribute方法用于定义类属性，Content属性用于设置元素内容。

元素的现有内容是通过异步GetChildContentAsync方法获得的，该方法返回一个TagHelperContent对象。这与TagHelperOutput.Content属性返回的对象相同，并允许使用相同的类型，通过表 25-6 中描述的方法来检查和更改元素的内容。

表 25-6　有用的 TagHelperContent 方法

| 名称 | 描述 |
|---|---|
| GetContent() | 此方法以字符串形式返回 HTML 元素的内容 |
| SetContent(text) | 此方法设置输出元素的内容。对字符串参数进行编码，以便安全地包含在 HTML 元素中 |
| SetHtmlContent(html) | 此方法设置输出元素的内容。假设字符串参数是安全编码的。应当谨慎使用 |
| Append(text) | 此方法安全地编码指定的字符串，并将其添加到输出元素的内容中 |
| AppendHtml(html) | 此方法将指定的字符串添加到输出元素的内容中，而不执行任何编码。应当谨慎使用 |
| Clear() | 此方法删除输出元素的内容 |

在代码清单 25-13 中，元素的现有内容通过 GetContent 元素读取，然后使用 SetHtmlContent 方法设置它。效果是在 tr 和 th 元素中对已转换元素中的现有内容换行。

重启 ASP.NET Core 并导航到 http://localhost:5000/home/index/1，就会看到标签助手的效果，如图 25-6 所示。

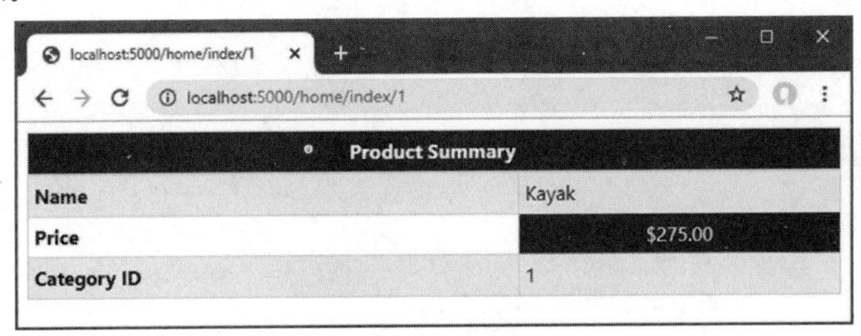


图 25-6　使用快捷元素

对于这个快捷元素：

```
...
<tablehead bg-color="dark">Product Summary</tablehead>
...
```

标签助手将其转换为以下元素：

```
...
<thead class="bg-dark text-white text-center">
    <tr>
        <th colspan="2">Product Summary</th>
    </tr>
</thead>
...
```

注意，转换后的元素不包括 bg-color 属性。与标签助手定义的属性匹配的属性将从输出元素中删除；如有必要，则必须显式地重新定义它们。

### 25.3.2 以编程方式创建元素

在生成新的 HTML 元素时，可以使用标准的 C#字符串格式来创建所需的内容，这就是在代码清单 25-13 中采用的方法。这样做很有效，但可能会很尴尬，需要密切注意以避免拼写错误。更健壮的方法是使用 TagBuilder 类，它是在 Microsoft.AspNetCore.Mvc.Rendering 名称空间中定义的，并允许以更结构化的方式创建元素。表 25-6 中描述的 TagHelperContent 方法接收 TagBuilder 对象，这将便于在标签助手中创建 HTML 内容，如代码清单 25-14 所示。

**代码清单 25-14 在 TagHelpers 文件夹的 TableHeadTagHelper.cs 文件中创建 HTML 元素**

```
using Microsoft.AspNetCore.Razor.TagHelpers;
using System.Threading.Tasks;
using Microsoft.AspNetCore.Mvc.Rendering;

namespace WebApp.TagHelpers {

    [HtmlTargetElement("tablehead")]
    public class TableHeadTagHelper: TagHelper {

        public string BgColor { get; set; } = "light";

        public override async Task ProcessAsync(TagHelperContext context,
                TagHelperOutput output) {

            output.TagName = "thead";
            output.TagMode = TagMode.StartTagAndEndTag;
            output.Attributes.SetAttribute("class",
                $"bg-{BgColor} text-white text-center");

            string content = (await output.GetChildContentAsync()).GetContent();

            TagBuilder header = new TagBuilder("th");
            header.Attributes["colspan"] = "2";
            header.InnerHtml.Append(content);

            TagBuilder row = new TagBuilder("tr");
            row.InnerHtml.AppendHtml(header);

            output.Content.SetHtmlContent(row);
        }
    }
}
```

这个示例使用 TagBuilder 对象创建每个新元素，并将它们组合成与代码清单 25-13 中基于字符串的版本相同的 HTML 结构。

### 25.3.3　追加、附加内容和元素

TagHelperOutput 类提供了四个属性，可以很容易地将新内容注入视图中，以便包围元素或元素的内容，如表 25-7 所示。

表 25-7　用于追加内容和元素的 TagHelperOutput 属性

| 名称 | 描述 |
| --- | --- |
| PreElement | 此属性用于在目标元素之前向视图中插入元素 |
| PostElement | 此属性用于在目标元素之后向视图中插入元素 |
| PreContent | 此属性用于在任何现有内容之前将内容插入目标元素 |
| PostContent | 此属性用于在任何现有内容之后将内容插入目标元素 |

接下来解释如何在目标元素周围和内部插入内容。

#### 1. 在输出元素周围插入内容

前两个 TagHelperOuput 属性是 PreElement 和 PostElement，它们用于在输出元素之前和之后向视图插入元素。为演示这些属性的用法，将一个名为 WrapperTagHelper.cs 的类文件添加到 WebApp/TagHelpers 文件夹中，其内容如代码清单 25-15 所示。

**代码清单 25-15　TagHelpers 文件夹中 WrapperTagHelper.cs 文件的内容**

```
using Microsoft.AspNetCore.Mvc.Rendering;
using Microsoft.AspNetCore.Razor.TagHelpers;

namespace WebApp.TagHelpers {

    [HtmlTargetElement("*", Attributes = "[wrap=true]")]
    public class ContentWrapperTagHelper: TagHelper {

        public override void Process(TagHelperContext context,
                TagHelperOutput output) {
            TagBuilder elem = new TagBuilder("div");
            elem.Attributes["class"] = "bg-primary text-white p-2 m-2";
            elem.InnerHtml.AppendHtml("Wrapper");

            output.PreElement.AppendHtml(elem);
            output.PostElement.AppendHtml(elem);
        }
    }
}
```

这个标签助手转换 wrap 属性值为 true 的元素，使用 PreElement 和 PostElement 属性在输出元素之前和之后添加 div 元素。代码清单 25-16 将一个元素添加到由标签助手转换的索引视图中。

**代码清单 25-16　在 Views/Index 文件夹的 Index.cshtml 文件中添加一个元素**

```
@model Product
@{
    Layout = "_SimpleLayout";
}

<div class="m-2" wrap="true">Inner Content</div>

<table class="table table-striped table-bordered table-sm">
    <tablehead bg-color="dark">Product Summary</tablehead>
    <tbody>
        <tr><th>Name</th><td>@Model.Name</td></tr>
        <tr>
            <th>Price</th>
            <td bg-color="dark">@Model.Price.ToString("c")</td>
        </tr>
        <tr><th>Category ID</th><td>@Model.CategoryId</td></tr>
    </tbody>
</table>
```

重启 ASP.NET Core 并使用浏览器请求 http://localhost:5000/home/index/1。响应包括转换后的元素，如图 25-7 所示。

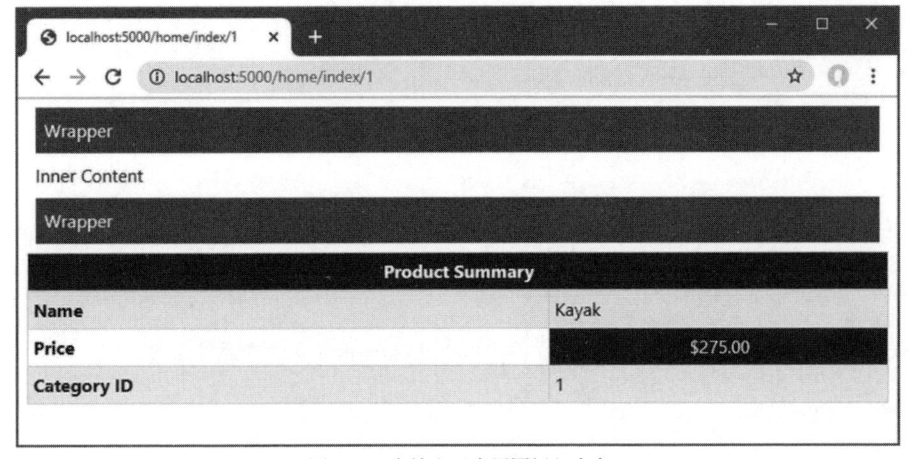


图 25-7　在输出元素周围插入内容

如果检查发送到浏览器的 HTML，就会看到这个元素：

```
...
<div class="m-2" wrap="true">Inner Content</div>
...
```

可将其转化为以下元素：

```
...
```

```
<div class="bg-primary text-white p-2 m-2">Wrapper</div>
<div class="m-2" wrap="true">Inner Content</div>
<div class="bg-primary text-white p-2 m-2">Wrapper</div>
...
```

注意，wrap 属性保留在输出元素上。这是因为没有在标签助手类中定义对应于此属性的属性。如果想防止属性包含在输出中，可在标签助手类中为它们定义一个属性(即使不使用属性值)。

### 2. 在输出元素中插入内容

PreContent 和 PostContent 属性用于在输出元素中围绕原始内容插入内容。为演示这个特性，将一个名为 HighlightTagHelper.cs 的类文件添加到 TagHelpers 文件夹中，并使用它定义如代码清单 25-17 所示的标签助手。

**代码清单 25-17　TagHelpers 文件夹中 HighlightTagHelper.cs 文件的内容**

```
using Microsoft.AspNetCore.Razor.TagHelpers;

namespace WebApp.TagHelpers {

    [HtmlTargetElement("*", Attributes = "[highlight=true]")]
    public class HighlightTagHelper: TagHelper {

        public override void Process(TagHelperContext context,
                TagHelperOutput output) {

            output.PreContent.SetHtmlContent("<b><i>");
            output.PostContent.SetHtmlContent("</i></b>");
        }
    }
}
```

这个标签助手在输出元素的内容周围插入 b 和 i 元素。代码清单 25-18 将 wrap 属性添加到 Index 视图中的一个表格单元格中。

**代码清单 25-18　在 Views/Home 文件夹的 Index.cshtml 文件中添加属性**

```
@model Product
@{
   Layout = "_SimpleLayout";
}

<div class="m-2" wrap="true">Inner Content</div>

<table class="table table-striped table-bordered table-sm">
    <tablehead bg-color="dark">Product Summary</tablehead>
    <tbody>
        <tr><th>Name</th><td highlight="true">@Model.Name</td></tr>
        <tr>
```

```
            <th>Price</th>
            <td bg-color="dark">@Model.Price.ToString("c")</td>
        </tr>
        <tr><th>Category ID</th><td>@Model.CategoryId</td></tr>
    </tbody>
</table>
```

重启 ASP.NET Core 并使用浏览器请求 http://localhost:5000/home/index/1。响应包括转换后的元素，如图 25-8 所示。

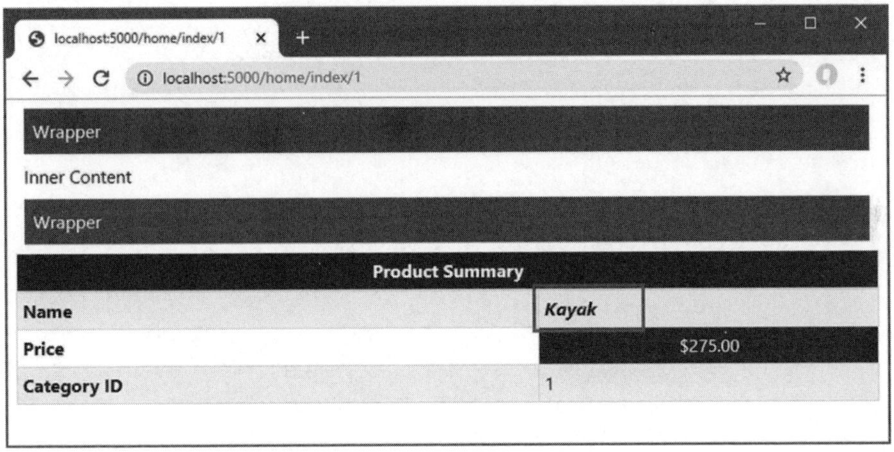

图 25-8　在元素中插入内容

检查发送到浏览器的 HTML，对于这个元素：

```
...
<td highlight="true">@Model.Name</td>
...
```

可以看到，上述元素已被转化为以下元素：

```
...
<td highlight="true"><b><i>Kayak</i></b></td>
...
```

## 25.3.4　获取视图上下文数据

标签助手的一个常见用途是转换元素，使它们包含当前请求或视图模型/页面模型的详细信息，这需要访问上下文数据。要创建这种类型的标签助手，将一个名为 RouteDataTagHelper.cs 的文件添加到 TagHelpers 文件夹，其内容如代码清单 25-19 所示。

**代码清单 25-19　WebApps/TagHelpers 文件夹中 RouteDataTagHelper.cs 文件的内容**

```
using Microsoft.AspNetCore.Mvc.Rendering;
using Microsoft.AspNetCore.Mvc.ViewFeatures;
using Microsoft.AspNetCore.Razor.TagHelpers;
```

```
using Microsoft.AspNetCore.Routing;

namespace WebApp.TagHelpers {

    [HtmlTargetElement("div", Attributes="[route-data=true]")]
    public class RouteDataTagHelper: TagHelper {

        [ViewContext]
        [HtmlAttributeNotBound]
        public ViewContext Context { get; set; }
        public override void Process(TagHelperContext context,
                TagHelperOutput output) {

            output.Attributes.SetAttribute("class", "bg-primary m-2 p-2");

            TagBuilder list = new TagBuilder("ul");
            list.Attributes["class"] = "list-group";
            RouteValueDictionary rd = Context.RouteData.Values;
            if (rd.Count > 0) {
                foreach (var kvp in rd) {
                    TagBuilder item = new TagBuilder("li");
                    item.Attributes["class"] = "list-group-item";
                    item.InnerHtml.Append($"{kvp.Key}: {kvp.Value}");
                    list.InnerHtml.AppendHtml(item);
                }
                output.Content.AppendHtml(list);
            } else {
                output.Content.Append("No route data");
            }
        }
    }
}
```

标签助手转换具有 route-data 属性(其值为 true)的 div 元素，并使用路由系统获得的段变量列表填充输出元素。

为获取路由数据，添加了一个名为 Context 的属性，并用两个属性装饰它，如下所示：

```
...
[ViewContext]
[HtmlAttributeNotBound]
public ViewContext Context { get; set; }
...
```

ViewContext 属性表示该属性的值应该在创建标签助手类的新实例时分配一个 ViewContext 对象，它提供了正在呈现的视图的详细信息，包括路由数据，如第 13 章所述。

如果在 div 元素上定义了匹配的属性，HTMLAttributeNotBound 属性将阻止为该属性分配值。这是一个很好的实践，在为其他开发人员编写标签助手时尤其如此。

■ **提示：** 标签助手可在它们的构造函数中声明对服务的依赖，这些依赖可通过第 14 章中描述的依赖注入特性来解决。

代码清单 25-20 向主控制器的 Index 视图添加了一个元素，这个元素由新的标签助手进行转换。

**代码清单 25-20　在 Views/Home 主文件夹的 Index.cshtml 文件中添加一个元素**

```
@model Product
@{
   Layout = "_SimpleLayout";
}

<div route-data="true"></div>

<table class="table table-striped table-bordered table-sm">
    <tablehead bg-color="dark">Product Summary</tablehead>
    <tbody>
        <tr><th>Name</th><td highlight="true">@Model.Name</td></tr>
        <tr>
            <th>Price</th>
            <td bg-color="dark">@Model.Price.ToString("c")</td>
        </tr>
        <tr><th>Category ID</th><td>@Model.CategoryId</td></tr>
    </tbody>
</table>
```

重启 ASP.NET Core 并使用浏览器请求 http://localhost:5000/home/index/1。响应将包括路由系统匹配的段变量的列表，如图 25-9 所示。

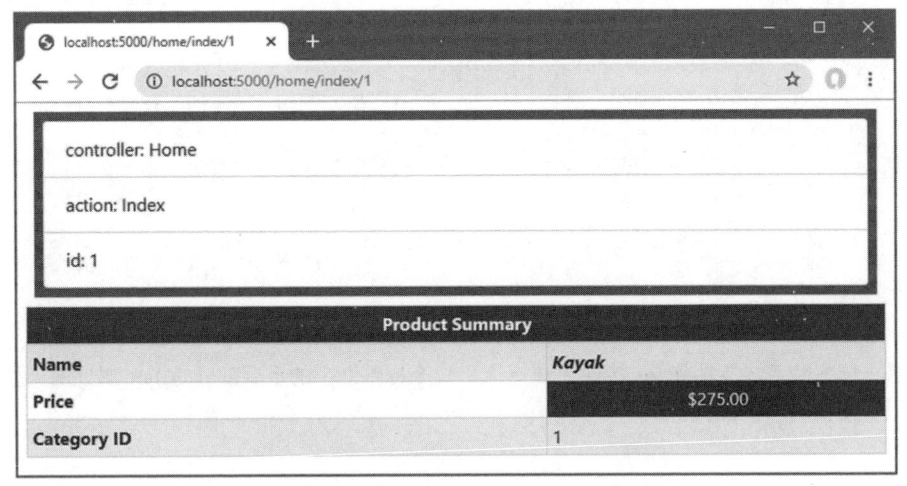


图 25-9　使用标签助手显示上下文数据

### 25.3.5　使用模型表达式

标签助手可以操作视图模型，裁剪它们执行的转换或它们创建的输出。要了解这个特性是如何工作的，可将一个名为 ModelRowTagHelper.cs 的类文件添加到 TagHelpers 文件夹中，如代码清单 25-21 所示。

**代码清单 25-21　TagHelpers 文件夹中 ModelRowTagHelper.cs 文件的内容**

```
using Microsoft.AspNetCore.Mvc.Rendering;
using Microsoft.AspNetCore.Mvc.ViewFeatures;
using Microsoft.AspNetCore.Razor.TagHelpers;

namespace WebApp.TagHelpers {

    [HtmlTargetElement("tr", Attributes = "for")]
    public class ModelRowTagHelper : TagHelper {

        public string Format { get; set; }
        public ModelExpression For { get; set; }
        public override void Process(TagHelperContext context,
                TagHelperOutput output) {

            output.TagMode = TagMode.StartTagAndEndTag;

            TagBuilder th = new TagBuilder("th");
            th.InnerHtml.Append(For.Name);
            output.Content.AppendHtml(th);

            TagBuilder td = new TagBuilder("td");
            if (Format != null && For.Metadata.ModelType == typeof(decimal)) {
                td.InnerHtml.Append(((decimal)For.Model).ToString(Format));
            } else {
                td.InnerHtml.Append(For.Model.ToString());
            }
            output.Content.AppendHtml(td);
        }
    }
}
```

这个标签助手转换具有 for 特性的 tr 元素。这个标签助手的重要部分是 For 属性的类型，该属性用于接收 for 特性的值。

```
...
public ModelExpression For { get; set; }
...
```

希望对视图模型的一部分进行操作时，将使用 ModelExpression 类，最简单的解释方法是向前跳转并显示如何在视图中应用标签助手，如代码清单 25-22 所示。

---

■ **注意：** ModelExpression 功能只能用于视图模型或页面模型，不能用于在视图中创建的变量，如@foreach 表达式。

---

**代码清单 25-22　在 Views/Home 文件夹的 Index.cshtml 文件中使用标签助手**

```
@model Product
@{
   Layout = "_SimpleLayout";
}

<div route-data="true"></div>

<table class="table table-striped table-bordered table-sm">
    <tablehead bg-color="dark">Product Summary</tablehead>
    <tbody>
        <tr for="Name" />
        <tr for="Price" format="c" />
        <tr for="CategoryId" />
    </tbody>
</table>
```

for 特性的值是视图模型类定义的属性的名称。在创建标签助手时，检测 For 属性的类型，并分配一个描述所选属性的 ModelExpression 对象。

这里不详细描述 ModelExpression 类，因为对类型的任何内省都会导致无尽的类和属性列表。此外，ASP.NET Core 提供了一组有用的内置标签助手，可以使用视图模型来转换元素，如第 26 章所述，这意味着不需要创建自己的标签助手。

对于示例标签助手，使用了三个值得描述的基本功能。第一个是获得模型属性的名称，这样就可以把它包含在输出元素中，如下：

```
...
th.InnerHtml.Append(For.Name);
...
```

Name 属性返回模型属性的名称。第二个功能是获取模型属性的类型，这样可以决定是否格式化值，如下所示：

```
...
if (Format != null && For.Metadata.ModelType == typeof(decimal)) {
...
```

第三个功能是获取属性的值，以便将其包含在响应中。

```
...
td.InnerHtml.Append(For.Model.ToString());
...
```

重启 ASP.NET Core 并使用浏览器请求 http://localhost:5000/home/index/2，将显示如图 25-10 所示的响应。

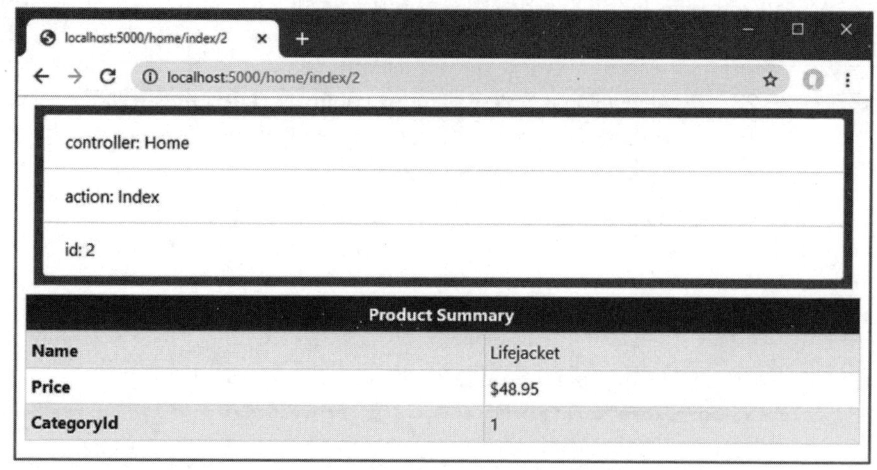


图 25-10　在标签助手中使用视图模型

### 使用页面模型

尽管选择属性的表达式必须考虑 Model 属性返回页面模型类的方式，但带有模型表达式的标签助手可应用于 Razor Pages。代码清单 25-23 将标签助手应用到 Editor Razor Pages，该页面模型定义了一个 Product 属性。

#### 代码清单 25-23　在 Pages 文件夹的 Editor.cshtml 文件中应用标签助手

```
@page "{id:long}"
@model EditorModel
@{
    Layout = null;
}

<!DOCTYPE html>
<html>
<head>
    <link href="/lib/twitter-bootstrap/css/bootstrap.min.css" rel="stylesheet" />
</head>
<body>
    <div class="bg-primary text-white text-center m-2 p-2">Editor</div>
    <div class="m-2">
        <table class="table table-sm table-striped table-bordered">
            <tbody>
                <tr for="Product.Name" />
                <tr for="Product.Price" format="c" />
            </tbody>
        </table>
        <form method="post">
```

```
            @Html.AntiForgeryToken()
            <div class="form-group">
                <label>Price</label>
                <input name="price" class="form-control"
                       value="@Model.Product.Price" />
            </div>
            <button class="btn btn-primary" type="submit">Submit</button>
        </form>
    </div>
</body>
</html>
```

for 属性的值通过 Product 属性选择嵌套属性，该属性为标签助手提供它需要的 ModelExpression。使用浏览器请求 http://localhost:5000/editor/1 以查看来自页面的响应，如图 25-11 的左侧所示。

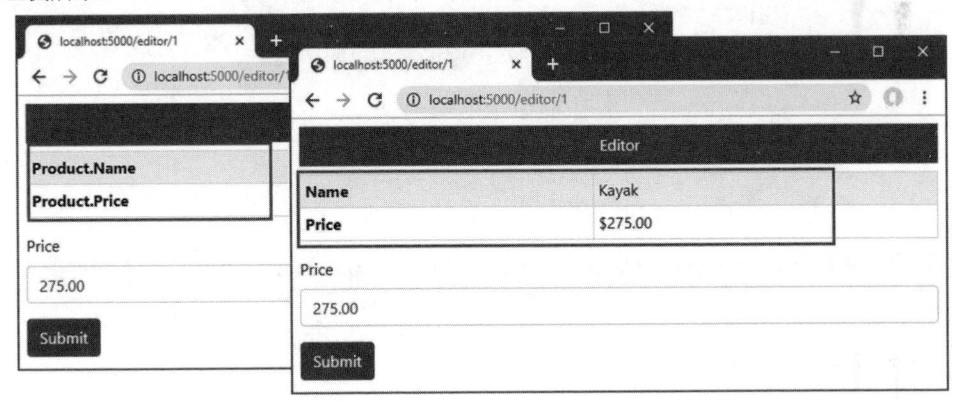


图 25-11　使用带有 Razor Pages 的模型表达式标签助手

该页面模型会产生效果；例如，ModelExpression.Name 属性将返回 Product.Name，而非 Name。代码清单 25-24 更新了标签助手，以便它只显示模型表达式名称的最后一部分。

■ **注意：**这个示例旨在突出页面模型对模型表达式的影响。与只显示名称的最后一部分不同，更灵活的方法是添加对另一个属性的支持，该属性允许根据需要覆盖显示值。

#### 代码清单 25-24　在 TagHelpers 文件夹的 ModelRowTagHelper.cs 文件中处理名称

```
using Microsoft.AspNetCore.Mvc.Rendering;
using Microsoft.AspNetCore.Mvc.ViewFeatures;
using Microsoft.AspNetCore.Razor.TagHelpers;
using System.Linq;

namespace WebApp.TagHelpers {

    [HtmlTargetElement("tr", Attributes = "for")]
    public class ModelRowTagHelper : TagHelper {
```

```
        public string Format { get; set; }
        public ModelExpression For { get; set; }

        public override void Process(TagHelperContext context,
                TagHelperOutput output) {

            output.TagMode = TagMode.StartTagAndEndTag;

            TagBuilder th = new TagBuilder("th");
            th.InnerHtml.Append(For.Name.Split(".").Last());
            output.Content.AppendHtml(th);

            TagBuilder td = new TagBuilder("td");
            if (Format != null && For.Metadata.ModelType == typeof(decimal)) {
                td.InnerHtml.Append(((decimal)For.Model).ToString(Format));
            } else {
                td.InnerHtml.Append(For.Model.ToString());
            }
            output.Content.AppendHtml(td);
        }
    }
}
```

重启 ASP.NET Core 并使用浏览器请求 http://localhost:5000/editor/1；修改后的响应如图 25-11 的右侧所示。

## 25.3.6　标签助手之间的协调

TagHelperContext.Items 属性提供一个字典，用于操作元素及其后代的标签助手。为了演示 Items 集合的使用，将一个名为 CoordinatingTagHelpers.cs 的类文件添加到 WebApp/TagHelpers 文件夹中，并添加如代码清单 25-25 所示的代码。

**代码清单 25-25　TagHelpers 文件夹中 CoordinatingTagHelpers.cs 文件的内容**

```
using Microsoft.AspNetCore.Razor.TagHelpers;

namespace WebApp.TagHelpers {

    [HtmlTargetElement("tr", Attributes = "theme")]
    public class RowTagHelper: TagHelper {

        public string Theme { get; set; }

        public override void Process(TagHelperContext context,
                TagHelperOutput output) {
            context.Items["theme"] = Theme;
        }
    }
```

```
    [HtmlTargetElement("th")]
    [HtmlTargetElement("td")]
    public class CellTagHelper : TagHelper {

        public override void Process(TagHelperContext context,
                TagHelperOutput output) {

            if (context.Items.ContainsKey("theme")) {
                output.Attributes.SetAttribute("class",
                    $"bg-{context.Items["theme"]} text-white");
            }
        }
    }
}
```

第一个标签助手对具有 theme 属性的 tr 元素进行操作。协调标签助手可转换它们自己的元素，但本示例只是将 theme 属性的值添加到 Items 字典中，以便对操作 tr 元素中包含的元素的标签助手可用。第二个标签助手对 th 和 td 元素进行操作，并使用 Items 字典中的 theme 值为其输出元素设置引导样式。

代码清单 25-26 向主控制器的 Index 视图添加了应用协调标签助手的元素。

---

■ **注意**：添加的 th 和 td 元素在代码清单 25-26 中转换，而不是依赖标签助手来生成它们。标签助手不应用于其他标签助手生成的元素，只影响视图中定义的元素。

---

**代码清单 25-26　在 Views/Home 文件夹的 Index.cshtml 文件中应用标签助手**

```
@model Product
@{
    Layout = "_SimpleLayout";
}

<table class="table table-striped table-bordered table-sm">
    <tablehead bg-color="dark">Product Summary</tablehead>
    <tbody>
        <tr theme="primary">
            <th>Name</th><td>@Model.Name</td>
        </tr>
        <tr theme="secondary">
            <th>Price</th><td>@Model.Price.ToString("c")</td>
        </tr>
        <tr theme="info">
            <th>Category</th><td>@Model.CategoryId</td>
        </tr>
    </tbody>
</table>
```

重启 ASP.NET Core 并使用浏览器请求 http://localhost:5000/home，这会生成如图 25-12 所示的响应。theme 元素的值已经从一个标签助手传递到另一个标签助手，并且应用颜色主题时不需要在每个被转换的元素上定义属性。

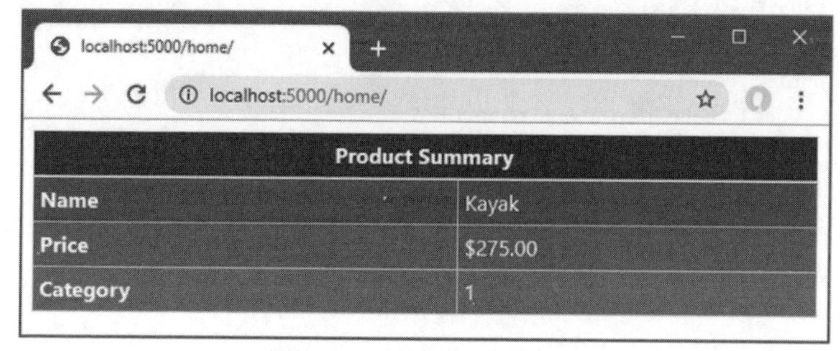


图 25-12　标签助手之间的协调

## 25.3.7　抑制输出元素

通过调用作为 Process 方法参数接收的 TagHelperOutput 对象上的 SuppressOuput 方法，可使用标签助手来防止元素包含在 HTML 响应中。在代码清单 25-27 中，向主控制器的 Index 视图添加了一个元素，只有在视图模型的 Price 属性超过指定值时才应该显示该元素。

**代码清单 25-27　在 Views/Home 文件夹的 Index.cshtml 文件中添加元素**

```
@model Product
@{
    Layout = "_SimpleLayout";
}

<div show-when-gt="500" for="Price">
    <h5 class="bg-danger text-white text-center p-2">
        Warning: Expensive Item
    </h5>
</div>

<table class="table table-striped table-bordered table-sm">
    <tablehead bg-color="dark">Product Summary</tablehead>
    <tbody>
        <tr theme="primary">
            <th>Name</th><td>@Model.Name</td>
        </tr>
        <tr theme="secondary">
            <th>Price</th><td>@Model.Price.ToString("c")</td>
        </tr>
        <tr theme="info">
            <th>Category</th><td>@Model.CategoryId</td>
        </tr>
    </tbody>
```

```
</table>
```

show-when-gt 属性指定了 div 元素应在其上显示的值，而 for 属性选择要检查的模型属性。要创建管理元素(包括响应)的标签助手，将一个名为 SelectiveTagHelper.cs 的类文件添加到 WebApp/TagHelpers 文件夹，如代码清单 25-28 所示。

#### 代码清单 25-28　TagHelpers 文件夹中 SelectiveTagHelper.cs 文件的内容

```
using Microsoft.AspNetCore.Mvc.ViewFeatures;
using Microsoft.AspNetCore.Razor.TagHelpers;

namespace WebApp.TagHelpers {

    [HtmlTargetElement("div", Attributes = "show-when-gt, for")]
    public class SelectiveTagHelper: TagHelper {

        public decimal ShowWhenGt { get; set; }
        public ModelExpression For { get; set; }

        public override void Process(TagHelperContext context,
                TagHelperOutput output) {

            if (For.Model.GetType() == typeof(decimal)
                    && (decimal)For.Model <= ShowWhenGt) {
                output.SuppressOutput();
            }
        }
    }
}
```

除非超过阈值，否则标签助手使用模型表达式访问属性并调用 SuppressOutput 方法。要查看效果，请重新启动 ASP.NET Core 并使用浏览器请求 http://localhost:5000/home/index/1 和 http://localhost:5000/home/index/5。第一个 URL 选择的产品的 Price 属性值小于阈值，因此该元素被抑制。第二个 URL 选择的产品的 Price 属性值超过了阈值，因此显示该元素。图 25-13 显示了这两种响应。

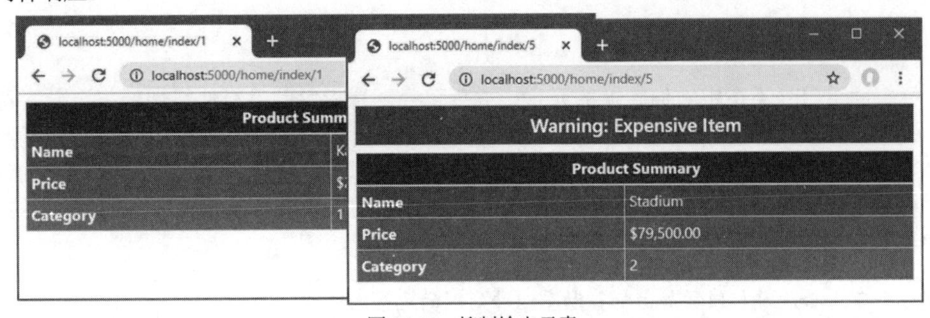


图 25-13　抑制输出元素

# 25.4　使用标签助手组件

标签助手组件提供了将标签助手用作服务的另一种方法。当需要设置标签助手来支持另一个服务或中间件组件时，该特性可能非常有用，这通常是同时具有客户端组件和服务器端组件的诊断工具或功能(如 Blazor)的情况。接下来将展示如何创建和应用标签助手组件。

## 25.4.1　创建标签助手组件

标签助手组件是从 TagHelperComponent 类派生的，该类提供了与前面示例中使用的 TagHelper 基类类似的 API。要创建标签助手组件，在 TagHelpers 文件夹中添加一个名为 TimeTagHelperComponent.cs 的类文件，内容如代码清单 25-29 所示。

**代码清单 25-29　TagHelpers 文件夹中 TimeTagHelperComponent.cs 文件的内容**

```
using Microsoft.AspNetCore.Mvc.Rendering;
using Microsoft.AspNetCore.Razor.TagHelpers;
using System;

namespace WebApp.TagHelpers {

    public class TimeTagHelperComponent: TagHelperComponent {

        public override void Process(TagHelperContext context,
                TagHelperOutput output) {

            string timestamp = DateTime.Now.ToLongTimeString();

            if (output.TagName == "body") {
                TagBuilder elem = new TagBuilder("div");
                elem.Attributes.Add("class", "bg-info text-white m-2 p-2");
                elem.InnerHtml.Append($"Time: {timestamp}");
                output.PreContent.AppendHtml(elem);
            }
        }
    }
}
```

标签助手组件不指定它们转换的元素，并为已经配置标签助手组件特性的每个元素调用 Process 方法。默认情况下，应用标签助手组件来转换 head 和 body 元素。这意味着标签助手组件类必须检查输出元素的 TagName 属性，以确保它们只执行预期的转换。代码清单 25-29 中的标签助手组件查找 body 元素，并使用 PreContent 属性在元素的其余内容之前插入包含时间戳的 div 元素。

**提示：**下一节展示如何增加由标签助手组件处理的元素的范围。

标签助手组件注册为实现 ITagHelperComponent 接口的服务，如代码清单 25-30 所示。

代码清单 25-30 在 WebApp 文件夹的 Startup.cs 文件中注册标签助手组件

```
using Microsoft.AspNetCore.Builder;
using Microsoft.Extensions.DependencyInjection;
using Microsoft.Extensions.Configuration;
using Microsoft.EntityFrameworkCore;
using WebApp.Models;
using Microsoft.AspNetCore.Razor.TagHelpers;
using WebApp.TagHelpers;

namespace WebApp {
    public class Startup {

        public Startup(IConfiguration config) {
           Configuration = config;
        }

        public IConfiguration Configuration { get; set; }

        public void ConfigureServices(IServiceCollection services) {
            services.AddDbContext<DataContext>(opts => {
                opts.UseSqlServer(Configuration[
                   "ConnectionStrings:ProductConnection"]);
                opts.EnableSensitiveDataLogging(true);
            });
            services.AddControllersWithViews().AddRazorRuntimeCompilation();
            services.AddRazorPages().AddRazorRuntimeCompilation();
            services.AddSingleton<CitiesData>();
            services.AddTransient<ITagHelperComponent, TimeTagHelperComponent>();
        }

        public void Configure(IApplicationBuilder app, DataContext context) {
            app.UseDeveloperExceptionPage();
            app.UseStaticFiles();
            app.UseRouting();
            app.UseEndpoints(endpoints => {
                endpoints.MapControllers();
                endpoints.MapDefaultControllerRoute();
                endpoints.MapRazorPages();
            });
            SeedData.SeedDatabase(context);
        }
    }
}
```

AddTransient 方法用于确保使用标签助手组件类的实例来处理每个请求。要查看标签助手组件的效果，请重新启动 ASP.NET Core 并使用浏览器请求 http://localhost:5000/home。该响应以及来自应用程序的其他所有 HTML 响应包含由标签助手组件生成的内容，如图 25-14 所示。

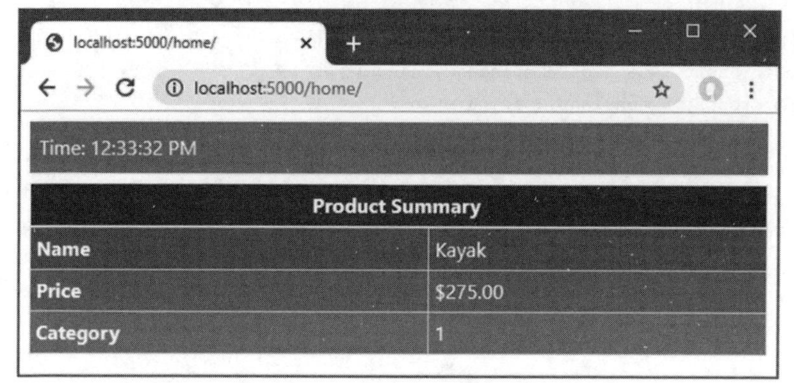


图 25-14　使用标签助手组件

## 25.4.2　展开标签助手的元素选择

默认情况下，标签助手组件只处理 head 和 body 元素，但可通过创建一个派生自 TagHelperComponentTagHelper 的类来选择其他元素。将类文件 TableFooterTagHelperComponent.cs 添加到 TagHelpers 文件夹中，并用它定义如代码清单 25-31 所示的类。

**代码清单 25-31　TagHelpers 文件夹中 TableFooterTagHelperComponent.cs 文件的内容**

```
using Microsoft.AspNetCore.Mvc.Razor.TagHelpers;
using Microsoft.AspNetCore.Mvc.Rendering;
using Microsoft.AspNetCore.Razor.TagHelpers;
using Microsoft.Extensions.Logging;

namespace WebApp.TagHelpers {

    [HtmlTargetElement("table")]
    public class TableFooterSelector: TagHelperComponentTagHelper {

        public TableFooterSelector(ITagHelperComponentManager mgr,
            ILoggerFactory log): base(mgr, log) { }
    }

    public class TableFooterTagHelperComponent: TagHelperComponent {

        public override void Process(TagHelperContext context,
                TagHelperOutput output) {

            if (output.TagName == "table") {
                TagBuilder cell = new TagBuilder("td");
                cell.Attributes.Add("colspan", "2");
                cell.Attributes.Add("class", "bg-dark text-white text-center");
                cell.InnerHtml.Append("Table Footer");
```

```
                TagBuilder row = new TagBuilder("tr");
                row.InnerHtml.AppendHtml(cell);
                TagBuilder footer = new TagBuilder("tfoot");
                footer.InnerHtml.AppendHtml(row);
                output.PostContent.AppendHtml(footer);
            }
        }
    }
}
```

TableFooterSelector 类派生自 TagHelperComponentTagHelper，并使用 HTMLTargetElement 属性装饰它，该属性扩展了由应用程序的标签助手组件处理的元素范围。在本例中，属性选择 table 元素。

在同一个文件中定义的 TableFooterTagHelperComponent 类是一个标签助手组件，它通过添加 tfoot 元素来转换表元素，tfoot 元素表示表的页脚。

---

▒ **警告：** 创建新的 TagHelperComponentTagHelper 时，所有标签助手组件将接收由 HtmlTargetAttribute 元素选择的元素。

---

标签助手组件必须注册为服务来接收转换的元素，但会自动发现、应用标签助手组件 tag helper(这是几年来我看到的最糟糕的命名之一)。代码清单 25-32 添加标签助手组件服务。

**代码清单 25-32　在 WebApp 文件夹的 Startup.cs 文件中注册标签助手组件**

```
...
public void ConfigureServices(IServiceCollection services) {
    services.AddDbContext<DataContext>(opts => {
        opts.UseSqlServer(Configuration[
           "ConnectionStrings:ProductConnection"]);
        opts.EnableSensitiveDataLogging(true);
    });
    services.AddControllersWithViews().AddRazorRuntimeCompilation();
    services.AddRazorPages().AddRazorRuntimeCompilation();
    services.AddSingleton<CitiesData>();
    services.AddTransient<ITagHelperComponent, TimeTagHelperComponent>();
    services.AddTransient<ITagHelperComponent, TableFooterTagHelperComponent>();
}
...
```

重启 ASP.NET Core 并使用浏览器请求呈现表的 URL，如 http://localhost:5000/home 或 http://localhost:5000/cities。每个表将包含一个表脚，如图 25-15 所示。

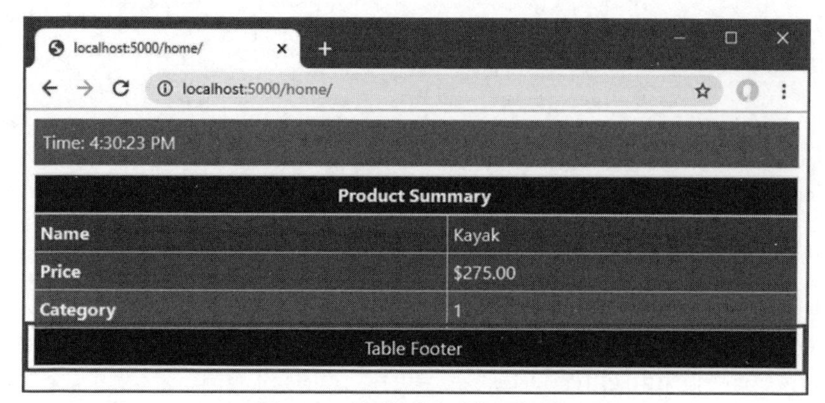


图 25-15 展开标签助手的元素选择

## 25.5 小结

本章解释了标签助手的工作原理以及它们在转换视图和页面中的 HTML 元素时的作用。展示了如何创建和应用标签助手，如何控制为转换选择的元素，以及如何使用高级特性来获得特定的结果。本章最后解释了标签助手组件特性(定义为服务)。第 26 章将描述 ASP.NET Core 提供的内置标签助手。

# 第 26 章

■■■

# 使用内置的标签助手

ASP.NET Core 提供了一组内置的标签助手，可以应用最常用的元素转换。本章解释那些处理锚、脚本、链接和图像元素的标签助手，以及用于缓存内容和基于环境选择内容的特性。第 27 章描述了支持 HTML 表单的标签助手。表 26-1 概述内置标签助手的相关知识。

表 26-1　内置标签助手的相关知识

| 问题 | 答案 |
| --- | --- |
| 它们是什么? | 内置的标签助手在 HTML 元素上执行通常需要的转换 |
| 它们为什么有用? | 使用内置的标签助手，意味着不必使用第 25 章中的技术来创建自定义助手 |
| 它们是如何使用的? | 使用标准 HTML 元素上的属性或通过自定义 HTML 元素应用标签助手 |
| 是否存在缺陷或限制? | 不，这些标签助手都经过良好测试，且易于使用。除非有特殊的需要，否则使用这些标签助手比自定义实现更可取 |
| 还有其他选择吗? | 这些标签助手是可选的，它们的使用不是必需的 |

表 26-2 总结了本章的内容。

表 26-2　本章内容摘要

| 问题 | 解决方案 | 代码清单 |
| --- | --- | --- |
| 创建指向端点的元素 | 使用锚元素标签助手属性 | 26-7～26-8 |
| 在响应中包含 JavaScript 文件 | 使用 JavaScript 标签助手属性 | 26-9～26-13 |
| 在响应中包含 CSS 文件 | 使用 CSS 标签助手属性 | 26-14～26-15 |
| 管理图像缓存 | 使用图像标签助手属性 | 26-16 |
| 缓存视图的部分 | 使用缓存标签助手 | 26-17～26-21 |
| 根据应用程序环境来改变内容 | 使用环境标签助手 | 26-22 |

## 26.1　准备工作

本章使用了第 25 章中的 WebApp 项目。为准备这一章，请注释掉在 Startup 类中注册标签组件助手的语句，如代码清单 26-1 所示。

**提示：**可从 https://github.com/apress/pro-asp.net-core-3 下载本章和本书中其他所有章节的示例项目。如果在运行示例时遇到问题，请参阅第 1 章以了解如何获得帮助。

代码清单 26-1　WebApp 文件夹中 Startup.cs 文件的内容

```
using Microsoft.AspNetCore.Builder;
using Microsoft.Extensions.DependencyInjection;
using Microsoft.Extensions.Configuration;
using Microsoft.EntityFrameworkCore;
using WebApp.Models;
//using Microsoft.AspNetCore.Razor.TagHelpers;
//using WebApp.TagHelpers;

namespace WebApp {
    public class Startup {

        public Startup(IConfiguration config) {
            Configuration = config;
        }

        public IConfiguration Configuration { get; set; }

        public void ConfigureServices(IServiceCollection services) {
            services.AddDbContext<DataContext>(opts => {
                opts.UseSqlServer(Configuration[
                    "ConnectionStrings:ProductConnection"]);
                opts.EnableSensitiveDataLogging(true);
            });
            services.AddControllersWithViews().AddRazorRuntimeCompilation();
            services.AddRazorPages().AddRazorRuntimeCompilation();
            services.AddSingleton<CitiesData>();
            //services.AddTransient<ITagHelperComponent, TimeTagHelperComponent>();
            //services.AddTransient<ITagHelperComponent,
            //        TableFooterTagHelperComponent>();
        }

        public void Configure(IApplicationBuilder app, DataContext context) {
            app.UseDeveloperExceptionPage();
            app.UseStaticFiles();
            app.UseRouting();
            app.UseEndpoints(endpoints => {
                endpoints.MapControllers();
                endpoints.MapDefaultControllerRoute();
                endpoints.MapRazorPages();
            });
            SeedData.SeedDatabase(context);
        }
```

```
        }
    }
```

接下来，在 Views/Home 文件夹中更新_RowPartial.cshtml 部分视图，执行如代码清单 26-2 所示的更改。

**代码清单 26-2　在 Views/Home 文件夹的_RowPartial.cshtml 文件中进行更改**

```
@model Product

<tr>
    <td>@Model.Name</td>
    <td>@Model.Price.ToString("c")</td>
    <td>@Model.CategoryId</td>
    <td>@Model.SupplierId</td>
    <td></td>
</tr>
```

添加如代码清单 26-3 所示的元素，在 Home 控制器的 List 视图中呈现的表内定义额外的列。

**代码清单 26-3　在 Views/Home 文件夹的 List.cshtml 文件中添加元素**

```
@model IEnumerable<Product>
@{ Layout = "_SimpleLayout"; }

<h6 class="bg-secondary text-white text-center m-2 p-2">Products</h6>
<div class="m-2">
    <table class="table table-sm table-striped table-bordered">
        <thead>
            <tr>
                <th>Name</th><th>Price</th>
                <th>Category</th><th>Supplier</th><th></th>
            </tr>
        </thead>
        <tbody>
            @foreach (Product p in Model) {
                <partial name="_RowPartial" model="p" />
            }
        </tbody>
    </table>
</div>
```

## 26.1.1　添加图像文件

本章中描述的一个标签助手为图像提供服务。创建 wwwroot/images 文件夹，并添加了一个名为 city.png 的图像文件。这是纽约市天际线的公共域全景图，如图 26-1 所示。

图 26-1　向项目添加图像

这个图像文件包含在本章的源代码中，可在本书的 GitHub 存储库中找到。如果不想下载示例项目，可以替换为自己的图像。

### 26.1.2　安装客户端包

本章中的一些示例演示了使用 JavaScript 文件的标签助手支持，对此使用 jQuery 包。使用 PowerShell 命令提示符在包含 WebApp.csproj 文件的项目文件夹中运行代码清单 26-4 所示的命令。如果使用的是 Visual Studio，可以选择 Project | Manage Client-Side Libraries 来选择 jQuery 包。

**代码清单 26-4　安装包**

```
libman install jquery@3.4.1 -d wwwroot/lib/jquery
```

### 26.1.3　删除数据库

打开一个新的 PowerShell 命令提示符，导航到包含 WebApp.csproj 文件的文件夹，并运行代码清单 26-5 所示的命令来删除数据库。

**代码清单 26-5　删除数据库**

```
dotnet ef database drop --force
```

### 26.1.4　运行示例应用程序

从 Debug 菜单中选择 Start Without Debugging 或 Run Without Debugging，或者使用 PowerShell 命令提示符运行代码清单 26-6 中所示的命令。

**代码清单 26-6　运行示例应用程序**

```
dotnet run
```

使用浏览器请求 http://localhost:5000/Home/list，这会显示一个产品列表，如图 26-2 所示。

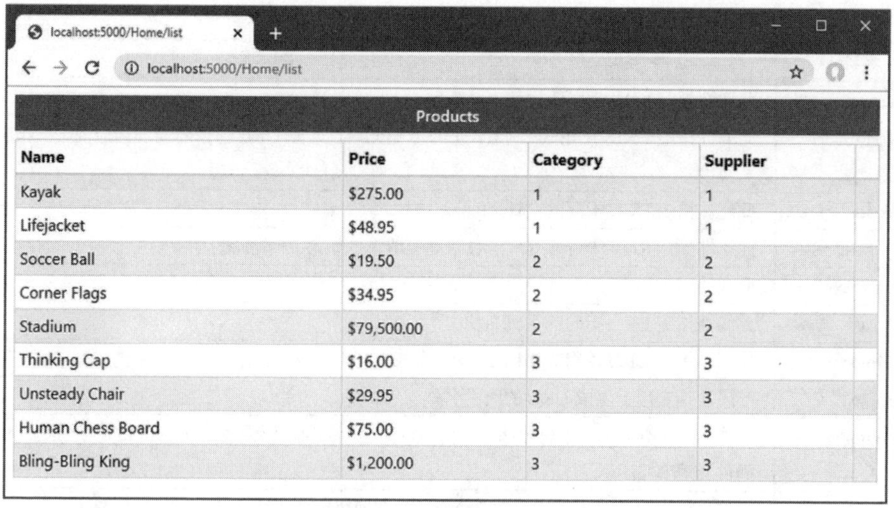

localhost:5000/Home/list

Products

| Name | Price | Category | Supplier |
|---|---|---|---|
| Kayak | $275.00 | 1 | 1 |
| Lifejacket | $48.95 | 1 | 1 |
| Soccer Ball | $19.50 | 2 | 2 |
| Corner Flags | $34.95 | 2 | 2 |
| Stadium | $79,500.00 | 2 | 2 |
| Thinking Cap | $16.00 | 3 | 3 |
| Unsteady Chair | $29.95 | 3 | 3 |
| Human Chess Board | $75.00 | 3 | 3 |
| Bling-Bling King | $1,200.00 | 3 | 3 |

图 26-2　运行示例应用程序

## 26.2　启用内置的标签助手

内置的标签助手都在 Microsoft.AspNetCore.Mvc.TagHelpers 名称空间中定义，启用它的方法是向单个视图或页面添加@addTagHelper 指令，或者像示例项目中那样，向视图导入文件添加该指令。下面是 View 文件夹中_ViewImports.cshtml 文件的必需指令，它为控制器视图启用内置的标签助手：

```
@using WebApp.Models
@addTagHelper *, Microsoft.AspNetCore.Mvc.TagHelpers
@using WebApp.Components
@addTagHelper *, WebApp
```

下面是 Pages 文件夹的_ViewImports.cshtml 文件中对应的指令，它启用了 Razor 页面(Razor Pages)的内置标签助手：

```
@namespace WebApp.Pages
@using WebApp.Models
@addTagHelper *, Microsoft.AspNetCore.Mvc.TagHelpers
@addTagHelper *, WebApp
```

在第 24 章中，这些指令添加到示例项目中，以启用视图组件特性。

## 26.3　改变锚元素

元素是在应用程序中导航和向应用程序发送 GET 请求的基本工具。AnchorTagHelper 类用于转换元素的 href 属性，以便它们指向使用路由系统生成的 URL，即不需要硬编码的 URL，路由配置中的更改将自动反映在应用程序的锚元素中。表 26-3 描述了 AnchorTagHelper 类支持的属性。

表 26-3　锚元素的内置标签助手属性

| 名称 | 描述 |
| --- | --- |
| asp-action | 此属性指定 URL 指向的操作方法 |
| asp-controller | 此属性指定 URL 指向的控制器。如果省略此属性，则 URL 将指向呈现当前视图的控制器或页面 |
| asp-page | 此属性指定 URL 指向的 Razor Pages |
| asp-page-handler | 此属性指定 Razor Pages 处理函数来处理请求，如第 23 章所述 |
| asp-fragment | 此属性用于指定 URL 片段(出现在#字符之后) |
| asp-host | 此属性指定 URL 指向的主机名称 |
| asp-protocol | 此属性指定 URL 使用的协议 |
| asp-route | 此属性指定用于生成 URL 的路由名称 |
| asp-route-* | 名称以 asp-route 开头的属性用于为 URL 指定附加值，以便使用 asp-route-id 属性为路由系统提供 id 段的值 |
| asp-all-route-data | 此属性提供用于路由的值作为单个值，而非使用单个属性 |

AnchorTagHelper 简单且可预测，很容易在使用应用程序路由配置的元素中生成 URL。代码清单 26-7 添加了一个锚元素，它使用表中的属性创建一个 URL，该 URL 针对主控制器定义的另一个操作。

**代码清单 26-7　在 Views/Home 文件夹的_RowPartial.cshtml 文件中转换元素**

```
@model Product

<tr>
    <td>@Model.Name</td>
    <td>@Model.Price.ToString("c")</td>
    <td>@Model.CategoryId</td>
    <td>@Model.SupplierId</td>
    <td>
        <a asp-action="index" asp-controller="home" asp-route-id="@Model.ProductId"
                class="btn btn-sm btn-info">
            Select
        </a>
    </td>
</tr>
```

asp-action 和 asp-controller 属性指定操作方法的名称和定义它的控制器。段变量的值是使用 asp-route-[name]属性定义的，这样，asp-route-id 属性为 id 段变量提供一个值，该值用于为 asp-action 属性选择的操作方法提供一个参数。

■ **提示：** 在代码清单 26-7 中添加到锚点元素的 class 属性应用了引导 CSS 框架样式，使元素具有按钮的外观。这不是使用标签助手的必要条件。

要查看锚元素转换，请使用浏览器请求 http://localhost:5000/home/list，这将生成如图 26-3 所示的响应。

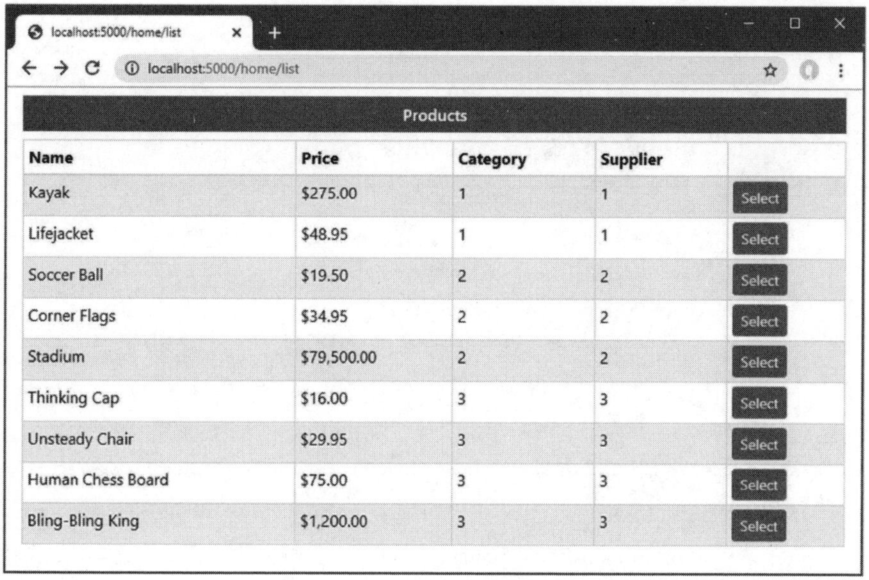

localhost:5000/home/list

Products

| Name | Price | Category | Supplier | |
|---|---|---|---|---|
| Kayak | $275.00 | 1 | 1 | Select |
| Lifejacket | $48.95 | 1 | 1 | Select |
| Soccer Ball | $19.50 | 2 | 2 | Select |
| Corner Flags | $34.95 | 2 | 2 | Select |
| Stadium | $79,500.00 | 2 | 2 | Select |
| Thinking Cap | $16.00 | 3 | 3 | Select |
| Unsteady Chair | $29.95 | 3 | 3 | Select |
| Human Chess Board | $75.00 | 3 | 3 | Select |
| Bling-Bling King | $1,200.00 | 3 | 3 | Select |

图 26-3　改变锚元素

如果检查 Select 锚元素，则每个 href 属性包括与之相关的 Product 对象的 ProductId 值，如下所示：

```
...
<a class="btn btn-sm btn-info" href="/Home/index/3">Select</a>
...
```

这种情况下，asp-route-id 属性提供的值意味着不能使用默认 URL，所以路由系统生成了 URL，包括控制器和操作名称部分，以及一个用来提供操作方法参数的段。在这两种情况下，由于只指定了一个操作方法，所以由标签助手创建的 URL 将指向呈现视图的控制器。单击锚元素将发送一个指向主控制器的 Index 方法的 HTTP GET 请求。

## 为 Razor Pages 使用锚元素

asp-page 属性用于指定 Razor Pages 作为锚元素的 href 属性的目标。页面的路径以/字符为前缀，@page 指令定义的路由段值使用 asp-route-[name]属性定义。代码清单 26-8 添加了一个锚元素，指向在 Pages/Suppliers 文件夹中定义的 List 页面。

■ **注意：** asp-page-handler 属性可用来指定处理请求的页面模型处理方法的名称。

**代码清单 26-8　在 Views/Home 文件夹的 List.cshtml 文件中指向 Razor Pages**

```
@model IEnumerable<Product>
@{
    Layout = "_SimpleLayout";
}

<h6 class="bg-secondary text-white text-center m-2 p-2">Products</h6>
```

```
<div class="m-2">
    <table class="table table-sm table-striped table-bordered">
        <thead>
            <tr>
                <th>Name</th><th>Price</th>
                <th>Category</th><th>Supplier</th><th></th>
            </tr>
        </thead>
        <tbody>
            @foreach (Product p in Model) {
                <partial name="_RowPartial" model="p" />
            }
        </tbody>
    </table>
    <a asp-page="/suppliers/list" class="btn btn-secondary">Suppliers</a>
</div>
```

使用浏览器请求 http://localhost:5000/home/list，就会看到样式为按钮的锚元素。如果检查发送到客户端的 HTML，会看到锚元素已转换，如下：

```
...
<a class="btn btn-secondary" href="/lists/suppliers">Suppliers</a>
...
```

href 属性中使用的这个 URL 反映了@page 指令，该指令用于覆盖此页中的默认路由约定。单击该元素，浏览器将显示 Razor Pages，如图 26-4 所示。

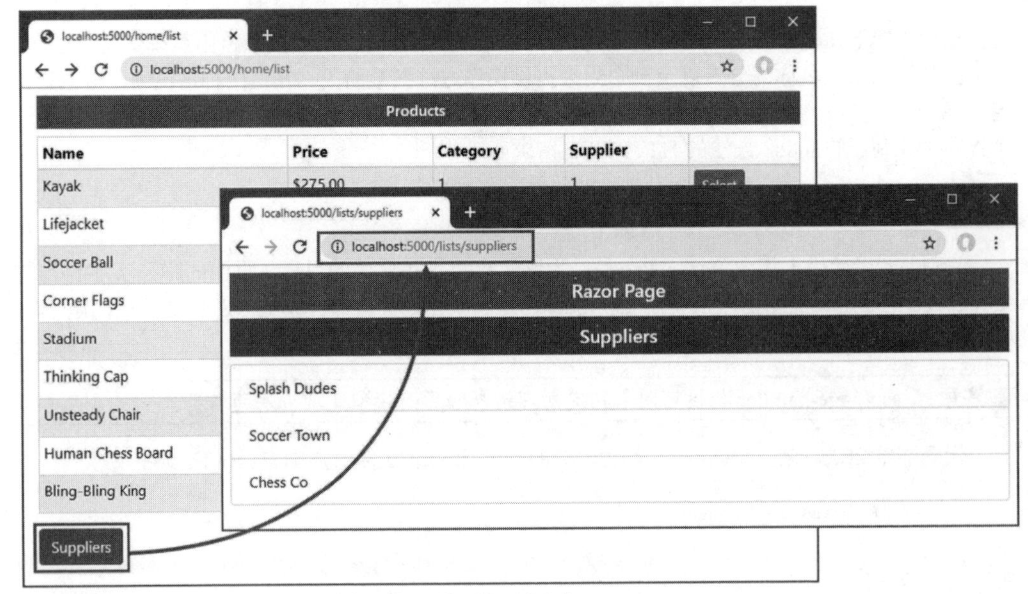


图 26-4　使用锚元素定位 Razor Pages

**生成 URL(而不是链接)**

标签助手只在锚元素中生成 URL。如果需要生成 URL，而不是链接，就可以使用 URL 属性，它在控制器、页面模型和视图中可用。这个属性返回一个实现 IUrlHelper 接口的对象，该接口提供了一组生成 URL 的方法和扩展方法。这是一个 Razor 片段，在视图中生成 URL:

```
...
<div>@Url.Page("/suppliers/list")</div>
...
```

此片段生成一个 div 元素，其内容是指向/Suppliers/List Razor Pages 的 URL。在控制器或页面模型类中使用相同的接口，例如下面的语句:

```
...
string url = Url.Action("List", "Home");
...
```

该语句生成一个 URL，指向主控制器上的 List 操作，并将其分配给名为 url 的字符串变量。

## 26.4 使用 JavaScript 和 CSS 标签助手

ASP.NET Core 提供了通过脚本和链接元素来管理 JavaScript 文件和 CSS 样式表的标签助手。如后面几节所述，这些标签助手功能强大且灵活，但需要密切关注，以免产生意外结果。

### 26.4.1 管理 JavaScript 文件

ScriptTagHelper 类是脚本元素的内置标签助手，可使用表 26-4 中描述的属性来管理视图中包含的 JavaScript 文件。

表 26-4 脚本元素的内置标签助手属性

| 名称 | 描述 |
| --- | --- |
| asp-src-include | 此属性用于指定包含在视图中的 JavaScript 文件 |
| asp-src-exclude | 此属性用于指定将从视图中排除的 JavaScript 文件 |
| asp-append-version | 此属性用于缓存破解，如“了解缓存破解”侧栏中所述 |
| asp-fallback-src | 此属性用于指定在内容传递网络出现问题时使用的回退 JavaScript 文件 |
| asp-fallback-src-include | 此属性用于选择在出现内容传递网络问题时使用的 JavaScript 文件 |
| asp-fallback-src-exclude | 此属性用于排除 JavaScript 文件，以便在出现内容传递网络问题时显示它们的用途 |
| asp-fallback-test | 此属性用于指定一个 JavaScript 片段，该 JavaScript 片段用于确定 JavaScript 代码是否已从内容传递网络中正确加载 |

#### 1. 选择 JavaScript 文件

asp-src-include 属性用于通过 Globbing 模式，在视图中包含 JavaScript 文件。Globbing 模式支持一组用于匹配文件的通配符，表 26-5 描述了最常见的 Globbing 模式。

表 26-5　常见的 Globbing 模式

| 模式 | 示例 | 描述 |
|---|---|---|
| ? | js/src?.js | 此模式匹配除/之外的任何单个字符。该示例匹配 js 目录中包含的任何名称为 src 后跟任何字符和.js 的文件，例如 js/src1.js 和 js/srcX.js，但不匹配 js/src123.js 或 js/mydir/src1.js |
| * | js/*.js | 此模式匹配除/之外的任意数量的字符。该示例匹配 js 目录中任何扩展名为.js 的文件，例如 js/src1.js 和 js/src123.js，但不匹配 js/mydir/src1.js |
| ** | js/**/*.js | 该模式匹配包括/在内的任意数量的字符。该示例匹配 js 目录或任何子目录中任何扩展名为.js 的文件，例如/js/src1.js 和/js/mydir/src1.js |

Globbing 是确保视图包含应用程序需要的 JavaScript 文件的一种有用方法，即使文件的确切路径发生了变化(通常发生在文件名中包含版本号或包添加了其他文件时)。

代码清单 26-9 使用 asp-src-include 属性来包含 wwwroot/lib/jquery 文件夹中的所有 JavaScript 文件，该文件夹是使用代码清单 26-4 中的命令安装的 jQuery 包的位置。

### 代码清单 26-9　在 Views/Shared 文件夹的_SimpleLayout.cshtml 文件中选择 JS 文件

```
<!DOCTYPE html>
<html>
<head>
    <title>@ViewBag.Title</title>
    <link href="/lib/twitter-bootstrap/css/bootstrap.min.css" rel="stylesheet" />
    <script asp-src-include="lib/jquery/**/*.js"></script>
</head>
<body>
    <div class="m-2">
       @RenderBody()
    </div>
</body>
</html>
```

在 wwwroot 文件夹评估模式，这里使用的模式对应于任何扩展名为.js 的文件，而不考虑它在 wwwroot 文件夹的位置；这意味着添加到项目中的任何 JavaScript 包都包含在发送给客户端的 HTML 中。

使用浏览器请求 http://localhost:5000/home/list，并检查发送到浏览器的 HTML。布局中的单一脚本元素已经转换为每个 JavaScript 文件的 script 元素，如下所示：

```
...
<head>
  <title></title>
  <link href="/lib/twitter-bootstrap/css/bootstrap.min.css" rel="stylesheet">
  <script src="/lib/jquery/core.js"></script>
  <script src="/lib/jquery/jquery.js"></script>
  <script src="/lib/jquery/jquery.min.js"></script>
  <script src="/lib/jquery/jquery.slim.js"></script>
```

```
<script src="/lib/jquery/jquery.slim.min.js"></script>

</head>
...
```

如果使用 Visual Studio，就可能没有意识到 jQuery 包包含这么多 JavaScript 文件，因为 Visual Studio 将它们隐藏在 Solution Explorer 中。要显示客户端包文件夹的全部内容，可以在 Solution Explorer 窗口中展开单独的嵌套条目，或者通过单击 Solution Explorer 窗口顶部的按钮来禁用文件嵌套，如图 26-5 所示(Visual Studio Code 不嵌套文件)。

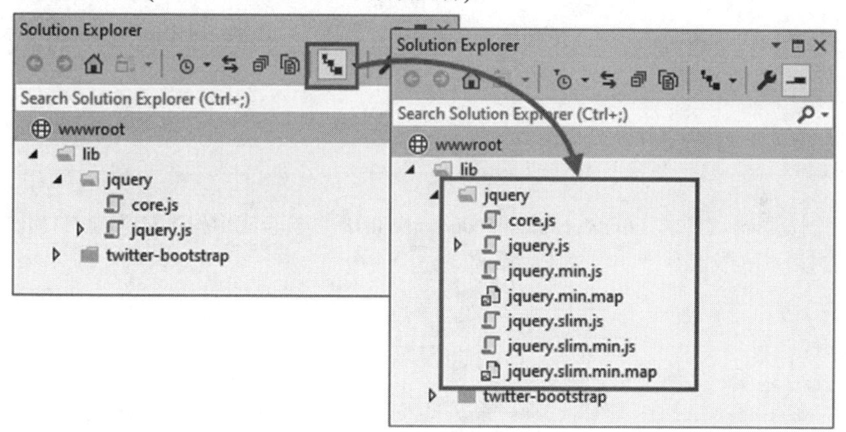


图 26-5 在 Visual Studio Solution Explorer 中禁用文件嵌套

**理解源映像**

JavasScript 文件被缩小，使它们更小，意味着它们可以更快地交付给客户端，使用更少的带宽。缩小过程将删除文件中的所有空白，并重命名函数和变量，这样有意义的名称(如 myHelpfullyNamedFunction)将用更少的字符表示，如 x1。当使用浏览器的 JavaScript 调试器跟踪已缩小代码中的问题时，像 x1 这样的名称几乎不可能跟踪整个代码的进度。

具有 map 扩展名的文件是源映射，浏览器使用源映射来帮助调试缩小的代码，它提供了缩小代码和开发人员可读的、未缩小的源文件之间的映射。打开浏览器的 F12 开发人员工具时，浏览器将自动请求源映射并使用它们来帮助调试应用程序的客户端代码。

### 2. 缩小 Globbing 模式

任何应用程序都不需要代码清单 26-9 中的模式的选择的所有文件。许多包都包含多个 JavaScript 文件，这些文件包含类似的内容，通常删除不太流行的特性以节省带宽。jQuery 包包括 jQuery.slim.js 文件，该文件包含与 jQuery.js 文件相同的代码，但没有处理异步 HTTP 请求和动画效果的特性(还有一个 core.js 文件，但它错误地包含在包中，应该忽略)。

这些文件中的每一个都对应于 min.js 文件扩展名，它表示一个缩小的文件。“缩小”通过删除所有空格、重命名函数和变量以使用更短的名称，来减小 JavaScript 文件的大小。

每个包只需要一个 JavaScript 文件，如果只需要缩小版本(大多数项目都是这样)，就可以限制 Globbing 模式匹配的文件集，如代码清单 26-10 所示。

**代码清单 26-10　在 Views/Shared 文件夹的_SimpleLayout.cshtml 文件中选择缩小的文件**

```
<!DOCTYPE html>
<html>
<head>
    <title>@ViewBag.Title</title>
    <link href="/lib/twitter-bootstrap/css/bootstrap.min.css" rel="stylesheet" />
    <script asp-src-include="lib/jquery**/*.min.js"></script>
</head>
<body>
    <div class="m-2">
        @RenderBody()
    </div>
</body>
</html>
```

使用浏览器再次请求 http://localhost:5000/home/list，并检查应用程序发送的 HTML。会看到只选择了缩小后的文件。

```
...
<head>
  <title></title>
  <link href="/lib/twitter-bootstrap/css/bootstrap.min.css" rel="stylesheet">
  <script src="/lib/jquery/jquery.min.js"></script>
  <script src="/lib/jquery/jquery.slim.min.js"></script>
</head>
...
```

缩小 JavaScript 文件的模式会有所帮助，但是浏览器最终仍然会得到 jQuery 的普通和精简版本，以及引导 JavaScript 文件的绑定和解绑定版本。为了进一步缩小选择范围，可以在模式中包括 slim，如代码清单 26-11 所示。

**代码清单 26-11　在 Views/Shared 文件夹的_SimpleLayout.cshtml 文件中缩小焦点**

```
<!DOCTYPE html>
<html>
<head>
    <title>@ViewBag.Title</title>
    <link href="/lib/twitter-bootstrap/css/bootstrap.min.css" rel="stylesheet" />
    <script asp-src-include="lib/jquery**/*slim.min.js"></script>
</head>
<body>
    <div class="m-2">
        @RenderBody()
    </div>
</body>
</html>
```

使用浏览器请求 http://localhost:5000/home/列表，并检查浏览器接收到的 HTML。脚本元素被转换，如下所示：

```
...
<head>
  <title></title>
  <link href="/lib/twitter-bootstrap/css/bootstrap.min.css" rel="stylesheet">
  <script src="/lib/jquery/jquery.slim.min.js"></script>
</head>
...
```

只有一个版本的 jQuery 文件发送到浏览器，同时保留了文件位置的灵活性。

### 3. 不包括文件

希望选择名称中包含特定术语(如 slim)的文件时，缩小 JavaScript 文件的模式将有所帮助。想要的文件没有这个术语时，例如想要缩小文件的完整版本时，它就没有帮助了。幸运的是，可使用 asp-src-exclude 属性从与 asp-src-include 属性匹配的列表中删除文件，如代码清单 26-12 所示。

**代码清单 26-12　在 Views/Shared 文件夹的_SimpleLayoutcshtml 文件中排除文件**

```
<!DOCTYPE html>
<html>
<head>
    <title>@ViewBag.Title</title>
    <link href="/lib/twitter-bootstrap/css/bootstrap.min.css" rel="stylesheet" />
    <script asp-src-include="/lib/jquery/**/*.min.js"
       asp-src-exclude="**.slim.**">
    </script>
</head>
<body>
    <div class="m-2">
       @RenderBody()
    </div>
</body>
</html>
```

如果使用浏览器请求 http://localhost:5000/home/list 并检查 HTML 响应，会看到 script 元素只链接到 jQuery 库的完整缩小版，如下：

```
...
<head>
    <title></title>
    <link href="/lib/twitter-bootstrap/css/bootstrap.min.css" rel="stylesheet">
    <script src="/lib/jquery/jquery.min.js"></script>
</head>
...
```

**了解缓存破解**

静态内容(如图像、CSS 样式表和 JavaScript 文件)通常会缓存，以阻止对很少更改的内容的请求到达应用服务器。缓存可以通过不同的方式完成：服务器可以告诉浏览器缓存内容，应用程序可以使用缓存服务器来补充应用服务器，或者可以使用内容分发网络来分发内容。并不是所有缓存都在控制之下。例如，大公司经常安装缓存以减少带宽需求，因为相当大比例的请求倾向于访问相同的站点或应用程序。

缓存的一个问题是，当部署静态文件时，客户端不会立即接收到新版本的静态文件，因为它们的请求仍然由以前缓存的内容提供服务。最终，缓存的内容将过期，而使用新内容；但在此期间，由应用程序控制器生成的动态内容与缓存所交付的静态内容不一致。这可能导致布局问题或意外的应用程序行为，具体取决于已更新的内容。

解决这个问题称为缓存破解(cache busting)。其思想是允许缓存处理静态内容，但立即反映在服务器上进行的任何更改。标签助手类通过向静态内容的 URL 添加查询字符串来支持缓存破解，静态内容包括作为版本号的校验和。例如，对于 JavaScript 文件，ScriptTagHelper 类通过 asp-append-version 属性支持缓存破解，如下所示：

```
...
<script asp-src-include="/lib/jquery/**/*.min.js"
   asp-src-exclude="**.slim.**" asp-append-version="true">
</script>
...
```

启用缓存破解功能会在发送到浏览器的 HTML 中生成这样的元素：

```
...
<script src="/lib/jquery/dist/jquery.min.js?v=3zRSQ1HF-ocUiVcdv9yKTXqM"></script>
...
```

标签助手使用相同的版本号，直到更改文件的内容(例如通过更新 JavaScript 库)，此时将计算不同的校验和。添加版本号意味着每次更改文件时，客户端将请求一个不同的 URL，缓存将此 URL 视为对无法满足先前缓存内容的新内容的请求，并将其传递给应用程序服务器。然后正常缓存内容，直到下一次更新，这会产生另一个不同版本的 URL。

#### 4. 使用内容发布网络

内容分发网络(CDN)用于将应用程序内容的请求转移到离用户更近的服务器。浏览器不是从服务器请求 JavaScript 文件，而是从解析到本地服务器的主机名请求它，这减少了加载文件所需的时间，并减少了必须为应用程序提供的带宽。如果有一个庞大的、分散在各地的用户群，那么签约一个 CDN 就有商业意义，但即使是最小最简单的应用程序也可以从使用大型技术公司运营的免费 CDN 中获益，这些 CDN 提供通用的 JavaScript 包，如 jQuery。

本章使用 CDNJS，它与库管理器工具在 ASP.NET Core 项目中安装客户端包所使用的 CDN 是相同的。可在 https://cdnjs.com 搜索软件包；对于 jQuery 3.4.1，即代码清单 26-4 中安装的包和版本，有 6 个 CDNJS URL。

- https://cdnjs.cloudflare.com/ajax/libs/jquery/3.4.1/jquery.js
- https://cdnjs.cloudflare.com/ajax/libs/jquery/3.4.1/jquery.min.js
- https://cdnjs.cloudflare.com/ajax/libs/jquery/3.4.1/jquery.min.map
- https://cdnjs.cloudflare.com/ajax/libs/jquery/3.4.1/jquery.slim.js

- https://cdnjs.cloudflare.com/ajax/libs/jquery/3.4.1/jquery.slim.min.js
- https://cdnjs.cloudflare.com/ajax/libs/jquery/3.4.1/jquery.slim.min.map

这些 URL 提供了常规 JavaScript 文件、缩小的 JavaScript 文件，以及完整版和精简版 jQuery 缩小文件的源映射(还有 core.js 文件的 URL，但如前所述，不会使用这个文件，并且会从未来的 jQuery 版本中删除)。

CDN 的问题在于它们不在组织的控制之下，这意味着它们可能会失败；应用程序继续运行，但由于 CDN 内容不可用而无法按预期工作。ScriptTagHelper 类提供了在客户端无法加载 CDN 内容时返回到本地文件的能力，如代码清单 26-13 所示。

**代码清单 26-13　在 Views/Shared 文件夹的_SimpleLayout.cshtml 文件中使用 CDN 回退**

```
<!DOCTYPE html>
<html>
<head>
    <title>@ViewBag.Title</title>
    <link href="/lib/twitter-bootstrap/css/bootstrap.min.css" rel="stylesheet" />
    <script src="https://cdnjs.cloudflare.com/ajax/libs/jquery/3.4.1/jquery.min.js"
        asp-fallback-src="/lib/jquery/jquery.min.js"
        asp-fallback-test="window.jQuery">
    </script>
</head>
<body>
    <div class="m-2">
       @RenderBody()
    </div>
</body>
</html>
```

src 属性用于指定 CDN URL。asp-fallback-src 属性用于指定一个本地文件，如果 CDN 不能传递常规 src 属性指定的文件，就使用该文件。为了弄清楚 CDN 是否在工作，asp-fallback-test 属性用来定义一个 JavaScript 片段，它将在浏览器中评估。如果片段计算为 false，将请求回退文件。

**提示**：可使用 asp-back-src-include 和 asp-back-src-exclude 属性来选择具有 Globbing 模式的本地文件。但是，如果 CDN script 元素选择单个文件，建议使用 asp-fallback-src 属性来选择相应的本地文件，如示例所示。

使用浏览器请求 http://localhost:5000/home/list，会看到 HTML 响应包含两个脚本元素，如下：

```
...
<head>
    <title></title>
    <link href="/lib/twitter-bootstrap/css/bootstrap.min.css" rel="stylesheet">
    <script src="https://cdnjs.cloudflare.com/ajax/libs/jquery/3.4.1/jquery.min.js">
</script>
    <script>
        (window.jQuery||document.write("\u003Cscript
           src=\u0022/lib/jquery/jquery.min.js\u0022\u003E\u003C/script\u003E"));
```

```
    </script>
</head>
...
```

第一个 Script 元素从 CDN 请求 JavaScript 文件。第二个脚本元素计算 asp-fallback-test 属性指定的 JavaScript 片段，该属性检查第一个脚本元素是否工作。如果片段的值为 true，就不会采取任何行动，因为 CDN 是有效的。如果片段的计算结果为 false，就向 HTML 文档添加一个新的 Script 元素，指示浏览器从回退 URL 中加载 JavaScript 文件。

重要的是测试回退设置，因为不会发现他们是否失败，直到 CDN 已经停止工作，用户不能访问应用程序为止。检查回退的最简单方法是将 src 属性指定的文件名称改为自己知道不存在的东西(这里给文件名追加 FAIL)，然后看看浏览器使用 F12 开发人员工具发出的网络请求。应该会看到 CDN 文件的错误，然后是对回退文件的请求。

**警告：** CDN 回退功能依赖于浏览器同步加载和执行脚本元素的内容，并按照它们定义的顺序加载和执行。有很多技术用来通过异步方式加快 JavaScript 的加载和执行过程，但是这些会导致在浏览器从 CDN 中检索文件，执行其内容之前，进行回退测试。即使在 CDN 工作得很完美，首先击败了 CDN 的使用，也导致了回退文件的请求。不要将异步脚本加载与 CDN 回退功能混在一起。

## 26.4.2　管理 CSS 样式表

LinkTagHelper 类是链接元素的内置标签助手，用于管理视图中包含的 CSS 样式表。这个标签助手支持表 26-6 中描述的属性。

表 26-6　链接元素的内置标签助手属性

| 名称 | 描述 |
| --- | --- |
| asp-href-include | 此属性用于为输出元素的 href 属性选择文件 |
| asp-href-exclude | 此属性用于从输出元素的 href 属性中排除文件 |
| asp-append-version | 此属性用于启用缓存破解 |
| asp-fallback-href | 此属性用于在 CDN 出现问题时指定一个回退文件 |
| asp-fallback-href-include | 此属性用于在出现 CDN 问题时选择要使用的文件 |
| asp-fallback-href-exclude | 此属性用于在存在 CDN 问题时从使用的文件集中排除文件 |
| asp-fallback-href-test-class | 此属性指定用于测试 CDN 的 CSS 类 |
| asp-fallback-href-test-property | 此属性指定用于测试 CDN 的 CSS 属性 |
| asp-fallback-href-test-value | 此属性指定用于测试 CDN 的 CSS 值 |

### 1. 选择样式表

LinkTagHelper 与 ScriptTagHelper 共享许多特性，包括支持 Globbing 模式来选择或排除 CSS 文件，这样它们就不必单独指定了。能够准确地选择 CSS 文件和选择 JavaScript 文件一样重要，因为样式表有常规版本和缩小版本，并且支持源映射。流行的 Bootstrap 包(本书一直使用它来设计 HTML 元素的样式)包括其在 wwwroot/lib/twitter-Bootstrap/CSS 文件夹中的 CSS 样式表。这些将在 Visual Studio Code 中可见，但只有在 Solution Explorer 中展开每个项，或者禁用嵌套，才能在 Visual Studio Solution Explorer 中查看它们，如图 26-6 所示。

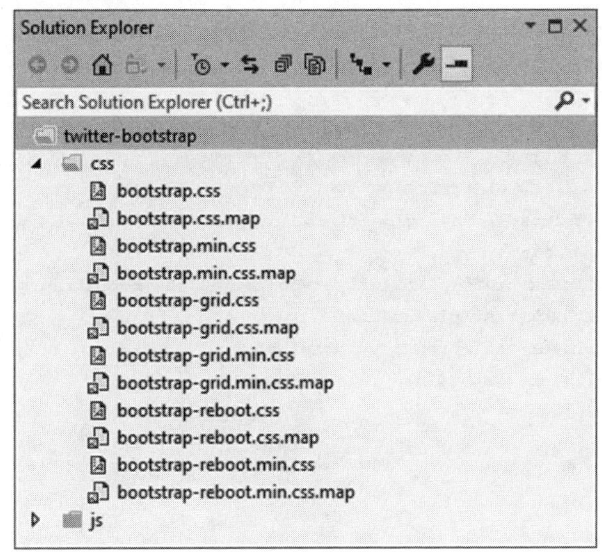


图 26-6　引导 CSS 文件

bootstrap.css 文件是常规样式表，bootstrap.min.css 文件是缩小版，bootstrap.css.map 文件是源映射。其他文件包含 CSS 特性的子集，以便在不使用这些特性的应用程序中节省带宽。

代码清单 26-14 将布局中的常规链接元素替换为使用 asp-href-include 和 asp-href-exclude 属性的元素(删除了 jQuery 的脚本元素，它不再是必需的)。

### 代码清单 26-14　在 Views/Shared 文件夹的_SimpleLayout.cshtml 文件中选择样式表

```
<!DOCTYPE html>
<html>
<head>
    <title>@ViewBag.Title</title>
    <link asp-href-include="/lib/twitter-bootstrap/css/*.min.css"
       asp-href-exclude="**/*-reboot*,**/*-grid*" rel="stylesheet" />
</head>
<body>
    <div class="m-2">
       @RenderBody()
    </div>
</body>
</html>
```

在选择 JavaScript 文件时同样需要注意细节，因为很容易为同一文件或不想要的文件的多个版本生成链接元素。

#### 2. 使用内容发布网络

LinkTag 助手类提供了一组属性，用于在 CDN 不可用时返回到本地内容，但测试样式表是否已加载的过程比测试 JavaScript 文件要复杂得多。代码清单 26-15 使用 CDNJS URL 作为引导 CSS

样式表。

代码清单 26-15　在 Views/Home 文件夹的_SimpleLayout.cshtml 文件中为 CSS 使用 CDN

```
<!DOCTYPE html>
<html>
<head>
    <title>@ViewBag.Title</title>
    <link href="https://cdnjs.cloudflare.com/ajax/libs/twitter-bootstrap/4.3.1/
css/bootstrap.min.css"
        asp-fallback-href="/lib/twitter-bootstrap/css/bootstrap.min.css"
        asp-fallback-test-class="btn"
        asp-fallback-test-property="display"
        asp-fallback-test-value="inline-block"
        rel="stylesheet" />
</head>
<body>
    <div class="m-2">
       @RenderBody()
    </div>
</body>
</html>
```

href 属性用于指定 CDN URL，这里使用 asp-fallback-href 属性来选择 CDN 不可用时要使用的文件。然而，为测试 CDN 是否工作，需要使用三种不同的属性，并了解正在使用的 CSS 样式表定义的 CSS 类。

使用浏览器请求 http://localhost:5000/home/list 并检查响应中的 HTML 元素。会看到布局中的 link 元素已经转换成三个独立元素，如下所示：

```
...
<head>
    <title></title>
    <link href="https://cdnjs.cloudflare.com/.../bootstrap.min.css" rel="stylesheet">
    <meta name="x-stylesheet-fallback-test" content="" class="btn">
    <script>
      ! function(a, b, c, d) {
        var e, f = document,
          g = f.getElementsByTagName("SCRIPT"),
          h = g[g.length1].previousElementSibling,
          i = f.defaultView && f.defaultView.getComputedStyle ?
          f.defaultView.getComputedStyle(h) : h.currentStyle;
        if (i && i[a] !== b)
          for (e = 0; e < c.length; e++)
            f.write('<link href="' + c[e] + '" ' + d + "/>")
      }("display", "inline-block", ["/lib/twitter-bootstrap/css/bootstrap.min.css"],
         "rel=\u0022stylesheet\u0022 ");
    </script>
```

```
</head>
...
```

为了使转换更容易理解，格式化 JavaScript 代码并缩短 URL。

第一个元素是一个常规链接，它的 href 属性指定 CDN 文件。第二个元素是 meta 元素，它指定来自视图中的 asp-fallback-test-class 属性的类。在代码清单中指定了 btn 类，这意味着像这样的元素添加到发送到浏览器的 HTML：

```
<meta name="x-stylesheet-fallback-test" content="" class="btn">
```

指定的 CSS 类必须在从 CDN 加载的样式表中定义。指定的 btn 类为引导按钮元素提供了基本格式。

asp-fallback-test-property 属性用于指定将 CSS 类应用到元素时设置的 CSS 属性，而 asp-fallback-test-value 属性用于指定为其设置的值。

由标签助手创建的 script 元素包含 JavaScript 代码，该代码将元素添加到指定的类中，然后测试 CSS 属性的值，以确定 CDN 样式表是否已加载。如果没有，则为回退文件创建 link 元素。Bootstrap btn 类将 display 属性设置为 inline-block，这样就可以测试浏览器能否从 CDN 加载 Bootstrap 样式表。

**提示：** 要想知道如何测试像 Bootstrap 这样的第三方包，最简单的方法是使用浏览器的 F12 开发工具。为了确定代码清单 26-15 中的测试，给 btn 类分配了一个元素，然后在浏览器中检查它，查看类更改的各个 CSS 属性。这比阅读长而复杂的样式表更容易。

## 26.5 处理图像元素

ImageTagHelper 类通过 img 元素的 src 属性为图像提供缓存破解，允许应用程序利用缓存，同时确保对图像的修改能够立即反映出来。ImageTagHelper 类在定义 asp-append-version 属性的 img 元素中操作，表 26-7 中描述了该属性，以便快速参考。

表 26-7 图像元素的内置标签助手属性

| 名称 | 描述 |
|---|---|
| asp-append-version | 此属性用于启用缓存破解，参见“了解缓存破解”侧栏 |

在代码清单 26-16 中，向城市天际线图像的共享布局中添加了一个 img 元素，这个城市天际线图像是本章开始时添加到项目中的。为简洁起见，还将 link 元素重置为使用本地文件。

**代码清单 26-16 在 Views/Shared 文件夹的_SimpleLayout.cshtml 文件中添加图像**

```
<!DOCTYPE html>
<html>
<head>
    <title>@ViewBag.Title</title>
    <link href="/lib/twitter-bootstrap/css/bootstrap.min.css" rel="stylesheet" />
</head>
<body>
```

```
    <div class="m-2">
        <img src="/images/city.png" asp-append-version="true" class="m-2" />
        @RenderBody()
    </div>
</body>
</html>
```

使用浏览器请求 http://localhost:5000/home/list，这将生成如图 26-7 所示的响应。

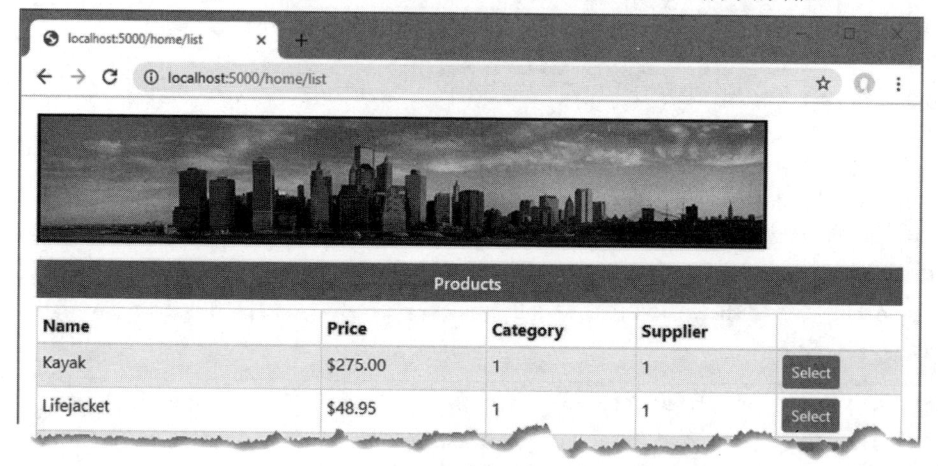


图 26-7　使用图像

检查 HTML 响应，会看到用于请求图像文件的 URL 包含一个版本校验和，如下所示：

```
...
<img src="/images/city.png?v=KaMNDSZFAJufRcRDpKh0K_IIPNc7E" class="m-2">
...
```

校验和的添加确保了对文件的任何更改都会通过任何缓存传递，从而避免陈旧的内容。

## 26.6　使用数据缓存

CacheTagHelper 类允许缓存内容片段，以加快视图或页面的呈现。缓存的内容使用 cache 元素表示，cache 元素使用表 26-8 中所示的属性配置。

> **注意：**缓存是重用部分内容的有用工具，这样它们就不必为每个请求生成。但是有效地使用缓存需要仔细的思考和计划。虽然缓存可以提高应用程序的性能，但会带来一些奇怪的效果，如用户收到陈旧内容，多个缓存包含不同版本的内容，更新被破坏的部署，因为从以前版本的应用程序中缓存的内容和新版本的内容混合在一起。除非有一个明确定义的性能问题需要解决，否则不要启用缓存，并确保理解缓存带来的影响。

表 26-8 缓存元素的内置标签助手属性

| 名称 | 描述 |
| --- | --- |
| enabled | 此 bool 属性用于控制是否缓存被缓存元素的内容。省略此属性将启用缓存 |
| expires-on | 此属性用于指定缓存内容过期的绝对时间，表示为 DateTime 值 |
| expires-after | 此属性用于指定缓存内容的相对过期时间，表示为 TimeSpan 值 |
| expires-sliding | 此属性用于指定缓存内容过期的时间段，即自上次使用它以来的时间段，并表示为 TimeSpan 值 |
| vly -by-header | 此属性用于指定请求头的名称，该请求头用于管理缓存内容的不同版本 |
| vary-by-query | 此属性用于指定查询字符串键的名称，该键用于管理缓存内容的不同版本 |
| vary-by-route | 此属性用于指定一个路由变量的名称，该变量用于管理缓存内容的不同版本 |
| vary-by-cookie | 此属性用于指定 cookie 的名称，该 cookie 用于管理缓存内容的不同版本 |
| vary-by-user | 此 bool 属性用于指定是否使用经过身份验证的用户的用户名来管理缓存内容的不同版本 |
| vary-by- content | 计算此属性，以提供一个用于管理不同版本内容的密钥 |
| priority | 此属性用于指定一个相对优先级，当内存缓存耗尽空间并清除未过期的缓存内容时，将考虑该优先级 |

代码清单 26-17 用包含时间戳的内容替换了上一节中的 img 元素。

### 代码清单 26-17 在 Views/Shared 文件夹的_SimpleLayout .cshtml 文件中缓存内容

```
<!DOCTYPE html>
<html>
<head>
    <title>@ViewBag.Title</title>
    <link href="/lib/twitter-bootstrap/css/bootstrap.min.css" rel="stylesheet" />
</head>
<body>
    <div class="m-2">
        <h6 class="bg-primary text-white m-2 p-2">
            Uncached timestamp: @DateTime.Now.ToLongTimeString()
        </h6>
        <cache>
            <h6 class="bg-primary text-white m-2 p-2">
                Cached timestamp: @DateTime.Now.ToLongTimeString()
            </h6>
        </cache>
        @RenderBody()
    </div>
</body>
</html>
```

cache 元素用于表示应该缓存的内容区域，该内容区域已应用于包含时间戳的 h6 元素之一。使用浏览器请求 http://localhost:5000/home/list，那么两个时间戳将是相同的。重新加载浏览器，会看到缓存的内容用于其中一个 h6 元素，时间戳没有变化，如图 26-8 所示。

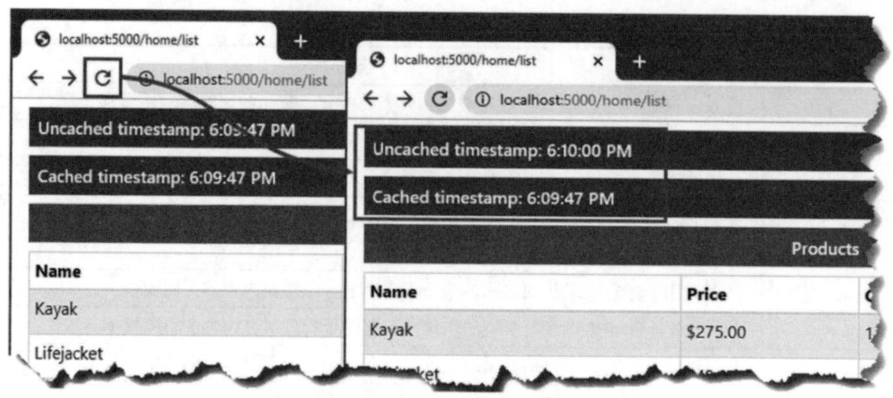


图 26-8　使用缓存标签助手

**对内容使用分布式缓存**

CacheTagHelper 类使用的缓存是基于内存的，这意味着它的容量受到可用 RAM 的限制，并且每个应用程序服务器维护一个单独的缓存。当可用容量不足时，内容将从缓存中弹出，当应用程序停止或重新启动时，整个内容将丢失。

distributed-cache 元素可用于在共享缓存中存储内容，从而确保所有应用程序服务器使用相同的数据，并且缓存在重新启动时仍然有效。distributed-cache 元素配置与缓存元素相同的属性，如表 26-8 所述。请参阅第 17 章，了解如何建立分布式缓存。

### 26.6.1　设置缓存到期时间

expires-*属性允许指定缓存内容何时过期，可以表示为绝对时间或相对于当前时间的时间，也可以指定缓存内容不被请求的持续时间。在代码清单 26-18 中，使用了 expires-after 属性来指定应该将内容缓存 15 秒。

**代码清单 26-18　在 Views/Shared 文件夹的_SimpleLayout.cshtml 文件中设置缓存过期时间**

```
<!DOCTYPE html>
<html>
<head>
    <title>@ViewBag.Title</title>
    <link href="/lib/twitter-bootstrap/css/bootstrap.min.css" rel="stylesheet" />
</head>
<body>
    <div class="m-2">
        <h6 class="bg-primary text-white m-2 p-2">
            Uncached timestamp: @DateTime.Now.ToLongTimeString()
        </h6>
        <cache expires-after="@TimeSpan.FromSeconds(15)">
            <h6 class="bg-primary text-white m-2 p-2">
                Cached timestamp: @DateTime.Now.ToLongTimeString()
            </h6>
        </cache>
```

```
        @RenderBody()
    </div>
</body>
</html>
```

使用浏览器请求 http://localhost:5000/home/list，然后重新加载页面。15 秒后，缓存的内容过期，并创建一个新的内容部分。

### 26.6.2 设置固定的过期点

可以使用 expireson-on 属性指定缓存内容的固定过期时间，该属性接收一个 DateTime 值，如代码清单 26-19 所示。

**代码清单 26-19 在 Views/Shared 文件夹的_SimpleLayout.cshtml 文件中设置缓存过期时间**

```
<!DOCTYPE html>
<html>
<head>
    <title>@ViewBag.Title</title>
    <link href="/lib/twitter-bootstrap/css/bootstrap.min.css" rel="stylesheet" />
</head>
<body>
    <div class="m-2">
        <h6 class="bg-primary text-white m-2 p-2">
           Uncached timestamp: @DateTime.Now.ToLongTimeString()
        </h6>
        <cache expires-on="@DateTime.Parse("2100-01-01")">
            <h6 class="bg-primary text-white m-2 p-2">
               Cached timestamp: @DateTime.Now.ToLongTimeString()
            </h6>
        </cache>
        @RenderBody()
    </div>
</body>
</html>
```

指定该数据应该缓存到 2100 年。这不是一种有用的缓存策略，因为应用程序很可能在下个世纪开始之前重新启动，但它确实说明了如何指定一个未来的固定时间点，而不是相对于缓存内容的时刻表示过期点。

### 26.6.3 设置最后使用的有效期

expires-sliding属性用于指定一个期限，如果内容还没有从缓存中检索，则在该期限之后内容将过期。在代码清单 26-20 中，指定了滑动到期时间为 10 秒。

**代码清单 26-20 在 Views/Shared 文件夹的_SimpleLayout.cshtml 文件中使用滑动到期时间**

```
<!DOCTYPE html>
<html>
```

```
<head>
    <title>@ViewBag.Title</title>
    <link href="/lib/twitter-bootstrap/css/bootstrap.min.css" rel="stylesheet" />
</head>
<body>
    <div class="m-2">
        <h6 class="bg-primary text-white m-2 p-2">
            Uncached timestamp: @DateTime.Now.ToLongTimeString()
        </h6>
        <cache expires-sliding="@TimeSpan.FromSeconds(10)">
            <h6 class="bg-primary text-white m-2 p-2">
                Cached timestamp: @DateTime.Now.ToLongTimeString()
            </h6>
        </cache>
        @RenderBody()
    </div>
</body>
</html>
```

通过请求 http://localhost:5000/home/list 并定期重新加载页面，可以看到 express-slide 属性的效果。如果在 10 秒内重新加载页面，就使用缓存的内容。如果等待超过 10 秒来重新加载页面，那么缓存的内容将被丢弃，视图组件用于生成新内容，过程将重新开始。

### 26.6.4　使用缓存的变化

默认情况下，所有请求接收相同的缓存内容。CacheTagHelper 类可以维护缓存内容的不同版本，并使用它们来满足不同类型的 HTTP 请求，使用名称以 vary-by 开头的属性指定。代码清单 26-21 显示了使用 vary-by-route 属性根据路由系统匹配的操作值来创建缓存变量。

**代码清单 26-21　在 Views/Shared 文件夹的_SimpleLayout.cshtml 文件中创建变体**

```
<!DOCTYPE html>
<html>
<head>
    <title>@ViewBag.Title</title>
    <link href="/lib/twitter-bootstrap/css/bootstrap.min.css" rel="stylesheet" />
</head>
<body>
    <div class="m-2">
        <h6 class="bg-primary text-white m-2 p-2">
            Uncached timestamp: @DateTime.Now.ToLongTimeString()
        </h6>
        <cache expires-sliding="@TimeSpan.FromSeconds(10)" vary-by-route="action">
            <h6 class="bg-primary text-white m-2 p-2">
                Cached timestamp: @DateTime.Now.ToLongTimeString()
            </h6>
        </cache>
        @RenderBody()
```

```
    </div>
</body>
</html>
```

如果使用两个浏览器选项卡来请求http://localhost:5000/home/index和http://localhost:5000/home/list，会看到每个窗口接收自己的缓存内容和过期时间，因为每个请求生成不同的操作路由值。

**提示：**如果使用 Razor Pages，就可以使用 Page 作为路由系统匹配的值来达到同样的效果。

## 26.7 使用宿主环境标签助手

EnvironmentTagHelper 类应用于自定义环境元素，并确定基于托管环境发送到浏览器的 HTML 中是否包含某个内容区域，参见第 15 章和第 16 章。environment 元素依赖于 names 属性，参见表 26-9。

表 26-9 environment 元素的内置标签助手属性

| 名称 | 描述 |
|---|---|
| names | 此属性用于指定以逗号分隔的主机环境名称列表，其中包含在 environment 元素中的内容将包含在发送给客户端的 HTML 中 |

在代码清单 26-22 中，向共享布局中添加了环境元素，包括视图中用于开发和生产托管环境的不同内容。

**代码清单 26-22 在 Views/Shared 文件夹的_SimpleLayout.cshtml 文件中使用环境**

```
<!DOCTYPE html>
<html>
<head>
    <title>@ViewBag.Title</title>
    <link href="/lib/twitter-bootstrap/css/bootstrap.min.css" rel="stylesheet" />
</head>
<body>
    <div class="m-2">
        <environment names="development">
            <h2 class="bg-info text-white m-2 p-2">This is Development</h2>
        </environment>
        <environment names="production">
            <h2 class="bg-danger text-white m-2 p-2">This is Production</h2>
        </environment>
        @RenderBody()
    </div>
</body>
</html>
```

environment 元素检查当前宿主环境名，并包含它所包含的内容或省略它(在发送给客户端的 HTML 中，environment 元素本身总是省略)。图 26-9 显示了开发和生产环境的输出(有关如何设置环境的细节，请参阅第 15 章)。

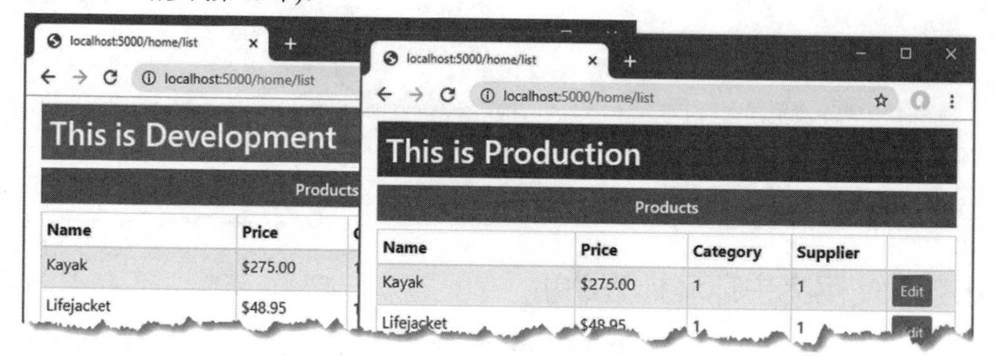


图 26-9　使用托管环境管理内容

## 26.8　小结

本章描述了基本的内置标签助手，并解释了它们是如何用于转换锚、链接、脚本和图像元素的。还解释了如何缓存内容部分以及如何根据应用程序的环境呈现内容。第 27 章将描述 ASP.NET Core 提供的用于处理 HTML 表单的标签助手。

# 第 27 章

# 使用表单标签助手

本章描述用于创建 HTML 表单的内置标签助手。这些标签助手确保表单提交到正确的操作或页面处理程序方法，并确保元素准确地表示特定的模型属性。表 27-1 概述表单标签助手的相关知识。

表 27-1　表单标签助手的相关知识

| 问题 | 答案 |
| --- | --- |
| 它们是什么? | 这些内置的标签助手转换 HTML 表单元素 |
| 它们为什么有用? | 这些标签助手确保 HTML 表单反映应用程序的路由配置和数据模型 |
| 它们是如何使用的? | 标签助手通过 asp-*属性应用于 HTML 元素 |
| 是否存在缺陷或限制? | 这些标签助手是可靠的、可预测的，并且没有严重的问题 |
| 还有其他选择吗? | 不必使用标签助手，如果愿意的话，也可以不使用它们来定义表单 |

表 27-2 总结了本章的内容。

表 27-2　本章内容摘要

| 问题 | 解决方案 | 代码清单 |
| --- | --- | --- |
| 指定如何提交表单 | 使用表单标签助手属性 | 27-10～27-13 |
| 转换 input 元素 | 使用输入标签助手属性 | 27-14～27-22 |
| 转换 label 元素 | 使用标签助手属性 | 27-23 |
| 填充 select 元素 | 使用选择标签助手属性 | 27-24～27-26 |
| 转换文本区域 | 使用文本区域标签助手属性 | 27-27 |
| 防止跨站点请求的伪造 | 启用防伪造功能 | 27-28～27-32 |

## 27.1　准备工作

本章使用了第 26 章中的 WebApp 项目。为准备这一章，替换 Views/Shared 文件夹的 _SimpleLayout.cshtml 文件中的内容，如代码清单 27-1 所示。

■ **提示：** 可从 https://github.com/apress/pro-asp.net-core-3 下载本章和本书中其他所有章节的示例项目。如果在运行示例时遇到问题，请参阅第 1 章以了解如何获得帮助。

**代码清单 27-1　Views/Shared 文件夹的_SimpleLayout.cshtml 文件中的内容**

```
<!DOCTYPE html>
<html>
<head>
    <title>@ViewBag.Title</title>
    <link href="/lib/twitter-bootstrap/css/bootstrap.min.css" rel="stylesheet" />
</head>
<body>
    <div class="m-2">
       @RenderBody()
    </div>
</body>
</html>
```

本章使用控制器视图和 Razor Pages 来呈现相似的内容。为了更容易区分控制器和页面，将代码清单 27-2 中所示的路由添加到 Startup 类中。

**代码清单 27-2　在 WebApp 文件夹的 Startup.cs 文件中添加一个路由**

```
using Microsoft.AspNetCore.Builder;
using Microsoft.Extensions.DependencyInjection;
using Microsoft.Extensions.Configuration;
using Microsoft.EntityFrameworkCore;
using WebApp.Models;

namespace WebApp {
    public class Startup {

        public Startup(IConfiguration config) {
           Configuration = config;
        }

        public IConfiguration Configuration { get; set; }

        public void ConfigureServices(IServiceCollection services) {
            services.AddDbContext<DataContext>(opts => {
                opts.UseSqlServer(Configuration[
                   "ConnectionStrings:ProductConnection"]);
                opts.EnableSensitiveDataLogging(true);
            });
            services.AddControllersWithViews().AddRazorRuntimeCompilation();
            services.AddRazorPages().AddRazorRuntimeCompilation();
            services.AddSingleton<CitiesData>();
        }
```

```
        public void Configure(IApplicationBuilder app, DataContext context) {
            app.UseDeveloperExceptionPage();
            app.UseStaticFiles();
            app.UseRouting();
            app.UseEndpoints(endpoints => {
                endpoints.MapControllers();
                endpoints.MapControllerRoute("forms",
                   "controllers/{controller=Home}/{action=Index}/{id?}");
                endpoints.MapDefaultControllerRoute();
                endpoints.MapRazorPages();
            });
            SeedData.SeedDatabase(context);
        }
    }
}
```

新路由引入了一个静态路径段，使 URL 明显地以控制器为目标。

### 27.1.1 删除数据库

打开一个新的 PowerShell 命令提示符，导航到包含 WebApp.csproj 文件的文件夹，并运行代码清单 27-3 所示的命令来删除数据库。

**代码清单 27-3 删除数据库**

```
dotnet ef database drop --force
```

### 27.1.2 运行示例应用程序

从 Debug 菜单中选择 Start Without Debugging 或 Run Without Debugging，或者使用 PowerShell 命令提示符运行代码清单 27-4 中所示的命令。

**代码清单 27-4 运行示例应用程序**

```
dotnet run
```

使用浏览器请求 http://localhost:5000/controllers/home/list，这会显示一个产品列表，如图 27-1 所示。

localhost:5000/controllers/home/list

Products

| Name | Price | Category | Supplier | |
|---|---|---|---|---|
| Kayak | $275.00 | 1 | 1 | Select |
| Lifejacket | $48.95 | 1 | 1 | Select |
| Soccer Ball | $19.50 | 2 | 2 | Select |
| Corner Flags | $34.95 | 2 | 2 | Select |
| Stadium | $79,500.00 | 2 | 2 | Select |
| Thinking Cap | $16.00 | 3 | 3 | Select |
| Unsteady Chair | $29.95 | 3 | 3 | Select |
| Human Chess Board | $75.00 | 3 | 3 | Select |
| Bling-Bling King | $1,200.00 | 3 | 3 | Select |

Suppliers

图 27-1　运行示例应用程序

## 27.2　理解表单处理模式

大多数 HTML 表单存在于定义良好的模式中，如图 27-2 所示。首先，浏览器发送一个 HTTP GET 请求，这会生成一个包含表单的 HTML 响应，使用户能够向应用程序提供数据。用户单击一个按钮，该按钮使用 HTTP POST 请求提交表单数据，这允许应用程序接收和处理用户的数据。处理完数据后，发送一个响应，该响应将浏览器重定向到一个 URL，该 URL 提供了对用户操作的确认。

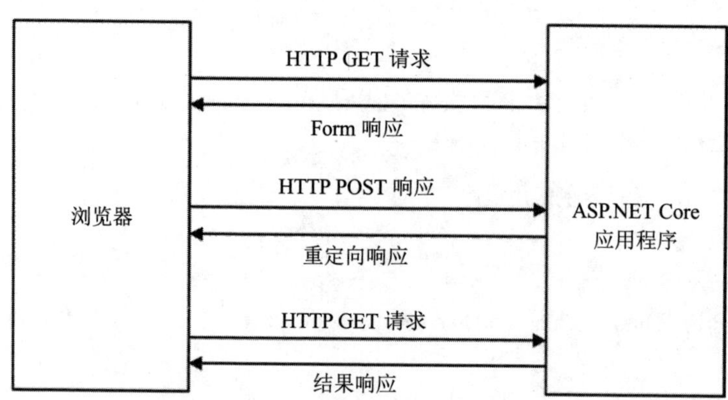


图 27-2　HTML Post/Redirect/Get 模式

这称为 Post/Redirect/Get 模式，这个重定向很重要，因为它意味着用户可以单击浏览器的 reload 按钮而不发送另一个 Post 请求，这可能会导致无意中重复操作。

接下来将展示如何使用控制器和 Razor Pages 遵循该模式。从模式的基本实现开始，然后使用标签助手和第 28 章中的模型绑定特性演示改进。

### 27.2.1 创建控制器来处理表单

处理表单的控制器是通过结合前面章节中描述的特性创建的。在 Controllers 文件夹中添加一个名为 FormController.cs的类文件，如代码清单 27-5 所示。

**代码清单 27-5 Controllers 文件夹中 FormController.cs 文件的内容**

```
using Microsoft.AspNetCore.Mvc;
using System.Linq;
using System.Threading.Tasks;
using WebApp.Models;

namespace WebApp.Controllers {

    public class FormController : Controller {
        private DataContext context;

        public FormController(DataContext dbContext) {
            context = dbContext;
        }

        public async Task<IActionResult> Index(long id = 1) {
            return View("Form", await context.Products.FindAsync(id));
        }

        [HttpPost]
        public IActionResult SubmitForm() {
            foreach (string key in Request.Form.Keys
                    .Where(k => !k.StartsWith("_"))) {
                TempData[key] = string.Join(", ", Request.Form[key]);
            }
            return RedirectToAction(nameof(Results));
        }
        public IActionResult Results() {
            return View(TempData);
        }
    }
}
```

Index 操作方法选择一个名为 Form 的视图，它向用户呈现一个 HTML 表单。当用户提交表单时，它由 SubmitForm 操作接收，SubmitForm 操作已经用 HttpPost 属性进行了修饰，因此它只能接收 HTTP POST 请求。这个操作方法处理通过 HttpRequest.Form 属性得到的 HTML 表单数据，以便可以使用临时数据特性存储它。临时数据特性可用于将数据从一个请求传递到另一个请求，但只能用于存储简单数据类型。每个表单数据值都以字符串数组的形式呈现，将其转换为以逗号分隔的字符串用于存储。浏览器重定向到 Results 操作方法，该方法选择默认视图并提供临时数据作为视图模型。

**提示：**只有名称不以下画线开头的表单数据值才会显示。原因参见本章后面的“使用防伪功能”一节。

为了给控制器提供视图，创建 Views/Form 文件夹，并添加一个名为 Form.cshtml 的 Razor 视图文件，内容如代码清单 27-6 所示。

#### 代码清单 27-6　Views /Form 文件夹中 Form.cshtml 文件的内容

```
@model Product
@{ Layout = "_SimpleLayout"; }

<h5 class="bg-primary text-white text-center p-2">HTML Form</h5>

<form action="/controllers/form/submitform" method="post">
    <div class="form-group">
        <label>Name</label>
        <input class="form-control" name="Name" value="@Model.Name" />
    </div>
    <button type="submit" class="btn btn-primary">Submit</button>
</form>
```

这个视图包含一个简单的 HTML 表单，配置为使用 POST 请求将数据提交给 SubmitForm 操作方法。表单包含一个 input 元素，该元素的值是使用 Razor 表达式设置的。接下来给 Views/Forms 文件夹添加一个名为 Results.cshtml 的 Razor 视图，内容如代码清单 27-7 所示。

#### 代码清单 27-7　Views/Forms 文件中 Results.cshtml 文件的内容

```
@model TempDataDictionary
@{ Layout = "_SimpleLayout"; }

<table class="table table-striped table-bordered table-sm">
    <thead>
        <tr class="bg-primary text-white text-center">
            <th colspan="2">Form Data</th>
        </tr>
    </thead>
    <tbody>
        @foreach (string key in Model.Keys) {
            <tr>
                <th>@key</th>
                <td>@Model[key]</td>
            </tr>
        }
    </tbody>
</table>
<a class="btn btn-primary" asp-action="Index">Return</a>
```

这个视图向用户显示表单数据。第 31 章展示如何以更有用的方式处理表单数据，但本章的重点是创建表单，查看表单中包含的数据就足够了。

重启 ASP.NET Core，并使用浏览器请求 http://localhost:5000/controllers/form 以查看 HTML 表单。在文本字段中输入一个值，并单击 Submit 发送一个 POST 请求，该请求将由 SubmitForm 操作处理。表单数据存储为临时数据，浏览器重定向，生成如图 27-3 所示的响应。

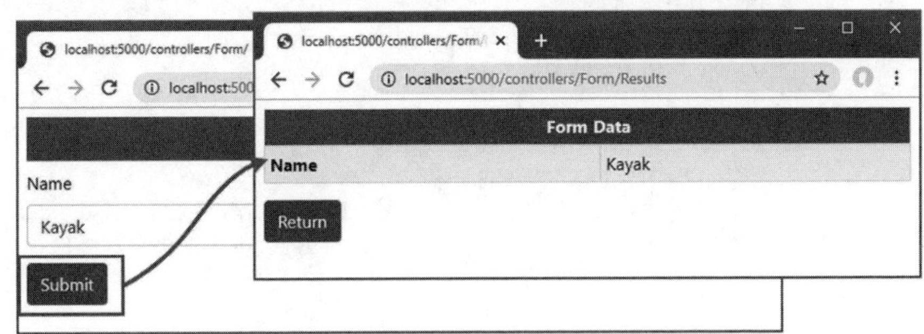


图 27-3 使用控制器来呈现和处理 HTML 表单

## 27.2.2 创建 Razor Pages 来处理表单

同样的模式可以使用 Razor Pages 实现。需要一个页面来呈现和处理表单数据，另一个页面显示结果。给 Pages 文件夹添加一个名为 FormHandler.cshtml 的 Razor Pages，内容如代码清单 27-8 所示。

#### 代码清单 27-8 Pages 文件夹中 FormHandler.cshtml 文件的内容

```
@page "/pages/form/{id:long?}"
@model FormHandlerModel
@using Microsoft.AspNetCore.Mvc.RazorPages

<div class="m-2">
    <h5 class="bg-primary text-white text-center p-2">HTML Form</h5>
    <form asp-page="FormHandler" method="post">
        <div class="form-group">
            <label>Name</label>
            <input class="form-control" name="Name" value="@Model.Product.Name" />
        </div>
        <button type="submit" class="btn btn-primary">Submit</button>
    </form>
</div>

@functions {

    [IgnoreAntiforgeryToken]
    public class FormHandlerModel : PageModel {
        private DataContext context;

        public FormHandlerModel(DataContext dbContext) {
            context = dbContext;
```

```
        }

        public Product Product { get; set; }

        public async Task OnGetAsync(long id = 1) {
            Product = await context.Products.FindAsync(id);
        }

        public IActionResult OnPost() {
            foreach (string key in Request.Form.Keys
                    .Where(k => !k.StartsWith("_"))) {
                TempData[key] = string.Join(", ", Request.Form[key]);
            }
            return RedirectToPage("FormResults");
        }
    }
}
```

OnGetAsync 处理程序方法从数据库中检索 Product，视图使用该产品设置 HTML 表单中输入元素的值。该表单配置为发送一个将由 OnPost 处理程序方法处理的 HTTP POST 请求。表单数据存储为临时数据，并向浏览器发送一个到 FormResults 表单的重定向。要创建浏览器将被重定向到的页面，向 Pages 文件夹添加一个名为 FormResults.cshtml 的 Razor Pages，内容如代码清单 27-9 所示。

**提示：** 代码清单 27-8 中的页面模型类使用 IgnoreAntiforgeryToken 属性修饰，该属性在“使用防伪功能”一节中描述。

**代码清单 27-9　Pages 文件夹中 FormResults.cshtml 文件的内容**

```
@page "/pages/results"

<div class="m-2">
    <table class="table table-striped table-bordered table-sm">
        <thead>
            <tr class="bg-primary text-white text-center">
                <th colspan="2">Form Data</th>
            </tr>
        </thead>
        <tbody>
            @foreach (string key in TempData.Keys) {
                <tr>
                    <th>@key</th>
                    <td>@TempData[key]</td>
                </tr>
            }
        </tbody>
    </table>
```

```
        <a class="btn btn-primary" asp-page="FormHandler">Return</a>
    </div>
```

这个页面不需要任何代码，它直接访问临时数据，并将其显示在表中。使用浏览器导航到 http://localhost:5000/pages/form，在文本字段中输入一个值，然后单击 Submit 按钮。表单数据由代码清单 27-9 中定义的 OnPost 方法处理，浏览器重定向到/pages/results，从而显示表单数据，如图 27-4 所示。

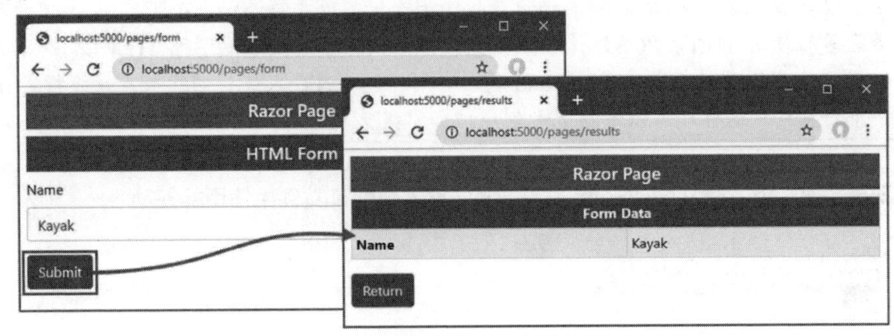


图 27-4 使用 Razor Pages 来呈现和处理 HTML 表单

# 27.3 使用标签助手改进 HTML 表单

上一节中的示例展示了处理 HTML 表单的基本机制，但是 ASP.NET Core 包含转换表单元素的标签助手。接下来描述标签助手并演示它们的用法。

## 27.3.1 使用表单元素

FormTagHelper 类是表单元素的内置标签助手类，用于管理 HTML 表单的配置，以便它们能够针对正确的操作或页面处理程序，而不需要硬编码 URL。这个标签助手支持表 27-3 中描述的属性。

表 27-3 表单元素的内置标签助手属性

| 名称 | 描述 |
| --- | --- |
| asp-controller | 此属性用于为操作属性 URL 指定路由系统的 controller 值。如果省略，那么使用呈现视图的控制器 |
| asp-action | 此属性用于为 action 属性 URL 的路由系统指定 action 值的操作方法。如果省略，就使用呈现视图的操作 |
| asp-page | 此属性用于指定 Razor Pages 的名称 |
| asp-page-handler | 此属性用于指定处理程序方法的名称，该处理程序方法用于处理请求。可在第 9 章的 SportsStore 应用程序中看到这个属性的示例 |
| asp-route-* | 名称以 asp-route 开头的属性用于为操作属性 URL 指定附加值，以便使用 asp-route-id 属性为路由系统提供 id 段的值 |
| asp-route | 此属性指定将用于为操作生成 URL 属性的路由名称 |
| asp-antiforgery | 这个属性控制是否将防伪信息添加到视图中，如“使用防伪功能”一节所述 |
| asp-fragmen | 此属性为生成的 URL 指定一个片段 |

### 设置表单目标

FormTagHelper 转换表单元素，这样它们就可以针对一个操作方法或 Razor Pages，而不需要硬编码的 URL。这个标签助手所支持的属性与第 26 章中描述的锚元素的工作方式相同，使用属性来提供值，帮助通过 ASP.NET Core 路由系统生成 URL。代码清单 27-10 修改表单视图中的 Form 元素，以应用标签助手。

---

■ **注意**：如果定义的表单元素没有 method 属性，那么标签助手将添加一个带有 post 值的元素，这意味着表单将使用 HTTP POST 请求来提交。如果因为希望浏览器遵循 HTML5 规范并使用 HTTP GET 请求发送表单而忽略了 method 属性，则可能导致令人惊讶的结果。始终指定 method 属性是一个好主意，这样就可以清楚地看到应该如何提交表单。

---

**代码清单 27-10　在 Views/Form 文件夹的 Form.cshtml 文件中使用标签助手**

```
@model Product
@{ Layout = "_SimpleLayout"; }

<h5 class="bg-primary text-white text-center p-2">HTML Form</h5>

<form asp-action="submitform" method="post">
    <div class="form-group">
        <label>Name</label>
        <input class="form-control" name="Name" value="@Model.Name" />
    </div>
    <button type="submit" class="btn btn-primary">Submit</button>
</form>
```

asp-action 属性用于指定将接收 HTTP 请求的操作的名称。路由系统用于生成 URL，就像第 26 章中描述的锚元素一样。代码清单 27-10 中没有使用 asp-controller 属性，这意味着呈现视图的控制器将在 URL 中使用。

asp-page 属性用于选择 Razor Pages 作为表单的目标，如代码清单 27-11 所示。

**代码清单 27-11　在 Pages 文件夹的 FormHandler.cshtml 文件中设置表单目标**

```
...
<div class="m-2">
    <h5 class="bg-primary text-white text-center p-2">HTML Form</h5>
    <form asp-page="FormHandler" method="post">
        <div class="form-group">
            <label>Name</label>
            <input class="form-control" name="Name" value="@Model.Product.Name" />
        </div>
        <button type="submit" class="btn btn-primary">Submit</button>
    </form>
</div>
...
```

使用浏览器导航到 http://localhost:5000/controllers/form，并检查浏览器接收到的 HTML；会看到标签助手添加为 form 元素的 action 属性，如下所示：

```
...
<form method="post" action="controllers/Form/submitform">
...
```

这与在创建视图时静态定义的 URL 相同，但它的优点是，对路由配置的更改将自动反映在表单 URL 中。请求 http://localhost:5000/pages/form，会看到表单元素已转换为针对页面 URL，如下所示：

```
...
<form method="post" action="/pages/form">
...
```

### 27.3.2 改变表单按钮

发送表单的按钮可以在表单元素之外定义。在这些情况下，按钮具有一个表单属性，该属性的值对应于它所关联的表单元素的 id 属性，以及一个 formaction 属性，该属性指定表单的目标 URL。

标签助手将通过 asp-action、asp-controller 或 asp-page 属性生成 formaction 属性，如代码清单 27-12 所示。

**代码清单 27-12 在 Views/Form 文件夹的 Form.cshtml 文件中转换按钮**

```
@model Product
@{ Layout = "_SimpleLayout"; }

<h5 class="bg-primary text-white text-center p-2">HTML Form</h5>

<form asp-action="submitform" method="post" id="htmlform">
    <div class="form-group">
        <label>Name</label>
        <input class="form-control" name="Name" value="@Model.Name" />
    </div>
    <button type="submit" class="btn btn-primary">Submit</button>
</form>

<button form="htmlform" asp-action="submitform" class="btn btn-primary mt-2">
   Sumit (Outside Form)
</button>
```

添加到表单元素的 id 属性的值被按钮用作 form 属性的值，它告诉浏览器，在单击按钮时提交哪个表单。表 27-3 中描述的属性用于标识表单的目标，标签助手在呈现视图时使用路由系统生成 URL。代码清单 27-13 对 Razor Pages 应用了相同的技术。

**代码清单 27-13　在 Pages 文件夹的 FormHandler.cshtml 文件中转换按钮**

```
...
<div class="m-2">
    <h5 class="bg-primary text-white text-center p-2">HTML Form</h5>
    <form asp-page="FormHandler" method="post" id="htmlform">
        <div class="form-group">
            <label>Name</label>
            <input class="form-control" name="Name" value="@Model.Product.Name" />
        </div>
        <button type="submit" class="btn btn-primary">Submit</button>
    </form>
    <button form="htmlform" asp-page="FormHandler" class="btn btn-primary mt-2">
       Sumit (Outside Form)
    </button>
</div>
...
```

使用浏览器请求 http://localhost:5000/controllers/form 或 http://localhost:5000/pages/form，并检查发送到浏览器的 HTML。会看到被转换的表单之外的按钮元素，如下所示：

```
...
<button form="htmlform" class="btn btn-primary mt-2"
        formaction="/controllers/Form/submitform">
    Sumit (Outside Form)
</button>
...
```

单击按钮就会提交表单，就像表单元素中定义的按钮一样，如图 27-5 所示。

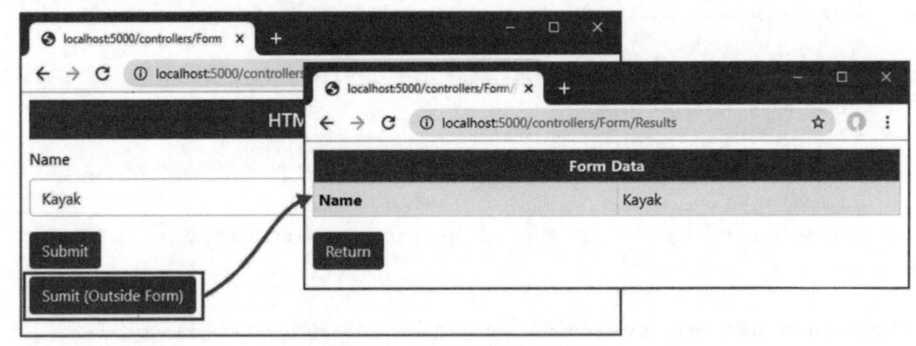


图 27-5　在表单元素外定义按钮

## 27.4　处理 input 元素

input(输入)元素是 HTML 表单的主干，它提供了用户向应用程序提供非结构化数据的主要方法。InputTagHelper 类用于转换 input 元素，以便它们使用表 27-4 中描述的属性，反映它们要收集的视图模型属性的数据类型和格式。

表 27-4 input 元素的内置标签助手属性

| 名称 | 描述 |
| --- | --- |
| asp-for | 这个属性用于指定 input 元素所表示的视图模型属性 |
| asp-format | 这个属性用于指定 input 元素所表示的视图模型属性的值的格式 |

asp-for 属性设置为视图模型属性的名称，然后用于设置 input 元素的 name、id、type 和 value 属性。代码清单 27-14 修改了控制器视图中的 input 元素，以使用 asp-for 属性。

**代码清单 27-14 在 Views/Form 文件夹的 Form.cshtml 文件中配置 input 元素**

```
@model Product
@{ Layout = "_SimpleLayout"; }

<h5 class="bg-primary text-white text-center p-2">HTML Form</h5>

<form asp-action="submitform" method="post" id="htmlform">
    <div class="form-group">
        <label>Name</label>
        <input class="form-control" asp-for="Name" />
    </div>
    <button type="submit" class="btn btn-primary">Submit</button>
</form>

<button form="htmlform" asp-action="submitform" class="btn btn-primary mt-2">
   Sumit (Outside Form)
</button>
```

这个标签助手使用一个模型表达式，如代码清单 27-14 所示，这就是为什么在指定 asp-for 属性的值时不使用@字符。如果查看在使用浏览器请求 http://localhost:5000/controllers/form 时应用程序返回的 HTML，可看到标签助手将 input 元素转换为如下形式：

```
...
<div class="form-group">
    <label>Name</label>
    <input class="form-control" type="text" id="Name" name="Name" value="Kayak">
</div>
...
```

id 和 name 属性的值是通过模型表达式获得的，从而确保在创建表单时不会引入拼写错误。其他属性更复杂，将在下面的部分中描述。

**在 Razor Pages 中选择模型属性**

这个标签助手的 asp-for 属性和本章中描述的其他标签助手可用于 Razor Pages，但是转换元素中的 name 和 id 属性值包括页面模型属性的名称。例如，这个元素通过页面模型的 Product 属性选择 Name 属性:

```
...
<input class="form-control" asp-for="Product.Name" />
...
```

转换后的元素将具有以下 id 和 name 属性:

```
...
<input class="form-control" type="text" id="Product_Name" name="Product.Name" >
...
```

当使用模型绑定特性来接收表单数据时，这种差异非常重要，如第 28 章所述。

### 27.4.1 转换 input 元素的类型属性

input 元素的 type 属性告诉浏览器如何显示元素，以及它应该如何限制用户输入的值。代码清单 27-14 中的 input 元素配置为 text 类型，这是默认的 input 元素类型，没有任何限制。代码清单 27-15 在表单中添加了另一个 input 元素，它为如何处理 type 属性提供更有用的演示。

**代码清单 27-15 在 Views/Form 文件夹的 Form.cshtml 文件中添加 input 元素**

```
@model Product
@{ Layout = "_SimpleLayout"; }

<h5 class="bg-primary text-white text-center p-2">HTML Form</h5>

<form asp-action="submitform" method="post" id="htmlform">
    <div class="form-group">
        <label>Id</label>
        <input class="form-control" asp-for="ProductId" />
    </div>
    <div class="form-group">
        <label>Name</label>
        <input class="form-control" asp-for="Name" />
    </div>
    <button type="submit" class="btn btn-primary">Submit</button>
</form>

<button form="htmlform" asp-action="submitform" class="btn btn-primary mt-2">
   Sumit (Outside Form)
</button>
```

新元素使用 asp-for 属性来选择视图模型的 ProductId 属性。使用浏览器请求 http:/localhost:5000/controllers/form，查看标签助手是如何转换元素的。

```
...
<div class="form-group">
    <label>Id</label>
    <input class="form-control" type="number" data-val="true"
        data-val-required="The ProductId field is required."
```

```
        id="ProductId" name="ProductId" value="1">
</div>
...
```

type 属性的值由 asp-for 属性指定的视图模型属性的类型决定。ProductId 属性的类型是 C# long 类型，它导致标签助手将 input 元素的 type 属性设置为 number，这将限制元素，使其只接收数字字符。data-val 和 data-val 所需的属性添加到 input 元素中，以帮助验证，参见第 29 章。表 27-5 描述了如何使用不同的 C#类型来设置 input 元素的 type 属性。

**注意：**浏览器如何解释 type 属性是有自由度的。并不是所有浏览器都响应 HTML5 规范中定义的所有 type 值，当它们响应时，它们的实现方式也有所不同。type 属性可以是表单中期望的数据类型，这是一个有用的提示，但是应该使用模型验证功能来确保用户提供可用的数据，如第 29 章所述。

表 27-5　属性类型及生成的 input 元素的 type 属性

| C#类型 | input 元素的 type 属性 |
| --- | --- |
| byte, sbyte, int, uint, short, ushort, long, ulong | number |
| float, double, decimal，如第 29 章所述 | text，具有模型验证的附加属性，如 checkbox |
| string | text |
| DateTime | datetime |

浮点、双精度和十进制类型将生成类型为 text 的 input 元素，因为并非所有浏览器都允许用于表示这种类型的合法值的全部字符。为向用户提供反馈，标签助手向 input 元素添加属性，这些属性与第 29 章中描述的验证功能一起使用。

通过显式地定义 input 元素上的 type 属性，可覆盖表 27-5 中所示的默认映射。标签助手不会覆盖定义的值，它允许指定 type 属性值。

这种方法的缺点是，必须在为给定模型属性生成 input 元素的所有视图中设置 type 属性。一种更优雅、可靠的方法是将表 27-6 中描述的一个属性应用于 C# model 类中的属性。

**提示：**如果模型属性不是表 27-5 中的类型之一，并且没有装饰属性，则标签助手将 input 元素的 type 属性设置为 text。

表 27-6　input 元素的 type 属性

| 属性 | input 元素的 type 属性 |
| --- | --- |
| [HiddenInput] | hidden |
| [Text] | text |
| [Phone] | tel |
| [Url] | url |
| [EmailAddress] | email |
| [DataType(DataType.Time)] | time |
| [DataType(DataType.Password)] | password |
| [DataType(DataType.Date)] | date |

## 27.4.2　格式化 input 元素值

当 action 方法为视图提供视图模型对象时，标签助手使用 asp-for 属性值来设置 input 元素的 value 属性。asp-format 属性用于指定数据值的格式。为演示默认格式，代码清单 27-16 向 Form 视图添加了一个新的 input 元素。

**代码清单 27-16　在 Views/Form 文件夹的 Form.cshtml 文件中添加元素**

```
@model Product
@{ Layout = "_SimpleLayout"; }

<h5 class="bg-primary text-white text-center p-2">HTML Form</h5>

<form asp-action="submitform" method="post" id="htmlform">
    <div class="form-group">
        <label>Id</label>
        <input class="form-control" asp-for="ProductId" />
    </div>
    <div class="form-group">
        <label>Name</label>
        <input class="form-control" asp-for="Name" />
    </div>
    <div class="form-group">
        <label>Price</label>
        <input class="form-control" asp-for="Price" />
    </div>
    <button type="submit" class="btn btn-primary">Submit</button>
</form>

<button form="htmlform" asp-action="submitform" class="btn btn-primary mt-2">
   Sumit (Outside Form)
</button>
```

使用浏览器导航到 http://localhost:5000/controllers/form/index/5 并检查浏览器接收到的 HTML。默认情况下，input 元素的值使用模型属性的值来设置，如下所示：

```
...
<input class="form-control" type="text" data-val="true"
    data-val-number="The field Price must be a number."
    data-val-required="The Price field is required."
    id="Price" name="Price" value="79500.00">
...
```

这种有两位小数的格式是数据库中存储值的方式。第 26 章使用 Column 属性选择 SQL 类型来存储 Price 值，如下所示：

```
...
[Column(TypeName = "decimal(8, 2)")]
```

```
public decimal Price { get; set; }
...
```

此类型指定的最大精度为 8 位，其中 2 位在小数点后。允许的最大值为 999 999.99，这足以容纳大多数在线商店的价格。asp-format 属性接收一个格式字符串，该字符串传递给标准的 C#字符串格式化系统，如代码清单 27-17 所示。

**代码清单 27-17　在 Views/Form 文件夹的 Form.cshtml 文件中格式化数据值**

```
@model Product
@{ Layout = "_SimpleLayout"; }

<h5 class="bg-primary text-white text-center p-2">HTML Form</h5>

<form asp-action="submitform" method="post" id="htmlform">
    <div class="form-group">
        <label>Id</label>
        <input class="form-control" asp-for="ProductId" />
    </div>
    <div class="form-group">
        <label>Name</label>
        <input class="form-control" asp-for="Name" />
    </div>
    <div class="form-group">
        <label>Price</label>
        <input class="form-control" asp-for="Price" asp-format="{0:#,###.00}" />
    </div>
    <button type="submit" class="btn btn-primary">Submit</button>
</form>

<button form="htmlform" asp-action="submitform" class="btn btn-primary mt-2">
   Sumit (Outside Form)
</button>
```

属性值是逐字使用的，这意味着必须包含大括号字符和 0:以及所需的格式。刷新浏览器，input 元素的值就已格式化，如下所示：

```
...
<input class="form-control" type="text" data-val="true"
    data-val-number="The field Price must be a number."
    data-val-required="The Price field is required."
    id="Price" name="Price" value="79,500.00">
...
```

使用此特性时应谨慎，因为必须确保应用程序的其余部分配置为支持使用的格式，并且创建的格式只包含 input 元素类型的合法字符。

### 通过模型类应用格式化

如果总是希望对模型属性使用相同的格式，就可以使用 DisplayFormat 属性来装饰 C#类，该属性在 System.ComponentModel.DataAnnotations 名称空间中定义。DisplayFormat 属性需要两个参数来格式化数据值：DataFormatString 参数指定格式化字符串，将 ApplyFormatInEditMode 设置为 true 意味着将值应用于编辑的元素(包括 input 元素)时应该使用格式化。代码清单 27-18 将该属性应用于 Product 类的 Price 属性，指定与前面示例不同的格式化字符串。

**代码清单 27-18　在 Models 文件夹的 Product.cs 文件中应用格式化属性**

```
using System.ComponentModel.DataAnnotations.Schema;
using System.ComponentModel.DataAnnotations;

namespace WebApp.Models {
    public class Product {

        public long ProductId { get; set; }

        public string Name { get; set; }

        [Column(TypeName = "decimal(8, 2)")]
        [DisplayFormat(DataFormatString = "{0:c2}", ApplyFormatInEditMode = true)]
        public decimal Price { get; set; }

        public long CategoryId { get; set; }
        public Category Category { get; set; }

        public long SupplierId { get; set; }
        public Supplier Supplier { get; set; }
    }
}
```

asp-format 属性优先于 DisplayFormat 属性，因此从视图中删除了该属性，如代码清单 27-19 所示。

**代码清单 27-19　在 Views/Form 文件夹的 Form.cshtml 文件中删除属性**

```
@model Product
@{ Layout = "_SimpleLayout"; }

<h5 class="bg-primary text-white text-center p-2">HTML Form</h5>

<form asp-action="submitform" method="post" id="htmlform">
    <div class="form-group">
        <label>Id</label>
        <input class="form-control" asp-for="ProductId" />
    </div>
    <div class="form-group">
```

```
        <label>Name</label>
        <input class="form-control" asp-for="Name" />
    </div>
    <div class="form-group">
        <label>Price</label>
        <input class="form-control" asp-for="Price" />
    </div>
    <button type="submit" class="btn btn-primary">Submit</button>
</form>

<button form="htmlform" asp-action="submitform" class="btn btn-primary mt-2">
    Sumit (Outside Form)
</button>
```

重启 ASP.NET Core 并使用浏览器请求 http://localhost:5000/controllers/Form/index/5，会看到由属性定义的格式化字符串已经应用，如图 27-6 所示。

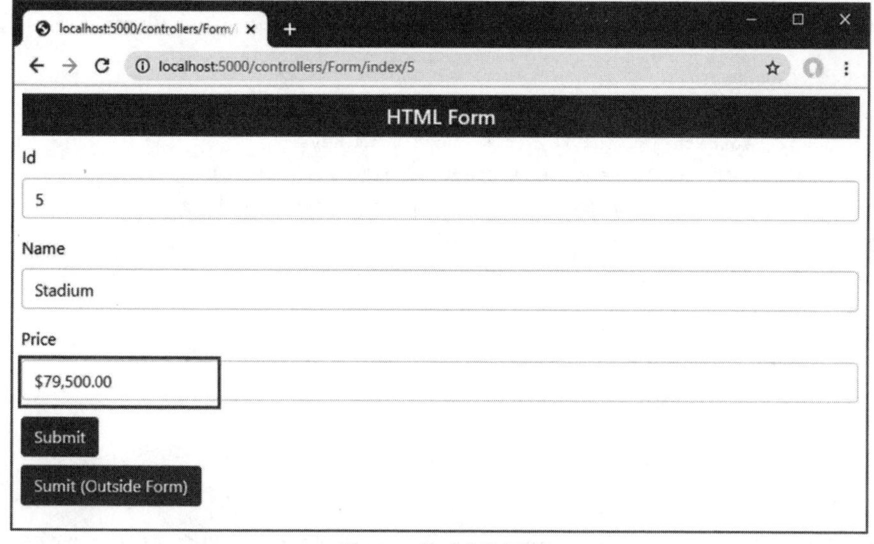


图 27-6　格式化数据值

选择这种格式是为了演示格式化属性的工作方式，但如前所述，务必确保应用程序能使用第 28 和 29 章中描述的模型绑定和验证功能来处理格式化的值。

### 27.4.3　在 input 元素中显示相关数据的值

在使用 Entity Framework Core 时，经常需要显示从相关数据中获得的数据值，使用 asp-for 属性很容易做到这一点，因为模型表达式允许选择嵌套的导航属性。首先，代码清单 27-20 在提供给视图的视图模型对象中包含相关数据。

**代码清单 27-20　在 Controllers 文件夹的 FormController.cs 文件中包含相关数据**

```
using Microsoft.AspNetCore.Mvc;
using System.Linq;
```

```
using System.Threading.Tasks;
using WebApp.Models;
using Microsoft.EntityFrameworkCore;

namespace WebApp.Controllers {

    public class FormController : Controller {
        private DataContext context;

        public FormController(DataContext dbContext) {
            context = dbContext;
        }

        public async Task<IActionResult> Index(long id = 1) {
            return View("Form", await context.Products.Include(p => p.Category)
                .Include(p => p.Supplier).FirstAsync(p => p.ProductId == id));
        }

        [HttpPost]
        public IActionResult SubmitForm() {
            foreach (string key in Request.Form.Keys
                    .Where(k => !k.StartsWith("_"))) {
                TempData[key] = string.Join(", ", Request.Form[key]);
            }
            return RedirectToAction(nameof(Results));
        }

        public IActionResult Results() {
            return View(TempData);
        }
    }
}
```

注意，不必担心处理相关数据中的循环引用，因为视图模型对象没有序列化。循环引用问题仅对 Web 服务控制器很重要。在代码清单 27-21 中，更新了 Form 视图，使其包含使用 asp-for 属性选择相关数据的 input 元素。

### 代码清单 27-21　在 Views/Form 文件夹的 Form.cshtml 文件中显示相关数据

```
@model Product
@{ Layout = "_SimpleLayout"; }

<h5 class="bg-primary text-white text-center p-2">HTML Form</h5>

<form asp-action="submitform" method="post" id="htmlform">
    <div class="form-group">
        <label>Id</label>
        <input class="form-control" asp-for="ProductId" />
```

```
        </div>
        <div class="form-group">
            <label>Name</label>
            <input class="form-control" asp-for="Name" />
        </div>
        <div class="form-group">
            <label>Price</label>
            <input class="form-control" asp-for="Price" />
        </div>
        <div class="form-group">
            <label>Category</label>
            <input class="form-control" asp-for="Category.Name" />
        </div>
        <div class="form-group">
            <label>Supplier</label>
            <input class="form-control" asp-for="Supplier.Name" />
        </div>
        <button type="submit" class="btn btn-primary">Submit</button>
    </form>

    <button form="htmlform" asp-action="submitform" class="btn btn-primary mt-2">
        Sumit (Outside Form)
    </button>
```

asp-for 属性的值是相对于视图模型对象表示的，可以包含嵌套属性，允许选择 Entity Framework Core 分配给 Category 和 Supplier 导航属性的相关对象的 Name 属性。Razor Pages 使用了相同的技术，只是属性是相对于页面模型对象表示的，如代码清单 27-22 所示。

### 代码清单 27-22　在 Pages 文件夹的 FormHandler.cshtml 文件中显示相关数据

```
@page "/pages/form/{id:long?}"
@model FormHandlerModel
@using Microsoft.AspNetCore.Mvc.RazorPages
@using Microsoft.EntityFrameworkCore

<div class="m-2">
    <h5 class="bg-primary text-white text-center p-2">HTML Form</h5>
    <form asp-page="FormHandler" method="post" id="htmlform">
        <div class="form-group">
            <label>Name</label>
            <input class="form-control" asp-for="Product.Name" />
        </div>
        <div class="form-group">
            <label>Price</label>
            <input class="form-control" asp-for="Product.Price" />
        </div>
        <div class="form-group">
            <label>Category</label>
```

```
            <input class="form-control" asp-for="Product.Category.Name" />
        </div>
        <div class="form-group">
            <label>Supplier</label>
            <input class="form-control" asp-for="Product.Supplier.Name" />
        </div>
        <button type="submit" class="btn btn-primary">Submit</button>
    </form>
    <button form="htmlform" asp-page="FormHandler" class="btn btn-primary mt-2">
       Sumit (Outside Form)
    </button>
</div>

@functions {

    [IgnoreAntiforgeryToken]
    public class FormHandlerModel : PageModel {
        private DataContext context;

        public FormHandlerModel(DataContext dbContext) {
            context = dbContext;
        }

        public Product Product { get; set; }

        public async Task OnGetAsync(long id = 1) {
            Product = await context.Products.Include(p => p.Category)
               .Include(p => p.Supplier).FirstAsync(p => p.ProductId == id);
        }
        public IActionResult OnPost() {
            foreach (string key in Request.Form.Keys
                    .Where(k => !k.StartsWith("_"))) {
                TempData[key] = string.Join(", ", Request.Form[key]);
            }
            return RedirectToPage("FormResults");
        }
    }
}
```

要查看效果，请重新启动 ASP.NET Core 以使对控制器的更改生效，并使用浏览器请求 http://localhost:5000/controller/form，这将生成如图 27-7 左侧所示的响应。使用浏览器请求 http://localhost:5000/pages/form，将看到 Razor Pages 使用的相同特性，如图 27-7 右侧所示。

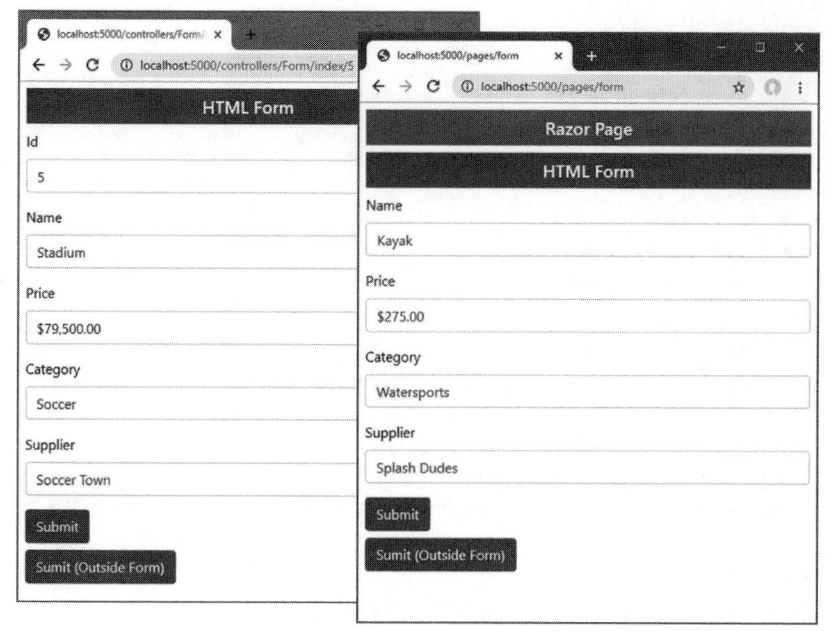


图 27-7　显示相关数据

## 27.5　使用 label 元素

LabelTagHelper 类用于转换 label(标签)元素，因此 for 属性的设置与用于转换 input 元素的方法一致。表 27-7 描述了标签助手支持的属性。

表 27-7　label 元素的内置标签助手属性

| 名称 | 描述 |
| --- | --- |
| asp-for | 此属性用于指定 label 元素所描述的视图模型属性 |

标签助手设置 label 元素的内容，以便它包含所选视图模型属性的名称。标签助手还设置 for 属性，该属性表示与特定 input 元素的关联。这将帮助依赖于屏幕阅读器的用户，并在单击关联的标签时使 input 元素获得焦点。

代码清单 27-23 将 asp-for 属性应用于表单视图，以将每个 label 元素与表示相同视图模型属性的 input 元素相关联。

#### 代码清单 27-23　在 Views/Form 文件夹的 Form.cshtml 文件中转换 label 元素

```
@model Product
@{ Layout = "_SimpleLayout"; }

<h5 class="bg-primary text-white text-center p-2">HTML Form</h5>

<form asp-action="submitform" method="post" id="htmlform">
    <div class="form-group">
        <label asp-for="ProductId"></label>
```

```
            <input class="form-control" asp-for="ProductId" />
        </div>
        <div class="form-group">
            <label asp-for="Name"></label>
            <input class="form-control" asp-for="Name" />
        </div>
        <div class="form-group">
            <label asp-for="Price"></label>
            <input class="form-control" asp-for="Price" />
        </div>
        <div class="form-group">
            <label asp-for="Category.Name">Category</label>
            <input class="form-control" asp-for="Category.Name" />
        </div>
        <div class="form-group">
            <label asp-for="Supplier.Name">Supplier</label>
            <input class="form-control" asp-for="Supplier.Name" />
        </div>
        <button type="submit" class="btn btn-primary">Submit</button>
    </form>

    <button form="htmlform" asp-action="submitform" class="btn btn-primary mt-2">
        Sumit (Outside Form)
    </button>
```

可通过自定义 label 元素来覆盖它的内容，这就是在代码清单 27-23 中对相关数据属性所做的工作。标签助手将把这两个 label 元素的内容设置为 Name，这不是一个有用的描述。定义元素内容意味着将应用 for 属性，但向用户显示一个更有用的名称。使用浏览器请求 http://localhost:5000/controllers/form，查看每个元素使用的名称，如图 27-8 所示。

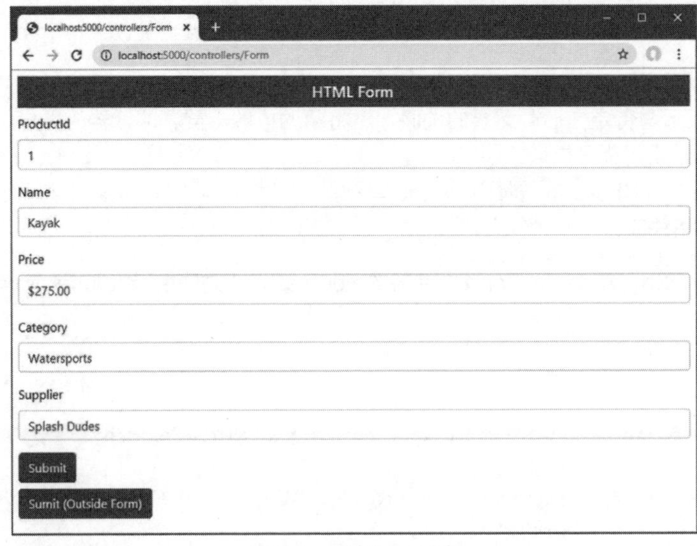


图 27-8　改变 label 元素

## 27.6 使用 select 和 option 元素

select 和 option 元素用于向用户提供一组固定的选择，而不是使用 input 元素可能提供的开发数据项。SelectTagHelper 负责转换 select 元素，并支持表 27-8 中描述的属性。

表 27-8 用于 select 元素的内置标签助手属性

| 名称 | 描述 |
| --- | --- |
| asp-for | 此属性用于指定 select 元素所表示的视图或页面模型属性 |
| asp-items | 此属性用于为 select 元素中包含的 option 元素指定值源 |

asp-for 属性设置 for 和 id 属性的值，以反映它接收到的模型属性。在代码清单 27-24 中，用 select 元素替换了类别的 input 元素，为用户提供了固定范围的值。

**代码清单 27-24 在 Views/Form 文件夹的 Form.cshtml 文件中使用 select 元素**

```
@model Product
@{ Layout = "_SimpleLayout"; }

<h5 class="bg-primary text-white text-center p-2">HTML Form</h5>

<form asp-action="submitform" method="post" id="htmlform">
    <div class="form-group">
        <label asp-for="ProductId"></label>
        <input class="form-control" asp-for="ProductId" />
    </div>
    <div class="form-group">
        <label asp-for="Name"></label>
        <input class="form-control" asp-for="Name" />
    </div>
    <div class="form-group">
        <label asp-for="Price"></label>
        <input class="form-control" asp-for="Price" />
    </div>
    <div class="form-group">
        <label asp-for="Category.Name">Category</label>
        <select class="form-control" asp-for="CategoryId">
            <option value="1">Watersports</option>
            <option value="2">Soccer</option>
            <option value="3">Chess</option>
        </select>
    </div>
    <div class="form-group">
        <label asp-for="Supplier.Name">Supplier</label>
        <input class="form-control" asp-for="Supplier.Name" />
    </div>
    <button type="submit" class="btn btn-primary">Submit</button>
```

```
</form>

<button form="htmlform" asp-action="submitform" class="btn btn-primary mt-2">
   Sumit (Outside Form)
</button>
```

用 option 元素手工填充 select 元素，这些 option 元素提供了一系列类别供用户选择。如果使用浏览器请求 http://localhost:5000/controllers/form/index/5 并检查 HTML 响应，会看到标签助手将 select 元素转换为如下形式：

```
...
<div class="form-group">
    <label for="Category_Name">Category</label>
    <select class="form-control" data-val="true"
          data-val-required="The CategoryId field is required."
          id="CategoryId" name="CategoryId">
        <option value="1">Watersports</option>
        <option value="2" selected="selected">Soccer</option>
        <option value="3">Chess</option>
    </select>
</div>
...
```

注意，所选的属性添加到 option 元素中，与视图模型的 CategoryId 值相对应，如下所示：

```
...
<option value="2" selected="selected">Soccer</option>
...
```

选择 option 元素的任务是由 OptionTagHelper 类执行的，该类通过 TagHelperContext. Items 集合接收来自 SelectTagHelper 的指令，参见第 25 章。结果是，select 元素显示与 Product 对象的 CategoryId 值相关联的类别名称。

```
</button>
```

### 填充 select 元素

为 select 元素显式地定义 option 元素在选择可能相同的值时是一个有用的方法，但在以下情形下无效：需要提供从数据模型中提取选项；在多个视图中需要相同的选项集，但不想手动维护重复的内容。

asp-items 属性用于为标签助手提供 SelectListItem 对象的列表序列，为这些对象生成 option 元素。代码清单 27-25 修改表单控制器的索引操作，以通过视图包向视图提供一系列 SelectListItem 对象。

**代码清单 27-25　在 Controllers 文件夹的 FormController.cs 文件中提供一个数据序列**

```
using Microsoft.AspNetCore.Mvc;
using System.Linq;
```

```
using System.Threading.Tasks;
using WebApp.Models;
using Microsoft.EntityFrameworkCore;
using Microsoft.AspNetCore.Mvc.Rendering;

namespace WebApp.Controllers {

    public class FormController : Controller {
        private DataContext context;

        public FormController(DataContext dbContext) {
            context = dbContext;
        }

        public async Task<IActionResult> Index(long id = 1) {
            ViewBag.Categories
                = new SelectList(context.Categories, "CategoryId", "Name");
            return View("Form", await context.Products.Include(p => p.Category)
                .Include(p => p.Supplier).FirstAsync(p => p.ProductId == id));
        }

        [HttpPost]
        public IActionResult SubmitForm() {
            foreach (string key in Request.Form.Keys
                    .Where(k => !k.StartsWith("_"))) {
                TempData[key] = string.Join(", ", Request.Form[key]);
            }
            return RedirectToAction(nameof(Results));
        }

        public IActionResult Results() {
            return View(TempData);
        }
    }
}
```

可直接创建 SelectListItem 对象，但是 ASP.NET Core 提供了 SelectList 类来适应现有的数据序列。在本例中，从数据库中获得的 Category 对象序列传递给 SelectList 构造函数，还传递了应该用作 option 元素的值和标签的属性名称。在代码清单 27-26 中，更新了 form 视图以使用 SelectList。

**代码清单 27-26　在 Views/Form 文件夹的 Form.cshtml 文件中使用 SelectList**

```
@model Product
@{ Layout = "_SimpleLayout"; }

<h5 class="bg-primary text-white text-center p-2">HTML Form</h5>

<form asp-action="submitform" method="post" id="htmlform">
```

```
    <div class="form-group">
        <label asp-for="ProductId"></label>
        <input class="form-control" asp-for="ProductId" />
    </div>
    <div class="form-group">
        <label asp-for="Name"></label>
        <input class="form-control" asp-for="Name" />
    </div>
    <div class="form-group">
        <label asp-for="Price"></label>
        <input class="form-control" asp-for="Price" />
    </div>
    <div class="form-group">
        <label asp-for="Category.Name">Category</label>
        <select class="form-control" asp-for="CategoryId"
            asp-items="@ViewBag.Categories">
        </select>
    </div>
    <div class="form-group">
        <label asp-for="Supplier.Name">Supplier</label>
        <input class="form-control" asp-for="Supplier.Name" />
    </div>
    <button type="submit" class="btn btn-primary">Submit</button>
</form>

<button form="htmlform" asp-action="submitform" class="btn btn-primary mt-2">
    Sumit (Outside Form)
</button>
```

重启ASP.NET Core，因此，对控制器的更改生效，并使用浏览器请求http://localhost:5000/controllers/form/index/5。呈现给用户的内容没有可见的变化，但是用于填充select元素的option元素是从数据库中生成的，如下所示：

```
...
<div class="form-group">
    <label for="Category_Name">Category</label>
    <select class="form-control" data-val="true"
            data-val-required="The CategoryId field is required."
            id="CategoryId" name="CategoryId">
        <option value="1">Watersports</option>
        <option selected="selected" value="2">Soccer</option>
        <option value="3">Chess</option>
    </select>
</div>
...
```

这种方法意味着提供给用户的选项将自动反映添加到数据库中的新类别。

## 27.7 处理文本区域

textarea 元素用于从用户那里请求大量文本，通常用于非结构化数据，如注释或观察。TextAreaTagHelper 负责转换 textarea 元素，并支持表 27-9 中描述的单个属性。

表 27-9 textarea 元素的内置标签助手属性

| 名称 | 描述 |
| --- | --- |
| asp-for | 这个属性用于指定 textarea 元素所表示的视图模型属性 |

TextAreaTagHelper 相对简单，为 asp-for 属性提供的值用于设置 textarea 元素上的 id 和 name 属性。asp-for 属性选择的属性值用作 textarea 元素的内容。代码清单 27-27 替换了 Supplier.Name属性的输入元素，带有已应用 asp-for 属性的文本区域。

**代码清单 27-27 在 Views/Form 文件夹的 Form.cshtml 文件中使用文本区域**

```
@model Product
@{ Layout = "_SimpleLayout"; }

<h5 class="bg-primary text-white text-center p-2">HTML Form</h5>

<form asp-action="submitform" method="post" id="htmlform">
    <div class="form-group">
        <label asp-for="ProductId"></label>
        <input class="form-control" asp-for="ProductId" />
    </div>
    <div class="form-group">
        <label asp-for="Name"></label>
        <input class="form-control" asp-for="Name" />
    </div>
    <div class="form-group">
        <label asp-for="Price"></label>
        <input class="form-control" asp-for="Price" />
    </div>
    <div class="form-group">
        <label asp-for="Category.Name">Category</label>
        <select class="form-control" asp-for="CategoryId"
           asp-items="@ViewBag.Categories">
        </select>
    </div>
    <div class="form-group">
        <label asp-for="Supplier.Name">Supplier</label>
        <textarea class="form-control" asp-for="Supplier.Name"></textarea>
    </div>
    <button type="submit" class="btn btn-primary">Submit</button>
</form>
```

```
<button form="htmlform" asp-action="submitform" class="btn btn-primary mt-2">
    Sumit (Outside Form)
</button>
```

使用浏览器请求 http://localhost:5000/controllers/form，并检查浏览器接收到的 HTML，以查看 textarea 元素的转换。

```
...
<div class="form-group">
    <label for="Supplier_Name">Supplier</label>
    <textarea class="form-control" id="Supplier_Name" name="Supplier.Name">
        Soccer Town
    </textarea>
</div>
...
```

TextAreaTagHelper 相对简单，但它提供了与本章描述的其他表单元素标签助手的一致性。

## 27.8　使用防伪功能

定义处理表单数据的控制器操作方法和页面处理程序方法时，过滤掉了名称以下画线开头的表单数据，如下所示：

```
...
[HttpPost]
public IActionResult SubmitForm() {
    foreach (string key in Request.Form.Keys
            .Where(k => !k.StartsWith("_"))) {
        TempData[key] = string.Join(", ", Request.Form[key]);
    }
    return RedirectToAction(nameof(Results));
}
...
```

应用这个过滤器来隐藏一个特性，以便将重点放在表单中 HTML 元素提供的值上。代码清单 27-28 从操作方法中删除了过滤器，以便从 HTML 表单接收到的所有数据都存储在临时数据中。

**代码清单 27-28　在 Controllers 文件夹的 FormController.cs 文件中删除过滤器**

```
...
[HttpPost]
public IActionResult SubmitForm() {
    foreach (string key in Request.Form.Keys) {
        TempData[key] = string.Join(", ", Request.Form[key]);
    }
    return RedirectToAction(nameof(Results));
}
...
```

重启 ASP.NET Core 并使用浏览器请求 http://localhost:5000/controllers。单击 Submit 按钮，将表单发送到应用程序，在结果中会看到一个新项，如图 27-9 所示。

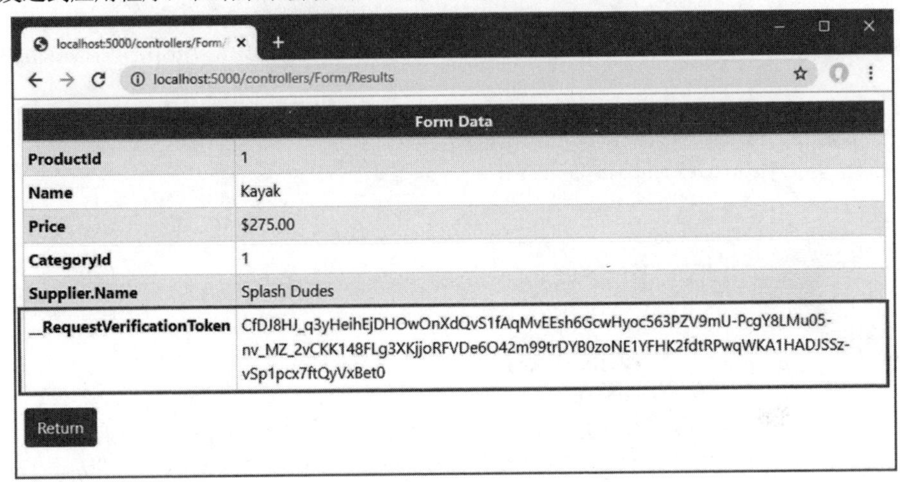


图 27-9　显示窗体数据

结果中显示的_RequestVerificationToken 表单值是一个安全特性，FormTagHelper 应用它来防止跨站点请求伪造。跨站点请求伪造(CSRF)通过利用用户请求的典型身份验证方式来利用 Web 应用程序。大多数 Web 应用程序——包括那些使用 ASP.NET Core 创建的应用程序——使用 cookie 来识别哪些请求与特定会话相关，通常与用户标识相关联。

CSRF(也称为 XSRF)依赖于用户在使用 Web 应用程序后访问恶意网站，并且没有显式地结束其会话。应用程序仍然认为用户的会话是活动的，并且浏览器存储的 cookie 还没有过期。恶意站点包含 JavaScript 代码，它会向应用程序发送表单请求，以便在未得到用户同意的情况下执行操作——操作的确切性质取决于被攻击的应用程序。由于 JavaScript 代码是由用户的浏览器执行的，因此对应用程序的请求包括会话 cookie，而应用程序在用户不知情或不同意的情况下执行操作。

■ **提示：** CSRF 的详细描述可在 http://en.wikipedia.org/wiki/Cross-site_request_forgery 找到。

如果 form 元素不包含 action 属性——因为它是由路由系统使用 asp-controller、asp-action 和 asp-page 属性生成的——那么 FormTagHelper 类将自动启用一个反 CSRF 特性，从而将一个安全令牌作为 cookie 添加到响应中。一个包含相同安全令牌的隐藏 input 元素添加到 HTML 表单中，如图 27-9 所示就是这个令牌。

## 27.8.1　在控制器中启用防伪功能

默认情况下，控制器接收 POST 请求(即使它们不包含所需的安全令牌)。要启用防伪功能，将一个属性应用于控制器类，如代码清单 27-29 所示。

**代码清单 27-29　在 Controllers 文件夹的 FormController.cs 文件中启用防伪功能**

```
using Microsoft.AspNetCore.Mvc;
using System.Linq;
using System.Threading.Tasks;
using WebApp.Models;
```

```
using Microsoft.EntityFrameworkCore;
using Microsoft.AspNetCore.Mvc.Rendering;

namespace WebApp.Controllers {

    [AutoValidateAntiforgeryToken]
    public class FormController : Controller {
        private DataContext context;

        public FormController(DataContext dbContext) {
            context = dbContext;
        }

        public async Task<IActionResult> Index(long id = 1) {
            ViewBag.Categories
                = new SelectList(context.Categories, "CategoryId", "Name");
            return View("Form", await context.Products.Include(p => p.Category)
                .Include(p => p.Supplier).FirstAsync(p => p.ProductId == id));
        }

        [HttpPost]
        public IActionResult SubmitForm() {
            foreach (string key in Request.Form.Keys) {
                TempData[key] = string.Join(", ", Request.Form[key]);
            }
            return RedirectToAction(nameof(Results));
        }

        public IActionResult Results() {
            return View(TempData);
        }
    }
}
```

并不是所有请求都需要一个防伪造令牌，AutoValidateAntiforgeryToken 确保对除 GET、HEAD、OPTIONS 和 TRACE 之外的所有 HTTP 方法执行检查。

**提示：**还有两个属性可以用来控制令牌验证。IgnoreValidationToken 属性抑制对操作方法或控制器的验证。ValidateAntiForgeryToken 属性执行相反的操作，并强制验证，即使对于那些通常不需要验证的请求(如 http get 请求)也是如此。建议使用 AutoValidateAntiforgeryToken 属性，如代码清单所示。

测试反 CSRF 特性有点棘手。请求包含表单的 URL(在本例中为 http://localhost:5000/controllers/forms)，然后使用浏览器的 F12 开发人员工具从表单中定位并删除隐藏的 input 元素(或更改元素的值)来实现这一点。填充并提交表单时，它丢失了所需数据的一部分，请求将失败。

### 27.8.2 在 Razor Pages 中启用防伪功能

在 Razor Pages 中默认启用了防伪造功能，这就是在创建 FormHandler 页面时将 IgnoreAntiforgeryToken 属性应用到代码清单 27-29 中的页面处理程序方法的原因。代码清单 27-30 删除了启用验证功能的属性。

**代码清单 27-30 在 Pages 文件夹的 FormHandler.cshtml 文件中启用请求验证**

```
@page "/pages/form/{id:long?}"
@model FormHandlerModel
@using Microsoft.AspNetCore.Mvc.RazorPages
@using Microsoft.EntityFrameworkCore

<div class="m-2">
    <h5 class="bg-primary text-white text-center p-2">HTML Form</h5>
    <form asp-page="FormHandler" method="post" id="htmlform">
        <div class="form-group">
            <label>Name</label>
            <input class="form-control" asp-for="Product.Name" />
        </div>

        <div class="form-group">
            <label>Price</label>
            <input class="form-control" asp-for="Product.Price" />
        </div>
        <div class="form-group">
            <label>Category</label>
            <input class="form-control" asp-for="Product.Category.Name" />
        </div>
        <div class="form-group">
            <label>Supplier</label>
            <input class="form-control" asp-for="Product.Supplier.Name" />
        </div>
        <button type="submit" class="btn btn-primary">Submit</button>
    </form>
    <button form="htmlform" asp-page="FormHandler" class="btn btn-primary mt-2">
       Sumit (Outside Form)
    </button>
</div>

@functions {

    //[IgnoreAntiforgeryToken]
    public class FormHandlerModel : PageModel {
        private DataContext context;

        public FormHandlerModel(DataContext dbContext) {
```

```
                context = dbContext;
            }

            public Product Product { get; set; }

            public async Task OnGetAsync(long id = 1) {
                Product = await context.Products.Include(p => p.Category)
                    .Include(p => p.Supplier).FirstAsync(p => p.ProductId == id);
            }

            public IActionResult OnPost() {
                foreach (string key in Request.Form.Keys
                        .Where(k => !k.StartsWith("_"))) {
                    TempData[key] = string.Join(", ", Request.Form[key]);
                }
                return RedirectToPage("FormResults");
            }
        }
    }
```

测试验证特性的方式与控制器相同，需要在向应用程序提交表单之前，使用浏览器的开发人员工具修改 HTML 文档。

### 27.8.3　使用 JavaScript 客户端防伪令牌

默认情况下，防伪功能依赖于 ASP.NET Core 应用程序能够在 HTML 表单中包含一个元素，当表单提交时浏览器将返回该元素。这对 JavaScript 客户端不起作用，因为 ASP.NET Core 应用程序提供的是数据而非 HTML，因此无法插入隐藏元素，并在以后的请求中接收它。

对于 Web 服务，防伪造令牌可作为 JavaScript 可读的 cookie 发送，JavaScript 客户端代码将读取该 cookie 并将其作为头包含在 POST 请求中。一些 JavaScript 框架(如 Angular)会自动检测 cookie 并在请求中包含一个头。对于其他框架和定制 JavaScript 代码，需要做额外的工作。

代码清单 27-31 显示了需要对 ASP.NET Core 应用程序进行的更改，来配置用于 JavaScript 客户端的防伪功能。

**代码清单 27-31　在 WebApp 文件夹的 Startup.cs 文件中配置防伪令牌**

```
using Microsoft.AspNetCore.Builder;
using Microsoft.Extensions.DependencyInjection;
using Microsoft.Extensions.Configuration;
using Microsoft.EntityFrameworkCore;
using WebApp.Models;
using Microsoft.AspNetCore.Antiforgery;
using Microsoft.AspNetCore.Http;

namespace WebApp {
    public class Startup {
```

```
public Startup(IConfiguration config) {
    Configuration = config;
}

public IConfiguration Configuration { get; set; }

public void ConfigureServices(IServiceCollection services) {
    services.AddDbContext<DataContext>(opts => {
        opts.UseSqlServer(Configuration[
            "ConnectionStrings:ProductConnection"]);
        opts.EnableSensitiveDataLogging(true);
    });
    services.AddControllersWithViews().AddRazorRuntimeCompilation();
    services.AddRazorPages().AddRazorRuntimeCompilation();
    services.AddSingleton<CitiesData>();

    services.Configure<AntiforgeryOptions>(opts => {
        opts.HeaderName = "X-XSRF-TOKEN";
    });
}

public void Configure(IApplicationBuilder app, DataContext context,
        IAntiforgery antiforgery) {
    app.UseRequestLocalization();

    app.UseDeveloperExceptionPage();
    app.UseStaticFiles();
    app.UseRouting();

    app.Use(async (context, next) => {
        if (!context.Request.Path.StartsWithSegments("/api")) {
            context.Response.Cookies.Append("XSRF-TOKEN",
                antiforgery.GetAndStoreTokens(context).RequestToken,
                new CookieOptions { HttpOnly = false });
        }
        await next();
    });

    app.UseEndpoints(endpoints => {
        endpoints.MapControllers();
        endpoints.MapControllerRoute("forms",
            "controllers/{controller=Home}/{action=Index}/{id?}");
        endpoints.MapDefaultControllerRoute();
        endpoints.MapRazorPages();
    });
    SeedData.SeedDatabase(context);
}
```

```
        }
    }
```

options 模式用于通过 AntiforgeryOptions 类配置防伪功能。HeaderName 属性用于通过接收的防伪令牌(在本例中是 X-XSRF-TOKEN)，来指定头的名称。

需要一个定制的中间件组件来设置 cookie，在本例中命名为 XSRF-TOKEN。cookie 的值是通过 IAntiForgery 服务获得的，并且必须将 HttpOnly 选项设置为 false，以便浏览器允许 JavaScript 代码读取 cookie。

---

**提示：**在本例中，使用了 Angular 支持的名称。其他框架遵循自己的约定，但通常可以配置为使用任意一组 cookie 和头名称。

---

要创建一个使用 cookie 和 header 的简单 JavaScript 客户端，给 Pages 文件夹添加一个 Razor Pages，内容如代码清单 27-32 所示。

**代码清单 27-32　Pages 文件夹中 JavaScriptForm.cshtml 文件的内容**

```
@page "/pages/jsform"

<script type="text/javascript">
    async function sendRequest() {
        const token = document.cookie
            .replace(/(?:(?:^|.*;\s*)XSRF-TOKEN\s*\=\s*([^;]*).*$)|^.*$/, "$1");

        let form = new FormData();
        form.append("name", "Paddle");
        form.append("price", 100);
        form.append("categoryId", 1);
        form.append("supplierId", 1);
        let response = await fetch("@Url.Page("FormHandler")", {
            method: "POST",
            headers: { "X-XSRF-TOKEN": token },
            body: form
        });
        document.getElementById("content").innerHTML = await response.text();
    }

    document.addEventListener("DOMContentLoaded",
        () => document.getElementById("submit").onclick = sendRequest);
</script>

<button class="btn btn-primary m-2" id="submit">Submit JavaScript Form</button>
<div id="content"></div>
```

Razor Pages 中的 JavaScript 代码通过向 FormHandler 发送一个 HTTP POST 请求来响应按钮的单击。XSRF-TOKEN cookie 的值被读取并包含在 X-XSRF-TOKEN 请求头中。来自 FormHandler 页面的响应是一个到 Results 页面的重定向，浏览器将自动跟随该页面。JavaScript 代码读取来自

Results 页面的响应，并将其插入元素中，以便向用户显示。要测试 JavaScript 代码，请使用浏览器请求 http://localhost:5000/pages/jsform 并单击按钮。JavaScript 代码将提交表单并显示响应，如图 27-10 所示。

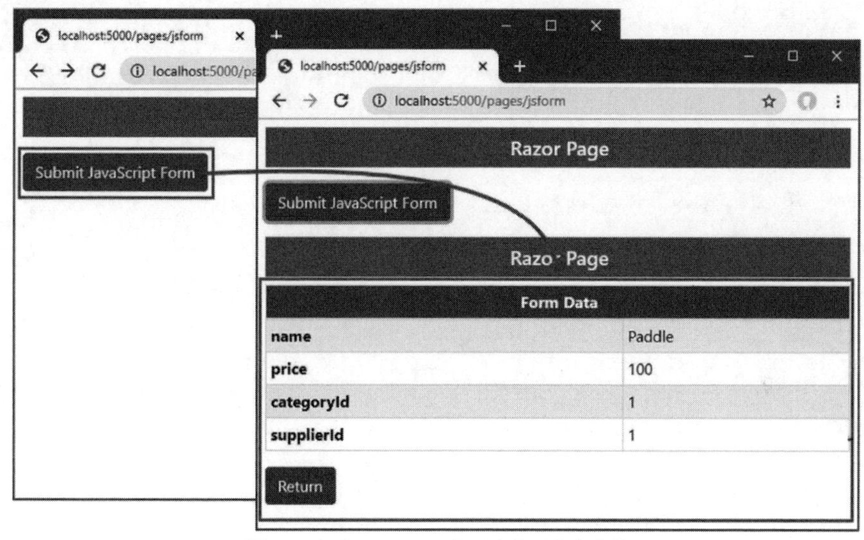


图 27-10 在 JavaScript 代码中使用安全令牌

## 27.9 小结

本章解释了 ASP.NET Core 提供的、创建 HTML 表单的功能。展示了如何使用标签助手来选择表单目标和关联的 input、textarea 和 select 元素。第 28 章描述模型绑定功能，它从请求中提取数据，以便在操作和处理程序方法中轻松地使用数据。

# 第28章

# 使用模型绑定

模型绑定是使用来自HTTP请求的值创建.NET对象的过程，以方便地访问操作方法和Razor Pages所需的数据。本章描述模型绑定系统的工作方式；显示它如何绑定简单类型、复杂类型和集合；并演示如何控制流程，以指定请求的哪一部分提供应用程序所需的数据值。表28-1概述模型绑定的相关知识。

表28-1　模型绑定的相关知识

| 问题 | 答案 |
| --- | --- |
| 它是什么? | 模型绑定是使用从HTTP请求获得的数据值，创建操作方法和页面处理程序所需的对象的过程 |
| 它为什么有用? | 模型绑定让控制器或页面处理程序使用C#类型来声明方法参数或属性，并自动从请求中接收数据，而不必直接检查、解析和处理数据 |
| 它是如何使用的? | 在最简单的形式中，方法声明参数或类可定义属性(其名称用于从HTTP请求中检索数据值)。可以通过将属性应用到方法参数或属性，来配置用于获取数据的请求部分 |
| 是否存在缺陷或限制? | 主要缺陷是从请求的错误部分获取数据。“理解模型绑定”一节中解释了搜索数据请求的方式，并且可以使用“指定模型绑定源”一节中描述的属性显式地指定搜索位置 |
| 还有其他选择吗? | 不必使用上下文对象进行模型绑定即可获得数据。然而，其结果是代码更加复杂，难以阅读和维护 |

表28-2总结了本章的内容。

表28-2　本章内容摘要

| 问题 | 解决方案 | 代码清单 |
| --- | --- | --- |
| 绑定基本类型 | 用基本类型定义方法参数 | 28-5～28-9 |
| 绑定复杂类型 | 用类类型定义方法参数 | 28-10 |
| 绑定到属性 | 使用BindProperty属性 | 28-11、28-12 |
| 绑定嵌套类型 | 确保表单值类型遵循句点标记 | 28-13～28-17 |
| 选择用于绑定集合的属性 | 使用Bind和BindNever属性 | 28-18、28-19 |
| 绑定集合 | 遵循序列绑定约定 | 28-20～28-25 |
| 指定用于绑定的源 | 使用其中一个源属性 | 28-26～28-31 |
| 手动执行绑定 | 使用TryUpdateModel方法 | 28-32 |

## 28.1 准备工作

本章使用了第 27 章中的 WebApp 项目。为准备本章，请用代码清单 28-1 所示的内容替换 Views/Form 文件夹中 Form.cshtml 文件的内容。

---

**提示：** 可以从 https://github.com/apress/pro-asp.net-core-3 下载本章和本书中其他所有章节的示例项目。如果在运行示例时遇到问题，请参阅第 1 章以了解如何获得帮助。

---

**代码清单 28-1 Views/Form 文件夹中 Form.cshtml 文件的内容**

```
@model Product
@{ Layout = "_SimpleLayout"; }

<h5 class="bg-primary text-white text-center p-2">HTML Form</h5>

<form asp-action="submitform" method="post" id="htmlform">
    <div class="form-group">
        <label asp-for="Name"></label>
        <input class="form-control" asp-for="Name" />
    </div>
    <div class="form-group">
        <label asp-for="Price"></label>
        <input class="form-control" asp-for="Price" />
    </div>
    <button type="submit" class="btn btn-primary">Submit</button>
</form>
```

接下来，注释掉已应用于 Product 模型类的 DisplayFormat 属性，如代码清单 28-2 所示。

**代码清单 28-2 在 Models 文件夹的 Product.cs 文件中删除一个属性**

```
using System.ComponentModel.DataAnnotations.Schema;
using System.ComponentModel.DataAnnotations;

namespace WebApp.Models {
    public class Product {

        public long ProductId { get; set; }

        public string Name { get; set; }

        [Column(TypeName = "decimal(8, 2)")]
        //[DisplayFormat(DataFormatString = "{0:c2}", ApplyFormatInEditMode = true)]
        public decimal Price { get; set; }

        public long CategoryId { get; set; }
        public Category Category { get; set; }
```

```
            public long SupplierId { get; set; }
            public Supplier Supplier { get; set; }
        }
    }
```

### 28.1.1 删除数据库

打开一个新的 PowerShell 命令提示符，导航到包含 WebApp.csproj 文件的文件夹，并运行代码清单 28-3 所示的命令来删除数据库。

**代码清单 28-3 删除数据库**

```
dotnet ef database drop --force
```

### 28.1.2 运行示例应用程序

从 Debug 菜单中选择 Start Without Debugging 或 Run Without Debugging，或者使用 PowerShell 命令提示符运行代码清单 28-4 所示的命令。

**代码清单 28-4 运行示例应用程序**

```
dotnet run
```

使用浏览器请求 http://localhost:5000/controllers/form，这将显示一个 HTML 表单。单击 Submit 按钮，将显示表单数据，如图 28-1 所示。

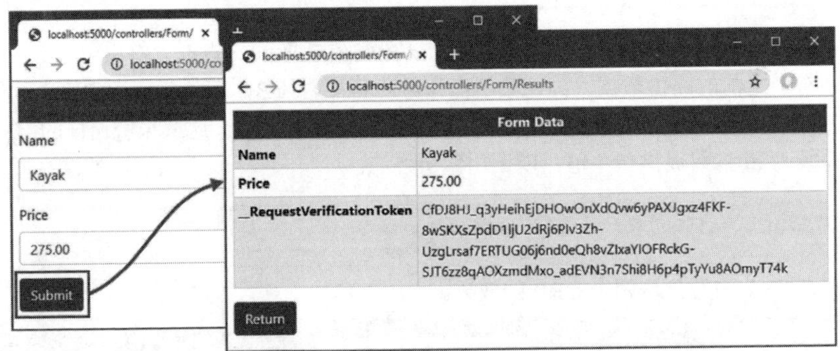


图 28-1 运行示例应用程序

## 28.2 理解模型绑定

模型绑定是 HTTP 请求和操作或页面处理程序方法之间的桥梁。大多数 ASP.NET Core 应用程序在某种程度上依赖于模型绑定，包括木章的示例应用程序。

通过使用浏览器请求 http://localhost:5000/controllers/form/index/5，可以看到模型绑定正在工作。这个 URL 包含想要查看的 Product 对象的 ProductId 属性值，如下所示：

```
http://localhost:5000/controllers/form/index/5
```

URL 的这一部分对应于控制器路由模式定义的 id 段变量，并与表单控制器的索引操作定义的参数名称相匹配:

```
...
public async Task<IActionResult> Index(long id = 1) {
...
```

在 MVC 框架调用操作方法之前需要为 id 参数设置一个值，而找到一个合适的值是模型绑定系统的职责。模型绑定系统依赖于模型绑定器，模型绑定器是负责从请求或应用程序的某个部分提供数据值的组件。默认的模型绑定在以下四个地方寻找数据值:

- 表单数据
- 请求主体(仅适用于用 ApiController 装饰的控制器)
- 路由段变量
- 查询字符串

按顺序检查每个数据源，直至找到参数的值为止。示例应用程序中没有表单数据，因此在那里不会找到任何值，并且表单控制器没有使用 ApiController 属性装饰，因此不会检查请求主体。下一步是检查路由数据，它包含一个名为 id 的段变量。这允许模型绑定系统提供一个值来允许调用索引操作方法。在找到合适的数据值后停止搜索，这意味着查询字符串不会搜索数据值。

**提示：**“指定模型绑定源”一节解释了如何使用属性来指定模型绑定数据的源。这允许指定从查询字符串等处获取数据值，即使路由数据中有合适的数据也是如此。

知道寻找数据值的顺序是很重要的，因为一个请求可以包含多个值，比如下面这个 URL:

```
http://localhost:5000/controllers/Form/Index/5?id=1
```

路由系统将处理请求，并将 URL 模板中的 id 段与值 3 匹配，查询字符串包含 id 值 1。由于搜索查询字符串之前的路由数据，因此索引操作方法将接收值 3，而忽略查询字符串的值。

另一方面，如果请求没有 id 段的 URL，则将检查查询字符串，这意味着这样的 URL 也允许模型绑定系统为 id 参数提供一个值，以调用索引方法。

```
http://localhost:5000/controllers/Form/Index?id=4
```

可以在图 28-2 中看到这两个 URL 的效果。

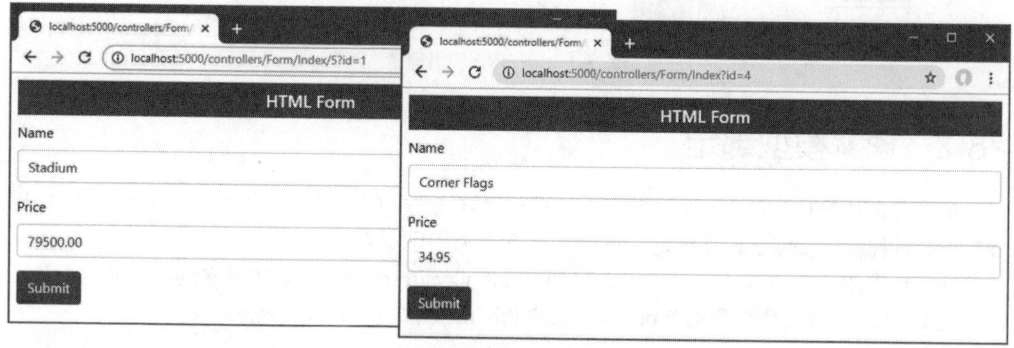


图 28-2　模型绑定数据源顺序的影响

# 28.3 绑定简单数据类型

请求数据值必须转换为C#值，这样它们才能用于调用操作或页面处理程序方法。简单类型是源自请求中的一项数据的值，该数据项可以从字符串中解析。这包括数值、bool 值、日期和字符串值。

用于简单类型的数据绑定很容易从请求中提取单个数据项，而不必通过上下文数据查找定义的位置。代码清单 28-5 向 Form 控制器方法定义的 SubmitForm 操作方法添加了参数，以便模型绑定器用于提供 name 和 price 值。

**代码清单 28-5 在 Controllers 文件夹的 FormController.cs 文件中添加方法参数**

```
using Microsoft.AspNetCore.Mvc;
using System.Linq;
using System.Threading.Tasks;
using WebApp.Models;
using Microsoft.EntityFrameworkCore;
using Microsoft.AspNetCore.Mvc.Rendering;

namespace WebApp.Controllers {

    [AutoValidateAntiforgeryToken]
    public class FormController : Controller {
        private DataContext context;

        public FormController(DataContext dbContext) {
            context = dbContext;
        }

        public async Task<IActionResult> Index(long id = 1) {
            ViewBag.Categories
                = new SelectList(context.Categories, "CategoryId", "Name");
            return View("Form", await context.Products.Include(p => p.Category)
                .Include(p => p.Supplier).FirstAsync(p => p.ProductId == id));
        }

        [HttpPost]
        public IActionResult SubmitForm(string name, decimal price) {
            TempData["name param"] = name;
            TempData["price param"] = price.ToString();
            return RedirectToAction(nameof(Results));
        }

        public IActionResult Results() {
            return View(TempData);
        }
    }
}
```

ASP.NET Core 接收将由 SubmitForm 操作方法处理的请求时，将使用模型绑定系统来获取 name 和 price 值。参数的使用简化了操作方法，并负责将请求数据转换为 C#数据类型，以便在调用操作方法之前将 Price 值转换为 C# decimal 类型(本例必须将十进制转换回字符串，以将其存储为临时数据。第 31 章演示处理表单数据的更有用方法)。重启 ASP.NET Core，对控制器的更改生效并请求 http://localhost:5000/controllers/Form。单击 Submit 按钮，会看到模型绑定特性从请求中提取的值，如图 28-3 所示。

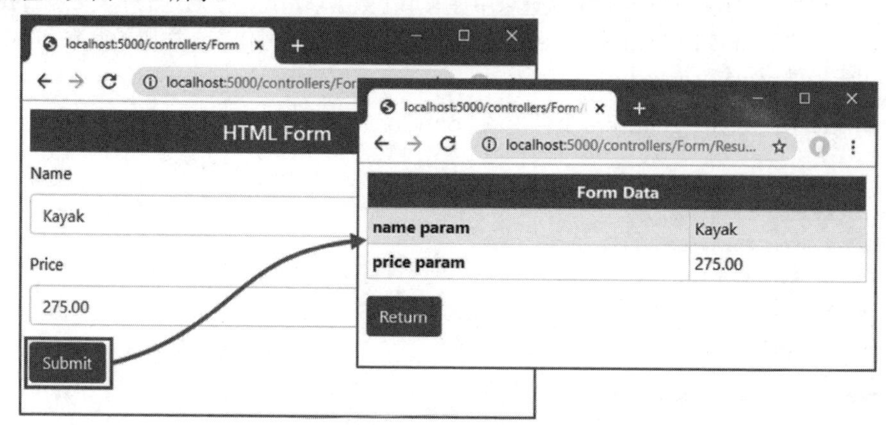


图 28-3　用于简单类型的模型绑定

## 28.3.1　绑定 Razor Pages 中的简单数据类型

Razor Pages 可以使用模型绑定，但是必须注意确保表单元素的 name 属性的值与处理程序方法参数的名称相匹配，如果 asp-for 属性用来选择嵌套属性，则可能不会出现这种情况。为了确保名称匹配，可以显式定义 name 属性，如代码清单 28-6 所示，这也简化了 HTML 表单，以便与控制器示例匹配。

### 代码清单 28-6　在 Pages 文件夹的 FormHandler.cshtml 文件中使用模型绑定

```
@page "/pages/form/{id:long?}"
@model FormHandlerModel
@using Microsoft.AspNetCore.Mvc.RazorPages
@using Microsoft.EntityFrameworkCore

<div class="m-2">
    <h5 class="bg-primary text-white text-center p-2">HTML Form</h5>
    <form asp-page="FormHandler" method="post" id="htmlform">
        <div class="form-group">
            <label>Name</label>
            <input class="form-control" asp-for="Product.Name" name="name"/>
        </div>
        <div class="form-group">
            <label>Price</label>
            <input class="form-control" asp-for="Product.Price" name="price" />
        </div>
```

```
        <button type="submit" class="btn btn-primary">Submit</button>
    </form>
</div>

@functions {

    public class FormHandlerModel : PageModel {
        private DataContext context;

        public FormHandlerModel(DataContext dbContext) {
            context = dbContext;
        }

        public Product Product { get; set; }

        public async Task OnGetAsync(long id = 1) {
            Product = await context.Products.Include(p => p.Category)
                .Include(p => p.Supplier).FirstAsync(p => p.ProductId == id);
        }

        public IActionResult OnPost(string name, decimal price) {
            TempData["name param"] = name;
            TempData["price param"] = price.ToString();
            return RedirectToPage("FormResults");
        }
    }
}
```

标签助手将输入元素的 name 属性设置为 Product.Name 和 Product.Price，阻止模型绑定与值匹配。显式地设置 name 属性将覆盖标签助手，并确保模型绑定过程正确工作。使用浏览器请求 http://localhost:5000/pages/form，并单击 Submit 按钮，会看到模型绑定器找到的值，如图 28-4 所示。

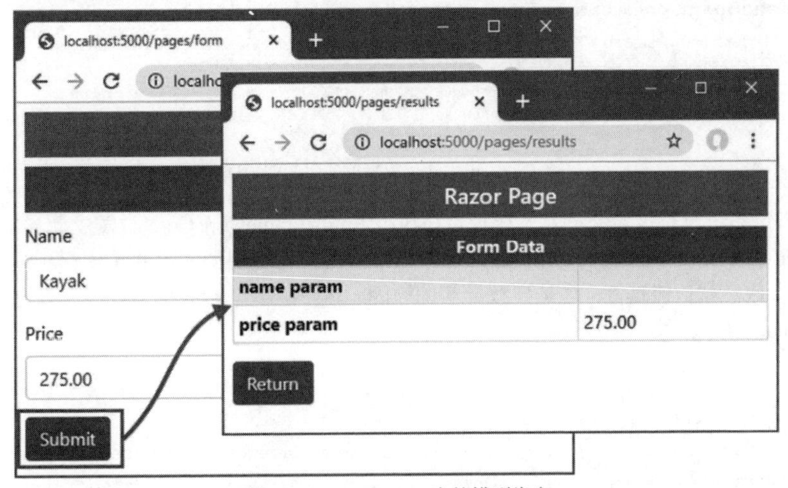


图 28-4 Razor Pages 中的模型绑定

## 28.3.2 理解默认绑定值

模型绑定是一个最佳的特性，这意味着模型绑定器将尝试获取方法参数的值，但是如果无法定位数据值，仍然会调用该方法。可以通过在 Form 控制器的 Index 操作方法中删除 id 参数的默认值来了解其工作原理，如代码清单 28-7 所示。

**代码清单 28-7 在 Controllers 文件夹的 FormController.cs 文件中删除一个参数**

```
...
public async Task<IActionResult> Index(long id) {
    ViewBag.Categories
        = new SelectList(context.Categories, "CategoryId", "Name");
    return View("Form", await context.Products.Include(p => p.Category)
        .Include(p => p.Supplier).FirstAsync(p => p.ProductId == id));
}
...
```

重启 ASP.NET Core，请求 http://localhost:5000/controllers/Form。URL 不包含模型绑定器可以用于 id 参数的值，也没有查询字符串或表单数据，但是仍然调用该方法，从而生成如图 28-5 所示的错误。

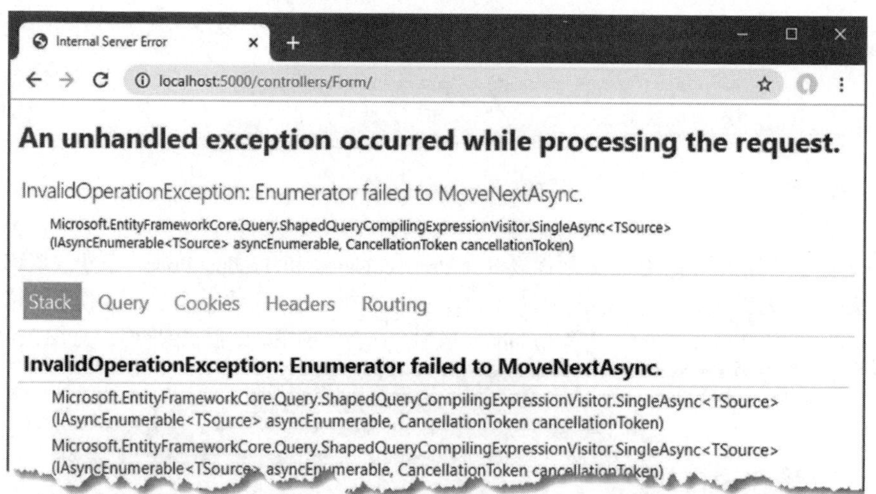


图 28-5 由于缺少数据值而引起的错误

模型绑定系统没有报告此异常。相反，在执行 Entity Framework Core 时发生此异常。MVC 框架必须为 id 参数提供一些值来调用 Index 操作方法，因此它使用默认值并希望得到最好的结果。对于 long 参数，默认值是 0，这是导致异常的原因。Index 操作方法使用 id 值作为键来查询数据库中的 Product 对象，如下所示：

```
...
public async Task<IActionResult> Index(long id) {
    ViewBag.Categories = new SelectList(context.Categories, "CategoryId", "Name");
    return View("Form", await context.Products.Include(p => p.Category)
```

```
        .Include(p => p.Supplier).FirstAsync(p => p.ProductId == id));
}
...
```

当模型绑定没有可用值时，操作方法尝试查询 id 为 0 的数据库。没有这样的对象，这会导致 Entity Framework Core 尝试处理结果时出现如图所示的错误。

必须编写应用程序来处理默认参数值，这可以通过几种方式实现。可以向控制器(如第 21 章所示)或页面(如第 23 章所示)使用的路由 URL 模式添加回退值。可以在操作或页面处理程序方法中定义参数时分配默认值，这是本书的这一部分所采用的方法。或者可以简单地编写容纳默认值而不导致错误的方法，如代码清单 28-8 所示。

**代码清单 28-8　在 Controllers 文件夹的 FormController.cs 文件中避免查询错误**

```
...
public async Task<IActionResult> Index(long id) {
    ViewBag.Categories = new SelectList(context.Categories, "CategoryId", "Name");
    return View("Form", await context.Products.Include(p => p.Category)
        .Include(p => p.Supplier).FirstOrDefaultAsync(p => p.ProductId == id));
}
...
```

Entity Framework Core 的 FirstOrDefaultAsync 方法在数据库中没有匹配对象时将返回 null，并且不会尝试加载相关数据。标签助手处理空值和显示空字段，通过重新启动 ASP.NET Core，请求 http://localhost:5000/controllers/Form，会生成如图 28-6 所示的结果。

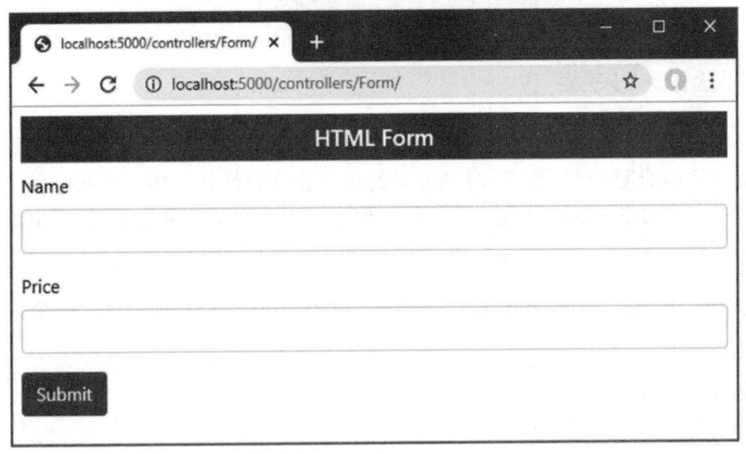


图 28-6　避免错误

一些应用程序需要区分丢失的值和用户提供的任何值。在这些情况下，可以使用可空的参数类型，如代码清单 28-9 所示。

**代码清单 28-9　在 Controllers 文件夹的 FormController.cs 文件中使用可空参数**

```
...
public async Task<IActionResult> Index(long? id) {
    ViewBag.Categories = new SelectList(context.Categories, "CategoryId", "Name");
```

```
        return View("Form", await context.Products.Include(p => p.Category)
            .Include(p => p.Supplier)
            .FirstOrDefaultAsync(p => id == null || p.ProductId == id));
    }
    ...
```

只有当请求不包含合适的值时，id 参数才为 null，这允许传递给 FirstOrDefaultAsync 方法的表达式在没有值时默认使用数据库中的第一个对象，并查询任何其他值。要查看效果，请重新启动 ASP.NET Core，并请求 http://localhost:5000/controllers/Form 和 http://localhost:5000/controllers/formation/index/0。第一个 URL 不包含 id 值，因此选择数据库中的第一个对象。第二个 URL 提供了一个 id 值 0，与数据库中的任何对象都不对应。图 28-7 显示了这两种结果。

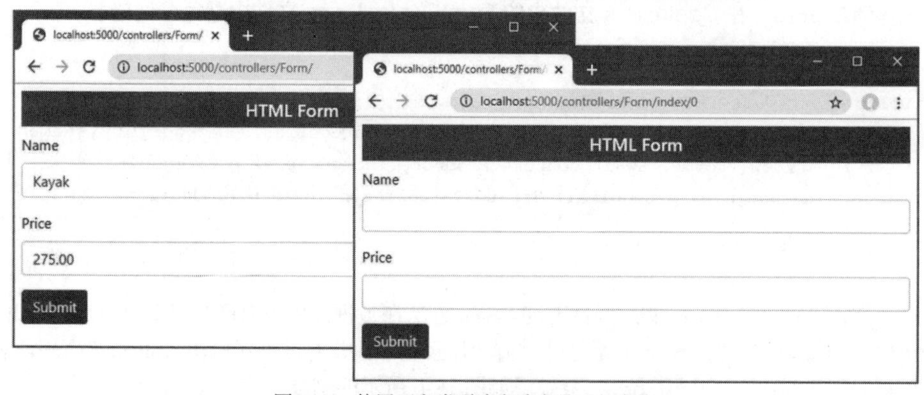


图 28-7　使用可空类型确定请求是否包含值

## 28.4　绑定复杂类型

模型绑定系统在处理复杂类型时非常出色，复杂类型是不能从单个字符串值解析的任何类型。模型绑定流程检查复杂类型，并对其定义的每个公共属性执行绑定流程。这意味着，可以使用绑定器创建完整的 Product 对象，而不是处理 name 和 price 等单独的值，如代码清单 28-10 所示。

**代码清单 28-10　在 Controllers 文件夹的 FormController.cs 文件中绑定一个复杂类型**

```
using Microsoft.AspNetCore.Mvc;
using System.Linq;
using System.Threading.Tasks;
using WebApp.Models;
using Microsoft.EntityFrameworkCore;
using Microsoft.AspNetCore.Mvc.Rendering;

namespace WebApp.Controllers {

    [AutoValidateAntiforgeryToken]
    public class FormController : Controller {
        private DataContext context;
```

```
        public FormController(DataContext dbContext) {
            context = dbContext;
        }

        public async Task<IActionResult> Index(long? id) {
            ViewBag.Categories
                = new SelectList(context.Categories, "CategoryId", "Name");
            return View("Form", await context.Products.Include(p => p.Category)
                .Include(p => p.Supplier)
                .FirstOrDefaultAsync(p => id == null || p.ProductId == id));
        }

        [HttpPost]
        public IActionResult SubmitForm(Product product) {
            TempData["product"] = System.Text.Json.JsonSerializer.Serialize(product);
            return RedirectToAction(nameof(Results));
        }

        public IActionResult Results() {
            return View(TempData);
        }
    }
}
```

代码清单更改 SubmitForm 操作方法，以便它定义 Product 参数。在调用操作方法之前，将创建一个新的 Product 对象，并将模型绑定流程应用于其每个公共属性。然后使用 Product 对象作为参数调用 SubmitForm 方法。

要查看模型绑定过程，请重新启动 ASP.NET Core，导航到 http://localhost:5000/controllers/Form，然后单击 Submit 按钮。模型绑定流程将从请求中提取数据值，并生成图 28-8 所示的结果。由模型绑定过程创建的 Product 对象序列化为 JSON 数据，这样就可以将其存储为临时数据，从而便于查看请求数据。

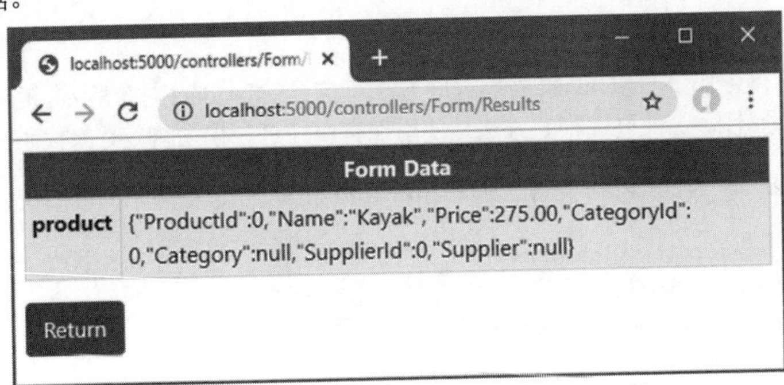


图 28-8　数据绑定复杂类型

复杂类型的数据绑定过程仍然是一项最佳功能，这意味着为 Product 类定义的每个公共属性寻找一个值，但缺少值不会阻止调用操作方法。相反，无法找到其值的属性将使用该属性类型的默认值。示例提供了 Name 和 Price 属性的值，但是 ProductId、CategoryId 和 SupplierId 属性为 0，而 Category 和 Supplier 属性为空。

### 28.4.1 绑定到属性

使用参数进行模型绑定不适合 Razor 页面(Razor Pages)开发风格，因为参数经常重复页面模型类定义的属性，如代码清单 28-11 所示。

**代码清单 28-11 在 Pages 文件夹的 FormHandler.cshtml 文件中绑定复杂类型**

```
...
@functions {

    public class FormHandlerModel : PageModel {
        private DataContext context;

        public FormHandlerModel(DataContext dbContext) {
            context = dbContext;
        }

        public Product Product { get; set; }

        public async Task OnGetAsync(long id = 1) {
            Product = await context.Products.Include(p => p.Category)
                .Include(p => p.Supplier).FirstAsync(p => p.ProductId == id);
        }

        public IActionResult OnPost(Product product) {
            TempData["product"] = System.Text.Json.JsonSerializer.Serialize(product);
            return RedirectToPage("FormResults");
        }
    }
}
...
```

这段代码可以工作，但是 OnPost 处理程序方法有自己的 Product 对象版本，镜像由 OnGetAsync 处理程序使用的属性。一种更优雅的方法是使用现有属性进行模型绑定，如代码清单 28-12 所示。

**代码清单 28-12 在 Pages 文件夹的 FormHandler.cshtml 文件中为模型绑定使用属性**

```
@page "/pages/form/{id:long?}"
@model FormHandlerModel
@using Microsoft.AspNetCore.Mvc.RazorPages
@using Microsoft.EntityFrameworkCore

<div class="m-2">
    <h5 class="bg-primary text-white text-center p-2">HTML Form</h5>
    <form asp-page="FormHandler" method="post" id="htmlform">
```

```
        <div class="form-group">
            <label>Name</label>
            <input class="form-control" asp-for="Product.Name" />
        </div>
        <div class="form-group">
            <label>Price</label>
            <input class="form-control" asp-for="Product.Price" />
        </div>
        <button type="submit" class="btn btn-primary">Submit</button>
    </form>
</div>

@functions {

    public class FormHandlerModel : PageModel {
        private DataContext context;

        public FormHandlerModel(DataContext dbContext) {
            context = dbContext;
        }

        [BindProperty]
        public Product Product { get; set; }

        public async Task OnGetAsync(long id = 1) {
            Product = await context.Products.Include(p => p.Category)
                .Include(p => p.Supplier).FirstAsync(p => p.ProductId == id);
        }

        public IActionResult OnPost() {
            TempData["product"] = System.Text.Json.JsonSerializer.Serialize(Product);
            return RedirectToPage("FormResults");
        }
    }
}
```

用 BindProperty 修饰属性表明它的属性应该服从模型绑定过程，这意味着 OnPost 处理程序方法可以在不声明参数的情况下获得它需要的数据。当使用 BindProperty 特性时，模型绑定器在定位数据值时使用属性名，因此不需要添加到输入元素的显式 name 特性。默认情况下，BindProperty 不会绑定 GET 请求的数据，但这可以通过将 BindProperty 特性的 SupportsGet 参数设置为 true 来更改。

---

■ **注意：**BindProperties 特性可应用于需要模型绑定过程的类，这些类需要对它们定义的所有 public 属性进行模型绑定，这比将 BindProperty 应用于许多单独的属性要方便得多。使用 BindNever 修饰属性，以将它们从模型绑定中排除。

---

### 28.4.2　绑定嵌套的复杂类型

如果使用复杂类型来定义受模型绑定约束的属性，则使用属性名作为前缀重复模型绑定过程。例如，Product 类定义 Category 属性，其类型是复杂的 Category 类型。代码清单 28-13 向 HTML 表单添加了一些元素，以向模型绑定器提供 Category 类定义的属性值。

**代码清单 28-13　在 Views/Form 文件夹的 Form.cshtml 文件中添加嵌套的表单元素**

```
@model Product
@{ Layout = "_SimpleLayout"; }

<h5 class="bg-primary text-white text-center p-2">HTML Form</h5>

<form asp-action="submitform" method="post" id="htmlform">
    <div class="form-group">
        <label asp-for="Name"></label>
        <input class="form-control" asp-for="Name" />
    </div>
    <div class="form-group">
        <label asp-for="Price"></label>
        <input class="form-control" asp-for="Price" />
    </div>
    <div class="form-group">
        <label>Category Name</label>
        <input class="form-control" name="Category.Name"
           value="@Model.Category.Name" />
    </div>
    <button type="submit" class="btn btn-primary">Submit</button>
</form>
```

name 特性组合了由句点分隔的属性名称。在本例中，元素用于给视图模型的 Category 属性指定的对象的 Name 属性，因此 Name 属性设置为 Category.Name。当应用 asp-for 特性时，输入元素标签助手将自动为 name 特性使用这种格式，如代码清单 28-14 所示。

**代码清单 28-14　在 Views/Form 文件夹的 Form.cshtml 文件中使用标签助手**

```
@model Product
@{ Layout = "_SimpleLayout"; }

<h5 class="bg-primary text-white text-center p-2">HTML Form</h5>

<form asp-action="submitform" method="post" id="htmlform">
    <div class="form-group">
        <label asp-for="Name"></label>
        <input class="form-control" asp-for="Name" />
    </div>
    <div class="form-group">
        <label asp-for="Price"></label>
```

```
        <input class="form-control" asp-for="Price" />
    </div>
    <div class="form-group">
        <label>Category Name</label>
        <input class="form-control" asp-for="Category.Name" />
    </div>
    <button type="submit" class="btn btn-primary">Submit</button>
</form>
```

标签助手是为嵌套属性创建元素的更可靠方法，可避免产生被模型绑定过程忽略的输入错误的风险。要查看新元素的效果，请求 http://localhost:5000/controllers/Form 并单击 Submit 按钮，生成如图 28-9 所示的响应。

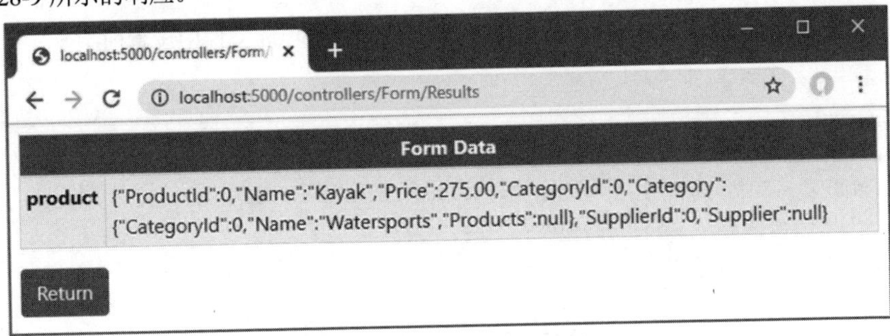


图 28-9　模型绑定嵌套的属性

在模型绑定过程中，创建一个新的 Category 对象，并将其分配给 Product 对象的 Category 属性。从图中可看到，模型绑定器定位 Category 对象的 Name 属性值，但是没有 CategoryId 属性的值；CategoryId 属性使用默认值。

### 为嵌套复杂类型指定自定义前缀

有些情况下，生成的 HTML 与一种类型的对象相关，但希望将其绑定到另一种类型。这意味着包含视图的前缀不会与模型绑定器所期望的结构相对应，数据也不会得到正确处理。代码清单 28-15 通过更改控制器的 SubmitForm 操作方法定义的参数类型来演示这个问题。

**代码清单 28-15　在 Controllers 文件夹的 FormController.cs 文件中更改参数**

```
...
[HttpPost]
public IActionResult SubmitForm(Category category) {
    TempData["category"] = System.Text.Json.JsonSerializer.Serialize(category);
    return RedirectToAction(nameof(Results));
}
...
```

新参数是一个 Category，但是模型绑定过程不能正确地挑选出数据值，即使表单视图发送的表单数据将包含 Category 对象的 Name 属性值也同样如此。相反，模型绑定器会找到 Product 对象的 Name 值并使用它，重新启动 ASP.NET Core，请求 http://localhost:5000/controllers/Form，并提交表单数据，这将生成第一个响应，如图 28-10 所示。

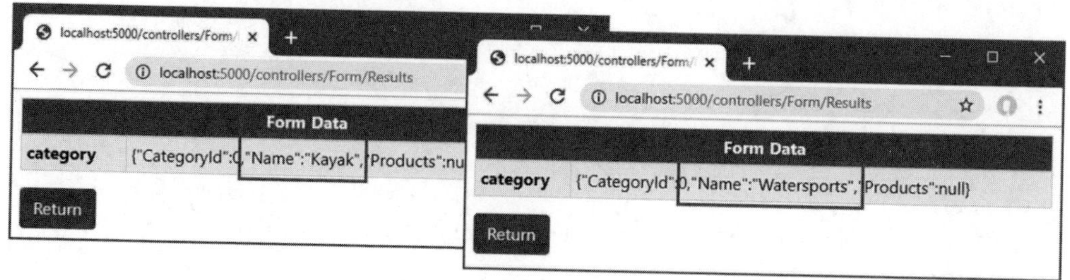


图 28-10　指定模型绑定前缀

通过将 Bind 属性应用到参数，并使用 Prefix 参数为模型绑定器指定一个前缀，可以解决这个问题，如代码清单 28-16 所示。

#### 代码清单 28-16　在 Controllers 文件夹的 FormController.cs 文件中设置前缀

```
...
[HttpPost]
public IActionResult SubmitForm([Bind(Prefix ="Category")] Category category) {
    TempData["category"] = System.Text.Json.JsonSerializer.Serialize(category);
    return RedirectToAction(nameof(Results));
}
...
```

虽然语法很笨拙，但该属性确保模型绑定器能够定位操作方法所需的数据。在这种情况下，将前缀设置为 Category 可确保使用正确的数据值来绑定 Category 参数。重启 ASP.NET Core，请求 http://localhost:5000/controllers/Form，并提交表单，表单将生成第二个响应，如图 28-10 所示。

在使用 BindProperty 特性时，用 Name 参数指定前缀，如代码清单 28-17 所示。

#### 代码清单 28-17　在 Pages 文件夹的 FormHandler.cshtml 文件中指定模型绑定前缀

```
@page "/pages/form/{id:long?}"
@model FormHandlerModel
@using Microsoft.AspNetCore.Mvc.RazorPages
@using Microsoft.EntityFrameworkCore

<div class="m-2">
    <h5 class="bg-primary text-white text-center p-2">HTML Form</h5>
    <form asp-page="FormHandler" method="post" id="htmlform">
        <div class="form-group">
            <label>Name</label>
            <input class="form-control" asp-for="Product.Name" />
        </div>
        <div class="form-group">
            <label>Price</label>
            <input class="form-control" asp-for="Product.Price" />
        </div>
        <div class="form-group">
```

```
            <label>Category Name</label>
            <input class="form-control" asp-for="Product.Category.Name" />
        </div>
        <button type="submit" class="btn btn-primary">Submit</button>
    </form>
</div>

@functions {

    public class FormHandlerModel : PageModel {
        private DataContext context;

        public FormHandlerModel(DataContext dbContext) {
            context = dbContext;
        }

        [BindProperty]
        public Product Product { get; set; }

        [BindProperty(Name = "Product.Category")]
        public Category Category { get; set; }

        public async Task OnGetAsync(long id = 1) {
            Product = await context.Products.Include(p => p.Category)
                .Include(p => p.Supplier).FirstAsync(p => p.ProductId == id);
        }

        public IActionResult OnPost() {
            TempData["product"] = System.Text.Json.JsonSerializer.Serialize(Product);
            TempData["category"]
                = System.Text.Json.JsonSerializer.Serialize(Category);
            return RedirectToPage("FormResults");
        }
    }
}
```

此代码清单添加了一个使用 asp-for 特性选择 Product.Category 属性的输入元素。页面处理程序类定义了一个用 BindProperty 特性修饰并用 Name 参数配置的 Category 属性。要查看模型绑定过程的结果，请使用浏览器请求 http://localhost:5000/pages/Form 并单击 Submit 按钮。模型绑定为这两个修饰属性查找值，从而生成如图 28-11 所示的响应。

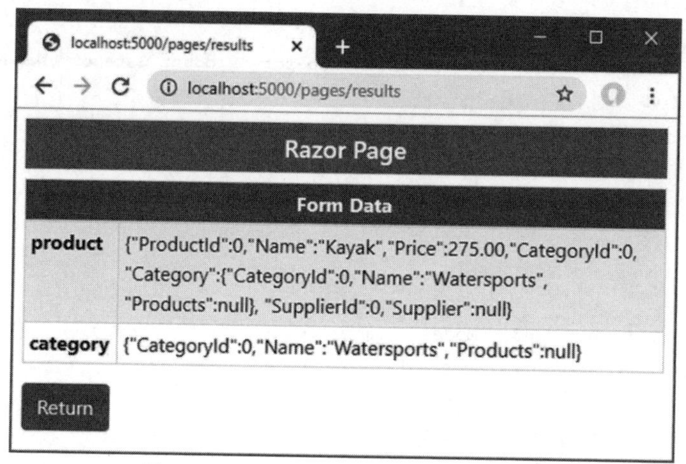


图 28-11　在 Razor Pages 中指定模型绑定前缀

### 28.4.3　选择性的绑定属性

一些模型类定义了一些敏感的属性，用户不应该为这些属性指定值。例如，用户可以更改 Product 对象的类别，但不应该更改价格。

可能会试图简单地创建视图，省略敏感属性的 HTML 元素，但这不能阻止恶意用户创建包含值的 HTTP 请求，这称为过度绑定攻击。为了防止模型绑定器使用敏感属性的值，可以指定应该绑定的属性列表，如代码清单 28-18 所示。

**代码清单 28-18　在 Controllers 文件夹的 FormController.cs 文件中有选择地绑定属性**

```
using Microsoft.AspNetCore.Mvc;
using System.Linq;
using System.Threading.Tasks;
using WebApp.Models;
using Microsoft.EntityFrameworkCore;
using Microsoft.AspNetCore.Mvc.Rendering;

namespace WebApp.Controllers {

    [AutoValidateAntiforgeryToken]
    public class FormController : Controller {
        private DataContext context;

        public FormController(DataContext dbContext) {
            context = dbContext;
        }

        public async Task<IActionResult> Index(long? id) {
            ViewBag.Categories
                = new SelectList(context.Categories, "CategoryId", "Name");
```

```
            return View("Form", await context.Products.Include(p => p.Category)
                .Include(p => p.Supplier)
                .FirstOrDefaultAsync(p => id == null || p.ProductId == id));
        }

        [HttpPost]
        public IActionResult SubmitForm([Bind("Name", "Category")] Product product) {
            TempData["name"] = product.Name;
            TempData["price"] = product.Price.ToString();
            TempData["category name"] = product.Category.Name;
            return RedirectToAction(nameof(Results));
        }

        public IActionResult Results() {
            return View(TempData);
        }
    }
}
```

为操作方法参数返回了 Product 类型，该参数用 Bind 特性修饰，以指定应该包含在模型绑定过程中的属性名称。这个示例告诉模型绑定特性寻找 Name 和 Category 属性的值，这将从流程中排除其他任何属性。重启 ASP.NET Core，导航到 http://localhost:5000/controller/form，并提交表单。即使浏览器将 Price 属性的值作为 HTTP POST 请求的一部分发送，也会被模型绑定器忽略，如图 28-12 所示。

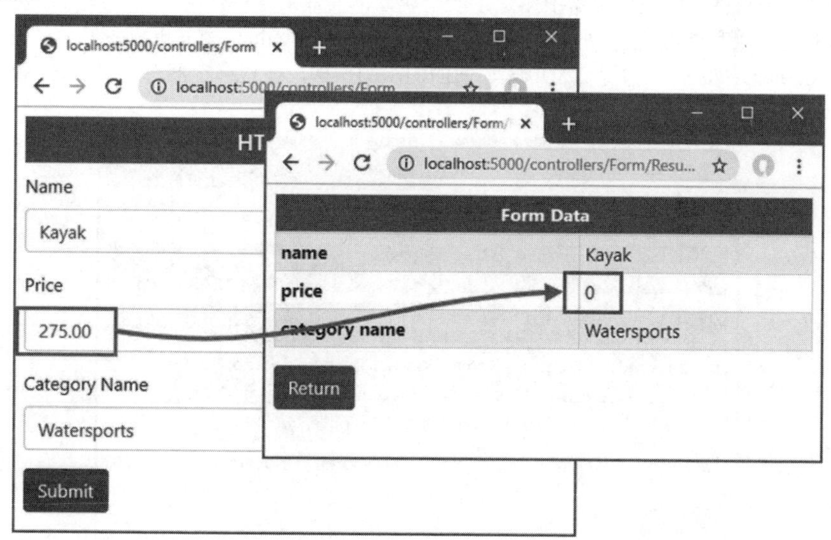


图 28-12　有选择性的绑定属性

### 有选择地绑定模型类

如果正在使用 Razor Pages，或者希望在整个应用程序中使用相同的一组属性进行模型绑定，那么可将 BindNever 特性直接应用到模型类，如代码清单 28-19 所示。

**代码清单 28-19 在 Models 文件夹的 Product.cs 文件中修饰属性**

```
using System.ComponentModel.DataAnnotations.Schema;
using System.ComponentModel.DataAnnotations;
using Microsoft.AspNetCore.Mvc.ModelBinding;

namespace WebApp.Models {
    public class Product {

        public long ProductId { get; set; }

        public string Name { get; set; }

        [Column(TypeName = "decimal(8, 2)")]
        [BindNever]
        public decimal Price { get; set; }

        public long CategoryId { get; set; }
        public Category Category { get; set; }

        public long SupplierId { get; set; }
        public Supplier Supplier { get; set; }
    }
}
```

BindNever 特性从模型绑定器中排除了一个属性，其效果与将它从上一节使用的列表中删除相同。要查看效果，请重新启动 ASP.NET Core，因此，对 Product 类的更改生效后，请求 http://localhost:5000/pages/form，并提交表单。与前面的示例一样，模型绑定器忽略了 Price 属性的值，如图 28-13 所示。

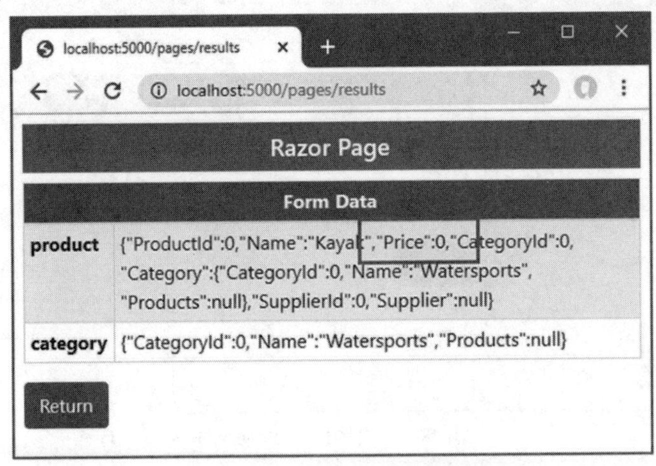


图 28-13 从模型绑定中排除属性

**提示：** 还有一个BindRequired属性，它告诉模型绑定过程，请求必须包含属性的值。如果请求没有必需的值，则会产生模型验证错误，如第29章所述。

# 28.5 绑定到数组和集合

模型绑定过程有一些很好的特性，可将请求数据绑定到数组和集合，参见下面的内容。

## 28.5.1 绑定到数组

默认模型绑定的一个优雅特性是它如何支持数组。要查看此特性的工作原理，给 Pages 文件夹添加一个名为 Bindings.cshtml 的 Razor Pages，内容如代码清单 28-20 所示。

**代码清单 28-20 Pages 文件夹中 Bindings.cshtml 文件的内容**

```
@page "/pages/bindings"
@model BindingsModel
@using Microsoft.AspNetCore.Mvc
@using Microsoft.AspNetCore.Mvc.RazorPages

<div class="container-fluid">
    <div class="row">
        <div class="col">
            <form asp-page="Bindings" method="post">
                <div class="form-group">
                    <label>Value #1</label>
                    <input class="form-control" name="Data" value="Item 1" />
                </div>
                <div class="form-group">
                    <label>Value #2</label>
                    <input class="form-control" name="Data" value="Item 2" />
                </div>
                <div class="form-group">
                        <label>Value #3</label>
                    <input class="form-control" name="Data" value="Item 3" />
                </div>
                <button type="submit" class="btn btn-primary">Submit</button>
                <a class="btn btn-secondary" asp-page="Bindings">Reset</a>
            </form>
        </div>
        <div class="col">
            <ul class="list-group">
                @foreach (string s in Model.Data.Where(s => s != null)) {
                   <li class="list-group-item">@s</li>
                }
            </ul>
        </div>
    </div>
</div>

@functions {
```

```
    public class BindingsModel : PageModel {

        [BindProperty(Name = "Data")]
        public string[] Data { get; set; } = Array.Empty<string>();
    }
}
```

数组的模型绑定需要将 name 特性设置为将提供数组值的所有元素的相同值。这个页面显示三个输入元素，它们的 name 特性值都是 Data。为让模型绑定器找到数组值，用 BindProperty 特性装饰了页面模型的 Data 属性，并使用了 Name 参数。

---

**提示：**代码清单 28-20 中的页面模型类没有定义处理程序方法。这很不寻常，但很有效，任何请求都不需要显式处理，因为请求只提供 Data 数组的值并显示它。

---

提交 HTML 表单时，将创建一个新数组，并使用来自所有三个输入元素的值填充该数组，这些值将显示给用户。要查看绑定过程，请求 http://localhost:5000/pages/bindings，编辑表单字段，然后单击 Submit 按钮。Data 数组的内容使用@foreach 表达式显示在列表中，如图 28-14 所示。

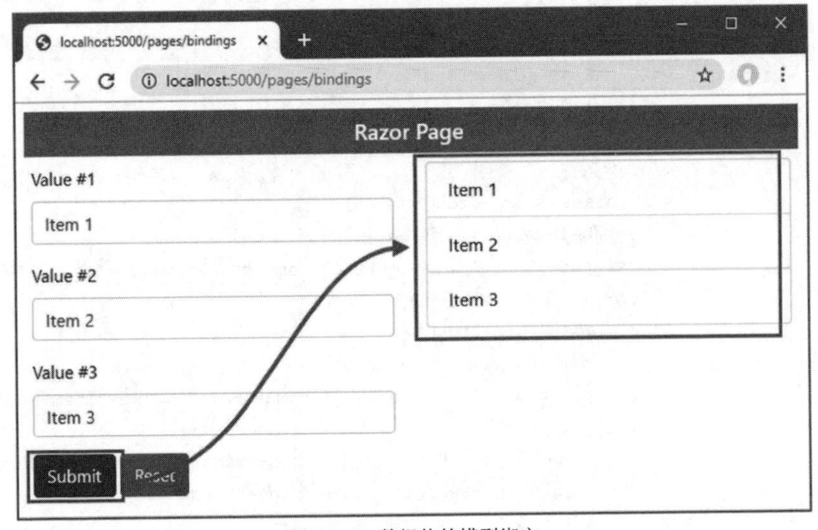


图 28-14　数组值的模型绑定

注意，在显示数组内容时，过滤掉了 null 值。

```
...
@foreach (string s in Model.Data.Where(s => s != null)) {
    <li class="list-group-item">@s</li>
}
...
```

空表单字段在数组中生成空值，这里不在结果中显示这些值。在第 29 章中，将展示如何确保为模型绑定属性提供了值。

### 指定数组值的索引位置

默认情况下，数组是按照从浏览器接收表单值的顺序填充的，这个顺序通常是定义 HTML 元素的顺序。如果需要覆盖默认值，则可以使用 name 特性指定数组中值的位置，如代码清单 28-21 所示。

**代码清单 28-21 在 Pages 文件夹的 Bindings.cshtml 文件中指定数组位置**

```
@page "/pages/bindings"
@model BindingsModel
@using Microsoft.AspNetCore.Mvc
@using Microsoft.AspNetCore.Mvc.RazorPages

<div class="container-fluid">
    <div class="row">
        <div class="col">
            <form asp-page="Bindings" method="post">
                <div class="form-group">
                    <label>Value #1</label>
                    <input class="form-control" name="Data[1]" value="Item 1" />
                </div>
                <div class="form-group">
                    <label>Value #2</label>
                    <input class="form-control" name="Data[0]" value="Item 2" />
                </div>
                <div class="form-group">
                        <label>Value #3</label>
                    <input class="form-control" name="Data[2]" value="Item 3" />
                </div>
                <button type="submit" class="btn btn-primary">Submit</button>
                <a class="btn btn-secondary" asp-page="Bindings">Reset</a>
            </form>
        </div>
        <div class="col">
            <ul class="list-group">
                @foreach (string s in Model.Data.Where(s => s != null)) {
                    <li class="list-group-item">@s</li>
                }
            </ul>
        </div>
    </div>
</div>

@functions {

    public class BindingsModel : PageModel {

        [BindProperty(Name = "Data")]
```

```
        public string[] Data { get; set; } = Array.Empty<string>();
    }
}
```

数组索引表示法用于指定值在数据绑定数组中的位置。使用浏览器请求 http:/localhost:5000/pages/bindings 并提交表单，将看到条目按照 name 特性指示的顺序出现，如图 28-15 所示。索引表示法必须应用于提供数组值的所有 HTML 元素，编号序列中不能有任何空白。

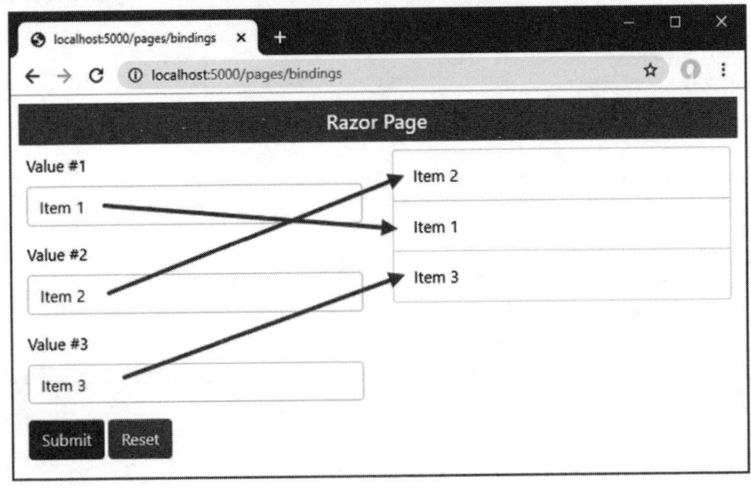


图 28-15　指定数组的位置

## 28.5.2　绑定到简单集合

模型绑定流程可以创建集合和数组。对于序列集合，例如列表和集合，只更改模型绑定器使用的属性或参数的类型，如代码清单 28-22 所示。

### 代码清单 28-22　在 Pages 文件夹的 Bindings.cshtml 文件中绑定到列表

```
@page "/pages/bindings"
@model BindingsModel
@using Microsoft.AspNetCore.Mvc
@using Microsoft.AspNetCore.Mvc.RazorPages

<div class="container-fluid">
    <div class="row">
        <div class="col">
            <form asp-page="Bindings" method="post">
                <div class="form-group">
                    <label>Value #1</label>
                    <input class="form-control" name="Data[1]" value="Item 1" />
                </div>
                <div class="form-group">
                    <label>Value #2</label>
                    <input class="form-control" name="Data[0]" value="Item 2" />
```

```
                </div>
                <div class="form-group">
                        <label>Value #3</label>
                    <input class="form-control" name="Data[2]" value="Item 3" />
                </div>
                <button type="submit" class="btn btn-primary">Submit</button>
                <a class="btn btn-secondary" asp-page="Bindings">Reset</a>
            </form>
        </div>
        <div class="col">
            <ul class="list-group">
                @foreach (string s in Model.Data.Where(s => s != null)) {
                    <li class="list-group-item">@s</li>
                }
            </ul>
        </div>
    </div>
</div>

@functions {

    public class BindingsModel : PageModel {

        [BindProperty(Name = "Data")]
        public SortedSet<string> Data { get; set; } = new SortedSet<string>();
    }
}
```

将 Data 属性的类型改为 SortedSet<string>。模型绑定过程用来自输入元素的值填充集合，这些值按字母顺序排序。在输入元素名称属性上保留了索引符号，但是它们没有任何作用，因为集合类按字母顺序对其值进行排序。要查看效果，请使用浏览器请求 http://localhost:5000/pages/bindings，编辑文本字段，然后单击 Submit 按钮。模型绑定过程用表单值填充排序的集合，表单值按顺序显示，如图 28-16 所示。

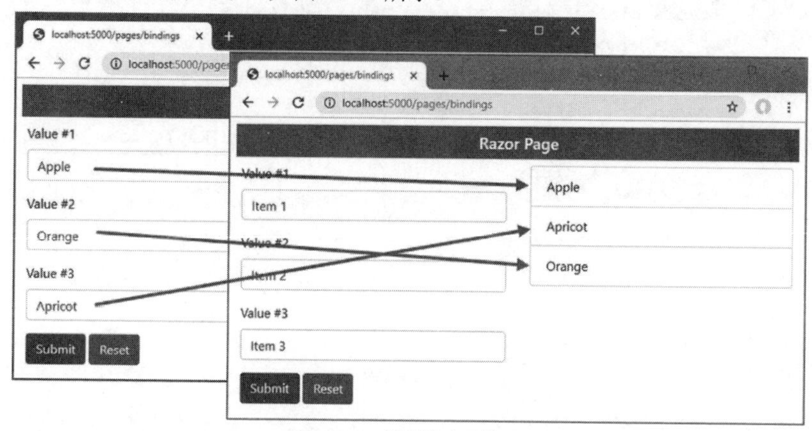


图 28-16 模型绑定到集合

### 28.5.3　绑定到字典

对于使用索引符号表示其 name 特性的元素，模型绑定器在绑定到字典时使用索引作为键，从而允许将一系列元素转换为键/值对，如代码清单 28-23 所示。

**代码清单 28-23　在 Pages 文件夹的 Bindings.cshtml 文件中绑定到字典**

```
@page "/pages/bindings"
@model BindingsModel
@using Microsoft.AspNetCore.Mvc
@using Microsoft.AspNetCore.Mvc.RazorPages

<div class="container-fluid">
    <div class="row">
        <div class="col">
            <form asp-page="Bindings" method="post">
                <div class="form-group">
                    <label>Value #1</label>
                    <input class="form-control" name="Data[first]" value="Item 1" />
                </div>
                <div class="form-group">
                    <label>Value #2</label>
                    <input class="form-control" name="Data[second]" value="Item 2" />
                </div>
                <div class="form-group">
                        <label>Value #3</label>
                    <input class="form-control" name="Data[third]" value="Item 3" />
                </div>
                <button type="submit" class="btn btn-primary">Submit</button>
                <a class="btn btn-secondary" asp-page="Bindings">Reset</a>
            </form>
        </div>
        <div class="col">
            <table class="table table-sm table-striped">
                <tbody>
                    @foreach (string key in Model.Data.Keys) {
                        <tr>
                            <th>@key</th><td>@Model.Data[key]</td>
                        </tr>
                    }
                </tbody>
            </table>
        </div>
    </div>
</div>

@functions {
```

```
    public class BindingsModel : PageModel {

        [BindProperty(Name = "Data")]
        public Dictionary<string, string> Data { get; set; }
           = new Dictionary<string, string>();
    }
}
```

为集合提供值的所有元素都必须共享一个公共前缀(在本例中为 Data)，后面跟着方括号中的键值。本例中的键是第一个、第二个和第三个字符串，用作用户在文本字段中提供的内容的键。要查看绑定过程，请求 http://localhost:5000/pages/bindings，编辑文本字段，并提交表单。表单数据的键和值显示在一个表中，如图 28-17 所示。

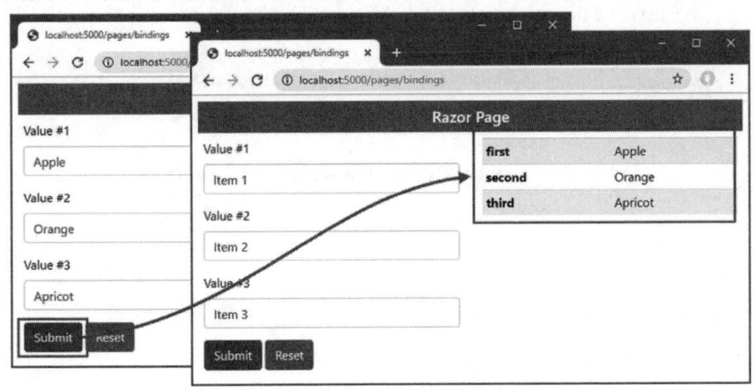


图 28-17 模型绑定到字典

## 28.5.4 绑定到复杂类型的集合

本节中的示例都是简单类型的集合，但同样的过程也可以用于复杂类型。为便于演示，代码清单 28-24 修改了 Razor Pages，以收集用于绑定到一个 Product 对象数组的详细信息。

### 代码清单 28-24 在 Pages 文件夹的 Bindings.cshtml 文件中绑定到复杂类型

```
@page "/pages/bindings"
@model BindingsModel
@using Microsoft.AspNetCore.Mvc
@using Microsoft.AspNetCore.Mvc.RazorPages

<div class="container-fluid">
    <div class="row">
        <div class="col">
            <form asp-page="Bindings" method="post">
                @for (int i = 0; i < 2; i++) {
                    <div class="form-group">
                        <label>Name #@i</label>
                        <input class="form-control" name="Data[@i].Name"
```

```
                            value="Product-@i" />
                    </div>
                    <div class="form-group">
                        <label>Price #@i</label>
                        <input class="form-control" name="Data[@i].Price"
                            value="@(100 + i)" />
                    </div>
                }
                <button type="submit" class="btn btn-primary">Submit</button>
                <a class="btn btn-secondary" asp-page="Bindings">Reset</a>
            </form>
        </div>
        <div class="col">
            <table class="table table-sm table-striped">
                <tbody>
                    <tr><th>Name</th><th>Price</th></tr>
                    @foreach (Product p in Model.Data) {
                        <tr>
                            <td>@p.Name</td><td>@p.Price</td>
                        </tr>
                    }
                </tbody>
            </table>
        </div>
    </div>
</div>

@functions {

    public class BindingsModel : PageModel {

        [BindProperty(Name = "Data")]
        public Product[] Data { get; set; } = Array.Empty<Product>();
    }
}
```

input 元素的 name 特性使用数组表示法，后跟句点，再跟它们所表示的复杂类型属性的名称。要定义 Name 和 Price 属性的元素，需要如下元素：

```
...
<input class="form-control" name="Data[0].Name" />
...
<input class="form-control" name="Data[0].Price" />
...
```

在绑定过程中，模型绑定器尝试定位目标类型定义的所有 public 属性的值，对表单数据中的每一组值重复此过程。

此示例依赖于 Product 类定义的 Price 属性的模型绑定，该属性在使用 BindNever 特性的绑定过程中排除。从属性中删除该特性，如代码清单 28-25 所示。

**代码清单 28-25　在 Models 文件夹的 Product.cs 文件中删除一个属性**

```
using System.ComponentModel.DataAnnotations.Schema;
using System.ComponentModel.DataAnnotations;
using Microsoft.AspNetCore.Mvc.ModelBinding;

namespace WebApp.Models {
    public class Product {

        public long ProductId { get; set; }

        public string Name { get; set; }

        [Column(TypeName = "decimal(8, 2)")]
        //[BindNever]
        public decimal Price { get; set; }

        public long CategoryId { get; set; }
        public Category Category { get; set; }

        public long SupplierId { get; set; }
        public Supplier Supplier { get; set; }
    }
}
```

重启 ASP.NET Core，对 Product 类的更改生效，并使用浏览器请求 http://localhost:5000/pages/bindings。在文本字段中输入名称和价格并提交表单，会看到根据表中显示的数据创建的 Product 对象的详细信息，如图 28-18 所示。

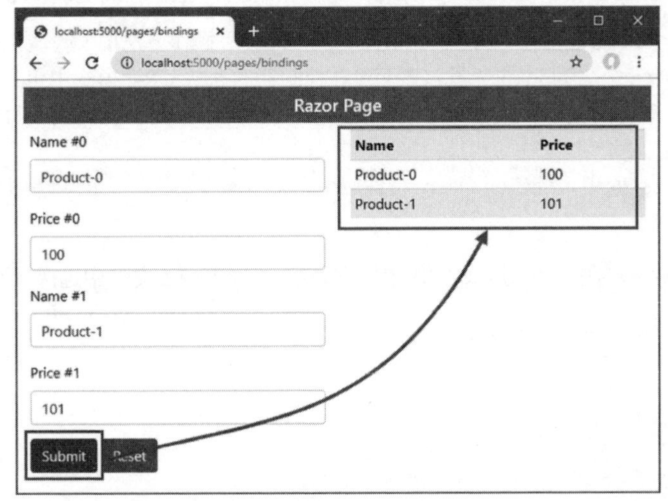


图 28-18　绑定到复杂类型的集合

## 28.6　指定模型绑定源

如本章开头所述，默认的模型绑定过程在四个地方查找数据：表单数据值、请求主体(仅针对 Web 服务控制器)、路由数据和请求查询字符串。

默认的搜索序列并不总是有用的，因为总是希望数据来自请求的特定部分，或者希望使用默认情况下没有搜索的数据源。模型绑定功能包括一组用于覆盖默认搜索行为的特性，如表 28-3 所示。

**提示：** 还有一个 FromService 特性，它不会从请求中获取值，而是使用第 14 章中描述的依赖注入功能。

表 28-3　模型绑定源特性

| 名称 | 描述 |
| --- | --- |
| FromForm | 该属性用于选择表单数据作为绑定数据的源。默认情况下，参数的名称用于定位表单值，但可以使用 Name 属性更改此值，该属性允许指定不同的名称 |
| FromRoute | 该属性用于选择作为绑定数据源的路由系统。默认情况下，参数的名称用于定位路由数据值，但可以使用 Name 属性进行更改，该属性允许指定不同的名称 |
| FromQuery | 该属性用于选择查询字符串作为绑定数据的源。默认情况下，参数的名称用于定位查询字符串值，但是可以使用 name 属性更改该值，该属性允许指定不同的查询字符串键 |
| FromHeader | 该属性用于选择一个请求头作为绑定数据的源。默认情况下，参数的名称用作标题名称，但是可以使用 Name 属性进行更改，该属性允许指定不同的标题名称 |
| FromBody | 该属性用于指定应该将请求体用作绑定数据的源，这在希望从未进行表单编码的请求(例如在提供 Web 服务的 API 控制器中)接收数据时是必需的 |

FromForm、FromRoute 和 FromQuery 属性允许指定从一个标准位置获取模型绑定数据，但不需要遵循正常的搜索顺序。本章前面使用了这个 URL：

```
http://localhost:5000/controllers/Form/Index/5?id=1
```

此 URL 包含两个可能的值，可用于表单控制器上的索引操作方法的 id 参数。路由系统把 URL 的最后一段分配给一个名为 id 的变量，该变量在控制器的默认 URL 模式中定义，查询字符串也包含一个 id 值。默认的搜索模式意味着将从路由数据中获取模型绑定数据，而查询字符串将被忽略。

在代码清单 28-26 中，将 FromQuery 特性应用于索引操作方法定义的 id 参数，该方法覆盖默认搜索序列。

#### 代码清单 28-26　在 Controllers 文件夹的 FormController.cs 文件中选择查询字符串

```
using Microsoft.AspNetCore.Mvc;
using System.Linq;
using System.Threading.Tasks;
using WebApp.Models;
using Microsoft.EntityFrameworkCore;
using Microsoft.AspNetCore.Mvc.Rendering;
```

```
namespace WebApp.Controllers {

    [AutoValidateAntiforgeryToken]
    public class FormController : Controller {
        private DataContext context;

        public FormController(DataContext dbContext) {
            context = dbContext;
        }

        public async Task<IActionResult> Index([FromQuery] long? id) {
            ViewBag.Categories
                = new SelectList(context.Categories, "CategoryId", "Name");
            return View("Form", await context.Products.Include(p => p.Category)
                .Include(p => p.Supplier)
                .FirstOrDefaultAsync(p => id == null || p.ProductId == id));
        }

        [HttpPost]
        public IActionResult SubmitForm([Bind("Name", "Category")] Product product) {
            TempData["name"] = product.Name;
            TempData["price"] = product.Price.ToString();
            TempData["category name"] = product.Category.Name;
            return RedirectToAction(nameof(Results));
        }

        public IActionResult Results() {
            return View(TempData);
        }
    }
}
```

该特性指定了模型绑定过程的源，重新启动 ASP.NET Core，并使用浏览器请求 http://localhost:5000/controllers/form/index/5?id=1，就可以看到该属性。不是使用路由系统匹配的值，而是使用查询字符串，生成如图 28-19 所示的响应。如果查询字符串不包含适合模型绑定过程的值，则不会使用其他位置。

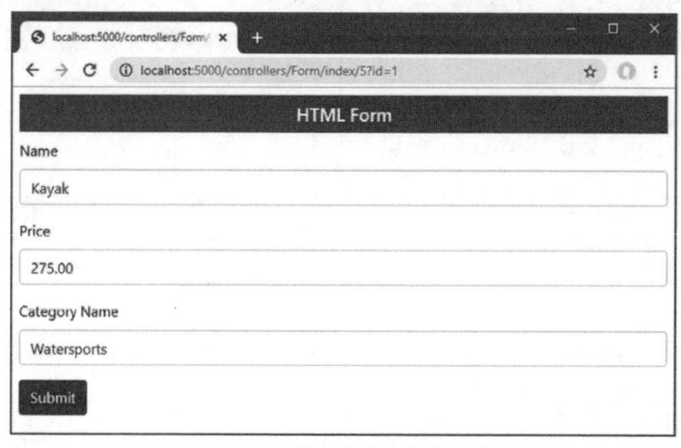


图 28-19 指定模型绑定数据源

**提示：** 在指定模型绑定源(如查询字符串)时，仍然可以绑定复杂类型。对于参数类型中的每个简单属性，模型绑定过程将查找同名的查询字符串键。

### 28.6.1　选择属性的绑定源

同样的属性可以用于由页面模型或控制器定义的模型绑定属性，如代码清单 28-27 所示。

**代码清单 28-27　在 Pages 文件夹的 Bindings.cshtml 文件中选择查询字符串**

```
...
@functions {

    public class BindingsModel : PageModel {

        //[BindProperty(Name = "Data")]
        [FromQuery(Name = "Data")]
        public Product[] Data { get; set; } = Array.Empty<Product>();
    }
}
...
```

FromQuery 属性的使用意味着，在模型绑定器创建 Product 数组时，使用查询字符串作为值的来源，可以通过请求 http://localhost:5000/pages/bindings?data[0].name=Skis&data[0].price=500 查看该数组，生成如图 28-20 所示的响应。

**注意：** 在这个例子中，使用了 get 请求，因为它允许方便地设置查询字符串。尽管在这样一个简单的例子中它是无害的，但在发送修改应用程序状态的 get 请求时必须小心。如前所述，对 get 请求进行更改可能会导致问题。

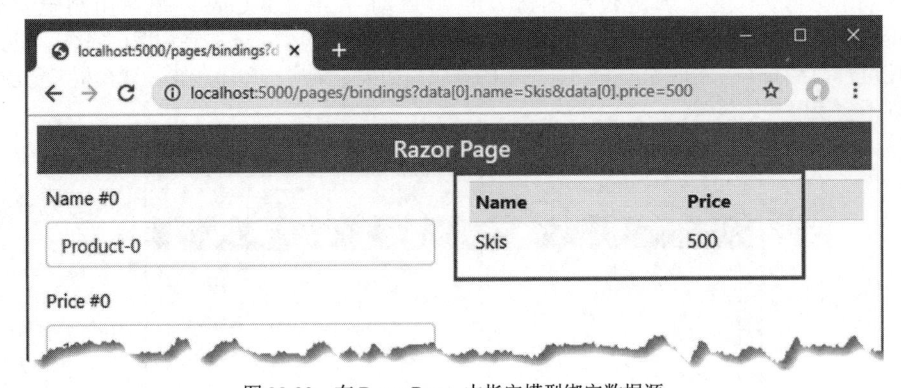


图 28-20　在 Razor Pages 中指定模型绑定数据源

**提示：** 尽管很少使用 FromHeader，但将 FromHeader 特性应用到模型类的属性，可以使用 Header 值绑定复杂类型。

### 28.6.2 使用标头进行模型绑定

FromHeader 属性允许将 HTTP 请求标头用作绑定数据的源。在代码清单 28-28 中，向表单控制器添加了一个简单的操作方法，该方法定义了一个参数，该参数将从标准 HTTP 请求头中进行模型绑定。

**代码清单 28-28 在 Controllers 文件夹的 FormController.cs 文件中，从 Header 进行模型绑定**

```
using Microsoft.AspNetCore.Mvc;
using System.Linq;
using System.Threading.Tasks;
using WebApp.Models;
using Microsoft.EntityFrameworkCore;
using Microsoft.AspNetCore.Mvc.Rendering;

namespace WebApp.Controllers {

    [AutoValidateAntiforgeryToken]
    public class FormController : Controller {
        private DataContext context;

        // ...other action methods omitted for brevity...

        public string Header([FromHeader]string accept) {
            return $"Header: {accept}";
        }
    }
}
```

Header 操作方法定义了一个 accept 参数，该参数的值从当前请求的 accept 头中获取，并作为方法结果返回。重启 ASP.NET Core，请求 http://localhost:5000/controllers/form/header，结果如下：

```
Header: text/html,application/xhtml+xml,application/xml;q=0.9,image/webp,
   image/apng,*/*;q=0.8,application/signed-exchange;v=b3
```

并不是所有的 HTTP 头名称都可以依赖操作方法参数的名称来轻松选择，因为模型绑定系统不会将 C#的命名约定转换为 HTTP 头使用的命名约定。在这些情况下，必须使用 Name 属性配置 FromHeader 特性，以指定 Header 的名称，如代码清单 28-29 所示。

**代码清单 28-29 在 Controllers 文件夹的 FormController.cs 文件中根据 Name 选择 Header**

```
using Microsoft.AspNetCore.Mvc;
using System.Linq;
using System.Threading.Tasks;
using WebApp.Models;
using Microsoft.EntityFrameworkCore;
using Microsoft.AspNetCore.Mvc.Rendering;
```

```
namespace WebApp.Controllers {

    [AutoValidateAntiforgeryToken]
    public class FormController : Controller {
        private DataContext context;

        // ...other action methods omitted for brevity...

        public string Header([FromHeader(Name = "Accept-Language")] string accept) {
            return $"Header: {accept}";
        }
    }
}
```

不能使用Accept-Language作为C#参数的名称，模型绑定器也不会自动将AcceptLanguage这样的名称转换为Accept-Language，以便它与标题匹配。相反，使用Name属性配置该特性，以便它与正确的标题匹配。如果重新启动ASP.NET Core，请求http://localhost:5000/controllers/form/header，会看到如下结果(结果因语言环境设置而异)：

```
Header: en-US;q=0.9,en;q=0.8
```

### 28.6.3 使用请求体作为绑定源

并不是客户端发送的所有数据都是表单数据，比如 JavaScript 客户端给 API 控制器发送 JSON 数据时。FromBody 属性指定应该对请求体进行解码，并将其用作模型绑定数据的来源。在代码清单 28-30 中，向表单控制器添加了一个新的操作方法，该方法带有一个用 FromBody 特性修饰的参数。

**提示：** 使用 ApiController 特性修饰的控制器不需要 FromBody 特性。

**代码清单 28-30 在 Controllers 文件夹的 FormController.cs 文件中添加一个操作方法**

```
using Microsoft.AspNetCore.Mvc;
using System.Linq;
using System.Threading.Tasks;
using WebApp.Models;
using Microsoft.EntityFrameworkCore;
using Microsoft.AspNetCore.Mvc.Rendering;

namespace WebApp.Controllers {

    [AutoValidateAntiforgeryToken]
    public class FormController : Controller {
        private DataContext context;

        public FormController(DataContext dbContext) {
            context = dbContext;
```

```
        }

        // ...other action methods omitted for brevity...

        [HttpPost]
        [IgnoreAntiforgeryToken]
        public Product Body([FromBody] Product model) {
            return model;
        }
    }
}
```

要测试模型绑定过程，请重新启动 ASP.NET Core，打开一个新的 PowerShell 命令提示符，运行代码清单 28-31 中的命令，向应用程序发送请求。

■ **注意**：在代码清单 28-30 的操作方法中添加了 IgnoreAntiforgeryToken，因为要发送的请求不包括防伪令牌，这一点在第 27 章中描述过。

**代码清单 28-31 发送请求**

```
Invoke-RestMethod http://localhost:5000/controllers/form/body -Method POST -Body
(@{ Name="Soccer Boots";
Price=89.99} | ConvertTo-Json) -ContentType "application/json"
```

JSON 编码的请求体用于对操作方法参数进行建模绑定，它生成以下响应：

```
productId  : 0
name       : Soccer Boots
price      : 89.99
categoryId : 0
category   :
supplierId : 0
supplier   :
```

## 28.7 手动模式绑定

当为操作或处理程序方法定义参数或应用 BindProperty 属性时，将自动应用模型绑定。如果始终如一地遵循名称约定，并且总是希望应用该过程，那么自动模型绑定可以很好地工作。如果需要控制绑定过程，或者希望有选择地执行绑定，那么可以手动执行模型绑定，如代码清单 28-32 所示。

**代码清单 28-32 在 Pages 文件夹的 Bindings.cshtml 文件中手动绑定**

```
@page "/pages/bindings"
@model BindingsModel
@using Microsoft.AspNetCore.Mvc
```

```
@using Microsoft.AspNetCore.Mvc.RazorPages

<div class="container-fluid">
    <div class="row">
        <div class="col">
            <form asp-page="Bindings" method="post">
                <div class="form-group">
                    <label>Name</label>
                    <input class="form-control" asp-for="Data.Name" />
                </div>
                <div class="form-group">
                    <label>Price</label>
                    <input class="form-control" asp-for="Data.Price"
                           value="@(Model.Data.Price + 1)" />
                </div>
                <div class="form-check m-2">
                    <input class="form-check-input" type="checkbox" name="bind"
                       value="true" checked />
                    <label class="form-check-label">Model Bind?</label>
                </div>
                <button type="submit" class="btn btn-primary">Submit</button>
                <a class="btn btn-secondary" asp-page="Bindings">Reset</a>
            </form>
        </div>
        <div class="col">
            <table class="table table-sm table-striped">
                <tbody>
                    <tr><th>Name</th><th>Price</th></tr>
                    <tr>
                        <td>@Model.Data.Name</td><td>@Model.Data.Price</td>
                    </tr>
                </tbody>
            </table>
        </div>
    </div>
</div>

@functions {

    public class BindingsModel : PageModel {

        public Product Data { get; set; }
            = new Product() { Name = "Skis", Price = 500 };

        public async Task OnPostAsync([FromForm] bool bind) {
            if (bind) {
                await TryUpdateModelAsync<Product>(Data,
```

```
                    "data", p => p.Name, p => p.Price);
            }
        }
    }
}
```

手动模型绑定使用 TryUpdateModelAsync 方法执行，该方法由 PageModel 和 ControllerBase 类提供，这意味着它对 Razor Pages 和 MVC 控制器都可用。

这个例子混合了自动和手动的模型绑定。OnPostAsync 方法使用自动模型绑定来接收其绑定参数的值，该参数已经用 FromForm 特性进行了修饰。如果参数的值为 true，则使用 TryUpdateModelAsync 方法应用模型绑定。TryUpdateModelAsync 方法的参数是将被模型绑定的对象、值的前缀和一系列选择将包含在流程中的属性的表达式，当然还有其他版本的 TryUpdateModelAsync 方法可用。

结果是，只有当用户选中代码清单 28-32 中添加到表单中的复选框时，才会执行 Data 属性的模型绑定过程。如果复选框未选中，则不会发生模型绑定，并且表单数据将被忽略。为了在使用模型绑定时更加明显，当呈现表单时，Price 属性的值会增加。要查看效果，请求 http://localhost:5000/pages/bindings，并分别在选中和取消选中复选框的情况下提交表单，如图 28-21 所示。

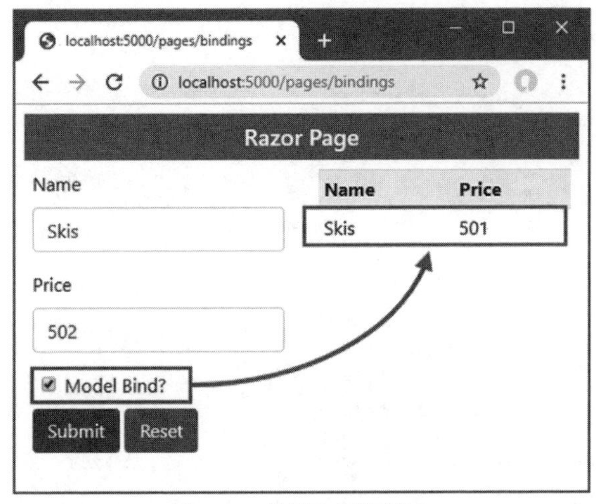


图 28-21 使用手动模型绑定

# 28.8 小结

本章介绍了模型绑定特性，展示了如何使用带有参数和属性的模型绑定，如何绑定简单和复杂类型，以及绑定到数组和集合所需的约定。还解释了如何控制请求的哪一部分用于模型绑定，以及如何控制何时执行模型绑定。第 29 章将描述 ASP.NET Core 提供的验证表单数据的功能。

# 第 29 章

# 使用模型验证

前一章展示了模型绑定流程如何从 HTTP 请求中创建对象。本章只简单地显示了应用程序接收到的数据。这是因为在对用户提供的数据进行检查以确保应用程序能够使用之前，不应该使用这些数据。实际情况是，用户经常会输入无效的、不能使用的数据，这就引出了本章的主题：模型验证。

模型验证是确保应用程序接收的数据适合绑定到模型的过程，如果不是这样，则向用户提供有助于解释问题的有用信息。

该过程的第一部分是检查接收到的数据，这是保持应用程序数据完整性的最重要方法之一。拒绝不能使用的数据可以防止应用程序中出现奇怪和不希望出现的状态。验证过程的第二部分是帮助用户纠正问题，这同样重要。如果没有纠正问题所需的反馈，用户会感到沮丧和困惑。在面向公众的应用程序中，这意味着用户将直接停止使用应用程序。在企业应用程序中，这意味着用户的工作流程将受到阻碍。这两种结果都不是理想的，但幸运的是，ASP.NET Core 为模型验证提供了广泛的支持。表 29-1 概述了模型验证的相关知识。

表 29-1　模型验证的相关知识

| 问题 | 答案 |
|---|---|
| 它是什么? | 模型验证是确保请求中提供的数据在应用程序中有效使用的过程 |
| 它为什么有用? | 用户并不总是输入有效数据，并且在应用程序中使用这些数据可能会产生意外和不希望出现的错误 |
| 它是如何使用的? | 控制器和 Razor Pages 检查验证过程的结果，标签助手用于在显示给用户的视图中包含验证反馈。可以在模型绑定过程中自动执行验证，并且可以通过自定义验证进行补充 |
| 是否存在缺陷或限制? | 重要的是测试验证代码的有效性，以确保它涵盖应用程序可以接收到的所有值 |
| 还有其他选择吗? | 模型验证是可选的，但是最好在使用模型绑定时使用它 |

表 29-2 总结了本章的内容。

表 29-2　本章内容摘要

| 问题 | 解决方案 | 代码清单 |
|---|---|---|
| 验证数据 | 手动使用 ModelState 特性或应用验证属性 | 29-5、29-13～29-20 |
| 显示验证消息 | 使用指定表单元素的类和验证标签助手 | 29-6～29-12 |
| 在提交表单之前验证数据 | 使用客户端和远程验证 | 29-21～29-25 |

## 29.1　准备工作

本章使用了第 28 章中的 WebApp 项目。为了准备这一章，更改表单控制器的表单视图的内容，使其包含 Product 类定义的每个属性(但不包括 Entity Framework Core 使用的导航属性)的输入元素，如代码清单 29-1 所示。

**提示：**可以从 https://github.com/apress/pro-asp.net-core-3 下载本章和本书中其他所有章节的示例项目。如果在运行示例时遇到问题，请参阅第 1 章以了解如何获得帮助。

### 代码清单 29-1　在 Views/Form 文件夹的 Form.cshtml 文件中更改元素

```
@model Product
@{ Layout = "_SimpleLayout"; }

<h5 class="bg-primary text-white text-center p-2">HTML Form</h5>

<form asp-action="submitform" method="post" id="htmlform">
    <div class="form-group">
        <label asp-for="Name"></label>
        <input class="form-control" asp-for="Name" />
    </div>
    <div class="form-group">
        <label asp-for="Price"></label>
        <input class="form-control" asp-for="Price" />
    </div>
    <div class="form-group">
        <label>CategoryId</label>
        <input class="form-control" asp-for="CategoryId" />
    </div>
    <div class="form-group">
        <label>SupplierId</label>
        <input class="form-control" asp-for="SupplierId" />
    </div>
    <button type="submit" class="btn btn-primary">Submit</button>
</form>
```

将 FormController.cs 文件的内容替换为代码清单 29-2 中所示的内容，这将允许显示代码清单 29-1 中定义的属性，并删除不再需要的模型绑定特性和操作方法。

### 代码清单 29-2　替换 Controllers 文件夹中 FormController.cs 文件的内容

```
using Microsoft.AspNetCore.Mvc;
using System.Linq;
using System.Threading.Tasks;
using WebApp.Models;
using Microsoft.EntityFrameworkCore;
```

```
namespace WebApp.Controllers {

    [AutoValidateAntiforgeryToken]
    public class FormController : Controller {
        private DataContext context;

        public FormController(DataContext dbContext) {
            context = dbContext;
        }

        public async Task<IActionResult> Index(long? id) {
            return View("Form", await context.Products
                .FirstOrDefaultAsync(p => id == null || p.ProductId == id));
        }

        [HttpPost]
        public IActionResult SubmitForm(Product product) {
            TempData["name"] = product.Name;
            TempData["price"] = product.Price.ToString();
            TempData["categoryId"] = product.CategoryId.ToString();
            TempData["supplierId"] = product.SupplierId.ToString();
            return RedirectToAction(nameof(Results));
        }

        public IActionResult Results() {
            return View(TempData);
        }
    }
}
```

## 29.2 删除数据库

打开一个新的 PowerShell 命令提示符，导航到包含 WebApp.csproj 文件的文件夹，并运行代码清单 29-3 所示的命令来删除数据库。

**代码清单 29-3 删除数据库**

```
dotnet ef database drop --force
```

## 29.3 运行示例应用程序

从 Debug 菜单中选择 Start Without Debugging 或 Run Without Debugging，或者使用 PowerShell 命令提示符运行代码清单 29-4 所示的命令。

**代码清单 29-4　运行示例应用程序**

```
dotnet run
```

使用浏览器请求 http://localhost:5000/controllers/Form，这将显示一个 HTML 表单。单击 Submit 按钮，将显示表单数据，如图 29-1 所示。

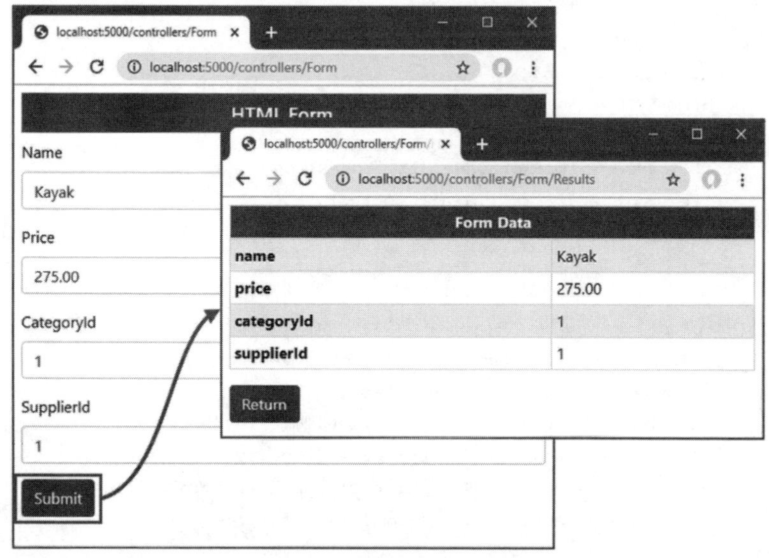


图 29-1　运行示例应用程序

## 29.4　理解对模型验证的需要

模型验证是应用程序对从客户端接收到的数据进行验证的过程。如果不进行验证，应用程序将尝试对接收到的任何数据进行操作，这可能导致立即出现的异常和意外行为，或随着数据库中充满了糟糕、不完整或恶意的数据而逐渐出现的长期问题。

目前，接收表单数据的操作和处理程序方法将接收用户提交的任何数据，这就是为什么示例只显示表单数据而没有将其存储在数据库中。

大多数数据值都有某种类型的约束。这可能包括要求提供一个值，要求值是特定类型，以及要求值在特定范围内。

例如，在安全地将产品对象存储到数据库之前，我需要确保用户提供了 Name、Price、CategoryId 和 SupplierId 属性的值。Name 值可以是任何有效的字符串，Price 属性必须是一个有效的货币金额，CategoryId 和 SupplierId 属性必须与数据库中现有的 Supplier 和 Category 产品相对应。下面演示如何通过检查应用程序接收到的数据，并在应用程序不能使用用户提交的数据时向用户提供反馈，从而使用模型验证来强制满足这些需求。

## 29.5　显式验证控制器中的数据

验证数据最直接的方法是在操作或处理程序方法中进行验证，如代码清单 29-5 所示，记录任何问题的细节，以便向用户显示。

代码清单 29-5 在 Controllers 文件夹的 FormController.cs 文件中显式验证数据

```
using Microsoft.AspNetCore.Mvc;
using System.Linq;
using System.Threading.Tasks;
using WebApp.Models;
using Microsoft.EntityFrameworkCore;
using Microsoft.AspNetCore.Mvc.ModelBinding;

namespace WebApp.Controllers {

    [AutoValidateAntiforgeryToken]
    public class FormController : Controller {
        private DataContext context;

        public FormController(DataContext dbContext) {
            context = dbContext;
        }

        public async Task<IActionResult> Index(long? id) {
            return View("Form", await context.Products
                .FirstOrDefaultAsync(p => id == null || p.ProductId == id));
        }

        [HttpPost]
        public IActionResult SubmitForm(Product product) {

            if (string.IsNullOrEmpty(product.Name)) {
                ModelState.AddModelError(nameof(Product.Name), "Enter a name");
            }

            if (ModelState.GetValidationState(nameof(Product.Price))
                    == ModelValidationState.Valid && product.Price < 1) {
                ModelState.AddModelError(nameof(Product.Price),
                    "Enter a positive price");
            }

            if (!context.Categories.Any(c => c.CategoryId == product.CategoryId)) {
                ModelState.AddModelError(nameof(Product.CategoryId),
                    "Enter an existing category ID");
            }

            if (!context.Suppliers.Any(s => s.SupplierId == product.SupplierId)) {
                ModelState.AddModelError(nameof(Product.SupplierId),
                    "Enter an existing supplier ID");
            }
```

```
            if (ModelState.IsValid) {
                TempData["name"] = product.Name;
                TempData["price"] = product.Price.ToString();
                TempData["categoryId"] = product.CategoryId.ToString();
                TempData["supplierId"] = product.SupplierId.ToString();
                return RedirectToAction(nameof(Results));
            } else {
                return View("Form");
            }
        }

        public IActionResult Results() {
            return View(TempData);
        }
    }
}
```

对于创建的 Product 参数的每个属性，检查用户提供的值，并记录使用从 ControllerBase 类继承的 ModelState 属性返回的 ModelStateDictionary 对象所发现的任何错误。

顾名思义，ModelStateDictionary 类是一个字典，用于跟踪模型对象状态的详细信息，重点关注验证错误。表 29-3 描述了最重要的 ModelStateDictionary 成员。

表 29-3　选择的 ModelStateDictionary 成员

| 名称 | 描述 |
|---|---|
| AddModelError(property, message) | 此方法用于记录指定属性的模型验证错误 |
| GetValidationState(property) | 此方法用于确定特定属性是否存在模型验证错误，相应属性表示为 ModelValidationState 枚举的值 |
| IsValid | 如果所有模型属性都有效，该属性返回 true，否则返回 false |
| Clear() | 此属性清除验证状态 |

作为使用 ModelStateDictionary 的一个示例，请考虑如何验证 Name 属性。

```
...
if (string.IsNullOrEmpty(product.Name)) {
    ModelState.AddModelError(nameof(Product.Name), "Enter a name");
}
...
```

Product 类的验证要求之一是确保用户为 Name 属性提供一个值，因此使用静态方法 string.IsNullOrEmpty 来测试模型绑定过程从请求中提取的属性值。如果 Name 属性是 null 或空字符串，那么说明应用程序不能使用该值，因此使用 ModelState.AddModelError 方法来注册一个验证错误，指定属性的名称(Name)和显示给用户的消息，以解释问题的性质(输入名称)。

在模型绑定过程中，还使用 ModelStateDictionary 来记录查找和为模型属性赋值的任何问题。GetValidationState 方法用于查看模型属性是否存在任何错误记录，这些错误可能来自模型绑定过程，也可能是因为 AddModelError 方法在操作方法的显式验证期间调用。GetValidationState 方法从 ModelValidationState 枚举中返回一个值，该枚举定义了表 29-4 中描述的值。

表 29-4 ModelValidationState 值

| 名称 | 描述 |
| --- | --- |
| Unvalidated | 未验证此值意味着没有对模型属性执行任何验证，这通常是因为请求中没有与属性名称对应的值 |
| Valid | 此值表示与该属性关联的请求值有效 |
| Invalid | 此值意味着与该属性关联的请求值无效，不应使用 |
| Skipped | 此值意味着没有处理模型属性，这通常意味着存在太多验证错误，没有必要继续执行验证检查 |

对于 Price 属性，检查模型绑定过程是否报告了一个问题，将浏览器发送的值解析为十进制值，如下所示：

```
...
if (ModelState.GetValidationState(nameof(Product.Price))
        == ModelValidationState.Valid && product.Price < 1) {
    ModelState.AddModelError(nameof(Product.Price), "Enter a positive price");
}
...
```

确保用户提供了一个等于或大于 1 的 Price 值、但是如果用户提供了一个模型绑定器不能转换成十进制的值，则没有记录一个值为零或负的错误。在执行自己的验证检查之前，使用 GetValidationState 方法来确定 Price 属性的验证状态。

在验证 Product 对象中的所有属性之后，检查 ModelState.IsValid 属性，以查看是否存在错误。如果在检查过程中调用 ModelState.AddModelError 方法，或模型绑定器在创建对象时出现任何问题，该方法就返回 true。

```
...
if (ModelState.IsValid) {
    TempData["name"] = product.Name;
    TempData["price"] = product.Price.ToString();
    TempData["categoryId"] = product.CategoryId.ToString();
    TempData["supplierId"] = product.SupplierId.ToString();
    return RedirectToAction(nameof(Results));
} else {
    return View("Form");
}
...
```

如果 IsValid 属性返回 true，则 Product 对象是有效的，在这种情况下，操作方法将浏览器重定向到 Results 操作，在该操作中显示经过验证的表单值。如果 IsValue 属性返回 false，则存在验证问题，这将通过调用 View 方法再次呈现表单视图来处理。

### 29.5.1 向用户显示验证错误

通过调用 View 方法来处理验证错误似乎有些奇怪，但是提供给视图的上下文数据包含模型验证错误的细节；标签助手使用这些细节来转换 input 元素。

要查看这是如何工作的，请重新启动 ASP.NET Core，因此，对控制器的更改生效，并使用浏览器请求 http:// localhost:5000/controllers/form。清除 Name 字段的内容，然后单击 Submit 按钮。

在浏览器显示的内容中没有任何可见的变化，但是如果检查 Name 字段的输入元素，将看到元素已转换。下面是表单提交前的 input 元素：

```
<input class="form-control" type="text" id="Name" name="Name" value="Kayak">
```

下面是表单提交后的 input 元素：

```
<input class="form-control input-validation-error" type="text" id="Name"
    name="Name" value="">
```

标签助手将值验证失败的元素添加到 input-validation-error 类中，然后可以对其进行样式化，以向用户突出显示问题。

可以通过在样式表中定义定制的 CSS 样式来实现这一点，但是如果想使用诸如 Bootstrap 的 CSS 库提供的内置验证样式，则需要做一些额外的工作。添加到 input 元素中的类的名称不能更改，这意味着需要一些 JavaScript 代码来在 ASP.NET Core 使用的名称和引导程序提供的 CSS 错误类之间进行映射。

**提示：**使用这样的 Javascript 代码可能十分尴尬，而且可能诱导使用自定义 CSS 样式，即使使用像 Bootstrap 这样的 CSS 库也是如此。然而，在 Bootstrap 中用于验证类的颜色可以通过使用主题或定制包并定义自己的样式来覆盖，这意味着必须确保对主题的任何更改都与对任何自定义样式的相应更改相匹配。理想情况下，Microsoft 使验证类名在 ASP.NET Core 的未来版本中可配置，但在此之前，使用 Javascript 应用 Bootstrap 样式是一种比创建自定义样式表更健壮的方法。

要定义 JavaScript 代码以便控制器和 Razor Pages 都可以使用它，可以使用 Visual Studio JavaScript 文件模板，给 Views/Shared 文件夹添加一个名为_Validation.cshtml 的文件，内容如代码清单 29-6 所示。Visual Studio Code 不需要模板，可以给 Views/Shared 文件夹添加一个名为_Validation.cshtml 的文件，如代码清单 29-6 所示。

**代码清单 29-6　Views/Shared 文件夹中_Validation.cshtml 文件的内容**

```
<script src="/lib/jquery/jquery.min.js"></script>
<script type="text/javascript">
    $(document).ready(function () {
        $("input.input-validation-error").addClass("is-invalid");
    });
</script>
```

使用新的文件作为局部视图，其中包含一个 script 元素来加载 jQuery 库，包含自定义脚本来定位 input 元素，该 input 元素是 input-validation-error 类的成员，并将它们添加到 is-invalid 类(Bootstrap 用于设置表单元素的错误颜色)。代码清单 29-7 使用部分标签助手将新的部分视图合并到 HTML 表单中，以便突出显示带有验证错误的字段。

**代码清单 29-7　在 Views/Form 文件夹的 Form.cshtml 文件中包含部分视图**

```
@model Product
@{ Layout = "_SimpleLayout"; }
```

```
<h5 class="bg-primary text-white text-center p-2">HTML Form</h5>

<partial name="_Validation" />

<form asp-action="submitform" method="post" id="htmlform">
    <div class="form-group">
        <label asp-for="Name"></label>
        <input class="form-control" asp-for="Name" />
    </div>
    <div class="form-group">
        <label asp-for="Price"></label>
        <input class="form-control" asp-for="Price" />
    </div>
    <div class="form-group">
        <label>CategoryId</label>
        <input class="form-control" asp-for="CategoryId" />
    </div>
    <div class="form-group">
        <label>SupplierId</label>
        <input class="form-control" asp-for="SupplierId" />
    </div>
    <button type="submit" class="btn btn-primary">Submit</button>
</form>
```

jQuery代码在浏览器解析完HTML文档中的所有元素后运行，其效果是突出显示已分配给input-validaton-error类的input元素。为查看效果，可以导航到http://localhost:5000/controllers/form，清除Name字段的内容，并提交表单，生成如图 29-2 所示的响应。

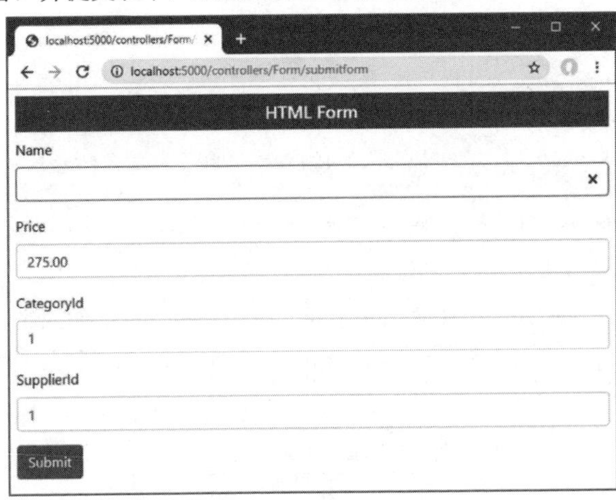


图 29-2　突出显示验证错误

只有当表单与可以被模型浏览器解析的数据一起提交，并通过操作方法中的显式验证检查时，用户才会看到结果视图。在此之前，提交表单将导致表单视图以突出显示的验证错误形式呈现。

### 29.5.2　显示验证消息

标签助手应用于输入元素的 CSS 类表明，表单字段存在问题，但没有告诉用户问题是什么。向用户提供更多信息需要使用不同的标签助手，从而将问题摘要添加到视图中，如代码清单 29-8 所示。

**代码清单 29-8　在 Views/Form 文件夹的 Form.cshtml 文件中显示摘要**

```
@model Product
@{ Layout = "_SimpleLayout"; }

<h5 class="bg-primary text-white text-center p-2">HTML Form</h5>

<partial name="_Validation" />

<form asp-action="submitform" method="post" id="htmlform">
    <div asp-validation-summary="All" class="text-danger"></div>
    <div class="form-group">
        <label asp-for="Name"></label>
        <input class="form-control" asp-for="Name" />
    </div>
    <div class="form-group">
        <label asp-for="Price"></label>
        <input class="form-control" asp-for="Price" />
    </div>
    <div class="form-group">
        <label>CategoryId</label>
        <input class="form-control" asp-for="CategoryId" />
    </div>
    <div class="form-group">
        <label>SupplierId</label>
        <input class="form-control" asp-for="SupplierId" />
    </div>
    <button type="submit" class="btn btn-primary">Submit</button>
</form>
```

ValidationSummaryTagHelper 类检测 div 元素上的 asp-validation-summary 特性，并通过添加描述已记录的任何验证错误的消息来响应。asp-validation-summary 特性的值来自 ValidationSummary 枚举，它定义了表 29-5 中所示的值。

表 29-5　ValidationSummary 值

| 名称 | 描述 |
|---|---|
| All | 该值用于显示已记录的所有验证错误 |
| ModelOnly | 该值用于仅显示整个模型的验证错误，不包括那些已经为单个属性记录的错误，如“显示模型级消息”一节所述 |
| None | 此值用于禁用标签助手，以便它不会转换 HTML 元素 |

显示错误消息可以帮助用户理解为什么不能处理表单。例如，尝试提交一个表单，其中的 Price 字段为负值(如-10)，或者该值不能转换为十进制(如 ten)。每个值都会导致不同的错误消息，如图 29-3 所示。

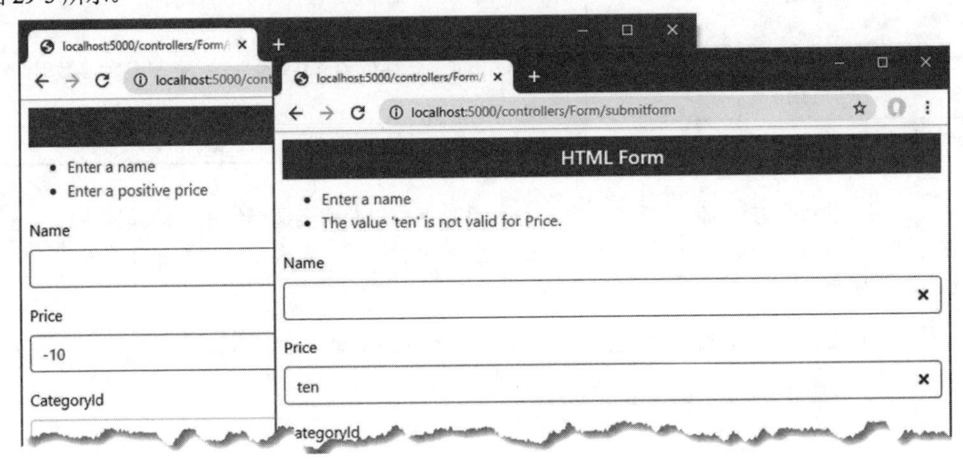


图 29-3 显示验证消息

### 配置默认的验证错误消息

当模型绑定流程尝试提供调用操作方法所需的数据值时，它会执行自己的验证，这就是为什么你会在 Price 值不能转换为小数时看到一条验证消息的原因。并不是模型绑定器生成的所有验证消息都对用户有帮助，可以通过清除 Price 字段并提交表单来看到这一点。空字段产生以下消息：

```
The value '' is invalid
```

当模型绑定过程找不到属性的值或找到了值但无法解析时，就将该消息添加到 ModelStateDictionary 中。在本例中，出现错误的原因是无法将表单数据中发送的空字符串解析为 Product 类的 Price 属性的十进制值。

模型绑定器有一组用于验证错误的预定义消息。可以使用 DefaultModelBindingMessageProvider 类定义的方法替换这些消息，如表 29-6 所述。

表 29-6 DefaultModelBindingMessageProvider 方法

| 名称 | 描述 |
|---|---|
| SetValueMustNotBeNullAccessor | 不可为空的模型属性值为空时，分配给此属性的函数用于生成验证错误消息 |
| SetMissingBindRequiredValueAccessor | 当请求不包含所需属性的值时，赋给此属性的函数用于生成验证错误消息 |
| SetMissingKeyOrValueAccessor | 当字典模型对象所需的数据包含空键或值时，赋给此属性的函数用于生成验证错误消息 |
| SetAttemptedValueIsInvalidAccessor | 当模型绑定系统不能将数据值转换为所需的 C#类型时，分配给此属性的函数用于生成验证错误消息 |

(续表)

| 名称 | 描述 |
| --- | --- |
| SetUnknownValueIsInvalidAccessor | 当模型绑定系统不能将数据值转换为已知的 C#类型时，分配给此属性的函数用于生成验证错误消息 |
| SetValueMustBeANumberAccessor | 当数据值不能解析为 C#数值类型时，分配给此属性的函数用于生成验证错误消息 |
| SetValueIsInvalidAccessor | 分配给此属性的函数用于生成回退验证错误消息，作为最后手段 |

表中描述的每个方法都接收一个函数，该函数被调用以获取要显示给用户的验证消息。这些方法通过 Startup 类中的 options 模式应用。在代码清单 29-9 中，替换了当值为 null 或无法转换时显示的默认消息。

### 代码清单 29-9　在 WebApp 文件夹的 Startup.cs 文件中更改验证消息

```
using Microsoft.AspNetCore.Builder;
using Microsoft.Extensions.DependencyInjection;
using Microsoft.Extensions.Configuration;
using Microsoft.EntityFrameworkCore;
using WebApp.Models;
using Microsoft.AspNetCore.Antiforgery;
using Microsoft.AspNetCore.Http;
using Microsoft.AspNetCore.Mvc;

namespace WebApp {
    public class Startup {

        public Startup(IConfiguration config) {
            Configuration = config;
        }

        public IConfiguration Configuration { get; set; }

        public void ConfigureServices(IServiceCollection services) {
            services.AddDbContext<DataContext>(opts => {
                opts.UseSqlServer(Configuration[
                    "ConnectionStrings:ProductConnection"]);
                opts.EnableSensitiveDataLogging(true);
            });

            services.AddControllersWithViews().AddRazorRuntimeCompilation();
            services.AddRazorPages().AddRazorRuntimeCompilation();
            services.AddSingleton<CitiesData>();

            services.Configure<AntiforgeryOptions>(opts => {
```

```
                opts.HeaderName = "X-XSRF-TOKEN";
            });

            services.Configure<MvcOptions>(opts => opts.ModelBindingMessageProvider
                .SetValueMustNotBeNullAccessor(value => "Please enter a value"));
        }

        public void Configure(IApplicationBuilder app, DataContext context,
                IAntiforgery antiforgery) {

            app.UseRequestLocalization();

            app.UseDeveloperExceptionPage();
            app.UseStaticFiles();
            app.UseRouting();

            app.Use(async (context, next) => {
                if (!context.Request.Path.StartsWithSegments("/api")) {
                    context.Response.Cookies.Append("XSRF-TOKEN",
                        antiforgery.GetAndStoreTokens(context).RequestToken,
                        new CookieOptions { HttpOnly = false });
                }
                await next();
            });

            app.UseEndpoints(endpoints => {
                endpoints.MapControllers();
                endpoints.MapControllerRoute("forms",
                    "controllers/{controller=Home}/{action=Index}/{id?}");
                endpoints.MapDefaultControllerRoute();
                endpoints.MapRazorPages();
            });
            SeedData.SeedDatabase(context);
        }
    }
}
```

指定的函数接收用户提供的值，尽管这在处理空值时不是特别有用。要查看自定义消息，请重新启动 ASP.NET Core，使用浏览器请求 http://localhost:5000/controllers/form，并提交带有空 Price 字段的表单。响应将包括自定义错误消息，如图 29-4 所示。

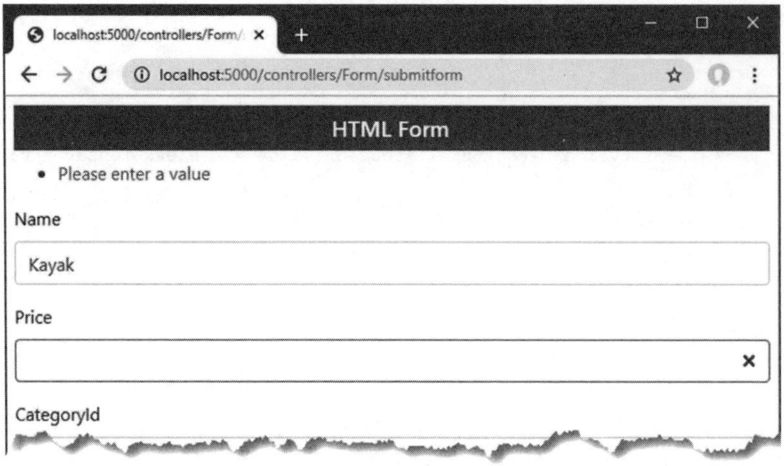

图 29-4　更改默认的验证消息

### 29.5.3　显示属性级的验证消息

尽管自定义错误消息比默认错误消息更有意义，但它仍然没有多大帮助，因为它不能清楚地指出问题与哪个字段相关。对于这类错误，更有用的是在包含问题数据的 HTML 元素旁边显示验证错误消息。这可以使用 ValidationMessageTag 标签助手来完成，它查找具有 asp-validation-for 特性的 span 元素，该特性用于指定应该显示错误消息的属性。

在代码清单 29-10 中，为表单中的每个 input 元素添加了属性级验证消息元素。

**代码清单 29-10　在 Views/Form 文件夹的 Form.cshtml 文件中添加属性级消息**

```
@model Product
@{ Layout = "_SimpleLayout"; }

<h5 class="bg-primary text-white text-center p-2">HTML Form</h5>

<partial name="_Validation" />

<form asp-action="submitform" method="post" id="htmlform">
    <div asp-validation-summary="All" class="text-danger"></div>
    <div class="form-group">
        <label asp-for="Name"></label>
        <div><span asp-validation-for="Name" class="text-danger"></span></div>
        <input class="form-control" asp-for="Name" />
    </div>
    <div class="form-group">
        <label asp-for="Price"></label>
        <div><span asp-validation-for="Price" class="text-danger"></span></div>
        <input class="form-control" asp-for="Price" />
    </div>
    <div class="form-group">
```

```
        <label>CategoryId</label>
        <div><span asp-validation-for="CategoryId" class="text-danger"></span></div>
        <input class="form-control" asp-for="CategoryId" />
    </div>
    <div class="form-group">
        <label>SupplierId</label>
        <div><span asp-validation-for="SupplierId" class="text-danger"></span></div>
        <input class="form-control" asp-for="SupplierId" />
    </div>
    <button type="submit" class="btn btn-primary">Submit</button>
</form>
```

由于 span 元素是内联显示的，所以在显示验证消息时必须非常小心，以使消息与哪个元素相关变得很明显。通过请求 http://localhost:5000/controllers/form，清除 Name 和 Price 字段并提交表单，可以看到新的验证消息的效果。响应如图 29-5 所示，在文本字段旁边包括验证消息。

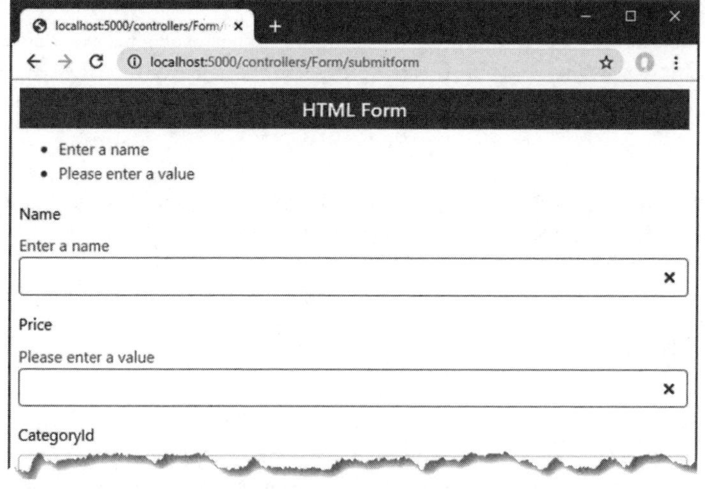


图 29-5　显示属性级的验证消息

## 29.5.4　显示模型级消息

验证摘要消息似乎是多余的，因为它重复了属性级消息。但是摘要有一个有用的技巧，它能显示应用于整个模型而不仅仅是单个属性的消息。这意味着可以报告由于单个属性的组合而产生的错误，否则就很难用属性级消息来表示这些错误。

在代码清单 29-11 中，添加了一个对 FormController.SubmitForm 操作的检查，该操作记录当名称值以 Small 开头且 Price 值超过 100 时的验证错误。

**代码清单 29-11　在 Controllers 文件夹的 FormController.cs 文件中执行模型级验证**

```
...
[HttpPost]
public IActionResult SubmitForm(Product product) {

    if (string.IsNullOrEmpty(product.Name)) {
```

```
        ModelState.AddModelError(nameof(Product.Name), "Enter a name");
    }

    if (ModelState.GetValidationState(nameof(Product.Price))
            == ModelValidationState.Valid && product.Price < 1) {
        ModelState.AddModelError(nameof(Product.Price), "Enter a positive price");
    }

    if (ModelState.GetValidationState(nameof(Product.Name))
            == ModelValidationState.Valid
            && ModelState.GetValidationState(nameof(Product.Price))
            == ModelValidationState.Valid
            && product.Name.ToLower().StartsWith("small") && product.Price > 100) {
        ModelState.AddModelError("", "Small products cannot cost more than $100");
    }

    if (!context.Categories.Any(c => c.CategoryId == product.CategoryId)) {
        ModelState.AddModelError(nameof(Product.CategoryId),
            "Enter an existing category ID");
    }

    if (!context.Suppliers.Any(s => s.SupplierId == product.SupplierId)) {
        ModelState.AddModelError(nameof(Product.SupplierId),
            "Enter an existing supplier ID");
    }

    if (ModelState.IsValid) {
        TempData["name"] = product.Name;
        TempData["price"] = product.Price.ToString();
        TempData["categoryId"] = product.CategoryId.ToString();
        TempData["supplierId"] = product.SupplierId.ToString();
        return RedirectToAction(nameof(Results));
    } else {
        return View("Form");
    }
}
...
```

如果用户输入以 small 开头的 Name 值和大于 100 的 Price 值，就会记录一个模型级验证错误。只有在单个属性值没有验证问题时，才检查这些值的组合，这可以确保用户不会得到冲突的消息。使用 AddModelError 记录与整个模型相关的验证错误，并将空字符串作为第一个参数。

代码清单 29-12 将 asp-validation-summary 属性的值更改为 ModelOnly，它排除了属性级的错误，这意味着摘要只显示应用于整个模型的错误。

**代码清单 29-12 在 Views/Form 文件夹的 Form.cshtml 文件中配置验证摘要**

```
@model Product
```

```
@{ Layout = "_SimpleLayout"; }

<h5 class="bg-primary text-white text-center p-2">HTML Form</h5>

<partial name="_Validation" />

<form asp-action="submitform" method="post" id="htmlform">
    <div asp-validation-summary="ModelOnly" class="text-danger"></div>
    <div class="form-group">
        <label asp-for="Name"></label>
        <div><span asp-validation-for="Name" class="text-danger"></span></div>
        <input class="form-control" asp-for="Name" />
    </div>
    <div class="form-group">
        <label asp-for="Price"></label>
        <div><span asp-validation-for="Price" class="text-danger"></span></div>
        <input class="form-control" asp-for="Price" />
    </div>
    <div class="form-group">
        <label>CategoryId</label>
        <div><span asp-validation-for="CategoryId" class="text-danger"></span></div>
        <input class="form-control" asp-for="CategoryId" />
    </div>
    <div class="form-group">
        <label>SupplierId</label>
        <div><span asp-validation-for="SupplierId" class="text-danger"></span></div>
        <input class="form-control" asp-for="SupplierId" />
    </div>
    <button type="submit" class="btn btn-primary">Submit</button>
</form>
```

重启 ASP.NET Core，请求 http://localhost:5000/controllers/Form。在 Name 栏中输入 Small Kayak，在 Price 字段输入 150，并提交表单。响应将包括模型级的错误消息，如图 29-6 所示。

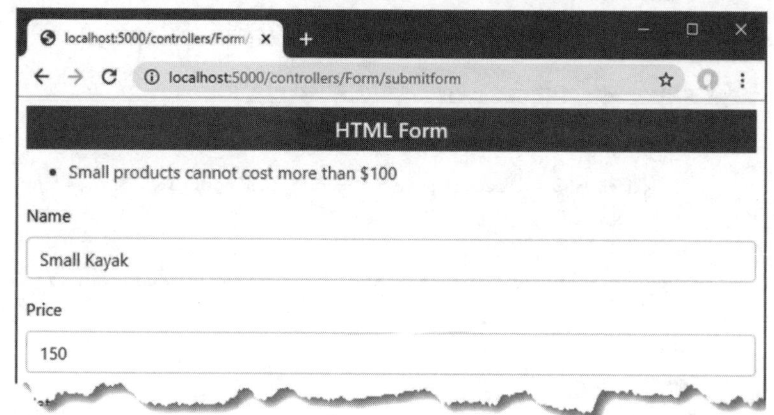


图 29-6　显示模型级验证消息

## 29.6　显式验证 Razor Pages 中的数据

Razor Pages 验证依赖于上一节介绍的控制器中使用的特性。代码清单 29-13 将在 FormHandler 页面显式地进行验证。

**代码清单 29-13　在 Pages 文件夹的 FormHandler.cshtml 文件中验证数据**

```
@page "/pages/form/{id:long?}"
@model FormHandlerModel
@using Microsoft.AspNetCore.Mvc.RazorPages
@using Microsoft.EntityFrameworkCore
@using Microsoft.AspNetCore.Mvc.ModelBinding

<partial name="_Validation" />

<div class="m-2">
    <h5 class="bg-primary text-white text-center p-2">HTML Form</h5>
    <form asp-page="FormHandler" method="post" id="htmlform">
        <div asp-validation-summary="ModelOnly" class="text-danger"></div>
        <div class="form-group">
            <label>Name</label>
            <div>
                <span asp-validation-for="Product.Name" class="text-danger">
                </span>
            </div>
            <input class="form-control" asp-for="Product.Name" />
        </div>
        <div class="form-group">
            <label>Price</label>
            <div>
                <span asp-validation-for="Product.Price" class="text-danger">
                </span>
            </div>
            <input class="form-control" asp-for="Product.Price" />
        </div>
        <div class="form-group">
            <label>CategoryId</label>
            <div>
                <span asp-validation-for="Product.CategoryId" class="text-danger">
                </span>
            </div>
            <input class="form-control" asp-for="Product.CategoryId" />
        </div>
        <div class="form-group">
            <label>SupplierId</label>
            <div>
                <span asp-validation-for="Product.SupplierId" class="text-danger">
```

```
                </span>
            </div>
            <input class="form-control" asp-for="Product.SupplierId" />
        </div>
        <button type="submit" class="btn btn-primary">Submit</button>
    </form>
</div>

@functions {

    public class FormHandlerModel : PageModel {
        private DataContext context;

        public FormHandlerModel(DataContext dbContext) {
            context = dbContext;
        }

        [BindProperty]
        public Product Product { get; set; }

        //[BindProperty(Name = "Product.Category")]
        //public Category Category { get; set; }

        public async Task OnGetAsync(long id = 1) {
            Product = await context.Products.FirstAsync(p => p.ProductId == id);
        }

        public IActionResult OnPost() {

            if (string.IsNullOrEmpty(Product.Name)) {
                ModelState.AddModelError("Product.Name", "Enter a name");
            }

            if (ModelState.GetValidationState("Product.Price")
                    == ModelValidationState.Valid && Product.Price < 1) {
                ModelState.AddModelError("Product.Price", "Enter a positive price");
            }

            if (ModelState.GetValidationState("Product.Name")
                    == ModelValidationState.Valid
                && ModelState.GetValidationState("Product.Price")
                    == ModelValidationState.Valid
                && Product.Name.ToLower().StartsWith("small")
                && Product.Price > 100) {
                    ModelState.AddModelError("",
                        "Small products cannot cost more than $100");
            }
```

```
                if (!context.Categories.Any(c => c.CategoryId == Product.CategoryId)) {
                    ModelState.AddModelError("Product.CategoryId",
                        "Enter an existing category ID");
                }

                if (!context.Suppliers.Any(s => s.SupplierId == Product.SupplierId)) {
                        ModelState.AddModelError("Product.SupplierId",
                            "Enter an existing supplier ID");
                }

                if (ModelState.IsValid) {
                    TempData["name"] = Product.Name;
                    TempData["price"] = Product.Price.ToString();
                    TempData["categoryId"] = Product.CategoryId.ToString();
                    TempData["supplierId"] = Product.SupplierId.ToString();
                    return RedirectToPage("FormResults");
                } else {
                    return Page();
                }
            }
        }
    }
```

PageModel 类定义了一个 ModelState 属性，它与在控制器中使用的该属性等价，并允许记录验证错误。验证过程是相同的，但是在记录错误时必须小心，以确保名称与 Razor Pages 使用的模式相匹配。当记录一个错误时，使用关键字 nameof 来选择与错误相关的属性，如下所示：

```
...
ModelState.AddModelError(nameof(Product.Name), "Enter a name");
...
```

这是一种常见约定，因为它防止输入问题导致错误被不当地记录下来。此表达式在 Razor Pages 中不起作用，在该页面中，必须针对 Product.Name(而不是 Name)记录错误，在 Razor Pages 中反射@Model 表达式会返回页面模型对象，如下所示：

```
...
ModelState.AddModelError("Product.Name", "Enter a name");
...
```

要测试验证过程，请使用浏览器请求 http://localhost:5000/pages/form 并提交表单，该表单包含空字段或不能转换为 Product 类所需的 C#类型的值。错误消息会像控制器一样显示出来，如图 29-7 所示(值 1、2 和 3 对 CategoryId 和 SupplierId 字段都有效)。

**提示：** 表 29-6 中描述的改变默认验证消息的方法会影响 Razor Pages 和控制器。

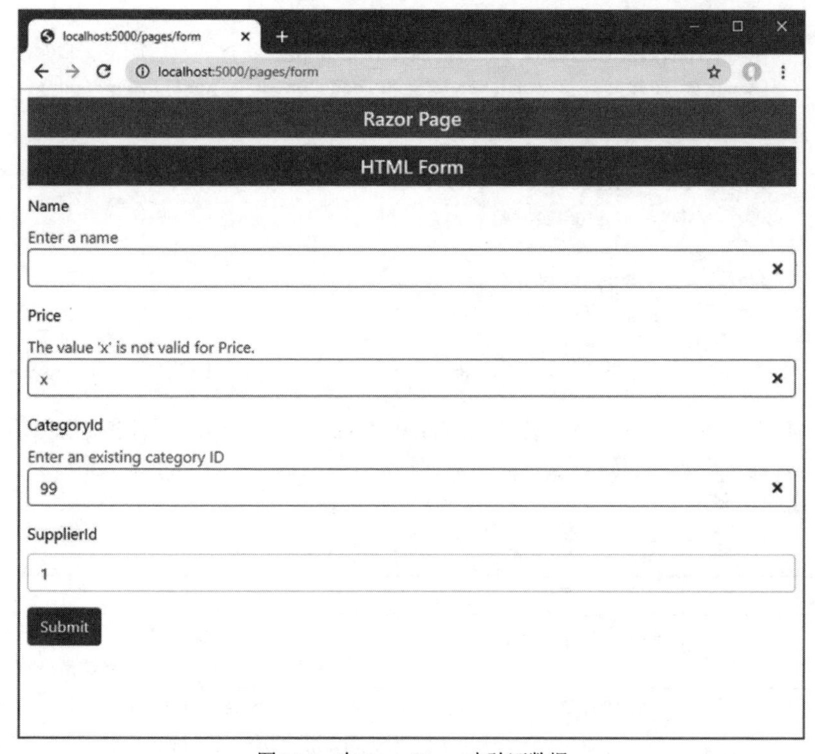


图 29-7　在 Razor Pages 中验证数据

## 29.7　使用元数据指定验证规则

将验证逻辑放入操作方法的一个问题是，它最终会在从用户接收数据的每个操作或处理程序方法中复制。为了帮助减少重复，验证过程支持使用属性在模型类中直接表达模型验证规则，确保无论使用哪种操作方法处理请求，都将应用相同的验证规则集。在代码清单 29-14 中，将属性应用于 Product 类，以描述 Name 和 Price 属性所需的验证。

**代码清单 29-14　在 Models 文件夹的 Product.cs 文件中应用验证属性**

```
using System.ComponentModel.DataAnnotations.Schema;
using System.ComponentModel.DataAnnotations;
using Microsoft.AspNetCore.Mvc.ModelBinding;

namespace WebApp.Models {
    public class Product {

        public long ProductId { get; set; }

        [Required]
        [Display(Name = "Name")]
        public string Name { get; set; }
```

```
        [Column(TypeName = "decimal(8, 2)")]
        [Required(ErrorMessage = "Please enter a price")]
        [Range(1, 999999, ErrorMessage = "Please enter a positive price")]
        public decimal Price { get; set; }

        public long CategoryId { get; set; }
        public Category Category { get; set; }

        public long SupplierId { get; set; }
        public Supplier Supplier { get; set; }
    }
}
```

在代码清单中使用了两个验证属性：Required 和 Range。Required 属性指定如果用户没有提交属性的值，则它是一个验证错误。Range 属性指定可接受值的子集。表 29-7 显示了一组可用的内置验证属性。

**表 29-7　内置的验证属性**

| 属性 | 示例 | 描述 |
| --- | --- | --- |
| Compare | [Compare "OtherProperty")] | 此属性确保属性必须具有相同的值，这在两次要求用户提供相同信息(如电子邮件地址或密码)时非常有用 |
| Range | [Range(10, 20)] | 该属性确保数值(或实现 IComparable 的任何属性类型)不超出指定的最小值和最大值的范围。若要仅在一侧指定边界，请使用 MinValue 或 MaxValue 常量 |
| RegularExpression | [RegularExpression("pattern")] | 此属性确保字符串值与指定的正则表达式模式匹配。注意，模式必须匹配用户提供的整个值，而不仅是其中的子字符串。默认情况下，它区分大小写，但是可以通过应用(?i)修饰符来使它不区分大小写——即[RegularExpression("(?i)mypattern")] |
| Required | [Required] | 该属性确保值不是空的，也不是只由空格组成的字符串。如果希望将空白视为有效，请使用[Required(AllowEmptyStrings = true)] |
| StringLength | [StringLength(10)] | 此属性确保字符串值不超过指定的最大长度值。还可以指定最小长度[StringLength(10, MinimumLength=2)] |

所有验证属性都支持通过设置 ErrorMessage 属性的值来自定义错误消息，例如：

```
...
[Column(TypeName = "decimal(8, 2)")]
[Required(ErrorMessage = "Please enter a price")]
[Range(1, 999999, ErrorMessage = "Please enter a positive price")]
public decimal Price { get; set; }
...
```

如果没有自定义的错误消息，那么使用默认消息，但是默认消息倾向于揭示对用户来说毫无意义的模型类细节，除非也使用 Display 属性，如下所示：

```
...
[Required]
[Display(Name = "Name")]
public string Name { get; set; }
...
```

Required 属性生成的默认消息反映了 Display 属性指定的名称，因此不会向用户显示属性的名称。

**验证工作**

在使用验证属性时，为获得所需的验证结果需要特别注意。例如，如果希望确保用户选中了一个复选框，则不能使用 Required 属性，因为当复选框未选中时，浏览器将发送 false 值，而该值将始终能通过 Required 属性应用的检查。相反，使用 Range 属性并将最小值和最大值指定为 true，如下所示：

```
...
[Range(typeof(bool), "true", "true", ErrorMessage="You must check the box")]
..
```

如果觉得这种解决方法不合适，那么可以创建自定义验证属性，如下一节所述。

在 Product 类上使用验证属性允许删除对 Name 和 Price 属性的显式验证检查，如代码清单 29-15 所示。

**代码清单 29-15　在 Controllers 文件夹的 FormController.cs 文件中删除显式验证**

```
...
[HttpPost]
public IActionResult SubmitForm(Product product) {

    //if (string.IsNullOrEmpty(product.Name)) {
    //    ModelState.AddModelError(nameof(Product.Name), "Enter a name");
    //}

    //if (ModelState.GetValidationState(nameof(Product.Price))
    //        == ModelValidationState.Valid && product.Price < 1) {
    //    ModelState.AddModelError(nameof(Product.Price), "Enter a positive price");
    //}

    if (ModelState.GetValidationState(nameof(Product.Name))
            == ModelValidationState.Valid
            && ModelState.GetValidationState(nameof(Product.Price))
            == ModelValidationState.Valid
            && product.Name.ToLower().StartsWith("small") && product.Price > 100) {
        ModelState.AddModelError("", "Small products cannot cost more than $100");
    }
```

```
        if (!context.Categories.Any(c => c.CategoryId == product.CategoryId)) {
            ModelState.AddModelError(nameof(Product.CategoryId),
                "Enter an existing category ID");
        }

        if (!context.Suppliers.Any(s => s.SupplierId == product.SupplierId)) {
            ModelState.AddModelError(nameof(Product.SupplierId),
                "Enter an existing supplier ID");
        }

        if (ModelState.IsValid) {
            TempData["name"] = product.Name;
            TempData["price"] = product.Price.ToString();
            TempData["categoryId"] = product.CategoryId.ToString();
            TempData["supplierId"] = product.SupplierId.ToString();
            return RedirectToAction(nameof(Results));
        } else {
            return View("Form");
        }
    }
...
```

验证属性是在调用操作方法之前应用的，这意味着在执行模型级验证时，仍然可以依赖模型状态来确定各个属性是否有效。要查看验证属性的作用，请重新启动 ASP.NET Core MVC，请求 http://localhost:5000/controllers/form，清除 Name 和 Price 字段，并提交表单。响应将包括属性生成的验证错误，如图 29-8 所示。

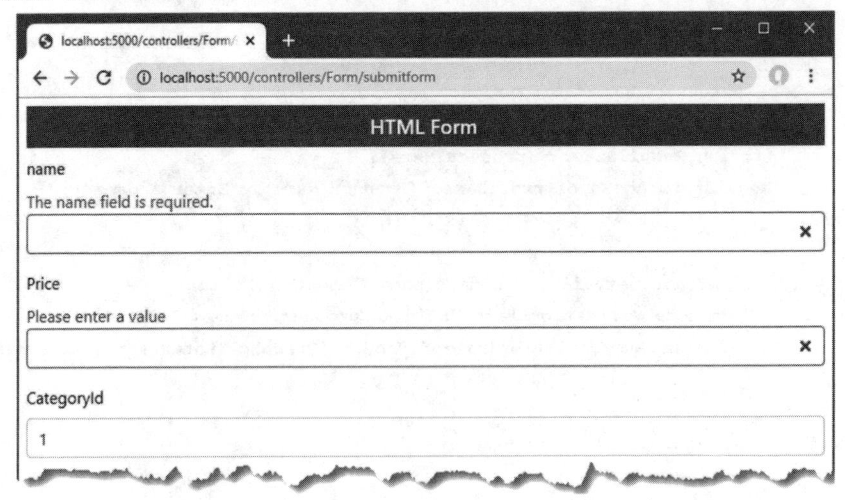


图 29-8　使用验证特性

**理解 Web 服务控制器验证**

用 ApiController 属性修饰的控制器不需要检查 ModelState.IsValid 属性。相反，只有在没有验证错误时才会调用操作方法，这意味着可以始终依赖通过模型绑定功能接收经过验证的对象。如果检测到任何验证错误，则终止请求，并向浏览器发送错误响应。

## 创建自定义属性的验证属性

可以通过创建扩展 ValidationAttribute 类的属性来扩展验证过程。为了演示，创建 WebApp/Validation 文件夹，并在其中添加了一个名为 PrimaryKeyAttribute.cs 的类文件。使用它来定义代码清单 29-16 所示的类。

**代码清单 29-16　Validation 文件夹中 PrimaryKeyAttribute.cs 文件的内容**

```
using Microsoft.EntityFrameworkCore;
using System;
using System.ComponentModel.DataAnnotations;

namespace WebApp.Validation {
    public class PrimaryKeyAttribute : ValidationAttribute {

        public Type ContextType { get; set; }

        public Type DataType { get; set; }

        protected override ValidationResult IsValid(object value,
                ValidationContext validationContext) {
            DbContext context
                = validationContext.GetService(ContextType) as DbContext;
            if (context.Find(DataType, value) == null) {
                return new ValidationResult(ErrorMessage
                    ?? "Enter an existing key value");
            } else {
                return ValidationResult.Success;
            }
        }
    }
}
```

自定义属性覆盖了 IsValid 方法和 ValidationContext 对象，该方法通过要检查的值来调用，ValidationContext 对象提供关于验证过程的上下文，并通过 GetService 方法提供对应用程序服务的访问。

在代码清单 29-16 中，自定义属性接收 Entity Framework Core 数据库上下文类的类型和模型类的类型。在 IsValid 方法中，属性获得上下文类的一个实例，并使用它查询数据库，以确定该值是否用作主键值。

**重新验证数据**

如果修改了从模型绑定器接收到的对象，那么可能需要再次执行验证过程。对于这些情况，使用 ModelState.Clear 方法清除任何现有的验证错误，并调用 TryValidateModel 方法。

定制验证属性也可以用于执行模型级验证。为便于演示，将一个名为 HraseAndPriceAttributere.cs 的类文件添加到 Validation 文件夹中，并使用它定义如代码清单 29-17 所示的类。

**代码清单 29-17　Validation 文件夹中 PhraseAndPriceAttribute.cs 文件的内容**

```
using System;
using System.ComponentModel.DataAnnotations;
using WebApp.Models;

namespace WebApp.Validation {
    public class PhraseAndPriceAttribute: ValidationAttribute {

        public string Phrase { get; set; }

        public string Price { get; set; }

        protected override ValidationResult IsValid(object value,
                ValidationContext validationContext) {
            Product product = value as Product;
            if (product != null
                && product.Name.StartsWith(Phrase,
                    StringComparison.OrdinalIgnoreCase)
                && product.Price > decimal.Parse(Price)) {
                    return new ValidationResult(ErrorMessage
                        ?? $"{Phrase} products cannot cost more than ${Price}");
            }
            return ValidationResult.Success;
        }
    }
}
```

该属性配置了 Phrase 和 Price 属性，这些属性在 IsValid 方法中用于检查模型对象的 Name 和 Price 属性。属性级的自定义验证属性直接应用于它们验证的属性，模型级属性应用于整个类，如代码清单 29-18 所示。

**代码清单 29-18　在 Models 文件夹的 Product.cs 文件中应用自定义验证属性**

```
using System.ComponentModel.DataAnnotations.Schema;
using System.ComponentModel.DataAnnotations;
using Microsoft.AspNetCore.Mvc.ModelBinding;
using WebApp.Validation;

namespace WebApp.Models {
```

```
    [PhraseAndPrice(Phrase ="Small", Price = "100")]
    public class Product {

        public long ProductId { get; set; }

        [Required]
        [Display(Name = "Name")]
        public string Name { get; set; }

        [Column(TypeName = "decimal(8, 2)")]
        [Required(ErrorMessage = "Please enter a price")]
        [Range(1, 999999, ErrorMessage = "Please enter a positive price")]
        public decimal Price { get; set; }

        [PrimaryKey(ContextType= typeof(DataContext), DataType = typeof(Category))]
        public long CategoryId { get; set; }
        public Category Category { get; set; }

        [PrimaryKey(ContextType = typeof(DataContext), DataType = typeof(Category))]
        public long SupplierId { get; set; }
        public Supplier Supplier { get; set; }
    }
}
```

自定义属性允许从表单控制器的操作方法中删除其余的显式验证语句，如代码清单 29-19 所示。

**代码清单 29-19　在 Controllers 文件夹的 FormController.cs 文件中删除显式验证**

```
using Microsoft.AspNetCore.Mvc;
using System.Linq;
using System.Threading.Tasks;
using WebApp.Models;
using Microsoft.EntityFrameworkCore;
using Microsoft.AspNetCore.Mvc.ModelBinding;

namespace WebApp.Controllers {

    [AutoValidateAntiforgeryToken]
    public class FormController : Controller {
        private DataContext context;

        public FormController(DataContext dbContext) {
            context = dbContext;
        }

        public async Task<IActionResult> Index(long? id) {
            return View("Form", await context.Products
```

```
                .FirstOrDefaultAsync(p => id == null || p.ProductId == id));
        }

        [HttpPost]
        public IActionResult SubmitForm(Product product) {
            if (ModelState.IsValid) {
                TempData["name"] = product.Name;
                TempData["price"] = product.Price.ToString();
                TempData["categoryId"] = product.CategoryId.ToString();
                TempData["supplierId"] = product.SupplierId.ToString();
                return RedirectToAction(nameof(Results));
            } else {
                return View("Form");
            }
        }

        public IActionResult Results() {
            return View(TempData);
        }
    }
}
```

验证属性会在调用操作方法之前自动应用，这意味着可以通过读取 ModelState. IsValid 属性来确定验证结果，同样的简化也可以应用到 Razor Pages，如代码清单 29-20 所示。

**代码清单 29-20　在 Pages 文件夹的 FormHandler.cshtml 文件中删除显式验证**

```
...
@functions {

    public class FormHandlerModel : PageModel {
        private DataContext context;

        public FormHandlerModel(DataContext dbContext) {
            context = dbContext;
        }

        [BindProperty]
        public Product Product { get; set; }

        public async Task OnGetAsync(long id = 1) {
            Product = await context.Products.FirstAsync(p => p.ProductId == id);
        }

        public IActionResult OnPost() {
            if (ModelState.IsValid) {
                TempData["name"] = Product.Name;
                TempData["price"] = Product.Price.ToString();
```

```
                TempData["categoryId"] = Product.CategoryId.ToString();
                TempData["supplierId"] = Product.SupplierId.ToString();
                return RedirectToPage("FormResults");
            } else {
                return Page();
            }
        }
    }
}
...
```

通过自定义属性表示验证将删除控制器和 Razor Pages 之间的代码重复，并确保在 Product 对象使用模型绑定的地方一致地应用验证。要测试验证属性，请重新启动 ASP.NET Core 并导航到 http://localhost:5000/controllers/form 或 http://localhost:5000/pages/form。清除表单字段或输入错误的键值，并提交表单。将显示由属性生成的错误消息，其中一些消息如图 29-9 所示(值 1、2 和 3 对 CategoryId 和 SupplierId 字段都有效)。

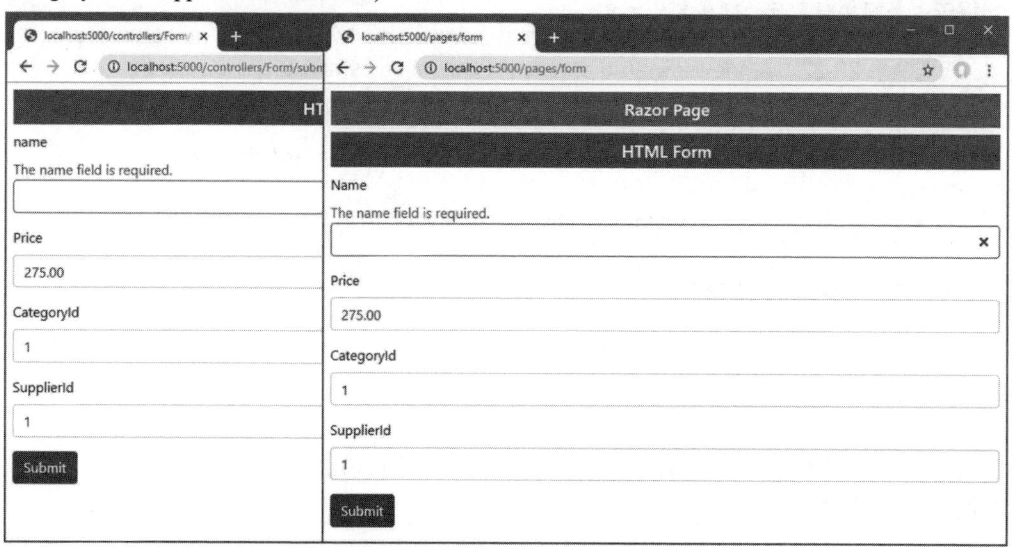


图 29-9 使用自定义验证属性

# 29.8 执行客户端验证

前面演示的验证技术都是服务器端验证的示例。这意味着用户向服务器提交数据，服务器验证数据并返回验证结果(要么成功处理数据，要么有一些需要纠正的错误)。

在 Web 应用程序中，用户通常希望立即得到验证反馈——而不必向服务器提交任何内容。这称为客户端验证，使用 JavaScript 实现。用户输入的数据在发送到服务器之前进行验证，从而为用户提供即时反馈和纠正任何问题的机会。

ASP.NET Core 支持不引人注目的客户端验证。术语“不引人注目”表示使用添加到 HTML 元素(由视图生成)中的属性来表达验证规则。这些属性由 Microsoft 分发的一个 JavaScript 库解释，该库反过来配置 jQuery 验证库，jQuery 验证库执行实际的验证工作。下面将展示内置验证支持是如何工作的，并演示如何扩展该功能，以提供定制的客户端验证。

第一步是安装处理验证的JavaScript包。打开一个新的PowerShell命令提示符，导航到WebApp项目文件夹，并运行代码清单 29-21 所示的命令。

**提示：** 核心 jQuery 命令是在第 26 章添加到项目中的。如果需要再次安装它，可运行以下命令：libman install jquery@3.4.1 -d wwwroot/lib/jquery。

**代码清单 29-21　安装验证包**

```
libman install jquery-validate@1.19.1 -d wwwroot/lib/jquery-validate
libman install jquery-validation-unobtrusive@3.2.11 -d wwwroot/lib/jquery-validation-
unobtrusive
```

安装包之后，在 Views/Shared 文件夹的_Validation.cshtml 文件中添加代码清单 29-22 中所示的元素，它提供了一种方便的方法，可将验证与应用程序中现有的 jQuery 代码一起引入。

**提示：** 必须按照显示的顺序定义元素。

**代码清单 29-22　在 Views/Shared 文件夹的_Validation.cshtml 文件中添加元素**

```
<script src="/lib/jquery/jquery.min.js"></script>
<script src="~/lib/jquery-validate/jquery.validate.min.js"></script>
<script
   src="~/lib/jquery-validation-unobtrusive/jquery.validate.unobtrusive.min.js">
</script>
<script type="text/javascript">
    $(document).ready(function () {
       $("input.input-validation-error").addClass("is-invalid");
    });
</script>
```

标签助手将 data-val*属性添加到描述字段验证约束的 input 元素中。下面是添加到 Name 字段的 input 元素的属性，例如：

```
...
<input class="form-control valid" type="text" data-val="true" data-val-required="The
name field is required."
id="Name" name="Name" value="Kayak" aria-describedby="Name-error"
aria-invalid="false">
...
```

不显眼的验证 JavaScript 代码会查找这些属性，并在用户尝试提交表单时在浏览器中执行验证。表单不会被提交，如果存在验证问题，将显示一个错误。只有解决全部验证问题，数据才会发送到应用程序。

JavaScript 代码查找具有 data-val 属性的元素，并在用户提交表单时在浏览器中执行本地验证，而不向服务器发送 HTTP 请求。通过使用 F12 工具运行应用程序并提交表单，可以看到效果，即使没有向服务器发送 HTTP 请求，也会显示验证错误消息。

**避免与浏览器验证冲突**

当前的一些 HTML5 浏览器支持基于应用于 input 元素的属性进行简单的客户端验证。例如，当用户尝试提交表单而没有提供值时，一个已应用了 reguired 属性的 input 元素将导致浏览器显示验证错误。

如果使用标签助手生成表单元素，就不会在浏览器验证中遇到任何问题，因为分配数据属性的元素会被浏览器忽略。

但是，如果不能完全控制应用程序中的标记，可能会遇到问题，这种情况在传递别处生成的内容时经常发生。结果是 jQuery 验证和浏览器验证都可以对表单进行操作，这只会让用户感到困惑。要避免此问题，可以向表单元素添加 novalidate 属性来禁用浏览器验证。

一个很好的客户端验证特性是在客户端和服务器上应用指定验证规则的相同属性。这意味着来自不支持 JavaScript 的浏览器的数据将与来自支持 JavaScript 的浏览器的数据进行相同的验证，而不需要任何额外的工作。

要测试客户端验证特性，请求 http://localhost:5000/controllers/Form 或 http://localhost:5000/pages/Form，清除 Name 字段，然后单击 Submit 按钮。

错误消息看起来与服务器端验证生成的消息类似，但是如果在字段中输入文本，将看到错误消息立即消失，因为 JavaScript 代码响应用户的交互，如图 29-10 所示。

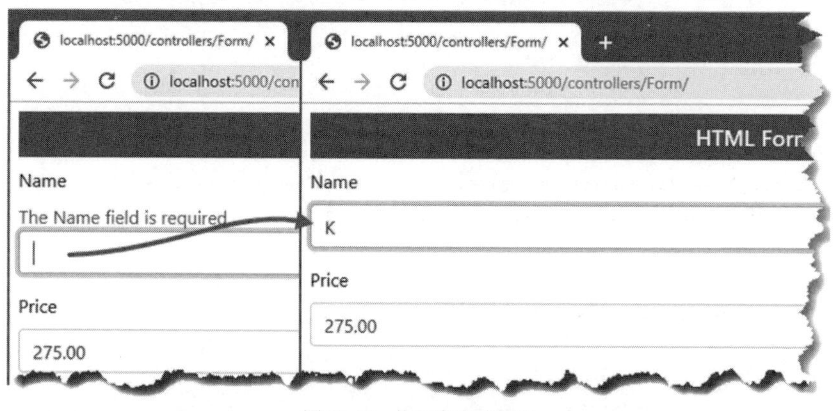


图 29-10　执行客户端验证

**扩展客户端验证**

客户端验证特性支持内置的属性级特性。该特性可以扩展，但需要熟练使用 Javascript，并且需要直接使用 jQuery 验证包。详细信息请参见 https://jqueryvalidation.org/documentation。

如果不想开始编写 Javascript 代码，那么可以遵循以下常见模式：使用客户端验证进行内置验证检查，使用服务器端验证进行自定义验证。

## 29.9 执行远程验证

远程验证模糊了客户端验证和服务器端验证之间的界线：验证检查由客户端 JavaScript 代码执行；通过向应用程序发送异步 HTTP 请求来测试用户在表单中输入的值，从而进行验证检查。

远程验证的一个常见示例是检查应用程序中是否有可用的用户名，当用户名必须唯一的时，用户提交数据，然后执行客户端验证。在此过程中，向服务器发出一个异步 HTTP 请求，以验证所请求的用户名。如果用户名已获取，则显示一个验证错误，以便用户可以输入另一个值。

这看起来像是常规的服务器端验证，但这种方法有一些好处。首先，只有一些属性进行远程验证；客户端验证的好处仍然适用于用户输入的其他所有数据值。其次，请求是相对轻量级的，并且关注于验证，而不是处理整个模型对象。

第三个区别是远程验证是在后台执行的。用户不必单击 Submit 按钮，然后等待新视图的呈现和返回。这可以提供响应更快的用户体验，特别是在浏览器和服务器之间的网络速度较慢的情况下。

也就是说，远程验证是一种折中。它在客户端验证和服务器端验证之间取得了平衡，但是它确实需要向应用服务器请求，而且它的验证速度不如普通的客户端验证快。

对于示例应用程序，使用远程验证来确保用户为 CategoryId 和 SupplierId 属性输入现有的键值。第一步是创建一个 Web 服务控制器，其操作方法将执行验证检查。将一个名为 ValidationController.cs 的类文件添加到 Controllers 文件夹，如代码清单 29-23 所示。

#### 代码清单 29-23　Controllers 文件夹中 ValidationController.cs 文件的内容

```
using Microsoft.AspNetCore.Mvc;
using WebApp.Models;

namespace WebApp.Controllers {

    [ApiController]
    [Route("api/[controller]")]
    public class ValidationController: ControllerBase {
        private DataContext dataContext;

        public ValidationController(DataContext context) {
            dataContext = context;
        }

        [HttpGet("categorykey")]
        public bool CategoryKey(string categoryId) {
            long keyVal;
            return long.TryParse(categoryId, out keyVal)
                && dataContext.Categories.Find(keyVal) != null;
        }

        [HttpGet("supplierkey")]
        public bool SupplierKey(string supplierId) {
            long keyVal;
            return long.TryParse(supplierId, out keyVal)
                && dataContext.Suppliers.Find(keyVal) != null;
        }
    }
}
```

验证操作方法必须定义一个参数，该参数的名称与它们要验证的字段相匹配，这允许模型绑定过程从请求查询字符串中提取要测试的值。操作方法的响应必须是 JSON，并且只能为 true 或 false，以指示一个值是否可接受。代码清单 29-23 中的操作方法接收候选值，并检查它们是否用作 Category 或 Supplier 对象的数据库键。

---

■ **提示：** 本可以利用模型绑定的优势，这样操作方法的参数将转换为 long 值，但这样做将意味着，如果用户输入的值不能转换为 long 类型，验证方法将不会调用。如果模型绑定器不能转换值，那么 MVC 框架就不能调用操作方法，验证也不能执行。通常，远程验证的最佳方法是在操作方法中接收字符串参数，并显式地执行任何类型转换、解析或模型绑定。

---

为了使用远程验证方法，将 Remote 属性应用到 Product 类中的 CategoryId 和 SupplierId 属性，如代码清单 29-24 所示。

### 代码清单 29-24　在 Models 文件夹的 Product.cs 文件中使用 Remote 属性

```
using System.ComponentModel.DataAnnotations.Schema;
using System.ComponentModel.DataAnnotations;
using Microsoft.AspNetCore.Mvc.ModelBinding;
using WebApp.Validation;
using Microsoft.AspNetCore.Mvc;

namespace WebApp.Models {

    [PhraseAndPrice(Phrase ="Small", Price = "100")]
    public class Product {

        public long ProductId { get; set; }

        [Required]
        [Display(Name = "Name")]
        public string Name { get; set; }

        [Column(TypeName = "decimal(8, 2)")]
        [Required(ErrorMessage = "Please enter a price")]
        [Range(1, 999999, ErrorMessage = "Please enter a positive price")]
        public decimal Price { get; set; }

        [PrimaryKey(ContextType= typeof(DataContext),
           DataType = typeof(Category))]
        [Remote("CategoryKey", "Validation", ErrorMessage = "Enter an existing key")]
        public long CategoryId { get; set; }
        public Category Category { get; set; }

        [PrimaryKey(ContextType = typeof(DataContext),
           DataType = typeof(Category))]
        [Remote("SupplierKey", "Validation", ErrorMessage = "Enter an existing key")]
        public long SupplierId { get; set; }
```

```
            public Supplier Supplier { get; set; }
        }
    }
```

Remote 属性的参数指定验证控制器的名称及其操作方法。还使用可选的 ErrorMessage 参数来指定验证失败时显示的错误消息。要查看远程验证，请重新启动 ASP.NET Core 并导航到 http://localhost:5000/controllers/form，输入一个无效的键值，然后提交表单。你将看到一条错误消息，在每次按下键之后，将验证输入元素的值，如图 29-11 所示(只有值 1、2 和 3 对 CategoryId 和 SupplierId 字段都有效)。

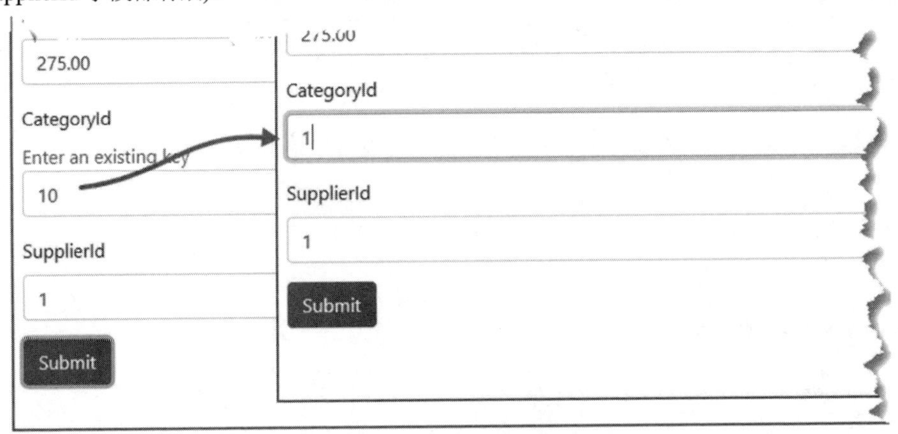


图 29-11　执行远程验证

---

■ **注意：** 验证操作方法将在用户首次提交表单时调用，在每次编辑数据时也会调用。对于文本输入元素，每次击键都将导致对服务器的调用。对于某些应用程序，这可能是相当多的请求，在指定应用程序在生产中需要的服务器容量和带宽时必须考虑到这一点。此外，对于验证开销较大的属性，可能选择不使用远程验证(示例反复查询数据库中的键值，这可能对所有应用程序或数据库都不合理)。

---

## 在 Razor Pages 中执行远程验证

远程验证可以在 Razor Pages 中工作，但是必须注意用于验证值的异步 HTTP 请求中使用的名称。对于上一节中的控制器示例，浏览器会向 URL 发送请求，如下所示：

```
http://localhost:5000/api/Validation/categorykey?CategoryId=1
```

但是对于示例 Razor Pages，URL 如下所示，反映了页面模型的使用：

```
http://localhost:5000/api/Validation/categorykey?Product.CategoryId=1
```

我喜欢通过向接收这两种请求的验证操作方法添加参数来解决这一差异，使用前面章节中描述的模型绑定特性很容易做到这一点，如代码清单 29-25 所示。

### 代码清单 29-25 在 Controllers 文件夹的 ValidationController.cs 文件中添加参数

```
using Microsoft.AspNetCore.Mvc;
using WebApp.Models;

namespace WebApp.Controllers {

    [ApiController]
    [Route("api/[controller]")]
    public class ValidationController: ControllerBase {
        private DataContext dataContext;

        public ValidationController(DataContext context) {
            dataContext = context;
        }

        [HttpGet("categorykey")]
        public bool CategoryKey(string categoryId, [FromQuery] KeyTarget target) {
            long keyVal;
            return long.TryParse(categoryId ?? target.CategoryId, out keyVal)
                && dataContext.Categories.Find(keyVal) != null;
        }

        [HttpGet("supplierkey")]
        public bool SupplierKey(string supplierId, [FromQuery] KeyTarget target) {
            long keyVal;
            return long.TryParse(supplierId ?? target.SupplierId, out keyVal)
                && dataContext.Suppliers.Find(keyVal) != null;
        }
    }
    [Bind(Prefix = "Product")]
    public class KeyTarget {
        public string CategoryId { get; set; }
        public string SupplierId{ get; set; }
    }
}
```

KeyTarget 类配置为绑定到请求的 Product 部分，其属性匹配这两种类型的远程验证请求。每个操作方法都有一个 KeyTarget 参数，如果没有接收到现有参数的值，就会使用这个参数。这允许相同的操作方法容纳两种类型的请求，为了查看效果，可重新启动 ASP.NET Core，导航到 http://localhost:5000/pages/form，输入一个不存在的键值，然后单击 Submit 按钮，这将生成如图 29-12 所示的响应。

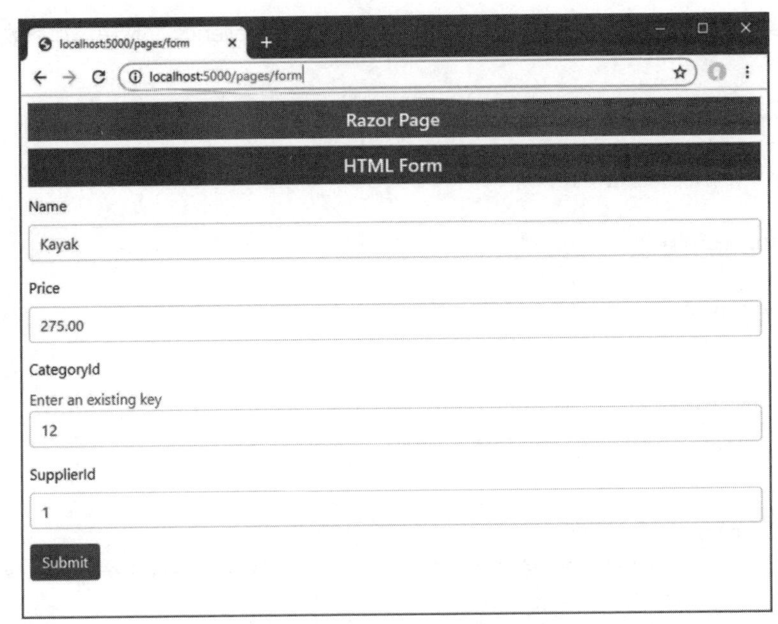


图 29-12　使用 Razor Pages 执行远程验证

## 29.10　小结

本章描述了 ASP.NET Core 数据验证功能。解释了如何显式地执行验证，如何使用属性来描述验证约束，以及如何验证单个属性和整个对象。演示了如何向用户显示验证消息，以及如何通过客户端和远程验证改进用户的验证体验。在第 30 章将描述 ASP.NET Core 过滤器功能。

# 第30章

# 使用过滤器

过滤器将额外的逻辑注入请求处理。过滤器类似于应用于单个端点的中间件，可以是操作或页面处理程序方法，它们提供了一种管理特定请求集的优雅方法。本章解释了过滤器是如何工作的，描述了ASP.NET Core支持的不同类型的过滤器，并演示自定义过滤器和ASP.NET Core提供的过滤器的使用。表30-1总结了本章的内容。

表30-1　本章内容摘要

| 问题 | 解决方案 | 代码清单 |
|---|---|---|
| 实现安全策略 | 使用授权过滤器 | 30-15～30-16 |
| 实现资源策略，如缓存 | 使用资源过滤器 | 30-17～30-19 |
| 更改操作方法的请求或响应 | 使用操作过滤器 | 30-20～30-23 |
| 更改页面处理程序方法的请求或响应 | 使用页面过滤器 | 30-24～30-26 |
| 检查或更改端点产生的结果 | 使用结果过滤器 | 30-27～30-29 |
| 检查或更改未捕获的异常 | 使用异常过滤器 | 30-30～30-31 |
| 改变过滤器生命周期 | 使用过滤器工厂或定义服务 | 30-32～30-35 |
| 在整个应用程序中应用过滤器 | 使用全局过滤器 | 30-36～30-37 |
| 改变应用过滤器的顺序 | 实现IOrderedFilter接口 | 30-38～30-42 |

## 30.1　准备工作

本章使用了第29章中的WebApp项目。为了准备这一章，打开一个新的PowerShell命令提示符，导航到WebApp项目文件夹，运行代码清单30-1所示的命令来删除不再需要的文件。

**代码清单30-1　从项目中删除文件**

```
Remove-Item -Path Controllers,Views,Pages -Recurse -Exclude _*,Shared
```

该命令删除控制器、视图和Razor Pages，留下共享布局、数据模型和配置文件。

■ **提示：**可从https://github.com/apress/pro-asp.net-core-3下载本章和本书中其他所有章节的示例项目。如果在运行示例时遇到问题，请参阅第1章以了解如何获得帮助。

创建 WebApp/Controllers 文件夹，并将一个名为 HomeController.cs 的类文件添加到 Controllers 文件夹，如代码清单 30-2 所示。

**代码清单 30-2　Controllers 文件夹中 HomeController.cs 文件的内容**

```
using Microsoft.AspNetCore.Mvc;

namespace WebApp.Controllers {

    public class HomeController : Controller {

        public IActionResult Index() {
            return View("Message",
               "This is the Index action on the Home controller");
        }
    }
}
```

操作方法呈现一个名为 Message 的视图，并传递一个字符串作为视图数据。添加一个名为 Message.cshtml 的 Razor 视图，内容如代码清单 30-3 所示。

**代码清单 30-3　Views/Shared 文件夹中 Message.cshtml 文件的内容**

```
@{ Layout = "_SimpleLayout"; }

@if (Model is string) {
   @Model
} else if (Model is IDictionary<string, string>) {
    var dict = Model as IDictionary<string, string>;
    <table class="table table-sm table-striped table-bordered">
        <thead><tr><th>Name</th><th>Value</th></tr></thead>
        <tbody>
            @foreach (var kvp in dict) {
               <tr><td>@kvp.Key</td><td>@kvp.Value</td></tr>
            }
        </tbody>
    </table>
}
```

在 Pages 文件夹中添加一个名为 Message.cshtml 的 Razor Pages，并添加如代码清单 30-4 所示的内容。

**代码清单 30-4　Pages 文件夹中 Message.cshtml 文件的内容**

```
@page "/pages/message"
@model MessageModel
@using Microsoft.AspNetCore.Mvc.RazorPages
@using System.Collections.Generic
```

```
@if (Model.Message is string) {
    @Model.Message
} else if (Model.Message is IDictionary<string, string>) {
    var dict = Model.Message as IDictionary<string, string>;
    <table class="table table-sm table-striped table-bordered">
        <thead><tr><th>Name</th><th>Value</th></tr></thead>
        <tbody>
            @foreach (var kvp in dict) {
                <tr><td>@kvp.Key</td><td>@kvp.Value</td></tr>
            }
        </tbody>
    </table>
}

@functions {
    public class MessageModel : PageModel {

        public object Message { get; set; } = "This is the Message Razor Page";
    }
}
```

### 30.1.1 启用 HTTPS 连接

本章中的一些示例需要使用 SSL。将代码清单 30-5 中所示的配置条目添加到 Properties 文件夹的 launchSettings.json 文件中，来启用 SSL 和 HTTPS，并将端口设置为 44350。

**代码清单 30-5　在 Properties 文件夹的 launchSettings.json 文件中启用 HTTPS**

```
{
  "iisSettings": {
    "windowsAuthentication": false,
    "anonymousAuthentication": true,
    "iisExpress": {
      "applicationUrl": "http://localhost:5000",
      "sslPort": 44350
    }
  },
  "profiles": {
    "IIS Express": {
      "commandName": "IISExpress",
      "launchBrowser": true,
      "environmentVariables": {
        "ASPNETCORE_ENVIRONMENT": "Development"
      }
    },
    "WebApp": {
      "commandName": "Project",
```

```
        "launchBrowser": true,
        "environmentVariables": {
          "ASPNETCORE_ENVIRONMENT": "Development"
        },
        "applicationUrl": "http://localhost:5000;https://localhost:44350"
      }
    }
  }
```

.NET Core 运行时包含一个用于 HTTPS 请求的测试证书。在 WebApp 文件夹中运行代码清单 30-6 所示的命令，以重新生成测试证书并信任它。

**代码清单 30-6　重新生成开发证书**

```
dotnet dev-certs https --clean
dotnet dev-certs https --trust
```

单击 Yes，提示删除已信任的现有证书；再次单击 Yes 表示信任新证书，如图 30-1 所示。

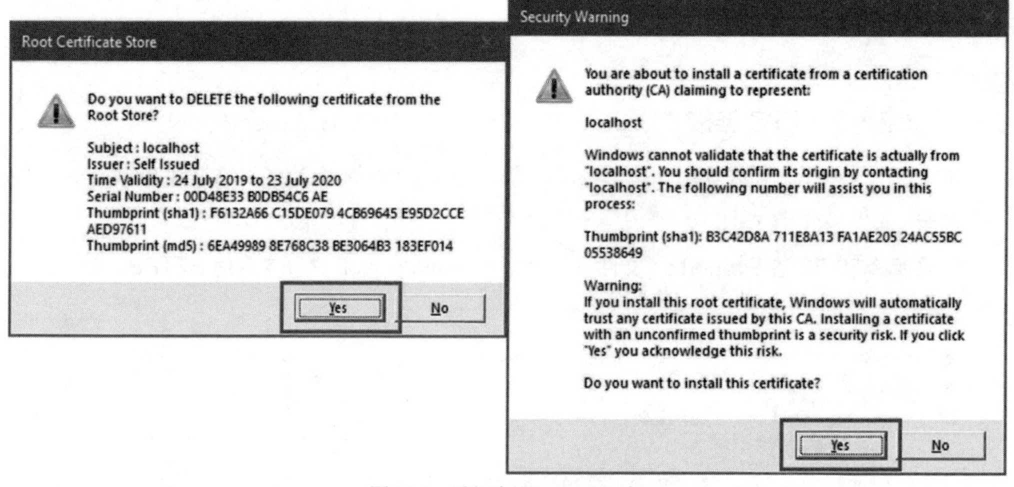


图 30-1　重新生成 HTTPS 证书

## 30.1.2　删除数据库

打开一个新的 PowerShell 命令提示符，导航到包含 WebApp.csproj 文件的文件夹，并运行代码清单 30-7 所示的命令来删除数据库。

**代码清单 30-7　删除数据库**

```
dotnet ef database drop --force
```

### 30.1.3 运行示例应用程序

从 Debug 菜单中选择 Start Without Debugging 或 Run Without Debugging，或者使用 PowerShell 命令提示符运行代码清单 30-8 所示的命令。

**代码清单 30-8 运行示例应用程序**

```
dotnet run
```

使用浏览器请求 http://localhost:5000 和 https://localhost:44350。两个 URL 都将由主控制器定义的索引操作处理，生成如图 30-2 所示的响应。

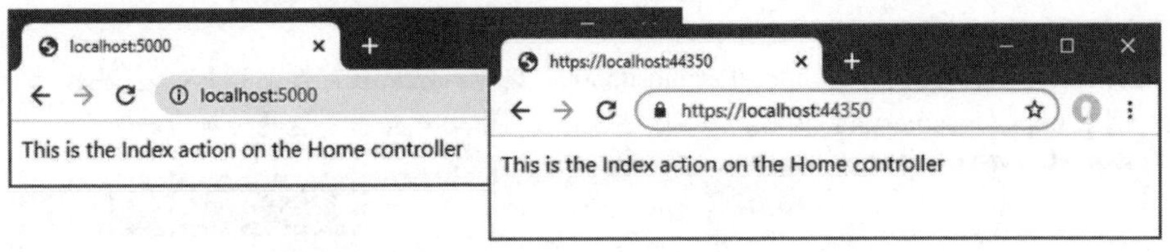


图 30-2 来自主控制器的响应

请求 http://localhost:5000/pages/message 和 https://localhost:44350/pages/message 以查看通过 HTTP和HTTPS传递的、来自Message Razor Pages的响应，如图 30-3 所示。

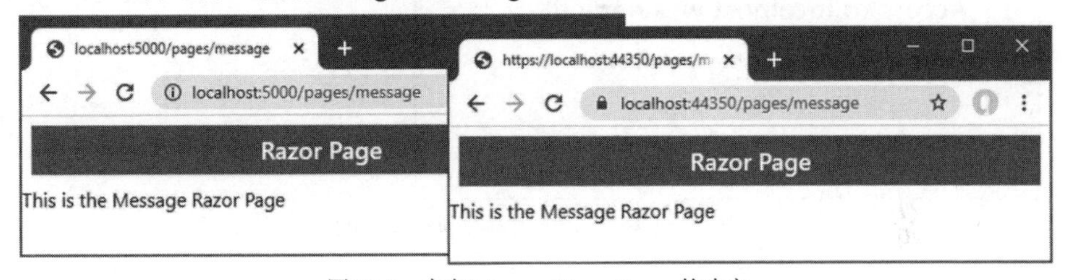


图 30-3 来自 Message Razor Pages 的响应

## 30.2 使用过滤器

对于将应用于中间件组件或操作方法的逻辑，可使用过滤器，在可以轻松重用的类中进行定义。

假设想对一些操作方法强制执行 HTTPS 请求。第 16 章展示了如何在中间件中通过读取 HttpRequest 对象的 IsHttps 属性来实现这一点。这种方法的问题是，中间件必须了解路由系统的配置，才能知道如何拦截针对特定操作方法的请求。更有针对性的方法是阅读操作方法中的 HttpRequest.IsHttps 属性。如代码清单 30-9 所示。

**代码清单 30-9 在 Controllers 文件夹的 HomeController.cs 文件中有选择地执行 HTTPS**

```
using Microsoft.AspNetCore.Mvc;
using Microsoft.AspNetCore.Http;

namespace WebApp.Controllers {
```

```
    public class HomeController : Controller {

        public IActionResult Index() {
            if (Request.IsHttps) {
                return View("Message",
                    "This is the Index action on the Home controller");
            } else {
                return new StatusCodeResult(StatusCodes.Status403Forbidden);
            }
        }
    }
}
```

重启 ASP.NET Core 并请求 http://localhost:5000。这个方法现在需要 HTTPS，会显示一个错误响应。请求 https://localhost:44350，会看到消息输出。图 30-4 显示了这两种响应。

图 30-4　在操作方法中强制执行 HTTPS

---

**提示：** 如果没有从本节的例子中得到想要的结果，请清除浏览器的历史记录。浏览器通常会拒绝向以前生成 HTTPS 错误的服务器发送请求，这是一种良好的安全实践，但在开发过程中可能会令人沮丧。

---

这种方法有效，但存在问题。第一个问题是操作方法包含的代码更多是关于实现安全策略的，而不是关于处理请求的。一个更严重的问题是，在操作方法中包含 HTTP 检测代码伸缩性不好，必须在控制器的每个操作方法中复制，如代码清单 30-10 所示。

#### 代码清单 30-10　在 Controllers 文件夹的 HomeController.cs 文件中添加操作方法

```
using Microsoft.AspNetCore.Mvc;
using Microsoft.AspNetCore.Http;

namespace WebApp.Controllers {

    public class HomeController : Controller {

        public IActionResult Index() {
```

```
            if (Request.IsHttps) {
                return View("Message",
                    "This is the Index action on the Home controller");
            } else {
                return new StatusCodeResult(StatusCodes.Status403Forbidden);
            }
        }

        public IActionResult Secure() {
            if (Request.IsHttps) {
                return View("Message",
                    "This is the Secure action on the Home controller");
            } else {
                return new StatusCodeResult(StatusCodes.Status403Forbidden);
            }
        }
    }
}
```

必须记得在想要使用 HTTPS 的每个控制器的每个操作方法中实现相同的检查。实现安全策略的代码是控制器的一个重要部分，它使控制器更难理解，而且迟早会忘记将其添加到新的操作方法中，从而在安全策略中创建一个漏洞。

这就是过滤器要解决的问题类型。代码清单 30-11 替换了对 HTTPS 的检查，并实现了一个过滤器。

### 代码清单 30-11　在 Controllers 文件夹的 HomeController.cs 文件中应用过滤器

```
using Microsoft.AspNetCore.Mvc;
using Microsoft.AspNetCore.Http;

namespace WebApp.Controllers {

    public class HomeController : Controller {

        [RequireHttps]
        public IActionResult Index() {
            return View("Message",
                "This is the Index action on the Home controller");
        }

        [RequireHttps]
        public IActionResult Secure() {
            return View("Message",
                "This is the Secure action on the Home controller");
        }
    }
}
```

RequireHttps 属性应用 ASP.NET Core 提供的一个内置过滤器。这个过滤器限制了对操作方法的访问，因此只支持 HTTPS 请求，允许从每个方法中删除安全代码，并专注于处理成功的请求。

---

■ **注意**：RequireHttps 过滤器的工作方式与自定义代码不一样。对于 GET 请求，RequireHttps 属性将客户端重定向到最初请求的 URL，但是它是通过使用 HTTPS 模式实现的，这样对 http://localhost:5000 的请求将重定向到 https://localhost:5000。这对于大多数已部署的应用程序来说是有意义的，但在开发期间没有意义，因为 HTTP 和 HTTPS 位于不同的本地端口。RequireHttpsAttribute 类定义了一个名为 HandleNonHttpsRequest 的受保护方法，可以重写该方法来更改行为。或者按“理解授权过滤器”一节的讲解从头重新创建原始功能。

---

仍然必须记住对每个操作方法应用 RequireHttps 属性，这意味着可能会忘记。但是过滤器有一个有用的技巧：将属性应用到控制器类与将它应用到每个单独的操作方法具有相同的效果，如代码清单 30-12 所示。

**代码清单 30-12　在 Controllers 文件夹的 HomeController.cs 文件中应用过滤器**

```
using Microsoft.AspNetCore.Mvc;
using Microsoft.AspNetCore.Http;

namespace WebApp.Controllers {

    [RequireHttps]
    public class HomeController : Controller {

        public IActionResult Index() {
            return View("Message",
                "This is the Index action on the Home controller");
        }

        public IActionResult Secure() {
            return View("Message",
                "This is the Secure action on the Home controller");
        }
    }
}
```

可以应用不同粒度级别的过滤器。如果希望限制对某些操作的访问，而不是对其他操作的访问，那么可仅对这些方法应用 RequireHttps 属性。如果希望保护所有的操作方法，包括将来添加到控制器中的任何操作方法，那么可将 RequireHttps 属性应用于该类。如果想对应用程序中的每个操作应用过滤器，那么可使用全局过滤器，参见本章后面的内容。

## 在 Razor Pages 中使用过滤器

也可以在 Razor Pages 中使用过滤器。例如，要在 Message Razor Pages 中实现仅限 HTTP 的策略，必须添加一个检查连接的处理程序方法，如代码清单 30-13 所示。

代码清单 30-13 在 Pages 文件夹的 Message.cshtml 文件中检查连接

```
@page "/pages/message"
@model MessageModel
@using Microsoft.AspNetCore.Mvc.RazorPages
@using System.Collections.Generic
@using Microsoft.AspNetCore.Http

@if (Model.Message is string) {
    @Model.Message
} else if (Model.Message is IDictionary<string, string>) {
    var dict = Model.Message as IDictionary<string, string>;
    <table class="table table-sm table-striped table-bordered">
        <thead><tr><th>Name</th><th>Value</th></tr></thead>
        <tbody>
            @foreach (var kvp in dict) {
                <tr><td>@kvp.Key</td><td>@kvp.Value</td></tr>
            }
        </tbody>
    </table>
}

@functions {

    public class MessageModel : PageModel {

        public object Message { get; set; } = "This is the Message Razor Page";

        public IActionResult OnGet() {
            if (!Request.IsHttps) {
                return new StatusCodeResult(StatusCodes.Status403Forbidden);
            } else {
                return Page();
            }
        }
    }
}
```

处理程序方法可以工作，但是它很笨拙，并且会出现与操作方法相同的问题。在 Razor Pages 中使用过滤器时，该属性可以应用于处理程序方法(如代码清单 30-14 所示)，也可以应用于整个类。

代码清单 30-14 在 Pages 文件夹的 Message.cshtml 文件中应用过滤器

```
@page "/pages/message"
@model MessageModel
@using Microsoft.AspNetCore.Mvc.RazorPages
@using System.Collections.Generic
@using Microsoft.AspNetCore.Http
```

```
@if (Model.Message is string) {
   @Model.Message
} else if (Model.Message is IDictionary<string, string>) {
    var dict = Model.Message as IDictionary<string, string>;
    <table class="table table-sm table-striped table-bordered">
        <thead><tr><th>Name</th><th>Value</th></tr></thead>
        <tbody>
            @foreach (var kvp in dict) {
               <tr><td>@kvp.Key</td><td>@kvp.Value</td></tr>
            }
        </tbody>
    </table>
}

@functions {

    [RequireHttps]
    public class MessageModel : PageModel {

       public object Message { get; set; } = "This is the Message Razor Page";
    }
}
```

如果请求 https://localhost:44350/pages/message，将看到一个正常的响应。如果请求常规 HTTP URL，如 http://localhost:5000/pages/messages，过滤器将重定向请求，你将看到一个错误(如前所述，RequireHttps 过滤器将浏览器重定向到示例应用程序中未启用的端口)。

## 30.3 理解过滤器

ASP.NET Core 支持不同类型的过滤器，每种过滤器用于不同的目的。表 30-2 描述了过滤器类别。

表 30-2　过滤器类型

| 名称 | 描述 |
| --- | --- |
| 授权过滤器 | 这种类型的过滤器用于应用应用程序的授权策略 |
| 资源过滤器 | 这种类型的过滤器用于拦截请求，通常用于实现诸如缓存的特性 |
| 操作过滤器 | 这种类型的过滤器用于在请求被操作方法接收之前修改请求，或者在生成请求之后修改操作结果。这种类型的过滤器只能应用于控制器和操作 |
| 页面过滤器 | 这种类型的过滤器用于在 Razor Pages 处理程序方法接收到请求之前修改请求，或者在生成操作结果之后修改操作结果。这种类型的过滤器只能应用于 Razor Pages |
| 结果过滤器 | 此类型的过滤器用于在操作执行前更改操作结果，或在执行后修改结果 |
| 异常过滤器 | 这种类型的过滤器用于处理操作方法或页面处理程序执行期间发生的异常 |

过滤器有自己的管道，并按照特定的顺序执行，如图 30-5 所示。

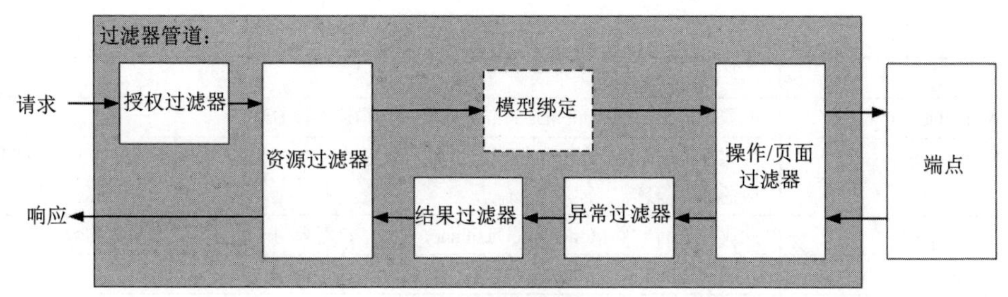


图 30-5 管道过滤器

过滤器可以短路过滤器管道，以阻止请求被转发到下一个过滤器。例如，如果用户未经身份验证，授权过滤器可以短路管道，并返回错误响应。资源、操作和页面过滤器能够在请求被端点处理之前和之后检查请求，允许这些类型的过滤器短路管道；在处理该请求之前更改该请求；或者改变反应(图 30-5 简化了过滤器的流程。页面过滤器在模型绑定过程之前和之后运行，如“理解页面过滤器”一节所述)。

每种类型的过滤器都是使用 ASP.NET Core 定义的接口实现的。它还提供了一些基类，可以方便地将某些类型的过滤器应用为属性。表 30-3 中显示了它们，以供快速参考。

表 30-3 过滤器类型、接口和属性基类

| 过滤器类型 | 接口 | 属性类 |
| --- | --- | --- |
| 授权过滤器 | IAuthorizationFilter、IAsyncAuthorizationFilter | 没有提供属性类 |
| 资源过滤器 | IResourceFilter、IAsyncResourceFilter | 没有提供属性类 |
| 操作过滤器 | IActionFilter、IAsyncActionFilter | ActionFilterAttribute |
| 页面过滤器 | IPageFilter、IAsyncPageFilter | 没有提供属性类 |
| 结果过滤器 | IResultFilter、IAsyncResultFilter、IAlwaysRunResultFilter、IAsyncAlwaysRunResultFilter | ResultFilterAttribute |
| 异常过滤器 | IexceptionFilter、IAsyncExceptionFilter | ExceptionFilterAttribute |

# 30.4 创建自定义过滤器

过滤器实现 IFilterMetadata 接口，它在 Microsoft.AspNetCore.Mvc.Filters 名称空间中。接口如下：

```
namespace Microsoft.AspNetCore.Mvc.Filters {
    public interface IFilterMetadata { }
}
```

该接口是空的，并且不需要过滤器来实现任何特定行为。这是因为上一节中描述的每个过滤器类别都以不同的方式工作。过滤器以 FilterContext 对象的形式提供上下文数据。为了方便起见，表 30-4 描述了 FilterContext 提供的属性。

表 30-4　FilterContext 属性

| 名称 | 描述 |
| --- | --- |
| ActionDescriptor | 此属性返回一个 ActionDescriptor 对象，它描述操作方法 |
| HttpContext | 此属性返回一个 HttpContext 对象，该对象提供 HTTP 请求和在返回时发送的 HTTP 响应的详细信息 |
| ModelState | 此属性返回一个 ModelStateDictionary 对象，该对象用于验证客户端发送的数据 |
| RouteData | 此属性返回一个 RouteData 对象，该对象描述路由系统处理请求的方式 |
| Filters | 此属性返回已应用于操作方法的过滤器列表，表示为 IList < IFilterMetadata > |

### 30.4.1　理解授权过滤器

授权过滤器用于实现应用程序的安全策略。授权过滤器在其他类型的过滤器和端点处理请求之前执行。以下是 IAuthorizationFilter 接口的定义：

```
namespace Microsoft.AspNetCore.Mvc.Filters {

    public interface IAuthorizationFilter : IFilterMetadata {

        void OnAuthorization(AuthorizationFilterContext context);
    }
}
```

调用 OnAuthorization 方法，为过滤器提供授权请求的机会。对于异步授权过滤器，下面是 IAsyncAuthorizationFilter 接口的定义：

```
using System.Threading.Tasks;

namespace Microsoft.AspNetCore.Mvc.Filters {

    public interface IAsyncAuthorizationFilter : IFilterMetadata {

        Task OnAuthorizationAsync(AuthorizationFilterContext context);
    }
}
```

调用 OnAuthorizationAsync 方法，以便过滤器可以对请求进行授权。无论使用哪个接口，过滤器都会通过 AuthorizationFilterContext 对象接收描述请求的上下文数据，该对象是从 FilterContext 类派生出来的，并添加了一个重要属性，如表 30-5 所示。

表 30-5　AuthorizationFilterContext 属性

| 名称 | 描述 |
| --- | --- |
| Result | 当请求不符合应用程序的授权策略时，授权过滤器会设置这个 IActionResult 属性。如果设置了此属性，则 ASP.NET Core 执行 IActionResult，而不是调用端点 |

### 创建授权过滤器

为了演示授权过滤器是如何工作的，在示例项目中创建 Filters 文件夹，并添加了一个名为 HttpsOnlyAttribute.cs 的文件，并使用它定义过滤器，如代码清单 30-15 所示。

**代码清单 30-15　Filters 文件夹中 HttpsOnlyAttribute.cs 文件的内容**

```
using System;
using Microsoft.AspNetCore.Http;
using Microsoft.AspNetCore.Mvc;
using Microsoft.AspNetCore.Mvc.Filters;

namespace WebApp.Filters {
    public class HttpsOnlyAttribute : Attribute, IAuthorizationFilter {

        public void OnAuthorization(AuthorizationFilterContext context) {
            if (!context.HttpContext.Request.IsHttps) {
                context.Result =
                    new StatusCodeResult(StatusCodes.Status403Forbidden);
            }
        }
    }
}
```

如果请求符合授权策略，授权过滤器就不执行任何操作，同时允许 ASP. NET Core 移到下一个过滤器，并最终执行端点。如果存在问题，过滤器将设置传递给 OnAuthorization 方法的 AuthorizationFilterContext 对象的 Result 属性。这将防止进一步执行操作，并提供返回给客户端的结果。在代码清单中，HttpsOnlyAttribute 类检查 HttpRequest 上下文对象的 IsHttps 属性，并设置 Result 属性，以在没有 HTTPS 的情况下发出请求时中断执行。授权过滤器可应用于控制器、操作方法和 Razor Pages。代码清单 30-16 将新的过滤器应用到主控制器。

**代码清单 30-16　在 Controllers 文件夹的 HomeController.cs 文件中应用自定义过滤器**

```
using Microsoft.AspNetCore.Mvc;
using Microsoft.AspNetCore.Http;
using WebApp.Filters;

namespace WebApp.Controllers {

    [HttpsOnly]
    public class HomeController : Controller {

        public IActionResult Index() {
            return View("Message",
                "This is the Index action on the Home controller");
        }

        public IActionResult Secure() {
```

```
            return View("Message",
                "This is the Secure action on the Home controller");
        }
    }
}
```

这个过滤器重新创建代码清单 30-10 的操作方法中包含的功能。在实际项目中，这不如进行像内置 RequireHttps 过滤器那样的重定向有用，因为用户不会理解 403 状态码的含义，但确实提供了一个关于授权过滤器如何工作的有用示例。重启 ASP.NET Core 并请求 http://localhost:5000，显示过滤器的效果。请求 https://localhost:44350，显示来自操作方法的响应，如图 30-6 所示。

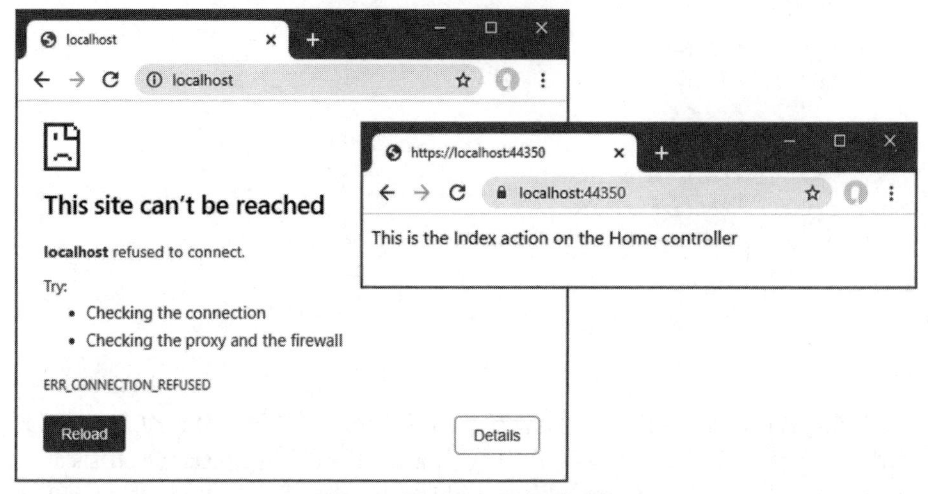


图 30-6 应用自定义的授权过滤器

## 30.4.2 理解资源过滤器

资源过滤器对每个请求执行两次 ASP.NET Core 模型绑定过程，在操作结果处理之前再生成结果。以下是 IResourceFilter 接口的定义：

```
namespace Microsoft.AspNetCore.Mvc.Filters {
    public interface IResourceFilter : IFilterMetadata {

        void OnResourceExecuting(ResourceExecutingContext context);

        void OnResourceExecuted(ResourceExecutedContext context);
    }
}
```

OnResourceExecuting 方法在处理请求时调用，而 OnResourceExecuted 方法在端点处理了请求之后、在操作结果执行之前调用。对于异步资源过滤器，IAsyncResourceFilter 接口的定义如下：

```
namespace Microsoft.AspNetCore.Mvc.Filters {
    public interface IAsyncResourceFilter : IFilterMetadata {
```

```
        Task OnResourceExecutionAsync(ResourceExecutingContext context,
            ResourceExecutionDelegate next);
    }
}
```

此接口定义了接收上下文对象和要调用的委托的单个方法。资源过滤器能够在调用委托之前检查请求，并在执行响应之前检查响应。OnResourceExecuting 方法使用 ResourceExecutingContext 类提供上下文；除了 FilterContext 类定义的属性外，该类还定义了表 30-6 中所示的属性。

表 30-6 由 ResourceExecutingContext 类定义的属性

| 名称 | 描述 |
| --- | --- |
| Result | 这个 IActionResult 属性用于提供一个导致管道短路的结果 |

OnResourceExecuted 方法使用 ResourceExecutedContext 类提供上下文，它定义了表 30-7 中所示的属性，以及 FilterContext 类定义的属性。

表 30-7 ResourceExecutedContext 类定义的属性

| 名称 | 描述 |
| --- | --- |
| Result | 这个 IActionResult 属性提供了用于生成响应的操作结果 |
| ValueProviderFactories | 这个属性返回一个 IList<IValueProviderFactory>，它允许访问为模型绑定过程提供值的对象 |

#### 1. 创建资源过滤器

资源过滤器通常用于可能使管道短路并尽早提供响应的地方，例如在实现数据缓存时。要创建一个简单的缓存过滤器，将一个名为 SimpleCacheAttribute.cs 的类文件添加到 Filters 文件夹，如代码清单 30-17 所示。

**过滤器和依赖注入**

作为属性应用的过滤器不能在其构造函数中声明依赖关系，除非它们实现 IFilterFactory 接口，并直接负责创建实例，如本章后面的“创建过滤器工厂”一节所述。

**代码清单 30-17 Filters 文件夹中 SimpleCacheAttribute.cs 文件的内容**

```
using Microsoft.AspNetCore.Http;
using Microsoft.AspNetCore.Mvc;
using Microsoft.AspNetCore.Mvc.Filters;
using System;
using System.Collections.Generic;

namespace WebApp.Filters {

    public class SimpleCacheAttribute : Attribute, IResourceFilter {
        private Dictionary<PathString, IActionResult> CachedResponses
            = new Dictionary<PathString, IActionResult>();
```

```
        public void OnResourceExecuting(ResourceExecutingContext context) {
            PathString path = context.HttpContext.Request.Path;
            if (CachedResponses.ContainsKey(path)) {
                context.Result = CachedResponses[path];
                CachedResponses.Remove(path);
            }
        }

        public void OnResourceExecuted(ResourceExecutedContext context) {
            CachedResponses.Add(context.HttpContext.Request.Path, context.Result);
        }
    }
}
```

这个过滤器不是一个特别有用的缓存，但是它显示了资源过滤器是如何工作的。OnResourceExecuting 方法通过将上下文对象的 Result 属性设置为先前缓存的操作结果，为过滤器提供了短路管道的机会。如果为 Result 属性分配了一个值，则过滤器管道将短路，并执行操作结果来为客户端生成响应。缓存的操作结果只使用一次，然后从缓存中丢弃。如果没有为 Result 属性分配值，则请求传递到管道中的下一个步骤，该步骤可能是另一个过滤器或端点。

OnResourceExecuted 方法为过滤器提供了当管道不短路时产生的操作结果。这种情况下，过滤器缓存操作结果，以便它可以用于后续请求。资源过滤器可以应用于控制器、操作方法和 Razor Pages。代码清单 30-18 将自定义资源过滤器应用到 Message Razor Pages，并添加一个时间戳，该时间戳将帮助确定何时缓存操作结果。

**代码清单 30-18 在 Pages 文件夹的 Message.cshtml 文件中应用资源过滤器**

```
@page "/pages/message"
@model MessageModel
@using Microsoft.AspNetCore.Mvc.RazorPages
@using System.Collections.Generic
@using Microsoft.AspNetCore.Http
@using WebApp.Filters
@if (Model.Message is string) {
    @Model.Message
} else if (Model.Message is IDictionary<string, string>) {
    var dict = Model.Message as IDictionary<string, string>;
    <table class="table table-sm table-striped table-bordered">
        <thead><tr><th>Name</th><th>Value</th></tr></thead>
        <tbody>
            @foreach (var kvp in dict) {
                <tr><td>@kvp.Key</td><td>@kvp.Value</td></tr>
            }
        </tbody>
    </table>
}
```

```
@functions {

    [RequireHttps]
    [SimpleCache]
    public class MessageModel : PageModel {

        public object Message { get; set; } =
        $"{DateTime.Now.ToLongTimeString()}: This is the Message Razor Page";
    }
}
```

要查看资源过滤器的效果，请重新启动 ASP.NET Core 并请求 https://localhost:44350/pages/message。因为这是对路径的第一个请求，所以不会缓存结果，请求将沿着管道转发。在处理响应时，资源过滤器将缓存操作结果供将来使用。重新加载浏览器以重复请求，会显示相同的时间戳，表明已使用了缓存的操作结果。缓存项在使用时被删除，这意味着重新加载浏览器将生成带有新时间戳的响应，如图 30-7 所示。

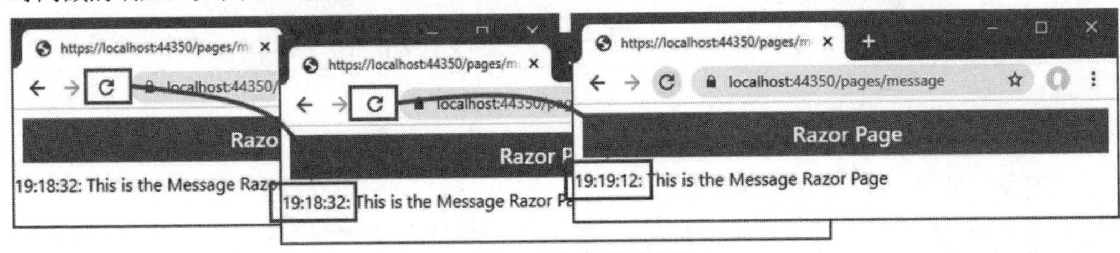


图 30-7 使用资源过滤器

### 2. 创建异步资源过滤器

异步资源过滤器的接口使用单个方法来接收用于沿着过滤器管道转发请求的委托。代码清单 30-19 重新实现了前一个示例中的缓存过滤器，以便它实现 IAsyncResourceFilter 接口。

**代码清单 30-19 在 Filters 文件夹的 SimpleCacheAttribute.cs 文件中创建异步过滤器**

```
using Microsoft.AspNetCore.Http;
using Microsoft.AspNetCore.Mvc;
using Microsoft.AspNetCore.Mvc.Filters;
using System;
using System.Collections.Generic;
using System.Threading.Tasks;

namespace WebApp.Filters {

    public class SimpleCacheAttribute : Attribute, IAsyncResourceFilter {
        private Dictionary<PathString, IActionResult> CachedResponses
            = new Dictionary<PathString, IActionResult>();

        public async Task OnResourceExecutionAsync(ResourceExecutingContext context,
                ResourceExecutionDelegate next) {
            PathString path = context.HttpContext.Request.Path;
```

```
                if (CachedResponses.ContainsKey(path)) {
                    context.Result = CachedResponses[path];
                    CachedResponses.Remove(path);
                } else {
                    ResourceExecutedContext execContext = await next();
                    CachedResponses.Add(context.HttpContext.Request.Path,
                        execContext.Result);
                }
            }
        }
    }
```

OnResourceExecutionAsync 方法接收一个 ResourceExecutingContext 对象，该对象用于确定管道是否可以短路。如果不能，则在没有参数的情况下调用委托，并在处理请求，沿着管道返回时，异步生成 ResourceExecutedContext 对象。重启 ASP.NET Core 并重复上一节中描述的请求，显示相同的缓存行为，如图 30-7 所示。

**注意：**不要混淆这两个上下文对象。由端点产生的操作结果仅在委托返回的上下文对象中可用。

### 30.4.3　理解操作过滤器

与资源过滤器一样，操作过滤器执行两次。区别在于操作过滤器在模型绑定过程之后执行，而资源过滤器在模型绑定之前执行。这意味着资源过滤器可以使管道短路，并将 ASP.NET Core 为处理请求所做的工作最小化。当需要模型绑定时，使用操作过滤器，这意味着它们用于诸如更改模型或强制验证的任务。操作过滤器只能应用于控制器和操作方法，这与资源过滤器不同，资源过滤器也可以用于 Razor Pages(与操作过滤器等价的 Razor Pages 是页面过滤器，参见“理解页面过滤器”一节)。下面是 IActionFilter 接口：

```
namespace Microsoft.AspNetCore.Mvc.Filters {

    public interface IActionFilter : IFilterMetadata {

        void OnActionExecuting(ActionExecutingContext context);

        void OnActionExecuted(ActionExecutedContext context);
    }
}
```

当操作过滤器应用到操作方法时，OnActionExecuting 方法会在调用操作方法之前调用，而 OnActionExecuted 方法会在调用操作方法之后调用。操作过滤器通过两个不同的上下文类提供上下文数据：用于 OnActionExecuting 方法的 ActionExecutingContext 和用于 OnActionExecuted 方法的 ActionExecutedContext。

ActionExecutingContext 类用于描述要调用的操作，它定义了表 30-8 中描述的属性，以及 FilterContext 属性。

表 30-8 ActionExecutingContext 属性

| 名称 | 描述 |
| --- | --- |
| Controller | 此属性返回要调用其操作方法的控制器(操作方法的详细信息可以通过从基类继承的 ActionDescriptor 属性获得) |
| ActionArguments | 此属性返回一个将传递给操作方法的参数字典，按名称索引。过滤器可以插入、删除或更改参数 |
| Result | 如果过滤器将 IActionResult 分配给此属性，则管道将被短路，操作结果用于生成对客户端的响应，而不调用操作方法 |

ActionExecutedContext 类用于表示已执行的操作，并定义表 30-9 中描述的属性，以及 FilterContext 属性。

表 30-9 ActionExecutedContext 属性

| 名称 | 描述 |
| --- | --- |
| Controller | 此属性返回将调用其操作方法的 Controller 对象 |
| Canceled | 如果另一个操作过滤器将操作结果分配给 ActionExecutingContext 对象的 Result 属性，从而使管道短路，则此 bool 属性设置为 true |
| Exception | 此属性包含操作方法抛出的任何异常 |
| ExceptionDispatchInfo | 此方法返回一个 ExceptionDispatchInfo 对象，该对象包含操作方法抛出的任何异常的堆栈跟踪细节 |
| ExceptionHandled | 将此属性设置为 true 表示过滤器已处理该异常，该异常将不再被进一步传播 |
| Result | 这个属性返回操作方法产生的 IActionResult。如有必要，过滤器可以更改或替换操作结果 |

异步操作过滤器是使用 IAsyncActionFilter 接口实现的。

```
namespace Microsoft.AspNetCore.Mvc.Filters {

    public interface IAsyncActionFilter : IFilterMetadata {

        Task OnActionExecutionAsync(ActionExecutingContext context,
           ActionExecutionDelegate next);
    }
}
```

该接口遵循与本章前面描述的 IAsyncResourceFilter 接口相同的模式。为 OnActionExecutionAsync 方法提供了一个 ActionExecutingContext 对象和一个委托。ActionExecutingContext 对象在请求被操作方法接收之前描述它。过滤器可以通过给 ActionExecutingContext.Result 属性赋值或通过调用委托将其传递来短路管道。委托异步地生成 ActionExecutedContext 对象，该对象描述来自操作方法的结果。

### 1. 创建操作过滤器

在 Filters 文件夹中添加一个名为 ChangeArgAttribute.cs 的类文件，并使用它来定义操作过滤器，如代码清单 30-20 所示。

**代码清单 30-20　Filters 文件夹中 ChangeArgAttribute.cs 文件的内容**

```
using Microsoft.AspNetCore.Mvc.Filters;
using System;
using System.Threading.Tasks;

namespace WebApp.Filters {
    public class ChangeArgAttribute : Attribute, IAsyncActionFilter {

        public async Task OnActionExecutionAsync(ActionExecutingContext context,
                ActionExecutionDelegate next) {

            if (context.ActionArguments.ContainsKey("message1")) {
                context.ActionArguments["message1"] = "New message";
            }
            await next();
        }
    }
}
```

过滤器查找名为 message1 的操作参数，并更改用于调用操作方法的值。用于操作方法参数的值由模型绑定过程决定。代码清单 30-21 向 Home 控制器添加了一个操作方法，并应用了新的过滤器。

**代码清单 30-21　在 Controllers 文件夹的 HomeController.cs 文件中应用过滤器**

```
using Microsoft.AspNetCore.Mvc;
using Microsoft.AspNetCore.Http;
using WebApp.Filters;

namespace WebApp.Controllers {

    [HttpsOnly]
    public class HomeController : Controller {

        public IActionResult Index() {
            return View("Message",
                "This is the Index action on the Home controller");
        }

        public IActionResult Secure() {
            return View("Message",
                "This is the Secure action on the Home controller");
        }

        [ChangeArg]
        public IActionResult Messages(string message1, string message2 = "None") {
            return View("Message", $"{message1}, {message2}");
```

```
            }
        }
    }
```

重启 ASP.NET Core，请求 https://localhost:44350/home/messages?message1=hello&message2 =world。模型绑定过程从查询字符串中定位操作方法定义的参数值。然后操作过滤器会修改其中一个值，生成如图 30-8 所示的响应。

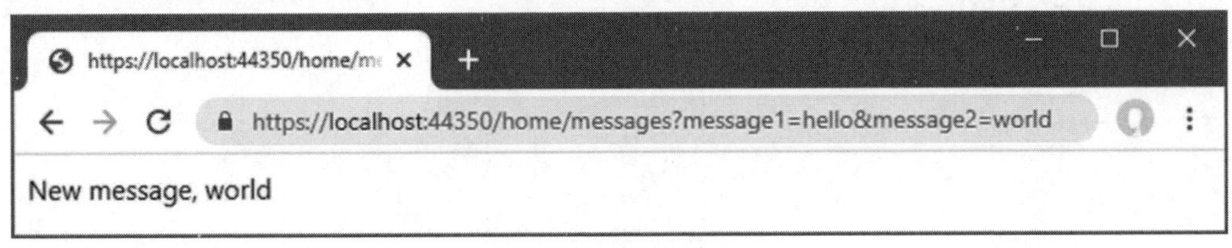


图 30-8　使用操作过滤器

### 2. 使用属性基类实现操作过滤器

操作属性也可以通过派生 ActionFilterAttribute 类来实现，该类扩展了 Attribute 并继承了 IActionFilter 和 IAsyncActionFilter 接口，以便实现类只覆盖它们需要的方法。在代码清单 30-22 中，重新实现了 ChangeArg 过滤器，使它派生自 ActionFilterAttribute。

**代码清单 30-22　在 Filters 文件夹的 ChangeArgsAttribute.cs 文件中使用 Filters 基类**

```
using Microsoft.AspNetCore.Mvc.Filters;
using System;
using System.Threading.Tasks;

namespace WebApp.Filters {
    public class ChangeArgAttribute : ActionFilterAttribute {

        public override async Task OnActionExecutionAsync(
            ActionExecutingContext context,
                ActionExecutionDelegate next) {

            if (context.ActionArguments.ContainsKey("message1")) {
                context.ActionArguments["message1"] = "New message";
            }
            await next();
        }
    }
}
```

此属性的行为方式与前面的实现完全相同，基类的使用取决于个人偏好。重启 ASP.NET Core，请求 https://localhost:44350/home/messages?message1=hello&message2=world，显示如图 30-8 所示的响应。

### 3. 使用控制器过滤方法

Controller 类是呈现 Razor 视图的控制器的基础，它实现了 IActionFilter 和 IAsyncActionFilter

接口，这意味着可以定义功能，并将其应用到由控制器和任何派生控制器定义的操作中。代码清单 30-23 在 HomeController 类中直接实现了 ChangeArg 过滤器功能。

**代码清单 30-23　在 Controllers 文件夹的 HomeController.cs 文件中操作过滤方法**

```
using Microsoft.AspNetCore.Mvc;
using Microsoft.AspNetCore.Http;
using WebApp.Filters;
using Microsoft.AspNetCore.Mvc.Filters;

namespace WebApp.Controllers {

    [HttpsOnly]
    public class HomeController : Controller {

        public IActionResult Index() {
            return View("Message",
                "This is the Index action on the Home controller");
        }

        public IActionResult Secure() {
            return View("Message",
                "This is the Secure action on the Home controller");
        }

        //[ChangeArg]
        public IActionResult Messages(string message1, string message2 = "None") {
            return View("Message", $"{message1}, {message2}");
        }

        public override void OnActionExecuting(ActionExecutingContext context) {
            if (context.ActionArguments.ContainsKey("message1")) {
                context.ActionArguments["message1"] = "New message";
            }
        }
    }
}
```

主控制器覆盖 OnActionExecuting 方法的控制器实现，并使用它来修改传递给执行方法的参数。

重启 ASP.NET Core，请求 https://localhost:44350/home/messages?message1=hello&message2=world，显示如图 30-8 所示的响应。

## 30.4.4　理解页面过滤器

页面过滤器是 Razor Pages 等效的操作过滤器。下面是 IPageFilter 接口，它是由同步页面过滤器实现的：

```
namespace Microsoft.AspNetCore.Mvc.Filters {

    public interface IPageFilter : IFilterMetadata {

        void OnPageHandlerSelected(PageHandlerSelectedContext context);

        void OnPageHandlerExecuting(PageHandlerExecutingContext context);

        void OnPageHandlerExecuted(PageHandlerExecutedContext context);
    }
}
```

在 ASP.NET Core 选择了页面处理程序方法后，但在执行模型绑定之前，调用 OnPageHandlerSelected 方法，这意味着处理程序方法的参数还没有确定。这个方法通过 PageHandlerSelectedContext 类接收上下文，除了 FilterContext 类定义的属性外，该类还定义了表 30-10 中所示的属性。此方法不能用于短路管道，但它可以更改将接收请求的处理程序方法。

表 30-10　PageHandlerSelectedContext 属性

| 名称 | 描述 |
| --- | --- |
| ActionDescriptor | 此属性返回 Razor Pages 的描述 |
| HandlerMethod | 此属性返回一个描述所选处理程序方法的 HandlerMethodDescriptor 对象 |
| HandlerInstance | 此属性返回处理请求的 Razor Pages 的实例 |

OnPageHandlerExecuting 方法在模型绑定过程完成之后，但在调用页面处理程序方法之前调用。这个方法通过 PageHandlerExecutingContext 类接收上下文，该类定义了表 30-11 中所示的属性。

表 30-11　PageHandlerExecutingContext 属性

| 名称 | 描述 |
| --- | --- |
| HandlerArguments | 此属性返回一个字典，其中包含按名称索引的页处理程序参数 |
| Result | 过滤器可以通过将 IActionResult 对象分配给此属性来短路管道 |

OnPageHandlerExecuted 方法在调用页面处理程序方法之后，但在处理操作结果以创建响应之前调用。这个方法通过 PageHandlerExecutedContext 类接收上下文，该类定义了表 30-12 中显示的属性以及 PageHandlerSelectedContext 属性。

表 30-12　PageHandlerExecutedContext 属性

| 名称 | 描述 |
| --- | --- |
| Canceled | 如果另一个过滤器短路了过滤器管道，则此属性返回 true |
| Exception | 如果页面处理程序方法引发异常，则此属性将返回异常 |
| ExceptionHandled | 此属性设置为 true，表示过滤器已处理由页面处理程序抛出的异常 |
| Result | 此属性返回用于为客户端创建响应的操作结果 |

异步页面过滤器是通过实现 IAsyncPageFilter 接口创建的，它的定义如下：

```
namespace Microsoft.AspNetCore.Mvc.Filters {
    public interface IAsyncPageFilter : IFilterMetadata {

        Task OnPageHandlerSelectionAsync(PageHandlerSelectedContext context);

        Task OnPageHandlerExecutionAsync(PageHandlerExecutingContext context,
            PageHandlerExecutionDelegate next);
    }
}
```

OnPageHandlerSelectionAsync 在选中处理程序方法之后调用，它等价于同步的 OnPageHandlerSelected 方法。为 OnPageHandlerExecutionAsync 提供了一个允许它短路管道的 PageHandlerExecutingContext 对象，以及一个被调用来传递请求的委托。委托产生一个 PageHandlerExecutedContext 对象，该对象可用于检查或更改处理程序方法产生的操作结果。

### 1. 创建页面过滤器

要创建页面过滤器，将一个名为 ChangePageArgs.cs 的类文件添加到 Filters 文件夹，并使用它定义如代码清单 30-24 所示的类。

**代码清单 30-24　Filters 文件夹中 ChangePageArgs.cs 文件的内容**

```
using Microsoft.AspNetCore.Mvc.Filters;
using System;

namespace WebApp.Filters {
    public class ChangePageArgs : Attribute, IPageFilter {

        public void OnPageHandlerSelected(PageHandlerSelectedContext context) {
            // do nothing
        }

        public void OnPageHandlerExecuting(PageHandlerExecutingContext context) {
            if (context.HandlerArguments.ContainsKey("message1")) {
                context.HandlerArguments["message1"] = "New message";
            }
        }

        public void OnPageHandlerExecuted(PageHandlerExecutedContext context) {
            // do nothing
        }
    }
}
```

代码清单 30-24 中的页面过滤器执行与上一节中创建的操作过滤器相同的任务。在代码清单 30-25 中，修改了 Message Razor Pages 以定义处理程序方法，并应用了页面过滤器。页面过滤器可以应用于单个处理程序方法，也可以应用于页面模型类(如代码清单中所示)，在这种情况下，过滤器用于所有处理程序方法(在代码清单 30-25 中还禁用了 SimpleCache 过滤器。资源过滤器可以与页面过滤器一起工作。禁用了这个过滤器，因为缓存响应会使一些示例更难理解)。

**代码清单 30-25　在 Pages 文件夹的 Message.cshtml 文件中使用页面过滤器**

```
@page "/pages/message"
@model MessageModel
@using Microsoft.AspNetCore.Mvc.RazorPages
@using System.Collections.Generic
@using Microsoft.AspNetCore.Http
@using WebApp.Filters

@if (Model.Message is string) {
   @Model.Message
} else if (Model.Message is IDictionary<string, string>) {
    var dict = Model.Message as IDictionary<string, string>;
    <table class="table table-sm table-striped table-bordered">
        <thead><tr><th>Name</th><th>Value</th></tr></thead>
        <tbody>
            @foreach (var kvp in dict) {
                <tr><td>@kvp.Key</td><td>@kvp.Value</td></tr>
            }
        </tbody>
    </table>
}

@functions {

    [RequireHttps]
    //[SimpleCache]
    [ChangePageArgs]
    public class MessageModel : PageModel {

        public object Message { get; set; } =
            $"{DateTime.Now.ToLongTimeString()}: This is the Message Razor Page";

        public void OnGet(string message1, string message2) {
            Message = $"{message1}, {message2}";
        }
    }
}
```

重启 ASP.NET Core，请求 https://localhost:44350/pages/message?message1=hello&message2=world。

页面过滤器替换 OnGet 处理程序方法的 message1 参数值，该方法生成如图 30-9 所示的响应。

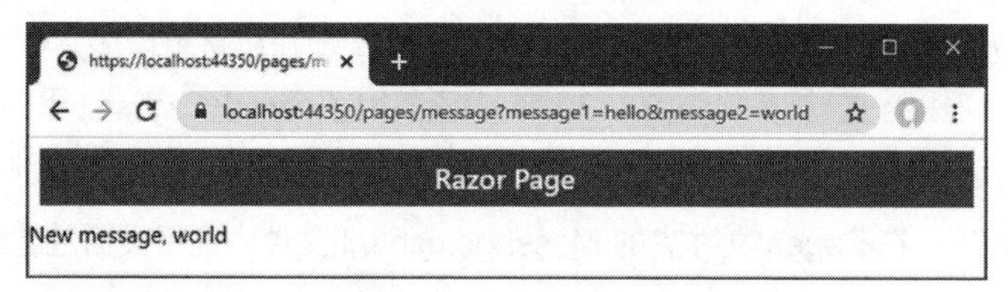


图 30-9　使用页面过滤器

### 2. 使用页面模型过滤方法

PageModel 类用作页面模型类的基础，它实现了 IPageFilter 和 IAsyncPageFilter 接口，这意味着可以直接向页面模型添加过滤功能，如代码清单 30-26 所示。

#### 代码清单 30-26　在 Pages 文件夹的 Message.cshtml 文件中使用 PageModel 过滤方法

```
@page "/pages/message"
@model MessageModel
@using Microsoft.AspNetCore.Mvc.RazorPages
@using System.Collections.Generic
@using Microsoft.AspNetCore.Http
@using WebApp.Filters
@using Microsoft.AspNetCore.Mvc.Filters

@if (Model.Message is string) {
    @Model.Message
} else if (Model.Message is IDictionary<string, string>) {
    var dict = Model.Message as IDictionary<string, string>;
    <table class="table table-sm table-striped table-bordered">
        <thead><tr><th>Name</th><th>Value</th></tr></thead>
        <tbody>
            @foreach (var kvp in dict) {
                <tr><td>@kvp.Key</td><td>@kvp.Value</td></tr>
            }
        </tbody>
    </table>
}

@functions {

    [RequireHttps]
    //[SimpleCache]
    //[ChangePageArgs]
    public class MessageModel : PageModel {

        public object Message { get; set; } =
            $"{DateTime.Now.ToLongTimeString()}: This is the Message Razor Page";

        public void OnGet(string message1, string message2) {
            Message = $"{message1}, {message2}";
```

```
        }

        public override void OnPageHandlerExecuting(
                PageHandlerExecutingContext context) {
            if (context.HandlerArguments.ContainsKey("message1")) {
                context.HandlerArguments["message1"] = "New message";
            }
        }
    }
}
```

请求 https://localhost:44350/pages/message?message1=hello&message2=world。代码清单 30-26 中的页面模型类实现的方法产生如图 30-9 所示的结果。

### 30.4.5 理解结果过滤器

结果过滤器在操作结果用于生成响应之前和之后执行，允许在端点处理响应之后修改响应。下面是 IResultFilter 接口的定义：

```
namespace Microsoft.AspNetCore.Mvc.Filters {
    public interface IResultFilter : IFilterMetadata {

        void OnResultExecuting(ResultExecutingContext context);

        void OnResultExecuted(ResultExecutedContext context);
    }
}
```

在端点生成操作结果后调用OnResultExecuting方法。该方法通过ResultExecutingContext类接收上下文，该类定义表 30-13 中描述的属性，以及FilterContext类定义的属性。

表 30-13 ResultExecutingContext 类属性

| 名称 | 描述 |
|---|---|
| Result | 此属性返回端点产生的操作结果 |
| ValueProviderFactories | 此属性返回 IList<IValueProviderFactory>，允许访问为模型绑定过程提供值的对象 |

OnResultExecuted方法在操作结果执行后调用，为客户端生成响应。这个方法通过ResultExecutedContext类接收上下文，该类除了从FilterContext类继承的属性外，还定义了表 30-14 中所示的属性。

表 30-14 ResultExecutedContext 类

| 名称 | 描述 |
|---|---|
| Canceled | 如果另一个过滤器短路了过滤器管道，则此属性返回 true |
| Controller | 此属性返回包含端点的对象 |
| Exception | 如果页面处理程序方法引发异常，则此属性返回异常 |
| ExceptionHandled | 此属性设置为 true，表示过滤器已处理由页面处理程序抛出的异常 |
| Result | 此属性返回用于为客户端创建响应的操作结果。此属性是只读的 |

异步结果过滤器实现 IAsyncResultFilter 接口，它的定义如下：

```
namespace Microsoft.AspNetCore.Mvc.Filters {

    public interface IAsyncResultFilter : IFilterMetadata {

        Task OnResultExecutionAsync(ResultExecutingContext context,
            ResultExecutionDelegate next);
    }
}
```

此接口遵循其他过滤器类型建立的模式。OnResultExecutionAsync 方法是通过上下文对象调用的，上下文对象的 Result 属性可用于更改响应，而委托将沿着管道转发响应。

#### 1. 理解始终运行的结果过滤器

实现 IResultFilter 和 IAsyncResultFilter 接口的过滤器仅在请求由端点正常处理时使用。如果另一个过滤器使管道短路或出现异常，则不使用它们。即使在管道短路时，需要检查或更改响应的过滤器，可以实现 IAlwaysRunResultFilter 或 IAsyncAlwaysRunResultFilter 接口。这些接口派生自 IResultFilter 和 IAsyncResultFilter，但没有定义新的特性。相反，ASP.NET Core 检测始终运行的接口，并始终应用过滤器。

#### 2. 创建结果过滤器

向 Filters 文件夹添加一个名为 ResultDiagnosticsAttribute.cs 的类文件，并使用它定义如代码清单 30-27 所示的过滤器。

**代码清单 30-27　Filters 文件夹中 ResultDiagnosticsAttribute.cs 文件的内容**

```
using Microsoft.AspNetCore.Mvc;
using Microsoft.AspNetCore.Mvc.Filters;
using Microsoft.AspNetCore.Mvc.ModelBinding;
using Microsoft.AspNetCore.Mvc.RazorPages;
using Microsoft.AspNetCore.Mvc.ViewFeatures;
using System;
using System.Collections.Generic;
using System.Threading.Tasks;

namespace WebApp.Filters {

    public class ResultDiagnosticsAttribute : Attribute, IAsyncResultFilter {

        public async Task OnResultExecutionAsync(
                ResultExecutingContext context, ResultExecutionDelegate next) {

            if (context.HttpContext.Request.Query.ContainsKey("diag")) {
                Dictionary<string, string> diagData =
                    new Dictionary<string, string> {
                        {"Result type", context.Result.GetType().Name }
```

```
            };
            if (context.Result is ViewResult vr) {
                diagData["View Name"] = vr.ViewName;
                diagData["Model Type"] = vr.ViewData.Model.GetType().Name;
                diagData["Model Data"] = vr.ViewData.Model.ToString();
            } else if (context.Result is PageResult pr) {
                diagData["Model Type"] = pr.Model.GetType().Name;
                diagData["Model Data"] = pr.ViewData.Model.ToString();
            }
            context.Result = new ViewResult() {
                ViewName = "/Views/Shared/Message.cshtml",
                ViewData = new ViewDataDictionary(
                                  new EmptyModelMetadataProvider(),
                                  new ModelStateDictionary()) {
                    Model = diagData
                }
            };
        }
        await next();
    }
  }
}
```

此过滤器检查请求是否包含名为 diag 的查询字符串参数，如果包含，则创建一个显示诊断信息的结果，而不是端点生成的输出。代码清单 30-27 中的过滤器使用主控制器或 Message Razor Pages 定义的操作。代码清单 30-28 将结果过滤器应用于 Home 控制器。

**提示：** 在代码清单 30-27 中创建操作结果时，为视图使用一个完全限定名。这样就避免了过滤器应用到 Razor Pages 时的问题，ASP.NET Core 试图以 Razor Pages 的形式执行新的结果，并抛出一个关于模型类型的异常。

### 代码清单 30-28 在 Controllers 文件夹的 HomeController.cs 文件中应用结果过滤器

```
using Microsoft.AspNetCore.Mvc;
using Microsoft.AspNetCore.Http;
using WebApp.Filters;
using Microsoft.AspNetCore.Mvc.Filters;

namespace WebApp.Controllers {

    [HttpsOnly]
    [ResultDiagnostics]
    public class HomeController : Controller {

        public IActionResult Index() {
            return View("Message",
                "This is the Index action on the Home controller");
```

```
        }

        public IActionResult Secure() {
            return View("Message",
                "This is the Secure action on the Home controller");
        }

        //[ChangeArg]
        public IActionResult Messages(string message1, string message2 = "None") {
            return View("Message", $"{message1}, {message2}");
        }

        public override void OnActionExecuting(ActionExecutingContext context) {
            if (context.ActionArguments.ContainsKey("message1")) {
                context.ActionArguments["message1"] = "New message";
            }
        }
    }
}
```

重启 ASP.NET Core，并请求 https://localhost:44350/?diag。过滤器将检测查询字符串参数，生成如图 30-10 所示的诊断信息。

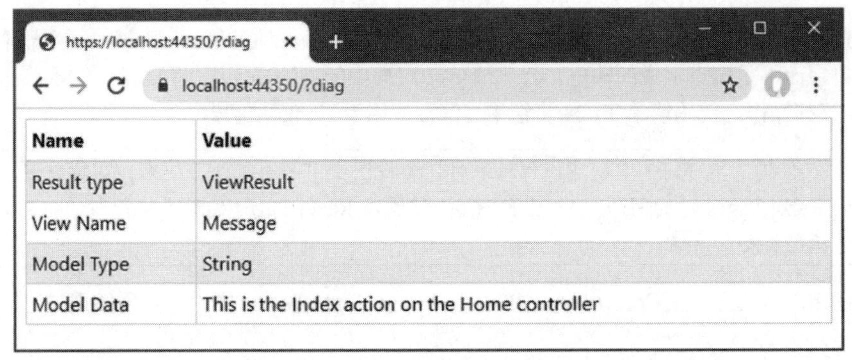


图 30-10　使用结果过滤器

### 3. 使用属性基类实现结果过滤器

ResultFilterAttribute 类派生自 Attribute 并实现 IResultFilter 和 IAsyncResultFilter 接口，可以用作结果过滤器的基类，如代码清单 30-29 所示。始终运行的接口没有属性基类。

#### 代码清单 30-29　在 Filters 文件夹的 ResultDiagnosticsAttribute.cs 文件中使用属性基类

```
using Microsoft.AspNetCore.Mvc;
using Microsoft.AspNetCore.Mvc.Filters;
using Microsoft.AspNetCore.Mvc.ModelBinding;
using Microsoft.AspNetCore.Mvc.RazorPages;
using Microsoft.AspNetCore.Mvc.ViewFeatures;
using System;
```

```
using System.Collections.Generic;
using System.Threading.Tasks;

namespace WebApp.Filters {

    public class ResultDiagnosticsAttribute : ResultFilterAttribute {

        public override async Task OnResultExecutionAsync(
               ResultExecutingContext context, ResultExecutionDelegate next) {

            if (context.HttpContext.Request.Query.ContainsKey("diag")) {
                Dictionary<string, string> diagData =
                    new Dictionary<string, string> {
                       {"Result type", context.Result.GetType().Name }
                    };
                if (context.Result is ViewResult vr) {
                    diagData["View Name"] = vr.ViewName;
                    diagData["Model Type"] = vr.ViewData.Model.GetType().Name;
                    diagData["Model Data"] = vr.ViewData.Model.ToString();
                } else if (context.Result is PageResult pr) {
                    diagData["Model Type"] = pr.Model.GetType().Name;
                    diagData["Model Data"] = pr.ViewData.Model.ToString();
                }
                context.Result = new ViewResult() {
                    ViewName = "/Views/Shared/Message.cshtml",
                    ViewData = new ViewDataDictionary(
                                        new EmptyModelMetadataProvider(),
                                        new ModelStateDictionary()) {
                        Model = diagData
                    }
                };
            }
            await next();
        }
    }
}
```

重启 ASP.NET Core，并请求 https://localhost:44350/?diag。过滤器将产生如图 30-10 所示的输出。

## 30.4.6 理解异常过滤器

使用异常过滤器，不必在每个操作方法中编写 try...catch 块，即可响应异常。异常过滤器可以应用于控制器类、操作方法、页面模型类或处理程序方法。当端点或已应用于端点的操作、页面和结果过滤器不处理异常时，将调用它们(操作、页面和结果过滤器可通过将其上下文对象的 ExceptionHandled 属性设置为 true，来处理未处理的异常)。异常过滤器实现 IExceptionFilter 接口，其定义如下：

```
namespace Microsoft.AspNetCore.Mvc.Filters {

    public interface IExceptionFilter : IFilterMetadata {

        void OnException(ExceptionContext context);
    }
}
```

如果遇到未处理的异常，则调用 OnException 方法。IAsyncExceptionFilter 接口可以用来创建异步异常过滤器。下面是异步接口的定义：

```
using System.Threading.Tasks;

namespace Microsoft.AspNetCore.Mvc.Filters {

    public interface IAsyncExceptionFilter : IFilterMetadata {

        Task OnExceptionAsync(ExceptionContext context);
    }
}
```

OnExceptionAsync 方法是 IExceptionFilter 接口中的 OnException 方法的异步对等物，当有未处理的异常时调用。对于这两个接口，上下文数据都是通过 ExceptionContext 类提供的，该类派生自 FilterContext，并定义了表 30-15 中所示的其他属性。

表 30-15　ExceptionContext 属性

| 名称 | 描述 |
|---|---|
| Exception | 此属性包含抛出的任何异常 |
| ExceptionHandled | 这个 bool 属性用于指示异常是否已处理 |
| Result | 此属性设置用于生成响应的 IActionResult |

## 30.4.7　创建异常过滤器

异常过滤器可以通过实现一个过滤器接口或从 ExceptionFilterAttribute 类派生来创建，该类派生自 Attribute 并实现了 IExceptionFilter 和 IAsyncException 过滤器。异常过滤器最常见的用途是为特定异常类型显示自定义错误页面，以便为用户提供比标准错误处理功能更有用的信息。

要创建异常过滤器，将一个名为 RangeExceptionAttribute.cs 的类文件添加到 Filters 文件夹，代码如代码清单 30-30 所示。

**代码清单 30-30　Filters 文件夹中 RangeExceptionAttribute.cs 文件的内容**

```
using Microsoft.AspNetCore.Mvc;
using Microsoft.AspNetCore.Mvc.Filters;
using Microsoft.AspNetCore.Mvc.ModelBinding;
using Microsoft.AspNetCore.Mvc.ViewFeatures;
```

```
using System;

namespace WebApp.Filters {
    public class RangeExceptionAttribute : ExceptionFilterAttribute {

        public override void OnException(ExceptionContext context) {
            if (context.Exception is ArgumentOutOfRangeException) {
                context.Result = new ViewResult() {
                    ViewName = "/Views/Shared/Message.cshtml",
                    ViewData = new ViewDataDictionary(
                        new EmptyModelMetadataProvider(),
                        new ModelStateDictionary()) {
                        Model = @"The data received by the
                                application cannot be processed"
                    }
                };
            }
        }
    }
}
```

此过滤器使用 ExceptionContext 对象来获取未处理异常的类型，如果类型是 ArgumentOutOfRangeException，则创建一个操作结果，向用户显示一条消息。代码清单 30-31 将一个操作方法添加到主控制器中，对其应用异常过滤器。

#### 代码清单 30-31　在 Controllers 文件夹的 HomeController.cs 文件中应用异常过滤器

```
using Microsoft.AspNetCore.Mvc;
using Microsoft.AspNetCore.Http;
using WebApp.Filters;
using Microsoft.AspNetCore.Mvc.Filters;
using System;

namespace WebApp.Controllers {

    [HttpsOnly]
    [ResultDiagnostics]
    public class HomeController : Controller {

        public IActionResult Index() {
            return View("Message",
                "This is the Index action on the Home controller");
        }

        public IActionResult Secure() {
            return View("Message",
                "This is the Secure action on the Home controller");
        }
```

```
        //[ChangeArg]
        public IActionResult Messages(string message1, string message2 = "None") {
            return View("Message", $"{message1}, {message2}");
        }

        public override void OnActionExecuting(ActionExecutingContext context) {
            if (context.ActionArguments.ContainsKey("message1")) {
                context.ActionArguments["message1"] = "New message";
            }
        }

        [RangeException]
        public ViewResult GenerateException(int? id) {
            if (id == null) {
                throw new ArgumentNullException(nameof(id));
            } else if (id > 10) {
                throw new ArgumentOutOfRangeException(nameof(id));
            } else {
                return View("Message", $"The value is {id}");
            }
        }
    }
}
```

GenerateException 操作方法依赖于默认的路由模式从请求 URL 接收一个可空的 int 值。如果没有匹配的 URL 段，操作方法抛出一个 ArgumentNullException；如果 int 的值大于 50，则抛出一个 ArgumentOutOfRangeException。如果有一个值并且处于范围内，那么操作方法返回 ViewResult。

重启 ASP.NET Core 并请求 https://localhost:44350/home/generateexception /100。最后一个段将超过 action 方法所期望的范围，该方法将抛出过滤器所处理的异常类型，生成如图 30-11 所示的结果。如果请求/Home/GenerateException，那么过滤器不会处理操作方法抛出的异常，而使用默认的错误处理。

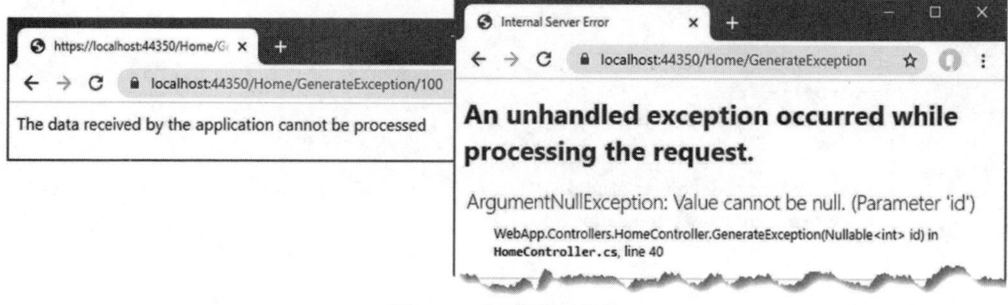


图 30-11　使用异常过滤器

## 30.5 管理过滤器生命周期

默认情况下，ASP.NET Core 管理它创建的过滤器对象，并在后续请求中重用它们。这并不总是期望的行为，接下来的部分描述控制如何创建过滤器的不同方法。要创建一个显示生命周期的过滤器，将一个名为 GuidResponseAttribute.cs 的类文件添加到 Filters 文件夹中，用它来定义如代码清单 30-32 所示的过滤器。

**代码清单 30-32 Filters 文件夹中 GuidResponseAttribute.cs 文件的内容**

```
using Microsoft.AspNetCore.Mvc;
using Microsoft.AspNetCore.Mvc.Filters;
using Microsoft.AspNetCore.Mvc.ModelBinding;
using Microsoft.AspNetCore.Mvc.ViewFeatures;
using System;
using System.Collections.Generic;
using System.Threading.Tasks;

namespace WebApp.Filters {

    [AttributeUsage(AttributeTargets.Method | AttributeTargets.Class,
       AllowMultiple = true)]
    public class GuidResponseAttribute : Attribute, IAsyncAlwaysRunResultFilter {
        private int counter = 0;
        private string guid = Guid.NewGuid().ToString();

        public async Task OnResultExecutionAsync(ResultExecutingContext context,
            ResultExecutionDelegate next) {

            Dictionary<string, string> resultData;
            if (context.Result is ViewResult vr
                && vr.ViewData.Model is Dictionary<string, string> data) {
                    resultData = data;
            } else {
                resultData = new Dictionary<string, string>();
                context.Result = new ViewResult() {
                    ViewName = "/Views/Shared/Message.cshtml",
                    ViewData = new ViewDataDictionary(
                                    new EmptyModelMetadataProvider(),
                                    new ModelStateDictionary()) {
                        Model = resultData
                    }
                };
            }
            while (resultData.ContainsKey($"Counter_{counter}")) {
                counter++;
            }
```

```
                resultData[$"Counter_{counter}"] = guid;
                await next();
            }
        }
    }
```

此结果过滤器将端点产生的操作结果替换为呈现消息视图并显示唯一 GUID 值的操作结果。对过滤器进行了配置，以便它可以多次应用于相同的目标，并且如果管道中较早的过滤器创建了合适的结果，它将添加新消息。代码清单 30-33 对主控制器应用了两次过滤器(为了简洁起见，我还删除了所有操作方法，只保留了一个)。

#### 代码清单 30-33 在 Controllers 文件夹的 HomeController.cs 文件中应用过滤器

```
using Microsoft.AspNetCore.Mvc;
using Microsoft.AspNetCore.Http;
using WebApp.Filters;
using Microsoft.AspNetCore.Mvc.Filters;
using System;

namespace WebApp.Controllers {

    [HttpsOnly]
    [ResultDiagnostics]
    [GuidResponse]
    [GuidResponse]
    public class HomeController : Controller {
        public IActionResult Index() {
            return View("Message",
                "This is the Index action on the Home controller");
        }
    }
}
```

要确认过滤器的重用，请重新启动 ASP.NET Core 并请求 https://localhost:44350/?diag。响应包含来自两个 GuidResponse 过滤器属性的 GUID 值。创建过滤器的两个实例来处理请求。重新加载浏览器，将看到显示的相同 GUID 值，这表明为处理第一个请求而创建的过滤器对象已重用(图 30-12)。

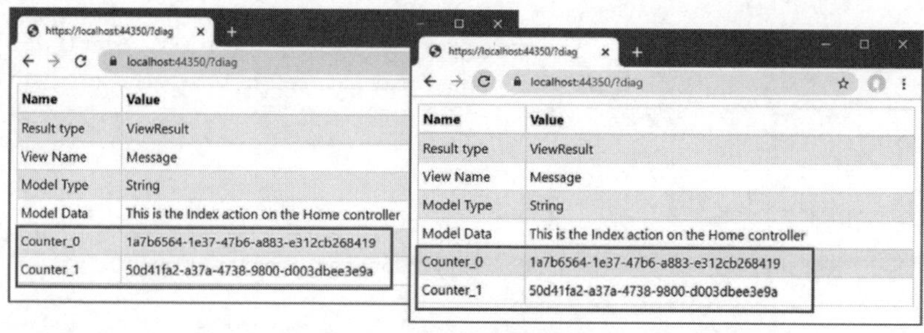


图 30-12 显示过滤器的重用

### 30.5.1 创建过滤器工厂

过滤器可以实现 IFilterFactory 接口来负责创建过滤器的实例，并指定是否可以重用这些实例。IFilterFactory 接口定义了表 30-16 中描述的成员。

表 30-16 IFilterFactory 成员

| 名称 | 描述 |
| --- | --- |
| IsReusable | 此 bool 属性指示是否可以重用过滤器的实例 |
| CreateInstance(serviceProvider) | 调用此方法来创建过滤器的新实例，并提供一个 IServiceProvider 对象 |

代码清单 30-34 实现了 IFilterFactory 接口，并为 IsReusable 属性返回 false，这将阻止过滤器的重用。

**代码清单 30-34 在 Filters 文件夹的 GuidResponseAttribute.cs 文件中实现接口**

```
using Microsoft.AspNetCore.Mvc;
using Microsoft.AspNetCore.Mvc.Filters;
using Microsoft.AspNetCore.Mvc.ModelBinding;
using Microsoft.AspNetCore.Mvc.ViewFeatures;
using System;
using System.Collections.Generic;
using System.Threading.Tasks;
using Microsoft.Extensions.DependencyInjection;

namespace WebApp.Filters {

    [AttributeUsage(AttributeTargets.Method | AttributeTargets.Class,
       AllowMultiple = true)]
    public class GuidResponseAttribute : Attribute,
            IAsyncAlwaysRunResultFilter, IFilterFactory {
        private int counter = 0;
        private string guid = Guid.NewGuid().ToString();

        public bool IsReusable => false;

        public IFilterMetadata CreateInstance(IServiceProvider serviceProvider) {
            return ActivatorUtilities
                .GetServiceOrCreateInstance<GuidResponseAttribute>(serviceProvider);
        }

        public async Task OnResultExecutionAsync(ResultExecutingContext context,
            ResultExecutionDelegate next) {

            Dictionary<string, string> resultData;
            if (context.Result is ViewResult vr
                && vr.ViewData.Model is Dictionary<string, string> data) {
```

```
                    resultData = data;
                } else {
                    resultData = new Dictionary<string, string>();
                    context.Result = new ViewResult() {
                        ViewName = "/Views/Shared/Message.cshtml",
                        ViewData = new ViewDataDictionary(
                                    new EmptyModelMetadataProvider(),
                                    new ModelStateDictionary()) {
                            Model = resultData
                        }
                    };
                }
                while (resultData.ContainsKey($"Counter_{counter}")) {
                    counter++;
                }
                resultData[$"Counter_{counter}"] = guid;
                await next();
            }
        }
    }
```

使用GetServiceOrCreateInstance方法创建新的过滤器对象，该方法由Microsoft.Extensions.DependencyInjection名称空间中的ActivatorUtilities类定义。尽管可以使用new关键字创建过滤器，但此方法解析通过过滤器的构造函数声明的、对服务的任何依赖关系。

要查看实现 IFilterFactory 接口的效果，请重新启动 ASP.NET Core 并请求 https://localhost:44350/?diag。重新加载浏览器，每次处理请求时，创建新的过滤器，并显示新的 GUID，如图 30-13 所示。

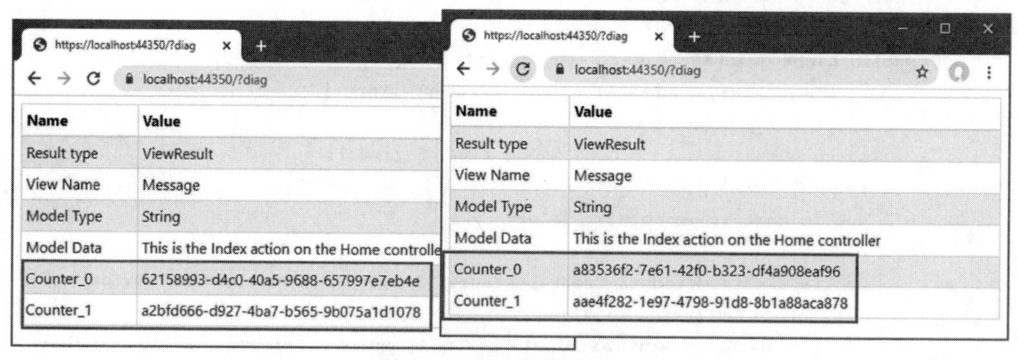


图 30-13　防止过滤器的重用

## 30.5.2　使用依赖注入范围来管理过滤器的生命周期

过滤器可以注册为服务，这允许通过依赖注入来控制它们的生命周期，第 14 章对此进行了描述。代码清单 30-35 将 GuidResponse 过滤器注册为范围限定的服务。

### 代码清单 30-35 在 WebApp 文件夹的 Startup.cs 文件中创建过滤器服务

```
using Microsoft.AspNetCore.Builder;
using Microsoft.Extensions.DependencyInjection;
using Microsoft.Extensions.Configuration;
using Microsoft.EntityFrameworkCore;
using WebApp.Models;
using Microsoft.AspNetCore.Antiforgery;
using Microsoft.AspNetCore.Http;
using Microsoft.AspNetCore.Mvc;
using WebApp.Filters;

namespace WebApp {
    public class Startup {

        public Startup(IConfiguration config) {
            Configuration = config;
        }

        public IConfiguration Configuration { get; set; }

        public void ConfigureServices(IServiceCollection services) {
            services.AddDbContext<DataContext>(opts => {
                opts.UseSqlServer(Configuration[
                    "ConnectionStrings:ProductConnection"]);
                opts.EnableSensitiveDataLogging(true);
            });
            services.AddControllersWithViews().AddRazorRuntimeCompilation();
            services.AddRazorPages().AddRazorRuntimeCompilation();
            services.AddSingleton<CitiesData>();

            services.Configure<AntiforgeryOptions>(opts => {
                opts.HeaderName = "X-XSRF-TOKEN";
            });
            services.Configure<MvcOptions>(opts => opts.ModelBindingMessageProvider
                .SetValueMustNotBeNullAccessor(value => "Please enter a value"));

            services.AddScoped<GuidResponseAttribute>();
        }

        public void Configure(IApplicationBuilder app, DataContext context,
                IAntiforgery antiforgery) {

            // ...statements omitted for brevity...
        }
    }
}
```

默认情况下，ASP.NET Core 为每个请求创建一个范围，这意味着为每个请求创建一个过滤器实例。要查看效果，请重新启动 ASP.NET Core，并请求 https://localhost:44350/?diag。应用于主控制器的两个属性都是使用过滤器的相同实例处理的，这意味着响应中的两个 GUID 是相同的。重新加载浏览器，创建一个新的范围，并使用一个新的过滤器对象，如图 30-14 所示。

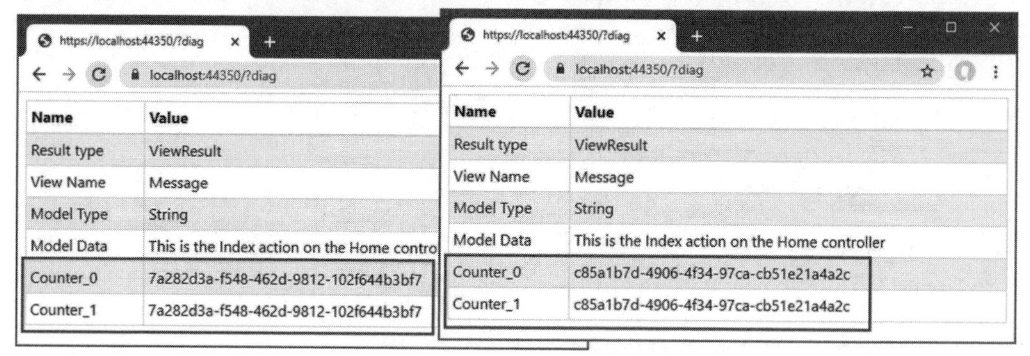


图 30-14　使用依赖注入来管理过滤器

**使用过滤器作为服务，而不使用 IFilterFactory 接口**

在本例中，生命周期中的更改立即生效，因为使用了 ActivatorUtilities.GetServiceOrCreateInstance 方法，用于在实现 IFilterFactory 接口时创建过滤器对象。此方法在调用请求类型的构造函数之前，检查是否有可用的服务。如果想使用过滤器作为服务，而不实现 IFilterFactory 和使用 ActivatorUtilities，可以使用 ServiceFilter 属性应用过滤器，如下所示：

```
...
[ServiceFilter(typeof(GuidResponseAttribute))]
...
```

ASP.NET Core 从服务中创建过滤器对象，并将其应用到请求中。以这种方式应用的过滤器不必从 Attribute 类派生。

## 30.6　创建全局过滤器

全局过滤器应用于 ASP.NET Core 处理的每一个请求，这意味着它们不必应用于单个控制器或 Razor Pages。任何过滤器都可以用作全局过滤器；然而，操作过滤器将仅应用于端点是操作方法的请求，而页面过滤器将仅应用于端点是 Razor Pages 的请求。

使用 Startup 类中的 options 模式设置全局过滤器，如代码清单 30-36 所示。

**代码清单 30-36　在 WebApp 文件夹的 Startup.cs 文件中创建全局过滤器**

```
...
public void ConfigureServices(IServiceCollection services) {
    services.AddDbContext<DataContext>(opts => {
        opts.UseSqlServer(Configuration[
            "ConnectionStrings:ProductConnection"]);
        opts.EnableSensitiveDataLogging(true);
    });
```

```
        services.AddControllersWithViews().AddRazorRuntimeCompilation();
        services.AddRazorPages().AddRazorRuntimeCompilation();
        services.AddSingleton<CitiesData>();

        services.Configure<AntiforgeryOptions>(opts => {
            opts.HeaderName = "X-XSRF-TOKEN";
        });

        services.Configure<MvcOptions>(opts => opts.ModelBindingMessageProvider
            .SetValueMustNotBeNullAccessor(value => "Please enter a value"));

        services.AddScoped<GuidResponseAttribute>();
        services.Configure<MvcOptions>(opts => opts.Filters.Add<HttpsOnlyAttribute>());
    }
    ...
```

MvcOptions.Filters 属性返回一个集合，其中添加了过滤器以全局应用它们，对于同样是服务的过滤器，可以使用 Add<T>方法或使用 AddService<T>方法。还有一个没有泛型类型参数的 Add 方法，可用于将特定对象注册为全局过滤器。

代码清单 30-36 中的语句注册了本章前面创建的 HttpsOnly 过滤器，这意味着它不再需要直接应用于单个控制器或 Razor Pages，因此代码清单 30-37 从主控制器中删除了这个过滤器。

---

■ **注意：** 在代码清单 30-37 中禁用了 GuidResponse 过滤器。这是一个始终运行的结果过滤器，将替换全局过滤器生成的结果。

---

#### 代码清单 30-37　在 Controllers 文件夹的 HomeController.cs 文件中删除过滤器

```
using Microsoft.AspNetCore.Mvc;
using Microsoft.AspNetCore.Http;
using WebApp.Filters;
using Microsoft.AspNetCore.Mvc.Filters;
using System;

namespace WebApp.Controllers {

    //[HttpsOnly]
    [ResultDiagnostics]
    //[GuidResponse]
    //[GuidResponse]

    public class HomeController : Controller {

        public IActionResult Index() {
            return View("Message",
                "This is the Index action on the Home controller");
        }
    }
```

```
}
```

重启 ASP.NET Core 并请求 http://localhost:5000 以确认正在应用 HTTPS-only 策略，即使该属性不再用于装饰控制器。全局授权过滤器将短路过滤器管道，并生成图 30-15 所示的响应。

图 30-15　使用全局过滤器

## 30.7　理解和改变过滤器的顺序

过滤器以特定顺序运行：授权、资源、操作(或页面)和结果。但是，如果有给定类型的多个过滤器，则应用它们的顺序由应用过滤器的范围驱动。

为了演示它是如何工作的，在 Filters 文件夹中添加一个名为 MessageAttribute.cs 的类文件，并使用它来定义如代码清单 30-38 所示的过滤器。

### 代码清单 30-38　Filters 文件夹中 MessageAttribute.cs 文件的内容

```
using Microsoft.AspNetCore.Mvc;
using Microsoft.AspNetCore.Mvc.Filters;
using Microsoft.AspNetCore.Mvc.ModelBinding;
using Microsoft.AspNetCore.Mvc.ViewFeatures;
using System;
using System.Collections.Generic;
using System.Threading.Tasks;

namespace WebApp.Filters {

    [AttributeUsage(AttributeTargets.Method | AttributeTargets.Class,
       AllowMultiple = true)]
    public class MessageAttribute : Attribute, IAsyncAlwaysRunResultFilter {
        private int counter = 0;
        private string msg;

        public MessageAttribute(string message) => msg = message;

        public async Task OnResultExecutionAsync(ResultExecutingContext context,
```

```
                ResultExecutionDelegate next) {
            Dictionary<string, string> resultData;
            if (context.Result is ViewResult vr
                && vr.ViewData.Model is Dictionary<string, string> data) {
                resultData = data;
            } else {
                resultData = new Dictionary<string, string>();
                context.Result = new ViewResult() {
                    ViewName = "/Views/Shared/Message.cshtml",
                    ViewData = new ViewDataDictionary(
                                    new EmptyModelMetadataProvider(),
                                    new ModelStateDictionary()) {
                        Model = resultData
                    }
                };
            }
            while (resultData.ContainsKey($"Message_{counter}")) {
                counter++;
            }
            resultData[$"Message_{counter}"] = msg;
            await next();
        }
    }
}
```

此结果过滤器使用前面示例中展示的技术来替换来自端点的结果，并允许多个过滤器构建一系列显示给用户的消息。代码清单 30-39 将消息过滤器的几个实例应用于主控制器。

**代码清单 30-39　在 Controllers 文件夹的 HomeController.cs 文件中应用过滤器**

```
using Microsoft.AspNetCore.Mvc;
using Microsoft.AspNetCore.Http;
using WebApp.Filters;
using Microsoft.AspNetCore.Mvc.Filters;
using System;

namespace WebApp.Controllers {

    [Message("This is the controller-scoped filter")]
    public class HomeController : Controller {

        [Message("This is the first action-scoped filter")]
        [Message("This is the second action-scoped filter")]
        public IActionResult Index() {
            return View("Message",
                "This is the Index action on the Home controller");
        }
    }
```

```
}
```

代码清单 30-40 全局注册了消息过滤器。

**代码清单 30-40　在 WebApp 文件夹的 Startup.cs 文件中创建全局过滤器**

```
...
public void ConfigureServices(IServiceCollection services) {
    services.AddDbContext<DataContext>(opts => {
        opts.UseSqlServer(Configuration[
            "ConnectionStrings:ProductConnection"]);
        opts.EnableSensitiveDataLogging(true);
    });
    services.AddControllersWithViews().AddRazorRuntimeCompilation();
    services.AddRazorPages().AddRazorRuntimeCompilation();
    services.AddSingleton<CitiesData>();

    services.Configure<AntiforgeryOptions>(opts => {
        opts.HeaderName = "X-XSRF-TOKEN";
    });

    services.Configure<MvcOptions>(opts => opts.ModelBindingMessageProvider
        .SetValueMustNotBeNullAccessor(value => "Please enter a value"));

    services.AddScoped<GuidResponseAttribute>();
    services.Configure<MvcOptions>(opts => {
        opts.Filters.Add<HttpsOnlyAttribute>();
        opts.Filters.Add(new MessageAttribute("This is the globally-scoped filter"));
    });
}
...
```

同一个过滤器有四个实例。要查看应用它们的顺序，请重新启动 ASP.NET Core，请求 https://localhost:44350，生成如图 30-16 所示的响应。

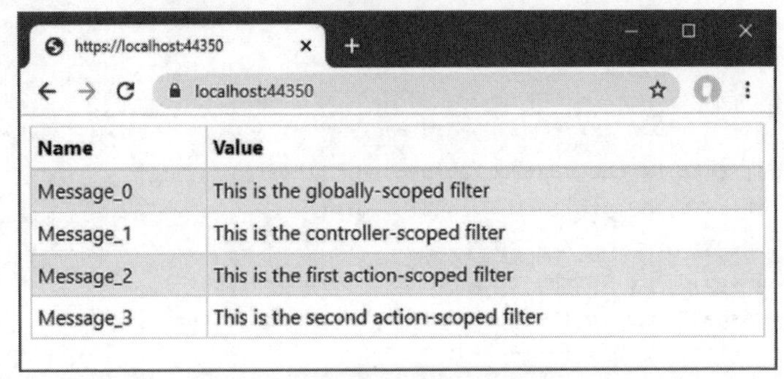


图 30-16　在不同范围内应用相同的过滤器

默认情况下，ASP.NET Core 运行全局过滤器，然后过滤器应用于控制器或页面模型类，最后过滤器应用于操作或处理程序方法。

## 改变过滤器顺序

默认的顺序可以通过实现 IOrderedFilter 接口来改变。ASP.NET Core 在确定过滤器如何排序时在该接口中寻找。接口的定义如下：

```
namespace Microsoft.AspNetCore.Mvc.Filters {

    public interface IOrderedFilter : IFilterMetadata {
        int Order { get; }
    }
}
```

Order 属性返回一个 int 值，低值的过滤器会在高 Order 值的过滤器之前应用。

代码清单 30-41 在消息过滤器中实现了接口，并定义了一个构造函数参数，该构造函数参数允许在应用过滤器时指定 Order 属性的值。

**代码清单 30-41 在 Filters 文件夹的 MessageAttribute.cs 文件中添加排序支持**

```
using Microsoft.AspNetCore.Mvc;
using Microsoft.AspNetCore.Mvc.Filters;
using Microsoft.AspNetCore.Mvc.ModelBinding;
using Microsoft.AspNetCore.Mvc.ViewFeatures;
using System;
using System.Collections.Generic;
using System.Threading.Tasks;

namespace WebApp.Filters {

    [AttributeUsage(AttributeTargets.Method | AttributeTargets.Class,
       AllowMultiple = true)]
    public class MessageAttribute : Attribute, IAsyncAlwaysRunResultFilter,
            IOrderedFilter {
        private int counter = 0;
        private string msg;

        public MessageAttribute(string message) => msg = message;

        public int Order { get; set; }

        public async Task OnResultExecutionAsync(ResultExecutingContext context,
                ResultExecutionDelegate next) {

            // ...statements omitted for brevity...
        }
    }
}
```

代码清单 30-42 使用构造函数参数来更改应用过滤器的顺序。

**代码清单 30-42　在 Controllers 文件夹的 HomeController.cs 文件中设置过滤器顺序**

```
using Microsoft.AspNetCore.Mvc;
using Microsoft.AspNetCore.Http;
using WebApp.Filters;
using Microsoft.AspNetCore.Mvc.Filters;
using System;

namespace WebApp.Controllers {

    [Message("This is the controller-scoped filter", Order = 10)]
    public class HomeController : Controller {

        [Message("This is the first action-scoped filter", Order = 1)]
        [Message("This is the second action-scoped filter", Order = -1)]
        public IActionResult Index() {
            return View("Message",
               "This is the Index action on the Home controller");
        }
    }
}
```

Order值可以是负数，这是一种有效方法，可以确保在任何具有默认顺序的全局过滤器之前应用过滤器(尽管也可在创建全局过滤器时设置顺序)。重启ASP.NET Core并请求https:/localhost:44350以查看新的过滤器顺序，如图 30-17 所示。

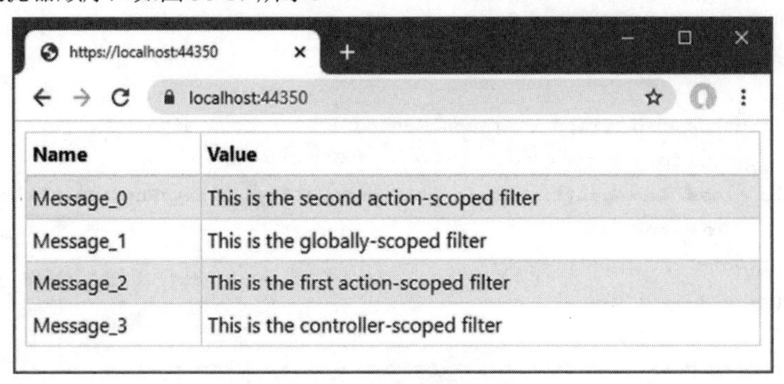


图 30-17　改变过滤器订单

# 30.8　小结

本章描述了 ASP.NET Core 过滤器特性，并解释了如何使用它来更改特定端点的请求和结果。描述了不同类型的过滤器，演示了如何创建和应用每种过滤器。还展示了如何管理过滤器的生命周期并控制它们的执行顺序。第 31 章展示如何结合本书这一部分中描述的功能来创建表单应用程序。

# 第 31 章

■■■

# 创建表单应用程序

前几章集中讨论了处理 HTML 表单一个方面的单个特性，有时很难看到它们如何组合在一起执行常见的任务。本章将介绍创建控制器、视图和 Razor Pages 的过程，这些页面支持具有创建、读取、更新和删除(CRUD)功能的应用程序。本章没有介绍新的功能，目标是演示如何将标签助手、模型绑定和模型验证等功能与 Entity Framework Core 结合使用。

## 31.1 准备工作

本章使用了第 30 章中的 WebApp 项目。为准备本章，请把 Controllers 文件夹中 HomeController.cs 文件的内容替换为代码清单 31-1 所示的代码。

■ **提示：**可以从 https://github.com/apress/pro-asp.net-core-3 下载本章和本书中其他所有章节的示例项目。如果在运行示例时遇到问题，请参阅第 1 章以了解如何获得帮助。

**代码清单 31-1　Controllers 文件夹中 HomeController.cs 文件的内容**

```
using Microsoft.AspNetCore.Mvc;
using Microsoft.EntityFrameworkCore;
using System.Collections.Generic;
using System.Threading.Tasks;
using WebApp.Models;

namespace WebApp.Controllers {

    [AutoValidateAntiforgeryToken]
    public class HomeController : Controller {
        private DataContext context;

        private IEnumerable<Category> Categories => context.Categories;
        private IEnumerable<Supplier> Suppliers => context.Suppliers;

        public HomeController(DataContext data) {
            context = data;
```

```
        }

        public IActionResult Index() {
            return View(context.Products.
               Include(p => p.Category).Include(p => p.Supplier));
        }
    }
}
```

创建如代码清单 31-2 所示的文件。

#### 代码清单 31-2　Views/Home 文件夹中 Index.cshtml 文件的内容

```
@model IEnumerable<Product>
@{ Layout = "_SimpleLayout"; }

<h4 class="bg-primary text-white text-center p-2">Products</h4>
<table class="table table-sm table-bordered table-striped">
    <thead>
        <tr>
           <th>ID</th><th>Name</th><th>Price</th><th>Category</th><th></th>
        </tr>
    </thead>
    <tbody>
        @foreach (Product p in Model) {
            <tr>
                <td>@p.ProductId</td>
                <td>@p.Name</td>
                <td>@p.Price</td>
                <td>@p.Category.Name</td>
                <td class="text-center">
                    <a asp-action="Details" asp-route-id="@p.ProductId"
                       class="btn btn-sm btn-info">Details</a>
                    <a asp-action="Edit" asp-route-id="@p.ProductId"
                       class="btn btn-sm btn-warning">Edit</a>
                    <a asp-action="Delete" asp-route-id="@p.ProductId"
                       class="btn btn-sm btn-danger">Delete</a>
                </td>
            </tr>
        }
    </tbody>
</table>
<a asp-action="Create" class="btn btn-primary">Create</a>
```

接下来，更新 Product 类，如代码清单 31-3 所示，以更改验证约束，删除模型级检查，并禁用远程验证。

### 代码清单 31-3　在 Models 文件夹的 Product.cs 文件中更改验证

```
using System.ComponentModel.DataAnnotations.Schema;
using System.ComponentModel.DataAnnotations;
using Microsoft.AspNetCore.Mvc.ModelBinding;
using WebApp.Validation;
using Microsoft.AspNetCore.Mvc;

namespace WebApp.Models {

    //[PhraseAndPrice(Phrase = "Small", Price = "100")]
    public class Product {

        public long ProductId { get; set; }

        [Required]
        [Display(Name = "Name")]
        public string Name { get; set; }
        [Column(TypeName = "decimal(8, 2)")]
        [Required(ErrorMessage = "Please enter a price")]
        [Range(1, 999999, ErrorMessage = "Please enter a positive price")]
        public decimal Price { get; set; }

        [PrimaryKey(ContextType = typeof(DataContext),
           DataType = typeof(Category))]
        //[Remote("CategoryKey", "Validation",
        //     ErrorMessage = "Enter an existing key")]
        public long CategoryId { get; set; }
        public Category Category { get; set; }

        [PrimaryKey(ContextType = typeof(DataContext),
           DataType = typeof(Category))]
        //[Remote("SupplierKey", "Validation",
        //    ErrorMessage = "Enter an existing key")]
        public long SupplierId { get; set; }
        public Supplier Supplier { get; set; }
    }
}
```

最后，在 Startup 类中禁用全局过滤器，如代码清单 31-4 所示。

### 代码清单 31-4　在 WebApp 文件夹的 Startup.cs 文件中禁用过滤器

```
...
public void ConfigureServices(IServiceCollection services) {
    services.AddDbContext<DataContext>(opts => {
        opts.UseSqlServer(Configuration[
           "ConnectionStrings:ProductConnection"]);
```

```
        opts.EnableSensitiveDataLogging(true);
    });
    services.AddControllersWithViews().AddRazorRuntimeCompilation();
    services.AddRazorPages().AddRazorRuntimeCompilation();
    services.AddSingleton<CitiesData>();

    services.Configure<AntiforgeryOptions>(opts => {
        opts.HeaderName = "X-XSRF-TOKEN";
    });

    services.Configure<MvcOptions>(opts => opts.ModelBindingMessageProvider
        .SetValueMustNotBeNullAccessor(value => "Please enter a value"));

    services.AddScoped<GuidResponseAttribute>();
    //services.Configure<MvcOptions>(opts => {
    // opts.Filters.Add<HttpsOnlyAttribute>();
    // opts.Filters.Add(new MessageAttribute(
    // "This is the globally-scoped filter"));
    //});
}
...
```

### 31.1.1　删除数据库

打开一个新的 PowerShell 命令提示符，导航到包含 WebApp.csproj 文件的文件夹，并运行代码清单 31-5 所示的命令来删除数据库。

**代码清单 31-5　删除数据库**

```
dotnet ef database drop --force
```

### 31.1.2　运行示例应用程序

从 Debug 菜单中选择 Start Without Debugging 或 Run Without Debugging，或者使用 PowerShell 命令提示符运行代码清单 31-6 中所示的命令。

**代码清单 31-6　运行示例应用程序**

```
dotnet run
```

使用浏览器请求 http://localhost:5000/controllers，这将显示一个产品列表，如图 31-1 所示。有一些锚定元素样式化为按钮，但直到添加了创建、编辑和删除对象的功能之后，这些元素才会工作。

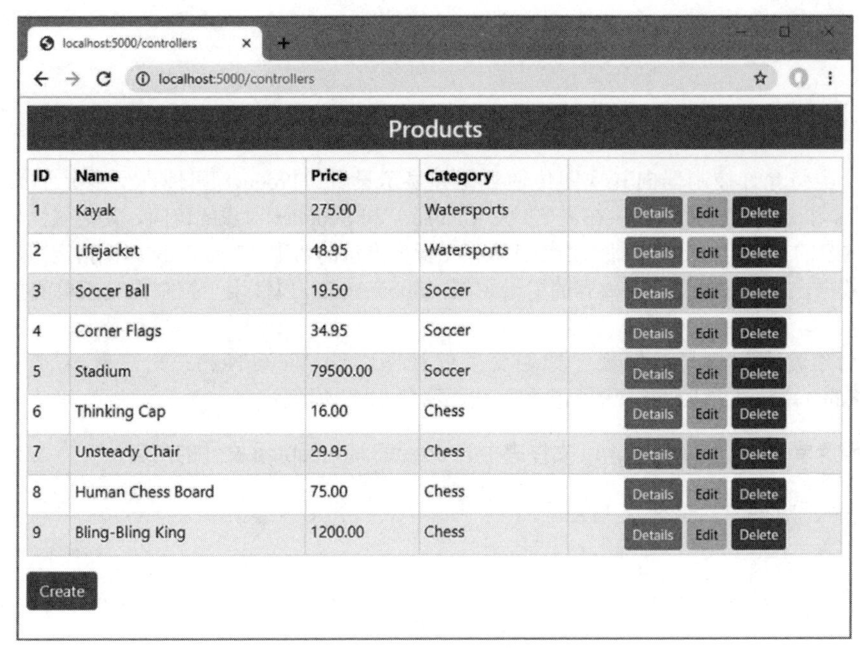


图 31-1 运行示例应用程序

# 31.2 创建 MVC 表单应用程序

在接下来的部分中，将展示如何使用 MVC 控制器和视图执行核心数据操作。本章的后面使用 Razor 页面(Razor Pages)创建相同的功能。

## 31.2.1 准备视图模型和视图

下面定义一个用于多个操作的表单，通过它的视图模型类进行配置。要创建视图模型类，向 Models 文件夹添加一个名为 ProductViewModel.cs 的类文件，并添加如代码清单 31-7 所示的代码。

**代码清单 31-7 Models 文件夹中 ProductViewModel.cs 文件的内容**

```
using System.Collections.Generic;
using System.Linq;

namespace WebApp.Models {

    public class ProductViewModel {
        public Product Product { get; set; }
        public string Action { get; set; } = "Create";
        public bool ReadOnly { get; set; } = false;
        public string Theme { get; set; } = "primary";
        public bool ShowAction { get; set; } = true;
        public IEnumerable<Category> Categories { get; set; }
            = Enumerable.Empty<Category>();
```

```
        public IEnumerable<Supplier> Suppliers { get; set; }
            = Enumerable.Empty<Supplier>();
    }
}
```

这个类将允许控制器向其视图传递数据和显示设置。Product 属性提供要显示的数据，Categories 和 Suppliers 属性提供在需要时对 Category 和 Supplier 对象的访问。其他属性配置给用户呈现内容的方式：Action 属性给当前任务指定操作方法的名称，ReadOnly 属性指定用户是否可以编辑数据，Theme 属性指定内容的引导主题，ShowAction 属性用于控制提交表单的按钮的可见性。

要创建允许用户与应用程序数据交互的视图，给 Views/Home 文件夹添加一个名为 ProductEditor.cshtml 的 Razor 视图，内容如代码清单 31-8 所示。

#### 代码清单 31-8　Views/Home 文件夹中 ProductEditor.cshtml 文件的内容

```
@model ProductViewModel
@{ Layout = "_SimpleLayout"; }

<partial name="_Validation" />

<h5 class="bg-@Model.Theme text-white text-center p-2">@Model.Action</h5>

<form asp-action="@Model.Action" method="post">
    <div class="form-group">
        <label asp-for="Product.ProductId"></label>
        <input class="form-control" asp-for="Product.ProductId" readonly />
    </div>
    <div class="form-group">
        <label asp-for="Product.Name"></label>
        <div>
            <span asp-validation-for="Product.Name" class="text-danger"></span>
        </div>
        <input class="form-control" asp-for="Product.Name"
               readonly="@Model.ReadOnly" />
    </div>
    <div class="form-group">
        <label asp-for="Product.Price"></label>
        <div>
            <span asp-validation-for="Product.Price" class="text-danger"></span>
        </div>
        <input class="form-control" asp-for="Product.Price"
               readonly="@Model.ReadOnly" />
    </div>
    <div class="form-group">
        <label asp-for="Product.CategoryId">Category</label>
        <div>
            <span asp-validation-for="Product.CategoryId" class="text-danger"></span>
```

```
            </div>
            <select asp-for="Product.CategoryId" class="form-control"
                    disabled="@Model.ReadOnly"
                    asp-items="@(new SelectList(Model.Categories,
                       "CategoryId", "Name"))">
                <option value="" disabled selected>Choose a Category</option>
            </select>
        </div>
        <div class="form-group">
            <label asp-for="Product.SupplierId">Supplier</label>
            <div>
               <span asp-validation-for="Product.SupplierId" class="text-danger"></span>
            </div>
            <select asp-for="Product.SupplierId" class="form-control"
                    disabled="@Model.ReadOnly"
                    asp-items="@(new SelectList(Model.Suppliers,
                       "SupplierId", "Name"))">
                <option value="" disabled selected>Choose a Supplier</option>
            </select>
        </div>
        @if (Model.ShowAction) {
            <button class="btn btn-@Model.Theme" type="submit">@Model.Action</button>
        }
        <a class="btn btn-secondary" asp-action="Index">Back</a>
    </form>
```

这个视图可能看起来很复杂，但它只结合了前面章节介绍的特性，一旦看到它的实际应用，它将变得更加清晰。这个视图的模型是一个 ProductViewModel 对象，它既提供了显示给用户的数据，也提供了有关数据应该如何显示的一些方向。

对于 Product 类定义的每个属性，视图包含一组元素：描述属性的 label 元素、允许对值进行编辑的 input 或 select 元素，以及显示验证消息的 span 元素。每个元素都配置了 asp-for 特性，确保标签助手为每个属性转换元素。这里有定义视图结构的 div 元素，所有元素都是用于样式化表单的引导 CSS 类的成员。

## 31.2.2 读取数据

最简单的操作是从数据库中读取数据并将其呈现给用户。在大多数应用程序中，这将允许用户看到列表视图中没有的额外细节。应用程序执行的每个任务都需要一组不同的 ProductViewModel 属性。为了管理这些组合，给 Models 文件夹添加一个名为 ViewModelFactory.cs 的类文件，如代码清单 31-9 所示。

**代码清单 31-9 Models 文件夹中 ViewModelFactory.cs 文件的内容**

```
using System.Collections.Generic;
using System.Linq;

namespace WebApp.Models {
```

```
    public static class ViewModelFactory {

        public static ProductViewModel Details(Product p) {
            return new ProductViewModel {
                Product = p, Action = "Details",
                ReadOnly = true, Theme = "info", ShowAction = false,
                Categories = p == null ? Enumerable.Empty<Category>()
                    : new List<Category> { p.Category },
                Suppliers = p == null ? Enumerable.Empty<Supplier>()
                    : new List<Supplier> { p.Supplier},
            };
        }
    }
}
```

Details 方法生成一个为查看对象而配置的 ProductViewModel 对象。当用户查看详细信息时，Category 和 Supplier 信息是只读的，这意味着只需要提供当前的 Category 和 Supplier 信息。

接下来，向使用 ViewModelFactory.Details 方法的主控制器添加一个操作方法，创建 ProductViewModel 对象，并使用 ProductEditor 视图将其显示给用户，如代码清单 31-10 所示。

#### 代码清单 31-10　在 Controllers 文件夹的 HomeController.cs 文件中添加一个操作方法

```
using Microsoft.AspNetCore.Mvc;
using Microsoft.EntityFrameworkCore;
using System.Collections.Generic;
using System.Threading.Tasks;
using WebApp.Models;

namespace WebApp.Controllers {

    [AutoValidateAntiforgeryToken]
    public class HomeController : Controller {
        private DataContext context;

        private IEnumerable<Category> Categories => context.Categories;
        private IEnumerable<Supplier> Suppliers => context.Suppliers;

        public HomeController(DataContext data) {
            context = data;
        }

        public IActionResult Index() {
            return View(context.Products.
                Include(p => p.Category).Include(p => p.Supplier));
        }
```

```
        public async Task<IActionResult> Details(long id) {
            Product p = await context.Products.
                Include(p => p.Category).Include(p => p.Supplier)
                .FirstOrDefaultAsync(p => p.ProductId == id);
            ProductViewModel model = ViewModelFactory.Details(p);
            return View("ProductEditor", model);
        }
    }
}
```

操作方法使用 id 参数(从路由数据中绑定模型)来查询数据库，并将 Product 对象传递给 ViewModelFactory.Details 方法。大多数操作都需要 Category 和 Supplier 数据，因此添加了提供对数据的直接访问的属性。

要测试 Details 特性，请重新启动 ASP.NET Core，请求 http://localhost:5000/controllers。单击其中一个 Details 按钮，将看到使用 ProductEditor 视图以只读形式显示选择的对象，如图 31-2 所示。

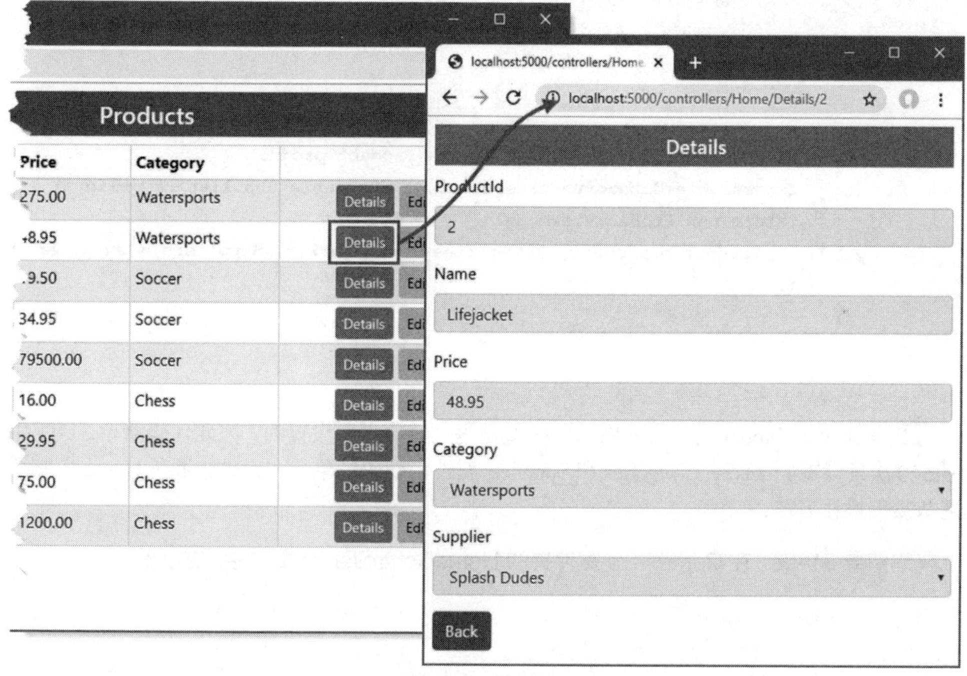


图 31-2 查看数据

如果用户导航到一个与数据库中的对象不对应的 URL，例如 http://localhost:5000/controllers/Home/Details/100，将显示一个空表单。

## 31.2.3 创建数据

创建数据依赖于模型绑定从请求中获取表单数据，并依赖验证来确保数据可以存储在数据库中。第一步是添加一个工厂方法，创建用于创建数据的视图模型对象，如代码清单 31-11 所示。

**代码清单 31-11　在 Models 文件夹的 ViewModelFactory.cs 文件中添加一个方法**

```
using System.Collections.Generic;
using System.Linq;

namespace WebApp.Models {

    public static class ViewModelFactory {

        public static ProductViewModel Details(Product p) {
            return new ProductViewModel {
                Product = p, Action = "Details",
                ReadOnly = true, Theme = "info", ShowAction = false,
                Categories = p == null ? Enumerable.Empty<Category>()
                    : new List<Category> { p.Category },
                Suppliers = p == null ? Enumerable.Empty<Supplier>()
                    : new List<Supplier> { p.Supplier},
            };
        }

        public static ProductViewModel Create(Product product,
            IEnumerable<Category> categories, IEnumerable<Supplier> suppliers) {
            return new ProductViewModel {
               Product = product, Categories = categories, Suppliers = suppliers
            };
        }
    }
}
```

为 ProductViewModel 属性所使用的默认值是为创建数据而设置的，因此代码清单 31-11 中的 Create 方法只设置 Product、Categories 和 Suppliers 属性。代码清单 31-12 添加了在 Home 控制器中创建数据的操作方法。

**代码清单 31-12　在 Controllers 文件夹的 HomeController.cs 文件中添加操作**

```
using Microsoft.AspNetCore.Mvc;
using Microsoft.EntityFrameworkCore;
using System.Collections.Generic;
using System.Threading.Tasks;
using WebApp.Models;

namespace WebApp.Controllers {

    [AutoValidateAntiforgeryToken]
    public class HomeController : Controller {
        private DataContext context;

        private IEnumerable<Category> Categories => context.Categories;
```

```
        private IEnumerable<Supplier> Suppliers => context.Suppliers;

        public HomeController(DataContext data) {
            context = data;
        }

        public IActionResult Index() {
            return View(context.Products.
                Include(p => p.Category).Include(p => p.Supplier));
        }

        public async Task<IActionResult> Details(long id) {
            Product p = await context.Products.
                Include(p => p.Category).Include(p => p.Supplier)
                .FirstOrDefaultAsync(p => p.ProductId == id);
            ProductViewModel model = ViewModelFactory.Details(p);
            return View("ProductEditor", model);
        }

        public IActionResult Create() {
            return View("ProductEditor",
                ViewModelFactory.Create(new Product(), Categories, Suppliers));
        }

        [HttpPost]
        public async Task<IActionResult> Create([FromForm] Product product) {
            if (ModelState.IsValid) {
                product.ProductId = default;
                product.Category = default;
                product.Supplier = default;
                context.Products.Add(product);
                await context.SaveChangesAsync();
                return RedirectToAction(nameof(Index));
            }
            return View("ProductEditor",
                ViewModelFactory.Create(product, Categories, Suppliers));
        }
    }
}
```

有两个 Create 方法，它们通过 HttpPost 属性和方法参数进行区分。HTTP GET 请求将由第一个方法处理，该方法选择 ProductEditor 视图并为其提供一个 ProductViewModel 对象。当用户提交表单时，它将被第二种方法接收，该方法依赖模型绑定来接收数据，并通过模型验证来确保数据有效。

如果数据通过验证，那么通过重置三个属性来准备存储在数据库中的对象，如下所示：

```
...
product.ProductId = default;
```

```
product.Category = default;
product.Supplier = default;
...
```

Entity Framework Core 配置数据库，以便在存储新数据时由数据库服务器分配主键。如果试图存储一个对象并提供一个不为零的 ProductId 值，将抛出一个异常。

重置了 Category 和 Supplier 属性，以防止存储对象时，Entity Framework Core 试图处理相关数据。Entity Framework Core 能够处理相关数据，但可能产生意外结果(本章后面的“创建新的相关数据对象”一节展示如何创建相关数据)。

注意，当验证失败时，使用参数调用 View 方法，如下所示：

```
...
return View("ProductEditor",
    ViewModelFactory.Create(product, Categories, Suppliers));
...
```

之所以这样做，是因为视图期望的视图模型对象与前面使用模型绑定从请求中提取的数据类型不同。相反，这里创建了一个新的视图模型对象，该对象包含模型绑定数据，并将其传递给 View 方法。

重启 ASP.NET Core，请求 http://localhost:5000/controllers，然后单击 Create。填写表单并单击 Create 按钮提交数据。新对象存储在数据库中，并在浏览器重定向到 Index 操作时显示出来，如图 31-3 所示。

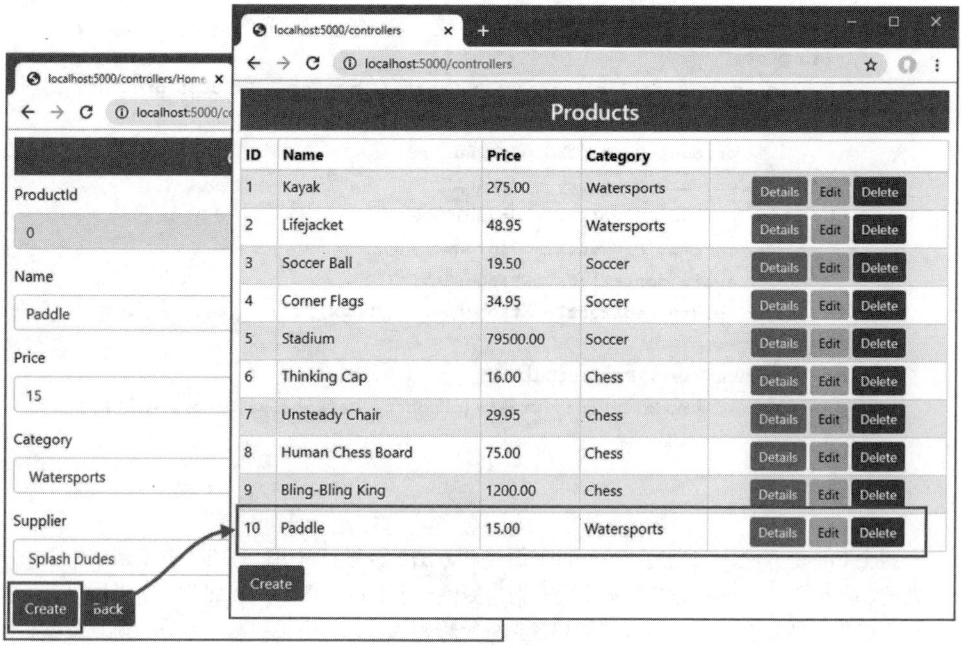

图 31-3　创建新对象

注意，select 元素允许用户使用类别和供应商名称来选择 CategoryId 和 SupplierId 属性的值，如下所示：

```
...
<select asp-for="Product.SupplierId" class="form-control" disabled="@Model.ReadOnly"
        asp-items="@(new SelectList(Model.Suppliers, "SupplierId", "Name"))">
    <option value="" disabled selected>Choose a Supplier</option>
</select>
...
```

第 30 章使用 input 元素来允许直接设置这些属性的值，但这是因为本章想演示不同类型的验证。在实际应用程序中，当应用程序已经拥有期望用户从中选择的数据时，最好为用户提供受限的选择。例如，在实际项目中，让用户输入有效的主键是没有意义的，因为应用程序可以很容易地为用户提供一个可供选择的键列表，如图 31-4 所示。

**提示：** “创建新的相关数据对象”一节展示了创建相关数据的不同技术。

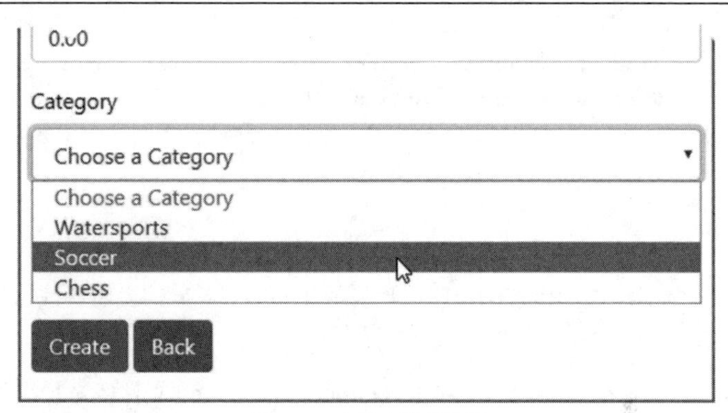


图 31-4　向用户提供选择

## 31.2.4　编辑数据

编辑数据的过程类似于创建数据。第一步是向视图模型工厂添加一个新方法，该方法将配置向用户显示数据的方式，如代码清单 31-13 所示。

### 代码清单 31-13　在 Models 文件夹的 ViewModelFactory.cs 文件中添加方法

```
using System.Collections.Generic;
using System.Linq;

namespace WebApp.Models {

    public static class ViewModelFactory {

        public static ProductViewModel Details(Product p) {
            return new ProductViewModel {
                Product = p, Action = "Details",
                ReadOnly = true, Theme = "info", ShowAction = false,
                Categories = p == null ? Enumerable.Empty<Category>()
                    : new List<Category> { p.Category },
```

```
                Suppliers = p == null ? Enumerable.Empty<Supplier>()
                    : new List<Supplier> { p.Supplier},
            };
        }

        public static ProductViewModel Create(Product product,
                IEnumerable<Category> categories, IEnumerable<Supplier> suppliers) {
            return new ProductViewModel {
                Product = product, Categories = categories, Suppliers = suppliers
            };
        }

        public static ProductViewModel Edit(Product product,
                IEnumerable<Category> categories, IEnumerable<Supplier> suppliers) {
            return new ProductViewModel {
                Product = product, Categories = categories, Suppliers = suppliers,
                Theme = "warning", Action = "Edit"
            };
        }
    }
}
```

下一步是向主控制器添加操作方法，向用户显示 Product 对象的当前属性，并接收用户所做的更改，如代码清单 31-14 所示。

**代码清单 31-14　在 Controllers 文件夹的 HomeController.cs 文件中添加操作方法**

```
using Microsoft.AspNetCore.Mvc;
using Microsoft.EntityFrameworkCore;
using System.Collections.Generic;
using System.Threading.Tasks;
using WebApp.Models;

namespace WebApp.Controllers {

    [AutoValidateAntiforgeryToken]
    public class HomeController : Controller {
        private DataContext context;

        private IEnumerable<Category> Categories => context.Categories;
        private IEnumerable<Supplier> Suppliers => context.Suppliers;

        public HomeController(DataContext data) {
            context = data;
        }

        // ...other action methods omitted for brevity...
```

```
        public async Task<IActionResult> Edit(long id) {
            Product p = await context.Products.FindAsync(id);
            ProductViewModel model = ViewModelFactory.Edit(p, Categories, Suppliers);
            return View("ProductEditor", model);
        }

        [HttpPost]
        public async Task<IActionResult> Edit([FromForm]Product product) {
            if (ModelState.IsValid) {
                product.Category = default;
                product.Supplier = default;
                context.Products.Update(product);
                await context.SaveChangesAsync();
                return RedirectToAction(nameof(Index));
            }
            return View("ProductEditor",
                ViewModelFactory.Edit(product, Categories, Suppliers));
        }
    }
}
```

要查看编辑功能的工作，请重新启动 ASP.NET Core，导航到 http://localhost:5000/controllers，并单击其中一个 Edit 按钮。更改一个或多个属性值并提交表单。更改将存储在数据库中，并在浏览器重定向到 Index 操作时显示的列表中反映出来，如图 31-5 所示。

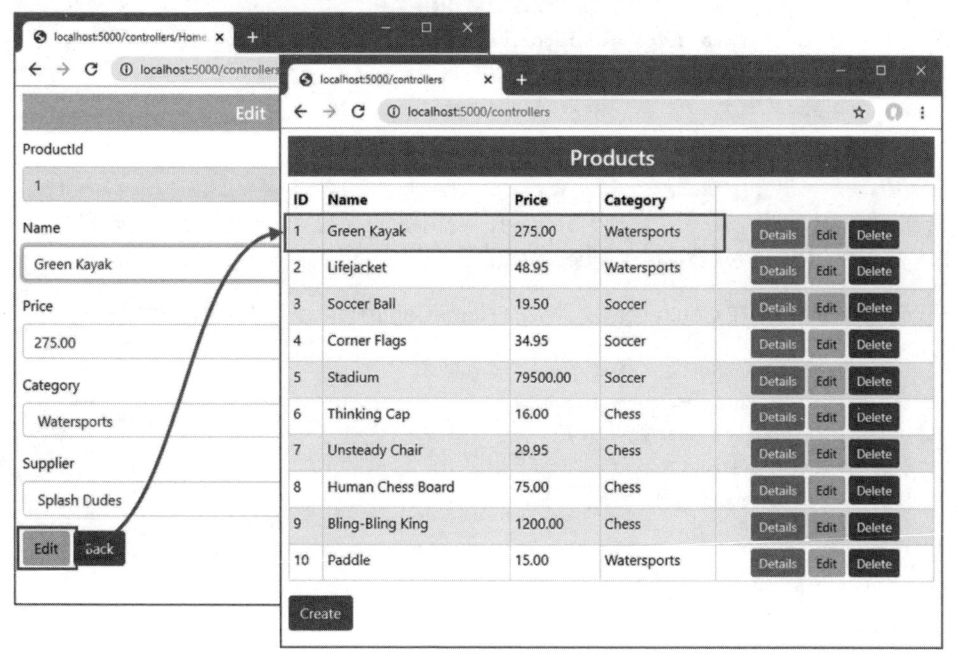


图 31-5　编辑产品

注意，ProductId 属性是不能更改的。应该避免尝试更改对象的主键，因为这会妨碍 Entity Framework Core 对其对象标识的理解。如果无法避免更改主键，那么最安全的方法是删除现有对象并存储一个新对象。

### 31.2.5　删除数据

最后一个基本操作是从数据库中删除对象。现在模式已经清晰了，第一步是添加一个方法来创建视图模型对象，以确定如何将数据呈现给用户，如代码清单 31-15 所示。

**代码清单 31-15　在 Models 文件夹的 ViewModelFactory.cs 文件中添加方法**

```
using System.Collections.Generic;
using System.Linq;

namespace WebApp.Models {

    public static class ViewModelFactory {

        // ...other methods omitted for brevity...

        public static ProductViewModel Delete(Product p,
                IEnumerable<Category> categories, IEnumerable<Supplier> suppliers) {
            return new ProductViewModel {
                Product = p, Action = "Delete",
                ReadOnly = true, Theme = "danger",
                Categories = categories, Suppliers = suppliers
            };
        }
    }
}
```

代码清单 31-16 将操作方法添加到 Home 控制器中，该控制器通过显示所选对象和 POST 请求来响应 GET 请求，以从数据库中删除该对象。

**代码清单 31-16　在 Controllers 文件夹的 HomeController.cs 文件中添加操作方法**

```
using Microsoft.AspNetCore.Mvc;
using Microsoft.EntityFrameworkCore;
using System.Collections.Generic;
using System.Threading.Tasks;
using WebApp.Models;

namespace WebApp.Controllers {

    [AutoValidateAntiforgeryToken]
    public class HomeController : Controller {
        private DataContext context;
```

```
        private IEnumerable<Category> Categories => context.Categories;
        private IEnumerable<Supplier> Suppliers => context.Suppliers;

        public HomeController(DataContext data) {
            context = data;
        }

        // ...other action methods removed for brevity...

        public async Task<IActionResult> Delete(long id) {
            ProductViewModel model = ViewModelFactory.Delete(
                await context.Products.FindAsync(id), Categories, Suppliers);
            return View("ProductEditor", model);
        }

        [HttpPost]
        public async Task<IActionResult> Delete(Product product) {
            context.Products.Remove(product);
            await context.SaveChangesAsync();
            return RedirectToAction(nameof(Index));
        }
    }
}
```

模型绑定过程从表单数据创建一个 Product 对象，该对象传递到 Entity Framework Core，以从数据库中删除。一旦数据从数据库中删除，浏览器将重定向到 Index 操作，如图 31-6 所示。

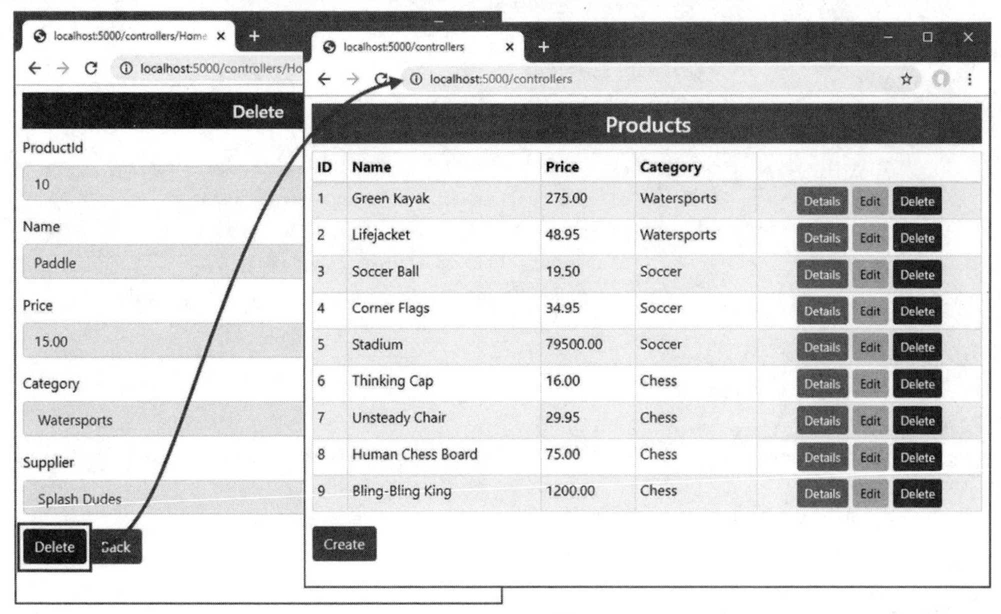


图 31-6　删除数据

## 31.3　创建 Razor Pages 表单应用程序

使用 Razor 表单依赖于与控制器示例类似的技术，尽管它被分解成更小的功能块。可以看到，主要困难是在不复制代码和标记的情况下保持 Razor Pages 的模块化特性。第一步是创建 Razor Pages，该页面将显示 Product 对象列表，并提供到其他操作的链接。给 Pages 文件夹添加一个名为 Index.cshtml 的 Razor Pages，内容如代码清单 31-17 所示。

**代码清单 31-17　Pages 文件夹中 Index.cshtml 文件的内容**

```
@page "/pages/{id:long?}"
@model IndexModel
@using Microsoft.AspNetCore.Mvc.RazorPages
@using Microsoft.EntityFrameworkCore

<div class="m-2">
    <h4 class="bg-primary text-white text-center p-2">Products</h4>
    <table class="table table-sm table-bordered table-striped">
        <thead>
            <tr>
               <th>ID</th><th>Name</th><th>Price</th><th>Category</th><th></th>
            </tr>
        </thead>
        <tbody>
            @foreach (Product p in Model.Products) {
                <tr>
                    <td>@p.ProductId</td>
                    <td>@p.Name</td>
                    <td>@p.Price</td>
                    <td>@p.Category.Name</td>
                    <td class="text-center">
                        <a asp-page="Details" asp-route-id="@p.ProductId"
                           class="btn btn-sm btn-info">Details</a>
                        <a asp-page="Edit" asp-route-id="@p.ProductId"
                           class="btn btn-sm btn-warning">Edit</a>
                        <a asp-page="Delete" asp-route-id="@p.ProductId"
                           class="btn btn-sm btn-danger">Delete</a>
                    </td>
                </tr>
            }
        </tbody>
    </table>
    <a asp-page="Create" class="btn btn-primary">Create</a>
</div>

@functions {
```

```
        public class IndexModel: PageModel {
            private DataContext context;

            public IndexModel(DataContext dbContext) {
                context = dbContext;
            }

            public IEnumerable<Product> Products { get; set; }

            public void OnGetAsync(long id = 1) {
                Products = context.Products
                    .Include(p => p.Category).Include(p => p.Supplier);
            }
        }
    }
```

页面的这个视图部分显示一个表，其中填充了页面模型从数据库中获得的 Product 对象的详细信息。使用浏览器请求 http://localhost:5000/pages，将显示如图 31-7 所示的响应。

除了 Product 对象的详细信息外，页面还显示导航到其他 Razor Pages 的锚元素，这些页面将在稍后讲述。

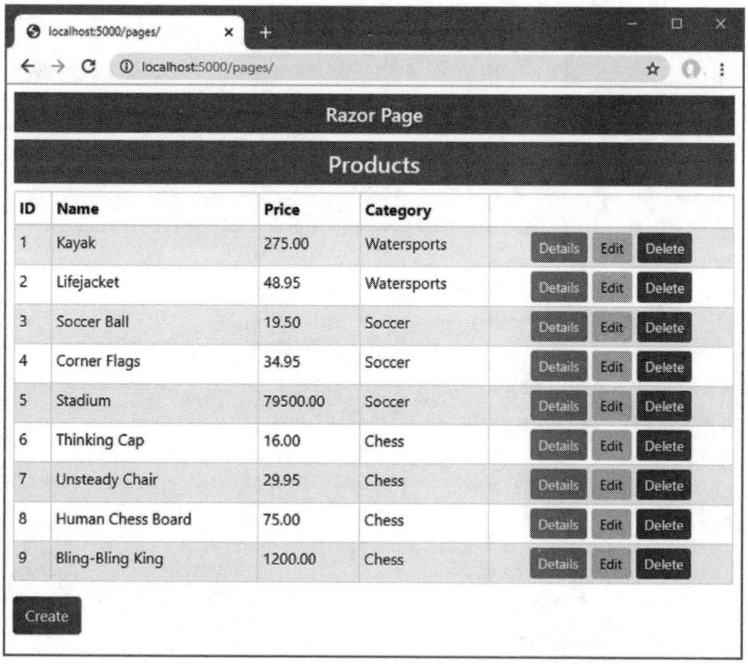

图 31-7 使用 Razor Pages 列出数据

### 31.3.1 创建常用功能

这里不在示例应用程序所需的每个页面中复制相同的 HTML 表单和支持代码。相反，定义一个部分视图来定义 HTML 表单，再定义一个基类，来定义页面模型类所需的公共代码。对于部分视图，给 Pages 文件夹添加一个名为_ProductEditor.cshtml 的 Razor 视图，内容如代码清单 31-18 所示。

**使用多个页面**

asp-page-handler 属性可用于指定处理程序方法的名称，该方法允许 Razor Pages 用于多个操作。我不喜欢这个特性，因为它的结果太接近标准的 MVC 控制器，并且破坏了我喜欢的 Razor Pages 开发的自包含性和模块化方面。

当然，我更喜欢本章采用的方法，即将公共内容合并到部分视图和共享基类中。两种方法都可以，建议都尝试一下，看看哪种更适用于自己的项目。

**代码清单 31-18　Pages 文件夹中_ProductEditor.cshtml 文件的内容**

```
@model ProductViewModel

<partial name="_Validation" />

<h5 class="bg-@Model.Theme text-white text-center p-2">@Model.Action</h5>

<form asp-page="@Model.Action" method="post">
    <div class="form-group">
        <label asp-for="Product.ProductId"></label>
        <input class="form-control" asp-for="Product.ProductId" readonly />
    </div>
    <div class="form-group">
        <label asp-for="Product.Name"></label>
        <div>
            <span asp-validation-for="Product.Name" class="text-danger"></span>
        </div>
        <input class="form-control" asp-for="Product.Name"
               readonly="@Model.ReadOnly" />
    </div>
    <div class="form-group">
        <label asp-for="Product.Price"></label>
        <div>
            <span asp-validation-for="Product.Price" class="text-danger"></span>
        </div>
        <input class="form-control" asp-for="Product.Price"
               readonly="@Model.ReadOnly" />
    </div>
    <div class="form-group">
        <label asp-for="Product.CategoryId">Category</label>
        <div>
            <span asp-validation-for="Product.CategoryId" class="text-danger"></span>
        </div>
        <select asp-for="Product.CategoryId" class="form-control"
                disabled="@Model.ReadOnly"
```

```
                asp-items="@(new SelectList(Model.Categories,
                    "CategoryId", "Name"))">
            <option value="" disabled selected>Choose a Category</option>
        </select>
    </div>
    <div class="form-group">
        <label asp-for="Product.SupplierId">Supplier</label>
        <div>
            <span asp-validation-for="Product.SupplierId" class="text-danger"></span>
        </div>
        <select asp-for="Product.SupplierId" class="form-control"
                disabled="@Model.ReadOnly"
                asp-items="@(new SelectList(Model.Suppliers,
                    "SupplierId", "Name"))">
            <option value="" disabled selected>Choose a Supplier</option>
        </select>
    </div>
    @if (Model.ShowAction) {
        <button class="btn btn-@Model.Theme" type="submit">@Model.Action</button>
    }
    <a class="btn btn-secondary" asp-page="Index">Back</a>
</form>
```

部分视图使用 ProductViewModel 类作为它的模型类型，并依赖内置的标签助手为 Product 类定义的属性显示 input 和 select 元素。这与本章前面使用的内容相同，只是将 asp-action 属性替换为 asp-page，以指定 form 和 anchor 元素的目标。

要定义页面模型基类，向 Pages 文件夹添加一个名为 EditorPageModel.cs 的类文件，并使用它定义如代码清单 31-19 所示的类。

### 代码清单 31-19　Pages 文件夹中 EditorPageModel.cs 文件的内容

```
using Microsoft.AspNetCore.Mvc.RazorPages;
using System.Collections.Generic;
using WebApp.Models;

namespace WebApp.Pages {

    public class EditorPageModel : PageModel {

        public EditorPageModel(DataContext dbContext) {
            DataContext = dbContext;
        }

        public DataContext DataContext { get; set; }

        public IEnumerable<Category> Categories => DataContext.Categories;
```

```
        public IEnumerable<Supplier> Suppliers => DataContext.Suppliers;

        public ProductViewModel ViewModel { get; set; }
    }
}
```

这个类定义的属性很简单，但是它们将帮助简化处理每个操作的 Razor Pages 的页面模型类。

本示例需要的所有 Razor Pages 都依赖相同的名称空间。将代码清单 31-20 中所示的表达式添加到 Pages 文件夹的_ViewImports.cshtml 文件中，以避免单个页面中的表达式重复。

**提示：**确保修改了 Pages 文件夹中的_ViewImports.cshtml 文件，而不是 Views 文件夹中的同名文件。

**代码清单 31-20　在 Pages 文件夹的_ViewImports.cshtml 文件中添加名称空间**

```
@namespace WebApp.Pages
@using WebApp.Models
@addTagHelper *, Microsoft.AspNetCore.Mvc.TagHelpers
@addTagHelper *, WebApp
@using Microsoft.AspNetCore.Mvc.RazorPages
@using Microsoft.EntityFrameworkCore
@using WebApp.Pages
@using System.Text.Json
@using Microsoft.AspNetCore.Http
```

## 31.3.2　为 CRUD 操作定义页面

有了部分视图和共享基类，处理各个操作的页面就十分简单了。给Pages文件夹添加一个名为Details.cshtml的Razor Pages，代码和内容如代码清单 31-21 所示。

**代码清单 31-21　Pages 文件夹中 Details.cshtml 文件的内容**

```
@page "/pages/details/{id}"
@model DetailsModel

<div class="m-2">
   <partial name="_ProductEditor" model="@Model.ViewModel" />
</div>

@functions {

    public class DetailsModel: EditorPageModel {

        public DetailsModel(DataContext dbContext): base(dbContext) {}

        public async Task OnGetAsync(long id) {
            Product p = await DataContext.Products.
                Include(p => p.Category).Include(p => p.Supplier)
```

```
                .FirstOrDefaultAsync(p => p.ProductId == id);
            ViewModel = ViewModelFactory.Details(p);
        }
    }
}
```

构造函数接收 Entity Framework Core 上下文对象，并将其传递给基类。处理程序方法通过查询数据库来响应请求，并使用响应创建一个使用 ViewModelFactory 类的 ProductViewModel 对象。

给 Pages 文件夹添加一个名为 Create.cshtml 的 Razor Pages，内容如代码清单 31-22 所示。

---

■ **提示：**使用部分视图意味着 asp-for 属性设置没有额外前缀的元素名称。这允许使用 FromForm 属性进行模型绑定，而不使用 Name 参数。

---

**代码清单 31-22　Pages 文件夹中 Create.cshtml 文件的内容**

```
@page "/pages/create"
@model CreateModel

<div class="m-2">
    <partial name="_ProductEditor" model="@Model.ViewModel" />
</div>

@functions {

    public class CreateModel: EditorPageModel {

        public CreateModel(DataContext dbContext): base(dbContext) {}

        public void OnGet() {
            ViewModel = ViewModelFactory.Create(new Product(),
                Categories, Suppliers);
        }

        public async Task<IActionResult> OnPostAsync([FromForm]Product product) {
            if (ModelState.IsValid) {
                product.ProductId = default;
                product.Category = default;
                product.Supplier = default;
                DataContext.Products.Add(product);
                await DataContext.SaveChangesAsync();
                return RedirectToPage(nameof(Index));
            }
            ViewModel = ViewModelFactory.Create(product, Categories, Suppliers);
            return Page();
        }
    }
}
```

给 Pages 文件夹添加一个名为 Edit.cshtml 的 Razor Pages，内容如代码清单 31-23 所示。

**代码清单 31-23　Pages 文件夹中 Edit.cshtml 文件的内容**

```
@page "/pages/edit/{id}"
@model EditModel

<div class="m-2">
   <partial name="_ProductEditor" model="@Model.ViewModel" />
</div>

@functions {

    public class EditModel: EditorPageModel {

        public EditModel(DataContext dbContext): base(dbContext) {}

        public async Task OnGetAsync(long id) {
            Product p = await this.DataContext.Products.FindAsync(id);
            ViewModel = ViewModelFactory.Edit(p, Categories, Suppliers);
        }

        public async Task<IActionResult> OnPostAsync([FromForm]Product product) {
            if (ModelState.IsValid) {
                product.Category = default;
                product.Supplier = default;
                DataContext.Products.Update(product);
                await DataContext.SaveChangesAsync();
                return RedirectToPage(nameof(Index));
            }
            ViewModel = ViewModelFactory.Edit(product, Categories, Suppliers);
            return Page();
        }
    }
}
```

给 Pages 文件夹添加一个名为 Delete.cshtml 的 Razor Pages，内容如代码清单 31-24 所示。

**代码清单 31-24　Pages 文件夹中 Delete.cshtml 文件的内容**

```
@page "/pages/delete/{id}"
@model DeleteModel

<div class="m-2">
   <partial name="_ProductEditor" model="@Model.ViewModel" />
</div>

@functions {
```

```
    public class DeleteModel: EditorPageModel {

        public DeleteModel(DataContext dbContext): base(dbContext) {}

        public async Task OnGetAsync(long id) {
            ViewModel = ViewModelFactory.Delete(
                await DataContext.Products.FindAsync(id), Categories, Suppliers);
        }

        public async Task<IActionResult> OnPostAsync([FromForm]Product product) {
            DataContext.Products.Remove(product);
            await DataContext.SaveChangesAsync();
            return RedirectToPage(nameof(Index));
        }
    }
}
```

重启 ASP.NET Core 并导航到 http://localhost:5000/pages，将能够单击链接，来查看、创建、编辑和删除数据，如图 31-8 所示。

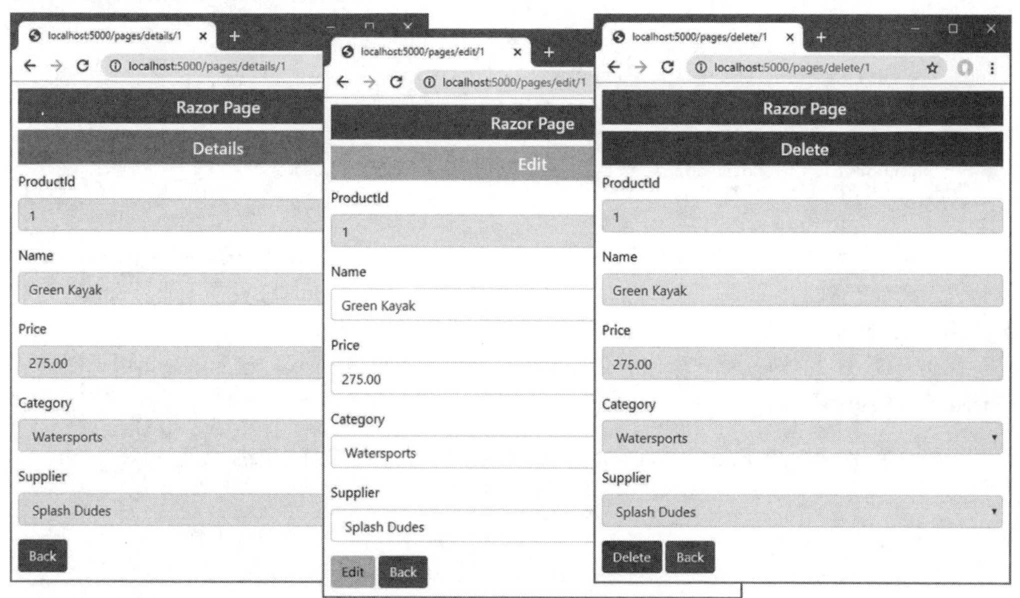


图 31-8 使用 Razor Pages

## 31.4 创建新的相关数据对象

一些应用程序需要允许用户创建新的相关数据，这样，例如，新的类别可以与该类别中的产品一起创建。有两种方法可以解决这个问题，如下一节所述。

### 31.4.1　在同一请求中提供相关数据

第一种方法是要求用户提供以相同形式创建相关数据所需的数据。对于示例应用程序，这意味着以用户输入 Product 对象值的相同形式来收集 Category 对象的详细信息。

对于简单数据类型，这可能是一种有用的方法。这种情况下，创建相关对象只需要少量数据，但对于具有许多属性的类型不太适合。

我更喜欢在各自的部分视图中定义相关数据类型的 HTML 元素。给 Pages 文件夹添加一个名为_CategoryEditor.cshtml 的 Razor 视图，内容如代码清单 31-25 所示。

**代码清单 31-25　Pages 文件夹中_CategoryEditor.cshtml 文件的内容**

```
@model Product
<script type="text/javascript">
    $(document).ready(() => {
        const catGroup = $("#categoryGroup").hide();
        $("select[name='Product.CategoryId']").on("change", (event) =>
            event.target.value === "-1" ? catGroup.show() : catGroup.hide());
    });
</script>

<div class="form-group bg-info p-1" id="categoryGroup">
    <label class="text-white" asp-for="Category.Name">
       New Category Name
    </label>
    <input class="form-control" asp-for="Category.Name" value="" />
</div>
```

Category 类型只需要一个属性，用户使用标准 input 元素提供该属性。部分视图中的 script 元素包含隐藏新元素的 jQuery 代码，直到用户选择为 Product.CategoryId属性设置-1 值的 option 元素为止(使用 JavaScript 完全是可选的，但它有助于强调新元素的用途)。

代码清单 31-26 向编辑器添加了部分视图，还显示 option 元素，该元素显示用于创建新 Category 对象的元素。

**代码清单 31-26　在 Pages 文件夹的_ProductEditor.cshtml 文件中添加元素**

```
...
<div class="form-group">
    <label asp-for="Product.CategoryId">Category</label>
    <div>
       <span asp-validation-for="Product.CategoryId" class="text-danger"></span>
    </div>
    <select asp-for="Product.CategoryId" class="form-control"
            disabled="@Model.ReadOnly" asp-items="@(new SelectList(Model.Categories,
                "CategoryId", "Name"))">
        <option value="-1">Create New Category...</option>
        <option value="" disabled selected>Choose a Category</option>
    </select>
```

```
</div>

<partial name="_CategoryEditor" for="Product" />
<div class="form-group">
    <label asp-for="Product.SupplierId">Supplier</label>
    <div><span asp-validation-for="Product.SupplierId" class="text-danger"></span></div>
    <select asp-for="Product.SupplierId" class="form-control" disabled="@Model.ReadOnly"
            asp-items="@(new SelectList(Model.Suppliers,
               "SupplierId", "Name"))">
        <option value="" disabled selected>Choose a Supplier</option>
    </select>
</div>
...
```

需要在多个页面中使用新功能，因此为了避免代码重复，添加了一个方法来处理页面模型基类中的相关数据，如代码清单 31-27 所示。

#### 代码清单 31-27 在 Pages 文件夹的 EditorPageModel.cs 文件中添加方法

```
using Microsoft.AspNetCore.Mvc.RazorPages;
using System.Collections.Generic;
using WebApp.Models;
using System.Threading.Tasks;

namespace WebApp.Pages {

    public class EditorPageModel : PageModel {

        public EditorPageModel(DataContext dbContext) {
            DataContext = dbContext;
        }

        public DataContext DataContext { get; set; }

        public IEnumerable<Category> Categories => DataContext.Categories;
        public IEnumerable<Supplier> Suppliers => DataContext.Suppliers;

        public ProductViewModel ViewModel { get; set; }

        protected async Task CheckNewCategory(Product product) {
            if (product.CategoryId == -1
                    && !string.IsNullOrEmpty(product.Category?.Name)) {
                DataContext.Categories.Add(product.Category);
                await DataContext.SaveChangesAsync();
                product.CategoryId = product.Category.CategoryId;
                ModelState.Clear();
                TryValidateModel(product);
            }
```

```
            }
        }
    }
```

新代码使用从用户接收到的数据创建一个 Category 对象，并将其存储在数据库中。数据库服务器为新对象分配一个主键，Entity Framework Core 使用该主键来更新 Category 对象。这允许更新 Product 对象的 CategoryId 属性，重新验证模型数据；分配给 CategoryId 属性的值将通过验证，因为它对应于新分配的键。要将新功能集成到 Create 页面中，请添加代码清单 31-28 所示的语句。

**代码清单 31-28　在 Pages 文件夹的 Create.cshtml 文件中添加语句**

```
...
public async Task<IActionResult> OnPostAsync([FromForm]Product product) {
    await CheckNewCategory(product);
    if (ModelState.IsValid) {
        product.ProductId = default;
        product.Category = default;
        product.Supplier = default;
        DataContext.Products.Add(product);
        await DataContext.SaveChangesAsync();
        return RedirectToPage(nameof(Index));
    }
    ViewModel = ViewModelFactory.Create(product, Categories, Suppliers);
    return Page();
}
...
```

向 Edit 页面中的处理程序方法添加相同的语句，如代码清单 31-29 所示。

**代码清单 31-29　在 Pages 文件夹的 Edit.cshtml 文件中添加语句**

```
...
public async Task<IActionResult> OnPostAsync([FromForm]Product product) {
    await CheckNewCategory(product);
    if (ModelState.IsValid) {
        product.Category = default;
        product.Supplier = default;
        DataContext.Products.Update(product);
        await DataContext.SaveChangesAsync();
        return RedirectToPage(nameof(Index));
    }
    ViewModel = ViewModelFactory.Edit(product, Categories, Suppliers);
    return Page();
}
...
```

重启 ASP.NETCore，这样，页面模型基类将重新编译，并使用浏览器请求 http://localhost:5000/pages/edit/1。单击 Category 元素并从选项列表中选择 Create New Category。在 input 元素中输入一

个新的类别名称，然后单击 Edit 按钮。当处理请求时，将一个新的 Category 对象存储在数据库中，并与 Product 对象相关联，如图 31-9 所示。

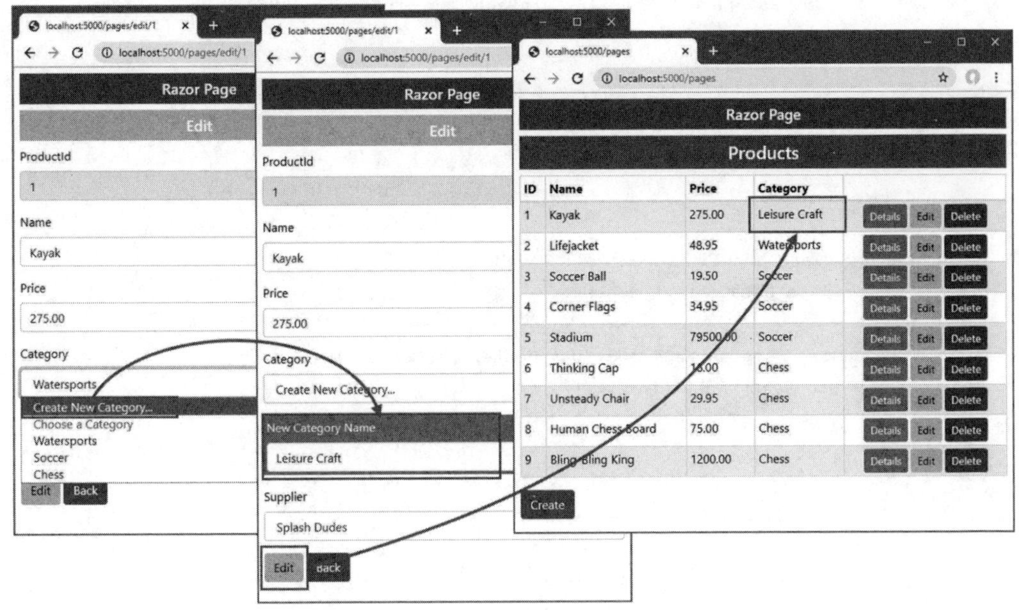

图 31-9 创建相关的数据

## 31.4.2 创建新数据

对于具有自己复杂创建过程的相关数据类型，向主表单添加元素可能让用户不知所措；更好的方法是从主表单导航到另一个控制器或页面，让用户创建新对象，然后返回来完成原始任务。下面演示用于创建 Supplier 对象的技术，尽管 Supplier 类型很简单，只需要用户提供两个值。

要创建允许用户创建 Supplier 对象的表单，请给 Pages 文件夹添加名为 SupplierBreakOut.cshtml 的 Razor Pages，内容如代码清单 31-30 所示。

**代码清单 31-30 Pages 文件夹中 SupplierBreakOut.cshtml 的内容**

```
@page "/pages/supplier"
@model SupplierPageModel

<div class="m-2">
    <h5 class="bg-secondary text-white text-center p-2">New Supplier</h5>
    <form asp-page="SupplierBreakOut" method="post">
        <div class="form-group">
            <label asp-for="Supplier.Name"></label>
            <input class="form-control" asp-for="Supplier.Name" />
        </div>
        <div class="form-group">
            <label asp-for="Supplier.City"></label>
            <input class="form-control" asp-for="Supplier.City" />
        </div>
```

```
            <button class="btn btn-secondary" type="submit">Create</button>
            <a class="btn btn-outline-secondary"
                    asp-page="@Model.ReturnPage" asp-route-id="@Model.ProductId">
                Cancel
            </a>
        </form>
    </div>

    @functions {

        public class SupplierPageModel: PageModel {
            private DataContext context;

            public SupplierPageModel(DataContext dbContext) {
                context = dbContext;
            }

            [BindProperty]
            public Supplier Supplier { get; set; }

            public string ReturnPage { get; set; }
            public string ProductId { get; set; }

            public void OnGet([FromQuery(Name="Product")] Product product,
                    string returnPage) {
                TempData["product"] = Serialize(product);
                TempData["returnAction"] = ReturnPage = returnPage;
                TempData["productId"] = ProductId = product.ProductId.ToString();
            }

            public async Task<IActionResult> OnPostAsync() {
                context.Suppliers.Add(Supplier);
                await context.SaveChangesAsync();
                Product product = Deserialize(TempData["product"] as string);
                product.SupplierId = Supplier.SupplierId;
                TempData["product"] = Serialize(product);
                string id = TempData["productId"] as string;
                return RedirectToPage(TempData["returnAction"] as string,
                    new { id = id });
            }

            private string Serialize(Product p) => JsonSerializer.Serialize(p);
            private Product Deserialize(string json) =>
                JsonSerializer.Deserialize<Product>(json);
        }
    }
```

用户使用 GET 请求导航到此页面，该请求包含用户提供的产品的详细信息和用户应该返回到的页面的名称。使用 temp 数据功能来存储该数据。

此页面向用户显示一个表单，其中包含创建新 Supplier 对象所需的 Name 和 City 属性字段。提交表单时，POST 处理程序方法存储一个新的 Supplier 对象，并使用数据库服务器分配的键来更新 Product 对象，然后将其再次存储为临时数据。用户重定向到他们到达的页面。

代码清单 31-31 向_ProductEditor 部分视图添加了元素，允许用户导航到新页面。

**代码清单 31-31　在 Pages 文件夹的_ProductEditor.cshtml 文件中添加元素**

```
...
<partial name="_CategoryEditor" for="Product" />

<div class="form-group">
    <label asp-for="Product.SupplierId">
        Supplier
        @if (!Model.ReadOnly) {
            <input type="hidden" name="returnPage" value="@Model.Action" />
            <button class="btn btn-sm btn-outline-primary ml-3"
                    asp-page="SupplierBreakOut" formmethod="get" formnovalidate>
                Create New Supplier
            </button>
        }
    </label>
    <div>
        <span asp-validation-for="Product.SupplierId" class="text-danger"></span>
    </div>
    <select asp-for="Product.SupplierId" class="form-control"
        disabled="@Model.ReadOnly" asp-items="@(new SelectList(Model.Suppliers,
                "SupplierId", "Name"))">
        <option value="" disabled selected>Choose a Supplier</option>
    </select>
</div>
...
```

新元素添加了一个隐藏的 input 元素(该元素捕获要返回的页面)，以及一个 button 元素(该元素使用 GET 请求将表单数据提交到 SupplierBreakOut 页面，这意味着表单值编码到查询字符串中，这是在代码清单 31-30 中使用 FromQuery 属性的原因)。代码清单 31-32 显示了 Create 页面所需的更改，以添加对检索临时数据和使用它填充 Product 表单的支持。

**代码清单 31-32　在 Pages 文件夹的 Create.cshtml 文件中检索数据**

```
@page "/pages/create"
@model CreateModel

<div class="m-2">
    <partial name="_ProductEditor" model="@Model.ViewModel" />
</div>
```

```
@functions {

    public class CreateModel: EditorPageModel {

        public CreateModel(DataContext dbContext): base(dbContext) {}

        public void OnGet() {
            Product p = TempData.ContainsKey("product")
                ? JsonSerializer.Deserialize<Product>(TempData["product"] as string)
                : new Product();
            ViewModel = ViewModelFactory.Create(p, Categories, Suppliers);
        }

        public async Task<IActionResult> OnPostAsync([FromForm]Product product) {
            await CheckNewCategory(product);
            if (ModelState.IsValid) {
                product.ProductId = default;
                product.Category = default;
                product.Supplier = default;
                DataContext.Products.Add(product);
                await DataContext.SaveChangesAsync();
                return RedirectToPage(nameof(Index));
            }
            ViewModel = ViewModelFactory.Create(product, Categories, Suppliers);
            return Page();
        }
    }
}
```

在 Edit 页面中需要进行类似的更改，如代码清单 31-33 所示(其他页面不需要更改，因为只有当用户能够创建或编辑 Product 数据时才需要中断)。

#### 代码清单 31-33　在 Pages 文件夹的 Edit.cshtml 文件中检索数据

```
@page "/pages/edit/{id}"
@model EditModel

<div class="m-2">
    <partial name="_ProductEditor" model="@Model.ViewModel" />
</div>

@functions {

    public class EditModel: EditorPageModel {

        public EditModel(DataContext dbContext): base(dbContext) {}
```

```
        public async Task OnGetAsync(long id) {
            Product p = TempData.ContainsKey("product")
                ? JsonSerializer.Deserialize<Product>(TempData["product"] as string)
                : await this.DataContext.Products.FindAsync(id);
            ViewModel = ViewModelFactory.Edit(p, Categories, Suppliers);
        }

        public async Task<IActionResult> OnPostAsync([FromForm]Product product) {
            await CheckNewCategory(product);
            if (ModelState.IsValid) {
                product.Category = default;
                product.Supplier = default;
                DataContext.Products.Update(product);
                await DataContext.SaveChangesAsync();
                return RedirectToPage(nameof(Index));
            }
            ViewModel = ViewModelFactory.Edit(product, Categories, Suppliers);
            return Page();
        }
    }
}
```

其效果是向用户显示 Create New Supplier 按钮，该按钮向浏览器发送到一个表单，该表单可用于创建 Supplier 对象。一旦 Supplier 存储在数据库中，浏览器就发送回原始页面，表单中填充用户输入的数据，并且 Supplier select 元素设置为新创建的对象，如图 31-10 所示。

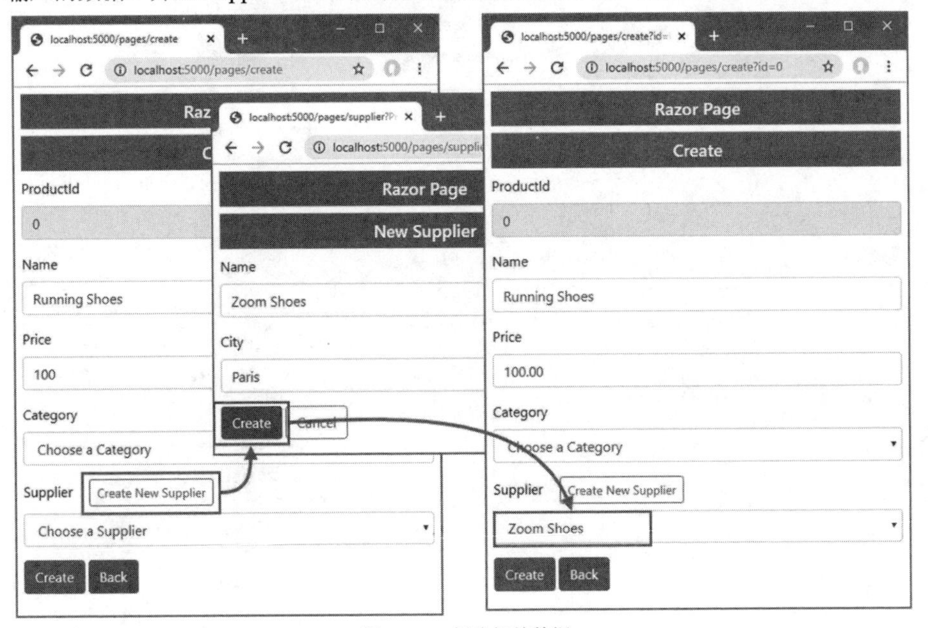


图 31-10　创建相关数据

## 31.5　小结

本章演示了如何将前面章节描述的功能与 Entity Framework Core 结合，来创建、读取、更新和删除数据。第Ⅳ部分将描述 ASP.NET Core 提供的一些高级功能。

# 第 IV 部分

# 高级 ASP.NET Core 功能

# 第 32 章

# 创建示例项目

本章将创建贯穿本书这一部分的示例项目。该项目包含一个使用简单控制器和 Razor Pages 显示的数据模型。

## 32.1 创建项目

从 Windows Start 菜单中打开一个新的 PowerShell 命令提示符，并运行代码清单 32-1 中所示的命令。

■ **提示：**可以从 https://github.com/apress/pro-asp.net-core-3 下载本章和本书中其他所有章节的示例项目。如果在运行示例时遇到问题，请参阅第 1 章，以了解如何获得帮助。

**代码清单 32-1 创建项目**

```
dotnet new globaljson --sdk-version 3.1.101 --output Advanced
dotnet new web --no-https --output Advanced --framework netcoreapp3.1
dotnet new sln -o Advanced

dotnet sln Advanced add Advanced
```

如果使用 Visual Studio，请打开 Advanced 文件夹中的 Advanced.sln 文件。选择 Project | Platform Properties，导航到 Debug 页面，并将 App URL 字段更改为 http://localhost:5000，如图 32-1 所示。这将更改用于接收 HTTP 请求的端口。选择 File | Save All 以保存配置变化。

如果使用的是 Visual Studio Code，请打开 Advanced 文件夹。当提示添加用于构建和调试项目的资源时，单击 Yes 按钮，如图 32-2 所示。

### 向项目中添加 NuGet 包

数据模型使用 Entity Framework Core 在 SQL Server LocalDB 数据库中存储和查询数据。要为 Entity Framework Core 添加 NuGet 包，使用 PowerShell 命令提示符在 Advanced 项目文件夹中运行代码清单 32-2 所示的命令。

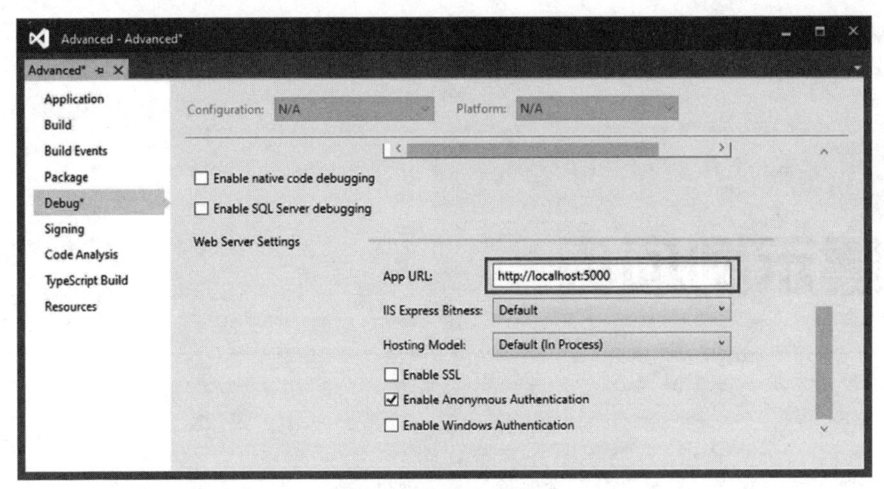


图 32-1　更改 HTTP 端口

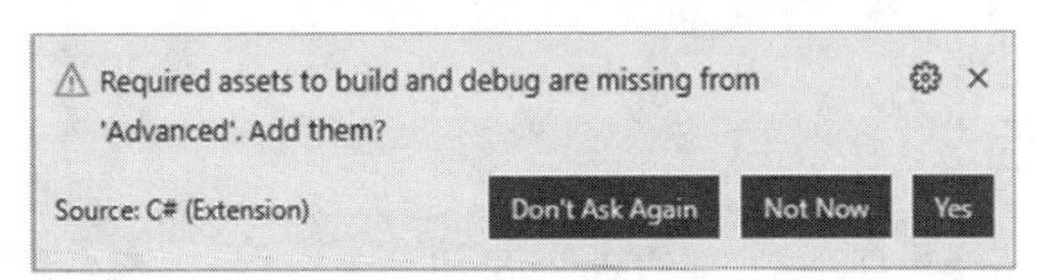


图 32-2　添加项目资源

**代码清单 32-2　向项目中添加包**

```
dotnet add package Microsoft.EntityFrameworkCore.Design --version 3.1.1
dotnet add package Microsoft.EntityFrameworkCore.SqlServer --version 3.1.1
```

如果使用的是 Visual Studio，可以通过选择 Project | Manage NuGet Packages 来添加包。请注意选择要添加到项目中的包的正确版本。

如果没有完成前面章节中的示例，就需要安装用于创建和管理 Entity Framework Core 迁移的全局工具包。运行代码清单 32-3 中所示的命令，删除包的任何现有版本，并安装本书所需的版本。

**代码清单 32-3　安装全局工具包**

```
dotne.t tool uninstall --global dotnet-ef
dotnet tool install --global dotnet-ef --version 3.1.1
```

## 32.2　添加数据模型

这个应用程序的数据模型由三个类组成，分别代表人员、工作部门和位置。创建一个 Models 文件夹，并向其中添加一个名为 Person.cs 的类文件，代码如代码清单 32-4 所示。

**代码清单 32-4　Models 文件夹中 Person.cs 文件的内容**

```
using System.Collections.Generic;
```

```
namespace Advanced.Models {

    public class Person {

        public long PersonId { get; set; }
        public string Firstname { get; set; }
        public string Surname { get; set; }
        public long DepartmentId { get; set; }
        public long LocationId { get; set; }

        public Department Department {get; set; }
        public Location Location { get; set; }
    }
}
```

将一个名为 Department.cs 的类文件添加到 Models 文件夹中，并使用它定义如代码清单 32-5 所示的类。

### 代码清单 32-5 Models 文件夹中 Department.cs 文件的内容

```
using System.Collections.Generic;

namespace Advanced.Models {
    public class Department {

        public long Departmentid { get; set; }
        public string Name { get; set; }

        public IEnumerable<Person> People { get; set; }
    }
}
```

在 Models 文件夹中添加一个名为 Location.cs 的类文件，并使用它来定义如代码清单 32-6 所示的类。

### 代码清单 32-6 Models 文件夹中 Location.cs 文件的内容

```
using System.Collections.Generic;

namespace Advanced.Models {
    public class Location {

        public long LocationId { get; set; }
        public string City { get; set; }
        public string State { get; set; }
        public IEnumerable<Person> People { get; set; }
    }
}
```

这三个数据模型类中的每个类都定义了一个关键属性，数据库在存储新对象时分配该键属性的值，并定义了外键属性(定义类之间关系)。这些是由导航属性补充的，这些导航属性与 Entity Framework Core 的 Include 方法一起使用，以将相关数据合并到查询中。

要创建提供对数据库的访问的 Entity Framework Core的上下文类，请将一个名为 DataContext.cs 的文件添加到 Models 文件夹，并添加如代码清单 32-7 所示的代码。

**代码清单 32-7　Models 文件夹中 DataContext.cs 文件的内容**

```
using Microsoft.EntityFrameworkCore;

namespace Advanced.Models {
    public class DataContext: DbContext {

        public DataContext(DbContextOptions<DataContext> opts)
            : base(opts) { }

        public DbSet<Person> People { get; set; }
        public DbSet<Department> Departments { get; set; }
        public DbSet<Location> Locations { get; set; }
    }
}
```

上下文类定义了一些属性，这些属性用于查询数据库中的 Person、Department 和 Location 数据。

### 32.2.1　准备种子数据

将一个名为 SeedData.cs 的类添加到 Models 文件夹中，并添加代码清单 32-8 所示的代码，以定义用于填充数据库的种子数据。

**代码清单 32-8　Models 文件夹中 SeedData.cs 文件的内容**

```
using Microsoft.EntityFrameworkCore;
using System.Linq;

namespace Advanced.Models {
    public static class SeedData {

        public static void SeedDatabase(DataContext context) {
            context.Database.Migrate();
            if (context.People.Count() == 0 && context.Departments.Count() == 0 &&
                context.Locations.Count() == 0) {

                Department d1 = new Department { Name = "Sales" };
                Department d2 = new Department { Name = "Development" };
                Department d3 = new Department { Name = "Support" };
                Department d4 = new Department { Name = "Facilities" };
```

```
        context.Departments.AddRange(d1, d2, d3, d4);
        context.SaveChanges();

        Location l1 = new Location { City = "Oakland", State = "CA" };
        Location l2 = new Location { City = "San Jose", State = "CA" };
        Location l3 = new Location { City = "New York", State = "NY" };
        context.Locations.AddRange(l1, l2, l3);

        context.People.AddRange(
            new Person {
                Firstname = "Francesca", Surname = "Jacobs",
                Department = d2, Location = l1
            },
            new Person {
                Firstname = "Charles", Surname = "Fuentes",
                Department = d2, Location = l3
            },
            new Person {
                Firstname = "Bright", Surname = "Becker",
                Department = d4, Location = l1
            },
            new Person {
                Firstname = "Murphy", Surname = "Lara",
                Department = d1, Location = l3
            },
            new Person {
                Firstname = "Beasley", Surname = "Hoffman",
                Department = d4, Location = l3
            },
            new Person {
                Firstname = "Marks", Surname = "Hays",
                Department = d4, Location = l1
            },
            new Person {
                Firstname = "Underwood", Surname = "Trujillo",
                Department = d2, Location = l1
            },
            new Person {
                Firstname = "Randall", Surname = "Lloyd",
                Department = d3, Location = l2
            },
            new Person {
                Firstname = "Guzman", Surname = "Case",
                Department = d2, Location = l2
            });
        context.SaveChanges();
    }
```

```
        }
    }
}
```

静态 SeedDatabase 方法确保所有未决迁移都已应用到数据库。如果数据库是空的，它将播种数据。Entity Framework Core 负责将对象映射到数据库的表中，并在存储数据时自动分配键属性。

### 32.2.2　配置 Entity Framework Core 服务和中间件

对代码清单32-9中所示的Startup类进行更改，它配置Entity Framework Core并设置DataContext服务，本书的这一部分将使用该服务访问数据库。

**代码清单 32-9　在 Advanced 文件夹的 Startup.cs 文件中准备服务和中间件**

```
using System;
using System.Collections.Generic;
using System.Linq;
using System.Threading.Tasks;
using Microsoft.AspNetCore.Builder;
using Microsoft.AspNetCore.Hosting;
using Microsoft.AspNetCore.Http;
using Microsoft.Extensions.DependencyInjection;
using Microsoft.Extensions.Hosting;
using Microsoft.Extensions.Configuration;
using Microsoft.EntityFrameworkCore;
using Advanced.Models;

namespace Advanced {
    public class Startup {

        public Startup(IConfiguration config) {
            Configuration = config;
        }

        public IConfiguration Configuration { get; set; }

        public void ConfigureServices(IServiceCollection services) {
            services.AddDbContext<DataContext>(opts => {
                opts.UseSqlServer(Configuration[
                    "ConnectionStrings:PeopleConnection"]);
                opts.EnableSensitiveDataLogging(true);
            });
        }

        public void Configure(IApplicationBuilder app, DataContext context) {

            app.UseDeveloperExceptionPage();
            app.UseRouting();
```

```
            app.UseEndpoints(endpoints => {
                endpoints.MapGet("/", async context => {
                    await context.Response.WriteAsync("Hello World!");
                });
            });

            SeedData.SeedDatabase(context);
        }
    }
}
```

要定义用于应用程序数据的连接字符串，在 appsettings.json 文件中添加如代码清单 32-10 所示的配置设置。连接字符串应该在一行中输入。

**代码清单 32-10　在 Advanced 文件夹的 appsettings.json 文件中定义连接字符串。**

```
{
    "Logging": {
        "LogLevel": {
            "Default": "Information",
            "Microsoft": "Warning",
            "Microsoft.Hosting.Lifetime": "Information",
            "Microsoft.EntityFrameworkCore": "Information"
        }
    },
    "AllowedHosts": "*",
    "ConnectionStrings": {
        "PeopleConnection": "Server=(localdb)
        \\MSSQLLocalDB;Database=People;MultipleActiveResultSets=True"
    }
}
```

除了连接字符串之外，代码清单 32-10 增加了 Entity Framework Core 的日志详细信息，以便记录发送到数据库的 SQL 查询。

### 32.2.3　创建和应用迁移

要创建建立数据库模式的迁移，使用 PowerShell 命令提示符在 Advanced 项目文件夹中运行代码清单 32-11 所示的命令。

**代码清单 32-11　创建 Entity Framework Core 迁移**

```
dotnet ef migrations add Initial
```

创建迁移后，使用代码清单 32-12 中所示的命令将其应用到数据库。

**代码清单 32-12　将迁移应用到数据库**

```
dotnet ef database update
```

应用程序显示的日志消息将显示发送到数据库的 SQL 命令。

> **注意：** 如果需要重置数据库，那么运行 dotnet ef database drop -force 命令，然后运行代码清单 32-12 中的命令。

## 32.3　添加引导 CSS 框架

按照前面章节中建立的模式，用引导 CSS 框架为示例应用程序生成的 HTML 元素设置样式。要安装引导包，在 Advanced 项目文件夹中运行代码清单 32-13 所示的命令。这些命令依赖于库管理器包。

**代码清单 32-13　安装引导 CSS 框架**

```
libman init -p cdnjs
libman install twitter-bootstrap@4.3.1 -d wwwroot/lib/twitter-bootstrap
```

如果使用的是 Visual Studio，可以通过在 Solution Explorer 中单击 Advanced项目并从弹出菜单中选择 Add | Client-Side Library 来安装客户端包。

## 32.4　配置服务和中间件

在这个项目中启用运行时 Razor 视图编译。在 Advanced 项目文件夹中运行代码清单 32-14 所示的命令，安装提供运行时编译服务的包。

**代码清单 32-14　向示例项目添加一个包**

```
dotnet add package Microsoft.AspNetCore.Mvc.Razor.RuntimeCompilation --version 3.1.1
```

本书这一部分中的示例应用程序同时使用 MVC 控制器和 Razor Pages 响应请求。将代码清单 32-15 中所示的语句添加到 Startup 类中，以配置应用程序使用的服务和中间件。

**代码清单 32-15　在 Advanced 文件夹的 Startup.cs 文件中添加服务和中间件**

```
using System;
using System.Collections.Generic;
using System.Linq;
using System.Threading.Tasks;
using Microsoft.AspNetCore.Builder;
using Microsoft.AspNetCore.Hosting;
using Microsoft.AspNetCore.Http;
using Microsoft.Extensions.DependencyInjection;
using Microsoft.Extensions.Hosting;
using Microsoft.Extensions.Configuration;
```

```
using Microsoft.EntityFrameworkCore;
using Advanced.Models;

namespace Advanced {
    public class Startup {

        public Startup(IConfiguration config) {
            Configuration = config;
        }

        public IConfiguration Configuration { get; set; }

        public void ConfigureServices(IServiceCollection services) {
            services.AddDbContext<DataContext>(opts => {
                opts.UseSqlServer(Configuration[
                    "ConnectionStrings:PeopleConnection"]);
                opts.EnableSensitiveDataLogging(true);
            });
            services.AddControllersWithViews().AddRazorRuntimeCompilation();
            services.AddRazorPages().AddRazorRuntimeCompilation();
        }

        public void Configure(IApplicationBuilder app, DataContext context) {

            app.UseDeveloperExceptionPage();
            app.UseStaticFiles();
            app.UseRouting();
            app.UseEndpoints(endpoints => {
                endpoints.MapControllerRoute("controllers",
                    "controllers/{controller=Home}/{action=Index}/{id?}");
                endpoints.MapDefaultControllerRoute();
                endpoints.MapRazorPages();
            });

            SeedData.SeedDatabase(context);
        }
    }
}
```

除了默认的控制器路由之外，还添加了一个匹配以控制器开始的 URL 路径的路由，这将使后续章节中的示例在控制器和 Razor Pages 之间切换时更容易理解。这与在前几章中采用的约定相同，把以/pages 开头的 URL 路径路由到 Razor 页面。

## 32.5 创建控制器和视图

要使用控制器显示应用程序的数据，请在 Advanced 项目文件夹中创建一个名为 Controllers

的文件夹，并向其中添加一个名为 HomeController.cs 的类文件，其内容如代码清单 32-16 所示。

### 代码清单 32-16　Controllers 文件夹中 HomeController.cs 文件的内容

```
using Advanced.Models;
using Microsoft.AspNetCore.Mvc;
using Microsoft.EntityFrameworkCore;
using System.Collections.Generic;
using System.Linq;

namespace Advanced.Controllers {
    public class HomeController : Controller {
        private DataContext context;

        public HomeController(DataContext dbContext) {
            context = dbContext;
        }

        public IActionResult Index([FromQuery] string selectedCity) {
            return View(new PeopleListViewModel {
                People = context.People
                    .Include(p => p.Department).Include(p => p.Location),
                Cities = context.Locations.Select(l => l.City).Distinct(),
                SelectedCity = selectedCity
            });
        }
    }

    public class PeopleListViewModel {
        public IEnumerable<Person> People { get; set; }
        public IEnumerable<string> Cities { get; set; }
        public string SelectedCity { get; set; }

        public string GetClass(string city) =>
            SelectedCity == city ? "bg-info text-white" : "";
    }
}
```

为给控制器提供视图，创建 Views/Home 文件夹，并添加一个名为 Index.cshtml 的 Razor 视图，内容如代码清单 32-17 所示。

### 代码清单 32-17　Views/Home 文件夹中的 Index.cshtml 文件的内容

```
@model PeopleListViewModel

<h4 class="bg-primary text-white text-center p-2">People</h4>

<table class="table table-sm table-bordered table-striped">
```

```
    <thead>
        <tr>
            <th>ID</th><th>Name</th><th>Dept</th><th>Location</th>
        </tr>
    </thead>
    <tbody>
        @foreach (Person p in Model.People) {
            <tr class="@Model.GetClass(p.Location.City)">
                <td>@p.PersonId</td>
                <td>@p.Surname, @p.Firstname</td>
                <td>@p.Department.Name</td>
                <td>@p.Location.City, @p.Location.State</td>
            </tr>
        }
    </tbody>
</table>

<form asp-action="Index" method="get">
    <div class="form-group">
        <label for="selectedCity">City</label>
        <select name="selectedCity" class="form-control">
            <option disabled selected>Select City</option>
            @foreach (string city in Model.Cities) {
                <option selected="@(city == Model.SelectedCity)">
                    @city
                </option>
            }
        </select>
    </div>
    <button class="btn btn-primary" type="submit">Select</button>
</form>
```

要启用标签助手并添加视图中默认可用的名称空间，请给 Views 文件夹添加一个名为 _ViewImports.cshtml 的 Razor 视图导入文件，内容如代码清单 32-18 所示。

#### 代码清单 32-18 Views 文件夹中_ViewImports.cshtml 文件的内容

```
@addTagHelper *, Microsoft.AspNetCore.Mvc.TagHelpers
@using Advanced.Models
@using Advanced.Controllers
```

要指定控制器视图的默认布局，给 Views 文件夹添加一个名为_ViewStart.cshtml 的 Razor 视图启动文件，内容如代码清单 32-19 所示。

#### 代码清单 32-19 Views 文件夹中_ViewStart.cshtml 文件的内容

```
@{
    Layout = "_Layout";
}
```

要创建布局，请创建Views/Shared文件夹，并向其中添加一个名为_Layout.cshtml的Razor布局，内容如代码清单32-20所示。

**代码清单 32-20　Views/Shared 文件夹中_Layout.cshtml 文件的内容**

```
<!DOCTYPE html>
<html>
<head>
    <title>@ViewBag.Title</title>
    <link href="/lib/twitter-bootstrap/css/bootstrap.min.css" rel="stylesheet" />
</head>
<body>
    <div class="m-2">
        @RenderBody()
    </div>
</body>
</html>
```

## 32.6　创建 Razor Pages

要使用 Razor Pages 显示应用程序的数据，请创建 Pages 文件夹，并向其中添加名为 Index.cshtml 的 Razor Pages，内容如代码清单 32-21 所示。

**代码清单 32-21　Pages 文件夹中 Index.cshtml 文件的内容**

```
@page "/pages"
@model IndexModel

<h4 class="bg-primary text-white text-center p-2">People</h4>

<table class="table table-sm table-bordered table-striped">
    <thead>
        <tr>
            <th>ID</th><th>Name</th><th>Dept</th><th>Location</th>
        </tr>
    </thead>
    <tbody>
        @foreach (Person p in Model.People) {
            <tr class="@Model.GetClass(p.Location.City)">
                <td>@p.PersonId</td>
                <td>@p.Surname, @p.Firstname</td>
                <td>@p.Department.Name</td>
                <td>@p.Location.City, @p.Location.State</td>
            </tr>
        }
    </tbody>
</table>
```

```
<form asp-page="Index" method="get">
    <div class="form-group">
        <label for="selectedCity">City</label>
        <select name="selectedCity" class="form-control">
            <option disabled selected>Select City</option>
            @foreach (string city in Model.Cities) {
                <option selected="@(city == Model.SelectedCity)">
                    @city
                </option>
            }
        </select>
    </div>
    <button class="btn btn-primary" type="submit">Select</button>
</form>

@functions {

    public class IndexModel: PageModel {
        private DataContext context;

        public IndexModel(DataContext dbContext) {
            context = dbContext;
        }

        public IEnumerable<Person> People { get; set; }

        public IEnumerable<string> Cities { get; set; }

        [FromQuery]
        public string SelectedCity { get; set; }

        public void OnGet() {
            People = context.People.Include(p => p.Department)
                .Include(p => p.Location);
            Cities = context.Locations.Select(l => l.City).Distinct();
        }

        public string GetClass(string city) =>
            SelectedCity == city ? "bg-info text-white" : "";
    }
}
```

要启用标签助手并在 Razor Pages 的 view 部分添加默认可用的名称空间，给 Pages 文件夹添加一个名为_ViewImports.cshtml 的 Razor view 导入文件，内容如代码清单 32-22 所示。

**代码清单 32-22　文件夹中_ViewImports.cshtml 文件的内容**

```
@addTagHelper *, Microsoft.AspNetCore.Mvc.TagHelpers
@using Advanced.Models
@using Microsoft.AspNetCore.Mvc.RazorPages
@using Microsoft.EntityFrameworkCore
```

要指定 Razor Pages 的默认布局，给 Pages 文件夹添加一个名为_ViewStart.cshtml 的 Razor 视图启动文件，内容如代码清单 32-23 所示。

**代码清单 32-23　Pages 文件夹中_ViewStart.cshtml 文件的内容**

```
@{
    Layout = "_Layout";
}
```

要创建布局，给 Pages 文件夹添加一个名为_Layout.cshtml 的 Razor 布局，内容如代码清单 32-24 所示。

**代码清单 32-24　Pages 文件夹中_Layout.cshtml 文件的内容**

```
<!DOCTYPE html>
<html>
<head>
    <title>@ViewBag.Title</title>
    <link href="/lib/twitter-bootstrap/css/bootstrap.min.css" rel="stylesheet" />
</head>
<body>
    <div class="m-2">
        <h5 class="bg-secondary text-white text-center p-2">Razor Page</h5>
        @RenderBody()
    </div>
</body>
</html>
```

## 32.7　运行示例应用程序

为启动应用程序，从 Debug 菜单中选择 Start Without Debugging 或 Run Without Debugging，或者在 Advanced 项目文件夹中运行代码清单 32-25 所示的命令。

**代码清单 32-25　运行示例应用程序**

```
dotnet run
```

使用浏览器请求 http://localhost:5000/controllers 和 http://localhost:5000/pages。使用 Select 元素选择一个城市，并单击 Select 按钮以突出显示表中的行，如图 32-3 所示。

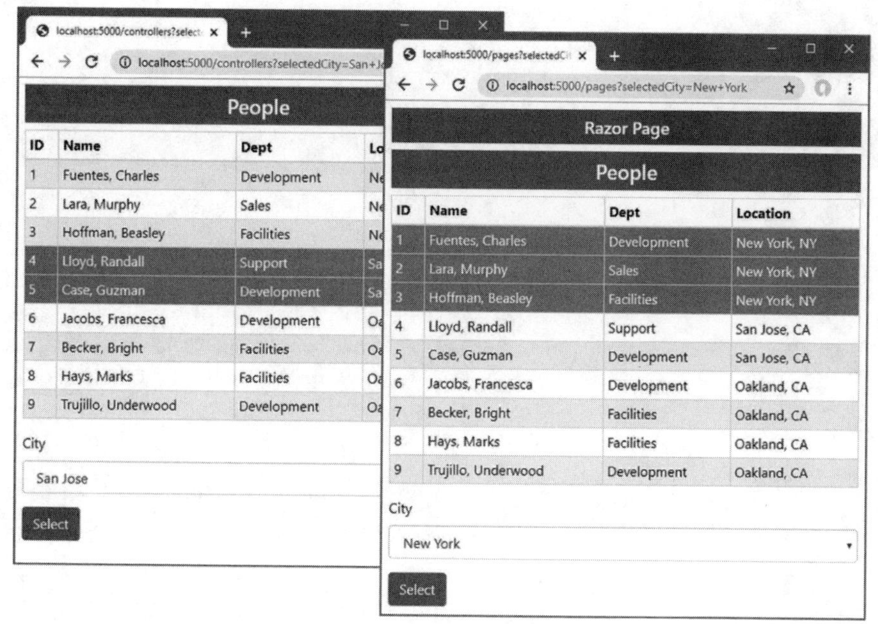

图 32-3 运行示例应用程序

# 32.8 小结

本章展示了如何创建贯穿全书这一部分的示例应用程序。项目是用空模板创建的，它包含一个依赖于 Entity Framework Core 的数据模型，并使用控制器和 Razor Pages 处理请求。第 33 章将介绍 Blazor，它是 ASP.NET Core 新增的一个功能。

# 第 33 章

# 使用 Blazor 服务器(第 1 部分)

Blazor 是 ASP.NET Core 新添加的一个功能，它向 Web 应用程序添加客户端交互性。Blazor 有两个变种，本章将重点介绍 Blazor 服务器。解释它解决的问题和它是如何工作的。展示如何配置 ASP.NET Core 应用程序来使用 Blazor 服务器，并描述使用 Razor 组件时可用的基本特性，这是 Blazor 服务器项目的构建块。第 34-36 章将描述更高级的 Blazor 服务器特性，第 37 章将描述 Blazor WebAssembly，这是 Blazor 的另一个变种。表 33-1 说明了 Blazor 服务器的情况。

表 33-1　Blazor 服务器的相关知识

| 问题 | 答案 |
| --- | --- |
| 它是什么? | Blazor 服务器使用 JavaScript 来接收浏览器事件，这些事件转发给 ASP.NET Core，使用 C#代码计算。事件对应用程序状态的影响发送回浏览器，并显示给用户 |
| 它为什么有用? | 与标准的 Web 应用程序相比，Blazor 服务器可以产生更丰富、响应更快的用户体验 |
| 它是如何使用的? | Blazor 服务器的构建块是 Razor 组件，它使用类似 Razor Pages 的语法。Razor 组件的视图部分包含特殊属性，这些属性指定应用程序如何响应用户交互 |
| 是否存在缺陷或限制? | Blazor 服务器依赖于与服务器的持久 HTTP 连接，并且当连接中断时不能正常工作。Blazor 服务器不受旧浏览器的支持 |
| 还有其他选择吗? | 本书第III部分中描述的特性可用于创建运行广泛但响应性较差的 Web 应用程序。也可以考虑客户端 JavaScript 框架，比如 Angular、React 或者 Vue.js |

表 33-2 总结了本章的内容。

表 33-2　本章内容摘要

| 问题 | 解决方案 | 代码清单 |
| --- | --- | --- |
| 配置 Blazor | 使用 AddServerSideBlazor 和 MapBlazorHub 方法设置所需的服务和中间件并配置 JavaScript 文件 | 33-3～33-6 |
| 创建 Blazor 组件 | 创建一个.blazor 文件，并使用它来定义代码和标记 | 33-7 |
| 应用组件 | 使用组件元素 | 33-8～33-9 |
| 处理事件 | 使用属性来指定处理事件的方法或表达式 | 33-10～33-15 |
| 创建与元素的双向关系 | 创建一个数据绑定 | 33-16～33-20 |
| 将代码与标记分开定义 | 使用代码隐藏类 | 33-21～33-23 |
| 定义没有声明式标记的组件 | 使用 Razor 组件类 | 33-24～33-25 |

## 33.1　准备工作

本章使用第 32 章中的 Advanced 项目。准备本章不需要任何更改。

---

**提示：** 可以从 https://github.com/apress/pro-asp.net-core-3下载本章和本书中其他所有章节的示例项目。如果在运行示例时遇到问题，请参阅第1章，以了解如何获得帮助。

---

打开一个新的 PowerShell 命令提示符，导航到包含 Advanced.csproj 文件的文件夹，并运行代码清单33-1所示的命令来删除数据库。

**代码清单33-1　删除数据库**

```
dotnet ef database drop --force
```

从 Debug 菜单中选择 Start Without Debugging 或 Run Without Debugging，或者使用 PowerShell 命令提示符运行代码清单 33-2 中所示的命令。

**代码清单 33-2　运行示例应用程序**

```
dotnet run
```

使用浏览器请求 http://localhost:5000/controllers，这将显示一个数据项列表。从下拉列表中选择一个城市，并单击 Select 按钮以突出显示元素，如图 33-1 所示。

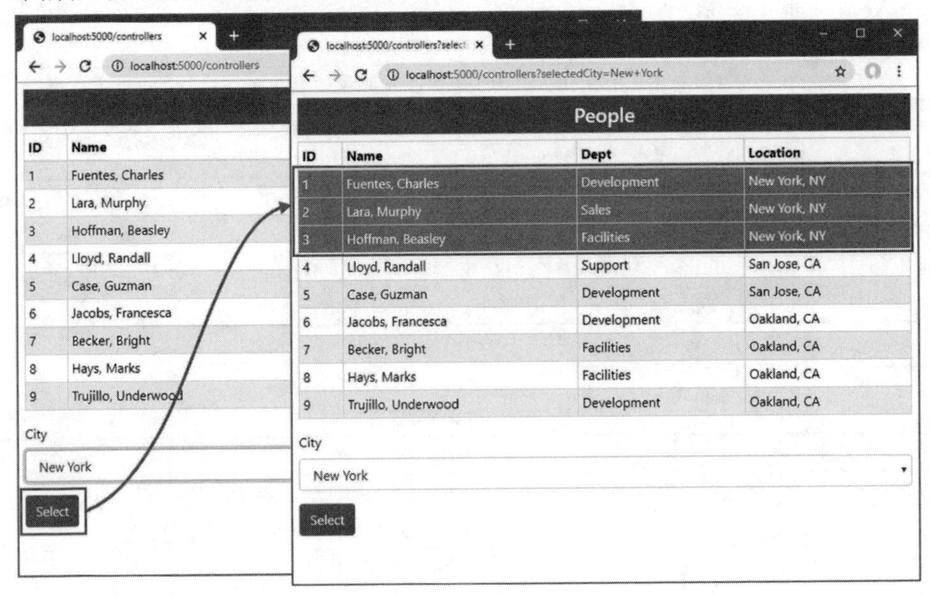


图 33-1　运行示例应用程序

## 33.2　理解 Blazor 服务器

考虑一下，选择一个城市并单击示例应用程序显示的 Select 按钮时会发生什么。浏览器发送一个提交表单的 HTTP GET 请求，由操作方法或处理程序方法接收，这取决于是使用控制器还是 Razor Pages。操作或处理程序呈现它的视图，视图向浏览器发送一个新的 HTML 文档，反映所选内容，如图 33-2 所示。

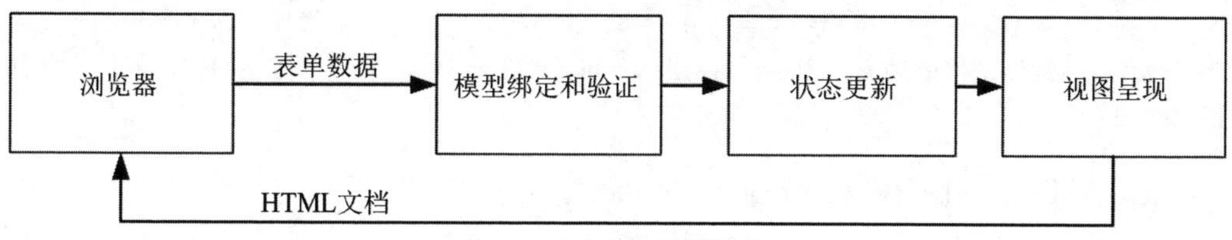


图 33-2　与示例应用程序交互

这种循环是有效的，但可能是低效的。每次单击 Submit 按钮时，浏览器都会向 ASP.NET Core 发送一个新的 HTTP 请求，每个请求包含一组完整的 HTTP 头，描述浏览器愿意接收的请求和响应的类型。在其响应中，服务器包含描述响应的 HTTP 头，并包含一个完整的 HTML 文档供浏览器显示。

在我的系统上，示例应用程序发送的数据量大约为 3KB，而且几乎所有数据都在请求之间重复。浏览器只希望告诉服务器选择了哪个城市，服务器只希望指示应该突出显示哪些表行；但是，每个 HTTP 请求都是自包含的，因此浏览器每次都必须解析一个完整的 HTML 文档。每个交互的根本问题都是一样的：发送请求并获得完整的 HTML 文档。

Blazor采取了不同的方法。发送到浏览器的HTML文档中包含一个JavaScript库。在执行JavaScript代码时，它将重新打开一个到服务器的HTTP连接，并使其保持打开状态，以便与用户交互。例如，当用户使用select元素选择一个值时，选择的详细信息发送到服务器，服务器只响应用于现有HTML的更改，如图 33-3 所示。

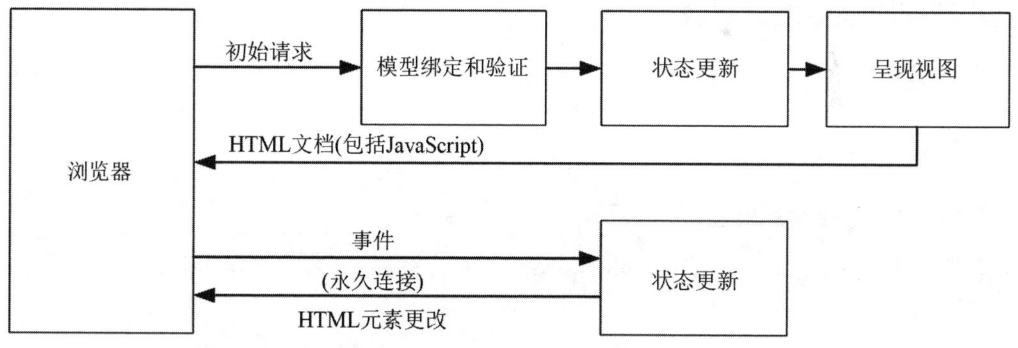


图 33-3　与 Blazor 交互

持久的 HTTP 连接将延迟最小化，仅用差异来响应会减少浏览器和服务器之间发送的数据量。

### 33.2.1　理解 Blazor 服务器的优势

Blazor 最大的吸引力在于它基于 C#编写的 Razor Pages。这意味着不需要学习新框架(如 Angular 或 React)和新语言(如 TypeScript 或 JavaScript)，就可提高效率和响应能力。Blazor 很好地集成到 ASP.NET Core 的其余部分中，构建在前面章节中描述的特性之上，这使得它易于使用(特别是与 Angular 这样的框架相比，后者的学习曲线令人眼花缭乱)。

### 33.2.2　理解 Blazor 服务器的缺点

Blazor 需要一个现代浏览器来建立和维护它的持久 HTTP 连接。而且，由于这种连接，如果连接丢失，使用 Blazor 的应用程序就会停止工作，这使得它们不适合离线使用，也就是不能依赖连接或者连接速度很慢的地方。Blazor WebAssembly(参见第 36 章)解决了这些问题，但是，它有自己的局限性。

### 33.2.3　在 Blazor 服务器和 Angular/React/Vue.js 之间选择

Blazor 和一个 JavaScript 框架之间的决定应该由开发团队的经验和用户对连接的期望来驱动。如果没有 JavaScript 专业知识，也没有使用过其中一个 JavaScript 框架，就应该使用 Blazor，但前提是可以依赖良好的连接和现代浏览器。这使 Blazor 成为业务应用程序的一个好选择，例如，可以预先确定浏览器的用户数量和网络质量。

如果有 JavaScript 经验，并编写面向公众的应用程序，就应该使用一个 JavaScript 框架，因为无法对浏览器或网络质量做出假设。选择哪种框架并不重要——我写过关于 Angular、React 和 View 的书，这些书都很出色。对于选择框架，我的建议是在每个框架中创建一个简单应用程序，然后选择一个开发模型最吸引人的框架。

如果编写面向公众的应用程序，而没有 JavaScript 经验，就有两种选择。最安全的选择是坚持使用前面的章节描述的 ASP.NET Core 特性，并接受它带来的低效率。这并不是一个糟糕的选择，仍然可以生成高质量的应用程序。更苛刻的选择是学习 TypeScript 或 JavaScript，还有 Angular、React 或者 Vue。但是不要低估掌握 JavaScript 所花费的时间或这些框架的复杂性。

## 33.3　从 Blazor 开始

最好的方式是直接从 Blazor 开始。接下来的部分将配置应用程序，以启用 Blazor，并重新创建控制器和 Razor Pages 提供的功能。之后讨论基础知识，解释 Razor 组件是如何工作的，以及它们提供的不同特性。

### 33.3.1　为 Blazor 服务器配置 ASP.NET Core

在使用 Blazor 之前需要做好准备工作。第一步是向 Startup 类添加服务和中间件，如代码清单 33-3 所示。

**代码清单 33-3　在 Advanced 文件夹的 Startup.cs 文件中添加服务和中间件**

```
using System;
using System.Collections.Generic;
using System.Linq;
using System.Threading.Tasks;
using Microsoft.AspNetCore.Builder;
using Microsoft.AspNetCore.Hosting;
using Microsoft.AspNetCore.Http;
using Microsoft.Extensions.DependencyInjection;
using Microsoft.Extensions.Hosting;
using Microsoft.Extensions.Configuration;
```

```
using Microsoft.EntityFrameworkCore;
using Advanced.Models;

namespace Advanced {
    public class Startup {

        public Startup(IConfiguration config) {
            Configuration = config;
        }

        public IConfiguration Configuration { get; set; }

        public void ConfigureServices(IServiceCollection services) {
            services.AddDbContext<DataContext>(opts => {
                opts.UseSqlServer(Configuration[
                    "ConnectionStrings:PeopleConnection"]);
                opts.EnableSensitiveDataLogging(true);
            });
            services.AddControllersWithViews().AddRazorRuntimeCompilation();
            services.AddRazorPages().AddRazorRuntimeCompilation();
            services.AddServerSideBlazor();
        }

        public void Configure(IApplicationBuilder app, DataContext context) {

            app.UseDeveloperExceptionPage();
            app.UseStaticFiles();
            app.UseRouting();

            app.UseEndpoints(endpoints => {
                endpoints.MapControllerRoute("controllers",
                    "controllers/{controller=Home}/{action=Index}/{id?}");
                endpoints.MapDefaultControllerRoute();
                endpoints.MapRazorPages();
                endpoints.MapBlazorHub();
            });

            SeedData.SeedDatabase(context);
        }
    }
}
```

MapBlazorHub 方法中的“Hub”与 SignalR 相关，它是 ASP.NET Core 的一部分，处理持久的 HTTP 请求。本书没有描述 SignalR，因为它很少直接使用，但是如果需要在客户端和服务器之间进行持续通信，那么它可能非常有用。详情请参见 https://docs.microsoft.com/en-gb/aspnet/core/signalr。

对于本书和大多数 ASP.NET Core 应用程序——知道 SignalR 用于管理 Blazor 所依赖的连接就

足够了。

### 1. 给布局添加 Blazor JavaScript 文件

Blazor 依赖 JavaScript 代码与 ASP.NET Core 服务器进行通信，在 Views/Shared 文件夹的_Layout.cshtml文件中添加代码清单 33-4 中所示的元素，将JavaScript文件添加到控制器视图使用的布局中。

**代码清单 33-4　在 Views/Shared 文件夹的_Layout.cshtml 文件中添加元素**

```
<!DOCTYPE html>
<html>
<head>
    <title>@ViewBag.Title</title>
    <link href="/lib/twitter-bootstrap/css/bootstrap.min.css" rel="stylesheet" />
    <base href="~/" />
</head>
<body>
    <div class="m-2">
        @RenderBody()
    </div>
    <script src="_framework/blazor.server.js"></script>
</body>
</html>
```

script 元素指定 JavaScript 文件的名称，对它的请求被添加到请求管道中的中间件拦截，如代码清单 33-3 所示，因此不需要额外的包将 JavaScript 代码添加到项目中。还必须添加基本元素以指定应用程序的根 URL。同样的元素必须添加到 Razor Pages 使用的布局中，如代码清单 33-5 所示。

**代码清单 33-5　在 Pages 文件夹的_Layout.cshtml 文件中添加元素**

```
<!DOCTYPE html>
<html>
<head>
    <title>@ViewBag.Title</title>
    <link href="/lib/twitter-bootstrap/css/bootstrap.min.css" rel="stylesheet" />
    <base href="~/" />
</head>
<body>
    <div class="m-2">
        <h5 class="bg-secondary text-white text-center p-2">Razor Page</h5>
        @RenderBody()
    </div>
    <script src="_framework/blazor.server.js"></script>
</body>
</html>
```

#### 2. 创建 Blazor 导入文件

Blazor 需要自己的导入文件来指定它使用的名称空间。很容易忘记将此文件添加到项目中，但是，如果没有它，Blazor 将悄无声息地失败。给 Advanced 文件夹添加一个名为_Imports.razor 的文件，内容如代码清单 33-6 所示。如果使用的是 Visual Studio，可以使用 Razor View Imports 模板来创建这个文件，但是要确保使用. Razor 文件扩展名。

**代码清单 33-6 Advanced 文件夹中_Imports.razor 文件的内容**

```
@using Microsoft.AspNetCore.Components
@using Microsoft.AspNetCore.Components.Forms
@using Microsoft.AspNetCore.Components.Routing
@using Microsoft.AspNetCore.Components.Web
@using Microsoft.JSInterop
@using Microsoft.EntityFrameworkCore
@using Advanced.Models
```

前五个@using 表达式用于 Blazor 所需的名称空间。最后两个表达式是为了便于在下面的示例中使用，因为它们允许使用 Entity Framework Core 和 Models 名称空间中的类。

### 33.3.2 创建 Blazor 组件

这里有一个术语上的冲突：技术是 Blazor，但关键的构建块称为 Razor 组件。Razor 组件是在扩展名为. Razor 的文件中定义的，并且必须以大写字母开头。组件可以在任何地方定义，但它们通常组合在一起，以帮助保持项目的组织性。在 Advanced 文件夹中创建一个 Blazor 文件夹，并添加一个名为 PeopleList. Razor 的 Razor 组件。内容如代码清单 33-7 所示。

**代码清单 33-7 Blazor 文件夹中 PeopleList. Razor 文件的内容**

```
<table class="table table-sm table-bordered table-striped">
    <thead>
        <tr>
            <th>ID</th><th>Name</th><th>Dept</th><th>Location</th>
        </tr>
    </thead>
    <tbody>
        @foreach (Person p in People) {
            <tr class="@GetClass(p.Location.City)">
                <td>@p.PersonId</td>
                <td>@p.Surname, @p.Firstname</td>
                <td>@p.Department.Name</td>
                <td>@p.Location.City, @p.Location.State</td>
            </tr>
        }
    </tbody>
</table>

<div class="form-group">
```

```
    <label for="city">City</label>
    <select name="city" class="form-control" @bind="SelectedCity">
        <option disabled selected>Select City</option>
        @foreach (string city in Cities) {
            <option value="@city" selected="@(city == SelectedCity)">
                @city
            </option>
        }
    </select>
</div>

@code {

    [Inject]
    public DataContext Context { get; set; }

    public IEnumerable<Person> People =>
        Context.People.Include(p => p.Department).Include(p => p.Location);

    public IEnumerable<string> Cities => Context.Locations.Select(l => l.City);

    public string SelectedCity { get; set; }

    public string GetClass(string city) =>
        SelectedCity == city ? "bg-info text-white" : "";
}
```

Razor 组件类似于 Razor Pages。视图部分依赖于在前面章节中看到的 Razor 特性，使用@表达式将数据值插入组件的 HTML 中，或者以序列化方式为对象生成元素，如下所示：

```
...
@foreach (string city in Cities) {
    <option value="@city" selected="@(city == SelectedCity)">
}
...
```

这个@foreach 表达式为 Cities 序列中的每个值生成 option 元素，与第 32 章中创建的控制器视图和 Razor Pages 中的等价表达式相同。

尽管 Razor 组件看起来很熟悉，但有一些重要的区别。第一个是没有页面模型类和@model 表达式。支持组件 HTML 的属性和方法直接在@code 表达式中定义，该表达式与 Razor Pages 的@functions 表达式对应。例如，要定义向视图部分提供 Person 对象的属性，只需要在@code 部分中定义一个 People 属性，如下所示：

```
...
public IEnumerable<Person> People =>
    Context.People.Include(p => p.Department).Include(p => p.Location);
...
```

而且，因为没有页面模型类，所以没有用于声明服务依赖关系的构造函数。相反，依赖注入会设置已被 Inject 属性修饰的属性值，如下所示：

```
...
[Inject]
public DataContext Context { get; set; }
...
```

最显著的区别是在 select 元素上使用了特殊属性。

```
...
<select name="city" class="form-control" @bind="SelectedCity">
   <option disabled selected>Select City</option>
...
```

这个 Blazor 属性在 select 元素的值和@code 部分中定义的 SelectedCity 属性之间创建一个数据绑定。

在 33.4.2 节中对数据绑定进行了更详细的描述，但是现在，当用户更改 select 元素的值时，SelectedCity 的值将被更新，知道这些就足够了。

### 使用 Razor 组件

Razor 组件作为 Razor Pages 或控制器视图的一部分交付给浏览器。代码清单 33-8 显示了如何在控制器视图中使用 Razor 组件。

**代码清单 33-8　在 Views/Home 文件夹的 Index.cshtml 文件中使用 Razor 组件**

```
@model PeopleListViewModel

<h4 class="bg-primary text-white text-center p-2">People</h4>

<component type="typeof(Advanced.Blazor.PeopleList)" render-mode="Server" />
```

Razor 组件是使用组件元素应用的，对于组件元素有一个标签助手。组件元素使用 type 和 render 模式属性配置。type 属性用于指定 Razor 组件。Razor 组件编译成类，就像控制器视图和 Razor Pages 一样。PeopleList 组件在 Advanced 项目的 Blazor 文件夹中定义，因此类型是 Advanced.Blazor.PeopleList，如下所示：

```
...
<component type="typeof(Advanced.Blazor.PeopleList)" render-mode="Server" />
...
```

render 模式属性用于选择组件如何使用 RenderMode 枚举中的值生成内容，该枚举如表 33-3 所示。

表 33-3 RenderMode 值

| 名称 | 描述 |
| --- | --- |
| Static | Razor 组件将其视图部分呈现为不支持客户端的静态 HTML |
| Server | HTML 文档连同组件的占位符一起发送到浏览器。组件显示的 HTML 通过持久 HTTP 连接发送到浏览器并显示给用户 |
| ServerPrerendered | 组件的视图部分包含在 HTML 中，并立即显示给用户。HTML 内容将通过持久 HTTP 连接再次发送 |

对于大多数应用程序，Server 选项是一个不错的选择。ServerPrerendered 在发送到浏览器的 HTML 文档中包含了 Razor 组件视图部分的静态再现。它充当占位符内容，这样在加载和执行 JavaScript 代码时，用户不会看到空的浏览器窗口。一旦建立了持久 HTTP 连接，占位符内容将被删除，替换为 Blazor 发送的动态版本。向用户显示静态内容的想法是好的，但它可以被混淆，因为 HTML 元素没有连接到应用程序的服务器端部分，与任何用户交互失效或一旦动态内容到来，HTML 元素就会被丢弃。

要查看 Blazor 的运行情况，请重启 ASP.NET Core，并使用浏览器请求 http://localhost:5000/controllers。在使用 Blazor 时不需要提交表单，因为一旦更改了 select 元素的值，数据绑定就会响应，如图 33-4 所示。

使用 select 元素时，选择的值将通过持久 HTTP 连接发送到 ASP.NET Core 服务器，它更新 Razor 组件的 SelectedCity 属性，并重新呈现 HTML 内容。将一组更新发送到 JavaScript 代码，后者更新表。

Razor 组件也可在 Razor Pages 中使用。将一个名为 Blazor.cshtml 的 Blazor 页面添加到 Pages 文件夹中，并添加如代码清单 33-9 所示的内容。

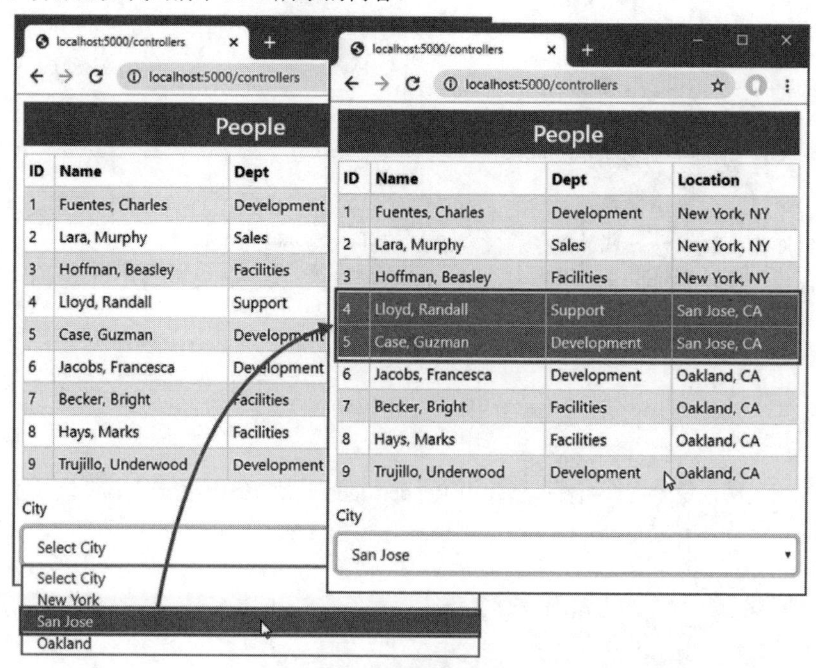


图 33-4 使用 Razor 组件

#### 代码清单 33-9　Pages 文件夹中 Blazor.cshtml 文件的内容

```
@page "/pages/blazor"

<script type="text/javascript">
    window.addEventListener("DOMContentLoaded", () => {
        document.getElementById("markElems").addEventListener("click", () => {
            document.querySelectorAll("td:first-child")
                .forEach(elem => {
                    elem.innerText = `M:${elem.innerText}`
                    elem.classList.add("border", "border-dark");
                });
        });
    });
</script>

<h4 class="bg-primary text-white text-center p-2">Blazor People</h4>

<button id="markElems" class="btn btn-outline-primary mb-2">Mark Elements</button>

<component type="typeof(Advanced.Blazor.PeopleList)" render-mode="Server" />
```

代码清单 33-9 中的 Razor Pages 包含额外的 JavaScript 代码，帮助演示只发送更改，而不是一个全新的 HTML 表。重启 ASP.NET Core，请求 http://localhost:5000/pages/blazor。单击 Mark Elements 按钮，ID 列中的单元格将更改为显示不同的内容和边框。现在使用 select 元素来选择一个不同的城市，表中的元素将被修改而没有删除，如图 33-5 所示。

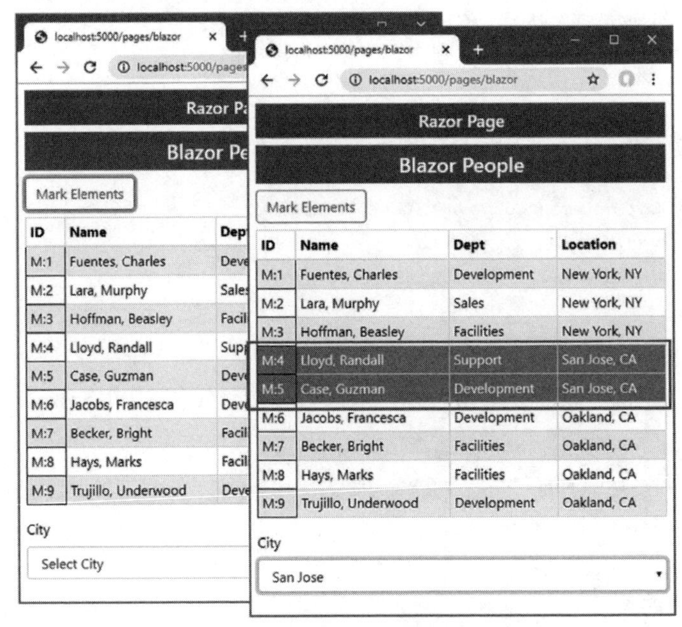


图 33-5　演示只使用更改

**理解 Blazor 连接消息**

停止 ASP.NET Core 时，会在浏览器窗口显示一个错误消息，它表明与服务器的连接已经丢失，并阻止用户与显示的组件交互。Blazor 将尝试重新连接，并恢复由于临时网络问题导致的中断，但当服务器停止或重新启动时，它将无法这样做，因为连接的上下文数据已经丢失；必须显式地请求一个新的 URL。

在连接消息中有一个默认的 reload 链接，但它指向网站的默认 URL，这对本书没有帮助，本书演示如何使用特定的 URL 来查看示例的效果。有关如何配置连接消息的详细信息，请参阅第 34 章。

## 33.4　理解 Razor 组件的基本特性

前面演示了 Blazor 的使用方法以及它的工作原理，现在是时候回到基本原理，介绍 Razor 组件提供的特性了。上一节中的示例展示了如何使用 Blazor 来复现标准的 ASP. NET Core 特性；实际上，Blazor 具有更广泛的可用特性。

### 33.4.1　理解 Blazor 事件和数据绑定

事件允许 Razor 组件响应用户的交互，Blazor 使用持久的 HTTP 连接将事件的详细信息发送到可以处理它的服务器。要查看 Blazor 事件的运行情况，为 Blazor 文件夹添加一个名为 Events.razor 的 Razor 组件，内容如代码清单 33-10 所示。

**代码清单 33-10　Blazor 文件夹中 Events.razor 的内容**

```
<div class="m-2 p-2 border">
    <button class="btn btn-primary" @onclick="IncrementCounter">Increment</button>
    <span class="p-2">Counter Value: @Counter</span>
</div>
@code {
    public int Counter { get; set; } = 1;

    public void IncrementCounter(MouseEventArgs e) {
        Counter++;
    }
}
```

向HTML元素添加属性，可注册事件处理程序，其中属性名称为@on，后面跟着事件名称。在这个例子中，为button元素生成的click事件设置了一个处理程序，如下所示：

```
...
<button class="btn btn-primary" @onclick="IncrementCounter">Increment</button>
...
```

赋给属性的值是在触发事件时调用的方法的名称。该方法可以定义一个可选参数，该参数要么是 EventArgs 类的实例，要么是 EventArgs 派生的类，提供关于事件的附加信息。

对于 onclick 事件，处理程序方法接收一个 MouseEventArgs 对象，该对象提供额外的细节，比如单击的屏幕坐标。表 33-4 列出了事件描述事件和使用它们的事件。

表 33-4　EventArgs 类及其所表示的事件

| 类 | 事件 |
|---|---|
| ChangeEventArgs | onchange、oninput |
| ClipboardEventArgs | oncopy、oncut、onpaste |
| DragEventArgs | ondrag、ondragend、ondragenter、ondragleave、ondragover、ondragstart、ondrop |
| ErrorEventArgs | onerror |
| FocusEventArgs | onblur、onfocus、onfocusin、onfocusout |
| KeyboardEventArgs | onkeydown、onkeypress、onkeyup |
| MouseEventArgs | onclick、oncontextmenu、ondblclick、onmousedown、onmousemove、onmouseout、onmouseover、onmouseup、onmousewheel、onwheel |
| PointerEventArgs | ongotpointercapture、onlostpointercapture、onpointercancel、onpointerdown、onpointerenter、onpointerleave、onpointermove、onpointerout、onpointerover、onpointerup |
| ProgressEventArgs | onabort、onload、onloadend、onloadstart、onprogress、ontimeout |
| TouchEventArgs | ontouchcancel、ontouchend、ontouchenter、ontouchleave、ontouchmove、ontouchstart |
| EventArgs | onactivate、onbeforeactivate、onbeforecopy、onbeforecut、onbeforedeactivate、onbeforepaste、oncanplay、oncanplaythrough、oncuechange、ondeactivate、ondurationchange、onemptied、onended、onfullscreenchange、onfullscreenerror、oninvalid、onloadeddata、onloadedmetadata、onpause、onplay、onplaying、onpointerlockchange、onpointerlockerror、onratechange、onreadystatechange、onreset、onscroll、onseeked、onseeking、onselect、onselectionchange、onselectstart、onstalled、onstop、onsubmit、onsuspend、ontimeupdate、onvolumechange、onwaiting |

Blazor JavaScript 代码在事件触发时接收事件，并通过持久 HTTP 连接将其转发给服务器。调用处理程序方法，并更新组件的状态。对组件的 view 部分生成的内容的任何更改都将发送回 JavaScript 代码，JavaScript 代码将更新浏览器显示的内容。

在本例中，单击事件将由 IncrementCounter 方法处理，该方法会更改 Counter 属性的值。Counter 属性的值包含在组件呈现的 HTML 中，因此 Blazor 将更改发送到浏览器，以便 JavaScript 代码可以更新显示给用户的 HTML 元素。要显示 Events 组件，请替换 Pages 文件夹中 Blazor.cshtml 文件的内容，如代码清单 33-11 所示。

#### 代码清单 33-11　在 Pages 文件夹的 Blazor.cshtml 文件中使用新的组件

```
@page "/pages/blazor"

<h4 class="bg-primary text-white text-center p-2">Events</h4>

<component type="typeof(Advanced.Blazor.Events)" render-mode="Server" />
```

代码清单 33-11 更改了组件元素的 type 属性，并删除了上一个示例中用于标记元素的定制 JavaScript 和按钮元素。重启 ASP.NET Core，请求 http://localhost:5000/pages/blazor 查看新组件。单击 Increment 按钮，Blazor JavaScript 代码将接收 Click 事件，并通过 IncrementCounter 方法发送到服务器进行处理，如图 33-6 所示。

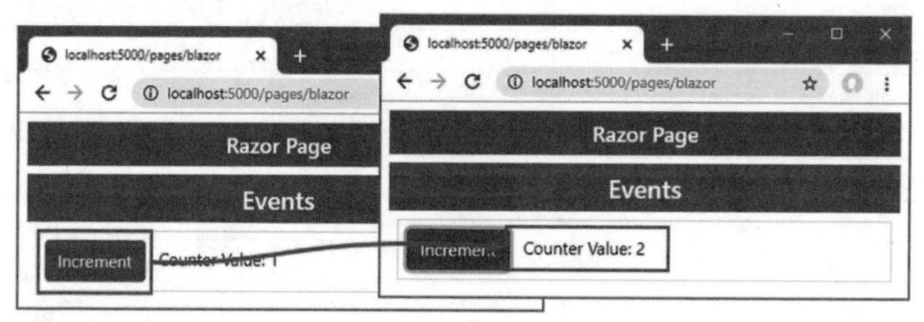


图 33-6　处理事件

### 1. 处理来自多个元素的事件

为避免代码重复，一个处理程序方法可以接收来自多个元素的元素，如代码清单 33-12 所示。

#### 代码清单 33-12　在 Blazor 文件夹的 Events.razor 文件中处理事件

```
<div class="m-2 p-2 border">
    <button class="btn btn-primary" @onclick="@(e => IncrementCounter(e, 0))">
        Increment Counter #1
    </button>
    <span class="p-2">Counter Value: @Counter[0]</span>
</div>

<div class="m-2 p-2 border">
    <button class="btn btn-primary" @onclick="@(e => IncrementCounter(e, 1))">
        Increment Counter #2
    </button>
    <span class="p-2">Counter Value: @Counter[1]</span>
</div>

@code {
    public int[] Counter { get; set; } = new int[] { 1, 1 };

    public void IncrementCounter(MouseEventArgs e, int index) {
        Counter[index]++;
    }
}
```

Blazor 事件属性可与 lambda 函数一起使用，后者接收 EventArgs 对象并调用带有附加参数的处理程序方法。本例向 IncrementCounter 方法添加了一个索引参数，该参数用于确定应该更新哪个计数器值。参数的值在@onclick 属性中定义，如下所示：

```
...
<button class="btn btn-primary" @onclick="@(e => IncrementCounter(e, 0))">
...
```

当以编程方式生成元素时，也可以使用这种技术，如代码清单 33-13 所示。本例使用@for 表达式来生成元素，并使用循环变量作为处理程序方法的参数。还从处理程序方法中删除了 EventArgs 参数，它没有使用。

### 2. 避免处理程序方法名 pitfall

在指定事件处理程序方法时，最常见的错误是包含括号，如下所示：

```
...
<button class="btn btn-primary" @onclick="IncrementCounter()">
...
```

它生成的错误消息将取决于事件处理程序方法。可能会看到一个警告，说明丢失了一个形参，或者 void 不能转换为 EventCallback。当指定一个处理程序方法时，必须只指定事件名，如下所示：

```
...
<button class="btn btn-primary" @onclick="IncrementCounter">
...
```

可以将方法名指定为 Razor 表达式，如下所示：

```
...
<button class="btn btn-primary" @onclick="@IncrementCounter">
...
```

一些开发人员发现这更容易解析，但结果是一样的。在使用 lambda 函数时，有一组不同的规则，必须在 Razor 表达式中定义，如下所示：

```
...
<button class="btn btn-primary" @onclick="@( ... )">
...
```

在 Razor 表达式中，lambda 函数的定义就像它在 C#类中的定义一样，这意味着定义参数，后跟箭头，再跟函数体，如下所示：

```
...
<button class="btn btn-primary" @onclick="@((e) => HandleEvent(e, local))">
...
```

如果不需要使用 EventArgs 对象，那么可以忽略 lambda 函数的参数，如下所示：

```
...
<button class="btn btn-primary" @onclick="@(() => IncrementCounter(local))">
...
```

开始使用 Blazor 时，会很快习惯这些规则，即使它们一开始看起来并不一致。

**代码清单 33-13　在 Blazor 文件夹的 Events.razor 文件中生成元素**

```
@for (int i = 0; i < ElementCount; i++) {
    int local = i;
    <div class="m-2 p-2 border">
        <button class="btn btn-primary" @onclick="@(() => IncrementCounter(local))">
            Increment Counter #@(i + 1)
        </button>
        <span class="p-2">Counter Value: @GetCounter(i)</span>
    </div>
}

@code {
    public int ElementCount { get; set; } = 4;

    public Dictionary<int, int> Counters { get; } = new Dictionary<int, int>();

    public int GetCounter(int index) =>
        Counters.ContainsKey(index) ? Counters[index] : 0;

    public void IncrementCounter(int index) =>
        Counters[index] = GetCounter(index) + 1;
}
```

要理解事件处理程序的重要一点是，直到服务器从浏览器接收到单击事件，才计算@onclick lambda 函数。这意味着必须注意不要使用循环变量 i 作为 IncrementCounter 方法的参数，因为它始终是循环生成的最终值，在本例中为 4。相反，必须在局部变量中捕获循环变量，如下所示：

```
...
int local = i;
...
```

然后本地变量用作属性中的事件处理程序方法的参数，如下所示：

```
...
<button class="btn btn-primary" @onclick="@(() => IncrementCounter(local))">
...
```

局部变量为每个生成的元素修复 lambda 函数的值。重启 ASP.NET Core 并使用浏览器请求 http://localhost:5000/pages/blazor，这将生成如图 33-7 所示的响应。所有 button 元素生成的 click 事件由相同的方法处理，但是 lambda 函数提供的参数确保更新了正确的计数器。

### 3. 不使用处理程序方法处理事件

简单的事件处理可直接在 lambda 函数中完成，而不需要使用处理程序方法，如代码清单 33-14 所示。

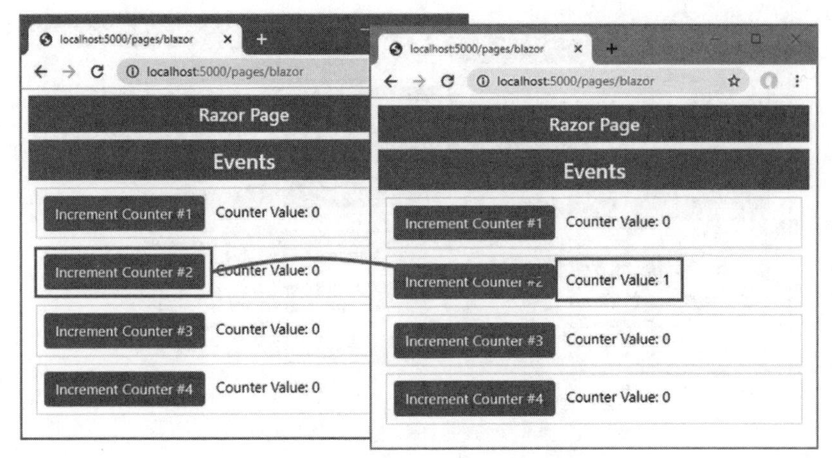


图 33-7　处理来自多个元素的事件

### 代码清单 33-14　在 Blazor 文件夹的 Events.razor 文件中处理事件

```
@for (int i = 0; i < ElementCount; i++) {
    int local = i;
    <div class="m-2 p-2 border">
        <button class="btn btn-primary" @onclick="@(() => IncrementCounter(local))">
            Increment Counter #@(i + 1)
        </button>
        <button class="btn btn-info" @onclick="@(() => Counters.Remove(local))">
            Reset
        </button>
        <span class="p-2">Counter Value: @GetCounter(i)</span>
    </div>
}

@code {
    public int ElementCount { get; set; } = 4;

    public Dictionary<int, int> Counters { get; } = new Dictionary<int, int>();

    public int GetCounter(int index) =>
        Counters.ContainsKey(index) ? Counters[index] : 0;

    public void IncrementCounter(int index) =>
        Counters[index] = GetCounter(index) + 1;
}
```

复杂的处理程序应该定义为方法，但这种方法对于简单的处理程序更简洁。重启 ASP.NET Core，请求 http://localhost:5000/pages/blazor。Reset 按钮从 Counters 集合中删除值，而不依赖于组件的@code 部分中的方法，如图 33-8 所示。

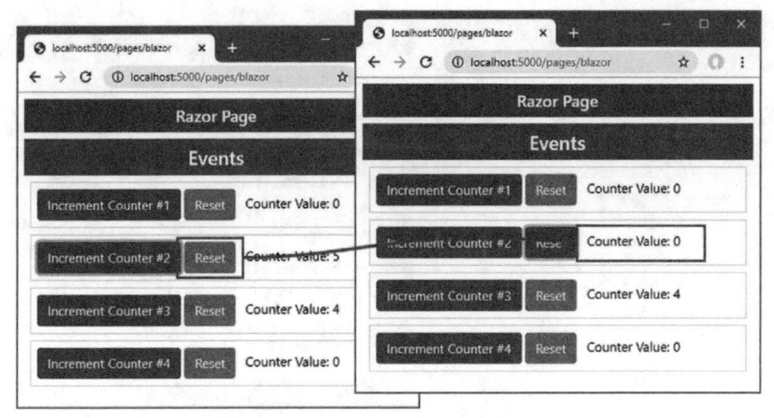


图 33-8　处理 lambda 表达式中的事件

### 4. 防止默认事件和事件传播

Blazor 提供了两个属性来改变浏览器中事件的默认行为，如表 33-5 所示。这些属性称为参数，其中事件的名称后跟一个冒号和一个关键字。

表 33-5　事件配置参数

| 名称 | 描述 |
| --- | --- |
| @on{event}:preventDefault | 这个参数决定了是否触发元素的默认事件 |
| @on{event}:stopPropagation | 这个参数决定一个事件是否传播到它的祖先元素 |

代码清单 33-15 演示了这些参数的作用以及它们为什么有用。

**代码清单 33-15　在 Blazor 文件夹的 Events.razor 文件中重写默认事件**

```
<form action="/pages/blazor" method="get">
    @for (int i = 0; i < ElementCount; i++) {
        int local = i;
        <div class="m-2 p-2 border">
            <button class="btn btn-primary"
                @onclick="@(() => IncrementCounter(local))"
                @onclick:preventDefault="EnableEventParams">
                    Increment Counter #@(i + 1)
            </button>
            <button class="btn btn-info" @onclick="@(() => Counters.Remove(local))">
                    Reset
            </button>
            <span class="p-2">Counter Value: @GetCounter(i)</span>
        </div>
    }
</form>
<div class="m-2" @onclick="@(() => IncrementCounter(1))">
    <button class="btn btn-primary" @onclick="@(() => IncrementCounter(0))"
        @onclick:stopPropagation="EnableEventParams">Propagation Test</button>
```

```
</div>

<div class="form-check m-2">
    <input class="form-check-input" type="checkbox"
          @onchange="@(() => EnableEventParams = !EnableEventParams)" />
    <label class="form-check-label">Enable Event Parameters</label>
</div>

@code {
    public int ElementCount { get; set; } = 4;

    public Dictionary<int, int> Counters { get; } = new Dictionary<int, int>();

    public int GetCounter(int index) =>
        Counters.ContainsKey(index) ? Counters[index] : 0;

    public void IncrementCounter(int index) =>
        Counters[index] = GetCounter(index) + 1;

    public bool EnableEventParams { get; set; } = false;
}
```

此示例创建了两种情况。这两种情况下，浏览器中事件的默认行为可能会导致问题。第一个是由添加表单元素引起的。默认情况下，表单中包含的按钮元素在被单击时将提交该表单，即使存在@onclick 属性。这意味着每当单击一个 Increment Counter 按钮时，浏览器就会将表单数据发送给 ASP.NET Core 服务器，它将响应 Blazor.cshtml Razor Pages 的内容。

第二个问题由一个元素演示，它的父元素也定义了一个事件处理程序，如下所示：

```
...
<div class="m-2" @onclick="@(() => IncrementCounter(1))">
    <button class="btn btn-primary" @onclick="@(() => IncrementCounter(0))"
...
```

事件在浏览器中经历一个定义良好的生命周期，包括沿着祖先元素链向上传递。在本例中，这意味着单击按钮将导致更新两个计数器，一次是由按钮元素的@onclick 处理程序更新，一次是由封闭 div 元素的@onclick 处理程序更新。

要查看这些问题，请重新启动 ASP.NET Core，请求 http://localhost:5000/pages/blazor。单击 Increment Counter 按钮；将看到表单被提交，页面被重新加载。单击 Propagation Test 按钮，将看到两个计数器被更新。图 33-9 显示了这两个问题。

代码清单 33-15 中的复选框切换应用表 33-5 中描述的参数的属性，结果是表单没有提交，只有按钮元素上的处理程序接收事件。要查看效果，请选中复选框，然后单击 Increment Counter 按钮和 Propagation Test 按钮，这会生成如图 33-10 所示的结果。

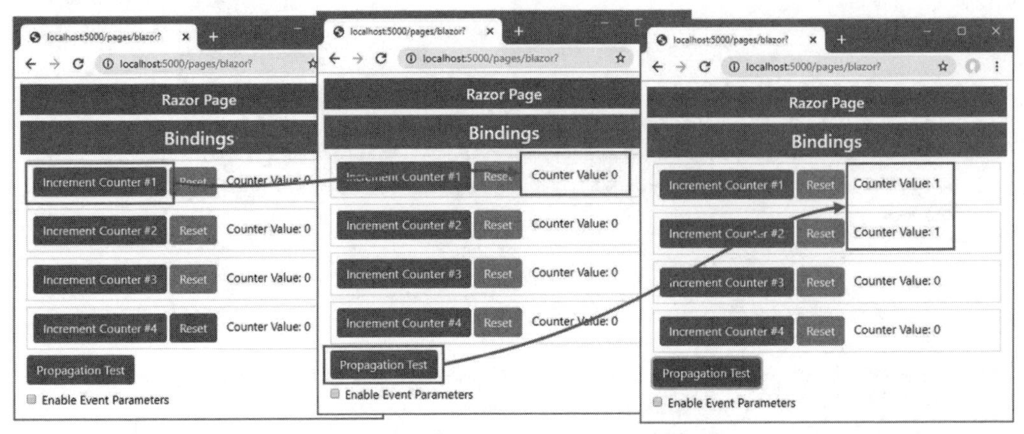


图 33-9　由浏览器中事件的默认行为引起的问题

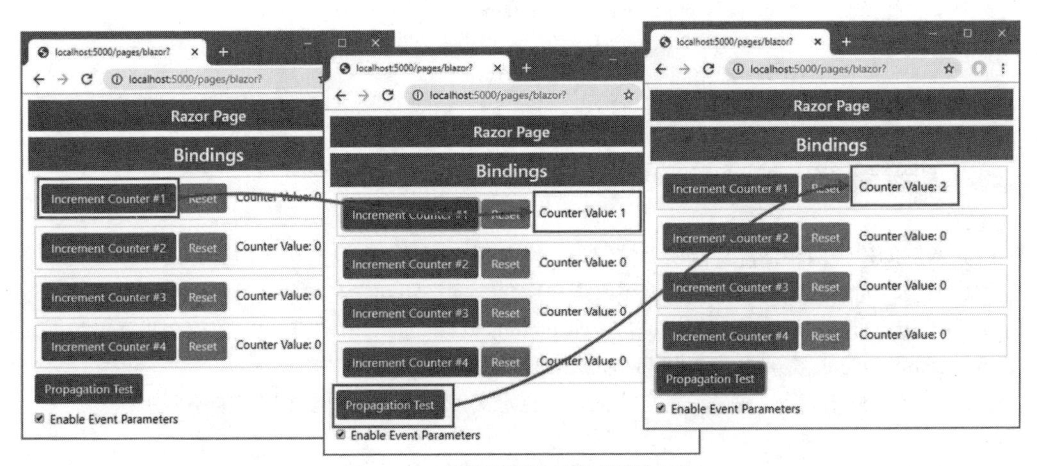


图 33-10　重写浏览器中事件的默认行为

## 33.4.2　使用数据绑定

事件处理程序和 Razor 表达式可用于创建 HTML 元素和 C#值之间的双向关系，这对于允许用户进行更改的元素(如输入和选择元素)非常有用。为 Blazor 文件夹添加名为 Bindings.razor 的 Razor 组件，内容如代码清单 33-16 所示。

**代码清单 33-16　Blazor 文件夹中 Bindings.razor 文件的内容**

```
<div class="form-group">
    <label>City:</label>
    <input class="form-control" value="@City" @onchange="UpdateCity" />
</div>
<div class="p-2 mb-2">City Value: @City</div>
<button class="btn btn-primary" @onclick="@(() => City = "Paris")">Paris</button>
<button class="btn btn-primary" @onclick="@(() => City = "Chicago")">Chicago</button>

@code {
```

```
    public string City { get; set; } = "London";

    public void UpdateCity(ChangeEventArgs e) {
        City = e.Value as string;
    }
}
```

@onchange 属性将 UpdateCity 方法注册为来自输入元素的更改事件的处理程序。这些事件是使用 ChangeEventArgs 类描述的，该类提供了一个 Value 属性。每次接收到更改事件时，都会使用输入元素的内容更新 City 属性。

input 元素的 value 属性创建了另一个方向的关系，这样当 City 属性的值发生变化时，元素的 value 属性也会发生变化，它会改变显示给用户的文本。要应用新的 Razor 组件，请更改 Razor Pages 中的组件属性，如代码清单 33-17 所示。

#### 代码清单 33-17　在 Pages 文件夹的 Blazor.cshtml 文件中使用 Razor 组件

```
@page "/pages/blazor"
<h4 class="bg-primary text-white text-center p-2">Events</h4>
<component type="typeof(Advanced.Blazor.Bindings)" render-mode="Server" />
```

要查看代码清单 33-16 中绑定定义的关系的两个部分，请重新启动 ASP.NET Core，导航到 http://localhost:5000/pages/blazor，并编辑输入元素的内容。更改事件只有在输入元素失去焦点时才会触发，因此一旦完成编辑，请按 Tab 键或单击输入元素的外部；将看到输入的、通过 div 元素中的 Razor 表达式显示出来的值，如图 33-11 的左边所示。单击其中一个按钮，City 属性更改为 Paris 或 Chicago，选择的值将由 div 元素和 input 元素同时显示，如图右侧所示。

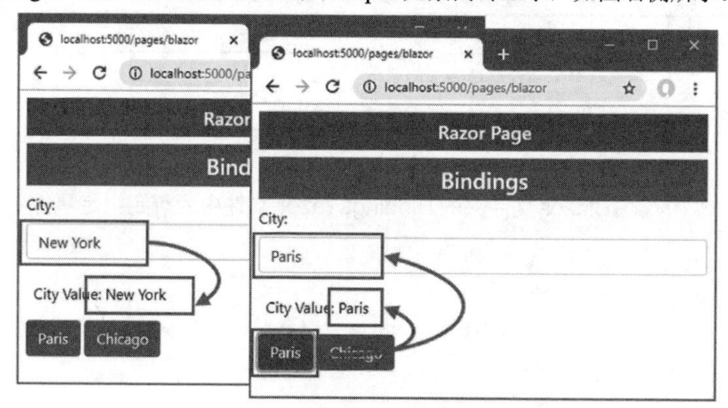


图 33-11　在元素和属性之间创建双向关系

涉及变更事件的双向关系可表示为数据绑定，数据绑定允许值和事件都用单个属性配置，如代码清单 33-18 所示。

#### 代码清单 33-18　在 Blazor 文件夹的 Bindings.razor 文件中使用数据绑定

```
<div class="form-group">
    <label>City:</label>
```

```
    <input class="form-control" @bind="City" />
</div>
<div class="p-2 mb-2">City Value: @City</div>
<button class="btn btn-primary" @onclick="@(() => City = "Paris")">Paris</button>
<button class="btn btn-primary" @onclick="@(() => City = "Chicago")">Chicago</button>

@code {

    public string City { get; set; } = "London";

    //public void UpdateCity(ChangeEventArgs e) {
    // City = e.Value as string;
    //}
}
```

@bind 属性用于指定在更改事件触发时更新的属性，以及在 Value 属性更改时更新的属性。代码清单 33-18 中的效果与代码清单 33-16 相同，但表达得更简洁，不需要处理程序方法或 lambda 函数来更新属性。

### 1. 更改绑定事件

默认情况下，绑定中使用更改事件，绑定为用户提供了合理的响应性，而不需要从服务器进行太多更新。可使用表 33-6 中描述的属性更改绑定中使用的事件。

这些属性替代了@bind，如代码清单 33-19 所示，但是只能用于用 ChangeEventArgs 类表示的事件。这意味着只能使用 onchange 和 oninput 事件，至少在当前版本中是这样。

表 33-6　用于指定事件的绑定属性

| 属性 | 描述 |
| --- | --- |
| @bind-value | 此属性用于选择数据绑定的属性 |
| @bind-value:event | 此属性用于选择数据绑定的事件 |

#### 代码清单 33-19　在 Blazor 文件夹的 Bindings.razor 文件中为绑定指定事件

```
<div class="form-group">
    <label>City:</label>
    <input class="form-control" @bind-value="City" @bind-value:event="oninput" />
</div>
<div class="p-2 mb-2">City Value: @City</div>
<button class="btn btn-primary" @onclick="@(() => City = "Paris")">Paris</button>
<button class="btn btn-primary" @onclick="@(() => City = "Chicago")">Chicago</button>

@code {
    public string City { get; set; } = "London";
}
```

这个属性组合为 City 属性创建了一个绑定，该属性在触发 oninput 事件时(每次击键后)更新，而不是只在输入元素失去焦点时更新。要查看效果，请重新启动 ASP.NET Core，导航到

http://localhost:5000/pages/blazor，并开始在 input 元素中输入。City 属性在每次击键之后更新，如图 33-12 所示。

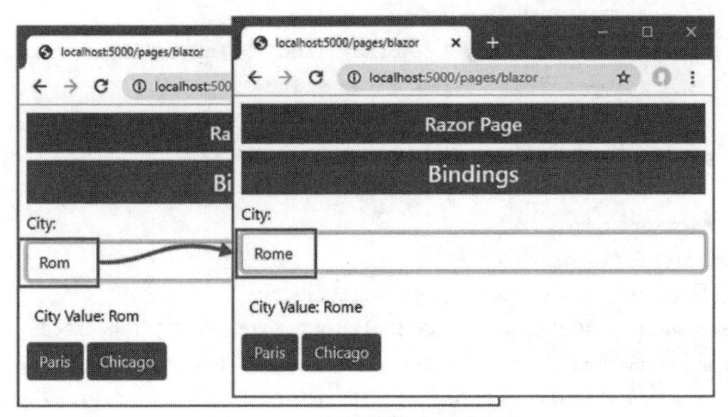


图 33-12 更改数据绑定中的事件

## 2. 创建 DateTime 绑定

Blazor 特别支持为 DateTime 属性创建绑定，允许使用特定区域性或格式字符串表示它们。使用表 33-7 中描述的参数应用该特性。

表 33-7 DateTime 参数

| 名称 | 描述 |
|---|---|
| @bind:culture | 此属性用于选择 CultureInfo 对象(用于格式化 DateTime 值) |
| @bind:format | 此属性用于指定一个数据格式化字符串(用于格式化 DateTime 值) |

■ **提示**：如果使用@bind-value和@bind-value:event属性来选择事件，则必须使用@bind-value:culture和@bind-value:format参数。

代码清单 33-20 显示了这些属性与 DateTime 属性的用法。

**注意**：这些示例中使用的格式化字符串在以下网址中描述：https://docs.microsoft.com/en-us/dotnet/api/system.datetime? = netcore-3.1。

### 代码清单 33-20 在 Blazor 文件夹的 Bindings.razor 文件中使用 DateTime 属性

```
@using System.Globalization

<div class="form-group">
    <label>City:</label>
    <input class="form-control" @bind-value="City" @bind-value:event="oninput" />
</div>
<div class="p-2 mb-2">City Value: @City</div>
<button class="btn btn-primary" @onclick="@(() => City = "Paris")">Paris</button>
<button class="btn btn-primary" @onclick="@(() => City = "Chicago")">Chicago</button>

<div class="form-group mt-2">
```

```
        <label>Time:</label>
        <input class="form-control my-1" @bind="Time" @bind:culture="Culture"
            @bind:format="MMM-dd" />
        <input class="form-control my-1" @bind="Time" @bind:culture="Culture" />
        <input class="form-control" type="date" @bind="Time" />
    </div>
    <div class="p-2 mb-2">Time Value: @Time</div>

    <div class="form-group">
        <label>Culture:</label>
        <select class="form-control" @bind="Culture">
            <option value="@CultureInfo.GetCultureInfo("en-us")">en-US</option>
            <option value="@CultureInfo.GetCultureInfo("en-gb")">en-GB</option>
            <option value="@CultureInfo.GetCultureInfo("fr-fr")">fr-FR</option>
        </select>
    </div>

    @code {
        public string City { get; set; } = "London";

        public DateTime Time { get; set; } = DateTime.Parse("2050/01/20 09:50");

        public CultureInfo Culture { get; set; } = CultureInfo.GetCultureInfo("en-us");
    }
```

有三个输入元素用于显示相同的 DataTime 值，其中两个使用表 33-7 中的属性配置。第一个元素配置了区域和格式字符串，如下所示：

```
...
<input class="form-control my-1" @bind="Time" @bind:culture="Culture"
    @bind:format="MMM-dd" />
...
```

使用在 select 元素中选择的区域以及显示缩写月份名称和数字日期的格式字符串来显示 DateTime 属性。第二个输入元素只指定区域，这意味着将使用默认的格式化字符串。

```
...
<input class="form-control my-1" @bind="Time" @bind:culture="Culture" />
...
```

要查看日期如何显示，请重新启动 ASP.NET Core，请求 http://localhost:5000/pages/blazor，并使用 select 元素来选择不同的区域设置。可用的设置表示美国英语、英国英语以及法国法语。图 33-13 显示了每个生成的格式。

本例中的初始语言环境是 en-US。当切换到 en-GB 时，月份和日期出现的顺序会发生变化。当切换到 fr-FR 时，缩写的月份名称会发生变化。

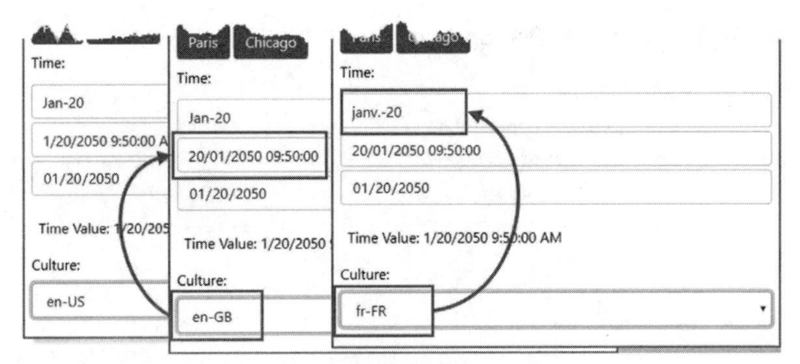


图 33-13　格式化 DateTime 值

**让浏览器格式化日期**

注意，代码清单 33-20 中第三个输入元素显示的值没有改变，无论选择的语言环境是什么。这个输入元素没有表 33-7 中描述的属性，但它的 type 属性设置为 Date，如下所示:

```
...
<input class="form-control" type="date" @bind="Time" />
...
```

将 type 属性设置为 date、datetime-local、month 或 time 时，不应指定区域或格式字符串，因为 Blazor 自动将日期值格式化为区域无关的格式，浏览器将其转换为用户的语言环境。图 33-11 显示了在 en-US 语言环境中如何格式化日期，但是用户将看到在他们的本地约定中表示的日期。

# 33.5　使用类文件定义组件

如果不喜欢 Razor 组件支持的代码和标记的混合，可以使用 C#类文件来定义一部分组件或全部组件。

## 33.5.1　使用代码隐藏类

Razor 组件的@code 部分可在单独的类文件中定义，称为代码隐藏类或代码隐藏文件。Razor 组件的代码隐藏类定义为部分类，其名称与为其提供代码的组件相同。

给 Blazor 文件夹添加名为 Split.razor 的 Razor 组件，内容如代码清单 33-21 所示。

**代码清单 33-21　Blazor 文件夹中 Split.razor 文件的内容**

```
<ul class="list-group">
    @foreach (string name in Names) {
        <li class="list-group-item">@name</li>
    }
</ul>
```

这个文件只包含 HTML 内容和 Razor 表达式，并呈现一个它希望通过 names 属性接收的名称列表。要为组件提供其代码，请将一个名为 Split.razor.cs 的类文件添加到 Blazor 文件夹中，并使用它定义如代码清单 33-22 所示的部分类。

### 代码清单 33-22　Blazor 文件夹中 Split.razor.cs 文件的内容

```
using Advanced.Models;
using Microsoft.AspNetCore.Components;
using System.Collections.Generic;
using System.Linq;

namespace Advanced.Blazor {

    public partial class Split {

        [Inject]
        public DataContext Context { get; set; }

        public IEnumerable<string> Names => Context.People.Select(p => p.Firstname);
    }
}
```

部分类必须在与其 Razor 组件相同的名称空间中定义，并且具有相同的名称。对于本例，这意味着名称空间是 Advanced.Blazor，类名是 Split。代码隐藏类不使用 Inject 属性定义构造函数和接收服务。代码清单 33-23 应用了新组件。

### 代码清单 33-23　在 Pages 文件夹的 Blazor.cshtml 文件中应用新的组件

```
@page "/pages/blazor"

<h4 class="bg-primary text-white text-center p-2">Code-Behind</h4>

<component type="typeof(Advanced.Blazor.Split)" render-mode="Server" />
```

重启 ASP.NET Core 并请求 http://localhost:5000/pages/blazor，将看到如图 33-14 所示的响应。

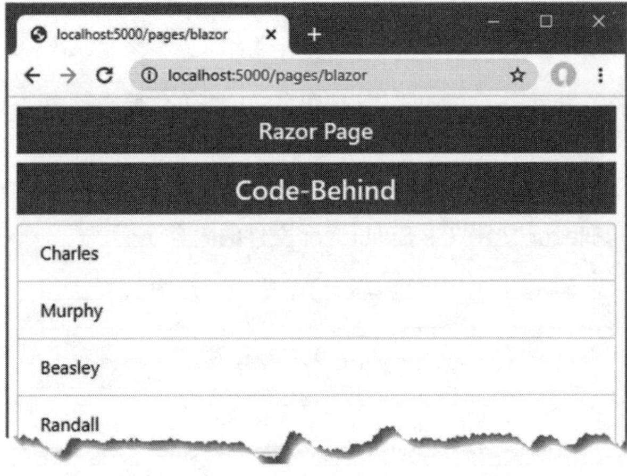


图 33-14　使用代码隐藏类定义 Razor 组件

### 33.5.2 定义 Razor 组件类

Razor 组件可以完全在类文件中定义，尽管这样做的表达性可能不如使用 Razor 表达式。将一个名为 CodeOnly.cs 的类文件添加到 Blazor 文件夹中，并使用它定义如代码清单 33-24 所示的类。

**代码清单 33-24　Blazor 文件夹中 CodeOnly.cs 文件的内容**

```
using Advanced.Models;
using Microsoft.AspNetCore.Components;
using Microsoft.AspNetCore.Components.Rendering;
using Microsoft.AspNetCore.Components.Web;
using System.Collections.Generic;
using System.Linq;

namespace Advanced.Blazor {

    public class CodeOnly : ComponentBase {

    [Inject]
    public DataContext Context { get; set; }

    public IEnumerable<string> Names => Context.People.Select(p => p.Firstname);

    public bool Ascending { get; set; } = false;

    protected override void BuildRenderTree(RenderTreeBuilder builder) {
        IEnumerable<string> data = Ascending
            ? Names.OrderBy(n => n) : Names.OrderByDescending(n => n);
            builder.OpenElement(1, "button");
            builder.AddAttribute(2, "class", "btn btn-primary mb-2");
            builder.AddAttribute(3, "onclick",
                EventCallback.Factory.Create<MouseEventArgs>(this,
                    () => Ascending = !Ascending));
            builder.AddContent(4, new MarkupString("Toggle"));
            builder.CloseElement();

            builder.OpenElement(5, "ul");
            builder.AddAttribute(6, "class", "list-group");
            foreach (string name in data) {
            builder.OpenElement(7, "li");
                builder.AddAttribute(8, "class", "list-group-item");
                builder.AddContent(9, new MarkupString(name));
                builder.CloseElement();
            }
            builder.CloseElement();
        }
    }
}
```

组件的基类是ComponentBase。通常表示为带注释的HTML元素的内容是通过覆盖BuildRenderTree方法并使用RenderTreeBuilder参数创建的。创建内容可能会很麻烦，因为每个元素都是使用多个代码语句创建和配置的，而且每个语句都必须有一个序列号，编译器使用这个序列号来匹配代码和内容。OpenElement方法启动一个新元素，该元素使用AddElement和AddContent方法配置，然后使用CloseElement方法完成。常规Razor组件中可用的所有特性都是可用的，包括事件和绑定，它们是通过向元素添加属性来设置的，就像在.Razor文件中定义的一样。代码清单33-24 中的组件显示排序后的名称列表，当单击按钮元素时，排序方向会改变。代码清单 33-25 应用了该组件，以便将其显示给用户。

**代码清单 33-25　在 Pages 文件夹的 Blazor.cshtml 上应用新的组件**

```
@page "/pages/blazor"

<h4 class="bg-primary text-white text-center p-2">Class Only</h4>

<component type="typeof(Advanced.Blazor.CodeOnly)" render-mode="Server" />
```

重启 ASP.NET Core，并请求 http://localhost:5000/pages/blazor，查看基于类的 Razor 组件生成的内容。当单击该按钮时，列表中名称的排序方向将发生改变，如图 33-15 所示。

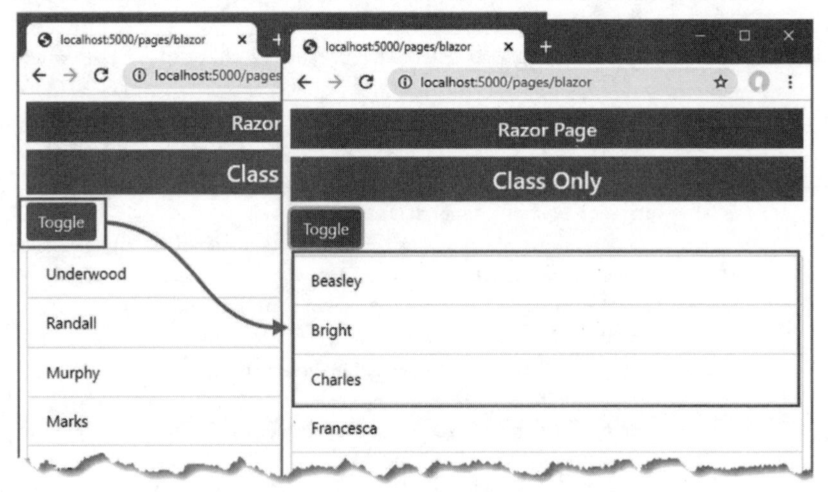


图 33-15　完全用代码定义组件

# 33.6　小结

本章介绍了 Blazor 服务器，解释了它解决的问题，并描述了它的优缺点。展示了如何配置 ASP.NET Core 应用程序来启用 Blazor 服务器，并展示了使用 Razor 组件时可用的基本特性，这些组件是 Blazor 的构建块。第 34 章将继续描述 Blazor 提供的特性。

# 第 34 章

# 使用 Blazor 服务器(第 2 部分)

本章继续描述 Blazor 服务器，重点介绍如何使用 Razor 组件来创建更复杂的特性。表 34-1 总结了本章的内容。

表 34-1 本章内容摘要

| 问题 | 解决方案 | 代码清单 |
| --- | --- | --- |
| 使用 Blazor 创建复杂的特性 | 组合组件来减少重复 | 34-3～34-4 |
| 配置组件 | 使用 Parameter 属性从属性中接收值 | 34-5～34-10 |
| 自定义事件和绑定 | 使用 EventCallbacks 来接收事件的处理程序，遵循创建绑定的约定 | 34-11～34-14 |
| 在组件中显示子内容 | 使用名为 ChildContent 的 RenderFragment | 34-15～34-16 |
| 创建模板 | 使用指定的 RenderFragment 属性 | 34-17～34-25 |
| 广泛地发布配置设置 | 使用级联参数 | 34-26～34-27 |
| 响应连接错误 | 使用连接元素和类 | 34-28～34-29 |
| 响应未处理的错误 | 使用 error 元素和类 | 34-30～34-31 |

## 34.1 准备工作

本章使用了第 33 章中的 Advanced 项目。准备本章不需要做任何更改。

■ **提示：**可以从 https://github.com/apress/pro-asp.net-core-3下载本章和本书中其他所有章节的示例项目。如果你在运行示例时遇到问题，请参阅第1章，以了解如何获得帮助。

打开新的 PowerShell 命令提示符，导航到包含 Advanced.csproj 文件的文件夹，并运行代码清单 34-1 所示的命令来删除数据库。

**代码清单 34-1 删除数据库**

```
dotnet ef database drop --force
```

从 Debug 菜单中选择 Start Without Debugging 或 Run Without Debugging，或者使用 PowerShell

命令提示符运行代码清单 34-2 中所示的命令。

**代码清单 34-2　运行示例应用程序**

```
dotnet run
```

使用浏览器请求http://localhost:5000/controllers，这将显示一个数据项列表。请求http://localhost:5000/pages/blazor，将看到第 33 章中演示数据绑定的组件。图 34-1 显示了这两种响应。

图 34-1　运行示例应用程序

## 34.2　结合组件

Blazor 组件可以组合起来创建更复杂的功能。接下来将展示如何使用多个组件，以及组件如何进行通信。首先，给 Blazor 文件夹添加一个名为 Razor 的组件，内容如代码清单 34-3 所示。

**代码清单 34-3　Blazor 文件夹中 SelectFilter.razor 文件的内容**

```
<div class="form-group">
    <label for="select-@Title">@Title</label>
    <select name="select-@Title" class="form-control" @bind="SelectedValue">
        <option disabled selected>Select @Title</option>
        @foreach (string val in Values) {
            <option value="@val" selected="@(val == SelectedValue)">
                @val
            </option>
        }
    </select>
</div>

@code {

    public IEnumerable<string> Values { get; set; } = Enumerable.Empty<string>();
```

```
    public string SelectedValue { get; set; }

    public string Title { get; set; } = "Placeholder";
}
```

该组件呈现一个select元素，该元素允许用户选择城市。在代码清单34-4中，应用了SelectFilter组件，替换现有的select元素。

### 代码清单 34-4 在 Blazor 文件夹的 PeopleList.razor 文件中应用组件

```
<table class="table table-sm table-bordered table-striped">
    <thead><tr><th>ID</th><th>Name</th><th>Dept</th><th>Location</th></tr></thead>
    <tbody>
        @foreach (Person p in People) {
            <tr class="@GetClass(p.Location.City)">
                <td>@p.PersonId</td>
                <td>@p.Surname, @p.Firstname</td>
                <td>@p.Department.Name</td>
                <td>@p.Location.City, @p.Location.State</td>
            </tr>
        }
    </tbody>
</table>

<SelectFilter />

@code {

    [Inject]
    public DataContext Context { get; set; }

    public IEnumerable<Person> People =>
        Context.People.Include(p => p.Department).Include(p => p.Location);

    public IEnumerable<string> Cities => Context.Locations.Select(l => l.City);

    public string SelectedCity { get; set; }

    public string GetClass(string city) =>
        SelectedCity == city ? "bg-info text-white" : "";
}
```

当组件添加到控制器视图或 Razor Pages 所呈现的内容中时，就会使用组件元素，如第 33 章所述。当一个组件添加到另一个组件所呈现的内容中时，该组件的名称将用作一个元素。本例向 PeopleList 组件所呈现的内容添加了 SelectFilter 组件，使用 SelectFilter 元素来实现。重要的是密切关注大小写，这必须准确匹配。

组合组件时，效果是一个组件将其部分布局的职责委托给另一个组件。本例删除了 PeopleList

组件用于向用户提供城市选择的 select 元素，并用 SelectFilter 组件替换它，后者将提供相同的功能。各组成部分构成父/子关系；PeopleList 组件是父组件，SelectFilter 组件是子组件。

在正确集成之前，还需要完成其他工作，但是可以看到，添加 SelectFilter 元素会通过重新启动 ASP.NET Core 来显示 SelectFilter 组件，并请求 http://localhost:5000/controllers，这会生成如图 34-2 所示的响应。

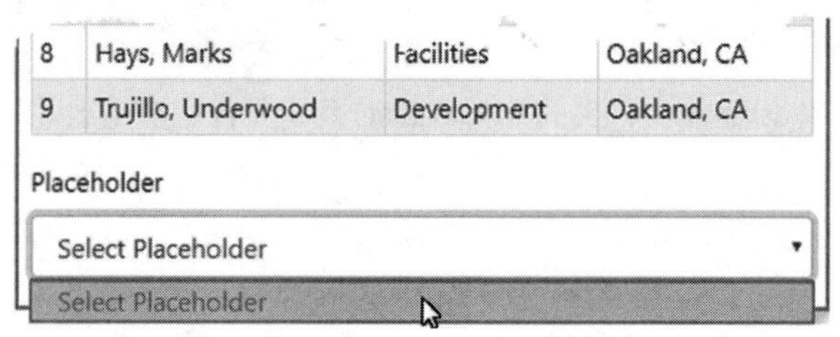


图 34-2　将一个组件添加到另一个组件所呈现的内容中

### 34.2.1　利用属性配置组件

使用 SelectList 组件的目标是创建一个可在整个应用程序中使用的通用特性，配置每次使用它时显示的值。Razor 组件是用属性配置的，这些属性添加到应用它们的 HTML 元素中。分配给 HTML 元素属性的值被分配给组件的 C#属性。Parameter 属性应用于组件允许配置的 C#属性，如代码清单 34-5 所示。

**代码清单 34-5　在 Blazor 文件夹的 SelectFilter.razor 文件中声明可配置属性**

```
<div class="form-group">
    <label for="select-@Title">@Title</label>
    <select name="select-@Title" class="form-control" @bind="SelectedValue">
        <option disabled selected>Select @Title</option>
        @foreach (string val in Values) {
            <option value="@val" selected="@(val == SelectedValue)">
                @val
            </option>
        }
    </select>
</div>

@code {
    [Parameter]
    public IEnumerable<string> Values { get; set; } = Enumerable.Empty<string>();

    public string SelectedValue { get; set; }

    [Parameter]
    public string Title { get; set; } = "Placeholder";
}
```

组件可以选择它们允许配置的属性。在本例中，Parameter 属性已应用于 SelectFilter 组件定义的两个属性。在代码清单 34-6 中，修改了 PeopleList 组件用于应用 SelectFilter 组件以添加配置属性的元素。

#### 代码清单 34-6 在 Blazor 文件夹的 PeopleList.razor 文件中配置组件

```
<table class="table table-sm table-bordered table-striped">
    <thead><tr><th>ID</th><th>Name</th><th>Dept</th><th>Location</th></tr></thead>
    <tbody>
        @foreach (Person p in People) {
            <tr class="@GetClass(p.Location.City)">
                <td>@p.PersonId</td>
                <td>@p.Surname, @p.Firstname</td>
                <td>@p.Department.Name</td>
                <td>@p.Location.City, @p.Location.State</td>
            </tr>
        }
    </tbody>
</table>

<SelectFilter values="@Cities" title="City" />

@code {

    [Inject]
    public DataContext Context { get; set; }

    public IEnumerable<Person> People =>
        Context.People.Include(p => p.Department).Include(p => p.Location);

    public IEnumerable<string> Cities => Context.Locations.Select(l => l.City);

    public string SelectedCity { get; set; }

    public string GetClass(string city) =>
        SelectedCity == city ? "bg-info text-white" : "";
}
```

对于每个应该配置的属性，将一个同名属性添加到父元素的 HTML 元素中。属性值可以是固定的值，比如分配给 title 属性的 City 字符串，也可以是 Razor 表达式，如@Cities(它将 City 属性中的对象序列分配给 values 属性)。

#### 1. 设置和接收批量配置设置

如果存在许多配置设置，特别是当组件接收这些值，以便将它们传递给子组件或常规 HTML 元素时，那么定义接收值的单个属性很容易出错。在这些情况下，可以指定一个属性来接收没有被其他属性匹配的任何属性值，然后可将其作为一个集合应用，如代码清单 34-7 所示。

#### 代码清单34-7 在Blazor文件夹的SelectFilter.razor文件中接收Bulk属性

```
<div class="form-group">
    <label for="select-@Title">@Title</label>
    <select name="select-@Title" class="form-control"
            @bind="SelectedValue" @attributes="Attrs">
        <option disabled selected>Select @Title</option>
        @foreach (string val in Values) {
            <option value="@val" selected="@(val == SelectedValue)">
                @val
            </option>
        }
    </select>
</div>

@code {

    [Parameter]
    public IEnumerable<string> Values { get; set; } = Enumerable.Empty<string>();

    public string SelectedValue { get; set; }

    [Parameter]
    public string Title { get; set; } = "Placeholder";

    [Parameter(CaptureUnmatchedValues = true)]
    public Dictionary<string, object> Attrs { get; set; }
}
```

将Parameter属性的CaptureUnmatchedValues参数设置为true，将标识一个属性作为未以其他方式匹配的属性的集合。属性的类型必须是Dictionary<string, object="">，这允许表示属性名称和值。

属性的类型是Dictionary<string，object >，可以使用@attribute表达式应用于元素，如下所示：

```
...
<select name="select-@Title" class="form-control" @bind="SelectedValue"
    @attributes="Attrs">
...
```

这称为属性splating，它允许一次性应用一组属性。代码清单34-7中更改的效果意味着SelectFilter组件接收Values和Title属性值，其他任何属性都分配给Attrs属性并传递给select元素。代码清单34-8添加了一些属性来演示这种效果。

#### 代码清单34-8 在Blazor文件夹的PeopleList.razor文件中添加元素属性

```
<table class="table table-sm table-bordered table-striped">
    <thead><tr><th>ID</th><th>Name</th><th>Dept</th><th>Location</th></tr></thead>
    <tbody>
```

```
        @foreach (Person p in People) {
            <tr class="@GetClass(p.Location.City)">
                <td>@p.PersonId</td>
                <td>@p.Surname, @p.Firstname</td>
                <td>@p.Department.Name</td>
                <td>@p.Location.City, @p.Location.State</td>
            </tr>
        }
    </tbody>
</table>

<SelectFilter values="@Cities" title="City" autofocus="true" name="city"
    required="true" />

@code {

    // ...statements omitted for brevity...
}
```

重启 ASP.NET Core，并导航到 http://localhost:5000/controllers。传递给 select 元素的属性不影响外观，但如果右击选择元素，并从弹出菜单中选择 Inspect，就将看到属性添加到 PeopleList 组件的 SelectFilter 元素中，SelectFilter 元素已添加到 SelectFilter 组件呈现的元素中，如下所示：

```
...
<select class="form-control" autofocus="true" name="city" required="true">
...
```

### 2. 在控制器视图或 Razor Pages 中配置组件

当使用组件元素应用组件时，属性还用于配置组件。在代码清单 34-9 中，向 PeopleList 组件添加了属性，该属性指定应该显示来自数据库的多少项，以及将传递给 SelectFilter 组件的字符串值。

**代码清单 34-9 在 Blazor 文件夹的 PeopleList.razor 文件中添加配置属性**

```
<table class="table table-sm table-bordered table-striped">
    <thead><tr><th>ID</th><th>Name</th><th>Dept</th><th>Location</th></tr></thead>
    <tbody>
        @foreach (Person p in People) {
            <tr class="@GetClass(p.Location.City)">
                <td>@p.PersonId</td>
                <td>@p.Surname, @p.Firstname</td>
                <td>@p.Department.Name</td>
                <td>@p.Location.City, @p.Location.State</td>
            </tr>
        }
    </tbody>
</table>
```

```
<SelectFilter values="@Cities" title="@SelectTitle" />

@code {

    [Inject]
    public DataContext Context { get; set; }

    public IEnumerable<Person> People => Context.People.Include(p => p.Department)
            .Include(p => p.Location).Take(ItemCount);

    public IEnumerable<string> Cities => Context.Locations.Select(l => l.City);

    public string SelectedCity { get; set; }

    public string GetClass(string city) =>
        SelectedCity == city ? "bg-info text-white" : "";

    [Parameter]
    public int ItemCount { get; set; } = 4;

    [Parameter]
    public string SelectTitle { get; set; }
}
```

C#属性的值是通过将名称以 param-开头、后跟属性名称的属性添加到组件元素来提供的，如代码清单 34-10 所示。

### 代码清单 34-10　在 Views/Home 文件夹的 Index.cshtm 文件中添加配置属性

```
@model PeopleListViewModel

<h4 class="bg-primary text-white text-center p-2">People</h4>

<component type="typeof(Advanced.Blazor.PeopleList)" render-mode="Server"
        param-itemcount="5" param-selecttitle="@("Location")" />
```

param-itemcount 属性为 ItemCount 属性提供一个值，param-selecttitle 属性为 SelectTitle 属性提供一个值。

在使用组件元素时，可以解析为数值或 bool 值的属性值处理为文字值而不是 Razor 表达式，这就是为什么能够将 ItemCount 属性的值指定为 4 的原因。其他值假定为 Razor 表达式而不是文字值，即使它们没有以@作为前缀。这意味着，为将 SelectTitle 属性的值指定为文字字符串，就需要一个 Razor 表达式，像这样:

```
...
<component type="typeof(Advanced.Blazor.PeopleList)" render-mode="Server"
        param-itemcount="5" param-selecttitle="@("Location")" />
...
```

要查看配置属性的效果，请重新启动 ASP.NET Core，并请求 http://localhost:5000/controllers，这将生成如图 34-3 所示的响应。

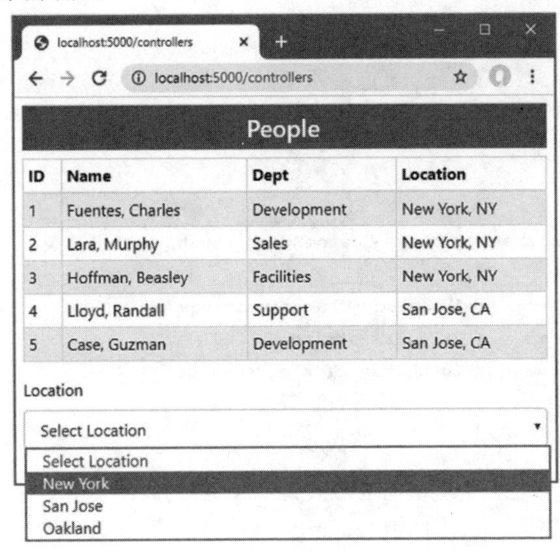

图 34-3　用属性配置组件

### 34.2.2　创建自定义事件和绑定

SelectFilter 组件从父组件接收数据值，但它无法指示用户何时进行选择。为此，需要创建一个自定义事件，父组件可为其注册一个处理程序方法，就像它为常规 HTML 元素中的事件所做的那样。代码清单 34-11 向 SelectFilter 组件添加了一个自定义事件。

**代码清单 34-11　在 Blazor 文件夹的 SelectFilter.razor 文件中创建事件**

```
<div class="form-group">
    <label for="select-@Title">@Title</label>
    <select name="select-@Title" class="form-control"
            @onchange="HandleSelect" value="@SelectedValue">
        <option disabled selected>Select @Title</option>
        @foreach (string val in Values) {
            <option value="@val" selected="@(val == SelectedValue)">
                @val
            </option>
        }
    </select>
</div>

@code {

    [Parameter]
    public IEnumerable<string> Values { get; set; } = Enumerable.Empty<string>();
```

```
    public string SelectedValue { get; set; }

    [Parameter]
    public string Title { get; set; } = "Placeholder";

    [Parameter(CaptureUnmatchedValues = true)]
    public Dictionary<string, object> Attrs { get; set; }

    [Parameter]
    public EventCallback<string> CustomEvent { get; set; }

    public async Task HandleSelect(ChangeEventArgs e) {
        SelectedValue = e.Value as string;
        await CustomEvent.InvokeAsync(SelectedValue);
    }
}
```

自定义事件是通过添加一个类型为 EventCallback<T>的属性来定义的。泛型类型参数是父类的事件处理程序接收的类型，在本例中为 string。前面更改了 select 元素，所以当 select 元素触发其 onchange 事件时，@onchange 属性会注册 HandleSelect 方法。

通过调用 EventCallback<T>.InvokeAsync 方法，HandleSelect 方法会更新 SelectedValue 属性，如下所示：

```
...
await CustomEvent.InvokeAsync(SelectedValue);
...
```

InvokeAsync方法的参数使用从ChangeEventArgs对象接收到的值(从select元素接收到的值)触发事件。代码清单 34-12 更改了PeopleList组件，以便它接收由SelectList组件发出的定制事件。

#### 代码清单 34-12　在 Blazor 文件夹的 PeopleList.razor 文件中处理事件

```
<table class="table table-sm table-bordered table-striped">
    <thead><tr><th>ID</th><th>Name</th><th>Dept</th><th>Location</th></tr></thead>
    <tbody>
        @foreach (Person p in People) {
            <tr class="@GetClass(p.Location.City)">
                <td>@p.PersonId</td>
                <td>@p.Surname, @p.Firstname</td>
                <td>@p.Department.Name</td>
                <td>@p.Location.City, @p.Location.State</td>
            </tr>
        }
    </tbody>
</table>

<SelectFilter values="@Cities" title="@SelectTitle" CustomEvent="@HandleCustom" />
```

```
@code {

    [Inject]
    public DataContext Context { get; set; }

    public IEnumerable<Person> People => Context.People.Include(p => p.Department)
            .Include(p => p.Location).Take(ItemCount);

    public IEnumerable<string> Cities => Context.Locations.Select(l => l.City);

    public string SelectedCity { get; set; }

    public string GetClass(string city) =>
        SelectedCity as string == city ? "bg-info text-white" : "";

    [Parameter]
    public int ItemCount { get; set; } = 4;

    [Parameter]
    public string SelectTitle { get; set; }

    public void HandleCustom(string newValue) {
        SelectedCity = newValue;
    }
}
```

要设置事件处理程序，需要向使用 EventCallback<T>属性名称应用子组件的元素添加一个属性。属性的值是一个 Razor 表达式，它选择接收类型为 T 的参数的方法。

重启 ASP.NET Core，请求 http://localhost:5000/controllers，并从城市列表中选择一个值。自定义事件完成父组件和子组件之间的关系。父元素通过其属性配置子元素，以指定呈现给用户的标题和数据值列表。子组件使用自定义事件来告诉父组件，当用户选择一个值时，允许父对象在其 HTML 表中突出显示相应的行，如图 34-4 所示。

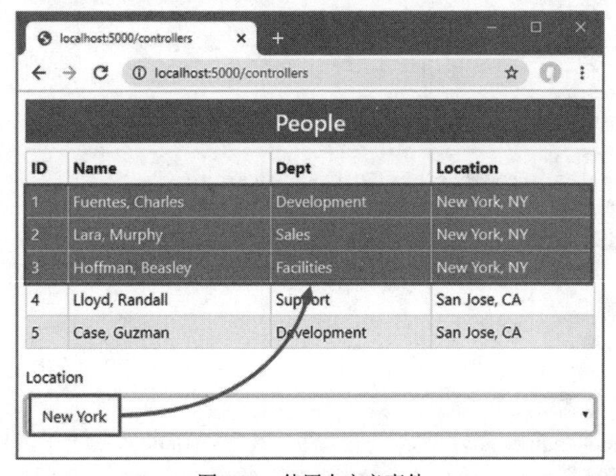


图 34-4 使用自定义事件

### 创建自定义绑定

父组件可以在子组件上创建绑定，前提是它定义了一对属性，其中一个属性被分配一个数据值，另一个属性是自定义事件。属性的名称很重要：事件属性的名称必须与data属性的名称加单词Changed相同。代码清单34-13更新了SelectFilter组件，以便它显示绑定所需的属性。

**代码清单34-13　在Blazor文件夹的SelectFilter.razor文件中准备自定义绑定**

```
<div class="form-group">
    <label for="select-@Title">@Title</label>
    <select name="select-@Title" class="form-control"
            @onchange="HandleSelect" value="@SelectedValue">
        <option disabled selected>Select @Title</option>
        @foreach (string val in Values) {
            <option value="@val" selected="@(val == SelectedValue)">
                @val
            </option>
        }
    </select>
</div>

@code {

    [Parameter]
    public IEnumerable<string> Values { get; set; } = Enumerable.Empty<string>();

    [Parameter]
    public string SelectedValue { get; set; }

    [Parameter]
    public string Title { get; set; } = "Placeholder";

    [Parameter(CaptureUnmatchedValues = true)]
    public Dictionary<string, object> Attrs { get; set; }

    [Parameter]
    public EventCallback<string> SelectedValueChanged { get; set; }

    public async Task HandleSelect(ChangeEventArgs e) {
        SelectedValue = e.Value as string;
        await SelectedValueChanged.InvokeAsync(SelectedValue);
    }
}
```

注意，Parameter属性必须同时应用于SelectedValue和SelectedValueChanged属性。如果省略其中任何一个属性，数据绑定将无法按预期工作。

父组件通过@bind-<name>属性绑定到子组件，其中<name>对应于子组件定义的属性。在本例中，子组件的属性名是SelectedValue，父组件可以使用@bind-SelectedValue创建绑定，如代码清单 34-14 所示。

**代码清单 34-14　在 Blazor 文件夹 PeopleList.razor 文件中使用自定义绑定**

```
<table class="table table-sm table-bordered table-striped">
    <thead><tr><th>ID</th><th>Name</th><th>Dept</th><th>Location</th></tr></thead>
    <tbody>
        @foreach (Person p in People) {
            <tr class="@GetClass(p.Location.City)">
                <td>@p.PersonId</td>
                <td>@p.Surname, @p.Firstname</td>
                <td>@p.Department.Name</td>
                <td>@p.Location.City, @p.Location.State</td>
            </tr>
        }
    </tbody>
</table>

<SelectFilter values="@Cities" title="@SelectTitle"
    @bind-SelectedValue="SelectedCity" />

<button class="btn btn-primary"
    @onclick="@(() => SelectedCity = "San Jose")">
        Change
</button>

@code {

    [Inject]
    public DataContext Context { get; set; }

    public IEnumerable<Person> People => Context.People.Include(p => p.Department)
            .Include(p => p.Location).Take(ItemCount);

    public IEnumerable<string> Cities => Context.Locations.Select(l => l.City);

    public string SelectedCity { get; set; }

    public string GetClass(string city) =>
        SelectedCity as string == city ? "bg-info text-white" : "";

    [Parameter]
    public int ItemCount { get; set; } = 4;

    [Parameter]
    public string SelectTitle { get; set; }
```

```
        //public void HandleCustom(string newValue) {
        // SelectedCity = newValue;
        //}
    }
```

重启 ASP.NET Core，请求 http://localhost:5000/controllers，并从城市列表中选择 New York。自定义绑定将导致 select 元素中选择的值通过表中的高亮显示反映出来。单击 Change 按钮以在另一个方向测试绑定，城市的更改将高亮显示，如图 34-5 所示。

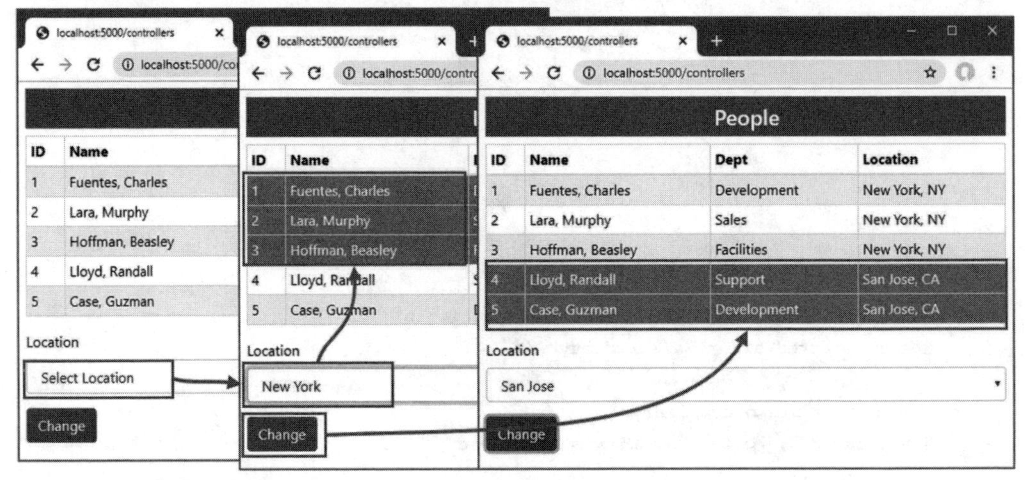


图 34-5　使用自定义绑定

# 34.3　在组件中显示子内容

显示子内容的组件充当父元素提供的元素的包装器。要查看子内容是如何管理的，请给 Blazor 文件夹添加名为 ThemeWrapper.razor 的 Razor 组件，内容如代码清单 34-15 所示。

### 代码清单 34-15　在 Blazor 文件夹中 ThemeWrapper.razor 的内容

```
<div class="p-2 bg-@Theme border text-white">
    <h5 class="text-center">@Title</h5>
    @ChildContent
</div>

@code {
    [Parameter]
    public string Theme { get; set; }

    [Parameter]
    public string Title { get; set; }

    [Parameter]
    public RenderFragment ChildContent { get; set; }
}
```

为接收子内容，组件定义了一个名为 ChildContent 的属性，它的类型是 RenderFragment，并且已经用 Parameter 属性进行了修饰。@ChildContent 表达式在组件的 HTML 输出中包含子内容。代码清单中的组件将其子内容包装在 div 元素中，div 元素使用引导主题颜色进行样式化，并显示标题。主题颜色的名称和标题的文本也作为参数接收。

### 限制元素的重用

当更新呈现给用户的内容时，Blazor会尽可能重用元素，因为创建新元素是一个相对昂贵的操作。在显示一系列值的元素时尤其如此，例如@for或@foreach表达式。如果序列发生变化，Blazor将重用它为旧数据值创建的元素来显示新数据。

如果对 Blazor 控制之外的元素进行了更改，比如对定制 JavaScript 代码的更改，则会导致问题。Blazor 不知道这些变化，当元素被重用时这些变化将会持续存在。尽管这种情况很少见，但可以通过使用@key 属性并提供一个表达式来限制元素的重用，该表达式将元素与序列中的一个数据值关联起来，如下所示:

```
...
@foreach (Person p in People) {
    <tr @key="p.PersonId" class="@GetClass(p.Location.City)">

        <td>@p.PersonId</td>
        <td>@p.Surname, @p.Firstname</td>
        <td>@p.Department.Name</td>
        <td>@p.Location.City, @p.Location.State</td>
    </tr>
}
...
```

Blazor 只在有具有相同键的数据项时才会重用元素。对于其他值，将创建新的元素。

子内容是通过在应用组件时在开始和结束标记之间添加HTML元素来定义的，如代码清单 34-16 所示。

#### 代码清单 34-16 在 Blazor 文件夹的 PeopleList.razor 文件中定义子内容

```
<table class="table table-sm table-bordered table-striped">
    <thead><tr><th>ID</th><th>Name</th><th>Dept</th><th>Location</th></tr></thead>
    <tbody>
        @foreach (Person p in People) {
            <tr class="@GetClass(p.Location.City)">
                <td>@p.PersonId</td>
                <td>@p.Surname, @p.Firstname</td>
                <td>@p.Department.Name</td>
                <td>@p.Location.City, @p.Location.State</td>
            </tr>
        }
    </tbody>
</table>
```

```
<ThemeWrapper Theme="info" Title="Location Selector">
    <SelectFilter values="@Cities" title="@SelectTitle"
        @bind-SelectedValue="SelectedCity" />
    <button class="btn btn-primary"
        @onclick="@(() => SelectedCity = "San Jose")">
            Change
    </button>
</ThemeWrapper>

@code {

    // ...statements omitted for brevity...
}
```

配置子内容不需要附加属性，子内容将被自动处理并分配给ChildContent属性。要查看ThemeWrapper组件如何显示它的子内容，请重启ASP.NET Core，请求http://localhost:5000/controllers。将看到选择主题的配置属性和用于生成响应的标题文本，如图 34-6 所示。

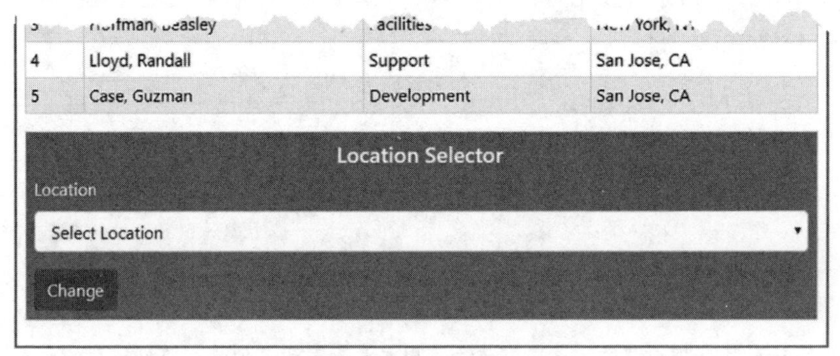


图 34-6　使用子内容

## 34.3.1　创建模板组件

模板组件为子内容的表示带来了更多的结构，允许显示内容的多个部分。模板组件可以很好地整合应用程序中使用的特性，以防止代码和内容的重复。

要了解其工作原理，请给 Blazor 文件夹添加一个名为 TableTemplate.razor 的 Razor 组件，内容如代码清单 34-17 所示。

### 代码清单 34-17　在 Blazor 文件夹中 TableTemplate.razor 文件的内容

```
<table class="table table-sm table-bordered table-striped">
    @if (Header != null) {
        <thead>@Header</thead>
    }
    <tbody>@Body</tbody>
</table>

@code {
```

```
    [Parameter]
    public RenderFragment Header { get; set; }

    [Parameter]
    public RenderFragment Body { get; set; }
}
```

组件为它支持的每个子内容区域定义一个 RenderFragment 属性。TableTemplate 组件定义了两个 RenderFragment 属性，命名为 Header 和 Body，它们代表了表格的内容部分。子内容的每个区域都是使用 Razor 表达式@Header 和@Body 呈现的，可以通过检查属性值是否为 null 来检查是否为特定部分提供了内容，该组件为 Header 部分执行此操作。

当使用模板组件时，每个区域的内容都包含在一个 HTML 元素中，该元素的标记与对应的 RenderFragment 属性的名称相匹配，如代码清单 34-18 所示。

#### 代码清单 34-18 在 Blazor 文件夹的 PeopleList.razor 文件中应用模板组件

```
<TableTemplate>
    <Header>
        <tr><th>ID</th><th>Name</th><th>Dept</th><th>Location</th></tr>
    </Header>
    <Body>
        @foreach (Person p in People) {
            <tr class="@GetClass(p.Location.City)">
                <td>@p.PersonId</td>
                <td>@p.Surname, @p.Firstname</td>
                <td>@p.Department.Name</td>
                <td>@p.Location.City, @p.Location.State</td>
            </tr>
        }
    </Body>
</TableTemplate>

<ThemeWrapper Theme="info" Title="Location Selector">
    <SelectFilter values="@Cities" title="@SelectTitle"
        @bind-SelectedValue="SelectedCity" />
    <button class="btn btn-primary"
        @onclick="@(() => SelectedCity = "San Jose")">
            Change
    </button>
</ThemeWrapper>

@code {

    // ...statements omitted for brevity...
}
```

子内容构造成与模板组件的属性、标题和主体对应的部分，这使得 TableTemplate 组件负责表结构，而 PeopleList 组件负责提供详细信息。重启 ASP.NET Core 并请求 http://localhost:5000/controllers，将看到模板组件生成的输出，如图 34-7 所示。

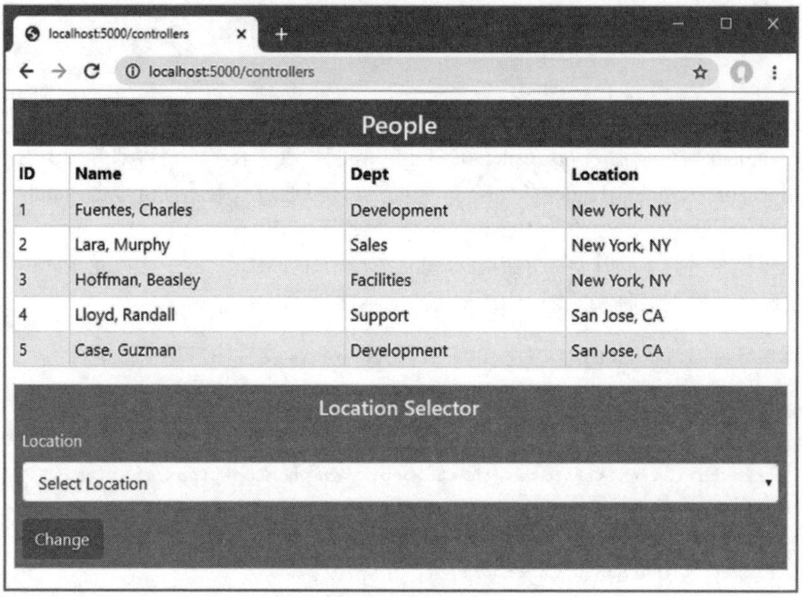

图 34-7　使用模板组件

## 34.3.2　在模板组件中使用泛型类型参数

上一节创建的模板组件很有用，因为它提供了表的一致表示，可在整个示例应用程序中使用它。但它也有局限性，因为它依赖父组件来负责为表主体生成行。模板组件对它呈现的内容没有任何洞察力，这意味着它不能对这些内容做任何事情，除了显示它。

模板组件可通过使用泛型类型参数来感知数据，泛型类型参数允许父组件提供一系列数据对象和用于表示这些对象的模板。模板组件负责为每个数据对象生成内容，因此可以提供更有用的功能。作为演示，下面将向模板组件添加对选择显示多少表行和选择表行的支持。第一步是向组件添加一个泛型类型参数，并使用它呈现表主体的内容，如代码清单 34-19 所示。

**代码清单 34-19　在 Blazor 文件夹的 TableTemplate.razor 文件中添加泛型类型参数**

```
@typeparam RowType

<table class="table table-sm table-bordered table-striped">
    @if (Header != null) {
        <thead>@Header</thead>
    }
    <tbody>
        @foreach (RowType item in RowData) {
            <tr>@RowTemplate(item)</tr>
        }
    </tbody>
```

```
</table>

@code {
    [Parameter]
    public RenderFragment Header { get; set; }

    [Parameter]
    public RenderFragment<RowType> RowTemplate{ get; set; }

    [Parameter]
    public IEnumerable<RowType> RowData { get; set; }
}
```

泛型类型参数是使用@typeparam属性指定的，本例为参数指定了名称RowType，因为引用组件将为其生成表行的数据类型。

组件处理的数据是通过添加一个属性来接收的，该属性的类型是泛型类型的对象序列。将属性命名为 RowData，其类型为 IEnumerable<RowType>。组件为每个对象显示的内容是使用RenderFragment<T> 属性接收的。将这个属性命名为 RowTemplate，它的类型是RenderFragment<RowType>，反映了为泛型类型参数选择的名称。

当组件通过 RenderFragment<T>属性接收到一个内容区段时，可通过调用该区段作为方法并使用该对象作为参数，来为单个对象呈现它，如下所示：

```
...
@foreach (RowType item in RowData) {
    <tr>@RowTemplate(item)</tr>
}
...
```

这段代码枚举 RowData 序列中的 RowType 对象，并为每个对象呈现通过 RowTemplate 属性接收到的内容部分。

#### 1. 使用通用模板组件

前面简化了 PeopleList 组件，使其仅使用模板组件生成 Person 对象表，并且删除了以前的特性，如代码清单 34-20 所示。

**代码清单 34-20　在 Blazor 文件夹的 PeopleList.razor 文件中使用通用模板组件**

```
<TableTemplate RowType="Person" RowData="People">
    <Header>
        <tr><th>ID</th><th>Name</th><th>Dept</th><th>Location</th></tr>
    </Header>
    <RowTemplate Context="p">
        <td>@p.PersonId</td>
        <td>@p.Surname, @p.Firstname</td>
        <td>@p.Department.Name</td>
        <td>@p.Location.City, @p.Location.State</td>
    </RowTemplate>
```

```
</TableTemplate>

@code {

    [Inject]
    public DataContext Context { get; set; }

    public IEnumerable<Person> People => Context.People
            .Include(p => p.Department)
            .Include(p => p.Location);
}
```

RowType 属性用于指定泛型类型参数的值。RowData 属性指定模板组件要处理的数据。

RowTemplate 元素表示为每个数据对象生成的元素。当为 RenderFragment<T>属性定义一个内容区段时，Context 属性用于为当前处理的对象分配一个名称。在本例中，使用 Context 属性将名称 p 分配给当前对象，然后在用于填充内容部分元素的 Razor 表达式中引用该对象。

总体效果是将模板组件配置为显示 Person 对象。该组件为每个 Person 生成一个表行，其中包含 td 元素，其内容是使用当前 Person 对象的属性设置的。

因为代码清单 34-20 删除了用 Parameter 属性装饰的属性，所以需要从应用 PepleList 组件的元素中删除相应的属性，如代码清单 34-21 所示。

### 代码清单 34-21　在 Views/Home 文件夹的 Index.cshtml 文件中删除属性

```
@model PeopleListViewModel

<h4 class="bg-primary text-white text-center p-2">People</h4>

<component type="typeof(Advanced.Blazor.PeopleList)" render-mode="Server" />
```

要查看通用模板组件，请重启 ASP.NET Core，请求 http://localhost:5000/controllers。PeopleList 组件提供的数据和内容部分已被 TableTemplate 组件用于生成图 34-8 所示的表。

localhost:5000/controllers

People

| ID | Name | Dept | Location |
|---|---|---|---|
| 1 | Fuentes, Charles | Development | New York, NY |
| 2 | Lara, Murphy | Sales | New York, NY |
| 3 | Hoffman, Beasley | Facilities | New York, NY |
| 4 | Lloyd, Randall | Support | San Jose, CA |
| 5 | Case, Guzman | Development | San Jose, CA |
| 6 | Jacobs, Francesca | Development | Oakland, CA |
| 7 | Becker, Bright | Facilities | Oakland, CA |
| 8 | Hays, Marks | Facilities | Oakland, CA |
| 9 | Trujillo, Underwood | Development | Oakland, CA |

图 34-8　使用通用模板组件

### 2. 向通用模板组件添加特性

这听起来像是倒退了一步，但可看到，让模板组件了解它处理的数据为添加特性奠定了基础，如代码清单 34-22 所示。

**代码清单 34-22　在 Blazor 文件夹的 TableTemplate.razor 文件中添加特性**

```
@typeparam RowType

<div class="container-fluid">
    <div class="row">
        <div class="col">
            <SelectFilter Title="@("Sort")" Values="@SortDirectionChoices"
                @bind-SelectedValue="SortDirectionSelection" />
        </div>
        <div class="col">
            <SelectFilter Title="@("Highlight")" Values="@HighlightChoices()"
                @bind-SelectedValue="HighlightSelection" />
        </div>
    </div>
</div>

<table class="table table-sm table-bordered table-striped">
    @if (Header != null) {
        <thead>@Header</thead>
    }
    <tbody>
        @foreach (RowType item in SortedData()) {
            <tr class="@IsHighlighted(item)">@RowTemplate(item)</tr>
        }
    </tbody>
</table>

@code {
    [Parameter]
    public RenderFragment Header { get; set; }

    [Parameter]
    public RenderFragment<RowType> RowTemplate{ get; set; }

    [Parameter]
    public IEnumerable<RowType> RowData { get; set; }

    [Parameter]
    public Func<RowType, string> Highlight { get; set; }

    public IEnumerable<string> HighlightChoices() =>
        RowData.Select(item => Highlight(item)).Distinct();
```

```
    public string HighlightSelection { get; set; }

    public string IsHighlighted(RowType item) =>
        Highlight(item) == HighlightSelection ? "bg-dark text-white": "";

    [Parameter]
    public Func<RowType, string> SortDirection { get; set; }

    public string[] SortDirectionChoices =
        new string[] { "Ascending", "Descending" };

    public string SortDirectionSelection{ get; set; } = "Ascending";

    public IEnumerable<RowType> SortedData() =>
        SortDirectionSelection == "Ascending"
            ? RowData.OrderBy(SortDirection)
            : RowData.OrderByDescending(SortDirection);
}
```

这些更改为用户提供了两个选择元素，这两个元素是使用本章前面创建的 SelectFilter 组件显示的。这些新元素允许用户按升序和降序对数据排序，并选择用于突出显示表行的值。父组件提供了额外参数，为模板组件函数提供功能，来选择用于排序和突出显示的属性，如代码清单 34-23 所示。

#### 代码清单 34-23　在 Blazor 文件夹的 PeopleList.razor 文件中配置模板组件特性

```
<TableTemplate RowType="Person" RowData="People"
        Highlight="@(p => p.Location.City)" SortDirection="@(p => p.Surname)">
    <Header>
        <tr><th>ID</th><th>Name</th><th>Dept</th><th>Location</th></tr>
    </Header>
    <RowTemplate Context="p">
        <td>@p.PersonId</td>
        <td>@p.Surname, @p.Firstname</td>
        <td>@p.Department.Name</td>
        <td>@p.Location.City, @p.Location.State</td>
    </RowTemplate>
</TableTemplate>

@code {

    [Inject]
    public DataContext Context { get; set; }

    public IEnumerable<Person> People => Context.People
            .Include(p => p.Department)
```

```
            .Include(p => p.Location);
    }
```

Highlight 属性为模板组件提供一个函数，该函数选择用于突出显示表行的属性，SortDirection 属性提供一个函数，该函数选择用于排序的属性。要查看效果，请重新启动 ASP.NET Core，请求 http://localhost:5000/controllers。响应将包含新的 select 元素，这些元素可用于更改排序顺序或选择要过滤的城市，如图 34-9 所示。

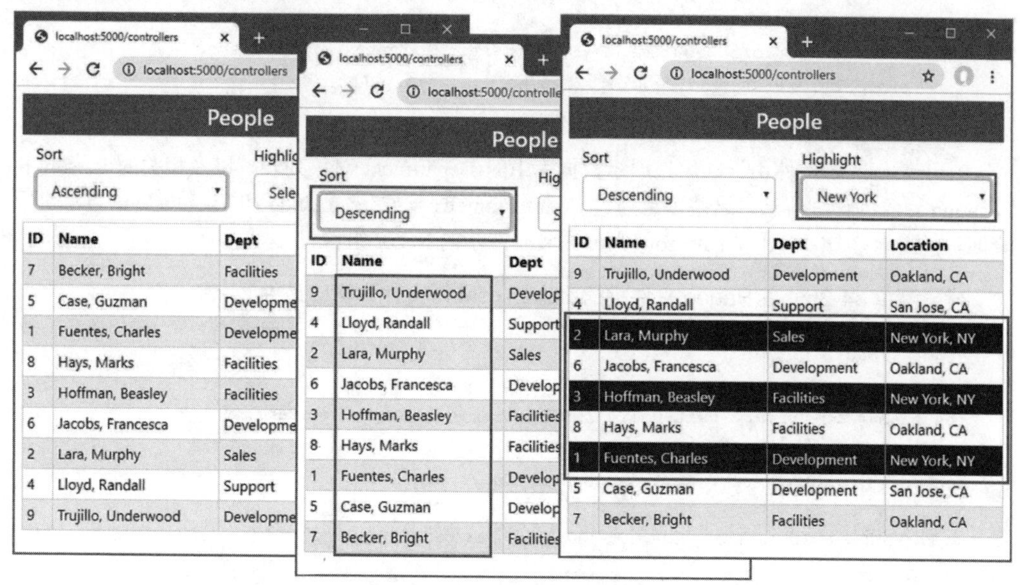

图 34-9 向模板组件添加特性

### 3. 重用通用模板组件

添加到模板组件的特性都依赖于泛型类型参数，允许组件修改它呈现的内容，而不绑定到特定的类。结果是一个组件，可用于在需要表的地方显示、排序和突出显示任何数据类型。给 Blazor 文件夹添加一个名为 DepartmentList.razor 的 Razor 组件，内容如代码清单 34-24 所示。

#### 代码清单 34-24 Blazor 文件夹中 DepartmentList.razor 文件的内容

```
<TableTemplate RowType="Department" RowData="Departments"
    Highlight="@(d => d.Name)"
    SortDirection="@(d => d.Name)">
    <Header>
        <tr><th>ID</th><th>Name</th><th>People</th><th>Locations</th></tr>
    </Header>
    <RowTemplate Context="d">
        <td>@d.Departmentid</td>
        <td>@d.Name</td>
        <td>@(String.Join(", ", d.People.Select(p => p.Surname)))</td>
        <td>
            @(String.Join(", ", d.People.Select(p => p.Location.City).Distinct()))
        </td>
```

```
    </RowTemplate>
</TableTemplate>

@code {

    [Inject]
    public DataContext Context { get; set; }

    public IEnumerable<Department> Departments => Context.Departments
            .Include(d => d.People).ThenInclude(p => p.Location);
}
```

TableTemplate 组件用于向用户展示数据库中的 Department 对象列表，以及相关的 Person 和 Location 对象的详细信息，这些信息通过 Entity Framework Core Include 和 ThenInclude 方法进行查询。代码清单 34-25 更改了 Razor Pages Blazor 显示的 Razor 组件。

**代码清单 34-25　在 Pages 文件夹的 Blazor.cshtml 文件中更改组件**

```
@page "/pages/blazor"

<h4 class="bg-primary text-white text-center p-2">Departments</h4>

<component type="typeof(Advanced.Blazor.DepartmentList)" render-mode="Server" />
```

重启 ASP.NET Core，请求 http://localhost:5000/pages/blazor。使用模板化组件显示响应，如图 34-10 所示。

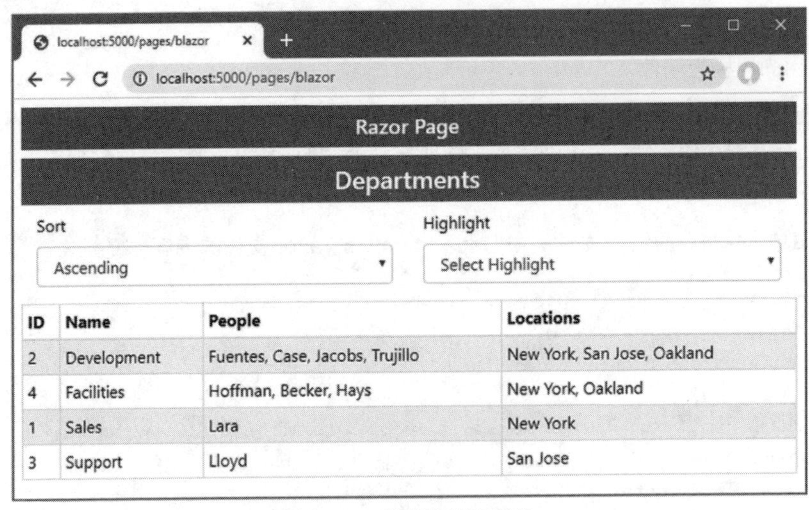


图 34-10　重用通用模板组件

### 34.3.3　级联参数

随着组件数量的增加，组件可向组件层次结构中的后代提供配置数据。这可以通过让链中的每个组件接收数据并将其传递给子组件来实现。但这很容易出错，并且需要每个组件都参与到这个过程中来，即使没有任何子组件使用它传递的数据。

Blazor 通过支持级联参数，提供了这个问题的解决方案，在级联参数中，组件提供的数据值可直接用于它的任何后代，而不由中间组件传递。级联参数是使用 CascadingValue 组件定义的，该组件用于包装部分内容，如代码清单 34-26 所示。

**代码清单 34-26 在 Blazor 文件夹的 DepartmentList.razor 文件中创建级联参数**

```
<CascadingValue Name="BgTheme" Value="Theme" IsFixed="false" >
    <TableTemplate RowType="Department" RowData="Departments"
        Highlight="@(d => d.Name)"
        SortDirection="@(d => d.Name)">
        <Header>
            <tr><th>ID</th><th>Name</th><th>People</th><th>Locations</th></tr>
        </Header>
        <RowTemplate Context="d">
            <td>@d.Departmentid</td>
            <td>@d.Name</td>
            <td>@(String.Join(", ", d.People.Select(p => p.Surname)))</td>
            <td>
                @(String.Join(", ",
                    d.People.Select(p => p.Location.City).Distinct()))
            </td>
        </RowTemplate>
    </TableTemplate>
</CascadingValue>

<SelectFilter Title="@("Theme")" Values="Themes" @bind-SelectedValue="Theme" />

@code {

    [Inject]
    public DataContext Context { get; set; }

    public IEnumerable<Department> Departments => Context.Departments
            .Include(d => d.People).ThenInclude(p => p.Location);
    public string Theme { get; set; } = "info";
    public string[] Themes = new string[] { "primary", "info", "success" };
}
```

CascadingValue 元素为它所包含的组件及其后代提供一个可用的值。Name 属性指定参数的名称，Value 属性指定值，isFixed 属性用于指定值是否会更改。代码清单 34-26 中使用了 CascadingValue 元素来创建名为 BgTheme 的级联参数，它的值由 SelectFilter 组件的一个实例设置，该组件向用户提供了一组引导 CSS 主题名。

---

■ 提示：每个CascadingValue元素创建一个级联参数。如果需要传递多个值，则可以嵌套CascadingValue或创建一个简单参数，通过字典提供多个设置。

---

使用 CascadingParameter 属性的组件直接接收级联参数，如代码清单 34-27 所示。

**代码清单 34-27　在 Blazor 文件夹的 SelectFilter.razor 文件中接收级联参数**

```
<div class="form-group p-2 bg-@Theme @TextColor()">
    <label for="select-@Title">@Title</label>
    <select name="select-@Title" class="form-control"
            @onchange="HandleSelect" value="@SelectedValue">
        <option disabled selected>Select @Title</option>
        @foreach (string val in Values) {
            <option value="@val" selected="@(val == SelectedValue)">
                @val
            </option>
        }
    </select>
</div>

@code (

    [Parameter]
    public IEnumerable<string> Values { get; set; } = Enumerable.Empty<string>();

    [Parameter]
    public string SelectedValue { get; set; }

    [Parameter]
    public string Title { get; set; } = "Placeholder";

    [Parameter(CaptureUnmatchedValues = true)]
    public Dictionary<string, object> Attrs { get; set; }

    [Parameter]
    public EventCallback<string> SelectedValueChanged { get; set; }

    public async Task HandleSelect(ChangeEventArgs e) {
        SelectedValue = e.Value as string;
        await SelectedValueChanged.InvokeAsync(SelectedValue);
    }

    [CascadingParameter(Name ="BgTheme")]
    public string Theme { get; set; }

    public string TextColor() => Theme == null ? "" : "text-white";
}
```

CascadingParameter 属性的 Name 参数用于指定级联参数的名称。代码清单 34-26 中定义的 BgTheme 参数由代码清单 34-27 中的 Theme 属性接收，并用于设置组件的背景。重启 ASP.NET Core

并请求 http://localhost:5000/pages/blazor，这会生成如图 34-11 所示的响应。

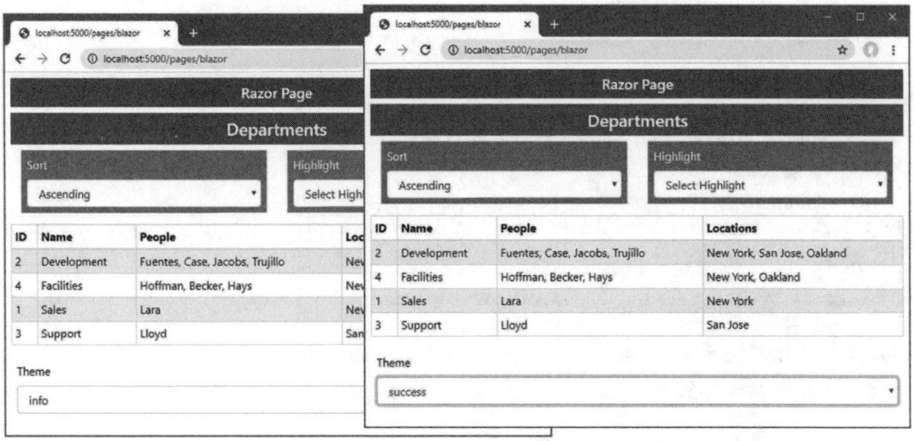

图 34-11 使用级联参数

本例中使用了三个 SelectFilter 组件实例，但其中只有两个在 CascadingValue 元素所包含的层次结构中。另一个实例在 CascadingValue 元素之外定义，不接收级联值。

# 34.4 处理错误

下面描述 Blazor 提供的、处理连接错误和未处理的应用程序错误的特性。

## 34.4.1 处理连接错误

Blazor 依赖于它在浏览器和 ASP.NET Core 服务器之间的持久 HTTP 连接。当连接中断时，应用程序无法正常工作，并显示一个模态错误消息，阻止用户与组件交互。

Blazor 允许通过定义具有特定 id 的元素来定制连接错误，如代码清单 34-28 所示。

**代码清单 34-28 在 Pages 文件夹的 Blazor.cshtml 文件中定义连接错误元素**

```
@page "/pages/blazor"

<h4 class="bg-primary text-white text-center p-2">Departments</h4>

<link rel="stylesheet" href="connectionErrors.css" />

<div id="components-reconnect-modal"
    class="h4 bg-dark text-white text-center my-2 p-2 components-reconnect-hide">
    Blazor Connection Lost
    <div class="reconnect">
        Trying to reconnect...
    </div>
    <div class="failed">
        Reconnection Failed.
        <button class="btn btn-light" onclick="window.Blazor.reconnect()">
```

```
            Reconnect
        </button>
    </div>
    <div class="rejected">
        Reconnection Rejected.
        <button class="btn btn-light" onclick="location.reload()">
            Reload
        </button>
    </div>
</div>
<component type="typeof(Advanced.Blazor.DepartmentList)" render-mode="Server" />
```

自定义错误元素的 id 属性必须是 componentes-reconnect-modal。当出现连接错误时，Blazor 将查找此元素并将其添加到表 34-2 中描述的四个类中的一个。

表 34-2　连接错误类

| 名称 | 描述 |
|---|---|
| components-reconnect-show | 当连接已经丢失且 Blazor 正在尝试重连接时，将元素添加到这个类中。错误消息应该显示给用户，并且应该防止与 Blazor 内容的交互 |
| components-reconnect-hide | 如果重新建立连接，则将元素添加到该类中。错误消息应该隐藏，并且应该允许交互 |
| components-reconnect-failed | 如果 Blazor 重连接失败，则将该元素添加到该类中。用户可以看到一个按钮，该按钮调用 window.Blazor.reconnect()来再次尝试重新连接 |
| components-reconnect-rejected | 如果 Blazor 能够到达服务器，但用户的连接状态已经丢失，则将元素添加到这个类中。这通常发生在服务器重新启动时。用户可以看到一个按钮，该按钮调用 location.reload()来重新加载应用程序并重试 |

元素最初没有添加到这些类中的任何一个，因此将其显式添加到 components-reconnect-hide 类中，以便在出现问题之前隐藏它。

希望针对重连接期间可能出现的每个条件向用户显示特定的消息。为此，添加了为每个条件显示消息的元素。为了管理它们的可见性，将一个名为 connectionError.css 的 CSS 样式表添加到 wwwroot 文件夹中，并用它定义如代码清单 34-29 所示的样式。

### 代码清单 34-29　wwwroot 文件夹中 connectionError.css 文件的内容

```
#components-reconnect-modal {
    position: fixed; top: 0; right: 0; bottom: 0;
    left: 0; z-index: 1000; overflow: hidden; opacity: 0.9;
}

.components-reconnect-hide { display: none; }
.components-reconnect-show { display: block; }

.components-reconnect-show > .reconnect { display: block; }
.components-reconnect-show > .failed,
```

```
.components-reconnect-show > .rejected {
   display: none;
}

.components-reconnect-failed > .failed {
   display: block;
}
.components-reconnect-failed > .reconnect,
.components-reconnect-failed > .rejected {
   display: none;
}

.components-reconnect-rejected > .rejected {
   display: block;
}
.components-reconnect-rejected > .reconnect,
.components-reconnect-rejected > .failed {
   display: none;
}
```

这些样式将 componentes-reconnect-modal 元素显示为一个模态项，其可见性由 componentes-reconnect-hide 和 componentes-reconnect-show 类决定。特定消息的可见性是基于表 34-2 中类的应用程序来切换的。

要查看效果，请重新启动 ASP.NET Core，请求 http://localhost:5000/pages/blazor。等到显示组件，就停止 ASP.NET Core 服务器。当 Blazor 尝试重新连接时，将看到一个初始错误消息。几秒钟后，将看到指示重连接失败的消息。

重启 ASP.NET Core，请求 http://localhost:5000/pages/blazor。等到显示组件，然后重新启动 ASP.NET Core。这一次 Blazor 能够连接到服务器，但是连接将被拒绝，因为服务器重启导致连接状态丢失。图 34-12 显示了这两个错误消息序列。

■ **提示：**仅用浏览器无法测试成功的连接恢复，因为无法中断持久的 HTTP 连接。所以这里使用了优秀的 Fiddler 代理(https://www.telerik.com/fiddler)，它允许在不停止 ASP.NET Core 服务器的情况下终止连接。

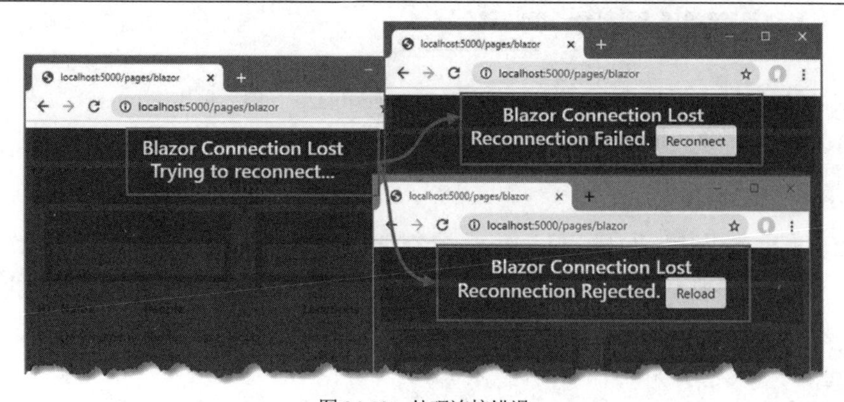


图 34-12　处理连接错误

### 34.4.2 处理未捕获的应用程序错误

Blazor不能很好地响应未捕获的应用程序错误，这些错误几乎总是被视为终端。要查看默认的错误行为，请将代码清单34-30中所示的元素添加到DepartmentList组件。

**代码清单34-30 在Blazor文件夹的DepartmentList.razor文件中添加元素**

```
<CascadingValue Name="BgTheme" Value="Theme" IsFixed="false" >
    <TableTemplate RowType="Department" RowData="Departments"
        Highlight="@(d => d.Name)"
        SortDirection="@(d => d.Name)">
        <Header>
            <tr><th>ID</th><th>Name</th><th>People</th><th>Locations</th></tr>
        </Header>
        <RowTemplate Context="d">
            <td>@d.Departmentid</td>
            <td>@d.Name</td>
            <td>@(String.Join(", ", d.People.Select(p => p.Surname)))</td>
            <td>
                @(String.Join(", ",
                    d.People.Select(p => p.Location.City).Distinct()))
            </td>
        </RowTemplate>
    </TableTemplate>
</CascadingValue>

<SelectFilter Title="@("Theme")" Values="Themes" @bind-SelectedValue="Theme" />

<button class="btn btn-danger" @onclick="@(() => throw new Exception())">
    Error
</button>

@code {

    // ...statements omitted for brevity...
}
```

重启ASP.NET Core，请求http://localhost:5000/pages/blazor，然后单击Error按钮。在浏览器中并无明显变化，但当单击按钮时在服务器上抛出的异常被证明是致命的：用户仍然可以使用select元素选择值，因为这些是由浏览器显示的，但响应选择的事件处理程序不再有效，应用程序本质上是停止了。

当出现未处理的应用程序错误时，Blazor会查找id为Blazor-error-ui的元素，并设置其CSS显示属性以阻止。代码清单34-31将具有此id的元素添加到Blazor.cshtml文件，该文件样式化为显示有用的消息。

**代码清单 34-31　在 Pages 文件夹的 Blazor.cshtml 文件中添加错误元素**

```
@page "/pages/blazor"

<h4 class="bg-primary text-white text-center p-2">Departments</h4>

<link rel="stylesheet" href="connectionErrors.css" />

<div id="components-reconnect-modal"
    class="h4 bg-dark text-white text-center my-2 p-2 components-reconnect-hide">
    Blazor Connection Lost
    <div class="reconnect">
        Trying to reconnect...
    </div>
    <div class="failed">
        Reconnection Failed.
        <button class="btn btn-light" onclick="window.Blazor.reconnect()">
            Reconnect
        </button>
    </div>
    <div class="rejected">
        Reconnection Rejected.
        <button class="btn btn-light" onclick="location.reload()">
            Reload
        </button>
    </div>
</div>

<div id="blazor-error-ui"
    class="text-center bg-danger h6 text-white p-2 fixed-top w-100"
    style="display:none">
    An error has occurred. This application will not respond until reloaded.
    <button class="btn btn-sm btn-primary" onclick="location.reload()">
        Reload
    </button>
</div>

<component type="typeof(Advanced.Blazor.DepartmentList)" render-mode="Server" />
```

当显示元素时，用户将看到一个警告和一个重新加载浏览器的按钮。要查看效果，请重新启动 ASP.NET Core，请求 http://localhost:5000/pages/blazor，然后单击 Error 按钮，它将显示如图 34-13 所示的消息。

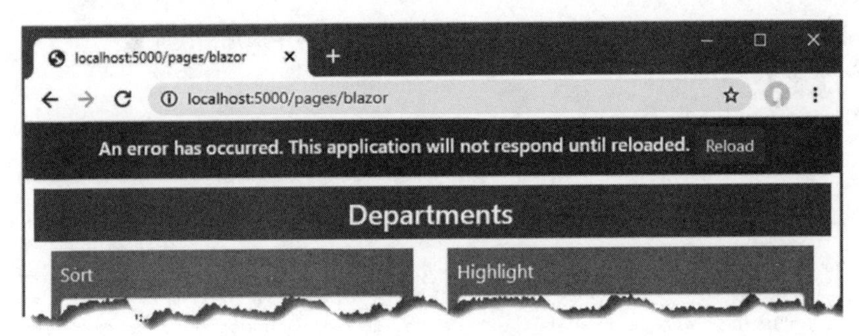


图 34-13　显示错误信息

## 34.5　小结

本章展示了如何组合 Razor 组件来创建更复杂的特性。展示了如何创建组件之间的父/子关系，如何利用属性配置组件，以及如何创建自定义事件，以在发生重要更改时发出信号。还展示了组件如何从父组件接收内容，以及如何使用模板组件一致地生成内容，模板组件可以用一个或多个泛型类型参数定义。在本章结束时，演示了 Blazor 应用程序如何对连接和应用程序错误做出反应。第 35 章将描述 Blazor 提供的高级特性。

# 第35章

# 高级Blazor特性

本章解释Blazor如何支持URL路由，以便通过一个请求显示多个组件。展示如何设置路由系统、如何定义路由以及如何在布局中创建公共内容。

本章还介绍了组件的生命周期，它允许组件积极地参与Blazor环境，这在开始使用URL路由特性时尤为重要。最后，本章解释了组件在前面章节描述的父/子关系之外的不同交互方式。表35-1对这些特性进行了具体分析。

表35-1　Blazor路由和组件生命周期的相关知识

| 问题 | 答案 |
| --- | --- |
| 它们是什么？ | 路由特性允许组件响应URL中的更改，而不需要新的HTTP连接。生命周期特性允许组件定义在执行应用程序时调用的方法，交互特性提供了在组件之间以及与其他JavaScript代码进行通信的有用方法 |
| 它们为什么有用？ | 这些特性允许创建利用Blazor架构的复杂应用程序 |
| 它们是如何使用的？ | URL路由使用内置组件设置，并使用@page指令配置。生命周期特性是通过覆盖组件的@code部分中的方法来使用的。根据组件与什么交互，交互特性以不同的方式使用 |
| 是否存在缺陷或限制？ | 这些是必须小心使用的高级特性，特别是在创建Blazor之外的交互时 |
| 还有其他选择吗？ | 本章中描述的所有特性都是可选的，但若没有它们，将很难创建复杂的应用程序 |

表35-2总结了本章的内容。

表35-2　本章内容摘要

| 问题 | 解决方案 | 代码清单 |
| --- | --- | --- |
| 基于当前URL选择组件 | 使用URL | 35-6～35-12 |
| 定义多个组件使用的内容 | 使用布局 | 35-13～35-14 |
| 响应组件生命周期的各个阶段 | 实现生命周期通知方法 | 35-15～35-17 |
| 协调多个组件的活动 | 保留@ref表达式的引用 | 35-18～35-19 |
| 协调Blazor之外的代码 | 使用互操作性特性 | 35-20～35-35 |

## 35.1　准备工作

本章使用了第 35 章中的 Advanced 项目。本章不需要任何更改。

**提示：** 可从 https://github.com/apress/pro-asp.netcore-3 下载本章和本书中所有其他章节的示例项目。如果在运行示例时遇到问题，请参阅第 1 章，以了解如何获得帮助。

打开一个新的 PowerShell 命令提示符，导航到包含 Advanced.csproj 文件的文件夹，并运行代码清单 35-1 所示的命令来删除数据库。

**代码清单 35-1　删除数据库**

```
dotnet ef database drop --force
```

从 Debug 菜单中选择 Start Without Debugging 或 Run Without Debugging，或者使用 PowerShell 命令提示符运行代码清单 35-2 所示的命令。

**代码清单 35-2　运行示例应用程序**

```
dotnet run
```

使用浏览器请求 http://localhost:5000/controllers，这将显示一个数据项列表。请求 http://localhost:5000/pages/blazor，将看到第 34 章中用于演示绑定的组件。图 35-1 显示了这两种响应。

图 35-1　运行示例应用程序

## 35.2　使用组件的路由

Blazor 支持根据 ASP.NET Core 路由系统选择要显示给用户的组件，使应用程序通过显示不同的 Razor 组件来响应 URL 中的变化。首先，给 Blazor 文件夹添加一个名为 Routed.razor 的 Razor 组件，内容如代码清单 35-3 所示。

**代码清单 35-3　Blazor 文件夹中 Routed.razor 文件的内容**

```
<Router AppAssembly="typeof(Startup).Assembly">
    <Found>
        <RouteView RouteData="@context" />
    </Found>
    <NotFound>
        <h4 class="bg-danger text-white text-center p-2">
            No Matching Route Found
        </h4>
    </NotFound>
</Router>
```

Router 组件包含在 ASP.NET Core 中，提供了 Blazor 和 ASP.NET Core 路由特性之间的链接。Router 是一个通用模板组件，它定义了 Found 和 NotFound 部分。

Router 组件需要 AppAssembly 属性，该属性指定要使用的.NET 程序集。对于大多数项目，这是当前的程序集，它的指定如下:

```
...
<Router AppAssembly="typeof(Startup).Assembly">
...
```

Router 组件的 Found 属性类型是 RenderFragment<RouteData>，它通过 RouteData 属性传递给 RouteView 组件，如下所示：

```
...
<Found>
    <RouteView RouteData="@context" />
</Found>
...
```

RouteView 组件负责显示与当前路由匹配的组件，以及通过布局显示公共内容(稍后将简要地进行解释)。NotFound 属性的类型是 RenderFragment，没有泛型类型参数，在当前路由找不到组件时显示内容部分。

## 35.2.1　准备 Blazor 页

单个组件可以显示在现有的控制器视图和 Blazor 页面中，如前几章所示。但是在使用组件路由时，最好创建一组与 Blazor 不同的 URL，因为 URL 被支持的方式是有限的，并且会导致复杂的解决方案。将一个名为_Host.cshtml 的 Razor Pages 添加到 Pages 文件夹，并添加如代码清单 35-4 所示的内容。

**代码清单 35-4　Pages 文件夹中_Host.cshtml 文件的内容**

```
@page "/"
@{ Layout = null; }

<!DOCTYPE html>
```

```
<html>
<head>
    <title>@ViewBag.Title</title>
    <link href="/lib/twitter-bootstrap/css/bootstrap.min.css" rel="stylesheet" />
    <base href="~/" />
</head>
<body>
    <div class="m-2">
        <component type="typeof(Advanced.Blazor.Routed)" render-mode="Server" />
    </div>
    <script src="_framework/blazor.server.js"></script>
</body>
</html>
```

这个页面包含一个应用代码清单 35-4 中定义的 Routed 组件的 component 元素和一个用于 Blazor JavaScript 代码的 script 元素。还有一个用于引导 CSS 样式表的 link 元素。修改示例应用程序的配置，以当请求与现有 URL 路由不匹配时使用_Host.cshtml 文件作为回退，如代码清单 35-5 所示。

**代码清单 35-5　添加回退**

```
...
public void Configure(IApplicationBuilder app, DataContext context) {

    app.UseDeveloperExceptionPage();
    app.UseStaticFiles();
    app.UseRouting();

    app.UseEndpoints(endpoints => {
        endpoints.MapControllerRoute("controllers",
            "controllers/{controller=Home}/{action=Index}/{id?}");
        endpoints.MapDefaultControllerRoute();
        endpoints.MapRazorPages();
        endpoints.MapBlazorHub();
        endpoints.MapFallbackToPage("/_Host");
    });

    SeedData.SeedDatabase(context);
}
...
```

MapFallbackToPage 方法将路由系统配置为使用_Host 页面，作为处理不匹配请求的最后手段。

## 35.2.2　向组件添加路由

组件声明应该使用@page 指令显示的 URL。代码清单 35-6 给 PeopleList 组件添加了@page 指令。

### 代码清单 35-6 在 Blazor 文件夹的 PeopleList.razor 文件中添加指令

```
@page "/people"

<TableTemplate RowType="Person" RowData="People"
      Highlight="@(p => p.Location.City)" SortDirection="@(p => p.Surname)">
   <Header>
      <tr><th>ID</th><th>Name</th><th>Dept</th><th>Location</th></tr>
   </Header>
   <RowTemplate Context="p">
      <td>@p.PersonId</td>
      <td>@p.Surname, @p.Firstname</td>
      <td>@p.Department.Name</td>
      <td>@p.Location.City, @p.Location.State</td>
   </RowTemplate>
</TableTemplate>

@code {

   [Inject]
   public DataContext Context { get; set; }

   public IEnumerable<Person> People => Context.People
         .Include(p => p.Department)
         .Include(p => p.Location);
}
```

代码清单 35-6 中的指令意味着为 http://localhost:5000/people URL 显示 PeopleList 组件。组件可使用多个@page 指令声明对多个路由的支持。代码清单 35-7 将@page 指令添加到 DepartmentList 组件以支持两个 URL。

### 代码清单 35-7 在 Blazor 文件夹的 DepartmentList.razor 文件中添加指令

```
@page "/departments"
@page "/depts"

<CascadingValue Name="BgTheme" Value="Theme" IsFixed="false" >
   <TableTemplate RowType="Department" RowData="Departments"
      Highlight="@(d => d.Name)"
      SortDirection="@(d => d.Name)">
      <Header>
         <tr><th>ID</th><th>Name</th><th>People</th><th>Locations</th></tr>
      </Header>
      <RowTemplate Context="d">
         <td>@d.Departmentid</td>
         <td>@d.Name</td>
         <td>@(String.Join(", ", d.People.Select(p => p.Surname)))</td>
         <td>
```

```
            @(String.Join(", ",
                d.People.Select(p => p.Location.City).Distinct()))
        </td>
      </RowTemplate>
    </TableTemplate>
</CascadingValue>

<SelectFilter Title="@("Theme")" Values="Themes" @bind-SelectedValue="Theme" />

<button class="btn btn-danger" @onclick="@(() => throw new Exception())">
    Error
</button>

@code {

    [Inject]
    public DataContext Context { get; set; }

    public IEnumerable<Department> Departments => Context.Departments
            .Include(d => d.People).ThenInclude(p => p.Location);

    public string Theme { get; set; } = "info";
    public string[] Themes = new string[] { "primary", "info", "success" };
}
```

第 13 章描述的大多数路由模式特征都可在@page表达式中使用，除了catchall段变量和可选段变量之外。使用两个@page表达式(其中一个带有一个段变量)，可重新创建可选的变量特性，如第 36 章所示，其中展示了如何使用Blazor实现CRUD应用程序。

要查看基本的 Razor 组件路由特性的工作情况，请重新启动 ASP.NET Core 并请求 http://localhost:5000/people 和 http://localhost:5000/dept。每个 URL 显示应用程序中的一个组件，如图 35-2 所示。

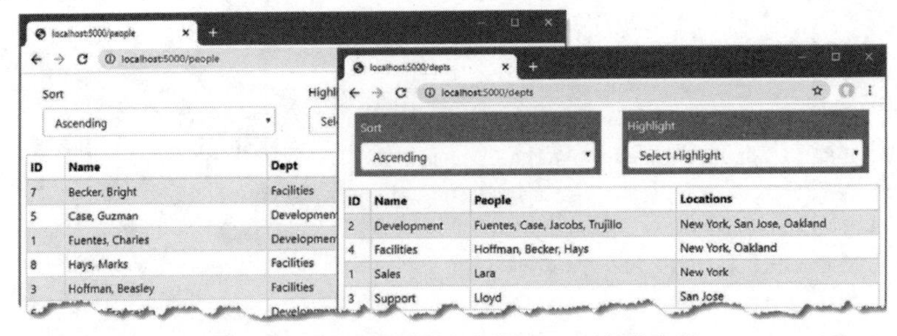

图 35-2　在示例应用程序中启用 Razor 组件路由

## 设置默认组件路由

代码清单 35-5 中的配置更改为 Startup 类中的请求设置了回退路由。应用程序的一个组件需要相应的路由来识别应该为应用程序的默认 URL http://localhost:5000 显示的组件，如代码清单 35-8 所示。

代码清单 35-8 在 Blazor 文件夹的 PeopleList.razor 文件中定义默认路由

```
@page "/"
@page "/people"

<TableTemplate RowType="Person" RowData="People"
        Highlight="@(p => p.Location.City)" SortDirection="@(p => p.Surname)">
    <Header>
        <tr><th>ID</th><th>Name</th><th>Dept</th><th>Location</th></tr>
    </Header>
    <RowTemplate Context="p">
        <td>@p.PersonId</td>
        <td>@p.Surname, @p.Firstname</td>
        <td>@p.Department.Name</td>
        <td>@p.Location.City, @p.Location.State</td>
    </RowTemplate>
</TableTemplate>

@code {

    [Inject]
    public DataContext Context { get; set; }

    public IEnumerable<Person> People => Context.People
            .Include(p => p.Department)
            .Include(p => p.Location);
}
```

重启 ASP.NET Core 并请求 http://localhost:5000，将看到 PeopleList 组件生成的内容，如图 35-3 所示。

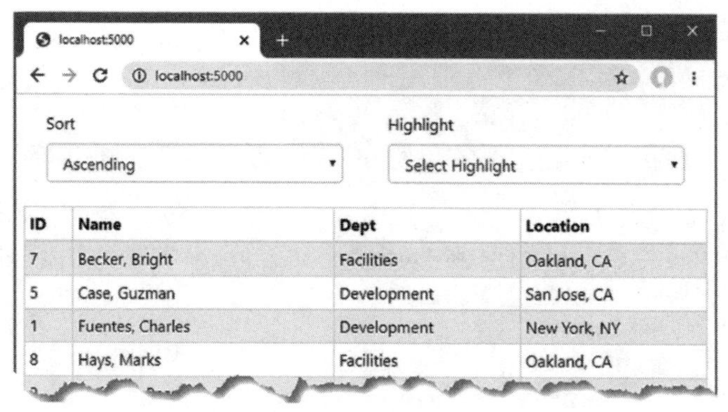


图 35-3 显示默认 URL 的组件

## 35.2.3 在路由组件之间导航

基本的路由配置已经就绪，但是为什么使用路由比前面章节中演示的独立组件更有优势还不明显。改进来自 NavLink 组件，它呈现连接到路由系统的锚元素。代码清单 35-9 将 NavLink 添加到 PeopleList 组件。

**代码清单 35-9　在 Blazor 文件夹的 PeopleList.razor 文件中添加导航**

```
@page "/"
@page "/people"

<TableTemplate RowType="Person" RowData="People"
        Highlight="@(p => p.Location.City)" SortDirection="@(p => p.Surname)">
    <Header>
        <tr><th>ID</th><th>Name</th><th>Dept</th><th>Location</th></tr>
    </Header>
    <RowTemplate Context="p">
        <td>@p.PersonId</td>
        <td>@p.Surname, @p.Firstname</td>
        <td>@p.Department.Name</td>
        <td>@p.Location.City, @p.Location.State</td>
    </RowTemplate>
</TableTemplate>

<NavLink class="btn btn-primary" href="/depts">Departments</NavLink>

@code {

    [Inject]
    public DataContext Context { get; set; }
    public IEnumerable<Person> People => Context.People
            .Include(p => p.Department)
            .Include(p => p.Location);
}
```

与在 ASP.NET Core 的其他部分中使用的锚元素不同，NavLink 组件是使用 URL 配置的，而不是组件、页面或动作名称。本例中的 NavLink 导航到 DepartmentList 组件的@page 指令所支持的 URL。

导航也可以通过编程方式执行，这在组件响应事件然后需要导航到不同的 URL 时非常有用，如代码清单 35-10 所示。

**代码清单 35-10　在 Blazor 文件夹的 DepartmentList.razor 文件中以编程方式导航**

```
@page "/departments"
@page "/depts"

<CascadingValue Name="BgTheme" Value="Theme" IsFixed="false" >
    <TableTemplate RowType="Department" RowData="Departments"
        Highlight="@(d => d.Name)"
        SortDirection="@(d => d.Name)">
        <Header>
            <tr><th>ID</th><th>Name</th><th>People</th><th>Locations</th></tr>
```

```
        </Header>
        <RowTemplate Context="d">
            <td>@d.Departmentid</td>
            <td>@d.Name</td>
            <td>@(String.Join(", ", d.People.Select(p => p.Surname)))</td>
            <td>
                @(String.Join(", ",
                    d.People.Select(p => p.Location.City).Distinct()))
            </td>
        </RowTemplate>
    </TableTemplate>
</CascadingValue>

<SelectFilter Title="@("Theme")" Values="Themes" @bind-SelectedValue="Theme" />

<button class="btn btn-primary" @onclick="HandleClick">People</button>

@code {

    [Inject]
    public DataContext Context { get; set; }

    public IEnumerable<Department> Departments => Context.Departments
            .Include(d => d.People).ThenInclude(p => p.Location);

    public string Theme { get; set; } = "info";
    public string[] Themes = new string[] { "primary", "info", "success" };

    [Inject]
    public NavigationManager NavManager { get; set; }

    public void HandleClick() => NavManager.NavigateTo("/people");
}
```

NavigationManager 类提供对导航的编程访问。表 35-3 描述了 NavigationManager 类提供的最重要成员。

表 35-3 有用的 NavigationManager 成员

| 名称 | 描述 |
|---|---|
| NavigateTo(url) | 此方法导航到指定的 URL，而不发送新的 HTTP 请求 |
| ToAbsoluteUri(path) | 此方法将一个相对路径转换为一个完整的 URL |
| ToBaseRelativePath(url) | 此方法从完整的 URL 中获取相对路径 |
| LocationChanged | 位置更改时触发此事件 |
| Uri | 此属性返回当前 URL |

NavigationManager 类作为服务提供，并通过 Inject 属性被 Razor 组件接收，该属性提供对第 14 章中描述的依赖注入特性的访问。

NavigationManager.NavigateTo 方法导航到一个 URL，并在本例中用于导航到/people URL，该 URL 由 PeopleList 组件处理。

要了解路由和导航为什么重要，请重新启动 ASP.NET Core，并请求 http://localhost:5000/people。单击 Departments 链接，该链接样式化为一个按钮，DepartmentList 组件将显示出来。单击 People 链接，将返回到 PeopleList 组件，如图 35-4 所示。

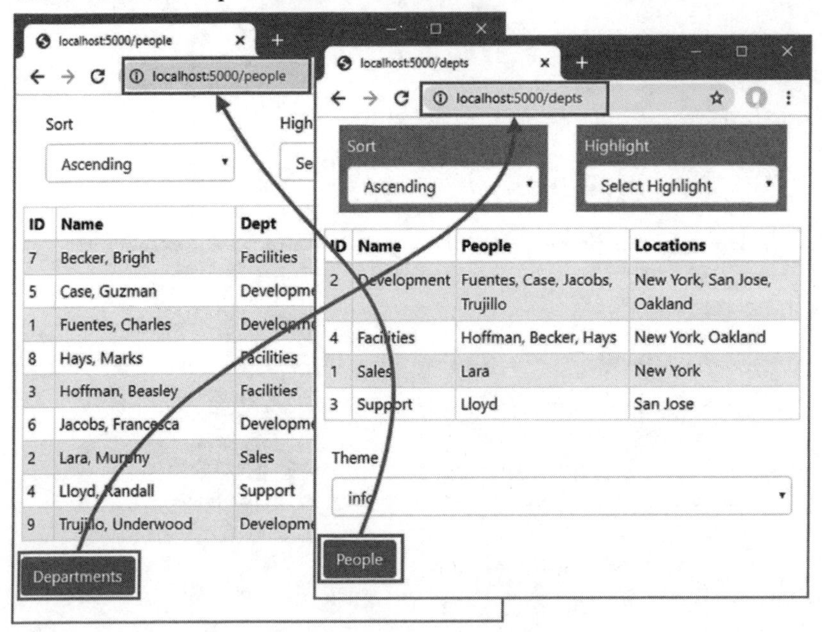

图 35-4　在路由组件之间导航

如果在 F12 开发人员工具打开的情况下执行这个序列，将看到从一个组件到下一个组件的转换不需要单独的 HTTP 请求，即使浏览器显示的 URL 发生了变化。Blazor 交付每个组件通过持久 HTTP 连接呈现的内容，持久 HTTP 连接是在显示第一个组件时建立的，并使用 JavaScript API 导航，而不加载新的 HTML 文档。

提示：NavigationManager.NavigateTo 方法接收一个可选参数，当为 true 时，强制浏览器发送新的 HTTP 请求，并重新加载 HTML 文档。

## 35.2.4 接收路由数据

组件可以通过使用 Parameter 属性修饰属性来接收段变量。为了便于演示，为 Blazor 文件夹添加一个名为 PersonDisplay.razor 的 Razor 组件，内容如代码清单 35-11 所示。

**代码清单 35-11　Blazor 文件夹中 PersonDisplay.razor 的内容**

```
@page "/person"
@page "/person/{id:long}"
```

```
<h5>Editor for Person: @Id</h5>

<NavLink class="btn btn-primary" href="/people">Return</NavLink>

@code {

    [Parameter]
    public long Id { get; set; }
}
```

直到在本章后面添加特性，这个组件除了显示它从路由数据中接收到的值之外，什么都不做。@page 表达式包括一个名为 id 的段变量，其类型指定为 long。通过定义具有相同名称的属性并使用 Parameter 属性修饰它，组件接收分配给段变量的值。

■ **提示**：如果在@page 表达式中没有为段变量指定类型，那么必须将属性的类型设置为 string。

代码清单 35-12 使用 NavLink 组件为 PeopleList 组件显示的每个 Person 对象创建导航链接。

**代码清单 35-12　在 Blazor 文件夹的 PeopleList.razor 文件中添加导航链接**

```
@page "/"
@page "/people"

<TableTemplate RowType="Person" RowData="People"
        Highlight="@(p => p.Location.City)" SortDirection="@(p => p.Surname)">
    <Header>
        <tr><th>ID</th><th>Name</th><th>Dept</th><th>Location</th>
            <td></td>
        </tr>
    </Header>
    <RowTemplate Context="p">
        <td>@p.PersonId</td>
        <td>@p.Surname, @p.Firstname</td>
        <td>@p.Department.Name</td>
        <td>@p.Location.City, @p.Location.State</td>
        <td>
            <NavLink class="btn btn-sm btn-info" href="@GetEditUrl(p.PersonId)">
                Edit
            </NavLink>
        </td>
    </RowTemplate>
</TableTemplate>

<NavLink class="btn btn-primary" href="/depts">Departments</NavLink>

@code {

    [Inject]
```

```
    public DataContext Context { get; set; }

    public IEnumerable<Person> People => Context.People
            .Include(p => p.Department)
            .Include(p => p.Location);

    public string GetEditUrl(long id) => $"/person/{id}";
}
```

Razor 组件不支持在属性值中混合静态内容和 Razor 表达式。相反，前面定义了 GetEditUrl 方法来为每个 Person 对象生成导航 URL，调用该方法来生成 NavLink href 属性的值。

重启 ASP.NET Core，请求 http://localhost:5000/people，然后单击其中一个 Edit 按钮。浏览器导航到新的 URL 而不重新加载 HTML 文档，并显示 PersonDisplay 组件生成的占位符内容，如图 35-5 所示，它显示了组件如何从路由系统中接收数据。

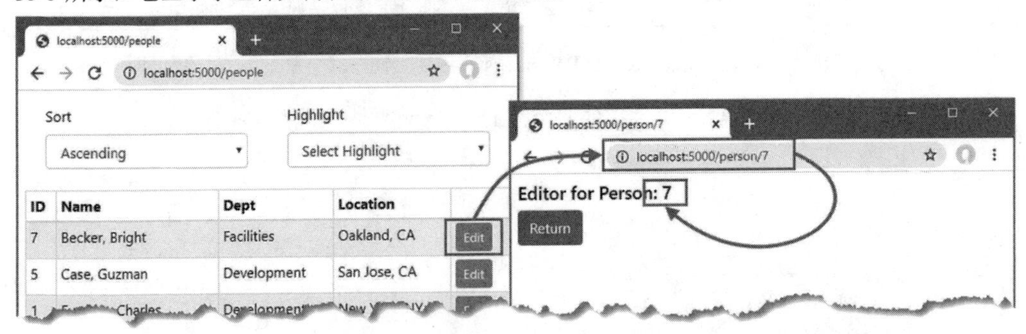


图 35-5　从 Razor 组件的路由系统接收数据

## 35.2.5　使用布局定义公共内容

布局是为 Razor 组件提供通用内容的模板组件。要创建布局，给 Blazor 文件夹添加一个名为 NavLayout.razor 的 Razor 组件，并添加如代码清单 35-13 所示的内容。

#### 代码清单 35-13　Blazor 文件夹中 NavLayout.razor 文件的内容

```
@inherits LayoutComponentBase

<div class="container-fluid">
    <div class="row">
        <div class="col-3">
            @foreach (string key in NavLinks.Keys) {
                <NavLink class="btn btn-outline-primary btn-block"
                        href="@NavLinks[key]"
                        ActiveClass="btn-primary text-white"
                        Match="NavLinkMatch.Prefix">
                    @key
                </NavLink>
            }
        </div>
```

```
        <div class="col">
            @Body
        </div>
    </div>
</div>

@code {

    public Dictionary<string, string> NavLinks
        = new Dictionary<string, string> {
            {"People", "/people" },
            {"Departments", "/depts" },
            {"Details", "/person" }
        };
}
```

布局使用@inherits 表达式指定 LayoutComponentBase 类作为从 Razor 组件中生成的类的基类。LayoutComponentBase 类定义了一个名为 Body 的 RenderFragment 类，用于指定布局中显示的公共内容中的组件内容。在本例中，布局组件创建了一个网格布局，该布局为应用程序中的每个组件显示一组 NavLink 组件。NavLink 组件配置了两个新属性，如表 35-4 所示。

表 35-4 NavLink 配置属性

| 名称 | 描述 |
| --- | --- |
| ActiveClass | 此属性指定一个或多个 CSS 类，当前 URL 与 href 属性值匹配时，由 NavLink 组件呈现的锚元素将添加到这些 CSS 类中 |
| Match | 此属性指定如何使用来自 NavLinkMatch 枚举的值将当前 URL 匹配到 href 属性。这些值是 Prefix(如果 href 匹配 URL 的开头，则认为是匹配的)和 All(要求整个 URL 相同) |

NavLink 组件配置为使用前缀匹配，并当有匹配时，添加它们呈现给引导程序 btn-primary 和 text-white 类的锚元素。

## 应用布局

有三种方法可以应用布局。组件可以使用@layout表达式选择自己的布局。父组件可通过将子组件包装到内置的LayoutView组件中，来为其子组件使用布局。通过设置RouteView组件的DefaultLayout属性，可将布局应用于所有组件，如代码清单 35-14 所示。

**代码清单 35-14 在 Blazor 文件夹的 Routed.razor 文件中应用布局**

```
<Router AppAssembly="typeof(Startup).Assembly">
    <Found>
        <RouteView RouteData="@context" DefaultLayout="typeof(NavLayout)" />
    </Found>
    <NotFound>
        <h4 class="bg-danger text-white text-center p-2">
            Not Matching Route Found
        </h4>
```

```
    </NotFound>
</Router>
```

重启 ASP.NET Core 并请求 http://localhost:5000/people。该布局与 PeopleList 组件呈现的内容一起显示。布局左侧的导航按钮可用于在应用程序中导航，如图 35-6 所示。

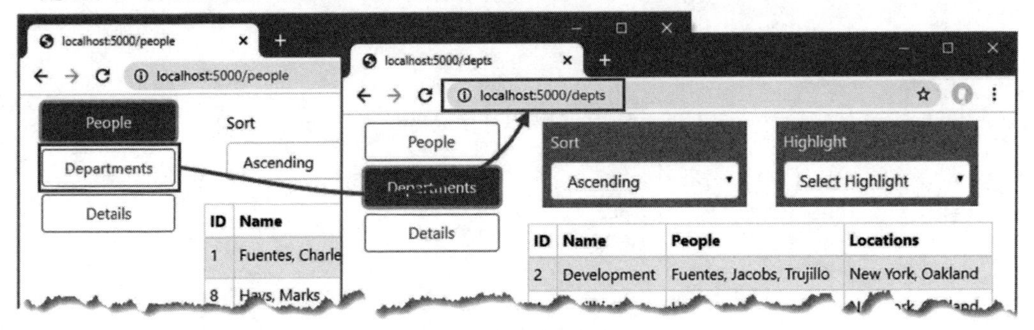


图 35-6　使用布局组件

■ **注意：**如果请求 http://localhost:5000，将看到 PeopleList 组件的内容，但相应的导航按钮不会突出显示。下一节将展示如何解决这个问题。

## 35.3　理解组件生命周期方法

Razor 组件具有定义良好的生命周期，用组件可以实现的方法来表示，以接收关键转换的通知。表 35-5 描述了生命周期方法。

表 35-5　Razor 组件的生命周期方法

| 名称 | 描述 |
|---|---|
| OnInitialized()<br>OnInitializedAsync() | 这些方法在组件第一次初始化时调用 |
| OnParametersSet()<br>OnParametersSetAsync() | 这些方法在应用通过 Parameter 属性修饰的属性值后调用 |
| ShouldRender() | 此方法在呈现组件的内容之前调用，以更新呈现给用户的内容。如果该方法返回 false，则组件的内容将不呈现，更新将被抑制。此方法不会取消组件的初始呈现 |
| OnAfterRender(first)<br>OnAfterRenderAsync(first) | 在呈现组件的内容之后调用此方法。当 Blazor 为组件执行初始渲染时，bool 参数为 true |

使用 OnInitialized 或 OnParameterSet 方法都可用于设置组件的初始状态。上一节定义的布局没有处理默认的 URL，因为 NavLink 组件只匹配一个 URL。对于 DepartmentList 组件也存在同样的问题，可以使用/departments 和/depts 路径请求该组件。

**理解路由组件的生命周期**

当使用 URL 路由时，可在 URL 更改时从显示中删除组件。组件可以实现 System.IDisposable 接口，Blazor 将在组件删除时调用该方法。

创建匹配多个 URL 的组件需要使用生命周期方法。要了解原因，请为 Blazor 文件夹添加一个名为 MultiNavLink.razor 的 Razor 组件，内容如代码清单 35-15 所示。

**代码清单 35-15 Blazor 文件夹中 MultiNavLink.razor 的内容**

```
<a class="@ComputedClass" @onclick="HandleClick" href="">
    @ChildContent
</a>

@code {

    [Inject]
    public NavigationManager NavManager { get; set; }

    [Parameter]
    public IEnumerable<string> Href { get; set; }

    [Parameter]
    public string Class { get; set; }

    [Parameter]
    public string ActiveClass { get; set; }

    [Parameter]
    public NavLinkMatch? Match { get; set; }

    public NavLinkMatch ComputedMatch { get =>
            Match ?? (Href.Count() == 1 ? NavLinkMatch.Prefix : NavLinkMatch.All); }

    [Parameter]
    public RenderFragment ChildContent { get; set; }

    public string ComputedClass { get; set; }

    public void HandleClick() {
        NavManager.NavigateTo(Href.First());
}

private void CheckMatch(string currentUrl) {
    string path = NavManager.ToBaseRelativePath(currentUrl);
        path = path.EndsWith("/") ? path.Substring(0, path.Length - 1) : path;
        bool match = Href.Any(href => ComputedMatch == NavLinkMatch.All
                ? path == href : path.StartsWith(href));
        ComputedClass = match ? $"{Class} {ActiveClass}" : Class;
    }

    protected override void OnParametersSet() {
```

```
            ComputedClass = Class;
            NavManager.LocationChanged += (sender, arg) => CheckMatch(arg.Location);
            Href = Href.Select(h => h.StartsWith("/") ? h.Substring(1) : h);
            CheckMatch(NavManager.Uri);
        }
    }
```

该组件的工作方式与常规NavLink相同，但接收一组路径来匹配。组件依赖于OnParametersSet生命周期方法，因为需要进行一些初始设置，而只有在为用 Parameter 属性修饰的属性分配了值之后，才能执行这些初始设置，如提取单个路径。

此组件通过侦听由 NavigationManager 类定义的 LocationChanged 事件来响应当前 URL 中的更改。事件的 Location 属性为组件提供当前 URL，该 URL 用于更改锚元素的类。代码清单 35-16 在布局中应用新组件。

**提示：** 在代码清单 35-14 中删除了 Match 属性。新组件支持这个属性，但默认情况下根据它通过 href 属性接收到的路径数量进行匹配。

**代码清单 35-16　在 Blazor 文件夹的 NavLayout.razor 文件中应用新组件**

```
@inherits LayoutComponentBase

<div class="container-fluid">
    <div class="row">
        <div class="col-3">
            @foreach (string key in NavLinks.Keys) {
                <MultiNavLink class="btn btn-outline-primary btn-block"
                        href="@NavLinks[key]" ActiveClass="btn-primary text-white">
                    @key
                </MultiNavLink>
            }
        </div>
        <div class="col">
            @Body
        </div>
    </div>
</div>

@code {

    public Dictionary<string, string[]> NavLinks
        = new Dictionary<string, string[]> {
            {"People", new string[] {"/people", "/" } },
            {"Departments", new string[] {"/depts", "/departments" } },
            {"Details", new string[] { "/person" } }
        };
}
```

重启 ASP.NET Core，并请求 http://localhost:5000 和 http://localhost:5000/departments。这两个 URL 都可以识别，相应的导航按钮高亮显示，如图 35-7 所示。

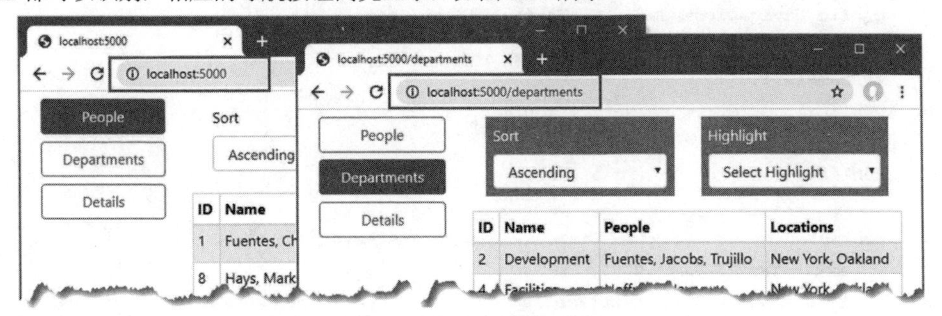


图 35-7 使用生命周期方法

### 对异步任务使用生命周期方法

生命周期方法对于执行可能在呈现组件的初始内容之后完成的任务也很有用，例如查询数据库。代码清单 35-17 替换了 PersonDisplay 组件中的占位符内容，并通过生命周期方法，使用接收到的值作为参数来查询数据库。

#### 代码清单 35-17 在 Blazor 文件夹的 PersonDisplay.razor 文件中查询数据

```
@page "/person"
@page "/person/{id:long}"

@if (Person == null) {
    <h5 class="bg-info text-white text-center p-2">Loading...</h5>
} else {
    <table class="table table-striped table-bordered">
        <tbody>
            <tr><th>Id</th><td>@Person.PersonId</td></tr>
            <tr><th>Surname</th><td>@Person.Surname</td></tr>
            <tr><th>Firstname</th><td>@Person.Firstname</td></tr>
        </tbody>
    </table>
}

<button class="btn btn-outline-primary" @onclick="@(() => HandleClick(false))">
    Previous
</button>
<button class="btn btn-outline-primary" @onclick="@(() => HandleClick(true))">
    Next
</button>

@code {

    [Inject]
    public DataContext Context { get; set; }
```

```
    [Inject]
    public NavigationManager NavManager { get; set; }

    [Parameter]
    public long Id { get; set; } = 0;

    public Person Person { get; set; }

    protected async override Task OnParametersSetAsync() {
        await Task.Delay(1000);
        Person = await Context.People
            .FirstOrDefaultAsync(p => p.PersonId == Id) ?? new Person();
    }

    public void HandleClick(bool increment) {
        Person = null;
        NavManager.NavigateTo($"/person/{(increment ? Id + 1 : Id -1)}");
    }
}
```

直到设置了参数值，组件才能查询数据库，因此在 OnParametersSetAsync 方法中获得 Person 属性的值。因为数据库是与 ASP.NET Core 服务器一起运行的，在查询数据库之前添加了一秒钟的延迟，以帮助强调组件的工作方式。

在查询完成前，Person 属性的值为 null，这时它是表示查询结果的对象，或者如果查询没有生成结果，它就是一个新的 Person 对象。当 Person 对象为空时，将显示加载消息。

重启ASP.NET Core，并请求http://localhost:5000。单击表中显示的一个Edit按钮，PersonDisplay 组件显示数据的摘要。单击Previous和Next按钮，以查询具有相邻主键值的对象，生成如图 35-8 所示的结果。

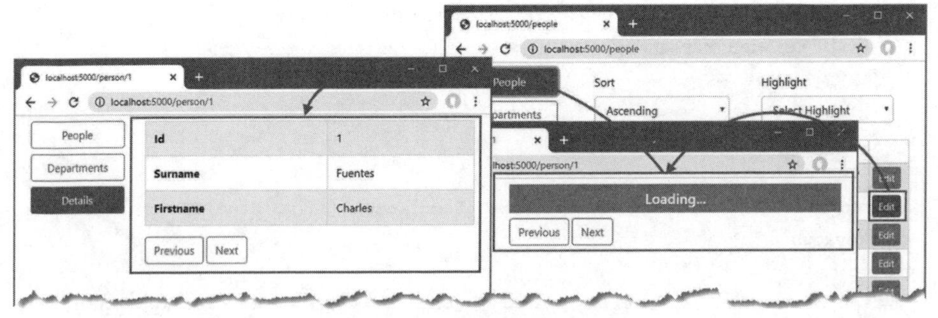


图 35-8　在组件中执行异步任务

注意，Blazor 在向用户显示内容之前，不会等待 OnParametersSetAsync 方法中执行的任务完成，这就是为什么当 Person 属性为 null 时，加载消息是有用的。一旦任务完成并为 Person 属性分配了一个值，组件的视图将自动重新呈现，更改将通过持久 HTTP 连接发送到浏览器，并显示给用户。

## 35.4 管理组件的交互

大多数组件通过参数和事件来工作，允许通过用户的交互来驱动应用程序中的更改。Blazor 还提供了用于管理与组件交互的高级选项，参见下面的讲解。

### 35.4.1 使用子组件的引用

父组件可以获得对子组件的引用，并可使用子组件定义的属性和方法。在准备过程中，代码清单 35-18 将禁用状态添加到 MultiNavLink 组件。

**代码清单 35-18 在 Blazor 文件夹的 MultiNavLink.razor 文件中添加特性**

```
<a class="@ComputedClass" @onclick="HandleClick" href="">
    @if (Enabled) {
        @ChildContent
    } else {
        @("Disabled")
    }
</a>

@code {

    [Inject]
    public NavigationManager NavManager { get; set; }

    [Parameter]
    public IEnumerable<string> Href { get; set; }

    [Parameter]
    public string Class { get; set; }

    [Parameter]
    public string ActiveClass { get; set; }

    [Parameter]
    public string DisabledClasses { get; set; }

    [Parameter]
    public NavLinkMatch? Match { get; set; }

    public NavLinkMatch ComputedMatch { get =>
        Match ?? (Href.Count() == 1 ? NavLinkMatch.Prefix : NavLinkMatch.All); }

    [Parameter]
    public RenderFragment ChildContent { get; set; }

    public string ComputedClass { get; set; }
```

```
    public void HandleClick() {
        NavManager.NavigateTo(Href.First());
    }

    private void CheckMatch(string currentUrl) {
        string path = NavManager.ToBaseRelativePath(currentUrl);
        path = path.EndsWith("/") ? path.Substring(0, path.Length - 1) : path;
        bool match = Href.Any(href => ComputedMatch == NavLinkMatch.All
                ? path == href : path.StartsWith(href));
        if (!Enabled) {
            ComputedClass = DisabledClasses;
        } else {
            ComputedClass = match ? $"{Class} {ActiveClass}" : Class;
        }
    }

    protected override void OnParametersSet() {
        ComputedClass = Class;
        NavManager.LocationChanged += (sender, arg) => CheckMatch(arg.Location);
        Href = Href.Select(h => h.StartsWith("/") ? h.Substring(1) : h);
        CheckMatch(NavManager.Uri);
    }

    private bool Enabled { get; set; } = true;

    public void SetEnabled(bool enabled) {
        Enabled = enabled;
        CheckMatch(NavManager.Uri);
    }
}
```

在代码清单 35-19 中，更新了共享布局，以便保留对 MultiNavLink 组件和一个按钮的引用，该按钮用于切换其 Enabled 属性值。

**代码清单 35-19　在 Blazor 文件夹的 NavLayout.razor 文件中保留引用**

```
@inherits LayoutComponentBase

<div class="container-fluid">
    <div class="row">
        <div class="col-3">
            @foreach (string key in NavLinks.Keys) {
                <MultiNavLink class="btn btn-outline-primary btn-block"
                        href="@NavLinks[key]"
                        ActiveClass="btn-primary text-white"
                        DisabledClasses="btn btn-dark text-light btn-block disabled"
                        @ref="Refs[key]">
```

```
                @key
            </MultiNavLink>
        }
        <button class="btn btn-secondary btn-block mt-5 " @onclick="ToggleLinks">
            Toggle Links
        </button>
    </div>
    <div class="col">
        @Body
    </div>
  </div>
</div>

@code {

    public Dictionary<string, string[]> NavLinks
        = new Dictionary<string, string[]> {
            {"People", new string[] {"/people", "/" } },
            {"Departments", new string[] {"/depts", "/departments" } },
            {"Details", new string[] { "/person" } }
        };

    public Dictionary<string, MultiNavLink> Refs
        = new Dictionary<string, MultiNavLink>();

    private bool LinksEnabled = true;

    public void ToggleLinks() {
        LinksEnabled = !LinksEnabled;
        foreach (MultiNavLink link in Refs.Values) {
            link.SetEnabled(LinksEnabled);
        }
    }
}
```

通过添加@ref 属性，并指定应该为组件分配的字段或属性的名称，可以创建对组件的引用。由于 MultiNavLink 组件是在由 Dictionary 驱动的@foreach 循环中创建的，因此保留引用的最简单方法也是在 Dictionary 中，如下所示：

```
...
<MultiNavLink class="btn btn-outline-primary btn-block"
    href="@NavLinks[key]" ActiveClass="btn-primary text-white"
    DisabledClasses="btn btn-dark text-light btn-block disabled"
    @ref="Refs[key]">
...
```

在创建每个 MultiNavLink 组件时，它添加到 Refs 字典中。Razor 组件编译成标准的 C#类，这意味着 MultiNavLink 组件的集合就是 MultiNavLink 对象的集合。

```
...
public Dictionary<string, MultiNavLink> Refs
        = new Dictionary<string, MultiNavLink>();
...
```

重启 ASP.NET Core，请求 http://localhost:5000，然后单击 Toggle Links 按钮。事件处理程序调用 ToggleLinks 方法，它为每个 MultiNavLink 组件设置 Enabled 属性的值，如图 35-9 所示。

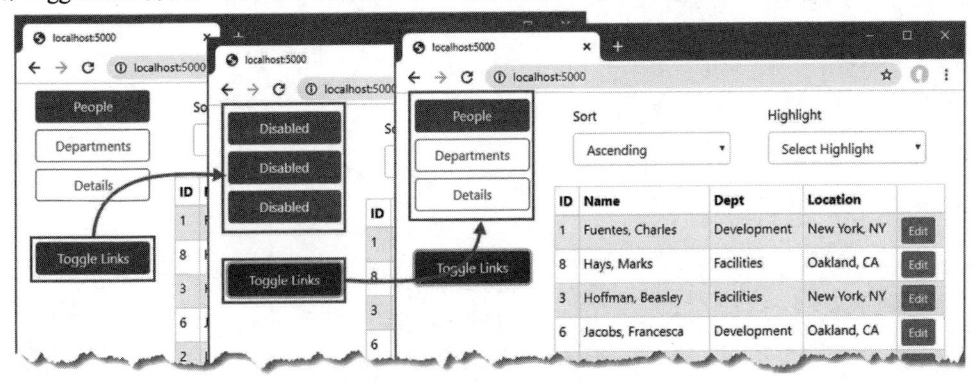


图 35-9　保留对组件的引用

■ **警告：** 引用只能在组件的内容呈现出来，调用了 OnAfterRender/OnAfterRenderAsync 生命周期方法之后使用。这使得引用非常适合在事件处理程序中使用，而不是在早期的生命周期方法中使用。

## 35.4.2　与来自其他代码的组件交互

组件可被 ASP.NET Core 应用程序中的其他代码使用，允许复杂项目的各个部分之间进行更丰富的交互。代码清单 35-20 修改了 MultiNavLink 组件中的方法，以便可以由 ASP.NET Core 应用程序的其他部分调用，来启用和禁用导航。

**代码清单 35-20　在 Blazor 文件夹的 MultiNavLink.razor 文件中替换一个方法**

```
<a class="@ComputedClass" @onclick="HandleClick" href="">
    @if (Enabled) {
        @ChildContent
    } else {
        @("Disabled")
    }
</a>

@code {

    // ...other properties and methods omitted for brevity...

    public void SetEnabled(bool enabled) {
```

```
            InvokeAsync(() => {
                Enabled = enabled;
                CheckMatch(NavManager.Uri);
                StateHasChanged();
            });
        }
    }
```

Razor 组件提供了在 Blazor 环境外部调用的代码中使用的两个方法，如表 35-6 所述。

表 35-6 Razor 组件的外部调用方法

| 名称 | 描述 |
| --- | --- |
| InvokeAsync(func) | 此方法用于在 Blazor 环境中执行一个函数 |
| StateHasChanged() | 当在正常生命周期之外发生更改时，将调用该方法，如下一节所示 |

InvokeAsync 方法用于调用 Blazor 环境中的一个函数，以确保正确地处理更改。当应用所有更改时，将调用 StateHasChanged 方法，触发 Blazor 更新并确保更改反映在组件的输出中。

要创建将在整个应用程序中可用的服务，请创建 Advanced/Services 文件夹，并将一个名为 ToggleService.cs 的类文件添加到其中，如代码清单 35-21 所示。

### 代码清单 35-21 Services 文件夹中 ToggleService.cs 文件的内容

```
using Advanced.Blazor;
using System.Collections.Generic;

namespace Advanced.Services {
    public class ToggleService {
        private List<MultiNavLink> components = new List<MultiNavLink>();
        private bool enabled = true;

        public void EnrolComponents(IEnumerable<MultiNavLink> comps) {
            components.AddRange(comps);
        }

        public bool ToggleComponents() {
            enabled = !enabled;
            components.ForEach(c => c.SetEnabled(enabled));
            return enabled;
        }
    }
}
```

此服务管理一个组件集合，并在调用其 ToggleComponents 方法时对所有组件调用 SetEnabled 方法。在这个服务中没有任何特定于 Blazor 的内容，它依赖于在编译 Razor 组件文件时生成的 C# 类。代码清单 35-22 更新了应用程序配置，将 ToggleService 类配置为单例服务。

**代码清单 35-22　在 Advanced 文件夹的 Startup.cs 文件中配置服务**

```
...
public void ConfigureServices(IServiceCollection services) {
    services.AddDbContext<DataContext>(opts => {
        opts.UseSqlServer(Configuration[
            "ConnectionStrings:PeopleConnection"]);
        opts.EnableSensitiveDataLogging(true);
    });
    services.AddControllersWithViews().AddRazorRuntimeCompilation();
    services.AddRazorPages().AddRazorRuntimeCompilation();
    services.AddServerSideBlazor();
    services.AddSingleton<Services.ToggleService>();
}
...
```

代码清单 35-23 更新了 Blazor 布局，以便保留对 MultiNavLink 组件的引用并向新服务注册。

**代码清单 35-23　在 Blazor 文件夹的 NavLayout.razor 文件中使用服务**

```
@inherits LayoutComponentBase
@using Advanced.Services

<div class="container-fluid">
    <div class="row">
        <div class="col-3">
            @foreach (string key in NavLinks.Keys) {
                <MultiNavLink class="btn btn-outline-primary btn-block"
                        href="@NavLinks[key]"
                        ActiveClass="btn-primary text-white"
                        DisabledClasses="btn btn-dark text-light btn-block disabled"
                        @ref="Refs[key]">
                    @key
                </MultiNavLink>
            }
            <button class="btn btn-secondary btn-block mt-5 " @onclick="ToggleLinks">
                Toggle Links
            </button>
        </div>
        <div class="col">
            @Body
        </div>
    </div>
</div>

@code {

    [Inject]
```

```
    public ToggleService Toggler { get; set; }

    public Dictionary<string, string[]> NavLinks
        = new Dictionary<string, string[]> {
            {"People", new string[] {"/people", "/" } },
            {"Departments", new string[] {"/depts", "/departments" } },
            {"Details", new string[] { "/person" } }
        };

    public Dictionary<string, MultiNavLink> Refs
        = new Dictionary<string, MultiNavLink>();

    protected override void OnAfterRender(bool firstRender) {
        if (firstRender) {
            Toggler.EnrolComponents(Refs.Values);
        }
    }

    public void ToggleLinks() {
        Toggler.ToggleComponents();
    }
}
```

如上一节所述，组件引用只有在内容呈现之后才可用。代码清单35-23使用 OnAfterRender 生命周期方法向服务注册组件引用，服务是通过依赖项注入接收的。

最后一步是从 ASP.NET Core 应用程序的不同部分使用服务。每次处理请求时，代码清单 35-24 向调用 ToggleService 的主控制器添加了一个简单的操作方法。

### 代码清单 35-24 在 Controllers 文件夹的 HomeController.cs 文件中添加一个操作方法

```
using Advanced.Models;
using Microsoft.AspNetCore.Mvc;
using Microsoft.EntityFrameworkCore;
using System.Collections.Generic;
using System.Linq;
using Advanced.Services;

namespace Advanced.Controllers {
    public class HomeController : Controller {
        private DataContext context;
        private ToggleService toggleService;

        public HomeController(DataContext dbContext, ToggleService ts) {
            context = dbContext;
            toggleService = ts;
        }
```

```
        public IActionResult Index([FromQuery] string selectedCity) {
            return View(new PeopleListViewModel {
                People = context.People
                    .Include(p => p.Department).Include(p => p.Location),
                Cities = context.Locations.Select(l => l.City).Distinct(),
                SelectedCity = selectedCity
            });
        }

        public string Toggle() => $"Enabled: { toggleService.ToggleComponents() }";
    }

    public class PeopleListViewModel {
        public IEnumerable<Person> People { get; set; }
        public IEnumerable<string> Cities { get; set; }
        public string SelectedCity { get; set; }

        public string GetClass(string city) =>
            SelectedCity == city ? "bg-info text-white" : "";
    }
}
```

重启ASP.NET Core并请求http://localhost:5000。打开一个单独的浏览器窗口并请求http:/localhost:5000/controllers/home/toggle。当ASP.NET Core应用程序处理第二个请求时。操作方法将使用服务来切换导航按钮的状态。每次请求/controllers/home/toggle时，导航按钮的状态将发生变化，如图35-10所示。

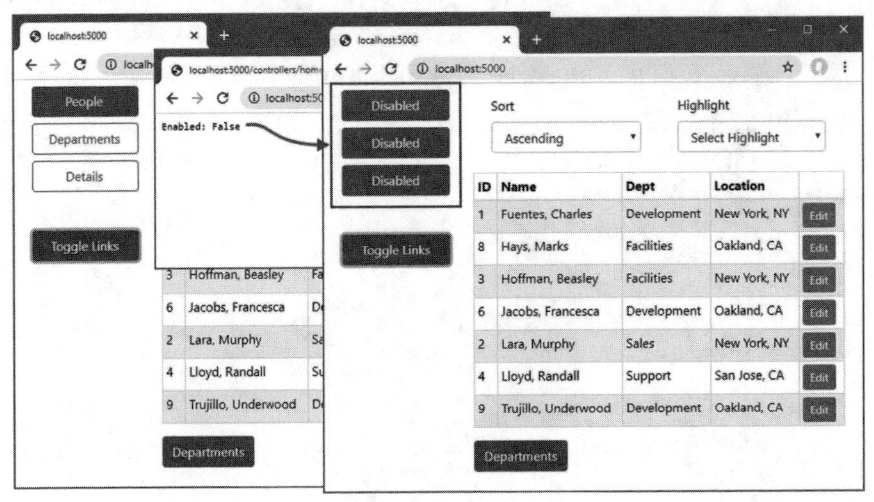


图 35-10 调用组件的方法

### 35.4.3 使用 JavaScript 与组件交互

Blazor 提供了一系列在 JavaScript 和服务器端 C#代码之间交互的工具，如下所述。

### 1. 从组件中调用 JavaScript 函数

为准备这些示例，在 wwwroot 文件夹中添加一个名为 interop.js 的 JavaScript 文件，并添加如代码清单 35-25 所示的代码。

**代码清单 35-25 wwwroot 文件夹中 interop.js 文件的内容**

```
function addTableRows(colCount) {
    let elem = document.querySelector("tbody");
    let row = document.createElement("tr");
    elem.append(row);
    for (let i = 0; i < colCount; i++) {
        let cell = document.createElement("td");
    cell.innerText = "New Elements"
        row.append(cell);
    }
}
```

JavaScript 代码使用浏览器提供的 API 来定位 tbody 元素，该元素表示表的主体，并添加一个新行，其中包含由函数参数指定的单元格数量。

要将 JavaScript 文件合并到应用程序中，请将代码清单 35-26 所示的元素添加到_Host Razor Pages，该页面配置为将 Blazor 应用程序交付到浏览器的回退页面。

**代码清单 35-26 在 Pages 文件夹的_Host.cshtml 文件中添加元素**

```
@page "/"
@{ Layout = null; }

<!DOCTYPE html>
<html>
<head>
    <title>@ViewBag.Title</title>
    <link href="/lib/twitter-bootstrap/css/bootstrap.min.css" rel="stylesheet" />
    <base href="~/" />
</head>
<body>
    <div class="m-2">
        <component type="typeof(Advanced.Blazor.Routed)" render-mode="Server" />
    </div>
    <script src="_framework/blazor.server.js"></script>
    <script src="~/interop.js"></script>
</body>
</html>
```

代码清单 35-27 修改了 PersonDisplay 组件，以便它呈现一个按钮，该按钮在触发 onclick 事件时调用 JavaScript 函数。还删除了之前为了演示组件生命周期方法的使用而添加的延迟。

**代码清单 35-27 在 Blazor 文件夹的 PersonDisplay.razor 文件中调用 JavaScript 函数**

```
@page "/person"
@page "/person/{id:long}"

@if (Person == null) {
    <h5 class="bg-info text-white text-center p-2">Loading...</h5>
} else {
    <table class="table table-striped table-bordered">
        <tbody>
            <tr><th>Id</th><td>@Person.PersonId</td></tr>
            <tr><th>Surname</th><td>@Person.Surname</td></tr>
            <tr><th>Firstname</th><td>@Person.Firstname</td></tr>
        </tbody>
    </table>
}

<button class="btn btn-outline-primary" @onclick="@HandleClick">
    Invoke JS Function
</button>

@code {

    [Inject]
    public DataContext Context { get; set; }

    [Inject]
    public NavigationManager NavManager { get; set; }

    [Inject]
    public IJSRuntime JSRuntime { get; set; }

    [Parameter]
    public long Id { get; set; } = 0;

    public Person Person { get; set; }

    protected async override Task OnParametersSetAsync() {
        //await Task.Delay(1000);
        Person = await Context.People
            .FirstOrDefaultAsync(p => p.PersonId == Id) ?? new Person();
    }

    public async Task HandleClick() {
        await JSRuntime.InvokeVoidAsync("addTableRows", 2);
    }
}
```

调用 JavaScript 函数是通过 IJSRuntime 接口完成的，该接口由组件通过依赖项注入接收。服务是作为 Blazor 配置的一部分自动创建的，并提供表 35-7 中描述的方法。

表 35-7　IJSRuntime 方法

| 名称 | 描述 |
| --- | --- |
| InvokeAsync<T>(name, args) | 此方法使用提供的参数调用指定的函数。结果类型由泛型类型参数指定 |
| InvokeVoidAsync(name, args) | 此方法调用一个不生成结果的函数 |

在代码清单 35-27 中，使用 InvokeVoidAsync 方法调用 JavaScript 函数 addTableRows，为函数参数提供一个值。重启 ASP.NET Core，导航到 http://localhost:5000/person/1，然后单击 Invoke JS Function 按钮。Blazor 将调用 JavaScript 函数，该函数将在表的末尾添加一行，如图 35-11 所示。

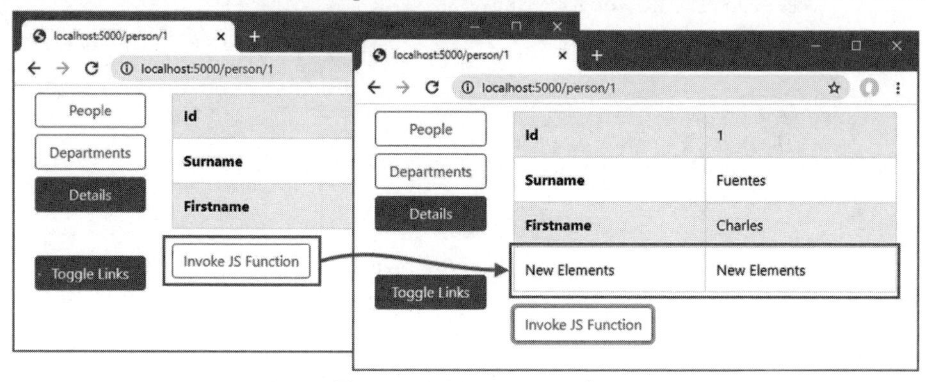


图 35-11　调用 JavaScript 函数

## 2. 保留对 HTML 元素的引用

Razor 组件可以保留对它们创建的 HTML 元素的引用，并将这些引用传递给 JavaScript 代码。代码清单 35-28 更改前一个示例中的 JavaScript 函数，使其对通过参数接收的 HTML 元素进行操作。

### 代码清单 35-28　在 wwwroot 文件夹的 interop.js 文件中定义参数

```
function addTableRows(colCount, elem) {
    //let elem = document.querySelector("tbody");
    let row = document.createElement("tr");
    elem.parentNode.insertBefore(row, elem);
    for (let i = 0; i < colCount; i++) {
        let cell = document.createElement("td");
        cell.innerText = "New Elements"
        row.append(cell);
    }
}
```

在代码清单 35-29 中，PersonDisplay 组件保留了对它创建的一个 HTML 元素的引用，并将其作为参数传递给 JavaScript 函数。

代码清单 35-29　在 Blazor 文件夹的 PersonDisplay.razor 文件中保留引用

```
@page "/person"
@page "/person/{id:long}"

@if (Person == null) {
    <h5 class="bg-info text-white text-center p-2">Loading...</h5>
} else {
    <table class="table table-striped table-bordered">
        <tbody>
            <tr><th>Id</th><td>@Person.PersonId</td></tr>
            <tr @ref="RowReference"><th>Surname</th><td>@Person.Surname</td></tr>
            <tr><th>Firstname</th><td>@Person.Firstname</td></tr>
        </tbody>
    </table>
}

<button class="btn btn-outline-primary" @onclick="@HandleClick">
    Invoke JS Function
</button>

@code {

    [Inject]
    public DataContext Context { get; set; }

    [Inject]
    public NavigationManager NavManager { get; set; }

    [Inject]
    public IJSRuntime JSRuntime { get; set; }

    [Parameter]
    public long Id { get; set; } = 0;

    public Person Person { get; set; }

    protected async override Task OnParametersSetAsync() {
        //await Task.Delay(1000);
        Person = await Context.People
            .FirstOrDefaultAsync(p => p.PersonId == Id) ?? new Person();
    }

    public ElementReference RowReference { get; set; }

    public async Task HandleClick() {
```

```
            await JSRuntime.InvokeVoidAsync("addTableRows", 2, RowReference);
        }
    }
```

@ref属性将HTML元素分配给一个属性，该属性的类型必须是ElementReference。重启ASP.NET Core，请求http://localhost:5000/person/1，然后单击Invoke JS Function按钮。ElementReference属性的值作为参数通过InvokeVoidAsync方法传递给JavaScript函数，生成如图35-12 所示的结果。

---

■ **注意：**对常规 HTML 元素的引用的唯一用途是将其传递给 Javascript 函数。使用前面章节中描述的绑定和事件特性与组件呈现的元素交互。

---

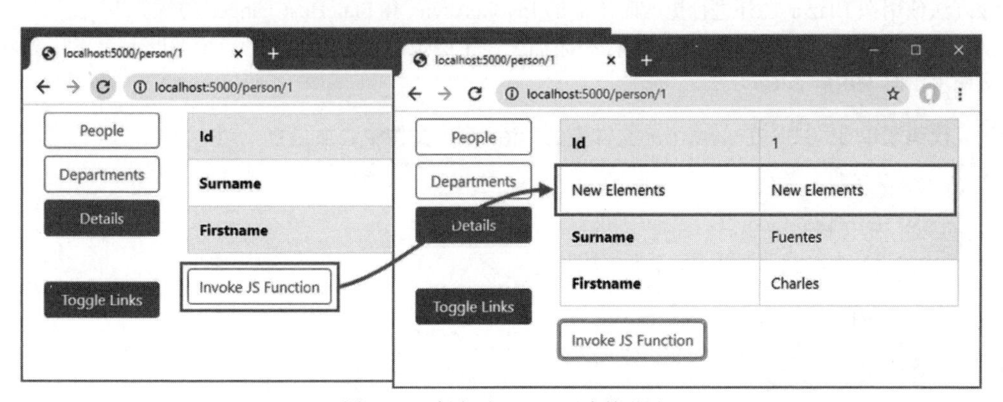


图 35-12 保留对 HTML 元素的引用

### 3. 从 JavaScript 调用组件方法

从 JavaScript 调用 C#方法的基本方法是使用静态方法。代码清单 35-30 将静态方法添加到更改启用状态的 MultiNavLink 组件。

#### 代码清单 35-30 在 Blazor 文件夹的 MultiNavLink.razor 文件中引入静态成员

```
<a class="@ComputedClass" @onclick="HandleClick" href="">
    @if (Enabled) {
        @ChildContent
    } else {
        @("Disabled")
    }
</a>

@code {

    // ...other methods and properties omitted for brevity...

    [JSInvokable]
    public static void ToggleEnabled() => ToggleEvent.Invoke(null, new EventArgs());
```

```
    private static event EventHandler ToggleEvent;

    protected override void OnInitialized() {
        ToggleEvent += (sender, args) => SetEnabled(!Enabled);
    }
}
```

静态方法必须使用 JSInvokable 属性进行修饰，然后才能从 JavaScript 代码中调用它们。使用静态方法的主要限制是难以更新单个组件，因此定义了组件的每个实例都会处理的一个静态事件。该事件命名为 ToggleEvent，它由将从 JavaScript 调用的静态方法触发。为了监听事件，使用了 OnInitialized 生命周期事件。当接收到事件时，组件的启用状态通过实例方法 SetEnabled 来切换，该方法使用在 Blazor 之外进行更改时所需的 InvokeAsync 和 StateHasChanged 方法。

代码清单 35-31 在 JavaScript 文件中添加了一个函数，该函数创建了一个 button 元素，该元素在单击时调用静态 C#方法。

#### 代码清单 35-31　在 wwwroot 文件夹的 interop.js 文件中添加函数

```
function addTableRows(colCount, elem) {
    //let elem = document.querySelector("tbody");
    let row = document.createElement("tr");
    elem.parentNode.insertBefore(row, elem);
    for (let i = 0; i < colCount; i++) {
        let cell = document.createElement("td");
        cell.innerText = "New Elements"
        row.append(cell);
    }
}

function createToggleButton() {
    let sibling = document.querySelector("button:last-of-type");
    let button = document.createElement("button");
    button.classList.add("btn", "btn-secondary", "btn-block");
    button.innerText = "JS Toggle";
    sibling.parentNode.insertBefore(button, sibling.nextSibling);
    button.onclick = () => DotNet.invokeMethodAsync("Advanced", "ToggleEnabled");
}
```

新函数定位一个现有 button 元素，并在其后面添加一个新按钮。当单击按钮时，调用组件方法，如下所示：

```
...
button.onclick = () => DotNet.invokeMethodAsync("Advanced", "ToggleEnabled");
...
```

注意用于 C#方法的 JavaScript 函数的大小写很重要：它是 DotNet，后面是句点，后面是 invokeMethodAsync，后面是小写的 i，参数是程序集的名称和静态方法的名称。组件的名称不是必需的。

代码清单 35-31 中函数寻找的按钮元素只有在 Blazor 为用户呈现了内容之后才可用。由于这个原因，代码清单 35-32 在 NavLayout 组件定义的 OnAfterRenderAsync 方法中添加了一条语句，仅在内容呈现后才调用 JavaScript 函数。NavLayout 组件是 MultiNavLink 组件的父组件，当调用静态方法时，MultiNavLink 组件会受到影响，并允许确保 JavaScript 函数只调用一次。

**代码清单 35-32　在 Blazor 文件夹的 NavLayout.razor 文件中调用 JavaScript 函数**

```
...
@code {

    [Inject]
    public IJSRuntime JSRuntime { get; set; }

    [Inject]
    public ToggleService Toggler { get; set; }

    public Dictionary<string, string[]> NavLinks
        = new Dictionary<string, string[]> {
            {"People", new string[] {"/people", "/" } },
            {"Departments", new string[] {"/depts", "/departments" } },
            {"Details", new string[] { "/person" } }
        };

    public Dictionary<string, MultiNavLink> Refs
        = new Dictionary<string, MultiNavLink>();

    protected async override Task OnAfterRenderAsync(bool firstRender) {
        if (firstRender) {
            Toggler.EnrolComponents(Refs.Values);
            await JSRuntime.InvokeVoidAsync("createToggleButton");
        }
    }

    public void ToggleLinks() {
        Toggler.ToggleComponents();
    }
}
...
```

重启 ASP.NET Core 并请求 http://localhost:5000。Blazor 呈现其内容后，调用 JavaScript 函数并创建一个新按钮。单击按钮将调用静态方法；该方法触发事件，来切换导航按钮的状态，并导致 Blazor 更新，如图 35-13 所示。

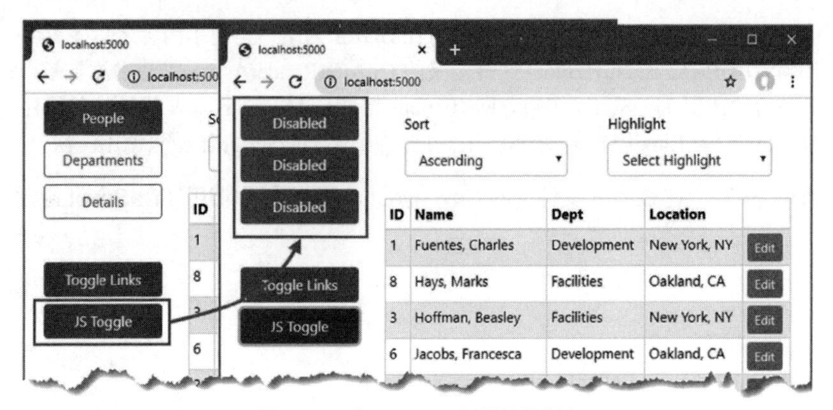


图 35-13　从 JavaScript 调用组件方法

#### 4. 从 JavaScript 函数中调用实例方法

上一个示例中的复杂性部分来自于对更新 Razor 组件对象的静态方法的响应。另一种方法是为 JavaScript 代码提供对实例方法的引用，然后可以直接调用该实例方法。

第一步是将 JSInvokable 属性添加到 JavaScript 代码将要调用的方法中。下面调用 ToggleService 类定义的 ToggleComponents 方法，如代码清单 35-33 所示。

**代码清单 35-33　在 Services 文件夹的 ToggleService.cs 文件中应用属性**

```
using Advanced.Blazor;
using System.Collections.Generic;
using Microsoft.JSInterop;

namespace Advanced.Services {
   public class ToggleService {
      private List<MultiNavLink> components = new List<MultiNavLink>();
      private bool enabled = true;

      public void EnrolComponents(IEnumerable<MultiNavLink> comps) {
         components.AddRange(comps);
      }

      [JSInvokable]
      public bool ToggleComponents() {
         enabled = !enabled;
         components.ForEach(c => c.SetEnabled(enabled));
         return enabled;
      }
   }
}
```

下一步是为 JavaScript 函数提供一个对象的引用，该对象的方法将被调用，如代码清单 35-34 所示。

### 代码清单 35-34　在 Blazor 文件夹的 NavLayout.razor 文件中提供实例

```
...
protected async override Task OnAfterRenderAsync(bool firstRender) {
    if (firstRender) {
        Toggler.EnrolComponents(Refs.Values);
        await JSRuntime.InvokeVoidAsync("createToggleButton",
            DotNetObjectReference.Create(Toggler));
    }
}
...
```

DotNetObjectReference.Create 方法创建对对象的引用，该引用使用 JSRuntime.InvokeVoidAsync 方法作为参数传递给 JavaScript 函数。最后一步是用 JavaScript 接收对象引用，并在单击按钮元素时调用其方法，如代码清单 35-35 所示。

### 代码清单 35-35　在 wwwroot 文件夹的 interop.js 文件中调用 C#方法

```
function addTableRows(colCount, elem) {
    //let elem = document.querySelector("tbody");
    let row = document.createElement("tr");
    elem.parentNode.insertBefore(row, elem);
    for (let i = 0; i < colCount; i++) {
        let cell = document.createElement("td");
        cell.innerText = "New Elements"
        row.append(cell);
    }
}

function createToggleButton(toggleServiceRef) {
    let sibling = document.querySelector("button:last-of-type");
    let button = document.createElement("button");
    button.classList.add("btn", "btn-secondary", "btn-block");
    button.innerText = "JS Toggle";
    sibling.parentNode.insertBefore(button, sibling.nextSibling);
    button.onclick = () => toggleServiceRef.invokeMethodAsync("ToggleComponents");
}
```

JavaScript 函数接收对 C#对象的引用作为参数，并使用 invokeMethodAsync 调用它的方法，将方法的名称指定为参数。也可提供方法的参数，但在本例中不是必需的。

重启 ASP.NET Core，请求 http://localhost:5000，然后单击 JS Toggle 按钮。结果如图 35-13 所示，但组件中的更改是通过 ToggleService 对象管理的。

## 35.5 小结

本章解释了如何将组件与路由结合起来，根据当前 URL 修改显示给用户的内容。描述了组件的生命周期以及在过程的每个阶段可以实现的方法，在本章的结尾，解释了从 Blazor 外部调用组件方法的不同方式，包括与 JavaScript 的互操作性。第 36 章将描述 Blazor 为 HTML 表单提供的特性。

# 第 36 章

# Blazor 表单和数据

本章将描述 Blazor 为处理 HTML 表单提供的特性，包括对数据验证的支持。还描述 Blazor 提供的内置组件，并展示了它们是如何使用的。本章还将解释 Blazor 模型如何在 Entity Framework Core 中导致意外结果，并展示如何解决这些问题。本章最后通过创建一个简单的表单应用程序来演示创建、读取、更新和删除数据(CRUD 操作)，并解释如何扩展 Blazor 表单特征来改善用户体验。表 36-1 概述了 Blazor 表单特征。

表 36-1　Blazor 表单特征的相关知识

| 问题 | 答案 |
| --- | --- |
| 它们是什么? | Blazor 提供一组内置组件，为用户提供一个易于验证的表单 |
| 它们为什么有用? | 表单仍然是 Web 应用程序的核心构建块之一，而这些组件提供了大多数项目所需的功能 |
| 它们是如何使用的? | EditForm 组件用作单个表单字段组件的父组件 |
| 是否存在缺陷或限制? | Entity Framework Core 和 Blazor 一起工作的方式可能会有问题，这些问题在使用表单时尤其明显 |
| 还有其他选择吗? | 可以创建自己的表单组件和验证特性，尽管本章描述的特性适用于大多数项目，而且如本章所述，可以轻松扩展 |

表 36-2 总结了本章的内容。

表 36-2　本章内容摘要

| 问题 | 解决方案 | 代码清单 |
| --- | --- | --- |
| 创建 HTML 表单 | 使用 EditForm 和 Input*组件 | 36-7～36-9、36-13 |
| 验证数据 | 使用标准验证属性和由 EditForm 组件触发的事件 | 36-10～36-12 |
| 舍弃未保存的数据 | 明确地释放数据或为组件创建新的作用域 | 36-14～36-16 |
| 避免重复查询数据库 | 显式管理查询的执行 | 36-17～36-19 |

## 36.1　准备工作

本章使用了第 35 章中的 Advanced 项目。为准备这一章，创建 Blazor/Forms 文件夹并添加一个名为 EmptyLayout.razor 的 Razor 组件。内容如代码清单 36-1 所示。本章使用这个组件作为主要的布局。

**提示：** 可从 https://github.com/apress/pro-asp.net-core-3 下载本章和本书中其他所有章节的示例项目。如果在运行示例时遇到问题，请参阅第 1 章以了解如何获得帮助。

**代码清单 36-1　Blazor/Forms 文件夹中 EmptyLayout.razor 文件的内容**

```
@inherits LayoutComponentBase
<div class="m-2">
    @Body
</div>
```

为 Blazor/Forms 文件夹添加一个名为 FormSpy.razor 的 Razor 组件，内容如代码清单 36-2 所示。这个组件用来显示表单元素和旁边正在编辑的值。

**代码清单 36-2　Blazor/Forms 文件夹中 FormSpy.razor 文件的内容**

```
<div class="container-fluid no-gutters">
    <div class="row">
        <div class="col">
            @ChildContent
        </div>
        <div class="col">
            <table class="table table-sm table-striped table-bordered">
                <thead>
                    <tr><th colspan="2" class="text-center">Data Summary</th></tr>
                </thead>
                <tbody>
                    <tr><th>ID</th><td>@PersonData?.PersonId</td></tr>
                    <tr><th>Firstname</th><td>@PersonData?.Firstname</td></tr>
                    <tr><th>Surname</th><td>@PersonData?.Surname</td></tr>
                    <tr><th>Dept ID</th><td>@PersonData?.DepartmentId</td></tr>
                    <tr><th>Location ID</th><td>@PersonData?.LocationId</td></tr>
                </tbody>
            </table>
        </div>
    </div>
</div>

@code {

    [Parameter]
    public RenderFragment ChildContent { get; set; }

    [Parameter]
    public Person PersonData { get; set; }
}
```

接下来，给 Blazor/Forms 文件夹添加一个名为 Editor.razor 的组件，并添加如代码清单 36-3 所示的内容。此组件将用于编辑现有的 Person 对象和创建新的 Person 对象。

---

■ **注意：**在阅读完本章的其余部分之前，不要在实际项目中使用 Editor 和 List 组件。本章已经包含一些常见的陷阱，将在后面解释。

---

**代码清单 36-3　Blazor/Forms 文件夹中 Editor.razor 文件的内容**

```
@page "/forms/edit/{id:long}"
@layout EmptyLayout

<h4 class="bg-primary text-center text-white p-2">Edit</h4>

<FormSpy PersonData="PersonData">
    <h4 class="text-center">Form Placeholder</h4>
    <div class="text-center">
        <NavLink class="btn btn-secondary" href="/forms">Back</NavLink>
    </div>
</FormSpy>

@code {

    [Inject]
    public NavigationManager NavManager { get; set; }

    [Inject]
    DataContext Context { get; set; }

    [Parameter]
    public long Id { get; set; }

    public Person PersonData { get; set; } = new Person();

    protected async override Task OnParametersSetAsync() {
        PersonData = await Context.People.FindAsync(Id);
    }
}
```

代码清单 36-3 中的组件使用@layout 表达式覆盖默认布局并选择 EmptyLayout。并排布局用于在占位符旁边显示 PersonTable 组件，在这里将添加表单。

最后，在 Blazor/Forms 文件夹中创建一个名为 List.razor 的组件，并添加如代码清单 36-4 所示的内容，以定义一个组件，该组件以表的形式向用户显示 Person 对象列表。

**代码清单 36-4　Blazor/Forms 文件夹中 List.razor 文件的内容**

```
@page "/forms"
@page "/forms/list"
```

```
@layout EmptyLayout

<h5 class="bg-primary text-white text-center p-2">People</h5>

<table class="table table-sm table-striped table-bordered">
    <thead>
        <tr>
            <th>ID</th><th>Name</th><th>Dept</th><th>Location</th><th></th>
        </tr>
    </thead>
    <tbody>
        @if (People.Count() == 0) {
            <tr><th colspan="5" class="p-4 text-center">Loading Data...</th></tr>
        } else {
            @foreach (Person p in People) {
                <tr>
                    <td>@p.PersonId</td>
                    <td>@p.Surname, @p.Firstname</td>
                    <td>@p.Department.Name</td>
                    <td>@p.Location.City</td>
                    <td>
                        <NavLink class="btn btn-sm btn-warning"
                            href="@GetEditUrl(p.PersonId)">
                            Edit
                        </NavLink>
                    </td>
                </tr>
            }
        }
    </tbody>
</table>

@code {

    [Inject]
    public DataContext Context { get; set; }

    public IEnumerable<Person> People { get; set; } = Enumerable.Empty<Person>();

    protected override void OnInitialized() {
        People = Context.People.Include(p => p.Department).Include(p => p.Location);
    }

    string GetEditUrl(long id) => $"/forms/edit/{id}";
}
```

### 删除数据库并运行应用程序

打开一个新的 PowerShell 命令提示符，导航到包含 Advanced.csproj 文件的文件夹，并运行代码清单 36-5 所示的命令来删除数据库。

**代码清单 36-5　删除数据库**

```
dotnet ef database drop --force
```

从 Debug 菜单中选择 Start Without Debugging 或 Run Without Debugging，或者使用 PowerShell 命令提示符运行代码清单 36-6 中所示的命令。

**代码清单 36-6　运行示例应用程序**

```
dotnet run
```

使用浏览器请求 http://localhost:5000/forms，这将生成一个数据表。单击其中一个 Edit 按钮，将看到表单占位符和显示所选 Person 对象当前属性值的摘要，如图 36-1 所示。

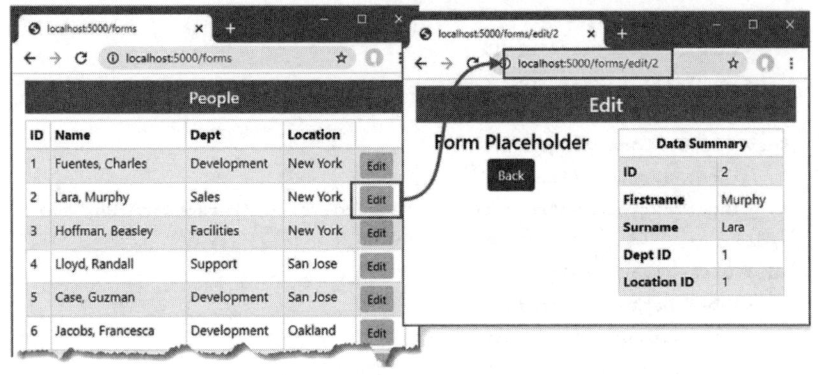


图 36-1　运行示例应用程序

## 36.2　使用 Blazor 表单组件

Blazor 提供了一组用于呈现表单元素的内置组件，确保在用户交互和集成验证之后更新服务器端组件属性。表 36-3 描述了 Blazor 提供的组件。

**表 36-3　Bazor 表单组件**

| 名称 | 描述 |
|---|---|
| EditForm | 此组件将呈现连接起来进行数据验证的表单元素 |
| InputText | 此组件呈现一个绑定到 C#字符串属性的输入元素 |
| InputCheckbox | 此组件呈现一个输入元素，它的类型属性是 checkbox，并且绑定到 C# bool 属性 |
| InputDate | 此组件呈现一个输入元素，该元素的类型属性为 date，并绑定到 C# DateTime 或 DateTimeOffset 属性 |
| InputNumber | 此组件呈现一个输入元素，其类型属性为 number，并绑定到 C# int、long、float、double 或 decimal 值 |
| InputTextArea | 此组件呈现一个绑定到 C#字符串属性的 textarea 组件 |

EditForm 组件必须用于任何其他组件才能工作。在代码清单 36-7 中添加了一个 EditForm 和两个 InputText 组件，来表示 Person 类定义的两个属性。

**代码清单 36-7　在 Blazor/Forms 文件夹的 Editor.razor 文件中使用表单组件**

```
@page "/forms/edit/{id:long}"
@layout EmptyLayout

<h4 class="bg-primary text-center text-white p-2">Edit</h4>

<FormSpy PersonData="PersonData">
    <EditForm Model="PersonData">
        <div class="form-group">
            <label>Person ID</label>
            <InputNumber class="form-control"
                @bind-Value="PersonData.PersonId" disabled />
        </div>
        <div class="form-group">
            <label>Firstname</label>
            <InputText class="form-control" @bind-Value="PersonData.Firstname" />
        </div>
        <div class="form-group">
            <label>Surname</label>
            <InputText class="form-control" @bind-Value="PersonData.Surname" />
        </div>
        <div class="form-group">
            <label>Dept ID</label>
            <InputNumber class="form-control"
                @bind-Value="PersonData.DepartmentId" />
        </div>
        <div class="text-center">
            <NavLink class="btn btn-secondary" href="/forms">Back</NavLink>
        </div>
    </EditForm>
</FormSpy>

@code {

    // ...statements omitted for brevity...
}
```

EditForm 组件呈现一个表单元素，并为“验证表单数据”一节描述的验证特性提供基础。Model 属性用于向 EditForm 提供表单用于编辑和验证的对象。

表 36-3 中名称以 Input 开头的组件用于显示单个模型属性的 input 或 textarea 元素。这些组件定义了一个名为 Value 的自定义绑定，该绑定使用@bind-Value 属性与模型属性关联。属性级组件必须与它们呈现给用户的属性类型相匹配。正是由于这个原因，这里使用 InputText 组件来处理 Person 类的 Firstname 和 Surname 属性，而 InputNumber 组件用于 PersonId 和 DepartmentId 属性。

如果使用的属性级组件具有错误类型的模型属性，那么当组件试图解析输入 HTML 元素中的值时，将收到一个错误。

重启 ASP.NET Core 并请求 http://localhost:5000/form/edit/2，将看到显示的三个输入元素。编辑值并通过按 Tab 键移动焦点，将在更新窗口的右侧看到汇总数据，如图 36-2 所示。内置的表单组件支持属性 splatting，这就是将 PersonId 属性的 disabled 特性(InputNumber 组件)应用到 input 元素的原因。

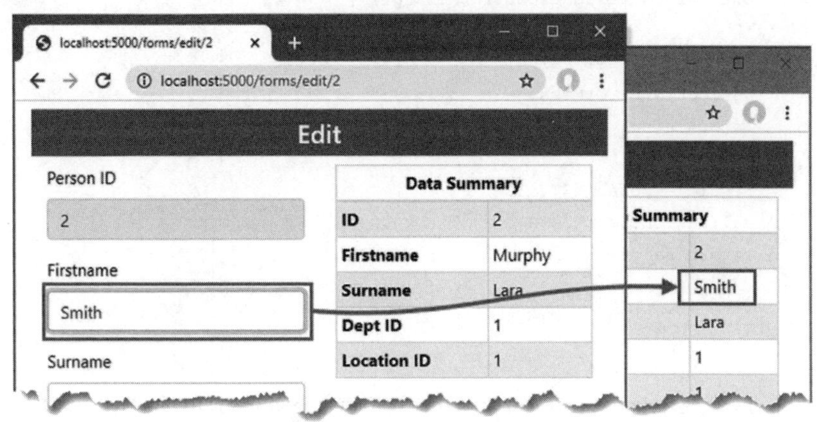


图 36-2 使用 Blazor 表单元素

## 36.2.1 创建自定义表单组件

Blazor 仅为 input 和 textarea 元素提供内置组件。幸运的是，创建一个集成到 Blazor 表单特性的自定义组件是一个简单的过程。在 Blazor/Forms 文件夹中添加一个名为 CustomSelect.razor 的 Razor 组件，并用它来定义代码清单 36-8 所示的组件。

### 代码清单 36-8 Blazor/Forms 文件夹中 CustomSelect.razor 文件的内容

```
@typeparam TValue
@inherits InputBase<TValue>

<select class="form-control @CssClass" value="@CurrentValueAsString"
        @onchange="@(ev => CurrentValueAsString = ev.Value as string)">
    @ChildContent
    @foreach (KeyValuePair<string, TValue> kvp in Values) {
        <option value="@kvp.Value">@kvp.Key</option>
    }
</select>

@code {

    [Parameter]
    public RenderFragment ChildContent { get; set; }

    [Parameter]
```

```
    public IDictionary<string, TValue> Values { get; set; }

    [Parameter]
    public Func<string, TValue> Parser { get; set; }

    protected override bool TryParseValueFromString(string value, out TValue result,
            out string validationErrorMessage) {
        try {
            result = Parser(value);
            validationErrorMessage = null;
            return true;
        } catch {
            result = default(TValue);
            validationErrorMessage = "The value is not valid";
            return false;
        }
    }
}
```

表单组件的基类是 InputBase<TValue>，其中通用类型参数是组件表示的模型属性类型。基类负责大部分工作，并提供 CurrentValueAsString 属性，该属性用于在用户选择新值时在事件处理程序中提供当前值，如下所示：

```
...
<select class="form-control @CssClass" value="@CurrentValueAsString"
        @onchange="@(ev => CurrentValueAsString = ev.Value as string)">
...
```

在准备数据验证的过程中，如下一节所述，该组件包括 CssClass 属性的值，CssClass 属性在 select 元素的 class 属性中，如下所示：

```
...
<select class="form-control @CssClass" value="@CurrentValueAsString"
        @onchange="@(ev => CurrentValueAsString = ev.Value as string)">
...
```

必须实现抽象的 TryParseValueFromString 方法，以便基类能够在 HTML 元素使用的字符串值和 C#模型属性的相应值之间进行映射。这里不想将自定义 select 元素实现为任何特定的 C#数据类型，因此使用@typeparam 表达式来定义通用类型参数。Values 属性用于接收将显示给用户的字典映射字符串值和用作 C#值的 TValue 值。该方法接收两个 out 参数，这些参数用于设置解析值，以及解析器验证错误消息，如果存在问题，该错误消息将显示给用户。由于正在使用泛型类型，因此 Parser 属性接收一个函数，调用该函数，以将字符串值解析为 TValue 值。

代码清单 36-9 应用了新的表单组件，因此用户可为 Person 类定义的 DepartmentId 和 LocationId 属性选择值。

**代码清单 36-9 在 Blazor/Forms 文件夹的 Editor.razor 文件中使用自定义表单元素**

```
@page "/forms/edit/{id:long}"
@layout EmptyLayout

<h4 class="bg-primary text-center text-white p-2">Edit</h4>

<FormSpy PersonData="PersonData">

    <EditForm Model="PersonData">
        <div class="form-group">
            <label>Firstname</label>
            <InputText class="form-control" @bind-Value="PersonData.Firstname" />
        </div>
        <div class="form-group">
            <label>Surname</label>
            <InputText class="form-control" @bind-Value="PersonData.Surname" />
        </div>
        <div class="form-group">
            <label>Dept ID</label>
            <CustomSelect TValue="long" Values="Departments"
                      Parser="@(str => long.Parse(str))"
                      @bind-Value="PersonData.DepartmentId">
                <option selected disabled value="0">Choose a Department</option>
            </CustomSelect>
        </div>
        <div class="form-group">
            <label>Location ID</label>
            <CustomSelect TValue="long" Values="Locations"
                      Parser="@(str => long.Parse(str))"
                      @bind-Value="PersonData.LocationId">
                <option selected disabled value="0">Choose a Location</option>
            </CustomSelect>
        </div>
        <div class="text-center">
            <NavLink class="btn btn-secondary" href="/forms">Back</NavLink>
        </div>
    </EditForm>
</FormSpy>

@code {

    [Inject]
    public NavigationManager NavManager { get; set; }

    [Inject]
    DataContext Context { get; set; }
```

```
    [Parameter]
    public long Id { get; set; }

    public Person PersonData { get; set; } = new Person();

    public IDictionary<string, long> Departments { get; set; }
        = new Dictionary<string, long>();

    public IDictionary<string, long> Locations { get; set; }
        = new Dictionary<string, long>();

    protected async override Task OnParametersSetAsync() {
        PersonData = await Context.People.FindAsync(Id);
        Departments = await Context.Departments
            .ToDictionaryAsync(d => d.Name, d => d.Departmentid);
        Locations = await Context.Locations
            .ToDictionaryAsync(l => $"{l.City}, {l.State}", l => l.LocationId);
    }
}
```

使用Entity Framework Core ToDictionaryAsync方法从Department和Location数据创建值和标签的集合，并使用它们配置CustomSelect组件。重启ASP.NET Core，请求http://localhost:5000/form/edit/2；将看到如图36-3所示的选择元素。当选择一个新值时，CustomSelect组件将更新CurrentValueAsString属性，这导致对TryParseValueFromString方法的调用，其结果用于更新Value绑定。

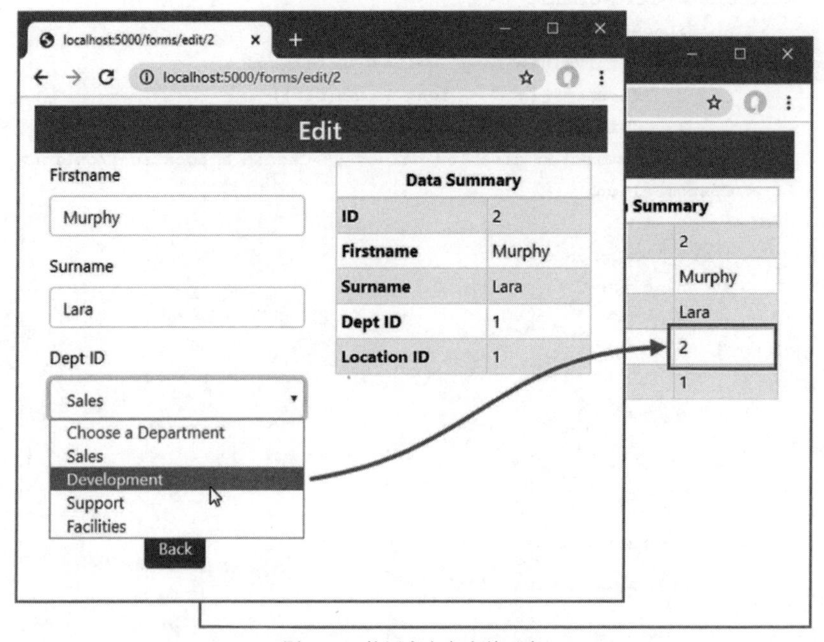


图 36-3　使用自定义表单元素

## 36.2.2 验证表单数据

Blazor 提供了使用标准属性执行验证的组件。表 36-4 描述了验证组件。

表 36-4 Blazor 验证组件

| 名称 | 描述 |
| --- | --- |
| DataAnnotationsValidator | 此组件将应用于模型类的验证属性集成到 Blazor 表单特性中 |
| ValidationMessage | 此组件显示单个属性的验证错误消息 |
| ValidationSummary | 此组件显示整个模型对象的验证错误消息 |

验证组件生成分配给类的元素，如表 36-5 所示，可以用 CSS 样式化这些元素，以吸引用户的注意。

表 36-5 Blazor 验证组件使用的类

| 名称 | 描述 |
| --- | --- |
| validation-errors | ValidationSummary 组件生成一个 ul 元素，该元素被分配给这个类，并且是验证消息摘要的顶级容器 |
| validation-message | ValidationSummary 组件使用为每个验证消息分配给这个类的 li 元素来填充它的 ul 元素。ValidationMessage 组件为这个类的属性级消息呈现一个分配给它的 div 元素 |

Blazor Input*组件将它们生成的 HTML 元素添加到表 36-6 描述的类中，以指示验证状态。这包括 InputBase<TValue>类，从这个类派生了 CustomSelect 组件，它也是代码清单 36-8 中 CssClass 属性的用途。

表 36-6 添加到表单元素的验证类

| 名称 | 描述 |
| --- | --- |
| modified | 一旦用户编辑了值，元素就会添加到这个类中 |
| valid | 如果包含的值通过验证，则将元素添加到该类中 |
| invalid | 如果元素包含的值验证失败，则将元素添加到该类中 |

这种组件和类的组合一开始可能会令人困惑，但关键是首先根据表 36-5 和表 36-6 中的类定义需要的 CSS 样式。将一个名为 blazorValidation.css 的 CSS 样式表添加到 wwwroot 文件夹中，其内容如代码清单 36-10 所示。

### 代码清单 36-10 wwwroot 文件夹中 blazorValidation.css 文件的内容

```
.validation-errors {
   background-color: rgb(220, 53, 69); color: white; padding: 8px;
   text-align: center; font-size: 16px; font-weight: 500;
}
div.validation-message { color: rgb(220, 53, 69); font-weight: 500 }
.modified.valid { border: solid 3px rgb(40, 167, 69); }
.modified.invalid { border: solid 3px rgb(220, 53, 69); }
```

这些样式将错误消息格式化为红色，并对单个表单元素应用红色或绿色边框。代码清单 36-11 导入 CSS 样式表并应用 Blazor 验证组件。

**代码清单 36-11　在 Blazor/Forms 文件夹的 Editor.razor 文件中应用验证组件**

```
@page "/forms/edit/{id:long}"
@layout EmptyLayout

<link href="/blazorValidation.css" rel="stylesheet" />
<h4 class="bg-primary text-center text-white p-2">Edit</h4>

<FormSpy PersonData="PersonData">
    <EditForm Model="PersonData">
        <DataAnnotationsValidator />
        <ValidationSummary />
        <div class="form-group">
            <label>Firstname</label>
            <ValidationMessage For="@(() => PersonData.Firstname)" />
            <InputText class="form-control" @bind-Value="PersonData.Firstname" />
        </div>
        <div class="form-group">
            <label>Surname</label>
            <ValidationMessage For="@(() => PersonData.Surname)" />
            <InputText class="form-control" @bind-Value="PersonData.Surname" />
        </div>
        <div class="form-group">
            <label>Dept ID</label>
            <ValidationMessage For="@(() => PersonData.DepartmentId)" />
            <CustomSelect TValue="long" Values="Departments"
                    Parser="@(str => long.Parse(str))"
                    @bind-Value="PersonData.DepartmentId">
                <option selected disabled value="0">Choose a Department</option>
            </CustomSelect>
        </div>
        <div class="form-group">
            <label>Location ID</label>
            <ValidationMessage For="@(() => PersonData.LocationId)" />
            <CustomSelect TValue="long" Values="Locations"
                    Parser="@(str => long.Parse(str))"
                    @bind-Value="PersonData.LocationId">
                <option selected disabled value="0">Choose a Location</option>
            </CustomSelect>
        </div>
        <div class="text-center">
            <NavLink class="btn btn-secondary" href="/forms">Back</NavLink>
        </div>
    </EditForm>
```

```
</FormSpy>

@code {

    // ...statements omitted for brevity...
}
```

DataAnnotationsValidator和ValidationSummary组件在应用时没有任何配置属性。ValidationMessage属性使用For属性配置，该属性接收一个函数，该函数返回组件所表示的属性。例如，下面是选择Firstname属性的表达式：

```
..s.
<ValidationMessage For="@(() => PersonData.Firstname)" />
...
```

该表达式不定义任何参数，并从用于 EditForm 组件中 Model 属性的对象(而不是模型类型)中选择属性。对于本例，这意味着表达式操作 PersonData 对象而不是 Person 类。

**提示：** Blazor 并不总是能够确定 ValidationMessage 组件的属性类型。如果接收到异常，则可以添加 TValue 属性来显式设置类型。例如，如果 ValidationMessage 表示的属性类型是 long，那么添加一个 TValue="long"属性。

启用数据验证的最后一步是将属性应用到模型类，如代码清单 36-12 所示。

**代码清单 36-12 在 Models 文件夹的 Person.cs 文件中应用验证属性**

```
using System.Collections.Generic;
using System.ComponentModel.DataAnnotations;

namespace Advanced.Models {

    public class Person {

        public long PersonId { get; set; }

        [Required(ErrorMessage = "A firstname is required")]
        [MinLength(3, ErrorMessage = "Firstnames must be 3 or more characters")]
        public string Firstname { get; set; }

        [Required(ErrorMessage = "A surname is required")]
        [MinLength(3, ErrorMessage = "Surnames must be 3 or more characters")]
        public string Surname { get; set; }

        [Required]
        [Range(1, long.MaxValue,
            ErrorMessage = "A department must be selected")]
        public long DepartmentId { get; set; }
```

```
        [Required]
        [Range(1, long.MaxValue,
            ErrorMessage = "A location must be selected")]
        public long LocationId { get; set; }

        public Department Department { get; set; }
        public Location Location { get; set; }
    }
}
```

要查看验证组件的效果，请重新启动 ASP.NET Core，请求 http://localhost:5000/form/edit/2。清除 Firstname 字段并通过按 Tab 键或单击另一个字段移动焦点。当焦点更改时，将执行验证，并显示错误消息。Editor 组件同时显示摘要消息和每个属性消息，因此相同的错误消息会显示两次。从 Surname 字段中删除除前两个字符以外的所有字符，当更改焦点时将显示第二条验证消息，如图 36-4 所示。也有对其他属性的验证支持，但是 select 元素不允许用户选择无效的有效值。如果更改了一个值，select 元素将用绿色边框装饰，以指示有效的选择，但是在演示如何使用表单组件创建新的数据对象之前，看不到无效的响应。

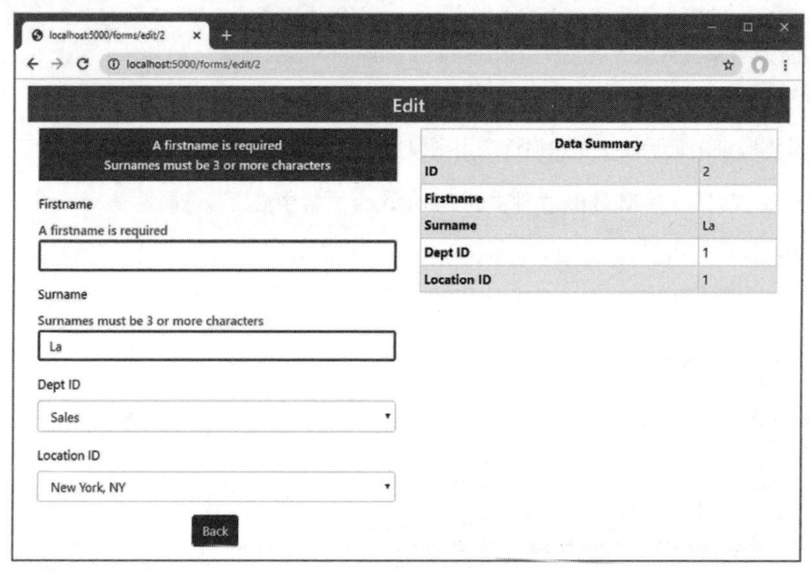


图 36-4　使用 Blazor 验证特性

## 36.2.3　处理表单事件

EditForm 组件定义了允许应用程序响应用户操作的事件，如表 36-7 所述。

表 36-7　EditForm 事件

| 名称 | 描述 |
| --- | --- |
| OnValidSubmit | 当提交表单且表单数据通过验证时触发此事件 |
| OnInvalidSubmit | 当提交表单且表单数据验证失败时触发此事件 |
| OnSubmit | 此事件在表单提交和验证执行之前触发 |

这些事件是通过在 EditForm 组件包含的内容中添加一个传统的提交按钮来触发的。EditForm 组件处理它所呈现的表单元素发送的 onsubmit 事件，应用验证，并触发表中描述的事件。代码清单 36-13 向 Editor 组件添加了一个 submit 按钮，并处理 EditForm 事件。

**代码清单 36-13 在 Blazor/Forms 文件夹的 Editor.razor 文件中处理 EditForm 事件**

```
@page "/forms/edit/{id:long}"
@layout EmptyLayout

<link href="/blazorValidation.css" rel="stylesheet" />

<h4 class="bg-primary text-center text-white p-2">Edit</h4>
<h6 class="bg-info text-center text-white p-2">@FormSubmitMessage</h6>

<FormSpy PersonData="PersonData">
   <EditForm Model="PersonData" OnValidSubmit="HandleValidSubmit"
         OnInvalidSubmit="HandleInvalidSubmit">
      <DataAnnotationsValidator />
      <ValidationSummary />
      <div class="form-group">
         <label>Firstname</label>
         <ValidationMessage For="@(() => PersonData.Firstname)" />
         <InputText class="form-control" @bind-Value="PersonData.Firstname" />
      </div>
      <div class="form-group">
         <label>Surname</label>
         <ValidationMessage For="@(() => PersonData.Surname)" />
         <InputText class="form-control" @bind-Value="PersonData.Surname" />
      </div>
      <div class="form-group">
         <label>Dept ID</label>
         <ValidationMessage For="@(() => PersonData.DepartmentId)" />
         <CustomSelect TValue="long" Values="Departments"
                  Parser="@(str => long.Parse(str))"
                  @bind-Value="PersonData.DepartmentId">
            <option selected disabled value="0">Choose a Department</option>
         </CustomSelect>
      </div>
      <div class="form-group">
         <label>Location ID</label>
         <ValidationMessage For="@(() => PersonData.LocationId)" />
         <CustomSelect TValue="long" Values="Locations"
                  Parser="@(str => long.Parse(str))"
                  @bind-Value="PersonData.LocationId">
            <option selected disabled value="0">Choose a Location</option>
         </CustomSelect>
      </div>
```

```
        <div class="text-center">
            <button type="submit" class="btn btn-primary">Submit</button>
            <NavLink class="btn btn-secondary" href="/forms">Back</NavLink>
        </div>
    </EditForm>
</FormSpy>

@code {

    // ...other statements omitted for brevity...

    public string FormSubmitMessage { get; set; } = "Form Data Not Submitted";
    public void HandleValidSubmit() => FormSubmitMessage = "Valid Data Submitted";
    public void HandleInvalidSubmit() =>
        FormSubmitMessage = "Invalid Data Submitted";
}
```

重启 ASP.NET Core，请求 http://localhost:5000/form/edit/2。清除 Firstname 字段，然后单击 Submit 按钮。除了验证错误之外，还将看到一条消息，指示提交的表单使用了无效数据。在字段中输入一个名称，再次单击 Submit，消息会更改，如图 36-5 所示。

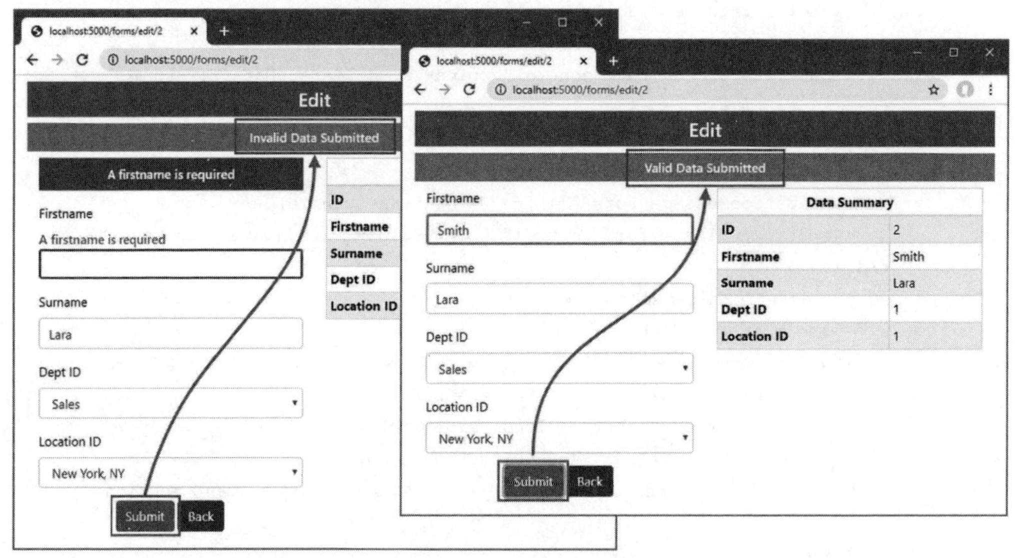


图 36-5　处理 EditForm 事件

## 36.3　使用 Entity Framework Core 与 Blazor

Blazor 模型改变了 Entity Framework Core 的行为方式，如果习惯于编写常规的 ASP.NET Core 应用程序，那么这可能会导致意想不到的结果。接下来将解释这些问题以及如何避免可能出现的问题。

### 36.3.1 理解 Entity Framework Core 上下文范围问题

要查看第一个问题，请求 http://localhost:5000/form/edit/2，清除 Firstname 字段，并将 Surname 字段的内容更改为 La。这两个值都没有通过验证，在表单元素之间移动时，将看到错误消息。单击 Back 按钮，将看到数据表反映了所做的更改，如图 36-6 所示，尽管这些更改是无效的。

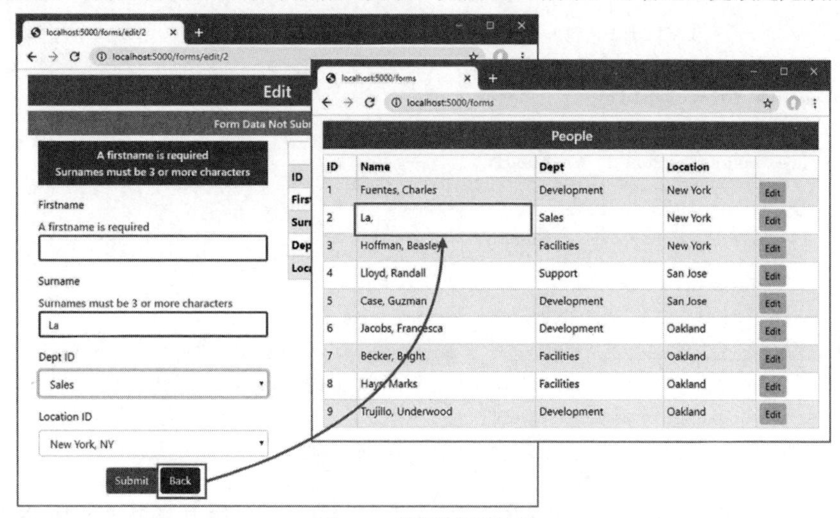


图 36-6 编辑数据的效果

在使用控制器或 Razor Pages 编写的传统 ASP.NET Core 应用程序中，单击按钮，触发一个新的 HTTP 请求。每个请求都隔离处理，并且每个请求都接收到它自己的 Entity Framework Core 上下文对象，该对象配置为范围限定的服务。结果是，在处理一个请求时，创建的数据只有在写入数据库后才会影响其他请求。

在 Blazor 应用程序中，路由系统响应 URL 更改，而不发送新的 HTTP 请求，这意味着只使用 Blazor 维护的到服务器的持久 HTTP 连接来显示多个组件。这将导致多个组件共享单个依赖注入范围，如图 36-7 所示；一个组件所做的更改将影响其他组件，即使这些更改没有写入数据库。

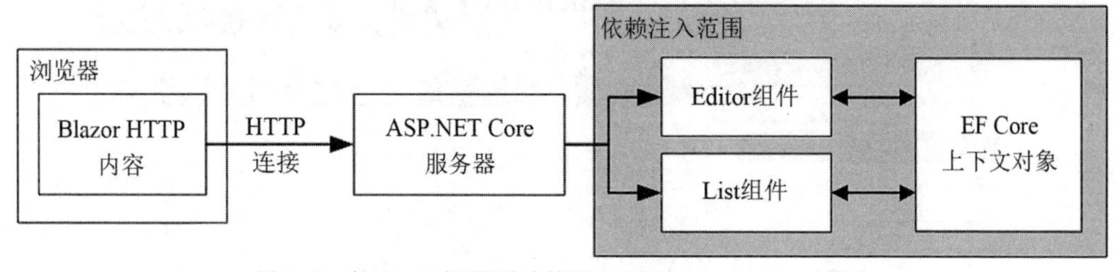


图 36-7 在 Blazor 应用程序中使用 Entity Framework Core 上下文

Entity Framework Core 试图提供帮助，这种方法允许在存储或丢弃之前，随着时间的推移执行复杂的数据操作。遗憾的是，就像第 35 章中描述的 Entity Framework Core 处理相关数据的有用方法一样，它也给那些期望组件像 ASP.NET Core 的其他组件一样处理数据的粗心开发人员挖了一个陷阱。

### 1. 丢弃未保存的数据更改

如果在组件之间共享上下文很有吸引力(对于某些应用程序来说是这样)，那么可采用这种方法，并确保组件在销毁时放弃任何更改，如代码清单 36-14 所示。

**代码清单 36-14　在 Blazor/Forms 文件夹的 Editor.razor 文件中丢弃未保存的数据更改**

```
@page "/forms/edit/{id:long}"
@layout EmptyLayout
@implements IDisposable

<!-- ...elements omitted for brevity... -->

@code {

    // ...statements omitted for brevity...

    public string FormSubmitMessage { get; set; } = "Form Data Not Submitted";
    public void HandleValidSubmit() => FormSubmitMessage = "Valid Data Submitted";
    public void HandleInvalidSubmit() =>
        FormSubmitMessage = "Invalid Data Submitted";

    public void Dispose() => Context.Entry(PersonData).State = EntityState.Detached;
}
```

如第 35 章所述，组件可以实现 System.IDisposable 接口，Dispose 方法在组件即将销毁时调用，在导航到另一个组件时会调用该方法。在代码清单 36-14 中，Dispose 方法的实现告诉 Entity Framework Core 忽略 PersonData 对象，这意味着它不会用于满足未来的请求。要查看效果，请重新启动 ASP.NET Core，请求 http://localhost:5000/form/edit/2，清除 Firstname 字段，然后单击 Back 按钮。当 Entity Framework Core 向列表组件提供其数据时，修改的 Person 对象将被忽略，如图 36-8 所示。

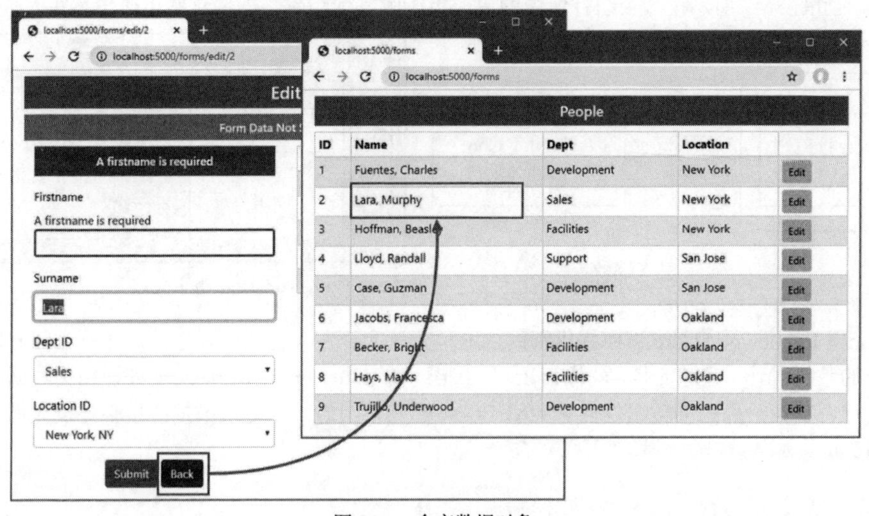


图 36-8　舍弃数据对象

### 2. 创建新的依赖注入范围

如果想要保留 ASP.NET Core 的其余部分使用的模型，就必须创建新的依赖注入范围。并让每个组件接收自己的 Entity Framework Core 上下文对象。这是通过使用@inherits 表达式将组件的基类设置为 OwningComponentBase 或 OwningComponentBase<T>来完成的。

OwningComponentCase 类定义了组件继承的 ScopedServices 属性，该属性提供了一个可用于获取服务的 IServiceProvider 对象，该服务在一个特定于组件的生命周期的作用域中创建，该范围不会与其他任何组件共享，如代码清单 36-15 所示。

**代码清单 36-15　在 Blazor/Forms 文件夹的 Editor.razor 文件中使用新的范围**

```
@page "/forms/edit/{id:long}"
@layout EmptyLayout
@inherits OwningComponentBase
@using Microsoft.Extensions.DependencyInjection

<link href="/blazorValidation.css" rel="stylesheet" />

<h4 class="bg-primary text-center text-white p-2">Edit</h4>
<h6 class="bg-info text-center text-white p-2">@FormSubmitMessage</h6>

<!-- ...elements omitted for brevity... -->

@code {

    [Inject]
    public NavigationManager NavManager { get; set; }

    //[Inject]
    DataContext Context => ScopedServices.GetService<DataContext>();

    [Parameter]
    public long Id { get; set; }

    // ...statements omitted for brevity...
    //public void Dispose() =>
    //      Context.Entry(PersonData).State = EntityState.Detached;
}
```

在代码清单中，注释掉了 Inject 属性，并通过获得 DataContext 服务来设置 Context 属性的值。Microsoft.Extensions.DependencyInjection 名称空间包含扩展方法，这样 IServiceProvider 对象更容易获取服务，如第 14 章所述。

**注意，** 更改基类不会影响使用 Inject 属性接收的服务，它仍然会在请求范围内获得。专用组件范围内需要的每个服务都必须通过 ScopedServices 属性获得，且 Inject 属性不应该应用于该属性。

OwningComponentBase < T >类定义了一个额外的便利属性，来访问范围类型 T 的服务，如果组件只需要范围内单一的服务，该属性就可以很有用，见代码清单36-16(尽管仍然可以通过ScopedServices属性获得更多服务)。

#### 代码清单36-16 在Blazor/Forms文件夹的Editor.razor文件中使用类型化基类

```
@page "/forms/edit/{id:long}"
@layout EmptyLayout
@inherits OwningComponentBase<DataContext>

<link href="/blazorValidation.css" rel="stylesheet" />

<h4 class="bg-primary text-center text-white p-2">Edit</h4>
<h6 class="bg-info text-center text-white p-2">@FormSubmitMessage</h6>

<!-- ...elements omitted for brevity... -->

@code {

    [Inject]
    public NavigationManager NavManager { get; set; }

    //[Inject]
    DataContext Context => Service;

    // ...statements omitted for brevity...
}
```

有作用域的服务可通过名为Service的属性使用。在这个例子中，指定DataContext作为基类的类型参数。

无论使用哪个基类，结果都是Editor组件有自己的依赖注入作用域和自己的DataContext对象。List组件没有修改，因此它将接收请求范围的DataContext对象，如图36-9所示。

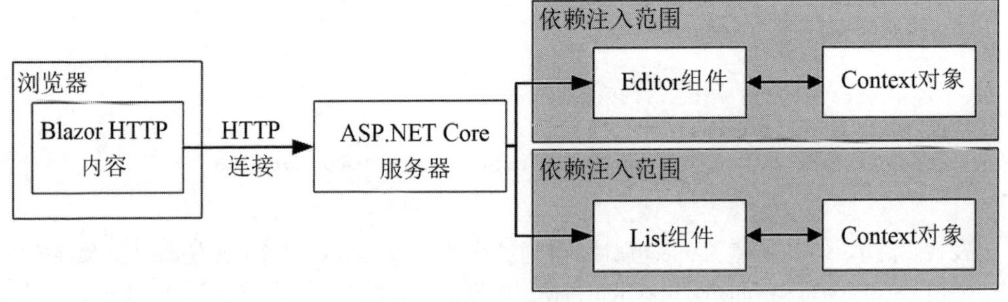


图36-9 为组件使用范围限定的服务

重启ASP.NET Core，导航到http://localhost:5000/form/edit/2，清除Firstname字段，然后单击Back按钮。Editor组件所做的更改不会保存到数据库中，并且由于Editor组件的数据上下文与List组件使用的数据上下文是分离的，因此编辑的数据将被丢弃，生成如图36-8所示的响应。

### 36.3.2 理解重复查询问题

Blazor尽可能高效地响应状态变化，但仍然必须呈现组件的内容，以确定应该发送到浏览器的变化。

Blazor 工作方式的一个后果是，它会导致发送到数据库的查询数量急剧增加。为了演示这个问题，代码清单 36-17 添加了一个按钮，该按钮向 List 组件增加一个计数器。

**代码清单 36-17 在 Blazor/Forms 文件夹的 List.razor 文件中添加按钮**

```
@page "/forms"
@page "/forms/list"
@layout EmptyLayout

<h5 class="bg-primary text-white text-center p-2">People</h5>

<table class="table table-sm table-striped table-bordered">
    <thead>
        <tr>
            <th>ID</th><th>Name</th><th>Dept</th><th>Location</th><th></th>
        </tr>
    </thead>
    <tbody>
        @if (People.Count() == 0) {
            <tr><th colspan="5" class="p-4 text-center">Loading Data...</th></tr>
        } else {
            @foreach (Person p in People) {
                <tr>
                    <td>@p.PersonId</td>
                    <td>@p.Surname, @p.Firstname</td>
                    <td>@p.Department.Name</td>
                    <td>@p.Location.City</td>
                    <td>
                        <NavLink class="btn btn-sm btn-warning"
                                href="@GetEditUrl(p.PersonId)">
                            Edit
                        </NavLink>
                    </td>
                </tr>
            }
        }
    </tbody>
</table>

<button class="btn btn-primary" @onclick="@(() => Counter++)">Increment</button>
<span class="h5">Counter: @Counter</span>

@code {
```

```
    [Inject]
    public DataContext Context { get; set; }

    public IEnumerable<Person> People { get; set; } = Enumerable.Empty<Person>();
    protected override void OnInitialized() {
        People = Context.People.Include(p => p.Department).Include(p => p.Location);
    }

    string GetEditUrl(long id) => $"/forms/edit/{id}";

    public int Counter { get; set; } = 0;
}
```

重启 ASP.NET Core 并请求 http://localhost:5000/forms。单击按钮并观察 ASP.NET Core 服务器的输出。每次单击该按钮时，都会调用事件处理程序，并向数据库发送一个新的数据库查询，生成如下日志消息：

```
...
info: Microsoft.EntityFrameworkCore.Database.Command[20101]
      Executed DbCommand (1ms) [Parameters=[], CommandType='Text',
      CommandTimeout='30']
  SELECT [p].[PersonId], [p].[DepartmentId], [p].[Firstname], [p].[LocationId],
      [p].[Surname], [d].[Departmentid], [d].[Name], [l].[LocationId], [l].[City],
      [l].[State]
    FROM [People] AS [p]
    INNER JOIN [Departments] AS [d] ON [p].[DepartmentId] = [d].[Departmentid]
    INNER JOIN [Locations] AS [l] ON [p].[LocationId] = [l].[LocationId]
info: Microsoft.EntityFrameworkCore.Database.Command[20101]
    Executed DbCommand (0ms) [Parameters=[], CommandType='Text',
    CommandTimeout='30']
  SELECT [p].[PersonId], [p].[DepartmentId], [p].[Firstname], [p].[LocationId],
      [p].[Surname], [d].[Departmentid], [d].[Name], [l].[LocationId], [l].[City],
      [l].[State]
    FROM [People] AS [p]
    INNER JOIN [Departments] AS [d] ON [p].[DepartmentId] = [d].[Departmentid]
    INNER JOIN [Locations] AS [l] ON [p].[LocationId] = [l].[LocationId]
...
```

每次呈现组件时，Entity Framework Core 都向数据库发送两个相同的请求，即使在没有执行数据操作的地方单击了 Increment 按钮，也是如此。

当使用 Entity Framework Core 时，就会出现这个问题，而 Blazor 则加重了这个问题。虽然常见的做法是将数据库查询分配给 IEnumerable<T>属性，这样做遮盖了 Entity Framework Core 的一个重要方面：即其 LINQ 表达式是查询表达式，而不是结果，每次读取属性，都会把一个新的查询发送到数据库。People 属性的值由 List 组件读取两次：由 Count 属性读取一次以确定数据是否已加载；由@foreach 表达式读取一次，生成 HTML 表的行。当用户单击 Increment 按钮时，Blazor 再次呈现 List 组件，以确定发生了什么变化，这将导致 People 属性被多读取两次，从而生成两个额外的数据库查询。

Blazor 和 Entity Framework Core 都在以它们应该的方式工作。Blazor 必须重新呈现组件的输出，以确定需要将哪些 HTML 更改发送到浏览器。只有在呈现元素并计算了所有 Razor 表达式之后，才知道单击按钮会产生什么效果。Entity Framework Core 在每次读取属性时，都执行其查询，以确保应用程序始终拥有新数据。

这些特性的组合带来两个问题。首先，将不必要的查询发送到数据库，这会增加应用程序所需的容量(尽管并非总是如此，因为数据库服务器擅长处理查询)。

另一个问题是，在用户进行不相关的交互之后，对数据库的更改将反映在呈现给用户的内容中。例如，如果另一个用户向数据库添加了 Person 对象，那么下次用户单击 Increment 按钮时，Person 对象将出现在表中。用户希望应用程序只反映他们的操作，而意外的变化会令人困惑、分心。

#### 管理组件中的查询

Blazor 和 Entity Framework Core 之间的交互对所有项目来说都不是问题，但是如果是的话，那么最好的方法是查询一次数据库，并且只对用户希望发生更新的操作再次进行查询。有些应用程序可能需要为用户提供显式选项来重新加载数据，特别是对于用户希望看到更新的应用程序，如代码清单 36-18 所示。

**代码清单 36-18　在 Blazor/Forms 文件夹的 List.razor 文件中控制查询**

```
@page "/forms"
@layout EmptyLayout

<h5 class="bg-primary text-white text-center p-2">People</h5>

<table class="table table-sm table-striped table-bordered">
    <thead>
        <tr>
            <th>ID</th><th>Name</th><th>Dept</th><th>Location</th><th></th>
        </tr>
    </thead>
    <tbody>
        @if (People.Count() == 0) {
            <tr><th colspan="5" class="p-4 text-center">Loading Data...</th></tr>
        } else {
            @foreach (Person p in People) {
                <tr>
                    <td>@p.PersonId</td>
                    <td>@p.Surname, @p.Firstname</td>
                    <td>@p.Department.Name</td>
                    <td>@p.Location.City</td>
                    <td></td>
                </tr>
            }
        }
    </tbody>
```

```
</table>

<button class="btn btn-danger" @onclick="UpdateData">Update</button>

<button class="btn btn-primary" @onclick="@(() => Counter++)">Increment</button>
<span class="h5">Counter: @Counter</span>

@code {

    [Inject]
    public DataContext Context { get; set; }

    public IEnumerable<Person> People { get; set; } = Enumerable.Empty<Person>();

    protected async override Task OnInitializedAsync() {
        await UpdateData();
    }

    private async Task UpdateData() =>
        People = await Context.People.Include(p => p.Department)
            .Include(p => p.Location).ToListAsync<Person>();

    public int Counter { get; set; } = 0;
}
```

UpdateData 方法执行相同的查询，但应用 ToListAsync 方法，该方法强制对 Entity Framework Core 查询进行评估。结果分配给 People 属性，可以重复读取，而不触发其他查询。为了让用户控制数据，添加了一个按钮，当单击 UpdateData 方法时，该按钮会调用该方法。重启 ASP.NET Core，请求 http://localhost:5000/forms，然后单击 Increment 按钮。监视 ASP.NET Core 服务器的输出，将看到只有在组件初始化时才进行查询。要显式触发查询，请单击 Update 按钮。

一些操作可能需要一个新的查询，这很容易执行。为了便于演示，代码清单 36-19 向 List 组件添加了一个排序操作，该操作是使用和不使用新查询实现的。

#### 代码清单 36-19　在 Blazor/Forms 文件夹的 List.razor 文件中添加操作

```
@page "/forms"
@page "/forms/list"
@layout EmptyLayout

<h5 class="bg-primary text-white text-center p-2">People</h5>

<table class="table table-sm table-striped table-bordered">
    <thead>
        <tr>
            <th>ID</th><th>Name</th><th>Dept</th><th>Location</th><th></th>
        </tr>
    </thead>
```

```
    <tbody>
        @if (People.Count() == 0) {
            <tr><th colspan="5" class="p-4 text-center">Loading Data...</th></tr>
        } else {
            @foreach (Person p in People) {
                <tr>
                    <td>@p.PersonId</td>
                    <td>@p.Surname, @p.Firstname</td>
                    <td>@p.Department.Name</td>
                    <td>@p.Location.City</td>
                    <td>
                        <NavLink class="btn btn-sm btn-warning"
                                href="@GetEditUrl(p.PersonId)">
                            Edit
                        </NavLink>
                    </td>
                </tr>
            }
        }
    </tbody>
</table>

<button class="btn btn-danger" @onclick="@(() => UpdateData())">Update</button>
<button class="btn btn-info" @onclick="SortWithQuery">Sort (With Query)</button>
<button class="btn btn-info" @onclick="SortWithoutQuery">Sort (No Query)</button>
<button class="btn btn-primary" @onclick="@(() => Counter++)">Increment</button>
<span class="h5">Counter: @Counter</span>

@code {

    [Inject]
    public DataContext Context { get; set; }

    public IEnumerable<Person> People { get; set; } = Enumerable.Empty<Person>();

    protected async override Task OnInitializedAsync() {
        await UpdateData();
    }

    private IQueryable<Person> Query => Context.People.Include(p => p.Department)
            .Include(p => p.Location);

    private async Task UpdateData(IQueryable<Person> query = null) =>
        People = await (query ?? Query).ToListAsync<Person>();

    public async Task SortWithQuery() {
        await UpdateData(Query.OrderBy(p => p.Surname));
```

```
    }

    public void SortWithoutQuery() {
        People = People.OrderBy(p => p.Firstname).ToList<Person>();
    }

    string GetEditUrl(long id) => $"/forms/edit/{id}";

    public int Counter { get; set; } = 0;
}
```

Entity Framework Core 查询表示为 IQueryable<T>对象，允许该查询在发送到数据库服务器之前与附加的 LINQ 方法组合。示例中的新操作都使用 LINQ OrderBy 方法，但其中一个将其应用于 IQueryable<T>，然后对其进行评估，以使用 ToListAsync 方法发送查询。另一个操作将 OrderBy 方法应用于现有结果数据，对其进行排序，而不发送新的查询。要查看这两个操作，请重新启动 ASP.NET Core，请求 http://localhost:5000/forms，并单击 Sort 按钮，如图 36-10 所示。当单击 Sort (With Query)按钮时，将看到一条日志消息，指示查询已发送到数据库。

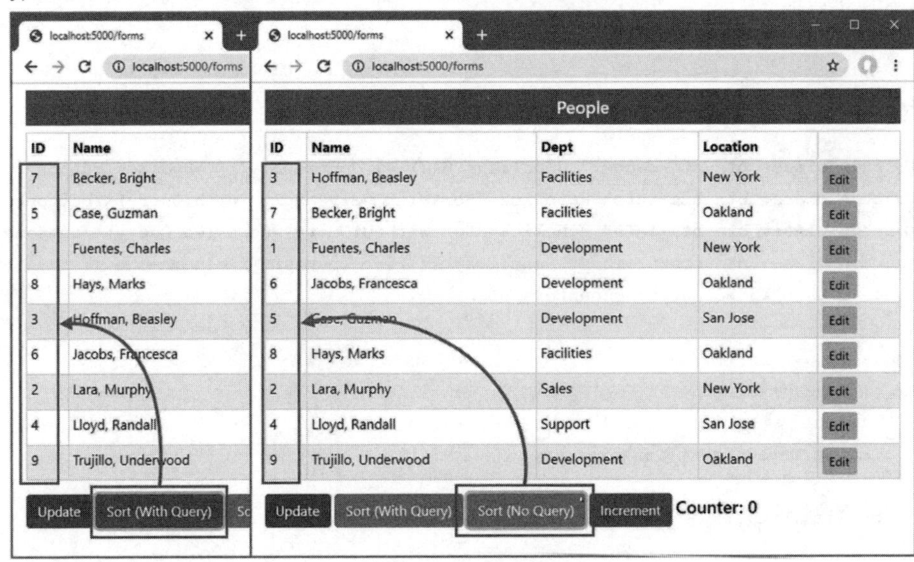


图 36-10　管理组件查询

## 避免重叠查询陷阱

可能会遇到一个异常，指出“在前一个操作完成之前，在此上下文中启动了第二个操作”。当子组件使用 OnParametersSetAsync 方法执行异步 Entity Framework Core 查询，并且父组件数据的更改触发查询完成前的第二次调用 OnParametersSetAsync 时，就会发生这种情况。第二个方法调用启动导致异常的重复查询。这个问题可以通过同步执行 Entity Framework Core 查询来解决。可以在代码清单 36-12 中看到一个示例，其中同步执行查询，因为父组件在接收到其数据时将触发更新。

# 36.4 执行创建、读取、更新和删除操作

为了展示如何将前面部分描述的特性组合在一起，下面创建一个简单的应用程序，允许用户对 Person 对象执行 CRUD 操作。

## 36.4.1 创建 List 组件

List 组件包含需要的基本功能。代码清单 36-20 删除了前面部分中不再需要的一些特性，并添加了允许用户导航到其他函数的按钮。

**代码清单 36-20 在 Blazor/Forms 文件夹的 List.razor 文件中准备组件**

```
@page "/forms"
@page "/forms/list"
@layout EmptyLayout
@inherits OwningComponentBase<DataContext>

<h5 class="bg-primary text-white text-center p-2">People</h5>

<table class="table table-sm table-striped table-bordered">
    <thead>
        <tr>
            <th>ID</th><th>Name</th><th>Dept</th><th>Location</th><th></th>
        </tr>
    </thead>
    <tbody>
        @if (People.Count() == 0) {
            <tr><th colspan="5" class="p-4 text-center">Loading Data...</th></tr>
        } else {
            @foreach (Person p in People) {
                <tr>
                    <td>@p.PersonId</td>
                    <td>@p.Surname, @p.Firstname</td>
                    <td>@p.Department.Name</td>
                    <td>@p.Location.City</td>
                    <td class="text-center">
                        <NavLink class="btn btn-sm btn-info"
                                href="@GetDetailsUrl(p.PersonId)">
                            Details
                        </NavLink>
                        <NavLink class="btn btn-sm btn-warning"
                                  href="@GetEditUrl(p.PersonId)">
                              Edit
                          </NavLink>
                          <button class="btn btn-sm btn-danger"
                              @onclick="@(() => HandleDelete(p))">
                          Delete
```

```
                    </button>
                </td>
            </tr>
        }
    }
    </tbody>
</table>

<NavLink class="btn btn-primary" href="/forms/create">Create</NavLink>

@code {

    public DataContext Context => Service;

    public IEnumerable<Person> People { get; set; } = Enumerable.Empty<Person>();

    protected async override Task OnInitializedAsync() {
        await UpdateData();
    }

    private IQueryable<Person> Query => Context.People.Include(p => p.Department)
            .Include(p => p.Location);

    private async Task UpdateData(IQueryable<Person> query = null) =>
        People = await (query ?? Query).ToListAsync<Person>();

    string GetEditUrl(long id) => $"/forms/edit/{id}";
    string GetDetailsUrl(long id) => $"/forms/details/{id}";

    public async Task HandleDelete(Person p) {
        Context.Remove(p);
        await Context.SaveChangesAsync();
        await UpdateData();
    }
}
```

对象的创建、查看和编辑操作导航到其他 URL，但是删除操作由 List 组件执行，注意在保存更改后重新加载数据，以将更改反映给用户。

### 36.4.2　创建 Details 组件

Details 组件显示数据的只读视图，它不需要 Blazor 表单特性，也不存在 Entity Framework Core 的任何问题。为 Blazor/Forms 文件夹添加一个名为 Details.razor 的 Blazor 组件，内容如代码清单 36-21 所示。

**代码清单 36-21　Blazor/Forms 文件夹中 Details.razor 文件的内容**

```
@page "/forms/details/{id:long}"
```

```
@layout EmptyLayout
@inherits OwningComponentBase<DataContext>

<h4 class="bg-info text-center text-white p-2">Details</h4>

<div class="form-group">
   <label>ID</label>
   <input class="form-control" value="@PersonData.PersonId" disabled />
</div>
<div class="form-group">
   <label>Firstname</label>
   <input class="form-control" value="@PersonData.Firstname" disabled />
</div>
<div class="form-group">
   <label>Surname</label>
   <input class="form-control" value="@PersonData.Surname" disabled />
</div>
<div class="form-group">
   <label>Department</label>
   <input class="form-control" value="@PersonData.Department?.Name" disabled />
</div>
<div class="form-group">
   <label>Location</label>
   <input class="form-control"
         value="@($"{PersonData.Location?.City}, {PersonData.Location?.State}")"
         disabled />
</div>
<div class="text-center">
   <NavLink class="btn btn-info" href="@EditUrl">Edit</NavLink>
   <NavLink class="btn btn-secondary" href="/forms">Back</NavLink>
</div>

@code {

   [Inject]
   public NavigationManager NavManager { get; set; }

   DataContext Context => Service;

   [Parameter]
   public long Id { get; set; }

   public Person PersonData { get; set; } = new Person();

   protected async override Task OnParametersSetAsync() {
      PersonData = await Context.People.Include(p => p.Department)
         .Include(p => p.Location).FirstOrDefaultAsync(p => p.PersonId == Id);
```

```
    }

    public string EditUrl => $"/forms/edit/{Id}";
}
```

此组件显示的所有输入元素都被禁用，这意味着不需要处理事件或处理用户输入。

### 36.4.3　创建 Editor 组件

其余特性将由 Editor 组件处理。代码清单 36-22 删除了前面示例中不再需要的特性，并添加了对创建和编辑对象的支持，包括持久化数据。

**代码清单 36-22　在 Forms/Blazor 文件夹的 Editor.razor 文件中添加应用程序特性**

```
@page "/forms/edit/{id:long}"
@page "/forms/create"
@layout EmptyLayout
@inherits OwningComponentBase<DataContext>

<link href="/blazorValidation.css" rel="stylesheet" />

<h4 class="bg-@Theme text-center text-white p-2">@Mode</h4>

<EditForm Model="PersonData" OnValidSubmit="HandleValidSubmit">
    <DataAnnotationsValidator />
    @if (Mode == "Edit") {
        <div class="form-group">
            <label>ID</label>
            <InputNumber class="form-control"
                @bind-Value="PersonData.PersonId" readonly />
        </div>
    }
    <div class="form-group">
        <label>Firstname</label>
        <ValidationMessage For="@(() => PersonData.Firstname)" />
        <InputText class="form-control" @bind-Value="PersonData.Firstname" />
    </div>
    <div class="form-group">
        <label>Surname</label>
        <ValidationMessage For="@(() => PersonData.Surname)" />
        <InputText class="form-control" @bind-Value="PersonData.Surname" />
    </div>
    <div class="form-group">
        <label>Deptartment</label>
        <ValidationMessage For="@(() => PersonData.DepartmentId)" />
        <CustomSelect TValue="long" Values="Departments"
        Parser="@(str => long.Parse(str))"
```

```
                @bind-Value="PersonData.DepartmentId">
            <option selected disabled value="0">Choose a Department</option>
        </CustomSelect>
    </div>
    <div class="form-group">
        <label>Location</label>
        <ValidationMessage For="@(() => PersonData.LocationId)" />
        <CustomSelect TValue="long" Values="Locations"
                      Parser="@(str => long.Parse(str))"
                      @bind-Value="PersonData.LocationId">
            <option selected disabled value="0">Choose a Location</option>
        </CustomSelect>
    </div>
    <div class="text-center">
        <button type="submit" class="btn btn-@Theme">Save</button>
        <NavLink class="btn btn-secondary" href="/forms">Back</NavLink>
    </div>
</EditForm>

@code {

    [Inject]
    public NavigationManager NavManager { get; set; }

    DataContext Context => Service;

    [Parameter]
    public long Id { get; set; }

    public Person PersonData { get; set; } = new Person();

    public IDictionary<string, long> Departments { get; set; }
        = new Dictionary<string, long>();

    public IDictionary<string, long> Locations { get; set; }
        = new Dictionary<string, long>();

    protected async override Task OnParametersSetAsync() {
        if (Mode == "Edit") {
            PersonData = await Context.People.FindAsync(Id);
        }
        Departments = await Context.Departments
            .ToDictionaryAsync(d => d.Name, d => d.Departmentid);
        Locations = await Context.Locations
            .ToDictionaryAsync(l => $"{l.City}, {l.State}", l => l.LocationId);
    }
```

```
        public string Theme => Id == 0 ? "primary" : "warning";
        public string Mode => Id == 0 ? "Create" : "Edit";

        public async Task HandleValidSubmit() {
            if (Mode== "Create") {
                Context.Add(PersonData);
            }
            await Context.SaveChangesAsync();
            NavManager.NavigateTo("/forms");
        }
    }
```

添加了对新 URL 的支持，并使用引导 CSS 主题来区分创建新对象和编辑现有对象。删除了验证摘要，以便只显示属性级别的验证消息，并添加了通过 Entity Framework Core 存储数据的支持。与使用控制器或 Razor Pages 创建的表单应用程序不同，本例不必处理模型绑定，因为 Blazor 直接处理 Entity Framework Core 从初始数据库查询生成的对象。重启 ASP.NET Core 并请求 http://localhost:5000/forms。将看到如图 36-11 所示的 Person 对象列表，单击 Create、Details、Edit 和 Delete 按钮，将允许处理数据库中的数据。

**提示：**如果需要重置数据库以撤消所做的更改，请打开命令提示符，并在 Advanced 项目文件夹中运行 dotnet ef database drop -force。当重新启动 ASP.NET Core 时，数据库将再次播种，显示如图所示的数据。

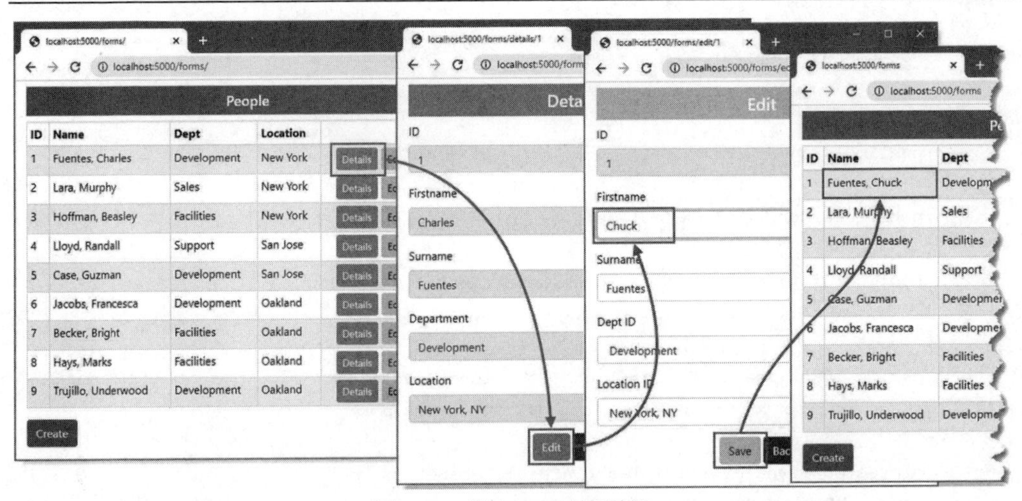

图 36-11　使用 Blazor 处理数据

## 36.5　扩展 Blazor 表单特性

Blazor 表单特性是有效的，但有缺陷，总是在新技术中发现这些缺陷。希望未来的发行版能够完善这个特性集，但与此同时，Blazor 更容易增强表单的工作方式。EditForm 组件定义了层叠式 EditContext 对象，该对象提供了对表单验证的访问，更容易通过表 36-8 中描述的事件、属性和方法创建自定义表单组件。

表 36-8 EditContext 特性

| 名称 | 描述 |
| --- | --- |
| OnFieldChanged | 修改任何表单字段时触发此事件 |
| OnValidationRequested | 当需要验证时触发此事件，可用于创建自定义验证过程 |
| OnValidationStateChanged | 当整个表单的验证状态发生变化时触发此事件 |
| Model | 此属性返回传递给 EditForm 组件的 Model 属性值 |
| Field(name) | 此方法用于获取描述单个字段的 FieldIdentifier 对象 |
| IsModified() | 如果修改任何表单字段，此方法返回 true |
| IsModified(field) | 如果修改了 FieldIdentifier 参数指定的字段，则此方法返回 true |
| GetValidationMessages() | 此方法返回一个序列，其中包含整个表单的验证错误消息 |
| GetValidationMessages(field) | 此方法使用从 Field 方法获得的 FieldIdentifer 对象返回一个序列，其中包含单个字段的验证错误消息 |
| MarkAsUnmodified() | 此方法将表单标记为未修改 |
| MarkAsUnmodified(field) | 此方法使用由 Field 方法获得的 FieldIdentifer 对象将特定的字段标记为未修改 |
| NotifyValidationStateChanged() | 此方法用于指示验证状态的更改 |
| NotifyFieldChanged(field) | 此方法使用从 Field 方法获取的 FieldIdentifer 来指示字段何时发生更改 |
| Validate() | 此方法对表单执行验证，如果所有表单字段通过验证，则返回 true，否则返回 false |

### 36.5.1 创建自定义验证约束

如果内置验证属性不够，可以创建应用自定义验证约束的组件。这种类型的组件不呈现自己的内容，它更容易定义为类。给 Blazor/Forms 文件夹添加名为 DeptStateValidator.cs 的类文件，并使用它定义如代码清单 36-23 所示的组件类。

**代码清单 36-23 Blazor/Forms 文件夹中 DeptStateValidator.cs 文件的内容**

```
using Advanced.Models;
using Microsoft.AspNetCore.Components;
using Microsoft.AspNetCore.Components.Forms;
using System.Collections.Generic;
using System.Linq;

namespace Advanced.Blazor.Forms {

    public class DeptStateValidator: OwningComponentBase<DataContext> {

        public DataContext Context => Service;

    [Parameter]
    public long DepartmentId { get; set; }
```

```
        [Parameter]
        public string State { get; set; }

        [CascadingParameter]
        public EditContext CurrentEditContext { get; set; }

        private string DeptName { get; set; }
        private IDictionary<long, string> LocationStates { get; set; }

        protected override void OnInitialized() {
            ValidationMessageStore store =
                new ValidationMessageStore(CurrentEditContext);
            CurrentEditContext.OnFieldChanged += (sender, args) => {
                string name = args.FieldIdentifier.FieldName;
                if (name == "DepartmentId" || name == "LocationId") {
                    Validate(CurrentEditContext.Model as Person, store);
                }
            };
        }

        protected override void OnParametersSet() {
            DeptName = Context.Departments.Find(DepartmentId).Name;
            LocationStates = Context.Locations
                .ToDictionary(l => l.LocationId, l => l.State);
        }

        private void Validate(Person model, ValidationMessageStore store) {

            if (model.DepartmentId == DepartmentId &&
                (!LocationStates.ContainsKey(model.LocationId) ||
                    LocationStates[model.LocationId] != State)) {
                store.Add(CurrentEditContext.Field("LocationId"),
                    $"{DeptName} staff must be in: {State}");
            } else {
                store.Clear();
            }
            CurrentEditContext.NotifyValidationStateChanged();
        }
    }
}
```

该组件对可以给部门所在的州实施限制，例如，只有在选择了开发部门时，加州的位置才是有效选项，而其他任何位置将产生验证错误。

组件有自己的范围限定的 DataContext 对象，它通过使用 OwningComponentBase<T>作为基类来接收该对象。父组件提供 DepartmentId 和 State 属性的值，它们用于执行验证规则。从 EditForm 组件接收级联 EditContext 属性，并提供对表 36-8 中描述的特性的访问。

初始化组件时，创建一个新的 ValidationMessageStore。该对象用于注册验证错误消息，并接收 EditContext 对象作为其构造函数参数，如下所示：

```
...
ValidationMessageStore store = new ValidationMessageStore(CurrentEditContext);
...
```

Blazor 负责处理添加到存储中的消息，定制验证组件只需要决定哪些消息是必需的，哪些消息是由 Validate 方法处理的。此方法检查 DepartmentId 和 LocationId 属性，以确保允许组合。如果有问题，则向存储区添加一条新的验证消息，如下所示：

```
...
store.Add(CurrentEditContext.Field("LocationId"),
   $"{DeptName} staff must be in: {State}");
...
```

Add 方法的参数是一个 FieldIdentifier，用于标识与错误相关的字段和验证消息。如果没有验证错误，则调用消息存储的 Clear 方法，该方法将确保不再显示之前由组件生成的任何陈旧消息。

Validation 方法由 OnFieldChanged 事件的处理程序调用，该事件允许组件在用户进行更改时响应。

```
...
CurrentEditContext.OnFieldChanged += (sender, args) => {
   string name = args.FieldIdentifier.FieldName;
   if (name == "DepartmentId" || name == "LocationId") {
      Validate(CurrentEditContext.Model as Person, store);
   }
};
...
```

处理程序接收一个 FieldChangeEventArgs 对象，该对象定义了一个 FieldIdentifer 属性，该属性指示哪个字段已经修改。代码清单 36-24 将新的验证应用到 Editor 组件。

**代码清单 36-24　在 Blazor/Forms 文件夹的 Editor.razor 文件中应用验证组件**

```
@page "/forms/edit/{id:long}"
@page "/forms/create"
@layout EmptyLayout
@inherits OwningComponentBase<DataContext>

<link href="/blazorValidation.css" rel="stylesheet" />

<h4 class="bg-@Theme text-center text-white p-2">@Mode</h4>

<EditForm Model="PersonData" OnValidSubmit="HandleValidSubmit">
   <DataAnnotationsValidator />
   <DeptStateValidator DepartmentId="2" State="CA" />
```

```
    <!-- ...elements omitted for brevity... -->

    </EditForm>

    @code {

    // ...statements omitted for brevity...
}
```

DepartmentId 和 State 属性指定了开发部门只能选择加州的位置的限制。重启 ASP.NET Core 请求 http://localhost:5000/form/edit/2。为 Department 字段选择 Development，就将看到一个验证错误，因为此人的位置是纽约。在选择加州的一个位置或更改部门之前，此错误将一直可见，如图 36-12 所示。

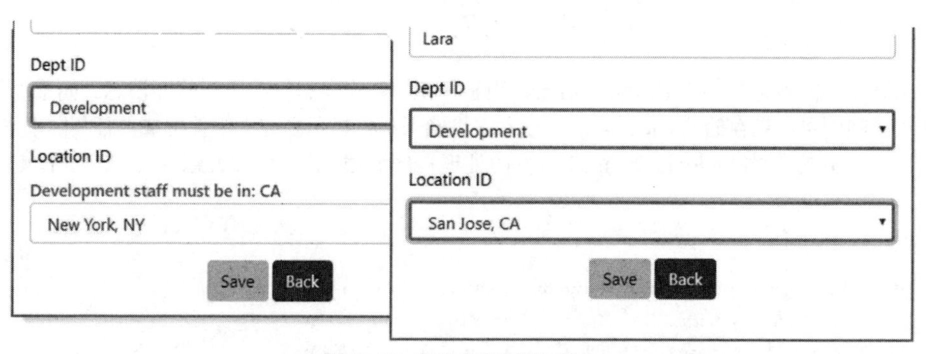


图 36-12　创建自定义验证组件

## 36.5.2　创建只验证提交按钮组件

为了完成本章，下面创建一个组件，它为表单呈现一个提交按钮，该按钮只有在数据有效时才启用。给 Forms/Blazor 文件夹添加一个名为 ValidButton.razor 的 Razor 组件，内容如代码清单 36-25 所示。

#### 代码清单 36-25　Forms/Blazor 文件夹中 ValidButton.razor 文件的内容

```
<button class="@ButtonClass" @attributes="Attributes" disabled="@Disabled">
    @ChildContent
</button>

@code {

    [Parameter]
    public RenderFragment ChildContent { get; set; }

    [Parameter]
    public string BtnTheme { get; set; }

    [Parameter]
```

```
    public string DisabledClass { get; set;} = "btn-outline-dark disabled";

    [Parameter(CaptureUnmatchedValues = true)]
    public IDictionary<string, object> Attributes { get; set; }

    [CascadingParameter]
    public EditContext CurrentEditContext { get; set; }

    public bool Disabled { get; set; }

    public string ButtonClass =>
        Disabled ? $"btn btn-{BtnTheme} {DisabledClass}" : $"btn btn-{BtnTheme}";

    protected override void OnInitialized() {
        SetButtonState();
        CurrentEditContext.OnValidationStateChanged +=
            (sender, args) => SetButtonState();
        CurrentEditContext.Validate();
    }

    public void SetButtonState() {
        Disabled = CurrentEditContext.GetValidationMessages().Any();
    }
}
```

此组件响应 OnValidationStateChanged 方法，该方法在表单的验证状态发生更改时触发。没有详细说明验证状态的 EditContext 属性，因此，查看是否存在任何验证问题的最佳方法是查看是否存在任何验证消息。如果存在，则存在验证问题。如果没有验证消息，则此表单有效。为了确保按钮状态正确显示，将调用 Validation 方法，以便在初始化组件时立即执行验证检查。

代码清单 36-26 使用新组件替换 Editor 组件中的常规按钮。

#### 代码清单 36-26　在 Blazor/Forms 文件夹的 Editor.razor 文件中应用组件

```
...
<div class="text-center">
    <ValidButton type="submit" BtnTheme="@Theme">Save</ValidButton>
    <NavLink class="btn btn-secondary" href="/forms">Back</NavLink>
</div>
...
```

重启 ASP.NET Core，请求 http://localhost:5000/forms/create;，就为每个表单元素显示验证消息，但 Save 按钮被禁用。一旦每个验证问题得到解决，该按钮将被启用，如图 36-13 所示。

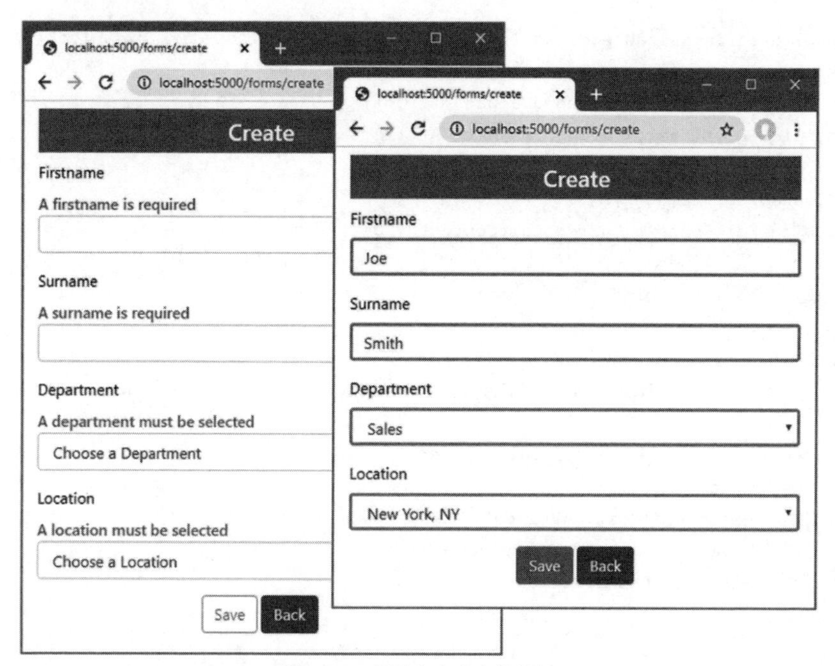


图 36-13　创建自定义表单按钮

# 36.6　小结

本章描述了 Blazor 的表单特性，并展示了如何使用它们创建验证数据的表单。还解释了 Entity Framework Core 和 Blazor 之间的交互如何导致意外结果，以及如何通过创建依赖注入范围和管理查询的执行方式来解决这些问题。第 37 章将描述 Blazor WebAssembly，它完全在浏览器中执行 Razor 组件。

# 第 37 章

# 使用 Blazor WebAssembly

本章演示了 Blazor WebAssembly 的使用，它是 Blazor 为 WebAssembly 编写的实现。

WebAssembly 是运行在浏览器中的虚拟机。高级语言编译成与底层语言无关的汇编程序格式，能以接近本机性能的方式执行。WebAssembly 提供对 JavaScript 应用程序可用的 API 的访问，这意味着 WebAssembly 应用程序可以访问域对象模型、使用层叠样式表、发起异步 HTTP 请求。

顾名思义，Blazor WebAssembly 是运行在 WebAssembly 虚拟机中的 Blazor 的一个实现。Blazor WebAssembly 打破了对服务器的依赖，并在浏览器中完全执行 Blazor 应用程序。其结果是一个真正的客户端应用程序，能够访问 Blazor 服务器的所有相同特性，但不需要持久的 HTTP 连接。

WebAssembly 和 Blazor WebAssembly 都还处于早期阶段，至少目前还存在一些严重的限制。WebAssembly 是一项新技术，只有最新的浏览器版本才支持它。如果项目需要支持旧浏览器——或者甚至是现代浏览器的旧版本，不能使用 WebAssembly。Blazor WebAssembly 仍处于预览阶段，这意味着当前的实现不受支持，存在一些缺陷，不支持生产使用。最终的版本可能会有所变化，当前的工具和运行时可能并不像 Blazor 服务器一样成熟。

即使允许预览状态存在，Blazor WebAssembly 也不是没有缺点。Blazor WebAssembly 依赖于为 WebAssembly 编写的.NET Core运行时的一个实现。要运行 Blazor WebAssembly 应用程序，浏览器必须下载.NETCore运行时、应用程序需要的.NET 程序集以及应用程序代码。在撰写本书时，这意味着浏览器必须下载大约 5MB 的数据，才能启动应用程序。希望生产版本会减少所需的数据量，但是这样的数据量会使 Blazor WebAssembly 不适合某些项目。Blazor WebAssembly 应用程序被限制在浏览器提供的一组 API 中，这意味着并不是所有.NET 特性都可以在 WebAssembly 应用程序中使用。与 Angular 这样的客户端框架相比，Blazor 并没有什么缺点，但它确实意味着像 Entity Framework Core 这样的特性是不可用的，因为浏览器限制 WebAssembly 应用程序只能发出 HTTP 请求。

不过，尽管 Blazor WebAssembly 有一些限制，但它仍然是一项令人兴奋的技术，它承诺能够使用 C#和 ASP.NET Core编写真正的客户端应用程序。不需要 JavaScript 框架。表 37-1 列出了 Blazor WebAssembly 的相关知识。

---

■ **警告：** 写这一章的原因是 Blazor WebAssembly 展示了 ASP.NET Core 的未来。但是 Blazor WebAssembly 还没有准备好投入生产，不应该在实际项目中使用它。

---

表 37-1 Blazor WebAssembly 的相关知识

| 问题 | 答案 |
| --- | --- |
| 它是什么? | Blazor WebAssembly 是 Blazor 的一个实现，它使用 WebAssembly 在浏览器中运行 |
| 它为什么有用? | Blazor WebAssembly 允许用 C#编写客户端应用程序，不必执行服务器端，也不必执行 Blazor 服务器所需的持久 HTTP 连接 |
| 它是如何使用的? | Blazor 组件添加到一个专门用于 Blazor WebAssembly 的项目中 |
| 是否存在缺陷或限制? | 并不是所有浏览器都支持 WebAssembly。需要更大的下载来提供浏览器所需的代码，并非所有 ASP.NET Core 功能都可用于 Blazor WebAssembly 组件 |
| 还有其他选择吗? | Blazor WebAssembly 是使用 ASP.NET Core 编写的真正的客户端应用程序的唯一组合。如果可以接受服务器端支持，可以使用 Blazor 服务器；否则应该使用 JavaScript 框架，如 Angular、React 或 Vue.js |

# 37.1 准备工作

本章使用了第 26 章中的 Advanced 项目。为了准备这一章，在 Controllers 文件夹中添加一个名为 DataController.cs 的类文件，并使用它来定义 Web 服务控制器，如代码清单 37-1 所示。

**提示：** 可以从 https://githubcom/apress/pro-asp.net-core-3 下载本章和本书中其他所有章节的示例项目。如果在运行示例时遇到问题，请参阅第 1 章以了解如何获得帮助。

**代码清单 37-1 Controller 文件夹中 DataController.cs 文件的内容**

```
using Advanced.Models;
using Microsoft.AspNetCore.Mvc;
using Microsoft.EntityFrameworkCore;
using System.Collections.Generic;
using System.Threading.Tasks;

namespace Advanced.Controllers {

    [ApiController]
    [Route("/api/people")]
    public class DataController : ControllerBase {
        private DataContext context;

        public DataController(DataContext ctx) {
            context = ctx;
        }

        [HttpGet]
        public IEnumerable<Person> GetAll() {
            IEnumerable<Person> people
                = context.People.Include(p => p.Department).Include(p => p.Location);
            foreach (Person p in people) {
```

```
                p.Department.People = null;
                p.Location.People = null;
            }
            return people;
        }

        [HttpGet("{id}")]
        public async Task<Person> GetDetails(long id) {
            Person p = await context.People.Include(p => p.Department)
                .Include(p => p.Location).FirstAsync(p => p.PersonId == id);
            p.Department.People = null;
            p.Location.People = null;
            return p;
        }

        [HttpPost]
        public async Task Save([FromBody]Person p) {
            await context.People.AddAsync(p);
            await context.SaveChangesAsync();
        }

        [HttpPut]
        public async Task Update([FromBody]Person p) {
            context.Update(p);
            await context.SaveChangesAsync();
        }

        [HttpDelete("{id}")]
        public async Task Delete(long id) {
            context.People.Remove(new Person() { PersonId = id });
            await context.SaveChangesAsync();
        }

        [HttpGet("/api/locations")]
        public IAsyncEnumerable<Location> GetLocations() => context.Locations;
        [HttpGet("/api/departments")]
        public IAsyncEnumerable<Department> GetDepts() => context.Departments;
    }
}
```

这个控制器提供了允许创建、读取、更新和删除 Person 对象的操作。还添加了返回 Location 和 Department 对象的操作。通常为每种类型的数据创建单独的控制器，但这些操作只在支持 Person 特性时需要，因此将所有操作合并到一个控制器中。

#### 删除数据库并运行应用程序

打开一个新的 PowerShell 命令提示符，导航到包含 Advanced.csproj 文件的文件夹，并运行代码清单 37-2 所示的命令来删除数据库。

**代码清单 37-2　删除数据库**

```
dotnet ef database drop --force
```

从 Debug 菜单中选择 Start Without Debugging 或 Run Without Debugging，或者使用 PowerShell 命令提示符运行代码清单 37-3 中所示的命令。

**代码清单 37-3　运行示例应用程序**

```
dotnet run
```

使用浏览器请求 http://localhost:5000/api/people，这将从数据库中生成 Person 对象的 JSON 表示，如图 37-1 所示。

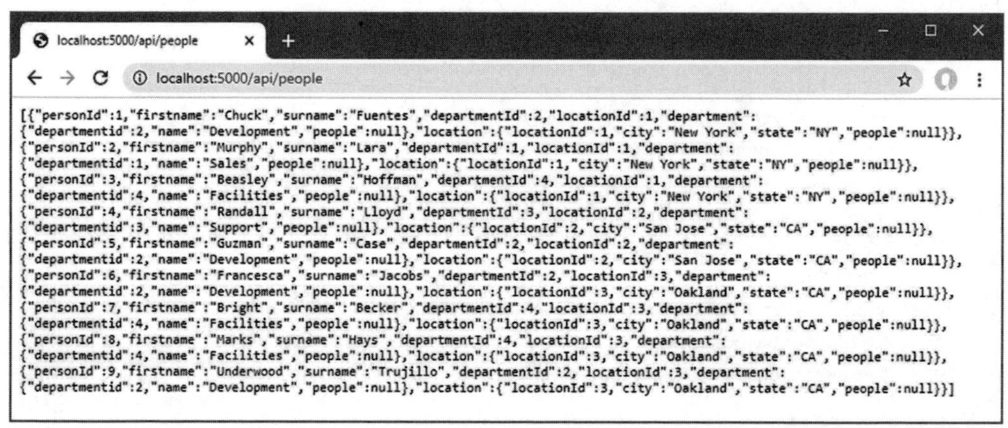


图 37-1　运行示例应用程序

# 37.2　设置 Blazor WebAssembly

Blazor WebAssembly 需要一个单独的项目，这样 Razor 组件就可以编译好，然后由浏览器执行。编译后的组件可以通过标准的 ASP.NET Core 服务器传递给浏览器，也可以通过 Web 服务提供数据。为了方便 Blazor WebAssembly 组件使用由 ASP.NET Core 服务器提供的数据。第三个项目需要包含它们之间共享的项。

这涉及创建这三项的过程，部分原因是要将一些现有的类从 Advanced 项目移到数据模型项目中。尽管可以使用 Visual Studio 向导执行其中的一些步骤，但这里使用命令行工具执行了这些步骤，以尽量减少错误。

**注意：** 如果在以下步骤中遇到问题，可以从本书的 GitHub 知识库下载这三项：https://github.com/apress/pro-asp.net-core-3。

## 37.2.1　创建共享项目

在开始之前，请确保关闭了 Visual Studio 或 Visual Studio Code。打开一个新的 PowerShell 命令提示符，并导航到 Advanced 项目文件夹，其中包含了 Advanced.csproj 文件，并运行代码清单

37-4 所示的命令。

**代码清单 37-4 为 Blazor 准备项目**

```
dotnet new classlib -o ../DataModel -f netstandard2.1
dotnet add ../DataModel package System.ComponentModel.Annotations --version 4.7.0
Move-Item -Path @("Models/Person.cs", "Models/Location.cs",
"Models/Department.cs") ../DataModel
```

这些命令创建一个名为 DataModel 的新项目。它们安装 System.ComponentModel.Annotation 包(其中包含用于数据验证的属性)并将数据模型类移动到新项目中。

## 37.2.2 创建 Blazor WebAssembly 项目

我通常喜欢从一个空项目开始，然后添加应用程序需要的包和配置文件，但 Blazor WebAssembly 仍处于试验阶段，它的项目结构和依赖关系很可能会改变；因此，在技术稳定之前，了解 Blazor WebAssembly 项目是如何工作的没有任何优势。相反，这里使用微软提供的模板创建 Blazor WebAssembly。Blazor WebAssembly 模板没有包含在标准. NET Core 模板库中，所以第一步是使用 PowerShell 命令提示符运行代码清单 37-5 所示的命令来安装模板。

■ **提示：** 只需要在第一次创建 Blazor WebAssembly 项目时运行此命令。安装好模板后，可以跳过这一步，跳到代码清单 37-6 中所示的命令。

**代码清单 37-5 安装 Blazor WebAssembly 项目模板**

```
dotnet new -i Microsoft.AspNetCore.Blazor.Templates::3.1.0-preview4.19579.2
```

这个命令的输出令人困惑，可能给人留下安装失败的印象。该命令显示已安装模板的列表，如果在列表中看到此条目，就知道该命令已成功：

```
...
Blazor WebAssembly App blazorwasm [C#] Web/Blazor/WebAssembly
...
```

接下来，使用 PowerShell 命令提示符在 Advanced 项目文件夹(包含 Advanced.csproj 文件的文件夹)中运行代码清单 37-6 所示的命令。

**代码清单 37-6 创建 Blazor WebAssembly 项目**

```
dotnet new blazorwasm -o ../BlazorWebAssembly
dotnet add ../BlazorWebAssembly reference ../DataModel
```

这些命令创建一个名为 BlazorWebAssembly 的项目，并添加对该数据模型项目的引用，它使 Person、Department 和 Location 类可用。

## 37.2.3 准备 ASP.NET Core 项目

使用 PowerShell 命令提示符在 Advanced 项目文件夹中运行代码清单 37-7 所示的命令。

**代码清单 37-7　准备 Advanced 项目**

```
dotnet add reference ../DataModel ../BlazorWebAssembly

dotnet add package Microsoft.AspNetCore.Blazor.Server --version 3.1.0-preview4.19579.2
```

这些命令创建对其他项目的引用，以便可以使用 Blazor WebAssembly 项目中的数据模型类和组件。

## 37.2.4　添加解决方案引用

在 Advanced 文件夹中运行代码清单 37-8 所示的命令，对新项目的引用添加到解决方案文件中。

**代码清单 37-8　添加解决方案引用**

```
dotnet sln add ../DataModel ../BlazorWebAssembly
```

## 37.2.5　打开项目

设置好这三个项目后，启动 Visual Studio 或 Visual Studio Code。如果使用 Visual Studio，请打开 Advanced 文件夹中的 Advanced.sln 文件。这三个项目都已打开，可以进行编辑，如图 37-2 所示。如果使用 Visual Studio Code，请打开包含所有三个项目的文件夹，如图 37-2 所示。

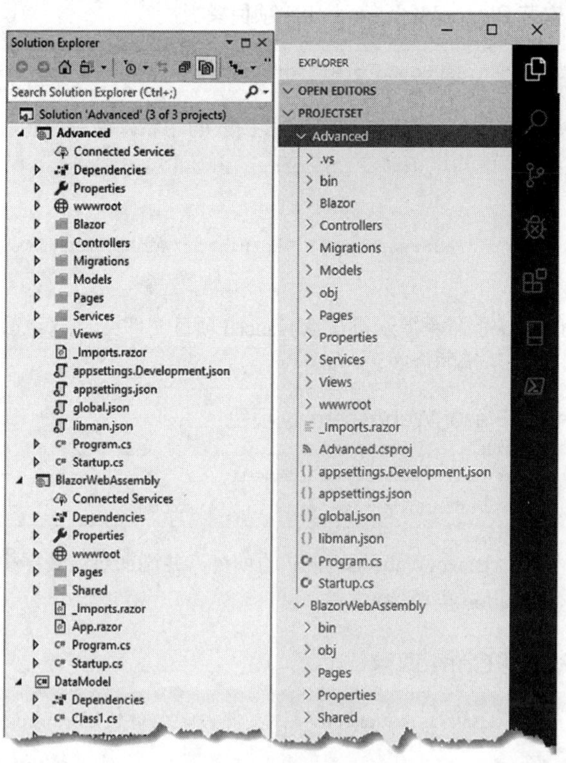


图 37-2　启动三个项目

## 37.2.6 完成 Blazor WebAssembly 配置

下一步是配置 ASP.NET Core 项目，这样它就可将 Blazor WebAssembly 项目的内容传递给客户端。将代码清单 37-9 所示的语句添加到 Advanced 文件夹的 Startup.cs 文件中。

---

■ **警告：** 密切关注当前编辑的文件是很重要的。具有相同名称的文件存在于多个项目中，如果不严格遵循示例，将无法得到正常工作的应用程序。Blazor 的未来版本可能更容易使用，但目前，细节还是十分重要的。

---

**代码清单 37-9 在 Advanced 项目的 Startup.cs 文件中配置应用程序**

```
using System;
using System.Collections.Generic;
using System.Linq;
using System.Threading.Tasks;
using Microsoft.AspNetCore.Builder;
using Microsoft.AspNetCore.Hosting;
using Microsoft.AspNetCore.Http;
using Microsoft.Extensions.DependencyInjection;
using Microsoft.Extensions.Hosting;
using Microsoft.Extensions.Configuration;
using Microsoft.EntityFrameworkCore;
using Advanced.Models;
using Microsoft.AspNetCore.ResponseCompression;

namespace Advanced {
   public class Startup {

      public Startup(IConfiguration config) {
         Configuration = config;
      }

      public IConfiguration Configuration { get; set; }

      public void ConfigureServices(IServiceCollection services) {
         services.AddDbContext<DataContext>(opts => {
            opts.UseSqlServer(Configuration[
               "ConnectionStrings:PeopleConnection"]);
            opts.EnableSensitiveDataLogging(true);
         });
         services.AddControllersWithViews().AddRazorRuntimeCompilation();
         services.AddRazorPages().AddRazorRuntimeCompilation();
         services.AddServerSideBlazor();
         services.AddSingleton<Services.ToggleService>();
         services.AddResponseCompression(opts => {
            opts.MimeTypes = ResponseCompressionDefaults.MimeTypes.Concat(
               new[] { "application/octet-stream" });
         });
```

```
        }

        public void Configure(IApplicationBuilder app, DataContext context) {

            app.UseDeveloperExceptionPage();
            app.UseStaticFiles();
            app.UseRouting();

            app.UseEndpoints(endpoints => {
                endpoints.MapControllerRoute("controllers",
                    "controllers/{controller=Home}/{action=Index}/{id?}");

                endpoints.MapDefaultControllerRoute();
                endpoints.MapRazorPages();
                endpoints.MapBlazorHub();

                endpoints.MapFallbackToClientSideBlazor<BlazorWebAssembly.Startup>
                    ("/webassembly/{*path:nonfile}", "index.html");

                endpoints.MapFallbackToPage("/_Host");
            });

            app.Map("/webassembly", opts =>
                opts.UseClientSideBlazorFiles<BlazorWebAssembly.Startup>());

            SeedData.SeedDatabase(context);
        }
    }
}
```

这些语句启用响应压缩，这是 Blazor WebAssembly 所需要的，并配置 ASP.NET Core 请求管道，使/webassembly 的请求由 Blazor WebAssembly 使用 BlazorWebAssembly 项目的内容来处理。

### 设置基本 URL

最后一步是修改 HTML 文件，该文件将用于响应对/webassembly URL 的请求。将代码清单 37-10 所示的更改应用到 BlazorWebAssembly 项目的 wwwroot 文件夹中的 index.html 文件。

---

■ **警告：** 请注意，在基本元素的 href 属性中，在 webassembly 之前和之后都有斜杠(/)字符。如果省略任何一个字符，Blazor WebAssembly 将无法工作。

---

**代码清单 37-10　在 BlazorWebAssembly 项目的 wwwroot 文件夹的 index.html 文件中设置 URL**

```
<!DOCTYPE html>
<html>

<head>
    <meta charset="utf-8" />
    <meta name="viewport" content="width=device-width" />
```

```
    <title>BlazorWebAssembly</title>
    <base href="/webassembly/" />
    <link href="css/bootstrap/bootstrap.min.css" rel="stylesheet" />
    <link href="css/site.css" rel="stylesheet" />
</head>

<body>
    <app>Loading...</app>

    <div id="blazor-error-ui">
        An unhandled error has occurred.
        <a href="" class="reload">Reload</a>
        <a class="dismiss">x</a>
    </div>
    <script src="_framework/blazor.webassembly.js"></script>
</body>

</html>
```

基本元素设置 URL，文档中的所有相对 URL 都是从该 URL 定义的，而且 Blazor WebAssembly 路由系统的正确操作需要该元素。

## 37.2.7 测试占位符组件

为了启动 ASP.NET Core，从 Debug 菜单中选择 Start Without Debugging 或 Run Without Debugging。如果喜欢使用命令提示符，可以在 Advanced 项目文件夹中运行代码清单 37-11 所示的命令。

**代码清单 37-11 运行示例应用程序**

```
dotnet run
```

使用浏览器请求 http://localhost:5000/webassembly，将看到由模板添加的占位符内容，该模板用于创建 BlazorWebAssembly 项目。

使用 PowerShell 命令提示符，在 Advanced 项目文件夹中运行以下命令。单击 Counter 和 Fetch data 链接，将显示不同的内容，如图 37-3 所示。

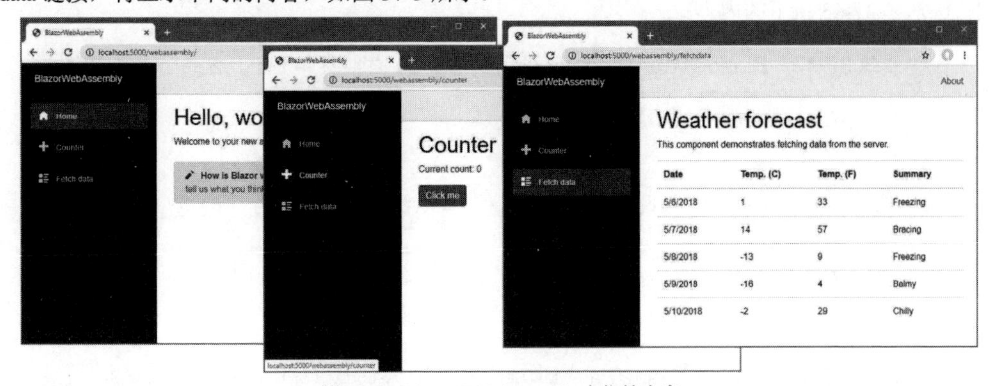


图 37-3 Blazor WebAssembly 占位符内容

## 37.3　创建 Blazor WebAssembly 组件

Blazor WebAssembly 使用与 Blazor Server 相同的方法，依赖组件作为应用程序的构建块，通过路由系统连接，并通过布局显示公共内容。本节将展示如何创建与 Blazor WebAssembly 一起工作的 Razor 组件，然后重新创建第 36 章中的简单表单应用程序。

### 37.3.1　导入数据模型名称空间

在本章中创建的组件都使用共享 DataModel 项目中的类。不是向每个组件添加@using 表达式，而是将数据模型类的名称空间添加到 BlazorWebAssembly 项目根文件夹中的_Imports.razor 文件，如代码清单 37-12 所示。

**代码清单 37-12　在 BlazorWebAssembly 项目的_Imports.razor 文件中添加名称空间**

```
@using System.Net.Http
@using Microsoft.AspNetCore.Components.Forms
@using Microsoft.AspNetCore.Components.Routing
@using Microsoft.AspNetCore.Components.Web
@using Microsoft.JSInterop
@using BlazorWebAssembly
@using BlazorWebAssembly.Shared
@using Advanced.Models
```

注意，尽管将模型类移到 DataModel 项目中，但指定了 Advanced.Models 名称空间。这是因为移动的类文件都有指定 Advanced.Models 的名称空间声明，这意味着移动文件不会改变类所在的名称空间。

### 37.3.2　创建组件

前面的章节在 Blazor 文件夹中定义了 Razor 组件，以保持新的内容与 ASP.NET Core 其他部分分离。BlazorWebAssembly 项目中只有 Blazor 内容，因此这里遵循项目模板采用的约定，使用 Pages 和 Shared 文件夹。

在 BlazorWebAssembly 项目的 Pages 文件夹中添加一个名为 List.razor 的 Razor 组件，并添加如代码清单 37-13 所示的内容。

**代码清单 37-13　BlazorWebAssembly 项目的 Pages 文件夹中 List.razor 文件的内容**

```
@page "/forms"
@page "/forms/list"

<h5 class="bg-primary text-white text-center p-2">People (WebAssembly)</h5>

<table class="table table-sm table-striped table-bordered">
    <thead>
        <tr>
            <th>ID</th><th>Name</th><th>Dept</th><th>Location</th><th></th>
        </tr>
```

```
    </thead>
    <tbody>
        @if (People.Count() == 0) {
            <tr><th colspan="5" class="p-4 text-center">Loading Data...</th></tr>
        } else {
            @foreach (Person p in People) {
                <tr>
                    <td>@p.PersonId</td>
                    <td>@p.Surname, @p.Firstname</td>
                    <td>@p.Department.Name</td>
                    <td>@p.Location.City</td>
                    <td class="text-center">
                        <NavLink class="btn btn-sm btn-info"
                                href="@GetDetailsUrl(p.PersonId)">
                            Details
                        </NavLink>
                        <NavLink class="btn btn-sm btn-warning"
                                href="@GetEditUrl(p.PersonId)">
                            Edit
                        </NavLink>
                        <button class="btn btn-sm btn-danger"
                                @onclick="@(() => HandleDelete(p))">
                            Delete
                        </button>
                    </td>
                </tr>
            }
        }
    </tbody>
</table>

<NavLink class="btn btn-primary" href="forms/create">Create</NavLink>

@code {

    [Inject]
    public HttpClient Http { get; set; }

    public Person[] People { get; set; } = Array.Empty<Person>();

    protected async override Task OnInitializedAsync() {
        await UpdateData();
    }

    private async Task UpdateData() {
        People = await Http.GetJsonAsync<Person[]>("/api/people");
    }
```

```
        string GetEditUrl(long id) => $"forms/edit/{id}";
        string GetDetailsUrl(long id) => $"forms/details/{id}";

        public async Task HandleDelete(Person p) {
            HttpResponseMessage resp =
                await Http.DeleteAsync($"/api/people/{p.PersonId}");
            if (resp.IsSuccessStatusCode) {
                await UpdateData();
            }
        }
    }
```

如果将此组件与第 36 章中的 Blazor 服务器相比较，会发现它们基本上是相同的。两种类型的 Blazor 使用相同的核心特性，这就是为什么内容使用相同的 Razor 指令，使用@onclick 属性处理事件，并对 C#语句使用相同的@code 部分。Blazor WebAssembly 组件编译成一个 C#类，就像它的 Blazor 服务器版本一样。当然，关键区别在于生成的 C#类是在浏览器中执行的——这就是与第 36 章中的组件不同的原因。

#### 1. 在 Blazor WebAssembly 组件中导航

请注意，用于导航的 URL 表示为没有前导斜杠字符，如下所示：

```
...
<NavLink class="btn btn-primary" href="forms/create">Create</NavLink>
...
```

应用程序的根 URL 是使用代码清单 37-13 中的基本元素指定的，使用相对 URL 可以确保相对根执行导航。这种情况下，相对的 forms/create URL 与 base 元素指定的/webassembly/root 结合在一起，导航是/webassembly/forms/create。包含前导斜杠将导航到/forms/create，这是在应用程序 Blazor WebAssembly 部分管理的 URL 集合之外的。仅需要对导航 URL 执行此更改。例如，用@page 指令指定的 URL 不会受到影响。

#### 2. 在 Blazor WebAssembly 组件中获取数据

最大的变化是 Blazor WebAssembly 不能使用 EntityFrameworkCore。尽管运行时可能能够执行 EntityFrameworkCore 类，但浏览器将 WebAssembly 应用程序限制为 HTTP 请求，从而阻止了 SQL 的使用。为获取数据，Blazor WebAssembly 应用程序需要使用 Web 服务，这就是为什么本章的开始部分将 API 控制器添加到 Advanced 项目中。

作为 Blazor WebAssembly 应用程序启动的一部分，为 HttpClient 类创建一个服务，组件可以使用标准的依赖注入特性接收该服务。List 组件通过修饰了 Inject 特性的属性接收 HttpClient 组件，如下所示：

```
...
[Inject]
public HttpClient Http { get; set; }
...
```

HttpClient 类提供了表 37-2 中描述的方法来发送 HTTP 请求。

表 37-2 HttpClient 类定义的方法

| 名称 | 描述 |
| --- | --- |
| GetAsync(url) | 此方法发送 HTTP GET 请求 |
| PostAsync(url, data) | 此方法发送 HTTP POST 请求 |
| PutAsync(url, data) | 此方法发送 HTTP PUT 请求 |
| PatchAync(url, data) | 此方法发送 HTTP PATCH 请求 |
| DeleteAsync(url) | 此方法发送一个 HTTP DELETE 请求 |
| SendAsync(request) | 此方法发送使用 HttpRequestMessage 对象配置的 HTTP |

表 37-2 中的方法返回一个 Task<HttpResponseMessage>结果，它描述了从 HTTP 服务器接收到的对异步请求的响应。表 37-3 显示了最有用的 HttpResponseMessage 属性。

表 37-3 有用的 HttpClient 属性

| 名称 | 描述 |
| --- | --- |
| Content | 此属性返回服务器返回的内容 |
| HttpResponseHeaders | 此属性返回响应头 |
| StatusCode | 此属性返回响应状态代码 |
| IsSuccessStatusCode | 如果响应状态码在 200 和 299 之间，则此属性返回 true，表示请求成功 |

当用户单击 Delete 按钮时，List 组件使用 DeleteAsync 方法来请求 Web 服务删除对象。

```
...
HttpResponseMessage resp =
    await Http.DeleteAsync($"/api/people/{p.PersonId}");
    if (resp.IsSuccessStatusCode) {
        await UpdateData();
    }
}
...
```

当不需要处理 Web 服务返回的数据时，这些方法非常有用，例如在本例中，只检查 DELETE 请求是否成功。请注意，仅在使用 HttpClient 服务时为请求 URL 指定路径，因为 Web 服务可使用与应用程序相同的方案、主机和端口。

对于 Web 服务返回数据的操作，表 37-4 中描述的 HttpClient 类的扩展方法更有用。这些方法将数据序列化为 JSON，这样就可将其发送到服务器，并将 JSON 响应解析为 C#对象。对于没有返回结果的请求，可以省略泛型类型参数。

表 37-4 HttpClient 扩展方法

| 名称 | 描述 |
| --- | --- |
| GetJsonAsync<T>(url) | 此方法发送 HTTP GET 请求并解析类型 T 的响应 |
| PostJsonAsync<T>(url, data) | 此方法发送 HTTP POST 请求并解析类型 T 的响应 |
| PutJsonAsync<T>(url, data) | 此方法发送 HTTP PUT 请求并解析类型 T 的响应 |
| SendJsonAsync<T>(method, url, data) | 此方法使用指定的方法发送 HTTP 请求，并解析类型 T 的响应 |

List 组件使用 GetJsonAsync<T>方法从 Web 服务请求数据。

```
...
private async Task UpdateData() {
    People = await Http.GetJsonAsync<Person[]>("/api/people");
}
...
```

将泛型类型参数设置为 Person[]，就告诉 HttpClient 将响应解析为一个 Person 对象数组。

**注意：**HttpClient 类不存在任何作用域或生命周期问题，只在调用表 37-2 或表 37-4 中描述的方法之一时发送请求。然而，需要考虑一下什么时候请求新数据。本例在删除对象之后请求 Web 服务，而不是简单地从组件初始化时请求的数据中删除对象。这可能并不适用于所有应用程序，因为它反映其他用户对数据库所做的任何更改。

### 37.3.3　创建布局

用于创建 Blazor WebAssembly 项目的模板包括显示占位符内容的导航特性的布局。本例不需要这些导航特性，所以第一步是创建一个新的布局。在 BlazorWebAssembly 项目的 Shared 文件夹中添加名为 EmptyLayout.razor 的 Razor 组件，内容如代码清单 37-14 所示。

**代码清单 37-14　BlazorWebAssembly 项目的 Shared 文件夹中的 EmptyLayout.razor 文件**

```
@inherits LayoutComponentBase

<div class="m-2">
    @Body
</div>
```

可以使用@layout 表达式来应用新的布局，就像第 36 章中做的那样，但是本例通过更改路由配置来使用这个布局作为默认布局，路由配置在 BlazorWebAssembly 项目的 App.razor 文件中定义，如代码清单 37-15 所示。

**代码清单 37-15　在 BlazorWebAssembly 项目的 App.razor 文件中应用布局**

```
<Router AppAssembly="@typeof(Program).Assembly">
    <Found Context="routeData">
        <RouteView RouteData="@routeData" DefaultLayout="@typeof(EmptyLayout)" />
    </Found>
    <NotFound>
        <LayoutView Layout="@typeof(EmptyLayout)">
            <p>Sorry, there's nothing at this address.</p>
        </LayoutView>
    </NotFound>
</Router>
```

第 35 章描述了 Router、RouteView、Found 和 NotFound 组件。

### 37.3.4 定义 CSS 样式

模板创建了 Blazor WebAssembly 项目，带有自己的 Bootstrap CSS 框架副本和附加的样式表，该样式表结合了配置 Blazor WebAssembly 错误和验证元素所需的样式，并管理应用程序的布局。替换 HTML 文件中的链接元素，如代码清单 37-16 所示，并将样式直接应用到错误元素。这样可以移除 Microsoft 布局使用的样式，并使用添加到 Advanced 项目中的引导 CSS 样式表。

**代码清单 37-16 在 BlazorWebAssembly 项目的 wwwroot 文件夹中修改 index.html 文件**

```
<!DOCTYPE html>
<html>

<head>
    <meta charset="utf-8" />
    <meta name="viewport" content="width=device-width" />
    <title>BlazorWebAssembly</title>
    <base href="/webassembly/" />
    <link href="/lib/twitter-bootstrap/css/bootstrap.min.css" rel="stylesheet" />
</head>

<body>
    <app>Loading...</app>

    <div id="blazor-error-ui"
              class="text-center bg-danger h6 text-white p-2 fixed-top w-100"
                 style="display:none">
        An unhandled error has occurred.
        <a href="" class="reload">Reload</a>
        <a class="dismiss">x</a>
    </div>
    <script src="_framework/blazor.webassembly.js"></script>
</body>

</html>
```

要查看新组件，请重新启动 ASP.NET Core，并请求 http://localhost:5000/webassembly/forms，这将生成如图 37-4 所示的响应。

Blazor WebAssembly 组件遵循标准的 Blazor 生命周期，并且组件显示它从 Web 服务接收到的数据。

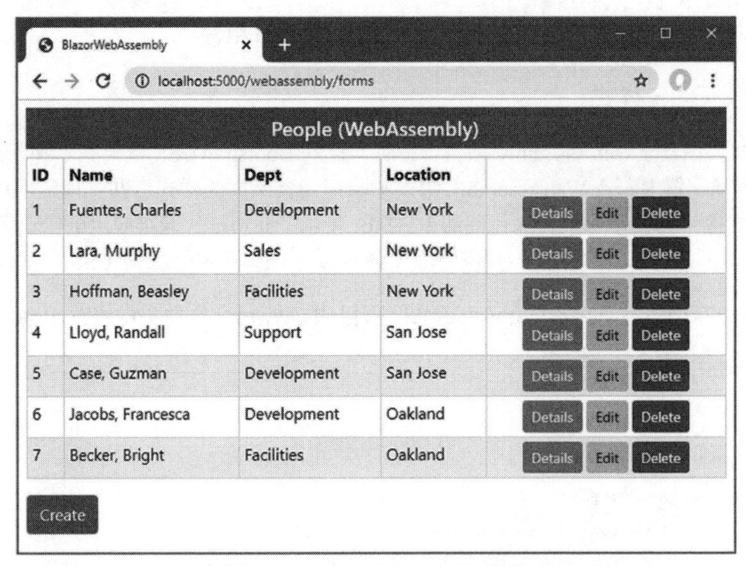


图 37-4　Blazor WebAssembly 组件

## 37.4　完成 Blazor WebAssembly 表单应用程序

当前只有List组件显示的Delete按钮有效。接下来通过创建附加组件来完成Blazor WebAssembly表单应用程序。

### 37.4.1　创建 Details 组件

在 BlazorWebAssembly 项目的 Pages 文件夹中添加名为 Details.razor 的 Razor 组件，内容如代码清单 37-17 所示。

**代码清单 37-17　在 BlazorWebAssembly 项目的 Pages 文件夹中 Details.razor 文件的内容**

```
@page "/forms/details/{id:long}"

<h4 class="bg-info text-center text-white p-2">Details (WebAssembly)</h4>
    <div class="form-group">
    <label>ID</label>
<input class="form-control" value="@PersonData.PersonId" disabled />
</div>
<div class="form-group">
    <label>Firstname</label>
    <input class="form-control" value="@PersonData.Firstname" disabled />
</div>
<div class="form-group">
    <label>Surname</label>
    <input class="form-control" value="@PersonData.Surname" disabled />
</div>
```

```
<div class="form-group">
    <label>Department</label>
    <input class="form-control" value="@PersonData.Department?.Name" disabled />
</div>
<div class="form-group">
    <label>Location</label>
    <input class="form-control"
           value="@($"{PersonData.Location?.City}, {PersonData.Location?.State}")"
           disabled />
</div>
<div class="text-center">
    <NavLink class="btn btn-info" href="@EditUrl">Edit</NavLink>
    <NavLink class="btn btn-secondary" href="forms">Back</NavLink>
</div>

@code {

    [Inject]
    public NavigationManager NavManager { get; set; }

    [Inject]
    public HttpClient Http { get; set; }

    [Parameter]
    public long Id { get; set; }

    public Person PersonData { get; set; } = new Person();

    protected async override Task OnParametersSetAsync() {
        PersonData = await Http.GetJsonAsync<Person>($"/api/people/{Id}");
    }

    public string EditUrl => $"forms/edit/{Id}";
}
```

Details 组件与 Blazor 服务器对应组件只有两个不同之处，遵循 List 组件建立的模式：数据通过 HttpClient 服务获得，导航目标使用相对 URL 表示。在所有其他方面，比如从路由数据中获取参数，Blazor WebAssembly 的工作方式与 Blazor Server 完全相同。

### 37.4.2 创建 Editor 组件

要完成表单应用程序，给 BlazorWebAssembly 项目的 Pages 文件夹添加一个名为 Editor.Razor 的 Razor 组件。内容如代码清单 37-18 所示。

**代码清单 37-18　BlazorWebAssembly 项目的 Pages 文件夹中 Editor.Razor 文件的内容**

```
@page "/forms/edit/{id:long}"
@page "/forms/create"
```

```
<link href="/blazorValidation.css" rel="stylesheet" />

<h4 class="bg-@Theme text-center text-white p-2">@Mode (WebAssembly)</h4>

<EditForm Model="PersonData" OnValidSubmit="HandleValidSubmit">
    <DataAnnotationsValidator />
    @if (Mode == "Edit") {
        <div class="form-group">
            <label>ID</label>
            <InputNumber class="form-control"
                @bind-Value="PersonData.PersonId" readonly />
        </div>
    }
    <div class="form-group">
        <label>Firstname</label>
        <ValidationMessage For="@(() => PersonData.Firstname)" />
        <InputText class="form-control" @bind-Value="PersonData.Firstname" />
    </div>
    <div class="form-group">
        <label>Surname</label>
        <ValidationMessage For="@(() => PersonData.Surname)" />
        <InputText class="form-control" @bind-Value="PersonData.Surname" />
    </div>
    <div class="form-group">
        <label>Department</label>
        <ValidationMessage For="@(() => PersonData.DepartmentId)" />
        <select @bind="PersonData.DepartmentId" class="form-control">
            <option selected disabled value="0">Choose a Department</option>
            @foreach (var kvp in Departments) {
                <option value="@kvp.Value">@kvp.Key</option>
            }
        </select>
    </div>
    <div class="form-group">
        <label>Location</label>
        <ValidationMessage For="@(() => PersonData.LocationId)" />
        <select @bind="PersonData.LocationId" class="form-control">
            <option selected disabled value="0">Choose a Location</option>
            @foreach (var kvp in Locations) {
                <option value="@kvp.Value">@kvp.Key</option>
            }
        </select>
    </div>
    <div class="text-center">
        <button type="submit" class="btn btn-@Theme">Save</button>
        <NavLink class="btn btn-secondary" href="forms">Back</NavLink>
```

```
    </div>
</EditForm>

@code {

    [Inject]
    public HttpClient Http { get; set; }

    [Inject]
    public NavigationManager NavManager { get; set; }

    [Parameter]
    public long Id { get; set; }

    public Person PersonData { get; set; } = new Person();

    public IDictionary<string, long> Departments { get; set; }
        = new Dictionary<string, long>();

    public IDictionary<string, long> Locations { get; set; }
        = new Dictionary<string, long>();

    protected async override Task OnParametersSetAsync() {
        if (Mode == "Edit") {
            PersonData = await Http.GetJsonAsync<Person>($"/api/people/{Id}");
        }
        Departments = (await Http.GetJsonAsync<Department[]>("/api/departments"))
            .ToDictionary(d => d.Name, d => d.Departmentid);
        Locations = (await Http.GetJsonAsync<Location[]>("/api/locations"))
            .ToDictionary(l => $"{l.City}, {l.State}", l => l.LocationId);
    }

    public string Theme => Id == 0 ? "primary" : "warning";
    public string Mode => Id == 0 ? "Create" : "Edit";

    public async Task HandleValidSubmit() {
        await Http.SendJsonAsync(Mode == "Create" ? HttpMethod.Post : HttpMethod.Put,
            "/api/people", PersonData);
        NavManager.NavigateTo("forms");
    }
}s
```

该组件使用第 36 章中描述的 Blazor 表单特性，但使用 HTTP 请求来读取本章开始时创建的 Web 服务，并在其中写入数据。GetJsonAsync<T>方法用于从 Web 服务读取数据，SendJsonAsync 方法用于在用户提交表单时发送 POST 或 PUT 请求。

注意，没有使用在第 36 章中创建的自定义选择组件或验证组件。在项目之间共享组件——尤其是在开发开始后引入 Blazor WebAssembly 时——是很尴尬的。希望这个过程在未来的版本中得到改进，但是本章没有使用这些特性。因此，当选择一个值时，select 元素不会触发验证，submit 按钮不会自动禁用，并且 department 和 location 的组合没有限制。

重启 ASP.NET Core，请求 http://localhost:5000/webassembly/forms，将看到 Blazor WebAssembly 版本的表单应用程序。单击表中第一项的 Details 按钮，将看到所选对象的字段。单击 Edit 按钮，将看到一个可编辑的表单。进行更改并单击 Save 按钮，更改将发送到 Web 服务并显示在数据表中，如图 37-5 所示。

图 37-5 完成的 Blazor WebAssembly 表单应用程序

## 37.5 小结

本章描述了 Blazor WebAssembly，展示了如何将其添加到项目中，并演示了它与前面章节中描述的 Blazor 服务器的相似之处(尽管并非完全相同)。第 38 章将解释如何使用 ASP.NET Core Identity 以确保应用程序的安全。

# 第 38 章

# 使用 ASP.NET Core Identity

ASP.NET Core Identity 是微软在 ASP.NET Core 应用程序中管理用户的 API，并支持将身份验证和授权集成到请求管道中。

ASP.NET Core Identity 是一个工具包，可以用来创建应用程序所需的授权和身份验证特性。对于诸如双因素身份验证、联合、单点登录和账户自助服务等特性，有无数的集成选项。有些选项仅在大型企业环境或使用云托管用户管理时有用。

ASP.NET Core Identity 已经发展成它自己的框架，它太大了，不可能在本书中详细介绍。相反，本章将重点放在 Identity API 与 Web 应用程序开发交叉的部分上，就像在 Entity Framework Core 上所做的那样。本章将展示如何给一个项目添加 ASP.NET Core Identity，并解释了如何使用 ASP.NET Core Identity API 创建工具，来执行基本的用户和角色管理。第 39 章将展示如何使用 ASP.NET Core Identity 验证用户和执行授权。表 38-1 描述了 ASP.NET Core Identity。

表 38-1　ASP.NET Core Identity

| 问题 | 答案 |
| --- | --- |
| 它是什么? | ASP.NET Core Identity 是一个用于管理用户的 API |
| 它为什么有用? | 大多数应用程序有一些不应该对所有用户都可用的特性。ASP.NET Core Identity 提供的特性允许用户验证自己并访问受限制的特性 |
| 它是如何使用的? | ASP.NET Core Identity 作为一个包添加到项目中，并使用 Entity Framework Core 将其数据存储在数据库中。对用户的管理是通过定义良好的 API 执行的，它的特性是作为属性应用的，如第 39 章所述 |
| 是否存在缺陷或限制? | ASP.NET Core Identity 很复杂，并提供了对广泛的身份验证、授权和管理模型的支持。理解所有选项可能会很困难，而且文档可能很少 |
| 还有其他选择吗? | 如果项目需要限制对特性的访问，ASP.NET Core Identity 没有合理的替代 |

表 38-2 总结了本章的内容。

表 38-2　本章内容摘要

| 问题 | 解决方案 | 代码清单 |
| --- | --- | --- |
| 为身份准备应用程序 | 创建上下文类并准备应用于数据库的迁移 | 38-4～38-7 |
| 管理用户账户 | 使用 UserManager<T>类 | 38-8～38-12、38-15、38-16 |

(续表)

| 问题 | 解决方案 | 代码清单 |
| --- | --- | --- |
| 设置用户名和密码策略 | 使用 options 模式配置标识 | 38-13～38-14 |
| 管理角色 | 使用 RoleManager<T>类来管理角色和 UserManager<T>类为用户分配角色 | 38-17～38-20 |

## 38.1　准备工作

本章使用第 37 章中的 Advanced、DataModel 和 BlazorWebAssembly 项目。如果使用的是 Visual Studio，请打开上一章创建的 Advanced.sln 文件，用于打开所有三个项目。如果使用的是 Visual Studio Code，请打开包含这三个项目的文件夹。

**提示：**可从https://github.com/apress/pro-asp.net-core-3 下载本章和本书中其他所有章节的示例项目。如果在运行示例时遇到问题，请参阅第 1 章以了解如何获得帮助。

打开一个新的 PowerShell 命令提示符，导航到包含 Advanced.csproj 文件的文件夹，并运行代码清单 38-1 所示的命令来删除数据库。

**代码清单 38-1　删除数据库**

```
dotnet ef database drop --force
```

从 Debug 菜单中选择 Start Without Debugging 或 Run Without Debugging，或使用 PowerShell 命令提示符运行代码清单 38-2 中所示的命令。

**代码清单 38-2　运行示例应用程序**

```
dotnet run
```

使用浏览器请求 http://localhost:5000，将生成如图 38-1 所示的响应。

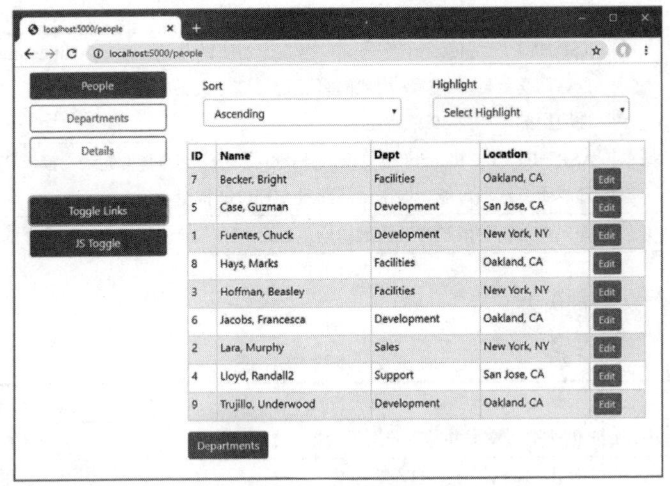


图 38-1　运行示例应用程序

## 38.2 为 ASP.NET Core Identity 准备项目

设置 ASP.NET Core Identity 的过程需要向项目添加一个包、配置应用程序和准备数据库。首先，在 Advanced 项目文件夹中使用 PowerShell 命令提示符运行代码清单 38-3 中所示的命令，该命令安装 ASP.NET Core Identity 包。如果使用的是 Visual Studio，就可以通过选择 Project | Manage NuGet Packages 来安装该包。

**代码清单 38-3 安装 ASP.NET Core Identity 包**

```
dotnet add package Microsoft.AspNetCore.Identity.EntityFrameworkCore --version 3.1.1
```

### 38.2.1 准备 ASP.NET Core Identity 数据库

ASP.NET Identity 需要一个数据库，该数据库通过 Entity Framework Core 进行管理。要创建提供对 Identity 数据访问的 Entity Framework Core 上下文类，将一个名为 IdentityContext.cs 的类文件添加到 Advanced/Models 文件夹中，代码如代码清单 38-4 所示。

**代码清单 38-4 Advanced 项目的 Models 文件夹中 IdentityContext.cs 文件的内容**

```
using Microsoft.AspNetCore.Identity;
using Microsoft.AspNetCore.Identity.EntityFrameworkCore;
using Microsoft.EntityFrameworkCore;

namespace Advanced.Models {
    public class IdentityContext: IdentityDbContext<IdentityUser> {

        public IdentityContext(DbContextOptions<IdentityContext> options)
            : base(options) { }
    }
}
```

ASP.NET Core Identity 包包括 IdentityDbContext<T>类，用于创建 Entity Framework Core 上下文类。泛型类型参数 T 用于指定表示数据库中的用户的类。可以创建自定义用户类，但是使用了名为 IdentityUser 的基本类，它提供了核心的标识特性。

**注意：**如果代码清单 38-4 中使用的类没有意义，不必担心。如果不熟悉 Entity Framework Core，那么建议将该类视为黑盒。一旦建立了 ASP.NET Core Identity 的构建块，就很少需要修改，可以将本章中的文件复制到自己的项目中。

### 38.2.2 配置数据库连接字符串

需要一个连接字符串来告诉 ASP.NET Core Identity，它应该存储数据的地方。在代码清单 38-5 中，向 appsettings.json 文件添加了一个连接字符串，以及用于应用程序数据的连接字符串。

代码清单 38-5　在 Advanced 项目的 appsettings.json 文件中添加连接字符串

```
{
   "Logging": {
      "LogLevel": {
         "Default": "Information",
         "Microsoft": "Warning",
         "Microsoft.Hosting.Lifetime": "Information",
         "Microsoft.EntityFrameworkCore": "Information"
      }
   },
   "AllowedHosts": "*",
   "ConnectionStrings": {
      "PeopleConnection": "Server=(localdb)
      \\MSSQLLocalDB;Database=People;MultipleActiveResultSets=True",
      "IdentityConnection": "Server=(localdb)
      \\MSSQLLocalDB;Database=Identity;MultipleActiveResultSets=True"
   }
}
```

连接字符串指定了一个名为 Identity 的 LocalDB 数据库。

**注意：**打印页面的宽度不允许对连接字符串进行合理的格式化，它必须出现在单个不间断的行中。将连接字符串添加到自己的项目时，请确保它在单行上。

### 38.2.3　配置应用程序

下一步是配置 ASP.NET Core，因此 Identity 数据库上下文设置为服务，如代码清单 38-6 所示。

代码清单 38-6　在 Advanced 项目的 Startup.cs 文件中配置 Identity

```
using System;
using System.Collections.Generic;
using System.Linq;
using System.Threading.Tasks;
using Microsoft.AspNetCore.Builder;
using Microsoft.AspNetCore.Hosting;
using Microsoft.AspNetCore.Http;
using Microsoft.Extensions.DependencyInjection;
using Microsoft.Extensions.Hosting;
using Microsoft.Extensions.Configuration;
using Microsoft.EntityFrameworkCore;
using Advanced.Models;
using Microsoft.AspNetCore.ResponseCompression;
using Microsoft.AspNetCore.Identity;

namespace Advanced {
   public class Startup {
```

```
    public Startup(IConfiguration config) {
        Configuration = config;
    }

    public IConfiguration Configuration { get; set; }

    public void ConfigureServices(IServiceCollection services) {
        services.AddDbContext<DataContext>(opts => {
            opts.UseSqlServer(Configuration[
                "ConnectionStrings:PeopleConnection"]);
            opts.EnableSensitiveDataLogging(true);
        });
        services.AddControllersWithViews().AddRazorRuntimeCompilation();
        services.AddRazorPages().AddRazorRuntimeCompilation();
        services.AddServerSideBlazor();
        services.AddSingleton<Services.ToggleService>();
        services.AddResponseCompression(opts => {
            opts.MimeTypes = ResponseCompressionDefaults.MimeTypes.Concat(
                new[] { "application/octet-stream" });
    });

    services.AddDbContext<IdentityContext>(opts =>
        opts.UseSqlServer(Configuration[
            "ConnectionStrings:IdentityConnection"]));
    services.AddIdentity<IdentityUser, IdentityRole>()
        .AddEntityFrameworkStores<IdentityContext>();
}

public void Configure(IApplicationBuilder app, DataContext context) {

    app.UseDeveloperExceptionPage();
    app.UseStaticFiles();
    app.UseRouting();

    app.UseEndpoints(endpoints => {
        endpoints.MapControllerRoute("controllers",
            "controllers/{controller=Home}/{action=Index}/{id?}");
        endpoints.MapDefaultControllerRoute();
        endpoints.MapRazorPages();
        endpoints.MapBlazorHub();

        endpoints.MapFallbackToClientSideBlazor<BlazorWebAssembly.Startup>
            ("/webassembly/{*path:nonfile}", "index.html");

        endpoints.MapFallbackToPage("/_Host");
    });
```

```
            app.Map("/webassembly", opts =>
                opts.UseClientSideBlazorFiles<BlazorWebAssembly.Startup>());

            SeedData.SeedDatabase(context);
        }
    }
}
```

### 38.2.4 创建和应用身份数据库迁移

剩下的步骤是创建 Entity Framework Core 数据库迁移，并将其应用于创建数据库。打开一个新的 PowerShell 窗口，导航到 Advanced 项目文件夹，然后运行代码清单 38-7 中所示的命令。

**代码清单 38-7 创建和应用数据库迁移**

```
dotnet ef migrations add --context IdentityContext Initial
dotnet ef database update --context IdentityContext
```

如前面的章节所述，Entity Framework Core 通过称为迁移的功能来管理对数据库架构的更改。既然项目中有两个数据库上下文类，那么 Entity Framework Core 工具需要使用--context 参数来确定使用哪个上下文类。代码清单 38-7 中的命令创建一个包含 ASP. NET Core Identity 架构的迁移，并将其应用于数据库。

**重置 ASP. NET Core Identity 数据库**

如果需要重置数据库，请在 Advanced 文件夹中运行 dotnet ef database drop --force --context IdentityContext 命令，然后运行 dotnet ef database update --context IdentityContext 命令。这将删除现有数据库并创建一个 new 和 empty 替换。不要在生产系统上使用这些命令，因为这样会删除用户凭证。如果需要重置主数据库，则运行 dotnet ef database drop --force --context DataContext 命令，后跟 dotnet ef database update --context DataContext 命令。

## 38.3 创建用户管理工具

这一节中创建通过 ASP.NET Core Identity 管理用户的工具。用户通过 UserManager<T>类进行管理，其中 T 是选择来表示数据库中的用户的类。在创建 Entity Framework Core 上下文类时，指定 IdentityUser 作为表示数据库中的用户的类。这是由 ASP.NET Core Identity 提供的内置类。它提供了大多数应用程序所需要的核心特性。表 38-3 描述了最有用的 IdentityUser 属性。IdentityUser 类还定义了其他属性，但这些是大多数应用程序需要的属性，也是本书使用的属性。

**搭建身份管理工具**

微软提供了一个工具，它为管理用户生成一组 Razor Pages。该工具将模板中的通用内容(称为脚手架)添加到项目中，然后根据应用程序对其进行定制。我不是脚手架或模板的粉丝，这也不例外。Microsoft Identity 模板经过了深思熟虑，但它们的用途有限，因为它们侧重于自我管理，允许用户创建账户、更改密码等，而不需要管理员干预。可以调整模板以限制用户执行的任务范围，但这些特性背后的前提是相同的。

如果编写的应用程序类型是用户管理他们自己的凭证，那么 scaffolding 选项可能值得考虑，并在 https://docs.microsoft.com/en-us/aspnet/core/security/authentication/scaffold-identity 上进行了描述。对于其他所有方法，应该使用 ASP.NET Core Identity 提供的用户管理 API。

表 38-3 有用的 IdentityUser 属性

| 名称 | 描述 |
| --- | --- |
| Id | 此属性包含用户的唯一 ID |
| UserName | 此属性返回用户的用户名 |
| Email | 此属性包含用户的电子邮件地址 |

表 38-4 描述了本节中用于管理用户的 UserManager<T>成员。

表 38-4 有用的 UserManager<T>成员

| 名称 | 描述 |
| --- | --- |
| Users | 此属性返回一个包含数据库中存储的用户的序列 |
| FindByIdAsync(id) | 此方法使用指定的 ID 查询数据库中的用户对象 |
| CreateAsync(user, password) | 此方法使用指定的密码在数据库中存储一个新用户 |
| UpdateAsync(user) | 此方法修改数据库中的现有用户 |
| DeleteAsync(user) | 此方法从数据库中删除指定的用户 |

## 38.3.1 准备用户管理工具

为准备创建管理工具，在Advanced 项目的 Pages 文件夹中，请将代码清单 38-8 所示的表达式添加到_ViewImports.cshtml 文件中。

**代码清单 38-8 在 Advanced 项目的 Pages 文件夹的_ViewImports.cshtml 文件中添加表达式**

```
@addTagHelper *, Microsoft.AspNetCore.Mvc.TagHelpers
@using Advanced.Models
@using Microsoft.AspNetCore.Mvc.RazorPages
@using Microsoft.EntityFrameworkCore
@using System.ComponentModel.DataAnnotations
@using Microsoft.AspNetCore.Identity
@using Advanced.Pages
```

接下来，在 Advanced 项目中创建 Pages/Users 文件夹，并向 Pages/Users 文件夹中添加一个名为_Layout.cshtml 的 Razor 布局，内容如代码清单 38-9 所示。

**代码清单 38-9 Advanced 项目的 Pages/Users 文件夹中的_Layout.cshtml 文件**

```
<!DOCTYPE html>
<html>
```

```
<head>
    <title>Identity</title>
    <link href="/lib/twitter-bootstrap/css/bootstrap.min.css" rel="stylesheet" />
</head>
<body>
    <div class="m-2">
        <h5 class="bg-info text-white text-center p-2">User Administration</h5>
        @RenderBody()
    </div>
</body>
</html>
```

将一个名为 AdminPageModel.cs 的类文件添加到 Pages 文件夹中，并使用它定义如代码清单 38-10 所示的类。

#### 代码清单 38-10　Advanced 项目的 Pages 文件夹中的 AdminPageModel.cs 文件

```
using Microsoft.AspNetCore.Mvc.RazorPages;
namespace Advanced.Pages {
    public class AdminPageModel: PageModel {

    }
}
```

这个类将是本节定义的页面模型类的基础。如第 39 章所述，在保护应用程序时，公共基类非常有用。

### 38.3.2　枚举用户账户

尽管数据库目前是空的，但下面首先创建一个枚举用户账户的 Razor Pages。在 Advanced 项目的 Pages/Users 文件夹中添加一个名为 List.cshtml 的 Razor Pages，内容如代码清单 38-11 所示。

#### 代码清单 38-11　Advanced 项目的 Pages/Users 文件夹中的 List.cshtml 文件的内容

```
@page
@model ListModel

<table class="table table-sm table-bordered">
    <tr><th>ID</th><th>Name</th><th>Email</th><th></th></tr>
    @if (Model.Users.Count() == 0) {
        <tr><td colspan="4" class="text-center">No User Accounts</td></tr>
    } else {
        foreach (IdentityUser user in Model.Users) {
            <tr>
                <td>@user.Id</td>
                <td>@user.UserName</td>
                <td>@user.Email</td>
                <td class="text-center">
                    <form asp-page="List" method="post">
```

```
                    <input type="hidden" name="Id" value="@user.Id" />
                    <a class="btn btn-sm btn-warning" asp-page="Editor"
                       asp-route-id="@user.Id" asp-route-mode="edit">Edit</a>
                    <button type="submit" class="btn btn-sm btn-danger">
                       Delete
                    </button>
                 </form>
              </td>
           </tr>
        }
    }
</table>

<a class="btn btn-primary" asp-page="create">Create</a>

@functions {

    public class ListModel : AdminPageModel {
        public UserManager<IdentityUser> UserManager;

        public ListModel(UserManager<IdentityUser> userManager) {
            UserManager = userManager;
        }

        public IEnumerable<IdentityUser> Users { get; set; }

        public void OnGet() {
            Users = UserManager.Users;
        }
    }
}
```

UserManager<IdentityUser>类设置为一个服务，以便可以通过依赖注入使用它。Users 属性返回一个 IdentityUser 对象集合，可用于枚举用户账户。这个 Razor Pages 将用户显示在一个表中，带有允许编辑或删除每个用户的按钮，尽管这在最初是不可见的，因为当没有用户对象显示时，占位符消息会显示出来。有一个按钮可以导航到名为 Create 的 Razor Pages，下一节将定义这个页面。

重启 ASP.NET 并请求 http://localhost:5000/users/list，以查看当前为空的数据表，如图 38-2 所示。

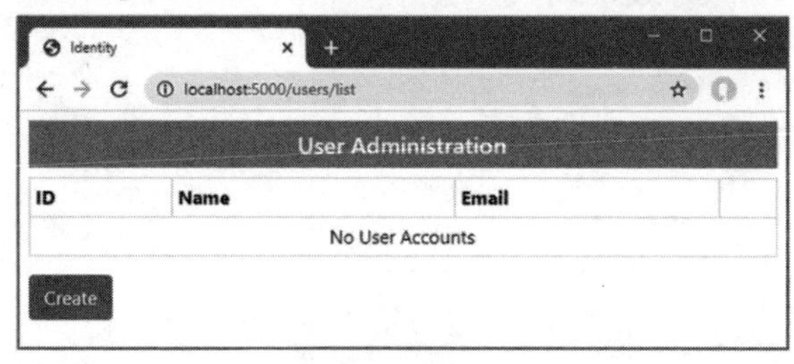


图 38-2 列举用户

### 38.3.3 创建用户

给Pages/Users文件夹添加一个名为Create.cshtml的Razor Pages，内容如代码清单38-12所示。

**代码清单38-12 Advanced项目的Pages/Users文件夹中的Create.cshtml文件内容**

```
@page
@model CreateModel

<h5 class="bg-primary text-white text-center p-2">Create User</h5>
<form method="post">
    <div asp-validation-summary="All" class="text-danger"></div>
    <div class="form-group">
        <label>User Name</label>
        <input name="UserName" class="form-control" value="@Model.UserName" />
    </div>
    <div class="form-group">
        <label>Email</label>
        <input name="Email" class="form-control" value="@Model.Email" />
    </div>
    <div class="form-group">
        <label>Password</label>
        <input name="Password" class="form-control" value="@Model.Password" />
    </div>
    <button type="submit" class="btn btn-primary">Submit</button>
    <a class="btn btn-secondary" asp-page="list">Back</a>
</form>

@functions {

    public class CreateModel : AdminPageModel {
        public UserManager<IdentityUser> UserManager;

        public CreateModel(UserManager<IdentityUser> usrManager) {
            UserManager = usrManager;
        }

        [BindProperty][Required]
        public string UserName { get; set; }

        [BindProperty][Required][EmailAddress]
        public string Email { get; set; }

        [BindProperty][Required]
        public string Password { get; set; }

        public async Task<IActionResult> OnPostAsync() {
```

```
            if (ModelState.IsValid) {
                IdentityUser user =
                    new IdentityUser { UserName = UserName, Email = Email };
                IdentityResult result =
                    await UserManager.CreateAsync(user, Password);
                if (result.Succeeded) {
                    return RedirectToPage("List");
                }
                foreach (IdentityError err in result.Errors) {
                    ModelState.AddModelError("", err.Description);
                }
            }
            return Page();
        }
    }
}
```

尽管 ASP.NET Core Identity 数据是使用 Entity Framework Core 存储的，但不需要直接使用数据库上下文类。相反，数据是通过 UserManager<T>类提供的方法来管理的。使用 CreateAsync 方法创建新用户，该方法接收一个 IdentityUser 对象和一个密码字符串作为参数。

这个 Razor Pages 定义了受制于模型绑定的三个属性。UserName 和 Email 属性用于配置 IdentityUser 对象，该对象与绑定到 Password 属性的值结合起来调用 CreateAsync 方法。这些属性配置了验证属性，验证属性确保提供了值，并且 Email 属性是格式化的电子邮件地址。

CreateAsync 方法的结果是一个 Task<IdentityResult>对象，它使用表 38-5 中描述的属性指示创建操作的结果。

表 38-5　由 IdentityResult 类定义的属性

| 名称 | 描述 |
| --- | --- |
| Succeeded | 如果操作成功，则返回 true |
| Errors | 返回一系列 IdentityError 对象，这些对象描述尝试操作时遇到的错误。每个 IdentityError 对象都提供一个 Description 属性来总结问题 |

检查 Succeeded 属性以确定是否在数据库中创建了一个新用户。如果 Succeeded 属性为 true，则将客户端重定向到 List 页面，以便显示用户列表，反映新添加的用户。

```
...
if (result.Succeeded) {
    return RedirectToPage("List");
}
foreach (IdentityError err in result.Errors) {
    ModelState.AddModelError("", err.Description);
}
...
```

如果 Succeeded 属性为 false，则枚举由 Errors 属性提供的 IdentityError 对象序列，并通过 ModelState.AddModelError 方法和 Description 属性来创建模型级验证错误。

要测试创建新用户账户的能力，请重新启动 ASP.NET Core，请求 http://localhost:5000/users/list。单击 Create 按钮并使用表 38-6 中所示的值填充表单。

**提示：**有一些域名保留用于测试，包括 example.com。可在 https://tools.ietf.org/html/rfc2606 上看到完整的列表。

表 38-6　用于创建示例用户的值

| 字段 | 描述 |
|---|---|
| Name | Joe |
| Email | joe@example.com |
| Password | Secret123$ |

输入值后，单击 Submit 按钮。ASP.NET Core Identity 在数据库中创建用户，浏览器重定向，如图 38-3 所示。你将看到一个不同的 ID 值，因为 ID 是为每个用户随机生成的。

**注意：**在 Password 字段中使用了一个常规的输入元素，以便更容易理解本章的示例。对于实际的项目，最好将输入元素的 type 属性设置为 password，这样输入的字符就看不到了。

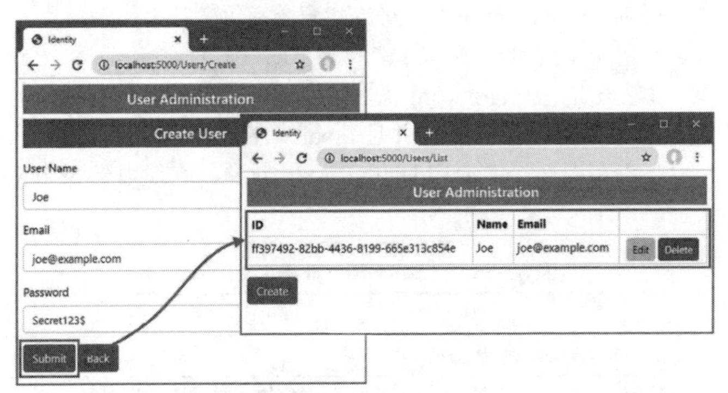


图 38-3　创建新用户

再次单击 Create 按钮，并使用表 38-6 中的值在表单中输入相同的详细信息。这一次，单击 Create 按钮时，将看到通过模型验证摘要报告的错误，如图 38-4 所示。这是 CreateAsync 方法生成的 IdentityResult 对象返回的错误示例。

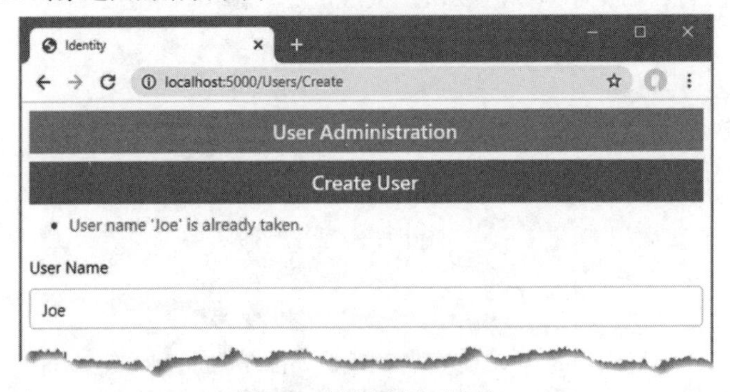


图 38-4　创建新用户时出错

### 1. 验证密码

最常见的需求之一(尤其是企业应用程序)是实施密码策略。通过导航到 http://localhost:5000/Users/Create 并使用表 38-7 中所示的数据填写表单，可以看到默认策略。

表 38-7 用于创建示例用户的值

| 字段 | 描述 |
|---|---|
| Name | Alice |
| Email | alice@example.com |
| Password | secret |

提交表单时，ASP.NET Core Identity 检查候选密码，如果与密码不匹配，则生成错误，如图 38-5 所示。

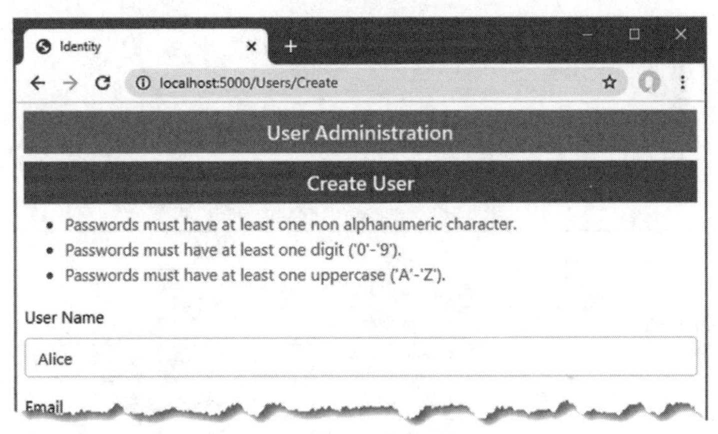


图 38-5 密码验证错误

密码验证规则使用 options 模式配置，如代码清单 38-13 所示。

#### 代码清单 38-13 在 Advanced 项目的 Startup.cs 文件中配置密码验证

```
...
public void ConfigureServices(IServiceCollection services) {
    services.AddDbContext<DataContext>(opts => {
        opts.UseSqlServer(Configuration[
            "ConnectionStrings:PeopleConnection"]);
        opts.EnableSensitiveDataLogging(true);
    });
    services.AddControllersWithViews().AddRazorRuntimeCompilation();
    services.AddRazorPages().AddRazorRuntimeCompilation();
    services.AddServerSideBlazor();
    services.AddSingleton<Services.ToggleService>();

    services.AddResponseCompression(opts => {
        opts.MimeTypes = ResponseCompressionDefaults.MimeTypes.Concat(
            new[] { "application/octet-stream" });
```

```
    });

    services.AddDbContext<IdentityContext>(opts =>
       opts.UseSqlServer(Configuration["ConnectionStrings:IdentityConnection"]));
    services.AddIdentity<IdentityUser, IdentityRole>()
       .AddEntityFrameworkStores<IdentityContext>();

    services.Configure<IdentityOptions>(opts => {
       opts.Password.RequiredLength = 6;
       opts.Password.RequireNonAlphanumeric = false;
       opts.Password.RequireLowercase = false;
       opts.Password.RequireUppercase = false;
       opts.Password.RequireDigit = false;
    });
 }
...
```

ASP.NET Core Identity 使用 IdentityOptions 类配置，其 Password 属性返回一个 PasswordOptions 类，使用表 38-8 中描述的属性配置密码验证。

表 38-8　PasswordOptions 属性

| 名称 | 描述 |
| --- | --- |
| RequiredLength | 这个 int 属性用于指定密码的最小长度 |
| RequireNonAlphanumeric | 将这个 bool 属性设置为 true，要求密码至少包含一个非字母或数字的字符 |
| RequireLowercase | 将这个 bool 属性设置为 true，要求密码至少包含一个小写字符 |
| RequireUppercase | 将这个 bool 属性设置为 true，要求密码至少包含一个大写字符 |
| RequireDigit | 将这个 bool 属性设置为 true，要求密码至少包含一个数字字符 |

在代码清单中，指定了密码的最小长度必须为 6 个字符，并禁用了其他约束。在实际项目中，不应该不仔细考虑就这样做，但是它允许进行有效的演示。重启 ASP.NET Core，请求 http://localhost:5000/user/create，并使用表 38-7 中的详细信息填写表单。单击 Submit 按钮时，密码将被新的验证规则接受，并创建一个新用户，如图 38-6 所示。

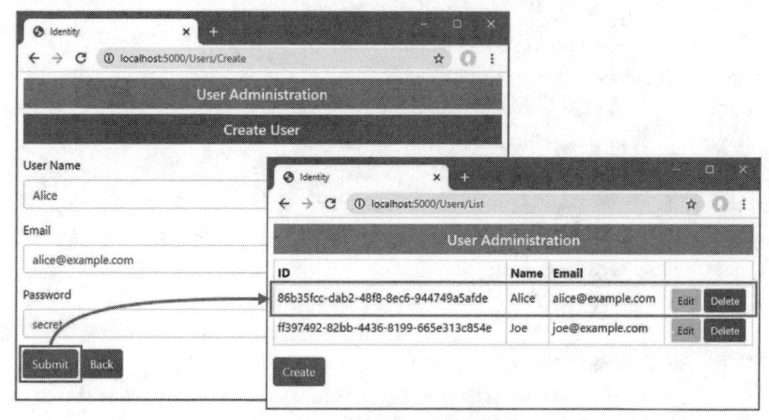


图 38-6　更改密码验证规则

### 2. 验证用户详细信息

在创建账户时，还对用户名和电子邮件地址执行验证。要查看如何应用验证，请求 http://localhost:5000/users/create，并使用表 38-9 中所示的值填写表单。

表 38-9　用于创建示例用户的值

| 字段 | 描述 |
| --- | --- |
| Name | Bob |
| Email | alice@example.com |
| Password | secret |

单击 Submit 按钮，将看到如图 38-7 所示的错误消息。

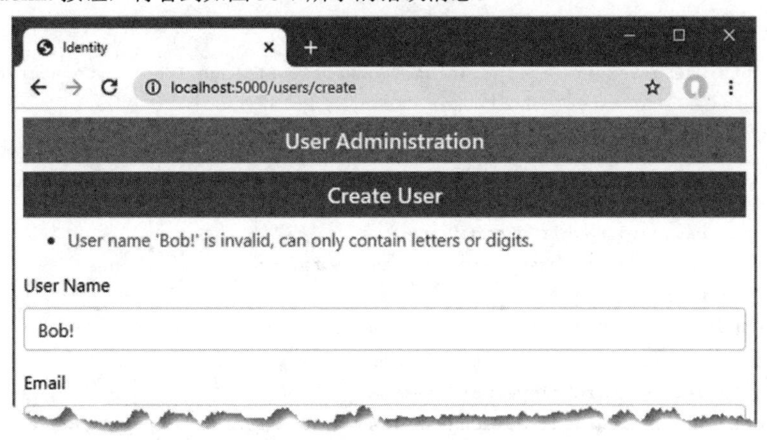


图 38-7　用户详细说明验证错误

使用 IdentityOptions 类定义的 User 属性，可以使用 options 模式配置验证。这个类返回一个 UserOptions 类，它的属性如表 38-10 所示。

表 38-10　UserOptions 属性

| 名称 | 描述 |
| --- | --- |
| AllowedUserNameCharacters | 这个字符串属性包含了用户名中可以使用的所有合法字符。默认值指定 a~z、A~Z 和 0~9 以及连字符、句点、下画线和@字符。此属性不是正则表达式，必须在字符串中显式指定每个合法字符 |
| RequireUniqueEmail | 将此 bool 属性设置为 true，要求新账户指定以前未使用的电子邮件地址 |

在代码清单 38-14 中，修改了应用程序的配置，要求使用唯一的电子邮件地址，用户名中只允许使用小写字母。

**代码清单 38-14　在 Advanced 项目的 Startup.cs 文件中更改用户的验证设置**

```
...
public void ConfigureServices(IServiceCollection services) {
    services.AddDbContext<DataContext>(opts => {
        opts.UseSqlServer(Configuration[
```

```
            "ConnectionStrings:PeopleConnection"]);
        opts.EnableSensitiveDataLogging(true);
    });
    services.AddControllersWithViews().AddRazorRuntimeCompilation();
    services.AddRazorPages().AddRazorRuntimeCompilation();
    services.AddServerSideBlazor();
    services.AddSingleton<Services.ToggleService>();

    services.AddResponseCompression(opts => {
        opts.MimeTypes = ResponseCompressionDefaults.MimeTypes.Concat(
            new[] { "application/octet-stream" });
    });

    services.AddDbContext<IdentityContext>(opts =>
        opts.UseSqlServer(Configuration["ConnectionStrings:IdentityConnection"]));
    services.AddIdentity<IdentityUser, IdentityRole>()
        .AddEntityFrameworkStores<IdentityContext>();

    services.Configure<IdentityOptions>(opts => {
        opts.Password.RequiredLength = 6;
        opts.Password.RequireNonAlphanumeric = false;
        opts.Password.RequireLowercase = false;
        opts.Password.RequireUppercase = false;
        opts.Password.RequireDigit = false;
        opts.User.RequireUniqueEmail = true;
        opts.User.AllowedUserNameCharacters = "abcdefghijklmnopqrstuvwxyz";
    });
}
...
```

重启 ASP.NET Core，请求 http://localhost:5000/user/create，并使用表 38-9 中的值填写表单。单击 Submit 按钮，将看到电子邮件地址现在会导致一个错误。用户名仍然包含非法字符，也被标记为错误，如图 38-8 所示。

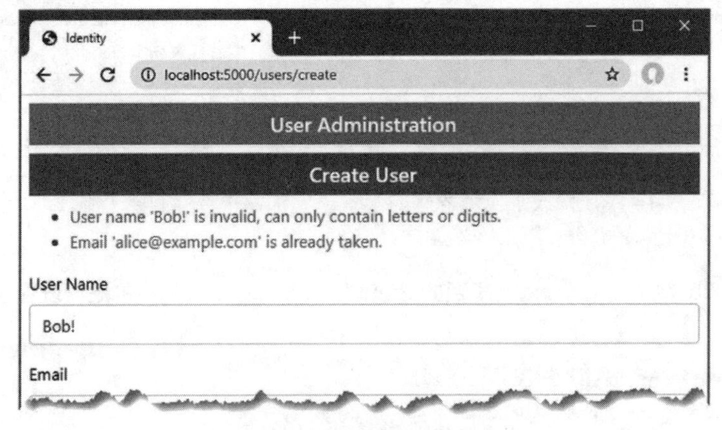


图 38-8　验证用户的详细信息

### 38.3.4 编辑用户

要添加对编辑用户的支持，请在 Advanced 项目的 Pages/Users 文件夹中添加一个名为 Editor.cshtml 的 Razor Pages，内容如代码清单 38-15 所示。

**代码清单 38-15 Advanced 项目的 Pages/Users 文件夹中 Editor.cshtml 文件的内容**

```
@page "{id}"
@model EditorModel

<h5 class="bg-warning text-white text-center p-2">Edit User</h5>
<form method="post">
    <div asp-validation-summary="All" class="text-danger"></div>
    <div class="form-group">
        <label>ID</label>
        <input name="Id" class="form-control" value="@Model.Id" disabled />
        <input name="Id" type="hidden" value="@Model.Id" />
    </div>
    <div class="form-group">
        <label>User Name</label>
        <input name="UserName" class="form-control" value="@Model.UserName" />
    </div>
    <div class="form-group">
        <label>Email</label>
        <input name="Email" class="form-control" value="@Model.Email" />
    </div>
    <div class="form-group">
        <label>New Password</label>
        <input name="Password" class="form-control" value="@Model.Password" />
    </div>
    <button type="submit" class="btn btn-warning">Submit</button>
    <a class="btn btn-secondary" asp-page="list">Back</a>
</form>

@functions {

    public class EditorModel : AdminPageModel {
        public UserManager<IdentityUser> UserManager;

        public EditorModel(UserManager<IdentityUser> usrManager) {
            UserManager = usrManager;
        }

        [BindProperty][Required]
        public string Id { get; set; }

        [BindProperty][Required]
```

```
        public string UserName { get; set; }

        [BindProperty][Required][EmailAddress]
        public string Email { get; set; }

        [BindProperty]
        public string Password { get; set; }

        public async Task OnGetAsync(string id) {
            IdentityUser user = await UserManager.FindByIdAsync(id);
            Id = user.Id; UserName = user.UserName; Email = user.Email;
        }

        public async Task<IActionResult> OnPostAsync() {
            if (ModelState.IsValid) {
                IdentityUser user = await UserManager.FindByIdAsync(Id);
                user.UserName = UserName;
                user.Email = Email;
                IdentityResult result = await UserManager.UpdateAsync(user);
                if (result.Succeeded && !String.IsNullOrEmpty(Password)) {
                    await UserManager.RemovePasswordAsync(user);
                    result = await UserManager.AddPasswordAsync(user, Password);
                }
                if (result.Succeeded) {
                    return RedirectToPage("List");
                }
                foreach (IdentityError err in result.Errors) {
                    ModelState.AddModelError("", err.Description);
                }

            }
            return Page();
        }
    }
}
```

Editor 页面使用 UserManager<T>.FindByIdAsync 方法查找用户，使用通过路由系统接收的 id 值查询数据库，并将 id 值作为 OnGetAsync 方法的参数接收。查询返回的 IdentityUser 对象的值用于填充页面的视图部分所显示的属性，确保在由于验证错误而重新显示页面时不会丢失这些值。

当用户提交表单时，FindByIdAsync 方法用于查询数据库中的 IdentityUser 对象，该对象将用表单中提供的用户名和电子邮件值更新。密码需要一种不同的方法，必须在分配新密码之前从用户对象中删除，如下所示：

```
...
await UserManager.RemovePasswordAsync(user);
result = await UserManager.AddPasswordAsync(user, Password);
...
```

只有当表单包含 Password 值，并且 UserName 和 Email 字段的更新成功时，Editor 页面才会更改密码。ASP.NET Core Identity中的错误以验证消息的形式显示，在成功更新后，浏览器重定向到 List 页面。请求 http://localhost:5000/users /List，单击 Joe 的 Edit 按钮，将 UserName 字段更改为 bob，全部使用小写字符。单击 Submit 按钮，将看到在用户列表中反映的更改，如图 38-9 所示。

---

■ **注意：** 如果单击 alice 账户的编辑按钮，然后不做修改就单击提交，会看到一个错误。这是因为账户是在更改验证策略之前创建的。ASP.NET Core Identity 对更新应用验证检查，这导致了一种奇怪的情况，即数据库中的数据可以读取和使用，但只有更改用户才能更新。

---

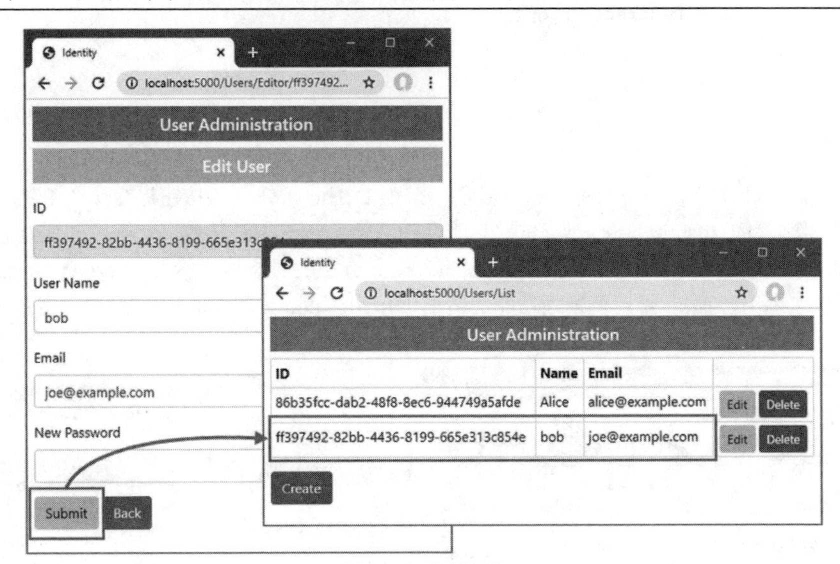


图 38-9　编辑用户

## 38.3.5　删除用户

基本用户管理应用程序需要的最后一个特性是删除用户的能力，如代码清单 38-16 所示。

**代码清单 38-16　在 Advanced 项目的 Pages/Users 文件夹的 List.cshtml 文件中删除用户**

```
...
@functions {

    public class ListModel : AdminPageModel {

        public UserManager<IdentityUser> UserManager;

        public ListModel(UserManager<IdentityUser> userManager) {
            UserManager = userManager;
        }

        public IEnumerable<IdentityUser> Users { get; set; }
```

```
        public void OnGet() {
            Users = UserManager.Users;
        }

        public async Task<IActionResult> OnPostAsync(string id) {
            IdentityUser user = await UserManager.FindByIdAsync(id);
            if (user != null) {
                await UserManager.DeleteAsync(user);
            }
            return RedirectToPage();
        }
    }
}
...
```

List 页面已经在数据表中为每个用户显示了一个 Delete 按钮，该按钮提交一个 POST 请求，其中包含要删除的 IdentityUser 对象的 Id 值。OnPostAsync 方法接收 Id 值，并通过它来使用 FindByIdAsync 方法查询标识，将返回给 DeleteAsync 方法的对象传递给 DeleteAsync 方法，该方法将其从数据库中删除。要检查删除功能，请求 http://localhost:5000/users/List 并单击 Alice 账户的 Delete。用户对象将被删除，如图 38-10 所示。

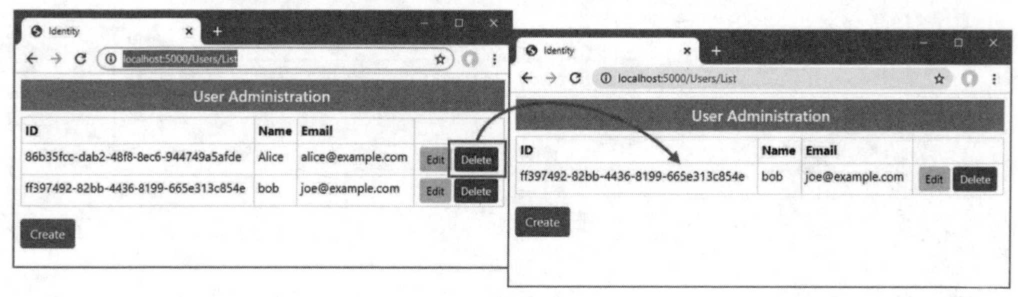


图 38-10　删除用户

## 38.4　创建角色管理工具

一些应用程序只执行两个级别的授权：允许经过身份验证的用户访问应用程序的所有特性，而未经身份验证的用户访问权限较少或无法访问。第 1 部分中的 SportsStore 应用程序采用了这种方法；一个用户经过身份验证后，就可以访问应用程序的所有特性，包括管理工具，而未经过身份验证的用户只能访问公共存储特性。

ASP.NET Core Identity 支持需要更细粒度授权的应用程序的角色。用户被分配到一个或多个角色，他们对这些角色的从属关系决定了哪些特性是可访问的。接下来将展示如何构建工具来创建和管理角色。

角色通过 RoleManager<T>类进行管理，其中 T 是数据库中角色的表示。在本章开头配置 ASP.NET Core Identity时，选择了 IdentityRole，它是 Identity 提供的用来描述角色的内置类，这意味着在这些示例中使用 RoleManager<IdentityRole>类。RoleManager<T>类定义了表 38-11 中所示的允许创建和管理角色的方法和属性。

表 38-11 由 RoleManager<T>类定义的成员

| 名称 | 描述 |
|---|---|
| CreateAsync(role) | 创建新角色 |
| DeleteAsync(role) | 删除指定的角色 |
| FindByIdAsync(id) | 通过 id 查找角色 |
| FindByNameAsync(name) | 根据角色的名称查找角色 |
| RoleExistsAsync(name) | 如果存在指定名称的角色，就返回 true |
| UpdateAsync(role) | 存储对指定角色的更改 |
| Roles | 返回已定义的角色的枚举 |

表 38-12 描述了 IdentityRole 类定义的关键属性。

表 38-12 有用的 IdentityRole 属性

| 名称 | 描述 |
|---|---|
| Id | 此属性包含角色的唯一 Id |
| Name | 此属性返回角色名称 |

虽然角色是通过 RoleManager<T>类管理的，但是角色的成员关系是通过表 38-13 中描述的 UserManager<T>提供的方法管理的。

表 38-13 用于管理角色成员关系的 UserManager<T>方法

| 名称 | 描述 |
|---|---|
| AddToRoleAsync(user, role) | 此方法将用户添加到角色 |
| RemoveFromRoleAsync(user, role) | 此方法从角色中删除用户 |
| GetRolesAsync(user) | 此方法返回用户所属的角色 |
| GetUsersInRoleAsync(role) | 此方法返回属于指定角色的用户 |
| IsInRoleAsync(user, role) | 如果用户是指定角色的成员，则此方法返回 true |

## 38.4.1 为角色管理工具做准备

要准备角色管理工具，请在 Advanced 项目中创建 Pages/Roles 文件夹，并向其中添加一个名为_Layout.cshtml 的 Razor 布局，内容如代码清单 38-17 所示。

**代码清单 38-17 Advanced 项目的 Pages/Roles 文件夹中的_Layout.cshtm 文件的内容**

```
<!DOCTYPE html>
<html>
<head>
    <title>Identity</title>
    <link href="/lib/twitter-bootstrap/css/bootstrap.min.css" rel="stylesheet" />
</head>
<body>
    <div class="m-2">
```

```
        <h5 class="bg-secondary text-white text-center p-2">Role Administration</h5>
        @RenderBody()
    </div>
</body>
</html>
```

这种布局确保用户和角色管理工具之间有明显的区别。

## 38.4.2 枚举和删除角色

在 Advanced 项目的 Pages/Roles 文件夹中添加一个名为 List.cshtml 的 Razor Pages，内容如代码清单 38-18 所示。

**代码清单 38-18 Advanced 项目的 Pages/Roles 文件夹中的 List.cshtml 文件的内容**

```
@page
@model ListModel
<table class="table table-sm table-bordered">
    <tr><th>ID</th><th>Name</th><th>Members</th><th></th></tr>
    @if (Model.Roles.Count() == 0) {
        <tr><td colspan="4" class="text-center">No Roles</td></tr>
    } else {
        foreach (IdentityRole role in Model.Roles) {
            <tr>
                <td>@role.Id</td>
                <td>@role.Name</td>
                <td>@(await Model.GetMembersString(role.Name))</td>
                <td class="text-center">
                    <form asp-page="List" method="post">
                        <input type="hidden" name="Id" value="@role.Id" />
                        <a class="btn btn-sm btn-warning" asp-page="Editor"
                            asp-route-id="@role.Id" asp-route-mode="edit">Edit</a>
                        <button type="submit" class="btn btn-sm btn-danger">
                            Delete
                        </button>
                    </form>
                </td>
            </tr>
        }
    }
</table>
<a class="btn btn-primary" asp-page="create">Create</a>

@functions {

    public class ListModel : AdminPageModel {
        public UserManager<IdentityUser> UserManager;
        public RoleManager<IdentityRole> RoleManager;
```

```
        public ListModel(UserManager<IdentityUser> userManager,
                RoleManager<IdentityRole> roleManager) {
            UserManager = userManager;
            RoleManager = roleManager;
        }

        public IEnumerable<IdentityRole> Roles { get; set; }

        public void OnGet() {
            Roles = RoleManager.Roles;
        }

        public async Task<string> GetMembersString(string role) {
            IEnumerable<IdentityUser> users
                = (await UserManager.GetUsersInRoleAsync(role));
            string result = users.Count() == 0
                ? "No members"
                : string.Join(", ", users.Take(3).Select(u => u.UserName).ToArray());
            return users.Count() > 3 ? $"{result}, (plus others)" : result;
        }

        public async Task<IActionResult> OnPostAsync(string id) {
            IdentityRole role = await RoleManager.FindByIdAsync(id);
            await RoleManager.DeleteAsync(role);
            return RedirectToPage();
        }
    }
}
```

对角色进行枚举，以及最多三个角色成员的名称，如果没有成员，则会显示一条占位符消息。还有一个 Create 按钮，每个角色都有 Edit 和 Delete 按钮，与在用户管理工具中使用的模式相同。

Delete 按钮向 Razor Pages 发送一个 POST 请求。OnPostAsync 方法使用 FindByIdAsync 方法检索角色对象，该对象传递给 DeleteAsync 方法，以从数据库中删除。

## 38.4.3 创建角色

在 Advanced 项目的 Pages/Roles 文件夹中添加一个名为 Create.cshtml 的 Razor Pages，内容如代码清单 38-19 所示。

**代码清单 38-19 Advanced 项目的 Pages/Roles 文件夹中 Create.cshtml 文件的内容**

```
@page
@model CreateModel

<h5 class="bg-primary text-white text-center p-2">Create Role</h5>
<form method="post">
    <div asp-validation-summary="All" class="text-danger"></div>
    <div class="form-group">
```

```
        <label>Role Name</label>
        <input name="Name" class="form-control" value="@Model.Name" />
    </div>
    <button type="submit" class="btn btn-primary">Submit</button>
    <a class="btn btn-secondary" asp-page="list">Back</a>
</form>

@functions {

    public class CreateModel : AdminPageModel {
        public RoleManager<IdentityRole> RoleManager;

        public CreateModel(UserManager<IdentityUser> userManager,
                RoleManager<IdentityRole> roleManager) {
            RoleManager = roleManager;
        }

        [BindProperty][Required]
        public string Name { get; set; }

        public async Task<IActionResult> OnPostAsync() {
            if (ModelState.IsValid) {
                IdentityRole role = new IdentityRole { Name = Name };
                IdentityResult result = await RoleManager.CreateAsync(role);
                if (result.Succeeded) {
                    return RedirectToPage("List");
                }
                foreach (IdentityError err in result.Errors) {
                    ModelState.AddModelError("", err.Description);
                }
            }
            return Page();
        }
    }
}
```

用户将看到一个包含输入元素的表单，用于指定新角色的名称。当提交表单时，OnPostAsync 方法创建一个新的 IdentityRole 对象并将其传递给 CreateAsync 方法。

### 38.4.4　分配角色从属关系

若要添加对管理角色成员资格的支持，请在 Advanced 项目的 Pages/Roles 文件夹中添加名为 Editor.cshtml 的 Razor Pages，内容如代码清单 38-20 所示。

**代码清单 38-20　Advanced 项目的 Pages/Roles 文件夹中的 Editor.cshtml 文件的内容**

```
@page "{id}"
@model EditorModel
```

```
<h5 class="bg-primary text-white text-center p-2">Edit Role: @Model.Role.Name</h5>

<form method="post">
    <input type="hidden" name="rolename" value="@Model.Role.Name" />
    <div asp-validation-summary="All" class="text-danger"></div>
    <h5 class="bg-secondary text-white p-2">Members</h5>
    <table class="table table-sm table-striped table-bordered">
        <thead><tr><th>User</th><th>Email</th><th></th></tr></thead>
        <tbody>
            @if ((await Model.Members()).Count() == 0) {
                <tr><td colspan="3" class="text-center">No members</td></tr>
            }
            @foreach (IdentityUser user in await Model.Members()) {
                <tr>
                    <td>@user.UserName</td>
                    <td>@user.Email</td>
                    <td>
                        <button asp-route-userid="@user.Id"
                                class="btn btn-primary btn-sm" type="submit">
                            Change
                        </button>
                    </td>
                </tr>
            }
        </tbody>
    </table>

    <h5 class="bg-secondary text-white p-2">Non-Members</h5>
    <table class="table table-sm table-striped table-bordered">
        <thead><tr><th>User</th><th>Email</th><th></th></tr></thead>
        <tbody>
            @if ((await Model.NonMembers()).Count() == 0) {
                <tr><td colspan="3" class="text-center">No non-members</td></tr>
            }
            @foreach (IdentityUser user in await Model.NonMembers()) {
                <tr>
                    <td>@user.UserName</td>
                    <td>@user.Email</td>
                    <td>
                        <button asp-route-userid="@user.Id"
                            class="btn btn-primary btn-sm" type="submit">
                            Change
                        </button>
                    </td>
                </tr>
            }
```

```
        </tbody>
    </table>
</form>

<a class="btn btn-secondary" asp-page="list">Back</a>

@functions {

    public class EditorModel : AdminPageModel {
        public UserManager<IdentityUser> UserManager;
        public RoleManager<IdentityRole> RoleManager;

        public EditorModel(UserManager<IdentityUser> userManager,
                RoleManager<IdentityRole> roleManager) {
            UserManager = userManager;
            RoleManager = roleManager;
        }

        public IdentityRole Role { get; set; }

        public Task<IList<IdentityUser>> Members() =>
                UserManager.GetUsersInRoleAsync(Role.Name);

        public async Task<IEnumerable<IdentityUser>> NonMembers() =>
                UserManager.Users.ToList().Except(await Members());

        public async Task OnGetAsync(string id) {
            Role = await RoleManager.FindByIdAsync(id);
        }

        public async Task<IActionResult> OnPostAsync(string userid,
                string rolename) {
            Role = await RoleManager.FindByNameAsync(rolename);
            IdentityUser user = await UserManager.FindByIdAsync(userid);
            IdentityResult result;
            if (await UserManager.IsInRoleAsync(user, rolename)) {
                result = await UserManager.RemoveFromRoleAsync(user, rolename);
            } else {
                result = await UserManager.AddToRoleAsync(user, rolename);
            }
            if (result.Succeeded) {
                return RedirectToPage();
            } else {
                foreach (IdentityError err in result.Errors) {
                    ModelState.AddModelError("", err.Description);
                }
                return Page();
```

```
                }
            }
        }
    }
```

向用户显示一个表，该表显示了作为角色成员的用户，并用一个表显示了非成员。每行包含一个提交表单的 Change 按钮。OnPostAsync 方法使用 UserManager.FindByIdAsync 方法从数据库中检索用户对象。IsInRoleAsync 方法用于确定用户是不是角色成员，而 AddToRoleAsync 和 RemoveFromRoleAsync 方法分别用于添加和删除用户。

重新启动 ASP.NET Core 并请求 http://localhost:5000/roles/list。该列表将为空，因为数据库中没有任何角色。单击 Create 按钮，在文本字段中输入 Admins，然后单击 Submit 按钮以创建新角色。创建角色后，单击 Edit 按钮，将看到可以添加到角色的用户列表。单击 Change 按钮会将用户移入和移出角色。单击 Back，列表将更新，以显示该角色成员的用户，如图 38-11 所示。

**警告：**当更改角色分配时，ASP.NET Core Identity 会重新验证用户的详细信息，如果尝试修改其详细信息与当前限制不匹配的用户，则会导致错误；当部署应用程序，数据库中已经包含在旧角色下创建的用户，并引入限制后会发生这种情况。因此，代码清单 38-20 中的 Razor Pages 会检查操作的结果，以在角色中添加或删除用户，并将所有错误显示为验证消息。

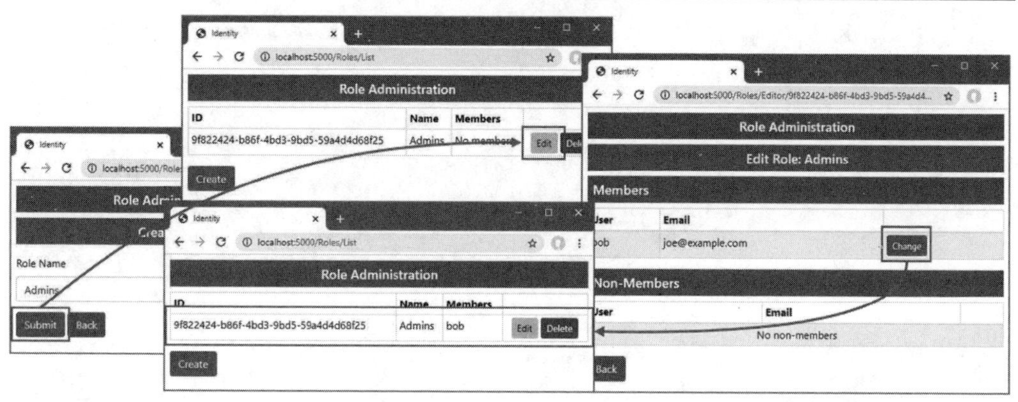


图 38-11 管理角色

## 38.5 小结

本章展示了如何给项目添加 ASP.NET Core Identity，并准备它的数据库来存储用户和角色。描述了基本的 ASP.NET Core Identity API，并展示了如何使用它创建管理用户和角色的工具。第 39 章将展示如何应用 ASP.NET Core Identity 来控制对控制器、Razor Pages、Blazor 应用程序和 Web 服务的访问。

# 第 39 章

# 应用ASP.NET Core Identity

本章将解释如何应用 ASP.NET Core Identity 对用户进行身份验证以及如何对应用程序的特性进行授权访问。将创建用户建立身份所需的特性，解释如何控制对端点的访问，并演示 Blazor 提供的安全特性。还将展示验证 Web 服务客户端的两种不同方法。表 39-1 总结了本章的内容。

表 39-1　本章内容摘要

| 问题 | 解决方案 | 代码清单 |
| --- | --- | --- |
| 验证用户的身份 | 使用 SignInManager<T>类验证用户提供的凭证，并使用内置中间件触发身份验证 | 39-3～39-18 |
| 限制对端点的访问 | 使用 Authorize 属性和内置的中间件来控制访问 | 39-9～39-13 |
| 限制对 Blazor 组件的访问 | 使用 Authorize 属性和内置的 Razor 组件来控制访问 | 39-14～39-17 |
| 限制对 Web 服务的访问 | 使用 cookie 身份验证或持有者令牌 | 39-18～39-30 |

## 准备工作

本章使用了第 38 章中的项目。为了准备这一章，将重置应用程序数据和 ASP.NET Core Identity 数据库并创建新的用户和角色。打开一个新的命令提示符，并在 Advanced 项目文件夹中运行代码清单 39-1 所示的命令，其中包含 Advanced.csproj 文件。这些命令将删除现有的数据库并重新创建它们。

■ **提示**：可从https://github.com/apress/pro-asp.net-core-3下载本章和本书中其他所有章节的示例项目。如果在运行示例时遇到问题，请参阅第1章以了解如何获得帮助。

**代码清单 39-1　重新创建项目数据库**

```
dotnet ef database drop --force --context DataContext
dotnet ef database drop --force --context IdentityContext
dotnet ef database update --context DataContext
dotnet ef database update --context IdentityContext
```

既然应用程序包含多个数据库上下文类，那么 Entity Framework Core 命令需要--context 参数，以选择应用命令的上下文。

从 Debug 菜单中选择 Start Without Debugging 或 Run Without Debugging，或者使用 PowerShell 命令提示符运行代码清单 39-2 中所示的命令。

**代码清单 39-2　运行示例应用程序**

```
dotnet run
```

使用浏览器请求 http://localhost:5000/home/index，这将生成如图 39-1 所示的响应。

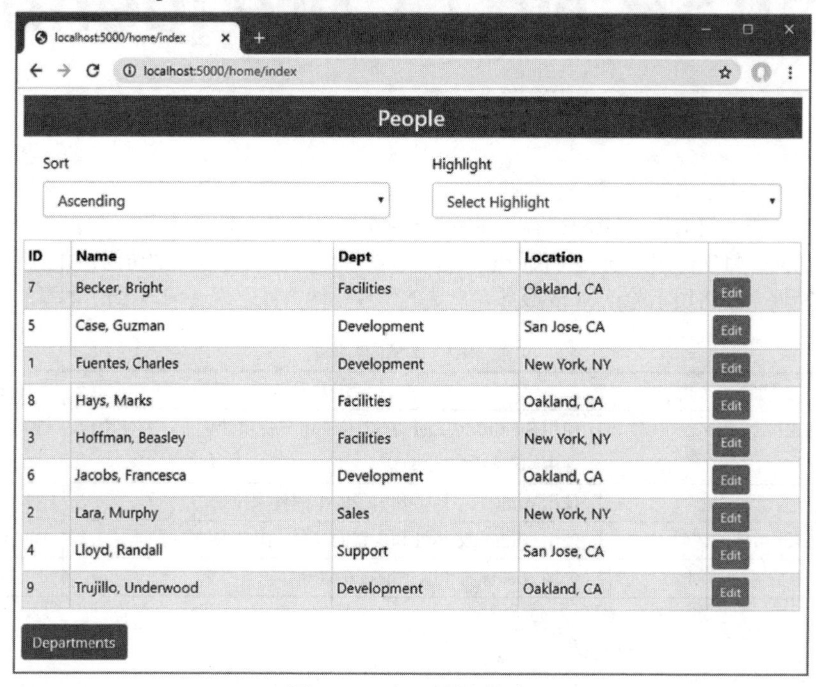


图 39-1　运行示例应用程序

当应用程序启动时，将自动重新播种主应用程序数据库。没有用于 ASP.NET Core Identity 数据库的种子数据。请求 http://localhost:5000/users/list 和 http://localhost:5000/roles/list，将看到图 39-2 中的响应，其中显示数据库是空的。

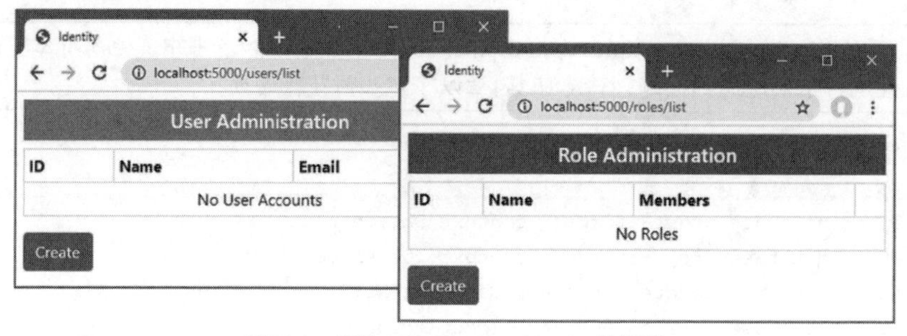


图 39-2　空的 ASP.NET Core Identity 数据库

# 39.1 验证用户的身份

接下来将展示如何向示例项目添加身份验证特性，以便用户可以向应用程序显示凭据并建立身份。

**身份验证与授权**

当使用 ASP.NET Core Identity 时，理解身份验证和授权之间的区别是很重要的。身份验证通常称为 AuthN，是建立用户身份的过程，用户通过向应用程序提交凭据来完成这一过程。在示例应用程序中，这些凭据是用户名和密码。用户名是公共信息，但密码只有用户知道；当提供正确的密码时，应用程序能够对用户进行身份验证。

授权通常称为 AuthZ，是基于用户身份授予对应用程序特性的访问权的过程。只有在用户经过身份验证后才能执行授权，因为应用程序必须知道用户的身份，然后才能决定用户是否有权使用特定功能。

## 39.1.1 创建登录特性

要实施安全策略，应用程序必须允许用户对自己进行身份验证，这是使用 ASP.NET Core Identity API 来完成的。创建 Pages/Account 文件夹，并向其中添加一个名为_Layout.cshtml 的 Razor Pages，内容如代码清单 39-3 所示。此布局将为身份验证特性提供通用内容。

**代码清单 39-3　Advanced 项目的 Pages/Account 文件夹中_Layout.cshtml 文件的内容**

```
<!DOCTYPE html>
<html>
<head>
    <title>Identity</title>
    <link href="/lib/twitter-bootstrap/css/bootstrap.min.css" rel="stylesheet" />
</head>
<body>
    <div class="m-2">
        @RenderBody()
    </div>
</body>
</html>
```

在 Advanced 项目的 Pages/Account 文件夹中添加一个名为 Login.cshtml 的 Razor Pages，内容如代码清单 39-4 所示。

**代码清单 39-4　Advanced 项目的 Pages/Account 文件夹中 Login.cshtml 文件的内容**

```
@page
@model LoginModel

<div class="bg-primary text-center text-white p-2"><h4>Log In</h4></div>

<div class="m-1 text-danger" asp-validation-summary="All"></div>
```

```
<form method="post">
    <input type="hidden" name="returnUrl" value="@Model.ReturnUrl" />
    <div class="form-group">
        <label>UserName</label>
        <input class="form-control" asp-for="UserName" />
    </div>
    <div class="form-group">
        <label>Password</label>
        <input asp-for="Password" type="password" class="form-control" />
    </div>
    <button class="btn btn-primary" type="submit">Log In</button>
</form>

@functions {

    public class LoginModel : PageModel {
        private SignInManager<IdentityUser> signInManager;

        public LoginModel(SignInManager<IdentityUser> signinMgr) {
            signInManager = signinMgr;
        }

        [BindProperty] [Required]
        public string UserName { get; set; }

        [BindProperty] [Required]
        public string Password { get; set; }

        [BindProperty(SupportsGet = true)]
        public string ReturnUrl { get; set; }

        public async Task<IActionResult> OnPostAsync() {
            if (ModelState.IsValid) {
                Microsoft.AspNetCore.Identity.SignInResult result =
                    await signInManager.PasswordSignInAsync(UserName, Password,
                            false, false);
                if (result.Succeeded) {
                    return Redirect(ReturnUrl ?? "/");
                }
                ModelState.AddModelError("", "Invalid username or password");
            }
            return Page();
        }
    }
}
```

ASP.NET Core Identity 提供 SigninManager<T>类来管理登录，其中泛型类型参数 T 是表示应用程序中的用户的类，在示例应用程序中是 IdentityUser。表 39-2 描述了本章使用的 SigninManager<T>成员。

表 39-2 有用的 SigninManager<T>成员

| 名称 | 描述 |
|---|---|
| PasswordSignInAsync(name, password, persist, lockout) | 此方法尝试使用指定的用户名和密码进行身份验证。persist 参数确定成功的身份验证是否生成一个在浏览器关闭后仍然存在的 cookie。lockout 参数决定在身份验证失败时是否锁定账户 |
| SignOutAsync() | 该方法注销用户 |

Razor Pages 为用户提供了一个表单，该表单收集用户名和密码，用于使用 PasswordSignInAsync 方法执行身份验证，如下所示：

```
...
Microsoft.AspNetCore.Identity.SignInResult result =
    await signInManager.PasswordSignInAsync(UserName, Password, false, false);
...
```

PasswordSignInAsync 方法的结果是一个 SignInResult 对象，它定义了 Suceeded 属性，如果身份验证成功，该属性为 true(在 Microsoft.AspNetCore.Mvc 名称空间中也定义了一个 SignInResult 类，这就是在代码清单中使用完全限定类名的原因)。

ASP.NET Core 应用程序中的身份验证通常是当用户试图访问一个需要授权的端点时引发的，如果身份验证成功，就按约定将用户返回端点，这就是为什么如果用户提供有效身份凭证，登录页面定义在重定向中使用的 ReturnUrl 属性的原因。

```
...
if (result.Succeeded) {
    return Redirect(ReturnUrl ?? "/");
}
...
```

如果用户没有提供有效凭据，则显示一条验证消息，并重新显示页面。

**保护认证 cookie**

身份验证 cookie 包含用户的身份，ASP.NET Core 相信，包含 cookie 的请求来自经过身份验证的用户。这意味着应该将 HTTPS 用于使用 ASP.NET Core Identity 的生产应用程序，以防止 cookie 被中间层拦截。有关在 ASP.NET Core 中启用 HTTPS 的详细信息，请参阅第 II 部分。

### 39.1.2 检查 ASP.NET Core Identity cookie

当用户经过身份验证时，向响应添加一个 cookie，以便后续请求可以标识为已经过身份验证。将一个名为 Details.cshtml 的 Razor Pages 添加到 Advanced 项目的 Pages/Account 文件夹，其内容如代码清单 39-5 所示，在 cookie 出现时显示它。

**代码清单 39-5　Advanced 项目的 Pages/Account 文件夹中 Details.cshtml 的内容**

```
@page
@model DetailsModel

<table class="table table-sm table-bordered">
    <tbody>
        @if (Model.Cookie == null) {
            <tr><th class="text-center">No Identity Cookie</th></tr>
        } else {
            <tr>
                <th>Cookie</th>
                <td class="text-break">@Model.Cookie</td>
            </tr>
        }
    </tbody>
</table>

@functions {

    public class DetailsModel : PageModel {

        public string Cookie { get; set; }

        public void OnGet() {
            Cookie = Request.Cookies[".AspNetCore.Identity.Application"];
        }
    }
}
```

用于 ASP.NET Core Identity cookie 的名称为.AspNetCore.Identity.Application。这个页面从请求中检索 cookie，如果没有 cookie，则显示它的值或占位符消息。

### 39.1.3　创建退出页面

让用户能够注销 cookie 是很重要的，这样用户就可以显式地删除 cookie，特别是在可以使用公共机器访问应用程序的情况下。在 Advanced 文件夹的 Pages/Account 文件夹中添加一个名为 Logout.cshtml 的 Razor Pages，其内容如代码清单 39-6 所示。

**代码清单 39-6　Advanced 项目的 Pages/Account 文件夹中 Logout.cshtml 文件的内容**

```
@page
@model LogoutModel

<div class="bg-primary text-center text-white p-2"><h4>Log Out</h4></div>
<div class="m-2">
    <h6>You are logged out</h6>
    <a asp-page="Login" class="btn btn-secondary">OK</a>
```

```
</div>

@functions {

    public class LogoutModel : PageModel {
        private SignInManager<IdentityUser> signInManager;

        public LogoutModel(SignInManager<IdentityUser> signInMgr) {
            signInManager = signInMgr;
        }

        public async Task OnGetAsync() {
            await signInManager.SignOutAsync();
        }
    }
}
```

此页面调用表 39-2 中描述的 SignOutAsync 方法来从应用程序中注销。ASP.NET Core Identity cookie 将删除，这样浏览器就不会在将来的请求中包含它(并使 cookie 失效，这样即使 cookie 再次使用，请求也不会被视为已验证)。

### 39.1.4 测试身份验证特性

重启 ASP.NET Core，请求 http://localhost:5000/users/list。单击 Create 按钮并使用表 39-3 中所示的数据填写表单。单击 Submit 按钮提交表单，并创建用户账户。

表 39-3 创建用户的数据值

| 字段 | 描述 |
|---|---|
| UserName | bob |
| Email | bob@example.com |
| Password | secret |

导航到 http://localhost:5000/account/login 并使用表 39-3 中的用户名和密码进行身份验证。没有指定返回 URL，一旦通过身份验证，就重定向到根 URL。请求 http://localhost:5000/account/details，将看到 ASP.NET Core Identity cookie。请求 http://localhost:5000/account/logout，退出应用程序，并返回 http://localhost:5000/account/details，以确认 cookie 已被删除，如图 39-3 所示。

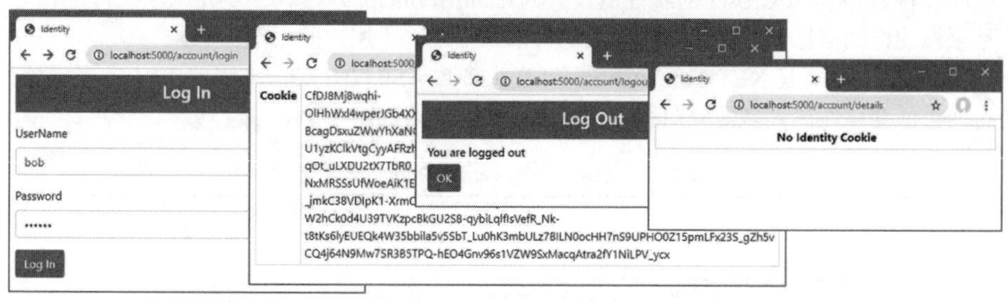


图 39-3 验证用户的身份

### 39.1.5　启用身份验证中间件

ASP.NET Core Identity 提供了一个中间件组件，它检测 SignInManager<T>类创建的 cookie，并使用经过身份验证的用户的详细信息填充 HttpContext 对象。这为端点提供了关于用户的详细信息，而不需要知道身份验证过程，也不需要直接处理由身份验证过程创建的 cookie。代码清单 39-7 将身份验证中间件添加到示例应用程序的请求管道中。

**代码清单 39-7　在 Advanced 文件夹的 Startup.cs 文件中启用中间件**

```
...
public void Configure(IApplicationBuilder app, DataContext context) {

    app.UseDeveloperExceptionPage();
    app.UseStaticFiles();
    app.UseRouting();

    app.UseAuthentication();
    app.UseEndpoints(endpoints => {
        endpoints.MapControllerRoute("controllers",
            "controllers/{controller=Home}/{action=Index}/{id?}");
        endpoints.MapDefaultControllerRoute();
        endpoints.MapRazorPages();
        endpoints.MapBlazorHub();

        endpoints.MapFallbackToClientSideBlazor<BlazorWebAssembly.Startup>
            ("/webassembly/{*path:nonfile}", "index.html");
        endpoints.MapFallbackToPage("/_Host");
    });

    app.Map("/webassembly", opts =>
        opts.UseClientSideBlazorFiles<BlazorWebAssembly.Startup>());

    SeedData.SeedDatabase(context);
}
...
```

中间件将 HttpContext.User 的属性值设置为 ClaimsPrincipal 对象。声明是关于用户的信息片段和信息来源的详细信息，提供了一种通用的方法来描述关于用户的已知信息。

ClaimsPrincipal 类是.NET Core 的一部分，在大多数 ASP.NET Core 应用程序中并不直接有用。但是有两个嵌套属性在大多数应用程序中是有用的，如表 39-4 所述。

**表 39-4　有用的嵌套 ClaimsPrincipal 属性**

| 名称 | 描述 |
|---|---|
| ClaimsPrincipal.Identity.Name | 此属性返回用户名，如果没有与请求关联的用户，则该用户名为 null |
| ClaimsPrincipal.Identity.IsAuthenticated | 如果与请求关联的用户已经过身份验证，则此属性返回 true |

通过 ClaimsPrincipal 对象提供的用户名可用于获取 ASP.NET Core Identity 用户对象，如代码清单 39-8 所示。

**代码清单 39-8　Advanced 项目的 Pages/Account 文件夹中 Details.cshtml 文件的详细信息**

```
@page
@model DetailsModel

<table class="table table-sm table-bordered">
    <tbody>
        @if (Model.IdentityUser == null) {
            <tr><th class="text-center">No User</th></tr>
        } else {
            <tr><th>Name</th><td>@Model.IdentityUser.UserName</td></tr>
            <tr><th>Email</th><td>@Model.IdentityUser.Email</td></tr>
        }
    </tbody>
</table>

@functions {

    public class DetailsModel : PageModel {
        private UserManager<IdentityUser> userManager;

        public DetailsModel(UserManager<IdentityUser> manager) {
            userManager = manager;
        }

        public IdentityUser IdentityUser { get; set; }

        public async Task OnGetAsync() {
            if (User.Identity.IsAuthenticated) {
                IdentityUser = await userManager.FindByNameAsync(User.Identity.Name);
            }
        }
    }
}
```

HttpContext.User 属性可以通过 PageModel 和 ControllerBase 类定义的便利属性 User 访问。此 Razor Pages 确认有一个经过身份验证的用户与该请求相关联，并获取描述该用户的 IdentityUser 对象。

重启 ASP.NET Core，请求 http://localhost:5000/account/login，并使用表 39-3 中的详细信息进行身份验证。请求 http://localhost:5000/account/details，将看到代码清单 39-7 中启用的 ASP.NET Core Identity 中间件如何处理 cookie，以将用户详细信息与请求关联起来，如图 39-4 所示。

**考虑双因素身份验证**

本章执行了单因素身份验证，用户可以使用预先知道的一段信息进行身份验证：密码。

ASP.NET Core Identity 也支持双因素身份验证，用户需要一些额外的东西，通常是在用户想要身份验证时提供给他们的东西。最常见的例子是来自硬件令牌或智能手机应用程序的值，或以电子邮件或文本消息发送的身份验证代码(严格地说，这两个因素可以是任何东西，包括指纹、虹膜扫描和声音识别，尽管这些选项对大多数 Web 应用程序来说很少是必需的)。

安全性得到了提高，因为攻击者需要知道用户的密码，并能够访问提供第二个因素的任何东西，例如电子邮件账户或手机。

书中没有展示双因素身份验证，有两个原因。首先，它需要大量的准备工作，比如建立分发第二因素电子邮件和文本的基础设施，以及实现验证逻辑，所有这些都超出了本书的讨论范围。

第二个原因是，双因素身份验证迫使用户记住，要通过一个额外的步骤来进行身份验证(比如记住他们的手机或在附近保留一个安全令牌)，这对于 Web 应用程序并不总是合适。十多年来，我从事不同的工作，随身带着这种或那种硬件令牌，我已经记不清有多少次因为把令牌忘在家里而无法登录雇主的系统。如果你正在考虑双因素身份验证，那么我建议你使用众多托管提供商之一，该提供商负责分发和管理第二个因素。

图 39-4　获取经过身份验证的用户的详细信息

## 39.2　对授权端点的访问

一旦应用程序具有身份验证功能，就可以使用用户身份来限制对端点的访问。接下来将解释启用授权的过程，并演示如何定义授权策略。

### 39.2.1　应用授权属性

Authorize 属性用于限制对端点的访问，并可应用于单个操作或页面处理程序方法，或应用于控制器或页面模型类，在这种情况下，策略应用于类定义的所有方法。我想限制对第 38 章中创建的用户和角色管理工具的访问。当需要为多个 Razor Pages 或控制器使用相同的授权策略时，最好定义一个通用基类，可将 Authorize 属性应用于该基类，因为这样可以确保不会意外地忽略该属性，并允许未经授权的访问。正是由于这个原因，第 38 章定义了 AdminPageModel 类，并使用它作为所有管理工具页面模型的基础。代码清单 39-9 将 Authorize 属性应用到 AdminPageModel 类以创建授权策略。

**代码清单 39-9　在 Advanced 项目的 Pages 文件夹中的 AdminPageModel.cs 文件内应用属性**

```
using Microsoft.AspNetCore.Mvc.RazorPages;
using Microsoft.AspNetCore.Authorization;

namespace Advanced.Pages {

    [Authorize(Roles="Admins")]
    public class AdminPageModel : PageModel {

    }
}
```

可在没有参数的情况下应用 Authorize 属性，从而限制任何经过身份验证的用户的访问权。Roles 参数用于进一步限制特定角色成员的用户的访问权限，特定角色表示为逗号分隔的列表。此代码清单中的属性限制分配给管理员角色的用户的访问权限。授权限制将继承，这意味着将该属性应用到基类，将限制对第 38 章中为管理用户和角色而创建的所有 Razor Pages 的访问。

▒ **注意：**如果想限制对控制器中大多数(而不是全部)操作方法的访问，那么可以为控制器类应用 Authorize 属性，把 AllowAnonymous 属性应用于需要通过验证访问的操作方法。

## 39.2.2　启用授权中间件

授权策略由中间件组件执行，该组件必须添加到应用程序的请求管道中，如代码清单 39-10 所示。

**代码清单 39-10　在 Advanced 项目的 Startup.cs 文件中添加中间件**

```
...
public void Configure(IApplicationBuilder app, DataContext context) {

    app.UseDeveloperExceptionPage();
    app.UseStaticFiles();
    app.UseRouting();

    app.UseAuthentication();
    app.UseAuthorization();

    app.UseEndpoints(endpoints => {
        endpoints.MapControllerRoute("controllers",
            "controllers/{controller=Home}/{action=Index}/{id?}");
        endpoints.MapDefaultControllerRoute();
        endpoints.MapRazorPages();
        endpoints.MapBlazorHub();
```

```
        endpoints.MapFallbackToClientSideBlazor<BlazorWebAssembly.Startup>
            ("/webassembly/{*path:nonfile}", "index.html");

        endpoints.MapFallbackToPage("/_Host");
    });

    app.Map("/webassembly", opts =>
        opts.UseClientSideBlazorFiles<BlazorWebAssembly.Startup>());

    SeedData.SeedDatabase(context);
}
...
```

必须在 UseRouting 和 UseEndpoints 方法之间以及调用 UseAuthentication 方法之后调用 UseAuthorization方法。这确保了授权组件可以在选择端点之后、在处理请求之前访问用户数据并检查授权策略。

### 39.2.3　创建被拒绝访问的端点

应用程序必须处理两种不同类型的授权失败。如果在请求受限制的端点时，没有用户经过身份验证，那么授权中间件将返回一个质询响应，这将触发对登录页面的重定向，以便用户可以显示他们的凭据并证明他们应该能够访问端点。

但是，如果经过身份验证的用户请求受限制的端点，但没有通过授权检查，则会生成一个拒绝访问的响应，以便应用程序可以向用户显示适当的警告。给 Advanced 文件夹的 Pages/Account 文件夹添加一个名为 AccessDenied.cshtml 的 Razor Pages，其内容如代码清单 39-11 所示。

**代码清单 39-11　Advanced 项目的 Pages/Account 文件夹中的 AccessDenied.cshtml 文件**

```
@page

<h4 class="bg-danger text-white text-center p-2">Access Denied</h4>

<div class="m-2">
    <h6>You are not authorized for this URL</h6>
    <a class="btn btn-outline-danger" href="/">OK</a>
    <a class="btn btn-outline-secondary" asp-page="Logout">Logout</a>
</div>
```

该页面向用户显示一条警告消息，并带有一个导航到根 URL 的按钮。在没有管理干预的情况下，用户通常很难解决授权失败问题，我倾向于使拒绝访问响应尽可能简单。

### 39.2.4　创建种子数据

在代码清单 39-9 中，限制了对用户和角色管理工具的访问，因此它们只能由 Admin 角色的用户访问。数据库中没有这样的角色，这就产生了一个问题：被管理工具锁定，因为不允许创建该角色的授权账户。

可在应用 Authorize 属性之前创建一个管理用户和角色，但是这样会使部署应用程序复杂化。

相反，下面为 ASP.NET Core Identity 创建种子数据，确保至少有一个账号可以用来访问用户和角色管理工具。在 Advanced 项目的 Models 文件夹中添加一个名为 IdentitySeedData.cs 的类文件，并使用它定义如代码清单 39-12 所示的类。

**代码清单 39-12　Advanced 项目的 Models 文件夹中 IdentitySeedData.cs 文件的内容**

```
using System;
using System.Threading.Tasks;
using Microsoft.AspNetCore.Identity;
using Microsoft.Extensions.Configuration;
using Microsoft.Extensions.DependencyInjection;

namespace Advanced.Models {
    public class IdentitySeedData {

        public static void CreateAdminAccount(IServiceProvider serviceProvider,
                IConfiguration configuration) {
            CreateAdminAccountAsync(serviceProvider, configuration).Wait();
    }

    public static async Task CreateAdminAccountAsync(IServiceProvider
            serviceProvider, IConfiguration configuration) {
        serviceProvider = serviceProvider.CreateScope().ServiceProvider;

        UserManager<IdentityUser> userManager =
            serviceProvider.GetRequiredService<UserManager<IdentityUser>>();
        RoleManager<IdentityRole> roleManager =
            serviceProvider.GetRequiredService<RoleManager<IdentityRole>>();

        string username = configuration["Data:AdminUser:Name"] ?? "admin";
        string email
            = configuration["Data:AdminUser:Email"] ?? "admin@example.com";
        string password = configuration["Data:AdminUser:Password"] ?? "secret";
        string role = configuration["Data:AdminUser:Role"] ?? "Admins";
            if (await userManager.FindByNameAsync(username) == null) {
                if (await roleManager.FindByNameAsync(role) == null) {
                    await roleManager.CreateAsync(new IdentityRole(role));
                }

                IdentityUser user = new IdentityUser {
                    UserName = username,
                    Email = email
                };

                IdentityResult result = await userManager
                    .CreateAsync(user, password);
                if (result.Succeeded) {
```

```
                await userManager.AddToRoleAsync(user, role);
            }
        }
    }
  }
}
```

UserManager<T>和 RoleManager<T>服务的作用域是确定的，这意味着需要在请求服务之前创建一个新的作用域，因为在应用程序启动时将完成播种。种子代码创建一个分配给角色的用户账户。种子数据的值使用回退值从应用程序的配置中读取，从而便于配置种子账户，而不需要修改代码。代码清单 39-13 向 Startup 类添加了一条语句，以便在应用程序启动时播种数据库。

---

**警告：**在代码文件或纯文本配置文件中设置密码，意味着在首次部署应用程序和初始化新数据库时，必须将更改默认账户密码作为部署过程的一部分。还可以使用用户机密特性将敏感数据保存在项目之外。

---

**代码清单 39-13 在 Advanced 项目的 Startup.cs 文件中播种标识**

```
...
public void Configure(IApplicationBuilder app, DataContext context) {

    app.UseDeveloperExceptionPage();
    app.UseStaticFiles();
    app.UseRouting();

    app.UseAuthentication();
    app.UseAuthorization();

    app.UseEndpoints(endpoints => {
        endpoints.MapControllerRoute("controllers",
            "controllers/{controller=Home}/{action=Index}/{id?}");
        endpoints.MapDefaultControllerRoute();
        endpoints.MapRazorPages();
        endpoints.MapBlazorHub();

        endpoints.MapFallbackToClientSideBlazor<BlazorWebAssembly.Startup>
            ("/webassembly/{*path:nonfile}", "index.html");

        endpoints.MapFallbackToPage("/_Host");
    });

    app.Map("/webassembly", opts =>
        opts.UseClientSideBlazorFiles<BlazorWebAssembly.Startup>());
    SeedData.SeedDatabase(context);
    IdentitySeedData.CreateAdminAccount(app.ApplicationServices, Configuration);
}
...
```

### 39.2.5 测试身份验证序列

重启 ASP.NET Core，并请求 http://localhost:5000/account/logout 以确保没有用户登录到应用程序。在没有登录的情况下，请求 http://localhost:5000/user/list。用于处理请求的端点将要求进行身份验证，并将显示登录提示，因为不存在经过验证的、与该请求关联的用户。使用用户名 bob 和密码 secret 进行身份验证。该用户没有访问受限制端点的权限，因此将显示拒绝访问响应，如图 39-5 所示。

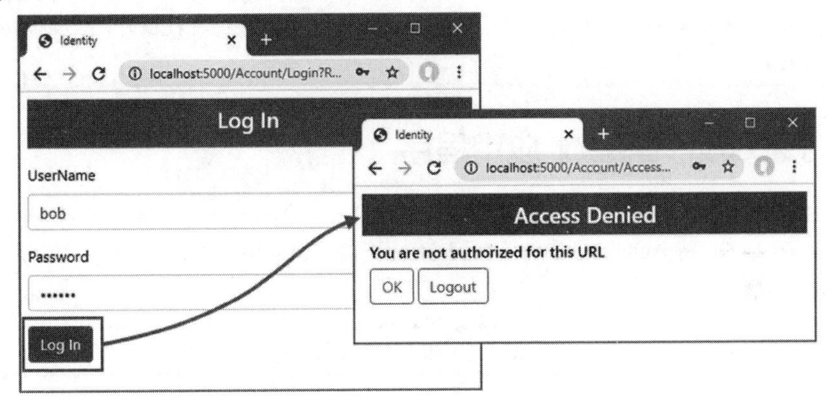


图 39-5　未经授权的用户

单击 Logout 按钮，再次请求 http://localhost:5000/users/list，这将导致显示登录提示符。使用用户名 admin 和密码 secret 进行身份验证。这是由种子数据创建的用户账户，是 Authorize 属性指定的角色的成员。用户通过授权检查，将显示所请求的 Razor Pages，如图 39-6 所示。

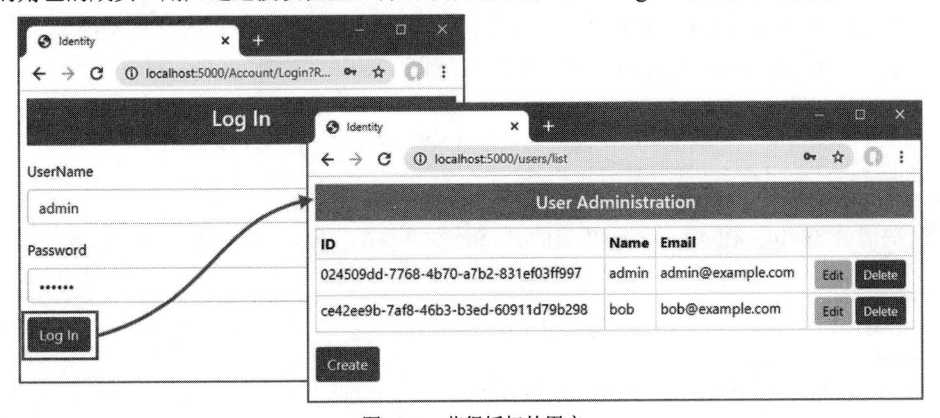


图 39-6　获得授权的用户

**更改授权 URL**

/Account/Login 和/Account/AccessDenied URL 是 ASP.NET Core 授权文件默认使用的。这些可以在 Startup 类中使用 options 模式进行更改，如下所示：

```
...
services.Configure<CookieAuthenticationOptions>(
        IdentityConstants.ApplicationScheme,
```

```
    opts => {
       opts.LoginPath = "/Authenticate";
       opts.AccessDeniedPath = "/NotAllowed";
    });
...
```

使用在 Microsoft.AspNetCore.Authentication.Cookies 名称空间中定义的 CookieAuthenticationOptions 类执行配置。LoginPath 属性用于指定当未经身份验证的用户试图访问受限制的端点时，将浏览器重定向到的路径。在经过身份验证的用户尝试访问受限制的端点且没有授权时，AccessDeniedPath 属性用于指定路径。

## 39.3 授权访问 Blazor 应用程序

保护 Blazor 应用程序的最简单方法是限制对作为入口点的 action 方法或 Razor Pages 的访问。在代码清单 39-14 中，将 Authorize 属性添加到_Host 页面的页面模型类中，它是示例项目中 Blazor 应用程序的入口点。

**理解 OAuth 和 IdentityServer**

如果阅读 Microsoft 文档，就需要使用调用的第三方服务器 IdentityServer (http://identityserver.io)对 Web 服务进行身份验证。

IdentityServer 是一个优质的开源包，提供身份验证和授权服务，并提供付费的附加组件和支持选项。IdentityServer 为 OAuth 提供支持，OAuth 是一种管理身份验证和授权的标准，并为一系列客户端框架提供包。

微软在包括 Web 服务身份验证的项目模板中使用了 IdentityServer。如果使用 Microsoft 提供的 ASP.NET Core 模板创建 Angular 或 React 项目，则身份验证是使用 IdentityServer 实现的。

身份验证很复杂，而且 IdentityServer 很难正确设置。我喜欢 IdentityServer，但它不是必需的，大多数项目都不需要。如果项目需要支持复杂的身份验证场景，IdentityServer 可能很有用，但我建议在必要时才使用第三方身份验证服务器。

**代码清单 39-14　在 Advanced 项目的 Pages 文件夹的_Host.cshtml 文件中应用属性**

```
@page "/"
@{ Layout = null; }
@model HostModel
@using Microsoft.AspNetCore.Authorization

<!DOCTYPE html>
<html>
<head>
    <title>@ViewBag.Title</title>
    <link href="/lib/twitter-bootstrap/css/bootstrap.min.css" rel="stylesheet" />
    <base href="~/" />
</head>
<body>
```

```
    <div class="m-2">
        <component type="typeof(Advanced.Blazor.Routed)" render-mode="Server" />
    </div>
    <script src="_framework/blazor.server.js"></script>
    <script src="~/interop.js"></script>
</body>
</html>

@functions {

    [Authorize]
    public class HostModel : PageModel {}
}
```

这可以防止未经身份验证的用户访问 Blazor 应用程序。请求 http://localhost:5000/account/logout，以确保浏览器没有身份验证 cookie，然后请求 http://localhost:5000。请求将由_Host 页面处理，但授权中间件将触发登录提示符的重定向。使用用户名 bob 和密码 secret 进行身份验证，将被授予访问 Blazor 应用程序的权限，如图 39-7 所示。

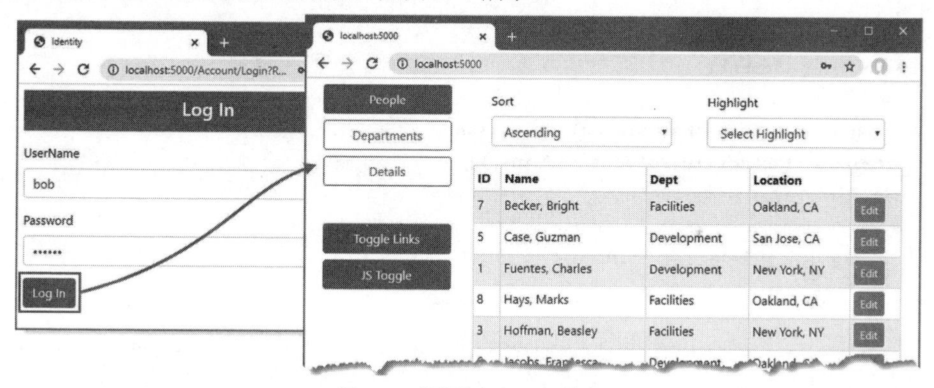


图 39-7 限制访问 Blazor 端点

## 39.3.1 在 Blazor 组件中执行授权

限制对端点的访问是一种有效的技术，但它对 Blazor 的所有功能应用相同级别的授权。对于需要更细粒度限制的应用程序，Blazor 提供了 AuthorizeRouteView 组件，当使用 URL 路由管理组件时，允许授权和未授权显示不同的内容。代码清单 39-15 将 AuthorizeRouteView 添加到示例应用程序中的路由组件。

### 代码清单 39-15 在 Advanced 项目的 Blazor 文件夹的 Routed.razor 文件中添加组件

```
@using Microsoft.AspNetCore.Components.Authorization

<Router AppAssembly="typeof(Startup).Assembly">
    <Found>
        <AuthorizeRouteView RouteData="@context" DefaultLayout="typeof(NavLayout)">
            <NotAuthorized Context="authContext">
```

```
                <h4 class="bg-danger text-white text-center p-2">Not Authorized </h4>
                <div class="text-center">
                    You may need to log in as a different user
                </div>
            </NotAuthorized>
        </AuthorizeRouteView>
    </Found>
    <NotFound>
        <h4 class="bg-danger text-white text-center p-2">
            Not Matching Route Found
        </h4>
    </NotFound>
</Router>
```

NotAuthorized 部分用于定义当用户试图访问受限制的资源时呈现给他们的内容。为演示这个特性，把对 DepartmentList 组件的访问限制为分配给 Admins 角色的用户，如代码清单 39-16 所示。

**代码清单 39-16　在 Advanced 项目的 Blazor 文件夹的 DepartmentList.cshtml 文件中限制访问**

```
@page "/departments"
@page "/depts"
@using Microsoft.AspNetCore.Authorization
@attribute [Authorize(Roles = "Admins")]
<CascadingValue Name="BgTheme" Value="Theme" IsFixed="false" >
    <TableTemplate RowType="Department" RowData="Departments"
        Highlight="@(d => d.Name)"
        SortDirection="@(d => d.Name)">
        <Header>
            <tr><th>ID</th><th>Name</th><th>People</th><th>Locations</th></tr>
        </Header>
        <RowTemplate Context="d">
            <td>@d.Departmentid</td>
            <td>@d.Name</td>
            <td>@(String.Join(", ", d.People.Select(p => p.Surname)))</td>
            <td>
                @(String.Join(", ",
                    d.People.Select(p => p.Location.City).Distinct()))
            </td>
        </RowTemplate>
    </TableTemplate>
</CascadingValue>

<SelectFilter Title="@("Theme")" Values="Themes" @bind-SelectedValue="Theme" />

<button class="btn btn-primary" @onclick="HandleClick">People</button>

@code {
```

```
    [Inject]
    public DataContext Context { get; set; }

    public IEnumerable<Department> Departments => Context.Departments
           .Include(d => d.People).ThenInclude(p => p.Location);

    public string Theme { get; set; } = "info";
    public string[] Themes = new string[] { "primary", "info", "success" };

    [Inject]
    public NavigationManager NavManager { get; set; }

    public void HandleClick() => NavManager.NavigateTo("/people");
}
```

使用@attribute 指令将 Authorize 属性应用到组件。重启 ASP.NET Core，并请求 http://localhost:5000/account/logout 以删除身份验证 cookie，然后请求 http://localhost:5000。出现提示时，使用用户名 bob 和密码 secret 进行身份验证，将看到 Blazor 应用程序，但是当单击 Departments 按钮时，将看到代码清单 39-15 中定义的授权内容，如图 39-8 所示。再次注销并以 admin 身份使用密码 secret 登录，就能使用受限组件。

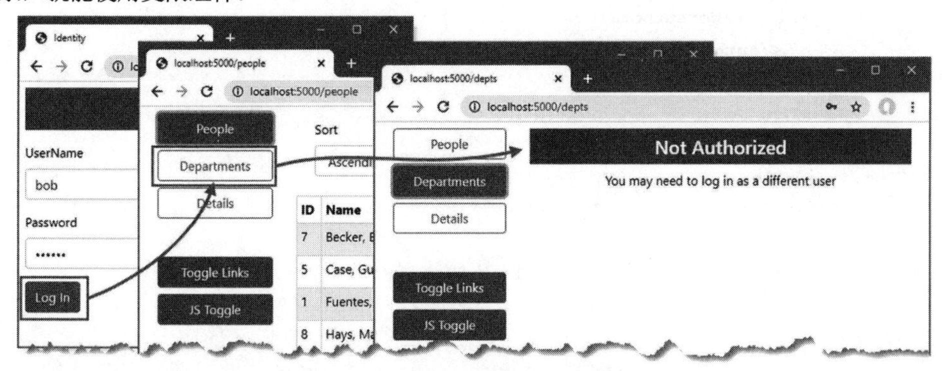


图 39-8 在 Blazor 应用程序中使用授权

## 39.3.2 向授权用户显示内容

AuthorizeView组件用于限制对组件呈现的内容部分的访问。在代码清单 39-17 中，更改了DepartmentList组件的授权，以便任何经过身份验证的用户都可以访问页面并使用AuthorizeView组件，这样表中Locations列的内容只显示给分配给Admins组的用户。

**代码清单 39-17 Advanced 项目的 Blazor 文件夹的 DepartmentList.razor 文件中的选择性内容**

```
@page "/departments"
@page "/depts"
@using Microsoft.AspNetCore.Authorization
@using Microsoft.AspNetCore.Components.Authorization
```

```
@attribute [Authorize]

<CascadingValue Name="BgTheme" Value="Theme" IsFixed="false" >
    <TableTemplate RowType="Department" RowData="Departments"
        Highlight="@(d => d.Name)"
        SortDirection="@(d => d.Name)">
        <Header>
            <tr><th>ID</th><th>Name</th><th>People</th><th>Locations</th></tr>
        </Header>
        <RowTemplate Context="d">
            <td>@d.Departmentid</td>
            <td>@d.Name</td>
            <td>@(String.Join(", ", d.People.Select(p => p.Surname)))</td>
            <td>
                <AuthorizeView Roles="Admins">
                    <Authorized>
                        @(String.Join(", ",
                            d.People.Select(p => p.Locations.City).Distinct()))
                    </Authorized>
                    <NotAuthorized>
                        (Not authorized)
                    </NotAuthorized>
                </AuthorizeView>
            </td>
        </RowTemplate>
    </TableTemplate>
</CascadingValue>
<SelectFilter Title="@("Theme")" Values="Themes" @bind-SelectedValue="Theme" />

<button class="btn btn-primary" @onclick="HandleClick">People</button>

@code {

// ...statements omitted for brevity...
}
```

AuthorizeView 组件配置了 Roles 属性，该属性接收以逗号分隔的已授权角色列表。Authorized 部分包含将显示给授权用户的内容。NotAuthorized 部分包含将显示给未授权用户的内容。

**提示：** 如果不需要向未经授权的用户显示内容，可省略 NotAuthorized 部分。

重启 ASP.NET Core，在请求 http://localhost:5000/depts 之前，使用密码 secret 进行 bob 身份验证。这个用户没有权限查看 Locations 列的内容，如图 39-9 所示。使用密码 secret 通过 admin 身份验证，并再次请求 http://localhost:5000/depts。这一次，用户是 Admins 角色的成员，并通过了授权检查，如图 39-9 所示。

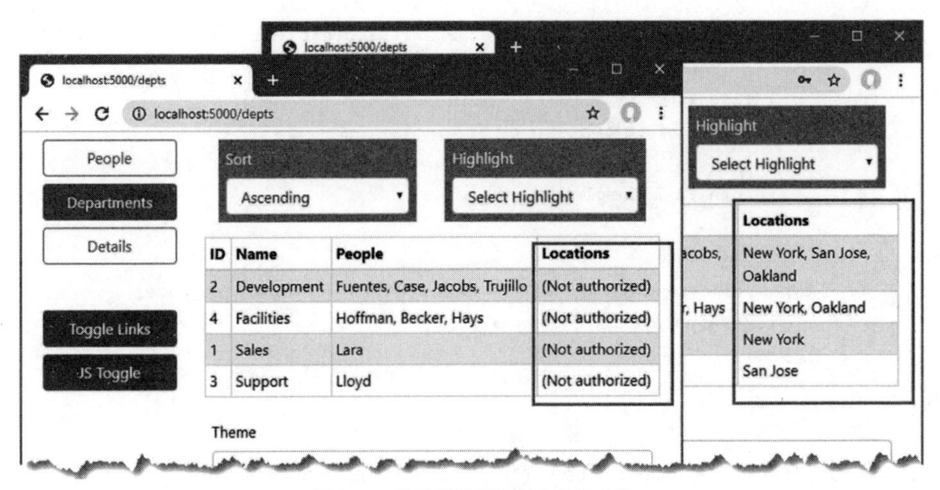


图 39-9　基于授权有选择地显示内容

## 39.4　对 Web 服务进行身份验证和授权

上一节中的授权过程依赖于能够将客户端重定向到允许用户输入其凭证的 URL。在向 Web 服务添加身份验证和授权时，需要一种不同的方法，因为无法向用户提供 HTML 表单来收集他们的凭据。添加对 Web 服务身份验证的支持的第一步是禁用重定向，以便客户端在尝试请求需要身份验证的端点时，收到 HTTP 错误响应。在 Advanced 文件夹中添加一个名为 CookieAuthenticationExtension.cs 的类文件，并用它定义扩展方法，如代码清单 39-18 所示。

### 代码清单 39-18　Advanced 文件夹中 CookieAuthenticationExtension.cs 文件的内容

```
using System;
using System.Collections.Generic;
using System.Linq;
using System.Linq.Expressions;
using System.Threading.Tasks;

namespace Microsoft.AspNetCore.Authentication.Cookies {
   public static class CookieAuthenticationExtensions {

      public static void DisableRedirectForPath(
         this CookieAuthenticationEvents events,
         Expression<Func<CookieAuthenticationEvents,
            Func<RedirectContext<CookieAuthenticationOptions>, Task>>> expr,
         string path, int statuscode) {

         string propertyName = ((MemberExpression)expr.Body).Member.Name;
         var oldHandler = expr.Compile().Invoke(events);

         Func<RedirectContext<CookieAuthenticationOptions>, Task> newHandler
               = context => {
```

```
                if (context.Request.Path.StartsWithSegments(path)) {
                    context.Response.StatusCode = statuscode;
                } else {
                    oldHandler(context);
                }
                return Task.CompletedTask;
            };

            typeof(CookieAuthenticationEvents).GetProperty(propertyName)
                .SetValue(events, newHandler);
        }
    }
}
```

这段代码很难理解。ASP.NET Core 提供了 CookieAuthenticationOptions 类，用于配置基于 cookie 的身份验证。CookieAuthenticationOptions.Events 属性返回一个 CookieAuthenticationEvents 对象，该对象用于设置由身份验证系统触发的事件的处理程序，包括当用户请求未经授权的内容时的重定向。代码清单 39-18 中的扩展方法将事件的默认处理程序替换为仅在请求不以指定路径字符串开始时执行重定向的处理程序。代码清单 39-19 使用扩展方法替换 OnRedirectToLogin 和 OnRedirectToAccessDenied 处理程序，这样当请求路径以/api 开始时，就不会执行重定向。

**代码清单 39-19　在 Advanced 文件夹的 Startup.cs 文件中防止重定向**

```
using System;
using System.Collections.Generic;
using System.Linq;
using System.Threading.Tasks;
using Microsoft.AspNetCore.Builder;
using Microsoft.AspNetCore.Hosting;
using Microsoft.AspNetCore.Http;
using Microsoft.Extensions.DependencyInjection;
using Microsoft.Extensions.Hosting;
using Microsoft.Extensions.Configuration;
using Microsoft.EntityFrameworkCore;
using Advanced.Models;
using Microsoft.AspNetCore.ResponseCompression;
using Microsoft.AspNetCore.Identity;
using Microsoft.AspNetCore.Authentication.Cookies;

namespace Advanced {
    public class Startup {

        public Startup(IConfiguration config) {
            Configuration = config;
        }

        public IConfiguration Configuration { get; set; }
```

```
public void ConfigureServices(IServiceCollection services) {
    services.AddDbContext<DataContext>(opts => {
        opts.UseSqlServer(Configuration[
            "ConnectionStrings:PeopleConnection"]);
        opts.EnableSensitiveDataLogging(true);
    });
    services.AddControllersWithViews().AddRazorRuntimeCompilation();
    services.AddRazorPages().AddRazorRuntimeCompilation();
    services.AddServerSideBlazor();
    services.AddSingleton<Services.ToggleService>();

    services.AddResponseCompression(opts => {
        opts.MimeTypes = ResponseCompressionDefaults.MimeTypes.Concat(
            new[] { "application/octet-stream" });
    });

    services.AddDbContext<IdentityContext>(opts =>
        opts.UseSqlServer(Configuration[
            "ConnectionStrings:IdentityConnection"]));
    services.AddIdentity<IdentityUser, IdentityRole>()
        .AddEntityFrameworkStores<IdentityContext>();

    services.Configure<IdentityOptions>(opts => {
        opts.Password.RequiredLength = 6;
        opts.Password.RequireNonAlphanumeric = false;
        opts.Password.RequireLowercase = false;
        opts.Password.RequireUppercase = false;
        opts.Password.RequireDigit = false;
        opts.User.RequireUniqueEmail = true;
        opts.User.AllowedUserNameCharacters = "abcdefghijklmnopqrstuvwxyz";
    });

    services.AddAuthentication(opts => {
        opts.DefaultScheme =
            CookieAuthenticationDefaults.AuthenticationScheme;
        opts.DefaultChallengeScheme =
            CookieAuthenticationDefaults.AuthenticationScheme;
    }).AddCookie(opts => {
        opts.Events.DisableRedirectForPath(e => e.OnRedirectToLogin,
            "/api", StatusCodes.Status401Unauthorized);
        opts.Events.DisableRedirectForPath(e => e.OnRedirectToAccessDenied,
            "/api", StatusCodes.Status403Forbidden);
    });
}
public void Configure(IApplicationBuilder app, DataContext context) {
```

```
            // ...statements omitted for brevity...
        }
    }
}
```

AddAuthentication 方法用于选择基于 cookie 的身份验证，并与 AddCookie 方法链接在一起，以替换可能触发重定向的事件处理程序。

### 39.4.1　构建简单的 JavaScript 客户端

为演示如何使用 Web 服务执行身份验证，下面创建一个简单的 JavaScript 客户端，它使用示例项目的 Data 控制器中的数据。

---

**提示：**不需要熟悉 JavaScript 来跟随本章这部分的例子。重要的是服务器端代码以及它支持客户端身份验证的方式，以便客户端能够访问 Web 服务。

---

使用代码清单 39-20 中所示的元素将名为 webclient.html 的 HTML 页面添加到 Advanced 项目的 wwwroot 文件夹中。

**代码清单 39-20　Advanced 项目的 wwwroot 文件夹中 webclient.html 文件的内容**

```
<!DOCTYPE html>
<html>
<head>
    <title>Web Service Authentication</title>
    <link href="/lib/twitter-bootstrap/css/bootstrap.min.css" rel="stylesheet" />
    <script type="text/javascript" src="webclient.js"></script>
</head>
<body>
    <div id="controls" class="m-2"></div>
    <div id="data" class="m-2 p-2">
        No data
    </div>
</body>
</html>
```

向 Advanced 项目的 wwwroot 文件夹中添加一个名为 webclient.js 的 JavaScript 文件，其内容如代码清单 39-21 所示。

**代码清单 39-21　Advanced 项目的 wwwroot 文件夹中 webclient.js 文件的内容**

```
const username = "bob";
const password = "secret";

window.addEventListener("DOMContentLoaded", () => {
    const controlDiv = document.getElementById("controls");
    createButton(controlDiv, "Get Data", getData);
    createButton(controlDiv, "Log In", login);
    createButton(controlDiv, "Log Out", logout);
```

```
});

function login() {
    // do nothing
}
function logout() {
    // do nothing
}

async function getData() {
    let response = await fetch("/api/people");
    if (response.ok) {
        let jsonData = await response.json();
        displayData(...jsonData.map(item => `${item.surname}, ${item.firstname}`));
    } else {
        displayData(`Error: ${response.status}: ${response.statusText}`);
    }
}

function displayData(...items) {
    const dataDiv = document.getElementById("data");
    dataDiv.innerHTML = "";
    items.forEach(item => {
        const itemDiv = document.createElement("div");
        itemDiv.innerText = item;
        itemDiv.style.wordWrap = "break-word";
        dataDiv.appendChild(itemDiv);
    })
}

function createButton(parent, label, handler) {
    const button = document.createElement("button");
    button.classList.add("btn", "btn-primary", "m-2");
    button.innerText = label;
    button.onclick = handler;
    parent.appendChild(button);
}
```

这段代码向用户展示了 Get Data、LogIn 和 LogOut 按钮。单击 Get Data 按钮使用 Fetch API 发送 HTTP 请求，处理 JSON 结果，并显示名称列表。其他按钮什么也不做，但是在后面的示例中通过它们来使用 JavaScript 代码中的硬连接凭证，对 ASP.NET Core 应用程序进行身份验证。

---

■ **警告：**这只是一个简单的客户端，来演示服务器端的验证功能。如果需要写一个 JavaScript 客户端，那么考虑像 Angular 或者 React 这样的框架。无论如何构建客户端，都不要在 JavaScript 文件中包含硬连接凭据。

---

请求 http://localhost:5000/webclient，然后单击 Get Data 按钮。JavaScript 客户端将向 Data 控制

器发送一个 HTTP 请求并显示结果，如图 39-10 所示。

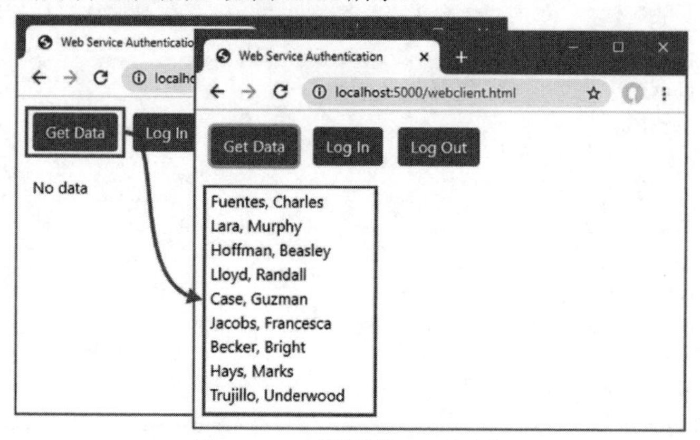


图 39-10　一个简单的 Web 客户端

## 39.4.2　限制对 Web 服务的访问

标准授权特性用于限制对 Web 服务端点的访问，在代码清单 39-22 中，给 DataController 类应用了 Authorize 属性。

**代码清单 39-22　在 Advanced 项目的 Controllers 文件夹的 DataController.cs 文件中应用属性**

```
using Advanced.Models;
using Microsoft.AspNetCore.Mvc;
using Microsoft.EntityFrameworkCore;
using System.Collections.Generic;
using System.Threading.Tasks;
using Microsoft.AspNetCore.Authorization;

namespace Advanced.Controllers {

    [ApiController]
    [Route("/api/people")]
    [Authorize]
    public class DataController : ControllerBase {
        private DataContext context;

        // ...methods omitted for brevity...
    }
}
```

重启 ASP.NET Core，并请求 http://localhost:5000/account/logout，以确保 JavaScript 客户端不使用上一个示例中的身份验证 cookie。请求 http://localhost:5000/webclient.html 以加载 JavaScript 客户端，并单击 Get Data 按钮发送 HTTP 请求。服务器将响应 401 Unauthorized，如图 39-11 所示。

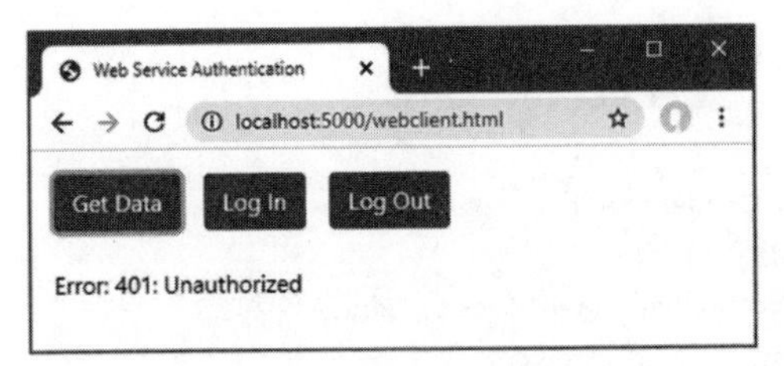


图 39-11　未经授权的请求

### 39.4.3　使用 cookie 验证

实现身份验证的最简单方法是依赖于上一节中演示过的标准 ASP.NET Core cookie。给 Advanced 项目的 Controllers 文件夹添加一个名为 ApiAccountController.cs 的类文件，并使用它来定义如代码清单 39-23 所示的控制器。

**代码清单 39-23　Advanced 项目 Controllers 文件夹中 ApiAccountController.cs 文件的内容**

```
using Microsoft.AspNetCore.Authorization;
using Microsoft.AspNetCore.Identity;
using Microsoft.AspNetCore.Mvc;
using System.ComponentModel.DataAnnotations;
using System.Threading.Tasks;

namespace Advanced.Controllers {

    [ApiController]
    [Route("/api/account")]
    public class ApiAccountController : ControllerBase {
        private SignInManager<IdentityUser> signinManager;

        public ApiAccountController(SignInManager<IdentityUser> mgr) {
            signinManager = mgr;
        }

        [HttpPost("login")]
        public async Task<IActionResult> Login([FromBody]Credentials creds) {
            Microsoft.AspNetCore.Identity.SignInResult result
                = await signinManager.PasswordSignInAsync(creds.Username,
                    creds.Password, false, false);
            if (result.Succeeded) {
                return Ok();
            }
            return Unauthorized();
        }

        [HttpPost("logout")]
        public async Task<IActionResult> Logout() {
            await signinManager.SignOutAsync();
```

```
            return Ok();
        }

        public class Credentials {
            [Required]
            public string Username { get; set; }
            [Required]
            public string Password { get; set; }
        }
    }
}
```

此 Web 服务控制器定义允许客户端登录和注销的操作。成功的身份验证请求的响应将包含一个 cookie，浏览器在 JavaScript 客户端发出的请求中自动包含该 cookie。

代码清单 39-24 向简单的 JavaScript 客户端添加了对使用代码清单 39-23 中定义的操作方法进行身份验证的支持。

**代码清单 39-24　在 Advanced 项目的 wwwroot 文件夹的 webclient.js 文件中添加身份验证**

```
const username = "bob";
const password = "secret";

window.addEventListener("DOMContentLoaded", () => {
    const controlDiv = document.getElementById("controls");
    createButton(controlDiv, "Get Data", getData);
    createButton(controlDiv, "Log In", login);
    createButton(controlDiv, "Log Out", logout);
});

async function login() {
    let response = await fetch("/api/account/login", {
        method: "POST",
        headers: { "Content-Type": "application/json" },
        body: JSON.stringify({ username: username, password: password })
    });
    if (response.ok) {
        displayData("Logged in");
    } else {
        displayData(`Error: ${response.status}: ${response.statusText}`);
    }
}

async function logout() {
    let response = await fetch("/api/account/logout", {
        method: "POST"
    });
    if (response.ok) {
```

```
            displayData("Logged out");
        } else {
            displayData(`Error: ${response.status}: ${response.statusText}`);
        }
    }

    async function getData() {
        let response = await fetch("/api/people");
        if (response.ok) {
            let jsonData = await response.json();
            displayData(...jsonData.map(item => `${item.surname}, ${item.firstname}`));
        } else {
            displayData(`Error: ${response.status}: ${response.statusText}`);
        }
    }
    function displayData(...items) {
        const dataDiv = document.getElementById("data");
        dataDiv.innerHTML = "";
        items.forEach(item => {
            const itemDiv = document.createElement("div");
            itemDiv.innerText = item;
            itemDiv.style.wordWrap = "break-word";
            dataDiv.appendChild(itemDiv);
        })
    }

    function createButton(parent, label, handler) {
        const button = document.createElement("button");
        button.classList.add("btn", "btn-primary", "m-2");
        button.innerText = label;
        button.onclick = handler;
        parent.appendChild(button);
    }
```

重启 ASP.NET Core，请求 http://localhost:5000/webclient.html，然后单击 Login In 按钮。等待确认身份验证的消息，然后单击 Get Data 按钮。浏览器包含身份验证 cookie，请求通过授权检查。单击 Log Out 按钮，然后再次单击 Get Data。没有使用 cookie，请求失败。图 39-12 显示了这两个请求。

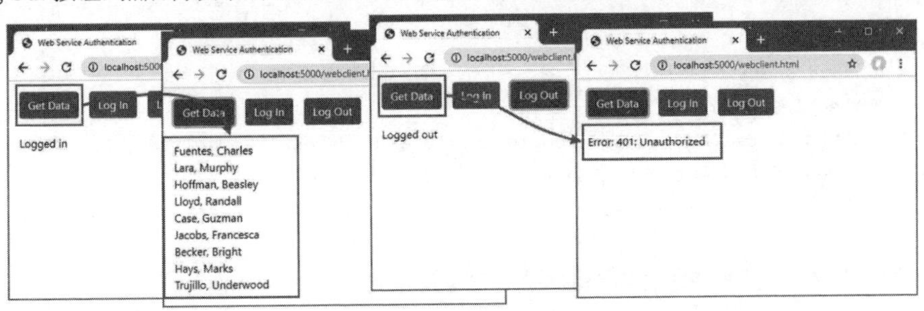


图 39-12 使用 cookie 验证

### 39.4.4 使用令牌认证

并不是所有Web服务都能够依赖cookie，因为并不是所有客户端都可以使用cookie。另一种方法是使用无记名令牌，它是客户端给定的字符串，包含在客户端发送给Web服务的请求中。客户端不理解这个令牌的含义(据说是不透明的)，只使用服务器提供的任何令牌。

下面演示使用JSON Web令牌(JWT)进行身份验证，该令牌为客户端提供了一个加密的令牌，其中包含经过身份验证的用户名。客户端无法解密或修改令牌，但当它包含在请求中时，ASP. NET Core服务器就解密令牌，并使用其中包含的名称作为用户身份。JWT格式在https://tools.ietf.org/html/rfc7519有详细描述。

> 注意：ASP. NET Core会相信任何包含令牌的请求都来自已验证用户。就像在使用cookie时一样，生产应用程序应该使用HTTPS来防止令牌被拦截和重用。

#### 准备应用程序

打开一个新的PowerShell命令提示符，导航到Advanced项目文件夹，运行代码清单39-25所示的命令，将JWT的包添加到项目中。

**代码清单39-25 安装NuGet包**

```
dotnet add package System.IdentityModel.Tokens.Jwt --version 5.6.0
dotnet add package Microsoft.AspNetCore.Authentication.JwtBearer --version 3.1.1
```

JWT需要一个用于加密和解密令牌的密钥。将代码清单39-26中所示的配置添加到appsettings.json文件中。如果在实际应用程序中使用JWT，请确保更改了密钥。

**代码清单39-26 在Advanced项目的appsettings.json文件中添加设置**

```
{
   "Logging": {
      "LogLevel": {
         "Default": "Information",
         "Microsoft": "Warning",
         "Microsoft.Hosting.Lifetime": "Information",
         "Microsoft.EntityFrameworkCore": "Information",
         "Microsoft.AspNetCore.Authentication": "Debug"
      }
   },
   "AllowedHosts": "*",
   "ConnectionStrings": {
      "PeopleConnection": "Server=(localdb)
      \\MSSQLLocalDB;Database=People;MultipleActiveResultSets=True",
      "IdentityConnection": "Server=(localdb)
      \\MSSQLLocalDB;Database=Identity;MultipleActiveResultSets=True"
   },
   "jwtSecret": "apress_jwt_secret"
}
```

### 39.4.5 创建令牌

客户端发送一个包含用户凭据的 HTTP 请求，并接收一个 JWT 作为响应。代码清单 39-27 向 ApiAccountController 添加了一个操作方法，该方法接收凭证、验证凭证并生成令牌。

**代码清单 39-27 在 Advanced 项目的 Controllers 文件夹中的 ApiAccountController.cs 文件中生成令牌**

```
using Microsoft.AspNetCore.Authorization;
using Microsoft.AspNetCore.Identity;
using Microsoft.AspNetCore.Mvc;
using System.ComponentModel.DataAnnotations;
using System.Threading.Tasks;
using Microsoft.IdentityModel.Tokens;
using System.IdentityModel.Tokens.Jwt;
using System.Text;
using System.Security.Claims;
using System;
using Microsoft.Extensions.Configuration;

namespace Advanced.Controllers {

    [ApiController]
    [Route("/api/account")]
    public class ApiAccountController : ControllerBase {
        private SignInManager<IdentityUser> signinManager;
        private UserManager<IdentityUser> userManager;
        private IConfiguration configuration;

        public ApiAccountController(SignInManager<IdentityUser> mgr,
                UserManager<IdentityUser> usermgr, IConfiguration config) {
            signinManager = mgr;
            userManager = usermgr;
            configuration = config;
        }

        [HttpPost("login")]
        public async Task<IActionResult> Login([FromBody]Credentials creds) {
            Microsoft.AspNetCore.Identity.SignInResult result
                = await signinManager.PasswordSignInAsync(creds.Username,
                    creds.Password, false, false);
            if (result.Succeeded) {
                return Ok();
            }
            return Unauthorized();
        }
```

```
[HttpPost("logout")]
public async Task<IActionResult> Logout() {
    await signinManager.SignOutAsync();
    return Ok();
}

[HttpPost("token")]
public async Task<IActionResult> Token([FromBody]Credentials creds) {
    if (await CheckPassword(creds)) {
        JwtSecurityTokenHandler handler = new JwtSecurityTokenHandler();
        byte[] secret = Encoding.ASCII.GetBytes(configuration["jwtSecret"]);
        SecurityTokenDescriptor descriptor = new SecurityTokenDescriptor {
            Subject = new ClaimsIdentity(new Claim[] {
                new Claim(ClaimTypes.Name, creds.Username)
            }),
            Expires = DateTime.UtcNow.AddHours(24),
            SigningCredentials = new SigningCredentials(
                new SymmetricSecurityKey(secret),
                    SecurityAlgorithms.HmacSha256Signature)
        };
        SecurityToken token = handler.CreateToken(descriptor);
        return Ok(new {
            success = true,
            token = handler.WriteToken(token)
        });
    }
    return Unauthorized();
}

    private async Task<bool> CheckPassword(Credentials creds) {
        IdentityUser user = await userManager.FindByNameAsync(creds.Username);
        if (user != null) {
            foreach (IPasswordValidator<IdentityUser> v in
                    userManager.PasswordValidators) {
                if ((await v.ValidateAsync(userManager, user,
                        creds.Password)).Succeeded) {
                    return true;
                }
            }
        }
        return false;
    }

    public class Credentials {
        [Required]
        public string Username { get; set; }
        [Required]
```

```
            public string Password { get; set; }
        }
    }
}
```

UserManager<T>类定义了 PasswordValidators 属性，该属性返回一系列实现 IPasswordValidator<T>接口的对象。当调用 Token 操作方法时，它将凭据传递给 CheckPassword 方法，该方法枚举 IPasswordValidator<T>对象，以调用每个对象上的 ValidateAsync 方法。如果密码被任何一个验证器验证，Token 方法将创建一个令牌。

JWT 规范定义了一个通用的令牌，并非仅限于在 HTTP 请求中标识用户，而且许多可用选项在本示例中并不是必要的。代码清单 39-27 中创建的令牌包含如下有效负载：

```
...
{
    "unique_name": "bob",
    "nbf": 1579765454,
    "exp": 1579851854,
    "iat": 1579765454
}
...
```

unique_name 属性包含用户名，用于对包含令牌的请求进行身份验证。其他有效负载属性是时间戳，这里不使用它。

有效负载使用代码清单 39-27 中定义的密钥加密，并作为 JSON 编码的响应返回给客户端，该响应如下所示：

```
...
{
    "success":true,
    "token":"eyJhbGciOiJIUzI1NiIsInR5cCI6IkpXVCJ9..."
}
...
```

这里只展示了标记的第一部分，因为它们是长字符串，而响应的结构才是重要的。客户端接收到令牌，并使用 Authorization 头将其包含在以后的请求中，如下所示：

```
...
Authorization: Bearer eyJhbGciOiJIUzI1NiIsInR5cCI6IkpXVCJ9
...
```

服务器接收令牌，使用密钥对其解密，并使用令牌有效负载中的 unique_name 属性值对请求进行验证。不执行进一步的验证，将使用负载中包含的任何用户名验证包含有效令牌的请求。

## 39.4.6 用令牌验证

下一步是配置应用程序以接收和验证令牌，如代码清单 39-28 所示。

**代码清单 39-28　在 Advanced 项目的 Startup.cs 文件中验证令牌**

```
using System;
using System.Collections.Generic;
using System.Linq;
using System.Threading.Tasks;
using Microsoft.AspNetCore.Builder;
using Microsoft.AspNetCore.Hosting;
using Microsoft.AspNetCore.Http;
using Microsoft.Extensions.DependencyInjection;
using Microsoft.Extensions.Hosting;
using Microsoft.Extensions.Configuration;
using Microsoft.EntityFrameworkCore;
using Advanced.Models;
using Microsoft.AspNetCore.ResponseCompression;
using Microsoft.AspNetCore.Identity;
using Microsoft.AspNetCore.Authentication.Cookies;
using Microsoft.IdentityModel.Tokens;
using System.Text;
using System.Security.Claims;
using Microsoft.AspNetCore.Authentication.JwtBearer;

namespace Advanced {
    public class Startup {

        public Startup(IConfiguration config) {
            Configuration = config;
        }

        public IConfiguration Configuration { get; set; }

        public void ConfigureServices(IServiceCollection services) {
            services.AddDbContext<DataContext>(opts => {
                opts.UseSqlServer(Configuration[
                    "ConnectionStrings:PeopleConnection"]);
                opts.EnableSensitiveDataLogging(true);
            });
            services.AddControllersWithViews().AddRazorRuntimeCompilation();
            services.AddRazorPages().AddRazorRuntimeCompilation();
            services.AddServerSideBlazor();
            services.AddSingleton<Services.ToggleService>();

            services.AddResponseCompression(opts => {
                opts.MimeTypes = ResponseCompressionDefaults.MimeTypes.Concat(
                    new[] { "application/octet-stream" });
            });
```

```
services.AddDbContext<IdentityContext>(opts =>
    opts.UseSqlServer(Configuration[
        "ConnectionStrings:IdentityConnection"]));

services.AddIdentity<IdentityUser, IdentityRole>()
    .AddEntityFrameworkStores<IdentityContext>();

services.Configure<IdentityOptions>(opts => {
    opts.Password.RequiredLength = 6;
    opts.Password.RequireNonAlphanumeric = false;
    opts.Password.RequireLowercase = false;
    opts.Password.RequireUppercase = false;
    opts.Password.RequireDigit = false;
    opts.User.RequireUniqueEmail = true;
    opts.User.AllowedUserNameCharacters = "abcdefghijklmnopqrstuvwxyz";
});

services.AddAuthentication(opts => {
    opts.DefaultScheme =
        CookieAuthenticationDefaults.AuthenticationScheme;
    opts.DefaultChallengeScheme =
        CookieAuthenticationDefaults.AuthenticationScheme;
}).AddCookie(opts => {
    opts.Events.DisableRedirectForPath(e => e.OnRedirectToLogin,
        "/api", StatusCodes.Status401Unauthorized);
    opts.Events.DisableRedirectForPath(e => e.OnRedirectToAccessDenied,
        "/api", StatusCodes.Status403Forbidden);
}).AddJwtBearer(opts => {
    opts.RequireHttpsMetadata = false;
    opts.SaveToken = true;
    opts.TokenValidationParameters = new TokenValidationParameters {
        ValidateIssuerSigningKey = true,
        IssuerSigningKey = new SymmetricSecurityKey(
            Encoding.ASCII.GetBytes(Configuration["jwtSecret"])),
        ValidateAudience = false,
        ValidateIssuer = false
    };
    opts.Events = new JwtBearerEvents {
        OnTokenValidated = async ctx => {
            var usrmgr = ctx.HttpContext.RequestServices
                .GetRequiredService<UserManager<IdentityUser>>();
            var signinmgr = ctx.HttpContext.RequestServices
                .GetRequiredService<SignInManager<IdentityUser>>();
            string username =
                ctx.Principal.FindFirst(ClaimTypes.Name)?.Value;
            IdentityUser idUser = await usrmgr.FindByNameAsync(username);
            ctx.Principal =
```

```
                        await signinmgr.CreateUserPrincipalAsync(idUser);
                    }
                };
            });
        }

        public void Configure(IApplicationBuilder app, DataContext context) {

            // ...statements omitted for brevity...
        }
    }
}
```

AddJwtBearer 身份验证系统添加对 JWT 的支持，并提供解密令牌所需的设置。此处为 OnTokenValidated 事件添加一个处理程序，该事件在验证令牌时触发，这样就可以查询用户数据库并将 IdentityUser 对象与请求关联起来。这充当 JWT 令牌和 ASP.NET Core Identity 数据之间的桥梁，确保像基于角色的授权这样的特性能够无缝地工作。

### 39.4.7 使用令牌限制访问

为允许使用令牌访问受限制的端点，修改应用于 Data 控制器的 Authorize 属性，如代码清单 39-29 所示。

**代码清单 39-29　在 Advanced 项目的 Controllers 文件夹中的 DataController.cs 文件中启用令牌**

```
using Advanced.Models;
using Microsoft.AspNetCore.Mvc;
using Microsoft.EntityFrameworkCore;
using System.Collections.Generic;
using System.Threading.Tasks;
using Microsoft.AspNetCore.Authorization;

namespace Advanced.Controllers {

    [ApiController]
    [Route("/api/people")]
    [Authorize(AuthenticationSchemes = "Identity.Application, Bearer")]
    public class DataController : ControllerBase {
        private DataContext context;

        // ...methods omitted for brevity...
    }
}
```

AuthenticationSchemes 参数指定身份验证类型，用于授权对控制器的访问。本例指定，可使用默认的 cookie 身份验证和新的 Bearer 令牌。

### 39.4.8 使用令牌请求数据

最后一步是更新 JavaScript 客户端，使其获得一个令牌，并将其包含在对数据的请求中，如代码清单 39-30 所示。

**代码清单 39-30 在 Advanced 项目的 wwwroot 文件夹的 webclient.js 文件中使用令牌**

```
const username = "bob";
const password = "secret";
let token;

window.addEventListener("DOMContentLoaded", () => {
    const controlDiv = document.getElementById("controls");
    createButton(controlDiv, "Get Data", getData);
    createButton(controlDiv, "Log In", login);
    createButton(controlDiv, "Log Out", logout);
});

async function login() {
    let response = await fetch("/api/account/token", {
        method: "POST",
        headers: { "Content-Type": "application/json" },
        body: JSON.stringify({ username: username, password: password })
    });
    if (response.ok) {
        token = (await response.json()).token;
        displayData("Logged in", token);
    } else {
        displayData(`Error: ${response.status}: ${response.statusText}`);
    }
}

async function logout() {
    token = "";
    displayData("Logged out");
}

async function getData() {
    let response = await fetch("/api/people", {
        headers: { "Authorization": `Bearer ${token}` }
    });
    if (response.ok) {
        let jsonData = await response.json();
        displayData(...jsonData.map(item => `${item.surname}, ${item.firstname}`));
    } else {
        displayData(`Error: ${response.status}: ${response.statusText}`);
    }
```

```
    }

    function displayData(...items) {
        const dataDiv = document.getElementById("data");
        dataDiv.innerHTML = "";
        items.forEach(item => {
            const itemDiv = document.createElement("div");
            itemDiv.innerText = item;
            itemDiv.style.wordWrap = "break-word";
            dataDiv.appendChild(itemDiv);
        })
    }

    function createButton(parent, label, handler) {
        const button = document.createElement("button");
        button.classList.add("btn", "btn-primary", "m-2");
        button.innerText = label;
        button.onclick = handler;
        parent.appendChild(button);
    }
```

客户端接收身份验证响应并分配令牌，以便它可以被设置 Authorization 头的 GetData 方法使用。注意，不需要注销请求，并且在用户单击 LogOut 按钮时，重置用于存储令牌的变量。

**注意：**在尝试测试令牌时，很容易以 cookie 进行身份验证。在测试此功能之前，请确保清除浏览器 cookie，以避免使用以前测试的 cookie。

重启 ASP.NET Core，请求 http://localhost:5000/webclient.html。单击 Log In 按钮，将生成并显示一个令牌。单击 Get Data 按钮，令牌就发送到服务器，并用于对用户进行身份验证，生成如图 39-13 所示的结果。

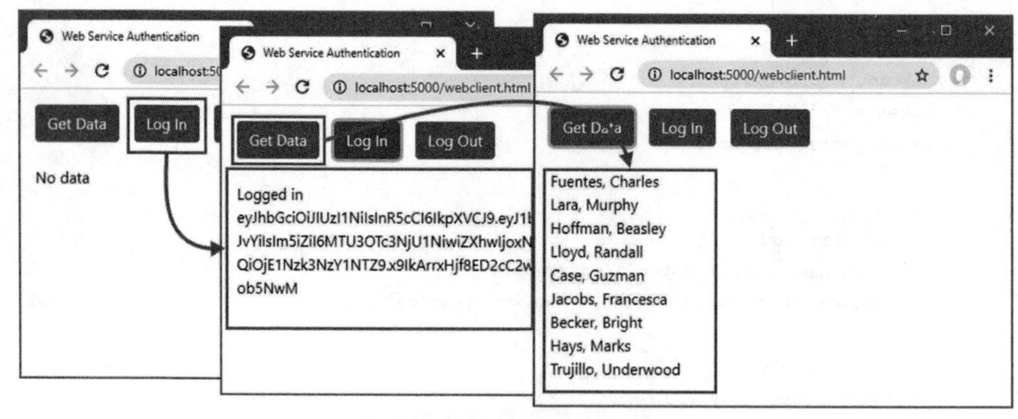


图 39-13　使用令牌进行身份验证

## 39.5 小结

本章展示了如何在 ASP.NET Core 应用程序中应用身份验证和授权，解释了对用户进行身份验证和限制端点访问的过程，解释了在 Blazor 应用程序中如何授权用户，并演示了如何使用 cookie 和 Bearer 令牌对 Web 服务客户端进行身份验证。

这就是关于 ASP.NET Core 的全部内容。希望你喜欢读这本书，祝你的 ASP.NET Core 项目获得成功。